2022 24th European Conference on Power Electronics and Applications
(EPE'22 ECCE Europe)

Hanover, Germany
5-9 September 2022

Pages 3362-4034

IEEE Catalog Number: CFP22850-POD
ISBN: 978-1-6654-8700-9

Copyright © 2022, The European Power Electronics and Drives Association
All Rights Reserved

*** *This is a print representation of what appears in the IEEE Digital Library. Some format issues inherent in the e-media version may also appear in this print version.*

IEEE Catalog Number: CFP22850-POD
ISBN (Print-On-Demand):   978-1-6654-8700-9
ISBN (Online):   978-9-0758-1539-9

Additional Copies of This Publication Are Available From:

Curran Associates, Inc
57 Morehouse Lane
Red Hook, NY 12571 USA
Phone: (845) 758-0400
Fax: (845) 758-2633
E-mail: curran@proceedings.com
Web: www.proceedings.com

2022 24th European Conference on Power Electronics and Applications (EPE'22 ECCE Europe)

Hanover, Germany
5-9 September 2022

Pages 3362-4034

IEEE Catalog Number: CFP22850-POD
ISBN: 978-1-6654-8700-9

TABLE OF CONTENTS

Dynamic Power Analysis of Inverter-Fed Drives Based on the Switching Period of the Power Electronics 1
Alexander Stock

Stability Analysis in an Inverter-Dominant Microgrid Facing In-Rush Current of an Induction Machine 11
Nastaran Fazli, David Hammes, Sidney Gierschner, Hans-Gunter Eckel

Self-Oscillating Capacitive Power Transfer with Multiple Receiver Capability and Coupling Path Adaption 22
Norbert Seliger

An Electrically Driven Gas Compressor for Hydrogen Refueling Stations with Active Power Smoothing 30
Alfred Rufer

Unsymmetrical Fault Behavior of PLL Based Grid-Connected Converters 39
Philipp Hackl, Ziqian Zhang, Robert Schuerhuber

Stability Assessment and Optimization of MMC Energy Balancing for Drive Applications at Standstill using an Averaging Approach 49
Qiuye Gui, Hendrik Fehr, Albrecht Gensior

Turn-On Losses Optimization for Medium Power SiC MOSFET Half-Bridge Module 59
Pham Ha Trieu To, Felix Kayser, Hans-Günter Eckel

Oscillation Damping in a 500kW Hybrid Si/SiC Three-Level ANPC Inverter with Decoupling Capacitor 70
Pham Ha Trieu To, Hans-Günter Eckel

Multi Busbar Sub-Module Modular Multilevel STATCOM with Partially Rated Energy Storage Configured in Sub-Stacks 80
Chuantong Hao, Wenhao Ma, Michael Merlin, Paul Judge, Stephen Finney

Three-Phase ZVS Inverter with Variable and Fixed Frequency Operation Based on GaN Semiconductors 88
Benedikt Kohlhepp, Michael Lutsch, Thomas Dürbaum

Influences of Conductor Positions and Fast Rising Impulse Voltages on the Line-End Coil Based on a Three-Phase High-Frequency Model 97
Ting Helmholdt-Zhu, Volker Grabs

Simulation Tool for Optimization of Digital Active Gate Drive Sequence using Genetic Algorithm 108
Hajime Takayama, Shuhei Fukunaga, Takashi Hikihara

Analysis of Balancing Algorithms for Quasi- Two/Three-Level Single Phase Operation of a Flying Capacitor Converter 115
Stefan Mersche, Markus Bayer, Kai Rickert, Marc Hiller

Instability in Active Balancing Control of Dc Bus Voltages in VSC Converters Interconnected via Multi-Winding Transformers 125
Duro Basic, Sami Siala

Online Learning-Based Islanding Detection Scheme for Grid-Connected Systems.. 135
    Mohammed Ali Khan, V S Bharath Kurukuru, Rupam Singh

Difference in the Design Process of LCL Filters for Grid Connected VSI When using SiC/GaN
Instead of Si Semiconductors 145
    Dennis Kampen, Lukas Fräger, Niklas Badenhop, Arthur Mambetow

Analysis and Design of a Resonant DC/DC Transformer in Modular Operation.. 152
    Abraham López, Manuel Arias, Pablo F. Miaja, Arturo Fernández

Predictive Braking Algorithm for Soft Starter Driven Induction Motors.. 160
    Hauke Nannen, Heiko Zatocil, Gerd Griepentrog

Ambient Electromagnetic Energy Harvesting Circuit using Rectennas Manufactured with
Stereolithography Resin 169
    Xuan Viet Linh Nguyen, Tony Gerges, Jacques Verdier, Philippe Lombard, Michel Cabrera,
    Bruno Allard, Jean-Marc Duchamp, Philippe Benech

Boost/Buck-Boost Based Grid Connected Solar PV Micro-Inverter with Reduced Number of
Switches and Having Power Decoupling Capability 178
    Arup Ratan Paul, Arghyadip Bhattacharya, Kishore Chatterjee

Operation and Selection of Multilevel Power Converters for Doubly Fed Induction Generator-
Based Wind Turbines 187
    Kapil Jha, Joseph Banda, Hridya I, Arvind Tiwari

A Detailed View on the Trapezoidal Operation for MMC Type Braking Chopper in Medium
Voltage Application... 195
    Patrick Hofstetter, Viktor Hofmann, Dennis Karwatzki

Influence of Operating Frequency on High-Power Medium-Voltage Medium-Frequency
Transformers 203
    Thomas B. Gradinger, Ralph M. Burkart, Marko Mogorovic

Output Power Characteristics of Isolated Secondary-Resonant SAB DC-DC Converter for Output
Voltage Variation 213
    Shota Yamashita, Kohei Budo, Takaharu Takeshita

Hardware and Control Design of a High Precision Modular Power Converter Based on GaN
Technology for Particle Accelerator Magnets 223
    Thomas Margreiter, Ivan De Cesaris, Maurizio Incurvati, Sebastien Pelletier, Martin
    Schiestl, Ronald Stärz

Battery Cycler to Generate Open Li-Ion Cell Aging Data and Models... 232
    Matthias Luh, Thomas Blank

Function Blocks of a Highly-Integrated All-In-GaN Power IC for DC-DC Conversion 242
    Michael Basler, Richard Reiner, Stefan Moench, Patrick Waltereit, Rüdiger Quay

Comparison of Redundancy Requirements for Modular Multilevel Converter Considering
Manufacturer Reliability Inputs and Mission Profile 251
    Diego Velazco, Guy Clerc, Emmanuel Boutleux, Francois Wallart

Impact of Insulation and Cooling on Performance Due to Reliability-Oriented Design of Electrical
Machines 261
    Lucas Vincent Hanisch, Jonas Franzki, Markus Henke

Long Switching Horizon Model Predictive Controller for High-Speed Integrated Modular Motor Drives 268

    Martin Schiestl, Maurizio Incurvati, Ronald Starz, Markus Schmid

Standalone Power Management System for Flexible Piezo Electric Nano Generators (PENG) Based on the Co-Polymer P(VDF:TrFE) 279

    Alexander Wölk, Mahmoud Shousha, Shashank Shekhawat Singh, Martin Haug, Lorandt Fölkel, Michael Brooks, Asier Alvarez, Andreas Petritz, Philipp Schäffner, Jonas Groten, Andreas Tschepp, Barbara Stadlober

Analysis and Estimation of Neutral-Point Voltage Balancing Ability of an Optimized Balancing Algorithm for Grid Connected Active-NPC Converter 289

    Joseph Banda, Kapil Jha, Hridya Ittamveettil, Arvind Kumar Tiwari, Fernando Ramirez

A Direct Model Predictive Control Strategy of Back-To-Back Modular Multilevel Converters using Arm Energy Estimation 297

    Akseli Hakkila, Antonios Antonopoulos, Petros Karamanakos

Study on Commutation Loop Inductance and Current Distribution to DC-Link Capacitors in a GaN Half-Bridge 307

    Benedikt Kohlhepp, Samuel Faber, Jeremias Kaiser, Thomas Dürbaum

Cooperative Control of Online Impedance Spectroscopy Monitoring Method and Maximum Power Point Tracking Method for Photovoltaic Panels 315

    Xin Wang, Zhixue Zheng, Michel Aillerie, Alexandre De Bernardinis, Jean–paul Sawicki, Marie-Cécile Péra, Daniel Hissel

Benefits of Switching from Si to SiC Modules with Further Converter Optimization 325

    Antxon Arrizabalaga, Mikel Mazuela, Iosu Aizpuru, June Urkizu, Jon Aztiria

On the Reduction of Output Capacitance in Two-Level Three Phase PFC Boost Rectifier for Pulsating Loads 335

    Tania C. Cano, Douglas Pedroso, Alberto Rodríguez, Ignacio Castro, Diego G. Lamar

Cognitive Insights into Metaheuristic Digital Twin Based Health Monitoring of DC-DC Converters 344

    Abdul Basit Mirza, Kushan Choksi, Sama Salehi Vala, Krishna Moorthy Radha, Madhu Sudhan Chinthavali, Fang Luo

A Three-Phase Isolated Secondary-Resonant Single-Active-Bridge DC-DC Converter with a Delta-Star Connected Transformer 351

    Atsushi Nishio, Kohei Budo, Mai Van Tuan, Takaharu Takeshita

A Novel Concept to Optimize Core Loss in Planar Magnetic Based on an Unbalanced-Flux-Approach 361

    Sobhi Barg, Kent Bertilsson, Grover Torrico

Model Reduction using Singular Perturbation Methods for a Microgrid Application 370

    Lasse Gnärig, Albrecht Gensior, Saioa Burutxaga Laza, Miguel Carrasco, Carsten Reincke-Collon

Drive Level Parameter Identification of an Induction Motor 380

    Andreas Bünte, Alex Hald, Andreas Kirsch

Impedance Stability of Single-Phase LCL Grid-Connected Voltage Source Inverters with Wideband Gap Devices Under Different Control Approaches 390

    Ramy Ali, Terence O'Donnell

Design and Modulation Optimization of an MMC Based Braking Chopper... 400
    Viktor Hofmann, Patrick Hofstetter

Modeling the Arrangement of Drill Holes for Orthogonal Biasing in Controllable Inductors for
Power Electronic Converters.. 411
    Jonas Pfeiffer, Christoph Drexler, Pierre Küster, Peter Zacharias, Michael Schmidhuber

A Sectorized FCS-MPC Transformerless SST for Power Transmission Application..................................... 421
    Gabriel Gaburro Bacheti, Renner Sartório Camargo, Emilio José Bueno, Marco Liserre,
    Lucas Frizera Encarnação

Inductance Estimation for Square-Shaped Multilayer Planar Windings 432
    Theofilos Papadopoulos, Antonios Antonopoulos

Cost and Efficiency Considerations in On-Board Chargers 442
    Marija Jankovic, Christian Felgemacher, Kevin Lenz, Aly Mashaly, Abdelmouneim
    Charkaoui

A Novel Combined Control of Ground Current and DC-Pole-To-Ground Voltage in Symmetrical
Monopole Modular Multilevel Converters for HVDC Applications.. 451
    Pablo Briff, Amit Kumar

A PFC Boost Converter with Reduced Switching Losses Operating at a Fixed Switching Frequency............ 459
    Burkhard Ulrich

Predictive Control of Power Electronics Autotransformer for Mitigating Three-Phase Grid Current
Unbalance in Railway Supply Systems.. 468
    Tabish Nazir Mir, Faysal Hardan, Masood Hajian, Tamer Kamel, Pietro Tricoli

Parameter Sensitivity of a MRAS-Based Sensorless Control for AFPMSM Considering Speed
Accuracy and Dynamic Response at Multiple Parameter Variations... 474
    Michael Brüns, Christian Rudolph, Tankred Müller

Synchronization Stability of a Grid Forming Converter Under the Effect of Current Limit in
Voltage Dips with VI Based Current Limiting Method: Analysis and Solution 484
    Siam Hasan Khan, Markel Zubiaga Lazkano, Pedro Izurza, Alain Sanchez-Ruiz, Javier Cañas
    Aceña, Joseba Arza

Analytic Calculation of Touch and Leakage Currents of Non-Isolated EV Chargers using a Fast
Common Mode Calculation Method and Non-Ideal Passive Component Models 493
    Christian Stutz, Sebastian Nielebock, Martin März

Triple-Phase-Shift Controlled Dual Active Bridge Converter with Variable Input Voltage in
Auxiliary Railway Supply 504
    Martin Scohier, Olivier Deblecker, Carlos Valderrama

Loss Characterization Methodology for Soft Magnetic Nano-Crystalline Tape Materials in Coupled
Inductors.. 514
    David Bohne, Valentin Wagner, Patrick Deck, Christian P. Dick

Substitution of Nanocrystalline Toroid by Laminated Ferrite Toroid in the Application of a
Common-Mode Choke 525
    Lukas Reißenweber, Fritz Wohlrath, Alexander Stadler

Direct Active Stabilization of the DC-Link in Voltage-Source Converters 534
    Matthieu Bertin, Mohamad Koteich

Hardware-In-The-Loop Control of a Modular Induction Motor Drive in Power Electronics Education........544
  Jens Peter Kaerst

Design and Efficiency Analysis of an LCL Capacitive Power Transfer System with Load-Independent ZPA........554
  Francesco Musolino, Ahmed Abdullah, Mario Pavone, Fabio Ferreyra, Paolo Crovetti

A Pulse Generator Based on Transmission Line Transformer for Insulation Aging Test........562
  Xiao Yu, Khanh-Hung Nguyen, Peter Zacharias

Design of a Single-Phase Common Mode and Differential Mode Inductor for Interleaved Converters........572
  Jonathan Robinson, Gopal Mondal, Stefan Hänsel, Matthias Neumeister

Steady-State Analysis and Comparison of SSFB, SDFB and DSFB MMC-Based STATCOM........582
  Mohamed Moez Belhaouane, Pierre Vermeerch, François Gruson, Pierre Rault, Sébastien Dennetiere, Xavier Guillaud

Current Distribution Control in Parallel Connected Power Converters with Continuous Output Voltage........593
  Sabrina Ulmer, Andreas Brunner, Philipp Czerwenka, Gernot Schullerus, Ertugrul Sönmez

Optimized Pulse Pattern with Half-Wave Symmetry for 5-Level Converter........604
  Jonas Weires, Pedro Leal Dos Santos, Steven Liu

Characterization of Si-IGBT Crosstalk with a Concentration on Power Circuit Parasitic Elements and the Device Operation Point........614
  Amir Azam Rajabian, Sadegh Mohsenzade, Javad Naghibi, Kamyar Mehran

Impact of Higher Current Harmonics on Component Current Stress and Conduction Losses of Half-Bridge-Series-Resonant-Converters in Discontinuous Conduction Mode for High-Power Applications........624
  Daniel Haake, Anton Grodnichev, Fabian Schnabel, Marco Jung

Control of a Zero-Voltage Switching Isolated Series-Resonant Power Circuit for Direct 3-Phase AC to DC Conversion........634
  Yusuf Kosesoy, Remco Bonten, Henk Huisman, Jan Schellekens

Design of a Robust Voltage Control for Inverters with LC Filter Based on the Internal Model Control........641
  Frederik Stallmann, Axel Mertens, Lukas Fräger

Influence of Power Semiconductor Device Variations on Pulse Shape of Nanosecond Pulses in a Solid-State Linear Transformer Driver........651
  Raffael Risch, Anliang Hu, Jürgen Biela

Optimal Design of Integrated Motor Drives - Comparison of Topologies (2L/3L/Modular), PWM Variants, and Switch Technologies (Si/SiC/GaN)........662
  Thilo Bringezu, Jürgen Biela

Distribution Transformer Voltage Control using a Single-Phase Matrix Converter........673
  Rui Wang, Henk Huisman, Korneel Wijnands

Influence of Carrier-Based PWM Techniques on the Common-Mode Voltage and Common-Mode Current of Six-Phase Full-Bridge Inverters........681
  Juris Arrozy, Esin Ilhan Caarls, Henk Huisman, Jorge L. Duarte, Lorenzo Ceccarelli

Mitigation of Dead-Time Effects on Transient DC Bias Elimination in Dual Active Bridge Link Current 689
MK Kharabela Mohanta, Dipankar De, Silpashree Sahu, Alberto Castellazzi

Generalized Automated Tool for Analysis and Design of Multiphase Coupled Inductor Buck Converters 698
Rana Asad Ali, Mahmoud Shousha, Martin Haug

Experimental Study of a Directly Oil-Cooled Electrical Machine for a Full-Electric Vehicle by using Low Viscosity Oil 709
Huihui Xu, Georg Tobias Götz, Shimin Zhang, Rik W. De Doncker

Development of a Family of High Voltage Gain Step-Up Multi-Port DC-DC Converters for Fuel Cell-Based Hybrid Vehicular Power Systems 719
Pouya Zolfi, Sina Vahid, Ayman El-Refaie

Bidirectional DC Circuit Breaker with Improved Performance During Commissioning and Reclosing 730
Aditya Pogulaguntla, Venkata Raghavendra I, Satish Naik Banavath, Andrii Chub, T Sreekanth, Harish Sarma Krishnamoorthy

Modeling Method for Conducted Noise Flowing in Power Lines of DC/DC Converter 739
Takato Hattori, Wataru Kitagawa, Takaharu Takeshita

High-Bandwidth Power Hardware-In-The-Loop for Motor and Battery Emulation at High Voltage Levels 749
Manuel Fischer, Philipp Kemper, Johannes Herbold, Daniel Epping, Frank Puschmann

Analysis and Discussion of Different Three-Phase dv/dt Filter Topologies and the Influences of Their Filter Parameters on Losses and EMC 758
Eric Fritze, Michael Meissner, Klaus F. Hoffmann, Kai-Uwe Rathjen, Stefan Dickmann, Oliver Woywode

State of Charge Prediction of Lithium-Ion Batteries Based on Artificial Neural Networks and Reduced Data 767
Sebastian Pohlmann, Ali Mashayekh, Dominic Karnehm, Manuel Kuder, Antje Gieraths, Thomas Weyh

Investigation for Condensation Test Condition of HVIGBT Modules 777
Kenji Hatori, Keiichi Nakamura, Wakana Noboru, Nils Soltau, Eugen Wiesner

Three Phase PV Inverter LCOE Optimization Considering Technological Choice 787
Morteza Tadbiri Nooshabadi, Jean-Luc Schanen, Shahrokh Farhangi, Hossein Iman-Eini

Square Wave Operation to Reduce Pulsating Power in Isolated MMC-Based Ultrafast Chargers 798
Ygor Pereira Marca, Maurice G. L. Roes, Jorge L. Duarte, Korneel Wijnands

Surge Current Protection for Railway Traction Applications 805
Michael Gleissner, Mark-M. Bakran

Impedance-Based Analysis of HVDC Converter Control for Robust Stability in AC Power Systems 814
André Schön, Andreas Lorenz, Rodrigo Alonso Alvarez Valenzuela

Class-E Push-Pull Resonance Converter with Load Variation Robustness for Industrial Induction Heating 825
Janus Dybdahl Meinert, Benjamin Futtrup Kjærsgaard, Thore Stig Aunsborg, Asger Bjorn Jorgensen, Stig Munk-Nielsen, Sune Bro Duun

Review of Power Converter Topologies for Electrochemical Impedance Spectroscopy of Lithium-Ion Batteries 833

Hamzeh Beiranvand, Julius M. Placzek, Marco Liserre, Giorgia Zampardi, Doriano Constantino Brogioli, Fabio La Mantia

Design and Experimental Validation of a Voltage Sensing-Current Cancellation Common Mode Linear Active Filter 843

B. Mohamed Nassurdine, PE Lévy, D. Labrousse, JL Schanen, X. Maynard, S. Carcouet

Partial Discharges of Insulated Wires Under Impulses from Wide Bandgap Power Electronics 854

Ting Helmholdt-Zhu, Vivien Grau, Urs Obernolte

Analysis of a Droop-Based Power Controller for Three-Phase Microgrids 865

Andrea Lauri, Hossein Abedini, Davide Biadene, Tommaso Caldognetto, Paolo Mattavelli

Efficiently Paralleling GaN-Transistors for High Current and High Frequency Applications using a Butterfly Layout 873

Martin Wattenberg, Oscar Lorenz, Juan Sanchez

Data-Driven Decentralized Volt/Var Control for Smart PV Inverters in Distribution Systems 883

Yizhou Lu, Qianwen Xu, Lars Nordström

Study of Current Ripple Generators for Accelerated Ageing of Capacitors 891

Robert Keilmann, Hendrik Schefer, Regine Mallwitz

Intra-Arm Balancing Control of Cascaded Multi-Port Converter for Whole Power Unbalance Conditions 902

Takumi Yasuda, Jun-Ichi Itoh

Investigation of Creepage Distances on Printed Circuit Boards for Avionic Applications 912

Hendrik Schefer, Zhongqing Xu, Tobias Kopp, Regine Mallwitz, Michael Kurrat

A 20 kW, 3-Level Flying Capacitor 1500 V Inverter with Characterized GaN Devices for Grid-Tie Applications 922

Van Sang Nguyen, Anthony Bier, Hajar Es-Seghier, Ulrich Soupremanien, Gérard Delette, Stephane Catellani

New Analytical Model for Calculating HF-Losses in Litz Wire Regions Located Outside the E/U-CoreWindow of Transformers 933

Qingchao Meng, Jürgen Biela

Fast and Accurate Soft-Switching and Hard-Switching Losses Estimation for Power Converter, Application to the Dual Active Bridge (DAB) Converter 944

Francois Boige, Nicolas Videau, Adel Ziani, Bruno Guerrero, Julien Laclaverie

Influence of an Electrical Machine on the Dimension and Packaging of Multi-Machine Systems 952

Thomas Stöckl, Hans-Georg Herzog

Design of a Serial Impingement Cooling Heatsink for a 30 kW PV String Inverter 960

Paul Bruyere, Guillaume Piquet Boisson, Gaëtan Perez

Online Junction Temperature Measurement of SiC-MOSFETs via Gate Impedance using the Gate-Signal Injection Method 971

David Hirning, Luca Bauer, Johannes Ruthardt, Jörg Haarer, Philipp Ziegler, Jörg Roth-Stielow

Powercycling Test Bench with Realistic Loss Distribution and Temperature Ripples 980
Till-Mathis Plötz, Jan Fuhrmann, Hans-Günter Eckel

Design, Implementation and Characterization of an Integrated Current Sensing in GaN HEMT
Device by using the Current-Mirroring Technique 990
*Van-Sang Nguyen, René Escoffier, Stéphane Catellani, Murielle Fayolle-Lecocq, Jérémy
Martin*

GaN-Based Modular Multilevel Converter for Low-Voltage Grid Enables High Efficiency 999
Philip Kiehnle, Patrick Himmelmann, Marc Hiller

Energy Management of Smart Homes with Electric Vehicles using Deep Reinforcement Learning............. 1006
Xavier Weiss, Qianwen Xu, Lars Nordström

Simple and Low-Computational Losses Modeling for Efficiency Enhancement of Differential
Inverters with High Accuracy at Different Modulation Schemes.. 1015
Ahmed Shawky, Mokhtar Aly, Emad M. Ahmed, Samir Kouro, José Rodriguez

Estimation of Battery Parameters in Cascaded Half-Bridge Converters with Reduced Voltage
Sensors 1025
Nima Tashakor, Bita Arabsalmanabadi, Elham Hosseini, Kamal Al-Haddad, Stefan Goetz

Method to Analyze the Influence of Switching Behavior in Hard Switching Half Bridge Topologies
for Traction Application 1036
Dominik Nehmer, Michael Gleissner, Lukas Bergmann, Mark-M. Bakran

Impact of Aluminum Casing on High-Frequency Transformer Leakage Inductance and AC
Resistance.. 1046
*Reda Bakri, Xavier Margueron, Wendell Da Cunha Alves, Xavier Cimetiere, Frédéric Gillon,
Antoine Bruyere, Lucian Vatamanu*

Neural Networks-Generalized Predictive Control for MIMO Grid-Connected Z-Source Inverter
Model 1056
Navid Salehi, Herminio Martinez-Garcia, Guillermo Velasco-Quesada

Voltage Estimation for Diode-Clamped MMCs Based on a Simplified Neural Network 1064
Nima Tashakor, Davood Keshavarzi, Shady Banana, Stefan Goetz

A Non-Cooperative Game-Theoretic Distributed Control Approach for Power Quality
Compensators 1074
*Claudio Burgos-Mellado, Victor Bucarey, Helmo K. Morales-Paredes, Diego Muñoz-
Carpintero*

A Comparative Analysis of Power Converter Topologies for Integration of Modular Batteries in
Electric Vehicles... 1083
*Alberto Cárcamo, Aitor Vázquez, Alberto Rodriguez, Diego G. Lamar, Marta M. Hernando,
Daniel Remón*

Design of a High-Dynamic Test Bench for Accelerated Dielectric Lifetime Testing with Adjustable
Voltage Slopes and Temperatures 1094
Hendrik Schefer, Lucas Hanisch, Tim-Hendrik Dietrich, Regine Mallwitz, Markus Henke

Novel Modulation Method for Common-Mode Noise Reduction in Solid-State Transformer Based
on ISOP Configuration 1104
Naoto Kikuchi, Hiroki Watanabe, Keisuke Kusaka, Jun-Ichi Itoh

Modular STATCOM for Compensation of Reactive Power and Voltage Asymmetry in Medium-Voltage Distribution Power Grids 1114

Josef Štengl, Tomáš Kormska, Jakub Talla, Zdenek Peroutka

Novel Method for Active Short Circuit (ASC) Tests of Power Module in Automotive Traction Application 1121

Tobias Appel, Arne Bieler

Short Circuit Performance and Current Limiting Mode of a Monolithically Integrated SiC Circuit Breaker for DC Applications Up to 800 V 1128

Norman Boettcher, Taro Takamori, Keiji Wada, Wataru Saito, Shin-Ichi Nishizawa, Tobias Erlbacher

Application of a HV Bipolar Square-Wave Voltage Generator for Qualification and Assessment of Energy Equipment 1137

Rico Fischer-Baeumer, Kai Gohrmann, Konrad Domes, Benjamin Sahan, Christian Staubach

A Decentralized and Communication-Free Control Algorithm of DC Microgrids for the Electrification of Rural Africa 1147

Lucas Richard, David Frey, Marie-Cecile Alvarez-Herault, Bertrand Raison

Universal Real-Time Model for Active Rectifiers in Versatile Totem-Pole PFC Configurations 1157

Axel Kiffe, Thorben Hoffstadt

Investigation of Core-Loss Mechanisms in Large-Scale Ferrite Cores for High-Frequency Applications 1167

Michael Baumann, Christoph Drexler, Jonas Pfeiffer, Jens Schueltzke, Erwin Lorenz, Michael Schmidhuber

Generation of Methodology for Making Benchmark Microgrids and Application in ESUSCON Microgrid 1177

Oscar Dorner, Patricio Mendoza-Araya

An Overview of Grid-Connection Requirements for Converters and Their Impact on Grid-Forming Control 1187

Paul Imgart, Mebtu Beza, Massimo Bongiorno, Jan R. Svensson

Modular Battery-Integrated Power Electronics-Modelling, Advantages, and Challenges 1197

Nima Tashakor, Jan Kacetl, Tomas Kacetl, Stefan Goetz

Design of Triple-Active Bridge Converter with Inherently Decoupled Power Flows 1207

Dong-Uk Kim, Byengjoo Byen, Byunghwang Jeong, Sungmin Kim

Application of a Multi-Winding Magnetic Component Characterization Method to Optimize Cross-Regulation Performances in DCM Flyback Converters 1216

Denis Motte-Michellon, Brahim Ramdane, Yves Lembeye, Bruno Cogitore

Application of an Electrostatic Machine in a Low-Voltage Microgrid 1226

Gabriel Ramos Huerta, Patricio Mendoza-Araya

Influences of Parasitic Capacitances in Wide Bandwidth Rogowski Coils for Commutation Current Measurement 1237

Philipp Ziegler, Tobias Festerling, Jorg Haarer, Philipp Marx, David Hirning, Jorg Roth-Stielow

Systematic Analysis of Oscillations in DC-Links of Fast Switching Power Electronics 1247

Tobias Fricke, Regine Mallwitz

EMI Mitigation Induced by an IGBT Driver Based on a Controlled Gate Current Profile 1256
Daniel S. Martinez-Padron, Nicolas Patin, Eric Monmasson

An Accurate and Fast Model of Three-Level Three-Phase Dual-Active Bridge Converters in Real-Time Simulation 1266
Ming Jia, Philipp Joebges, Rik W. De Doncker

A Calorimetric and Electrical Method for Measuring Loss Energies of Half-Bridges 1277
Jörg Haarer, Mattea Eckstein, Philipp Ziegler, Philipp Marx, David Hirning, Jörg Roth-Stielow

Condition Monitoring Approach of a SiC Power Semiconductor using Turn-Off Delay with an Integration in a SiC Driver 1286
Victor Golev, Ulf Schümann, Rando Raßmann, Jan Bockholt

Measurement Results of Multilevel Hysteresis Control for Paralleled Two-Level Converters 1294
Magdalena Gierschner, Yves Hein, Hans-Günter Eckel, Christian Heien

Design and Development of a Short-Circuit Test Bench for Low-Voltage Direct Current Protection Devices 1300
Simon Ravyts, Thomas Vandenbussche, Koen Stul, Jan Cappelle

A Novel Modified-TOGI Based PLL for the Three-Phase Unbalanced and Distorted Grid Conditions 1309
Khanh-Hung Nguyen, Ahmad Ali Nazeri, Xiao Yu, Peter Zacharias

Comparison of Two and Three-Level AC-DC Rectifier Semiconductor Losses with SiC MOSFETs Considering Reverse Conduction 1319
Guangyao Yu, Thiago Batista Soeiro, Jianning Dong, Pavol Bauer

Measurement Method for Simple Determination of Sinusoidal Large Signal Losses in Inductive Components 1328
Peter Zacharias, Alejandro Aganza-Torres

A Novel Technique for the Suppression of the Displacement Current Through Power Module Base-Plate Capacitance 1336
Mahmoud Saeidi, Ahmad Ali Nazeri, Rufad Zilic, Peter Zacharias

Analysis and Implementation of Effective Placement of EMC Capacitors for WBG Modules 1343
Mahmoud Saeidi, Ahmad Ali Nazeri, Firas Jenhani, Peter Zacharias

Power Hardware-In-The-Loop Verification of a Cold Load Pickup Scenario for a Bottom-Up Black Start of an Inverter-Dominated Microgrid 1350
Mina Mirzadeh, Robin Strunk, Tobias Erckrath, Axel Mertens

Detection of Incipient Inter-Turn Short-Circuit Faults by Artificial Intelligence Classifiers 1361
Osman Örgüt, Ilker Sahin, Ece Olcay Günes

Modeling the Impact of Grid-Forming E-STATCOMs on Inter-Area System Oscillations 1371
A. Bolzoni, N. Johansson, J. P. Hasler

Combining Schwarz-Christoffel Mappings and Biot-Savart Law to Calculate the High-Frequency Current Distribution Inside a Single Slot 1381
Torben Fricke, Phil Leon Pickert, Babette Schwarz, Bernd Ponick

Standardised Switching Cell Building Block for Converter Design Optimisation with Detailed
Electro-Thermal Model 1391
Georgios Papadopoulos, Jürgen Biela

Design Procedure for Transformer-Based Solid-State Pulse Modulators with Damping Network 1402
Spyridon Stathis, Juergen Biela

DC Bias Impact on Magnetic Core Losses at High Frequency 1413
Bima Nugraha Sanusi, Ziwei Ouyang

Investigation of the Short-Circuit Type II Safe Operating Area of IGBTs....................................... 1424
Madhu Lakshman Mysore, Mohamed Alaluss, Abhishek Maitra, Thomas Basler, Roman
Baburske, Franz-Josef Niedernostheide, Hans-Joachim Schulze

Single Transformer, MMC Based MV Power Electronic Traction Transformer 1434
Simon Fuchs, Simon Beck, Jürgen Biela

A New Power MOSFET Technology Achieves a Further Milestone in Efficiency 1445
Ralf Siemieniec, Michael Hutzler, Cesar Braz, Tomasz Naeve, Elias Pree, Heimo Hofer,
Ingmar Neumann, David Laforet

Experimental Evaluation of Battery Impedance and Submodule Loss Distribution for Battery
Integrated Modular Multilevel Converters 1456
Arvind Balachandran, Tomas Jonsson, Lars Eriksson, Anders Larsson

Constant DC Power Infeed Grid Forming with Improved Ability to Ride-Through Unbalanced
Low-Voltage Faults.. 1466
Tayssir Hassan, Malte Eggers, Huoming Yang, Peter Teske, Sibylle Dieckerhoff

Constrained Long-Horizon Direct Model Predictive Control for Grid-Connected Converters with
LCL Filters 1476
Mattia Rossi, Petros Karamanakos, Francesco Castelli-Dezza

Performance Evaluation of SiC-Based Isolated Bidirectional DC/DC Converters for Electric
Vehicle Charging.. 1486
Kaushik Naresh Kumar, Rafal Miskiewicz, Przemyslaw Trochimiuk, Jacek Rabkowski,
Dimosthenis Peftitsis

Impact of Threshold Voltage Shifting on Junction Temperature Sensing in GaN HEMTs........... 1497
Burhan Etoz, Jose Ortiz Gonzalez, Arkadeep Deb, Saeed Jahdi, Olayiwola Alatise

Comparison of Power Cycling Results of Discrete GaN Cascodes for Automotive Power
Electronics with High Temperature Swings 1506
Florian Lippold, Philipp Hauenschild, Regine Mallwitz

Current Distortion Study for Hybrid Multi-Level Grid Inverter with Active Neutral-Point-Clamped
4-Leg Topology 1515
Jonas Steffen, Matthias Klee, Fabian Schnabel, Axel Seibel, Marco Jung

Dynamic Maximum Power Point Tracking Method Including Detection of Varying Partial Shading
Conditions for Photovoltaic Systems 1525
Rosalie Rouphael, Nezha Maamri, Jean-Paul Gaubert

Novel Operation Mode of the Modular Multilevel Matrix Converter Based on a Dimensioning
Algorithm 1533
Rebecca Dierks, Axel Mertens

On the Cosmic Ray Influence on the Electronics Design of a High Altitude Electric Aircraft 1543
    Philippe Morey, Mauro Carpita

DC-Bus Control Considerations of Asymmetrical Multilevel Inverters with Embedded Buck-Boost Converter 1551
    Theodoros P. Mouselinos, Emmanuel C. Tatakis

A Seamless Modulation Strategy for Step-Up/Down Partial Power Processing Converter (SUD-P3C) 1561
    Chao Liu, Zhe Zhang, Ziwei Ouyang, Jiasheng Huang, Michael A. E. Andersen, Tiberiu Gabriel Zsurzsan

Performances Analysis of Non-Model-Based Speed Estimation Algorithms for Motor Drives 1569
    Gaetano Turrisi, Luigi Danilo Tornello, Giacomo Scelba, Giulio De Donato, Giuseppe Scarcella

A Method to Design Power Control System of Wayside Energy Storage System for Energy Saving in DC-Electrified Railway 1580
    Kota Sato, Keiichiro Kondo, Hiroyasu Kobayashi, Makoto Chida

A Reconfigurable Single-Stage Three-Phase Electric Vehicle DC Fast Charger Compatible with Both 400V and 800V Automotive Battery Packs 1590
    Mojtaba Forouzesh, Yan-Fei Liu, Paresh C. Sen

Efficiency Improvement of Single-Stage AC-DC LLC Converter using a Line Cycle Synchronous Rectifier (SR) Driving Strategy 1601
    Mojtaba Forouzesh, Yan-Fei Liu, Paresh C. Sen

Influence of DC Supply Voltage Unbalances on the Performance of ARCP Inverters 1611
    Gholamreza Tabrizi, Sebastian Sprunck, Marco Jung

Grid-Forming Control for Enhanced Microgrid Interconnection 1620
    Tobias Erckrath, Christian Bendfeld, Peter Unruh, Axel Seibel, Marco Jung

Low Phase Shift Filter for Current Sensing Based on the Difference Between AC Machine Models with and Without Iron Losses 1631
    Niklas Himker, Marcel Krümpelmann, Axel Mertens

Design and Analysis of a Voltage Clamping Active Delay Control Method for Series Connected SiC MOSFETs 1641
    Rui Wang, Asger Bjørn Jørgensen, Hongbo Zhao, Stig Munk-Nielsen

Practical Implementation of a Concept for In-Situ Detection of Humidity-Related Degradation of IGBT Modules 1649
    Benedikt Kostka, Axel Mertens

Design for Enhanced Noise Immunity of PCB Coils Used for Sensing Current Through Power Devices 1658
    Aamir Rafiq, Sumit Pramanick

Measurement Principle for Measuring High Frequency Bearing Currents in Electric Machines and Drive Systems 1665
    Benjamin Knebusch, Lennart Junemann, Pauline Holtje, Axel Mertens, Bernd Ponick

Climatically Induced Insulation Degradation in Power Semiconductor Modules of Wind Turbines 1674
    Timo Lichtenstein, Sören Fröhling, Bernd Tegtmeier, Katharina Fischer

Comparison of Magnetic Noise Compensation Techniques for Dual Three-Phase Electrically Excited Synchronous Machines.. 1684
Jonas Henkenjohann, Jan Andresen, Axel Mertens

PCB Technology Comparison Enabling a 900V SiC MOSFET Half Bridge Design for Automotive Traction Inverters 1692
Matthias Spieler, Che-Wei Chang, Ayman El-Refaie, Muhammad H Alvi, Dong Dong, Rolando Burgos

Desaturated Turn-Off of Low-Saturation IGBTs with Clamping Method to Reduce Turn-Off Energy Losses.. 1703
Vishwas Acharya Nayampalli, Hans-Günter Eckel

Impact of Bond Wire Configuration on the Power Cycling Capability of Discrete SiC-MOSFET Devices 1713
Patrick Heimler, Nick Thönelt, Josef Lutz, Thomas Basler

A Low-Leakage, Low-Loss Magnetic Transformer Structure for High-Frequency Applications.................. 1722
Allen Nguyen, Ajinkya Phanse, Michael Solomentsev, Alex J. Hanson

Temperature Distribution of an IGBT Chip During Repetitive Switching Events Under Consideration of Front-Side Ageing.. 1733
Christian Bäumler, Bo Zhang, Maximilian Goller, Xing Liu, Thomas Basler

Boosting Pilot-Diode Reverse-Conducting IGBTs Turn-ON and Reverse-Recovery Losses with a Simple Gate-Control Technique... 1744
Daniel Lexow, Hans-Günter Eckel

Modeling of an Interleaved DC-DC Boost Converter for a Direct Model Predictive Control Strategy.. 1754
Thomas Effenberger, Hannes Böorngen, Eyke Liegmann, Michael Hoerner, Petros Karamanakos, Ralph Kennel

Static Analysis and Control Strategies of the Single Active Bridge Converter 1765
Alexis A. Gómez, Alberto Rodríguez, Marta M. Hernando, Diego G. Lamar, Javier Sebastián, Ibán Ayarzaguena, Jose Manuel Bermejo, Igor Larrazabal, David Ortega, Francisco Vázquez

Multi-Port Inductive Power Transfer System Considering Charging Auxiliary Battery in EVs................... 1776
Zhuoqi Zhang, Ryosuke Ota, Ryohei Okada, Nobukazu Hoshi

Influence of IGBT and Diode Parameters on the Current Sharing and Switching-Waveform Characteristics of Parallel-Connected Power Modules.. 1785
Y. Ando, J. Sakai, K. Hatori, N. Soltau, E. Wiesner

Innovative Driving Scheme for Electrical Generators in More Electric Aircrafts Employing Series Active Filtering... 1796
Nena Apostolidou, Nick Papanikolaou

Field-Measurement Based Hygrothermal Modelling of the Converter-Cabinet Climate in Wind Turbines... 1804
Katharina Fischer, Katherina Gohler

A Multi-Mode Control Based Asymmetrical Dual-Active-Bridge Series-Resonant DC-DC Converter (DABSRC) 1815
M. Yaqoob, Grover Torrico, Wang Shuqin

Extended Balancing and Dimensioning of Capacitors in MMC Double Submodules 1824
Ali Sharaf Addin, Christopher Dahmen, Thomas Brückner

Saliency Extraction and Torque Sharing Estimation of Dual Motor Drive using Special Current
Sensor Configuration 1834
E. Rodriguez Montero, M. Vogelsberger, T. Wolbank

Soft-Switching Converter for Inductive Power Transfer System with Double-Sided LCC Resonant
Network 1844
Ryohei Okada, Ryosuke Ota, Nobukazu Hoshi

Ultra Low Loss - MMC Submodules Favorable for SiC-FET Enabling High Functional Safety 1855
Christopher Dahmen, Rainer Marquardt

Control of an Active Gate Driver for an Electric Vehicle Traction Inverter using Artificial Neural
Networks 1865
Julius Wiesemann, Jacob Dumtzlaff, Axel Mertens

Cascaded H-Bridge Converter Designs for Future Short-Range All-Electric Aircraft Propulsion 1875
Maximilian Hagedorn, Malte Lorenz, Axel Mertens

Overview and Evaluation of Energy Balancing Techniques for MMCs with Various Input and
Output Frequencies 1885
Gyanendra Kumar Sah, Michael Schütt, Hans-Günter Eckel

Comparative Lifetime Estimations for IGBT Modules in Wind Turbine Converters 1895
Christian Neumann, Hans-Gunter Eckel

Single-Phase, Five-Level Inverter with SPWM-Based Neutral Point Voltage Balancing Scheme 1906
Dmytro Kondratenko, Arkadiusz Lewicki, Charles Odeh

Magnetic Core Evaluation Kit for the Comparison of Core Losses 1914
Wilmar Martinez, Xiaobing Shen, Siqi Lin, Jens Friebe

Multi-Objective Optimization of Modular Multilevel Converter Systems 1923
Nikolaus Patzelt, Christian Schlegel, Michail Vasiladiotis

Sizing of Hybrid Energy Storage System for Residential PV Applications 1933
Xiangqiang Wu, Zhongting Tang, Tamas Kerekes

DC Bias Currents in Full-Bridge DC-DC Converters in Context of WBG Semiconductors and High
Switching Frequencies 1939
Niklas Badenhop, Lukas Fräger, Dennis Kampen, Sascha Langfermann, Michael Owzareck

Parameter Tuning Method for Class Φ_2 Converters for High-Frequency Wireless Power Transfer
Applications 1947
Yining Liu, Prasad Jayathurathnage, Jorma Kyyrä

Inductor Design Optimization using FEA Supervised Machine Learning 1955
D. Cajander, I. Viarouge, P. Viarouge, D. Aguglia

Enabling Large-Scaled MMC EMT-RMS Co-Simulation by Data Exchange in the Loop (DXiL) 1966
Xiong Xiao, Soham Choudhury, Martin Coumont, Jutta Hanson

Advanced Low-Voltage System-In-Package Half-Bridge MOSFET with Added Protection Features 1975
S. Musumeci, V. Barba, F. Scrimizzi, C. Mistretta

Evaluation of Common-Mode Leakage Current of Aalborg-Type Transformerless PV Inverters.............. 1985
Georgios I. Orfanoudakis, Eftychios Koutroulis, Georgios Foteinopoulos, Weimin Wu

Multi-Frequency Traction-To-Auxiliary Integrated EV Drivetrain: Eliminating the Need for an
Auxiliary Power Module 1995
Caniggia Viana, Mehanathan Pathmanathan, Peter W. Lehn

Potentials to Improve the Post-Fault Performance of a Fault-Tolerant Inverter System in Electrified
Aircraft Propulsion System 2003
Yongtao Cao, Leon Fauth, Jens Friebe, Axel Mertens

Model Predictive Control-Enabled Fault Ride Through Operation Strategy for High Power Wind
Turbine 2011
Pedro Catalán, Yanbo Wang, Zhe Chen, Joseba Arza

A Theoretical Comparison of Different Virtual Synchronous Generator Implementations on
Inverters 2021
Patrick Körner, Andrea Reindl, Hans Meier, Michael Niemetz

Linear Flux-Switching Machine Design - A Multiobjective Optimization 2030
Hendrik Marks, Henning Schillingmann, Sridhar Balasubramanian, Markus Henke

Single-Arm MMC-Based Converter for Transformerless Rail Interties.. 2038
Simon Beck, Simon Fuchs, Jürgen Biela

Medium Voltage Diode Rectifier Design for High Step-Up DC-DC Converter 2049
Pierre Le Métayer, Cyril Buttay, Drazen Dujic, Piotr Dworakowski

Fast Switching Planar Inductance Current Source ZETA Converter with Integrated Common Mode
Filter 2058
Benjamin H. Zacher, Christian Schumann

System Level Simulation of Moisture Propagation and Effects in Wind Power Converters.......... 2066
Johannes C. Wenzel, Axel Mertens

PWM-Based Optimization-Free Active Voltage-Balancing Control of 7-Level Active Neutral-
Point-Clamped Flying-Capacitor Multicell Inverters 2073
Vahid Dargahi

Model Predictive Power Sharing Algorithm for Fuel Cell Integration in a Dual Inverter Electric
Vehicle Drivetrain 2084
Mehanathan Pathmanathan, Caniggia Viana, Sukhjit Singh, Peter W. Lehn

Comparative Evaluation of the 5-Phase Vienna and the 5-Phase PWM Rectifiers Under DC
Voltage Control 2092
A. Dieng

Modelling and Control of a 50kW SiC-Based Isolated DAB Converter for Off-Board Chargers of
Electric Vehicles... 2101
*Haaris Rasool, Manh Tuan Tran, Sajib Chakraborty, Joeri Van Mierlo, Thomas Geury,
Mohamed El Baghdadi, Omar Hegazy*

Impact of Cyber Attacks on Cost Oriented Power Routing Schemes in Microgrids..................... 2110
Kirti Gupta, Subham Sahoo, Bijaya Ketan Panigrahi, Frede Blaabjerg

Response of IGBT Chip Characteristics Due to Critical Stress.. 2119
Kohei Yamauchi, Rik W. De Doncker

Mega-Hertz High-Power WPT System with Parallel-Connected Inverters using Current Balance Circuit 2127

Masamichi Yamaguchi, Keisuke Kusaka, Jun-Ichi Itoh

Investigation and Mitigation of Common-Mode Voltage in Four-Level NPC Converters Modulated by Redundant Level Modulation 2136

Jun Wang, Wei Xu, Xibo Yuan, Lihong Xie

Ferrite Optimization for a Three-Phase Wireless Power Transfer System for Electric Vehicles 2145

Shuang Nie, Mehanathan Pathmanathan, Peter W. Lehn

Frequency and Modulation Index Related Effects in Continuous and Discontinuous Modulated Y-Inverter for Motor-Drive Applications 2156

Hamzeh J. Jaber, Alberto Castellazzi

Performance Evaluation of Sinusoidal-Flux Reluctance Machine for Improving Power Density with Reduced Torque and Input-Current Ripples 2164

Kiwa Nagayasu, Masaki Iida, Kazuhiro Umetani, Mastaka Ishihara, Eiji Hiraki

Power Hardware-In-The-Loop Test of Low-Voltage Battery for a Plug-In Hybrid Electric Vehicle 2175

Ronan German, Florian Tournez, Alain Bouscayrol, Aurelien Lievre, Betty Lemaire-Semail

Stability Analysis of DFIG System Connected with High-Frequency Capacitive Grid Based on Closed-Loop Current Control and Direct Power Control 2182

Bin Hu, Heng Nian, Subham Sahoo, Frede Blaabjerg, Yaqian Zhang, Zixiao Xu

Full-Bridge Modular Multilevel Converter for the Four-Quadrant Supply of High Power Magnets in Particle Accelerators 2189

Manuel Colmenero, Ricardo Vidal-Albalate, Francisco R. Blanquez, Ramon Blasco-Gimenez

Deep Neural Network for Magnetic Core Loss Estimation using the MagNet Experimental Database 2197

Xiaobing Shen, Hans Wouters, Wilmar Martinez

Hybrid Circuit Board Structure for Power Electronics 2205

Gerrit Braun, Deniz-Heinz Moldenhauer

Active Control of Gear Mesh Vibration using a Permanent-Magnet Synchronous Motor and Simultaneous Equation Method 2211

Dominik Reitmeier

Research Laboratory for Testing Grid Connected Devices Under Grid Voltage / Grid Impedance Variations and Microgrid Conditions 2219

Swen Bosch, Jochen Staiger, Heinrich Steinhart

Reducing the Impact of Skin Effect Induced Measurement Errors in M-Shunts by Deliberate Field Coupling 2230

Hauke Lutzen, Jonas Müller, Vladimir Polezhaev, Till Huesgen, Nando Kaminski

Grid Forming Control for HVDC Systems: Opportunities and Challenges 2241

Adil Abdalrahman, Ying-Jiang Häfner, Malaya Kumar Sahu, Khirod Kumar Nayak, Ashkan Nami

A Highly Integrated and Modular High Speed Electric Drive for Lightweight Electric Mountain Bikes 2251

Matthias Hofer, Mario Nikowitz, Manfred Schrödl

Performance Enhancement of Power Conditioning Systems in More Electric Aircrafts 2257
Nick Rigogiannis, Nick Papanikolaou, Yongheng Yang

Steady State Simulations of a Hybrid HVAC/HVDC Network using OS Based ARM Devices 2266
Ioan Catalin Damian, Mircea Eremia

Experimental Comparison of FPGA-Implemented Model Predictive Voltage Control to Cascaded
Proportional Resonant Control for a Three-Phase Four-Wire Three-Level Grid-Forming Inverter of
250 kVA 2276
Jarren Lange, Dominik Schmies, Karl Stephan Stille, Joachim Böcker, Oliver Wallscheid

Experimental Study of Interleaved Y-Inverter Performance 2285
Yusuke Endo, Masataka Minami, Hamzeh J. Jaber, Alberto Castellazzi

Design of a GaN-Based Reconfigurable Resonant Converter for High Frequency On-Board
Charger of Battery Electric Vehicles 2293
*Manh Tuan Tran, Haaris Rasool, Dai Duong Tran, Mohamed El Baghdadi, Philippe Lataire,
Omar Hegazy*

Transient Liquid Phase Bond Reliability Evaluation of Die-Attach for Power Module Packaging 2301
Laxma R. Billa, Yangang Wang, Thomas Grant, Xiang Li, Harley Neal, Muhammad Morshed

Experimental Evaluation on Observer-Based Delay-Compensating Active Damping for LC-Filters 2308
Michael Schütt, Hans-Günter Eckel

Influence of Static Rotor Imbalance on the Roller Bearing Damage Due to Inverter-Induced
Bearing Currents 2316
Martin Weicker, Omid Safdarzadeh, Andreas Binder

Novel Current Balancing Method for HF Interleaved Converters with Reduced Control Effort 2327
Christian Beckemeier, Jens Friebe

dV/dt-Based Filter Design for Motor Inverters with Continuous Output Voltage 2334
Sabrina Ulmer, Stevan Bugarski, Gernot Schullerus, Ertugrul Sönmez

Evaluation of Core Losses in Transformers for Three-Phase Multi-Level DAB Converters 2344
Babak Khanzadeh, Yuriy Serdyuk, Torbjörn Thiringer

A Quasi-Offline Condition Monitoring Method of DC-Link Capacitor Banks in Accelerator Power
Converters 2355
*Timm Felix Baumann, Konstantinos Papastergiou, Raul Murillo Garcia, Dimosthenis
Peftitsis*

Minimizing Voltage Stress in Auxiliary Resonant Commutated Pole Inverters using Saturable
Inductors 2366
Markus Zocher, Norbert Grass, Ralph Kennel

Adaptive Dead-Time Control in a Resonant Wireless Power Transfer System 2375
Tim Krigar, Martin Pfost

Multilevel Battery Converter with Cascaded H-Bridges on Cell Level-Battery Management System
Or a Renewed Attempt for Power Electronic Building Blocks? 2383
*Max Rothenburger, Markus Horn, Xiao Yu, Gerold Schulze, Koenraad Muyllaert, Peter
Zacharias, Ludwig Brabetz, Hartmut Hillmer*

Design and Potential of EMI cm Chokes with Integrated DM Inductance 2392
Mohammad Ali, Rehnuma Bushra, Jens Friebe, Axel Mertens

Implementation Options of a Fully SiC Buck-CSI for Advanced Motor Drive Application........................ 2402
Yonghwa Lee, Alberto Castellazzi

Optimized Control Scheme to Achieve ZVS for the Complete Pre-Charging Phase of
Supercapacitors with a 500 kHz SiC- And GaN-Based Dual Active Bridge 2413
Patrick Lenzen, Martin Pfost

Fault Blocking Capability in the DC-MMC with Reduced Number of Sub-Modules................................... 2422
J. D. Páez, F. Morel, S. Bacha, P. Dworakowski

An Open-Source FEM Magnetic Toolbox for Calculating Electric and Thermal Behavior of Power
Electronic Magnetic Components 2432
Nikolas Förster, Jonas Hölscher, Till Piepenbrock, Philipp Rehlaender, Oliver Wallscheid,
Frank Schafmeister, Joachim Böcker

Comparison of Dual-Active-Bridge-Based Topologies for Single-Phase Single-Stage EV On-Board
Chargers 2441
Daniel Gaona, Denis Pauls, Eduardo Facanha De Oliveira

Design Concepts for Medium Voltage DC Networks Supplying the Future Circular Collider (FCC)........... 2451
Manuel Colmenero, Francisco R. Blanquez, Ramon Blasco-Gimenez

A Novel Dual CC-CV Output Wireless EV Charger with Minimal Dependency on Both Coil
Coupling and Load Variation 2462
Subhranil Barman, Kishore Chatterjee

A High-Performance EMI Filter Based on Laminated Ferrite Ring Cores 2470
Marcin Kacki, Marek S. Rylko, John G. Hayes, Charles R. Sullivan

Investigation of the Static Performance and Avalanche Reliability of High Voltage 4H-SiC
Merged-PiN-Schottky Diodes 2477
Chengjun Shen, Saeed Jahdi, Phil Mellor, Juefei Yang, Erfan Bashar, Jose Ortiz-Gonzalez,
Olayiwola Alatise

On Chain-Link Based Multi-Port Converters Able to Connect HVDC and MVDC to AC
Transmission Network... 2486
Daniele Falchi, Oriol Gomis-Bellmunt, Eduardo Prieto-Araujo, Olivier Despouys

Voltage Control Scheme for Multilevel Interfacing PV Application: Real-Time MRAC-Based
Approach 2496
Mohammad Sadegh Orfi Yeganeh, Mehdi Rahmani, Nenad Mijatovic, Tomislav Dragicevic,
Frede Blaabjerg, Pooya Davari

Control Principles for Island Operation and Black Start by Offshore Wind Farms Integrating Grid-
Forming Converters... 2504
Daniela Pagnani, Lukasz Kocewiak, Jesper Hjerrild, Frede Blaabjerg, Claus Leth Bak

Experimental Study of the Reduction and Removal of Turn-On Snubber for IGCT Based MMC
Submodule using Fast Silicon Diodes 2515
Arthur Boutry, Cyril Buttay, Besar Asllani, Bruno Lefebvre, Eric Vagnon, Dong Dong

Characterisation of a Ferrite-Polymer Based Magnetic Material 2526
Johan Le Leslé, Guillaume Lefevre, Julien Morand, Rémi Perrin, Pierre-Yves Pichon,
Guillaume Regnat

Model Predictive-Based Control Technique for Fault Ride-Through Capability of VSG-Based Grid-Forming Converter 2537
  Mobina Pouresmaeil, Amir Sepehr, Basit Ali Khan, Jafar Adabi, Edris Pouresmaeil

Grounding Points in HV/MV Hybrid Transformer Auxiliary Converters 2544
  Adrian Wiemer, Jürgen Biela

Non-Parasitic Induced Transient Overvoltage in ANPC Topology Due to Critical Switching Sequences 2554
  Michael Geiss, Robert Kragl, Jürgen Thoma, Benjamin Volzer

Open-Delta SBC: A New Converter Topology with Low Number of Sub-Modules for MV Applications 2564
  D. Lanzarotto, P. B Steckler, K. Vershinin, F. Morel

Characterising the Effect of an Inverter on the Regulation of the AC Voltage using a Frequency Response Identification Technique 2574
  Mohamed Aldarmon, Joan Marc Rodriguez, Adria Junyent-Ferre

Artificial-Intelligence Based DC-DC Converter Efficiency Modelling and Parameters Optimization 2581
  Fanghao Tian, Diego Bernal Cobaleda, Wilmar Martinez

Analysis of the Loss Distribution of a 6 kW Two Stage Power Supply for 600 V DC Applications 2588
  Lukas Fräger, Sascha Langfermann, Michael Owzareck, Dennis Kampen, Jens Friebe

Study on the Gate Loop Design and Its Impact on Switching Characteristics of GaN Transistors 2596
  Xiaomeng Geng, Carsten Kuring, Oliver Hilt, Mihaela Wolf, Joachim Würfl, Sibylle Dieckerhoff

Analysis of Current Sharing in the Parallel Connection of GaN Transistors 2607
  Frederik Stalleicken, Sibylle Dieckerhoff, Karsten Handt, Sebastian Nielebock

Verification of GaN-HEMT Spice Models using an S-Parameters Approach 2618
  Alonso Gutierrez, Nasri Said, Emmanuel Marcault, Mathieu Gavelle

Power Loss Modelling of GaN HEMT-Based 3L-ANPC Three-Phase Inverter for Different PWM Techniques 2628
  Salvatore Mita, Arjun Sujeeth, Giuseppe Aiello, Dario Patti, Francesco Gennaro, Giacomo Scelba, Mario Cacciato

Generalized Core and Winding Area Ratio - Trends for Inductors and Transformers in Power Electronics with High Switching Frequencies 2638
  Siqi Lin, Leon Fauth, Wilmar Martnez, Jens Friebe

Active Substrate Termination of Discrete and Monolithic Bidirectional GaN HEMTs in a T-Type Inverter 2644
  Carsten Kuring, Yannic Lange, Xiaomeng Geng, Oliver Hilt, Mihaela Wolf, Joachim Würfl, Sibylle Dieckerhoff

Transformer Design Optimization and Comparison for a DC-DC Converter Used in PV Micro-Inverters 2655
  Tobias Manthey, Meriem Khader, Jens Friebe

Automated Gate Impedance Network Design for SiC MOSFETs using SPICE Solver Interfaced with MATLAB Environment 2661
  Pawel Piotr Kubulus, Szymon Michal Beczkowski, Stig Munk-Nielsen, Asger Bjørn Jørgensen

An Improved Multi-Loop Resonant and Plug-In Repetitive Control Schemes for Three-Phase Stand-Alone PWM Inverter Supplying Non-Linear Loads 2670
Ahmad Ali Nazeri, Peter Zacharias

High Switching Frequency Operation of a Single-Phase Five-Level Hybrid Active Neutral Point Clamped Inverter with a Model Predictive Control Approach 2682
Mohammad Najjar, Mahdi Shahparasti, Rasool Heydari, Morten Nymand

Design of Planar Coupled Inductor Applied to Zero-Current Switching Clamped Current Converter 2689
Vinicius Freire Bezerra, Tobias Manthey, Montiê Alves Vitorino, Jens Friebe

Characterization of Online Junction Temperature of the SiC Power MOSFET by Combination of Four TSEPs using Neural Network 2698
Kanuj Sharma, Simon Kamm, Kevin Muñoz Barón, Ingmar Kallfass

Novel Extended Robust Disturbance Observer for Improved Cogging Force Compensation in Permanent Magnet Linear Motors 2706
Franz Luckert, Axel Mertens

Improvement of a Self-Powered Gate Driver Power Supply 2715
Mariana Raya, Oriol Aviñó, Sergio Busquets-Monge, Xavier Perpiñá, Miquel Vellvehi, Xavier Jordà

Optimization and Scaling of a Compact High-Power IGCT Capacitor Charger Based on Simulation and Measurements with a 300 kW/3.3 kV Demonstrator 2726
Felix Haag, Fabian Albrecht, Volker Brommer, Oliver Liebfried, Klaus F. Hoffmann

Multilayer Busbars for Medium Voltage ANPC Converter Dedicated to Battery Energy Storage Systems 2736
Mamadou Lamine Beye, Luc Bimmel, Anthony Bier, Jérémy Martin

A Simulation Model for SiC MOSFET Switching Transients Controlled by an Adaptive Gate Driver with the Capability of Reducing Switching Losses and EMI Across the Full Operating Range 2744
Zheming Li, Robert W. Maier, Mark-M. Bakran, Franz-J. Niedernostheide, Daniel Domes

Phase-Shift Modulation for Flying-Capacitor DC-DC Converters 2754
Philipp Rehlaender, Frank Schafmeister, Joachim Böcker

An EV Integrated Isolated DC Charger using a Six-Phase Synchronous Machine 2763
Sukhjit S Ghumman, Mehanathan Pathmanathan, Peter W Lehn

Configurable ISOP-IPOP DC-DC Converter for Universal Solid-State Transformer 2773
Pramod Apte, Jens Friebe, Lukas Fräger

Using System-On-Chip Boards for the Deployment of Controller for Verification and Prototyping 2780
Adeel Jamal, Gerd Griepentrog

Utilizing the Reactive Current Control Capability of an MMC-Fed AC/DC Converter for Volt-Second Balancing in Medium Frequency Transformers 2788
Kaveh Pouresmaeil, Maurice Roes, Jorge Duarte, Korneel Wijnands, Nico Baars, George Papafotiou

Cost Comparison for Different PV-Battery System Architectures Including Power Converter Reliability 2795
Martijn Deckers, Leander Van Cappellen, Glenn Emmers, Fereshteh Poormohammadi, Johan Driesen

Insulation Design and Analysis of a Medium Voltage Planar PCB-Based Power Bus Considering Interconnects and Ancillary Circuit Integration 2806
  Joshua Stewart, Rolando Burgos, Dushan Boroyevich

Modular Multilevel Converter Control with using a General Space Vector PWM Method in Medium Voltage Hydro Power Application............ 2813
  Chengjun Tang, Torbjörn Thiringer

A Technical Overview of Single-Stage Three-Port DC-DC-AC Converters 2824
  Sebastian Neira, Zoe Blatsi, Michael M. C. Merlin, Javier Pereda

Common-Mode EMI Noise Modeling of Three-Level T-Type Inverter for Adjustable Speed Drive Systems............ 2835
  Vefa Karakasli, Abdelmoumin Allioua, Gerd Griepentrog

A Condition Monitoring Scheme for Semiconductor Devices in Modular Multilevel Converters with Cascaded H-Bridge Submodules 2843
  Mohsen Asoodar, Mehrdad Nahalparvari, Christer Danielsson, Hans-Peter Nee

Particular Requirements on Drive Inverters for Safe and Robust Operation on an Open Industrial DC Grid............ 2852
  Simon Puls, Jan-Niklas Koch, Martin Ehlich, Holger Borcherding

Investigation About Operation and Performance of Gate Drivers for Power Electronics Converters for Cryogenic Temperatures............ 2860
  Mustafeez-Ul-Hassan, Yuxuan Wu, Vyacheslav Solovyov, Fang Luo

Synchronization Angle Determination in DVCSFO of DFIM Naval Propulsion............ 2869
  Youssef Drimizi, Maria Pietrzak-David, Pascal Maussion

Power Control of LCR-DAB Converter with Phase Shift in Fixed Switching Frequency 2877
  Seung-Hyuk Baek, Jaehong Lee, Seung-Hwan Lee, Sungmin Kim

A Simplified Braking Method for Direct Matrix Converter-Fed PMSM Drives with Consideration of Avoiding Regenerative Energy 2885
  Jun Xie, Dustin Henneberg, Martin Suberski, Thomas Ellinger, Uwe Radel, Jürgen Petzoldt

Inverter-Machine Parametric Co-Design for Energy Efficient Electric Drives............ 2893
  Jaedon Kwak, Alberto Castellazzi

Bidirectional Cuk Converter in Partial-Power Architecture with Current Mode Control for Battery Energy Storage System in Electric Vehicles 2903
  J. S. Artal-Sevil, J. Anzola, V. Ballestín-Bernad, I. Aizpuru

Design Space Exploration for a Capacitive 36V, 4A, 4:1 DCDC Converter with GaN Switches using a Performance-Cost-Matrix Including Uncommon Topologies............ 2912
  Adrian Gehl, Malte Kempchen, Simon Disselkamp, Markus Olbrich, Bernhard Wicht

A Fast Control for a Three-Switch Multi-Input DC-DC Converter............ 2919
  Simone Cosso, Andrea Formentini, Mario Marchesoni, Massimiliano Passalacqua, Luis Vaccaro

Impact on the Torque and on the Copper Losses Under Fault-Tolerant Control of 5-Phase PMSG............ 2930
  A. Dieng

Weighting Factor Design for FS-MPC in VSCs: A Brain Emotional Learning-Based Approach 2939
Mohammad Sadegh Orfi Yeganeh, Arman Oshnoei, Saeed Peyghami, Nenad Mijatovic, Tomislav Dragicevic, Frede Blaabjerg

A Strategy for Smooth Microgrid Transitions Without Phase Misalignment and Voltage Mismatch 2948
Gabriel Silva Rocha, Amiron Wolff Dos Santos Serra, Cesar Augusto Santana Castelo Branco, Hercules Araujo Oliveira, Jose Gomes De Matos, Luiz Antonio De Souza Ribeiro

Subtle Design and Performance Comparison of WF-FSM and DC-VRM for Large-Scale Direct-Drive Wind Power Generation 2958
Udochukwu B. Akuru, Maarten J. Kamper, Zi-Qiang Zhu

Analysis and Implementation of Different Non-Isolated Partial-Power Processing Architectures Based on the Cuk Converter 2967
J. S. Artal-Sevil, J. Anzola, V. Ballestín-Bernad, J. L. Bernal-Agustín

GaN HEMT and SiC Diode Commutation Cell Based Dual-Buck Single-Phase Inverter with Premagnetized Inductors and Negative Gate Driver Turn-Off Voltage 2977
Tobias Brinker, Hendrik Gräber, Jens Friebe

Determination of Optimal Associated Discrete Circuit Switch Model Parameters for Real-Time Simulation of Dual-Active Bridge Converters 2985
Marija Stevic, Ravinder Venugopal

Integrated Motor Drive: A Multidisciplinary Approach 2996
Betty Lemaire-Semail, Nadir Idir, Eric Semail, Souad Harmand

Hardware in the Loop Test of an Electric Aircraft Powertrain 3005
Sebastian Mönninghoff, Moritz Scholjegerdes, Kay Hameyer

A Multi-Port Smart Transformer for Green Airport Electrification 3014
Giampaolo Buticchi, Giovanni De Carne, Thiago Pereira, Kangan Wang, Xiang Gao, Jiajun Yang, Youngjong Ko, Zhixiang Zou, Marco Liserre

Improvement of EMI Filter Attenuation using Shielding 3022
Mohammad Ali, Rehnuma Bushra, Jens Friebe, Axel Mertens

Implementation of Onsite Junction Temperature Estimation for a SiC MOSFET Module for Condition Monitoring 3031
Farzad Hosseinabadi, Shahid Jaman, Sachin Kumar Bhoi, Md. Mahamudul Hasan, Sajib Chakraborty, Mohamed El Baghdadi, Omar Hegazy

Energy Storage Systems for Airborne Wind Generators 3037
Bakr Bagaber, Axel Mertens

Design Interactions of AC- And DC-Side Filters for Traction Drives with SiC Inverters 3048
Hedieh Movagharnejad, Benjamin Knebusch, Axel Mertens, Bernd Ponick

Investigation of an Interleaved Current-Fed Single Active Bridge DC-DC Converter for PV Applications 3059
Lucas Vinícius De Araújo Gomes, Tobias Manthey, Montiê Alves Vitorino, Jens Friebe

Real-Time Thermal Characterization of Power Semiconductors using a PSO-Based Digital Twin Approach 3067
Johannes Kuprat, Yoann Pascal, Marco Liserre

Self-Sensing Design and Control for an Induction Machine with an Additional Short-Circuited Rotor Coil 3075
Stefan Luecke, Axel Mertens

Calculating the Tractive Power and Power Conversion Efficiency of Battery Electric Vehicles using a Global Navigation Satellite System and a Road Elevation Database 3084
Shinichi Domae, Alberto Castellazzi, Hamzeh J. Jaber, Tenghui Dong, Taketsune Nakamura

PCB Layer Optimization of Planar Medium Frequency Transformer for On-Board EV Chargers 3092
Fabian Groon, Hamzeh Beiranvand, Thiago Pereira, Görkem Can, Marco Liserre

Fault Current Capability Assessment of Low-Voltage Side Inverters in Smart-Transformers 3101
Thiago Pereira, Luis Camurca, Francisco Santos, Marco Liserre

Adaptive Resonant-Valley Switching for a GaN HEMT Direct AC-AC Auxiliary Resonant Commutated Pole Converter 3112
Kyle Steyn, Johan Beukes

The Variation of Core Loss in High-Frequency Transformers Under Different Load Conditions 3120
Navid Rasekh, Jun Wang, Xibo Yuan

A Complete PFC Inductor Design for Lighting Equipment Applications 3130
Wai Keung Mo, Kasper M. Paasch, Thomas Ebel

Automatic Generation Control-Based Charging/Discharging Strategy for EV Fleets to Enhance the Stability of a Vehicle-To-Weak Grid System 3140
Majid Mehrasa, Mehrdad Gholami, Reza Razi, Khaled Hajar, Antoine Labonne, Ahmad Hably, Seddik Bacha

Model-Based Converter Control for the Emulation of a Wind Turbine Drive Train 3149
Alexander Ernst, Wilfried Holzke, Dawid Koczy, Nando Kaminski, Bernd Orlik

A Novel Grid-Demanded Power Point Tracking (GPPT) Control Method for Wind Turbines to Preserve Grid Stability with High Wind Energy Penetration 3159
David Matthies, Alexander Ernst, Henning Sauerland, René Reimann, Wilfried Holzke, Bernd Orlik

Extension and Implementation of a Model-Based Lifetime Monitoring System with Parallel Calculation of Multiple Power Semiconductors 3169
Steffen Menzel, Wilfried Holzke, Michael Hanf, Holger Groke, Bernd Orlik, Nando Kaminski

Smart Charging Strategy for Electric Vehicles using an Optimized Fuzzy Logic System 3179
M. Gholami, M. Mehrasa, R. Razi, K. Hajar, A. Hably, S. Bacha, A. Labonne

Analysis and Discussion of a Concept for an Adjustable Inductance Based on an Impact of an Orthogonal Magnetic Field 3188
Guido Schierle, Michael Meissner, Klaus F. Hoffmann

A Field Programmable and Dynamic Configurable Power Electronic Converter Concept 3198
Bjarte Hoff

DAB Converter Discrete ADRC Control into Real-Time CHIL Simulation of a MVDC/LVDC Power Grid 3206
Alessio Clerici, Riccardo Chiumeo, Diego Raggini, Alessandro Veroni

SNNFT: Sequential Neural Network-Fuzzy Thermal Early Warning System for Lithium-Ion Batteries.. 3215
    Marui Li, Chaoyu Dong, Yunfei Mu, Qian Xiao, Jingming Cao, Hongjie Jia

Fine-Grained Dynamics Representation and Stability Analysis for MMC-Based Hybrid AC/DC Power Systems 3225
    Jingming Cao, Chaoyu Dong, Qian Xiao, Marui Li, Xiaodan Yu, Hongjie Jia

Adaptive Pontryagin's Minimum Principle-Inspired Supervised-Learning-Based Energy Management for Hybrid Trains Powered by Fuel Cells and Batteries 3235
    Hujun Peng, Feifei Li, Zhu Chen, Kai Deng, Sebina Jeschke, Kay Hameyer

A Case Study of Pole-Phase Changing Induction Machine Performance 3246
    Konstantina Bitsi, Sjoerd G. Bosga

New Topology of Superconducting Fault Current Limiter with Bypass Resistor 3254
    D. Baimel, Eli Barbi, S. Bronstein, N. Baimel, A. Kuperman

A Pre- And Discharge Unit for Capacitive DC-Links Based on a Dual-Switch Bidirectional Flyback Converter 3262
    Madlen Hoffmann, Martin März

Control and Integration of a Multiphase Brushless Wounded Synchronous Motor Drive 3272
    Remi Perrin, Guilherme Bueno-Mariani

A Way Forward to Achieve Interoperability in Multi-Vendor HVDC Systems 3282
    Adil Abdalrahman, Ying-Jiang Häfner, Philippe Maibach, Christoph Haederli

Model Predicitve Position Control of Electrical Drives on an Industrial PC 3292
    Fabian Karau, Michael Leuer

Bidirectional Active EMC Filter for Industrial Power Converters... 3301
    Bernhard Wunsch, Stanislav Skibin, Ville Forsstrom

A General Method to Measure Parasitic Capacitance of Transformer using Guarding Technique 3309
    Shaokang Luan, Stig Munk-Nielsen, Bruce Wakelin, Magnus Hortans, Jan Schupp, Hongbo Zhao

Inductance Analysis of Electric Machines by Classical and Numerical Methods.......................... 3318
    J. J. Germishuizen, T. J. E. Miller

Dynamic Wireless Power Transfer DWPT Time Domain Model: Xyz Position and Speed Coupling Effect 3327
    Iosu Aizpuru, Eneko Agirrezabala, Mikel Mazuela, Unai Iraola, Estanis Oyarbide, Carlos Bernal

Dynamic Average Small Signal Model of the SAB Converter 3336
    Alexis A. Gómez, Alberto Rodríguez, Marta M. Hernando, Diego G. Lamar, Javier Sebastián, Ibán Ayarzaguena, Jose Manuel Bermejo, Igor Larrazabal, David Ortega, Francisco Vázquez

Algorithm for Optimal Selection of Drive Motor Transmission Combination................................ 3344
    Santiago Ramos Garces, Dries Jacques, Stijn Derammelaere, Simon Houwen, Nick Van Oosterwyck, Bart Vanwalleghem

Evaluation of Drain-Source Voltage in Switch Transient Time Intervals as Gate Oxide Degradation Precursor of SiC Power MOSFETs.. 3353
    Javad Naghibi, Sadegh Mohsenzade, Kamyar Mehran, Martin P. Foster

Active Output LLC Converter Topology 3362
  Hannes Börngen, Eyke Liegmann, Sriram Jagannath, Ralph Kennel

Short Circuit Type II and III Behavior of 1.2 kV Power SiC-MOSFETs............................ 3373
  Xing Liu, Xupeng Li, Thomas Basler

Analog MPPT Comparison for Interplanetary Small Satellites Missions 3382
  C. Torres, A. Garrigós, J. M. Blanes, P. Casado, D. Marroquí, C. Orts

Feasibility Assessment of Variable-Speed Generator Set Concepts with Focus on Rating of Power
Electronic Equipment 3391
  Hendrik Fehr, Albrecht Gensior, Andreas Möckel, Frank Atzler, Tilo Roß, Carsten Reincke-Collon

Bus Voltage Regulation using Sequentially Switched ZVZCS Converters for Spacecraft Power
Systems.. 3401
  A. Garrigós, C. Orts, D. Marroquí, J. M. Blanes, C. Torres, P. Casado

A Standardized and Modular Power Electronics Platform for Academic Research on Advanced
Grid-Connected Converter Control and Microgrids 3411
  Frank S. R., Schulz D., Stefanski L., Schwendemann R., Hiller M.

Gate Input Capacitance Characterization for Power MOSFETs using Turn-On and Turn-Off
Switching Waveforms 3420
  Yota Nishitani, Michiko Inoue, Takashi Sato, Michihiro Shintani

AC Battery: Modular Layout with Cell-Level Degradation Control 3429
  Claudio Burgos-Mellado, Marcos Orchard, Diego Muñoz-Carpintero, Tomislav Dragicevic, Lorenzo Reyes-Chamorro, Jacqueline Llanos

Analysis of Test Methods for Measurement of Leakage and Magnetising Inductances in Integrated
Transformers 3440
  Sajad A. Ansari, Jonathan N. Davidson, Martin P. Foster, David A. Stone

A Topology-Morphing Series Resonant Converter for Photovoltaic Module Applications............ 3450
  Grigorios Sergentanis, Liliana De Lillo, Lee Empringham, C. Mark Johnson

A Novel Parameter for the Evaluation of Protective Circuits for IGBT Explosion Protection in
Submodules of MMC 3460
  Christoph Junghans, Hans-Guenter Eckel

Sub-Modules Switching Algorithms for Dual Active Bridge Modular Multilevel Converters to
Optimize Capacitor Voltage Deviation Versus Power Efficiency................................. 3470
  Peizhou Xia, Chuantong Hao, Stephen Finney, Michael Merlin

Systematic Adaptive Robust State Feedback Control for Active Front-End Rectifiers 3480
  Aidar Zhetessov, Giri Venkataramanan

An Optimized Compensation Strategy of Direct Matrix Converter-Fed PMSM Drives with Field
Weakening Under Unbalanced Supply Conditions 3491
  Jun Xie, Dustin Henneberg, Martin Suberski, Manuel Kusebauch, Uwe Rädel, Jürgen Petzoldt

Double Inverter Concept for High-Speed Drives Without Motor Filters 3501
  Henning Kasten, Stephan Beineke, Matthias Bachmann

A Universal Single Stage Current-Fed Bidirectional Converter with Both AC and DC Input Power Source Compatibility.. 3511
Manish Kumar, Sumit Pramanick, Bijaya Ketan Panigrahi

Optimization of Electric Vehicle Charge Scheduling with Consideration of Battery Degradation............... 3518
Raka Jovanovic, Sertac Bayhan, Islam Safak Bayram

Onboard ESU Sizing and Dynamic IPT Charging Scenarios for a Tramway Application........................... 3529
Endika Bilbao Muruaga, Irma Villar, Florian Legay, Pierre Prenleloup, Jean-François Reynaud

Investigations on the Active Reduction of Common Mode Noise with Opposing Noise Sources................. 3536
Philipp Marx, Felix Seybold, Philipp Ziegler, David Hirning, Jörg Roth-Stielow

Knowledge Based Grey Box Modeling of Inaccessible Circuits for System EMC-Simulation in Time Domain.. 3545
Jan-Philipp Roche, Jens Friebe, Oliver Niggemann

Novel Quasi-Direct Rotor Position Estimator for Permanent Magnet Synchronous Machines Based on the Back-Electromotive Force using Current Oversampling... 3555
Georg Lindemann, Viktor Willich, Axel Mertens

Design Considerations for Fast On-State Voltage Measurement Circuits.. 3565
Mathias C. J. Weiser, Manuel Rueß, Ingmar Kallfass

Analytical, FEM and Experimental Study of the Influence of the Airgap Size in Different Types of Ferrite Cores... 3574
Asier Arruti, Francisco Jose Perez-Cebolla, Jon Anzola, Iosu Aizpuru, Mikel Mazuela

Design Method of a High Frequency GaN-Based Half-Bridge with Bottom-Side Cooled Transistors using Multi-PCB Assembly... 3582
Loris Pace, Florian Chevalier, Thierry Duquesne, Nadir Idir

A 30 kW Dynamic Wireless Inductive Charging System for EVs.. 3590
Zariff Meira Gomes, José Renes Pinheiro, Gilney Damm, Karim Kadem, Hassan Moussa

Dynamic Control of the Switching Behavior of SiC MOSFETs in Converter Operation 3599
Jochen Henn, Laurids Schmitz, Rik W. De Doncker

A Series Resonant Balancing Converter for Bipolar DC Grids on Ships... 3607
Sachin Yadav, Zian Qin, Pavol Bauer

A V2G-Enabled Seven-Level Buck PFC Rectifier for EV Charging Application 3615
Anekant Jain, Ritika Agarwal, Krishna Kumar Gupta, Sanjay K. Jain

Experimental Demonstration of a 2.2kW Active-Clamp Converter for High-Current Wide-Voltage-Transfer Ratio Applications 3625
Philipp Rehlaender, Bastian Korthauer, Frank Schafmeister, Joachim Böcker

A Simplified Model for the Battery Ageing Potential Under Highly Rippled Load 3636
Tomáš Kacetl, Jan Kacetl, Nima Tashakor, Stefan Goetz

System Modeling and Design of a Hybrid Renewable Energy System for a Cable Network Head-End Station in Rural Area.. 3646
Tobias Schillinger, Thomas Schuhmann, Martin Eckart

Comparison of System-Level Availability in Industrial Grids 3655
  G. Emmers, J. Driesen

Ageing Mitigation and Loss Control in Reconfigurable Batteries in Series-Level Setups............................. 3665
  Tomáš Kacetl, Jan Kacetl, Nima Tashakor, Stefan Goetz

Characterization of Conventional and Advanced Current Measurement Techniques Suitable for
WBG Semiconductor Devices.. 3676
  Severin Klever, André Thönnessen, Rik W. De Doncker

Zero-Sequence Voltage Reduces DC-Link Capacitor Demand in Cascaded H-Bridge Converters for
Large-Scale Electrolyzers by 40% 3686
  Roland Unruh, Frank Schafmeister, Joachim Böcker

Thermal Behavior Impact on the Electric Motor Shape Multi-Objective Optimization............................... 3696
  Aissam Riad Meddour, Anthony Babin, Nassim Rizoug, Christopher Vagg, Richard Burke,
  Laid Degaa

Modelling Approaches of Power Systems Considering Grid-Connected Converters and Renewable
Generation Dynamics 3704
  Jaume Girona-Badia, Vinícius Albernaz Lacerda, Eduardo Prieto-Araujo, Oriol Gomis-
  Bellmunt, Stephan Kusche, Florian Pöschke, Horst Schulte

Efficiency and Lifetime Analysis of Several Airborne Wind Energy Electrical Drive Concepts 3711
  Bakr Bagaber, Daniel Heide, Bernd Ponick, Axel Mertens

Design and Performance Analysis of Single-Phase Axial Flux Permanent Magnet Motor for
Coaxial Cascade 3722
  Chu Wang, Xiaowei Hu, Xiaoya Wang, Weiwei Geng, Qiang Li, Jingning Hou

Comparison of Pulse Current Capability of Different Switches for Modular Multilevel Converter-
Based Arbitrary Wave Shape Generator Used for Dielectric Testing of High Voltage Grid Assets............. 3729
  Dhanashree Ashok Ganeshpure, Ajeeth Phrassanna Soundararajan, Thiago Batista Soeiro,
  Mohamad Ghaffarian Niasar, Peter Vaessen, Pavol Bauer

Accurate Modeling of IGBT-Based Converters in PLECS 3740
  Anne Von Hoegen, Philipp Tillmann, Tetsuya Kojima, Rik W. De Doncker

Novel Analytical Method for Estimating the Junction-To-Top Thermal Resistance of Power
MOSFETs.. 3750
  José Miguel Sanz-Alcaine, Francisco Jose Perez-Cebolla, Carlos Bernal-Ruiz, Asier Arruti,
  Iosu Aizpuru

DC-Side Impedance for Handling Interoperability of Multi-Vendor Multi-Terminal HVDC
Systems... 3757
  Ashkan Nami, Adil Abdalrahman, Ying-Jiang Häfner, Malaya Kumar Sahu, Khirod Kumar
  Nayak

Utilizing the Electroluminescence of SiC MOSFETs as Degradation Sensitive Optical Parameter 3766
  Lukas A. Ruppert, Michael Laumen, Rik W. De Doncker

Characterization of GaN-On-AlN/SiC Transistors Towards Monolithic Integrability 3775
  Nick Wieczorek, Xiaomeng Geng, Carsten Kuring, Oliver Hilt, Frank Brunner, Mihaela Wolf,
  Joachim Würfl, Sibylle Dieckerhoff

Optimal Frequency for Dynamic Wireless Power Transfer 3786
  Mincui Liang, Khalil El Khamlichi Drissi, Christophe Pasquier

A Wide-Input-Voltage-Range 50W Series-Capacitor Buck Converter with Ancillary Voltage Bus
for Fast Transient Response in 48V PoL Applications.. 3796
Nameer Khan, James Xu, Gerard Villar Piqué, John Pigott, Henk Jan Bergveld, Alaa El
Sherif, Olivier Trescases

Four-Level Boost Inverter Based on ANPC Topology with Switched-Capacitor Branch............................ 3804
Robert Stala, Adam Penczek, Stanislaw Piróg, Aleksander Skala, Andrzej Mondzik, Zbigniew
Waradzyn, Krishna Kumar Gupta, Pallavee Bhatnagar, Sanjay K. Jain, Kasinath Jena

Comparative Evaluation of Partially-Rated Energy Storage Integration Topologies for High
Voltage Modular Multilevel Converters.. 3813
Zoe Blatsi, Sebastian Neira, Stephen Finney, Michael M. C. Merlin

Influence of Current Collapse Due to V_{ds} Bias Effect on GaN-HEMTs I_d-V_{ds} Characteristics in
Saturation Region 3822
Xuyang Lu, Arnaud Videt, Ke Li, Soroush Faramehr, Petar Igic, Nadir Idir

Deep-Learning Fault Detection and Classification on a UAV Propulsion System 3831
Pierre-Yves Brulin, Fouad Khenfri, Nassim Rizoug

A Compact Solid State Transformer for Replacing Conventional Medium Power Transformer in
Weight-Critical Applications.. 3838
Leon Fauth, Felix Willer, Jens Friebe

Comparative Study of Single-Phase and Three-Phase DAB for EV Charging Application......................... 3846
Nicola Blasuttigh, Hamzeh Beiranvand, Thiago Pereira, Marco Liserre

Dynamic Load Emulation for Automotive Power IC Robustness Validation 3855
Alexander Ulbing, Daniel Kostynski, Markus Sievers

DAB Frequency Decoupling Control with Current Minimization 3862
Simon Uicich, Jean-Yves Gauthier, Xuefang Lin-Shi, Bruno Allard, Arnaud Plat

Design and Performance Analysis of a Modified Proportional Multi-Resonant (PMR) Controller
for Three-Phase Voltage-Source Inverters 3871
Ahmad Ali Nazeri, Mahmoud Saeidi, Peter Zacharias

Proposition and Comparison of Several Solutions for High Induced Voltage Across Inactive
Transmitting Coils in a Series-Series Compensation DIPT System 3883
Wassim Kabbara, Tanguy Phulpin, Mohamed Bensetti, Antoine Caillierez, Serge Loudot,
Daniel Sadarnac

Modeling and Measuring the Bearing Capacitance of Radially Loaded Bearings 3893
Stefan Quabeck, Daniel C. Rodriguez, Rik W. De Doncker

Comprehensive Control of Matrix Converters in On-Board Electric Drive Applications............................ 3903
Galina Mirzaeva

Power System Simulation Tool for Quick Benchmarking of Innovative MVDC Grids in E-Mobility
Applications... 3910
Daniel Siemaszko, Philippe Noisette

An Artificial Intelligence Pipeline for Critical Equipment Thermal Conditioning System Design 3920
Raik Orbay, Athanasios Tzanakis, Inko Marcaide, Jonas Löfgren, Torbjörn Thiringer,
Thomas Bernichon

Aspects of Stability Issues of HVAC/HVDC Coupled Grids.. 3928
    Gianni Bakhos, Kosei Shinoda, Juan-Carlos Gonzalez-Torres, Abdelkrim Benchaib, Luigi Vanfretti, Seddik Bacha

Measurement of Coss-V Characteristic of the 1.7kV/900A SiC Power Module and Estimation of the Channel Current... 3938
    Jacek Rabkowski, Fernando Gonzalez-Hernando, Mariusz Zdanowski, Irma Villar, Uxue Larrañaga

In-Slot Cooling of Electrical Machines using Traditional Techniques and Additive Manufacturing 3947
    Ahmed Hembel, Gokhan Cakal, Bulent Sarlioglu

Comparison of High-Power 2-Level and 3-Level Converters in Terms of Power Density, Costs and Performance.. 3957
    Ludwig Schlegel, Wilfried Hofmann

Autonomous Characterization of Lithium-Ion Battery Model Parameters Utilizing a Mathematical Optimization Methodology 3966
    Hamzeh Beiranvand, Helge Krüger, Sandra Hansen, Marco Liserre, Christian Werlig, Andreas Würsig

SOC Governed Algorithm for an EV Cascaded H-Bridge Connected to a DC Charger 3975
    Giulia Tresca, Andrea Formentini, Filippo Gemma, Federico Lusardi, Riccardo Leuzzi, Pericle Zanchetta

Shaping the Transition from Si-Based Power Devices to SiC MOSFETs and GaN HEMTs 3984
    Gerald Deboy

Reinventing Batteries Through Nanotechnology 3986
    Yi Cui

Advancing GaN Power ICs: Efficiency, Reliability & Autonomy.. 3987
    Dan Kinzer

Electrification Strategy of Volkswagen Group.. 3989
    Alexander Krick

Make it Fly — the Future of Sustainable Aviation.. 3991
    Tanja Neuland

The Instrumental but Extremely Challenging Role of Hydrogen Towards a Decarbonized Society 3992
    Stefan Linder

Short Circuit Behavior of Dual Three-Phase Permanent Magnet Synchronous Motors with Different Mutual Inductance in Electric Propulsion Application 3993
    Yinghui Yang, Georg Möhlenkamp

Hybrid Silicon-SiC Inverter – Combining the Best of Both Worlds 4003
    Hans-Günter Eckel, Felix Kayser, Pham Ha Trieu To

Robustness of SiC Trench MOSFETs 4004
    Christian Felgemacher

3D Predictive Fatigue Modeling of Power Modules 4005
    Ben Samples, Brandon Passmore

Heterogeneous Integration of Power Conversion using Power Supply on Chip and Power Supply in Package 4006
Cian Ó Mathúna, Seamus O'Driscoll

Driving Innovations for Power Electronics with Integratable and Sustainable Magnetics 4008
Matt Wilkowski

Impact of Package Technology on the Switching Behavior of High-Voltage GaN FETs 4011
Sebastian Klötzer

Impact of Power Electronics on Battery Operation 4012
Dirk Uwe Sauer

Trends in Power Electronics and Batteries for Electrified Vehicle Infrastructure 4013
Torsten Leifert

Impact of High Frequency Current Pulses on Battery Ageing 4014
Julia Kowal

Aircraft Electrification – System-Level Potentials for Aviation Decarbonization 4015
Kathrin Ebner, Antoine Habersetzer, Arne Seitz

About Power Electronics Challenges in Aviation 4016
Marco Bohllaender

Development of Electric Motors for Aircraft Applications 4017
Simon Wolfstädter

Powertrain Trends in Electric Trucks 4018
Luciana C. Afonso

Modulation Strategy Impact of BEV Inverters on the Voltage Ripple and the High-Voltage Traction System Stability 4019
Cornelius Rettner

Zero Emission Trucks & Bodies 4020
Martin Glaser

Integrating Offshore Wind & Hydrogen - An Operator's View 4021
Florian Gremme

Status Quo and Future Prospects of Power Electronic Solutions for Electrolysis Plants 4022
Sven Schumann

Modular Power Supply System for Large Scale Water Electrolyzers 4023
Ralf Juchem, Klaus Rigbers

Properties of a Lithium-Ion Battery as a Partner of Power Electronics 4025
Alexander Blömeke, Katharina Lilith Quade, Dominik Jöst, Weihan Li, Florian Ringbeck, Dirk Uwe Sauer

Author Index

Active output LLC converter topology

Hannes Börngen, Eyke Liegmann, Sriram Jagannath, Ralph Kennel

Chair of High-Power Converter Systems , Technical University of Munich
Arcisstr. 21, 80333 Munich, Germany

Email: hannes.boerngen@tum.de

Keywords

≪DC-DC converter≫, ≪Resonant converter≫, ≪ZVS converters≫, ≪Soft switching≫.

Abstract

This paper presents an "active output" LLC topology, i.e. a variant of the load-resonant LLC converter with *asynchronous* rectification stage. In this topology, the output switches actively block and thus change the apparent parameters of the resonant tank. Advantages, such as an increased design space for converter optimisation, are gained.

Introduction

The state-of-the-art LLC converter, Fig. 1, is a dc-dc converter belonging to a group of so called load-resonant converters. Advantages of the LLC topology are the low component count, mainly a resonant tank comprising two inductances, L_s and L_p, and one capacitance, C_s, and the fact that the topology and its modes of operation lend themselves to soft switching. For majority charge carrier devices, zero-voltage switching (ZVS) transitions for the primary-side switches (S_1–S_4) are most beneficial to achieve. On the output side, diodes or switches emulating diode behaviour ($S_{o,1}$, $S_{o,2}$) are used to rectify the transformer's secondary-side current. The secondary-side switches usually achieve soft switching when switching from *off* to *on* state. These characteristics prime the LLC topology as a compact and high efficiency solution for a broad range of applications.

To mitigate a major issue of the LLC, namely the influence of the ratio of the resonant inductances on the achievable voltage transfer ratio *and* efficiency, a novel topology and control approach will be presented. The contribution can be summarised as making the resonant tank's behaviour controllable via adding reverse blocking switches on the output side and in this way realising an active output.

In literature, different means can be found that achieve an adaptive damping-factor, namely secondary-side phase-shift modulation (SSPSM) [1], secondary-side pulse width modulation (SSPWM) [2], and

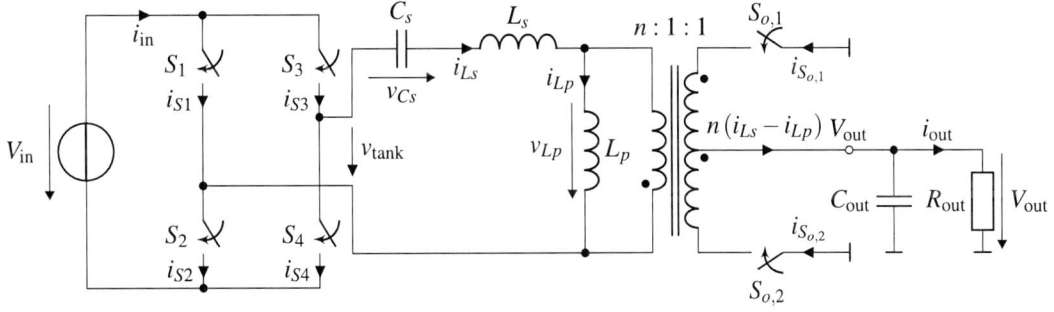

Fig. 1: State-of-the-art LLC converter topology.

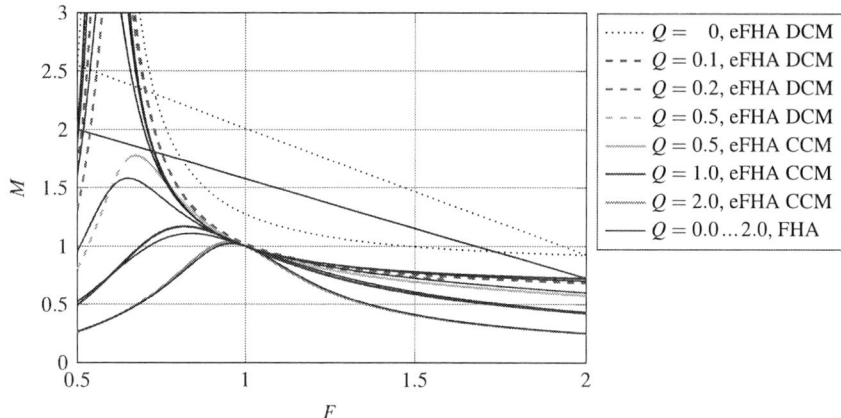

Fig. 2: Voltage transfer ratio of the frequency-modulated LLC; FHA, coded via solid black lines —, eFHA, coded via *coloured* lines, where discontinuous conduction mode (DCM) is coded via dashed lines, e.g. ---, and continuous conduction mode (CCM) is coded via solid lines, e.g. —.

a whole family of secondary-side "active boost rectifier" topologies [3]. In the case of SSPSM, the advantage of having voltage regulation without a change in switching frequency of the primary-side switches is employed to increase the hold up time for a fixed frequency LLC converter. For SSPWM, an adaptive voltage doubler topology on the output side is implemented for a fixed frequency LLC and is controlled to achieve the voltage regulation of the output side.

State-of-the-art LLC modulation techniques

In the following, two state-of-the-art modulation schemes for the LLC converter will be presented: frequency modulation and phase-shift modulation.

Frequency modulation

One major drawback of the LLC topology is the limited range of the output to input voltage ratio, which is linked to the fact that the flow of power is usually controlled via frequency modulation (fm). The voltage transfer function, $V_{\text{out}}/V_{\text{in}}$, of the frequency-modulated LLC can, by means of a first harmonic approximation (FHA) [4], be calculated as

$$M_{\text{LLC,fm}} = \frac{1}{n\sqrt{\left[1 + \kappa\left(1 - \frac{1}{F^2}\right)\right]^2 + \frac{\pi^4 Q^2}{64}\left(F - \frac{1}{F}\right)^2}} \tag{1}$$

with κ equalling the ratio of the resonant inductances, L_s/L_p, n being the transformer's transfer ratio, and with the control parameter F, the normalised switching frequency. The normalised switching frequency is defined as $f_{\text{sw}}/f_{s,0}$, where $f_{s,0}$ is the series resonance frequency $1/2\pi\sqrt{L_s C_s}$ and f_{sw} is the reciprocal of the duration of the switching period, T_{sw}. The quality factor Q of the resonant circuit is defined as $Z_0 P_{\text{out}}/n^2 V_{\text{out}}^2$, where Z_0 is $\sqrt{L_s/C_s}$. Thus, Q is also a measure of the transferred power.

The voltage transfer ratio of the frequency-modulated LLC is shown in Fig. 2, with Q as a parameter. In the plot, two different means of calculating the transfer ratio are employed, the aforementioned FHA and the extended FHA (eFHA) [5]. When switching with frequencies higher than F equal to 1 (*load independent point*), the resonant tank always exhibits inductive behaviour, which is necessary to achieve ZVS of the input switches. Below the *load independent point*, the resonant tank can show inductive behaviour as well, if the operating point is on the right-hand side of the maximum of the respective curve. For a fixed voltage transfer ratio, a reduction in transferred power is achieved with higher switching frequencies and is bounded by a lower limit, which depends on the ratio L_s/L_p and is as such fixed in the design stage. This behaviour can be visualised by the no-load curve, i.e. Q is 0, ..., where the voltage transfer ratio tends towards a non-zero value for higher switching frequencies. Changing the inductance

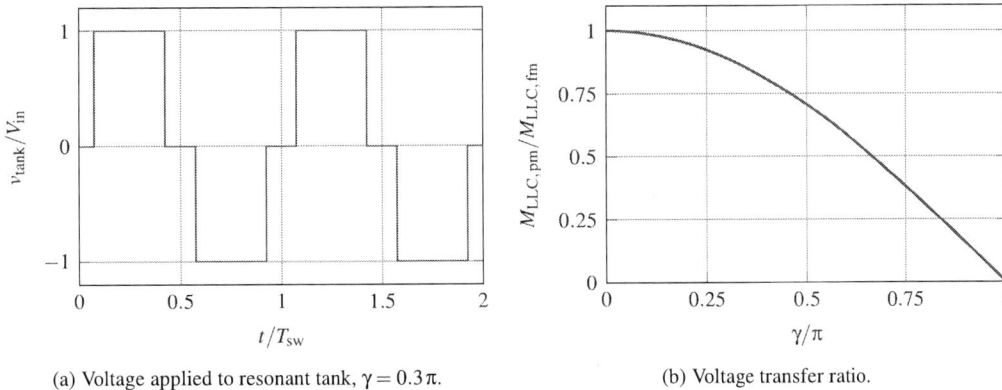

(a) Voltage applied to resonant tank, $\gamma = 0.3\,\pi$.

(b) Voltage transfer ratio.

Fig. 3: Phase-shift modulation of the LLC.

ratio in order to increase the range of achievable voltage ratios, e.g. by decreasing the value of L_p, is detrimental to the efficiency of the converter. How this constraint can be taken into account for the design of an LLC has been covered in [6].

Primary-side phase-shift modulation

In recent years, different methods have been investigated to mitigate this disadvantageous characteristic of the frequency-modulated LLC. One approach is to employ a different modulation technique, namely a phase-shift modulation (pm) for the primary-side switches [7], [8]. By this, the voltage applied to the terminals of the resonant tank can be clamped to 0 V in addition to positive and negative voltages. The voltage applied to the resonant tank depends on the phase-shift, γ, and is:

$$
v_{\text{tank}}\left(t\right) = \begin{cases} 0 & 0 & < \ t \bmod T_{\text{sw}} & \leq \frac{T_{\text{sw}}}{2} \frac{\gamma}{2\pi} \\ V_{\text{in}} & \frac{T_{\text{sw}}}{2} \frac{\gamma}{2\pi} & < \ t \bmod T_{\text{sw}} & \leq \frac{T_{\text{sw}}}{2} \left(1 - \frac{\gamma}{2\pi}\right) \\ 0 & \frac{T_{\text{sw}}}{2} \left(1 - \frac{\gamma}{2\pi}\right) & < \ t \bmod T_{\text{sw}} & \leq \frac{T_{\text{sw}}}{2} \left(1 + \frac{\gamma}{2\pi}\right) \\ -V_{\text{in}} & \frac{T_{\text{sw}}}{2} \left(1 + \frac{\gamma}{2\pi}\right) & < \ t \bmod T_{\text{sw}} & \leq \frac{T_{\text{sw}}}{2} \left(2 - \frac{\gamma}{2\pi}\right) \\ 0 & \frac{T_{\text{sw}}}{2} \left(2 - \frac{\gamma}{2\pi}\right) & < \ t \bmod T_{\text{sw}} & \leq T_{\text{sw}}. \end{cases}
\tag{2}
$$

An exemplary tank voltage is drawn in Fig. 3a. When using a similar approach to the FHA we can yield an analytical formula for the phase-shift modulated LLC converter. I.e. the time-domain voltage leads to a voltage gain $M_{\text{LLC,ph}}$ that exhibits the same behaviour as the frequency-modulated LLC, $M_{\text{LLC,fm}}$, with an additional multiplicative term as:

$$
M_{\text{LLC,pm}} = \frac{\sqrt{2}}{2} \sqrt{\cos\gamma + 1}\, M_{\text{LLC,fm}},
\tag{3}
$$

which is shown in Fig. 3b. It is apparent that the drawback of a design-determined lower output voltage is fixed, as the control input γ allows achieving output voltages down to 0 V.

Some literature can be found on the phase-shift LLC [7], [8], but to the knowledge of the authors, no in-depth investigation has been published that discloses a design optimisation strategy. The major disadvantage when designing the phase-shifted LLC is that a considerable phase-shift must be used for nominal input voltage, since the minimum input voltage should be covered by operation with zero phase-shift. As with the phase-shifted full-bridge converter, an asymmetry exists with regard to ZVS [9], the so called "active-to-passive", *strong*, and "passive-to-active", *weak*, transition of the input full bridge. This leads to a constrained design space in which κ has to be chosen large for ZVS, which in turn leads to low efficiency caused by an increase in circulating currents. To mitigate this issue, a topology and modulation scheme will be presented in the following section.

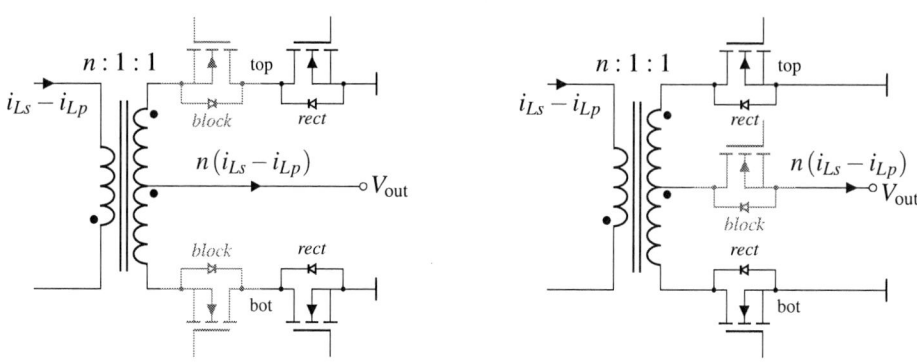

(a) Proposed active output configuration.　　　　(b) Active output alternative configuration.

Fig. 4: Output configurations of an active output LLC converter; proposed changes in red.

Proposed active output LLC converter

The main advantage of the proposed solution comes into play if a wide voltage transfer range is required, as a boosting of the output voltage can be achieved at low switching frequencies, as well as a step-down at higher switching frequencies. Another benefit lies in the fact that the resonant tank can be forced to consist of C_s, L_s, and L_p instead of solely C_s and L_s. The additional energy stored in the resonant tank enables a wider design range where ZVS of the primary-side switches can be achieved. Through this, the problems arising from the *weak leg* of the phase-shift modulation can largely be avoided, including the rather constrained suitable range of κ.

Topology and blocking switch implementation

State-of-the-art LLC converters use a *synchronous rectification*, i.e. an active switch that emulates diode behaviour. For the implementation of this, it comes in handy that most switches possess an anti-parallel diode and conduct current in the reverse direction when turned on. Both rectification schemes, using diodes and synchronous rectification, are in the following referred to as *passive rectification*.

In order to be able to actively block the output current path, normal, reverse conducting power switches cannot be used. Two approaches can be taken here; the development of reverse blocking switches or adding anti-series connected reverse conducting switches. One possible configuration of the proposed active output topology is presented in Fig. 4a, where the combination of switches act as a reverse blocking switch. This reverse blocking is activated as long as the gate signal of the respective blocking switch (*block*) is low. If the gate signal of the blocking switch is high, the respective rectifying switch (*rect*) can be used as a synchronous rectifier. It is possible to reconfigure the output in a manner that only *one* blocking switch is sufficient, as shown in Fig. 4b. This would entail the additional advantage that the *blocking* switch gate signal can be referenced to a constant voltage, namely V_{out}. The drawback of the latter solution is a more complicated timing of the gate drive pattern.

In Fig. 5, the active output LLC with the proposed blocking output configuration is shown in its entirety. The voltages across the blocking switches, v_{block}, have been defined as they would be generally defined for active switches, such as MOSFETs. In contrast, the voltages across the rectifying switches, v_{rect}, have been defined as more akin to the case where diodes are used.

Timing definitions

The gate signals of the blocking switch or, more precisely, the extensions of the blocking time are the control input for the active output LLC's secondary side, alongside with the switching frequency and the phase-shift modulation of the input full bridge. To illustrate the behaviour of the proposed topology, key voltage and current waveforms of an active output LLC are shown in Fig. 6. Additionally, the gate signals of the blocking switches are plotted.

In the diagrammed operating point, the active output LLC operates in DCM, where the bottom *rectifying* switch is "naturally" not conducting from ca. $0.245\,T_{sw}$ to $0.5\,T_{sw}$. The duration thereof is $T_{\alpha,1} = \alpha_1/2\pi\,T_{sw}$.

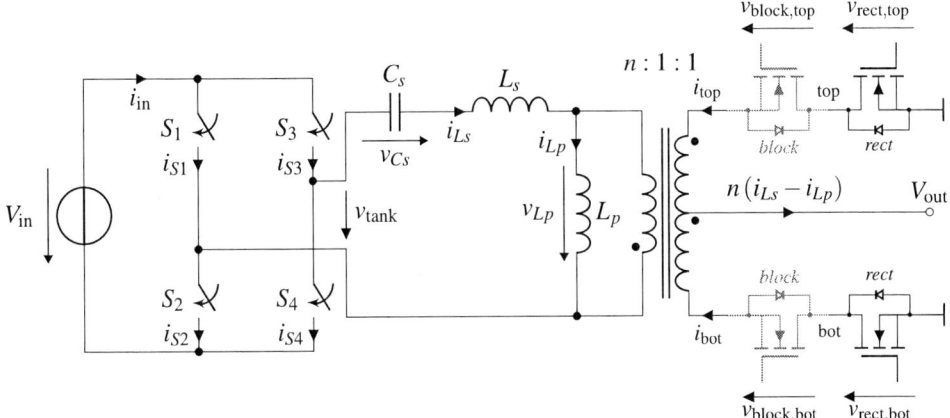

Fig. 5: Active output LLC converter topology; proposed changes in red.

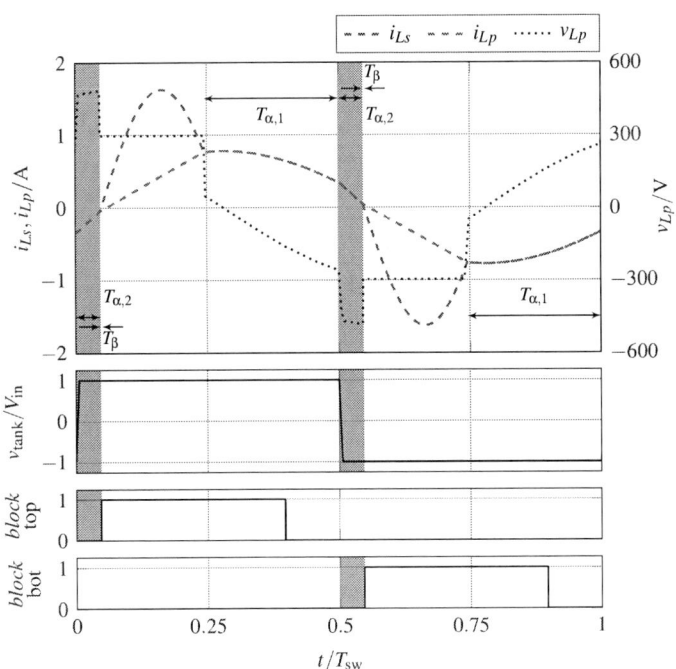

Fig. 6: Resonant tank behaviour and timing definitions of the active output LLC; $\alpha_2 = 0.08\,\pi$, $F = 0.465$, $\kappa = 0.166$, $Q = 0.17$. The output is *naturally* blocking during α_1 and is *actively* blocked during α_2. The currents i_{Ls} and i_{Lp} are equal during α and the energy stored in both inductances aids in achieving ZVS transitions.

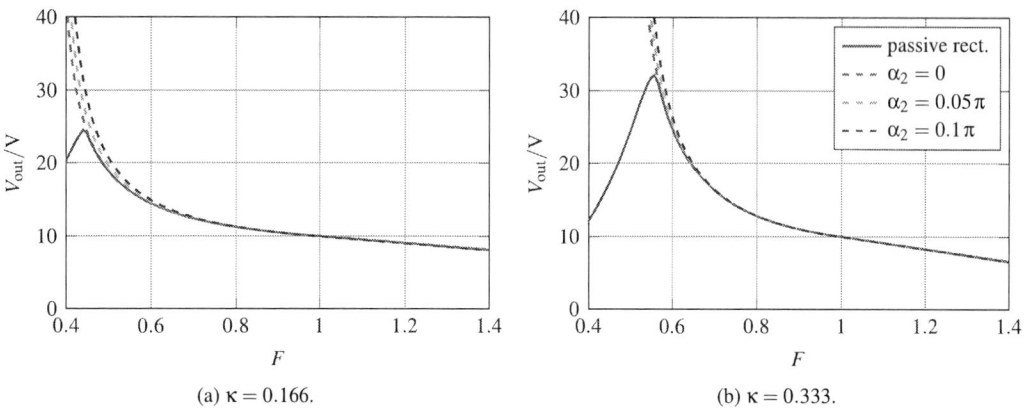

(a) $\kappa = 0.166$.   (b) $\kappa = 0.333$.

Fig. 7: Frequency-domain behaviour with improved boost capabilities; $Q = 0.17$, $V_{in} = 120\,\text{V}$.

From $0.5\,T_{sw}$ until $0.54\,T_{sw}$ the bottom output switch is blocking and thus forces L_s to be equal to L_p. The time spans during which one of the output switches is actively blocking are shaded in grey. The duration, $T_{\alpha,2} = \alpha_2/2\pi\,T_{sw}$, is the previously mentioned control input of the active output LLC, where α_2 denotes the delay from the, if applicable, *weak* half bridge transition to the switch *on* of the output blocking switch. While i_{Ls} is equal to i_{Lp}, both L_s and L_p take part in the resonance with C_s. The energy stored in both L_s and L_p supports the charging of the primary-side switch output capacitances C_{oss} and thus helps to achieve a ZVS transition.

As soon as the respective output blocking switch gate signal is *high*, the output current rises and i_{Ls} is not equal to i_{Lp} anymore. In the example in Fig. 6, i_{Ls} drops quickly below $0\,\text{A}$ after turning on of the respective output blocking switch, the duration of this time span is T_β.

Frequency-domain investigation

The previous section showed that the actively blocking output changes the waveforms of the resonant tank significantly. The effects in the frequency domain are discussed in the following. To this end, Fig. 7 depicts the output voltages plotted against the switching frequency F for two for different resonant inductance ratios, κ. In Fig. 7a we see the behaviour for a value of κ that is equal to 0.166, in Fig. 7b for a value of κ being equal to 0.333. From both plots it is apparent that for a normalised switching frequency of 1 we are at (or in fact rather close to) the load independent point, and both subplots exhibit the same voltage transfer ratio of $1/n = 1/12$. In addition, it is evident that the influence of the additional blocking time is marginal for the frequency range close to and above the load independent point if no phase-shit is used. The voltage transfer ratio with primary-side phase-shift will be covered in a following section.

It is important to note, that a significant increase of the voltage transfer ratio can be observed at switching frequencies close to the maximum voltage transfer ratio of the respective *passive rectification* curve (——). This characteristic can be used to achieve a higher voltage transfer ratio, without having to increase κ to reach the desired output voltage levels. Comparing the behaviour of the output blocking curves in Fig. 7a with the *passive rectification* curve in Fig. 7b, it is apparent that the same output voltages can be achieved with a κ that is 50 % smaller. This augments the design space of the LLC significantly, as a decrease in κ comes with an increase in efficiency.

Time-domain investigation

In the following, focus is put on the time-domain behaviour of the active output LLC, to understand how the proposed topology influences the resonant tank's waveforms. In the first subsection, the operating mode with switching frequencies below the series resonance is investigated. To achieve higher output voltages, the proposed LLC topology is operated solely with frequency modulation. Subsequently, the behaviour of the topology with additional phase-shift modulation of the input full bridge is scrutinised.

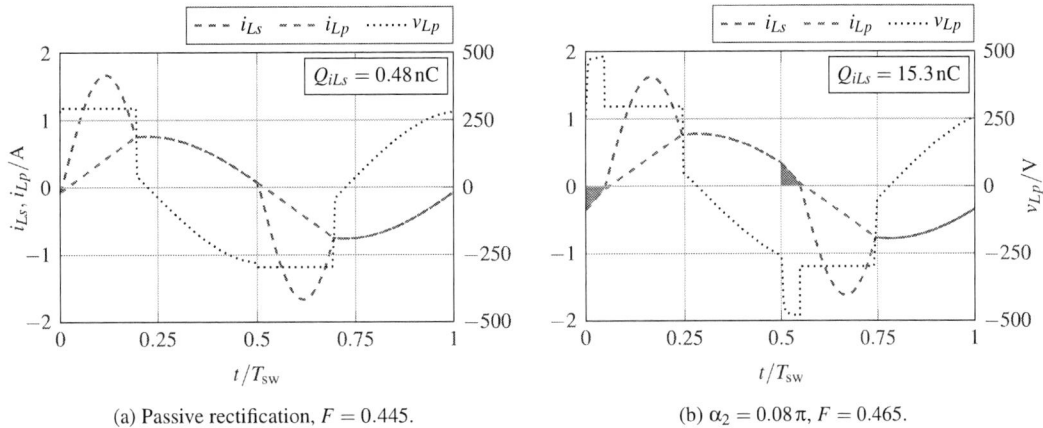

(a) Passive rectification, $F = 0.445$.

(b) $\alpha_2 = 0.08\,\pi$, $F = 0.465$.

Fig. 8: ZVS charge of passive rectification and active output LLC, with a 30-fold increase in case of active rectification; $\kappa = 0.166$, $Q = 0.17$, $V_{\text{in}} = 120\,\text{V}$.

Behaviour below series resonance frequency

For low input voltages, the LLC is operated close to its lower resonance frequency $1/2\pi\sqrt{C_s(L_s+L_p)}$, i.e. close to the maximum voltage transfer ratio. Here, one important characteristic is the increased boost capability to achieve the desired output voltage. Additionally, with the extended blocking time comes an increase in the available energy for the ZVS transition.

This improved ZVS transition is based on two different effects that are caused by the active blocking. Firstly, we delay the tank's current response to the applied voltage, which is directly related to α_2. This can be seen as making the resonant tank more inductive. Secondly, even without the delay of the tank's current with respect to the input voltage, we force i_{Ls} to drop more slowly, by connecting L_s essentially in series with L_p.

To show how pronounced the combination of both effects is, Fig. 8 depicts two operating points for an output voltage of 25 V. In Fig. 8a, the waveforms of the resonant tank's inductive components are drawn for a standard LLC. At t equal to $0.5\,T_{\text{sw}}$, a switching transition of the primary-side full bridge takes place, which can be seen at the small, step like change of v_{Lp}, the voltage across L_p. It is apparent that at this time instance the currents of both, L_s and L_p, have decreased close to 0 A. The charge that flows through L_s until the current has fully dropped to 0 A, i.e. the charge for the ZVS transition, is ca. 0.48 nC. In Fig. 8b, the same waveforms are drawn for the proposed topology. As before, at $0.5\,T_{\text{sw}}$ a switching transition of the primary-side full bridge takes place, which can be seen at the steep change of the voltage across L_p. At this time instance the currents of both, L_s and L_p, are still significantly larger. The charge through L_s is 15.3 nC, visualised by a dark grey area, and is thus a factor of thirty times larger compared to the standard solution and shows how the ZVS behaviour is greatly improved.

Behaviour with primary-side phase-shift

For high input voltages, the LLC is operated *at* or *above* its series resonance frequency and, additionally, with a significant primary-side phase-shift. In the following, only the operation at F equal to 1 will be presented as both regions exhibit similar characteristics.

To make the interaction between the active output LLC's control input α_2 and the primary-side phase-shift γ evident, Fig. 9a is plotted. We see that both an increase in α_2 and γ leads to a step-down of the output voltage. If the output is controlled to 20 V, a primary-side phase-shift of $0.6\,\pi$ (●) will be employed for a standard output. The charge Q_{ZVS} present for the *weak leg* ZVS-transition, Fig. 9b, is 2.2 nC (●) in this operating point. In case of the active output scheme with an α_2 of $0.1\,\pi$, a significantly smaller primary-side phase-shift of $0.5\,\pi$ (▲) is sufficient. At this operating point, Q_{ZVS} is larger with 9.2 nC (▲).

If we assume that a charge of 14 nC (indicated by the black dashed line) is necessary for the ZVS tran-

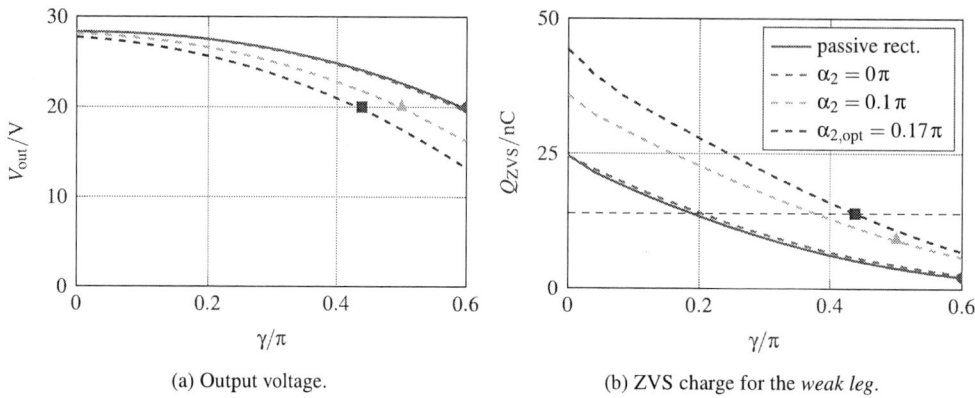

(a) Output voltage.

(b) ZVS charge for the *weak leg*.

Fig. 9: Behaviour with primary-side phase-shift γ, dependency of increase of ZVS charge with parameter α_2; $F = 1$, $\kappa = 0.166$, $Q = 0.25$, $V_{in} = 340$ V.

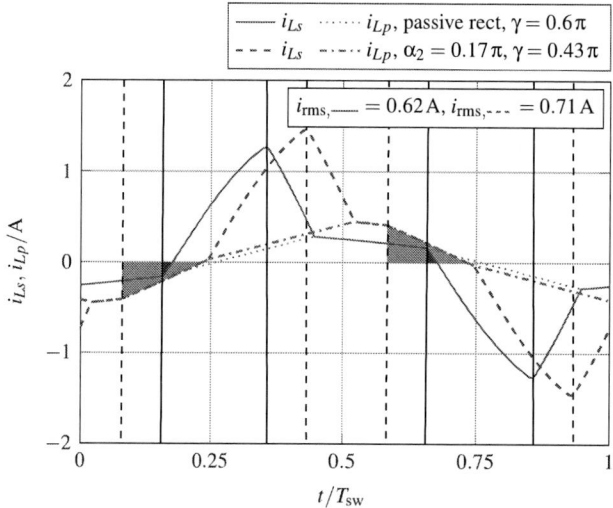

Fig. 10: Resonant tank current waveforms with improvement of ZVS charge visualised by grey shaded area; $F = 1$, $\kappa = 0.166$, $Q = 0.25$, $V_{in} = 340$ V.

sition in the selected operating point, the optimal α_2, with the lowest $i_{Ls,rms}$, is 0.17π (■). Fig. 10 shows this optimal operating point as well as the waveforms of the corresponding passive rectification operating point. The instances where the input bridges change state are indicated by solid black lines for the passive rectification case, and with dashed black lines for the active rectification case. We clearly see the larger charge present for the active output case (light ▨▨ versus dark shaded area ■■). Furthermore, there is a change of the rms value of i_{Ls}, that is comparatively small, as the larger primary-side phase-shift for the passive rectification scheme increases the rms value of the current as well. For the passive rectification scheme $i_{Ls,rms}$ is 0.62 A, for the active output scheme 0.71 A. Even though the rms value and the losses tied to this metric increase, achieving ZVS (and with that avoiding switching losses altogether) has an overall benefit with regard to the total losses and thus the efficiency. Therefore, the active output modulation scheme can improve the ZVS behaviour at the series resonance frequency without a drastic change of rms losses.

The same reasoning holds true if the LLC is operated above its series resonance frequency. In this case, there are three different means of achieving the desired output voltage: the switching frequency, the primary-side phase-shift, and the active output modulation.

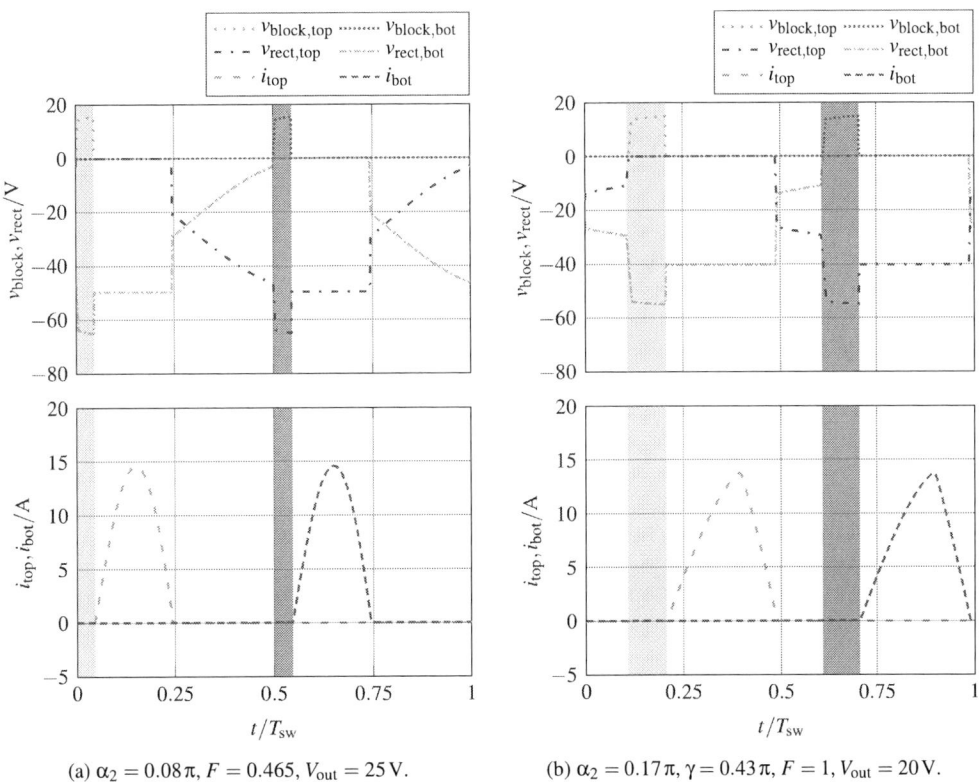

(a) $\alpha_2 = 0.08\,\pi$, $F = 0.465$, $V_{\text{out}} = 25\,\text{V}$.

(b) $\alpha_2 = 0.17\,\pi$, $\gamma = 0.43\,\pi$, $F = 1$, $V_{\text{out}} = 20\,\text{V}$.

Fig. 11: Voltages and currents of output switches for active rectification and two distinct operating points.

Secondary-side switch requirements

One aspect, that should be elaborated on, are the requirements for the secondary-side *blocking* as well as the *rectifying* switches. For this, Fig. 11 depicts the voltages and currents of the secondary-side switches for two operating points. In the top plots of Fig. 11, the voltages across the blocking *and* the rectifying switches, v_{block} and v_{rect}, are presented. In the bottom plots of the same figure, the currents through the switches, i_{top} and i_{bot} are shown[1]. To compare the passive and active rectification *intra* operating point—in contrast to *inter* operating points—the drawings in Fig. 12 are provided. As before, the time spans during which one of the output switches is actively blocking are shaded in grey. Additionally, some curves are drawn slightly transparent to improve the overall legibility.

The first operating point—$\alpha_2 = 0.08\,\pi$, $F = 0.465$—is chosen to be representative for switching frequencies below the *load independent point* and is shown in Fig. 11a and Fig. 12. This is the same operating point that was utilised for Fig. 8b. The second operating point—$\alpha_2 = 0.17\,\pi$, $\gamma = 0.43\,\pi$, $F = 1$—is chosen to be representative for operation with a significant primary-side phase-shift and is shown in Fig. 11b. Here, the same operating point was used as in Fig. 10.

In both operating points, the output currents exhibit a similar rms value if compared to their corresponding *passive* rectification case. For example, for the operating point in Fig. 11a, the rms value of the output currents, i.e. $i_{\text{top}} + i_{\text{bot}}$, is 6.59 A and 6.53 A, for the *passive* and the *active* rectification case, respectively. In Fig. 12a these currents are explicitly compared between the operating schemes. The same metric for the operating point shown in Fig. 11b also exhibits no strong dependency if comparing to the *passive* rectification case and, furthermore, is similar to the values of the first operating point. Accordingly, an increase in rms losses can be expected to mainly scale with the on-resistance of the additional blocking switch.

---

[1]The definitions of the voltages and the currents can be found in Fig. 5.

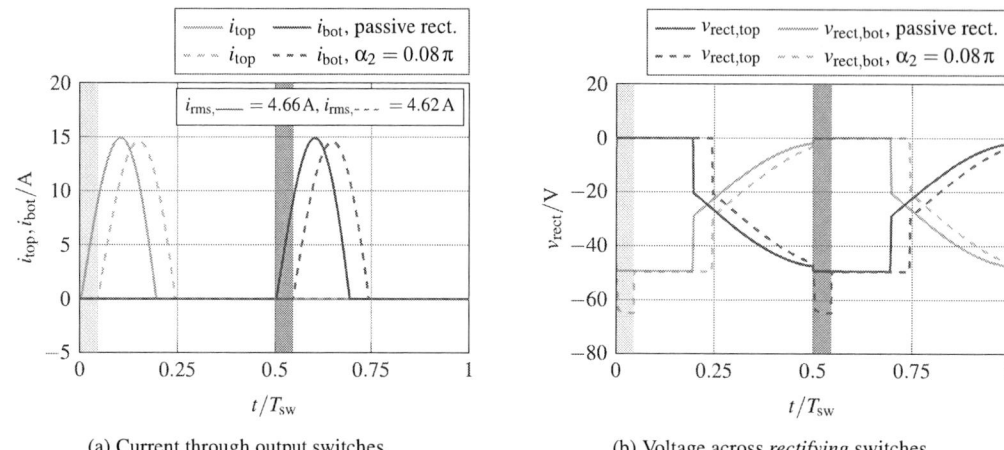

(a) Current through output switches.

(b) Voltage across *rectifying* switches.

Fig. 12: Output switch voltage and current waveforms for passive and active rectification; $F = 0.445$ (——, passive rectification), $F = 0.465$ with $\alpha_2 = 0.08\,\pi$ (- - -, active rectification), $V_{\mathrm{out}} = 25\,\mathrm{V}$.

On the other hand, the off-state voltages of the switches show a significantly different characteristic if the *active* rectification is used. In the respective grey-shaded areas in Fig. 11, the voltages across the *blocking* switches are about 15 V. The same voltage is visible in v_{rect} as an *additional* component on top of the corresponding *passive* rectification waveform, which is, in the ideal case, limited to $2\,V_{\mathrm{out}}$. This *additional* voltage is—in general—dependent on the input voltage, V_{in}, and transfer ratio of the transformer, n. To make the differences more apparent, Fig. 12b is provided, in which the *rectifying* switch voltage of the passive rectification is juxtaposed with the voltage of the active rectification. Consequently, the *rectifying* switches have to withstand an approximately 30 % higher voltage in the chosen operating points. Here, a trade-off becomes apparent between the primary-side ZVS and circulating currents, on one hand, and the blocking capabilities of the output switches, on the other hand.

Conclusion

In this paper, the modification of adding an active output to the load-resonant LLC converter was proposed. Based on simulation results, the converter's behaviour was presented in the time and frequency domain. The main advantages of the active output LLC are the increased boost as well as the ZVS capabilities for low switching frequencies, which will enable higher efficiencies compared to a standard LLC converter. For higher switching frequencies, a combination of frequency and phase-shift modulation was presented, which also improves the ZVS behaviour.

These advantages do not come at the expense of the key benefits of the LLC converter, namely the low component count and high efficiency, which is mainly accountable to ZVS operation, as the presented topology inherits these from the state-of-the art LLC topology. One drawback of the proposed topology is the increased control and drive effort on the secondary side. Additionally, it has to be ensured that the blocking switches do not switch off the *inductive* output current, which could result in the destruction of the respective switch. Furthermore, it is apparent that without *true* reverse blocking switches, at least one *additional* switch has to be placed in the output path, resulting in an increase in conduction losses.

We showed that the active output LLC, paired with a suitable modulation strategy, offers an additional control input that allows supplementary boosting or stepping-down of the output voltage, if required in the operating point. This increases the design space of the LLC and its transformer. The combination of both modulation strategies decouples the resonant tank's and transformer's design parameters, Q and κ—at least partially—from the achievable voltage transfer range, and thus enables optimal designs with higher efficiency.

To prove the validity of the presented topology, a prototype sub-100 W converter has been built. The results will be presented in a follow-up publication.

References

[1] H. Wu, T. Mu, X. Gao, and Y. Xing, "A secondary-side phase-shift-controlled LLC resonant converter with reduced conduction loss at normal operation for hold-up time compensation application," *IEEE Trans. Power Electron.*, vol. 30, no. 10, pp. 5352–5357, 2015.

[2] H. Wang and Z. Li, "A PWM LLC type resonant converter adapted to wide output range in PEV charging applications," *IEEE Trans. Power Electron.*, vol. 33, no. 5, pp. 3791–3801, 2018.

[3] H. Wu, Y. Lu, T. Mu, and Y. Xing, "A family of soft-switching dc–dc converters based on a phase-shift-controlled active boost rectifier," *IEEE Trans. Power Electron.*, vol. 30, no. 2, pp. 657–667, 2015.

[4] R. Steigerwald, "A comparison of half-bridge resonant converter topologies," *IEEE Trans. Power Electron.*, vol. 3, no. 2, pp. 174–182, 1988.

[5] M. P. Foster, C. R. Gould, A. J. Gilbert, D. A. Stone, and C. M. Bingham, "Analysis of CLL voltage-output resonant converters using describing functions," *IEEE Trans. Power Electron.*, vol. 23, no. 4, pp. 1772–1781, 2008.

[6] T. Duerbaum, "First harmonic approximation including design constraints," in *Intern. Telecom. Energy Conf.*, San Francisco, CA, Oct. 1998, pp. 321–328.

[7] W. Liu, B. Wang, W. Yao, Z. Lu, and X. Xu, "Steady-state analysis of the phase shift modulated LLC resonant converter," in *Proc. IEEE Energy Convers. Congr. Expo.*, Milwaukee, WI, Sep. 2016, pp. 1–5.

[8] N. Kollipara, M. K. Kazimierczuk, A. Reatti, and F. Corti, "Phase control and power optimization of LLC converter," in *Intern. Symp. Circuits and Systems*, Sapporo, Japan, May 2019, pp. 1–5.

[9] R. W. Erickson and D. Maksimović, *Fundamentals of Power Electronics*, 2nd. Norwell, MA: Kluwer Academic, 2001.

Short Circuit Type II and III Behavior of 1.2 kV Power SiC-MOSFETs

Xing Liu[1], Xupeng Li[1], Thomas Basler[1]

[1]Chemnitz University of Technology, Chair of Power Electronics
Reichenhainer Str. 70, 09126, Chemnitz, Germany
Tel.: +49 371 531-30465
E-Mail: xing.liu@etit.tu-chemnitz.de

Keywords

«Silicon Carbide (SiC)», «MOSFET», «Short Circuit», « Robustness» »

Abstract

The application relevant short circuit (SC) behavior of 1.2 kV power SiC-MOSFETs during the forward conduction mode (type II) and the body-diode conduction (type III) is experimentally investigated in this paper. Compared with Si-IGBTs, the SC type II and III behavior are less critical from the perspective of short-circuit energy. The reason is due to the smaller ratio between the Miller-capacitance and gate-source capacitance and further the more significant short-channel effect or Drain Induced Barrier Lowering (DIBL) effect for SiC-MOSFETs. Moreover, the influence of the operation temperature and gate trapping effect on the SC behavior are discussed.

Introduction

Silicon-Carbide (SiC) MOSFETs have experienced significant development over the recent years, and their level of majority is now sufficient for product development of converters for various applications for medium-power classes [1]. It ranges from motor drives for different kinds of electron vehicles to various industrial applications. The major advantages of SiC devices are their lower on-state resistance, higher operating temperature capability, and faster-switching speed compared with Silicon (Si) IGBTs. However, due to the higher current density and the thinner base region, the thermal capacitance of the SiC-MOSFETs chip is considerably smaller compared with Si counterparts. Therefore, the behavior and robustness of SiC-MOSFETs under the short circuit (SC) condition have to be crucially considered and investigated. The first type of SC, also called Hard Switching Fault (HSF), has been studied in numerous publications [2]-[4]. For 1.2 kV voltage classes, there are two main failure modes that have been found: thermal runaway and failure in the gate oxide [3]. However, minor research is currently focused on the application relevant SC type II and III characteristics of SiC-MOSFETs, especially for the SC III, where the SC event occurs under the conduction of the body diode (BD) mode. Unlike Si-IGBT applications, where a separate free-wheeling diode chip has to be used for an inductive load. SiC-MOSFETs can conduct reversely because of the integrated body diode. To reduce the conduction losses during the BD conduction, a positive gate voltage can be applied to open the channel and make the MOSFET work in reverse conduction mode. During this time, if another blocked device on the same bridge arm losses the blocking capability or the load is shorted accidentally, the reverse-conducting MOSFET will directly enter the short circuit type III mode.

In this paper, the low-inductive SC type II and III behavior for 1.2 kV SiC-MOSFETs will be experimentally investigated. Different gate structures from three manufacturers will be compared. The effects of different negative gate voltages, gate resistances, conduction current and operating temperatures will be analyzed. For SC type III, the bipolar effect of the BD and the charge-carrier trapping at the SiO_2/SiC interface will be considered.

Test setup and conditions

Three different gate structures have been selected as Devices Under Test (DUTs) for this investigation. The general information is summarized in Table I. A half-bridge based test setup, and the used pulse pattern to trigger the different SC events have already been introduced in our previous studies [5]. The only difference is the utilized auxiliary switch (aux. switch): two 30 mΩ SiC-MOSFETs (IMZ120R030M1H) were used in parallel as the high side (HS), see also Fig. 6. The drive voltage for aux. switch was set to -5/18 V, and a 0 Ω external gate resistance was implemented to achieve a fast turn-ON. This can simulate the SC behavior in the laboratory environment as close as possible to the reality. The lumped parasitic inductance of the SC loop is around 50 nH.

Table I: DUT information

Abbreviation	Name	Manufacturer/Gate structure	$R_{DS,ON}$	$R_{G,int}$	Packaging
M1	IMZ120R060M1H	Infineon/Asymmetrical Trench	60 mΩ	6 Ω	
M2	SCT3080KR	ROHM/Double Trench	80 mΩ	12 Ω	TO-247-4
M3	NTH4L080N120SC1	ON Semiconductor/Planar	80 mΩ	1.7 Ω	

For the SC type III measurements, the pulse is described later in detail to consider the bipolar and trapping effects. The DC-link voltage was set to 800 V, the applied negative gate voltage $V_{GS,OFF}$ was changed between different levels to investigate the trapping effect causing the gate threshold voltage $V_{G,th}$ shift/hysteresis. The positive gate voltage $V_{GS,ON}$ was fixed to 15 V to guarantee a certain short circuit capability. All DUTs are packaged in a 4-pin TO-247 discrete housing, and the sense-source is used as the measurement reference point to minimize the measurement error caused by the packaging parasitic inductance due to the fast current commutation.

The Short Circuit Type II Behavior of SiC-MOSFETs

The SC type I and II waveforms for M1 are shown in Fig. 1 (a). The duration of the SC event is limited to 1 μs to avoid possible destruction. The OFF-state gate voltage was set to 0 V to prevent the negative $V_{G,th}$ shift. Before the SC II event, the MOSFET was carrying a 13 A load current, which is the application-near current according to the datasheet of M1. The most considerable difference compared with the SC I is the desaturation process of the drain-source voltage V_{DS}. Due to the desaturation process, a displacement current flows through the Miller-capacitance C_{GD} from drain to gate and further charges the gate-source capacitance C_{GS} [6]. Hence, a gate voltage overshoot can be observed in the SC II measurement just after around 0.1 μs. As a consequence of the V_{GS} overshoot, a higher current peak is recorded as well. After approximately 0.27 μs, the SC II behavior is analogous to the SC type I. Although the SC II current peak is higher, the short circuit energy E_{SC} for both SC types is still comparable.

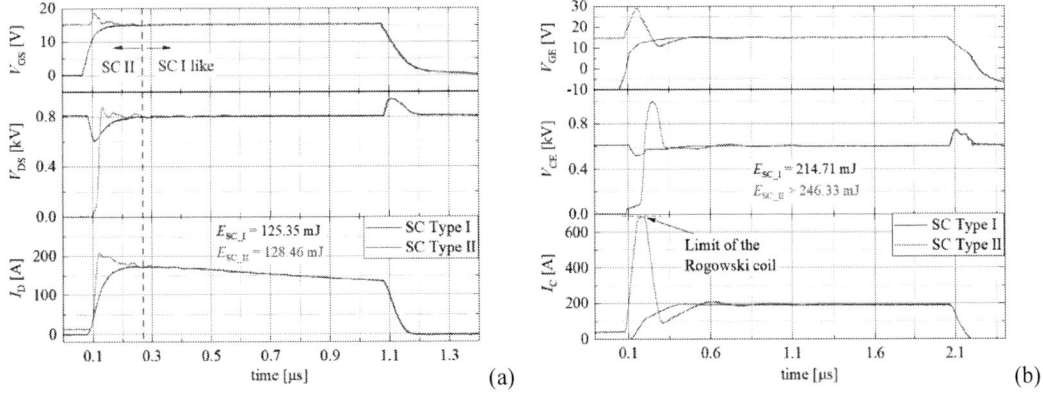

Fig. 1: (a) SC type I and II behavior comparison for M1, $V_{DS} = 800$ V, $R_{G,ON} = 7.5$ Ω, $V_{GS,OFF} = 0$ V, $V_{GS,ON} = 15$ V, room temperature; (b) SC II for Si-IGBT, $V_{CE} = 600$ V, $R_{G,ON} = 7.5$ Ω, $V_{GS,OFF} = -10$ V, $V_{GS,ON} = 15$ V

For comparison, the SC II behavior of a 1.2 kV, 40 A rated IGBT (IKY40N120CH3) is shown in Fig. 1 (b), a more significant Miller-capacitor feedback can be detected. For both measurements, no clamping circuit for gate voltage was utilized. The measured gate voltage overshoot of IGBT reaches 30 V, which leads to a current peak of more than 700 A. This current value already exceeds the measurement range of the used 600 A Rogowski coil. In addition to higher currents, the induced voltage peak from negative dI_C/dt is another possible cause of failure. Because the avalanche breakdown robustness of Si-IGBTs is weaker compared with SiC-MOSFETs. The reason for this less critical SC II behavior for SiC-MOSFET M1 should be the low ratio between the charge from the Miller-capacitance Q_{GD} and the gate-source capacitance C_{GS}. In Fig. 2, this ratio is expressed versus the drain-source voltage and is based on the static voltage-dependent capacitances (C-V) measurement. With the increased DC-link voltage (here expressed via V_{DS}), the Miller-effect becomes more and more pronounced. It can be seen from Fig. 2 that the ratio for SiC-MOSFETs is remarkably lower compared with IGBTs from the comparable current/voltage class except for M2 from ROHM. At 800 V, the ratio for M1 and M3 is around 5.5 V, respectively. For M2, the ratio is even higher than for a similar IGBT.

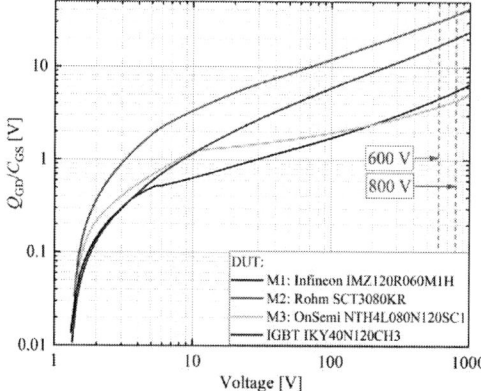

Fig. 2: Ratio between Q_{GD} and C_{GS} for SiC-MOSFETs and a comparable IGBT based on the static C-V measurement

However, it has to be emphasized that this ratio is just a rough estimation for the intensity of the Miller-effect under the SC II condition. Two essential factors are not considered for this estimation. The first is the gate-voltage-dependent gate capacitor (MOS-capacitor), and the second is the gate-drive loop impedance, which will be discussed in the following. Nevertheless, as a consequence of the much smaller C_{GD}, the Miller-feedback is in general weaker for SiC -MOSFETs under the SC II condition.

The effect of the OFF-state gate voltage $V_{GS,OFF}$, conduction current, and the turn-ON gate resistance $R_{G,ON}$ on the SC II behavior for M1 is shown in Fig. 3. Because the main consideration in SC II is in the desaturation stage, the subsequent SC characteristics are similar to SC I. Therefore, the following test results only exhibit the SC behavior around the desaturation stage.

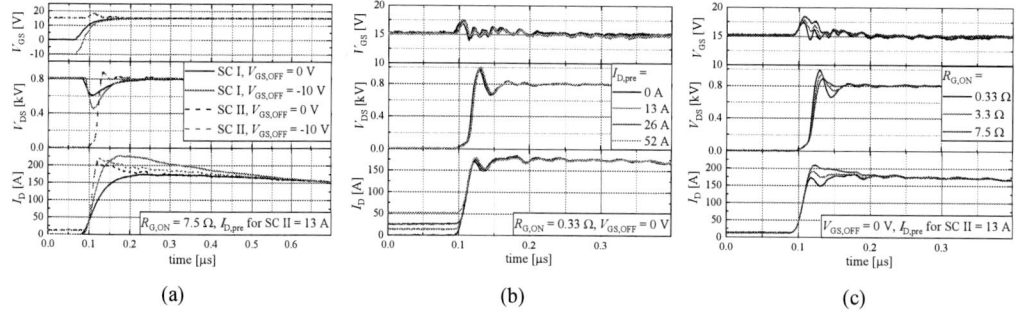

Fig. 3: Short circuit behavior for M1 with the effect of (a) $V_{GS,OFF}$ for SC I and SC II; (b) conducting current $I_{D,pre}$; (c) $R_{G,ON}$

At lower $V_{GS,OFF}$ during OFF-state, more positive charges from the p-region can be trapped at the SiO_2/SiC interface, due to the significantly higher trap density compared to the Si/SiO_2 interface [1]. This positive trapped charge acts similar like a positive gate voltage and leads to a dynamically negative shift of $V_{G,th}$. Therefore, the SC current peak becomes higher with lower $V_{GS,OFF}$. However, this effect is less pronounced under the SC II condition. The difference of the current peak caused by the $V_{GS,OFF}$ is around 55 A and 13 A for SC I and SC II, respectively. Because during the conduction phase before SC II, a positive gate voltage has already been applied, which pushes away the trapped positive charges. Meanwhile, a neutralizing process between channel electron current (load current) and trapped charges takes place. Both effects simultaneously attenuate the negative shift of $V_{G,th}$. Therefore, the influence of $V_{GS,OFF}$ on the SC II current peak is not as significant as that of SC I. In general, this effect depends on the applied turn-ON gate voltage $V_{GS,ON}$, the conduction current, conduction duration, and the operating temperature.

The influence of the conduction current is given in Fig. 3 (b). The pre-conducted current $I_{D,pre}$ was increased from 0 A to 4 times application-near current (52 A), $V_{GS,OFF}$ was set to 0 V to prevent the influence of the trapping effect for this measurement. There is almost no visible difference in the SC behavior that can be found. As a unipolar device, the channel current determines the total current flowing through the MOSFET, if the operating temperature is not extremely high, which could trigger the parasitic npn-transistor. With simplification that neglects the JEFET effect, different currents can be represented with different electron drift velocities in the channel:

$$J_D = J_n = n_{ch} \cdot q \cdot v_n = n_{ch} \cdot q \cdot \mu_n \cdot E_{ch} \tag{1}$$

where n_{ch} is the electron density in the channel, E_{ch} is the electrical field strength across the channel. As the conducting current increases, the device will move earlier from the ohmic-region into the current saturation region. But there is no plasma sweep-out process like in bipolar devices during desaturation, e.g. in IGBTs. Therefore, different currents hardly affect the short-circuit characteristics of the device, especially the current and voltage peaks. In Fig. 3 (c), the influence of $R_{G,ON}$ on the SC behavior is shown. The total capacitive charge from C_{GD} during the desaturation flows to the C_{GS} and is discharged by the gate loop simultaneously. The discharge process of overcharged C_{GS} depends on the output stage topology of the Gate Drive Unit (GDU) and mainly the gate resistance (internal and external) for a low inductive gate connection, which is described in [5]. Before SC II and III, C_{GS} is already fully charged to the static ON-state voltage and the GDU holds the turn-ON signal. Therefore, only $R_{G,ON}$ is involved in the gate loop for a source and sink separated GDU output stage. With increased $R_{G,ON}$, the GDU and MOSFET are more de-coupled, more capacitive charge flows into C_{GS} and leads to a higher V_{GS} overshoot and accordingly a higher current peak. Meanwhile, the induced voltage peak caused by the negative dI_D/dt becomes smaller. This can also be seen for M2 and M3 as shown in Fig. 4 [7].

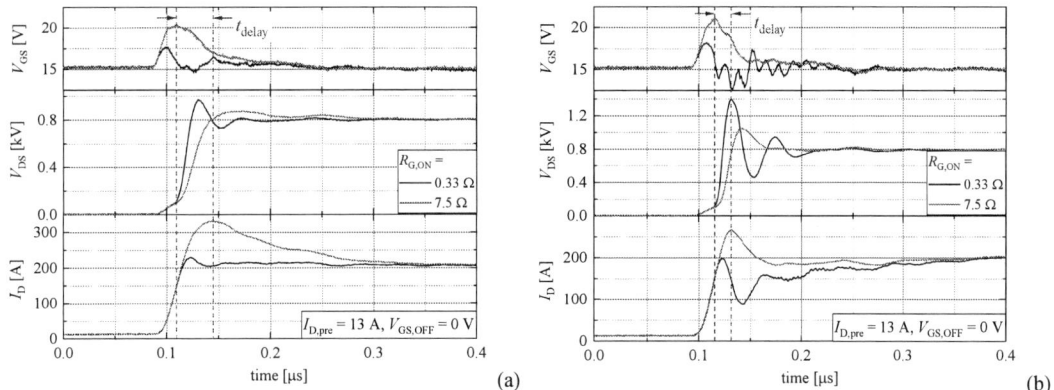

Fig. 4: SC II measurement with different $R_{G,ON}$, (a) M2 with double trench gate, (b) M3 with planar

As can be seen from Fig. 3 and 4, the gate impedance has the most impact on the SC II behavior. A trade-off between current peak and voltage peak has to be made. The relation between current peak and voltage peak according to the gate resistance is explained in [7]. Further, the different gate structures have a distinct response for varied $R_{G,ON}$. The SC II behavior of M2 with double trench gate is relatively soft even with very small (0.33 Ω) external $R_{G,ON}$. In contrast, M3 has an excessive voltage peak around 1.4 kV with the same condition due to the steep negative dI_D/dt. This voltage peak is close to the measured static avalanche breakdown voltage. Hence, such a voltage peak could force the DUT into avalanche mode and lead to potential destruction. The reason for this phenomenon may be due to different gate structures and their capacitive conditions. On the other hand, the integrated internal gate resistor $R_{G,int}$ (see Table I) on the chip, which is only 1.6 Ω for M3, while for M2 it is 12 Ω.

One interesting point that can be found in Fig. 4 is the time offset between the V_{GS} peak and the current peak for all three DUTs marked as t_{delay} for $R_{G,ON} = 7.5$ Ω for M2 and M3. At the early phase of the SC II (from the starting of desaturation to SC I like operation), the increased chip temperature should not be high enough to trigger already the parasitic npn-transistor. Therefore, the total current is still the electron current across the channel, which is controlled by the applied V_{GS}. Hence, a higher gate voltage should cause a higher current peak according to the saturation current equation for MOSFETs:

$$I_{D,sat} = \frac{W \cdot \mu_n \cdot C_{OX}}{L} \cdot \left(V_{GS,ON} - V_{G,th}\right)^2 \tag{2}$$

where W and L are channel width and channel length, respectively. C_{OX} is the gate-oxide capacitance. Two main factors cause this time offset t_{delay}: first, a measurement error is caused by the chip integrated $R_{G,int}$ and the gate inductance from the internal gate bond wire. Hence, a precise gate voltage measurement is almost not possible at the outside of the device packaging. The measured value is lower than the real chip V_{GS} if the gate current flows out of the gate. Besides this measurement error, the significant short-channel effect or Drain Induced Barrier Lowering (DIBL) effect plays a major role in this offset. The $V_{G,th}$ depends on the applied drain-source voltage V_{DS}. After the onset of the SC II, the current increases rapidly to the saturation value for the applied V_{GS}, V_{DS} rises slowly in this interval and holds on a low value. At the time point before fast desaturation, V_{GS} reaches the maximum, but the current does not. During the desaturation process (interval t_{delay} for $R_{G,ON} = 7.5$ Ω), the increased V_{DS} leads to a continuously decreased $V_{G,th}$. Hence, the SC current further increases after the V_{GS} peak according to Eq. (2). In other words, although the V_{GS} reaches the maximum, the SC current value is limited due to the high $V_{G,th}$ at low V_{DS} condition. Therefore, SC II behavior of SiC-MOSFETs is relatively soft compared to IGBTs. In addition, the negative temperature-dependency/coefficient of $V_{G,th}$ is also more pronounced due to the higher temperature rise rate of SiC-MOSFETs in SC operation [8]. This could lead to a further increase in the current value. At the current peak, the dI_D/dt is equal to zero, and V_{DS} reaches to the DC-link voltage.

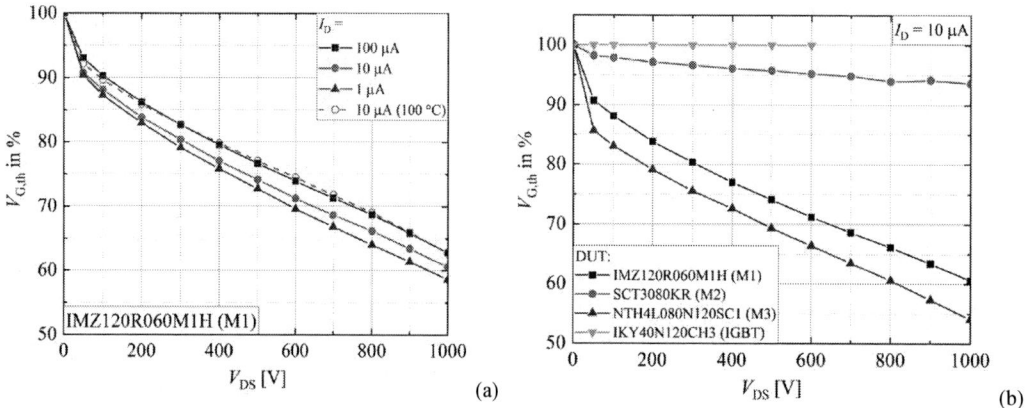

Fig. 5: (a) current and temperature dependency of the DIBL-effect for M1; (b) DIBL-effect for different DUTs

Fig. 5 shows the measurement of the DIBL effect. $V_{G,th}$ is extracted at I_D which is 10 µA to avoid self-heating. The current and temperature dependency of the DIBL effect for M1 is shown in Fig. 5 (a), the difference on the slope is negligible. It can be seen from Fig. 5 (b) that when the voltage increases to a specific value, (around 50 V), the threshold voltage shows a linear reduction with the increase of V_{DS}. This is valid for the asymmetric trench gate DUT (M1) and planar gate DUT (M3), where the DIBL effect is significant. $V_{G,th}$ is decreased about 34% and 40% at $V_{DS} = 800$ V, respectively. For the double trench gate structure (M2), the DIBL effect is weaker (5% at 800 V), and for the IGBT no DIBL-effect can be observed until 600 V. The reason for this phenomenon is on the one hand due to the lower electric field strength at the front-side p-region and n-base junction for IGBTs. On the other hand, it may be due to the particular electric field shielding structure, which reduces the influence of the electric field on the channel region, such as for the M2 double trench gate.

There are two concerns that cannot be explained with the performed measurements. First, for M2, although it has the biggest Q_{GD}/C_{GS} ratio, the corresponding SC II characteristic is not like that of IGBTs, which has a very high current peak. Moreover, attributed to its weak DIBL effect, $V_{G,th}$ should be hardly affected by V_{DS}. However, a large time shift between the V_{GS} and I_D (t_{delay}) peaks can still be observed in Fig. 4 (a). One possible reason for this behavior is the large internal gate resistance caused measurement error. The second point is that for the M3 device, as shown in Fig. 2, the Q_{GD}/C_{GS} ratio is the minimum, the V_{GS} overshoot due to the Miller-effect alone should not be higher than 5 V. However, the measured peak gate voltage is about 21 V under the condition of $R_{G,ON} = 7.5$ Ω. Meanwhile, V_{DS} at this time is only around 100 V, see Fig. 4 (b). Considering the presence of $R_{G,int}$, the actual V_{GS} should be even higher than the measured value. Therefore, a hypothesis can be constructed here: the gate voltage overshoot in SC II should not only be generated by the Miller-effect but there should be other unrecognized effects that dynamically affect V_{GS} in the process of current rising. Furthermore, under SC II condition, the current could be limited by other factors besides the gate voltage, such as the low channel electron mobility and the electric field applied across the channel. These open questions will be further investigated with the help of semiconductor device simulation in the future.

Short Circuit Type III Behavior of SiC-MOSFETs

In this section, the SC type III behavior is discussed, the utilized pulse pattern is explained in Fig. 6. In time interval t_1, the high-side auxiliary switch is turned-ON, the load current flows through the HS-switch and the load inductor L_{load}. The freewheeling current through the body diode $I_{D,BD}$ can be adjusted by changing the time of this period. In t_2, the HS-switch is turned-OFF and the BD starts to carry the load current. Depending on the applied gate voltage, the Body Diode (BD) can be operated in bipolar mode or the current flows unipolar via the channel. When the applied $V_{GS,OFF}$ is low enough, the channel will be completely closed. In this case, the current flows entirely through the BD in bipolar mode. When $V_{GS,OFF}$ is not low enough or equals zero, the current will be shared by BD and the channel due to the body-effect [9]. In this circumstance, the injection of holes from p-region is small.

To demonstrate the effect of plasma on the SC III behavior, $V_{GS,OFF}$ = -10 V was selected for M1, and -5 V for M2 and M3 during t_2. The duration of t_2 is fixed to 5 µs, to ensure that the BD enters a stable working state and the plasma distribution is stable as well. In interval t_{detrap}, the gate voltage of the DUT shifts to +15 V, the channel is opened. After this time point, the MOSFET works in the reverse conduction mode. This time period is defined as t_{detrap}, because the negative shift of the $V_{G,th}$ in t_2 is gradually compensated during this time. $V_{G,th}$ gradually increases with the increase of t_{detrap}. After t_{detrap}, the auxiliary switch turns-ON again to trigger SC III. The reverse recovery process of the BD starts, the current commutates to the channel and continues to rise to the peak value. The voltage and current waveforms are similar to SC type II. The minimum value of t_{detrap} was set to 50 ns. Because if t_{detrap} is 0, the DUT is turned-ON at the same time as the auxiliary switch. The turn-ON process of the aux. switch will affect the SC III characteristics of the DUT.

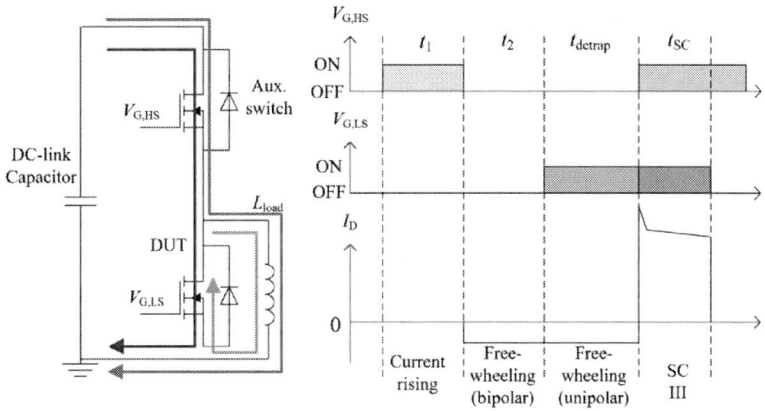

Fig. 6: SC type III test setup and pulse pattern

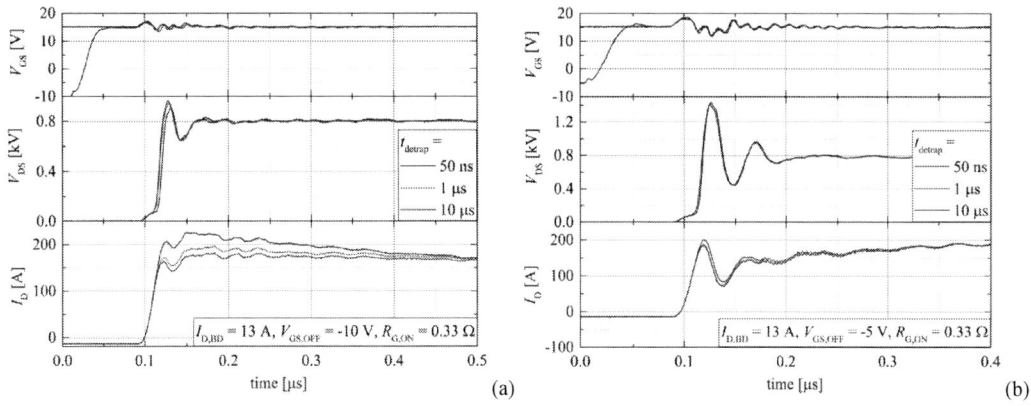

Fig. 7: SC type III behavior with different t_{detrap}: (a) M1, $V_{GS,OFF}$ = -10 V; (b) M3, $V_{GS,OFF}$ = -5 V

Fig. 7 (a) shows the effect of the de-trapping time t_{detrap} in the SC III behavior of M1. Since -10 V was applied to the DUT during t_2, therefore, the current flows entirely through the body-diode. When t_{detrap} is reduced to 50 ns, the dynamic $V_{G,th}$ of the DUT becomes the smallest, the hysteresis effect is most pronounced. Hence, the corresponding SC current peak value is the maximum. This current peak decreases with increasing t_{detrap}. The results for M2 are similar to M1, but a distinct result was measured for M3, as shown in Fig. 7 (b). The difference of the current peak is smaller and the peak value remains the same for t_{detrap} = 1 μs and 10 μs. This should be firstly related to a smaller $V_{G,th}$ hysteresis due to the lowest interface trap density of M3, which can be confirmed with a sub-threshold sweep in Fig. 8. The second reason could be the applied negative gate voltage (-5 V), which is not fully close the channel due to the body-effect. Hence, a small electron current can already flow through the channel and neutralize the trapped positive charge during t_2. Nevertheless, with this test setup, the negative gate voltage during t_2 can be further decreased to close the channel completely. But the DC-link voltage has to be reduced to avoid the possible critical electrical field in the oxide layer. t_{detrap} can be freely adjusted until the current peak reaches a stable value, which means that the trapped positive charge is fully neutralized. With this method, the time constant of the de-trapping phase for specific current and temperature can be estimated. For instance, the time constant of the de-trapping phase for M3 under given conditions is smaller than 200 ns.

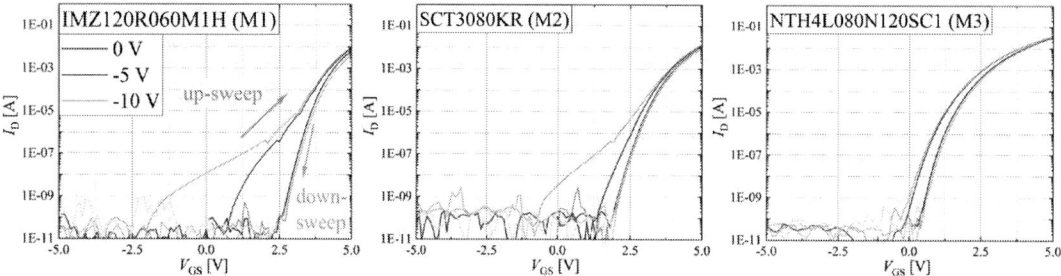

Fig. 8: Sub-threshold sweep from V_{GS} = -10, -5 and 0 V to V_{GS} = 15 V (up-sweep) and vice versa (down-sweep), V_{DS} = 0.1 V, I_D is limited to 500 mA

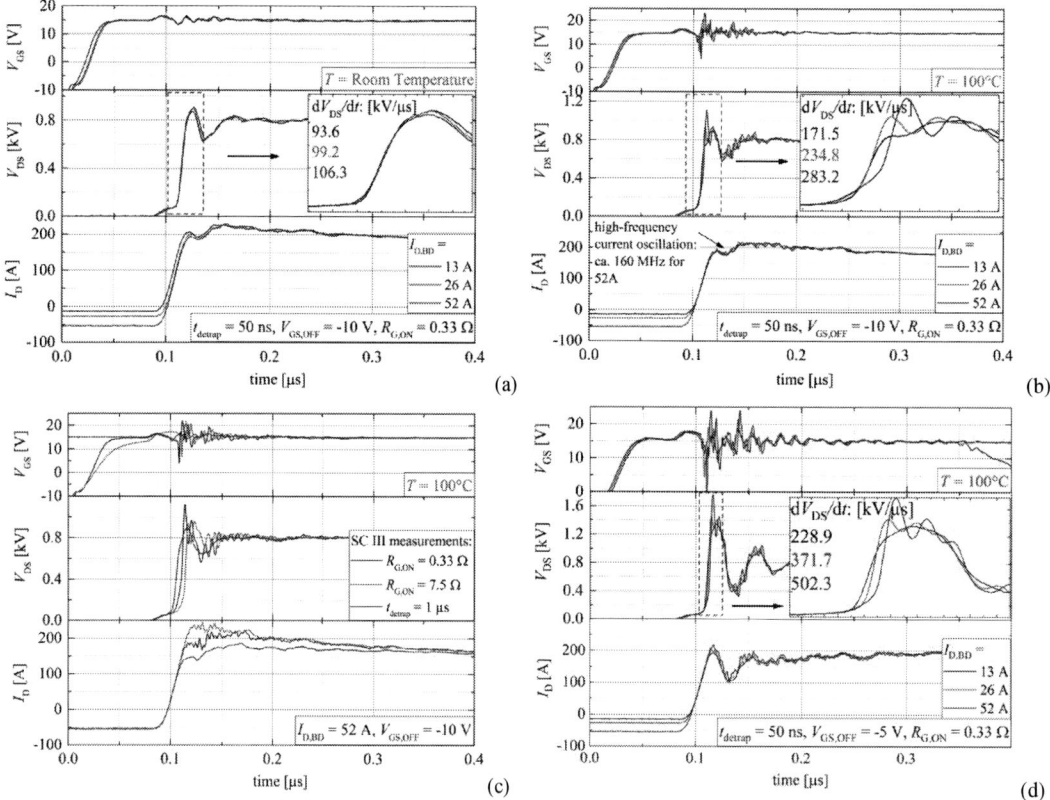

Fig. 9: (a) SC III behavior with different body diode current for M1 at room temperature; (b) at 100 °C. (c) Influence of the $R_{G,ON}$ and de-trapping time t_{detrap}, $I_{D,BD}$ = 52 A at 100 °C, t_{detrap} = 50 ns if not declared otherwise. (d) SC III behavior with different $I_{D,BD}$ for M3 at 100 °C, for $I_{D,BD}$ = 52 A, only 300 ns SC pulse was given to avoid a possible destruction.

The effect of different BD currents $I_{D,BD}$ on the SC III characteristics of M1 at room temperature is shown in Fig. 9 (a). The maximum dV_{DS}/dt during the desaturation is given in the figure as well. It can be seen that $I_{D,BD}$ value is hardly affecting the SC III characteristics. On the one hand, due to the wide bandgap of SiC material, p-doping cannot be fully ionized at room temperature [1]. Consequently, the emitter efficiency of the BD is quite low, despite the fact that the BD is working in the bipolar mode during t_2, which causes a specific density of plasma inside the MOSFET. But the plasma density is extremely low; although the current value is high. On the other hand, the minority carrier lifetime of SiC devices is much lower compared to Si-counterparts at room temperature [1]. The stored charge carriers are probably fully recombining in the t_{detrap} stage before SC occurs. Therefore, basically no influence from $I_{D,BD}$ on the SC III characteristics can be observed at room temperature.

However, when the temperature of the device increases, the minority carrier lifetime and the p-emitter efficiency of the BD increase simultaneously. The plasma may not fully be recombined after t_{detrap} and needs to be swept out during the reverse recovery process in SC III. Therefore, the value of $I_{D,BD}$ will affect the SC III characteristics. As shown in Fig. 9 (b) at 100°C, the peak value of the SC current decreases slightly due to the decreased electron mobility. But the gate voltage and drain-source voltage oscillate strongly. The oscillations become more severe as the current increases. The maximum dV_{DS}/dt at $I_{D,BD}$ = 52 A increases strongly from 106.3 kV/µs to 283.2 kV/µs. The frequency of this oscillation is approximately 160 MHz. The oscillations and V_{DS} spikes can be damped with a larger gate resistor. Besides, if t_{detrap} extends, the plasma could already recombine completely before SC III, and will hence no longer affect the SC characteristics as shown in Fig. 9 (c) where the blue current peak with t_{detrap} = 1 µs is the smallest.

For M2 with a double trench gate, the effect of the temperature on the SC behavior is not apparent. This is firstly due to the large 12 Ω internal gate resistor, which damps the oscillations. On the other hand, the selected $V_{GS,OFF}$ was -5 V during the t_2 phase. This voltage is already lower than the V_{GS} surge value specified in the datasheet (-4 V), but it may still not totally close the channel due to the body-effect. Hence, a part of the current flows through the channel during t_2. The plasma density in the MOSFET could be negligibly low before the short circuit. For M3 with a planar gate in Fig. 9 (d), the effect of the temperature is analogous to M1. However, due to the smaller $R_{G,int}$, the V_{DS} oscillation and the maximum dV_{DS}/dt of the device are even more pronounced than that of M1. For $I_{D,BD}$ = 13 A, the induced voltage peak already reaches 1350 V, which is above the rated blocking voltage. At 26 A and 52 A, it reaches 1420 V and 1764 V, respectively. In this case, the device will inevitably enter the avalanche mode.

In practical applications, the current and temperature are determined by the MOSFETs operating conditions, and SC can occur at any time of t_{detrap}. Therefore, to realize a better short circuit robustness, the external gate resistance should be carefully considered.

Conclusion

In general, due to the smaller Miller-capacitance and the significant DIBL effect of SiC-MOSFETs, the current overshoot of SC II is much smaller than that of IGBTs. Therefore, from the perspective of short-circuit energy, the SC II behavior of SiC-MOSFETs is not more critical than SC type I. The choice of the gate resistor should be carefully considered because it affects the current peak and induced voltage peak, which is similar to IGBTs.

The SC type III characteristics at room temperature are similar to SC II. The low plasma density at room temperature hardly influences the reverse recovery of the body diode under SC condition. When the temperature increases, the plasma effect on the reverse recovery of the BD can be observed from high-frequency oscillation and high induced voltage peak. However, a destruction could not be observed during the reverse-recovery phase, even at very high dV/dt. With a small R_G in combination with a short de-trapping time (the time difference between the application of positive gate voltage after bipolar conduction and the occurrence of SC), the device could enter the avalanche mode after the onset of SC III.

The reasons for the oscillation in the SC type III and the reverse recovery process of the body diode under short circuit condition will be further studied with the help of semiconductor device simulation in the future.

References

[1] T. Kimoto, J.A. Cooper: Fundamentals of Silicon Carbide Technology: Growth, Characterization, Devices, and Applications, John Wiley & Sons Singapore Pte. Ltd, 2014

[2] A. Castellazzi, T. Funaki, T. Kimoto and T. Hikihara, "Short-circuit tests on SiC power MOSFETs," PEDS 2013, pp. 1297-1300

[3] G. Romano et al., "A Comprehensive Study of Short-Circuit Ruggedness of Silicon Carbide Power MOSFETs," in IEEE Journal of Emerging and Selected Topics in Power Electronics, vol. 4, no. 3, , Sept. 2016pp. 978-987

[4] Z. Wang et al., "Temperature-Dependent Short-Circuit Capability of Silicon Carbide Power MOSFETs," in IEEE TPEL, Feb. 2016, pp. 1555-1566

[5] X. Liu, J. Kowalsky, C. Herrmann, C. Bäumler and J. Lutz, "The Influence of the Gate Driver and Common-Source Inductance on the Short-Circuit Behavior of IGBT Modules and Protection," in IEEE Transactions on Power Electronics, vol. 35, no. 10, pp. 10789-10798, Oct. 2020

[6] J. Lutz and T. Basler, "Short-circuit ruggedness of high-voltage IGBTs," 28th International Conference on Microelectronics Proceedings, 2012, pp. 243-250

[7] X. Liu et al., "Influence of the gate resistance on the short circuit type II & III behavior of IGBT modules and protection," PCIM Europe digital days 2020, pp. 1-9.

[8] A. Tsibizov et al., "Accurate Temperature Estimation of SiC Power mosfets Under Extreme Operating Conditions," in IEEE TPEL, Feb. 2020, pp. 1855-1865

[9] Dolny, Sapp, Elbanhaway and Wheatley, "The influence of body effect and threshold voltage reduction on trench MOSFET body diode characteristics," ISPSD 2004, pp. 217-220

Analog MPPT Comparison for Interplanetary Small Satellites Missions

C. Torres, A. Garrigós, J. M. Blanes, P. Casado, D. Marroquí, C. Orts.

Miguel Hernández University of Elche

Industrial Electronics Group

Elche, Spain

E-Mail: c.torres@umh.es

URL: http://www.umh.es

Acknowledgment

This work has been supported in part by the Generalitat Valenciana and the European Social Fund through the Subvención Para la Contratación de Personal Investigador de Carácter Predeoctoral under grant reference ACIF2020/154

Keywords

MPPT, Aerospace, Battery charger, Photovoltaic, P&O MPPT.

Abstract

This paper describes and compares two different Maximum Power Point Tracking (MPPT) techniques devised for micro satellite Solar Array Regulation (SAR). An Analog Global MPPT is introduced and compared with an Analog Oscillating MPPT. Advantages and limitations are discussed. Experimental validation has been carried out using a Power Conditioning Unit and different I/V curves.

Introduction

The use of small satellites with attitude control and ability to perform orbital jumps is nowadays an increasingly widespread practice, which allows interplanetary or deep space missions at a lower cost than traditional space missions with larger spacecrafts. For this reason, an increasing number of companies and scientific research groups are working on the development and evolution of the different subsystems that make them up.

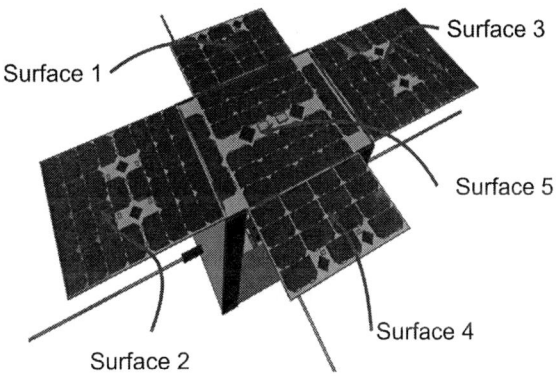

Fig. 1: Rendered image of the proposed solar cell distribution in deployable and body-mounted panels.

In this context, a power subsystem for a deep space mission with a variable distance between 1.5 AU (astronomical unit) and 2 AU from the Sun has been proposed in [1]. That platform has a volume of 36 U, a size of 30 x 30 x 40 cm and a weight around 50 kg including fuel and payload. To power the various subsystems the satellite has a total amount of 120 solar cells model 3G30C from the manufacturer Azur Space distributed in six solar arrays (SA). Each SA is distributed in two parallel connected strings made up of ten series-connected solar cells. The solar cells are accommodated in four deployable panels plus a body-mounted panel, as represented in Fig. 1.

The Power Conditioning Unit (PCU) is responsible for extracting power from the six SAs, charge the batteries and supply the rest of subsystems. The main element of the PCU is the Solar Array Regulator (SAR), which in this case, is divided into six independent phases with their outputs connected in parallel, see Fig. 2. Each phase includes a synchronous Buck converter powered by one of the six available SA's, an error amplifier that regulates the end of charge battery voltage or extracts maximum power from the photovoltaic source otherwise. A detailed explanation of the SAR control loop operation is described in [2]. Additionally, as a means of protection, the SAR has three overvoltage monitors that disconnect the Buck converters from the SA's if the battery voltage exceeds a pre-set threshold. The SAR has been designed following European Space Agency (ESA) guidelines to assure single point failure free (SPFF) operation

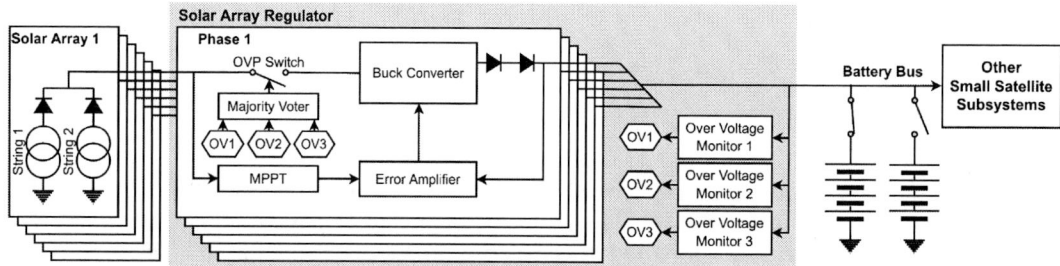

Fig. 2: Block diagram of the proposed PCU.

In extreme operating conditions, such in interplanetary missions, it is expected to have very different solar cell I/V curves in the same solar array. This may be due to different factors, some of them are summarised in Table I.

Table I: Effects causing mismatches between various solar cells.

Factor	*Causes and effects*
Δ Irradiance	Variations in irradiance affect the amount of current supplied by each solar cell. Further, the arrangement of the solar cells on different facets of the satellite can cause them not to receive irradiance homogeneously or to be affected by the shadow of some other element. This effect can be aggravated if the deployment of the panels is not fully successful.
Δ Temperature	Variations in temperature will affect the voltage and current provided by the solar cells. Temperature variations may be due, for example, to the location of the cell on the satellite. In the proposed satellite, cells located on the body-mounted panel could experience higher temperature than cells located on the rest of the facets of the satellite.
Cell Mismatch	Due to the manufacturing process, solar cells may present variations of up 10% in current and open circuit voltage V_{OC}.
Aging	Aging, mainly due to operation in highly radiation environments, produces a degradation in both voltage and current. Aging is independent in each cell.
Micrometeorites	Impacts produced by micrometeorites can damage the cells producing different degradation effects.

The aforementioned factors can create complex P/V curves such as those shown in Fig. 3. Fig 3 shows simulated I/V and P/V curves of a single SA at 1.5 AU with different degradation effects.

As can be seen, degradation, shadowing and other effects create P/V curves with several local maximum power points (MPP). This can cause that some oscillation-based Maximum Power Point Trackers (MPPTs) to get trapped oscillating around a local maximum, making impossible to extract the maximum available power from the solar panel.

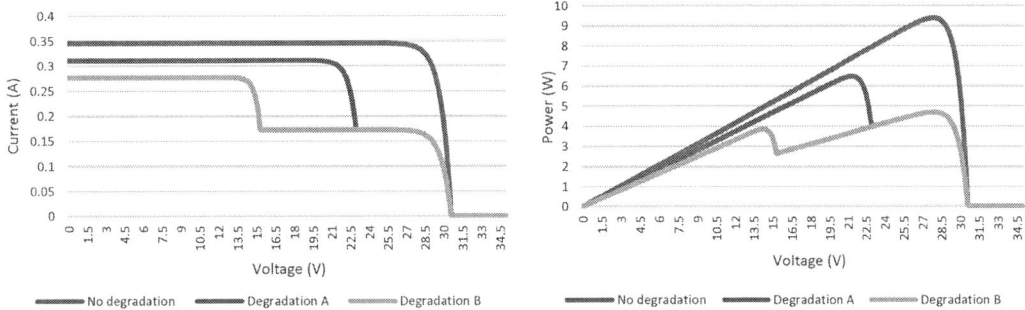

Fig. 3: Representation of curves of SA in different conditions. On the left I/V curve. On the right P/V curve.

In this work, the design of an analog oscillation-based MPPT and an Analog Global MPPT (AGMPPT) are presented. Both designs are compared, and their performance is measured under different I/V curves scenarios. Finally, the results obtained are discussed and the conclusions are presented.

Analog oscillation-based MPPT

The first of the two methods proposed in this work is an oscillating MPPT. This method, first proposed in 1960s [3], has been adapted and widely adopted in many space missions due to its simplicity and robustness [4]. This MPPT requires measurement of SA voltage and current. Fig. 4 shows the diagram of the proposed MPPT.

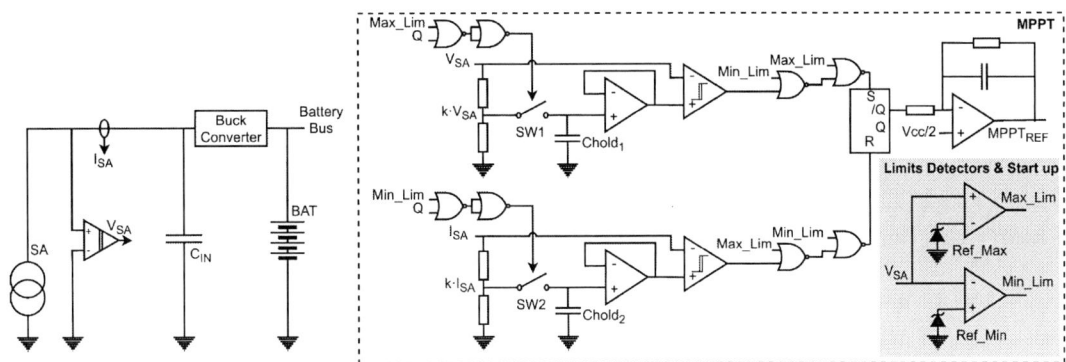

Fig. 4: Diagram of the proposed SAR using the oscillating-based MPPT method.

Initially, the MPPT starts at OC (open circuit) of the SA. In this moment, the output voltage of the SA is captured and the MPPT reference decreases causing an increase in the duty cycle of the DC-DC converter. As the duty cycle of the dc-dc converter increases the SA operating point moves towards to SC (short circuit), until the voltage V_{SA} decrease a certain (ΔV) amount determined by $k \cdot V_{SA}$. At that moment, RS flip-flop's SET signal becomes active, output Q changes its state and I_{SA} is recorded. As a result, the MPPT reference increases, the duty cycle of the converter is reduced, and the input voltage moves towards OC until I_{SA} decrease a certain (ΔI) amount determined by $k \cdot I_{SA}$. Then, RS flip-flop's RESET signal turns on, the MPPT reference is inverted, and the cycle starts again. This process is repeated continuously, causing the MPPT to oscillate around the Maximum Power Point (MPP) of the SA, as explained in detail in [3] and [4]. The MPPT reference signal is generated by integrating the flip-flop's output signal. Additionally, the MPPT has maximum and minimum voltage limit detectors, that forces the MPPT to work within a defined voltage range and ensure system starts.

Analog Global MPPT

The second method analysed is an Analog Global MPPT. The block diagram of the proposed AGMPPT is shown in Fig. 5. The AGMPPT uses the input capacitor of the Buck converter to scan the solar array and to save the global MPP (GMPP). The operation principle is as follows, first, the buck converter is turned-off and the input capacitor discharged (CLK1 signal). Once CLK1 is released, the I/V curve is scanned as the capacitor charges. During I/V curve scanning, the AGMPPT detects the global MPP using a peak detector. To measure the power (P), an analog PWM multiplier computes $I_{SA} \cdot V_{SA}$. The V_{SA} voltage corresponding to the global MPP of the SA is stored to be used later as a reference for the Buck converter during MPPT period. Once the scan is complete, the Buck converter is turned-on again. The converter remains on and working in the global MPP during the MPPT *on*-period. When the *on*-period (MPPT period) ends, the input capacitor is discharged again and the scanning process is repeated, thus obtaining a new global MPP for the buck converter. A more detailed explanation of the operating principle of this AGMPPT is provided in [5].

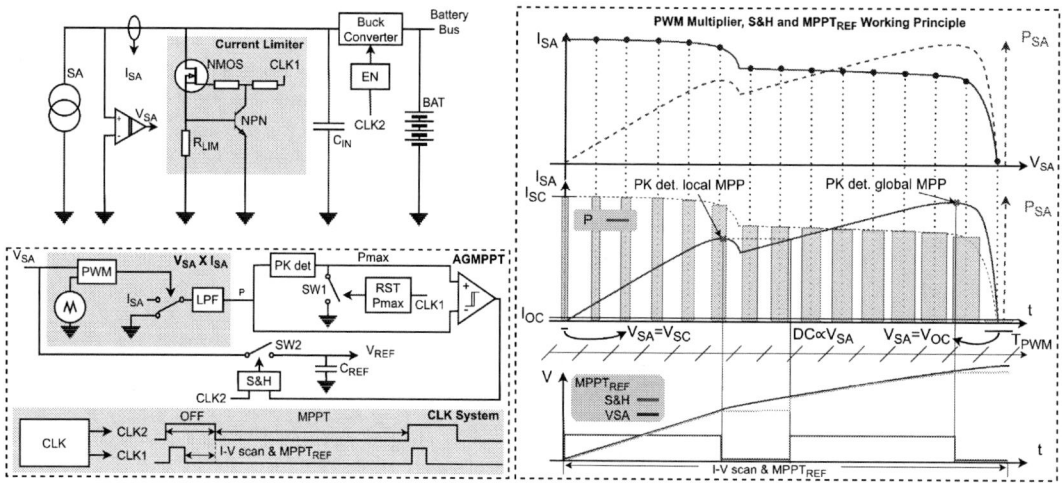

Fig. 5: Diagram of the proposed SAR using the AGMPPT method.

Theoretical comparison of the oscillating-MPPT vs the AGMPPT

Oscillating, perturb and observe and other MPPT techniques have been widely used due to its high efficiency and good performance. However, all of them have a major drawback, they cannot discriminate between local and global MPP. This can be a serious problem in certain I/V curves, as could be the case of the curve "Degradation A" shown in Fig. 3. In that case, the MPPT starts working from V_{OC} and move towards the I_{SC} point on the I/V curve, getting trapped oscillating around a local MPP.

The AGMPPT has been proposed to overcome this issue, since the full I/V curve is scanned. This AGMPPT requires a scan time during which the converter remains off, affecting its average efficiency. To minimize this issue, a compromise between the *off* time (t_{OFF}) and the MPPT-*on* time (t_{ON}) must be reached. The AGMPPT period (T) is equal to $t_{OFF} + t_{ON}$.

Another critical aspect that must be taken into consideration is the implementation of the analog multiplier. Simple and low consumption device is required to maximise efficiency, but also accuracy and fast response is essential, otherwise it is not possible to perform a correct scan of the I/V curve and, consequently, an erroneous MPP point would be used. Since monolithic space-qualified analog multipliers are scarce and complex, a simple analog PWM multiplier is used.

Regarding the implementation of both MPPT, Table II indicates the number of components, classified by groups, that are required for the implementation of each of the designs. It is observed that the number of components is similar in both MPPTs, so it is assumed that the area required for the implementation is also similar. As a main difference in terms of components is that oscillating-MPPT does not require any timer (i. e. 555 IC), although, it uses two more comparators than the AGMPPT. Therefore, the implementation cost of both systems is similar.

Table II: Comparison of the number of components required in each implementation.

Component	*N° of components*		*Component*	*N° of components*		*Component*	*N° of components*	
	Osc. MPPT	*AGMPPT*		*Osc. MPPT*	*AGMPPT*		*Osc. MPPT*	*AGMPPT*
Resistors	27	33	Comparators	4	2	555 IC	0	2
Capacitors	23	25	Diodes/Zener	3	2	OpAmp	4	4
Logic gates	8	9	Switches	2	3	I sensor	1	1
BJT Trans.	0	2	Flip-flops	1	0	MOSFET	0	2
Total n° of MPPT components:	**73**					**Total n° of AGMPPT components:**	**85**	

Implementation of both MPPTs and experimental tests

To verify that both MPPTs behave as expected, the complete SAR is implemented with the oscillating-MPPT integrated, and the AGMPPT implemented on a separate PCB. To optimize the period (T) of the AGMPPT, several experimental efficiency measurements have been carried out, obtaining a maximum efficiency for T = 2 s and t_{Off} = 45 ms. The developed prototypes are shown in Fig. 6.

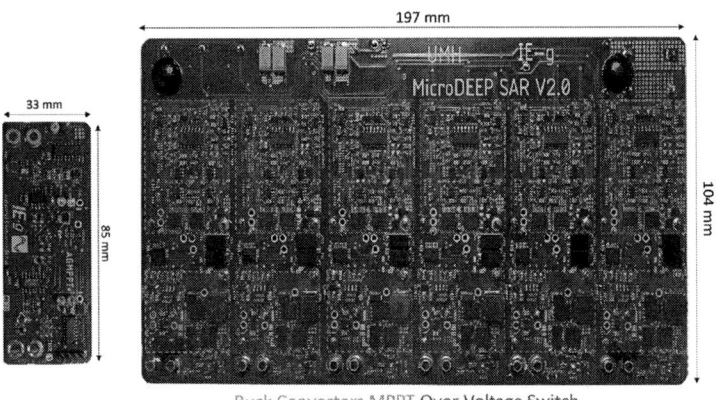

Fig. 6: AGMPPT is shown on the left. Full SAR with MPPT is shown on the right.

Tests have been performed using two different scenarios corresponding to a solar array in configuration 10S2P at a sun distance of 1.5 AU (610w/m^2, -30 °C) and 2 AU (342 w/m^2, -54,3 °C) respectively. For each scenario, three I/V curves are considered. First curve considers an ideal I/V curve with a single MPP. Second (Degradation 1) and third (Degradation 2) cases consider I/V curves with two local MPPs. These curves are simulated using two Agilent E4351B SA simulators connected in parallel. The curves, as well as the working point set by each MPPT in each of the tests, are shown in Fig. 7.

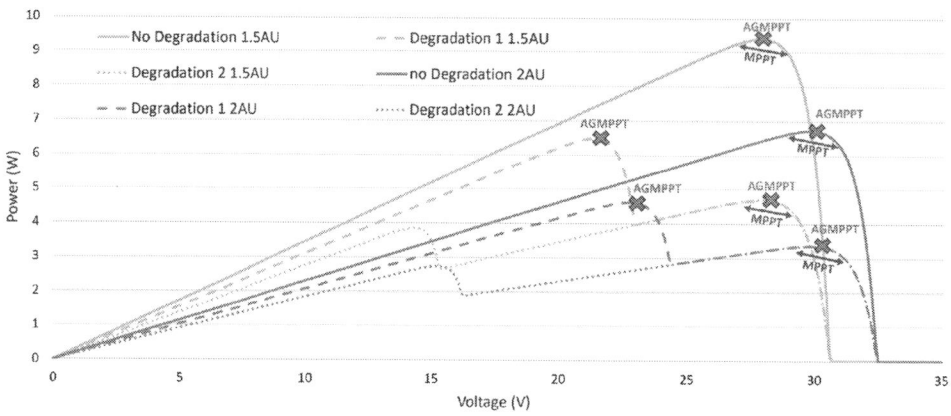

Fig. 7: Different simulated curves and working points obtained by both MPPT

To carry out the tests, an ITECH IT-M3412 battery simulator connected to the SAR output has been used. All tests have been carried out at a fixed output voltage of 16V. Measurement equipment is the following: Tektronix MDO3104 oscilloscope, Tektronix TPP1000 voltage probes and Tektronix TCPA0030A current probes. Efficiency measurements is performed by calculating the average power extracted from the SA divided by the total power available in the SA at MPP.

The results show that the oscillating-MPPT works correctly when the global MPP is the closest to the V_{OC} point. These are the "No Degradation" and "Degradation 2" cases for curves located both 1.5 AU and 2 AU. In cases where there is a local MPP between the global MPP and the V_{OC} point, it is observed that the oscillating MPPT is trapped working in that local MPP. That is the case "Degradation 1" at both 1.5 AU and 2 AU.

On the other hand, the AGMPPT scans properly the entire SA I/V curve, being able to identify its global MPP and working on it in all cases. Next figures show the behaviour of each of the MPPTs for some of the tested curves.

Fig. 8 shows the operation of the oscillating-MPPT with "Degradation 1" curve at a sun distance of 1.5 AU. Initially the MPPT starts at the V_{OC} point, as shown by the V_{SA} signal. The MPPT takes about 0.5s to start up because the control loop is slow and is initially saturated. Once steady-state is achieved, it oscillates around the 28 V point on the I/V curve, which, as shown in Fig. 7, corresponds to the voltage around the local MPP of SA. From this moment on, the MPPT is permanently working around the local MPP, thus wasting part of the power of the SA. Additionally, Fig. 8 also shows the VSA voltage, the ISA current, the power extracted from the SA and the MPPT reference signal.

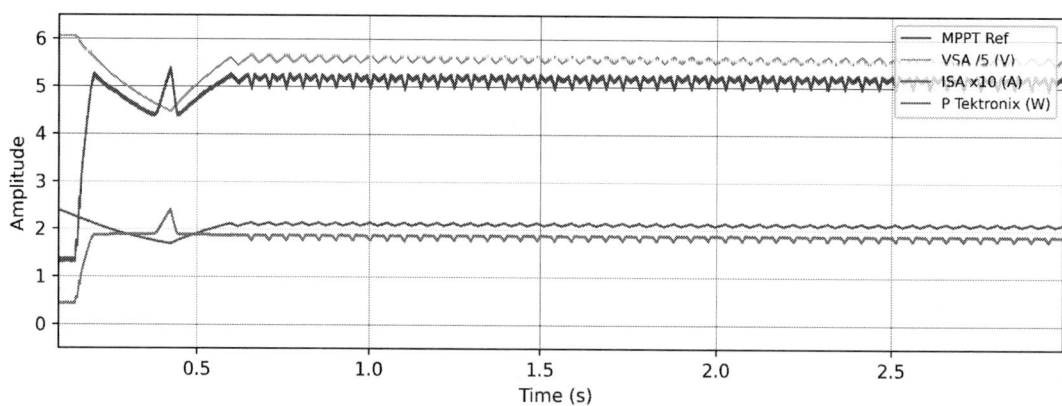

Fig. 8: "Degradation 1" curve, distance from sun of 1.5 AU. oscillating-MPPT.

On the other hand, Fig. 9 shows the behaviour of the AGMPPT in these same operating conditions and P/V curve. Between t = 0 and t = 0.046s, the capacitor is discharged, and the I/V curve is scanned. It is observed that power measured in the AGMPPT multiplier (P AGMPPT) corresponds accurately to the power measured in the oscilloscope (P Tektronix). It can be also observed that peak power detector identifies the global MPP of the I/V curve and stores the V_{SA} voltage value, which is used later as a reference for the Buck converter. At t = 0.046s, the Buck converter turns on and works in the global MPP, taking the previously stored value as a reference. The converter remains working at the global MPP for the next 2 seconds. Then, it will turn off again to repeat the scanning process. It should be noted that the AGMPPT has been able to work at the global MPP of the P/V curve, obtaining an efficiency greater than 96%, while the oscillating one has been stuck at the local MPP, reducing its efficiency to 75%.

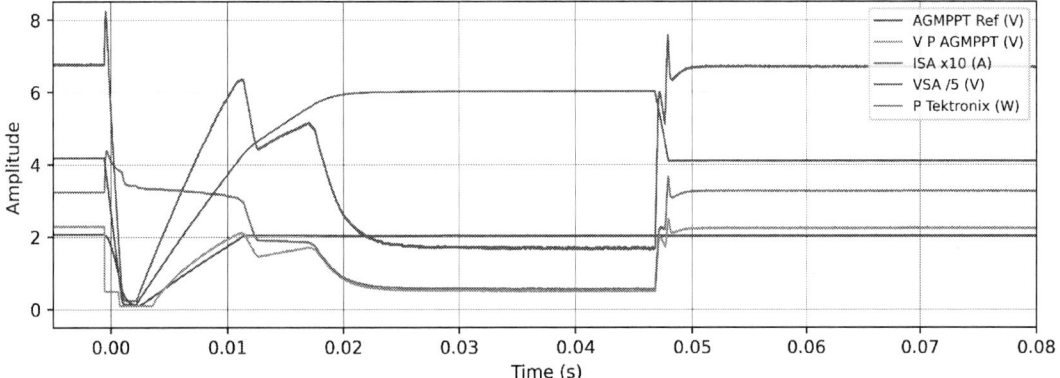

Fig. 9: "Degradation 1" curve, distance from sun of 1.5 AU. AGMPPT.

A second batch of measurements consider the 1.5AU, "Degradation 2" I/V curves. Fig. 10 shows the result obtained with the oscillating-MPPT. The MPPT starts at the V_{OC} point of the SA I/V curve. Once the system has started, the operating point moves to the next maximum found when tracking the I/V curve from the V_{OC} point to the I_{SC} point. In this case, that local maximum found corresponds to the global MPP, so now the oscillating-MPPT operates in the global MPP. The V_{SA} and I_{SA} signals show how the MPPT is oscillating around the MPP. Additionally, the power extracted from the SA (P Tektronix) and the MPPT reference signal (MPPT Ref) are shown. It should be noted that in this case the efficiency is better than 99%.

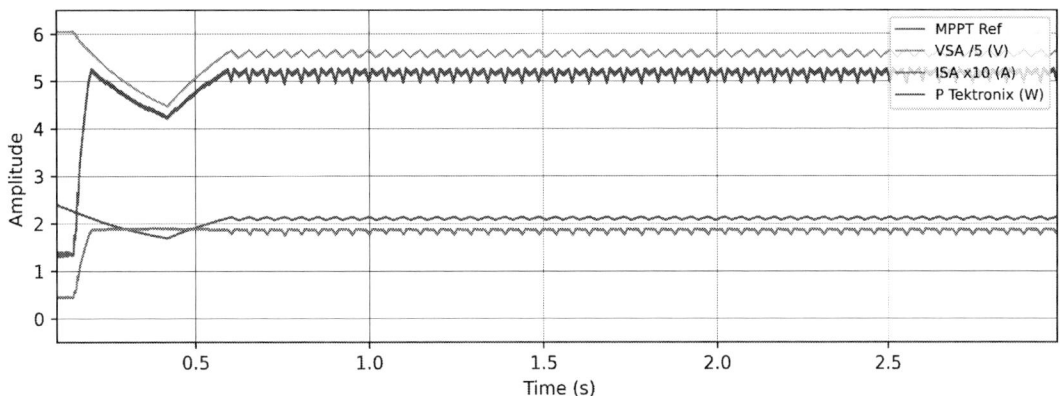

Fig. 10: "Degradation 2" curve, distance from sun of 1.5 AU. oscillating-MPPT.

The last test has been carried out using the AGMPPT with the previous I/V curve. The results obtained are shown in Fig. 11. It is shown that AGMPPT scans the I/V curve accurately, retaining at each moment the voltage V_{SA} corresponding to the MPP of the section of the curve already scanned. Once the scanning process is finished, the Buck converter is activated, setting the I/V point of the curve corresponding to the detected global MPP as its working point. The converter will remain working in the global MPP for 2 seconds. Then, the scanning process will start again. It should be noted that the efficiency obtained by the AGMPPT in this case is better than 97.60%, being slightly below the efficiency obtained with the oscillating MPPT under these same conditions.

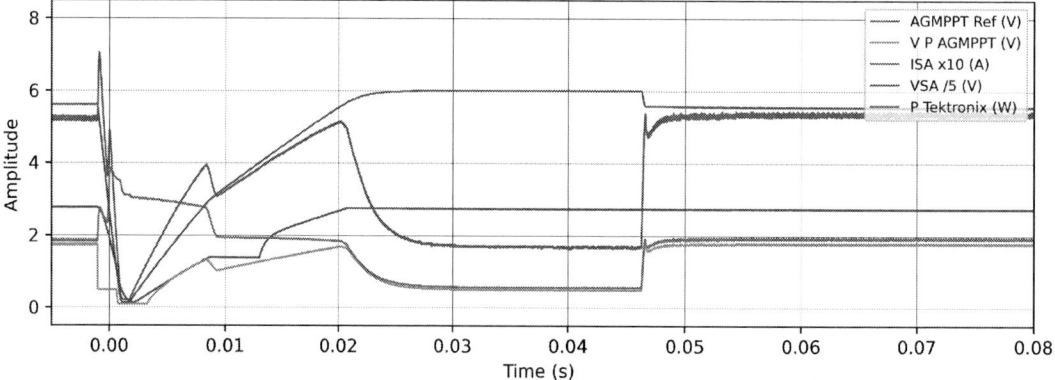

Fig. 11: "Degradation 2" curve, distance from sun of 1.5 AU. AGMPPT.

Experimental validation continues repeating these tests for the different curves, obtaining similar results. The following section shows the efficiency results obtained in each test.

Results

The data related to the efficiency of the MPPTs obtained experimentally in the various tests are shown in Table III. It is observed that when the curves do not have local MPP, both MPPTs work correctly, obtaining efficiencies between 96.01% and 98.82%. It is observed that in this case the efficiency of the oscillating-MPPT is slightly better than that of the AGMPPT. In the cases in which the curves present local MPP, but these are not between the global MPPT and the V_{OC} point (Degradation 2), the results are similar. The oscillating-MPPT has obtained a maximum efficiency of 99.06% while the AGMPPT has obtained a maximum efficiency of 97.65%.

On the other hand, in the cases in which the SA I/V curve presented local MPP between the global MPP and the V_{OC} point (Degradation 1), it is observed that the efficiency of the AGMPPT has remained between 96.02% and 96.07%. However, the efficiency obtained by the oscillating-MPPT has decreased to values between 71.90% and 75.04%. It is in this type of curves where the AGMPPT shows a much better performance than the oscillating-MPPT.

Table III: Experimental results obtained.

	SA MPP Power (W)		*Average Power Output (W)*				*Efficiency (%)*			
			AGMPPT		*Osc. MPPT*		*AGMPPT*		*Osc. MPPT*	
Sun Distance (AU):	*1.5*	*2.0*	*1.5*	*2.0*	*1.5*	*2.0*	*1.5*	*2.0*	*1.5*	*2.0*
No Degradation	9.684	6.76	9.36	6.49	9.564	6.68	96.65	96.01	98.76	98.81
Degradation 1	6.45	4.63	6.1936	4.448	4.84	3.329	96.03	96.07	75.04	71.90
Degradation 2	4.92	3.4	4.773	3.32	4.829	3.368	97.01	97.65	98.15	99.06

In general, both MPPTs have worked correctly, although the oscillating-MPPT has obtained a slightly higher efficiency than the AGMPPT in most cases. However, in some curves the oscillating-MPPT has presented efficiency losses of almost 30%, while the AGMPPT has always remained above 96%.

Analysing these results and considering that the cost of implementation and the area required by both models is very similar, it can be concluded that the use of the AGMPPT is advantageous compared to the use of the oscillating-MPPT in applications in which complex P/V are expected and SA power optimization is critical. This is because the AGMPPT ensures high performance with any kind of I/V curve with a penalty of 1% of efficiency reduction compared to traditional methods. If complex I/V curves are not anticipated, oscillating-MPPT might be more indicated.

Conclusion

In this work, two MPPT analog methods have been presented, an oscillating-MPPT and an AGMPPT, for high performance small satellites. The application to which both MPPTs are intended has been presented, the electrical diagrams have been shown and the operating principle of both has been explained. Subsequently, a theoretical comparison is made. This comparison presents the main advantages of each design, as well as the bill of materials that allows a fair estimation of the area and the cost of implementing one design compared to the other. Then, both designs have been implemented and tests have been carried out under various conditions with the aim of validating the correct operation of the designs. The results of the tests show that both MPPTs behave as expected. Oscillating-MPPT performs slightly better than AGMPPT if the I/V curve does not present local MPP between the global MPP and the V_{OC}. However, it cannot operate properly if the first MPP found is a local MPP. On the other hand, the AGMPPT has high efficiency in all cases, which makes it ideal for use in applications where reliability is a critical design aspect and high performance is sought.

References

[1] J. A. Carrasco *et al.*, "Micro-platform Power System for Scientific Deep Space Exploration," in *12th European Space Power Conference*, 2019.

[2] A. Garrigós, J. L. Lizan, J. M. Blanes, and R. Gutierrez, "Exploring the Use of the LT3480 (RH3480) Circuit as Low-Power, Low-Voltage Solar Array Regulator," in *10th European Space Power Conference*, 2014, ESA SP-719.

[3] A. F. Boehringer, "Self-Adapting dc Converter for Solar Spacecraft Power Supply," *IEEE Trans. Aerosp. Electron. Syst.*, vol. AES-4, no. 1, pp. 102–111, 1968, doi: 10.1109/TAES.1968.5408938.

[4] W. Denzinger, "Electrical power system of globalstar," *Proc. 4th Eur. Sp. Power Conf.*, pp. 171–174, 1995.

[5] A. Garrigós, D. Marroquí, J. M. Blanes, R. Gutiérrez, M. Compadre, and C. Clark, "An analog global maximum power point tracking for photovoltaic systems: Application to nanospacecrafts," *2017 19th Eur. Conf. Power Electron. Appl. EPE 2017 ECCE Eur.*, vol. 2017-Janua, pp. 1–9, 2017, doi: 10.23919/EPE17ECCEEurope.2017.8098986.

Feasibility assessment of variable-speed generator set concepts with focus on rating of power electronic equipment

Hendrik Fehr[1], Albrecht Gensior[1], Andreas Möckel[1], Frank Atzler[2], Tilo Roß[2],
Carsten Reincke-Collon[3]

[1]Technische Universität Ilmenau, Germany; [2]Technische Universität Dresden, Germany;
[3]Aggreko Deutschland GmbH, Germany

hendrik.fehr[a]; albrecht.gensior[a]; andreas.moeckel[a]; frank.atzler[b]; tilo.ross[b];
carsten.reincke-collon@aggreko.com
[a]@tu-ilmenau.de; [b]@tu-dresden.de

Keywords

≪Generation of electrical energy≫, ≪Microgrid≫, ≪Adjustable speed drive≫, ≪Doubly-Fed Induction Generator≫, ≪Non-standard electrical machine≫, ≪Power flow≫

Abstract

Four different engine-generator concepts with variable-speed operation of the internal combustion engine are analyzed regarding relevant performance criteria. The topologies under investigation have an internal energy storage in order to mitigate the impact of load steps. The performance criteria address the required rating of the power converters in stationary operation, after load steps and to deliver short-circuit current.

Introduction

Mobile generator sets are mainly used for the supply of electrical power in off-grid applications, e.g. at construction sites, as well as for grid support parallel to weak grid access points. Furthermore, they are used for emergency power supply.

The state-of-the-art solution consists of a Diesel engine coupled to a synchronous generator running at constant speed, as depicted in Fig. 1a. In practical applications, where the generator set has to supply loads with a high starting current, e.g. asynchronous machines, or nonlinear loads, e.g. diode rectifiers, the rated power of the generator set is chosen up to three times the rated power of the loads [1,2]. This setup has been proven to be very robust and reliable, however large and not energy-efficient:

- Since the generator set is designed for the peak power, most of the time it runs only at part load, typically below 20% of the rated power. In order to illustrate this, Fig. 1b shows the histogram of the power demanded from a generator set in a typical application from a construction site. Such a load profile implies that the Diesel engine consumes more fuel for the same amount of electrical energy compared to an operation in the engine map range of optimal fuel consumption. This range is located typically at approximately half the rated engine speed and 70% to 80% load. The engine can also be designed specifically for lowest fuel consumption in a particular map range, but this is always also a function of the desired rated power.
- Since the speed of the generator set is proportional to the grid frequency, it is neither possible to run the engine at higher nor at lower speeds. Higher speeds may be beneficial in order to gain more power from the same engine, lower speeds may help to reduce fuel consumption at lower loads.

This motivates to investigate alternative concepts. Similar as in wind energy conversion systems, there are two main concepts that have been proposed [3]:

- First of which uses a synchronous machine running at variable speed which is connected to a back-to-back converter. The DC link can optionally be supported by an energy storage [4]. In low-power

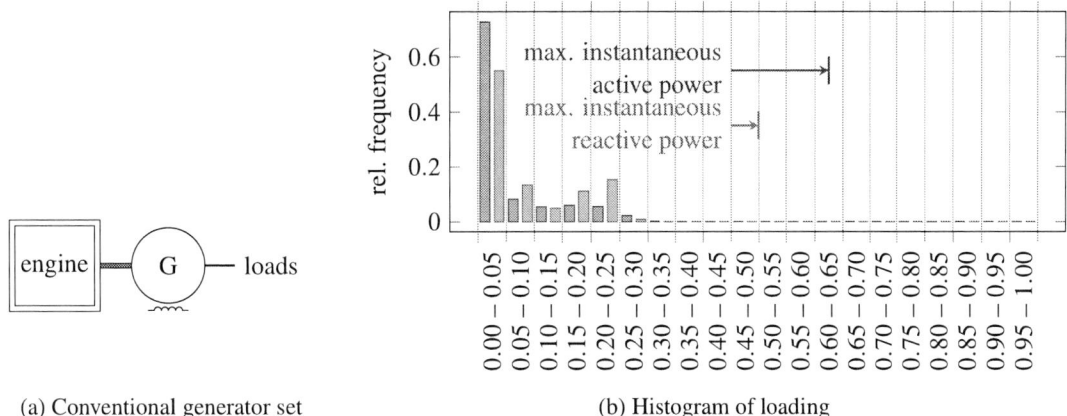

(a) Conventional generator set   (b) Histogram of loading

Fig. 1: Block diagram of the conventional fixed speed concept (a) and histogram (b) of the loading of such a genset with a rated power of 500 kVA (400 kW active power) in a crane application: the data is taken from 3-minute-average measurements of active power (blue) and reactive power (red). The abscissa shows the relative loading referring to the rated active power.

    applications, the machine side front end can be realized by a simple diode rectifier with a boost converter as in [4, 5]. Otherwise, an active front-end is proposed [6]. The concept is also attractive for full or hybrid ship propulsion systems [7, 8].

- The second concept employs a doubly-fed induction generator (DFIG) as in [9,10]. In this concept, a partial power rating of the power electronics converters is intended. Here as well, the DC link may be equipped with an energy storage.

Both concepts are gearless and increase the efficiency especially when operated at partial load. A further configuration employs an electric Continuously Variable Transmission (e-CVT) which is a special gear box with three mechanical ports. In [11, 12], it has been proposed for ship propulsion systems. Although being applicable for the present system as well, it is considered to be out of scope here in order to stay with a gearless solution.

Crucial for an industrial application is the rating of the power converters because they are considered to be the most expensive part. For low-power applications, selectivity in the protection system may not be required. This means that a short circuit may shut down the whole system. However, in applications for higher power or emergency supplies this is not acceptable and the power system must be able to sustain the short circuit current in order to trigger the protection devices, e.g. fuses. This is the weak point of both concepts: For the fully-rated conversion, the grid-side converter (GC) must be rated to more than rated current. In case of the DFIG with partially rated power converters, the control of the system via the machine-side converter (MC) is lost during the fault due to the clamping action of the crow-bar circuit [13, 14] that protects the MC from overvoltage.

This is why, another solution is proposed that preserves a synchronous machine directly coupled to the grid. Thus, short-circuit currents can be delivered without stressing the power electronics converters with this task. In order to adapt the speed of the internal combustion engine (ICE) to the fixed speed of the synchronous machine, a cascaded machine (CM) with two rotors as in [15] is inserted between both components, see Fig. 2. It exposes three energy ports: the two rotating shafts both of which carry the same torque at different speeds (except for the accelerating and frictional torque) and an electrical port accessible via three-phase slip rings which provide the slip power to a power electronic converter.

In order to be able to withstand load steps, in particular when the ICE is running at low speed and needs to be accelerated before being able to take over the full load, an energy storage is required. This can be realized as an electrical one connected to the DC link or a flywheel and allows for a set of combinations. The essential ones are presented below. The goal of the present paper is to evaluate the concepts in terms of efficiency and the power rating of the ICE, the electrical machines and the power converters. For this

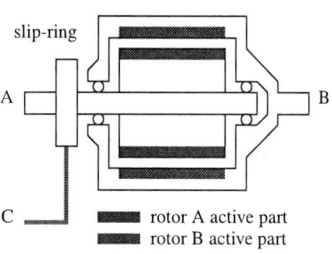

Fig. 2: Principle diagram of a cascaded machine, depicted without outer housing. The construction allows independent rotational speeds at the shaft ends A and B whose interacting torque is controlled by the three-phase electric machine that is formed by the active parts at both sides of the air gap. The speed difference and the interacting torque determine the corresponding electric power that is exchanged via the slip-ring system at the connection C. For a considerable operating range, the electric power is smaller than the mechanical power, offering the benefit of a partial power converter.

investigation, a power plant in the medium power range is considered.

The paper is organized as follows: In the next section, the requirements and performance criteria are discussed. This is followed by a section describing the concepts investigated before a comparison is done in a further section. The last section gives a conclusion.

Requirements and performance criteria

Focusing on stand-alone applications in the medium to high power range (30 kVA to 500 kVA rated power), one single generator set has to supply multiple electrical devices. This implies the following:

- The grid provided by the generator set has to supply the (possibly stepwise changing) load, which is one of the main reasons for the overdimensioning according to the state-of-the-art.
- The protection system of the grid must provide selectivity, i.e. a short circuit must trigger only the corresponding breaking element in order to minimize the number of loads affected. Therefore, the system must be capable of supplying a sufficiently large current for a few seconds.
- The efficiency of the whole system should be as high as possible.
- Targeting cost-effective mobile applications, the total weight and the rating of the power components must be as low as possible.

In view of the above expectations, potential variable speed concepts should combine the individual strengths of each system component and aim at a mutual compensation of possible weaknesses. Here, the following comparison criteria are used:

- Sizing of the grid facing components considering the short circuit current requirement.
- Power flow sharing among the components in stationary operation.
- Power flow sharing among the components after load steps.

Since the considered power range can be handled with a low-voltage solution, the power electronic converters will be of the two-level or three-level voltage source converter type, because these topologies are expected to provide the most economic solution. Moreover, these topologies allow a straight-forward extension from three phases to four phases to allow feeding of single phase loads via the GC.

Engine-generator concepts

Concept A: Fully-rated conversion   A straight-forward concept for variable speed operation is given by the block diagram in Fig. 3a. The conventional concept is extended by a MC and a GC operating in a back-to-back configuration in order to unlink the engine speed from the grid frequency. The energy storage, e.g. a super capacitor or battery, is connected to the DC link.

Each energy conversion stage must provide full system power capacity. In principle, a diode rectifier suffices as the MC. However, an active rectifier is assumed in the following, since it does not limit the choice of the generator type, can boost the generator voltage and allows motor-operation to quickly accelerate the ICE to the new operating point after load steps. For large load steps, the generator accelerates the ICE with the help of the energy storage which also takes over the load via the GC. Both converters will have a considerable cost footprint since each must handle the full power while the GC must even be rated for the short circuit current, which is the main drawback of this concept. The advantages are a low

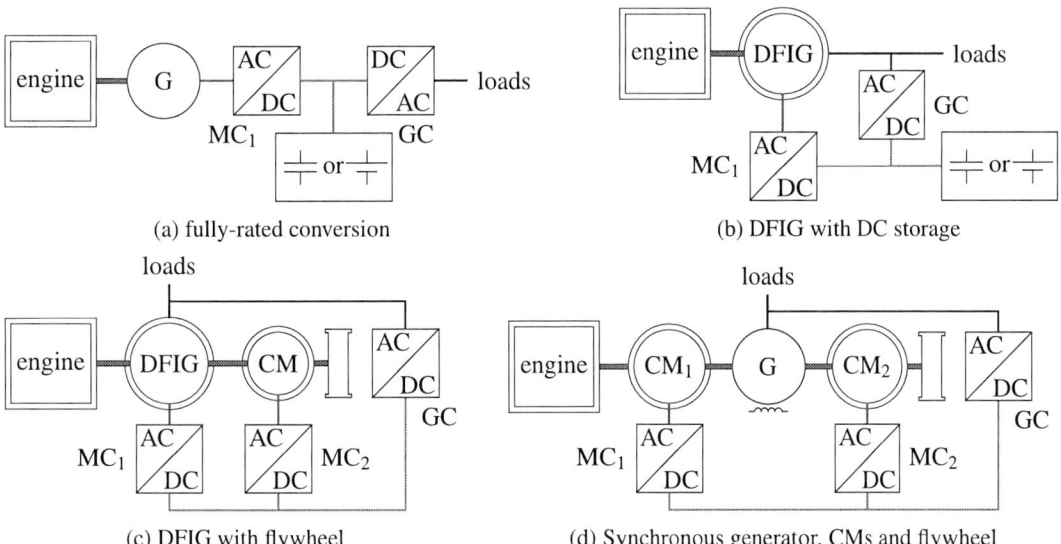

(a) fully-rated conversion

(b) DFIG with DC storage

(c) DFIG with flywheel

(d) Synchronous generator, CMs and flywheel

Fig. 3: Block diagrams of variable speed concepts: (a) fully-rated conversion allowing various generator types; (b) and (c) partial power conversion based on a DFIG; (d) partial power conversion based on a CM and synchronous generator. Concepts (a) and (b) use a super-capacitor- or battery-based storage connected to the DC link, while concepts (c) and (d) use a flywheel-based storage, coupled to the shaft.

number of main components, a simple mechanical system, and more design freedom for the optimization of the generator, since it is decoupled from the grid such that the respective requirements do not apply anymore.

Concept B: Doubly-fed induction generator with DC storage DFIGs offer two parallel paths for the energy flow, allowing a partial power rating of both, the machine and the power electronic conversion stages in a useful speed range. The block diagram is depicted in Fig. 3b. During high-load operation, the shaft speed is higher than synchronous speed and the slip power is transmitted via the MC and GC to the load. During low-load operation, the shaft speed is lower than synchronous speed while the corresponding slip power is fed into the rotor implying a circular energy flow involving the DFIG and both converters. Although the DFIG can help to relieve the GC from a large part of the short circuit current, the control of the system is impaired during grid faults, because of the crow-bar operation in the rotor circuit. Matching the MC and the standstill voltage of the DFIG in order to spare protective measures during grid faults is unfeasible because it impedes the partial power rating. Similar to the fully-rated conversion, the energy storage is connected to the DC link of the converters.

Concept C: Doubly-fed induction generator with flywheel A flywheel is used in order to avoid motor operation of the DFIG during load steps with the aim to reduce the power requirements of the GC. This is depicted in the block diagram in Fig. 3c. In order to control the torque and the speed range of the flywheel, a CM is used. During stationary operation, the flywheel maintains a higher speed than the main shaft. During a load step, the flywheel momentum is released to accelerate the ICE via the shaft and to supply the load via the DFIG. Moreover, a part of the momentum is extracted as electric power via the slip rings and provided to the DC link to be used by the MC or GC. Since a large part of the flywheels momentum is transferred as mechanical power, the respective power electronics converter only handles a fraction of that.

Compared to the previous concept, the GC can be smaller, however, this concept faces the same challenges during grid faults: control of the system is severely restricted due to the crow-bar operation.

Concept D: Synchronous generator with electromechanical transmission with flywheel In order to increase the short circuit current capability while maintaining full system control during faults, this concept employs a DC excited synchronous generator to take over the major part of the load during normal operation and to provide the short circuit current during faults. The block diagram is depicted in Fig. 3d. Variable speed operation is achieved by CM_1 and the energy storage is based on a flywheel, coupled via CM_2 to the main shaft, as in the previous concept.

In this concept, partial power rating for the converters and the synchronous generator is maintained during all relevant modes of operation. For high-load condition, the speed of the ICE is higher than the main shaft speed and the slip power of CM_1 is provided via the power electronics converters to the load, implying power sharing between the machines and the converters. During a load step, the flywheel feeds the load via the generator and the GC while it also accelerates the ICE via both CMs. Since the energy flow uses parallel electrical and mechanical paths during stationary and dynamic operation, this arrangement enables partial power rating for both, power converters and electric machines. Its main drawback is the increased number of main components. However, the rated power of each conversion stage is lower than the full system power.

Comparison

Sizing of the grid facing components considering the short circuit current requirement

In order to design a selective protection scheme, a three-phase short circuit current of three times the full load current for several seconds can typically be expected from a conventional generator set [16]. Assuming the same for the variable-speed systems, the GC of Concept A must be rated to three times the generator power, because the short circuit duration is much longer than the semiconductor thermal time constants, implying stationary operation of the converter. In the DFIG-based Concepts B and C, the electric machine and the GC contribute to the short circuit current, raising the hope of a reduced GC rating if the control challenges are solved that arise from the crow-bar operation during the fault. Compared to the other concepts, the lowest short circuit current requirement can be expected for the GC of Concept D, since the synchronous generator is able to reliably provide the majority of fault current by leveraging its thermal overload capability.

Power flow sharing among the components in stationary operation

In order to compare the power flow for each of the concepts, a minimum set of assumptions has to be made. Here, an application with a grid frequency of 50 Hz is assumed. The electrical machines have four poles which leads to a synchronous speed of $n_{\text{sync}} = 1500$ rpm. The speed of the ICE is $n_E \in [\check{n}_E, \hat{n}_E]$, with $\check{n}_E = 1000$ rpm and $\hat{n}_E = 3000$ rpm, where the lower or upper end is used when the power demand is low or high, respectively. The efficiency of the power conversion stages in a steady-state operation is denoted as $\eta_E = 0.4$, $\eta_G = \eta_{\text{CM}} = \eta_C = 0.95$ for the ICE, the generator, the CM and the power electronics converters, respectively. These values are assumed to be realistic for industrial applications above 30 kVA.

The DFIG and the CM are modeled as three-port devices as depicted in Fig. 4a including the direction definitions of the port powers P_1, P_2, and P_3. When P_1 is positive in stationary operation, the relation

$$P_1 \eta = P_2 + P_3 \tag{1}$$

is used to consider an energy conversion efficiency $0 < \eta \le 1$. The port powers of the DFIG and the CM are defined in Figs. 4b and 4c, respectively and they correspond to the powers in (1) as given by Table I. The slip of the DFIG and the CM are denoted by s and s_{CM}, respectively, and their definitions read

$$s = \frac{n_{\text{sync}} - n_E}{n_{\text{sync}}}, \qquad s_{\text{CM}} = \frac{n_B - n_A}{n_B}, \tag{2}$$

with the mechanical speeds n_A, n_B of the CM. In a similar manner, the stationary operation of the power electronics converters, the ICE, and the synchronous generator can be modeled with their respective port

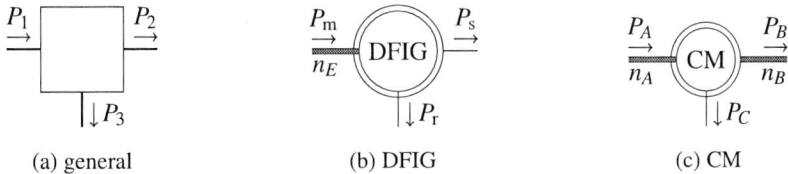

(a) general   (b) DFIG   (c) CM

Fig. 4: Three-port representations: general (a); DFIG (b); CM (c). The power relations are given by (1) after inserting the corresponding definitions from Table I.

Table I: Port power and efficiency definitions for the component modeling with (1). The general three-port system with the power directions are indicated in Fig. 4a. The slip of the DFIG and the CM are given by (2). In case of non-ideal conversion, i.e. $\eta < 1$, the models are valid for $P_1 > 0$.

	general	DFIG	CM	ICE	MC	GC	generator
power at port 1	P_1	P_m	P_A	P_fuel	$P_\mathrm{MC,AC}$	$P_\mathrm{GC,DC}$	$P_\mathrm{gen,m}$
power at port 2	P_2	P_s	P_B	$P_\mathrm{E,m}$	$P_\mathrm{MC,DC}$	$P_\mathrm{GC,AC}$	$P_\mathrm{gen,s}$
power at port 3	P_3	$P_\mathrm{r} = -sP_\mathrm{s}$	$P_C = -s_\mathrm{CM}P_B$	0	0	0	0
efficiency	η	η_G	η_CM	η_E	η_C	η_C	η_G

powers and conversion efficiencies, as indicated in the other columns of Table I. The third port power is not needed in these cases, since they are two-port devices, i.e. $P_3 = 0$.

In the following analysis, constant conversion efficiencies have been assumed for the components and their dependency on the DFIG- or CM-slip is neglected. In order to obtain a mathematical representation of each concept, the respective block diagrams can be translated to equations in the port powers from Table I. This results in

$$P_\mathrm{load} = P_\mathrm{GC,AC} \qquad P_\mathrm{GC,DC} = P_\mathrm{MC_1,DC} \qquad P_\mathrm{MC_1,AC} = P_\mathrm{gen,s} \qquad P_\mathrm{gen,m} = P_\mathrm{E,m} \qquad (3)$$

for Concept A,

$$P_\mathrm{load} = P_\mathrm{GC,AC} + P_\mathrm{s} \qquad P_\mathrm{GC,DC} = P_\mathrm{MC_1,DC} \qquad P_\mathrm{MC_1,AC} = P_\mathrm{r} \qquad P_\mathrm{m} = P_\mathrm{E,m} \qquad (4)$$

for Concepts B and C, and

$$P_\mathrm{load} = P_\mathrm{GC,AC} + P_\mathrm{gen,s} \qquad P_\mathrm{gen,m} = P_B \qquad P_\mathrm{GC,DC} = P_\mathrm{MC_1,DC} \qquad P_\mathrm{MC_1,AC} = P_C \qquad P_A = P_\mathrm{E,m} \qquad (5)$$

for Concept D. During stationary operation at high load, the ICE runs at $\hat{n}_E$, implying operation above synchronous speed for the DFIG- or CM-based concepts B, C, and D. For this mode of operation, the models in Table I are valid and the required fuel power P_fuel for a given load P_load can be determined by successive elimination and substitution of the relations in (3), (4), and (5) for Concept A, the Concepts B and C, and Concept D, respectively, until only P_load and P_fuel remain. This can be achieved by applying the general model (1) with the definitions from Table I and results in

$$P_\mathrm{fuel} = P_\mathrm{load} \left(\eta_E \eta_G \eta_C^2\right)^{-1} \qquad\qquad\qquad = 2.92\,P_\mathrm{load} \quad \text{for Concept A} \qquad (6)$$

$$P_\mathrm{fuel} = P_\mathrm{load} \left(1 - \hat{s}\right) \left(\eta_E \eta_G \left(1 - \hat{s}\eta_C^2\right)\right)^{-1} \qquad = 2.77\,P_\mathrm{load} \quad \text{for Concepts B, C} \qquad (7)$$

$$P_\mathrm{fuel} = P_\mathrm{load} \left(1 - \hat{s}_\mathrm{CM_1}\right) \left(\eta_E \eta_\mathrm{CM} \eta_G \left(1 - \hat{s}_\mathrm{CM_1} \eta_C^2\right)\right)^{-1} = 2.84\,P_\mathrm{load} \quad \text{for Concept D} \qquad (8)$$

with the slip $\hat{s} = \hat{s}_\mathrm{CM_1} = \frac{n_\mathrm{sync} - \hat{n}_E}{n_\mathrm{sync}}$. From the intermediate expressions, the port powers of each component can be obtained leading to the Sankey diagrams in Fig. 5 that depict the corresponding energy flow for the Concepts A to D. As a common basis for comparison, the load power P_load is the same in each diagram.

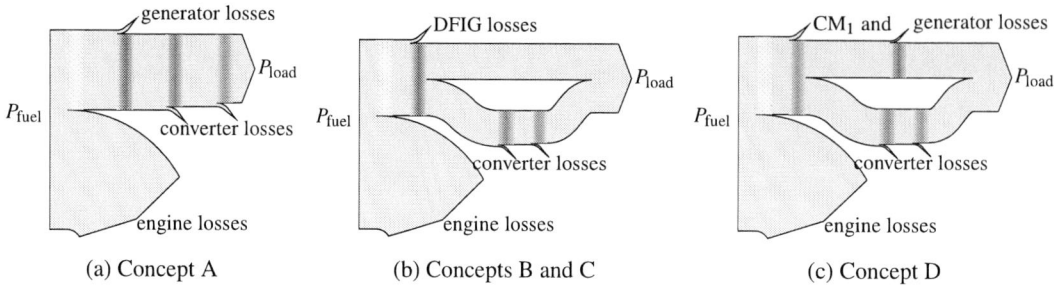

(a) Concept A   (b) Concepts B and C   (c) Concept D

Fig. 5: Sankey diagrams for the Concepts A to D during stationary operation for high-load, implying operation above synchronous speed for the DFIG or CM. The load is the same for each diagram in order to visually compare the power carried by each component. The output power of the respective component is indicated by color: □ ICE, ▨ generator or DFIG, ▨ CM_1, ▨ GC, ▨ MC_1.

The relevant power for the rating of the respective power electronics converters are given by

$$P_{GC,AC} = P_{load} \qquad\qquad P_{MC_1,DC} = P_{load}/\eta_C \qquad\qquad\qquad \text{for Concept A} \quad (9)$$

$$P_{GC,AC} = P_{load}\hat{s}/(\hat{s} - \eta_C^{-2}) \qquad P_{MC_1,DC} = P_{load}\hat{s}/(\eta_C\hat{s} - \eta_C^{-1}) \qquad \text{for Concepts B, C} \quad (10)$$

$$P_{GC,AC} = P_{load}/(1 - \eta_G\eta_C^{-2}\hat{s}_{CM_1}^{-1}) \quad P_{MC_1,DC} = P_{load}/(\eta_C - \eta_G\eta_C^{-1}\hat{s}_{CM_1}^{-1}) \qquad \text{for Concept D.} \quad (11)$$

Compared to the fully-rated Concept A, the converter powers of the other concepts are smaller, since the DFIG of Concepts B and C or the CM_1 of Concept D allow a fractional rating of the power converters.

Power flow sharing among the components at the start of load adaptation

As outlined above, the generator set must be able to deal with large load changes. In order to keep the energy storage small, the ICE must change operating points quickly which means that it may need to be accelerated. The worst-case scenario applies directly after a large load step during a low-load condition, i.e. where the internal combustion engine needs to be accelerated from low to required engine speed before it is able to take over the load. For the sake of simplicity, the contribution of the combustion process to the acceleration is neglected, leaving the respective energy storage as the sole source. Since the focus lies on the distribution of the power between the components, the losses are neglected because their impact is considered to be low. Thus, the power flow sharing at the start of load adaptation can be obtained with the modeling scheme from the previous section as well, since the models become valid for a potentially reverse power flow. In this case, the dc storage for concepts A, B and the flywheel for concepts C, D need to be considered, since they provide the energy. With P_{load} and P_{acc} being the load power after the load step and the power dedicated to the acceleration of the ICE, the powers relevant for the rating of the respective power electronics converters are:

$$P_{GC,AC} = P_{load} \qquad\qquad P_{MC_1,AC} = P_{acc} \qquad\qquad \text{for Concept A} \quad (12)$$

$$P_{GC,AC} = P_{acc}/(1 - \check{s}) + P_{load} \qquad P_{MC_1,DC} = P_{acc}\check{s}/(1 - \check{s}) \qquad \text{for Concept B,} \quad (13)$$

where $\check{s} = \frac{n_{sync} - \check{n}_E}{n_{sync}}$ denotes the slip at the start of load adaptation. For the flywheel-based concepts, the powers depend on the slips $\hat{s}_{CM} = \frac{\hat{n}_{FW} - \check{n}_E}{\hat{n}_{FW}}$, $\check{s}_{CM_1} = \frac{n_{sync} - \check{n}_E}{n_{sync}}$, and $\hat{s}_{CM_2} = \frac{\hat{n}_{FW} - n_{sync}}{\hat{n}_{FW}}$ of the CMs, leading to

$$P_{GC,AC} = P_{MC_2,DC} + P_{MC_1,DC} \tag{14}$$

$$P_{MC_1,DC} = \check{s}\left[P_{acc}\hat{s}_{CM} + P_{load}(\hat{s}_{CM} - 1)\right](1 - \check{s})^{-1} \tag{15}$$

$$P_{MC_2,DC} = (P_{load} + P_{acc})\hat{s}_{CM} \tag{16}$$

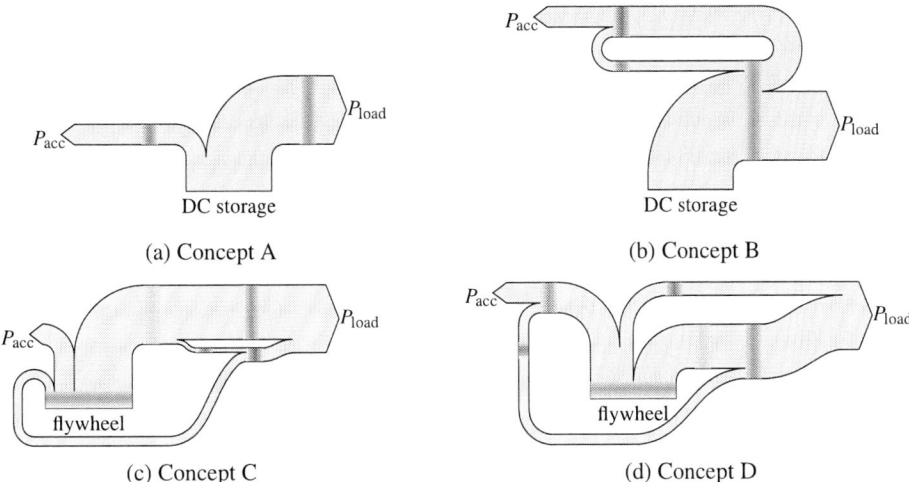

(a) Concept A (b) Concept B

(c) Concept C (d) Concept D

Fig. 6: Sankey diagrams for the Concepts A to D at the instant after a large load step during low-load condition, implying operation below synchronous speed for the DFIG and CM. The load P_{load} is the same for each diagram as well as the accelerating power P_{acc} that is used to quickly reach the new operating point of the ICE. The power of the respective component is indicated by color: ▨ GC, ▨ MC$_1$, ☐ MC$_2$, ▨ generator or DFIG, ▨ CM or CM$_1$, ▨ CM$_2$.

for Concept C, and for Concept D to

$$P_{\text{GC,AC}} = P_{\text{MC}_1,\text{DC}} + P_{\text{MC}_2,\text{DC}} \tag{17}$$

$$P_{\text{MC}_1,\text{DC}} = -P_{\text{acc}}\check{s}_{\text{CM}_1}(\check{s}_{\text{CM}_1} - 1)^{-1} \tag{18}$$

$$P_{\text{MC}_2,\text{DC}} = (P_{\text{load}} + P_{\text{acc}})\hat{s}_{\text{CM}_2}. \tag{19}$$

Similar as above, the Sankey diagrams in Fig. 6 illustrate the power flow and the required rating of the converters. In each diagram, the load P_{load} and the accelerating power $P_{\text{acc}} = 0.3P_{\text{load}}$ are the same, relying on the DC storage for the Concepts A and B or on the flywheel for the Concepts C and D. For all concepts, the speed of the engine is assumed to be at its lower end $\check{n}_E$ and for Concepts C and D, the flywheel is running at its upper speed range end $\hat{n}_{\text{FW}} = 3000\,\text{rpm}$. Similar as in Fig. 5 for the stationary operation, the required power capability of the power electronics and the electric machines can be compared for this key operating condition. The GC of Concept A continues to carry the full system power, while the MC acts as inverter, carrying the accelerating power and the generator losses since the latter runs in motor operation, as shown in Fig. 6a. The GC of Concept B handles even larger power than the GC of the fully-rated concept, because it takes over the full load plus the accelerating power and even the rotor power which the DFIG requires, since it runs below synchronous speed. Consequently, the Concept B is unsuitable for accelerating the internal combustion engine, since the energy storage connection at the DC link implies a suboptimal energy flow. The flywheel-based Concepts C and D do not suffer from this problem and thus, are better suited for accelerating the ICE, as visible from Figs. 6c and 6d. The flywheel-provided powers in Concepts C and D are similar to the battery or super-cap powers in Concepts A and B, however, the flywheel converter power in Concept C is larger than in Concept D, since the respective speed difference to the main shaft is larger. In contrast to Concept B, the rotor power of the DFIG in Concept C advantageously contributes to feeding the load, because the circular energy flow of Concept B is avoided in generator operation below synchronous speed. However, similar to Concept B, Concept C suffers from a problematic power flow pattern when the flywheel speed is lower than the main shaft speed while the latter is still lower than synchronous speed. This rotational speed constellation requires a power flow into both, the CM and the DFIG rotor, which must be provided via the DFIG stator and the GC. Avoiding this operation mode increases the system design effort and control challenges for Concept C. Concept D is better in this regard, since the speed difference of the flywheel

Table II: Comparison of the rating of the converters in relation to P_{load} for stationary and transient operation as well as for providing short circuit current. The values in the transient and stationary rows correspond to Figs. 5 and 6, respectively. For the transient case, the losses have been neglected. The maximum power determining the required rating is highlighted in boldface.

	Concept A		Concept B		Concept C			Concept D		
	GC	MC_1	GC	MC_1	GC	MC_1	MC_2	GC	MC_1	MC_2
stationary	1.00	**1.05**	0.47	**0.50**	0.47	**0.50**		0.48	**0.51**	
transient	1.00	0.30	**1.45**	0.15	**0.80**	0.06	**0.87**	**0.80**	0.15	**0.65**
short circuit	**3.00**									

to the main shaft decreases much slower during discharging operation than for Concept C, because the main shaft speed is maintained at synchronous speed.

Results

The quantitative results of the comparison are collected in Table II and can be summarized as follows:

Concept A has the lowest number of main components. In order to supply short-circuit current, the GC must be rated much higher than for nominal load. However, for acceleration of the ICE, the power flow of the MC is reversed but does not require a higher rating of the converter. Nevertheless, the required rating of the GC seems to be the largest drawback of the concept.

Concept B suffers from an inconvenient loading of the GC when it is necessary to accelerate the ICE. Thus, maintaining this feature demands a higher than nominal grid converter rating. However, together with Concept C, it seems to be the most efficient configuration in a stationary operation at full load.

Concept C avoids the main drawback of Concept B such that a partial-load rating of the GC can be achieved. However, as for Concept B, the controllability of the system is lost in case of short circuits.

Concept D avoids the main drawbacks of the other concepts with a slightly worse efficiency compared to Concepts B and C but a better efficiency than Concept A, making it highly attractive for further research.

Conclusion

Conventional fixed-speed generator sets suffer from low efficiency in case of highly variable load scenarios, such as e.g. construction sites, due to the long periods of low power operation and occasional peak load demand. Variable-speed operation of the ICE improves the conversion efficiency during low-load operation, allowing for an overall fuel saving compared to the conventional fixed-speed concept. The present paper compares different concepts of gearless generator sets that allow for a variable-speed operation of the ICE and feature an integrated energy storage to increase the resiliency with respect to load steps. Furthermore, since all configurations have a power electronic converter directly coupled to the grid, they are, in principle, capable of dealing with nonlinear loads. The paper compares four concepts two of which use an electrical storage while the other two use a flywheel. The investigations show that one of the concepts outperforms the other ones because it is the only one which satisfies the following key criteria: 1) the rating of the power electronics converters is below the rated power of the system, 2) it is possible to accelerate the ICE externally after a load step, 3) the system is able to deliver short-circuit current in order to trigger protection devices without losing the controllability. As a result, the paper adds one more promising concept to the set of variable-speed generator set configurations that have been investigated in recent research.

However, there remain open questions that have not been targeted yet. The dimensioning of the energy storage, the ability to supply reactive power as well as more detailed investigations on the efficiency obtained in a realistic scenario are left for future research.

References

[1] D. Kruger and R. Van Maaren. Sizing generators for motor starting—a practical guide to understanding how motor-starting loads affect generator performance. Technical report, Kohler, 2018.

[2] A. Thakur. Managing emergency generators with nonlinear loads. Technical report, Kohler, 2017.

[3] W. Koczara and G. Iwanski. Fuel saving variable speed generating set. In *2009 International Conference on Clean Electrical Power*, pages 22–28, Capri, Italy, Aug. 2009.

[4] J. Hamilton, M. Negnevitsky, and X. Wang. The potential of variable speed diesel application in increasing renewable energy source penetration. In *2nd International Conference on Energy and Power*, pages 558–565, Sydney, Australia, Dec. 2018.

[5] J. Leuchter, et al. Dynamic behavior modeling and verification of advanced electrical-generator set concept. *IEEE Transactions on Industrial Electronics*, 56(1):266–279, 2009.

[6] W. Koczara and T. Balkowiec. Smart microgrid grid power quality improvement and reduction of fuel consumption by application of adjustable speed generation system. In *2015 International Conference on Clean Electrical Power (ICCEP)*, pages 743–748, 2015.

[7] Z. Jin, et al. Hierarchical control design for a shipboard power system with DC distribution and energy storage aboard future more-electric ships. *IEEE Transactions on Industrial Informatics*, 14(2):703–714, 2018.

[8] D. Park and M. Zadeh. Dynamic modeling, stability analysis, and power management of shipboard DC hybrid power systems. *IEEE Transactions on Transportation Electrification*, 8(1):225–238, 2022.

[9] T. Waris and C. V. Nayar. Variable speed constant frequency diesel power conversion system using doubly fed induction generator (DFIG). In *2008 IEEE Power Electronics Specialists Conference*, pages 2728–2734, Rhodes, Greece, Aug. 2008.

[10] G. Carrasco, et al. Control of a four-leg converter for the operation of a DFIG feeding stand-alone unbalanced loads. *IEEE Transactions on Industrial Electronics*, 62(7):4630–4640, 2015.

[11] C. Rossi, et al. Power split e-CVT solution for combined ship propulsion and electric energy generation. In *Electrical Systems for Aircraft, Railway and Ship Propulsion*, 2010.

[12] J. Ch. Dermentzoglou and J. M. Prousalidis. Emulation of a system with a power split device for hybrid propulsion of ships. In *2018 XIII International Conference on Electrical Machines (ICEM)*, pages 2529–2534, 2018.

[13] M. Rodriquez, et al. Crowbar control algorithms for doubly fed induction generator during voltage dips. In *2005 European Conference on Power Electronics and Applications*, 2005.

[14] A. Al-Quteimat, C. Niewienda, and U. Schäfer. Low voltage ride through of doubly fed induction generator in wind power generation using crowbar solution. In *2017 International Conference on Optimization of Electrical and Electronic Equipment (OPTIM) 2017 Intl Aegean Conference on Electrical Machines and Power Electronics (ACEMP)*, pages 667–674, 2017.

[15] W. Qiang, T. Bäckström, and C. Sadarangani. A novel drive strategy for hybrid electric vehicles. In *IEMDC 2001. IEEE International Electric Machines and Drives Conference (Cat. No.01EX485)*, pages 79–81, 2001.

[16] A. Rosa. *Projektierung von Ersatzstromaggregaten*. VDE-Schriftenreihe Normen verständlich. VDE Verlag, 3. edition, 2018. in German.

Bus voltage regulation using sequentially switched ZVZCS converters for spacecraft power systems

A. Garrigós, C. Orts, D. Marroquí, J. M. Blanes, C. Torres, P. Casado
MIGUEL HERNANDEZ UNIVERSITY OF ELCHE – IEG
Avda. de la Universidad s/n. Edificio Torrevaillo, 1D5
Elche, Spain
Tel.: +34 / 96 665.88.92.
Fax: +34 / 96 665.84.97.
E-Mail: augarsir@umh.es
URL: http://www.umh.es

Keywords

«ZVZCS», «Space», «Photovoltaic», «Power conditioning», «Paralleling», «Current-source DC-DC».

Abstract

This work proposes a regulation technique for photovoltaic sources power bus based on quasi resonant zero-voltage, zero-current power switching cells controlled in a sequential manner to provide tight bus voltage regulation, fast transient response, and simple control loop design. This method has been devised for high-power bus regulated satellites following European Space Agency standards.

Acronyms

BC(D)R – Battery Charge (Discharge) Regulator
DCX – DC Transformer
EMI – Electromagnetic Interference
EP – Electric Propulsion
ESA – European Space Agency
IPOS – Input Parallel Output Series
MEA – Main Error Amplifier
MPP – Maximum Power Point
PCU – Power Conditioning Unit
SAR – Solar Array Regulator
SAS – Solar Array Section
SMART – Standard Multiple Application Regulator Topology
S^3R – Sequential Switching Shunt Regulator
S^3MPR – Sequential Switching Shunt Maximum Power Regulator
S^3ZVZCS – Sequential Switching Shunt Zero Voltage Zero Current Switching
$S^3ZVZCSMPR$ – Sequential Switching Shunt Zero Voltage Zero Current Switching Maximum Power Regulator
S^4R – Sequential Switching Shunt Series Regulator
S^4MPR – Sequential Switching Shunt Series Maximum Power Regulator
S^4ZVZCS – Sequential Switching Shunt Series Zero Voltage Zero Current Switching
$S^4ZVZCSMPR$ – Sequential Switching Shunt Series Zero Voltage Zero Current Switching Maximum Power Regulator
TWTA – Travelling Wave Tube Amplifier
ZCS – Zero Current Switching
ZVS – Zero Voltage Switching
ZVZCS – Zero Voltage Zero Current Switching

Introduction

Today the typical bus voltage for high-power satellites is 100 V. As spacecraft power is scaled up to tens of kW, mainly due to electric propulsion and more demanding payload, a 100 V power bus leads to high mass and losses in the DC harness, connectors, and equipment components. An increase in bus voltage to 300 V or more would reduce power losses, mass and would increase the electrical efficiency.

Considering the ESA standard [1], a 25kW power conditioning unit (currently the highest bus power capability of the largest European telecommunications platforms) will impose the following design constraints in 100V and 300V bus voltage distribution, please refer to table 1. Clearly, 300V bus voltage distribution will bring multiple benefits.

Table I: Estimated values for bus regulated architecture, 25kW – ECSS-S-ST-20C

Parameter	Bus: 100V	Bus: 300V	Comments
Bus current	250A	83.3A	Harness, connectors, distribution and protection switches specifications relaxed for 300V.
Maximum output impedance (Z_{Omax}) - § 5.7.2.o	8mΩ	72mΩ	$Z_O < 10$mΩ imposes a difficult bus bar design for 100V.
Bus capacitance (C_{BUS}) - § 5.7.2.o	8mF	880uF	Bus capacitance improvement (mass and volume) around 30% for 300V.

The most common electrical architecture for large satellites is the one represented in figure 1 (left) and it is composed of a direct-energy transfer solar array regulator, known as Sequential Switching Shunt Regulator (S^3R), a Battery Discharge Regulator (BDR) and a Battery Charge Regulator (BCR), being all of them implemented by several modules. The main error amplifier (MEA) is used to control the three main regulators and provide bus voltage regulation [2]. The control of the S^3R to achieve bus voltage regulation, is the sequential activation and deactivation of individual solar array sections based on hysteretic control [3].

To achieve 300V bus regulation with an S^3R electrical architecture [4], 300V-S^3R, high-voltage photovoltaic strings are required which in turn represent many challenges in the solar array design:

- Solar array arcing due to differential charging of the different materials on the solar array surface.
- Slip rings, design for 300V requires sealed assembly and arc mitigation system.
- String voltage equalization of solar cells. The larger the number of solar cells in series the more complicated the voltage distribution between cells. This is aggravated if the solar array section is shunted.
- Parasitic elements, solar array section capacitance (C_{SAS}) has a critical effect in 300V S^3R regulator in terms of energy dissipation and bus voltage performance, both in static and transient response.
- Qualification and cost, the solar array is the most expensive part of the satellite platform.

The proposed electrical architecture, represented in figure 1 (right), only changes the Solar Array Regulator (SAR) of the original Power Conditioning Unit (PCU) concept. Obviously, an important number of benefits are derived from this approach:

- Solar array design is decoupled from bus voltage distribution, higher but also lower bus voltage might be realized from a standard solar array.
- PCU design could be optimized in terms of efficiency, mass and volume while keeping existing and well-established control techniques and power conversion topologies.

- Modularization is easily achieved, for instance, higher bus voltage distribution could be easily achieved connecting several ZVZCS converters in input-parallel and output-series (IPOS) connection.

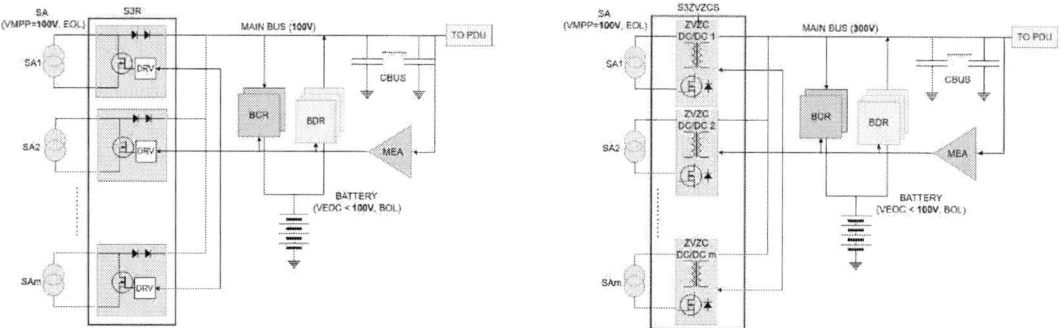

Fig. 1: Electrical Power Conditioning Unit architectures for large satellites. Left side: S³R bus regulated - 100V. Right side: S³ZVZCS – 300V.

In the space sector, high-voltage power supplies are common for Travelling Wave Tubes Amplifiers (TWTA's) and Electrical Propulsion (EP) systems, and therefore, there is always a special interest in reliable, high-efficiency, high-voltage conversion ratio converters. A widely spread ZVZCS technique for DCX applications, used by many European space manufacturers and ESA itself, is the one detailed in [5-7], which has been revisited recently [8, 9] applied to low-voltage, low-power, and high-voltage, high-power applications, respectively.

This ZVZCS switching technique, which can be applied to push-pull, half-bridge, full-bridge or any other dual-ended topology has the following benefits:

- all the parasitic elements are used in a resonant manner (efficiency > 97%).
- all power semiconductors are operated in ZVS and ZCS when turned on and off.
- ZVS and ZCS (neglecting magnetizing current) are load independent in a wide range.
- very simple low loss gate drive (especially in push-pull version).
- good power semiconductor utilization (with high duty cycles).
- operation at fixed frequency and duty cycle.
- reduced number of components.
- very low EMI.

Obviously, the main drawback is the absence of regulation capability, which has been also an important topic of research, and different approaches have been described in the literature to achieve output voltage regulation. Focusing on European space sector, a widely adopted solution by ESA and many manufacturers is the two-stage approach, typically a buck converter (pre-regulator) followed by a ZVZCS push-pull (DCX), known as SMART [10]. This approach, although it is widespread today, has an efficiency penalty due to the cascade connection. Other techniques, as parallel power processing and variable transformer turns ratio were introduced at that time [11,12] to alleviate that problem. This topic is still relevant and different approaches, like [13], are being proposed.

In the case of a segregated power source, such as it happens with the satellite solar array, another approach for output voltage regulation is possible. Essentially, some DCX converters are permanently providing power to the bus, while others DCX are disconnected. Only one DCX is turning on and off to eventually regulate the output voltage. This is the natural evolution of the sequential hysteretic control S³R, and as consequence it has been named here, S³ZVZCS.

The rest of the work is organized as follows. Section 2 describes the ZVZC conversion (focusing on the push-pull converter) and the output voltage regulation. Section 3 covers the design and simulation of 300V bus distribution and section 4 briefly discusses the results and issues related to space-grade implementation.

Photovoltaic power regulation using the S³ZVZCS technique

In essence, a DCX is an unregulated DC-DC converter that converts voltage and current with a fixed ratio, n, ($V_o=nV_{in}$ and $I_{in}=nI_o$). Typically, the DCX converter family considered in this work [5-7] can maintain ZVZC conditions in a wide load range. The S³ZVZCS power cells using a push-pull and a half-bridge structure are represented in figure 2. The solar array section (SAS) is modelled by the equivalent single-diode model and the parasitic capacitance (C_{SAS}) and the harness inductance (L_h). The photovoltaic source has some interesting characteristics when used with a ZVZCS converter:

- The power source exhibits inherent current limited characteristics (I_{SAS}).
- The maximum input voltage is limited to the open-circuit solar array section voltage (V_{OC}).
- The converter is current fed, so the solar array harness inductance can be part of it.

The shunt section consists in a power transistor with a current limiting circuit. The main function is to short the solar array section when its energy is not required while provides controlled energy discharge of the parasitic elements. M_1 and M_2 could also be controlled to perform the same operation, however considering a separate circuit allows optimization of MOSFET selection for each function, low-frequency, <10kHz, and linear operation (shunt) and high-frequency operation (ZVZCS converter). Further, instead of shunting M_1 and M_2, these can be left open to disconnect the solar array section from the bus; however, this method has some drawbacks related to high voltage at very low temperatures, e.g. during eclipse exits. A key difference with the two-stage approach, S³R cascaded by a ZVZCS converter, is that the S³R diode function is performed by the DCX output diodes.

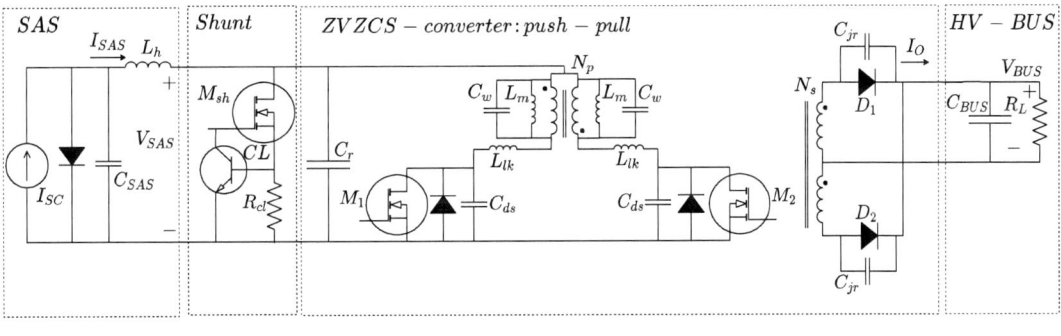

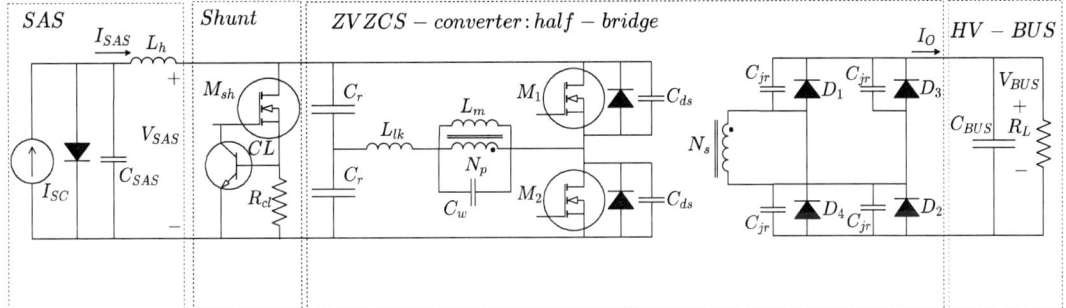

Fig. 2: S³ZVZCS power cells. Top side: push-pull DCX. Bottom side: half-bridge DCX.

The ZVZCS converter operates at fixed switching frequency and duty cycle. It has two operating states, ON state (one of the power MOSFETs is turned-on) and GAP state (both power MOSFETs are turned-off), with equivalent circuits shown in figure 3. During the ON state, a resonant switch current appears due to the resonant tank formed by C_r (resonant capacitor) and L_{lk} (transformer leakage inductance). During the GAP state, the magnetizing current charges and discharges the equivalent capacitances. The idealized MOSFET waveforms are shown in figure 3 (right). For a detailed description of the converter operation and design equations, please refer to [7,8].

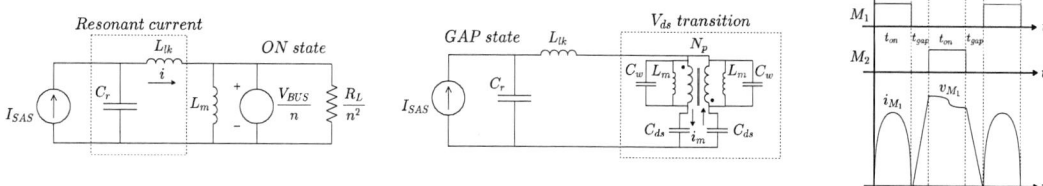

Fig. 3: ZVZCS-push-pull: equivalent circuits (ON and gap) and idealized waveforms.

When supplied from a photovoltaic source, as represented in figure 4, the converter can maintain ZVS and ZCS in a wide range of load variation (between operating points A and B), and it is not sensitive to irradiance deviations and temperature changes. For load lines exceeding the operating point B, the transferred power becomes negligible since the SAS current becomes magnetizing current and it is employed to charge and discharge parasitic capacitors on each cycle. On the other side, if the load line crosses below operating point A, the reflected input voltage becomes low enough that the voltage ripple in the resonant capacitor goes to zero. Between these two extremes, ZVZCS conditions are guaranteed, and for maximum power transfer, the load line should cross the Maximum Power Point (MPP).

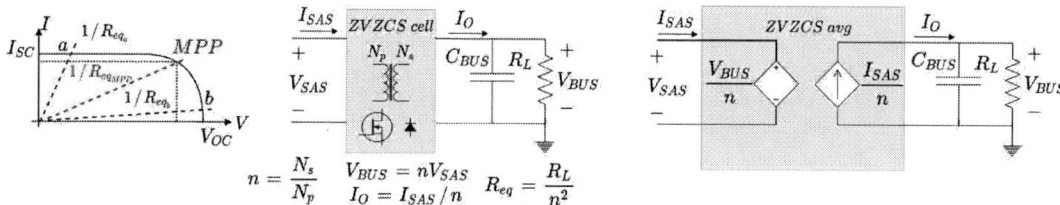

Fig. 4: ZVZCS-push-pull: connection to a photovoltaic source and equivalent DC circuit.

The DCX can be modelled as an output current source, whose value is the SAS current divided by the transformer turns ratio (n), that feeds the bus capacitance, see figure 4. At fixed duty cycle, the output current of a power cell is not controllable and mainly depends on irradiance, temperature and the load. However, in an arrangement of several power cells connected in parallel at the output, the average bus current (sum of all output currents) and, therefore the output voltage, can be controlled by connecting and disconnecting different power cells. In case of using a hysteretic-sequential control, in steady-state conditions, only one power cell will switch on and off while the others remain either fully-on or fully-off . An averaged linear model of the complete power regulator considers a voltage-controlled current-source, whose gain is simply the output current of a DCX divided by the hysteresis voltage [14]. Please refer to figure 5 for the S³ZVZCS configuration, non-linear power stage model and its linearization.

As a result of using sequential-hysteresis control, a variable low frequency signal (in the kHz range) defines the bus voltage ripple between two limits. A much higher frequency component appears at the switching frequency (or multiple of switching frequency if DCX interleaving is considered), but its amplitude remains very small because is filtered by the bus capacitance. Since the power stage is conductance-modelled, a simple Proportional-Integral error amplifier (MEA) is sufficient to guarantee bus voltage regulation.

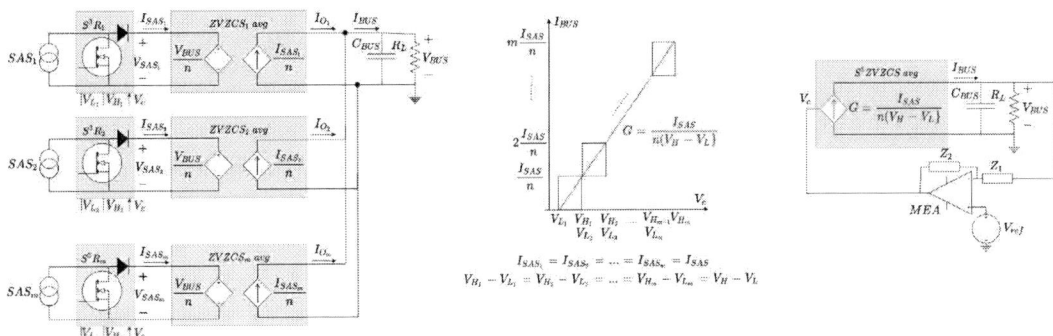

Fig. 5: S³ZVZCS-regulator. Left side: Equivalent circuit S³R and ZVZCS averaged model. Middle: Non-linear power regulator model. Right side: S³ZVZCS linearized model.

An implementation of the S³ZVZCS using a push-pull power cells is shown in figure 6.

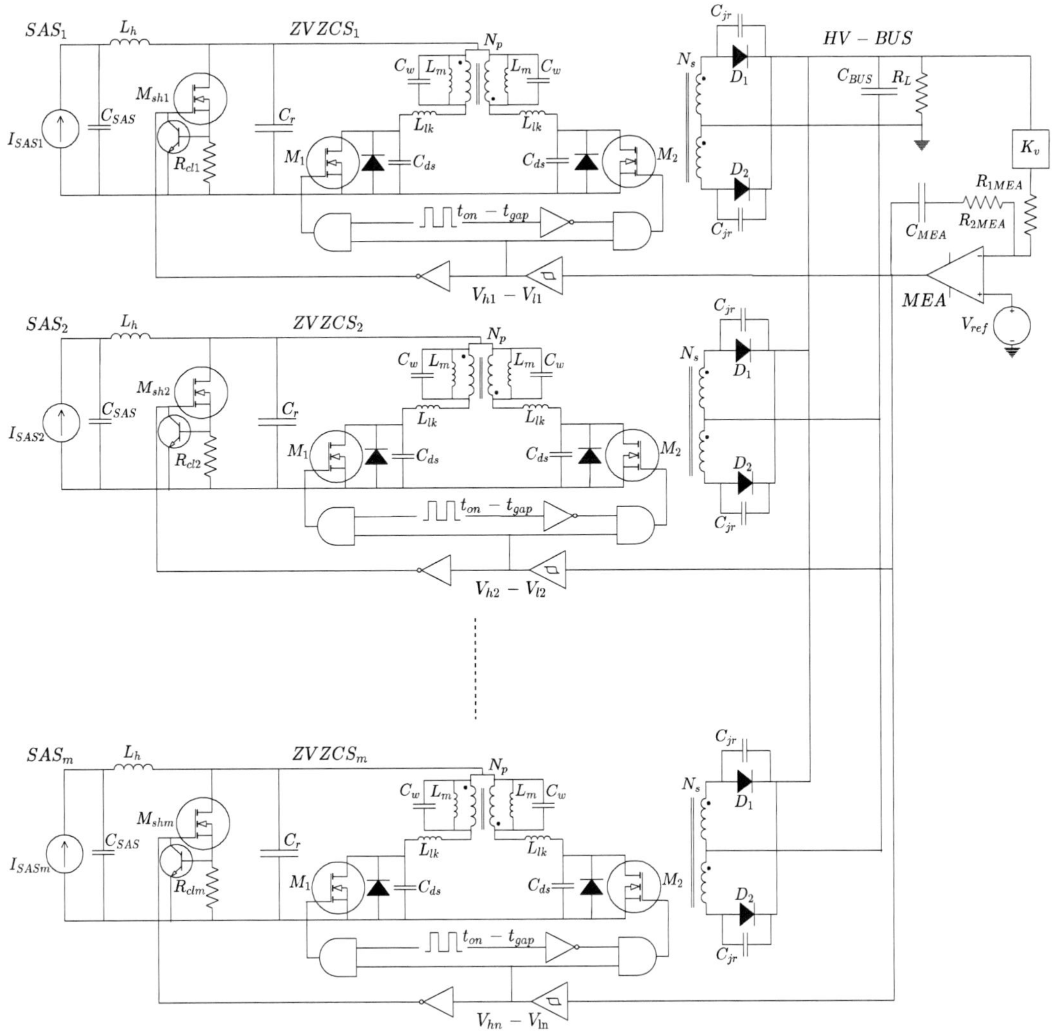

Fig. 6: S³ZVZCS-regulator block diagram: push-pull implementation.

Design and simulation of a S^{3}ZVZCS for 300V satellite bus regulation

The design of a three-power cell S^{3}ZVCS push-pull regulator has been carried out (three is the minimum number of cells required to observe the static, dynamic and potential *double sectioning* issues) to validate the proposed power regulation technique. The main characteristics of the regulator have been gathered in table 2.

Table II: 300V S^{3}ZVZCS main parameters

Parameter	Value	Comment
Solar array section - SAS		
V_{SAS}	100V	Operating conditions (1367W/m2; T=60°C)
V_{OC}	120V	Operating conditions (1367W/m2; T=60°C)
I_{SAS}	5A	Operating conditions (1367W/m2; T=60°C)
C_{SAS}	2uF	Estimated value
L_h	33uH	Estimated value. It also considers S^{3}R inductor used for current limitation.
Transformer ZVZCS		
Core	ETD49	Material 3C90
N_p	8	Magnetizing current approx. 15% of SAS current
N_s	24	V_{BUS} = 300V
L_m	307.5uH	Measured value
L_{lk}	1uH	Measured value
Resonant circuit ZVZCS		
C_r	500nF	ON state (L_{lk}-C_r resonant circuit)
C_{ds}	1.4nF	Estimated capacitance during GAP interval
f_s	104.5kHz	MOSFET Si (100kHz – 150kHz)
D	0.418	ON state power transfer optimization
Bus capacitance – ECSS-S-ST-20C		
C_{BUS}	250uF	Zo mask impedance (§ 5.7.2.o)
MEA - ECSS-S-ST-20C		
V_{ref}	6.4V	Temperature compensated voltage reference
K_v	0.0213	V_{BUS}=300V
R_{1MEA}	5kΩ	Zo mask (§ 5.7.2.o). A_{min}=20
R_{2MEA}	100kΩ	Zo mask (§ 5.7.2.o). A_{min}=20
C_{MEA}	6.8nF	Zo mask (§ 5.7.2.o). f_{zero}=234Hz
Hysteresis comparator - ECSS-S-ST-20C		
V_H-V_L	1V	Zo mask (§ 5.7.2.o).
Shunt current limitation		
R_{lim}	0.05Ω	I shunt max=14A

Figure 7 shows the bus steady-state voltage ripple and voltage response under a step load of 50% of the bus power. Before t=10ms, ZVZCS1 is maintaining the bus voltage regulation. After t=10ms, ZVZCS1 and ZVZCS2 are fully on to provide power to the bus and ZVZCS3 performs bus regulation by hysteresis control action. It is clearly observed that bus voltage ripple and bus voltage transient response only depend on the hysteresis-sequential control loop, while DCX high-frequency ripple (red trace) is filtered by the bus capacitance. Figure 8 is a zoomed view around t=13ms, where bus voltage, DCX-MOSFET voltage and DCX-MOSFET current are shown. Superimposed ripple on the bus voltage due to DCX and zero-voltage and zero-current switching of the power cell are clearly observed.

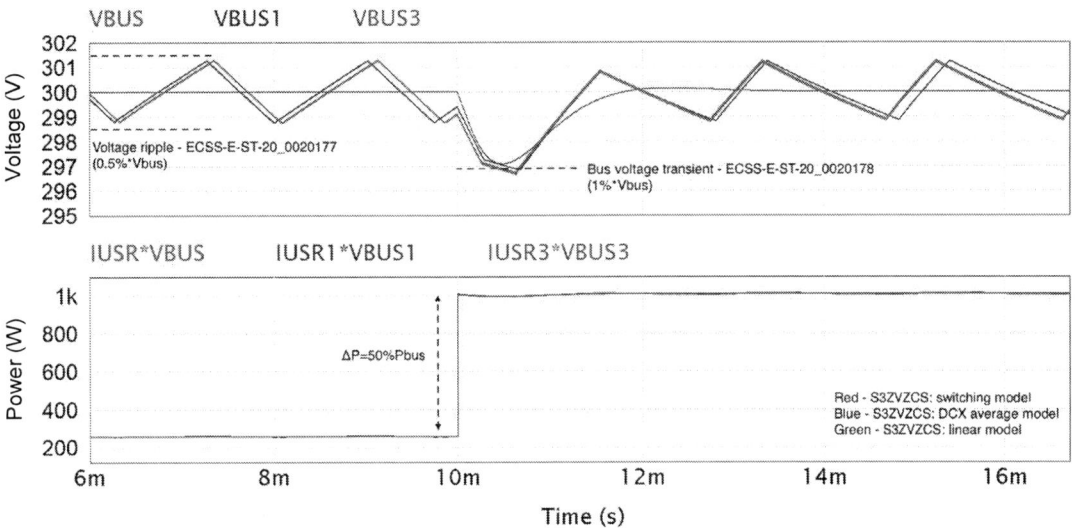

Fig. 7: S³ZVZCS-regulator transient response. Top figure: bus voltage. Bottom figure: Power step load

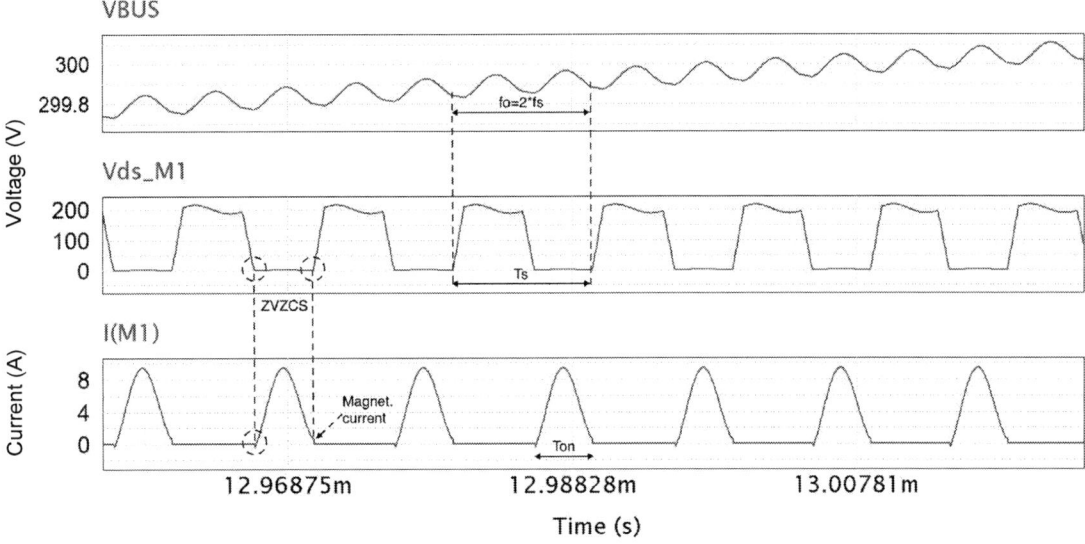

Fig. 8: S³ZVZCS-switching frequency detail. Top figure: bus voltage. Middle figure: DCX-MOSFET drain-source voltage. Bottom figure: DCX-MOSFET current.

Results discussion

A space-qualified implementation of the proposed concept has different issues to be dealt with. Some of them are briefly discussed below.

a) <u>Power semiconductors</u>. A critical evaluation of available technologies and potential needs is required to assess the viability of the presented approach. The increase of voltage and switching frequency ask for, as many other power electronics applications, wide bandgap power devices. Today, up to 650V/60A E-mode GaN space-qualified power transistors and drivers are available for DCX primary switches, allowing the implementation of any kind of the available topologies, e.g. push-pull, half-bridge or full-bridge. 1.2kV - SiC space-qualified power diodes are also a good starting point for rectifiers, but their availability is limited. However, there is an interest for 1.2kV SiC power diodes in other space applications, like TWTAs, and technology is already mature in other high reliable terrestrial applications, like automotive or aeronautical. Finally, shunt transistor requirements are covered by current space-qualified Si MOSFET technology and no major issues are anticipated.

b) <u>Passive components</u>. WBG power devices will allow an increase of switching frequency up to MHz range, and recent studies have been demonstrated operation up to 1MHz in similar applications [13]. Soft ferrites and space-qualified polyester PET film capacitors are already available for energy storage and tuning purposes.

c) <u>Analog vs. digital control</u>. Fully analog control implementation appears to be the simplest approach since all the elements involved in the S³ZVZCS have flown as separate subsystems. Dynamic interleaving between DCX converters will minimize RMS bus current and it could be implemented with analog space-qualified timing circuits. However, additional benefits would be derived from a digital implementation of the DCX control. Dynamic and real-time reconfiguration of t_{on} and t_{gap}, to compensate any variation of the resonant conditions, can be more easily achieved using a digital approach, as well as dynamic interleaving and other high-level monitoring and control options. Obviously, this is at the expense of additional complexity.

d) <u>Electrical power architectures variations</u>. The proposed concept could be easily adapted to other satellite electrical architectures, S⁴R [14], S³MPR [15] and S⁴MPR. In particular, the evolution of the S⁴R architecture, either the MPP version, S⁴ZVZCSMPR or the full regulated, S⁴ZVZCS, has the advantage of removing the double power processing step (SA Low Voltage – BUS High Voltage – BAT Low Voltage) as represented in figure 10.

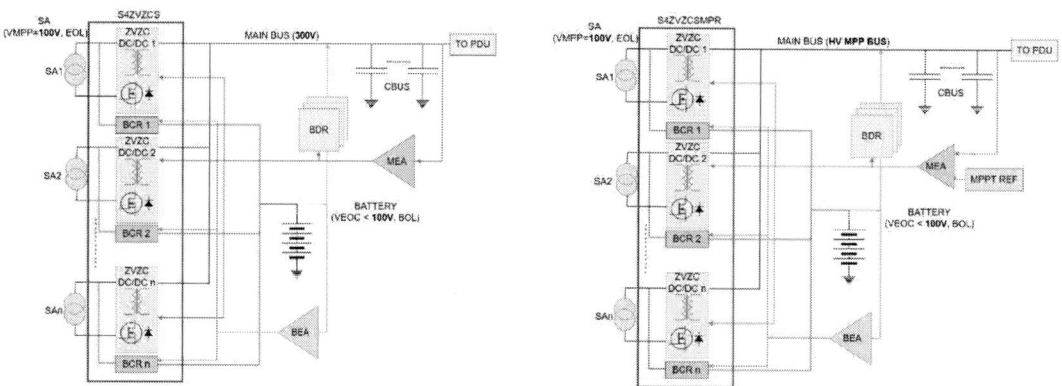

Fig. 9: S⁴ZVZCS electrical architecture (left side). S⁴ZVZCSMPR electrical architecture (right side).

e) <u>Scalability</u>. Another interesting point regards to DCX output serialization using a IPOS configuration (Input-Parallel Output-Series configuration). This would allow for a higher bus voltage, a reduction in the transformer turns ratio, and a voltage stress reduction in the secondary side power semiconductors. This also open new applications, like power conditioning for solar satellites, being in line with other recent studies [16].

Conclusion

Two well-known space power conversion techniques have been combined to propose a novel solar array regulation technique for high-voltage satellite power bus. Although, DC transformer function can adopt many forms, the proposed ZVZCS technique is well suited for a photovoltaic source since it operates as almost an ideal current source at the voltage regulation point. In addition, this method is applicable to push-pull, half-bridge, full-bridge or any dual-ended topology which provides a large degree of flexibility for designers. Design and simulation of the proposed method illustrate the benefits for a 300V bus voltage distribution in high-power satellites. The concept can be extrapolated to other bus regulated architectures, like S⁴R, S³MPR or S⁴MPR.

References

[1] -, "ECSS-E-ST-20C Rev.1 (15 Oct 2019): Space engineering – Electrical and Electronic"

[2] A. Capel, D. O'Sullivan, and J. C. Marpinard, "High-power conditioning for space applications," *Proceedings of the IEEE*, vol. 76, no. 4, Apr. 1988.

[3] D. O'Sullivan, A. H. Weinberg, "The sequential switching shunt regulator (S3R)," in *Third ESTEC Spacecraft Power Conditioning Semina*r, Sept. 1977, pp. 123-131.

[4] J. B. de Boissieu, *et al*, "High voltage electrical power system architecture optimized for electrical propulsion and high-power payload," in *ESA 12ᵗʰ European Space Power Conference*, 2019.

[5] A. H. Weinberg, "DC to DC converter using quasi-resonance," US Patent 4959765.

[6] A. H. Weinberg, L. Ghislanzoni, "A new zero voltage and zero current power switching technique," in *20ᵗʰ IEEE Power Electronics Specialists Conference*, 1989, pp. 909-919.

[7] A. H. Weinberg, L. Ghislanzoni, "A new zero voltage and zero current power switching technique," *IEEE Transactions on Power Electronics*, vol. 7. no. 4, Oct. 1992, pp. 655-665.

[8] W. Qin, X. Wu, J. Zhang, "A family of DC transformer (DCX) topologies based on new ZVZCS cells with DC resonant capacitance," IEEE Transactions on Power Electronics, vol. 32, no. 4, Apr. 2017, pp. 2822-2834.

[9] Q. Zhu, L. Wang, L. Zhang, A. Q. Huang, "A 10kV DC transformer (DCX) based on current fed SRC and 15kV SiC MOSFETs", in *IEEE Applied Power Electronics Conference*, 2018, pp. 149-155.

[10] D. O'Sullivan, M. M. Alfonso, "Rationale behind the SMART regulator," in *4ᵗʰ European Space Power Conference*, 1995, pp. 47-54.

[11] L. Ghislanzoni, "Parallel power regulation of a constant frequency, ZV-ZC switching resonant push-pull," in *ESA 2ⁿᵈ European Space Power Conference*, 1991, pp. 191-198.

[12] A. H. Weinberg, D. O'Sullivan, J. A. Carrasco, "Variable transformer turns ratio regulator {TR2} for a DC/DC converter or inverter," in *ESA 3ʳᵈ European Space Power Conference*, 1993, pp. 33-37.

[13] C. Wang, M. Li, Z. Ouyang, G. Wang, "Resonant push-pull converter with flyback regulator for MHz high step-up power conversion," IEEE Transactions on Industrial Electronics, vol. 68, no. 2, Feb. 2021, pp. 1178-1187.

[14] A. Garrigós, J. A. Carrasco, J. M. Blanes, E. Sanchis, "Modeling the Sequential Switching Shunt Series Regulator," *IEEE Power Electronics Letters*, vol. 3, no. 1, march 2005, pp. 7-13.

[15] A. Garrigós, J. M. Blanes, J. A. Carrasco, A. H. Weinberg, E. Maset, E. Sanchis-Kilders, J. B. Ejea, A. Ferreres, "The Sequential Switching Shunt Maximum Power Regulator and its application in the Electric Propulsion System of a spacecraft," in *IEEE Power Electronics Specialists Conference*, 2007, pp. 1374-1379.

[16] L. Wang, D. Zhang, J. Duan, J. Li, "Design and research of high voltage power conversion system for space solar power station," in *IEEE International Power Electronics and Application Conference and Exposition*, 2018.

A standardized and modular power electronics platform for academic research on advanced grid-connected converter control and microgrids

Frank S.R., Schulz D., Stefanski L., Schwendemann R., Hiller M.
Karlsruhe Institute of Technology (KIT), Institute of Electrical Engineering (ETI)
Kaiserstraße 12
Karlsruhe, Germany
Phone: +49 721 608 42465
Email: s.frank@kit.edu
URL: http://www.kit.edu

Acknowledgments

This work was supported by the KIT Future Fields Stage 2 funding program.

Keywords

≪Microgrid≫, ≪Standardization≫, ≪Hardware≫, ≪Grid-forming converters≫, ≪Grid-connected converter≫, ≪Measurements≫, ≪Compensation≫, ≪Harmonic Injection≫, ≪Active damping≫, ≪Smart grids≫, ≪Active Front End≫, ≪LCL≫, ≪Isolated Converter≫

Abstract

This paper introduces a multifunctional converter platform rated at 30 kW. Individual units allow research on advanced grid-connected converter control, while their interconnection enables isolated microgrid investigations. The standardized and modular design allows simple reconfiguration of the system for different setups, for which multiple measurements are presented.

Introduction

As the electrical grid evolves from a system supported by power generation using synchronous generators to rising levels of renewable energy and thus power electronics interfaced power sources, the design, control and coordination of grid-connected power converters are crucial for this ongoing power system evolution.

The microgrid paradigm [1] has established itself as the main approach to operate multiple power converters in a concerted manner. Microgrids consist of multiple converters that allow decentralized grid control, coordination of local energy balance and possible islanding in case of failure of the main grid. Three main aspects are particularly relevant when considering microgrids for research purposes. First, they enable the investigation of a concerted operation of grid-connected converters in a full converter-based microgrid with the above mentioned advantages. Second, an isolated microgrid can be used to emulate arbitrary grid conditions, such as distorted or weak grids, as well as voltage and frequency deviations. Stricter grid connection standards defined for example in [2] and [3] require some degree of grid support in reaction to such deviations, while also specifying necessary Fault-Ride-Through (FRT) capabilities. These can be implemented and examined in the stand-alone microgrid. The isolated microgrid is also beneficial to investigate converters and their controls operating in a grid-forming mode. This includes the necessary synchronization and load sharing between grid-forming units, as well as appropriate short-circuit current limiting. Third, a microgrid provides a testbed for converter-dominated grids facing challenges such as a lack of inertia, overcurrent protection and instability phenomena arising from converter interaction [4].

In literature, several power electronics platforms or testbeds are introduced, developed by various universities and research institutions. In [5], a digital-physical hybrid real-time simulation platform is proposed, [6] presents a platform based on an OPAL-RT simulator and different converters. In [7], [8] and [9] microgrid testbeds are presented to study the interconnected operation of various distributed generators and energy storage. However, most of these platforms utilize commercial power electronics or signal processing subsystems. The disadvantage of this approach is that the system is not cost effective, expensive to maintain or upgrade and modification might not be possible [6]. To avoid these problems and gain full control over the complete system it is therefore expedient to design and construct a platform with self-developed fundamental building blocks.

This paper presents a Standardized Power Electronic Converter Cabinet (SPECC) as building block for a microgrid to take advantage of the aforementioned research benefits. Apart from the microgrid application, the SPECC is intended to offer a platform for general grid-connected converter investigations, including harmonic current mitigation resulting from distorted grid voltages. Due to more restrictive grid codes as well as more demanding size and cost requirements of the converters, LCL line filters are typically employed to reduce current harmonics near the switching frequency. However, possible instabilities due to the filters resonance must be addressed with damping techniques, especially when multiple grid-connected converters with LCL filters are nearby and may affect each other [10].

The developed SPECC-based converter platform provides a testbed for all these investigations under real-world conditions. Its modular design together with the completely self-developed hard- and software enables simple reconfiguration and extension and will be presented in the following section. Thereafter, the standardized measurement acquisition and signal processing system is detailed together with its benefits in the next section. Finally, possible control strategies and associated measurements are shown in the last two sections.

Topology and Hardware Platform

Due to their simple design, low need for semiconductor switches and well-established control, two-level converters are widespread and sufficient for most applications. Therefore, the topology of one SPECC, shown in Fig. 1, is based on standard Two-Level Voltage Source Converters (2L-VSC). It consists of two AC/DC 2L-VSCs at both AC-sides, shown in red. The first DC link (DC1) can be either directly connected to the secondary side (DC2) of a Dual Active Bridge converter (DAB), depicted in blue, or a DC/DC converter, shown in green, can be switched in between. The DC link of the second AC/DC 2L-VSC (DC3) is directly connected to the DABs primary side. The DAB ensures a galvanic isolation between its primary and secondary DC side that allows the realization of galvanically isolated microgrids. The optional DC/DC converter can be used to variably adjust the DABs secondary DC voltage to reach its most efficient operating point or to provide a connection for a DC grid with a specific voltage.

The DAB has been internally developed and used for research before [11], [12]. It consists of two full-bridge units, equipped with Infineon FF23MR12W1M1B11 Silicon Carbide (SiC)-Metal-Oxide Field-Effect Transistor (MOSFET) half-bridge modules, coupled by a Medium Frequency (MF)-transformer. This setup allows a bidirectional power transmission of 30 kW. For the AC/DC 2L-VSC and the DC/DC converter the internally developed 'combi single board converter (combi-EPSR)' introduced in [13] is used. The converter can be either equipped with Cree/Wolfspeed CCS050M12CM SiC-MOSFET or Infineon FS75R12K Silicon (Si)-Insulated-Gate Bipolar Transistor (IGBT) semiconductor modules. In the following, a setup with one SiC-MOSFET combi-EPSR (AC/DC converter 1) switched with $f_{sw,SiC} = 50$ kHz and one Si-IGBT combi-EPSR (AC/DC converter 2) switched with $f_{sw,IGBT} = 10$ kHz is used, to evaluate the different behaviour of these technologies in further research. For this reason, two different LCL line filters are used with their parameters given in table I.

For Electromagnetic Compatibility (EMC) reasons two serial connected common mode chokes WE744839010400 with 10.5 mH common mode inductance and 40 A rated current are integrated on both sides between the LCL filters and the combi-EPSRs. Besides the two three-phase AC terminals (AC1 and AC2), the DC terminals of both DAB sides (DC2 and DC3) as well as the output of the

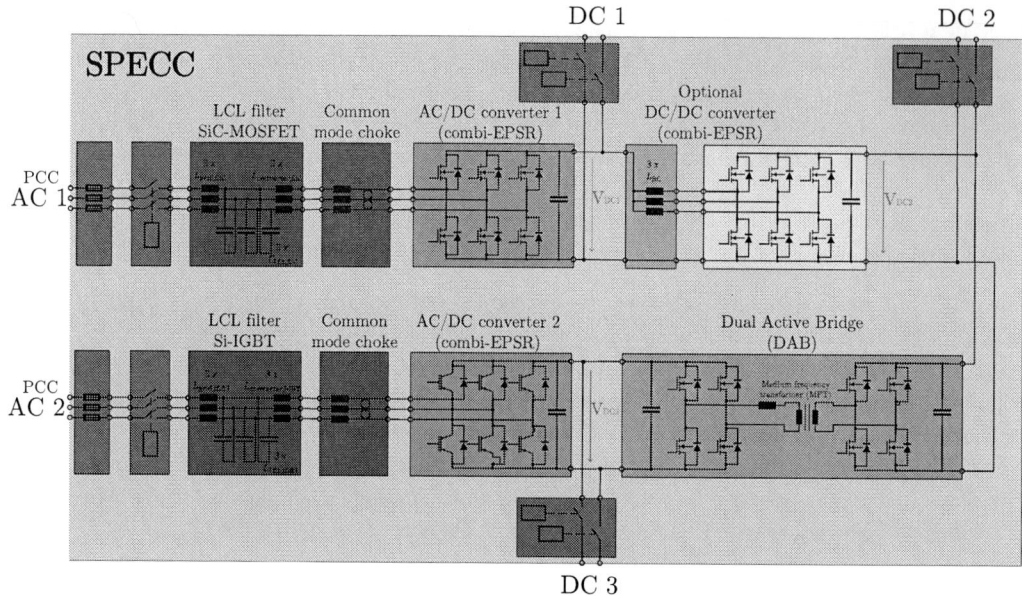

Fig. 1: Topology of one Standardized Power Electronic Converter Cabinet (SPECC)

Table I: Characteristics of the LCL line filters.

LCL filter type	Grid-sided inductance	Converter-sided inductance	Capacitance	Resonance frequency	Current rms	Dimensions	Mass
	L_{grid}	$L_{converter}$	C_{LCL}	f_{res}	$I_{ph,rms}$	$l \times w \times h$	
Si-IGBT LCL	$3 \times 150\,\mu H$	$3 \times 1\,mH$	$3 \times 6\,\mu F$ (delta)	$3.28\,kHz$	$45\,A$	$260 \times 275 \times 173\,mm$	$17.5\,kg$
SiC-MOSFET LCL	$3 \times 50\,\mu H$	$3 \times 100\,\mu H$	$3 \times 4.7\,\mu F$ (delta)	$7.34\,kHz$	$45\,A$	$253 \times 196 \times 128\,mm$	$5.7\,kg$

DC/DC converter (DC1) are accessible to offer the possibility for DC grid connections. The setup allows DC voltages up to 800 V. First measurements are done without the optional DC/DC converter, resulting in $V_{DC2} = V_{DC3}$. The whole setup is pictured in Fig. 2. All power electronics units (DAB and combi-EPSRs) are realized with double-height Eurocard Printed Circuit Boards (PCBs) [14], that are mounted in 19" racks in the lower region of the cabinet. The Central System-on-Chip (SoC)-based signal processing and Control Unit (CCU) and a status panel are placed spatially separated in the upper region. A computer functioning as Human Machine Interface (HMI) and interface to other SPECCs is also placed in this region. It can be operated using a touchscreen or a remote desktop connection. This arrangement ensures the separation of different function units and voltages levels a well as EMC. The mid part of the cabinet offers space for optional project-specific power electronics.

Sensors and Signal Processing

To enable the investigation of advanced control methods for grid-connected converters, each SPECC is equipped with a large number of sensors to measure the voltages and currents of each power electronics unit (DAB and combi-EPSR) as well as the grid currents at the AC terminals i_{grid} and the capacitor voltages v_c of the LCL filter.

Since a SPECC employs undamped LCL filters for both the IGBT and SiC AC/DC converter, the possibility of resonances needs to be considered in the controller design. Therefore, additional measurements such as capacitor voltages or capacitor currents are necessary. These measurements can be used to dampen LCL oscillations using appropriate feedback structures. With the measured quantities shown

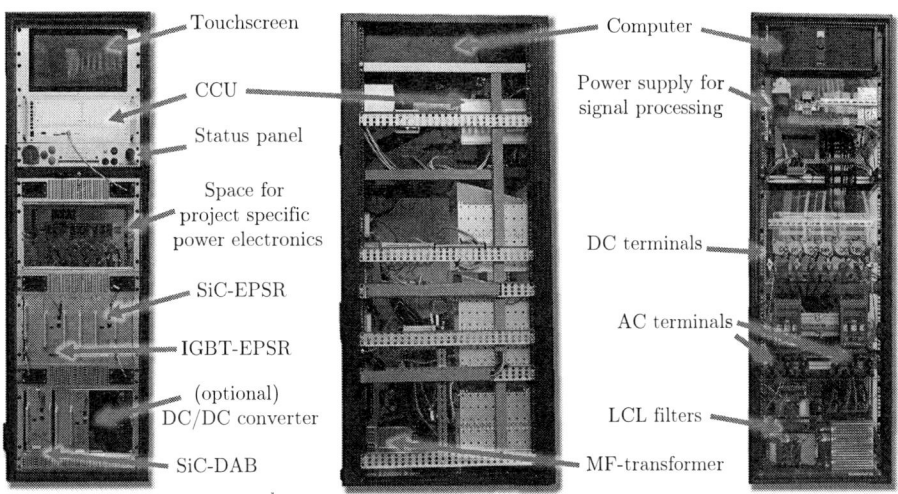

Fig. 2: Hardware setup of the Standardized Power Electronic Converter Cabinet (SPECC)

Table II: Measured Voltages and Currents

Physical quantity	Grid current	Capacitor volt. LCL	AC phase cur. EPSR	DC cur. DAB	AC cur. DAB	Phase volt. EPSR	DC volt. EPSR	DC volt. DAB
	i_{grid}	v_C	$i^\star_{\text{ph,EPSR}}$	$i^\star_{\text{DC,DAB}}$	$i^\star_{\text{AC,DAB}}$	$v^\star_{\text{ph,EPSR}}$	$v^\star_{\text{DC,EPSR}}$	$v^\star_{\text{DC,DAB}}$
Measuring range	$\pm102.4\,\text{A}$	$\pm1000\,\text{V}$	$\pm80\,\text{A}$	$\pm160\,\text{A}$	$\pm163.8\,\text{A}$	$\pm600\,\text{V}$	$0..1000\,\text{V}$	$0..1100\,\text{V}$
Resolution	$3.1\,\frac{\text{mA}}{\text{bit}}$ (16 bit)	$30.5\,\frac{\text{mV}}{\text{bit}}$ (16 bit)	$39\,\frac{\text{mA}}{\text{bit}}$ (12 bit)	$78\,\frac{\text{mA}}{\text{bit}}$ (12 bit)	$80\,\frac{\text{mA}}{\text{bit}}$ (12 bit)	$293\,\frac{\text{mV}}{\text{bit}}$ (12 bit)	$488\,\frac{\text{mV}}{\text{bit}}$ (12 bit)	$537\,\frac{\text{mV}}{\text{bit}}$ (12 bit)
Sample rate	5 MSPS	5 MSPS	650 kSPS	650 kSPS	1 MSPS	650 kSPS	650 kSPS	650 kSPS

in Tab. II, both the capacitor voltages v_C and capacitor currents i_C are available. In the current state of development, the capacitor currents are used for the active damping approach, which are calculated using

$$i_C = i_{\text{grid}} - i_{\text{ph,EPSR}}, \tag{1}$$

with the measured grid currents i_{grid} and phase currents $i_{\text{ph,EPSR}}$ of the respective three-phase converter.

The large number of accessible measurements allows the implementation of various control strategies and ensures a save operation of each power electronics unit. Besides, the temperatures of the LCL-filters, MF-transformer and semiconductor modules can be observed, guaranteeing global supervision and safe operation of the whole system.

The signal processing tasks necessary within the SPECC are distributed on multiple layers and processing units, which is shown in Fig. 3. The core of the signal processing architecture is the CCU depicted in Fig. 4, that is based on a *ZYNQ7030* SoC, which consists of two ARM processor cores together with an Field Programmable Gate Array (FPGA). The CCU has eight slots for miscellaneous extension cards, e.g. for fast analogue signal sampling, fiber wire communication and as interface to peripheral components. A detailed description is given in [15]. The FPGA allows a fast reaction to error states, the implementation of low-latency control algorithms and low-level interfacing of the previous mentioned extension cards. The first ARM core is tasked with handling the network communication over TCP/IP. The second ARM core is used to execute control algorithms, to coordinate the overall system using state machines and to perform a safe and controlled system shutdown in case a subsystem reports a fault.

The supervising control and data acquisition tasks are performed by a LabView-based Monitor Control Tool (MCT) on a desktop computer. This computer is located in the same network as the CCU(s), such that both measurement data and control commands can be sent over ethernet connections. Since each SPECC uses the identical signal processing system, the central MCT instance can facilitate communi-

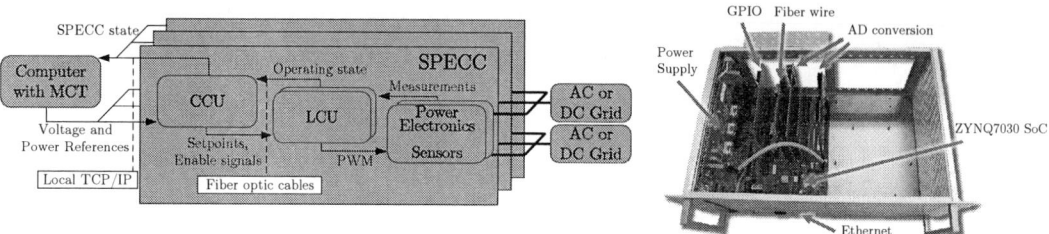

Fig. 3: Hierarchy and interconnections of the signal processing structure

Fig. 4: CCU, based on a *ZYNQ7030* SoC

cation between individual SPECCs. This enables the coordination of multiple SPECCs as required for cooperative operation such as in a microgrid setup.

The lowest layer of the signal processing structure shown in Fig. 3 is formed by Local Control Units (LCU). These are circuit boards equipped with an Artix-7 FPGA that provides signal processing capabilities with minimum signal paths and latency for the associated power electronics. Functionalities implemented on the LCUs include fault detection, for instance overcurrents, overvoltages and overtemperatures as well as high-bandwidth control loops and modulation for the connected semiconductor switches. In a SPECC, both AC/DC converters, the DAB and the optional DC/DC converter each possess a dedicated LCU. The quantities marked with "*" in Tab. II are measured, digitized and processed directly on the respective power electronics unit, the other quantities are centrally digitized on the CCU.

Control

A SPECC employs a distributed control architecture that mirrors the signal processing hierarchy presented in Fig. 3. The main design principle for this architecture is the ability of each individual power electronics unit (DAB and combi-EPSR) to control its outputs according to setpoints specified by higher-level control units. For this purpose, the control algorithms can either be executed on the associated LCU or alternatively on the CCU. An overview of several control methods for the operation within the SPECC that are implemented so far is given in Fig. 5.

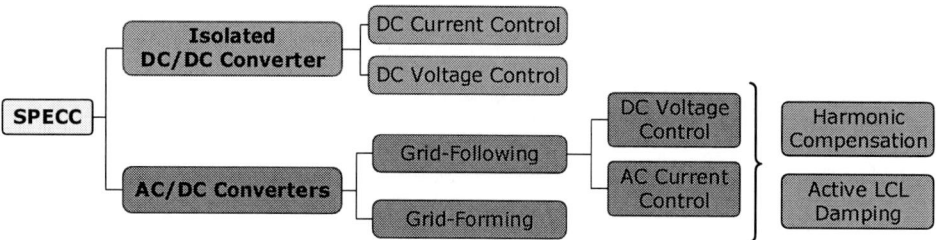

Fig. 5: Overview of control methods used within the SPECC

The interconnection, coordination and interaction of converter systems can be investigated together with the reaction of the converter system to different grid situations. At the same time, the control methods themselves are research subjects. The SPECC provides a platform for the investigation of possible control strategies including resonant controllers, harmonic compensation techniques and active LCL damping approaches.

The control of the single-phase DAB is based on the phase shift φ between the primary-side and secondary-side H-bridge voltage. In [16], the transferred power for single-phase-shift modulation and a MF-transformer with a turns ratio of $n = 1 : 1$ is derived from the characteristic DAB current and voltage waveforms to

$$P_0 = \frac{V_S^2}{\omega L_\sigma} \cdot d \cdot \varphi \cdot \left(1 - \frac{\varphi}{\pi}\right). \tag{2}$$

Ideal semiconductors and an instantaneous current commutation are assumed. For the presented topology with $P_0 = P_{DC}$, $V_S = V_{DC3}$, $d = V_{DC2}/V_{DC3}$ and $\omega = 2\pi \cdot f_{sw}$ follows for both power directions:

$$P_{DC} = \frac{V_{DC2} V_{DC3}}{2\pi f_{sw} L_\sigma} \cdot \varphi \cdot \left(1 - \frac{|\varphi|}{\pi} \right),\tag{3}$$

with the DC voltages on the primary and secondary side V_{DC3} and V_{DC2}, the switching frequency f_{sw} and the leakage inductance L_σ. The specified DC power or current setpoint can be achieved by adjusting the phase shift φ accordingly. Alternatively, the DAB can be operated in a voltage controlled mode. In this case, the phase shift φ is the output variable of a PI controller, which acts on the error between the voltage setpoint and voltage measurement.

The control strategies for the AC/DC converters are mainly distinguished between grid-forming and grid-following operation. First, the control options for the grid-following mode are described. In this case, the three-phase grid currents are the controlled quantities. Additionally, the active current setpoint may be specified by a superimposed control loop that regulates the voltage of the DC link. This control structure is shown in Fig. 6. The current setpoints in the rotating dq-reference frame are transformed into the stationary $\alpha\beta$ frame using the grid angle γ obtained by a Phase-Locked Loop (PLL). The error in the grid currents is then fed into a dampened Proportional-Resonant (PR) controller tuned to the grid frequency of 50 Hz. Additional resonant paths tuned to multiples of the grid frequency can be connected in parallel to the fundamental PR controller as harmonic compensators to attenuate harmonic currents [17]. The resulting controller is described by

$$G_{PR,\,HC}(s) = K_p + \frac{2K_i s}{s^2 + 2\omega_c s + \omega_0} + \sum_{h=7,11,13,17} \frac{2K_{ih}\omega_{ch} s}{s^2 + 2\omega_{ch} s + (h\omega_0)^2},\tag{4}$$

where K_p is the proportional gain, K_i the resonant gain, ω_0 the grid frequency and ω_c the damping coefficient. Parameters with the additional subscript h denote identical quantities for the aforementioned harmonic compensator terms. The measured grid voltages can be fed forwarded to the controller output. In order to dampen potential oscillations arising from the LCL grid filters, the LCL capacitor currents are also measured and can be utilized for a proportional active damping feedback with a gain of K_{AD}. Both the outer current loop gain K_p and the active damping gain K_{AD} are tuned to ensure stable system poles using root locus analysis [10]. Finally, the calculated output voltage is translated into power electronics switching states by zero sequence Pulse-Width Modulation (PWM).

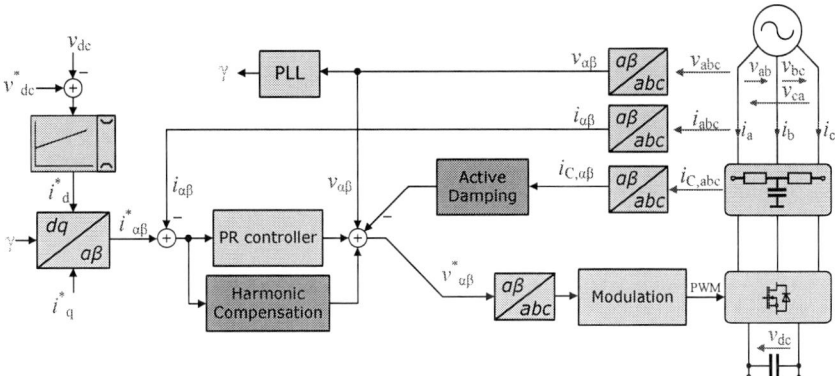

Fig. 6: Control structure for grid-connected AFE with harmonic compensation and active damping

The SPECC is also intended to be used as a research platform for the investigation of grid-forming controls. In this mode, the the three-phase grid voltages at the Point of Common Coupling (PCC) are the control objective. In case the converter system is connected to an existing grid with a voltage system present, the grid-forming control needs to synchronize to the grid angle and adapt its voltage setpoints using droop controls. Alternatively, in the islanded microgrid case where no other voltage source is connected to the grid, the control needs to be able to regulate the grid voltage on its own. In both cases, either the DC link connectors or the secondary AC connection of the SPECC is used to supply the necessary power for the operation as a grid-forming converter. The overall control setup for the grid-forming operation with a power supply from the laboratory grid is depicted in Fig. 7 with the control objectives shown in black. The same figure shows an alternative application of the SPECC operating as two constant power loads in an existing DC grid.

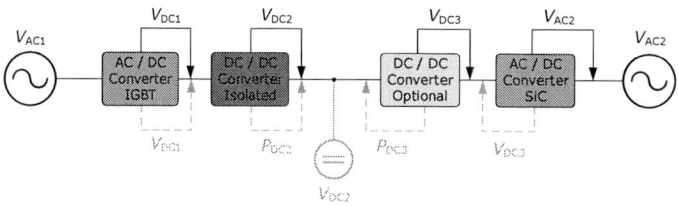

Fig. 7: Control setup within one SPECC. Control objectives shown in black are for grid-forming operation, gray quantitites below are control objectives when operating as constant power load in a DC grid

Measurements

This section will present first measurements taken with the SPECC. First, the application as a platform for the investigation of control methods for grid-connected power converters is demonstrated. Fig. 8 shows the measurements of the AC/DC converter operating as Active Front End (AFE) in a grid-following mode controlling its DC link voltage. Depicted are the phase currents i_a, i_b and i_c, phase voltage v_a, as well as the DC current i_{DC} and DC voltage v_{DC}. In fig. 8(a), the DC voltage setpoint is changed from $v_{DC} = 600\,\text{V}$ to $700\,\text{V}$ at $t = 16\,\text{ms}$. Fig. 8(b) shows multiple load steps. The DC load current as disturbance variable is first reduced to zero at $t = 5\,\text{ms}$ and then set to $i_{DC} = 12\,\text{A}$ at $t = 10\,\text{ms}$. In this application, the SPECC serves to verify the design of both the DC voltage controller and AC current controller, including their optimized controller parameters, real-world performance and controller robustness.

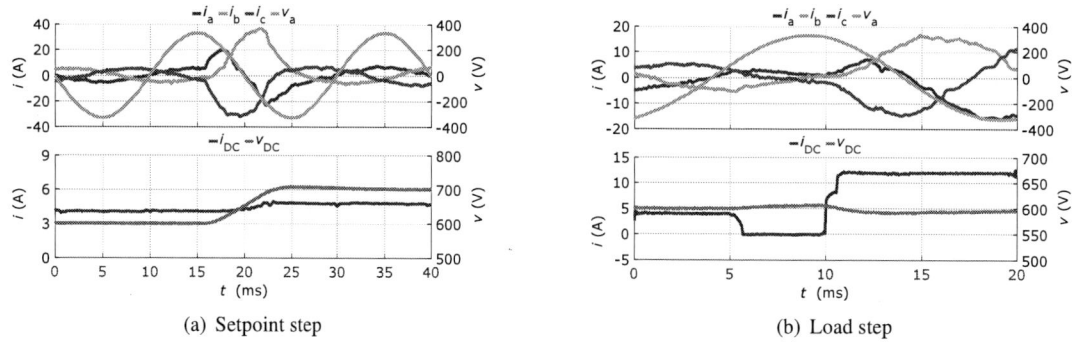

(a) Setpoint step

(b) Load step

Fig. 8: Measurements of AFE DC voltage control with step responses

Apart from the conventional PI and PR controllers under investigation, the SPECC is also able to provide a testbed for more advanced controller structures. One such extension is the addition of harmonic current compensators as described in the control section. The harmonic currents are caused by the harmonic distortions of the grid voltages. This grid may either be the laboratory grid, which already presents distorted grid conditions, or alternatively the isolated microgrid with adjustable voltage distortions. Fig. 9 and 10 show the effect of compensators for the 7th, 11th, 13th and 17th current harmonic, with the grid-side phase currents i_{on} and i_{off} for activated and deactivated harmonic compensation. The Total Harmonic Distortion (THD) is reduced from 0.58 % to 0.15 %.

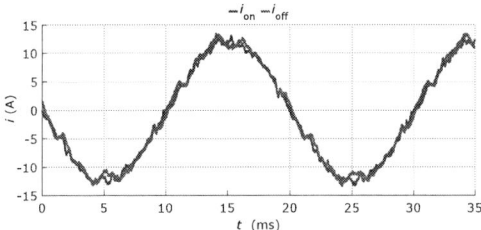

Fig. 9: Measurement of harmonic compensator effect on grid current waveform

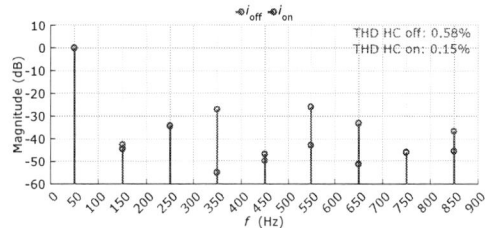

Fig. 10: Measurement of harmonic compensator effect on grid current spectrum

With no passive damping used in the LCL filters on both the SiC and IGBT side, the SPECC allows the examination of active resonance damping approaches described in the previous section. In Fig. 11 the effect of active damping on the capacitor current is illustrated. Depicted are the phase current i_a and line-to-line voltage v_{ab}. At $t = 17\,\text{ms}$ the active damping is disabled. Due to the high bandwidth of the current controller, an oscillation occurs immediately and the system must be switched off.

For the grid-forming operation, first measurements were taken for the most basic case in which only one converter is tasked with regulating the grid voltage, which is depicted in Fig. 12. The grid-forming converter is connected to an ohmic load of $R_{LL} = 146\,\Omega$ and its setpoint increased from $v^*_{LL,1} = 200\,\text{V}$ to $v^*_{LL,2} = 250\,\text{V}$.

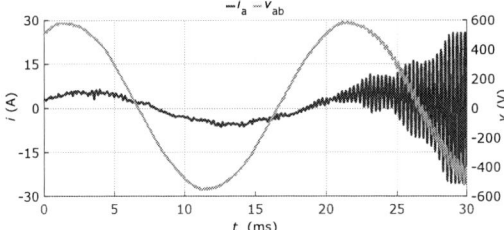

Fig. 11: Measurement of active damping effect

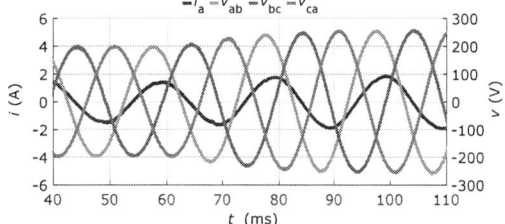

Fig. 12: Measurement of grid forming voltage step

Conclusion and Future Work

This paper introduces a standardized and modular converter platform based on so-called SPECCs. The SPECC, consisting of two AC/DC converters, an isolated DC/DC converter and a standardized signal processing system is well-suited for numerous research applications in the field of grid-connected converter control and microgrids. Measurements demonstrate the function of harmonic compensation and active damping controls. Finally, a basic grid-forming control is also verified with an ohmic load. The shown measurements form the basis for further research on islanded microgrids and the behavior of their converters.

Future work includes studies on the active damping of LCL filters such as capacitor voltage feedback, replacement of capacitor measurements with estimators and state-space control. Apart from harmonic voltage distortions, other typical grid disturbances such as voltage dips, unbalances and changes of grid impedance and their effects on the control stability need to be considered. Furthermore, the coupling of several SPECCs on both the AC and DC side to form AC and DC power electronics based microgrids will be realized. These can then be used to investigate the operation of such isolated microgrids, grid-forming converter controls, converter interactions and possible instabilities in power electronics grids.

References

[1] N. Hatziargyriou, Ed., Microgrid: architectures and control. Chichester, West Sussex, United Kingdom: Wiley, 2013

[2] DIN VDE V 0124-100:2020-06, "Grid integration of generator plants - Low voltage - Test requirements for generator units to be connected to and operated in parallel with low-voltage distribution grids"

[3] VDE-AR-N 4105:2018-11, "Generation Plants on the Low-Voltage Grid - Minimum Technical Requirements for Connection and Parallel Operation of Generation Plants on the Low-Voltage Grid"

[4] Q. Peng, Q. Jiang, Y. Yang, T. Liu, H. Wang, F. Blaabjerg, "On the Stability of Power Electronics-Dominated Systems: Challenges and Potential Solutions", IEEE Trans. on Ind. Applicat. 55, 2019, pp. 7657–7670.

[5] C. Mao et al., "A 400-V/50-kVA Digital–Physical Hybrid Real-Time Simulation Platform for Power Systems," in IEEE Transactions on Industrial Electronics, vol. 65, no. 5, pp. 3666-3676, May 2018

[6] W. Liu, J. -M. Kim, C. Wang, W. -S. Im, L. Liu and H. Xu, "Power Converters Based Advanced Experimental Platform for Integrated Study of Power and Controls" in IEEE Transactions on Industrial Informatics, vol. 14, no. 11, pp. 4940-4952, Nov. 2018

[7] G. Turner, J. P. Kelley, C. L. Storm, D. A. Wetz and W. -J. Lee, "Design and Active Control of a Microgrid Testbed" in IEEE Transactions on Smart Grid, vol. 6, no. 1, pp. 73-81, Jan. 2015

[8] C. Wang et al., "A Highly Integrated and Reconfigurable Microgrid Testbed with Hybrid Distributed Energy Sources" in IEEE Transactions on Smart Grid, vol. 7, no. 1, pp. 451-459, Jan. 2016.

[9] O. A. Mohammed, M. A. Nayeem and A. K. Kaviani, "A laboratory based microgrid and distributed generation infrastructure for studying connectivity issues to operational power systems" IEEE PES General Meeting, 2010, pp. 1-6

[10] J. Dannehl, C. Wessels, F.W. Fuchs, "Limitations of Voltage-Oriented PI Current Control of Grid-Connected PWM Rectifiers With *LCL* Filters", IEEE Transactions on Industrial Electronics 56, 2009, pp. 380–388.

[11] N. Menger, F. Sommer, T. Merz and M. Hiller, "Transient Power Control Algorithm for a Dual Active Bridge", 2021 23rd European Conference on Power Electronics and Applications (EPE'21 ECCE Europe), 2021, pp. 1-8.

[12] F. Sommer, N. Menger, T. Merz and M. Hiller, "Accurate Time Domain Zero Voltage Switching Analysis of a Dual Active Bridge with Triple Phase Shift", 2021 23rd European Conference on Power Electronics and Applications (EPE'21 ECCE Europe), 2021, pp. 1-9.

[13] R. Schwendemann, S. Decker, M. Hiller and M. Braun, "A Modular Converter- and Signal-Processing-Platform for Academic Research in the Field of Power Electronics", 2018 International Power Electronics Conference (IPEC-Niigata 2018 -ECCE Asia), 2018, pp. 3074-3080

[14] DIN EN 60297-3-101:2005-06, Construction methods for electronic devices - Dimensions of the 482,6-mm-(19-in-)construction - Part 3-101: Subracks and assemblies (IEC 60297-3-101:2004); German version EN 60297-3-101:2004

[15] B. Schmitz-Rode, L. Stefanski, R. Schwendemann, S. Decker, S. Mersche, P. Kiehnle, P. Himmelmann A. Liske and M. Hiller, "A modular signal processing platform for grid and motor control, HIL and PHIL applications", 2022 International Power Electronics Conference (IPEC), 2022, preprint

[16] R. W. De Doncker, D. M. Divan and M. H. Kheraluwala, "A three-phase soft-switched high power density DC/DC converter for high power applications", Conference Record of the 1988 IEEE Industry Applications Society Annual Meeting, 1988, pp. 796-805 vol.1

[17] R. Teodorescu, F. Blaabjerg, M. Liserre, P.C. Loh, "Proportional-resonant controllers and filters for grid-connected voltage-source converters" in IEE Proceedings - Electric Power Applications 153, 2006, pp. 750–762

Gate Input Capacitance Characterization for Power MOSFETs Using Turn-on and Turn-off Switching Waveforms

Yota Nishitani[1], Michiko Inoue[1], Takashi Sato[2], and Michihiro Shintani[3]

[1]Nara Institute of Science and Technology
8916-5 Takayama-cho, Ikoma, 630-0192, Japan
Email: nishitani.yota.nt2@is.naist.jp

[2]Kyoto University
Yoshida-honmachi, Sakyo-ku, Kyoto, 606-8501, Japan
Email: takashi@i.kyoto-u.ac.jp

[3]Kyoto Institute of Technology
Matsugasaki, Sakyo-ku, Kyoto, 606-8585 Japan
Email: shintani@kit.ac.jp

Acknowledgments

This work was supported by JST-OPERA Program Grant Number JPMJOP1841, Japan.

Keywords

≪MOSFET≫, ≪Simulation≫, ≪Capacitors≫, ≪Transient analysis≫

Abstract

We propose a novel method for characterizing the input capacitance of power metal-oxide semiconductor field-effect transistors (MOSFETs). In contrast with the conventional method, our switching-based characterization extracts gate-source and gate-drain capacitances in a single setup, without partial differentiation. Characterization using both turn-on and turn-off switching waveforms improved the simulation accuracy and reduced the switching timing error by a factor of more than 2.5x.

1 Introduction

Improving the conversion efficiency is the primary goal in the design of power converters, and power metal-oxide-semiconductor field-effect transistors (MOSFETs) are the key devices for improving the conversion efficiency. In recent years, the demand for more efficient power converters has been increasing to realize a low-carbon society. Power MOSFETs using wide-bandgap semiconductors such as silicon carbide (SiC) and gallium nitride (GaN) have exhibited potential for dramatically improving the conversion efficiency [1, 2]. SiC and GaN power MOSFETs can operate at higher switching speeds, with lower losses, and at higher temperatures than conventionally used Silicon (Si) MOSFETs. To take full advantage of these characteristics, circuit simulations play a crucial role in the design of power converters [3].

The accuracy of a circuit simulation is highly dependent on the accuracy of the device models, which represent the device characteristics. The device model of power MOSFETs represents the current and capacitance characteristics. In particular, the gate input capacitance characteristic significantly affects the switching waveform in transient analysis; thus, the characterization of the gate input capacitance is particularly crucial. The gate input capacitance is the sum of the capacitances between the gate-source electrodes (C_{gs}) and drain-gate electrodes (C_{dg}). In general, the input capacitance of a power MOSFET is measured using an LCR meter. During the measurement, the LCR meter provides an operating bias voltage and applies a small signal to measure the impedance between each electrode [4, 5]. Hereafter, this method is called small-signal capacitance (SSC) measurement. However, the operating voltage of power MOSFETs changes dynamically during switching. Consequently, the MOSFETs exhibit operating conditions that differ significantly from the fixed voltages in the SSC measurements. In fact, it has been reported that switching waveforms cannot be accurately represented by device models characterized via SSC measurements [6].

To address this issue, an input capacitance characterization using the turn-on switching waveforms of a double-pulse tester circuit has been proposed [7]. In this method, the input capacitance is characterized by the trajectories of the gate-charge and gate-voltage space during switching. The capacitance model developed using this method can accurately reproduce transient waveforms. Partial differentiation was performed to differentiate between C_{gs} and C_{dg}. To numerically differentiate the measured data, multiple measurements were performed while changing the load current of the double-pulse tester. Therefore, in the region where C_{gs} and C_{dg} are charged simultaneously, it is important to select the bias voltage to perform partial differentiation. In addition, the charge-voltage trajectories during turn-on and turn-off operations may not be consistent. Capacitances obtained using only the turn-on waveforms may not accurately reproduce turn-off waveforms.

This study proposes a novel input capacitance measurement method that is based on [7]. The proposed method takes advantage of the fact that C_{gs} can be treated as a constant, regardless of the gate bias voltage. The drain-gate charge Q_{dg} associated with C_{dg} is obtained by subtracting the gate-source charge Q_{gs} from the total gate charge obtained by the gate current measurement. The drain-gate capacitance C_{dg} can be directly calculated using the obtained Q_{dg}, without partial differentiation. Because partial differentiation is eliminated, the input capacitance is characterized by a switching waveform with a single-load current condition. In addition, separate modeling of turn-on and turn-off behaviors becomes possible, enabling more accurate transient analysis.

The remainder of this paper is organized as follows. In Section 2, we review related studies and explain the motivation for this study. Section 3 describes the proposed input capacitance measurement method. In Section 4, the input capacitance of a commercial SiC power MOSFET is measured using the proposed and conventional methods, and the differences in the measurement results are discussed. Then, the effectiveness of the proposed method was verified by performing simulations using a boost converter. Finally, Section 5 concludes the paper.

2 Switching-based capacitance characterization

This section briefly reviews the approach to obtain the switching-based (SB) capacitance [7]. While widely adopted SSC measurements use a predetermined operating condition, the conventional SB method measures the input capacitance under an actual switching operation with varying bias voltages.

The double-pulse tester circuit shown in Fig. 1 was used for the conventional SB method. When the power MOSFET is turned on, the gate charge Q_g is expressed as

$$Q_g = Q_{gs} - Q_{dg}. \tag{1}$$

By measuring the gate current I_g during turn-on operation, Q_g can be calculated as

$$Q_g = \int I_g dt = \int \frac{V_{in} - V_{gs}}{R_g} dt, \tag{2}$$

where V_{in} is the output voltage of the pulse generator and V_{gs} is the gate-source voltage. Q_g was calculated by measuring the potential difference in the external gate resistor R_g. In the measurement, a relatively large R_g is chosen to suppress ringing owing to the parasitic capacitance and inductance as well as to

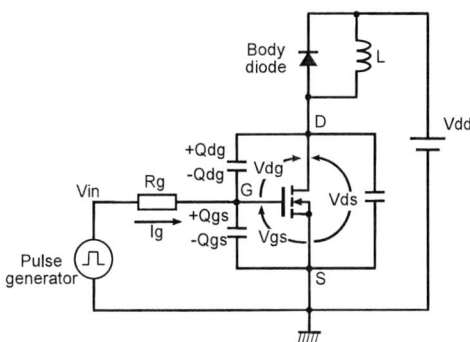

Fig. 1: Double-pulse tester circuit.

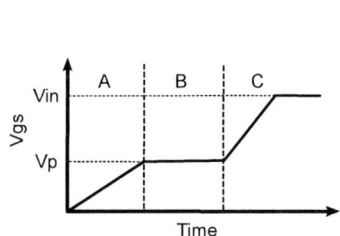

Fig. 2: Gate voltage during turn-on.

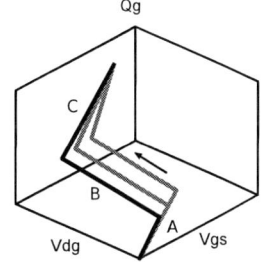

Fig. 3: Turn-on trajectories for different load currents in a $V_{gs}V_{dg}Q_g$ space.

reduce the influence of the intrinsic gate resistance of the power MOSFET. By definition,

$$C_{gs} = \frac{dQ_{gs}}{dV_{gs}} \tag{3}$$

$$C_{dg} = \frac{dQ_{dg}}{dV_{dg}}. \tag{4}$$

Note that Q_g is measurable only in this setup. In order to separate Q_g into Q_{gs} and Q_{dg}, the authors in [7] made an assumption that "Q_{gs} depends only on V_{gs} and Q_{dg} depends only on V_{dg}." Then, C_{gs} and C_{dg} are derived as the partial differentiations of the gate-source voltage V_{gs} and drain-gate voltage, respectively V_{dg}:

$$\frac{\partial Q_g}{\partial V_{gs}} = \frac{\partial \left(Q_{gs} - Q_{dg} \right)}{\partial V_{gs}} = \frac{\partial Q_{gs}}{\partial V_{gs}} = C_{gs} \tag{5}$$

$$\frac{\partial Q_g}{\partial V_{dg}} = \frac{\partial \left(Q_{gs} - Q_{dg} \right)}{\partial V_{dg}} = -\frac{\partial Q_{dg}}{\partial V_{dg}} = -C_{dg}. \tag{6}$$

In this way, C_{gs} and C_{dg} are determined separately.

Here, we consider the gate charge accumulation during the turn-on period. A typical turn-on waveform of V_{gs} is illustrated in Fig. 2. In period "A," C_{gs} is charged primarily until V_{gs} reaches the plateau voltage V_p. The period "B" is called the "Miller plateau," where V_{gs} takes a constant value, and only C_{dg} is charged. In period "C," both C_{gs} and C_{dg} are charged until V_{gs} reaches V_{in}.

Fig. 3 shows the trajectories of the switching waveform plotted in a $V_{gs}V_{dg}Q_g$ space, where "A," "B," and "C" correspond to the periods shown in Fig. 2. In the double-pulse tester, because the switching waveforms differ depending on the load current, multiple trajectories are drawn, as shown in Fig. 3, collectively forming a plane in the $V_{gs}V_{dg}Q_g$ space. The slope along the V_{gs} axis represents C_{gs}, and that

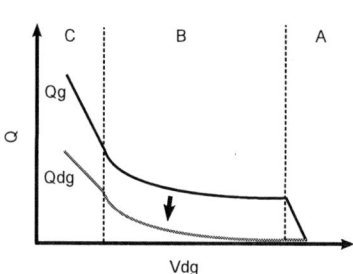

Fig. 4: Gate charge as a function of V_{dg}.

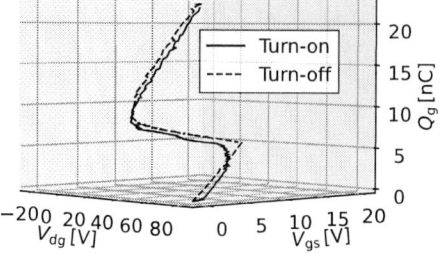

Fig. 5: Switching trajectories during turn-on and turn-off.

along the V_{dg} axis is C_{dg}. These values were obtained by performing measurements with different load currents.

Although the conventional SB method captures capacitances during switching operations more accurately than SSC measurements, the following issues exist:

- Multiple measurements with different load currents are required to form a plane in the $V_{gs}V_{dg}Q_g$ space.
- Determining the bias point to calculate partial differentiation is non-trivial.
- Only turn-on trajectories are used, and turn-on and turn-off characteristics may not match.

3 Switching-based charge-subtraction capacitance characterization

Here, we propose a novel input capacitance characterization method that resolves the aforementioned issues in the SB method. Hereafter, we refer to the proposed method as switching-based, charge-subtraction (SB-CS) capacitance characterization. The SB-CS method uses the same measurement setting as the conventional switching-based, partial differentiation (SB-PD) method. A distinct feature of the SB-CS method is the separation of Q_g to calculate C_{gs} and C_{dg}. We propose to calculate Q_{dg} by subtracting Q_{gs} from Q_g and regarding C_{gs} as invariant. In addition, while the conventional method uses only turn-on waveforms, the SB-CS method uses both turn-on and turn-off waveforms.

The SB-CS capacitance characterization is as follows: First, C_{gs} was calculated from the turn-on trajectory. As explained, charge accumulates mainly in C_{gs} in period "A." Strictly speaking, charge is still transferred to C_{dg} in period "A," but it is sufficiently small compared to C_{gs}. Hence, $Q_g = Q_{gs}$ holds true in this period. Using this period, C_{gs} can be derived as

$$\frac{dQ_g}{dV_{gs}} = \frac{dQ_{gs}}{dV_{gs}} = C_{gs}. \tag{7}$$

Here, we assume that "C_{gs} is invariant and voltage independent." This is a reasonable assumption because the voltage dependency of C_{gs} is much smaller than that of C_{ds} and C_{dg} in vertical-power MOSFETs [3, 8]. With this assumption, the C_{gs} value obtained using period "A" is used to calculate C_{dg}.

C_{dg} is then calculated as the residual, i.e., Q_{dg} is written as

$$-Q_{dg} = Q_g - Q_{gs} = Q_g - C_{gs} \cdot V_{gs}. \tag{8}$$

Fig. 4 shows the charge trajectory from the V_{dg} direction in the $V_{gs}V_{dg}Q_g$ space. Q_{dg} was obtained by subtracting Q_{gs} from Q_g, and the gradient of the Q_{dg} curve was defined as C_{dg}.

Because the SB-CS method does not rely on partial differentiation, multiple measurements with varying load currents are not required. As we verify later through experiments, the input capacitance can be determined with a single shot under a typical load current condition. Obviously, biased point selection for partial differentiation is no longer necessary.

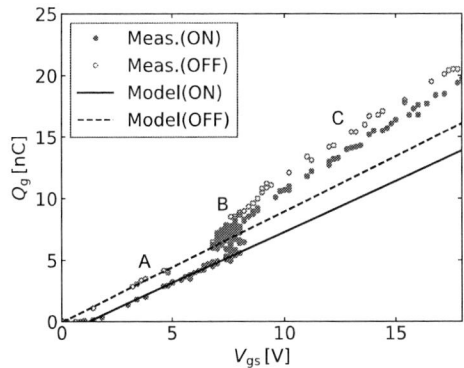

Fig. 6: Measured Q_g-V_{gs} characteristics and its model.

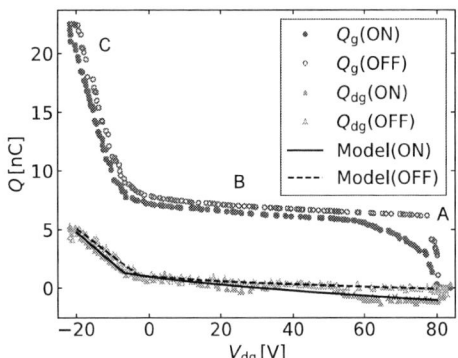

Fig. 7: Q_g-V_{dg} and Q_{dg}-V_{dg} characteristics and Q_{dg} model.

We also propose the use of both the turn-on and turn-off waveforms. Fig. 5 shows the measured waveforms of a commercial SiC power MOSFET (Infineon Technologies AG, IMW120R090M1H, 1200 V, 26 A [9]). The turn-on and turn-off Q_g trajectories are not identical. For such cases, we model the capacitance separately for the turn-on and turn-off transitions. As an example of model implementation, the turn-on capacitance models are written as follows [7]:

$$C_{gs} = \mathbf{CGSO_{on}} \tag{9}$$

$$C_{dg} = \begin{cases} \mathbf{CDGO_{on}} \left(V_{dg} + \mathbf{VJ_{on}}\right)^{-\frac{1}{2}} & (V_{dg} > -0.5\mathbf{VJ_{on}}) \\ \mathbf{COXD_{on}} & (V_{dg} \leq -0.5\mathbf{VJ_{on}}), \end{cases} \tag{10}$$

where the model parameters are in bold. For the turn-off model, the subscript **on** is replaced with **off**. These model parameters can be determined separately from the turn-on and turn-off charge characteristics, respectively. At the point where the C_{dg} function switches, the two functions are implemented to connect smoothly.

4 Measurements and evaluations

The SB-CS method was evaluated using a commercially available SiC power MOSFET (Infineon Technologies AG, IMW120R090M1H, 1200 V, 26 A [9]) as the device under test (DUT). The device models were implemented using Verilog-A [10], and transient analyses were performed using a circuit simulator [11]. For comparison, we applied the SB-PD method [7] to the same DUT.

Capacitance modeling

The double-pulse tester shown in Fig. 1 was used to measure Q_g, where $R_g = 1\,\mathrm{k\Omega}$, $V_{dd} = 80\,\mathrm{V}$, $L = 220\,\mu\mathrm{H}$. A MOSFET body diode of the same type as the DUT was used as the freewheeling diode. $V_{in} = 20\,\mathrm{V}$ was applied using a pulse generator, and the load current was set to 2 A. Fig. 5 shows the measured turn-on and turn-off trajectories. The capacitance characteristics were modeled using Eqs. (9) and (10), and the model parameters were fitted to reproduce these characteristics using simulated annealing [12]. The current characteristics were also modeled to match the drain current measured by the curve tracer using a threshold voltage model [13]. The body diode characteristics were also modeled using the traditional p-n junction diode model [14].

First, to obtain C_{gs}, we used Fig. 6, in which Fig. 5 is viewed from the positive direction of V_{dg}. In Fig. 6, the symbols represent the Q_g-V_{gs} characteristics obtained from the measurement. Closed and open circles indicate turn-on and turn-off, respectively. The solid and broken lines correspond to the fitting results for turn-on and turn-off, respectively. The capacitances of (9) and (10) were integrated and used for fitting as charges. The gate charge Q_g in the period "A" was used for C_{gs} fitting.

Table I: Comparison of model parameters for input capacitance

Parameter	on	off	Conv.
CGSO [pF]	827	899	929
CDGO [pF]	157	78.6	139
COXD [pF]	265	235	381
VJ [V]	13.5	6.05	11.3

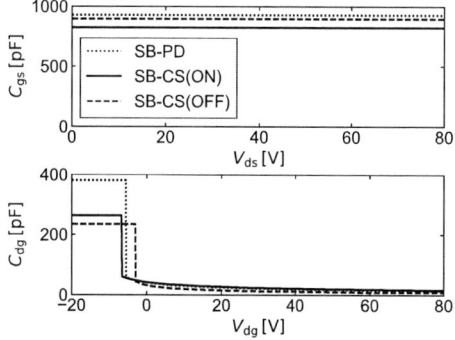

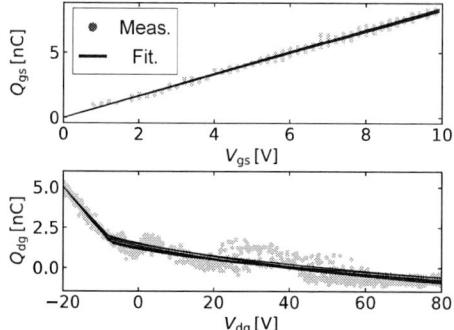

Fig. 8: Comparison of characterized input capacitances.

Fig. 9: Load current dependence of charge characteristics.

Fig. 7 shows the measured Q_g-V_{dg} characteristics. The circle and triangle symbols represent Q_g and Q_{dg}, respectively, which were obtained using Eq.(8). The open symbols correspond to the turn-off waveforms. The solid and broken lines represent the fitting results for the turn-on and turn-off waveforms, respectively. It can be observed that the slope of Q_{dg} is significantly reduced when $V_{dg} < 0$ compared to Q_g. Furthermore, those slopes were different for the turn-on and turn-off waveforms, indicating different plateau period lengths.

Comparison of capacitance model parameters

We compared the capacitance models obtained using the SB-CS method with the conventional SB-PD method [7]. For the SB-PD method, the input capacitance was calculated using the turn-on waveform.

Table I lists the model parameters obtained using the SB-CS and SB-PD methods. The capacitance characteristics of C_{gs} and C_{dg}, which were obtained using these parameters, are shown in Fig. 8. C_{gs} obtained by the SB-CS method was almost the same as that obtained by the SB-PD method. However, for C_{dg}, a large difference is observed for $V_{dg} < 0$. This is because of the value of **COXD**; that of the SB-PD method is approximately 1.5 times larger than that of the SB-CS method. In addition, C_{gs} in the region of $V_{dg} < 0$ is approximately three times smaller than C_{dg}. This may be because the slope of C_{dg} could not be determined accurately when taking partial derivatives.

To confirm that the SB-CS method can be used to accurately determine the input capacitance regardless of the load current condition, the change in the charge characteristics was evaluated for various load currents. Fig. 9 shows the measured Q_{gs} and Q_{dg}, as well as the fitting result obtained using the SB-CS method. The load current was varied from 1 A to 7 A in increments of 1 A. The dotted symbols show the measurement of each load current, and the solid lines show the fitting results. The figure shows that the fitting results are almost the same regardless of the load current, indicating that a consistent capacitance characteristic can be obtained at any load .

Transient analysis

To investigate the effectiveness of the capacitance model with a practical circuit, a boost converter was fabricated, and the transient analysis waveforms and measurement results were compared. Transient analyses were performed using two models characterized by the turn-on and turn-off waveforms. The

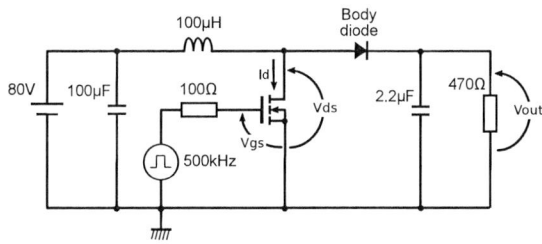

Fig. 10: Circuit diagram of a boost converter.

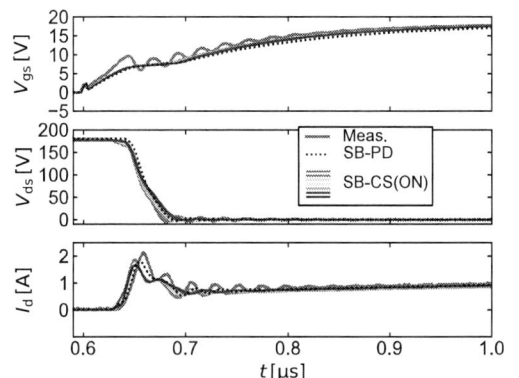

Fig. 11: Simulated and measured turn-on waveforms.

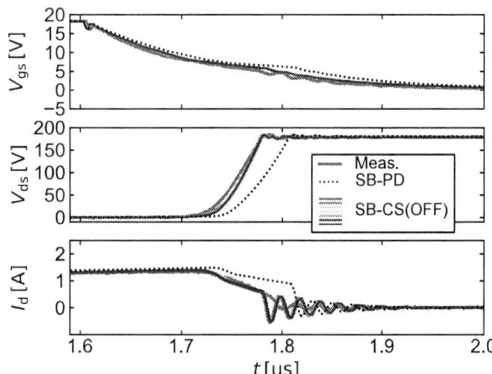

Fig. 12: Simulated and measure turn-off waveforms.

parasitic components of each element, such as the printed circuit board and MOSFET package obtained using an impedance analyzer, were considered in the simulation.

Fig. 10 shows a schematic of the boost converter circuit. The DUT was IMW120R090M1H, and the freewheeling diode was the body diode of the same device as the DUT. A square waveform with an amplitude of 18 V amplitude is applied using a pulse generator. The operating frequency was 500 kHz and the duty ratio was 50%. The measurement and simulation results of the turn-on and turn-off waveforms are shown in Figs. 11 and 12, respectively, where the gray solid line indicates the actual measurement. The colored solid lines and dotted lines are the simulation results obtained using the SB-CS and SB-PD methods, respectively. For the SB-CS method, seven lines are shown in different colors for load currents of 1– 7 A.

In Fig. 11, the V_{gs} waveform of the SB-CS method after the plateau period is in good agreement with the measurements. From the plateau voltage, it can be concluded that the C_{dg} parameter obtained using the SB-CS method is valid. In the SB-PD method, a larger delay of V_{gs} is observed owing to the overestimation of **COXD**. In addition, in the SB-CS method, the simulation results obtained using the capacitance parameters obtained by changing the load current were almost the same. From the transient simulation results, we again confirmed that an arbitrary load current can be used to determine the capacitance parameters.

The SB-CS method reproduced the measured waveform well even for the turn-off waveform, as shown in Fig. 12. In the case of turn-off waveforms, the accuracy of **COXD** affects that of the transient analysis more significantly as the bias condition of the C_{dg}-V_{dg} characteristic transitions from negative to positive values. Therefore, a larger error was observed over the entire switching waveform for the SB-PD method than for the turn-on analysis. Furthermore, from Fig. 12, we can confirm that the load current of the proposed method does not significantly affect the results of the transient analyses.

To evaluate the simulation accuracy quantitatively, we compared the timing errors between the measure-

Table II: Timing error between measured and simulation

	V_{gs}		V_{ds}		I_{d}	
	SB-PD	SB-CS	SB-PD	SB-CS	SB-PD	SB-CS
Turn-on [ns]	59.8	23.8	2.4	−2.6	−1.4	−5.4
Turn-off [ns]	6.2	2.2	26.4	3.4	34.2	−6.8

ment and simulation at V_{gs} =15 V. The timing error in the capacitance model determined from the 2 A load current is presented in Table II. For V_{gs}, the SB-CS method reduced timing error by approximately 2.5 times for the turn-on periods and 2.8 times for the turn-off periods. By reducing the timing error of V_{gs}, the timing errors of V_{ds} and I_{d} were also improved by 7.8 and 5.0 times for V_{ds} and I_{d}, respectively, at V_{ds} =100 V and I_{d} =1 A.

5 Conclusion

In this paper, we proposed a novel input capacitance measurement method using switching waveforms. The conventional method uses the turn-on waveforms with different load current conditions and required partial differentiation for separating C_{gs} and C_{dg}. However, in the proposed method, C_{dg} is obtained as the difference between Q_{g} and Q_{gs} by taking advantage of the fact that C_{gs} is nearly bias voltage independent. In addition, the proposed method requires a single measurement and the selection of bias points for partial differentiation is unnecessary. Furthermore, because the trajectory of Q_{g} is different between the turn-on and turn-off waveforms, separate modeling of the turn-on and turn-off waveforms was performed to improve the accuracy of the transient analysis. The evaluation using a SiC power MOSFET proved that the proposed method can be used to extract consistent input capacitance. We confirmed that the model obtained using the proposed method accurately simulates the measured waveforms of a boost converter and reduces the timing error of V_{gs} in the transient analysis by 2.5x smaller compared with the conventional method.

References

[1] B. Jayant Baliga, *Fundamentals of Power Semiconductor Devices*. Springer, 2008.

[2] T. Kimoto and J. A. Cooper, *Fundamentals of Silicon Carbide Technology: Growth, Characterization, Devices and Applications*. Wiley, 2014.

[3] M. Shintani, Y. Nakamura, K. Oishi, M. Hiromoto, T. Hikihara, and T. Sato, "Surface-potential-based silicon carbide power MOSFET model for circuit simulation," *IEEE Transactions on Power Electronics*, vol. 33, no. 12, pp. 10 774–10 783, 2018.

[4] T. Funaki, N. Phankong, T. Kimoto, and T. Hikihara, "Measuring terminal capacitance and its voltage dependency for high-voltage power devices," *IEEE Transactions on Power Electronics*, vol. 24, no. 6, pp. 1486–1493, 2009.

[5] *B1505A Power Device Analyzer/Curve Tracer*, Keysight Technologies, Inc., 2015.

[6] V. Hoch, J. Petzoldt, A. Schlogl, H. Jacobs, and G. Deboy, "Dynamic characterization of high voltage power MOSFETs for behavior simulation models," in *Proceedings of IEEE European Conference on Power Electronics and Applications*, 2009, pp. 1–10.

[7] K. Oishi, M. Shintani, M. Hiromoto, and T. Sato, "Input capacitance determination of power MOSFETs from switching trajectories," in *Proceedings of IEEE International Conference of Microelectronic Test Structures*, 2017, p. 4.4.

[8] N. Phankong, T. Funaki, and T. Hikihara, "A static and dynamic model for a silicon carbide power MOSFET," in *Proceedings of IEEE European Conference on Power Electronics and Applications*, 2009, pp. 1–10.

[9] *IMW120R090M1H datasheet*, Infineon Technologies AG, 12 2020.

[10] C. C. McAndrew, G. J. Coram, K. K. Gullapalli, J. R. Jones, L. W. Nagel, A. S. Roy, J. Roy-chowdhury, A. J. Scholten, G. D. J. Smit, X. Wang, and S. Yoshitomi, "Best practices for compact modeling in Verilog-A," *IEEE Journal of the Electron Devices Society*, vol. 3, no. 5, pp. 383–396, 2015.

[11] *HSPICE User Guide: Basic Simulation and Analysis Version P-2019.06*, Synopsys, Inc., 2019.

[12] S. Kirkpatrick, C. D. Gelatt, Jr., and M. P. Vecchi, "Optimization by simulated annealing," *Science*, vol. 220, no. 4598, pp. 671–680, 1983.

[13] T. Sakurai and A. R. Newton, "A simple MOSFET model for circuit analysis," *IEEE Transactions on Electron Devices*, vol. 38, no. 4, pp. 887–894, 1991.

[14] G. Massobrio and P. Antognetti, *Semiconductor Device Modeling with SPICE*. McGraw-Hill, 1993.

AC Battery: Modular Layout with Cell-level Degradation Control

Claudio Burgos-Mellado[1], Marcos Orchard[2], Diego Muñoz-Carpintero[1],
Tomislav Dragičević[3], Lorenzo Reyes-Chamorro[4], Jacqueline Llanos[5]

[1]Universidad de O'Higgins, Rancagua, Chile
[2]University of Chile, Santiago, Chile
[3]Technical University of Denmark, Copenhagen, Denmark
[4]Universidad Austral de Chile, INVENT UACh, Valdivia, Chile
[5]Universidad de las Fuerzas Armadas ESPE, Sangolqui, Ecuador
Email: claudio.burgos@uoh.cl

Acknowledgments

This work was supported in part by "Agencia Nacional Investigacion y Desarrollo" (ANID) under grants: ANID/FONDECYT de Iniciación/11220989, ANID/FONDECYT de Iniciación/11221230 and ANID/PAI/PAI7719002.

Keywords

≪Battery≫, ≪Consensus-based cooperative control≫, ≪Degradation≫, ≪Energy storage≫, ≪Lifetime≫, ≪Modular Multilevel Converters≫

Abstract

This paper proposes a three-phase AC battery based on the modular multilevel converter (M2C). The AC battery concept allows plug-and-play combinatorial integration of diverse battery cells with different characteristics such as nominal voltage, state of charge (*SoC*), and degradation levels. The resulting modular and reconfigurable battery pack can cost-effectively cover various applications, from electrified vehicles to stationary storage. To this end, in each sub-module (*SM*) of the M2C, battery cells (or battery modules composed of one or more battery cells) are connected to a single capacitor, thus enabling a cell-to-cell battery integration and control. In this scenario, the traditional battery management system (BMS) can be replaced by control schemes at the converter level, aiming to equalise critical parameters associated with battery cells (or battery modules). In this context, the degradation level of battery cells integrated by the proposed AC battery is a critical parameter to be considered. Indeed, it is likely that the integrated battery cells by the AC battery have different degradation levels. Therefore, methods for managing their degradation state are required to equalise the lifespan of battery cells within the AC battery. Based on that, this paper proposes a consensus-based distributed control scheme to manage the degradation process of battery cells (or battery modules composed of one or more battery cells) within the AC battery. Simulation results corroborate the effectiveness of the proposal.

Introduction

In recent years, the use of electric vehicles (EVs) has increased worldwide. This has promoted policies in several countries to adopt environment-friendly transport means, targeting the reduction of greenhouse gases. Recently, some European countries have announced that fossil fuel-based vehicles will be forbidden in only a few years. Thus, the demand for EVs will continue to increase in the coming years [1]. It is noteworthy that a critical component of EVs is the battery energy storage system (BESS). For electromobility applications, a common practice is to use the BESS until its capacity reaches approximately 80%

of its nominal value. After that, the current recommendation is that the BESS is discarded and replaced by a new one. This situation generates opportunities and challenges for efficient and novel solutions that could reuse degraded BESSs as second-life battery systems in smart grids, modern power systems, or less demanding electromobility applications.

The integration of second-life batteries has several challenges as batteries may have different manufacturers, chemistry, voltage, capacity, degradation levels, lifespan, etc. [2]. In this sense, the modular multilevel converter (M2C) has been proposed as a prominent solution that meets all the requirements needed to integrate heterogeneous BESS coming from EVs [3, 4, 5, 6]. Indeed, the M2C allows fulfilling crucial tasks such as (i) distributing battery cells (or battery modules composed of one or more battery cells) on its sub-modules (*SMs*), providing modularity for battery integration, (ii) equalising the state of charge (*SoC*) of battery cells (or battery modules) placed on the M2C *SMs* thanks to the circulating currents, and (iii) lifetime prolongation of battery cells by using suitable modulation techniques and/or proper control of their usage policies [7]. Indeed, in [8, 9, 10], it is argued that pulsed charging/discharging (which is an intrinsic M2C property) can prolong the battery life. In addition, in [7], it is demonstrated that the degradation and lifespan of BESS can be managed, among other criteria, through the operational policies for charging and discharging the battery. In particular, the *SoC* range is constrained to pre-defined ranges during the discharging process. For instance, the degradation of a battery discharged between 100% and 0% of *SoC* is much higher than the same battery being discharged, with an operational policy limiting the *SoC* between 100% and 75% (see [7] for more information).

The use of the M2C for integrating second-life batteries has been proposed in [3, 4, 5, 6]. These references show that the M2C allows a modular integration of battery cells coming from EV applications. Also, control schemes for regulating the *SoC* of batteries placed on the M2C *SMs* were proposed, showing promising results. However, these research efforts have the following drawbacks: (i) they do not provide modularity in terms of computational burden for the M2C control system, as they are based on a centralised control scheme (where a central controller manages the whole system), and (ii) they do not consider active degradation control of the battery cells (or battery modules composed of one or more battery cells) integrated by the M2C.

Based on the research gaps discussed above, this paper proposes the AC battery concept based on the M2C shown in Fig. 1. This topology enables the integration of battery cells on the *SMs* of the M2C. It uses local controllers (*LCs*) placed on *SMs* to implement a distributed control architecture, providing modularity in both battery cells integration and computing capacity required to implement the control system of the proposed AC battery. In addition, it is demonstrated that the degradation level of battery cells within the AC battery can be controlled using a proposed consensus-based distributed control scheme augmented by a degradation management system (DMS). Note that the DMS generates control actions (*dms* signals in Fig. 2) for the proposed consensus algorithm. Additionally, the DMS should be implemented in the cloud to avoid the need for placing control platforms with high computational capabilities in the M2C.

Control Architecture of the Proposed AC Battery

Fig. 2 shows a generic cluster of N *SMs* that comprises the M2C-based AC battery. As observed (on the left), battery cells with different characteristics can be integrated into the cluster through the *SMs*. In each *SM*, an *LC* is placed, which is in charge of driving the respective semiconductor devices and calculating the main operational parameters of the battery cell (or battery modules composed of one or more battery cells) connected to the *SM* (e.g., degradation level, temperature, cycle number). Then, this local information is exchanged with neighbour *LCs* via a distributed communication network (see Fig. 2), enabling the implementation of consensus-based distributed control schemes to regulate one or more parameters of battery cells. As shown in Fig. 2, the AC battery concept allows for integrating battery cells (or battery modules) with different characteristics and specifications. Indeed, each *SM* may be using a battery cell or a battery module with unequal states of charge (*SoCs*), power ratings, and degradation levels. Focusing on the latter, it is noteworthy that the degradation level of the entire AC battery is often determined by the most degraded battery cell (or battery modules) that composes it. In this regard, it is

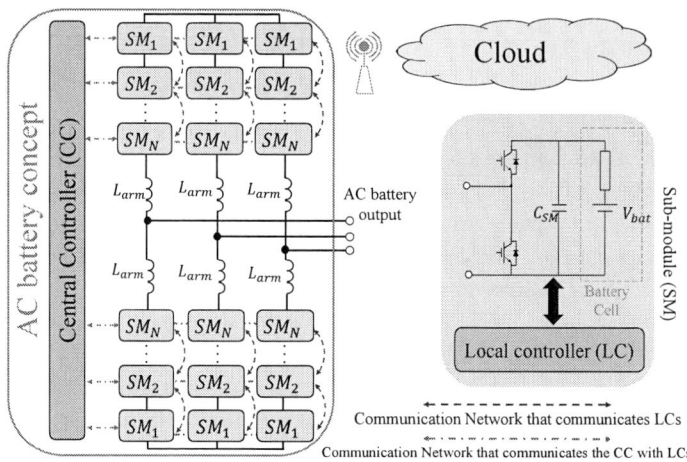

Fig. 1: Proposed modular layout for integrating second-life battery cells (and/or battery modules).

paramount to develop control schemes to manage the degradation that each battery cell/module undergoes, thus improving the lifespan of the entire AC battery. To the authors' best knowledge, this is the first work that proposes the integration of battery cells in a modular layout with cell-level degradation control. In this sense, Fig. 2 shows the proposed AC battery system from where it is possible to distinguish three main points: (i) a central controller (CC), (ii) local controllers (*LCs*) operating in a consensus-based distributed control architecture, and (iii) the cloud, where a degradation management system (DMS) is being run. It must be highlighted that the CC performs the power control of the entire AC battery. In contrast, *LCs* regulate the degradation level of the battery cells (or battery modules) within the AC battery. This is done by the proposed consensus algorithm introduced in the next section. Finally, the DMS aims to manage the M2C to achieve battery-cell-degradation control by generating control actions for the *LCs* to accomplish that control objective. In this sense, it should be pointed out that the DMS is similar to the well-known battery management system (BMS), widely used in battery systems, but in this case, the DMS is tailored for battery degradation. In particular, the DMS is in charge of generating operational policies for charging and discharging each battery cell (in terms of *DoD*) within the AC battery to regulate their degradation levels. In this paper, point (ii) is addressed: it is discussed in the next section. The DMC mentioned in (iii) is ongoing research and will be addressed in future publications.

Proposed Distributed Control Scheme for Degradation Control

Fig. 3 illustrates the proposed control scheme implemented on the *i*th *LC* (for the rest of the *LCs*, the procedure is analogous). The main aspects of the implementation are discussed below.

Degradation model

Battery degradation is the process in which capacity fades over time. It is also characterised by an increase of the internal impedance, which may dangerously increase operation temperature, eventually making a battery replacement necessary. Battery degradation depends on many factors such as current rate, temperature, number of cycles and their depths of discharge (*DoDs*) [7]. In particular, the *DoD* has an enormous influence on battery lifespan [11, 12, 13]: deep discharges reduce battery lifetimes. Because of this, battery degradation can be indirectly controlled by regulating the *DoD*. In particular, in this paper, we focus on reducing the *DoD*-related degradation by regulating the discharging policies on battery cells [7] (or battery modules composed of one or more battery cells).

To properly analyse and reduce the *DoD*-related degradation, a model that relates the *DoD* policies of the battery cells with their long term degradation is needed. In this work, a simplified version of the model of [7] is used. A cycle is defined as the period of time since a discharging period starts until a charging period ends (Fig. 4 shows the cycle considered in this work). Let $Q_i(k)$ be the capacity of battery cell *i*th

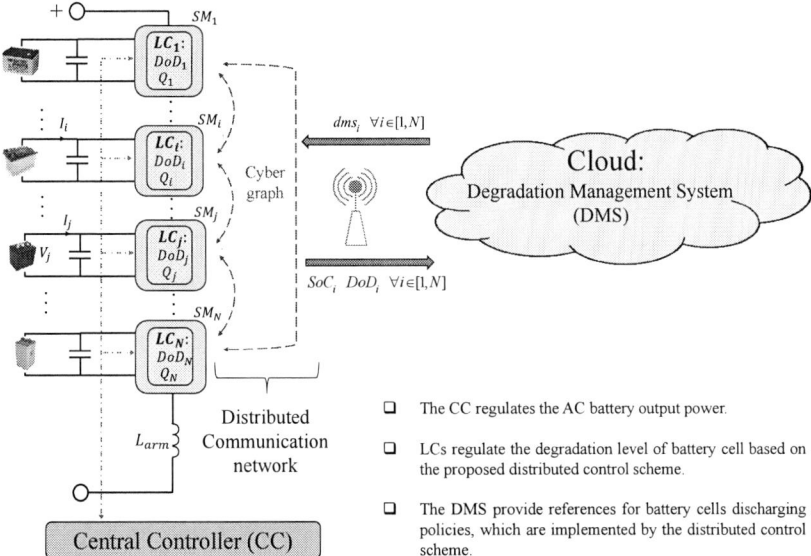

Fig. 2: Proposed control architecture for regulating the degradation level of battery cells that composed the AC battery. (One cluster of the M2C is shown)

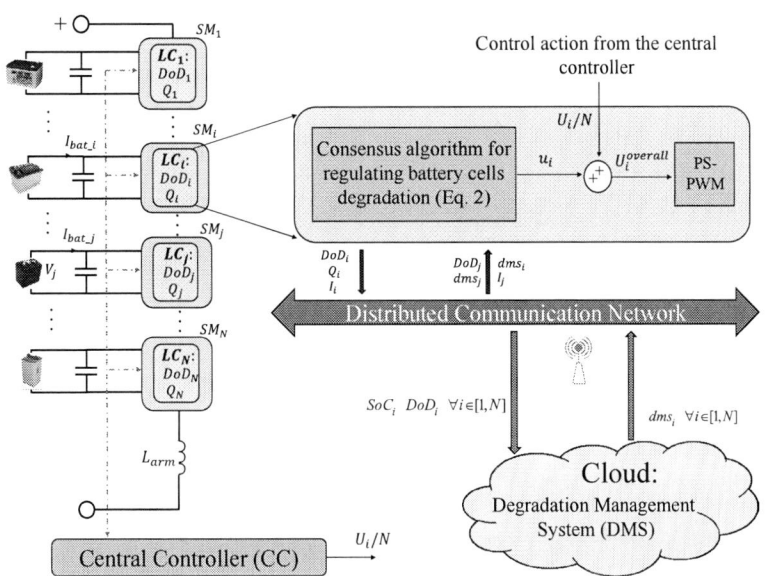

Fig. 3: Implementation of the proposed distributed control scheme for controlling the degradation of battery cells.

at discharge cycle k. Then, at the next cycle $k+1$, the capacity evolves as:

$$Q_i(k+1) = \eta_i(DoD_i(k)) \cdot Q_i(k), \qquad (1)$$

where $\eta_i(DoD_i(k))$ is the Coulombic efficiency of the battery cell. This model is thus fully defined by having η_i as a function of $DoD_i(k)$. One usually will have a table for the values of η_i for different $DoD_i(k)$, and the remaining values (those in between) can be found by interpolation. It must be pointed out that these Coulombic efficiencies can be calculated using data provided by the manufacturer that characterises the battery lifespan. Thus, once the cycle is identified, the model characterises cycle k in terms of its $DoD_i(k)$. Then the Coulombic efficiency is found, and finally the capacity for time $k+1$, $Q_i(k+1)$ is obtained.

Note that the simplification of this model is that it is assumed that η_i depends only on $DoD_i(k)$, whereas in [7] it depends both on $DoD_i(k)$ and the SoC swing (which is the net difference between the highest and the lowest SoC value within a cycle), and in other works [8] it is also affected by other battery parameters such as SR, temperature, and cycles. These effects will be considered in future publications.

Proposed distributed control scheme for controlling battery cells degradation

The proposed control scheme is based on the consensus theory and looks to indirectly regulate battery cell degradation by managing the battery cell discharge policy. Indeed, as is shown in (1), the degradation in the ith battery cell is a function of the DoD that undergoes during a given operational cycle. Also, a battery cell with an operating policy that drives it to high $DoDs$ will degrade faster than the same battery being managed with an operational approach with lower $DoDs$: This situation motivates the proposed control scheme for battery cell degradation. For instance, let us consider that both the ith and jth battery cells illustrated in Fig. 3 have the same characteristics, but due to their different usage profiles before being recycled to be part of the AC battery, their current capacities are: $Q_i(k) < Q_j(k)$. In this case, for regulating their degradation levels, it is preferable that both battery cells follow the criterion: $DoD_i(k) < DoD_j(k)$ while the AC battery is operating. By doing that, and according to (1), while the AC battery is being used, internally, the jth battery cell is being more degraded than the battery cell ith. To achieve this, the consensus algorithm (2) is proposed. To implement (2), the cyber graph illustrated in Fig. 2 is modelled as an undirected graph $\mathbb{G} = (\mathfrak{N}, \xi, A)$ among the local controllers (LCs) $\mathfrak{N} = \{1, ..., N\}$, where ξ is the set of communication links and A is a non-negative $N \times N$ weighted adjacency matrix. The elements of A are $h_{ij} = h_{ji} \geq 0$, with $h_{ij} \geq 0$ if and only if $\{i, j\} \in \xi$ [14]. In this context, the consensus can be reached via a feedback loop by applying the protocol u_i given by (2). This control is distributed because it only depends on the immediate neighbours $j \in \mathfrak{N}(i)$ of node i in the graph topology. Finally, it must be pointed out that the proposed control scheme needs to know the current degradation level of battery cells. For the simulation work presented here, these values are assumed as known. For real applications, this information can be obtained from experimental tests applied to the battery cells before their integration into the AC battery (such as capacity tests and electrochemical impedance spectroscopy [15]).

Note that algorithm (2) regulates the DoD of battery cells by manipulating their output currents I_{bat} (see Fig. 3). The term g_i holds the dynamic response of the controller, h_{ij} are the elements of the adjacency matrix A, and the terms dms_i and dms_j are generated by the degradation management system (DMS) to set the usage policies on battery cells (according to some criteria). As was discussed earlier in this paper, the development of the DMP is not addressed here; it is ongoing research and will be published further. However, simulation work discusses and validates some case studies for these terms in the following section.

$$u_i = -\frac{1}{g_i} \sum_{j \in \mathfrak{N}(i)} h_{ij} \cdot \left(\frac{I_{bat\_i}}{dms_i} - \frac{I_{bat\_j}}{dms_j} \right) \qquad (2)$$

It must be highlighted that the overall control action $U_i^{overall}$ for the ith SM, sent to the modulation stage, comprises two parts, as shown in Fig. 3: U_i is produced by a central controller for regulating active and reactive powers of the proposed AC battery, and u_i is given by the proposed consensus-based distributed scheme (see Fig. 3) for the regulation of battery cells degradation. In this case, u_i is generated by (2), whereas U_i is generated by the central controller reported in [16].

Finally, as seen in Fig. 3, the implementation of the proposed distributed control for regulating the degradation of batteries cells requires that each LC compute the following variables (in each battery cell cycle): (i) cycle length calculation, (ii) DoD in the identified cycle and (iii) battery cell current. Once this information is calculated (considering the cycle k), the degraded capacity of the ith battery cell, after that cycle, i.e., in the cycle $k+1$, is given by (1).

Simulation Work: Proposal Validation

In this section, the performance of the proposed distributed control scheme for controlling the degradation level of battery cells inside the AC battery is validated via simulation work. To this end, the M2C-based AC battery shown in Fig. 1 is simulated using PLECS software with the parameters listed in Table I. Note in this table that the coulombic efficiencies were arbitrarily selected (following real patterns for these efficiencies [7, 13]) to validate the methodology proposed in this paper and achieve a fast degradation ratio (allowing a sensible simulation time). On the other hand, the central controller (see Fig. 1), in charge of active and reactive power regulation, is implemented in the $\Sigma\Delta\alpha\beta$ reference frame (see [16]), and the battery cells degradation is achieved through the distributed control scheme (2), which implements the discharging policy given by the DMS (dms_i and dms_j in (2)). This section demonstrates that the proposed distributed control scheme illustrated in Figs. 1-3 can control the degradation level of the battery cells that composes the AC battery. In particular, it is shown that the proposed approach can extend the lifespan of the AC battery (and battery cells that compose the AC battery) by properly setting the control actions generated by the DMS (dms signals in (2)). For now, these control actions are chosen empirically for two case studies. Both case studies consider that the AC battery is connected to an AC grid, operating with the discharging and charging profile illustrated in Fig. 4. As observed, the AC battery is being discharged in each cycle by power steps of 18kW. Then, it is charged with a power step of the same magnitude (18kW). It must be highlighted that this pattern, along with the small capacities for the battery cells (see Table II and Table III), were selected only for validation purposes as it allows a sensible number of cycles and in an achievable simulation time. Finally, note that the pattern displayed in Fig. 4 is repeated until one or more battery cells reach their usable life. The characteristics of the case studies considered in this paper and the simulation results are presented as follows.

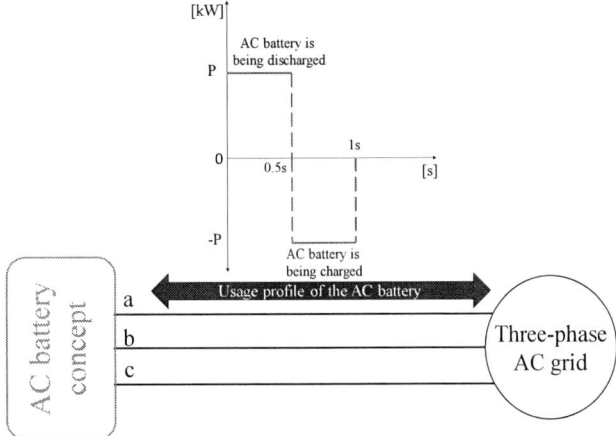

Fig. 4: Main characteristics of the discharging/charging cycle used for operating the AC battery.

Table I: AC battery parameters.

Parameter	Value
Arm inductance (L_{arm})	5mH
SM capacitance (C_{SM})	6mF
No. *SM* per arm (N)	3
Switching frequency (PS-PWM)	8kHz
Active Power reference (P)	18kW
Reactive Power reference (Q)	0Var
Consensus gain (k_i)	1
Coulombic efficiencies as a function of the *DoD* [$\eta(DoD)$] (considered in this work)	$\eta(0.45) = 99\%$ $\eta(0.65) = 98\%$ $\eta(0.85) = 97\%$
Battery cells voltage (V_{bat})	260V

Table II: Parameter of battery cells considered in the case study 1. (Only battery cells related to the upper arm of the phase "a" in the M2C are studied)

	Nominal Capacity Q_n [As]	Current Capacity	Control actions for the DMS considered in this scenario
Battery cell 1	3	$0.8Q_n$	$dms_1 = 0.8$
Battery cell 2	3	$0.6Q_n$	$dms_2 = 0.6$
Battery cell 3	3	$0.4Q_n$	$dms_3 = 0.4$

Scenario 1

First, for the sake of clarity, the degradation of battery cells associated with the upper arm of phase "a" in the M2C is considered. In this case, it is assumed that all the battery cells composing that arm have the same characteristics (same manufacturer, nominal capacity and nominal voltage). However, due to their use in electromobility applications, their current condition in terms of current capacity (degradation level) is different, as shown in Table II. As seen, battery cell 1 is the less degraded battery, while battery cell 3 is the most degraded one. In this condition, to achieve regulation of those battery cells in terms of their degradation level, the proposed control scheme displayed from Fig. 1 to Fig. 3 and given by (2) is run considering that the control actions generated by the DMS (dms_1, dms_2 and dms_3 in Table II) are set by reason of their current capacities (see Table II). By doing this, battery cell 1 will be operated with a more aggressive policy, in terms of *DoD*, than battery cell 3, while the AC battery is operating. This will equalise the degradation level of these battery cells during the AC battery operation.

Fig. 5(a) shows the evolution of capacities, calculated as (1), for the three battery cells considered in this case study, when the proposed control scheme is not enabled. As seen, for the parameters considered in this test and the discharging/charging profile shown in Fig. 4, it is found that the AC battery can operate until around 55 cycles. After that, the AC battery cannot continue working as battery cell 3 has reached its end of life, limiting the operation of the entire AC battery. In contrast, when the AC battery is working with the proposed distributed control scheme for controlling the degradation of battery cells, it can operate for around 70 cycles, as shown in Fig. 5(b). This result demonstrates that the proposed control scheme can effectively extend the lifespan of the whole AC battery by a proper degradation control of the battery cells that compose it. It must be pointed out that this degradation control is achieved by changing the discharging policy of battery cells, as is illustrated in Fig. 6 after 20 seconds. As observed in that figure, before the 20s, the proposal was not working, and all battery cells were operating with the same policy: they are discharged until a *DoD* of 0.68. Then, after the 20s, when the proposed control scheme is activated, their operating policy is changed, and they operate with a policy proportional (in terms of *DoD*) to their degradation level. By doing this, regulation of their degradation level can be achieved, therefore, increasing the lifespan of the proposed AC battery. Finally, it is worth remembering that the

SoC is calculated, as shown in Fig. 8.

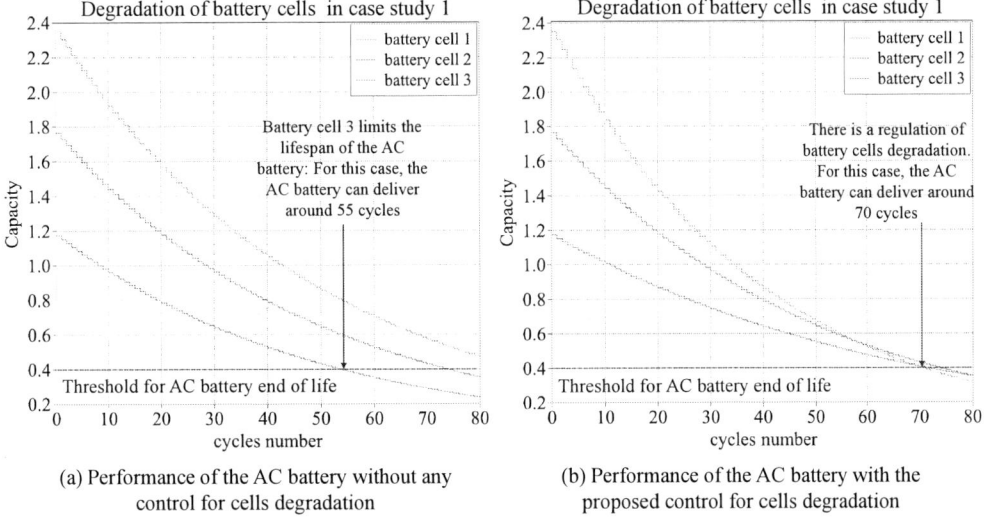

(a) Performance of the AC battery without any control for cells degradation

(b) Performance of the AC battery with the proposed control for cells degradation

Fig. 5: Capacity of battery cells in the upper arm of phase "a" during case study 1: (a) the AC battery is operating without the proposal for degradation control of battery cells, (b) the AC battery is operating with the proposal for degradation control of battery cells: By averaging degradations across cells, prolong the lifetime of the AC battery.

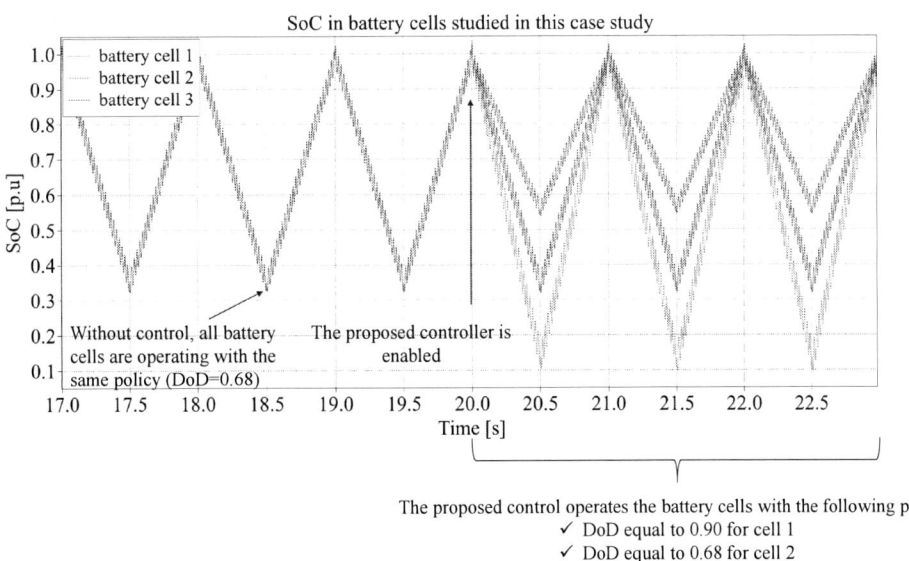

The proposed control operates the battery cells with the following policy:
✓ DoD equal to 0.90 for cell 1
✓ DoD equal to 0.68 for cell 2
✓ DoD equal to 0.45 for cell 3

Fig. 6: *SoC* of battery cells in the studied arm during case study 1: before the 20s, the AC battery is operating without the proposal for degradation control of battery cells; after the 20s, the AC battery is running with the proposed control scheme for degradation control of battery cells. (In this figure, six discharging/charging cycles are shown)

Scenario 2

Similar to case 1, in this case, battery cells related to the upper arm of phase "a" are considered. This case study assumes that battery cells in that arm have the same electrical characteristics but different chemistry. Because of that, battery cell 1 is much cheaper than battery cells 2 and 3. In this condition, the proposed control scheme for degradation purposes could be driven by the DMS considering the

Table III: Parameter of battery cells considered in the case study 2. (Only battery cells related to the upper arm of the phase "a" in the M2C are studied)

	Battery cell cost (a)	Current Capacity	Control actions for the DMS considered in this scenario
Battery cell 1	$0.05a$	$0.6Q_n$	$dms_1 = 20$
Battery cell 2	a	$0.6Q_n$	$dms_2 = 1$
Battery cell 3	a	$0.6Q_n$	$dms_3 = 1$

following control objective: degrade battery cell 1 as much as possible as it is cheaper than others. Thus, the lifespan of battery cells 2 and 3 is prolonged. In this test, the control actions generated by the DMS (dms_1, dms_2 and dms_3) are set for an inverse reason of their costs, as shown in Table III. By doing this, battery cell 1 will be operated with a more aggressive policy, in terms of DoD, than others battery cells. This will prolong the lifespan of battery cells 2 and 3 (the more expensive ones) by highly degrading battery cell 1: This is not an issue in this case, as it is assumed that this battery cell is the cheapest, and there is plenty of availability on the market for buying new ones and replacing it. The results of this test are illustrated in Fig. 7 and Fig. 8. From Fig. 7(a), it is concluded that all the battery cells are degraded in the same ratio without any degradation control, considering this case study's parameters. As was discussed above, from an economic point of view, this situation is not the best because it is preferable to degrade the cheapest battery cell than the expensive ones, to prolong the lifespan of the latter ones. The proposed distributed control scheme achieves this aim, as shown in Fig. 7(b), showing its effectiveness. Additionally, Fig. 8 shows the operating policy used before the activation of the proposal (before 20 seconds) and the operating policy of battery cells when the proposal is working (after 20 seconds). As observed, in this latter situation, battery cell 1 (the cheapest one) has an aggressive discharging policy (DoD close to 0.9), increasing its degradation ratio. Finally, it must be pointed out that, as discussed in [8, 13, 17], managing degradation ratios in a battery system composed of different types of batteries can bring economic benefits, as is the case discussed in this case study. Finally, it is worth remembering that the SoC is calculated, as shown in Fig. 8.

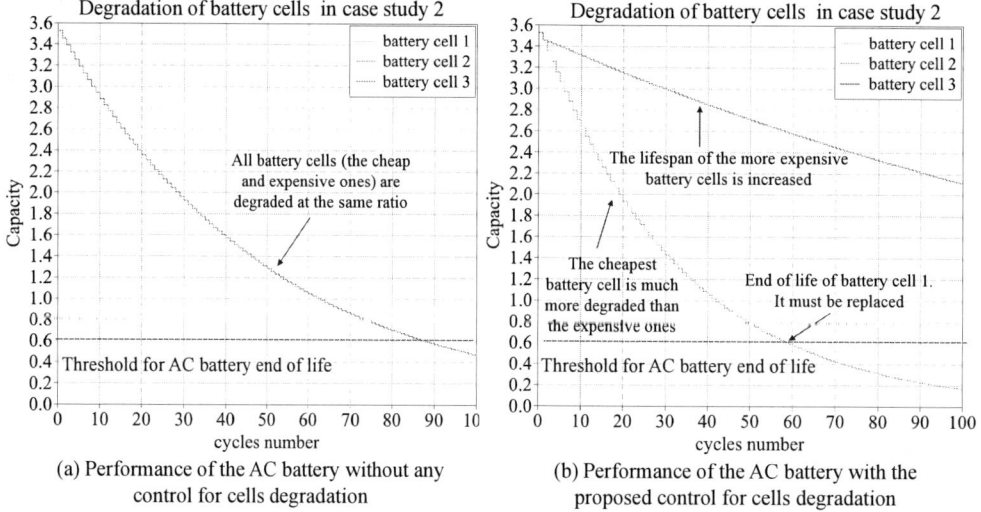

(a) Performance of the AC battery without any control for cells degradation

(b) Performance of the AC battery with the proposed control for cells degradation

Fig. 7: Capacity of battery cells in the upper arm of phase "a" during case study 2: (a) the AC battery is operating without the proposal for degradation control of battery cells, (b) the AC battery is operating with the proposal for degradation control of battery cells.

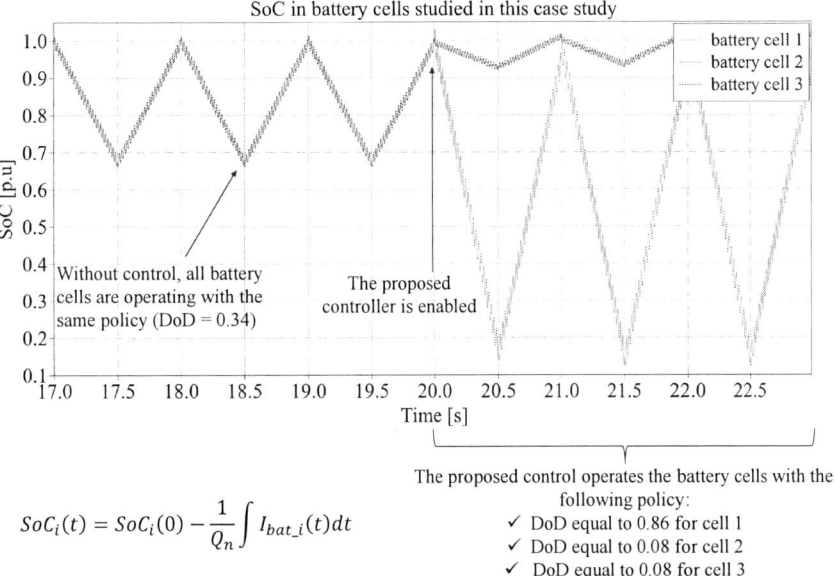

$$SoC_i(t) = SoC_i(0) - \frac{1}{Q_n} \int I_{bat\_i}(t)dt$$

The proposed control operates the battery cells with the following policy:
- ✓ DoD equal to 0.86 for cell 1
- ✓ DoD equal to 0.08 for cell 2
- ✓ DoD equal to 0.08 for cell 3

Fig. 8: *SoC* of battery cells in the upper arm of phase "a" during case study 1: before the 20s, the AC battery is operating without the proposal for degradation control of battery cells; after the 20s, the AC battery is running with the proposal for degradation control of battery cells. (In this figure, six discharging/charging cycles are shown)

Conclusions

This paper proposed an AC battery concept based on the M2C that facilitates the use of battery cells (or battery modules composed of one or more battery cells) in second-life applications. The proposed control scheme demonstrated that it can control the degradation level of battery cells that compose the AC battery, bringing economic benefits to the entire AC battery system. Finally, this work demonstrates that the degradation of battery cells can be achieved by the proposed degradation management system (DMS). Note that the case studies and their corresponding parameters were selected to prove the proposed architecture for controlling the degradation of the battery cells (or battery modules composed of one or more battery cells) within the AC battery. Future publications will consider the DMS development and its validation, considering the real operating conditions.

References

[1] N. Mukherjee and D. Strickland, "Analysis and Comparative Study of Different Converter Modes in Modular Second-Life Hybrid Battery Energy Storage Systems," IEEE Journal of Emerging and Selected Topics in Power Electronics, vol. 4, no. 2, pp. 547 - 563, 2016.

[2] N. Mukherjee and D. Strickland, "Control of Second-Life Hybrid Battery Energy Storage System Based on Modular Boost-Multilevel Buck Converter," IEEE Transactions on Industrial Electronics, vol. 62, no. 2, pp. 1034 - 1046, 2015.

[3] G. Liang, H. Dehghani Tafti , G. G. Farivar , J. Pou, C. D. Townsend , G. Konstantinou and S. Ceballos, "Analytical Derivation of Intersubmodule Active Power Disparity Limits in Modular Multilevel Converter-Based Battery Energy Storage Systems," IEEE Transactions on Power Electronics, vol. 36, no. 3, pp. 2864 - 2874, 2021.

[4] M. Quraan, T. Yeo and P. Tricoli, "Design and Control of Modular Multilevel Converters for Battery Electric Vehicles," IEEE Transactions on Power Electronics, vol. 31, no. 1, pp. 507 - 517, 2016.

[5] M. Quraan, P. Tricoli, S. D'Arco and L. Piegari, "Efficiency Assessment of Modular Multilevel Converters for Battery Electric Vehicles," IEEE Transactions on Power Electronics, vol. 32, no. 3, pp. 2041 - 2051, 2017.

[6] X. Yang, Y. Xue, B. Chen, Y. Mu, Z. Lin, T. Q. Zheng and S. Igarashi, "Reverse-blocking modular multilevel converter for battery energy storage systems," Journal of Modern Power Systems and Clean Energy, vol. 5, no. 4, pp. 652 - 662, 2017.

[7] A. Perez, R. Moreno, R. Moreira, M. Orchard and G. Strbac, "Effect of Battery Degradation on Multi-Service Portfolios of Energy Storage," IEEE Transactions on Sustainable Energy, vol. 7, no. 4, pp. 1718 - 1729, 2016.

[8] D. R. R. Kannan and M. H. Weatherspoon, "The effect of pulse charging on commercial lithium nickel cobalt oxide (NMC) cathode lithium-ion batteries," Journal of Power Sources, vol. 479, pp. 1-8, 2020.

[9] H. Lv, X. Huang and Y. Liu, "Analysis on pulse charging–discharging strategies for improving capacity retention rates of lithium-ion batteries," Ionics, vol. 26, p. 1749–1770, 2020.

[10] M. Abdel-Monem, K. Trad, N. Omar, O. Hegazy, P. Van den Bossche and J. Van Mierlo, "Influence analysis of static and dynamic fast-charging current profileson ageing performance of commercial lithium-ion batteries," Energy, vol. 120, pp. 179-191, 2017.

[11] L. Serrao, S. Onori, A. Sciarretta, Y. Guezennec and G. Rizzoni, "Optimal energy management of hybrid electric vehicles including battery aging," in Proceedings of the 2011 American Control Conference, San Francisco, CA, USA, 2011.

[12] V. Marano, S. Onori, Y. Guezennec, G. Rizzoni and N. Madella, "Lithium-ion batteries life estimation for plug-in hybrid electric vehicles," in 2009 IEEE Vehicle Power and Propulsion Conference, Dearborn, MI, USA, 2009.

[13] A. Perez, "Effect of temperature-dependent degradation models for lithium-ion energy storage devices on optimized multiservice portfolio strategies," PhD thesis, University of Chile, Santiago, Chile, 2018.

[14] J. W. Simpson-Porco, Q. Shafiee, F. Dörfler, J. C. Vasquez, J. M. Guerrero and F. Bullo, "Secondary Frequency and Voltage Control of Islanded Microgrids via Distributed Averaging," IEEE Transactions on Industrial Electronics, vol. 62, no. 11, pp. 7025 - 7038, 2015.

[15] M. Kwiecien, J. Badeda, M. Huck, K. Komut, D. Duman and D. Uwe Sauer, "Determination of SoH of Lead-Acid Batteries by Electrochemical Impedance Spectroscopy," Applied Sciences, vol. 8, 2018.

[16] F. Donoso, R. Cardenas, M. Espinoza, J. Clare, A. Mora and A. Watson, "Experimental Validation of a Nested Control System to Balance the Cell Capacitor Voltages in Hybrid MMCs," IEEE Access, vol. 9, pp. 21965 - 21985, 2021.

[17] D. Jimenez, "Gestión óptima de la energía de una nano-red para minimizar la degradación de un pack modular de baterías de ion-litio," Master thesis, University of Chile, Santiago, Chile, 2018.

Analysis of Test Methods for Measurement of Leakage and Magnetising Inductances in Integrated Transformers

Sajad A. Ansari, Jonathan N. Davidson, Martin P. Foster and David A. Stone
THE UNIVERSITY OF SHEFFIELD
Department of Electronic and Electrical Engineering, The University of Sheffield
Sheffield, UK
Tel.: +44 114 222 5355
E-Mail: SArabAnsari1@sheffield.ac.uk
URL: http://www.sheffield.ac.uk/

Keywords

«Integrated Transformer», «Resonant converter», «Leakage inductance», «Impedance analysis», «Measurement».

Abstract

The attention in integrated transformers has increased recently. The operation of the converters with integrated transformers depends on their components' value significantly. Therefore, a sensitivity analysis is provided to find the most precise measurement method for characterization of integrated transformers, especially when they have different primary and secondary leakage inductances. The theoretical analysis is verified by simulation and experimental results.

Introduction

The pulse-width-modulated converters, such as boost converter, cannot provide high efficiency at high switching frequency due to high switching losses. However, resonant converters can provide high efficiency and power density since they benefit from soft switching capability and magnetic integration [1-6].

The attention in integrated transformers has increased in recent years and they have been developed to be used in many resonant converters for different applications [7-9]. The integrated transformers usually have a high leakage inductance either in the primary or secondary side or in both sides. This leakage inductance may be used as the series inductor of the converters in which there is an inductor in series with the isolated transformer [10, 11]. For example, the schematic of the LLC resonant converter while it has three separate magnetic components, including series (L_S) and parallel (L_P) inductors and transformer, is shown in Fig. 1(a). As shown, the series and parallel inductors of the LLC converter can be integrated into the transformer to reduce the number of passive components. In other words, as presented in Fig. 1(b), the magnetising inductance (L_m) of the transformer can be used as the parallel inductor and the leakage inductance (L_{lk}) of the transformer can be used as the series inductor to enhance the power density and efficiency of the converter [12].

The operation of the resonant converters highly depends on their resonant components. Therefore, the leakage and magnetising inductances of the integrated transformers need to be quantified precisely to guarantee the proper operation of the converter [13-15].

A lot of research has been done on the calculation and measurement of the leakage and magnetising inductances of the transformers in recent years [16, 17]. The standard test approach has been largely used to characterize a transformer. This method uses open-circuit and short-circuit tests and assumes that the leakage inductances of the primary and secondary windings are identical and very small compared to the magnetising inductance [18]. This method may be used for the characterization of the integrated transformers with similar primary and secondary leakage inductances, e.g., shunt-inserted integrated planar transformers. However, in some topologies of the integrated transformers, not only the

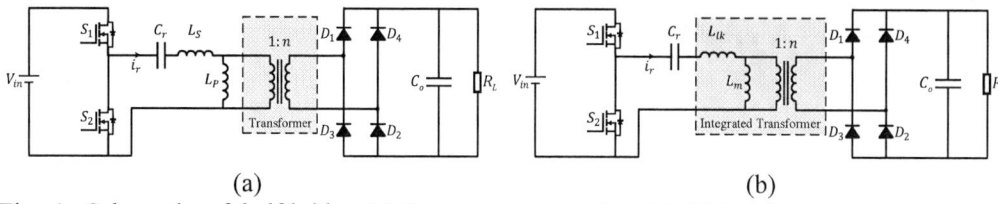

(a)   (b)

Fig. 1: Schematic of half-bridge LLC resonant converter. (a) Using three separate magnetic components. (b) Using an integrated transformer.

primary and secondary leakage inductances may not be identical but also the leakage inductance has a noticeable value in comparison with the magnetising inductance.

This issue with the standard test method can be addressed by adding a third test. In this approach, named as extended test method, the primary and secondary leakage inductances can be measured separately even when they have high and different values. This method is effective for transformers with low resistance, e.g., low-frequency transformers. However, this approach may not be precise for high-frequency integrated transformers since there is significant resistance in the windings of the transformer [18].

The series-coupling test is introduced to overcome the issue of high resistance of the windings [19]. The inductances measured in the series-coupling tests are the algebraic addition and subtraction of inductances and are independent of the winding resistances and parallel paths, facilitating direct measurement by the LCR meter. However, this approach can be very sensitive to measurement errors.

The attention in integrated transformers increases daily and new topologies of the integrated transformers have been proposed with different primary and secondary leakage inductances for utilization in unidirectional and bidirectional resonant converters. As pointed out earlier too, the inaccurate quantifying of the leakage and magnetising inductances can affect the converter operation, optimization, efficiency, modelling and control system. The methods of transformer characterization were introduced in the 1990s and they have not been proposed for the recently introduced integrated transformers, and investigation of their accuracy for characterization of the new integrated transformers is a gap in the literature. Therefore, in this paper, different methods of transformer characterization are investigated to find the most accurate method for characterization of the integrated transformers.

A set of six common transformer tests are considered for the transformer characterization and a measurement analysis is provided for the different combination of this set when three of the tests is only taken. In addition, the sensitivity of the different combinations to measurement errors is investigated. The most accurate combination is finally proposed, and its accuracy is confirmed by finite element analysis and experimental implementation.

Integrated transformer modelling

Fig. 2(a) shows the modelling of a lossless transformer. As shown, the relationship between the voltages and currents of the transformer may be obtained as (1) [12, 20].

$$\begin{bmatrix} v_{\mathrm{P}} \\ v_{\mathrm{S}} \end{bmatrix} = \begin{bmatrix} L_{\mathrm{PP}} & L_{\mathrm{PS}} \\ L_{\mathrm{SP}} & L_{\mathrm{SS}} \end{bmatrix} \frac{\mathrm{d}}{\mathrm{d}t} \begin{bmatrix} i_{\mathrm{P}} \\ i_{\mathrm{S}} \end{bmatrix} \tag{1}$$

where L_{PP} and L_{SS} are primary and secondary self-inductances and L_{PS} and L_{SP} are mutual inductances and i_{P} and i_{S} are primary and secondary currents, respectively.

In Fig. 2(b), another modelling for a lossless transformer is presented which is based on primary magnetising inductance, $L_{\mathrm{m_P}}$, primary, $L_{\mathrm{lk_P}}$, and secondary, $L_{\mathrm{lk_S}}$, leakage inductances and transformer turns ratio, n. The relationship between the voltages and currents of this model may be obtained by (2).

(a)   (b)

Fig. 2: Transformer models. (a) Based on self and mutual inductances. (b) Based on magnetising and leakage inductances.

$$\begin{bmatrix} v_P \\ v_S \end{bmatrix} = \begin{bmatrix} L_{\mathrm{lk}_P} + L_{\mathrm{m}_P} & \dfrac{N_S}{N_P} L_{\mathrm{m}_P} \\[2ex] \dfrac{N_S}{N_P} L_{\mathrm{m}} & L_{\mathrm{lk}_S} + \dfrac{N_S^2}{N_P^2} L_{\mathrm{m}_P} \end{bmatrix} \dfrac{\mathrm{d}}{\mathrm{d}t} \begin{bmatrix} i_P \\ i_S \end{bmatrix} \tag{2}$$

where N_P and N_S are the primary and secondary turns numbers, respectively and the turns ratio, n, is equal to $\dfrac{N_S}{N_P}$. The primary magnetising inductance may be obtained by (3).

$$L_{\mathrm{m}_P} = \frac{N_P}{N_S} L_{\mathrm{PS}} \tag{3}$$

The mutual inductance may be defined by (4).

$$L_{\mathrm{PS}} = \frac{N_S}{I_P} \phi_{\mathrm{PS}} \tag{4}$$

where ϕ_{PS} is the mutual flux produced by the magnetic field of the current in the primary winding that links with the secondary winding and I_P is the RMS of primary current. From (3) and (4), the primary magnetising inductance may be obtained as (5).

$$L_{\mathrm{m}_P} = \frac{N_P}{I_P} \phi_{\mathrm{PS}} \tag{5}$$

From (5), the primary magnetising inductance of a transformer can be obtained after the calculation of the mutual flux of its windings. The primary self-inductance can be obtained by (6).

$$L_{\mathrm{PP}} = \frac{N_P^2}{\mathcal{R}_{\mathrm{T}_P}} \tag{6}$$

where $\mathcal{R}_{\mathrm{T}_P}$ is the core reluctance (evaluated from the primary side). The primary leakage inductance may be calculated by (7) [12, 20].

$$L_{\mathrm{lk}_P} = L_{\mathrm{PP}} - L_{\mathrm{m}_P} \tag{7}$$

From a similar approach, the secondary leakage inductance can be obtained by (8).

$$L_{\mathrm{lk}_S} = L_{\mathrm{SS}} - L_{\mathrm{m}_S} \tag{8}$$

where L_{m_S} is the secondary magnetising inductance and it may be obtained by (9).

$$L_{\mathrm{m}_S} = \frac{N_S}{N_P} L_{\mathrm{SP}} = \left(\frac{N_S}{N_P}\right)^2 L_{\mathrm{m}_P} \tag{9}$$

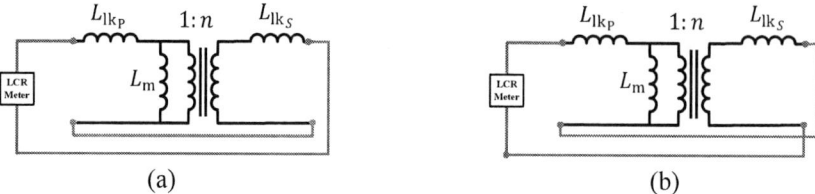

(a)   (b)

Fig. 3: Transformer test set-up. (a) Differentially-coupling. (b) Cumulative-coupling.

and the secondary self-inductance can be obtained by (10).

$$L_{SS} = \frac{N_S^2}{\mathcal{R}_{T_S}}$$ (10)

where $\mathcal{R}_{T_S}$ is the core reluctance (evaluated from the secondary side).

Measurement tests

In general, there are six well-known transformer tests that can be used for the characterization of a transformer. Different transformer characterization approaches usually use either two or three of these tests to characterize a transformer [16-19]. These transformer tests are presented as follows.

Test A: The transformer configuration of this test is presented in Fig. 3(a). As shown, a common current goes through the positive terminal of the primary winding and then into the negative terminal of the secondary winding. According to Fig. 3(a), the differential-coupling inductance, L_D, measured by the LCR meter can be given as (11).

$$L_D = L_{lk_P} + (1 - n)L_{m_P} - n(1 - n)L_{m_P} + L_{lk_S}$$ (11)

Test B: The transformer configuration of this test is presented in Fig. 3(b). As shown, a common current goes through the positive terminal of the primary winding and then into again the positive terminal of the secondary winding. According to Fig. 3(b), the cumulative-coupling inductance, L_C, measured by the LCR meter can be given as (12).

$$L_C = L_{lk_P} + (1 + n)L_{m_P} + n(1 + n)L_{m_P} + L_{lk_S}$$ (12)

Test C: The third test can be performed by measuring the inductance from the primary side while the secondary side is left open. According to Fig. 2(b), the open-circuit inductance measured from the primary side can be given as (13).

$$L_{P_{SOC}} = L_{PP} = L_{lk_P} + L_{m_P}$$ (13)

Test D: The fourth test can be performed by measuring the inductance from the secondary side while the primary side is left open. According to Fig. 2(b), the open-circuit inductance measured from the secondary side can be given as (14).

$$L_{S_{POC}} = L_{SS} = L_{lk_S} + L_{m_S} = L_{lk_S} + n^2 L_{m_P}$$ (14)

Test E: The fifth test can be performed by measuring the inductance from the primary side while the secondary side is shorted. According to Fig. 2(b), the short-circuit inductance measured from the primary side can be given as (15).

$$L_{P_{SSC}} = L_{lk_P} + \frac{L_{m_P} L_{lk_S}}{n^2 L_{m_P} + L_{lk_S}}$$ (15)

Table I: Different combinations for transformer characterization

Combination	Equations	Combination	Equations
ABC	(11), (12), (13)	BCD	(12), (13), (14)
ABD	(11), (12), (14)	BCE	(12), (13), (15)
ABE	(11), (12), (15)	BCF	(12), (13), (16)
ABF	(11), (12), (16)	BDE	(12), (14), (15)
ACD	(11), (13), (14)	BDF	(12), (14), (16)
ACE	(11), (13), (15)	BEF	(12), (15), (16)
ACF	(11), (13), (16)	CDE	(13), (14), (15)
ADE	(11), (14), (15)	CDF	(13), (14), (16)
ADF	(11), (14), (16)	CEF	(13), (15), (16)
AEF	(11), (15), (16)	DEF	(14), (15), (16)

Test F: The sixth test can be performed by measuring the inductance from the secondary side while the primary side is shorted. According to Fig. 2(b), the short-circuit inductance measured from the secondary side can be given as (16).

$$L_{S_{PSC}} = L_{lk_S} + \frac{n^2 L_{m_P} L_{lk_P}}{L_{m_P} + L_{lk_P}} \tag{16}$$

Transformer characterization methods

The standard open- and short-circuit (SOS) approach (test C and test E) can be used to find leakage and magnetising inductance of a transformer. In this approach, it is assumed that the primary and secondary leakage inductances are identical and winding resistance is negligible. Therefore, this approach cannot be accurate when the primary and secondary leakage inductances are different and there is high winding resistance.

The limitation of SOS can be addressed by having a third test. This approach, which is named as extended open- and short-circuit (EOS) method, uses tests C, D and E to characterize a transformer. Even though the primary and secondary leakage inductances can be different in this approach, a large winding resistance relative to magnetising inductance can distort the measurement.

The series-coupling (SC) approach uses tests A, B and C to characterize a transformer. Since the inductances measured in the series-coupling tests are the algebraic addition and subtraction of inductances and are independent of winding resistances and parallel paths, this approach can be used for transformers with high winding resistance. However, this approach is very sensitive to measurement errors.

Amongst various transformer characterization methods, the SOS, EOS and SC approaches are very popular. These approaches have been developed to be used mainly for conventional transformers. Therefore, their performance for the characterization of integrated transformers needs to be investigated.

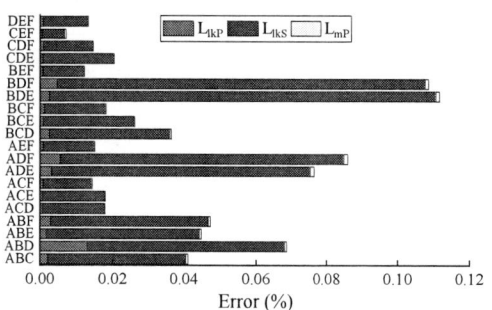
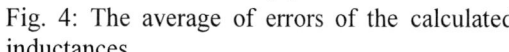
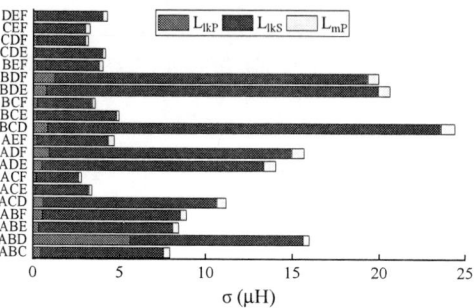

Fig. 4: The average of errors of the calculated inductances.

Fig. 5: The standard deviation of the calculated inductances.

The SOS test is not a good candidate for the characterization of the integrated transformers since the integrated transformers usually have different primary and secondary leakage inductances.

There are six tests and the number of combinations of them when a sample size of three is chosen is twenty and EOS and SC approaches are only two of these different combinations. The other eighteen combinations can also characterize a transformer. Hence, the other eighteen approaches need to be also investigated to find the best combination for the characterization of an integrated transformer.

Investigation of transformer characterization methods

As mentioned earlier, there are six common tests for a transformer and the number of combinations of them when a sample size of three is chosen is twenty. In this section, all twenty combinations are investigated and the best combination and the best way to characterize an integrated transformer is presented.

The different combination of six tests with a sample size of three is presented in Table I. Each combination is named according to the tests that are used in it. For example, the combination ABC uses tests A, B and C. The combination ABC is the same as the SC approach and the combination CDE is the same as the EOS approach. Therefore, the six transformer tests, presented earlier, can provide other eighteen combinations that can be used to characterize a transformer.

An air-gap is added to the integrated transformers to store more energy in their magnetising inductance and integrated transformers are developed to have high primary or/and secondary leakage inductances. An important application of the integrated transformers is in resonant converters. The operation of the resonant converters is very sensitive to their resonant components and inaccurate characterization of their transformer can affect the converter operation, optimization, efficiency, modelling and control system. Therefore, it is very important to characterize an integrated transformer precisely.

The measurement errors in the transformer tests are inevitable and these errors lead to the inaccurate characterization of a transformer. Therefore, in the first step of the investigation of the different combinations, the sensitivity of each combination to the measurement errors is investigated.

The sensitivity analysis has been done in MATLAB for a transformer with characteristics presented in Table II. In the first step, the ideal measurement for each test is calculated by substituting the values presented in Table II in (11)-(16). In the second step, 10000 random errors are applied to each measurement test by using "randn function" in MATLAB while the maximum error is 0.2%. Finally, the primary and secondary leakage inductances and magnetising inductance are calculated for each combination while there are 10000 random errors in each measurement.

The average of the calculated inductances with applying the 10000 random errors to the measurements is calculated for each combination and is presented in Fig. 4. The difference between the average of

Table II: Transformer characteristics

Magnetising inductance (L_{m_P})	Primary leakage inductance (L_{lk_P})	Secondary leakage inductance (L_{lk_S})	Turns ratio (1:n)
300 μH	120 μH	110 μH	1 : 5

Table III: The maximum error of the calculated inductances for each combination

Combination		ABC	ABD	ABE	ABF	ACD	ACE	ACF	ADE	ADF	AEF
Error (%)	L_{m_P}	0.386	0.386	0.38	0.38	0.61	0.17	0.18	0.69	0.74	0.34
	L_{lk_P}	1.518	19.24	1.00	1.75	1.80	0.39	0.54	1.65	3.02	0.72
	L_{lk_S}	25.15	38.19	27.3	28.1	34.0	11.2	9.04	45.4	49.5	15.6
Combination		BCD	BCE	BCF	BDE	BDF	BEF	CDE	CDF	CEF	DEF
Error (%)	L_{m_P}	0.787	0.169	0.18	0.68	0.64	0.23	0.18	0.17	0.28	0.26
	L_{lk_P}	2.631	0.580	0.71	2.37	3.89	0.65	0.48	0.61	0.54	0.68
	L_{lk_S}	77.59	16.88	11.7	67.6	63.9	13.7	13.4	9.78	10.6	14.4

errors of different combinations is negligible and there is no advantage for any combination when the average of errors is considered.

The maximum error of the calculated inductances for each combination after applying the 10000 random errors to the measurements is presented in Table III, low errors are written in green and the lowest error for each inductance is underlined. As shown, the combination BCE has the lowest sensitivity to the measurement errors for magnetising inductance measurement and the combinations ACE, ACF, BCF, CDE and CDF have also lower sensitivity compared to others. The combination ACE has the lowest sensitivity to the measurement errors for the primary leakage inductance measurement and the combinations ACF, CDE and CEF have also lower sensitivity compared to others. The combination ACF has the lowest sensitivity to measurement errors for secondary leakage inductance measurement and the combinations ACE and CDF have also lower sensitivity compared to others. Therefore, from Table III, it can be concluded that the combination ACF may be the best combination in terms of sensitivity to the measurement errors. In addition, to obtain the most accurate answer, the magnetising inductance and primary and secondary leakage inductance can be obtained from the combinations BCE, ACE and ACF, respectively.

The standard deviation of the calculated inductances for each combination after applying the 10000 random errors to the measurements is presented in Fig. 5. As shown, the combination ACF has the lowest standard deviation and therefore has the lowest sensitivity in general. The combination CDF is the second rank, but it has higher sensitivity for every inductance compared to the combination ACF.

The histogram of the calculated inductances after applying the 10000 random errors to the measurements and after dividing by their precise value (presented in Table II) is shown for the combinations ABC (conventional SC), CDE (conventional EOS), and ACF in Fig. 6. As shown, in general, the calculated inductances from the combination ACF are closer to the right values and are less sensitive to measurement errors.

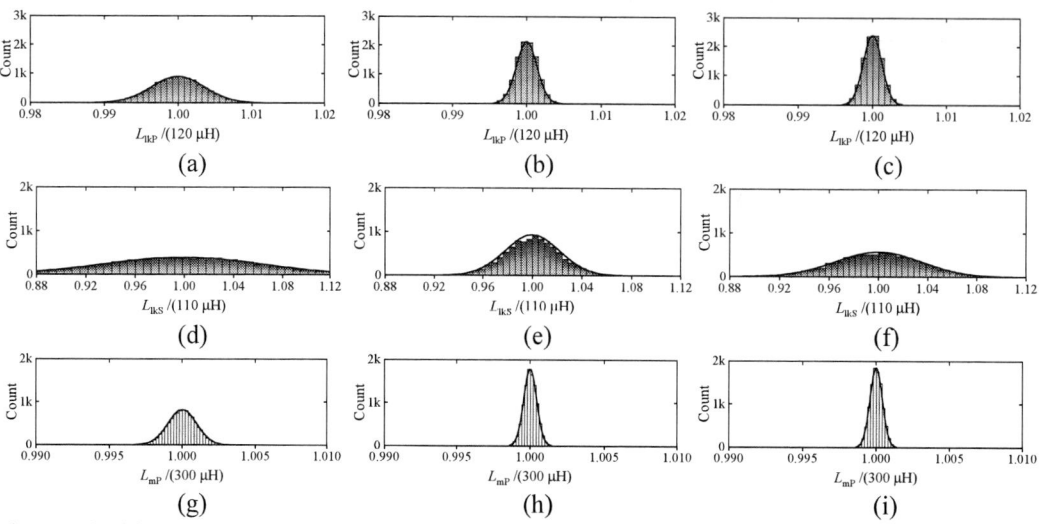

Fig. 6: The histogram of the calculated inductances. (a) L_{lk_P} from ABC. (b) L_{lk_P} from ACF. (c) L_{lk_P} from CDE. (d) L_{lk_S} from ABC. (e) L_{lk_S} from ACF. (f) L_{lk_S} from CDE. (g) L_{m_P} from ABC. (h) L_{m_P} from ACF. (i) L_{m_P} from CDE.

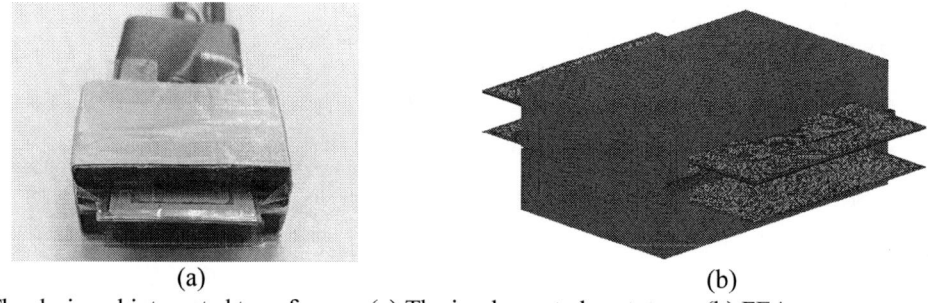

Fig. 7: The designed integrated transformer. (a) The implemented prototype. (b) FEA.

In summary, the combination ACF is less sensitive to measurement errors and can be used when primary and secondary leakage inductance are different. In addition, the combination ACF benefits from one of the series-coupling tests (measurement is independent of winding resistances). Therefore, the combination ACF is a suitable method to characterize the integrated transformers.

Experimental verification

To verify the theoretical analysis, an integrated transformer is implemented in the laboratory and simulated in Ansys Maxwell as presented in Figs. 7(a) and (b), respectively. The characteristic of the implemented integrated transformer is presented in Table IV. The topology of the chosen integrated transformer is similar to the topology presented in [10, 15]. This integrated transformer is based on a planar transformer in which a magnetic shunt is inserted between two E-cores. This integrated transformer is designed to be used in the LLC resonant converters and therefore it can provide high leakage inductance and can be designed for most of the required leakage inductance values.

The designed integrated transformer is characterized by finite-element analysis (FEA) and its primary and secondary leakage inductances and magnetising inductance are presented in Table IV. In addition, the implemented transformer is characterized by the combination ACF which is also presented in Table IV. As shown, the primary and secondary leakage inductances and magnetising inductance can be measured precisely by using the combination ACF.

Table IV: Transformer characteristics verification

	Primary leakage inductance (L_{lk_P})	Secondary leakage inductance (L_{lk_S})	Magnetising inductance (L_{m_P})
Designed	8 µH	0.03 µH	28.5 µH
FEA	8.2 µH	0.04 µH	28.85 µH
Measurement	8.3 µH	0.05 µH	29.5 µH

Conclusion

The attention in integrated transformers has increased recently. The operation of the converters with integrated transformers depends on their components' value. However, the transformer characterization methods are introduced for the conventional transformers and may not be accurate if they apply to integrated transformers. In this paper, different transformer characterization methods are investigated. The advantages and disadvantages of different characterization approaches for the integrated transformer characterization are presented. The six well-known transformer tests that can be used to characterize a transformer are also presented. It is shown that the popular characterization methods such as EOS and SC use three of these tests. However, there are other eighteen combinations for these tests with a sample size of three. Therefore, to find the best combination to characterize an integrated transformer, all twenty combinations are investigated. It is shown that the best combination uses an open-circuit test, a short-circuit test and a series-coupling test and it is compared with the conventional EOS and SC approaches. In addition, the accuracy of the best combination is verified by finite-element analysis and experimental implementation.

References

[1] S. A. Ansari and J. S. Moghani, "A novel high voltage gain noncoupled inductor SEPIC converter," *IEEE Transactions on Industrial Electronics,* vol. 66, no. 9, pp. 7099-7108, 2018.

[2] S. A. Ansari and J. S. Moghani, "Soft switching flyback inverter for photovoltaic AC module applications," *IET Renewable Power Generation,* vol. 13, no. 13, pp. 2347-2355, 2019.

[3] S. Arab Ansari, J. S. Moghani, and M. Mohammadi, "Analysis and implementation of a new zero current switching flyback inverter," *International Journal of Circuit Theory and Applications,* vol. 47, no. 1, pp. 103-132, 2019.

[4] A. Mirzaee, S. Arab Ansari, and J. Shokrollahi Moghani, "Single switch quadratic boost converter with continuous input current for high voltage applications," *International Journal of Circuit Theory and Applications,* vol. 48, no. 4, pp. 587-602, 2020.

[5] S. A. Ansari, J. N. Davidson, and M. P. Foster, "Evaluation of silicon MOSFETs and GaN HEMTs in soft-switched and hard-switched DC-DC boost converters for domestic PV applications," *IET Power Electronics,* vol. 14, no. 5, pp. 1032-1043, 2021.

[6] A. Mizani, S. A. Ansari, A. Shoulaie, J. N. Davidson, and M. P. Foster, "Single-active switch high-voltage gain DC–DC converter using a non-coupled inductor," *IET Power Electronics,* vol. 14, no. 3, pp. 492-502, 2021.

[7] M. D'Antonio, S. Chakraborty, and A. Khaligh, "Planar Transformer with Asymmetric Integrated Leakage Inductance Using Horizontal Air Gap," *IEEE Transactions on Power Electronics,* vol. 36, no. 12, pp. 14014 - 14028, 2021, doi: 10.1109/TPEL.2021.3089606.

[8] S. A. Ansari, J. N. Davidson, and M. P. Foster, "Analysis, design and modelling of two fully-integrated transformers with segmental magnetic shunt for LLC resonant converters," in *IECON 2020 The 46th Annual Conference of the IEEE Industrial Electronics Society,* 2020: IEEE, pp. 1273-1278.

[9] S. A. Ansari, J. Davidson, and M. Foster, "Inserted-shunt Integrated Planar Transformer with Low Secondary Leakage Inductance for LLC Resonant Converters," *IEEE Transactions on Industrial Electronics,* 2022.

[10] S. A. Ansari, J. N. Davidson, and M. P. Foster, "Fully-Integrated Solid Shunt Planar Transformer for LLC Resonant Converters," *IEEE Open Journal of Power Electronics,* vol. 3, pp. 26 - 35, 2021, doi: 10.1109/OJPEL.2021.3137016.

[11] S. A. Ansari, J. N. Davidson, M. P. Foster, and D. A. Stone, "Design and analysis of a Fully-integrated planar transformer for LCLC resonant converters," in *2021 23rd European Conference on Power Electronics and Applications (EPE'21 ECCE Europe)*, 2021: IEEE, pp. P. 1-P. 8.

[12] M. Li, Z. Ouyang, and M. A. Andersen, "High-frequency LLC resonant converter with magnetic shunt integrated planar transformer," *IEEE Transactions on Power Electronics,* vol. 34, no. 3, pp. 2405-2415, 2018.

[13] G. Spiazzi and S. Buso, "Effect of a split transformer leakage inductance in the LLC converter with integrated magnetics," in *2013 Brazilian Power Electronics Conference*, 2013: IEEE, pp. 135-140.

[14] H.-S. Choi, "AN4151. Half-bridge LLC resonant converter design using Fairchild Power Switch (FPS)," *Fairchild semiconductor,* 2007.

[15] S. A. Ansari, J. Davidson, and M. Foster, "Fully-integrated planar transformer with a segmental shunt for LLC resonant converters," *IEEE Transactions on Industrial Electronics,* 2021.

[16] J. Wang, A. F. Witulski, J. L. Vollin, T. Phelps, and G. Cardwell, "Derivation, calculation and measurement of parameters for a multi-winding transformer electrical model," in *APEC'99. Fourteenth Annual Applied Power Electronics Conference and Exposition. 1999 Conference Proceedings (Cat. No. 99CH36285)*, 1999, vol. 1: IEEE, pp. 220-226.

[17] W. G. Hurley and D. J. Wilcox, "Calculation of leakage inductance in transformer windings," *IEEE Transactions on Power electronics,* vol. 9, no. 1, pp. 121-126, 1994.

[18] J. G. Hayes, N. o'Donovan, M. G. Egan, and T. O'Donnell, "Inductance characterization of high-leakage transformers," in *Eighteenth Annual IEEE Applied Power Electronics Conference and Exposition, 2003. APEC'03.*, 2003, vol. 2: IEEE, pp. 1150-1156.

[19] M. Honda, *The Impedance Measurement Handbook: A Guide to Measurement Technology and Techniques. Hauptw.* Hewlett-Packard Company, 1990.

[20] S. De Simone, C. Adragna, and C. Spini, "Design guideline for magnetic integration in LLC resonant converters," in *2008 International Symposium on Power Electronics, Electrical Drives, Automation and Motion*, 2008: IEEE, pp. 950-957.

A Topology-Morphing Series Resonant Converter for Photovoltaic Module Applications

Grigorios Sergentanis, Liliana de Lillo, Lee Empringham, C. Mark Johnson
The University of Nottingham
University Park, NG7 2RD
Nottingham, UK
Tel.: +44 / (115) – 8468840.
E-Mail: grigorios.sergentanis@nottingham.ac.uk
URL: http://www.nottingham.ac.uk

Acknowledgements

This work was supported by the Engineering and Physical Sciences Research Council (grant number EP/S024069/1). The work was conducted within the Centre for Doctoral Training in Sustainable Electric Propulsion.

Keywords

«Resonant Converter», «Boost», «Photovoltaic», «DC-DC converter», «Efficiency»

Abstract

Residential solar photovoltaic (PV) installations frequently use power optimizers to increase their energy production. In this application, the ability to regulate a wide range of voltage with high efficiency is highly desirable. Thus, this paper proposes a novel hybrid-controlled series resonant converter (SRC) for photovoltaic power optimizers. The converter utilizes the advantage of GaN devices, which have improved switching transition times compared to Si devices, hence providing a lower switching loss. Regulation is achieved with fixed-frequency PWM control on the secondary side, while ZVS and ZCS of the devices are achieved with the proposed resonant tank design. The proposed converter maintains high efficiency over a wide voltage range, making the PV system shade-tolerant while keeping the number of switching devices low. The paper presents the operating principles, the design methodology, and simulation results. The results show a high efficiency over a wide voltage range, as well as a wide load range.

Introduction

Due to the nonlinear nature of photovoltaic (PV) panels, to harvest the maximum amount of energy, a maximum power point tracking (MPPT) algorithm is required [1]. This algorithm measures the PV side voltage and current and then gives a reference input voltage for the interfacing converter. This control is normally enacted on a string or array of PV panels, but with the rise of residential installations, module-level power electronics (MPLE) have become a viable option in the form of a dc power optimizer or a microinverter. A PV installation using MPLE has the benefits of high scalability, decreased magnetics size, zero mismatch loss between panels, and increased reliability [2]. DC power optimizers may also be used in building-integrated PV systems, where using a DC microgrid is financially attractive [3]. One of the difficulties in implementing a DC power optimizer is the need to accommodate a wide range of PV voltages, as environmental conditions like insolation, temperature and shading may vary the optimum voltage greatly.

In this work, the authors propose a DC power optimizer that can be used as the first stage in a microinverter, or as a standalone solution that may be connected to a central inverter or DC microgrid. The proposed converter, as seen in Fig. 1, consists of the highly efficient series resonant converter (SRC), used to provide galvanic isolation and voltage step-up via its transformer. On the rectifying full bridge, the lower diodes are replaced by active switches. These switches are used to achieve input voltage regulation as well as topology morphing. Under typical conditions, the two switches are phase-shift modulated to provide the equivalent function as a boost converter integrated into the SRC. In heavily shaded conditions, where the desired voltage can be quite low, one of the switches is set constantly ON, morphing the rectifying side into a Greinacher voltage doubling circuit, while the other switch is PWM modulated to control the input voltage. The concept of integrating a boost converter with an SRC has been successfully explored by researchers [4], [5], but their regulation range is limited due to duty cycle constraints. The concept of using a topology morphing rectifier to increase the regulation range is also well known [6], but comes at the cost of increased semiconductor devices. The novelty of this research is those two ideas are combined in a converter that does not require additional devices to operate. Another advantage of this converter is that it operates with a fixed switching frequency, allowing more freedom in the transformer design compared with frequency-modulated resonant converters. The latter, such as the LLC converter, have an inherent disadvantage by requiring specific ratios of resonant inductance to magnetizing inductance, limiting the achievable efficiencies [7]. Furthermore, researchers have developed other attractive topologies in the field of micro converters, such as the quasi-Z source SRC converter [8] and converters with switched capacitor/inductor cells [9], however, these solutions require increased magnetic component count, which may pose a problem in terms of volumetric density, if trying to integrate the DC optimizer with the PV panel.

Topology of the Converter

As seen in Fig. 1, the PV panel is connected directly to the input, and an input capacitance is inserted in parallel to stabilize the input voltage to suppress oscillations around the MPP and thereby improve the panel's utilization ratio [10]. The full-bridge created by $S_1 - S_4$ provides the AC voltage to be fed to the transformer. The transformer itself is designed so that its magnetizing inductance provides zero voltage switching (ZVS) on the primary side. The resonant tank, comprised of L_r and C_r is designed so that their resonant frequency is slightly higher than the converter's operating frequency, guaranteeing zero current switching (ZCS) at the secondary bridge (D_1, D_2, S_5, S_6). The resonant inductance may be integrated with the transformer design, or it may be an external inductor in series with the transformer's leakage inductance, depending on the designing switching frequency and the designer's priorities. C_{out} is assumed to be much larger compared to C_r during the converter analysis, and the output of the converter can be a fixed voltage DC microgrid or the DC link of a two-stage inverter system. In either case, the output voltage is assumed constant, as even in the latter scenario, the control of the DC-link voltage is normally encased in the inverter circuit [11]. The converter may operate as a DC transformer (DCX) when the normalized voltage gain is equal to one, and in the boosting mode or the Greinacher mode when the MPP voltage is reduced.

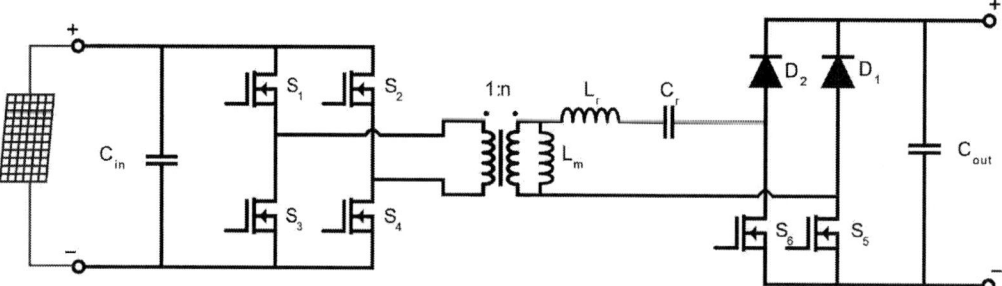

Fig. 1: The proposed converter

DCX operation

At the nominal input voltage, the converter operates as an SRC. This is the highest efficiency state, as energy is transferred from the source to the output for nearly all the operating period. As seen in Fig. 2(a), the current flowing through the resonant branch is sinusoidal. The switching frequency is slightly lower than the resonant tank's resonant frequency, hence ZCS is guaranteed on D_1, D_2, S_5, S_6 at time t_1. The primary bridge of $S_1 - S_4$ can achieve ZVS turn-on, if the transformer is designed to provide enough magnetizing current during the dead-time interval. For example, during the time $t_1 - t_2$, the voltage transitions on the primary bridge are shown pictorially in Fig. 2(b). The condition to achieve ZVS may be formulated mathematically as:

$$L_m \leq \frac{n^2 t_{dt}}{8 f_s C_{oss}} \tag{1}$$

With t_{dt} being the dead-time, and C_{oss} the charge equivalent output capacitance of the FETs on the primary bridge. An additional dead-time interval may be added on S_5, S_6 at time t_2, to allow them to also achieve ZVS turn on. Plus, the primary bridge switches turn-off with low losses, as they are conducting only the reflected magnetizing current at the time t_1.

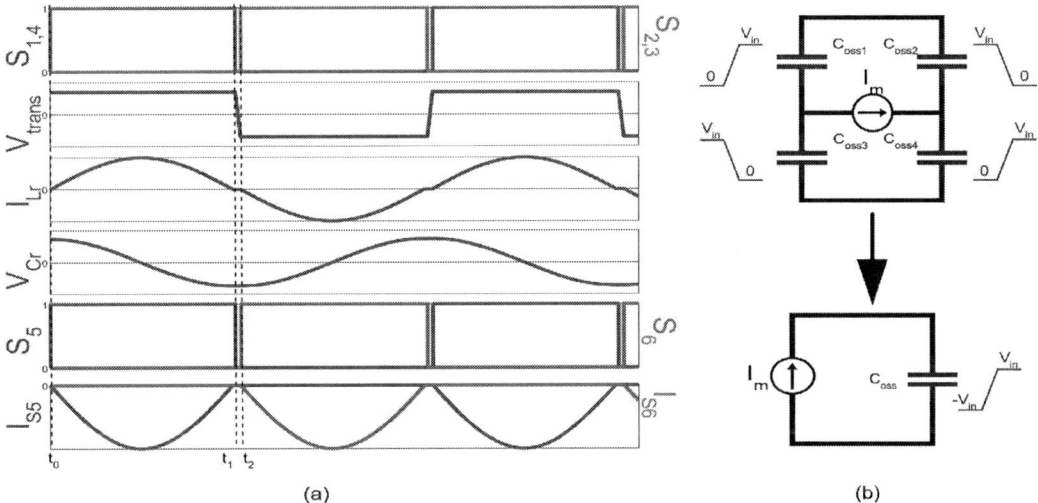

(a)   (b)

Fig. 2: Operation waveforms at the DCX mode (a), and the equivalent circuit of the full-bridge during the dead-time period (b)

Boosting Operation

When the desired input voltage is lower than the nominal, the converter operates in its boosting mode. This is achieved by adding extra ON time to the switches S_5, S_6 in the form of phase-shifting their respective pulses, as is seen in Fig 3(a). During the overlap of the pulses, the resonant tank is shorted, and L_r is rapidly charged. This results in the secondary side acting as a boost converter operating at double the switching frequency. Afterwards, at $t_1 - t_2$, the current is sinusoidal according to the resonant tank's frequency, hence it will reach zero before time t_3, allowing ZCS for the diodes D_1, D_2. Similar to the DCX mode, in the primary bridge ZVS is achieved by utilizing the dead-time interval, and low turn-off currents are guaranteed as the current in the resonant branch is zero in advance of time t_3. The secondary side switches can still achieve ZVS by adding an additional dead-time delay.

After performing Kirchhoff's laws for each state of the converter, the state plane diagram of the resonant tank can be created, and the converter's gain may be computed geometrically. For the time period $t_0 - t_1$, the following equation describes the trajectory of the resonant tank:

$$(v_{cr} - nV_{in})^2 + (Z_r i_{Lr})^2 = (nV_{in} + \Delta v_{cr})^2 \tag{2}$$

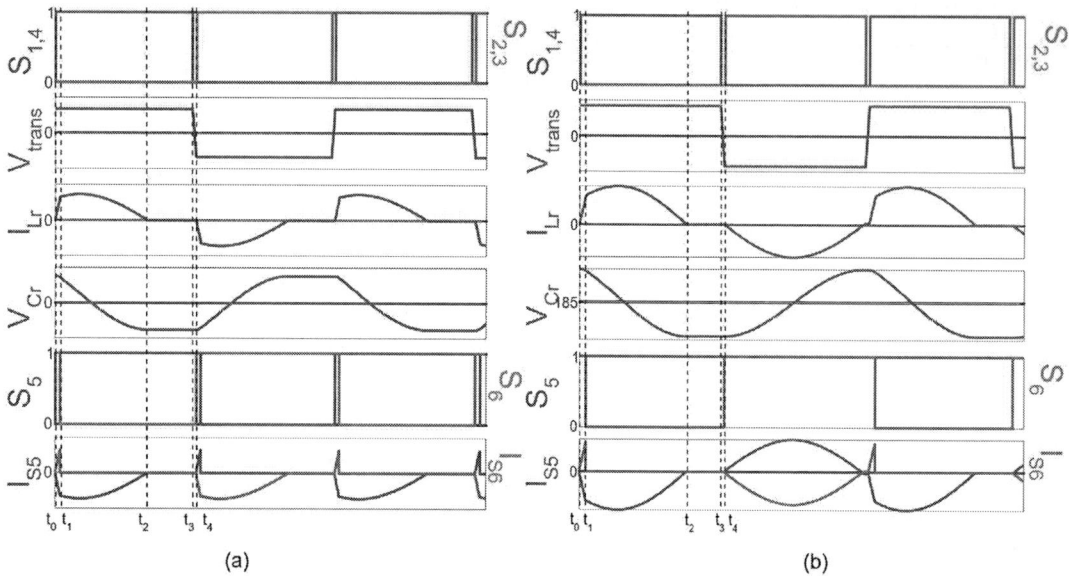

Fig 3: Operation of the converter in the boosting mode (a), and in the Greinacher mode (b).

At the time period $t_1 - t_2$, the respective equation is:

$$(v_{cr} - nV_{in} + V_{out})^2 + (Z_r i_{Lr})^2 = (V_{out} - nV_{in} + \Delta v_{cr})^2 \tag{3}$$

Where,

$$Z_r = \sqrt{L_r/C_r} \tag{4}$$

And by assuming that the converter is operating at 100% efficiency, so that all the power is passing via the resonant tank, the voltage swing at the capacitor is

$$\Delta v_{cr} = \frac{P_o T_s}{4 n V_{in} C_r} \tag{5}$$

Finally, solving equations (2) and (3) at their intersection which is at the time $t_1 = d \cdot T_s$, the converter gain can be retrieved. The state plane trajectories are plotted in Fig. 4(a).

$$\frac{V_{out}}{nV_{in}} = M(d) = \frac{P_o T_s + \sqrt{(P_o T_s)^2 + 4 P_o T_s V_{out}^2 C_r (1 - cos^2(\omega_r dT_s))}}{P_o T_s (cos(\omega_r dT_s) + 1)} \tag{6}$$

Greinacher Operation

In the Greinacher operation, S_5 is kept continuously in the ON state. In this state, the resonant capacitor voltage is DC biased, thereby extending the voltage gain of the converter. Like the other states, the primary bridge achieves ZVS turn-on and low current at their turn off times. S_6 can still achieve ZVS turn-on with the addition of an extra dead-time, and ZCS is achieved for D_2, while D_1 is not conducting any current during this operation.

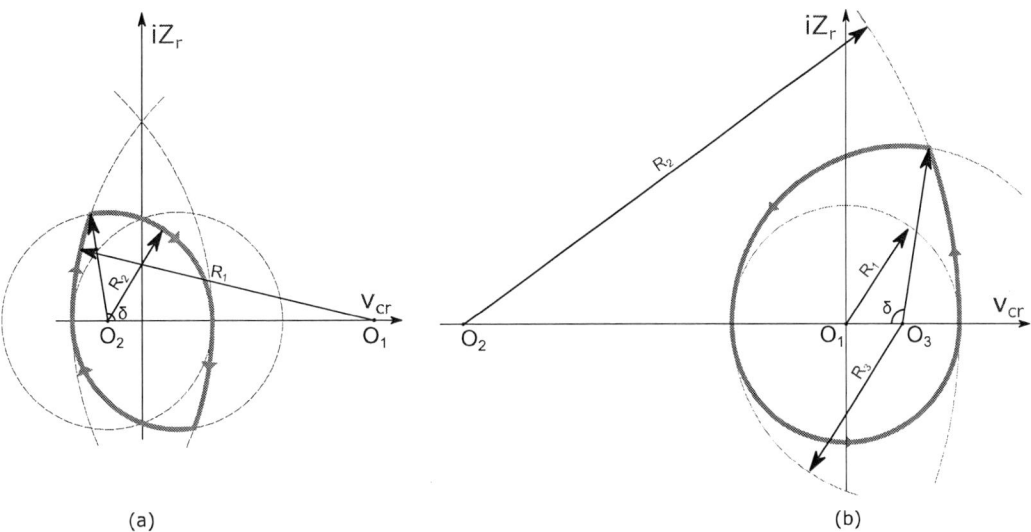

(a)   (b)

Fig. 4: State plane diagram of the resonant tank in the boosting mode (a) and at the Greinacher mode (b)

The DC biasing of the resonant capacitor using the charge balance equation is found to be equal to nV_{in}. Based on this fact, the gain of this mode can be found by solving for the state plane trajectories for the periods $t_0 - t_1$, and $t_1 - t_2$, and solving for their intersection point.

$$(v_{cr} + 2nV_{in})^2 + (Z_r i_{Lr})^2 = (2nV_{in} + \Delta v_{cr})^2 \tag{7}$$

$$(v_{cr} + 2nV_{in} - V_{out})^2 + (Z_r i_{Lr})^2 = (V_{out} - 2nV_{in} + \Delta v_{cr})^2 \tag{8}$$

$$M(d) = 2 \cdot \frac{nP_o T_s + \sqrt{(nP_o T_s)^2 + 2n^2 P_o T_s V_{out}^2 C_r (1 - \cos^2(\omega_r d T_s))}}{nP_o T_s (1 + \cos(\omega_r d T_s))} \tag{9}$$

Design of the Converter

To select an input capacitance, the criteria are the volume occupied by the capacitors, plus that the energy yield from the PV panel that must be kept high. A low capacitance would have a high voltage swing, therefore creating oscillations centered at the MPP, reducing the energy harvested from the panel. The desired capacitor value can be determined by the capacitor's charge equation, by setting the desired voltage swing and integrating the input current over half the switching period.

$$C_{in} = Q/\Delta v_{in} \tag{10}$$

To select the devices $S_1 - S_4$, apart from selecting a device with the required voltage rating, another consideration must be their performance. These devices will have ZVS turn-on and will be turning off with a moderate current, therefore the switching losses will be low, allowing for a potential silicon design. However, as silicon devices have increased parasitic capacitances, GaN FETs EPC2021 will be used to provide a design with a lower dead-time, and a lower gate driving loss.

For the design of the transformer turns ratio, it is designed to be close to the nominal voltage of the panel, which is taken to be a value typical for residential PV panels, equal to 35V, and the output high voltage side is taken as 380V.

$$n = V_{out}/V_{in,nom} \tag{11}$$

With the turn ratio decided, it is possible to design the transformer for maximum efficiency. The transformer core flux swing is equal to

$$\Delta B = \frac{V_{in} T_s}{4 n_{pri} A_c} \tag{12}$$

And since the excitation of the transformer terminals is a square wave with minimal periods of zero voltage, the original Steinmetz equation gives an accurate result for the core loss [12].

$$P_{core} = k \cdot f^a \cdot B^\beta \tag{13}$$

For the core loss calculation, the ferrite N95 was used as it possesses a relatively flat loss to temperature curve [13], in the form factor of RM12. It is assumed the proximity effects on the wiring resistance will be minimal, as Litz wire will be used on the prototype, as well as interleaving to reduce the proximity effect in the primary side of the transformer, and a thinly stranded Litz wire is used on the secondary. Interleaving also brings the benefit of low leakage inductance, ameliorating the voltage stress that the primary devices will be under. Fig. 5 shows that the optimum number of primary turns is five, and for the secondary side, it is chosen that there will be 54 turns, to provide interleaving on the primary side, and give a total voltage step-up of $n = 10.8$.

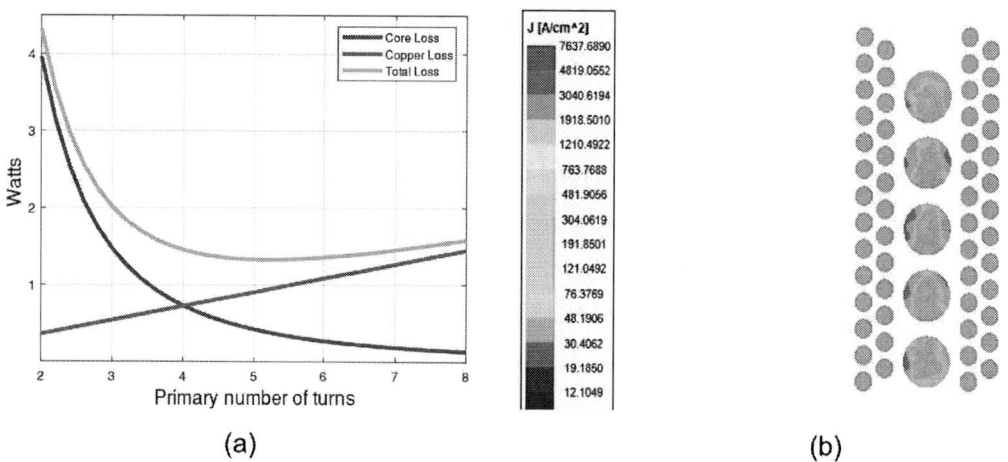

(a)   (b)

Fig. 5: Power losses of the transformer for different winding numbers(a), and the effects of balanced interleaving on a solid copper wire (b). It is seen that the current density does not gravitate towards a particular side in the middle (primary) conductors.

For the design of the air gap, utilizing equation (1), and setting the switching frequency equal to 140kHz gives

$$L_m \leq 54812 \cdot t_{dt} \tag{14}$$

And the transformer's magnetizing inductance may be controlled by adding an air gap approximated as

$$l_g = \frac{\mu_0 A_c n_{sec}}{L_m} \tag{15}$$

An air gap of $0.7mm$ was selected, which should give an approximate magnetizing inductance of $L_m = 658\mu H$, according to equation (15). The design is verified by Ansys Maxwell 3D simulation and experimentally. The transformer has been constructed with enameled copper wire to test the resulting magnetizing and leakage inductances. The results were retrieved via the Keysight E4990A impedance analyzer with an AC voltage excitation at the switching frequency. By replacing the measured magnetizing inductance in Table I with equation (14), it is found that $t_{dt} > 13ns$. C_{oss} was calculated by integrating the $C_{oss} - V$ graph provided by EPC for their EPC2021 eFET from 0 to 35V and dividing by 35.

Table I: Simulated and measured transformer values

	Simulation	Experiment
$L_{leak,sec}$	9.42 μH	9.46 μH
$L_{m,sec}$	749.60 μH	701.19 μH
Coupling factor	99.3%	99.2%

Concerning the design of the resonant tank, a higher resonant inductor correlates with lower DCM periods, and therefore lower RMS currents and higher efficiency. However, DCM must be achieved in all operating regions to benefit from ZCS on the diodes and low current turn off on $S_1 - S_4$. This condition can be formulated geometrically from state plane diagrams (Fig. 4).

$$T_s/2 \geq dT_s + \delta/\omega_r \tag{16}$$

For the boosting operation, this equals to:

$$T_s/2 \geq dT_s + \sin^{-1}\left(R_1 \sin(\omega_r dT_s)\right)/\omega_r \tag{17}$$

And for the Greinacher operation:

$$T_s/2 \geq dT_s + \sin^{-1}(R_2 \sin(\omega_r dT_s))/\omega_r \tag{18}$$

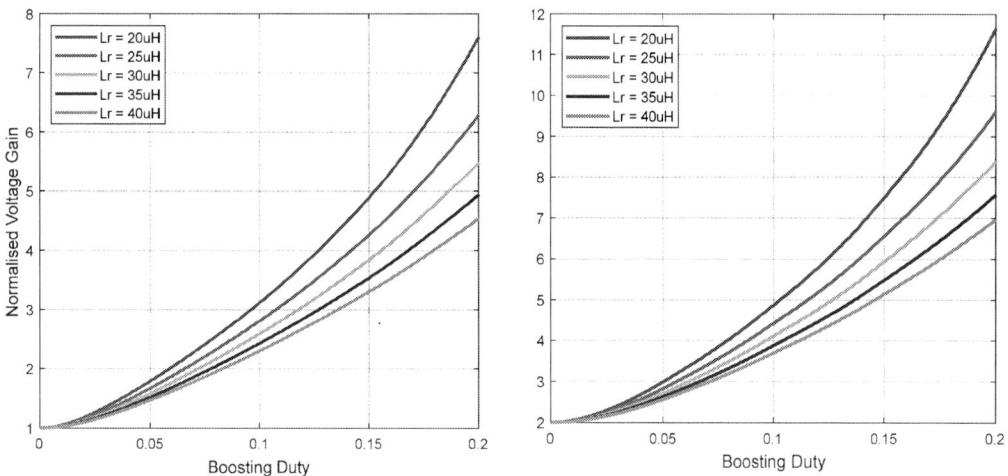

Fig. 6: Widening of duty ratio in the boosting operation (left) and in Greinacher operation (right) at full load.

Nevertheless, an increased inductor also results in a widened duty cycle required for boosting the voltage, thereby increasing the turn-off loss of S_5, S_6 and the RMS currents converter wide. For this reason, the required voltage gain range must be set, and a duty cycle constraint set to select the resonant inductance. A voltage gain of 1-4 will be sufficient in controlling the PV panel to its MPP even in case of extreme shading where two bypass diodes are conducting, and the panel's output voltage is derated by two thirds. Another restriction on the resonant tank is that the resonant capacitor may not exceed the output voltage, otherwise the converter will not operate as required. The limiting factor is greater in the Greinacher mode, as the capacitor is DC biased with nV_{in}. The capacitor should also be selected so that it does not have a large capacitance derating at DC biasing operation.

$$\frac{P_o T_s}{4nV_{in}C_r} + nV_{in} < V_{out} \tag{19}$$

Based on the above discussion, the resonant inductance is selected as $L_r = 40\mu H$, and the resonant capacitance is selected based on the resonance frequency being slightly higher than the switching frequency.

$$C_r \approx \frac{1}{L_r \omega_s} \tag{20}$$

For the design of S_5, S_6, the devices are selected based on their voltage rating, and their ability to perform a fast switch-off transition. For the diodes, similarly, the key metrics are the low conduction losses and sufficient voltage rating.

Simulation Results

The proposed converter was tested in PLECS 4.5.8, with the solar panel used being the FuturaSun FU 300M. The PV panel is modelled as a LUT with the simplified model presented by Bellini et al [14]. The converter is running at $f_s = 140kHz$. It is shown in Fig. 7 that the converter controls the PV panel voltage on the full designed range, and it reaches the desired voltage point in less than 5ms. One PI block is used to control the duty cycle, and a voltage sensor is attached to the panel to control the input voltage, as well as to change the control scheme at $M = 2$, which occurs at $V_{ref} = 17.5\,V$. At the transition point, the PI block is damped to avoid large currents due to the rapid duty cycle change.

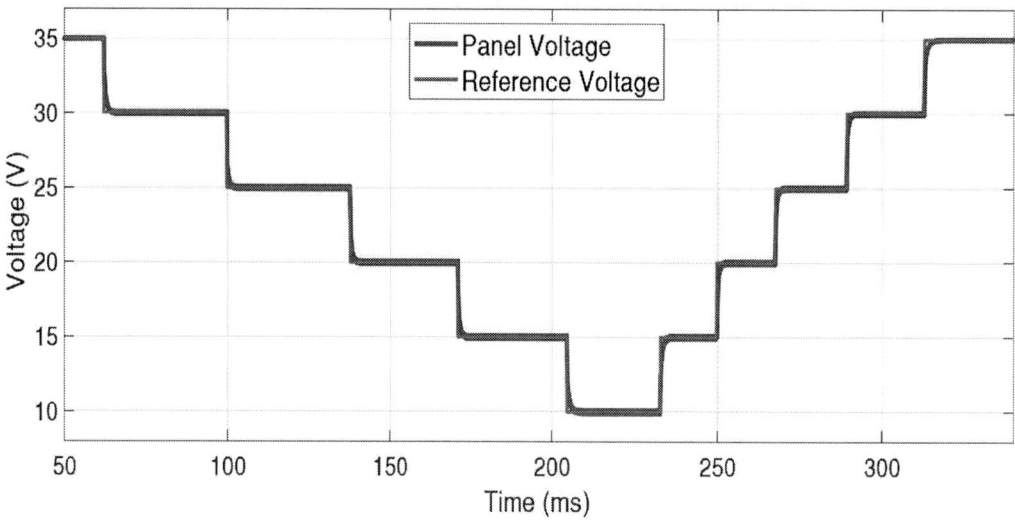

Fig. 7: PLECS model of the proposed topology. FU 300M is interfaced with a 380V DC bus.

Furthermore, the system was modelled using the PLECS thermal and magnetics domains, to estimate the converter's efficiency. The parameters used are summarized in Table II.

To calculate the losses, the data provided in the respective manufacturers' datasheet were used. For the switching losses, double pulse tests were run in LTSpice using manufacturer-provided SPICE models to estimate the switching transition time. The efficiency in different operating modes is shown in Fig. 8. In the partial shading condition, the power given by the PV will be less than the nominal, hence from this figure, it is seen that the converter maintains high efficiency of over 90% over the controlled voltage range. The loss distribution is shown in Fig. 9 to demonstrate the impact of the increase in the gain ratio. The majority of the losses when the gain is increased arise from the copper losses in the transformer and inductor, as well as the turn off loss in S_5, S_6.

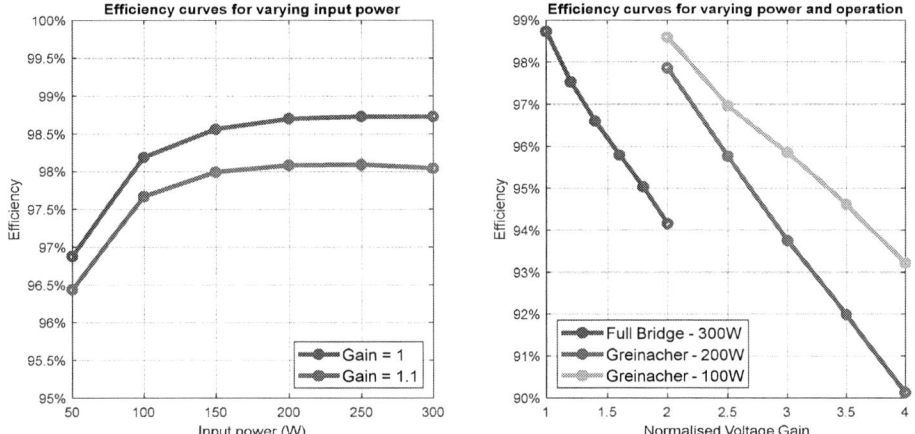

Fig. 8: Efficiency curves for varying PV power (left) and varying voltage gain (right).

Table II: Simulated components list

C_{in}:	$80\mu F$	$D_1 - D_2$:	C3D02060E
$S_1 - S_4$:	EPC 2021	L_r:	$40\mu H$
$S_5 - S_6$:	NV 6115	C_r:	$35nF$
Transformer Primary/Secondary coils:	5 turns: 435/40AWG/ 54 turns 300/46AWG	External Inductor:	14 turns 300/46AWG
Transformer Core:	N95/RM12, 0.7mm gap	Inductor Core:	N95/RM8, 0.5mm gap

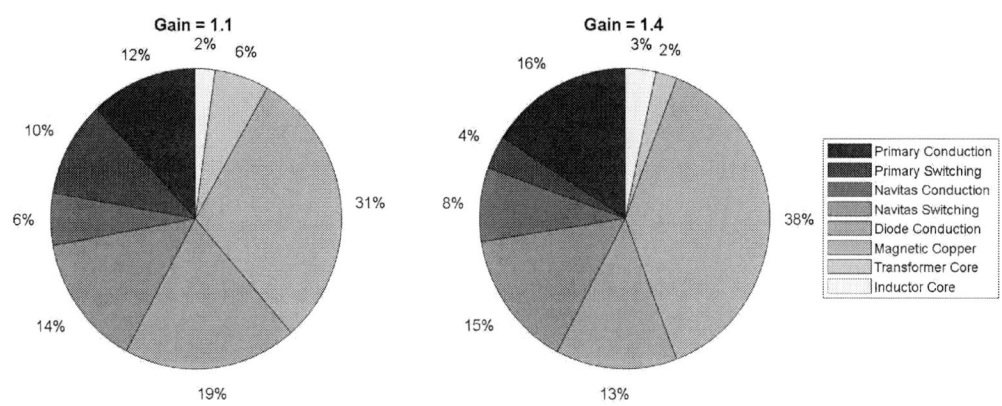

Fig. 9: Loss distribution at gain 1.1 (left) and gain set to 1.4 (right) with power set to the nominal 300W.

Conclusion

A highly efficient isolated DC-DC converter was presented for the PV power optimizer application. The proposed converter can operate in a wide input voltage range, making it an excellent candidate for shade-tolerant PV power production. The design benefits from zero-voltage switching and zero-current switching based on its resonant design, but it is operating at a fixed frequency, thus curtailing the limitations frequency-modulated resonant converters have. Based on its low switching losses and low magnetics count, it is also a good candidate for high power density design, enabling future integration with PV modules directly.

References

[1] B. Subudhi and R. Pradhan, "A comparative study on maximum power point tracking techniques for photovoltaic power systems," IEEE Transactions on Sustainable Energy, vol. 4, no. 1, pp. 89–98, 2013, doi: 10.1109/TSTE.2012.2202294.

[2] K. Alluhaybi, I. Batarseh, and H. Hu, "Comprehensive Review and Comparison of Single-Phase Grid-Tied Photovoltaic Microinverters," IEEE Journal of Emerging and Selected Topics in Power Electronics, vol. 8, no. 2, pp. 1310–1329, Jun. 2020, doi: 10.1109/JESTPE.2019.2900413.

[3] A. Chub, D. Vinnikov, O. Korkh, M. Malinowski, and S. Kouro, "Ultra-Wide Voltage Gain Range Microconverter for Integration of Silicon and Thin-Film Photovoltaic Modules in DC Microgrids," IEEE Transactions on Power Electronics, pp. 1–1, 2021, doi: 10.1109/TPEL.2021.3084918.

[4] T. Labella, W. Yu, J. S. Lai, M. Senesky, and D. Anderson, "A bidirectional-switch-based wide-input range high-efficiency isolated resonant converter for photovoltaic applications," IEEE Transactions on Power Electronics, vol. 29, no. 7, pp. 3473–3484, Jul. 2014, doi: 10.1109/TPEL.2013.2282258.

[5] X. Zhao, C. W. Chen, and J. S. Lai, "A High-Efficiency Active-Boost-Rectifier-Based Converter with a Novel Double-Pulse Duty Cycle Modulation for PV to DC Microgrid Applications," IEEE Transactions on Power Electronics, vol. 34, no. 8, pp. 7462–7473, Aug. 2019, doi: 10.1109/TPEL.2018.2878225.

[6] A. Chub, D. Vinnikov, O. Korkh, T. Jalakas, and G. Demidova, "Wide-Range Operation of High Step-Up DC-DC Converters with Multimode Rectifiers," Electronics (Basel), vol. 10, no. 8, p. 914, Apr. 2021, doi: 10.3390/electronics10080914.

[7] Y. Wei, Q. Luo, and H. Alan Mantooth, "A Novel LLC Converter with Topology Morphing Control for Wide Input Voltage Range Application," IEEE Journal of Emerging and Selected Topics in Power Electronics, pp. 1–1, 2020, doi: 10.1109/JESTPE.2020.3044207.

[8] D. Vinnikov, A. Chub, E. Liivik, and I. Roasto, "High-Performance Quasi-Z-Source Series Resonant DC-DC Converter for Photovoltaic Module-Level Power Electronics Applications," IEEE Transactions on Power Electronics, vol. 32, no. 5, pp. 3634–3650, May 2017, doi: 10.1109/TPEL.2016.2591726.

[9] H. M. Maheri, S. S. Vala, A. B. Mirza, E. Babaei, and D. Vinnikov, "A Novel Extendable High Gain Step up DC-DC Converter," 2021 IEEE 62nd International Scientific Conference on Power and Electrical Engineering of Riga Technical University (RTUCON), pp. 1–6, Nov. 2021, doi: 10.1109/RTUCON53541.2021.9711745.

[10] S. B. Kjaer, J. K. Pedersen, and F. Blaabjerg, "A review of single-phase grid-connected inverters for photovoltaic modules," IEEE Transactions on Industry Applications, vol. 41, no. 5, pp. 1292–1306, Sep. 2005, doi: 10.1109/TIA.2005.853371.

[11] Y. Shen, H. Wang, Z. Shen, Y. Yang, and F. Blaabjerg, "A 1-MHz Series Resonant DC-DC Converter with a Dual-Mode Rectifier for PV Microinverters," IEEE Transactions on Power Electronics, vol. 34, no. 7, pp. 6544–6564, Jul. 2019, doi: 10.1109/TPEL.2018.2876346.

[12] J. Mühlethaler, J. Biela, J. W. Kolar, and A. Ecklebe, "Improved core-loss calculation for magnetic components employed in power electronic systems," IEEE Transactions on Power Electronics, vol. 27, no. 2, pp. 964–973, 2012, doi: 10.1109/TPEL.2011.2162252.

[13] "Ferrites and accessories SIFERRIT material N95," 2017.

[14] A. Bellini, S. Bifaretti, V. Iacovone, and C. Cornaro, "Simplified model of a photovoltaic module," in Applied Electronics, 2009, pp. 47–51.

A novel parameter for the evaluation of protective circuits for IGBT explosion protection in submodules of MMC

Christoph Junghans, Hans-Guenter Eckel
UNIVERSITY OF ROSTOCK
Albert-Einstein-Str. 2
18059 Rostock, Germany
christoph.junghans@startmail.com
www.iee.uni-rostock.de

Keywords

«Modular Multilevel Converters (MMC)», «HVDC», «Short circuit», «Faults», «IGBT»

Abstract

Faults in high-power converters can cause sudden release of the energy stored in the charged intermediate circuit capacitor, destroying power semiconductors and causing secondary damage, such as the release of debris, the deformation of busbars and the emission of interference fields. This paper examines the effectiveness of certain protective circuits for the electrical explosion protection of IGBT modules in submodules of Modular Multilevel Converters (MMC) for HVDC applications and introduces a new parameter for the characterization of IGBT explosions. The protective circuits are arranged within the intermediate circuit of the submodule and are intended to reduce the surge current in the event of a fault and accordingly to dissipate or divert the energy of the storage capacitor and thus relieve the IGBT modules. The examinations indicate, that the connection of RL-combination and bypass thyristor can be recommended as protective circuit for IGBT explosion protection and surge current reduction used in MMC submodules with low switching frequencies. The newly introduced Explosion Integral XI improves the evaluation of electrical explosion protection measures. The parameter takes into account the thermal and magnetic mechanisms that cause the explosion of IGBT modules.

Introduction

Today MMC for high power HVDC transmission [1] surpass power levels of 1 GW and are constantly being further developed to increase the performance parameters. Occasional submodule faults in MMC are considered in the design and occur for different reasons (cosmic radiation, vibrations, thermal aging, fabrication defect, etc.). The submodules of the overall converter are designed in such a way that they are grouped out of the network with the help of a fast bypass switch in the event of a fault. The converter can then continue to work with redundant modules until a routine maintenance. The failure of one submodule must therefore not adversely affect other submodules or system components. A short circuit in the power semiconductors of the submodule induces the release of the energy stored in the intermediate circuit capacitor, which can cause surge currents and IGBT explosions to occur. These effects worsen with higher capacitor energy due to increasing MMC performance parameters. The electromagnetic as well as the mechanical effects of a fault must be restrained by appropriate technical measures like advanced semiconductor packages, reinforced semiconductor cells [2], interference shielding and additional protective circuits. The issue is quite important for the ongoing industrial converter evolution and widely ignored in the academic MMC sphere.

Two effects are responsible for the explosion of IGBT modules, which occur in combination during a surge current event and have a different degree of impact depending on the voltage class and module structure. On the one hand, it is the Lorentz force, which acts on the internal conductor structure of the IGBT module, on the other hand, it is the gas pressure that builds up as a result of electric arcs in the IGBT module. The arcs also cause the evaporation and decomposition of silicone gel that seals the internal electric structure.

The effect of the magnetic force component is considered in [3] for an IGBT module of the same package size as in this paper. However, the energy available in this study of the module explosion is relatively small at 10 kJ and the effect of gas formation is not taken into account. For the measurements, IGBT modules in the so-called IHV housing (IGBT High Voltage) with a base plate in the dimensions 140 mm x 190 mm are used. The IGBT modules are connected as a half-bridge and arranged in a closed and mechanically reinforced semiconductor cell, that is a modified version of a commercial MMC submodule semiconductor cell.

Different parameters can be used to evaluate the effectiveness of protective circuits in the intermediate circuit of a submodule. In this paper, the peak value of the fault current, the I²t-value, the energy conversion inside the IGBT module, the mechanical warpage of the IGBT heat sink and the novel parameter Explosion Integral XI are used. This paper is based on excerpts of [4]. Extensive experimental results, researched detailed design as well as simulation of the transient electrical and mechanical behaviour (e.g. interference fields, deformation of busbars with Ansys Workbench) can also be found there.

1. MMC-submodule fault and type of protective circuits

The MMC submodule considered here consists of the storage capacitor C, the IGBT half-bridge T1/D1 + T2/D2 and the bypass switch BPS (Fig. 1a). For the examination, the serious fault case is assumed, in which the majority of the energy of the storage capacitor is released in the IGBT module T1/D1. This occurs when the IGBT modules of the submodule are no longer switching (e.g. due to driver failure or signal interruption) and the storage capacitor C is reaching an unacceptably high voltage. If this case is detected by the monitoring electronics, the submodule is grouped out by closing the high-speed bypass switch BPS [5]. The freewheel diode D1 of the upper IGBT module T1/D1 is hard switched off and destroyed by the reverse recovery current. In this way, the charged intermediate circuit capacitor is short-circuited so that the surge current flows via T1/D1 and the bypass switch (red arrows, Fig. 1a).

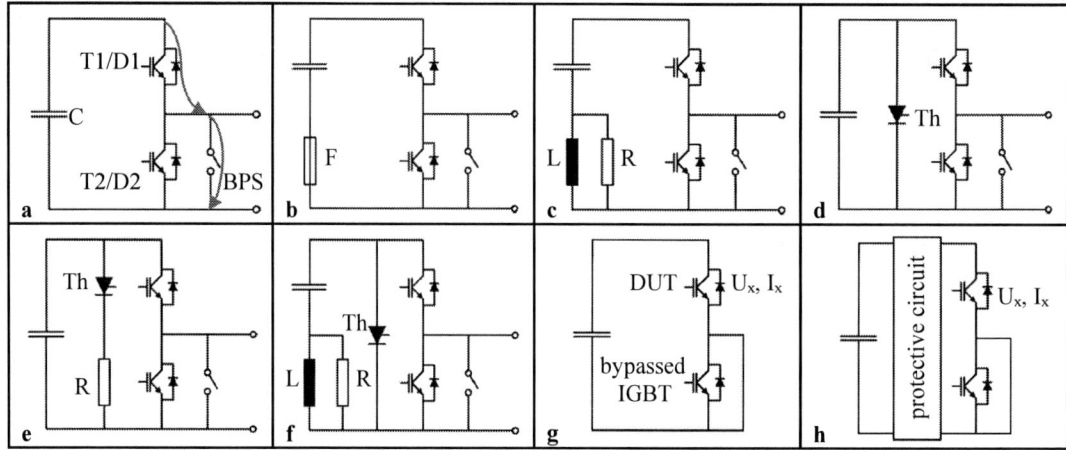

Fig. 1 a-h: Schematics of submodule, protective circuits and experimental setup

The IGBT modules CM900HC90 (CM900) and CM1200HC90 (CM1200) from the manufacturer Mitsubishi are used in the experiments. The modules belong to the voltage class 4.5 kV and the rated current is 900 A and accordingly 1200 A. The experiments show that the CM900 builds up more voltage during a failure and that the explosion effect is greater than that of the CM1200. This is a consequence of the unequal internal structure and the different nominal current ratings. For the examination of the fault event, the storage capacitor has a capacitance of 13 mF and is charged to a voltage of 4 kV, so that an energy of 104 kJ is available. The alternating current in the intermediate circuit under normal operating conditions in this design is defined to 1 kA$_{rms}$, even if this would be too high for the IGBT modules used. The switching frequency of the exemplary submodule is rated at 150 Hz.

Based on the energy content and the nominal current of the intermediate circuit used, according to the current state of the art, a commercially used submodule would each be equipped with two IGBT modules for T1/D1 and T2/D2 connected in parallel. In the event of a fault, it cannot be assumed that both modules connected in parallel fail at the same time and that the released energy is evenly distributed among the modules. The examinations with only one IGBT module used therefore not only show a case of particularly high module stress, but also anticipate a further increase in the power density of IGBT modules due to technical progress. With a comparable intermediate circuit, only one module of the reviewed housing category could be sufficient in the future as a power semiconductor for the formation of a half-bridge consisting of a total of two IGBT modules.

The examined protective circuits in the intermediate circuit are high-speed fuses (F, Fig. 1b), the series connection of a parallel RL-combination (R+L, Fig. 1c), a thyristor (Th, Fig. 1d) bypassing the IGBT half-bridge without or with a discharge resistor (Th+R, Fig. 1e) and an integration of RL-combination and bypass thyristor (Th+R+L, Fig. 1f). Benchmark measurements without protective circuits (Fig. 1g) are used for comparison. The experiments are started by switching on T1 ("0" on the diagrams x-axis), as the experimental effort for destroying D1 by hard switching off is too great. After switching on T1, the IGBT desaturates for a few microseconds and is then destroyed. The IGBT module is mounted in a semiconductor cell on an aluminium alloy heat sink, which is 20 mm thick and pierced with cooling channels for water cooling. The IGBT module explosion induces recoil as well as an increase in pressure in the semiconductor cell, causing the heat sink to warp. The semiconductor cell is a modified commercially used package containing two IGBT modules as a half-bridge. One of the IGBT modules is bridged with busbars, so that only the failure of a single module (DUT, T1/D1) is subject to consideration (U_x+I_x, Fig. 1g+1h). Hereinafter selected tests are presented, an overview of the conducted experiments is given in Table I.

2. Details on the measurements

2.1 Benchmark tests (Fig. 1g)

The benchmark tests are conducted without protective circuits, voltage and current curves of the IGBT modules CM900 and CM1200 are shown in Fig. 2. U_x is the voltage drop of the semiconductor cell (IGBT module + internal busbars), I_t is the surge current, 9 or 12 stands for CM900 and CM1200 respectively. Fig. 5a shows a damaged semiconductor cell (the two IGBT modules are mounted inside) of a benchmark test with the IGBT CM900, the upper deformed heat sink is the one of the DUT, the lower one is the least deformed heat sink of the bridged IGBT module T2/D2.

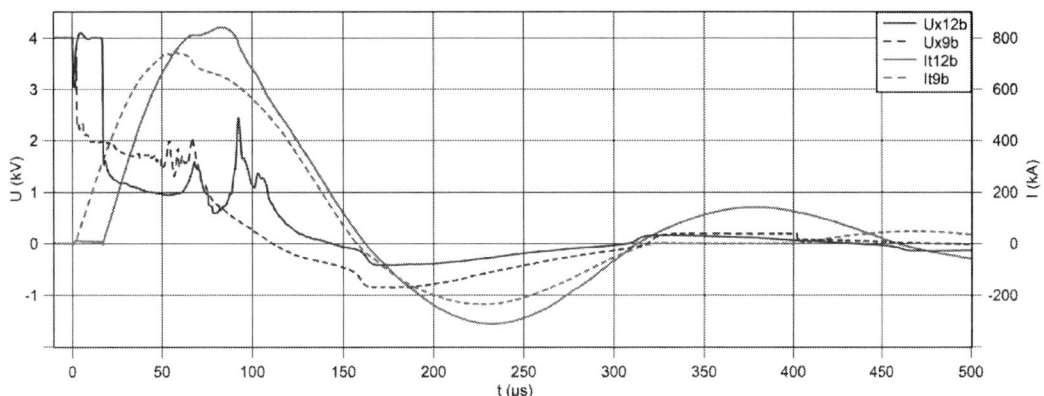

Fig. 2: Benchmark tests of the IGBT modules CM900 and CM1200

2.2 High-speed fuses (Fig. 1b)

The examination of high-speed fuses as a well-known variant of IGBT explosion protection [6, 7] is carried out with two parallel fuses of the type PC123UD25C500TF from the manufacturer Mersen/Ferraz with a nominal voltage of 2.5 kV_{rms} (3.6 kV_{peak}) and a rated current of 500 A each. The design depends on the nominal AC current of the intermediate circuit and the appropriate voltage drop to be built up by the fuses in a surge current event. For a practical design, the fuses should have a higher rated current in order to maintain safety reserves. The IGBT module CM900 is destroyed and the housings of the fuses crack (Fig. 5b). Nevertheless, the purpose of IGBT explosion protection is fulfilled with this protective circuit, as the heat sink is not deformed. The commercial availability of types of high-speed fuses suitable for the current and voltage values in MMC submodules is limited. The diagram in Fig. 3 shows the curves of the voltage drop of the IGBT semiconductor cell U_{x9f} and of the fuses U_{f9f} as well as the total fault current I_{t9f}.

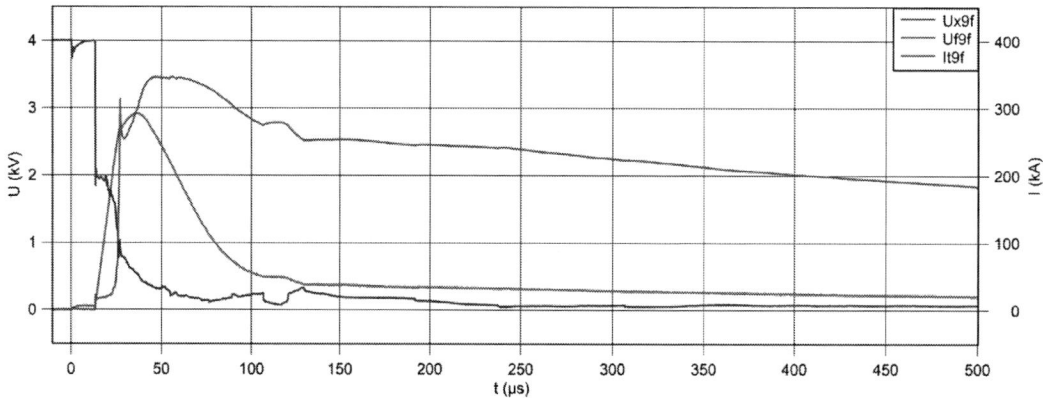

Fig. 3: Test of submodule with high-speed fuses and CM900

2.3 RL-combination (Fig. 1c)

The combination of a resistor and an inductor connected in parallel (RL-combination) in the intermediate circuit of a submodule is known from [8]. The RL-combination causes losses in normal operation which increase with the switching frequency. The RL-combination is suitable for surge current limitation during a fault for semiconductors in disc housings and is examined here for IGBT module explosion protection.

The newly introduced RL-transformer [9] (so-called Hochenergie-Stossstrombegrenzer) as a compact derivative of the RL-combination with two primary windings and one secondary winding would also be suitable. The electrical behaviour is identical to the RL-combination. The short-circuited secondary winding is placed within the primary windings and crosses them. The secondary winding represents the inductively coupled parallel resistor and is made of stainless steel (V2A). To avoid saturation effects, the RL-transformer has no core material.

In the RL-combination with a design at the values 6 mΩ and 430 nH as well as with aforementioned nominal AC current and switching frequency, losses of approx. 120 W including the skin effect are to be expected in normal operation of the submodule. The inductor consists of one winding of a formed busbar (Fig. 5c) with a cross-section of 100 mm x 5 mm and a diameter of about 340 mm. The resistor is made up of a folded stainless steel (V2A) sheet with insulating intermediate layers of synthetic mica sheets (Fig. 5d). The IGBT module CM1200 is completely destroyed in the test, the heat sink warpage is at 3 mm. The diagram in Fig. 4 shows the curves of the voltage drop of the semiconductor cell U_{x12rl} and of the RL-combination U_{rl12rl} as well as the total current I_{t12rl}, the current of the inductor I_{12rl} and the current of the parallel resistor I_{r12rl}.

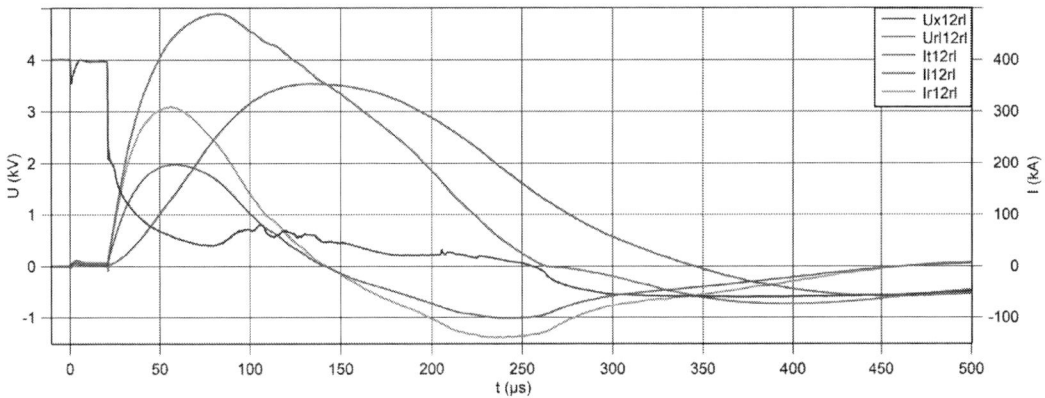

Fig. 4: Voltages and currents when testing the RL-combination with the IGBT CM1200

Fig. 5a-i: Photographs of the experiments

2.4 Bypass thyristor parallel with the semiconductor cell (Fig. 1d)

A bypass thyristor is intended to short-circuit the intermediate circuit and thus relieve the IGBT module in case of a failure [10]. The thyristor in a disc housing is turned on approximately one microsecond after the failure of the IGBT module by a controller that detects the subsequent current increase as the IGBT is destroyed after desaturation. The module housing of the CM900 is slightly damaged after the experiment. No deformation of the heat sink can be detected. The amplitude of the fault current of the IGBT module is limited to 65 kA.

The diagram in Fig. 6 shows the voltage drop U_{x9bp} of the semiconductor cell, the total surge current I_{t9bp} and the fault current of the IGBT module I_{x9bp}. The total surge current with a peak value of 1300 kA is greater than in the benchmark test, the thyristor disc housing burst due to high internal pressure. The effect on 5 mm x 120 mm busbars is shown in Fig. 5e.

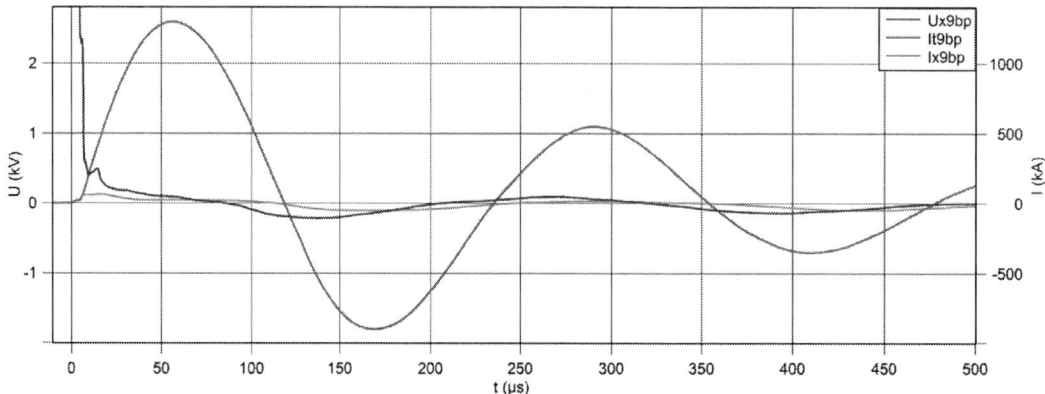

Fig. 6: Current and voltage curves when tested the bypass thyristor with the CM900 IGBT module

2.5 Bypass thyristor with discharge resistor (Fig. 1e)

Parallel to the IGBT semiconductor cell, a series connection of thyristor and discharge resistor is arranged, which is a simplified arrangement of [11] without inductance. The thyristor is turned on when the IGBT module fails. Then the thyristor and discharge resistor take over a part of the surge current and thus partly relieve the IGBT module. The resistor thermally dissipates a part of the energy stored in the intermediate circuit capacitor. The rating of the resistor value is based on the balancing between the maximum permissible surge current in the intermediate circuit and the distribution of energy and current to both paths.

The discharge resistor is dimensioned to a value of 1.2 milliohms for the transfer of the larger proportion of current and is designed as a folded stainless steel (V2A) sheet resistor. As with the bypass thyristor without discharge resistor, this protective circuit does not cause any additional losses during normal submodule operation. The measurements are conducted with the CM900 and CM1200 IGBT modules. The IGBT modules are destroyed (CM900 in Fig. 5g) and the heat sink is warped 1 mm. The busbars between the bypass and the capacitor bank are less widened (Fig. 5f) by I²t-value than the same place when using a thyristor without discharge resistor (Fig. 5e). Fig. 7 shows the measurements with a CM1200 IGBT module. U_{x12bpr} is the voltage of the semiconductor cell, I_{t12bpr} stands for the total surge current of the intermediate circuit, I_{x12bpr} is the fault current of the IGBT module and $I_{th12bpr}$ represents the current of the bypass consisting of thyristor and resistor.

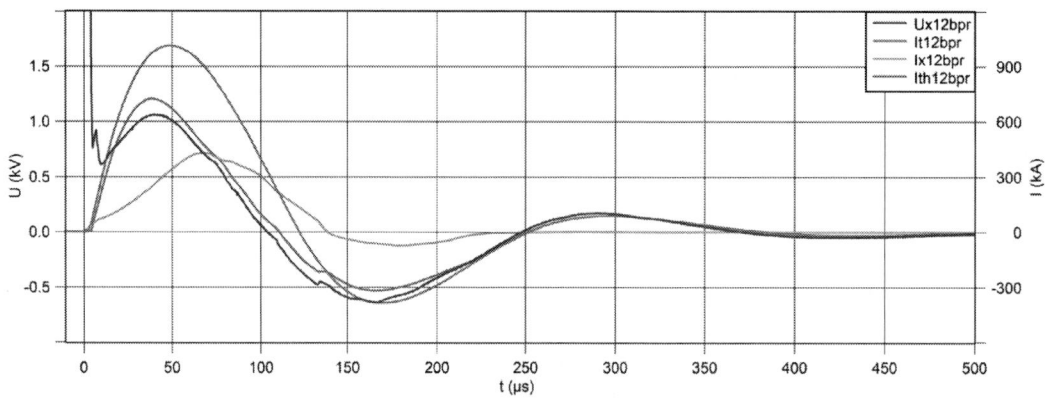

Fig. 7: Measurements on bypass thyristor with discharge resistor and IGBT CM1200

2.6 RL-combination joined with bypass thyristor (Fig. 1f)

RL-combination or RL-transformer can be connected to the bypass thyristor, thus combining IGBT explosion protection and surge current limitation [12]. The RL-combination with a resistance of R = 6 mΩ and an inductance of L = 430 nH is accommodated with the bypass thyristor in a clamping device located between the semiconductor cell and the capacitor bank. Fig. 5h shows the setup after the test, the busbars between capacitor bank an protective circuit are barely bended.

The housing of the IGBT module CM900 remains largely intact (Fig. 5i). The heat sink of the IGBT module has no deformation. Fig. 8 shows the voltages of IGBT semiconductor cell U_{x9bprl} and RL-combination $U_{rl9bprl}$ as well as the total surge current I_{t9bprl} of the intermediate circuit, the fault current I_{x9bprl} of the IGBT module and the current curve $I_{th9bprl}$ of the bypass thyristor.

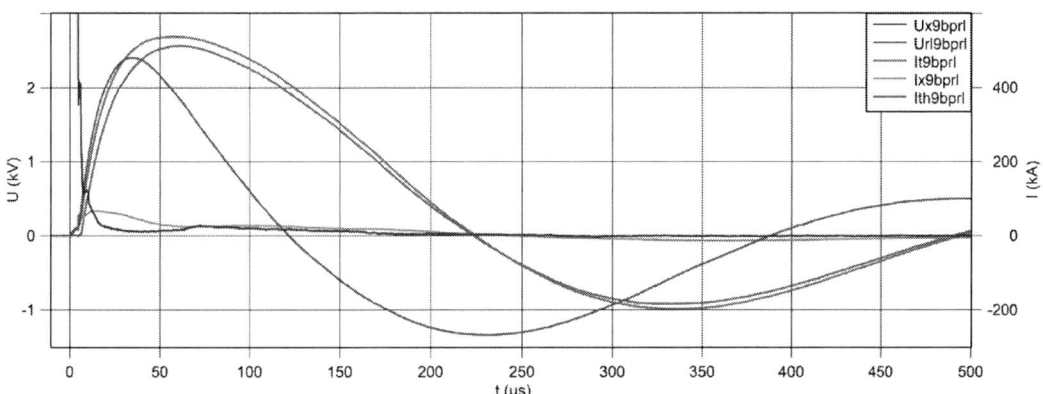

Fig. 8: Currents and voltages of joined RL-combination and bypass thyristor with IGBT CM900

3. The Explosion Integral XI

The quantification of the explosion effect of IGBT modules based on the mechanical impact on the semiconductor cell is associated with high material effort and error-prone. Usually most of the components must be renewed for each additional measurement. For instance, this includes deformed, damaged or weakened heat sinks, screws, mounting assemblies, busbars, reinforced plastics and - of course - the destroyed IGBT modules. The Explosion Integral XI should therefore make it possible to induce from electrical measurements the effectiveness of protective circuits for IGBT explosion protection and thus to reduce experimental expenses.

The I²t-value and the energy E converted during an IGBT failure are not in a fixed relationship to each other even with identical IGBT modules, but depend on the slope of current rise, the level of the surge current and the duration of the surge current event. This is due to the fact that the energy conversion of the IGBT module depends not only on the surge current, but also on the arc voltage, which develops after the mechanical or thermal destruction of the internal electrical connecting components of the IGBT module. Therefore, not one of the parameters alone is sufficient for the characterization of the explosion behaviour of IGBT modules. For the better evaluation of protective circuits, it is therefore proposed to use a characteristic parameter that includes both the magnetic component from the I²t-value, as well as the thermal component from the energy E. Thus, the two integral expressions for I²t and E can be combined to form the Explosion Integral XI.

The formation of the Explosion Integral XI is not a derivation but a definition:

I²t-value: $\qquad I^2t = \int i^2 dt$ $\hfill (1)$

Energy: $\qquad E = \int uidt$ $\hfill (2)$

Explosion Integral: $\qquad XI = \int \sqrt{i^2}\, uidt = \int |i| uidt$ $\hfill (3)$

Unit XI: $\qquad [XI] = A^2Vs = AJ$ $\hfill (4)$

The current "i" and the voltage drop "u" are the electrical quantities of the IGBT module or rather the semiconductor cell during the fault event. The absolute value formation of one share of the fault current ensures that the energy temporarily stored in the stray inductance of the semiconductor cells' busbars is not affecting the Explosion Integral XI. This makes it possible that the measurement of the voltage does not need to be taken directly on the terminals of the IGBT module. For the calculation of the Explosion Integral XI, it is also conceivable to weight the proportions of the magnetic and thermal components differently, since both effects can have an unequal share of the IGBT explosion behavior.

4. Results

Fig. 9 shows the fault currents in the intermediate circuit for all tests with the IGBT CM900: benchmark test (It9b), high-speed fuse (It9f), RL-combination (It9rl), bypass thyristor (It9bp), bypass thyristor with resistor (It9bpr) and the connection of RL-combination with bypass thyristor (It9bprl).

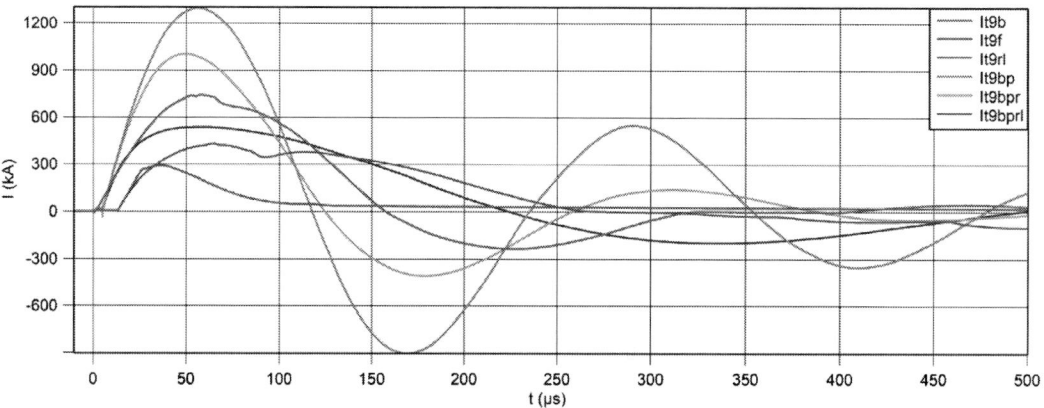

Fig. 9: Overview of fault currents in the intermediate circuit with IGBT module CM900

The IGBT current curves with the module CM900 in which the heat sink was warped by 1 mm or less are shown in Fig. 10: high-speed fuse (Ix9f), bypass thyristor (Ix9bp), bypass thyristor with resistor (Ix9bpr) and the connection of RL-combination with bypass thyristor (Ix9bprl).

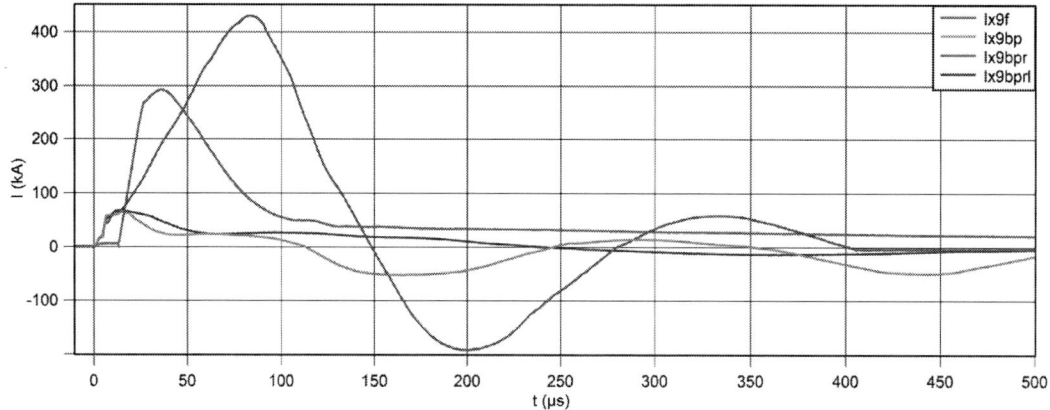

Fig. 10: CM900 IGBT module current of experiments with heat sink deformation by 1 mm or less

Table I: Overview of the measurement results

Protective circuit	IGBT	I_{tmax}	I^2t_t	I_{xmax}	I^2t_x	E_x	E_{pc}	Δ_{hs}	XI	Remarks
#	Unit	kA	MA²s	kA	MA²s	kJ	kJ	mm	GAJ	
Benchmark										
1	CM900	746	46	746	46	87	-	14	49	
2	CM1200	843	55	843	55	78	-	12	49	
High-speed fuse										
3	CM900	300	4	300	4	10	83	0	2	
RL-combination										
4	CM900	434	23	434	23	55	41	8	18	Folded V2A-resistor
5	CM1200	491	26	491	26	46	51	3	14	Folded V2A-resistor
6	CM1200	505	28	505	28	41	56	3	14	Graphite disc resistor
Bypass thyristor										
7	CM900	1301	173	65	0.5	2	-	0	0.08	Thyristor disc burst
Bypass thyristor + discharge resistor										
8	CM900	1004	71	431	13	20	43	1	5	
9	CM1200	1016	70	431	11	20	44	1	6	
Bypass thyristor + RL-combination										
10	CM900	540	38	68	0.2	1	82	0	0.04	

Table I gives an overview of all measurement results: intermediate circuit surge current peak value I_{tmax}, intermediate circuit I^2t_t-value, DUT (IGBT module) surge current peak value Ixmax, DUT I^2t_x-value, DUT energy conversion E_x, protective circuit energy conversion E_{pc} (without losses of bypass thyristor, if any), deformation of heat sink Δ_{hs} and Explosion Integral XI. The bound of integration for I^2t-value, energy and Explosion Integral is about one millisecond from switching on T1.

Fig. 11 indicates the relation of the heat sink warpage Δ_{hs} to energy conversion E_x, I^2t_x-value and Explosion Integral XI. Regression lines (subindex "r") are derived from the respective measuring points. In order to evaluate the significance of E_x, I^2t_x and XI for the relationship with Δ_{hs}, the determination coefficient R^2 of the measuring points for the respective regression line is determined. The Explosion Integral XI has the highest determination coefficient with a value of $R^2_{XI} = 0.94$, followed by the energy conversion E_x with $R^2_{Ex} = 0.90$ and the I^2t_x-value with a $R^2_{I^2tx} = 0.83$.

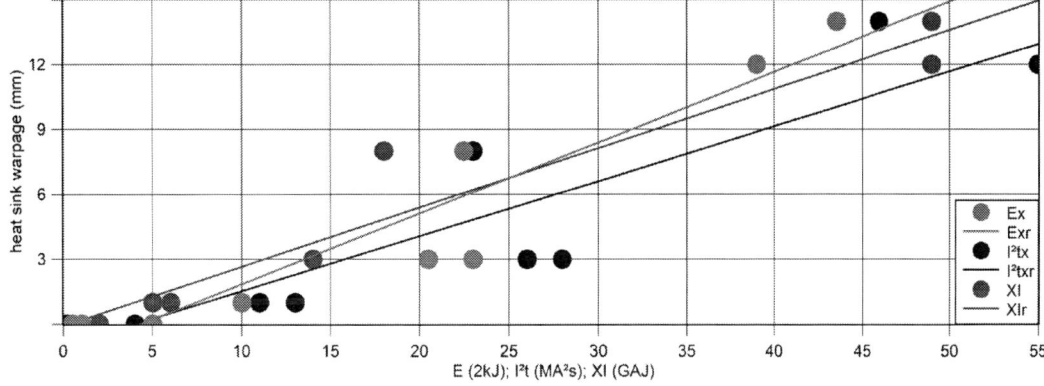

Fig. 11: heat sink warpage Δ_{hs} in relation to energy conversion E_x, I^2t_x-value and Explosion Integral XI

Conclusion

The connection of RL-combination and bypass thyristor can be recommended as protective circuit for IGBT module explosion protection and surge current reduction in MMC submodules with low switching frequencies. The parallel bypass thyristor reduces the voltage drop over the IGBT module to a minimum to avoid an explosion. RL-combination or RL-transformer may be designed for optimal trade-off of surge current reduction in the intermediate circuit and additional losses in normal operation. RL-combination or RL-transformer are freely scalable and can be manufactured from commercially available standard materials.

The newly introduced Explosion Integral XI is suitable for the evaluation of the effectiveness of protective circuits for IGBT explosion protection because it combines the magnetic and thermal components of the fault event. Based on the results of the examinations, it can be assumed that the Explosion Integral XI as a parameter for describing the explosion behaviour of IGBT modules is superior to the I^2t-value and the energy E. For further verification, examinations should be carried out with other IGBT module types and differing converter characteristics.

References

[1] R. Marquardt, A. Lesnicar: „An Innovative Modular Multilevel Converter Topology Suitable for a Wide Power Range", IEEE Bologna Power Tech Conference Proceedings, ISBN: 0-7803-7967-5, 2003

[2] M. Billmann, D. Malipaard, H. Gambach: „Explosion Proof Housings for IGBT Module Based High Power Inverters in HVDC Transmission Application", PCIM Conference, 2009

[3] D. Li, F. Qi, M. Packwood, A. Islam, L. Coulbeck, X. Li, Y. Wang, H. Luo, X. Dai, G. Liu: „Explosion Mechanism Investigation of High Power IGBT Module", 19th Interntl. Conf. on Thermal, Mechanical and Multi-Physics Simulation and Experiments in Microelectronics and Microsystems (EuroSimE), 2018

[4] C. Junghans: "Zur Beherrschung von Stoßstromereignissen bei Fehlerfällen in Submodulen von Modularen Mehrpunktumrichtern", Dissertation, Universität Rostock, 2022

[5] J. Bürger, J. Dorn, T. Kübel, R. Renz, A. Stelzer, A. Zenkner: „Kurzschlussvorrichtung mit pyrotechnischer Auslösung", Patent WO/2009/092621A1, Siemens AG, 2009

[6] F. Iov, F. Abrahamsen, F. Blaabjerg, K. Ries, H. Rasmussen, P. Bjornaa: „Fusing IGBT-based Inverters", PCIM Conference, 2001

[7] D. Braun, D. Pixler, P. LeMay: „IGBT Module Rupture Categorization and Testing", IEEE IAS '97, ISBN: 07803-4070-1/97, 1997

[8] H.-G. Eckel, H. Gambach, F. Schremmer, M. Wahle: „Submodul mit Stromstossbegrenzung", Patent WO/2014/090272A1, Siemens AG, 2014

[9] H.-G. Eckel, C. Junghans, D. Schmitt: „Hochenergie-Stossstrombegrenzer", Patent WO/2017/097351, Siemens AG, 2017

[10] M. Billmann, J. Dorn, H. Gambach, M. Wahle: „Kurzschlussstromentlastung für Submodul eines Modularen Mehrstufenumrichters (MMC)", Patent WO/2013/044961A1, Siemens AG, 2013

[11] A. Zuckerberger, F. Bolgiani, J. Eckerle: „Spulenbauelement zur Verwendung in einer Crowbar-Schaltung", Patent EP2109122A1, ABB AG, 2009

[12] H.-G. Eckel, C. Junghans, D. Schmitt: „Modul für Modularen Mehrpunktumrichter mit Kurzschließer und Kondensatorstrombegrenzung", Patent WO/2018/149493, Siemens AG, 2018

Sub-Modules Switching Algorithms for Dual Active Bridge Modular Multilevel Converters to Optimize Capacitor Voltage Deviation versus Power Efficiency

Peizhou Xia, Chuantong Hao, Stephen Finney, Michael Merlin
The University of Edinburgh
Sanderson Building Robert Stevenson Road The King's Buildings Edinburgh EH9 3FB
Edinburgh, the United Kingdom
Email: P.Xia@sms.ed.ac.uk

Acknowledgments

This project is supported by the University of Edinburgh, and China Scholarships Council (CSC).

Keywords

≪Dual Active Bridge (DAB) DC-DC converters≫, ≪Electromagnetic Interference (EMI)≫, ≪Modular Multilevel Converter (MMC)≫, ≪Soft switching≫, ≪Switching losses≫.

Abstract

This article firstly presents a detailed analysis and improvement for the dual active bridge employing MMC cells (MMC-DAB converter). The presence of MMC cells can effectively address the concern that medium or high voltage level is beyond voltage rating of semi-conductor devices in conventional DAB converter. However, the introduction of MMC cells results in significant losses originated from resonance between line inductor and sub-module capacitor. Besides, the non-negligible switching loss is another factor of efficiency drop. In order to solve the additional losses and improve the overall conversion efficiency, this article proposes three innovative switching algorithms, where manipulating cell states to stop current oscillation and enable zero voltage switching (ZVS). Simulated results for a medium voltage level MMC-DAB converter is provided to verify the efficiency gain under optimal switching algorithms. Moreover, a 2-level MMC-DAB converter is implemented and the experimental result shows the proficiency of the converter.

Introduction

Isolated Bi-directional DC-DC converters (IBDC) are a key building block for DC power systems, due to their advantages of easy power flow regulation and reduced power conversion stages [1]. Various topologies of IBDC converters have been discussed in recent researches, such as dual-flyback converter [2], flyback-forward converter [3], dual-half-bridge converter [4], and Dual-Active-Bridge (DAB) converter. Among all these topologies, DAB-IBDC has the advantages of high power capacity, bi-directional energy flow, simple control strategy, and inherent soft switching [5]. These merits make DAB converter a preferred IBDC topology in many applications such as solid-state transformer [6], micro-grid [7], and electric vehicle [8].

DAB Converter Employing Modular Multilevel Converter (MMC) Cells

Utility scale DC power networks will require DC-DC converters capable of operating beyond the ratings of available power semiconductor devices. In order to address these concerns, DAB converters based around Modular Multi-Level topologies are introduced [9]-[10]. As shown in Fig. 1, By replacing the power devices of a single phase DAB converter with series-connected sub-modules. The harmful dv/dt stress is reduced by introducing intermediate voltage steps, where the output voltage of one bridge can be modulated as a trapezoidal waveform. The analysed MMC converter has a capacitor connected in series with an IGBT, which can act as a soft-voltage clamps for the converter. When the MMC cells are employed into DAB converter, all cells in upper or lower arm are utilizing half-bridge module, denoted as "quasi two-level converter" (Q2LC) [11].

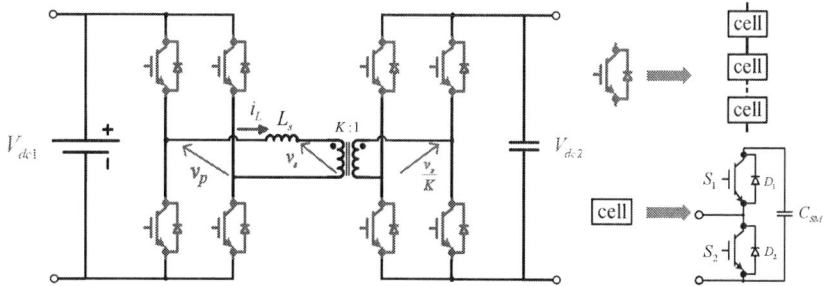

Fig. 1: The schematic of conventional DAB converter which applies MMC cells.

Fig. 2 depicts the schematic of MMC-DAB converter which is used in this article. The conventional H-Bridge connection of DAB is replaced by one bridge leg with centralized ground for simplifying set-up. Meanwhile, voltages across the leakage inductance L_s of AC transformer are denoted as v_p and v_s respectively, where trapezoidal modulation is used to create quasi two-level voltage waveform. The phase shift between v_p and v_s can be controlled to achieve output regulation, where controlling strategies such as linear quadratic regulator (LQR) can be added [12]. Moreover, N represents the number of cells within one arm, and there will be $N+1$ levels on aggregate for stack voltage generation. T_w indicates the transition dwell time which can be short enough compared to the main switching period T_s.

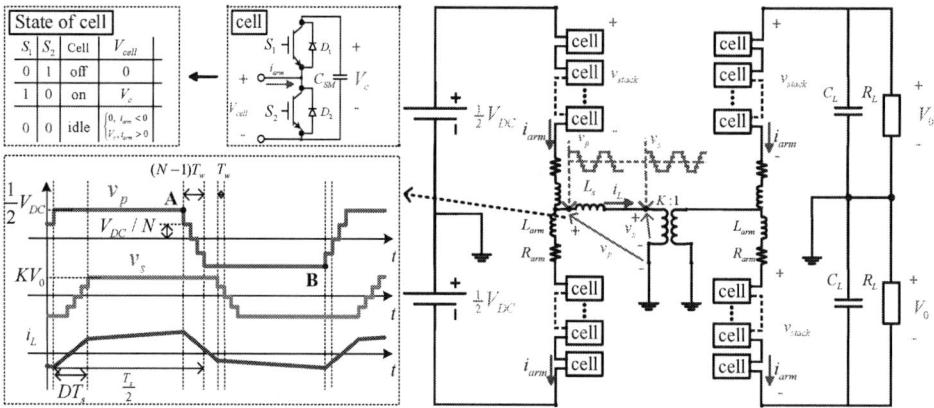

Fig. 2: Schematic of single phase MMC-DAB converter.

Loss Classification in MMC-DAB Converter

However, the introduction of MMC can result in extra losses, which can be classified as conduction losses and switching losses. The energy trapped in LC resonance is a dominant factor in efficiency reduction. Each time when a new cell being switched in, the sinusoidal charging current will cause oscillation in capacitor voltage which will draw more energy from sources. Subsequently, this extra energy will be transferred and trapped between cell capacitor and line inductor, and finally cause power dissipation on line resistance. This resonance can also result in electromagnetic interference (EMI) [13]-[14], which could have negative effect on the performances of semiconductor devices and also be harmful to circuits. In this article, an alternative switching algorithm is proposed (Algorithm 1) to control oscillation and reduce losses associated with quasi-two level switching of MMC stacks.

Additionally, during each switching transient, there will be a significant overlap between switch current and voltage due to the high-stand voltage of sub-module capacitor [15]. The switching losses can be relatively large under high load current and high switching frequency. In order to address this problem, the second algorithm (Algorithm 2), is proposed which exploit zero voltage switch (ZVS) to reduce switching losses. Moreover, Algorithm 3 is proposed by combining Algorithm 1& 2.

Operating principle of MMC-DAB

As T_d is small enough and negligible compared to switching period, quasi two-level voltage waveforms are generated across AC transformer. Thus, the DAB characteristics of MMC-DAB converter, such as output voltage and leakage inductor current, can be derived directly using DAB analysis. In a typical DAB converter, if the duty ratio D is conventionally constrained between 0 and 0.5, the power will be transmitted from primary to secondary side. Two conditions of AC current have been depicted as shown in Fig. 3. This article will analyse forward power flow. Moreover, the polarity of I_1 determines whether the cells in the upper arm of secondary bridge will experience positive (inductive loading) or negative (capacitive loading) current. The inductor current is analysed in DAB converter [12],

$$
\begin{aligned}
I_1 &= \frac{v_p - v_s - 2Dv_p}{4Lf_s} - \frac{N-1}{2L_s}(v_p + v_s)T_w & I_2 &= -\frac{2v_sD - v_s + v_p}{4Lf_s} + \frac{N-1}{2L_s}(v_p + v_s)T_w \\
I_3 &= \frac{2Dv_p + v_s - v_p}{4Lf_s} + \frac{N-1}{2L_s}(-v_p + v_s)T_w & I_4 &= \frac{2v_sD - v_s + v_p}{4Lf_s} - \frac{N-1}{2L_s}(v_p - v_s)T_w
\end{aligned}
\tag{1}
$$

where $v_p = V_{dc}/2$ and $v_s = KV_0$ are the voltages across leakage inductance L_s, K is the turns ratio of AC transformer, D represents the duty ratio. The power and the output voltage will be,

$$
\begin{cases}
P_{sec} = P_{pri} = \dfrac{KV_{DC}V_0D(1-D)}{4L_sf_s} \\[4mm]
V_0 = \dfrac{KV_{DC}D(1-D)R_L}{L_sf_sC_LR_L + 2L_sf_s}
\end{cases}
\tag{2}
$$

From (2), we can see that the output voltage can be regulated by controlling phase shift D.

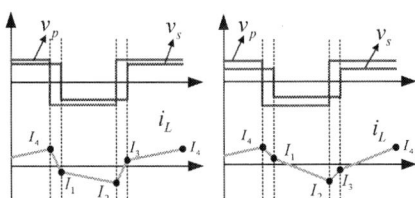

Fig. 3: AC current waveform under high (left) and low (right) modulation index.

Voltage balancing

Achieve sub-module capacitor voltage V_c balance is a premise of MMC successful operation [14]. As the cell which turns on under positive current first sees the greatest charge, the cells with lower voltage should be turned on before those with higher voltage in order to achieve voltage balancing. Thus, each individual cell voltage can be balanced by firstly applying conventional sorting algorithms, where sub-module capacitor voltage V_c are sorted in ascending or descending order based on arm current polarity. Then the cell capacitor is brought into conduction following the order [11]. Other advanced balancing techniques are researched such as rotating gate signals in turn [16] and in [14] which solves the challenge that current polarity may reverse.

In this article, conventional balancing control is sufficient enough as medium voltage level and small quantity of sub-modules are mainly considered. Suppose $D \in [0, 0.5]$ which indicates forward power flow, the switched-in arm of primary bridge would always meet positive current as $I_4 > 0$. Positive (charging) arm current illustrates that for primary bridge, V_c should be sorted in the ascending order, where cell with the lowest V_c will be firstly switched in, and cell with the highest V_c will be the last one.

However, for secondary bridge, it depends on polarity of I_1. When $I_1 < 0$, cells should still be switched in from low V_c to high V_c. For negative (discharging) current where $I_1 > 0$, cells will be sorted in the descending order. The cell capacitor with the highest V_c will be firstly brought into conduction while the cell with the lowest V_c will be lastly switched in. This article mainly analyse the condition of $I_1 < 0$, where

$$I_1 < 0 \rightarrow \frac{v_p - v_s - 2Dv_p}{4Lf_s} - \frac{N-1}{2L_s}(v_p + v_s)T_w < 0 \tag{3}$$

Component sizing

The choice of sub-module capacitance C_{PM} is of great importance in MMC-DAB converter, since it determines the voltage overshoot which could be potentially harmful for semiconductor devices [17]. Thus, we consider the worst case for voltage overshoot, where phase shift reaches maximum, $D = 0.5$, exhibiting the highest current. Meanwhile, the capacitor voltage rise is closely related to the energy released from leakage inductor L_s. Thus, we have,

$$C_{PM} = \frac{2N}{\ell V_{DC}} \int_0^{(N-1)T_w} \left(I_4 + \frac{1}{L_s} \int_0^t (v_p - v_s)dt \right) dt \tag{4}$$

where $\ell = \delta V / V_{cn}$, δV is the voltage overshoot, and V_{cn} is the nominal cell voltage. As shown in Fig 2, according to the values of v_p, v_s, we can derive the designed equation for C_{PM},

$$C_{PM} \geq \begin{cases} \dfrac{N(N-1)T_w}{12\ell L_s}\left(3DT_s - \dfrac{3T_s}{2} + \dfrac{3v_sT_s}{2v_p} - \dfrac{2v_s}{v_p}(N-1)T_w\right) & , v_p \leq v_s \\[3mm] \dfrac{N(N-1)T_w}{12\ell L_s}(3DT_s - 2(N-1)T_w) & , v_p = v_s \\[3mm] \dfrac{N(N-1)T_w}{12\ell L_s}\left(\dfrac{T_s}{2} - \dfrac{T_sv_s}{2v_p} - (N-1)T_w + \dfrac{Dv_sT_s}{v_p} - \dfrac{v_s}{v_p}(N-1)T_w\right) & , v_p > v_s \end{cases} \tag{5}$$

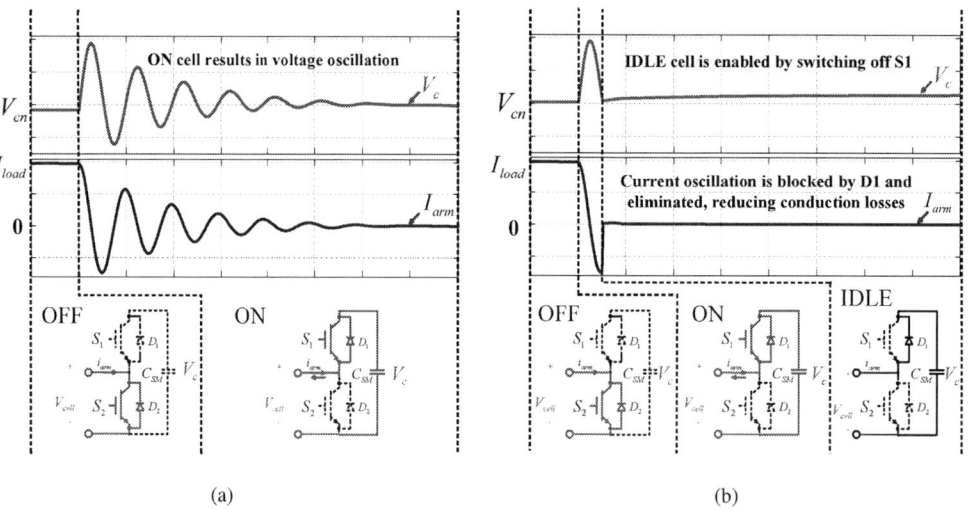

(a)   (b)

Fig. 4: The capacitor voltage V_c and arm current i_{arm} of MMC-DAB converter regarding different cell states under (a)Normal operation, (b)Algorithm 1.

Optimal Switching Algorithms for Gating Signals

In regular MMC, the cell output voltage can be regulated to V_{cn} or 0 by adjusting states of two devices. During "ON" state, S_1 turns on while S_2 turns off, leading to $V_{cell} = V_{cn}$. While during "OFF" state, the

S_1 turns off and S_2 turns on result in $V_{cell} = 0$. Conventionally the gating signals of two IGBT devices will be complementary, always keeping the aggregate number of "ON" cells and "OFF" cells to be N, where $N_{on} + N_{off} = N$.

MMC Mode (Algorithm 1)

Based on regular MMC, "Idle" state is enabled, where both of S_1 and S_2 are switched off, and the discharging current of submodule capacitor C_{PM} can be blocked by the diode D_1. As a result, LC resonance is forced to stop, and cell capacitor voltage V_c will remain constant. The switch off operation for S_1 should only happen during discharging process, otherwise the current could still charge the capacitor through diode D_1. Fig. 4 demonstrates the corresponding cell states for one switching instance under normal operation and Algorithm 1. There will be voltage oscillation caused by LC resonance under normal operation, where cell state is always "ON". When Algorithm 1 is applied, S_1 will be switched off when V_c oscillates back to nominal voltage V_{cn} after the first oscillation cycle. The energy stored in the inductance is returned to the DC supply when S_1 is turned off.

DAB Mode (Algorithm 2)

In order to achieve ZVS in MMC-DAB, S_1 can be turned on during all operating time, allowing C_{PM} to be fully charged and discharged. In this mode, C_{PM} is equivalent to the snubber capacitance in conventional DAB converter. Fig. 5(a) illustrates the power flow within the stack. For turn off transient, as $V_c = 0$ there will be no overlap between device voltage and current. For turn on transient, the S_{2L} of complementary arm will be switched off first, which results in power transfer between two sub-module capacitors. Thus the cell capacitor will be fully discharged, and anti-parallel diode conducting, resulting in zero voltage switching. Based on different delay time T_d between S_{2U} and S_{2L}, there will be a minimum load current for C_{PM} to be fast charged and discharged. According to Fig. 5(a),

$$V_1 + V_2 + L_{arm}\frac{dI_1}{dt} - L_{arm}\frac{dI_2}{dt} = V_{DC}; \quad I_1 + I_2 = I_{load}; \quad C_{equi}\frac{dV_1}{dt} = I_1; \quad C_{equi}\frac{dV_2}{dt} = -I_2 \tag{6}$$

where $C_{equi} = C_{PM}/N$, and $\omega_d = 1/\sqrt{L_{arm}C_{equi}}$. Then the discharging equation would be

$$V_2 = \frac{I_{load}}{2\omega_d C_{equi}}\sin(\omega_d t) - \frac{I_{load}t}{2C_{equi}} + V_{cn} \tag{7}$$

The derivation of required minimum load current, or duty ratio D_{min}, would be

$$\begin{cases} D_{min} = \dfrac{-4C_{equi}V_{cn}}{v_p\left(\omega_d\sin\left(\frac{T_d}{\omega_d}\right) - T_d\right)} + \dfrac{-v_p + v_s}{2v_p} & \text{For primary bridge} \\[4mm] D_{min} = \dfrac{-4C_{equi}V_{cn}}{v_p\left(\omega_d\sin\left(\frac{T_d}{\omega_d}\right) - T_d\right)} + \dfrac{v_p - v_s}{2v_p} & \text{For secondary bridge} \end{cases} \tag{8}$$

Additionally, there should be constraint on T_d for certain load current, to avoid reverse current, where C_{PM} could be charged again. Based on (1), the value of $T_{d\max}$ can be derived by letting $i_L = 0$,

$$i_L = \frac{2Dv_s + v_p - v_s}{4L_s f_s} - \frac{(N-1)T_w}{2L_s}(v_p + v_s) - \frac{v_p + v_s}{L_s}T_d = 0$$

$$T_{d\max} = -\frac{(N-1)T_w}{2} + \frac{2Dv_s + v_p - v_s}{4f_s(v_p + v_s)} \tag{9}$$

Based on (8) and (9), Fig. 5(b) shows the ZVS range for varying chosen delay time T_d.

Algorithm 3

By combining Algorithm 1& 2, the improved optimal gating signals in DAB mode (Algorithm 3) will switch off S_1 when capacitor voltage returns to V_{cn}. Fig 6 contrasts the current flow (red arrow) of cells

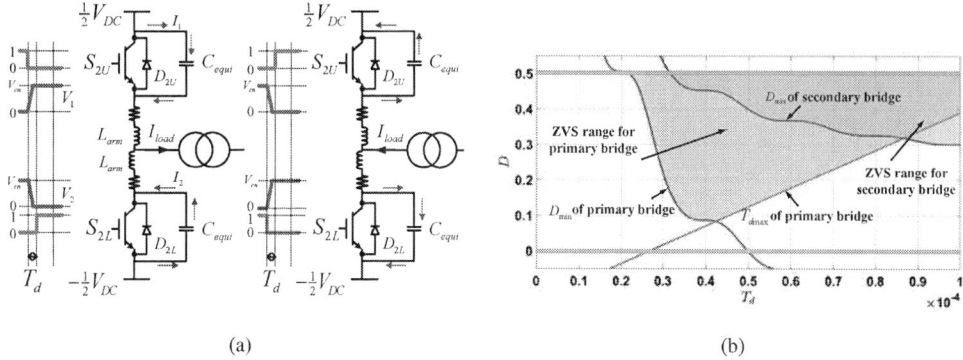

(a)   (b)

Fig. 5: When Algorithm 2 is enabled, the power flow diagram during dead-band between upper S_{2U} and lower S_{2L}, with charging current (red solid line) and discharging current (blue solid line). (b)ZVS range of primary (blue area) and secondary bridge (grey area) under varying T_d.

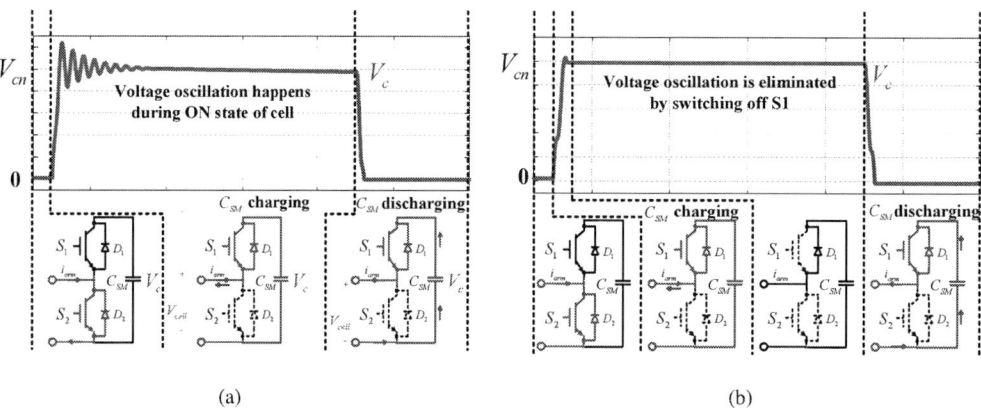

(a)   (b)

Fig. 6: The capacitor voltage V_c of MMC-DAB converter regarding different cell states under (a)Algorithm 2, (b)Algorithm 3.

under Algorithm 2 and 3. As noted from Fig. 6(b), the switching off of S_1 can block discharging current, and thus the voltage variation caused by LC resonance will be eliminated. Voltage balancing algorithm becomes unnecessary under Algorithm 2& 3, as V_c will be fully discharged at the end of each switching cycle. Fig. 7 gives specific flowchart of applying each optimal switching algorithms.

Simulated Results

To begin with, a DAB converter employing MMC cells has been set up in Simulink, which operates under the voltage range of -2 kV to 2 kV. Based on (4) and (5), the value of C_{PM} can be determined, where the detailed parameters are shown in Table I.

Table I: Designed circuit parameters of MMC-DAB converter and optimal algorithms.

Power rating	Values	Circuit parameter	Values
DC source voltage (V_{dc})	4kV	Number of cells per arm (N)	4
Output voltage (V_0)	2kV	Submodule capacitance (C_{SM})	60 μF
Sub-module nominal voltage (V_{cn})	1kV	Arm impedance (R_{arm})	10 mΩ
Duty ratio D	0.4	Arm inductance (L_{arm})	2 μH
Dwell time (T_w)	5 μs	Leakage inductance L_s	1mH
Operating frequency (f_s)	1 kHz		

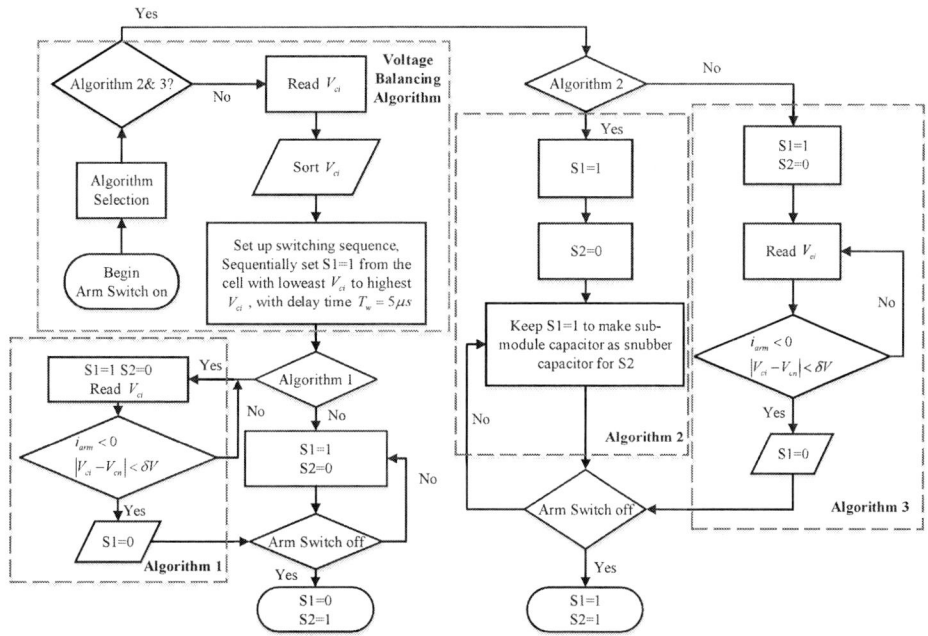

Fig. 7: The flowchart of applying proposed optimal switching algorithms in MMC-DAB converter.

Output Characteristics under Optimal Gating Signals

Fig. 8 contrasts the output and cell capacitor voltage under trapezoidal modulated MMC complementary gating signals and optimal gating signals in MMC mode (Algorithm 1). It indicates that there will be no more voltage oscillation in output waveform after one cycle of resonance. One thing has to be noticed is that not all capacitor voltage is returning to the voltage level within the statutory limit. This is because if one cell is switched to Idle, negative charging current for all cells will be blocked since they are connected in series. Although some cell capacitor voltages are not returned to the statutory limit, voltage balancing control can guarantee the stability of cell capacitor voltage.

From Fig. 8(b), we can notice that there will be a spike in v_p. When S_1 is turned off, the negative current can conduct D_2 then result in $V_{cell} = 0$. The overall conversion efficiency will not be affected, where the efficiency gain is still achieved by applying Algorithm 1 as it has eliminated the power loss from resonance.

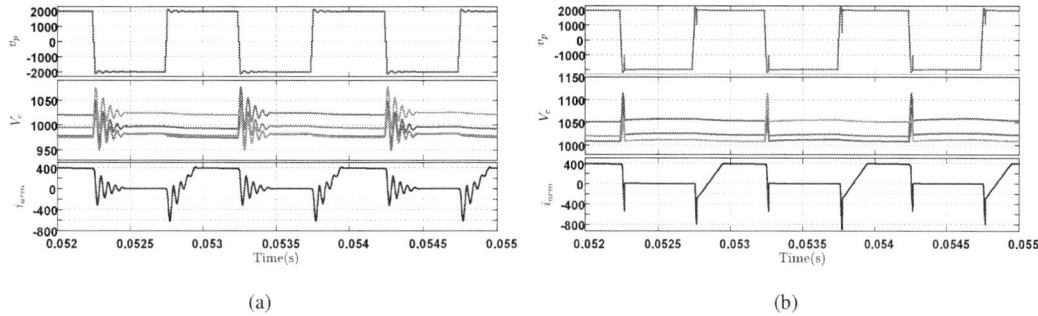

(a)   (b)

Fig. 8: Submodule capacitor voltage and arm current of MMC-DAB converter under (a)normal operation, (b)optimal switching Algorithm 1.

Fig. 9 contrasts the output and cell capacitor voltage operated under Algorithm 2 and 3, which is under DAB mode. The cell capacitor will be charged and fully discharged every switching cycle. The submodule voltage can sill be stabilized although there is no voltage balancing control. Under Algorithm 3, the LC damping has been eliminated by switching off S_1 as V_c approaches V_{cn}.

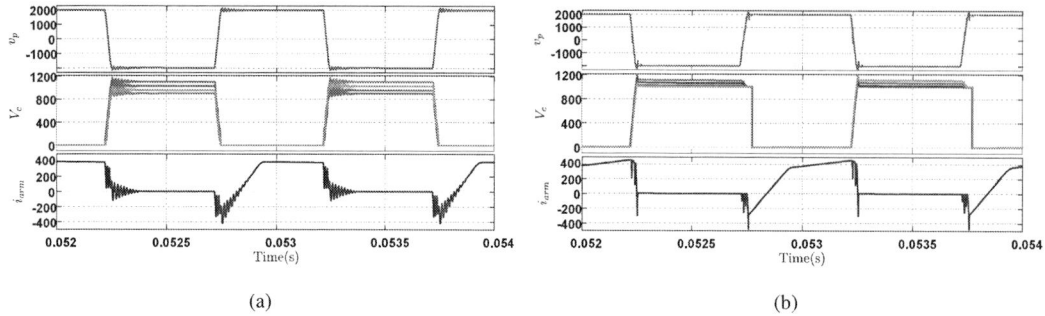

(a)   (b)

Fig. 9: Submodule capacitor voltage and arm current of MMC-DAB converter under (a)Algorithm 2, (b)Algorithm 3.

Switching loss analysis

A simulink-simscape model was set up to calculate the switching losses of semiconductor devices. Fig. 10 gives simulated result of the switching transient under normal operation and Algorithm 2. For normal operation, during one switching cycle, there will be two hard switches, especially the turn off transient of S_2 would result in 95.72mJ losses. Under Algorithm 2, C_{PM} is utilized as snubber capacitor for the turning off transient of S_2, and switching losses have been significantly reduced as ZVS is achieved. The total switching loss of the converter under Algorithm 1 will be similar to that of normal operation, but the turn off process of S_1 causes more energy consumption as there is negative current flowing through the device during switching off. The turn off switching losses of S_1 increase from 11.78mJ to 101.37mJ. Nevertheless, the application of Algorithm 1 would not contribute to significant switching losses since only 1 S_1 needs to be switched off among the whole arm under Algorithm 1.

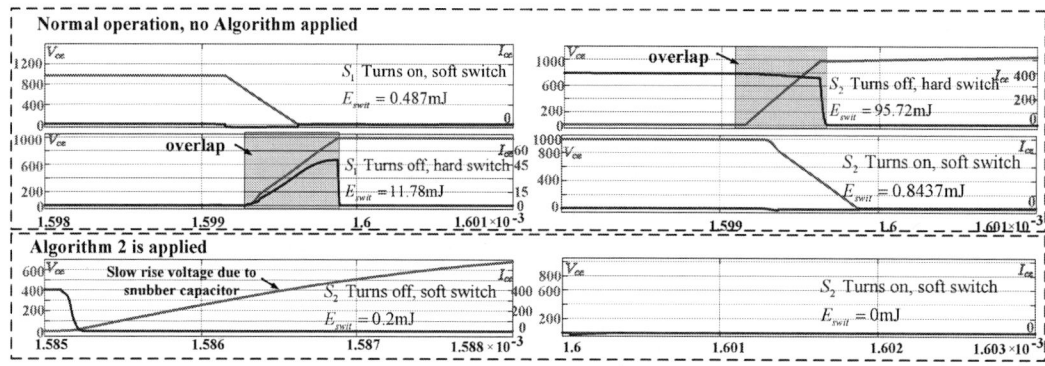

Fig. 10: The switching characteristics of S_1 and S_2 under normal operation and switching algorithm 2.

Table II gives detailed power flow and losses during one entire switching cycle for MMC-DAB converter. Under the proposed switching algorithms, the overall conversion efficiency has been improved. Although efficiency is improved under Algorithm 1, the output spike may potentially affect output stability. The effect could be mitigated when more cells are applied. More efficiency gain can be achieved for Algorithm 1 when intense resonance occurs, such as with higher L_{arm}. The system under Algorithm 2& 3 have higher conversion efficiency. However, there is limited operating range for adopting these two algorithms as they both require high load current and careful design of delay time T_d, to achieve rapid charging and discharging, subsequently ZVS. The employment of Algorithm 3 does not achieve efficiency gain as expected, since it has additional switching on and off transients for S_1.

Table II: Power losses of MMC-DAB converter under different operating switching algorithms.

Algorithms	Input power P_{in}	Conduction losses P_{cond}	Switching losses P_{swit}	Total losses P_{tot}	Efficiency η
Normal Operation	478kW	23.11kW	3.47kW	26.58kW	94.4%
Algorithm 1	477kW	17.7kW	3.83kW	21.53kW	95.5%
Algorithm 2	477kW	19.14kW	8W	19.15kW	95.9%
Algorithm 3	498kW	18.5kW	4.3kW	22.8kW	95.4%

Experimental Test

In order to verify the simulated result, a DAB converter employing MMC cells needs to be built. It remains impractical to build an MMC-DAB converter in industrial level, which typically rates at hundreds of Megawatts within a traditional laboratory. Thus, a scaled-down version of MMC-DAB converter was set up, where the proposed algorithms can be tested. The designed sub-module is in half-bridge operation, and rated at 12V. When 4 series-connected sub-modules of each stack are enabled, $\pm$ 24V square wave will be generated. C_{PM} is chosen as 4.7μF and L_{arm} is relatively large at 27μH to create enough oscillation which can be later tested and eliminated. f_s is chosen at 2kHz and L_s is set to 560μH. The allowed maximum i_L is 3A. Fig. 11(a) shows the circuit diagram of the experimental test, where only 1 cell is enabled in each arm. Thus, 2-level voltage waveforms with phase shift as shown in Fig. 11(b) across leakage inductance L_s are generated under normal operation. As only one cell of the arm is enabled, the v_p and v_s depict at $\pm24/4 = \pm6$V, with phase shift $D = 0.25$. The voltage oscillation is not obvious in this test indicates that mosfets with superior performance (specifically lower $R_{ds(on)}$) are required.

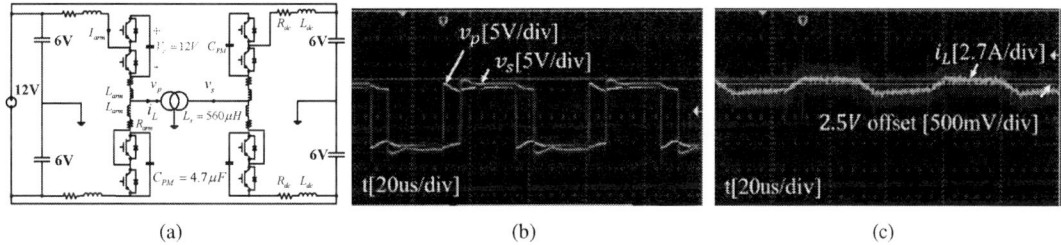

(a)   (b)   (c)

Fig. 11: (a)The circuit diagram of experimental test for 2-level MMC-DAB converter. Under normal operation, (b)the voltages across L_s, and (c)the current flowing through L_s, which is i_L.

The quasi 2-level voltages contribute to the typical current waveform of DAB, as shown in Fig. 11(c). The highest inductor current from the diagram is 670mA, which matches the derivation of (1). Besides, the current shows symmetrical characteristics which verifies that the MMC-DAB converter is under stable operation, and with power transfer from primary to secondary bridge. Generally, the result verifies the fundamental operation of DAB converter employing MMC cells. The later work will be focusing on the experiment of converters with full cells and optimal algorithms enabled.

Conclusion

This article firstly reviews the analysis and design of DAB converter employing MMC cells. Then, three advanced gating signals are proposed in order to mitigate the significant losses caused by LC resonance, and the switching losses. The overall conversion efficiency under these optimal gating signals are tested in Simulink, verifying the efficiency improvement brought by these algorithms. As for experiment, a scaled down MMC-DAB converter was established, DAB converter with 2 voltage levels has been tested, which can confirm the correct operation of DAB converter employing MMC structures. The future work will enable full MMC cells to generate trapezoidal waveform, and verify the efficiency improvement when optimal algorithms are applied.

References

[1] H. Wen, W. Xiao, and B. Su, "Nonactive power loss minimization in a bidirectional isolated DC–DC converter for distributed power systems," IEEE Trans. Ind. Electron., vol. 61, no. 12, pp. 6822–6831, Dec. 2014.

[2] J.-W. Yang and H.-L. Do, "Soft-switching dual-flyback DC–DC converter with improved efficiency and reduced output ripple current," IEEE Trans. Ind. Electron., vol. 64, no. 5, pp. 3587–3594, May 2017.

[3] Y. Hu, R. Zeng, W. Cao, J. Zhang and S. J. Finney, "Design of a Modular, High Step-Up Ratio DC–DC Converter for HVDC Applications Integrating Offshore Wind Power," in IEEE Transactions on Industrial Electronics, vol. 63, no. 4, pp. 2190-2202, April 2016.

[4] B. Han, C. Bai, J. S. Lee, and M. Kim, "Repetitive controller of capacitor-less current-fed dual-half-bridge converter for grid-connected fuel cell system," IEEE Trans. Ind. Electron., vol. 65, no 10, pp. 7841–7855, Oct. 2018.

[5] B. Zhao, Q. Song, W. Liu, and Y. Sun, "Overview of dual-active-bridge isolated bidirectional dc-dc converter for high-frequency-link power-conversion system," IEEE Transactions on Power Electronics, vol. 29, no. 8, pp. 4091-4106, Aug 2014

[6] H. Shi et al., "Minimum-backflow-power scheme of DAB-based solidstate transformer with extended-phase-shift control," IEEE Trans. Ind. Appl., vol. 54, no. 4, pp. 3483–3496, Jul. 2018.

[7] Q. Ye, R. Mo, and H. Li, "Low-frequency resonance suppression of a dual-active-bridge DC/DC converter enabled DC microgrid," IEEE J. Emerg. Sel. Topics Power Electron., vol. 5, no. 3, pp. 982–994, Sep. 2017.

[8] L. Gill, T. Ikari, T. Kai, B. Li, K. Ngo and D. Dong, "Medium Voltage Dual Active Bridge Using 3.3 kV SiC MOSFETs for EV Charging Application," 2019 IEEE Energy Conversion Congress and Exposition (ECCE), 2019, pp. 1237-1244.

[9] T. Lüth, M. Merlin and T. Green, "A DC/DC converter suitable for HVDC applications with large step-ratios," 2014 IEEE Energy Conversion Congress and Exposition (ECCE), 2014, pp. 5331-5338.

[10] B. Zhao, Q. Song, J. Li, X. Xu, and W. Liu, "Comparative analysis of multilevel-high-frequency-link and multilevel-dc-link dc-dc transformers based on mmc and dual-active bridge for mvdc application," IEEE Transactions on Power Electronics, vol. 33, no. 3, pp. 2035-2049, March 2018

[11] I. Gowaid, G. P. Adam, S. Ahmed, D. Holliday, and B. W. Williams, "Analysis and design of a modular multilevel converter with trapezoidal modulation for medium and high voltage dc-dc transformers," IEEE Transactions on Power Electronics, vol. 30, no. 10, pp. 5439-5457, 2015.

[12] P. Xia, H. Shi, H. Wen, Q. Bu, Y. Hu and Y. Yang, "Robust LMI-LQR Control for Dual-Active-Bridge DC–DC Converters With High Parameter Uncertainties," in IEEE Transactions on Transportation Electrification, vol. 6, no. 1, pp. 131-145, March 2020.

[13] Y. Zhong, N. Roscoe, D. Holliday, T. C. Lim and S. J. Finney, "High-Efficiency mosfet-Based MMC Design for LVDC Distribution Systems," in IEEE Transactions on Industry Applications, vol. 54, no. 1, pp. 321-334, Jan.-Feb. 2018.

[14] S. Shao, M. Jiang, J. Zhang and X. Wu, "A Capacitor Voltage Balancing Method for a Modular Multilevel DC Transformer for DC Distribution System," in IEEE Transactions on Power Electronics, vol. 33, no. 4, pp. 3002-3011, April 2018.

[15] Z. Lu, L. Lin, X. Wang and C. Xu, "LLC-MMC Resonant DC-DC Converter: Modulation Method and Capacitor Voltage Balance Control Strategy," 2020 IEEE Applied Power Electronics Conference and Exposition (APEC), 2020, pp. 2056-2061.

[16] B. Zhao, Q. Song, J. Li, Y. Wang and W. Liu, "High-Frequency-Link Modulation Methodology of DC–DC Transformer Based on Modular Multilevel Converter for HVDC Application: Comprehensive Analysis and Experimental Verification," in IEEE Transactions on Power Electronics, vol. 32, no. 5, pp. 3413-3424, May 2017.

[17] M. M. C. Merlin, T. C. Green, P. D. Mitcheson, F. J. Moreno, K. J. Dyke and D. R. Trainer, "Cell capacitor sizing in modular multilevel converters and hybrid topologies," 2014 16th European Conference on Power Electronics and Applications, 2014, pp. 1-10.

Systematic Adaptive Robust State Feedback Control for Active Front-End Rectifiers

Aidar Zhetessov, Giri Venkataramanan
UNIVERSITY OF WISCONSIN-MADISON
1415 Engineering Dr.
Madison, WI 53706, United States
Phone: +1 (608) 262-3934
Email: zhetessov@wisc.edu, giri@engr.wisc.com
URL: https://wempec.wisc.edu/

Acknowledgments

The authors gratefully acknowledge the members of the Wisconsin Electric Machines and Power Electronics Consortium (WEMPEC) for supporting this work.

Keywords

≪Active Front-End≫, ≪State-space model≫, ≪Small signal stability≫, ≪Adaptive control≫, ≪Robustness≫.

Abstract

This paper presents the systematic use of physics-based frequency-domain techniques to design the adaptive and robust state-feedback controller for an AFE rectifiers. The proposed design approach ensures robustness and stability of the controller and the converter, independent of the size of the DC-link capacitance. This will enable the optimization of the power circuit from the perspective of efficiency, power density, cost, etc. with the widest possible design space. Furthermore, the control structure is operating-point adaptive, optimally robust and computationally efficient due to closed-form controller representation. Detailed analytical development of the controller is presented in the paper along with a proof for robustness verified using simulations that support the theoretical findings.

1 Introduction

Three-phase six-switch Active Front-End (AFE) rectifiers are the active ac-dc power electronic converters that can be used in various applications to regulate the output dc voltage, keep high ac-side power factor, compensate for the reactive power, or even do primary grid-forming support [1]-[3]. There is an ongoing industrial trend towards constrained multi-objective optimization of power converters, including the AFE rectifiers [4]. One of the major enablers of this trend is the recent advent of wide band-gap (WBG) semiconductor technology [5]. With higher switching frequencies of WBG devices the passive component values can be reduced while preserving the acceptable current/voltage ripple during steady-state operation, thus opening the path towards further passives reduction/optimization [5]. This paper is focused on examining the relationship between the dynamic properties of control of the converter and the size of the reactive components.

While various modulation/control methods have been developed for the AFE rectifiers [2], [7]-[9], conventional cascaded dq-frame PI regulator design has been widely adopted [2], [6]. Furthermore, it is widely known and acknowledged that AFE rectifiers feature a right half plane (RHP) zero arising from its parent boost converter topology. In order to prevent any dynamic instabilities caused by the non-minimal phase response due to the RHP zero over the complete load range, constraints on passive components sizing were proposed in [6], as reviewed in brief detail in Section 2.

It should be noted that in [6] rather conservative regulator-induced constraints were imposed over and beyond the current/voltage ripple limits on passives even for relatively low switching frequencies. Thus, even at low switching frequencies, the potential for miniaturization of passive components was not properly utilized due to the overly conservative constraints. When higher frequency operation is enabled by WBG devices, such regulator-induced constraints on passive component values would constitute even greater barrier on the path towards system optimization. The goal of the approach presented in this paper is to overcome such constraints through a systematic design of the commonly used cascaded controller structure with an inner current regulator and an outer voltage regulator, illustrated further in Section 3.

While this work preserves the classical control structure it employs various "physics-based" regulator modifications [10]-[12] as detailed in Section 3. Section 4 presents eigenvalue migration of the system with load variation illustrating the designed system stability and adaptability that overcome the dynamic constraints on passive components, along with detailed simulations. Although the evolution of controller parameters is carried out in frequency domain, a parallel viewpoint is taken in Section 5 using state-space modeling, that provides guarantees on overall system robustness, adaptability, transient performance, and computational efficiency. The paper concludes with a section summarizing the contributions.

2 Preliminaries: AFE Rectifier Modeling & Dynamic Constraint

First, we revisit the model of the grid-connected three-phase AFE rectifier, realized as a standard six-switch two-level three-phase topology. An example schematic of the rectifier with resistive load is shown in Fig. 1. Note that in this control-oriented work the AC-side L-filter instead of LCL-filter was considered to focus on control instead of modeling/EMI. The initial nonlinear plant model and electrical parameters are taken from [6]. Table I summarizes the respective specifications, including the minimum DC-link capacitance derived in [6].

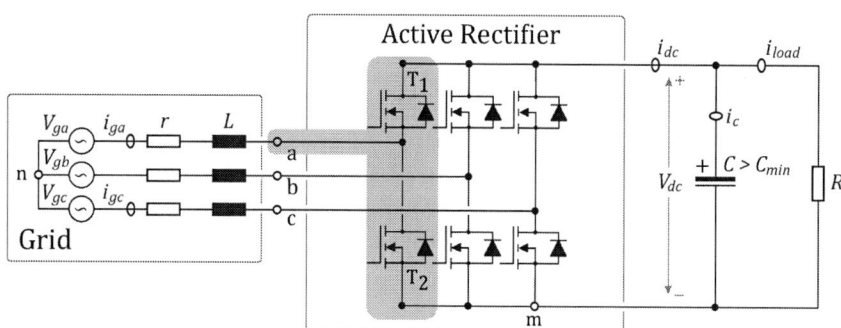

Fig. 1: Active Front End Rectifier Schematic Diagram

Table I: System Specifications and Operating Point Parameters [6]

Parameter	Smbl	Value	Unit	Parameter	Smbl	Value	Unit
AC Input LL RMS Volt.	V_{gLL}	230	Vrms	Sw. Frequency	f_{sw}	10	kHz
DC-bus Voltage	V_{dc}	400	V	Current Ctrl BW*	f_i^*	1	kHz
Rated (Max) Power	P	25	kW	Voltage Ctrl BW*	f_v^*	100	Hz
Rated Load Resist.	R	6.4	Ω	Op. Point d-Crnt	I_{gd}	88.96	A
Grid Frequency	f_o	60	Hz	Op. Point q-Crnt	I_{gq}	0	A
Input Inductance	L	0.34	mH	Op. Point d-Duty	M_d	0.4684	-
Minimum DC Cap [6]	C_{min}	505	μF	Op. Point q-Duty	M_q	-0.0285	-
L Series Resistance	r	5	mΩ	Cmmn-Mode Duty	M_{CM}	0.5	-

After perturbing and linearizing that nonlinear model in dq-frame one can arrive at the following steady-state operating point relations (1) and state-space plant model (2):

$$M_d V_{dc} = V_{gd} - I_{gd} r, \qquad M_q V_{dc} = -\omega_0 L I_{gd}, \qquad P = \frac{V_{dc}^2}{R} = \frac{3}{2}\left(M_d V_{dc} I_{gd}\right), \qquad I_{gq} = 0 \qquad (1)$$

$$\underbrace{\begin{bmatrix} \dot{\tilde{i}}_{gd} \\ \dot{\tilde{i}}_{gq} \\ \dot{\tilde{v}}_{dc} \end{bmatrix}}_{\dot{x}} = \underbrace{\begin{bmatrix} -\frac{r}{L} & \omega_0 & -\frac{M_d}{L} \\ -\omega_0 & -\frac{r}{L} & -\frac{M_q}{L} \\ \frac{3M_d}{2C} & \frac{3M_q}{2C} & -\frac{1}{CR} \end{bmatrix}}_{A} \underbrace{\begin{bmatrix} \tilde{i}_{gd} \\ \tilde{i}_{gq} \\ \tilde{v}_{dc} \end{bmatrix}}_{x} + \underbrace{\begin{bmatrix} -\frac{V_{dc}}{L} & 0 \\ 0 & -\frac{V_{dc}}{L} \\ \frac{3I_{gd}}{2C} & \frac{3I_{gq}}{2C} \end{bmatrix}}_{B_1} \underbrace{\begin{bmatrix} \tilde{m}_d \\ \tilde{m}_q \end{bmatrix}}_{u_1} + \underbrace{\begin{bmatrix} \frac{1}{L} & 0 \\ 0 & \frac{1}{L} \\ 0 & 0 \end{bmatrix}}_{B_2} \underbrace{\begin{bmatrix} \tilde{v}_{gd} \\ \tilde{v}_{gq} \end{bmatrix}}_{u_2}$$

$$\underbrace{\begin{bmatrix} \tilde{v}_{dc} \\ \tilde{i}_{gq} \end{bmatrix}}_{y} = \underbrace{\begin{bmatrix} 0 & 0 & 1 \\ 0 & 1 & 0 \end{bmatrix}}_{C} \underbrace{\begin{bmatrix} \tilde{i}_{gd} \\ \tilde{i}_{gq} \\ \tilde{v}_{dc} \end{bmatrix}}_{x}, \qquad \mathbf{D = 0} \tag{2}$$

Here the state variables were selected to be small-signal grid dq-frame currents $\tilde{i}_{gd}, \tilde{i}_{gq}$ and DC voltage $\tilde{v}_{dc}$. The input B matrix was split in two parts to differentiate between manipulated inputs (small-signal modulation indices $\tilde{m}_d, \tilde{m}_q$) and disturbances (small-signal grid voltage disturbances $\tilde{v}_{gd}, \tilde{v}_{gq}$). M_d, M_q, I_{gd}, $I_{gq} = 0$, V_{dc}, V_{gd} and $V_{gq} = 0$ are the operating point large-signal parameters, around which the linearization was performed. At nominal conditions and in dq-frame aligned with grid v_{gd} voltage, these steady state parameters are evaluated in Table I. The common-mode duty cycle M_{CM} reflects the steady state voltage between the grid star point and the lower rail of the rectifier. For a standard PWM modulation assumed in this work, the mentioned voltage is kept constant at $V_{CM} = M_{CM}V_{dc} = 0.5V_{dc}$. This model serves as the case study for the rest of the work. For now, the underlying assumptions of the model are: undisturbed grid voltages, perfect synchronization with grid voltages via PLL, availability of state measurements, perfect knowledge of system parameters, and constant linearized operating point (OP). To be sure, many of these assumptions will be relaxed in the next sections while we evaluate system adaptability and robustness.

Regarding the dynamic constraint, [6] has found the small-signal stability constraint on DC-link capacitance for an AFE rectifier regulated by a conventional cascaded PI control. It turns out that, provided the AC-side inductance $L \geq L_{min}$ (current THD requirement) and maximum power throughput that corresponds to a minimum load resistance R_{min}, the DC-link capacitance should be greater than a minimum value for small-signal stability over the entire load range. That minimum value was derived as follows: $C \geq C_{min} = 20 \cdot L / [M_{d(R_{min})}^2 R_{min}^2]$. Often, in practical designs, the minimum capacitance could be dictated by other requirements such as maximum voltage ripple, ESR, loss, ride-through, etc. However, with the advent of WBG devices, the aforementioned constraints may not be the barriers anymore [5], opening the room for further reduction in capacitor sizing. In fact, as we will see next, even for relatively low switching frequency [6] the minimum capacitance is already dictated by the control-inducted constraint above instead of maximum voltage ripple. As f_{sw} goes up this will become even more true, since from voltage ripple standpoint higher f_{sw} requires even smaller C for the same ripple. The following section takes a systematic design approach for the regulator design using various decoupling techniques in order to remove the aforementioned control-induced constraint.

3 Physics-based Controller Design in Frequency Domain

State Feedback (SFB) control method pertains to the linear systems theory and time-domain state-space framework. The state-space framework, although being extremely powerful, might sometimes obscure the physical insight of the control action behind various theorems, eigenpairs, system norms and numerical methods (see next sections). On the other hand, a "physically insightful" controller design techniques have been developed in frequency domain with feedback of state variables, notably for electric machine drives [10]-[12]. This section employs such techniques for the conventional cascaded controller design [6] for the small-signal Frequency Domain (FD) SFB controller. Although the controller design occurs in frequency domain, it is possible to obtain the equivalent state-space representation of the system in order to benefit from both frameworks.

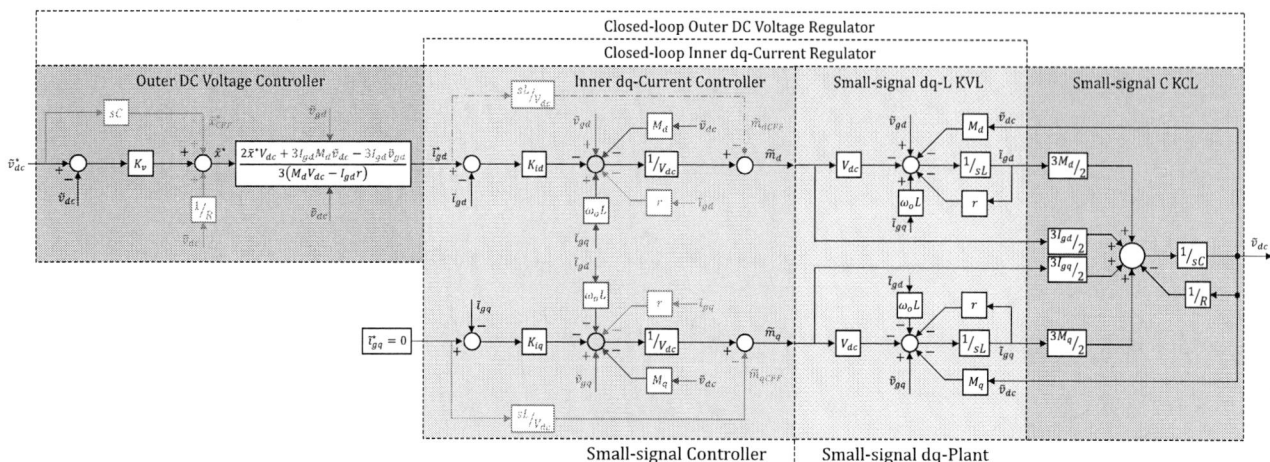

Fig. 2: Block diagram illustrating the augmentations to the classical cascaded controller for the AFE converter using the physics-based techniques: Disturbance Input (DID) / State Feedback (SFBD) / Virtual Zero Reference (VZRD) Decoupling, Command Feedforward (CFF).

The classical cascaded controller (in black) with "physics-based" modifications (in color) for the AFE plant is depicted in the closed-loop block diagram of Fig. 2. In the gray regions on the right one can find the detailed block diagram of the AFE plant small-signal model (2). Unlike the conventional lumped transfer function (TF) representation, the plant block diagram is broken down to its smallest building blocks that provide physical insight. Blue regions on the left are the controller blocks with similar details. The integrator gains were removed from the conventional controller in black due to the SFB nature of the control obviating the need for an integrator [10]. Classical structure uses d- and q-manipulated inputs to regulate d- and q-current components respectively. The q-current reference is zero for unity power factor, while the d-current reference is dictated by the outer cascaded DC voltage controller.

Four frequency domain techniques are used to bring the classical controller to the (optimal) SFB form: Disturbance Input Decoupling (DID), State Feedback Decoupling (SFBD), Virtual Zero Reference Decoupling (VZRD), and the Command Feedforward (CFF). DID attempts to cancel the external disturbances by measuring and applying the same and opposite signals through manipulated inputs. In the AFE case, grid voltage small-signal disturbances could be decoupled as shown in Fig. 2. SFBD attempts to cancel the influence of all other states on the given state dynamics by measuring and applying the same and opposite signals through manipulated inputs. For example, from the plant in Fig. 2 it can be seen that the $\tilde{i}_{gd}$ dynamics are influenced by the $\tilde{v}_{dc}$ through M_d and the appropriate SFBD in blue decouples this effect. VZRD can be regarded as the subset of SFBD, which accounts for the influence of the state itself on its own dynamics. Together, DID, SFBD, and VZRD, starting from the inner-loop towards outer-loop dynamics, simplify the plant such that the respective loop-gain dynamics become first-order and second-order integrators respectively. Those integrator plants can be controlled with near-zero steady state errors using only the proportional gain, hence integrator in controls is redundant. For example, focusing only on the $\tilde{i}_{gd}$ dynamics, DID, SFBD, and VZRD cancel all the influences on $\tilde{i}_{gd}$, leaving simple $1/sL$ as the (sub)plant to be controlled. By choosing the gain $K_{id} = \omega_i L$, one simply places the closed-loop pole of the inner $\tilde{i}_{gd}$ loop to the desired location ω_i. Finally, CFF accounts for potentially non-constant commands into the respective loops and feeds those forward to the manipulated inputs to ensure the best command tracking (no phase lag and signal attenuation). Inner d-axis CFF may not be easily realizable due to differentiation of measurements coming from outer loop.

Once all the controller modifications of classical PI regulator are made, one can tune the respective cascaded loops to the desired bandwidths that are listed in Table I. Q-axis current loop gain:

$$T_q(s) = \frac{\tilde{i}_{gq}}{(\tilde{i}_{gq}^* - \tilde{i}_{gq})} = \frac{K_{iq}}{sL} \quad \rightarrow \quad K_{iq} = 2\pi f_i L = \omega_i L \tag{3}$$

D-axis cascaded closed-loop TF and gains (assuming perfect decoupling between inner and outer loops):

$$G_d(s) = \frac{\tilde{v}_{dc}}{\tilde{v}_{dc}^*} = \frac{1 - s\frac{I_{gd}L}{M_d V_{dc}}}{\frac{s^2 CL}{K_{id}K_v} + \frac{sC}{K_v} + 1} = \frac{1 - \frac{s}{\omega_z}}{\frac{s^2}{\omega_i \omega_v} + \frac{s}{\omega_i || \omega_v} + 1} \rightarrow K_{id} = (\omega_i + \omega_v)L, \; K_v = C(\omega_i || \omega_v) \quad (4)$$

Without the assumption above, the TF and gains become OP-dependent ($r \approx 0$, see Table I):

$$G_d(s) = \frac{\tilde{v}_{dc}}{\tilde{v}_{dc}^*} = \frac{1 - s\frac{I_{gd}L}{M_d V_{dc}}}{s^2 \frac{CL}{K_{id}K_v} + s\left(\frac{2L}{K_{id}K_v R} + \frac{2I_{gd}L}{K_v M_d R V_{dc}} + \frac{C}{K_v} - \frac{I_{gd}L}{M_d V_{dc}}\right) + 1} = \frac{1 - \frac{s}{\omega_z}}{\frac{s^2}{\omega_i \omega_v} + \frac{s}{\omega_i || \omega_v} + 1}$$

$$\rightarrow K_{id} = \frac{CM_d R V_{dc}(\omega_i + \omega_v) + CI_{gd}LR\omega_i\omega_v - 2M_d V_{dc}}{2I_{gd} + CM_d R V_{dc}L^{-1}}, \; K_v = LC\omega_i\omega_v K_{id}^{-1}, \; \omega_z = \frac{M_d V_{dc}}{LI_{gd}} \quad (5)$$

Finally, multiplying out all the gains for $\tilde{v}_{dc}^* = 0$ in Fig. 2, the manipulated inputs become:

$$\tilde{m}_d = k_{11}\tilde{i}_{gd} + k_{12}\tilde{i}_{gq} + k_{13}\tilde{v}_{dc}, \quad \tilde{m}_q = k_{21}\tilde{i}_{gd} + k_{22}\tilde{i}_{gq} + k_{23}\tilde{v}_{dc} \quad (6)$$

Where the gains k_{ij}, based on the controller gains (5) and Table I specs, are as follows:

$$k_{11} = \frac{K_{id} - r}{V_{dc}} = 5.3545 mA^{-1}, \qquad k_{12} = \frac{\omega_o L}{V_{dc}} = 0.3204 mA^{-1}$$

$$k_{13} = -\frac{M_d}{V_{dc}} + \frac{2K_{id}V_{dc}[K_v - R^{-1}] - 3K_{id}I_{gd}M_d}{3V_{dc}[M_d V_{dc} - I_{gd}r]} = -1.1458 mV^{-1} \quad (7)$$

$$k_{21} = -\frac{\omega_o L}{V_{dc}} = -0.3204 mA^{-1}, \; k_{22} = \frac{K_{iq} - r}{V_{dc}} = 5.3407 mA^{-1}, \; k_{23} = -\frac{M_q}{V_{dc}} = 0.0711 mV^{-1}$$

Given the state-space SFB form ($\dot{\mathbf{x}} = \mathbf{A}\mathbf{x} + \mathbf{B_1}\mathbf{K_{fd}}\mathbf{x}$), the state-space form of FD controller $\mathbf{K_{fd}}$ is:

$$\mathbf{K_{fd}} = \begin{bmatrix} 5.3545 & 0.3204 & -1.1458 \\ -0.3204 & 5.3407 & 0.0711 \end{bmatrix} \cdot 10^{-3} \quad (8)$$

4 Regulator Evaluation

To evaluate the SFB regulator design above, first, it should be evaluated whether the regulator relaxed the dynamic constraint on DC-link capacitance [6]. For that, one can study only the d-axis closed-loop (CL) system (4-5) - thanks to dq-axes decoupling and stabilizing the q-axis dynamics at the origin. From (5) one can derive the equivalent loop-gain TF, similar to (3):

$$T_d(s) = \frac{\tilde{v}_{dc}}{(\tilde{v}_{dc}^* - \tilde{v}_{dc})} = \frac{\left(\frac{2L}{K_{id}K_v R} + \frac{2I_{gd}L}{K_v M_d R V_{dc}} + \frac{C}{K_v}\right)^{-1}\left(1 - s\frac{I_{gd}L}{M_d V_{dc}}\right)}{s\left[1 + s\left(\frac{2I_{gd}K_{id}}{CM_d R V_{dc}} + \frac{2}{CR} + \frac{K_{id}}{L}\right)^{-1}\right]} \stackrel{(5)}{=} \frac{\left(\omega_z || \omega_i || \omega_v\right)\left(1 - \frac{s}{\omega_z}\right)}{s\left[1 + \frac{s}{\omega_i\omega_v}\left(\omega_i || \omega_v || \omega_z\right)\right]} \quad (9)$$

From (9) it is seen that two control gains K_{id}, K_v can be used to place the $T_d(s)$ crossover frequency and the open-loop pole to the locations corresponding to the CL voltage/current bandwidths of ω_v, ω_i respectively. Comparing (9) to equations (8)-(10) in [6], one can see that decoupling techniques removed the uncontrolled LHP pole and the integrator zero from the loop gain $T_d(s)$. Recalling that the dynamic constraint arose from the necessary frequency window between the uncontrolled LHP pole and the RHP zero, removing that pole essentially removes the constraint on DC-link capacitance. The RHP zero, being the inherent attribute of the boost-type topologies, is not affected by decoupling. Thus, it is still uncontrollable and still sets the upper bound on the CL bandwidth of the outer-loop voltage regulation.

To further evaluate the designed SFB regulator, closed-loop eigenvalue migration plots and transient simulations can be studied over the load range and for various DC-link capacitance values. For the eigenvalues, the closed-loop system matrices have been derived for both conventional [6] and SFB regulators. The eigenvalues of these matrices were evaluated and plotted (Fig. 3) over the range of DC-link capacitance values and load resistances. The capacitance was swept from the C_{min} value to $0.3C_{min}$, while the load was swept from the rated power $P = 25$ kW to 5 kW (Table I, Fig. 3).

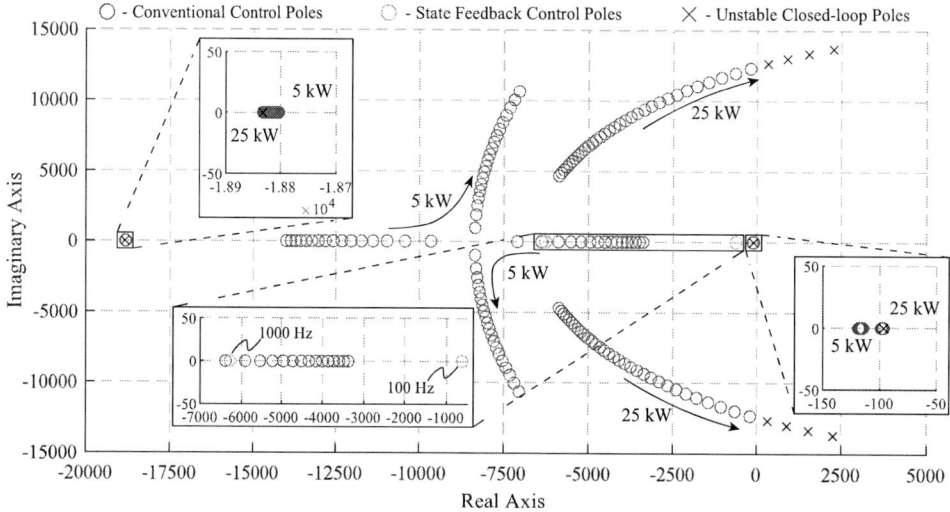

Fig. 3: Effect of DC-link ($C_{min} < C < 0.3C_{min}$) on the closed-loop stability for two control structures/loads.

The conventional system (without any of the decoupling) in Fig. 3 has four distinct CL poles, shown in blue. For 5kW, $C = C_{min}$ conditions, all these poles are along the real axis. When the capacitance is reduced, the poles at the lowest frequency and the highest frequency are relatively stationary. On the other hand, the two of the mid-frequency poles approach each other on real axis and become a complex conjugate pair after meeting each other as the capacitance decreases. For the 25kW case, this pair of poles starts out to be complex, ultimately leading to instability (RHP poles) at $C = 0.3C_{min}$, as predicted by [6]. On the other hand, the SFB system has only two distinct CL poles, that are shown in green. This is expected, as the desired CL bandwidths for both current loops were selected to be the same - 1 kHz (6.28 krad/s). The third pole of the voltage loop is located at 100 Hz (628 rad/s). Notably, the CL eigenvalues with the proposed controller using (7) are stationary irrespective of capacitance and/or load variation, provided that capacitance and load values are known.

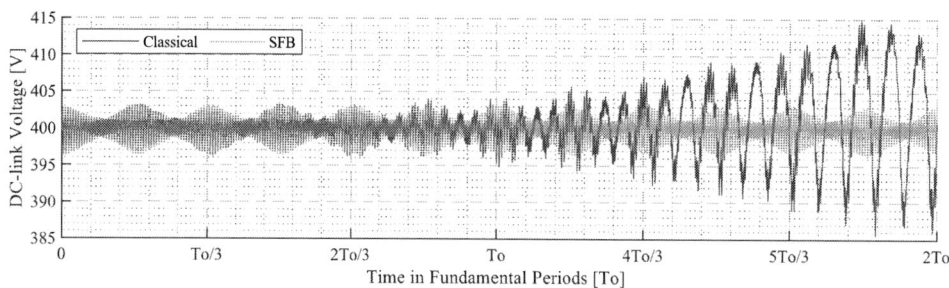

Fig. 4: Switched model simulation results for DC-link voltage at $C = 0.5C_{min}$ and Classical [6]/SFB controls.

A detailed simulation of the system using the switched nonlinear model was developed to verify the approach presented in the paper. Fig. 4 illustrates the dc bus voltage for conventional control (blue) and the proposed control (gray), illustrating the effective stabilization of the unstable operating condition. Looking at the gray waveform, note how small the voltage ripple envelope is - only $\pm 3V$ peak, which is less than 1% for a dc voltage of 400V. Recalling from Table I that simulated $f_{sw} = 10kHz$, one can

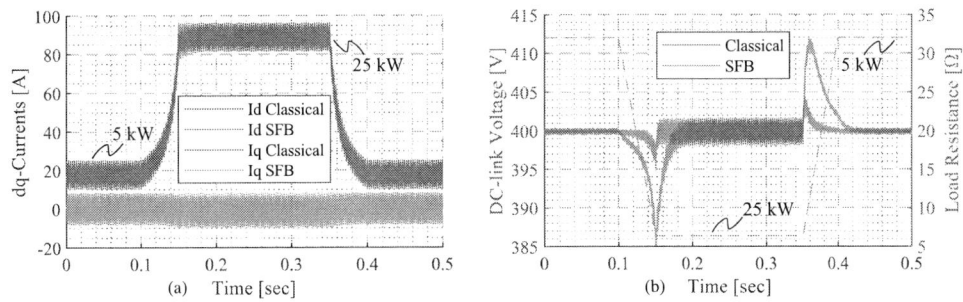

Fig. 5: Transient simulation results for (a) dq-axes currents, and (b) DC-link voltage for Classical and SFB controls. The R_{load} changes between 32Ω (5 kW) and 6.4Ω (25 kW). Capacitance value: $C = C_{min}$.

get an idea how large the simulated capacitance of $0.5C_{min}$ is. Indeed, even at such a low switching frequency the voltage ripple is still very small, implying unnecessarily large DC-link capacitance value. From voltage ripple perspective $0.5C_{min}$ is too large, yet from stability perspective we see that classical CL system (blue) is unstable, requiring even more capacitance for stability. As f_{sw} increases, this gap between voltage ripple and stability constraints is set to increase even further. Thus, one can see the importance of a proper control system - it "unleashes" the advantages of high switching frequency in passive component minimization (power density, efficiency, cost etc).

In addition to small-signal stability at reduced capacitance values, the designed SFB controller improves the transient system performance even for larger C values, without compromising the stability of conventional control. Fig. 5 shows large-signal transient simulation scenario where the load is rapidly varied between 5 and 25 kW. The capacitance is kept at $C = C_{min}$. It can be seen that for all three state variables of interest (i_{gd}, i_{gq}, v_{dc}) the transient overshoot, settling time, and command tracking are improved.

5   Insights on Robustness & Adaptability

So far the work addressed the SFB controller designed in frequency domain (FD). However, as it was mentioned, the SFB control originally pertains to the State-Space (SS) framework. The SS framework has the controller design tools that, provided controllability of the system, can place the closed-loop (CL) eigenvalues to any desired locations. In addition, if there are degrees of freedom (DOFs) in feedback matrix $\mathbf{K_{ss}}$, the tools can use those to achieve the desired characteristics. For instance, the numerical algorithm behind Matlab PLACE command uses the DOFs to minimize the upper bound on sensitivities of CL poles to small perturbations in $\mathbf{A}$ and $\mathbf{B_1}$, thus maximizing small-signal robustness [13]-[14].

An interesting insight is found when applying Matlab PLACE tool to place the CL eigenvalues of the AFE rectifier system to $\Lambda = [\lambda_1, \lambda_2, \lambda_3] = [2\pi \cdot 1000, 2\pi \cdot 1000, 2\pi \cdot 100]$ (Table I):

$$\dot{\mathbf{x}} = \mathbf{A}\mathbf{x} + \mathbf{B_1} \underbrace{(-\mathbf{K_{ss}}\mathbf{x})}_{\mathbf{u_1}} \ \rightarrow \ \mathbf{K_{ss}} = \text{PLACE}(\mathbf{A}, \mathbf{B_1}, \Lambda) = \begin{bmatrix} 5.3545 & 0.3204 & -1.1458 \\ -0.3204 & 5.3407 & 0.0711 \end{bmatrix} \cdot 10^{-3} \quad (10)$$

Here $\mathbf{K_{ss}} \in \mathbb{R}^{2 \times 3}$ is a SS SFB controller matrix that moves the CL eigenvalues of the state dynamics to the desired locations. Comparing the controllers in (8) and (10), one can see that they are identical for the operating point (OP) at hand. Other OPs (other load resistances) were also evaluated, and the observed alignment holds true for those OPs as well. Therefore, at least for the AFE case, FD SFB controller gives identical results as the SS SFB controller generated numerically using PLACE tool. Method [13] maximizes small-signal robustness using the control DOFs. Moreover, the numerical method minimizes the upper bounds on the norm of the feedback matrix and transient response (minimal control effort and transient peaks), and maximizes the lower bound on stability margin. For the proofs of these statements refer to [13]-[14]. Since the FD SFB controller is identical to the SS SFB one, it retains all the advantages above, while also being in an explicit closed-form.

Gaining a bit more insight into the small-signal robustness of "physics-based" SFB controls, [14] derives the upper bound on CL eigenvalue sensitivity as:

$$\frac{d\lambda_{i\in\{1..n\}}}{d\varepsilon} \leq n \cdot \kappa_2(\mathbf{X}) \equiv n \cdot ||\mathbf{X}||_2 \cdot ||\mathbf{X}^{-1}||_2, \text{ where } \mathbf{X} \in \mathbb{C}^{n\times n}\text{- matrix of CL right eigenvectors} \quad (11)$$

Here $\kappa_2(\mathbf{X})$ is the *spectral condition number* of the CL matrix $\mathbf{A_{cl}} = \mathbf{A} + \mathbf{B_1K}$, ε is the small-signal perturbation of any entry in $\mathbf{A_{cl}}$, n is the number of states. $\kappa_2(\mathbf{X})$ is always greater or equal to 1 and it gets to its unique minimum (=1) when $\mathbf{A_{cl}}$ is a normal matrix (all of its CL normalized eigenvectors $\mathbf{X}$ are orthogonal to each other - orthonormal eigenvector condition). From physics perspective, the CL eigenvectors $\mathbf{X}$ are orthogonal when the state dynamics are decoupled from each other and that is exactly what the "physics-based" techniques aim for - they do physics-informed decoupling actions for plant simplification and control. Thus, it turns out that "physics-based" decoupling techniques can result in an optimally robust CL SFB system, at least for the AFE rectifiers.

Theoretical insight above relates only to the small-signal robustness, or CL eigenvalue sensitivity to parameter variations (11). To assess the effect of greater parameter variations, switched model simulations and/or Lyapunov functions [16] can be used. For simulations, consider the same transient scenario as in Fig. 5, but with $L = 2 \times L$ and $C = 0.8 \times C_{min}$ as an example of unmodeled large parameter variation in the plant. Fig. 6a shows the transient waveforms of the system states in such scenario. It is seen that classical PI CL system (gray) becomes unstable at high power (low load resistance), while the SFB CL system (blue) preserves stability. Note that lack of integrator in control leads to a very small steady-state error in DC-link voltage and grid currents even for significant parameter variations. Slow integrators can be added to the outer loops to eliminate steady state error completely.

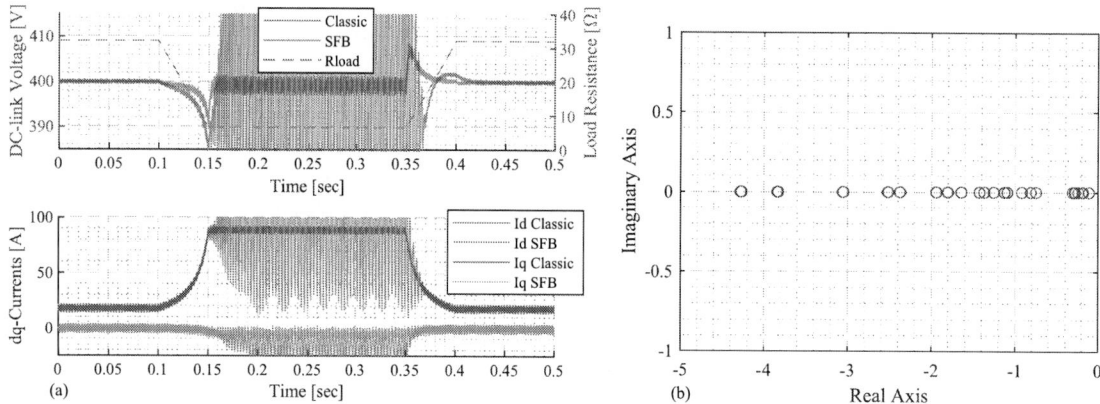

Fig. 6: Transient robustness simulation (a) for dq-axes currents and DC-link voltage for Classic and SFB controls. The R_{load} is swept 32Ω (5 kW) - 6.4Ω (25 kW). Parameter variation: $C = 0.8 \cdot C_{min}$, $L = 2 \cdot L$. Eigenvalue sweep (b) of the Lyapunov derivative matrix $\mathbf{A_{cl}}^\top(t)\mathbf{P} + \mathbf{PA_{cl}}(t)$ for $0.5 \cdot L \to 2 \cdot L, 0.5 \cdot r \to 2 \cdot r$ parameter variations.

As of Lyapunov functions, one can use some results for linear time-varying (LTV) systems to evaluate the system robustness at any given OP more precisely. For a CL LTV systems with certain bounds on parameter variations in $\mathbf{A_{cl}}(t)$, one can conclude asymptotic stability of the system irrespective of the temporal nature of parameter variations within the defined bounds, if there exists a positive-definite Lyapunov function $V(\mathbf{x})$ with a negative-definite time derivative for all possible $\mathbf{A_{cl}}(t)$ [16].

Consider $2\times$ parameter variations in L and r due to thermal/saturation/proximity effects for example. Since L, r enter the system matrices $\mathbf{A}$ and $\mathbf{B_1}$, entries of those will change, affecting $\mathbf{A_{cl}}(t) = \mathbf{A}(t) + \mathbf{B_1}(t)\mathbf{K}$. As those variations are not modeled, the controller matrix will not change. Robust asymptotic stability of the system in such conditions can be concluded if there exists $\mathbf{P} = \mathbf{P}^\top > 0$ such that:

$$V(\mathbf{x}) = \mathbf{x}^\top \mathbf{P}\mathbf{x} > 0, \quad \dot{V}(\mathbf{x}) = \mathbf{x}^\top \left(\mathbf{A_{cl}}^\top(t)\mathbf{P} + \mathbf{PA_{cl}}(t)\right)\mathbf{x} < 0, \quad \forall \mathbf{x} \neq 0, \quad \forall t \quad (12)$$

Inspired by solution of $\dot{V}(\mathbf{x}) = -\mathbf{x}^\top \mathbf{I}\mathbf{x}$, one can get the following energy-like Lyapunov matrix $\mathbf{P}$:

$$\mathbf{P} = \begin{bmatrix} \frac{L}{2} & 0 & \frac{\sqrt{LC}}{4}; & 0 & \frac{L}{2} & 0; & \frac{\sqrt{LC}}{4} & 0 & \frac{C}{2} \end{bmatrix} = \mathbf{P}^\top > 0 \tag{13}$$

The range of parameters of interest $(0.5 \cdot L \to 2 \cdot L, 0.5 \cdot r \to 2 \cdot r)$ was swept with high resolution to evaluate the negative-definiteness of $\mathbf{A_{cl}}^\top(t)\mathbf{P} + \mathbf{PA_{cl}}(t)$ in each case by checking the respective eigenvalues. From Fig. 6b one can see that all eigenvalues throughout the sweep are strictly negative, implying robust asymptotic stability of the system at the peak power OP in the presence of large L, r parameter variations.

Moving on to adaptability, it is enabled by the fact that, unlike the output of numerical optimization, the designed regulator is in the closed-form (7). Studying (7) and the control gains (3), (5) therein, it is seen that the regulator depends on AFE parameter estimations (L, C, r), PLL output ω_o, and OP large-signal quasi-steady-state parameters $(M_{dq}, I_{gdq}, V_{dc}, R = V_{dc}/I_{load}, V_{gdq})$. The CL robustness to AFE parameter variations has already been shown above. ω_o is the measurement coming from the PLL. The OP parameters can be obtained from measurements and OP relations (1) - see the example below.

For example, $V_{dc} = V_{dc}^*$, I_{gd} can be obtained from low-pass filtering the i_{gd} measurement. The low-pass filter frequency used in this work was $f_v^*/10$ to separate the high-frequency ripple content from the low-frequency quasi-steady-state OP part ($i_{gd} = I_{gd} + \tilde{i}_{gd}$). Similarly one can obtain $I_{load}, R, V_{gdq}, \tilde{v}_{gdq}, \tilde{v}_{dc}$. There are no measurements for M_{dq}, but those can be calculated from (1). The combined manipulated input (duty cycles) would then become: $m_{dq} = M_{dq} + \tilde{m}_{dq}$ - large-signal part M_{dq} and small-signal part $\tilde{m}_{dq}$ (Fig. 2). Note how the manipulated input comprises large- and small-signal parts.

Having shown how the OP parameters can be obtained from measurements in real time, one can now see how the SFB regulator is made adaptive. Indeed, since the SFB control gains (7) explicitly depend on measurable OP parameters, SFB gains will change with OP variation - load step, AC voltage change etc.

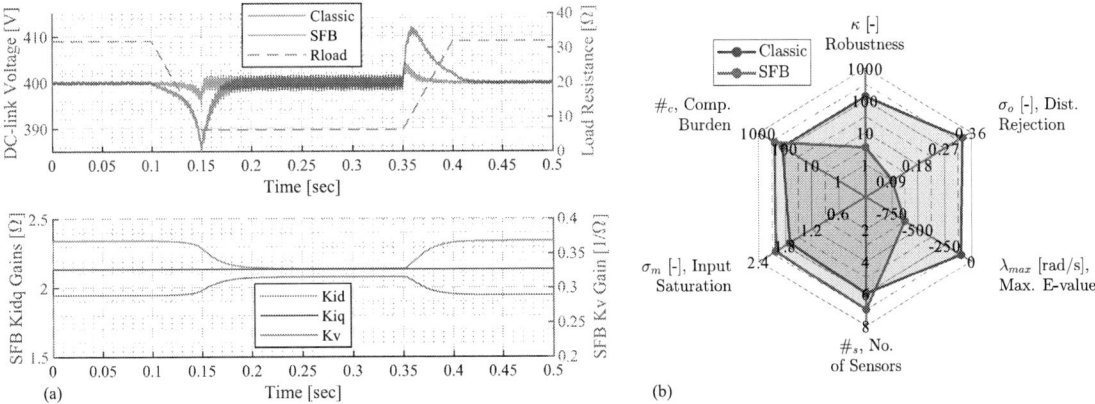

Fig. 7: Transient adaptability simulation (a) for DC-link voltage for Classical and SFB controls. The R_{load} changes between 32Ω (5 kW) and 6.4Ω (25 kW). SFB control gains K_{id}, K_{iq}, K_v adapt to OP variation. Comparative evaluation (b) of the control structures using spider plot of performance metrics. Smaller hexagon - better performance.

To support the claims on adaptability, Fig. 7a shows the same dc voltage transient as in Fig. 5. Also, it plots the control gains that were used in real time during the transient. The gains are derived from (3), (5) using OP parameter measurements. Indeed, it is seen that the control gains do change based on OP. Moreover, adaptive SFB control results in a better transient performance - less over/undershoot, settling time etc., in addition to aforementioned robustness and reduced capacitance capability.

With the obtained robustness and adaptability, the systematically designed SFB AFE regulator at hand retains the same CL eigenvalues at any OP, does not overload the small-signal controller with large-signal measurements, while also preserving optimal performance at any given OP without numerical optimization. Hence the name of the paper - systematic adaptive robust SFB control for AFE rectifiers.

To highlight various aspects of the designed SFB versus classical regulator, Fig. 7b depicts the comparative evaluation spider diagram. The diagram aims to evaluate and compare various small-signal performance metrics of controllers to explicitly see the pros and cons of SFB/classical controllers applied to the same AFE rectifier plant given in Table I. Generally, the smaller the diagram area the better the overall controller performance.

Six axes indicate the performance metrics evaluated. κ - measure of small-signal robustness, CL eigenvalue sensitivity of the normalized (unitless) CL matrix $\mathbf{A_{cl}}$ (11). σ_o, σ_m - measures of small-signal disturbance rejection and small-signal input saturation, disturbance-to-output and disturbance-to-manipulated input normalized (unitless) peak singular values, also known as $\mathcal{H}_\infty$-norms [15]. λ_{max} - measure of small-signal transient convergence, maximum (the least negative) real-part CL eigenvalue. $\#_s, \#_c$ - measures of sensor requirement and computational burden, number of sensors and number of multiplication/addition operations made by controller in simulation within one switching period.

It can be seen from Fig. 7b that the designed SFB controller has much better small-signal robustness, disturbance rejection and transient convergence. All these come at the cost of sightly increased computational burden, input saturation and one more sensor (load current). With these one can claim superiority of SFB over the classical controller for the nominal AFE plant. The SFB advantage grows even further when capacitance reduction and adaptability are taken into account.

6   Conclusion

With recent proliferation of WBG semiconductors and higher switching frequencies, there is a trend towards passives' miniaturization in power converters. Although higher switching frequencies enable even smaller passive components to provide satisfactory ripple values during steady state, more subtle dynamic stability constraints still put a lower bound on passive component values even for low switching frequencies, as it was shown for AFE rectifiers with cascaded control. This work attempts to remove the aforementioned dynamic constraints for three-phase six-switch two-level AFE rectifiers by modifying the conventional control through physics-based frequency-domain techniques.

It turns out from the analysis and simulations that within the bandwidth, dictated by the RHP zero of the rectifier, the reported dynamic constraint on AFE DC-link capacitance was removed by the designed regulator - smaller DC-link capacitance values can be used compared to previous literature.

Interesting observation leading to numerous conclusions is that analytically-designed state-feedback controller of this paper is the same as the one obtained numerically in [13]. The controllers are precisely the same over the full load range. Moreover, the numerically obtained controller guarantees maximum robustness for closed-loop eigenvalues, minimum upper bound on the norm of the feedback matrix (minimal control action) and transient response, as well as maximum lower bound on stability margin. Equality of the analytical and numerical controllers implies 1) that the analytically designed SFB controller retains all the guarantees of the numerical solution; and 2) that utilized analytical controller design techniques might constitute a generalized design methodology for a robust power electronics controllers based on physical insight. Validating the second implication above requires more studies in the future.

Finally, the work proposed how to make the designed robust controller operating point-aware by explicitly splitting the measurements into the operating point large-signal part and the ripply small-signal part using low-pass filters. Combined with the fact that the designed controller is analytical (in the closed-form), the system can achieve robust behavior in an operating-point-adaptive fashion without numerical optimization for every operating point. This approach to adaptability is not constrained to the AFE, implying that it also might constitute a generalized adaptability methodology for the linearized nonlinear systems. Again, validating this implication requires a deeper dive into control systems theory.

References

[1] R. Teodorescu, M. Liserre and P. Rodriguez, Grid converters for photovoltaic and wind power systems. Chichester, U.K.: Wiley, 2011, pp. 1-4.

[2] Y. Yin et al., "Adaptive Control for Three-Phase Power Converters With Disturbance Rejection Performance," in IEEE Transactions on Systems, Man, and Cybernetics: Systems, vol. 51, no. 2, pp. 674-685, Feb. 2021.

[3] Q.-C. Zhong and Z. Lyu, "Droop-Controlled Rectifiers That Continuously Take Part in Grid Regulation," IEEE Transactions on Industrial Electronics, vol. 66, no. 8, pp. 6516-6526, 2019.

[4] J. W. Kolar and T. Friedli, "The essence of three-phase pfc rectifier systems - part i," IEEE Transactions on Power Electronics, vol. 28, no. 1, pp. 176-198, 2013.

[5] H. A. Mantooth, M. D. Glover and P. Shepherd, "Wide Bandgap Technologies and Their Implications on Miniaturizing Power Electronic Systems," in IEEE Journal of Emerging and Selected Topics in Power Electronics, vol. 2, no. 3, pp. 374-385, Sept. 2014.

[6] B. Shi, G. Venkataramanan, and N. Sharma, "Design consideration for reactive elements and control parameters for three phase boost rectifiers," in IEEE International Conference on Electric Machines and Drives, 2005, pp. 1757–1764.

[7] M. P. Kazmierkowski and L. Malesani, "Current control techniques for three-phase voltage-source PWM converters: a survey," in IEEE Transactions on Industrial Electronics, vol. 45, no. 5, pp. 691-703, Oct. 1998.

[8] B. Singh, B. N. Singh, A. Chandra, K. Al-Haddad, A. Pandey and D. P. Kothari, "A review of three-phase improved power quality AC-DC converters," in IEEE Transactions on Industrial Electronics, vol. 51, no. 3, pp. 641-660, June 2004.

[9] Hengchun Mao, C. Y. Lee, D. Boroyevich and S. Hiti, "Review of high-performance three-phase power-factor correction circuits," in IEEE Transactions on Industrial Electronics, vol. 44, no. 4, pp. 437-446, Aug. 1997.

[10] M. J. Ryan, W. E. Brumsickle and R. D. Lorenz, "Control topology options for single-phase UPS inverters," in IEEE Transactions on Industry Applications, vol. 33, no. 2, pp. 493-501, March-April 1997.

[11] F. B. del Blanco, M. W. Degner and R. D. Lorenz, "Dynamic analysis of current regulators for AC motors using complex vectors," in IEEE Transactions on Industry Applications, vol. 35, no. 6, pp. 1424-1432, Nov.-Dec. 1999.

[12] F. Briz, M. W. Degner and R. D. Lorenz, "Analysis and design of current regulators using complex vectors," in IEEE Transactions on Industry Applications, vol. 36, no. 3, pp. 817-825, May-June 2000.

[13] J. Kautsky, N. K. Nichols, and P. Van Dooren, "Robust pole assignment in linear state feedback," International Journal of Control, vol. 41, pp. 1129–1155, 1985.

[14] J. Wilkinson, Algebraic eigenvalue problem. Oxford: Clarendon Press, 1965, pp. 87-90.

[15] S. Skogestad and I. Postlethwaite, Multivariable Feedback Control: Analysis and Design. Chichester, U.K.: Wiley, 2010, pp. 537-538.

[16] K. Gu, M. A. Zohdy and N. K. Loh, "Necessary and sufficient conditions of quadratic stability of uncertain linear systems," in IEEE Transactions on Automatic Control, vol. 35, no. 5, pp. 601-604, May 1990, doi: 10.1109/9.53534.

An Optimized Compensation Strategy of Direct Matrix Converter-Fed PMSM Drives with Field Weakening under Unbalanced Supply Conditions

Jun Xie, Dustin Henneberg, Martin Suberski, Manuel Kusebauch,
Uwe Rädel and Jürgen Petzoldt
Technische Universität Ilmenau
Power Electronics and Control Group
Ilmenau, Germany
Phone: +49 367769-1553
Fax: +49 367769-1469
Email: jun.xie@tu-ilmenau.de
URL: http://www.tu-ilmenau.de

Keywords

≪Direct matrix converter≫, ≪AC-AC converter≫, ≪Unbalanced AC grid≫, ≪Permanent magnet motor≫, ≪Field Oriented Control≫

Abstract

This paper presents an optimized compensation strategy for direct matrix converter (DMC) - fed permanent magnet synchronous motor (PMSM) drives with special consideration of the modification in field weakening control under unbalanced supply conditions. The imbalance of input voltage conditions in a three wire system are analysed and considered to be composed of positive and negative sequence components. To determine unbalance factor of the grid and decompose the positive and negative sequence components, the dual second order generalized integrator (DSOGI) and positive sequence calculator (PSC) algorithm are introduced. In order to achieve balanced output while optimizing the input current to be sinusoidal, the modulation index based on direct modulation method and the field weakening regulator are then adjusted according to unbalance factor of the input voltage conditions. The principle of the proposed compensation strategy is explained in detail. Simulation results are used to verify this proposed strategy. A low voltage laboratory platform consisting of DMC and servo motor is implemented and controlled using hybrid hardware concept based on field-programmable gate array (FPGA) and digital signal processor (DSP). Experimental studies on the laboratory prototype confirm the feasibility and effectiveness of the proposed method.

Introduction

The three-phase DMC is an alternative topology of direct AC-AC conversion without any bulky energy storage elements in an intermediate link [1,2]. This topology currently attracts many research interests in power electronics due to its compact size, bidirectional energy-flow capability and longtime durability, especially in adjustable speed drives applications.

On the other hand, because of the absence of DC-link energy storage components, the input and output sides of DMC are directly coupled. Any distortion, imbalance or voltage sag in input voltage conditions will reflect to the load side instantly, which results in harmonics in input currents and output voltages [3, 7]. Furthermore, as discussed in [7, 9, 10], the imbalance in input voltage conditions also reduces amplitude of the maximum achievable balanced output voltage, which is verified in the presented strategy as well.

Several methods have been proposed to mitigate the impact of unbalanced input voltage conditions under respective control strategies for DMC. In [3,4], the harmonic content in input current is analytically evaluated and minimised by controlling the input current displacement angle using direct space vector modulation (SVM) technique. In [7], the method with special consideration when desired output voltage greater than maximum attainable balanced output voltage is discussed, by means of solving optimization problems. In [8], a modified indirect SVM technique is used to determine the modulation index. However, the limitation of balanced output voltage is left undiscussed. In [5], a similar compensation strategy is proposed by using positive and negative sequence components of the input voltage under direct modulation technique. However, the method of how to decompose the positive and negative sequence components is not mentioned. In [6], a compensation method based on fuzzy logic control is discussed, which increases the complexity of calculating the modulation index.

In this paper, an optimized compensation strategy for DMC-fed PMSM drives with special consideration of the modification in field weakening control under unbalanced input voltage conditions is proposed. The modulation index m based on direct modulation method is determined by the desired modulation factor m_m and the compensation factor m_c. The desired modulation factor m_m is derived based on motor demands from the closed cascade control loops using field oriented control (FOC) method. The compensation factor m_c is calculated using amplitude of the instantaneous input voltages and the designed input phase displacement angle. Compared with previous research, the main features of the proposed strategy are as follows:

- It provides an optimized compensation method as a simple but practical solution to not only eliminate impact of unbalanced input voltages on output side of DMC, but also optimized the input current quality to be sinusoidal.
- It employs a method to decompose the measured instantaneous input voltage into positive and negative sequence components by introducing the algorithm of DSOGI with PSC, which are widely used in synchronisation of power systems [11–13].
- It dynamically adjusts the upper limit of the modulation factor m_m and the field weakening regulator according to unbalance factor of the input voltage conditions. While increasing the voltage transfer ratio under field weakening as much as possible, the modulation index m is ensured not to exceed the limitation of 1.

The rest of this paper is organized as follows. In section II, a brief introduction of the system topology together with the control scheme is presented. The proposed strategy is subsequently addressed in section III. Simulation results verifying the performance of the proposed strategy are demonstrated in section IV. In section V, experimental studies on the laboratory prototype confirm the feasibility and effectiveness of the proposed compensation method. Finally, this study is summarized in section VI.

System topology and control scheme

The system topology of three-phase to three-phase DMC-fed PMSM drives is shown in Fig. 1, which consists of power supply, input LC filter, DMC switching array, PMSM drive system and a protection circuit.

A. Topology of DMC

The DMC is composed of nine bidirectional switches (BDS) as a 3×3 matrix, which connects the grid to the PMSM drives. Each BDS should have the capability to block voltage and conduct current in both directions. There are many configurations for the realization [17]. One possibility could be two power semiconductor switches in back-to-back arrangement, which is shown in Fig. 1. The power semiconductor switch could be anti-parallel insulated gate bipolar transistor (IGBT) or metal–oxide–semiconductor field-effect transistor (MOSFET). The input LC-Filter is necessary to reduce high-frequency current harmonics and voltage fluctuation. It also removes the zero sequence component of the grid voltage. Many topologies of input filter for DMC have been proposed [15]. In this study, the damped LC filter is employed, in which the filter capacitor C_f is parallel with a RC damping circuit. It is worth noticing that, if the grid impedance $(R_N + jX_{L_N})$ can be precisely determined, it is possible to use grid inductance L_N

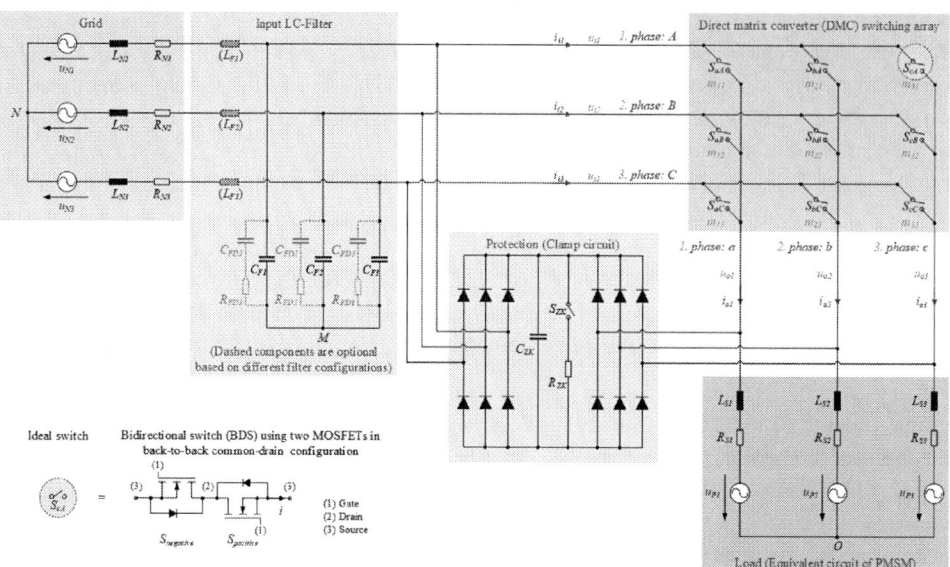

Fig. 1: System topology of three-phase to three-phase DMC-fed PMSM drives

instead of the normal filter inductance L_f within the permissible resonance frequency range. The output filter is commonly neglected due to the inductive nature of PMSM drive system. The clamping circuit is frequently implemented as an overvoltage protection circuit, which makes up the deficiencies of DMC for absence of passive free-wheeling paths [16]. It consists of 12 fast-recovery diodes and connects the input and output of DMC using double three-phase diode bridge (B6). The double B6 diode bridge are connected by a DC-link capacitor C_{ZK}, which is determined proportional to the energy stored in PMSM drives [16]. A switchable resistor R_{ZK} is typically in parallel to the DC-link capacitor to share the engergy and prevent overvoltage, which also allows to further reduce the size of the DC-link capacitor.

B. Model of PMSM drive system

Since no output filter is required, the motor voltages and currents are actually the output voltages $\vec{u}_o$ and currents $\vec{i}_o$ of DMC. The continuous-time model of the PMSM is given in dq-coordinate in (1):

$$
\begin{cases} u_d = R_s i_d - \omega_e \cdot \psi_q + \frac{d\psi_d}{dt} \\ u_q = R_s i_q + \omega_e \cdot \psi_d + \frac{d\psi_q}{dt} \end{cases} \quad , \quad \begin{cases} \psi_d = L_d i_d + \psi_m \\ \psi_q = L_q i_q \end{cases} \tag{1}
$$

The electromagnetic torque m_i and mechanical dynamics of PMSM can be estimated in (2):

$$
m_i = \frac{3}{2} P_p (\psi_d i_q - \psi_q i_d) \quad , \quad m_i - m_l - m_r = \frac{1}{J} \frac{d\omega_m}{dt} \quad , \quad m_r = k_r \omega_m \quad , \quad \omega_e = P_p \cdot \omega_m \tag{2}
$$

where u_d and u_q, i_d and i_q, ψ_d and ψ_q, L_d and L_q represent stator voltages $\vec{u}_o$, stator currents $\vec{i}_o$, magnet flux and stator inductance in dq-coordinate, respectively; ψ_m represents the magnet flux linkage of PMSM; R_s is the resistance of stator; P_p is the number of pole pairs; J is the moment of inertia; ω_e and ω_m are the electrical and mechanical angular frequency of PMSM, respectively; m_l is the load moment and m_r is the frictional moment, which is proportional to the mechanische angular frequency with the friction coefficient k_r.

C. Decompose positive and negative sequence components using DSOGI-PSC algorithm

To obtain the positive and negative sequence components under unbalanced input voltage conditions, the DSOGI-PSC algorithm is introduced. This strategy is presented in [14, 18] in detail and the main content is summarized as follows with slight modifications.

The DSOGI-PSC algorithm requires two components in quadrature. To achieve this, Clark transformation is used to transfer the input voltage $\vec{u}_{i,abc}$ to $\alpha\beta$-coordinate $\vec{u}_{i,\alpha\beta}$. For each voltage component $u_{i,\alpha}$ and $u_{i,\beta}$, a single SOGI structure is employed, which leads to the output voltage signals $u'_{i,\alpha}$ and $u'_{i,\beta}$. The transfer function of voltage $u'_{i,\alpha\beta}$ to $u_{i,\alpha\beta}$ is described in (3) as $D(s)$, which corresponds to a band-pass filter (BPF). The voltage signals $u'_{i,\alpha\beta}$ are then lagged by $90°$, which are described as $q \cdot u'_{i,\alpha}$ and $q \cdot u'_{i,\beta}$, where q is $e^{-j(\pi/2)}$. In this way, a pair of quadrature signals $u'_{i,\alpha\beta}$ and $q \cdot u'_{i,\alpha\beta}$ are formed. The transfer function of voltage $q \cdot u'_{i,\alpha\beta}$ to $u_{i,\alpha\beta}$ is described in (3) as $Q(s)$, which corresponds to a low-pass filter (LPF).

$$D(s) = \frac{u'_{i,\alpha\beta}}{u_{i,\alpha\beta}}(s) = \frac{k\omega' s}{s^2 + k\omega' s + \omega'^2} \quad , \quad Q(s) = \frac{q \cdot u'_{i,\alpha\beta}}{u_{i,\alpha\beta}}(s) = \frac{k\omega'^2}{s^2 + k\omega' s + \omega'^2} \tag{3}$$

where ω' is the angular frequency of input voltage, which is estimated from the synchronous reference frame (SRF) phase locked loop (PLL) structure and k is the gain factor of SOGI. The quadrature signals $u'_{i,\alpha\beta}$ and $q \cdot u'_{i,\alpha\beta}$ are then decomposed into positive $u_{p,\alpha\beta}$ and negative $u_{n,\alpha\beta}$ sequence components in the PSC structure, which is expressed in (4).

$$\begin{bmatrix} u_{p,\alpha} \\ u_{p,\beta} \\ u_{n,\alpha} \\ u_{n,\beta} \end{bmatrix} = \frac{1}{2} \begin{bmatrix} 1 & -q \\ q & 1 \\ 1 & q \\ -q & 1 \end{bmatrix} \cdot \begin{bmatrix} u'_{i,\alpha} \\ u'_{i,\beta} \end{bmatrix} \quad , \quad q = e^{-j(\pi/2)} \quad , \quad \begin{bmatrix} u_{i,\alpha} \\ u_{i,\beta} \end{bmatrix} = \begin{bmatrix} u_{p,\alpha} \\ u_{p,\beta} \end{bmatrix} + \begin{bmatrix} u_{n,\alpha} \\ u_{n,\beta} \end{bmatrix} \tag{4}$$

The block diagram of DSOGI-PSC-SRF-PLL is shown in Fig. 2:

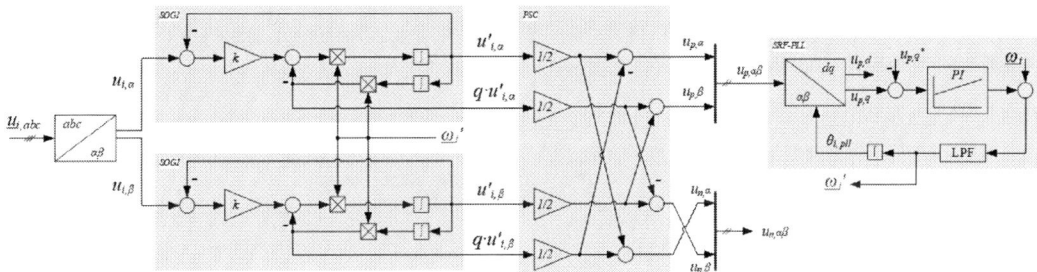

Fig. 2: Block diagram of the DSOGI-PSC-SRF-PLL algorithm [14, 18]

Proposed control strategy

The control scheme for the DMC-fed PMSM drive system with field weakening control is shown in Fig. 3. The PMSM drive system is controlled based on FOC algorithm in closed cascade control loops. The DMC is controlled using direct modulation strategy, namely the optimized Venturini's method, because of its convenience and straightforward of understanding the basics of DMC.

In outer loop, the motor speed n is estimated and regulated by proportional-integral (PI) controller. The output of speed controller is the reference value of torque producing current i_q^*. In inner loop, the magnetizing current i_d^* is set to zero before field weakening and is set to be negative during field weakening in order to further increase the motor speed. In the mean while, as the magnetizing current i_d^* increases, the reference value of torque producing current i_q^* is restricted by the maximum load current I_{max} in (5).

$$|i_q^*| \leq \sqrt{(I_{max})^2 - (i_d^*)^2} \quad , \quad u_d^2 + u_q^2 \leq \hat{U}_{o,max}^2 \tag{5}$$

Suppose the positive $\vec{u}_{i,p}$, negative $\vec{u}_{i,n}$ sequence components of the input voltages and the unbalance

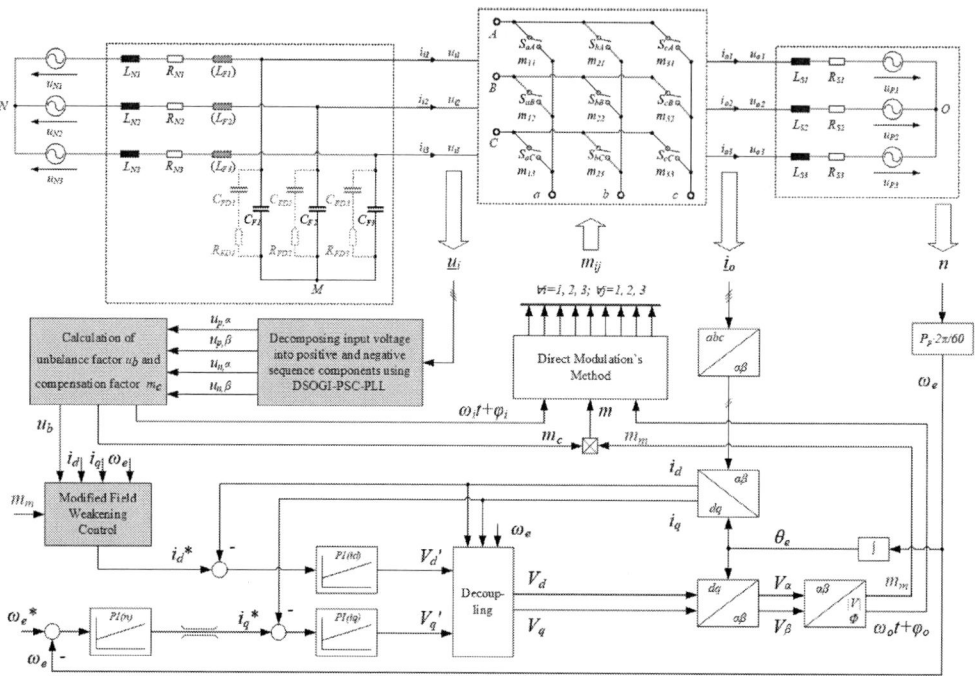

Fig. 3: Control scheme of DMC-fed PMSM drive system with field weakening

faktor u_b are defined in (6):

$$\vec{u}_{i,p} = \begin{bmatrix} \hat{U}_p cos(\omega_i t + \phi_p) \\ \hat{U}_p cos(\omega_i t - \frac{2\pi}{3} + \phi_p) \\ \hat{U}_p cos(\omega_i t + \frac{2\pi}{3} + \phi_p) \end{bmatrix} \quad , \quad \vec{u}_{i,n} = \begin{bmatrix} \hat{U}_n cos(\omega_i t + \phi_n) \\ \hat{U}_n cos(\omega_i t + \frac{2\pi}{3} + \phi_n) \\ \hat{U}_n cos(\omega_i t - \frac{2\pi}{3} + \phi_n) \end{bmatrix} \quad , \quad u_b = \frac{\hat{U}_n}{\hat{U}_p} \tag{6}$$

where ω_i is the angular frequency of input voltage, $\hat{U}_p$ and $\hat{U}_n$ are the amplitude of positive and negative sequence components of input voltage. ϕ_p and ϕ_n are the phase displacement angle of the positive and negative sequence components respectively.

The unbalanced input voltage $\vec{u}_i$ is considered to be sum of positive and negative sequence components as described in (7) using Fortescue's theorem. The zero sequence component is neglected in a three wire system.

$$\vec{u}_i = \vec{u}_{i,p} + \vec{u}_{i,n} = \begin{bmatrix} \hat{U}_p cos(\omega_i t + \phi_p) \\ \hat{U}_p cos(\omega_i t - \frac{2\pi}{3} + \phi_p) \\ \hat{U}_p cos(\omega_i t + \frac{2\pi}{3} + \phi_p) \end{bmatrix} + \begin{bmatrix} \hat{U}_n cos(\omega_i t + \phi_n) \\ \hat{U}_n cos(\omega_i t + \frac{2\pi}{3} + \phi_n) \\ \hat{U}_n cos(\omega_i t - \frac{2\pi}{3} + \phi_n) \end{bmatrix} \tag{7}$$

The amplitude $\hat{U}_i$ of input voltage $\vec{u}_i$ can be expressed using instantaneous input voltages after Clarke and Cartesian-Polar transformation, which can be further described using $\vec{u}_{i,p}$ and $\vec{u}_{i,n}$ as shown in (8).

$$\hat{U}_i = \sqrt{\frac{2}{3}(u_{i1}^2 + u_{i2}^2 + u_{i3}^2)} = \sqrt{\hat{U}_p^2 + \hat{U}_n^2 + 2\hat{U}_p\hat{U}_n cos(2\omega_i t + \phi_p + \phi_n)} \tag{8}$$

The phase angle θ_{ui} of input voltage $\vec{u}_i$ is shown in (9):

$$\theta_{ui} = tan^{-1}\left(\frac{\hat{U}_p sin(\omega_i t + \phi_p) - \hat{U}_n sin(\omega_i t + \phi_n)}{\hat{U}_p cos(\omega_i t + \phi_p) + \hat{U}_n cos(\omega_i t + \phi_n)}\right) \tag{9}$$

In average value model of DMC, the instantaneous output voltage $\underline{u}_o$ can be expressed as (10):

$$\underline{u}_o = \hat{U}_o e^{j\omega_o t} = \hat{U}_i \cdot \frac{\sqrt{3}}{2} \cdot m \cdot cos(\phi_i) e^{j\omega_o t} \quad , \quad 0 \leq m \leq 1 \tag{10}$$

where ω_o is the angular frequency of output voltage, $cos(\phi_i)$ is the input power factor with ϕ_i the input phase displacement angle, $m = \frac{q}{\sqrt{3}/2}$ is the modulation index with a range from 0 to 1 and q the corresponding voltage transfer ratio between output and input side of DMC. Introducing (8) in (10), leads to the following equation (11):

$$\underline{u}_o = \sqrt{\hat{U}_p^2 + \hat{U}_n^2 + 2\hat{U}_p \hat{U}_n cos(2\omega_i t + \phi_p + \phi_n)} \cdot \frac{\sqrt{3}}{2} \cdot m \cdot cos(\phi_i) e^{j\omega_o t} \tag{11}$$

Equation (11) shows, under unbalanced input voltage conditions, the amplitude of the input voltage $\hat{U}_i$ oscillates with twice the input frequency due to presence of the negative sequence component $u_{i,n}$, which results in harmonics to the output side of DMC.

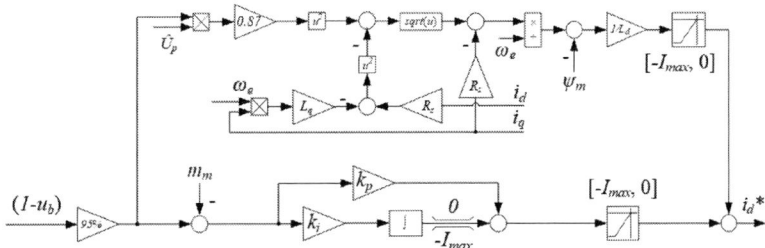

Fig. 4: Block diagram of the modified field weakening control

In order to eliminate the oscillation in the output of DMC, the simplest method is to modify the modulation index m through two factors m_m and m_c in such a way that:

- the modulation factor m_m keeps the modulation depth according to the desired motor demands.
- the compensation factor m_c compensates the oscillation of input voltage amplitude $\hat{U}_i$ while keeping the positive sequence component $u_{i,p}$ as maximal achievable balanced output voltage.
- the input phase displacement angle ϕ_i is determined to be the phase angle of voltage $(u_{i,p} - u_{i,n})$, so that the amplitude of input current oscillates exactly in the opposite direction of input voltage, meanwhile keeps the input power factor $cos(\phi_i)$ to be unity every time when amplitude of input voltage reaches $(\hat{U}_p + \hat{U}_n)$ or $(\hat{U}_p - \hat{U}_n)$ and ensures the input current to be sinusoidal.
- the input power factor $cos(\phi_i)$ is then divided in compensation factor m_c to eliminate the oscillation and keep the amplitude of output voltage relative constant.

To achieve this, the modulation index m is expressed in (12):

$$m = m_m \cdot m_c = m_m \cdot \left(\frac{\hat{U}_p}{\sqrt{\hat{U}_p^2 + \hat{U}_n^2 + 2\hat{U}_p \hat{U}_n cos(2\omega_i t + \phi_p + \phi_n)}} \cdot \frac{1}{cos(\phi_i)} \right) \leq 1 \tag{12}$$

Since the modulation index m can't exceed 1, which is the intrinsic limitation, the range of the compensation factor m_c is $[\frac{\hat{U}_p}{\hat{U}_p + \hat{U}_n}, \frac{\hat{U}_p}{\hat{U}_p - \hat{U}_n}]$ or $[\frac{1}{1+u_b}, \frac{1}{1-u_b}]$. Thus, m_m should be smaller than the minimal value of $\frac{1}{m_c}$, which is $(1 - u_b)$.

Introducing (12) in (11), the equivalent equation of balanced output voltage of DMC can be expressed in (13).

$$\underline{u}_o = \hat{U}_p \frac{\sqrt{3}}{2} \cdot m_m \, e^{j\omega_o t} \quad , \quad 0 \leq m_m \leq (1 - u_b) \tag{13}$$

Equation (13) shows that under unbalanced input voltage conditions, the balanced output voltage can be achieved using the proposed compensation strategy, but the maximal modulation depth is reduced to $(1 - u_b)$, compared with 1 under balanced supply conditions. The amplitude of maximal output voltage $\hat{U}_{o,max}$ is then $\frac{\sqrt{3}}{2}\hat{U}_p(1 - u_b)$ or $\frac{\sqrt{3}}{2}(\hat{U}_p - \hat{U}_n)$.

Because the maximal achievable balanced output voltage is reduced, the field weakening control of PMSM drive system needs to be adjusted as well in order to get rid of any over modulation according to the reduced maximal modulation depth.

The block diagram of a modified field weakening control with feedforward is described in Fig. 4. The field weakening control is based on PI controller. A feedforward loop is employed to help increase the dynamic response and determine the reference magnetizing current i_d^* as shown in (14), which is derived from model of PMSM as described in (1) together with the voltage limitation shown in (5).

$$i_d^* = \frac{1}{L_d}\left(\frac{\sqrt{\left(\frac{\sqrt{3}}{2}\hat{U}_p(1 - u_b)\right)^2 - (R_s i_d - \omega_e L_q i_q)^2} - R_s i_q}{\omega_e} - \psi_m\right) \tag{14}$$

In the block diagram, the gain of 95% is determined from experience for ensurance of the controllability.

Simulation results

Simulation results in Fig. 5 are used to verify the effectiveness of the proposed method using MATLAB. The parameters of the system are shown in Table I.

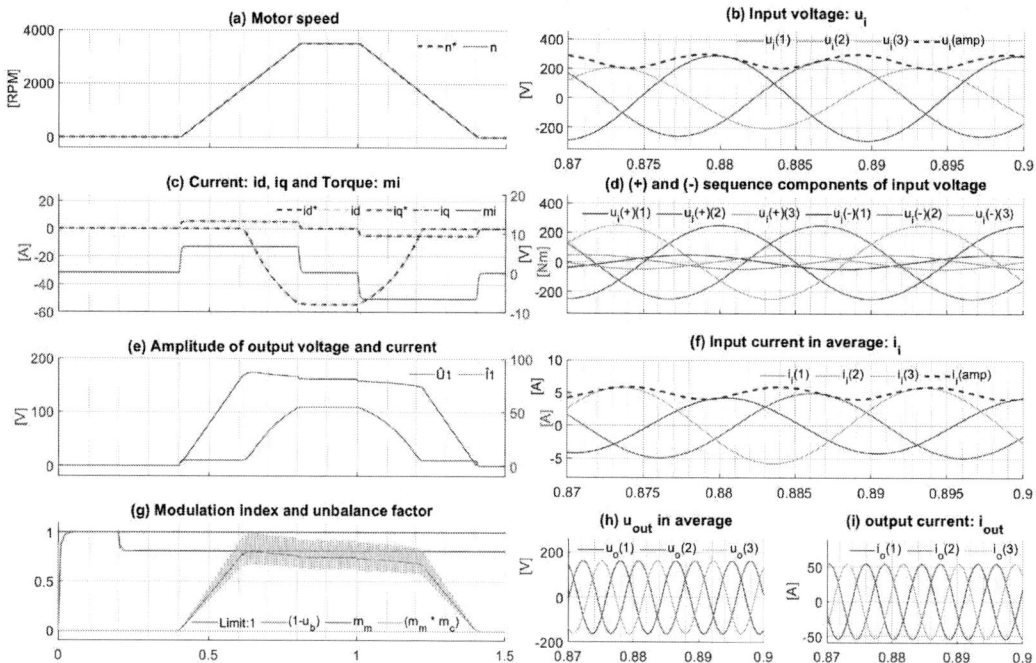

Fig. 5: Simulation results: (a) Motor speed. (b) Input Voltages. (c) Output currents in dq-Coordinate and the Torque. (d) Estimated positive and negative sequence components. (e) Amplitude of output voltage and current. (f) Input Current in average. (g) Modulation index and unbalance factor. (h) Output voltage in average. (i) Output current.

Table I: System parameter in simulation

Symbol	Value	Symbol	Value
$U_{N0,RMS}$	$230V$	f_N	$50Hz$
P_p	2	L_d, L_q	$4mH$
R_s	0.4Ω	J	$0.007kg \cdot m^2$
f_p	$20kHz$	I_{max}	$70A$
ψ_m	$0.4V/(rad \cdot s^{-1})$	k	$\sqrt{2}$

In the simulation, the imbalanced grid voltages $\vec{u}_{N,imbalanced}$ are defined in (15):

$$\vec{u}_{N,imbalanced} = \begin{bmatrix} 230\sqrt{2}\,\cos(100\pi t) \\ 230\sqrt{2} \cdot 80\%\,\cos(100\pi t - \frac{2\pi}{3}) \\ 230\sqrt{2} \cdot 50\%\,\cos(100\pi t + \frac{2\pi}{3}) \end{bmatrix} \tag{15}$$

Through input LC-Filter, the zero sequence component of the grid voltage is removed. The input voltages of DMC are then the sum of positive and negative sequence components as shown in (16).

$$\vec{u}_{N,imbalanced} - \frac{1}{3}\sum(u_{N1} + u_{N2} + u_{N3}) = \vec{u}_i = \vec{u}_{i,p} + \vec{u}_{i,n} \tag{16}$$

The input voltages $\vec{u}_i$ start to be unbalanced from $t = 0.2s$. It can be noticed in (g) that the upper limit of maximal available modulation depth m_m reduced from 1 to be $1 - u_b$. It takes in this simulation environment around $60ms$ for the DSOGI structure to estimate these positive and negative sequence components of input voltage ($\vec{u}_{i,p}$ and $\vec{u}_{i,n}$) based on the configuration set in DSOGI and the estimated result is shown in (d).

The oscillation of input voltage amplitude with doubled grid frequency is shown in (b). The quality of input current is optimized to be sinusoidal in (f) by adjusting the input phase displacement angle ϕ_i. By employing the proposed compensation factor m_c to desired modulation depth m_m, the total modulation index m is not exceeding the intrinsic limitation 1 of DMC as shown in (g) and the output voltage and current is balanced, which are shown in (h) and (i).

The PMSM drive is accelerated by $t = 0.5s$ from static state to be $3500RPM$. After keeping the maximal speed for short and then braked to be zero speed. The field weakening region starts by around $1795RPM$, where the magnetizing current i_d starts to be negative to further increase the motor speed. By employing the proposed field weakening control, the torque producing current i_q^* and motor torque keep relative constant as shown in (c).

Experimental results

A low voltage laboratory platform consisting of a DMC and a servo motor is implemented. A hybrid hardware architecture based on FPGA and DSP is employed to control the demonstrator and the experimental results are shown in Fig. 6 (a).

The unbalanced input voltages are generated by a three phase DC-link voltage source inverter (VSI) as shown in (b), which consists of a positive sequence modulator with $50Hz$ and a negative sequence modulator with frequency $-50Hz$.

The output voltages and currents are shown in (c) and (d) and are measured at the operation point of motor speed by $-500RPM$.

As can be seen in the measurement results, the output voltage and current are sinusoidal and balanced, which verified the effectiveness of the proposed compensation method.

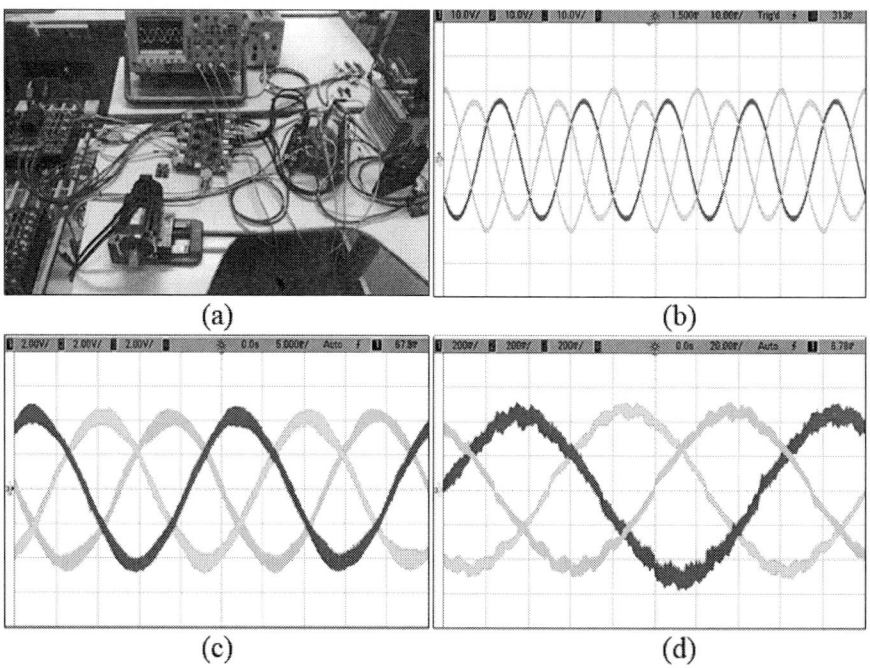

Fig. 6: (a) Demonstrator. (b) Unbalanced input voltage. (c) output voltage. (d) output current.

Conclusion

In this paper, an optimized compensation strategy for DMC-fed PMSM drives with special consideration of field weakening control under unbalanced input voltage conditions is presented. By employing the presented method, the impact of unbalanced input voltages is eliminated and the output voltages and currents are balanced. Meanwhile, the quality of input currents is optimized to be sinusoidal. The impacts on maximal achievable balanced output voltage and the upper limitation in field weakening regulator are discussed. Simulation and experiment results confirm the validity of the proposed strategy.

References

[1] Alesina, A. and Venturini, M.: Solid-state power conversion: A Fourier analysis approach to generalized transformer synthesis, IEEE Transactions on Circuits and Systems, Vol 28 no 4, pp. 319-330, 1981.

[2] Alesina, A. and Venturini, M.G.B.: Analysis and design of optimum-amplitude nine-switch direct AC-AC converters, IEEE Transactions on Power Electronics, Vol 4 no 1, pp. 101-112, 1989.

[3] Casadei, D. and Serra, G. and Tani, A.: A general approach for the analysis of the input power quality in matrix converters, IEEE Transactions on Power Electronics, Vol 13 no 5, pp. 882-891, 1998.

[4] Casadei, D. and Serra, G. and Tani, A.: Reduction of the input current harmonic content in matrix converters under input/output unbalance, IEEE Transactions on Industrial Electronics, Vol 45 no 3, pp. 401-411, 1998.

[5] Dastfan, A. and Haghshenas, M.: Design and Simulation of A Power Supply Based on A Matrix Converter Under Unbalanced Input Voltage, 41st IUPEC, Vol 2, pp. 569-573, 2006.

[6] Karaca, H. and Akkaya, R. and Dogan, H.: A novel compensation method based on fuzzy logic control for matrix converter under distorted input voltage conditions, 18th ICEM, pp. 1-5, 2008.

[7] Dasika, J. D. and Saeedifard, M.: A modulation strategy to control the Matrix Converter under unbalanced input voltage conditions, IEEE APEC, pp. 1556-1563, 2015.

[8] Patel, P. and Mulla, M. A.: Modified Space Vector Modulated Three-phase to Three-phase Matrix Converter Under Unbalanced Supply Conditions, 8th IICPE, pp. 1-6, 2018.

[9] Wei, L. and Matsushita, Y. and Lipo, T.A.: Investigation of dual-bridge matrix converter operating under unbalanced source voltages, IEEE PESC '03., Vol 3, pp. 1293-1298, 2003.

[10] Satish, T. and Mohapatra, K.K. and Mohan, N.: Modulation methods based on a novel carrier-based PWM scheme for matrix converter operation under unbalanced input voltages, APEC '06., pp. 127-132, 2006.

[11] Patil, K. R. and Patel, H. H.: Modified dual second-order generalised integrator FLL for synchronization of a distributed generator to a weak grid, EEEIC '16., pp. 1-5, 2016.

[12] Zhang, C. and Føyen, S. and Suul, J. A. and Molinas, M.: Modeling and Analysis of SOGI-PLL/FLL-Based Synchronization Units: Stability Impacts of Different Frequency-Feedback Paths, IEEE Transactions on Energy Conversion, Vol 36 no 3, pp. 2047-2058, 2021.

[13] R. Izah and S. Subiyanto and D. Prastiyanto: Improvement of DSOGI PLL Synchronization Algorithm with Filter on Three-Phase Grid-connected Photovoltaic System, Jurnal Elektronika dan Telekomunikasi, Vol 18 no 1, pp. 35-45, 2018.

[14] Q. Cheng and F. Tan and J. Gao and Y. Zhang and D. Yu: The separation of positive and negative sequence component based on SOGI and cascade DSC and its application at unbalanced PWM rectifier, IEEE 29th CCDC, pp. 5804-5808, 2017.

[15] She, Hongwu and Lin, Hua and Wang, Xingwei and Yue, Limin: Damped input filter design of matrix converter, PEDS, pp. 672-677, 2009.

[16] Costa, L. A. and Fan, B. and Burgos, R. and Boroyevich, D. and Chen, W. and Blasko, V.: The Fast Overvoltage Protection Consideration and Design for SiC-based Matrix Converters, APEC, pp. 1567-1574, 2020.

[17] Wheeler, P.W. and Rodriguez, J. and Clare, J.C. and Empringham, L. and Weinstein, A.: Matrix converters: a technology review, IEEE Transactions on Industrial Electronics, Vol 49 no 2, pp. 276-288, 2002.

[18] P. Rodríguez, A. Luna, R. S. Munoz-Aguilar, I. Etxeberria-Otadui, R. Teodorescu, and F. Blaabjerg: A stationary reference frame grid synchronization system for three-phase grid-connected power converters under adverse grid conditions, IEEE Transactions on Power Electronics, vol. 27, no. 1, pp. 99–112, 2012.

Double inverter concept for high-speed drives without motor filters

Henning Kasten, Stephan Beineke, Matthias Bachmann
KEBA Industrial Automation Germany GmbH
Gewerbestraße 5-9
35633 Lahnau/Germany
Tel.: +49 6441 966-0.
Fax: +49 6441 966-137
E-Mail: Henning.Kasten@keba.com
URL: www.keba.com

Keywords

Multi-level inverters, Interleaved converters, Multiphase converter, High-speed drive, Harmonics

Abstract

It is known that high-speed machines have high power and loss densities [1]. The additional losses in the rotor caused by harmonics are particularly critical, as they act in the center of the machine. This paper shows an approach to reach small additional losses without a filter between inverter and machine.

Introduction

High-Speed drives become an important key technology to gain sustainable solutions in many applications. Water treatment, processes in food industry or high-performance pumps and blowers [2] require high rotational speeds. Direct drives, build with Permanent Magnet Synchronous Machines (PMSM) and equipped with magnetic bearings, avoid mechanical transmissions. Thus, they offer much higher energy efficiency and less wear resulting in lower operation and maintenance costs compared to conventional drives systems. Also regenerative energy systems require these high speed drives; they drive kinetic energy storages or they are used as turbo generators in many processes, like ORC (organic Rankine cycle) processes for e.g. waste heat recovery, biomass or geothermal power plant. However, rotor losses are critical and should be considered carefully. They are one reason why these drives cannot be operated by standard inverters without additional measures. One standard measure for reducing rotor losses is to use standard two-level inverters with medium switching frequencies in combination with *LC* motor filters to smooth the current ripple. Because these filters are cost intensive, heavy and can have negative impact on robustness and performance, they should be avoided. For development of high-speed drives without such motor filter some specific effects have to be understood, leading to a special design of the drive system, that considers the interaction of inverter hardware and software and the electrical machines, as discussed in the following. Especially the effects of the switched output voltage of an inverter on the electrical machine are manifold. The high rates of voltage rise lead to overvoltage at the winding, inhomogeneous voltage distribution in the winding, bearing currents and electromagnetic interference. However, these problems will not be considered here. The focus is on the understanding of the additional losses in the rotor, which are excited by harmonic waves and oscillations. The harmonic oscillations are caused by the inverter. The spectrum of the output voltage contains numerous harmonics in addition to the desired fundamental oscillation. These lead to a current with harmonics, whereby harmonics near twice the switching frequency appear particularly dominant in the spectrum. Finally, the winding generates pulsating fields and rotating fields from the harmonics, which have speeds different from the mechanical speed of the rotor. In contrast to the fundamental wave field circumventing with the rotor, the harmonic fields generate time-varying inductions in the rotor. Due to the electrical conductivity of the solid components in the rotor, eddy currents are excited causing corresponding additional losses. In addition to the converter, the winding itself also generates harmonics, even if they are supplied by sinusoidal currents. The reason for this is that the coils are mounted in slots. The resulting staircase-shaped distribution of the magnetomotive force generates corresponding harmonics. Another source of harmonics are the slot openings. This changes the magnetic resistance at the circumference, so that the generated fields are modulated. The next section discusses the reduction of harmonic losses by using suitable inverter topologies. Then it will be shown how the losses in the electrical machine can be reduced by supplying them with several inverters. The knowledge gained is then verified by experimental measurements. Finally, a suitable inverter with further possible areas of application is presented.

Methods for reducing the harmonic content at the inverter output

Increasing the switching frequency: This method is often seen as the first option for reducing additional losses. In the power range up to a few 10 kW, converters have been developed that operate with switching frequencies of up to 100 kHz [2]. With the new but cost-intensive wide bandgap (WBG) semiconductor materials SiC and GaN, these high switching frequencies can be achieved even for higher power levels [3]. Note, that in nearly all applications high-speed drives, as considered here, operate in quasi-stationary mode. This is different from the requirements in servo drives and one reason, why such developments are not the cost optimized and best solution. Other reasons for this are:

- The speed is limited to approx. 45,000 rpm for power ratings above 100 kW. This results in fundamental frequencies of max. 1500 Hz. An increase of the switching frequency beyond 20 kHz reduces the losses only slightly.
- Inverters with conventional silicon semiconductors generate very high losses at the high switching frequencies, are large and correspondingly expensive.
- Inverters with WBG semiconductors are much more cost-intensive than a standard inverter with corresponding sinusoidal filters. The greater control dynamics of the filterless variant is not required in the applications.

Under the present conditions, inverters built with ordinary silicon semiconductors and switching frequencies between 10 kHz ... 25 kHz seem to be an ideal compromise between requirements from the application and costs. In the following, the influence of the frequency ratio on the harmonic losses is to be shown with measurements on motors with relatively low power ratings. Fig. 1 shows the measured harmonic losses of a standard asynchronous machine and a fast rotating permanent magnet synchronous machine. The measurements of the asynchronous machine are taken from [5]. Table 1 shows the machine data.

Table 1: Data of the used machines

Parameter	Asynchronous Machine	Synchronous Machine
Rated power	7.5 kW	35 kW
Rated speed	2900 rpm	54 000 rpm
Rated frequency	50 Hz	1800 Hz

If we look at the harmonic loss curves, we see that the loss proportion is much smaller for permanently excited high-speed machines. Above frequency ratios of 25, hardly any reductions are possible. With standard asynchronous machines, the harmonic losses can be decreased significantly up to frequency ratios around 150. The difference in loss reduction can be suspected from the fact that in the asynchronous machine a large part of the additional losses occurs in the cage winding. In the synchronous machines considered here, however, this is not existing. Due to the small improvements and the problems arising in the inverter to achieve high frequency ratios at high fundamental frequencies, a large increase in the switching frequency does not make sense.

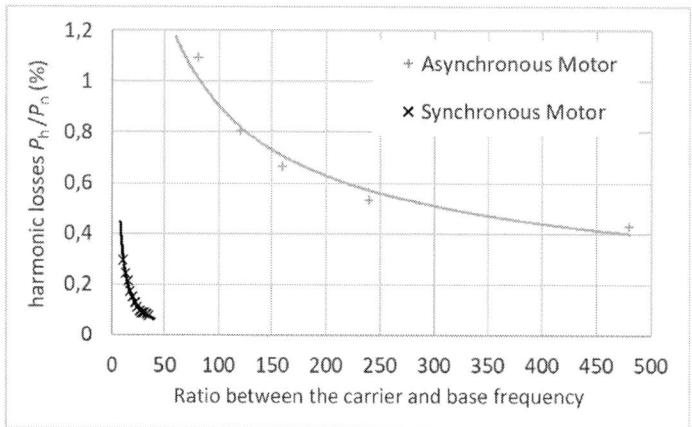

Fig. 1: Harmonic losses normalized to the rated power as a function of the switching frequency

Multi-level inverters: Another possibility to reduce the harmonic content is to increase the number of switching levels. Since the number of components used increases strongly with the number of switching levels, only three-level converters can currently be used effectively in the low-voltage range. In the meantime, there are many semiconductor modules available for this inverter topology, so that the module costs have been greatly reduced in recent years. The switching voltages at the inverter output are only half as large as those of the standard two-level inverter. This means that at the same switching frequency, the current ripple and the harmonic fields excited in the machine are also halved. According to [4, 6], the magnetic induction is squared in the magnetic losses. Therefore, by using inverters with three-level technology, the additional losses in the rotor and stator caused by the converter are reduced by factor 4. This fact was also confirmed by measurements [5]. The loss reduction through three-level inverters is very strong and sufficient for many applications [16].

Interleaved inverter

Use of coupling inductors: An interleaved inverter is built by connecting n inverters of the same type. Fig. 2 shows such a setup with two inverters. The fundamental oscillations are given out synchronously, but the carrier signals C_1 and C_2 of the modulation have phase offsets by $2\pi/n$ to each other. Fig. 3 shows the pulse patterns of two interleaved switching sub-inverters. The output currents i_1 and i_2 show the typical ripple of an inverter feeding an inductor. However, the sum results in a current that contains smaller harmonic amplitudes with double the frequency. The harmonic components of each sub-inverter are therefore not reduced by interleaving. The harmonics, which are no longer present in the total current, flow as circulating currents. The limitation of these harmonics must be done with correspondingly large chokes L_1 and L_2 [7]. Fig. 4 shows an example of the spectrum of the output voltage of an interleaved inverter with two sub-inverters.

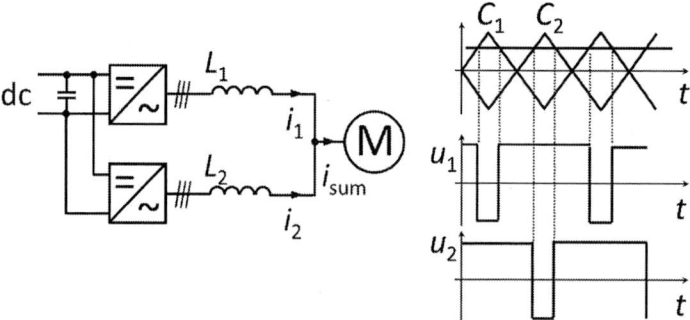

Fig. 2: Interleaved inverter

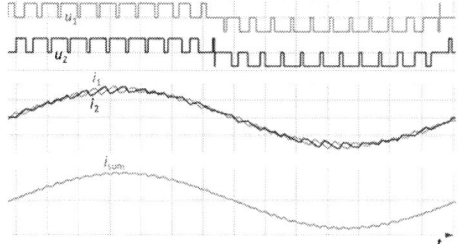 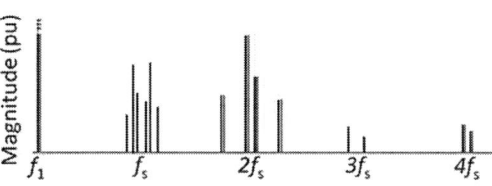

Fig. 3: Inverter outputs and their sum current one

Fig. 4: Frequency spectrum of the output voltage of inverter (black) and the sum of both inverters (red)

The harmonics and their sidebands with multiples of 1, 3, 5, ... of the switching frequency, which are present when only one inverter is working, can be suppressed. However, the particularly dominant harmonics with twice the switching frequency cannot be influenced. Because the density of the spectrum decreases with $1/n$ through n interleaved switching inverters, it can be assumed that the harmonic losses decrease with the same ratio.

Direct connection of the winding: If the armature winding is divided into several galvanically separated partial windings, a correspondingly large number of interleaved converters can be connected directly to them. The separate wiring between the inverter and the machine with several three-phase cables has further additional advantages:
- At high currents and frequencies, the current displacement losses in the cables are no longer negligible. A division into several three-phase cables reduces the current displacement. For this reason, [8] also recommends such cabling.
- The balancing of the currents is achieved by the positive temperature coefficient of the conductors. If the cables are connected to separate winding sections, the current is distributed more homogeneously among the cables.

In this configuration the circular currents are now limited by the inductances of the individual partial windings. Fig. 5 shows this situation for a winding with two parts. First, the circuit currents are limited by the leakage inductances of the two winding parts. Furthermore, the coupling of the winding parts by the main flux is not complete, so that the main inductance is modelled with 3 equivalent elements. L_{m1} and L_{m2} stand for the leakage inductances not chained to the main flux. L_m stands for the main inductance linked to the main flux. The effect of the winding resistances on the circulating currents can be neglected.

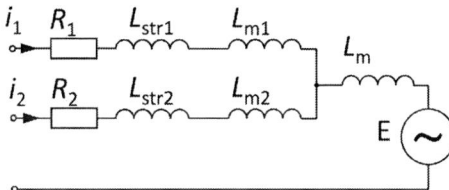

Fig. 5: Simplified equivalent circuit diagram for a synchronous machine fed by 2 inverters

In the following it will be shown how the winding construction influences the coupling of the winding parts. For simplicity, only windings with two partial windings are considered. In general, however, these statements can also be applied to several sub-windings.
Case a) The coils of the partial windings lie on top of each other and in the same slots:
In this case, very small stray inductances of the partial windings result. The magnetic coupling is at a maximum. As a result, very high circulating currents are generated through the two windings and inverters. Usually the inductances are too small, so that additional external inductances are necessary. The construction of such windings is difficult, as there are always two coils to be insulated, one on top of the other in the winding. However, the suppression of the harmonic content is maximum with this type of winding. It can be assumed that the same results are achieved as with interleaved converters connected by coupling chokes.

Case b) Partial windings with coil groups shifted by 360°:
Fig. 6 shows the position of the two coil groups of a phase in a 4-pole machine. From the indicated field lines it can be seen that the coil groups are not chained via the main flux. This means that there are no circulating currents between the two inverters. Due to the lack of coupling, however, the full harmonic spectrum of an inverter reaches the rotor, so that no reduction in losses can be expected with this arrangement.

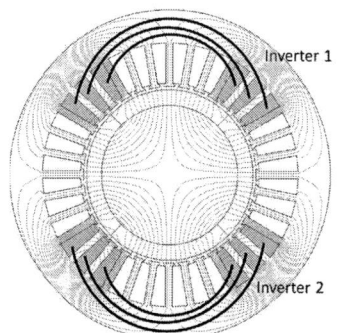

Fig. 6: Position of the coil groups shifted by 360° in a 4-pole machine

Case c) Partial windings with coil groups shifted by 180°:
Fig. 7 shows the placement of the coil groups. Since here, too, each slot magnetomotive force is generated by both inverters, the harmonics, eliminated by interleaved operation can also be suppressed. The end windings that are not directly on top of each other increase the leakage inductance, so that the circular currents in this case are smaller compared to case a (coils are on top of each other). However, to suppress winding harmonics, chorded coils are often used so that the coil groups do not lie exactly in the same slots. Fig. 8 shows such a possible position. The field shown results from the circular current. As can be seen, the circular current forms fluxes that only partially reach the rotor. For this reason, the suppression of harmonics is not as good as in an interleaved inverter with coupling chokes or in the case of windings where the two parts of the winding are placed in the same slots. On the other hand, in this case there is a considerable inductance that limits the circulating currents, so that additional chokes can be omitted. As a rule, if the suppression of the additional losses caused by the inverter is sufficient, this variant with chorded coils is forced in an interleaved concept, since no additional chokes are necessary.

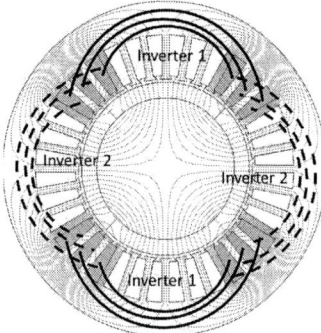 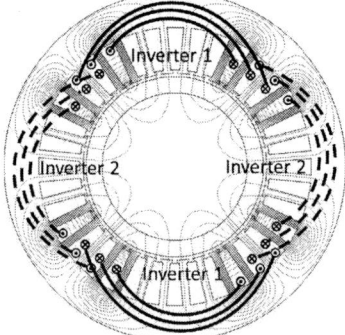

Fig. 7: Position of the coil groups shifted by 180° in a 4-pole machine

Fig. 8: Position of the coil groups shifted by 180° with chorded coils in a 4-pole machine

High phase windings

Similar to the interleaved concept, several 3-phase inverters feeding a corresponding number of galvanically separated winding parts are used here. In contrast to interleaved system, in which the carrier frequencies are shifted, the phase position between the fundamental oscillations is shifted here. As a result, the winding parts must also be shifted arranged in the stator. This known method [9-11] has the following advantages:

- Increase of the fundamental winding factor, so that 8-12 % winding losses can be reduced with the same material usage.
- The lower winding losses reduce the winding temperature and indirectly also the critical rotor temperatures.
- Reduction of the harmonic content of the winding, so that the losses in the rotor decrease.
- The rotor losses generated by harmonics oscillations can be well suppressed by using shielding sleeves [12]. However, this solution produces higher winding harmonic losses. If such an arrangement is to be used effectively, low-harmonic windings must be used accordingly. This goal can be achieved well by using high phase windings.

The ideal phase shift α_i between the 3-phase systems depends on the number n of winding systems used:

$$\alpha_i = \frac{60°}{n}.$$

The number n of systems influences the harmonic content, which can be described by the coefficient of differential leakage [12, 13]. The influence of the number of systems is shown in the example of a 2-pole winding with 72 slots (Table 2). It is noticeable that the fundamental winding factor can be increased when two three-phase systems are used. A further increase in the number of winding systems, on the other hand, neither increases the fundamental winding factor, nor reduces the coefficient of the differential leakage. Investigations in[14] show that the winding losses decrease quadratically to the fundamental winding factor.

Table 2: Properties of the high phase windings

number of 3-phase systems	1	2	3	4
phase shift (deg)	-	30	20	15
nuber of slots per pole and phase	12	6	4	3
coil pitch (slots)	30	33	34	35
fundamental winding factor ξ_1	0.9227	0.9805	0.9915	0.9965
$(\xi_{1,3}/\xi_1)^2$	1	0.886	0.866	0.8573
differential leakage factor (%)	0.0928	0.065	0.0639	0.0638

With two shifted three-phase systems, the winding losses can be reduced by approx. 11 %. If four systems are used, the winding losses are reduced to approx. 14 %. As can be seen, an increase to more than two three-phase systems hardly results in any advantages. In addition, such high-phase systems often require an impracticably high number of slots. Apart from the loss considerations, there is another advantage with these windings. Due to the high frequency and power, the number of coil turns are very small, so that an adjustment to the ideal number of turns is not possible. Furthermore, the required large conductor cross-sections tend to cause large current displacement losses, which can only be avoided by twisted strands with very thin individual wires. This results in a poor slot fill factor [15]. Both problems can be reduced if the number of parallel armature paths is increased. It is known that three-phase windings can have a maximum of as many parallel armature paths as this winding has poles. In the case of windings constructed by n systems, this applies to each system. Accordingly, such windings can be constructed by $2p·n$ parallel interacting armature paths. In addition, the 5th and 7th harmonics can be suppressed so that the generated torque pulsates less. Typically, such harmonics occur in V/f controlled operation. This can disturb the start-up. Due to the effective suppression when using two inverters shifted by $30°_{el}$, the start-up can take place much less critically. This mode of operation is no more up-to-date today. Nevertheless, it is used for sensorless control in the lower speed range when there are only small differences between the d- and q-axis inductances in synchronous machines.

Measurement results

All measurements were done on permanently excited 300 kW machines, which can be operated as motors or generators. The tests were made in the speed range between 18 000 rpm and 22 000 rpm.

Reduction of losses through 3-level inverter and interleaved operation:
The power loss reduction was measured in no-load operation, as this can be determined directly via the power consumption. To measure the power consumption, a power meter with sufficient bandwidth was used between the inverter and the electric machine operated as a motor. Table 3 shows the losses measured in this way. Comparing the two interleaved measurements, it can be seen that the harmonic losses can be reduced to approx. ¼ by the 3-level control. The loss reduction is much weaker when comparing non-interleaved and interleaved operation in 3-level mode. The loss reduction is less than ½. In chapter II it was derived that the losses decrease proportionally with the number of inverters in interleaved operation. In the measurement, two inverters were used. However, the currents were not added by a coupling inductor. Both converters fed one winding part each. Their windings are shifted by 180°. As described above, the power loss reduction is less than ½, in this case 0.4. Fig. 9 shows the influence of interleaved operation on the rotor temperatures under load with nominal torque. These comparative measurements were also carried out with 3-level inverters. The rotor temperatures were measured at 6 axial points during operation. Data transmission was carried out using near-field telemetry. Here, too, it was shown that the rotor temperatures could only be reduced slightly by approx. 3 K through the interleaved drive.

Table 3: Measured no load losses

Mode	Total losses (W)	Harmonic current losses (W)
2-Level, interleaved	3583	437
3-Level, not interleaved	3330	184
3-Level, interleaved	3255	109

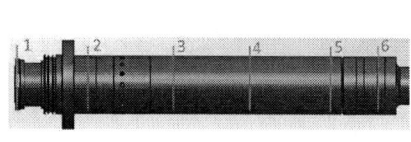
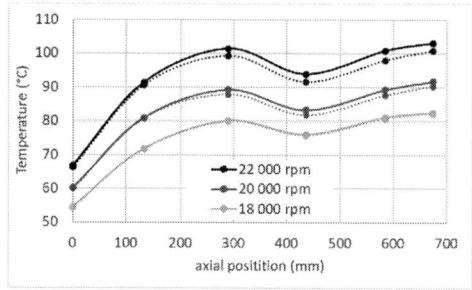

Fig. 9: measured rotor temperatures (interleaved operation is shown with dotted lines)

Winding with two winding parts shifted by 30°:
Compared to the 3-phase standard winding, the measured cold resistance has increased by factor 1.08. The reason for this is that the end windings are 1 slot pitch longer. Since the winding factor also increased from 0.925 to 0.983, the winding losses can be slightly reduced:

$$\frac{P_{cu,6\sim}}{P_{cu,3\sim}} = 1.08 \cdot \left(\frac{0.925}{0.983}\right)^2 = 0.96 \, .$$

The smaller winding cross-section results in smaller current displacement losses, which, however, could not be determined by measurements. However, the total losses at 300 kW power could be measured. Table 4 shows the correspondingly determined values for this and the usual 3-phase winding. The total losses could be reduced by approx. 10 % at 18 000 rpm. At 22 000 rpm, more field weakening current has to be provided due to the higher winding factor, so that the loss reduction is lower.

Table 4: measured total losses at 300 kW mechanical power

Winding type	18 000 rpm	22 000 rpm
3-phase standard	5.04 kW	4.74 kW
30° shifted winding parts	4.51 kW	4.56 kW

Inverter concept

Based on the research results discussed above, a multifunctional inverter has been designed as shown in Figure 10. It contains two 3-level inverters operating on one DC link. The two inverters can operate synchronized for driving one motor (interleaved operation, generation of two phase-shifted three-phase systems). However, it is also possible for the two converters to perform different control tasks independently of each other. The inverter is characterized by the following additional features:

- The connectable DC link: The inverter can also be fed or fed back directly into the DC link.
- Braking resistors for safe braking even in the event of a mains failure.
- Input rectifier with thyristors. These can ramp up the DC link voltage in a controlled process. A start via charging resistors with additional external relay is not necessary.
- Integrated power supply unit for magnetic bearings, which is internally fed from the DC link. This means that even in the event of mains failures, the magnetic bearing electronics can work as long as the machine is rotating. In this case the DC link voltage is maintained by braking operation.
- The output power of the two inverters is fed out to four 3-phase terminals. This allows the power to be distributed over several thin cable cross-sections. The thinner cable cross-sections are easier to install and avoid high current displacement at high output frequencies required by high speed drives.
- All 12 phase connections are monitored with current sensors, so asymmetries and cable breaks can be detected.

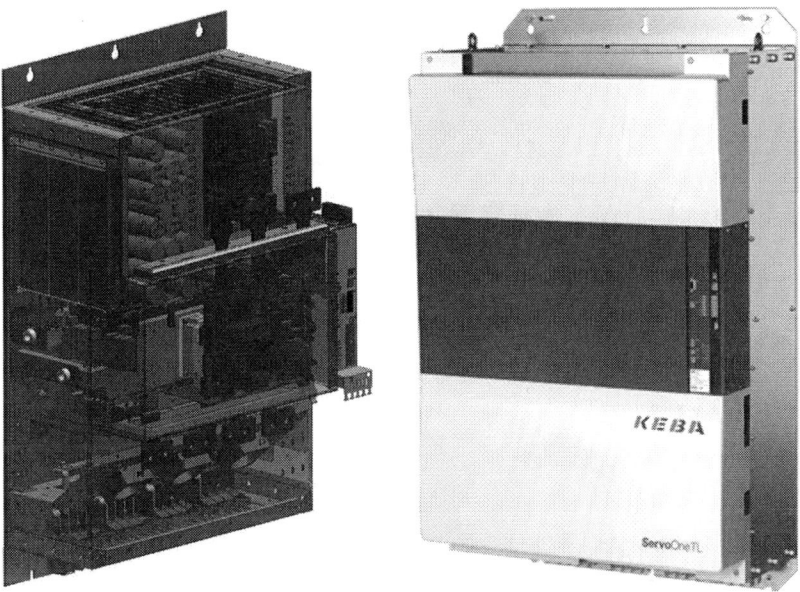

Fig. 10: Double inverter: Basic layout (left) and view of the finished unit (right)

In addition to the high-speed applications, that benefits from the reduction of additional losses caused by the inverter, the modular double inverter can be used in other applications:

- Operation on 2 independently running motors (multi-axis drive)

- Operation as active front end (boost converter) with higher DC link voltages
- Inverters that generate three-phase systems with a multiple of 60°: This does not reduce the losses in the electrical machine. However, common mode is completely avoided, so that the center of the DC link no longer oscillates at 3 times the fundamental frequency. The modulation level can thus be increased at low fundamental frequencies.
- Operation as regenerative converter: One output stage works as an active rectifier. To reduce the output currents and cable cross-sections, it is suitable to use higher DC link voltages in the range between 700 V and 800 V. The other output stage operates as an inverter on the electrical machine. Due to the possible synchronization of both output stages, the undesired overshooting common-mode voltage at the machine terminals can be actively damped [17].

Conclusion

If high-speed machines with high power are fed by an inverter, the harmonics generated cause a problem. Since typical motor filters are very large and expensive, they have to be avoided and other approaches for reducing the harmonic content should be evaluated. In this paper, the approaches of increasing the switching frequency, increasing the number of switching levels, and interleaved converters have been discussed. Like the harmonic oscillations, the harmonic waves generated by the winding also have an impact. Both effects cause an increase in losses and temperature in the most critical component of the electrical machine, the rotor. A further method for loss reduction has been proposed, which makes use of several converters generating phase-shifted three-phase systems. In this method, the winding losses and the losses generated by harmonics are reduced. As a result from this analysis, a multifunctional double converter has been designed. By housing two converters in one unit, each equipped with an own controller, the double inverter provide all operation modes and include most methods, which have been proposed for the reduction of the additional losses caused by the converter. The flexible and modular concept of the inverter allows its use in even more applications than just high-speed drives.

References

[1] L. Schwager, A. Tuysuz, C. Zwyssig and J. W. Kolar, "Modeling and comparison of machine and converter losses for PWM and PAM in high-speed drives", *IEEE Trans. Ind. Appl.,* vol. 50, no. 2, pp. 995-1006, Mar./Apr. 2012.

[2] S. Beineke, L. Hebing, A. Bünte, "High-Speed Drive with Three-Level Inverter for Vacuum Pumps, Laser Cooling and High-Speed Cutting", EPE 2003

[3] Shirabe, M. Swamy, J. Kang, M. Hisatsune, Y. Wu, D. Kebort, et al., "Advantages of High Frequency PWM in AC Motor Drive Applications", *IEEE Energy Conversion Congress and Exposition (ECCE),* 2012.

[4] G. Müller, K. Vogt, B. Ponick, and G. Müller, "Berechnung elektrischer Maschinen" Wiley, 2012.

[5] M. Schweizer, J.W. Kolar, "High Efficiency Drive System with 3-Level T-Type Inverter", *Proceedings of the 14th IEEE International Power Electronics and Motion Control Conference (ECCE Europe 2011)*, Birmingham

[6] M. K. Bradley, W. Cao, J. Clare, and P. Wheeler, "Predicting inverter-induced harmonic loss by improved harmonic injection", *IEEE Transactions on Power Electronics*, vol. 23, no. 5, pp. 2619–2624, 2008.

[7] J. Endres, "Hochdynamischer Stromrichter in Hybridstruktur", Diss. Technischen Universität Ilmenau, 2017

[8] IEC TS 60034-25 Rotating electrical machines-Part 25: Guidance for design and performance of a.c. motors specifically designed for converter supply, 2007

[9] R. Gregor, F. Barrero, S. Toral, M.J. Durán, "Realization of an Asynchronous Six-Phase Induction Motor Drive Test-Rig", Renewable Energy & Power Quality Journal, Vol. 1, No.6, March 2008

[10] H. Kasten and W. Hofmann, "Electrical machines with higher efficiency through combined star-delta windings", *Electric Machines Drives Conference (IEMDC) 2011 IEEE International*, pp. 1374-1379, May 2011.

[11] M. B. Slimene, "Performance analysis of six-phase induction machine-multilevel inverter with arbitrary displacement", *Electrical engineering & electromechanics*, 2020, no. 4, pp. 12-16.

[12] H. Kasten, G. Che, S. Beineke, "PM-machines with an additional shielding sleeve to reduce the rotor losses", *Energietechnische Gesellschaft, ETG-Fb. 164*, Elektromechanische Antriebssysteme, 2021

[12] Klima, V, On the Theorem of the Sum of Squares of Winding Factors Invariance, *Acta Technica* (1979) H 3, pp 365

[13] M. Caruso, A. O. Di Tommaso, F. Genduso, R. Miceli, G. R. Galluzzo, "A General Mathematical Formulation for the Determination of Differential Leakage Factors in Electrical Machines with Symmetrical and Asymmetrical Full or Dead-coil Multiphase Windings" *IEEE Transactions on Industry Application*, 2018

[14] H. Kasten, "Verbesserung der Betriebseigenschaften elektrischer Maschinen durch den Einsatz kombinierter Wicklungen", Diss. Technischen Universität Dresden, 2015

[15] H. Kasten and W. Hofmann, "Optimal number of strands in electrical medium-frequency machines," *The XIX International Conference on Electrical Machines - ICEM 2010*, pp. 1-6

[16] G. Kucera, "Drei-Level-Frequenzumrichter reduziert die Motorverluste", Leistungselektronik, 17.11.2020, https://www.leistungselektronik.de/drei-level-frequenzumrichter-reduziert-die-motorverluste-a-1043366/

[17] M. Schmitt, A. Ackva, "Active damping of common-mode oscillations in electric drive systems using direct current control," *2017 19th European Conference on Power Electronics and Applications (EPE'17 ECCE Europe)*, pp. P.1-P.9

A Universal Single Stage Current-fed Bidirectional Converter with both AC and DC Input Power Source Compatibility

Manish Kumar, Sumit Pramanick and Bijaya Ketan Panigrahi
Department of Electrical Engineering, Indian Institute of Technology, Delhi
Email: manish.kumar@ee.iitd.ac.in

Acknowledgments

This work was supported in part by the Ministry of Education under the project titled "Smart Infrastructure for an Electric Vehicle Ecosystem."

Keywords

≪On-board Charger≫, ≪Isolated converters≫, ≪Current Source Converter (CSC)≫, ≪Photovoltaic≫, ≪MPPT≫,.

Abstract

This article demonstrates a universal, single power conversion stage (1-S) current-fed bidirectional converter based on-board electric vehicle (EV) charger (OBC). The converter can be connected to either residential ac utility gird or solar photovoltaic (PV) array. Additionally, when connected to ac grid, the converter maintains unity power factor (UPF) and when connected to PV array, the power is extracted at the maximum power point (MPP). A laboratory prototype of 1.5 kW is developed to validate the theoretical analysis and can be connected to 230 V, 50 Hz ac mains voltage or 100-300 V PV array. The battery voltage range is of 300-400 V.

1 Introduction

Advancements in battery technology, charging infrastructures, and efficient drive trains, electric vehicles (EVs) have grown in popularity and customer acceptability. As shown in Fig. 1a, EVs power systems generally comprises of drivetrain, on-board charger (OBC), auxiliary power module (APM) and motor drive inverter [1]. OBC is an integral part of EV as it allows the customer to charge their vehicle overnight using a residential ac utility grid.

General design criteria for OBCs is to meet power quality requirements on the grid side, as defined by IEEE 519 standards [2]. In addition, galvanic isolation between the power source and the vehicle battery is required for safety requirements as defined by Underwriters Laboratories (UL) 2202 standards [3]. Furthermore, owing to space constraints in EVs, OBCs should have high volumetric and electrical efficiency. These design requirements are met by OBCs with two power conversion stages (2-S). The first stage is a power factor correction (PFC), while the second stage is an isolated dc-dc converter [4,5]. A 2-S based OBCs have increased component counts, resulting in higher losses.

To overcome the drawbacks of 2-S based OBCs, several single power conversion stage (1-S) OBCs have been explored. In [6, 7], a 1-S bridgeless ac–dc converter with power factor correction for EV charging has been reported. However, these converters lack galvanic isolation, as required by safety standards. [8] showed a 1-S half-bridge matrix convert with combined frequency and phase shift modulation and [9] proposed an optimal zero voltage switching (ZVS) modulation for DAB-based converter, but requires an additional line frequency synchronous rectifier on the grid side. Additionally, the above mentioned

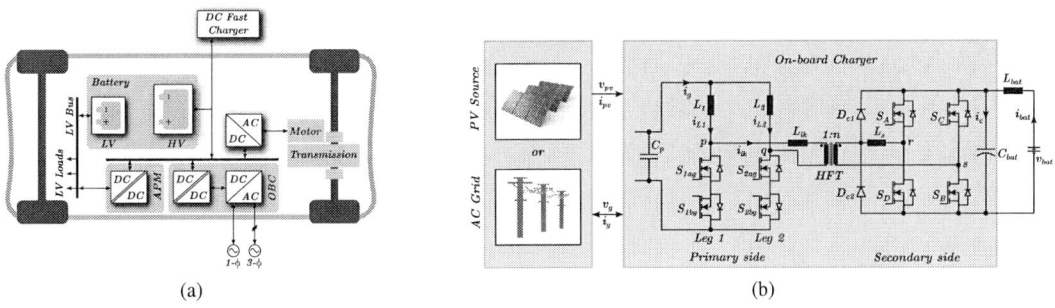

Fig. 1: (a) EV power system architecture. (b) L-L type current-fed half bridge bidirectional converter.

converters can only be connected to ac residential grid and therefore are classified as isolated ac-dc power converters. The charger developed in [10, 11] connects only to solar photovoltaic (PV) array to charge the battery.

This paper demonstrates a 1-S L-L type current-fed half-bridge (CFHB) bidirectional converter with both ac and dc input power source compatibility. The CFHB is a good alternative to voltage-fed converters due to reduced input current ripple, efficient performance over a wide input range, and easier input current controllability [12]. Furthermore, the converter can be connected to a residential ac utility grid and draw power at unity power factor (UPF). The converter can also be connected to a PV array in the absence of ac grid and extract power at the maximum power point (MPP).

The paper is organized as follows: section 2 provides the detailed description of the converter along with the operation. Experimental results are presented and discussed in section 3. Lastly, conclusions are given in section 4.

2 Converter Description and Operation

2.1 Converter description

Fig. 1b depicts the schematic of a 1-S galvanically isolated current-fed half bridge bidirectional converter. On the primary side, it consists of a current-fed half bridge converter with bidirectional switches, and on the secondary side, it consists of a voltage-fed full bridge converter connected via a high frequency transformer (HFT). The two boost inductors are L_1 and L_2. The HFT's leakage inductance is L_{lk}. To obtain the necessary total series inductance $L_t = L_{lk} + L_s/n^2$, an external series inductor L_s is added on the secondary side. The primary and secondary side filter capacitors are denoted by C_p and C_{bat} respectively. D_{c1} and D_{c2} are the diode clamps that prevents parasitic ringing across the primary side switches.

2.2 Converter Operation

The 1-S converter can be connected to either a residential ac utility grid or solar PV array. The switching scheme when connected to ac utility grid is based on improved discontinuous current phase shift modulation (IDCPSM) and is described as follows. Primary side switches S_{1ag} and S_{2ag} are switched at a duty ratio of $d_1^{(k)} > 0.5$ in the positive half cycle of grid voltage and is continuously on in the negative half cycle. Similarly, S_{1bg} and S_{2bg} are switched at a duty ratio of $d_1^{(k)}$ in the negative half cycle of grid voltage and is continuously on in the positive half cycle. Secondary side switches S_A and S_D are switched at a fixed duty ratio of 0.5. S_B and S_C are switched at a variable duty ratio of $d_2^{(k)}$. The switching scheme when connected to solar PV array remains same and depending on the connection polarity, the corresponding primary side switching combination is applied. Fig. 2 shows the operation waveform of the converter in a grid cycle along with the switching interval waveform shown in the zoomed section.

Following assumption are made for simplifying the analysis of the converter.

- When connected to a domestic ac grid, the converter operates at UPF.

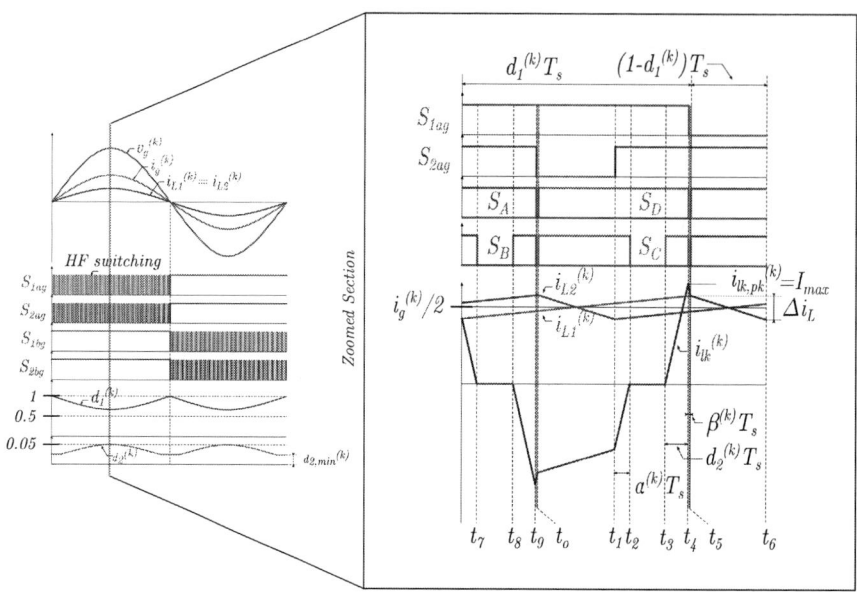

Fig. 2: Operational waveform of the converter in a grid cycle.

- In a switching interval k, the input grid voltage v_g and grid current i_g are assumed to be constant and is given by (1) and (2) respectively.

- For high switching operation, the effect of parasitic capacitance and device output capacitance are ignored.

$$v_g^{(k)}(kT_s) = V_m sin(\omega kT_s) \tag{1}$$

$$i_g^{(k)}(kT_s) = \frac{2P_o}{V_m} sin(\omega kT_s) \tag{2}$$

Primary side duty cycle is given by

$$d_1^{(k)} = \frac{V_{bat} - n|v_g^{(k)}(kT_s)|}{V_{bat}} \tag{3}$$

Secondary side duty cycle is given by

$$d_2^{(k)} = \frac{nL_t|i_{lk,pk}^{(k)}|}{V_{bat}T_s} \tag{4}$$

where, $\omega = 2\pi/t_g$, t_g is the time period of a grid cycle, T_s is the time period of a switching cycle.

2.3 AC utility grid operation

The OBC when connected to ac utility grid, a dual control variable approach proposed in [13] has been adopted. In this control structure, the primary side duty $d_1^{(k)}$ is modulated to maintain UPF. In addition, to minimize the peak circulating current flowing through the HFT, secondary side duty $d_2^{(k)}$ is modulated across the grid cycle of ac mains voltage. The nature of primary and secondary side duty in a grid cycle is depicted in Fig. 2. Furthermore, throughout the operation, ZCS turn-off is achieved for primary side switches and ZVS turn-on is achieved for secondary side switches. This results in very low conduction and switching losses.

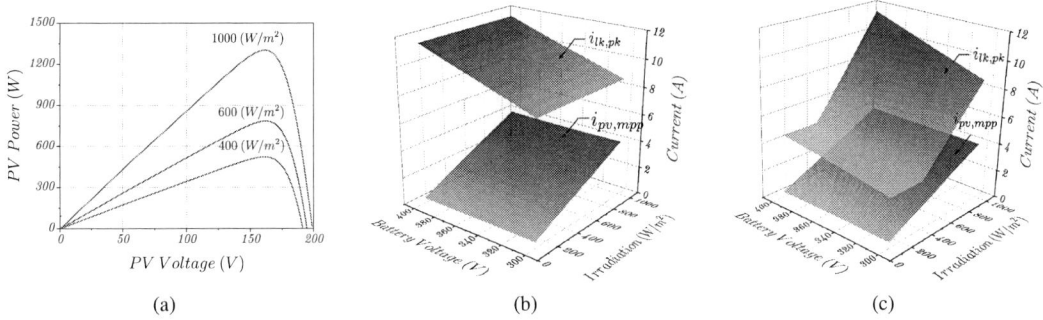

Fig. 3: (a) PV curve for different irradiation conditions, (b) variation of transformer winding peak current $i_{lk,pk}$ and MPP PV array current $i_{pv,mpp}$ for different battery voltage and irradiation at fixed secondary side duty and (c) varying secondary side duty

2.4 PV array operation

The OBC when connected to PV array, a similar dual control variable approach is developed. This control structure allows for MPPT by modulating the primary side duty $d_1^{(k)}$ using adaptive incremental conductance algorithm [14]. The secondary side duty $d_2^{(k)}$ can be kept fixed to 0.05 for active current commutation as proposed in [15]. However, the MPP of the PV array reduces under low irradiation conditions as depicted by the PV curve in Fig. 3a. Therefore, from (4) it can be concluded that for a fixed battery voltage, the peak value of transformer winding current would remain fixed irrespective of the irradiation condition. This means that for low irradiation conditions, the circulating current through the transformer winding would be substantially high as depicted in Fig. 3b, thereby contributing to higher losses. In order to minimize the circulating at low irradiation conditions, the secondary side duty $d_2^{(k)}$ is varied in this control structure. This can be seen observed in Fig. 3c that for varying irradiation, the peak current through the transformer winding is closely following the MPP PV array current, thereby substantially reducing the losses in the converter.

2.4 Current expressions

For detailed loss analysis, the section provides this current expressions flowing through the various components of the converter when connected to ac grid. Same expressions are used when connected to PV array by assuming $t_g = T_s$

- Primary side HFT RMS current

$$I_{lk,RMS} = \sqrt{\frac{T_s}{t_g} \sum_{k=1}^{t_g/T_s} \frac{(i_g^{(k)})^2}{2} \left(1 - d_1^{(k)} + \frac{4d_2^{(k)}}{3} + \frac{\alpha^{(k)}}{3} + \frac{7\beta^{(k)}}{3} \right)} \tag{5}$$

- Primary side switch RMS current

$$I_{S_{pri,RMS}} = \sqrt{\frac{T_s}{t_g} \sum_{k=1}^{t_g/T_s} \left(\frac{i_g^{(k)}}{2} \right)^2 \left(\frac{11}{3} - \frac{10d_1^{(k)}}{3} + 4d_2^{(k)} + \frac{7\alpha^{(k)}}{3} + \frac{17\beta^{(k)}}{3} \right)} \tag{6}$$

- Primary side switch body diode average current

$$I_{D_{pri,avg}} = \frac{T_s}{t_g} \sum_{k=1}^{t_g/T_s} \frac{i_g^{(k)}}{2} \frac{\beta^{(k)}}{2} \tag{7}$$

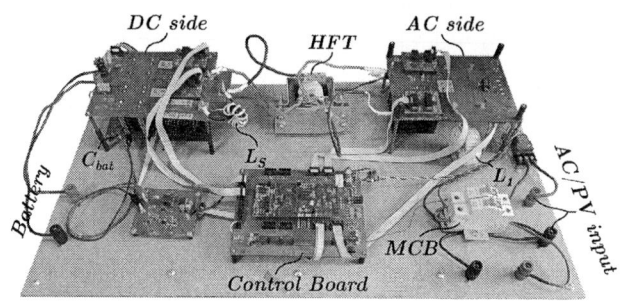

Fig. 4: Laboratory prototype of L-L type current fed half bridge bidirectional converter.

Table I: Experimental Setup Parameters

Converter Parameters		PV Panel Parameters	
Descriptions	**Specifications**	**Descriptions**	**Specifications**
Power Rating (P_o)	1.5 kW	Maximum Power ($P_{pv,MPP}$)	0.325 kW
Input Grid Voltage (V_g)	230 V	MPP Voltage ($v_{pv,MPP}$)	40 V
Battery Voltage (V_{bat})	300...400 V	Open Circuit Voltage ($v_{pv,OC}$)	50 V
Transformer Turns ratio (n) and Leak Inductance (L_{lk})	0.37 and 6.5 μH	No. of Panel in series	4
Boost Inductor (L_1) and (L_2)	1.5 mH	Model	Kyocera Solar KD325GX
Primary side filter Capacitor (C_p)	4.7 μF		
Secondary side filter Capacitor (C_{bat})	100 μF		
Switching Frequency (f_{sw})	100 kHz		
Primary side Mosfets (S_{xyg})	C2M0080170P		
Secondary side Mosfets (S_p)	UF3C065030K4S		

- Secondary side switch RMS current

$$I_{S_{sec,RMS}} = \sqrt{\frac{T_s}{t_g} \sum_{k=1}^{t_g/T_s} \left(\frac{i_g^{(k)}}{2n} \right)^2 \left(1 - d_1^{(k)} + \frac{4d_2^{(k)}}{3} + \frac{\alpha^{(k)}}{3} + \frac{7\beta^{(k)}}{3} \right)} \tag{8}$$

3 Experimental Results and Discussion

To evaluate the converter's performance with both ac and dc input power sources, a 1.5-kW laboratory prototype is developed as shown in Fig. 4. The ac input source is 230V, 50Hz, and the dc input is via a PV emulator. The parameters of the experimental setup are listed in Table I. A TI TMS320F28397D launchpad is used to implement the control scheme, and a Xilinx XC6SLX4-based FPGA board is used to generate the gate pulse combinational logic.

3.1 Experimental waveform

The measured ac grid voltage v_g, grid current i_g, battery current i_{bat} and primary side HFT winding current i_{lk} for few grid cycle is shown in Fig. 5a. It is evident from the figure that converter is operating at 0.999 power factor and $P_o = 1.5\,kW$. The primary side HFT winding current has a sinusoidal envelope with a frequency that is twice that of the ac mains voltage. The measured primary side v_{pq} and secondary side v_{rs} HFT voltage and primary side HFT winding current i_{lk} for a few switching interval is shown in Fig. 5b.

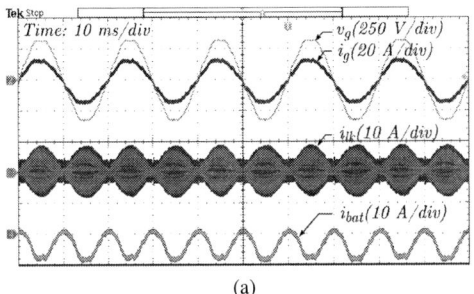
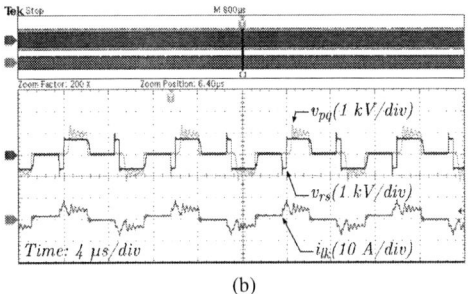

(a)

(b)

Fig. 5: (a) Measured grid voltage v_g, grid current i_g, battery current i_{bat} and primary side HFT winding current i_{lk} ($P_o = 1.5\,kW, v_g = 230\,V_{rms}$ and $V_{bat} = 345\,V$), (b) measured primary side v_{pq} and secondary side v_{rs} HFT voltage and primary side HFT winding current i_{lk} in a few switching interval.

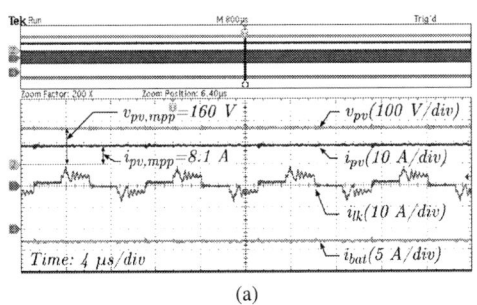
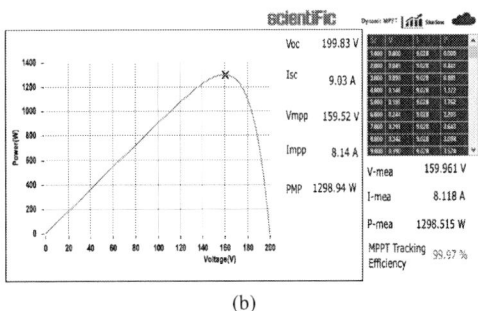

(a)

(b)

Fig. 6: (a) Measured PV voltage v_{pv}, PV current i_{pv}, battery current i_{bat} and primary side HFT winding current i_{lk} ($P_{pv,mpp} = 1.3\,kW$ and $V_{bat} = 345\,V$), (b) PV emulator software showing MPP tracking efficiency.

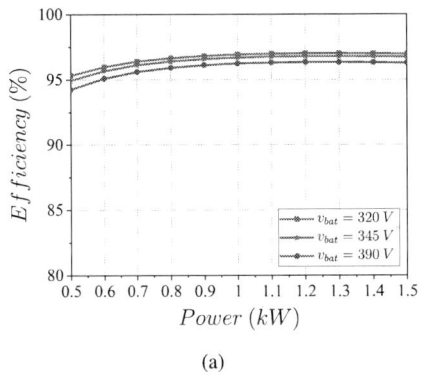
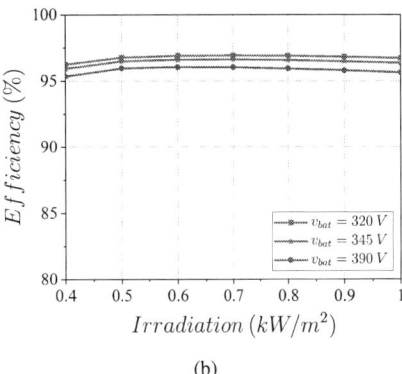

(a)

(b)

Fig. 7: Efficiency curve for different battery voltage V_{bat} (a) grid connected varying loading conditions (b) PV connected varying irradiation conditions.

It can be clearly observed from Fig. 6a showing the measured PV voltage v_{pv}, PV current i_{pv}, battery current i_{bat} and primary side HFT winding current i_{lk}, that the converter is extracting power from the PV array at MPP i.e. $P_{pv,mpp} = 1.3\,kW$. This can be also seen in Fig. 6b showing the MPP tracking efficiency of 99.97% from PV emulator software.

3.1 Efficiency

Fig. 7a and 7b shows the theoretical efficiency cure for different battery voltage V_{bat} when connected to ac grid and PV array respectively. A peak efficiency of 97.01% is observed when operated in grid connected mode and peak efficiency of 96.92% and almost a flat efficiency profile across irradiation conditions is seen when connected to PV array.

4 Conclusion

A universal, 1-S current-fed bidirectional converter as an OBC has been demonstrated in this article. The converter is compatible with both ac and dc input power source i.e. it can be connected to either residential ac utility gird or PV array. The converter draws power at UPF from the ac grid and the power is extracted at the MPP when connected to PV array. The IDCPSM switching scheme has been modified to minimize circulating current across varying irradiation conditions. Peak efficiency of 97.01% when connected to ac grid and 96.92% when connected to PV array is achieved.

References

[1] A. Khaligh and M. D'Antonio, "Global Trends in High-Power On-Board Chargers for Electric Vehicles," IEEE Trans. Veh. Technol., vol. 68, no. 4, pp. 3306–3324, Apr. 2019.

[2] "IEEE Recommended Practice and Requirements for Harmonic Control in Electric Power Sys- tems," IEEE Std 519-2014 Revis. IEEE Std 519-1992, pp. 1–29, Jun. 2014.

[3] "UL-2202StandardforElectricVehicle(EV)ChargingSystemEquipment—StandardsCatalog," https://standardscatalog.ul.com/standards/en/standard 2202 2.

[4] J. Lu, K. Bai, A. R. Taylor, G. Liu, A. Brown, P. M. Johnson, and M. McAmmond, "A Modular-Designed Three-Phase High-Efficiency High-Power-Density EV Battery Charger Using Dual/Triple-Phase-Shift Control," IEEE Trans. Power Electron., vol. 33, no. 9, pp. 8091–8100, Sep. 2018.

[5] B.-K. Lee, J.-P. Kim, S.-G. Kim, and J.-Y. Lee, "An Isolated/Bidirectional PWM Resonant Con- verter for V2G(H) EV On-Board Charger," IEEE Trans. Veh. Technol., vol. 66, no. 9, pp. 7741– 7750, Sep. 2017.

[6] B. R. Ananthapadmanabha, R. Maurya, and S. R. Arya, "Improved Power Quality Switched In- ductor Cuk Converter for Battery Charging Applications," IEEE Trans. Power Electron., vol. 33, no. 11, pp. 9412–9423, Nov. 2018.

[7] J. Gupta, R. Kushwaha, and B. Singh, "Improved Power Quality Transformerless Single-Stage Bridgeless Converter Based Charger for Light Electric Vehicles," IEEE Trans. Power Electron., vol. 36, no. 7, pp. 7716–7724, Jul. 2021.

[8] F. Jauch and J. Biela, "Combined Phase-Shift and Frequency Modulation of a Dual-Active-Bridge AC–DC Converter With PFC," IEEE Trans. Power Electron., vol. 31, no. 12, pp. 8387–8397, Dec. 2016.

[9] J. Everts, F. Krismer, J. V. den Keybus, J. Driesen, and J. W. Kolar, "Optimal ZVS Modulation of Single-Phase Single-Stage Bidirectional DAB AC–DC Converters," IEEE Trans. Power Electron., vol. 29, no. 8, pp. 3954–3970, Aug. 2014.

[10] J. Traube, F. Lu, D. Maksimovic, J. Mossoba, M. Kromer, P. Faill, S. Katz, B. Borowy, S. Nichols, and L. Casey, "Mitigation of Solar Irradiance Intermittency in Photovoltaic Power Systems With Integrated Electric-Vehicle Charging Functionality," IEEE Trans. Power Electron., vol. 28, no. 6, pp. 3058–3067, Jun. 2013.

[11] S.Biswas,L.Huang,V.Vaidya,K.Ravichandran,N.Mohan,andS.V.Dhople,"UniversalCurrent- Mode Control Schemes to Charge Li-Ion Batteries Under DC/PV Source," IEEE Trans. Circuits Syst. Regul. Pap., vol. 63, no. 9, pp. 1531–1542, Sep. 2016.

[12] K. Gnanasambandam, A. K. Rathore, A. Edpuganti, D. Srinivasan, and J. Rodriguez, "Current-Fed Multi-level Converters: An Overview of Circuit Topologies, Modulation Techniques, and Applica- tions," IEEE Trans. Power Electron., vol. 32, no. 5, pp. 3382–3401, May 2017.

[13] M. Kumar, S. Pramanick, and B. K. Panigrahi, "Reduction in Circulating Current With Improved Secondary Side Modulation in Isolated Current-Fed Half Bridge AC–DC Converter," IEEE Trans. Power Electron., vol. 37, no. 5, pp. 5625–5636, May 2022.

[14] E. Kim, M. Warner, and I. Bhattacharya, "Adaptive Step Size Incremental Conductance Based Maximum Power Point Tracking (MPPT)," in 2020 47th IEEE Photovoltaic Specialists Conference (PVSC), Jun. 2020, pp. 2335–2339.

[15] U. R. Prasanna, A. K. Rathore, and S. K. Mazumder, "Novel Zero-Current-Switching Current-Fed Half-Bridge Isolated DC/DC Converter for Fuel-Cell-Based Applications," IEEE Trans. Ind. Appl., vol. 49, no. 4, pp. 1658–1668, Jul. 2013.

Optimization of electric vehicle charge scheduling with consideration of battery degradation

Raka Jovanovic and Sertac Bayhan
Hamad bin Khalifa University
Doha, Qatar
Email:{rjovanovic,sbayhan}@hbku.edu.qa,

Islam Safak Bayram
University of Strathclyde
Glasgow, United Kingdom
Email: safak.bayram@strath.ac.uk

Keywords

≪Charge scheduling≫, ≪Optimization algorithm≫, ≪Vehicle-to-Grid≫

Abstract

In this work, we explore the potential of exploiting the demand-flexibility of electric vehicles (EVs) for flattening the electricity duck curve that emerge as a result of growing solar power production. The focus is on vehicle-to-grid technology in which smart charging allows bidirectional energy flow between EVs and the utility grid. The main objective of this study is to evaluate the impact of V2G technologies on battery degradation. To do that a mathematical model is developed in the form of a mixed integer linear program (MILP). In the MILP the battery degradation is modeled based on charge/discharge cycles using the rising edge method for which appropriate constraints are provided. The proposed method is used to asses the relation between battery degradation and the level of flattening of the duck curve that can be achieved in V2G systems at park and ride facilities. The conducted computational experiments, based on real world data, show that the additional degradation caused by battery discharge in such systems can be substantial and can reach close to 9% of battery degradation in standard V2G systems.

1 Introduction

In recent years, there has been an astonishing increase in the use of solar energy for electricity production. The main reasons for this growth are the need for lowering carbon emissions and improvements to air quality. This growth creates operational issues for the power grid operators. The main cause for this is the imbalance between peak electricity demand and renewable energy production during afternoons and mornings resulting in the "duck curve" issue. This results in the need for power system operators to ramp-up or ramp-down their production capacity, which leads to financial losses [1, 2]. With the further increase of solar electricity generation it is expected that the "duck curve" issue will grow further.

At the same time, many governments have incentivised the use of electric vehicles (EV) to meet net-zero emission goals. This has resulted in a fast growth of EVs adoption, and it is expected that in the near decades, EVs will become a primary mode of ground transportation. The demand flexibility of EV charging has a high potential to alleviate the "duck curve" related issues. To be more precise, the use of smart charging of large groups of vehicles can minimizes the ramp-up or ramp-down requirements of non-renewable power generation. Such use of EV charging, in combination with residential and industrial demand response programs, as well as energy storage technologies can significantly decrease the costs of power grid operators related to integration of renewable generation [3].

The majority of EV charging is done by one of three types of chargers. Firstly, by fast, Level 3 (50+ kW) chargers, that can fill a typical EV battery in less than 30 minutes. Level 1 (2-3 kW or up to 16 amps) chargers are generally used at residential premises and not well suited for demand side management due to low charge rates and the fact they are generally used for overnight charging when no electricity is generated from solar power. In case of Level 3 chargers, the charging time is short and it is not practical for shifting demand. On the other hand, smart charging of such systems generally lowers the charging power, and decreases the main benefit of fast charging.

The most suitable type of chargers for smoothing "duck curves" are Level 2 ones (5-7 kW) since they are mostly used at parking lots (e.g. workplace, university, etc.) where EVs are parked for several hours [4, 5]. In recent years, extensive research has been conducted on optimizing the scheduling of EV charging with Level 2 chargers at parking lots [6–8]. The positive aspects of EV smart charging have been analyzed in the context of workplace parking [9–12], commercial [13] parking lots and park-and-ride (PR) facilities [14, 15]. The majority of this research is dedicated to systems that only allow EVs to receive power, commonly called grid-to-vehicle systems (G2V). Recently, more research effort has been dedicated to the evaluation of systems in which EVs can also provide energy from their batteries to the grid, known as vehicle-to-grid systems (V2G) [16, 17]. When analyzing such systems, it is necessary to consider battery degradation, since a significant part of an EV value is related to its battery. The problem is that modeling battery degradation is in essence nonlinear and results in complex optimization problems which are frequently solved using different types of gradient based [18, 19], heuristic [20] and metaheuristic [21] methods.

On the other hand, the use of linear or quadratic mixed-integer programs for modeling such systems can provide significant benefits. In [22] a mixed integer linear program (MILP) is developed to analyze financial costs of battery degradation in V2G systems, that is solved using iterative solution method. This paper focuses on defining a MILP for optimizing V2G systems with consideration of battery degradation that can be solved using standard solvers. To be more precise, the fact that battery degradation is related to charge/discharge cycles is a convex function is exploited to create a mixed integer linear program (MILP) for the problem of interest using piece-wise linear approximation. The developed mathematical model is used to evaluate the relation between battery degradation and the smoothing of the "duck curve". To be exact, this analysis is conducted based on real world data for PR facilities through a case study.

The paper is organized as follows. The following section provides an outline of the system being optimized and the corresponding mathematical model. Section 3 is dedicated to the MILP for the problem of interest. The next section focuses on the conducted experiments and their analyses. The paper is finalized with concluding remarks that are presented in the last section.

2 Mathematical model

In this section, the mathematical model for optimizing smart scheduling of EV charging with the goal of minimizing the "duck curve" is presented. The model considers a V2G system where multiple vehicles visit a charging facility using Level 2 chargers. The model analyses a setting in which bidirectional charging is allowed and it is possible for EVs to inject power back to the electrical grid from their batteries. This type of smart scheduling is also observed from the aspect of battery degradation induced by the use of EV batteries to provide energy to the grid. The objective of the model is to find an optimal charging schedule for all the EVs that are present in the station at the same time. To be more precise, the goal is to find the amount of power each EV receives or discharges at each time period.

In the following section, an outline of the model is provided along with the process to transform the problem into a MILP. Note that the practical problem of smart scheduling of EVs is an online one, in the sense that the information (EV charge levels etc.) is not initially available. The goal of this work is to evaluate the upper bounds of the benefits such systems can provide, because of this it is assumed that all the information is available to the optimization algorithm. In that sense, we follow the approach, based on neural networks proposed in [15], shows how optimal solutions acquired using a mixed integer program for smart EV scheduling can be used for finding near optimal solutions in an online setting.

2 Scheduling setup

The proposed model starts from the following assumptions. The scheduling is performed over a time window $\mathcal{T}$ which is divided into a set of periods $\{1, \ldots, T\}$. Each time period $t \in \mathcal{T}$ has a parameter

q_t equal to the base electricity consumption minus the solar generation. Due to technological limits, all chargers in the system have the same maximal charging power c^{max} (in kW). This value provides information on how much power an EV can receive in one time periods. In addition, parameter d^{max} indicates the maximal power an EV can provide to the grid in one time period. It is assumed that a number of EVs $i \in \mathcal{E}$, $E = \{1, \ldots, M\}$ visit the charging station and that for each EV i the arrival time a_i and departure time l_i are known in advance. For each EV i, its battery capacity f_i and state of charge at arrival $0 \le b_i^a \le f_i$ are known. It is assumed that an EV is connected to a charger during the time it is at the station and can receive/discharge power during its stay. A battery of an EV can only be charged until its full capacity or discharged to a minimal allowed state of charge (SoC). It is assumed that the maximal charging power of the system is equal to the total charging potential of all the chargers in the station. Each EV i requests a certain amount of charge r_i, and the charging station must provide it with at least αr_i power during its visit to the station, where α is a predetermined constant. The flattening of the "duck curve" is done by minimizing the change in total load (sum of q_t and energy used for charging all the EVs) in successive time periods. A more detailed description of the objective function is provided in the next section that is dedicated to the MILP.

2   Battery degradation

There are several sources of EV battery degradation related to charging. Some major ones include battery temperature, charge/discharge cycles, and extreme states of charge, as reported in [22]. Battery temperature is generally related to weather and the use of fast chargers, and does not have a major impact in case of Level 2 chargers. Therefore, the focus is on charge/discharge cycles, since they are directly related to the use of EV batteries by the station operator for providing power to the grid. There are several approaches for cycle counting, the most commonly used methods are based on half-cycle, rising edge, rain flow, and max-edge analysis [23]. Aforementioned methods analyze the peaks of the corresponding load curve (or stress curve). A graphical illustration of these methods can be seen in Fig 1. The rainflow method is the most effective as it manages to track the amplitudes of short and long period cycles. The issue with this approach is that it does not have a closed form expression and, to the best of the authors' knowledge, does not a have a formulation suitable for linear programming. Because of this, models for battery degradation that use the rainflow method for cycle counting are generally optimized using metaheuristics such as particle swarm optimization [21]. It should be noted that there are also approaches to model the charge/discharge cycle-based battery degradation cost as convex function [18]. Consequently, gradient based methods have been developed for the related optimization problems [18, 19]. On the other hand, although other methods provide cycle counts that have a higher error than the

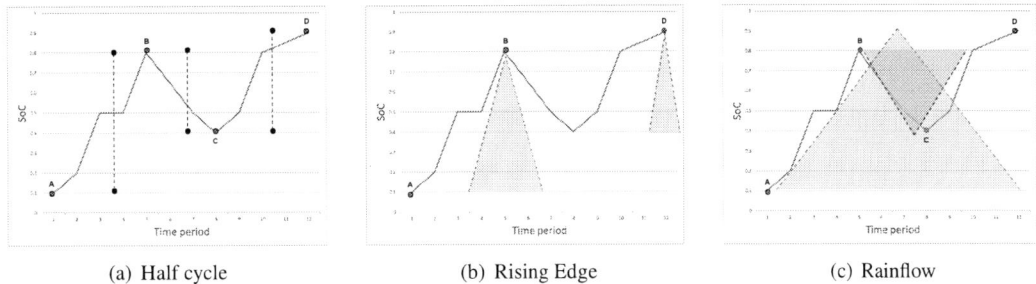

| (a) Half cycle | (b) Rising Edge | (c) Rainflow |

Fig. 1: Overview of different methods of cycle counting. Lines present half cycles and triangles full cycles. The amplitude of the cycles is given in the y axis.

rainflow method, they are commonly used since they can provide some insight in the problem of interest. Some comparison of the quality of cycle counts found by different methods can be found in [23]. In the proposed model, the selected approach for cycle counting is the rising edge method. The main reason is that the rising edge method can focus on the cycles that are induced by the battery discharges done by the charging station operator. Note that in this case, the method is applied to the load curve symmetric to the x-axis, so it could be called "falling edge" method. In a sense, it tracks the damage given to the battery resulting from the use of an EV's battery by the charging station operator. In the next sections, it will be shown how rising edge based cycle counting can be included in a mixed integer linear program.

3 Mixed integer program

In this section, the details of the MILP for the proposed mathematical model are given. In practice, the MILP formulation is divided into the following components. The first one focuses on the charge/discharge power of EVs related to minimizing the duck curve. The second part of the model focuses on peak tracking and their use for evaluating battery degradation.

The model uses the following set of decision variables. For each EV $i \in \mathcal{E}$ and time period $t \in \mathcal{T}$ a real decision variables c_{it} is used to indicate the amount power an EV i receives at time period t. In the same way, decision variables d_{it}, defined for all $i \in \mathcal{E}$ and $t \in \mathcal{T}$, are used to indicate the amount of power that an EV i provides to the grid at time period t. Next, let us define real variables b_{it}, for all $i \in \mathcal{E}$ and $t \in \mathcal{T}$, equal to the SoC of EV i at time t. With the intention of having a simpler formulation, instead of parameters for arrival (a_i) and departure times (d_i) for an EV i, a set of binary parameters v_{it}, for all $i \in \mathcal{E}$ and $t \in \mathcal{T}$, are defined to indicate if EV i is at the charging station at time period t. The model

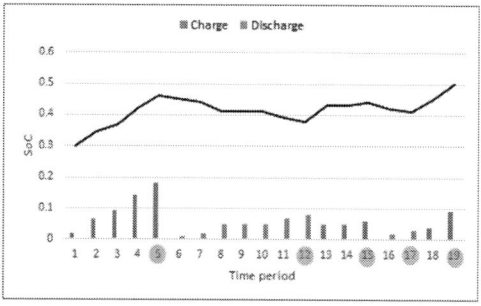

Fig. 2: Illustration of charge/discharge windows based on SoC. The color blue/orange is used for time period is in a charge/discharge window.

uses auxiliary real variables l_t, defined for all $t \in \mathcal{T}$, to store the information on the total load related to charging/discharging of the EVs currently at the station. Next, a set of auxiliary real variables C_t, defined for $t \in \mathcal{T}_0$ where $\mathcal{T}_0 = \mathcal{T} \setminus \{1\}$, is used the track the difference in total load (sum of base load and EV related load) between consecutive time periods.

Using these variables, the constraints for modeling an EV charging facility and the objective function for minimizing load ramp-up/ramp-down requirements can be specified. As in previously published research [12, 24], the minimization of ramp-up requirements is done based on the sum of the quadratic change in total load C_t overall the time periods. Due to the higher computational complexity of the proposed MIP, compared to the ones in [12, 24], instead of using the sum of C_t^2 in the objective function, its linear piece-wise approximation is used, as

$$\text{Minimize} \sum_{t \in \mathcal{T}_0} lpa(C_t^2). \tag{1}$$

In Eq. (1), the notation $lpa(\cdot)$ is used for the linear piece-wise approximation of a function, which is a quadratic one. The next step in defining the proposed MILP, is specifying the constraints related to the SoC, charge and discharge powers for EVs and tracking the total load difference. The constraints are given in Eqs. (2)-(10) as shown below.

$$0 \leq d_{it} \leq d^{max} \qquad i \in \mathcal{E}, t_e \in \mathcal{T} \tag{2}$$

$$0 \leq c_{it} \leq c^{max} \qquad i \in \mathcal{E}, t_e \in \mathcal{T} \tag{3}$$

$$b_i^{min} \leq b_{it} \leq f_i \qquad i \in \mathcal{E}, t \in \mathcal{T} \tag{4}$$

$$\tag{5}$$

$$b_{i1} = b_i^a \qquad i \in \mathcal{E} \tag{6}$$

$$b_{it+1} = b_{it} + c_{it} - d_{it} \qquad i \in \mathcal{E}, t \in \mathcal{T}_0 \tag{7}$$

$$b_{tT} - b_i^a \geq \alpha r_i \qquad i \in \mathcal{E} \tag{8}$$

$$l_t = \sum_{i \in \mathcal{E}} (c_{it} - d_{it}) \qquad t \in \mathcal{T} \tag{9}$$

$$C_t = l_t + q_t - l_{t-1} - q_{t-1} \qquad t \in \mathcal{T}_0 \tag{10}$$

The Eqs. (2),(3) are used to specify the bounds on the amount of power an EV can receive or discharge at each time period. The constraints given in (4) guarantee that the SoC of each EV i at each time period t is bound by the battery capacity and minimal allowed state of charge. Next, the constraints given in Eqs. (6),(7) are used to guarantee that at each time period the b_{it} is equal to the correct SoC for EV i at time period t. The Eq. 8 guarantees that each EV receives at least the amount of charge promised by the charging station operator. The Eq. (9) is used to track the total load (l_t) related to EVs at the station at time period t. The Eq.(10) uses this value to get the changes in total load (C_t) between two consecutive time periods

It should be noted that two sets of non-negative decision variables for charging and discharging of EV batteries are used to make it possible to track peaks and valleys of the SoC curve for each EV. The idea for the constraints used to track peaks and valleys is to track continues charging or discharging windows. A discharging window consists of all time periods from the start of discharging until the first charging period (not included), see Fig. 2. Note that periods in which there is no charge or discharge will be considered being in the same window as the previous period. Now, peaks occur at time periods that are at the ends of charging windows and valleys occur at ends of discharging periods.

The following text describes the part of the MILP dedicated to the tracking of peaks and valleys of individual EVs SoC and how they are used for cycle counting. Firstly, let us define a auxiliary binary variables c_{it}^w and d_{it}^w, defined for $i \in \mathcal{E}$ and $t \in \mathcal{T}$, indicating if time period t for EV i is inside a charging or discharging window, respectively. Next, let us define a non-negative real variables d_{it}^s, defined for $i \in \mathcal{E}$ and $t \in \mathcal{T}$, providing the information on how much power has an EV discharged since the beginning of the current discharge window. The last set of auxiliary variables needed for peak tracking are d_{it}^p, which are equal to the d_{it}^s in case time period t corresponds to a peak and zero otherwise. The constraints specifying the behavior of the mentioned variables are given in Eqs. (11)-(28).

$$0 \leq c_{it}^w \leq 1, \qquad i \in \mathcal{E}, t \in \mathcal{T} \tag{11}$$

$$0 \leq d_{it}^w \leq 1, \qquad i \in \mathcal{E}, t \in \mathcal{T} \tag{12}$$

$$c_{it}^w + d_{it}^w = v_{it}, \qquad i \in \mathcal{E}, t \in \mathcal{T} \tag{13}$$

$$c_{it} \leq M c_{it}^w, \qquad i \in \mathcal{E}, t \in \mathcal{T} \tag{14}$$

$$d_{it} \leq M d_{it}^w, \qquad i \in \mathcal{E}, t \in \mathcal{T} \tag{15}$$

$$c_{it+1}^w \geq c_{it}^w - d_{it}^w, \qquad i \in \mathcal{E}, t \in \mathcal{T} \tag{16}$$

$$d_{it+1}^w \geq d_{it}^w - c_{it}^w, \qquad i \in \mathcal{E}, t \in \mathcal{T} \tag{17}$$

$$d_{i1}^s = d_{i1}, \qquad i \in \mathcal{E} \tag{18}$$

$$d_{it}^s \geq 0, \qquad i \in \mathcal{E}, t \in \mathcal{T} \tag{19}$$

$$d_{it}^s \leq M(1 - c_{it}^a), \qquad i \in \mathcal{E}, t \in \mathcal{T} \tag{20}$$

$$d_{it}^s \geq d_{it-1}^s + d_{it} - M c_{it}^a, \qquad i \in \mathcal{E}, t \in \mathcal{T} \tag{21}$$

$$d_{it}^s \leq d_{it-1}^s + d_{it} + M c_{it}^a, \qquad i \in \mathcal{E}, t \in \mathcal{T} \tag{22}$$

$$d_{it}^p \geq 0, \qquad i \in \mathcal{E}, t \in \mathcal{T} \tag{23}$$

$$d_{it}^p \leq v_{it}, \qquad i \in \mathcal{E}, t \in \mathcal{T} \tag{24}$$

$$d_{it}^p \leq d_{it}^s, \qquad i \in \mathcal{E}, t \in \mathcal{T} \tag{25}$$

$$d_{it}^p \geq d_{it}^s - (1 - c_{it+1}^a + c_{it}^a)M, \qquad i \in \mathcal{E}, t \in \mathcal{T} \tag{26}$$

$$d_{it}^p \geq d_{min}^p - (1 - c_{it+1}^a + c_{it}^a)M, \qquad i \in \mathcal{E}, t \in \mathcal{T} \tag{27}$$

$$d_{it}^p \leq Mc_{it+1}^a, \qquad i \in \mathcal{E}, t \in \mathcal{T} \tag{28}$$

Eq. (13) states that for an EV i at time period t, is either in a charge or discharge window if it is at the charging station. Eqs. (14) and (15) guarantee that an EV can charge or discharge power only inside a charge or discharge window, respectively. In Eqs. (14) and (15), the notation M is used for a sufficiently large value in the standard way for large M constraints. The same notation is used later in the text. The constraints given in Eq. (16) guarantee that a charging window can only end with a discharge period. The Eq. (17) provides analogous constraints for discharge windows.

The next group of constraints is dedicated to the total amount of power that has been discharged during a discharge window. The constraints given in Eq. (18) are used to initialize the amount of discharge at the first time period. Eq. (20) states that in case of a charge period the total discharge is equal to 0. The constraints given in Eqs. (21),(22) are used for tracking the value of the total discharge in a discharge window between consecutive time periods. The Eqs. (21) and (22) state that the total amount of discharge in a discharge window, is equal to the total discharge in the previous time period increased by the current discharge.

The last group of constraints given in Eqs. (23)-(28) are related to recognizing the amount of total discharge at the end of a discharge window. The constraints given in Eq. (24) state that the total peak discharge is equal to 0, when an EV is not at the charging station. Eq. (25) states that the peak discharge at time period t is less than or equal to the total discharge in the discharge window at time period t. Constraints given in Eqs. (26) guarantees that the peak discharge is larger or equal to the total discharge at the end of a discharge window. The Eq. (27) guarantees that there is a minimal amount of discharge d_{min}^p that must happen in a discharge window. Although these constraints are not necessary to represent the real world systems, we have observed that there inclusion greatly decreases the computational cost of solving the MILP. The constraints given in (28) guarantee that a peak discharge at time period t can only be greater than 0 in case the following time period is in a charge window.

In the proposed MILP, as in [19], the battery degradation is modeled using the life loss from a single cycle of depth u measured in terms of (normalized) changes in the SoC. To be more precise, it is assumed that if a battery cell is repetitively cycled with depth u, then it can operate $\frac{1}{\Phi(u)}$ number of cycles before reaching its end of life. The degradation function $\Phi(u)$ is normalized between 0 and 1 with respect to the total battery life. In the proposed model the discharge cycle depth is calculated based on the peak discharge. As suggested in [19], the function $\Phi(u)$ is well approximated for electro-chemical batteries using the following formula

$$\Phi(u) = 0.524 \times 10^{-4} u^{2.03}. \tag{29}$$

Now, using the function $\Phi(u)$ the objective related to battery degradation can be included in the model in the following way

$$\text{Minimize} \sum_{i \in \mathcal{E}} \sum_{t \in \mathcal{T}} lpa(\Phi(d_{it}^p)) \tag{30}$$

Eq. (30) uses the rising edge approach for analyzing battery discharge cycles using rising edges. Note that the rising edges correspond to the"decreasing edges" of an EV SOC. The total battery degradation is equal to the sum of degradation for all peak discharges for all EVs over all the time periods. Note that there is only one peak discharge for each window. It is important to point out that the proposed objective function, in practice, only considers the battery degradation induced by the discharge of batteries induced by the charging station operator. In the proposed model, due to computational complexity of the model instead of the function $\Phi(u)$ its piece-wise linear approximation $lpa(\Phi(u))$ is used.

4 Computational Experiments

In this section, the results of the conducted computational experiments are presented. The proposed MILP has been implemented using OPL in IBM ILOG CPLEX Optimization Studio Version: 12.6.1.0, and executed using the default solver settings. The computational experiments have been performed on a personal computer running Windows 10 having an Intel(R)Xeon(R) Gold 6244 CPU @3.60 GHz processor with 128 GB memory. The goal of the conducted computational experiments is to evaluate the potential of reducing the duck curve issue using a V2G system, in the sense of lowering ramping requirements in relation to battery degradation within an EV smart scheduling system. This is done based on real world data, with a special focus on park and ride (PR) facilities. In this section, firstly the method for incorporating such data in the model is presented. Next, details on the use of the proposed model are given and an analysis through a case study is provided.

4 Generation of data for experiments

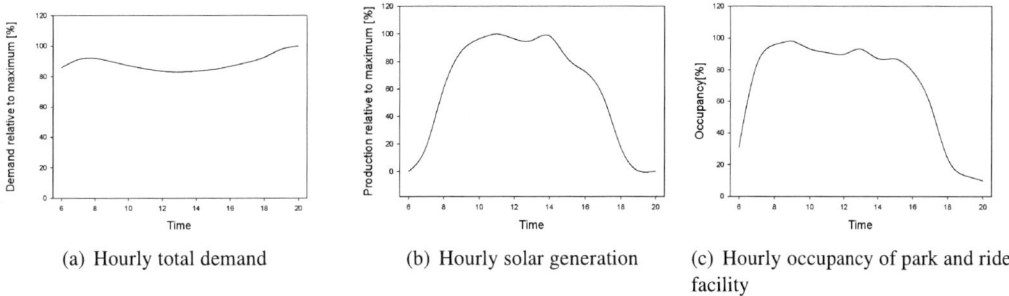

(a) Hourly total demand   (b) Hourly solar generation   (c) Hourly occupancy of park and ride facility

Fig. 3: Graphical illustration of data used for generating test instances taken from [25]. All the values are given in a normalized form and given in percentage.

The first group of parameters that need to be incorporated into the model are related to EV visits to the PR facilities. The data set that is used for this task is the hourly utilization rate of PR facilities taken from [25] (see Fig. 3(c) for a graphical illustration). This data is combined with metro passenger behavior statistics taken from [26]. The battery types of EVs have been randomly generated based on data related EV sales taken from [27] and corresponding battery sizes [28]. The SoC of individual batteries is randomly generated form the range 15% - 80% using a uniform distribution. This information is used to generate problem instances in the same way as in [14]. An extension of the proposed model, compared to [14], is the inclusion of requested amount of charge for EVs visiting the station. The requested charge is selected randomly, using a uniform distribution, 50% to 100% of battery capacity that can be maximally charged during the EVs stay at the facility.

The intention of the proposed model is to evaluate the relation of battery degradation and reductions in ramping requirements for varying percentages of solar generation and EV penetration levels. In the conducted case studies, the information on total demand and solar generation are taken from the publicly available data for California [29] and presented in Figs. 3(a) and 3(b). Using this data, separate problem instances have been generated for different portions of power used for EV charging and solar generation. To be more precise, it is assumed that the model optimizes the scheduling of N EVs and that the total requested power is E. Next, the base load and solar generation are scaled to be a specific proportion of this value. In other words, the normalized values given in Fig. 3 are multiplied by constants to get the correct proportion for EV consumption E, base load L and solar production S.

In generating the problem instances, the time period between 6:00 and 20:00 has been used to mimic actual operational times of PR parking lots. Each time period in the model corresponded to 15 minutes, which means the value of parameter T is 56. In the generated problem instances, a fixed number $N = 10$ is used for the number of EVs visiting the PR facility during the day. The potential of scheduling EV charging to minimize issues related to the "duck curve" are evaluated for settings where 15%, and 30% of the maximal load is produced from solar generation. The assumed power of the chargers is 4 kW, which translates to $c_{max} = 1$ inside the model. The same value of maximal discharge power is used for $d_{max} = 1$. The chosen minimal amount of charge received by an EV was equivalent to 90% of the requested charge, hence $\alpha = 0.9$. The maximal amount of energy used for EV charging was 2.5% and 5.0% of the total energy consumption. The minimal total amount of discharge during a discharge window was 1.5 kWh.

4 Computational experiment setup

The objectives of the conducted research are to evaluate several aspects of the use of V2G systems. Firstly, to observe the level of decrease in ramping requirements of smart charge scheduling in V2G systems compared to G2V. Next, to evaluate the level of additional battery degradation that occurs in V2G compared to G2V systems. Finally, to see what is the maximal level of decrease in ramping requirements that can be achieved for different levels of battery degradation. To achieve this, the proposed model is used in the following way. A MILP using the objective function given in Eq. (1) is used to find the minimal value of the objective function $Ramp_{min}$ related to ramping requirements in V2G system. Note that this model only needs constraints given in Eqs. (2) - (10). The next step is finding the minimal level of battery degradation needed to achieve this level of decrease in ramping requirements. This is achieved, by minimizing the objective function given in Eq. (30) for the MILP given in constraints Eqs. (2)-(28) with the following additional constraint.

$$\sum_{t \in \mathcal{T}_0} lpa(C_t^2) \leq Ramp_{min} \tag{31}$$

In this way, the value of the objective function $DegBat_{min}$ related to minimal battery degradation for the maximal decrease in ramping requirements is acquired. The next step, is to evaluate the decrease in ramping requirements for different levels of battery degradation. Different levels, for $i \in \{0,...,n\}$., of battery degradation $DegBat_i$ are specified using the following equation

$$DegBat_i = \frac{i}{n} DegBat_{min}. \tag{32}$$

For each value of $DegBat_i$ the maximal possible decrease in ramping requirements is calculated using a MILP with the objective function given in Eq. (1) with constraints given in Eqs. (2) - (28) with the addition of the following constraint.

$$\sum_{i \in \mathcal{E}} \sum_{t \in \mathcal{T}} lpa(\Phi(d_{it}^p)) \leq DegBat_i \tag{33}$$

Eq. (33) guarantees that the total battery degradation is less than $DegBat_i$. The used approach is equivalent to the standard ε-constraint method [30] for finding the Pareto Front of a bi-objective problem.

4 Case Study

The presented approach is used for a case study on the effect of smart charging in V2G systems at PR facilities. The method for incorporating real-world data has been provided in the previous section. The first set of results is given in Fig. 4 (left sub-figures), each of them provides information for settings with different levels of solar generation and EV adoption. Each of this figures shows the base load curve (base demand minus solar generation) over time. In addition, these figures contain load curves in which additional power is used for smart scheduling of EVs charging in G2V and V2G systems. In case of V2G two levels of battery degradation are considered. The first observation that can be drawn from these figures is that the V2G systems have a much higher impact on the load curve in the morning where there is a need for a high level of ramp-down (lowering of electricity production) than in the evening where there is need for ramp-up compared to G2V systems. One reason for this is that in the evening, there is a significant decrease of EVs at the PR facility due to drivers leaving after finishing their work day. The second reason is that at this period, it is necessary to have EVs that have received a higher amount of charge than requested that could potentially be discharged which is not possible for most of the EVs. Because of this, the focus of the quantitative analysis of V2G systems focuses on the morning period.

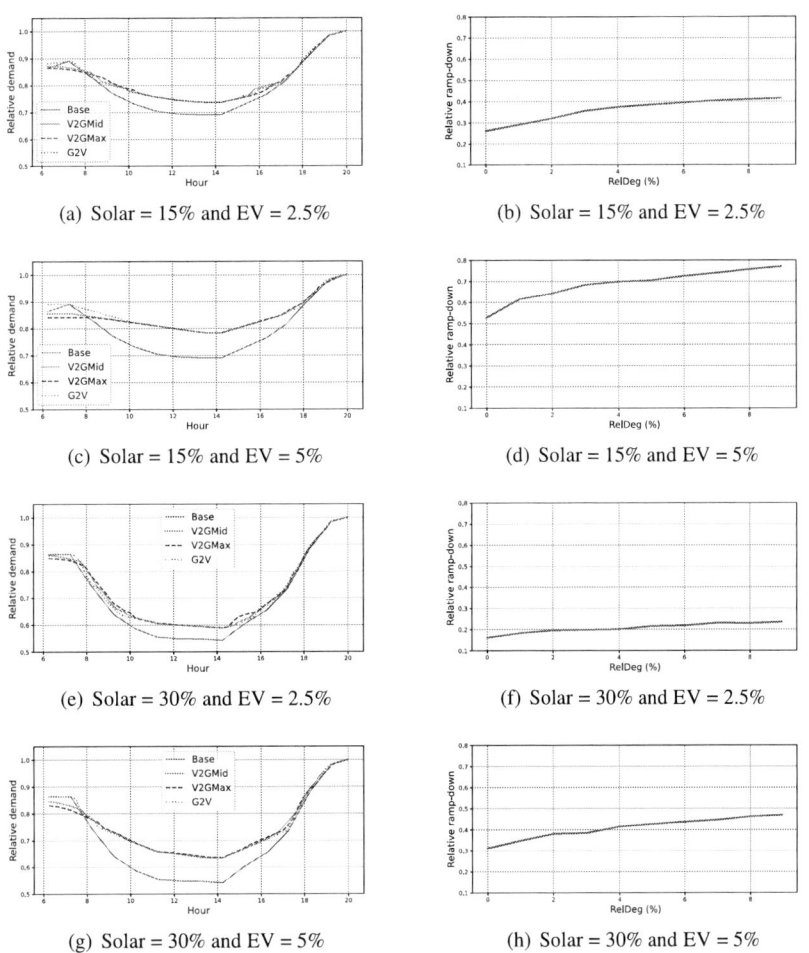

Fig. 4: Illustration of the effect of smart charging in V2G systems at PR facilities on the "duck curve". Left side shows load curves for different levels of battery degradation. "V2GMid" is used for battery degradation corresponding to the value $i = \frac{n}{2}$ in Eq.32. The right side shows the relation between relative battery degradation and relative ramp-down requirements. The values "Solar" and "EV" correspond to the percentage of the total energy consumption satisfied from solar or used by EVs.

The objective functions given in Eq. (1) and Eq. (30) are abstractions that are not suitable for understanding the relation of battery degradation and ramp-down requirements. Because of this the following measures are used to evaluate this relation. Firstly, the ramp-down requirements are observed using the difference between the maximal and minimal load in the time period between 7:15 and 11:30 divided by the time difference between them. The positive effect of the use of smart scheduling of EVs in G2V and V2G systems on ramp-down requirements is evaluated relative to the ramp-down requirements needed for the base load.

The battery degradation of V2G systems is analyzed in relation to the level of battery degradation in G2V systems. To be more precise, the additional battery degradation that occurs in V2G systems due to battery discharges relative to systems in which no discharge is allowed is observed. Formally, for each of the degradation level $DegBat_i$, the following equations are used to as a measure.

$$Deg_{Charge} \quad = \quad \sum_{e \in \mathcal{E}} \Phi(\frac{b_{iT} - b_i^a}{f_i}) \tag{34}$$

$$Deg_{Discharge} \quad = \quad \sum_{i \in \mathcal{E}} \sum_{t \in \mathcal{T}} \Phi(d_{it}^p) \tag{35}$$

$$RelDeg \quad = \quad \frac{Deg_{Discharge}}{Deg_{Charge}} \tag{36}$$

The Eq. (34) provide the battery degradation for all the EVs in the system based on the amount of charge each vehicle received. To be exact, for each EV i the amount of received charge equal to the SoC at arrival (b_i^a) minus the state of charge at departure (b_{iT}). The degradation is calculated using function Φ, so it is necessary to normalize the amount of received charge by the battery size. The battery degradation related to discharges is given in Eq. (35), and is the same as the objective function (33). Finally, the relative additional battery degradation is given in Eq. (36) as the total discharge degradation divided by the total charge related degradation.

The results on the relation of change in ramp-down requirements and battery degradation can be observed for different levels of EV adoption and solar generation in Fig. 4 (right sub-figures). Note that in these figures additional battery degradation of 0% corresponds to a G2V system. From these results, it is evident that the V2G systems provides a substantial additional decrease in ramp-down requirements. In case of solar generation of 15% G2V systems provide a decrease in ramp-down requirements of around 25% and 55% compared to 40% and 75% in case of V2G systems when EV charging consumes 2.5% and 5% of the total load, respectively. In case of solar generation of 30%, same values for G2V are 15% and 30% and grow to 25% and 45% in case of V2G, when EV charging consumes 2.5% and 5% of the total load, respectively. This maximal decrease in case of V2G system comes at a significant additional battery degradation which is close to 9% for all of the settings. It is important to point out that more than half of the decrease in ramp-down requirements can be achieved with an additional battery degradation of around 3% for the tested instances. These results indicate that although the use of smart charging in V2G systems for flattening the "duck curve" is effective it can result in significant additional cost for EV drivers. This negative effect can be substantially lowered if the optimization of the EV charge scheduling considers battery degradation.

5 Conclusion

In this paper, a mathematical model for analyzing the smart scheduling of EV charging in V2G system has been presented. The focus is on evaluating the relation of battery degradation and the potential for flattening of the "duck curve" in such systems. The proposed model is a MILP in which battery degradation is observed through charge/discharge cycles. It introduces a set of new constraints corresponding to the rising edge method which is highly suitable for evaluating battery degradation related to battery discharge induced by the charging station operator. The developed model is used to evaluate the potential of V2G systems at PR facilities based on real-world data. The conducted computational experiments indicate that, based on the new model, such systems are more suitable for lowering ramp-down requirements in the morning period than ramp-up requirements in the evening. It has been shown that smart scheduling of EV charging at such facilities can result in a substantial additional cost to EV drivers in case battery degradation is not considered.

References

[1] M. Obi and R. Bass, "Trends and challenges of grid-connected photovoltaic systems–a review," *Renewable and Sustainable Energy Reviews*, vol. 58, pp. 1082–1094, 2016.

[2] P. Denholm, M. O'Connell, G. Brinkman, and J. Jorgenson, "Overgeneration from solar energy in california. a field guide to the duck chart," National Renewable Energy Lab.(NREL), Golden, CO (United States), Tech. Rep., 2015.

[3] M. A. Zehir, A. Batman, and M. Bagriyanik, "Review and comparison of demand response options for more effective use of renewable energy at consumer level," *Renewable and Sustainable Energy Reviews*, vol. 56, pp. 631–642, 2016.

[4] D. Meyer and J. Wang, "Integrating ultra-fast charging stations within the power grids of smart cities: a review," *IET Smart Grid*, vol. 1, no. 1, pp. 3–10, 2018.

[5] I. Rahman, P. M. Vasant, B. S. M. Singh, M. Abdullah-Al-Wadud, and N. Adnan, "Review of recent trends in optimization techniques for plug-in hybrid, and electric vehicle charging infrastructures," *Renewable and Sustainable Energy Reviews*, vol. 58, pp. 1039–1047, 2016.

[6] X. Wu, "Role of workplace charging opportunities on adoption of plug-in electric vehicles–analysis based on gps-based longitudinal travel data," *Energy Policy*, vol. 114, pp. 367–379, 2018.

[7] B. Ferguson, V. Nagaraj, E. C. Kara, and M. Alizadeh, "Optimal planning of workplace electric vehicle charging infrastructure with smart charging opportunities," in *2018 21st International Conference on Intelligent Transportation Systems (ITSC)*. IEEE, 2018, pp. 1149–1154.

[8] R. S. Levinson and T. H. West, "Impact of convenient away-from-home charging infrastructure," *Transportation Research Part D: Transport and Environment*, vol. 65, pp. 288–299, 2018.

[9] Y. Zhang and L. Cai, "Dynamic charging scheduling for ev parking lots with photovoltaic power system," *IEEE Access*, vol. 6, pp. 56 995–57 005, 2018.

[10] R. Dhawan and S. Prabhakar Karthikeyan, "An efficient EV fleet management for charging at workplace using solar energy," in *2018 National Power Engineering Conference (NPEC)*, March 2018, pp. 1–5.

[11] Z. Wei, Y. Li, Y. Zhang, and L. Cai, "Intelligent parking garage EV charging scheduling considering battery charging characteristic," *IEEE Transactions on Industrial Electronics*, vol. 65, no. 3, pp. 2806–2816, March 2018.

[12] R. Jovanovic, S. Bayhan, and I. S. Bayram, "A multiobjective analysis of the potential of scheduling electrical vehicle charging for flattening the duck curve," *Journal of Computational Science*, vol. 48, p. 101262, 2021.

[13] K. Jhala, B. Natarajan, A. Pahwa, and L. Erickson, "Coordinated electric vehicle charging for commercial parking lot with renewable energy sources," *Electric Power Components and Systems*, vol. 45, no. 3, pp. 344–353, 2017.

[14] R. Jovanovic and I. S. Bayram, "Scheduling electric vehicle charging at park-and-ride facilities to flatten duck curves," in *2019 IEEE Vehicle Power and Propulsion Conference (VPPC)*. IEEE, 2019, pp. 1–5.

[15] R. Jovanovic, S. Bayhan, and I. S. Bayram, "An online model for scheduling electric vehicle charging at park-and-ride facilities for flattening solar duck curves," in *2020 International Joint Conference on Neural Networks (IJCNN)*, 2020, pp. 1–8.

[16] M. A. Ortega-Vazquez, "Optimal scheduling of electric vehicle charging and vehicle-to-grid services at household level including battery degradation and price uncertainty," *IET Generation, Transmission & Distribution*, vol. 8, no. 6, pp. 1007–1016, 2014.

[17] K. Ginigeme and Z. Wang, "Distributed optimal vehicle-to-grid approaches with consideration of battery degradation cost under real-time pricing," *IEEE Access*, vol. 8, pp. 5225–5235, 2020.

[18] Y. Shi, B. Xu, Y. Tan, and B. Zhang, "A convex cycle-based degradation model for battery energy storage planning and operation," in *2018 Annual American Control Conference (ACC)*, 2018, pp. 4590–4596.

[19] Y. Shi, B. Xu, Y. Tan, D. Kirschen, and B. Zhang, "Optimal battery control under cycle aging mechanisms in pay for performance settings," *IEEE Transactions on Automatic Control*, vol. 64, no. 6, pp. 2324–2339, 2019.

[20] Z. Wei, Y. Li, and L. Cai, "Electric vehicle charging scheme for a park-and-charge system considering battery degradation costs," *IEEE Transactions on Intelligent Vehicles*, vol. 3, no. 3, pp. 361–373, 2018.

[21] Q. Yang, J. Li, W. Cao, S. Li, J. Lin, D. Huo, and H. He, "An improved vehicle to the grid method with battery longevity management in a microgrid application," *Energy*, vol. 198, p. 117374, 2020.

[22] H. Farzin, M. Fotuhi-Firuzabad, and M. Moeini-Aghtaie, "A practical scheme to involve degradation cost of lithium-ion batteries in vehicle-to-grid applications," *IEEE Transactions on Sustainable Energy*, vol. 7, no. 4, pp. 1730–1738, 2016.

[23] K. Mainka, M. Thoben, and O. Schilling, "Lifetime calculation for power modules, application and theory of models and counting methods," in *Proceedings of the 2011 14th European Conference on Power Electronics and Applications*. IEEE, 2011, pp. 1–8.

[24] Z. J. Lee, T. Li, and S. H. Low, "ACN-data: Analysis and applications of an open EV charging dataset," in *Proceedings of the Tenth ACM International Conference on Future Energy Systems*, 2019, pp. 139–149.

[25] L. K Cherrington, J. Brooks, J. Cardenas, Z. Elgart, L. David Galicia, T. Hansen, K. Miller, M. J. Walk, P. Ryus, C. Semler, and K. Coffel, "Decision-making toolbox to plan and manage park-and-ride facilities for public transportation: Guidebook on planning and managing park-and-ride," Tech. Rep., 01 2017.

[26] J. Neff, "A profile of public transportation passenger demographics and travel characteristics reported in on-board surveys," *American Public Transportation Association and others*, 2007.

[27] InsideEVs. (2019) Monthly plug-in ev sales scorecard. [Online]. Available: https://insideevs.com/monthly-plug-in-sales-scorecard/

[28] Wikipedia. (2022) Electric vehicle battery. Accessed 2022-03-04. [Online]. Available: https://en.wikipedia.org

[29] California ISO. (2022) California ISO. [Online]. Available: http://www.caiso.com

[30] G. Mavrotas, "Effective implementation of the ε-constraint method in multi-objective mathematical programming problems," *Applied Mathematics and Computation*, vol. 213, no. 2, pp. 455 – 465, 2009.

Onboard ESU Sizing and Dynamic IPT Charging Scenarios for a Tramway Application

Endika Bilbao Muruaga*, Irma Villar*, Florian Legay**, Pierre Prenleloup**, Jean-François Reynaud***

* Senior Researcher, Team Leader	** Battery Simulation Leader,	*** Project Manager
Ikerlan Technology Research Centre	Railways Market Manager	(Energy Storage Expert)
Basque Research and Technology	*Saft SAS*	*CAF Power & Automation*
Alliance (BRTA)	Bordeaux, Levallois-Perret,	San Sebastián, Spain
Arrasate-Mondragon, Spain	France	

Keywords

≪Energy Storage Unit≫, ≪Sizing≫, ≪Batteries≫, ≪Tramway≫, ≪Wireless Power Transfer≫, ≪Energy Management≫, ≪Traction≫.

Abstract

In this paper a battery based Energy Storage Unit (ESU) sizing methodology is proposed for tramway applications. At the same time, the required recharging system is also defined. For that, a tramway is modeled to define power and energy requirements. Then, the methodology is applied to obtain a set of solutions for LTO (power oriented) and sLFP (energy oriented) cell technologies, as well as, static and dynamic wireless recharging systems. The objective is to size the required system using batteries and its associated recharging system as an ESU for 15-year operation.

Introduction

Onboard Energy Storage Units (ESUs) in tramways is a current trend with multiple purposes. By employing these systems, peak power demanded and delivered from/to the overhead lines (catenary) can be reduced, improving the efficiency, and minimizing the voltage drops in the main grid [1, 2]. Furthermore, the use of ESU permits a certain range of self-sufficient operation of the tramway, providing driving autonomy to the vehicle. This has already been employed by different tramway manufacturers for catenary free operation in different locations. In these cases, the ESU is sized to fully supply the tramway power demand until the next recharging station [3, 4, 5]. On the other hand, wireless dynamic charging systems [6, 7, 8] will potentially enable ESU size reduction, as well as its degradation extending the lifetime (i.e. system powering for accelerating periods).

The objective of this paper is to present a methodology for the optimal sizing of ESU onboard a tramway considering different wireless charging stations along a city profile in a catenary free operation. For that, the tramway will be modeled, obtaining kinematics equations of the electromechanical system. Then, these expressions will be simulated with a real application data to obtain power and energy requirements. Afterwards, ESU sizing methodology will be proposed and tested with the real case-study and two battery technologies (LTO, sLFP). Apart from the ESU sizing common information inputs (i.e. recharging power, locations, degradation level), the wireless dynamic charging system influence will be evaluated with degrees of freedom of power level and length. All these potential solutions will be analysed trying to obtain the most suitable ones for the application, getting a compromise between all good solutions and each objective.

Tramway Modeling

In this section the tramway modeling will be presented. The elements of an ESU powered tramway are shown in Fig. 1. The traction controller is in charge controlling and managing the power flows. On the load side, the main elements are air-conditioner and lighting, grouped as auxiliary loads, and motor responsible for moving the tramway. On the power source side, the catenary and ESU (batteries). Note that, motor and ESU are bidirectional elements, therefore, they can act as loads and power sources depending on the circumstances. There are two main operation modes, on-grid and off-grid, in other words catenary connected or disconnected, respectively. Table I describes system configuration and assumptions according to the operation mode. As it can be seen, the ESU will be recharged in stations while the catenary is connected, and discharged during the tramway movement.

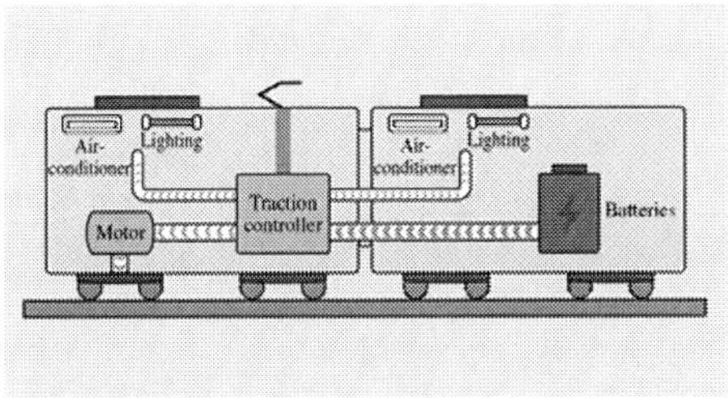

Table I: Tramway operation modes.

Element	On-Grid	Off-Grid
ESU	Recharging (no grid injection)	Powering (auxiliary loads, motor)
Motor	Zero (tramway stopped)	Consuming (traction) Regenerating (braking)
Auxiliary Loads	Consuming	Consuming
Catenary	Powering (auxiliary loads, ESU)	Zero (tramway disconnected)

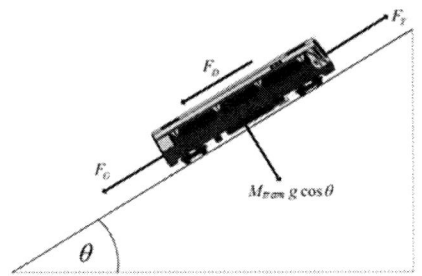

Fig. 1: Tramway main power sources and loads diagram.

The motor power profiles for a trip are defined by the tramway movement equations. Figure 2 shows all the forces acting in the system [9, 10, 11], where the kinetics are defined in (1), (2), (3). The tramway mass (M_{tram}) acceleration (a) is defined by the traction force (F_T) produced by motors, drag force (F_D) due to velocity and gravity force (F_G) due to elevation (θ). The tramway traction should be evaluated in electrical kW, which can be obtained with the speed value (v). The traction power (P_T) is expressed in equation (4). Note that, the rolling resistance has not been considered in this analysis due to neglectable values in terms of power and energy requirements.

$$M_{tram} \cdot a = F_T - F_D - F_G \tag{1}$$

$$F_D = A + B \cdot v + C \cdot v^2 \tag{2}$$

$$F_G = M_{tram} \cdot g \cdot \sin\theta \tag{3}$$

Fig. 2: Free body diagram of the tramway.

$$P_T = F_T \cdot v = ((M_{tram} \cdot a) + (A + B \cdot v + C \cdot v^2) + (M_{tram} \cdot g \cdot \sin\theta)) \cdot v \tag{4}$$

The simulation parameters are shown in Table II based on tramway manufacturer data, for the case-study presented in Fig. 3. Note that, these results will represent the power and energy requirements of one coach. Usually, tramways are composed by several units. For this analysis, we assume there is one ESU per coach, being independent units. This approach helps to have a scalable solution, where total requirements can be obtained multiplying the simulation results by the number of coaches of the full tramway. Regarding auxiliary loads, it will be assumed a constant power consumption. Auxiliary loads are mainly composed by HVAC and lighting devices. Therefore, in a short trip the value is constant.

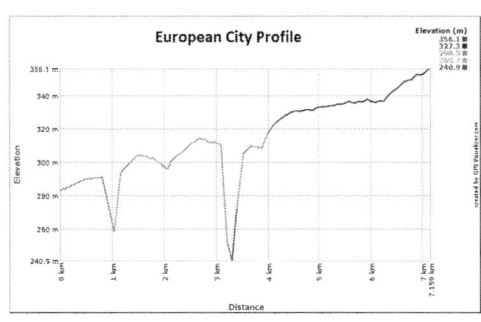

Fig. 3: Tramway route elevation profile.

Table II: Simulation parameters.

Parameter	Description	Value
M_{tram}	Tramway mass (coach+passengers)	12000kg
A	Mass coeff.	$2.965kg$
B	Track coeff.	0.46
C	Aerodynamic coeff.	0.005
g	Gravitational accel.	$9.81m/s^2$

Fig. 4a presents the speed profile. It represents a tramway completing a full round trip, go and back six times per day. Fig. 4b presents the power profile for the full trip. Maximum acceleration power is 447kW and minimum braking power is -387kW. The sign criterion is positive for acceleration power values, and negative for braking situations. It has been assumed that in braking mode the energy will be stored by the ESU and used afterwards, increasing system efficiency. Also mention that this power has been delivered through the catenary, being the coach continuously connected to the grid. Finally, the electrical requirements can be represented as power versus energy requirements. This graph illustrates visually the limits of the ESU sizing, in other words, what are the bounds of the sizing study, as well as what are the power requirements for each SOC of the ESU. Fig. 4d shows the results for the city tramway. This is because the easiest solution to power the tramway is to install an ESU covering all the requirements of all the stations (-387kW+447kW and 61.2kWh). That solution is not optimal due to the over sizing and lower efficiency, which would require an installation of a huge ESU in every single coach. The objective is to reduce this sizing, satisfying tramway requirements, while the efficiency is only slightly decreased.

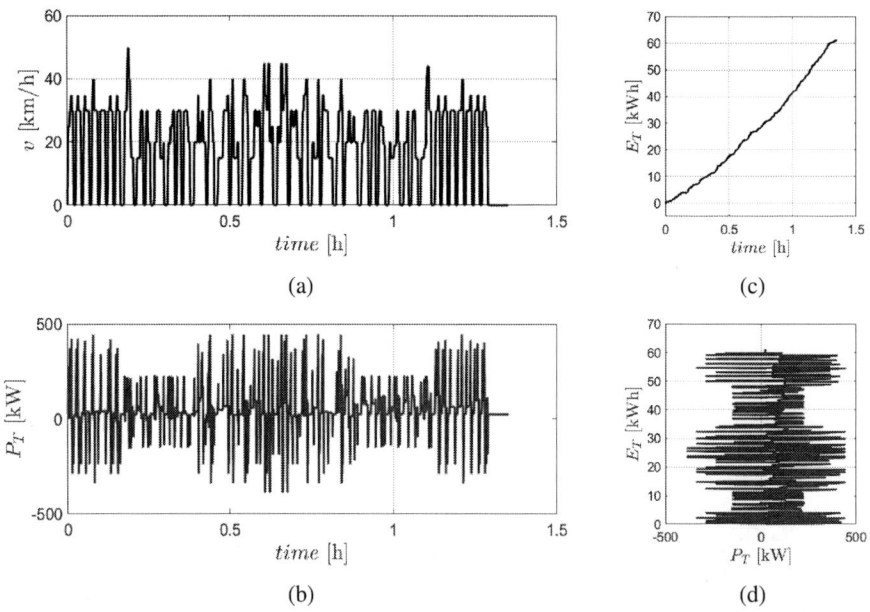

Fig. 4: Case-study: (a) speed profile, (b) power profile, (c) energy profile, (d) power vs. energy diagram.

ESU Systems Sizing and Associated Recharging System Definition

In this section the ESU sizing methodology will be presented, applied to the real case-study and the results analyzed.

ESU Sizing Methodology

The methodology is presented in Fig. 5. Steps 1 was completed in the previous section. Step 2 defines the recharging strategy (i.e. power level, recharging locations, etc.). Step 3 and step 4 are composed by the first round of simulations considering the State Of Charge (SOC) control, series/parallel cells configuration and ESU lifetime at the Beginning Of Life (BOL). The second round of simulations is carried out in step 5, considering the ESU End Of Life (EOL). Finally (step 6), the results are validated and system configuration evaluated in terms of mass and size. It has to be mentioned, the lifetime and system performance will be defined by the capacity and internal resistance degradation, decreasing the storage capacity and increasing thermal requirements. The sizing criteria is presented in Table III.

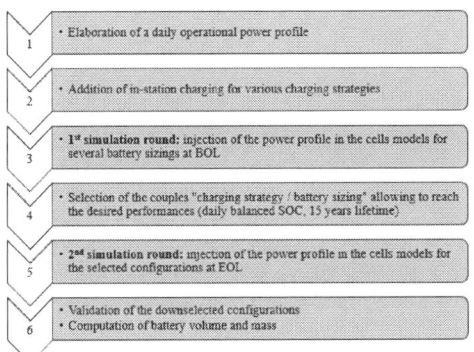

Fig. 5: Sizing methodology.

Table III: Sizing criteria.

Parameter	Value
Capacity Ageing	40% degradation (15-year lifetime)
Resistance Ageing	150% increasing (15-year lifetime)
Balanced SOC	1% tolerance (daily)
Balanced Temperature	1% tolerance (daily)

Real Case-Study Approach

For this application two battery technologies are considered (LTO and sLFP), see Fig. 6 and Fig. 7 and their main characteristics (Table IV). Both cells are currently used for railway applications. LTO ones are more oriented to power applications than sLFP ones. However, both ones are suitable for the application for the power levels obtained in the modeling section. For each cell technology, the number of cells in series was chosen to allow to reach a standard railway traction voltage (600V minimum), being 360S for LTO (612V-972V) and 216S for sLFP (606V-832V). Finally, several parallel configurations will be evaluated, for LTO (4P, 6P, 8P, 10P) and for sLFP (4P, 6P, 8P, 10P, 12P, 14P).

Fig. 7: SAFT LP31MTi cell (LTO).

Fig. 6: SAFT VL45EFe cell (sLFP), VL47EFe precursor.

Table IV: Cell Characteristics.

Cell Name	Format	Capacity (Ah)	Energy (Wh)
LP31MTi	Prismatic	31	70
VL47EFe	Cylindrical	47	132

Regarding the recharging system, the operational cycle selected for the study is repeated 6 times per day (8h working and 16h resting approximately). And this profile will be applied 320 days per year. Recharging configurations are presented in Table V and Table VI, showing all the configuration inputs for the sizing. The static charger is used when the tramway is stopped (only station catenary). On the other side, the dynamic wireless system will support acceleration processes (right after the station). Both systems can be installed in the same station. The number of charging stations indicates what is the total of recharging points in the route, from no charging points (0 stations) to all the stations (42 stations).

Table V: Static wired charging system.

Charging Stations	Power Level	Night Charge
0, 7, 14, 21, 28, 35, 42	100kW 200kW 300kW	100kW (optional recharge)

Table VI: Dynamic wireless charging system.

Charging Stations	Power Level	Path Length
0, 7, 14, 21, 28, 35, 42	100kW 200kW 300kW	15m 30m

Results

Once the methodology is applied with discrete input values and ESU constraints, a family of solutions is obtained. Therefore, an analysis is required to evaluate most representative and valuable results in terms of ESU size (series/parallel cells) and required recharging system (dynamic wireless and static systems). This analysis has been completed for each technology individually, and then compared.

Starting with LTO cells, the results are shown in Fig. 8. In Fig. 8a for each sizing the minimum number of stations and recharging power requirements (dynimac and static systems) is presented. As it can be seen, if a wireless dynamic recharging system is introduced there are two effects. The ESU size or number of stations can be reduced. There are two interesting solutions for this application. If 28 stations with 200kW of static recharging systems are selected, just to add a 100kW dynamic wireless recharging systems reduce the required ESU from 10P360S to 6P360S. On the other hand, there are no solutions for only static recharging systems at 100kW, but adding a wireless system at 100kW allows to have a feasible 42 stations solution.

Secondly, the wireless recharging system length also determines optimal solutions. In this case (Fig. 8b), if the length is increased the ESU size can be reduced. However, the most interesting conclusion is that longer solutions make it possible to have lower number of stations. As an example, if the wireless system length is increased up to 30m with an static system of 100kW, required stations are reduced from 35 stations down to 28 or 21 stations (depending on the ESU size), not being feasible before for 15m solutions.

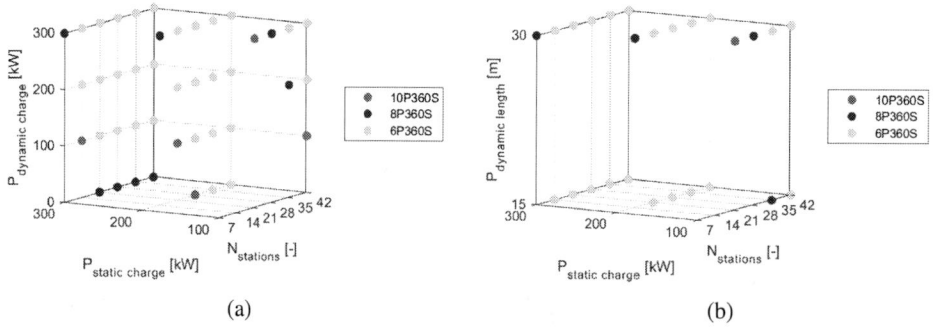

Fig. 8: LTO cells sizing results: (a) wireless charger power, (b) wireless charger path length.

Regarding sLFP cell solutions presented in Fig. 9, the conclusion is the same. Among all ESU sizing solutions, there is a clear trend where adding wireless dynamic recharging system reduces significantly the ESU size. For example, if a 14 stations and 300kW of static recharging system is selected, the required ESU size is reduced from 12P360S down to 8P360S only adding a 100kW wireless recharging systems.

In this case and compared to LTO cells technology, there are ESU solutions for all the wireless recharging systems lengths. Then, if the length is increased the ESU size is reduced. As an example, in a 21 stations and 100kW static system solution, the ESU required size is reduced from 12P360S down to 8P360S.

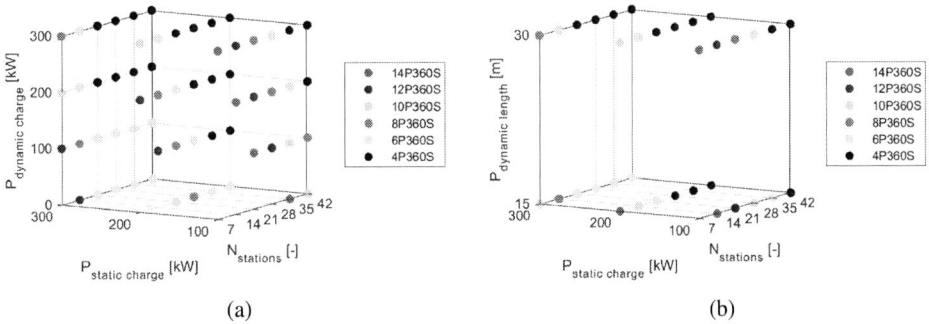

(a)   (b)

Fig. 9: sFP cells sizing results: (a) wireless charger power, (b) wireless charger length.

Finally, if both ESU cell technologies are compared the sLFP has an advantage in terms of optimizing the infrastructure. It is possible to perform the profile with a smaller number of charging stations and/or less powerful ones, which is expected because the tram profile studied is more energy-oriented. In terms of volume and weight onboard the tramway, the LTO has an advantage due to its prismatic format which allows to reach a better integration factor. Note that the volumes and weights presented here would be impacted by the thermal management needs of the current ESU. Having smaller batteries (be it sLFP or LTO) would also mean a smaller cooling system, minimizing the impact on the tramway architecture. Depending on the objectives at system level, sLFP cell will minimize infrastructure costs and LTO will minimize impact on the tramway (e.g., retrofit of an existing tramway). It is interesting to note that, if eventually power needs to be increased, LTO would become a more favored solution as it is a power-oriented cell and can accommodate very fast charge or discharge sequences.

In summary, this methodology enables to obtain a family of feasible ESU solutions suitable for the case-study, but not being possible to chose a single solution. For this selection more factors must be considered in further steps (i.e. engineering, budgeting, infrastructure limitations, etc.). However, this methodology defines the best ESU sizing for the number of stations and recharging systems (static and dynamic). Note that, a small 100kW wireless dynamic recharging system allows to reduce the number of stations and ESU size, as it has been concluded.

Conclusions

In this paper, a tramway has been analytically modeled in order to obtain the power and energy requirements for a real application. The objective is to size the required system using batteries and its associated recharging system as an ESU for 15-year operation. For that, it has been proposed an ESU sizing methodology where the constraints are defined by the ESU cells safe operation (capacity/resistance ageing and SOC/temperature balance).

This methodology has been applied to a real case-study using LTO and sLFP cells for the tramway. Recharging system is composed by static and dynamic wireless systems with different power levels, lengths and number of stations. A set of solutions has been obtained for both ESU technologies, being both suitable for the application. sLFP (energy oriented) cells have an advantage in terms of optimizing the infrastructure and LTO (power oriented) cells for a better integration. Mention, small wireless dynamic recharging system allows to reduce number of stations and ESU size, being a key for the final solution.

Acknowledgment

This project has received funding from the Shift2Rail Joint Undertaking under the European Union's Horizon 2020 research and innovation programme under grant agreement No 101015423. The content of this publication does not reflect the official opinion of the European Union. Responsibility for the information and views expressed in the report lies entirely with the author(s).

References

[1] L. Mir, I. Etxeberria-Otadui, I. P. de Arenaza, I. Sarasola, and T. Nieva, "A supercapacitor based light rail vehicle: system design and operations modes," in *2009 IEEE Energy Conversion Congress and Exposition*, 2009, pp. 1632–1639.

[2] J. J. Mwambeleko and T. Kulworawanichpong, "Battery and accelerating-catenary hybrid system for light rail vehicles and trams," in *2017 International Electrical Engineering Congress (iEECON)*, 2017, pp. 1–4.

[3] A. Rujas, I. Villar, I. Etxeberria-Otadui, U. Larrañaga, and T. Nieva, "Design and experimental validation of a silicon carbide 100kw battery charger operating at 60khz," in *2014 IEEE 15th Workshop on Control and Modeling for Power Electronics (COMPEL)*, 2014, pp. 1–7.

[4] V. I. Herrera, H. Gaztañaga, A. Milo, A. Saez-de Ibarra, I. Etxeberria-Otadui, and T. Nieva, "Optimal energy management and sizing of a battery–supercapacitor-based light rail vehicle with a multiobjective approach," *IEEE Transactions on Industry Applications*, vol. 52, no. 4, pp. 3367–3377, 2016.

[5] J. J. Mwambeleko, U. Leeton, and T. Kulworawanichpong, "Effect of partial charging at intermediate stations in reducing the required battery pack capacity for a battery powered tram," in *2016 IEEE/SICE International Symposium on System Integration (SII)*, 2016, pp. 19–24.

[6] I. Villar, A. Garcia-Bediaga, U. Iruretagoyena, R. Arregi, and P. Estevez, "Design and experimental validation of a 50kw ipt for railway traction applications," in *2018 IEEE Energy Conversion Congress and Exposition (ECCE)*, 2018, pp. 1177–1183.

[7] A. Avila, A. Garcia-Bediaga, U. Iruretagoyena, I. Villar, and A. Rujas, "Comparative evaluation of front- and back-end pfc ipt systems for a contactless battery charger," *IEEE Transactions on Industry Applications*, vol. 54, no. 5, pp. 4842–4850, 2018.

[8] F. Gonzalez-Hernando, U. Iruretagoyena, M. Arias, and I. Villar, "Dynamic ipt system with lumped coils for railway application," in *2017 19th European Conference on Power Electronics and Applications (EPE'17 ECCE Europe)*, 2017, pp. P.1–P.9.

[9] P. Li and M. Mitchell, *Chapter 2 - Railway Industry Overview*. AREMA, 2003.

[10] Z. Tian, N. Zhao, S. Hillmansen, S. Su, and C. Wen, "Traction power substation load analysis with various train operating styles and substation fault modes," *Energies*, vol. 13, no. 11, 2020. [Online]. Available: https://www.mdpi.com/1996-1073/13/11/2788

[11] K. Mongkoldee, U. Leeton, and T. Kulworawanichpong, "Single train movement modelling and simulation with rail potential consideration," in *2016 IEEE/SICE International Symposium on System Integration (SII)*, 2016, pp. 7–12.

Investigations on the Active Reduction of Common Mode Noise with Opposing Noise Sources

Philipp Marx, Felix Seybold, Philipp Ziegler, David Hirning, Jörg Roth-Stielow
INSTITUTE FOR POWER ELECTRONICS AND ELECTRICAL DRIVES
UNIVERSITY OF STUTTGART
Pfaffenwaldring 47
70569 Stuttgart, Germany
Tel.: +49 / (0) –71168567373
Fax: +49 / (0) – 71168567378
E-Mail: Philipp.Marx@ilea.uni-stuttgart.de
URL: http://www.ilea.uni-stuttgart.de

Keywords

«Active filter», «EMC/EMI», «impedance analysis», «filtering», «frequency domain analysis»

Abstract

Conventional filters needed for the electromagnetic compatibility lower the system power density by consuming additional space in power electronic setups. To increase the power density, alternatives for these filters are needed. If the power electronic setup offers the possibility to synchronize two opposing switching transitions, the common mode noise can be reduced without additional filters. In this case two opposing noise currents reduce one another by superimposition. For power electronic setups which do not offer the possibility to synchronize switching transitions, this paper presents an approach which uses additional half bridges as opposing noise sources connected to a replicated grounding impedance to create the opposing common mode noise. The main goal of the opposing noise source is to create similar potential changes as the power electronic setup over the replicated grounding impedance. The replicated grounding impedance emulates the grounding impedance of an ohmic inductive load with discrete elements. The achieved reduction of the common mode noise with this approach is evaluated with measurements.

Introduction

The usage of wide band gap transistors in industrial and automotive applications is increasing. Their characteristics such as higher blocking voltages and increasing switching transitions slopes compared to conventional silicon transistors promise advantages in efficiency and power density. On the other hand, increasing switching slopes and smaller power electronic designs lead to increasing challenges regarding electromagnetic interference (EMI). The components used in conventional passive EMI filters get bigger and heavier so that they lower the gained benefits. Hence, active EMI filters to omit the conventional EMI filters are examined.

Basic idea of active EMI filters is to inject an inverted noise signal to suppress the disturbances by superimposition. Therefore, the noise is measured with a sensing system, inverted and adapted by an analog or digital system and injected in a feedback or feedforward loop into the circuit [1–4]. Because the signals have to be adapted and inverted with an analog or digital system which has a limited bandwidth, the bandwidth of the injected signal is also limited. In addition, the sensing and injection circuits need space, especially if transformers are used.

To overcome these drawbacks an alternative way to damp the noise is to synchronize the timing of two opposing switching transitions [5–11]. Because the noise is generated by the same noise source, there are theoretically no limits regarding the bandwidth. For this purpose, it is mandatory for the power electronic setup to offer the possibility to synchronize switching transitions and identical disturbance propagation paths. If these requirements are met, the disturbances can be damped without any additional components. However, in various power electronic topologies, the adaption of the timing of the switching transitions is not possible, the propagation paths differ from each other or only one transistor is used. In these cases, the damping of the generated noise is not possible with this attempt.

To generate inverted noise signals, in [12] additional half bridges are used to generate switching transitions. The half bridges are connected with capacitances to ground. In this paper, different realization options to generate the switching transitions for such an approach are investigated. The designs attempt to save costs and space in comparison to the use of the same design as the load system. Moreover, the grounding impedances are investigated to provide an accurate emulation of the parasitic propagation path of the noise. A replication with discrete elements as a grounding impedance is built with the knowledge of the parasitic impedances. Furthermore, a possibility to adapt the timing of the generated switching transitions is presented. The potential of the approach is presented with measurements on a test setup. The different designs are compared to each other in terms of cost, space and reduction of the disturbances. Furthermore, the discrete elements of the replication are varied to show the influence of tolerances on the approach.

Approach

The origins of common mode noise are changing electrical potentials due to switching transitions in power electronic systems. The changing potential φ_1 across the parasitic impedances of the load to the ground potential result in the common mode current i_1. The magnitude and frequency components of the common mode currents are dependent on the derivative of the potential change and the parasitic grounding impedance.

It is possible to reduce the common mode disturbances of the load system, if a second potential change φ_2 with an opposing direction across a similar grounding impedance occurs at the same time. The second potential change generates an opposing common mode current i_2, which reduces the other common mode current by superimposition. The sum of both currents is theoretically zero. This principle is depicted in Fig. 1.

In this approach, the second potential change is generated by an additional opposing noise source which is exclusively used for the active common mode reduction, so that it conducts no load current. The opposing noise source is connected to an impedance network of discrete elements, which is connected to the ground potential. The system components of the approach are depicted in Fig. 2. The impedance network emulates the parasitic impedances of the load to form the

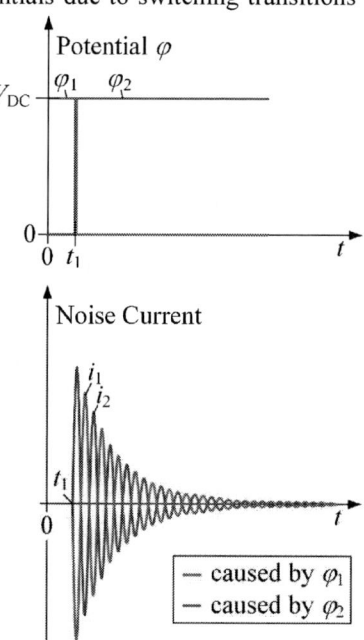

Fig. 1: Principle of the common mode reduction

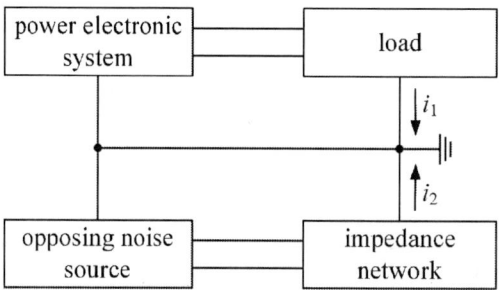

Fig. 2: System components of the approach

inverse temporal course of the common mode current i_1. Therefore, the impedance network should have the same impedance than the original parasitic common mode propagation path of the load connected to the power electronic system. To build up an accurate impedance network, the setup and its common mode propagation paths are evaluated. Based on this evaluation, the opposing noise source and the impedance network are designed to create an opposing noise current i_2.

Overall structure

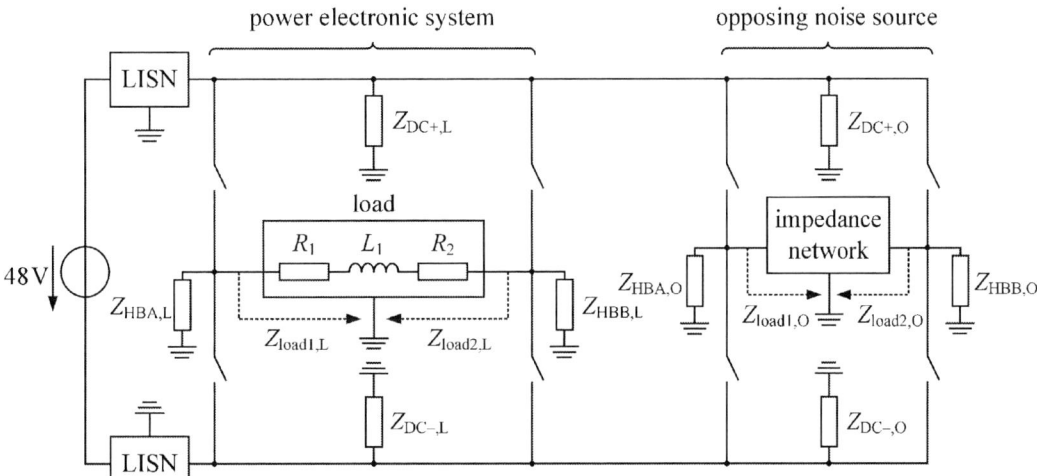

Fig. 3: Overall structure

The overall structure of the approach is depicted in Fig. 3. As the power electronic system, a full bridge converter connected to an ohmic inductive load is used. The design of the full bridge converter of the power electronic system is shown in Fig. 4. The full bridge converter consists of four transistors which are located in a TO263 package. The gate signals of the transistors are generated on another PCB which can be attached by connectors. The transistors are cooled through the PCB which is mounted with an isolation pad on a grounded heat sink. All electrically conductive areas of the full bridge converter form a parasitic impedance to the grounded heat sink. In detail, the impedance between the dc link terminals to ground $Z_{DC+,L}$ and $Z_{DC-,L}$, as well as the grounding impedances at both load terminals $Z_{HBA,L}$ and $Z_{HBB,L}$ are evaluated.

A conceivable field of application of the approach are power electronic converters for electrical machines. Hence, the inductor L_1 of the load is located in a machine housing to ensure realistic parasitic impedances. The resistors R_1 and R_2 are used to emulate the induced voltage of an electrical machine. They are placed at the two inductor terminals to create symmetrical parasitic impedances at both inductor terminals. $Z_{load1,L}$ and $Z_{load2,L}$ represent the grounding impedances of the load.

To create the same switching pattern as the power electronic system, also a full bridge converter is used as the opposing noise source. The opposing noise source itself forms parasitic grounding impedances $Z_{HBA,O}$, $Z_{HBB,O}$, $Z_{DC+,O}$, $Z_{DC-,O}$. The opposing noise source is connected to an impedance network. The grounding impedances $Z_{load1,O}$ and $Z_{load2,O}$ are formed with discrete elements by using the impedance network.

Opposing Noise Source

The main goal of the opposing noise source is to generate an inverse changing potential over the impedance network. Furthermore, the design of the opposing noise source attempts to form the same parasitic impedances to ground as the power electronic system. Therefore, three different designs are realized and evaluated.

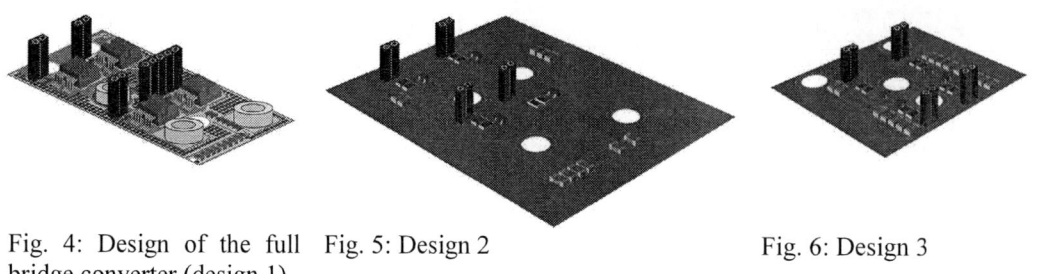

Fig. 4: Design of the full bridge converter (design 1) Fig. 5: Design 2   Fig. 6: Design 3

Design 1 uses the same power electronic setup for the opposing noise source as the load setup in Fig. 4. In that case, the copper areas are the same and therefore, the same parasitic impedances are formed. Although the switching transition slopes are dependent on the load current, with the use of same transistors, the derivative of the potential change is more similar as with different transistors.

Design 2 in Fig. 5 tries to omit the expensive power transistors and uses cheap signal transistors instead. This is possible as the filter full bridge converter only needs to carry the disturbance current. On the other hand, the use of different transistors influences the switching transition slope. The copper areas on the additional power electronic setup have the same size as the load power electronic. Additional discrete capacitances can be placed on the ground plate to adjust the parasitic impedance. By using the same copper area, the size of the additional noise source is equal to the power electronic setup and therefore not minimal.

In design 3, shown in Fig. 6, the goal is to minimize the opposing noise source. Therefore, the parasitic impedance is emulated only with discrete elements. Similar to design 2, cheap signal transistors are used to generate a changing potential. With this design the size of the opposing noise source is the smallest, but the emulated impedance has to be accurate.

To design the right discrete elements of design 2 and 3, the grounding impedances $Z_{DC+,L}$, $Z_{DC-,L}$, $Z_{HBA,L}$ and $Z_{HBB,L}$ of the power electronic system are measured with an impedance analyzer (Omicron Lab. Bode 100). The measurements are depicted in Fig. 7. The impedance courses of all impedances of the power electronic system show capacitive behavior in the considered frequency range. For this reason, only capacitors are used to adapt the grounding impedances $Z_{DC+,O}$, $Z_{DC-,O}$, $Z_{HBA,O}$ and $Z_{HBB,O}$ with discrete elements. With an iterative process the values of the required capacitances C_{DC-}, C_{DC+}, C_{HBA} and C_{HBB}, depicted in Tab. I, are found. Using these values, the impedance courses of design 2 and 3 indicated with O2 and O3, match with the impedance courses of the full bridge converter.

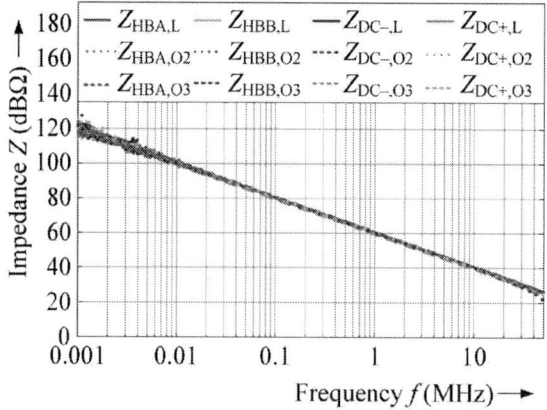

Fig. 7: Grounding impedances of the full bridge converter

Table I: Values of the used capacitances

	C_{DC+}	C_{DC-}	C_{HBA}	C_{HBB}
Design 2	66 pF	68 pF	55 pF	61 pF
Design 3	90 pF	95 pF	100 pF	96 pF

Design of the Impedance Network

To find a matching replication, the ohmic inductive load connected to the full bridge converter is evaluated. The connected ohmic inductive load can be described with the impedance circuit in Fig. 8 [10]. Two resistors are connected symmetrically to both terminals of the inductor. The two resistors are considered with its resistive part $R_{add,1}$ and $R_{add,2}$, and its parasitic inductive part $L_{add,1}$ and $L_{add,2}$. The inductor is located in a grounded machine housing. In particular the copper wires of the inductor are coiled around a stator tooth to ensure realistic parasitic impedances.

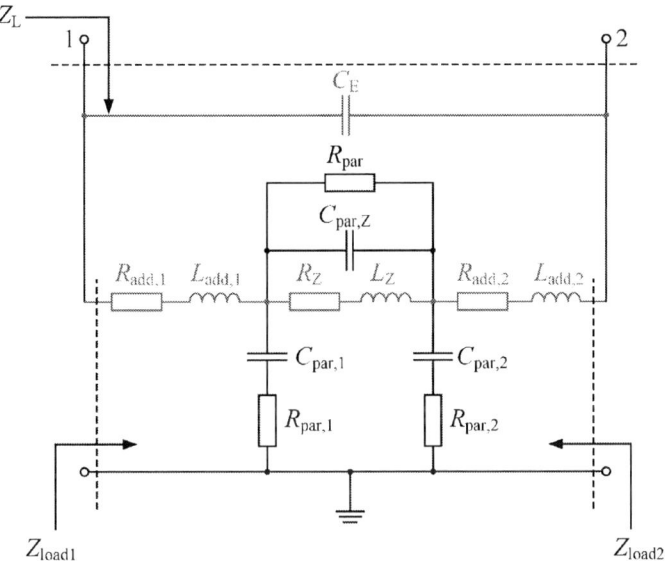

Fig. 8: Impedance circuit of the load

The impedance between the terminals of the inductor is modelled with R_Z and L_Z for frequencies below the resonance frequency and R_{par} and $C_{par,Z}$ are needed to describe the impedance in the frequency range above the resonance frequency. Between the windings of the inductor and the grounded machine housing a parasitic impedance is formed. Because the windings are coiled in several layers, the distance and therefore the effective parasitic impedance between the windings at both inductor terminals differ from each other. These grounding impedances are considered with $C_{par,1}$ and $R_{par,1}$ at one inductor terminal as well as with $C_{par,2}$ and $R_{par,2}$ at the other inductor terminal.

To generate the opposing common mode noise, which is propagating through the grounding impedances, an impedance network with discrete elements is designed. Therefore, the impedance courses of the parasitic impedances $Z_{load1,L}$ and $Z_{load2,L}$ of the load system need to be replicated as well as possible. The main focus is the frequency range close to the resonance frequency at around 6 MHz because in this frequency range the impedance is the lowest and vice versa the disturbance current is the biggest. The resonance frequencies f_1 for Z_{load1} and f_2 for Z_{load2} can be calculated with:

$$f_1 = \frac{1}{2\pi}\sqrt{\frac{1}{(L_{add1}+L_{add2})\cdot(C_{par,1}+\dfrac{C_{par,Z}\cdot C_{par,2}}{C_{par,Z}+C_{par,2}})}} \tag{1}$$

$$f_2 = \frac{1}{2\pi}\sqrt{\frac{1}{(L_{add1}+L_{add2})\cdot(C_{par,2}+\dfrac{C_{par,Z}\cdot C_{par,1}}{C_{par,Z}+C_{par,1}})}} \tag{2}$$

Both grounding paths are linked with the parasitic capacitance $C_{par,Z}$ in the considered frequency range, also seen in (1) and (2). The inductance L_Z and the resistance R_Z have a negligibly small impact on the grounding impedances Z_{load1} and Z_{load2}. Because the impedance Z_L between the inductors terminals is insignificant for the approach, those two elements are neglected in the discrete impedance network to save space. Furthermore, the impedance network is connected to the same resistors as the load. Hence, the elements $R_{add,1}$, $R_{add,2}$, $L_{add,1}$, $L_{add,2}$ and C_E are already existing and are not replicated with the impedance network. The impedance network is built up with the values in Tab. II and measured with an impedance analyzer. The measurements of the impedance Z_L, depicted in Fig. 9, show big differences between the parasitic and replicated course. Due to the missing elements L_Z and R_Z the differences are as expected, but as mentioned not important for the approach.

Table II: Values of the impedance network

$C_{par,Z}$	$R_{par,Z}$	$R_{par,1}$	$R_{par,2}$	$C_{par,1}$	$C_{par,2}$
45 pF	2.8 kΩ	0 Ω	0 Ω	52 pF	51 pF

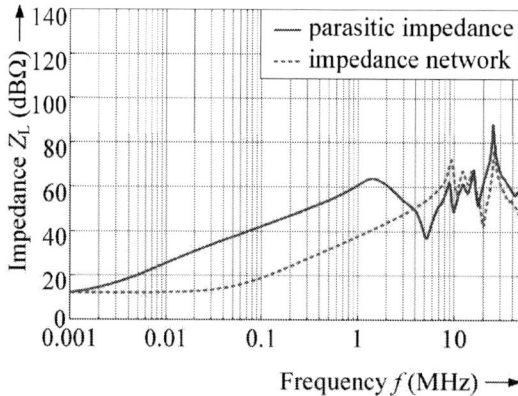

Fig. 9: Impedance between the inductor terminals

The measured impedance courses of $Z_{load1,O}$ and $Z_{load2,O}$ are shown in Fig. 10. Despite the missing elements L_Z and R_Z both impedance courses show a good replication of the original impedance course in the expanded view. Between 400 kHz and 1.5 MHz the courses slightly differ, due to an extra resonance. Also in the high frequency range, the replication is not accurate. The detailed view shows a good replication of $Z_{load1,L}$ with $Z_{load1,O}$ at the resonance frequency, but the replicated impedance $Z_{load2,O}$ differs slightly from the measured parasitic impedance $Z_{load2,L}$.

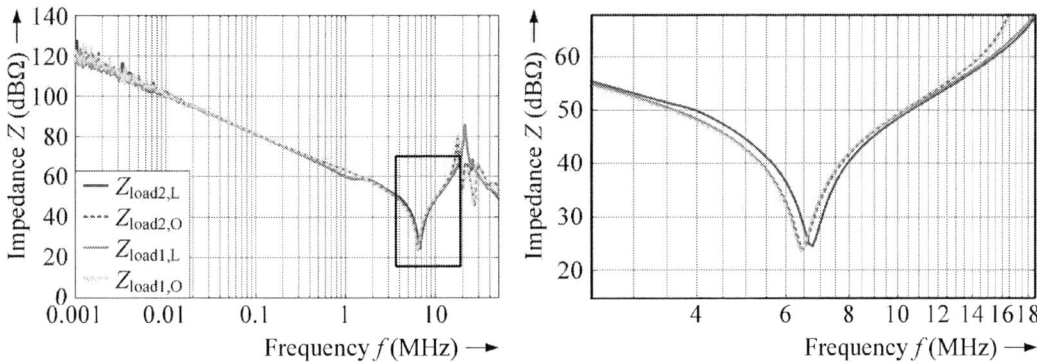

Fig. 10: Grounding impedances – expanded view (left) and detailed view (right)

Timing

For the best reduction of the disturbance currents, the changing potentials in both systems have to occur at the same time. In the opposing noise source the current direction in the power electronic system has to be considered. Depending on the current direction, the commutation type at the half bridges is either hard or soft. If the commutation type is hard, the potential change occurs before the dead time. If the commutation type is soft, the changing potential occurs after the dead time. Hence, the gate signals of the opposing noise source have to be shifted by the dead time dependent on the load current direction. Furthermore, to improve the reduction of the common mode noise, differences in the propagation delay of the gate signals should be adjusted. The gate signals are generated with a rapid prototyping system and afterwards processed with a FPGA. For large time shifts in terms of the dead time, delay chains are used to shift the gate signal in multiples of the FPGA clock. For small time shifts, a phase locked loop is used, which provides a precise timing in steps down to 120 ps. With the combination of both options the timing can be adjusted in a wide time range in a fine resolution. In this approach, the best timing is set manual.

Measurements and Results

The overall structure in Fig. 3 is built up on an electrically conductive ground plane. To provide a defined input impedance, two line stabilization networks (LISN) are connected to every DC-link terminal. The noise current generates the disturbance voltages v_{DC+} and v_{DC-} at both 50 Ω termination resistors of the LISNs. With a power splitter, the common mode voltage v_{CM} which is measured with a spectrum analyzer (Tektronix RSA306B), can be extracted. The measurements are performed at a DC-link voltage of 48 V at a load current of 5 A. The PWM frequency is set to 20 kHz with a duty cycle of 0.5, so the time between the switching transitions of the full bridge converter is at its maximum. In this case, the affection of two consecutive noise signals is at its minimum.

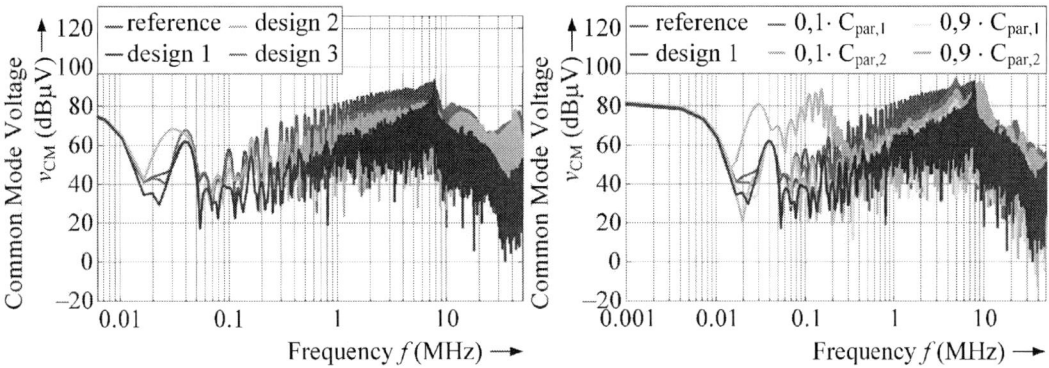

Fig. 11: Common mode voltage with different designs

Fig 12: Common mode voltage with different values of the impedance network

The measurement results of the common mode voltage v_{CM} for the different designs of the opposing noise source and of a reference system are shown in Fig. 11. As a reference system, only the noise source connected to the load is operated. So the course depicted in blue shows the common mode voltage v_{CM} without any filtering. In all other colors the power electronic system and the opposing noise source are operated. In purple, the opposing noise source consists of the same power electronic design as the power electronic system but is connected to the discrete impedance network. In this case, the common mode voltage can be reduced over a wide frequency range. From 50 kHz to 5 MHz the reduction is more than 10 dB, at some frequencies the reduction is even higher than 20 dB. At around 8 MHz the common mode voltage peaks and no reduction to the reference system is measured. This shows that with the impedance network an opposing noise current with the same frequency spectrum can be shaped. The small differences of the impedance courses around the resonance frequency, where the noise current is the highest, have a big impact on the common mode voltage.

For the measurement results in yellow, design 2, which uses signal transistors and the same PCB areas as the load power electronic setup, is used. The reduction int the entire frequency range is lower than in the first setup. In the frequency area at the resonance frequencies the course is similar. Still, the reduction of the common mode voltage is more than 5 dB over a large frequency range. On the other hand, the common mode voltage rises at frequencies higher than 10 MHz.

With the use of design 3 as the opposing noise source, shown in red, the course of the common mode voltage is similar, but the reduction is even lower than with the second setup.

The lowered reduction of the common mode voltage with design 2 and 3 show that the signal transistors have an impact on the approach. Reasons can be differences in the grounding impedances and differences in the switching transitions slopes.

To evaluate how accurate the discrete elements have to be, the values of the discrete elements are changed. Design 1 is used as the opposing noise source due to the best results. The grounding capacitances $C_{par,1}$ and $C_{par,2}$ are reduced with 10% and 90% of its value. The measurements are shown in Fig. 12. If the differences are small the course of the common mode voltage is almost the same. With large differences, the reduction of the common mode voltage decreases. Furthermore, new resonance frequencies are formed which leads to higher common mode voltages at these frequencies. This shows that small differences due to construction tolerances in the impedance network are tolerable. Large differences reduce the effectiveness of the common mode reduction. In some frequencies the common mode is even reduced. Furthermore, this evaluation shows that the values of the impedance network were suitable in the first place.

Conclusion

With an additional noise source connected to an impedance network, it is possible to reduce the common mode noise in power electronic setups which do not offer the possibility to synchronize switching transitions. Three different designs are evaluated as an opposing noise source which provide changing potentials. The parasitic impedances of the designs are adapted to the load power electronic setup. Using the same design as the power electronic system provides the best results. By using cheaper signal transistors, the common mode noise can still be reduced, but the reduction is smaller due to differences In the grounding impedances and in the switching transition slopes. A space-reduced design, which emulates the grounding impedances with capacitances, lowers the reduction of the common mode noise. Future work will investigate adjustments of the switching transition slopes to improve the results.

References

[1] M. Biskoping, M. Rosekeit, and R. W. De Doncker, "Active EMI-filter using the gate-drivers power supply," in 2015 IEEE 11th International Conference on Power Electronics and Drive Systems, 2015, pp. 449–455.

[2] I. Takahashi, A. Ogata, H. Kanazawa, and A. Hiruma, "Active EMI filter for switching noise of high frequency inverters," in Proceedings of Power Conversion Conference - PCC '97, 1997, pp. 331-334.

[3] D. Shin et al., "Analysis and Design Guide of Active EMI Filter in a Compact Package for Reduction of Common-Mode Conducted Emissions," IEEE Transactions on Electromagnetic Compatibility, vol. 57, no. 4, pp. 660–671, 2015, doi: 10.1109/TEMC.2015.2401001.

[4]   S. Ogasawara, H. Ayano, and H. Akagi, "An active circuit for cancellation of common-mode voltage generated by a PWM inverter," IEEE Trans. Power Electron., vol. 13, no. 5, pp. 835–841, 1998, doi: 10.1109/63.712285.

[5]   M. Zehelein, J. Portik, M. Nitzsche, P. Marx, and J. Roth-Stielow, "Reduction of the Leakage Currents by Switching Transition Synchronization for a Four-Switch Buck-Boost Converter," in 2019 10th International Conference on Power Electronics and ECCE Asia (ICPE 2019 - ECCE Asia), 2019, pp. 2217–2223.

[6]   M. Zehelein, J. Ruthardt, M. Nitzsche, T. Tymosch, and J. Roth-Stielow, "Leakage current reduction for a double-leg boost converter by switching transition synchronisation," The Journal of Engineering, vol. 2019, no. 17, pp. 3789–3792, 2019, doi: 10.1049/joe.2018.8128.

[7]   Youngjin Baek, Gwigeun Park, Dongmin Park, Honnyong Cha, and Heung-Geun Kim, "Common-Mode Current Reduction with Synchronized PWM Strategy in Two-Inverter Air-Conditioning Systems," Journal of Power Electronics, vol. 19, no. 6, pp. 1582–1590, 2019. [Online]. Available: https://jpels.org/digital-library/22272

[8]   A. von Jouanne, H. Zhang, and A. K. Wallace, "An evaluation of mitigation techniques for bearing currents, EMI and overvoltages in ASD applications," IEEE Transactions on Industry Applications, vol. 34, no. 5, pp. 1113–1122, 1998, doi: 10.1109/28.720452.

[9]   J. Bertelmann, M. Beltle, S. Tenbohlen, and R. Eidher, "Minimierung der Gleichtaktstörung in elektrischen Lenkungssystemen durch gegenphasiges Takten der Leistungshalbleiter," in VDE Automotive meets Electronics 2019.

[10]   P. Marx et al., "Common Mode Voltage Cancellation in Integrated Modular Motor Drives," in 2021 23rd European Conference on Power Electronics and Applications (EPE'21 ECCE Europe), 2021, P.1-P.9.

[11]   P. Marx, J. Assenheimer, P. Ziegler, J. Haarer, and J. Roth-Stielow, "Reduction of Common Mode Disturbances in Parallel Modules of Integrated Modular Motor Drives," in The 2022 International Power Electronics Conference (IPEC-Himeji 2022 -ECCE Asia-), in press.

[12]   S. Cordes and F. Klotz, "Active common mode cancellation," in 2018 IEEE International Symposium on Electromagnetic Compatibility and 2018 IEEE Asia-Pacific Symposium on Electromagnetic Compatibility (EMC/APEMC), Singapore, May. 2018 - May. 2018, pp. 127–130.

Knowledge Based Grey Box Modeling of Inaccessible Circuits for System EMC-Simulation in Time Domain

Jan-Philipp Roche	Jens Friebe	Oliver Niggemann
KEB Automation KG	Leibniz University Hannover	Helmut-Schmidt-University
R&D Electronics	Institute for Drive Systems	Institute of Automation Technology
Südstraße 38	and Power Electronics	Holstenhofweg 85
32683 Barntrup, Germany	Welfengarten 1	22043 Hamburg, Germany
jan-philipp.roche@keb.de	30167 Hannover, Germany	oliver.niggemann@hsu-hh.de
https://www.keb.de	friebe@ial.uni-hannover.de	https://www.hsu-hh.de/imb/
	https://www.ial.uni-hannover.de	

Acknowledgment

Jens Friebe would like to acknowledge the funding by the Ministry of Science and Culture of Lower Saxony and the Volkswagen Foundation. The authors are responsible for the content of this publication.

Keywords

≪Neural network≫, ≪Passive Filters≫, ≪Modelling≫, ≪Discrete-time≫, ≪EMC/EMI≫

Abstract

Time domain simulations are important to efficiently optimize function and EMC of electrical circuits in one setup together. Knowledge based grey box modeling is a promising approach for modeling inaccessible circuits which enables simulations of whole electrical systems. Grey box modeling combines the advantages of white and black box modeling. This work examines and evaluates the application possibilities in the field of EMC considerations. Furthermore, perspectives for future work are given.

Introduction

In a product development process, electromagnetic compatibility (EMC) is just as important as the function of the electrical circuit [1]. Prototype-based design iterations can be highly time-consuming, so simulations are aimed. A time domain simulation is suitable to cover aspects of EMC and circuit function in one single simulation setup. The circuit function has to be monitored and preserved during EMC optimization process. Electromagnetic emissions (EME) are part of the EMC and in the application focus of this work. As being the initial source for EME, voltage and current spectra are used as modeling evaluation criteria in the following considerations. These spectra are calculated by fast Fourier transform (FFT) from time domain results. EME results are not presented, because they can only be determined for a whole system including other circuits, the environment and measurement equipment, not for a single component. So only the component modeling capability of low and high frequency, small and large signals at the same time will be examined by spectra of voltages and currents.

Models of electrical components have to be built if they are not available or not sufficient for the simulation of function and EME-relevant spectra. In general, there are two different modeling approaches: White and black box modeling. Pure white box modeling by physical insight can be quite complicated and costly. Pure black box modeling suffers from other disadvantages: Previously known model parts are not included, so available knowledge is neglected. Furthermore, the learned behaviour can be limited to the used training data, extrapolation for example demands for new approaches [2] and the amount of required training data can be a practical problem. Grey box modeling combines the advantages of these two general modeling approaches: Previous knowledge and physical insight can be implemented. Very complex or unknown behaviour can be represented by trained black box model parts. Compared to a pure black box model, less training data is needed and the extrapolation possibilities are better [3]. Furthermore, some components are not accessible, so measurement and characterization of internal parts are not possible. The grey box modeling approach is applied to model components with limited insight. So EME considerations by simulations with inaccessible components are thus aimed to be made possible by knowledge based grey box modeling. In Fig. 1, this topic is shown in context of a development process. For developing such an approach, nonlinear passive filter circuits are useful application examples. They are widely spread in power electronics applications, are important for the function and EME of electrical circuits and can have strong nonlinearities. Furthermore, they can have a significant time dependent behaviour due to their energy storage elements. Some characteristics are previously known and can be directly implemented in the model. The remaining amount of the overall behaviour can be trained to and represented by a black box model part. The knowledge based grey box modeling approach based on equivalent circuits is introduced. The modeling results are compared to an existing pure neural network model implementation of the same filter application example [6]. In future work, the grey box approach should be applied to an integrated DC-to-DC-converter.

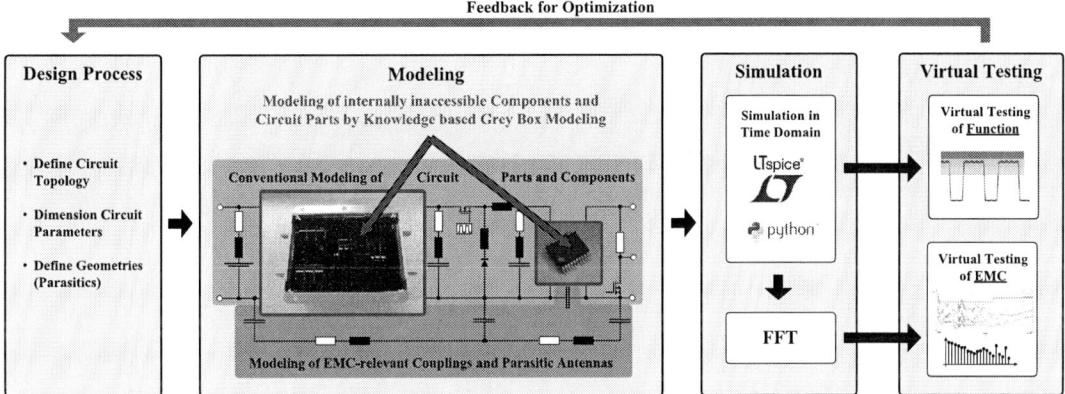

Fig. 1: Modeling of internally inaccessible components and circuit parts by knowledge based grey box modeling in context of a development process (LTspice- and Python-Logo from [4] [5])

The filters are simple and well known examples to explain, implement and verify the knowledge based grey box modeling approach. They are used as a placeholder for much more complex components and thus are assumed to be inaccessible. Only the rated component values without nonlinearities are known. The goal is to achieve the modeling of existing inaccessible components to enable the EMC simulation of a whole system. It is not intended to apply knowledge based grey box modeling for filter design optimization.

The research questions (RQ) addressed in this paper are:
- Are knowledge based grey box models suitable to model partly known and internally inaccessible circuits for time domain simulations of EME-relevant voltage and current spectra (by easily understandable and representative example of nonlinear filters)? (RQ1)
- Do knowledge based grey box models of partly known and internally inaccessible circuits lead to better modeling results than pure neural network models (from just as little training data)? (RQ2)
- Is there a simulation speed advantage? (RQ3)

State of the Art

In this section, different filter modeling approaches are introduced to show the difference between the general modeling approaches. Filter modeling is used here as a well known example. Conventional filter modeling is briefly summarized without any claim of completeness. Neural network (filter) modeling is presented by literature overview. The used nonlinear filter application examples and neural network modeling methods in the mentioned reference paper [6] are addressed after that. Finally, knowledge based grey box modeling with neural networks is explained.

Conventional Filter Modeling

A conventional filter modeling approach is the usage of an equivalent circuit (EC). An EC is a simplification which represents the major physical dependencies. The complexity of the circuit structure determines the grade of simplification. Including parasitics and nonlinearities, the EC model has to represent the filter behavior across all required operating points and frequencies. Depending on the application, there can be a huge amount of operating points to be linearized by small signal measurements. Such nonlinearities like a current dependent inductance and a voltage dependent capacitance are described in the LTspice documentation [7]. LTspice is one of many suitable possibilities to simulate EC models and used here exemplary. Ideas for specific modeling procedures can be found in literature, e.g. in [1] [8]. Every component feature has to be known for applying ECs. Even frequency dependent inductances can occur. So all relevant internal components have to be accessible for measurements with e.g. vector network and impedance analyzers. The biasing for nonlinear component measuring can be a practical problem, especially by limited accessibility. On the one hand, the usage of finite element programs is limited by available component information. On the other hand, finite element programs can be too costly and extensive for a wide industrial application.

Neural Network (Filter) Modeling

In literature, neural network modeling is applied to different kinds of circuits. A waveguide and a microstrip low-pass filter is modeled by a recurrent neural network (RNN) [9]. Nonlinear circuits in time domain (CMOS-Receiver, integrated power amplifier and integrated CMOS-Logic) are modeled by a deep RNN [10]. Long short-term memory (LSTM) networks are applied to power amplifiers [11]. Different kinds of RNN are applied for fast simulation of high-speed channels [12]. It may be useful to also apply these modeling approaches from the field of high frequency, low power communications technology to power electronics and its EME. In [6], a LSTM network is applied for modeling of nonlinear filters in discrete time domain. This is the pure neural network reference example for comparison. The nonlinear filter examples, the training details and the results will be summarized later.

Knowledge Based Grey Box Modeling with Neural Networks

There are several approaches concerning knowledge based grey box modeling with neural networks. In [13], existing RF and microwave knowledge is combined with neural networks. This includes the integration of knowledge into neural network structures and the combination of circuit models with neural networks. The last approach can be realized e.g. by the source difference method (SDM), the prior knowledge input method (PKI) and space-mapped neural models (SMN). In the SDM, the neural network models the difference between the original and the equivalent circuit model behaviour. Previously unknown behaviour is added to the EC model by the neural network. In the PKI, the neural network is used to map the EC output to the desired, original output behaviour. In SMN, the neural network is used to map the original problem input-space to a coarse model (e.g. equivalent circuit). There, all desired model behaviour has to be considered by the chosen coarse model structure. Additional output mapping is proposed in [14].

Further approaches, like physics-informed neural networks (PINN), neural networks with transfer functions (neuro-TF) [3] and the combination of equivalent circuits, neural networks and state-space-equations (EC-SSE-NN) [16] and state-space neural networks [15] are not considered here. State-Space neural networks are time discrete state-space equations with neural network functions instead of typical matrices [15]. The modeling capability of neural networks is combined with well-known properties of state-space equations such as stability criteria and internal state estimations.

Solution Approach

In this section, the solution approach using knowledge based neural grey box models is presented. The chosen types are based on equivalent circuits, because of the transparent implementation of previous knowledge. Due to the author's best knowledge, they are applied for the first time to strongly nonlinear and discretely constructed filter circuits in discrete time domain. The filters are only simple application examples to introduce the approach. In future work, it should be applied to more complex circuits. The desired goal of such an approach is a smaller modeling error compared to a pure neural network modeling with the same little amount of training data. This data is available from [6] and applied in the following. State-space neural networks are also a promising approach but proposed for future work.

Equivalent Circuit Based Neural Grey Box Models

The combination of equivalent circuits (white box) and neural networks (black box) is chosen due to the straightforward, versatile and transparent implementation of previous knowledge. In Fig. 2, the finished grey box model structures of the SDM (Fig. 2(a)), the PKI (Fig. 2(b)) and SMN (Fig. 2(c)) are shown. Unknown component behaviour is trained and represented in different ways by the grey box model structures in Fig. 2. In SDM, the difference between the original and the white box model behaviour is represented directly by the black box model part. In PKI, the output of the known white box model part is modified until it fits the original component behaviour. It is a kind of output mapping. Furthermore, the black box part uses the input values as additional knowledge. In SMN, the input values of the white box are mapped by a black box model part so that the white box model output fits the real component behaviour. It is a kind of input mapping.

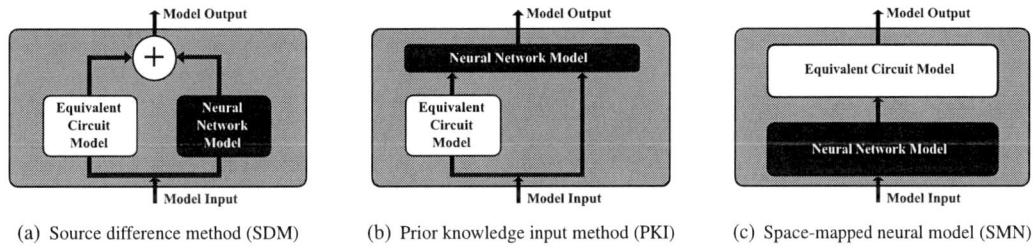

(a) Source difference method (SDM)   (b) Prior knowledge input method (PKI)   (c) Space-mapped neural model (SMN)

Fig. 2: Finished model structures following [13]

The basic training principle for the SDM following [13] is shown in Fig. 3. The training principles for the PKI and SMN have to be adapted to the respective model structure.

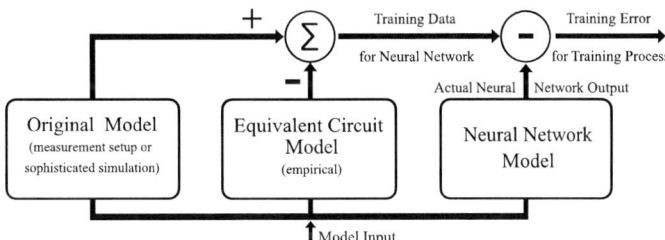

Fig. 3: Basic training principle of source difference method (SDM) following [13]

Experiments

The proposed grey box models are applied to the nonlinear filter examples of [6]. These filters are suitable and simple application examples to introduce the method of knowledge based grey box modeling. First, the application examples are briefly reintroduced and the training of the pure neural network model from [6] is summarized. After that, the training of the proposed equivalent circuit based neural grey box model is presented.

Application Examples

The used application examples (AE) from [6] are briefly reintroduced in the following. Three different nonlinear filters are used as AE. Two of them are imaginary examples (Fig. 4(a), Fig. 4(b)), which are simulated with LTspice to generate more AEs. One AE is assembled and measured at the laboratory (Fig. 4(c)), which is the preferred and intended application of the used modeling methods. These three low pass filters can reduce high-frequency (HF) noise of a signal or supply voltage. They contain nonlinear, voltage dependent capacitances and current dependent inductances.

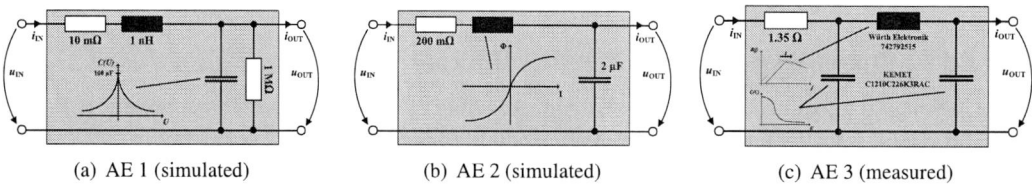

| (a) AE 1 (simulated) | (b) AE 2 (simulated) | (c) AE 3 (measured) |

Fig. 4: Nonlinear filter application examples (AE)

It is assumed that the AEs are inaccessible and only the rated component values without nonlinearities are known. In Fig. 4(a), a capacitance value of $100\,\mu F$ without voltage dependency is assumed to be known. In Fig. 4(b), the inductance value is assumed to be $300\,\mu H$ (no current dependency). In Fig. 4(c), constant values of $22\,\mu F$ and $1.6\,\mu H$ are chosen for the white box model part. The rest of the white box models is identical to Fig. 4(a), Fig. 4(b) and Fig. 4(c).

These AEs are only used to develop and apply a method for modeling existing and inaccessible components based on measurements and previous knowledge. These modeling methods are not intended to be used for the component design itself or for converting a physical model into a neural network model.

Neural Network Reference with LSTM

In [6], LSTM networks are trained with input and output voltage and current time series of these three AE. The time series are given an equidistant time base and they are fed to the neural network by tensors. These tensors result from the sliding window and its sequence length for modeling the dependency on the past time series. One dataset of each AE is reserved for testing the trained neural network. The other datasets (e.g. six) are used for training the neural network in an alternating procedure for many epochs (e.g. 1000). The detailed training scheme can be found in [6]. The used neural network structure and the respective inputs and outputs are shown in Fig. 5(a) and Fig. 5(b). The applied training (hyper)parameters can be found in Tab. I.

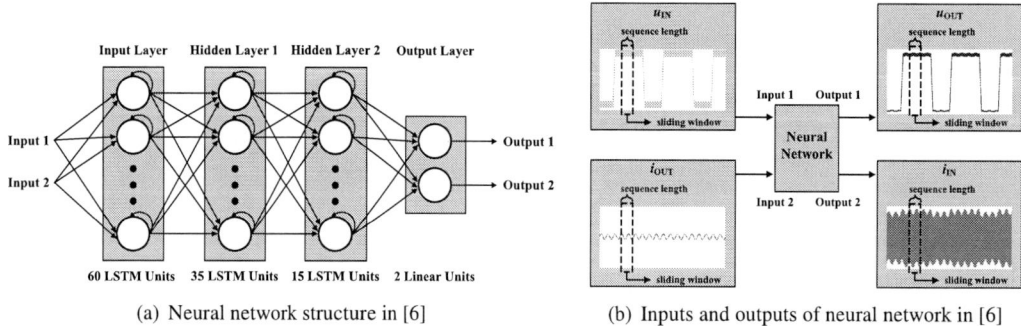

| (a) Neural network structure in [6] | (b) Inputs and outputs of neural network in [6] |

Fig. 5: Neural network in [6]

More details concerning training procedure can be found in [6]. In Tab. II, used time series characteristics examples are shown. The test results of the trained pure neural network are the reference for the comparisons to the proposed knowledge based grey box models.

TABLE I: Training (hyper)parameters in [6]

Parameter	Value
Input features	u_{IN} and i_{OUT}
Output features	u_{OUT} and i_{IN}
Optimizer	Adam
Loss function	MSE (sum of u- and i-loss)
Learning rate	0.001
Scaler	MinMaxScaler
Sequence length	3
Batch size	all batches
Epochs	1000

TABLE II: Time series characteristic examples in [6]

No.	u_{IN}	i_{OUT}
1	DC + LF-Sine	DC
2	DC + LF-Sine + HF-Sine	DC
3	DC + LF-Sine	LF-Sine with DC
4	DC + LF-Sine + HF-Sine	LF-Sine with DC
5	DC + LF-Sine + HF-Sine	Trapezoidal
6	DC + LF-Sine + HF-Sine	Trapezoidal + LF-Sine with DC
7	DC + Sine Sweep + HF-Sine	Trapezoidal
...	...	...

Training of EC based grey box models

The black box model parts in the proposed grey box models are the same in structure and size as the neural network in [6]. The training parameters and data characteristics are shown in Tab. I and Tab. II. They are chosen to be the same for a good comparability to the pure neural network model approach from [6]. Only six datasets were used for training as well. Due to the different application in form of the grey box models, an adjustment can be necessary for an optimal modeling result (future work). Only the sampling rate was increased to 100 Mpts/s to increase the frequency range. The required simulation time with a neural network (part) depends directly on the desired frequency range. The thus required sampling rate of the equidistant time series data determines the required simulation time. So it can be shorter or longer than the simulation time of the non-equidistant LTspice simulation.

In the SDM, the difference between the real and the white box model behaviour is trained to the neural network. In the PKI, the white box model outputs (prior knowledge) and the original inputs are fed to the neural network. It is trained to fit the original component behaviour. The only difference to the previous network structures is the number of input features of four instead of two caused by the additional prior knowledge input.

The training of the neural network in the SMN in Fig. 2 does not work if the equivalent circuit model is located outside the Python environment (e.g. in LTspice). It is possible to integrate LTspice in the training loop by calling it in batch mode and to pass parameters. But the gradients needed for the backpropagation in the training process are no more available after the external simulation. A possible solution can be an implementation of a kind of SPICE-simulation in Python, e.g. with a solver like [17] which allows backpropagation through ordinary differential equation for deep learning applications (future work).

Results and Discussion

The trained models are tested with time series that have been excluded from the training process. So the test data is not known by the model. First, the AE1 from Fig. 4(a) is used to compare the different modeling approaches. Additionally, one chosen modeling approach is applied to all three application examples.

In Fig. 6(a), Fig. 7(a) and Fig. 8(a), the model output time domain results for u_{OUT} and i_{IN} of the different modeling approaches are compared to the original time series. The general shape of u_{OUT} is modeled well by all three approaches. The superimposed high frequency signal is modeled slightly to high and misshapen by the pure neural network approach. The amplitude of the high frequency signal is modeled way to low by the SDM approach. It fits a bit better by the PKI approach. The shape of i_{IN} does not fit for the pure neural network results. The results from the SDM and the PKI approach do fit better. Quantified details are discussed based on the results in the frequency domain.

In Fig. 6(b), Fig. 7(b) and Fig. 8(b), the model output frequency domain results for u_{OUT} and i_{IN} of the different modeling approaches are compared to the simulated results in the frequency domain (original data). The results are presented as magnitudes of the respective voltages and currents. EME results (e.g. quasi peak in dB) are not presented, because they can only be achieved for a whole electrical system including the EME test equipment characteristics. The modeling of an electrical component is only a small part to make a simulation of a whole system realizable (see Fig. 1).

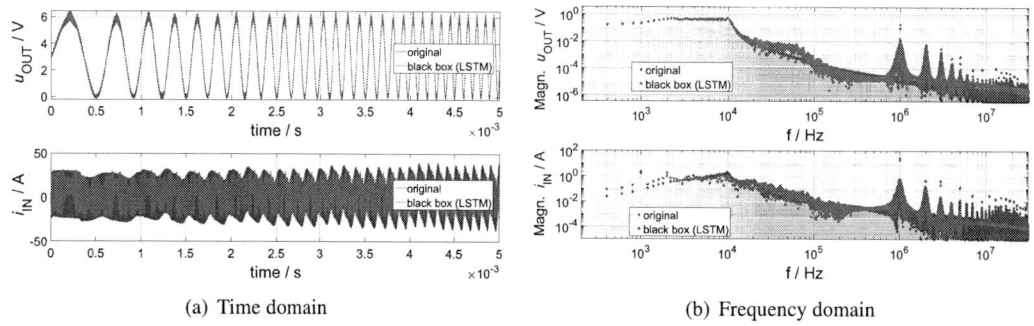

(a) Time domain   (b) Frequency domain

Fig. 6: Pure neural network model results for AE1

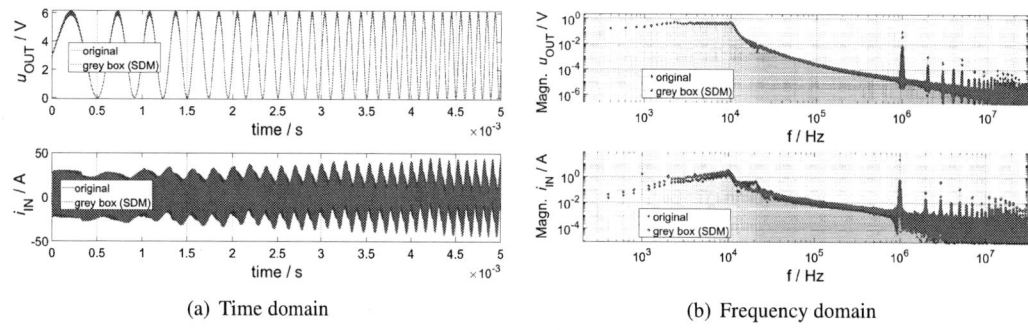

(a) Time domain   (b) Frequency domain

Fig. 7: Source difference method (SDM) model results for AE1

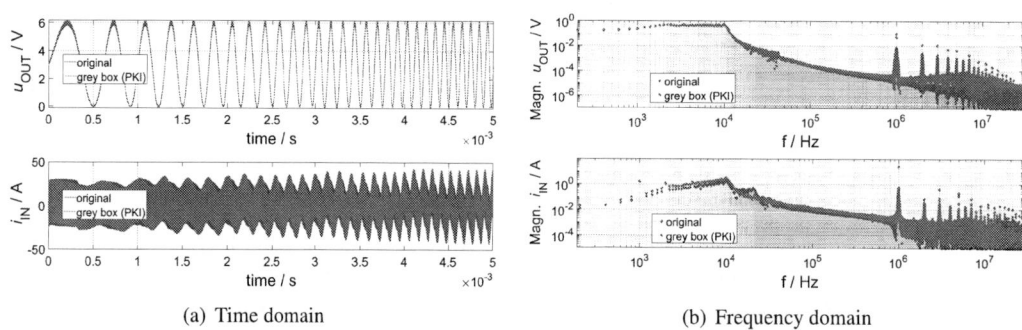

(a) Time domain   (b) Frequency domain

Fig. 8: Prior knowledge input (PKI) model results for AE1

The results for the pure neural network approach from [6] show a congruent fit for u_{OUT} at low frequencies (less than 1 dB deviation for most points below 10 kHz). But there are fluctuating deviations between 20 kHz and 300 kHz of up to about 15 dB. The magnitudes of the 1 MHz component and its harmonics are shown in Tab. III. They are presented in dBμV to provide a typical scaling in the field of EMC. The respective sidebands of the model in Fig. 6(b) are much to widespread. And there is a general large deviation above 5 MHz. The ability of the used neural network to model such very small high frequency amplitudes inside a large signal can be the limit here. The results characteristic of i_{IN} is very similar. But the deviation in the low frequency range is bigger (about 10 dB below 2 kHz).

TABLE III: Magnitudes of 1 MHz component and its harmonics in dBμV and dBμA for AE1
(green: up to 3 dB deviation; yellow: between 3 dB and 15 dB deviation; red: more than 15 dB deviation)

Frequency	u_{OUT} (simulated)	u_{OUT} (LSTM)	u_{OUT} (SDM)	u_{OUT} (PKI)	i_{IN} (simulated)	i_{IN} (LSTM)	i_{IN} (SDM)	i_{IN} (PKI)
1 MHz	99.57 dBμV	105.00 dBμV	91.59 dBμV	93.84 dBμV	148.03 dBμA	145.33 dBμA	147.70 dBμA	147.69 dBμA
2 MHz	60.26 dBμV	69.74 dBμV	32.26 dBμV	81.58 dBμV	96.12 dBμA	112.46 dBμA	82.47 dBμA	96.61 dBμA
3 MHz	51.59 dBμV	69.40 dBμV	49.40 dBμV	81.02 dBμV	111.82 dBμA	105.85 dBμA	112.93 dBμA	112.34 dBμA
4 MHz	40.00 dBμV	54.65 dBμV	28.04 dBμV	74.15 dBμV	77.70 dBμA	133.73 dBμA	79.31 dBμA	80.00 dBμA
5 MHz	56.90 dBμV	47.23 dBμV	48.30 dBμV	72.81 dBμV	101.87 dBμA	75.96 dBμA	101.36 dBμA	105.71 dBμA
6 MHz	16.65 dBμV	30.85 dBμV	14.57 dBμV	56.83 dBμV	75.82 dBμA	55.86 dBμA	73.50 dBμA	83.52 dBμA
7 MHz	37.50 dBμV	28.30 dBμV	25.32 dBμV	65.25 dBμV	89.25 dBμA	57.91 dBμA	89.25 dBμA	96.52 dBμA

The outputs of the SDM model in Fig. 7(b) show better results than the pure neural network approach. The components of u_{OUT} in general do fit better across the whole frequency range. The magnitudes of the 1 MHz component and its harmonics are shown in Tab. III. The respective spreading is fitting better than in Fig. 6(b). The general fitting above 5 MHz is better. The results characteristic of u_{OUT} and i_{IN} are very similar. The deviation in the low frequency range is lower, only about a third as in Fig. 6(b).

For u_{OUT}, the PKI model in Fig. 8(b) shows a very good fit up to about 20 kHz. There seems to be a local modeling problem around 20 kHz. The general fit is good, but becomes worse in the MHz range. In general, the magnitudes above 2 MHz are to high. The magnitudes of the 1 MHz component and its harmonics are shown in Tab. III. The results characteristic of i_{IN} is very similar to the characteristic of u_{OUT}. Across all approaches, there is a modeling problem for i_{IN} around 20 kHz.

The SDM is chosen for AE2 and AE3, because it shows the best results for AE1 compared to the other approaches. The time domain results for AE2 are shown in Fig. 9. There are large deviations for u_{OUT} and i_{IN} between the original and modeled results. The modeling for AE2 is as difficult as in [6]. A possible reasons can be the complexity of the example or the used training parameters. But the comparison to all other results suggests that the used training data is insufficient. A presentation of the frequency domain results is omitted.

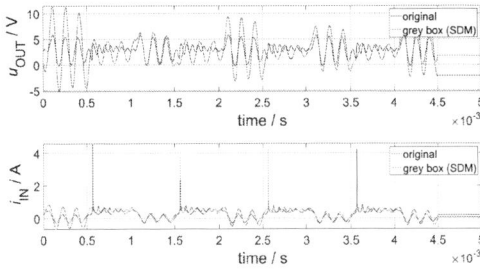

Fig. 9: Source difference method (SDM) model time domain results for AE2

In Fig. 10(a), the model output time domain results for u_{OUT} and i_{IN} are shown for AE3. The general shape of u_{OUT} is quite similar compared to the original time series and distinctive details are recognisable. But the general amplitude is about 13% smaller and there is an offset of about 10%. The general shape and the details of i_{IN} is also similar compared to the original time series. The general amplitude is about 15% smaller but there is no visible offset.

The respective frequency domain results for u_{OUT} and i_{IN} of AE3 are compared and shown in Fig. 10(b). A separate consideration of noise floor and dominant amplitudes is made for analysing the results systematically. The components of u_{OUT} show a generally increasing deviation with rising frequency. In contrast to that, the noise floor of i_{IN} is modeled constantly too high.

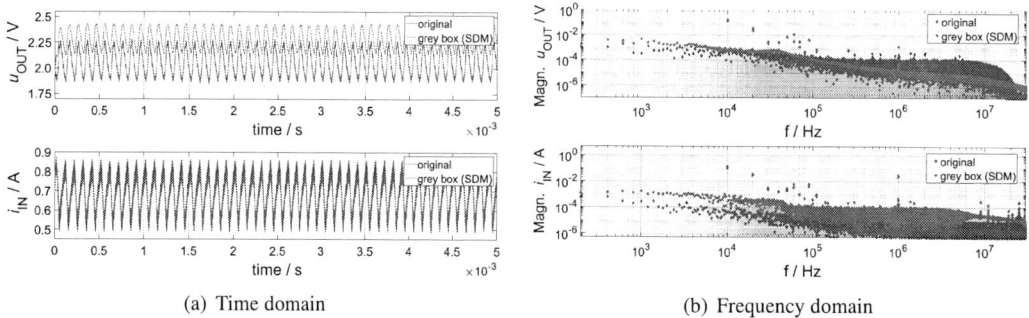

(a) Time domain   (b) Frequency domain

Fig. 10: Source difference method (SDM) model results for AE3

The magnitudes of selected, dominant frequency components are shown in Tab. IV. As an overall result, half of the compared magnitudes are inside a 3 dB deviation range. The rest shows more or less larger deviations.

TABLE IV: Magnitudes of selected, dominant frequency components in dBμV and dBμA for AE3
(green: up to 3 dB deviation; yellow: between 3 dB and 15 dB deviation; red: more than 15 dB deviation)

Frequency	u_{OUT} (measured)	u_{OUT} (SDM)	i_{IN} (measured)	i_{IN} (SDM)
10 kHz	104.61 dBμV	102.59 dBμV	102.53 dBμA	99.54 dBμA
20 kHz	90.51 dBμV	86.36 dBμV	90.47 dBμA	88.94 dBμA
30 kHz	66.67 dBμV	69.04 dBμV	69.75 dBμA	67.38 dBμA
40 kHz	68.09 dBμV	57.00 dBμV	63.97 dBμA	38.59 dBμA
50 kHz	77.20 dBμV	77.08 dBμV	68.63 dBμA	66.01 dBμA
60 kHz	80.81 dBμV	69.43 dBμV	75.37 dBμA	70.65 dBμA
70 kHz	68.97 dBμV	60.87 dBμV	59.40 dBμA	40.86 dBμA
80 kHz	49.95 dBμV	53.38 dBμV	53.80 dBμA	42.28 dBμA
90 kHz	73.12 dBμV	71.86 dBμV	75.40 dBμA	71.55 dBμA
100 kHz	45.93 dBμV	47.23 dBμV	51.57 dBμA	45.51 dBμA
110 kHz	53.66 dBμV	56.81 dBμV	59.50 dBμA	57.11 dBμA
1000 kHz	31.27 dBμV	61.41 dBμV	89.22 dBμA	86.97 dBμA

Conclusion and Future Work

Knowledge based grey box models are generally suitable to model partly known and internally inaccessible circuits for time domain simulations of EME-relevant voltage and current spectra (RQ1). They lead to better modeling results than pure neural network models from just as little training data (RQ2). In this work, only six datasets have been used for the training of every model. The requested amount of training data is an important aspect in industrial applications. A simulation speed advantage of knowledge based grey box models is not existing in general (RQ3).

For future work, it is advised to apply such knowledge based grey box modeling approaches to more complex components with also unknown model parts. The advantages of knowledge based grey box models are assumed to be determined more clearly. So e.g. an integrated DC-to-DC-converter can be integrated into a whole system EMC simulation which would not have been possible before. The used nonlinear filters served as simple and well known examples for comparing the applied methods. For that, they are assumed to be inaccessible. Potential for future improvements is offered by hyperparameter optimizing for every application example, by usage of more training data, by using a higher order filter model with heavy parasitics and by a combination of input and output mapping [14]. The space-mapped neural network approach and state-space neural network should be implemented to examine their potential.

References

[1] T. Williams, EMC for Product Designers, fourth edition ed. Oxford: Elsevier Ltd., 2007.

[2] W. Na, W. Liu, L. Zhu, F. Feng, J. Ma and Q. -J. Zhang, "Advanced Extrapolation Technique for Neural-Based Microwave Modeling and Design," in IEEE Transactions on Microwave Theory and Techniques, vol. 66, no. 10, pp. 4397-4418, Oct. 2018, doi: 10.1109/TMTT.2018.2854163.

[3] F. Feng, W. Na, J. Jin, W. Zhang and Q. -J. Zhang, "ANNs for Fast Parameterized EM Modeling: The State of the Art in Machine Learning for Design Automation of Passive Microwave Structures," in IEEE Microwave Magazine, vol. 22, no. 10, pp. 37-50, Oct. 2021, doi: 10.1109/MMM.2021.3095990.

[4] https://911electronic.com/de/ltspice-tutorial/ltspice-logo/

[5] https://www.python.org/community/logos/

[6] J.-P. Roche, J. Friebe and O. Niggemann, "Neural Network Modeling of Nonlinear Filters for EMC Simulation in Discrete Time Domain," IECON 2021 – 47th Annual Conference of the IEEE Industrial Electronics Society, 2021, pp. 1-7, doi: 10.1109/IECON48115.2021.9589226.

[7] G. Brocard and M. Engelhardt, Simulation in LTSpice IV: Handbuch, Methoden und Anwendungen, 1st ed. K¨unzelsau: Swiridoff Verlag, 2013.

[8] M. L. Heldwein, "Emc filtering of three-phase pwm converters," Ph.D. dissertation, ETH Zurich, Zurich, 2008, diss. ETH No. 17554.

[9] H. Sharma and Q. Zhang, "Transient electromagnetic modeling using recurrent neural networks," in IEEE MTT-S International Microwave Symposium Digest, 2005., 2005, pp. 1597–1600.

[10] Z. Naghibi, S. A. Sadrossadat, and S. Safari, "Time-domain modeling of nonlinear circuits using deep recurrent neural network technique," AEU - International Journal of Electronics and Communications, vol. 100, pp. 66–74, 2019. [Online]. Available: https://www.sciencedirect.com/science/article/pii/S1434841118318235

[11] P. Chen, S. Alsahali, A. Alt, J. Lees, and P. J. Tasker, "Behavioral modeling of gan power amplifiers using long short-term memory networks," in 2018 International Workshop on Integrated Nonlinear Microwave and Millimetre-wave Circuits (INMMIC), 2018, pp. 1–3.

[12] T. Nguyen, T. Lu, K. Wu, and J. Schutt-Aine, "Fast transient simulation of high-speed channels using recurrent neural network," 2019.

[13] Q. J. Zhang and K. C. Gupta, Neural Networks for RF and Microwave Design. Norwood, MA: Artech House, 2000.

[14] Q. Zhang, W. Na, M. Li, Y. Lan, Q. Ding and G. Wu, "Knowledge-based Neural Models for Modelling High-Frequency Electronics Circuits," 2019 6th International Conference on Systems and Informatics (ICSAI), 2019, pp. 1589-1593, doi: 10.1109/ICSAI48974.2019.9010157.

[15] W. Kirchgässner, O. Wallscheid and J. Böcker, "Thermal Neural Networks: Lumped-Parameter Thermal Modeling With State-Space Machine Learning", arXiv:2103.16323.v2, 2021.

[16] X. Ding, V. Devabhaktuni, B. Chattaraj, M. Yagoub, M. Deo, J. Xu, and Q. J. Zhang, "Neural-network approaches to electromagnetic-based modeling of passive components and their applications to high-frequency and high-speed nonlinear circuit optimization," IEEE Transactions on Microwave Theory and Techniques, vol. 52, no. 1, pp. 436–449, 2004.

[17] "PyTorch Implementation of Differentiable ODE Solvers", https://github.com/rtqichen/torchdiffeq (Accessed: 26 March 2022)

Novel Quasi-Direct Rotor Position Estimator for Permanent Magnet Synchronous Machines based on the Back-Electromotive Force using Current Oversampling

Georg Lindemann, Viktor Willich and Axel Mertens
Leibniz University Hannover
Institute for Drive Systems and Power Electronics
Welfengarten 1
Hannover, Germany
Email: georg.lindemann@ial.uni-hannover.de
URL: https://www.ial.uni-hannover.de

Acknowledgments

This work was supported by the German Research Foundation (DFG) - Project 329209868.

Keywords

≪Self-sensing control≫, ≪Current derivative≫, ≪Field programmable gate array (FPGA)≫, ≪Permanent magnet motor≫.

Abstract

This paper presents an approach for rotor position estimation in permanent magnet synchronous machines (PMSMs) based on back-electromotive force (EMF). By directly evaluating the voltage equation and using current oversampling, highly responsive dynamics can be achieved. Online optimisation ensures an analytically adjustable, quasi-direct calculation of the rotor position.

Introduction

In the past, different methods for rotor position estimation in PMSMs have been presented [1]. The estimation methods can be divided into methods for low speeds and standstill and methods for higher speeds. The methods used for higher speeds are based on the EMF of the machine. The proposed EMF-based method should first be classified according to the current state of research. Conventional methods are based on the evaluation of the current measurement and voltage setpoints of the converter once per pulse width modulation (PWM) period. Direct calculation of the EMF in the stator reference frame is possible, but additional filters are needed for estimated position and speed [2]. The evaluation of the EMF in the estimated rotor reference frame takes advantage of handling direct quantities. For interior PMSMs, the approach of extended EMF has been established, which considers the fact that the EMF is dependent not only on the induced voltage from the permanent magnet (PM) but also to the stator inductances [3]. The extended EMF also considers the positional information given by the magnetic saliency. Despite the additional considerations of this approach, the magnetic saturation effects and non-linearity of the machine are neglected. The proposed method bridges this gap as it is derived by using the non-linear differential equation of the PMSM to estimate the rotor position with a high accuracy at high-saturation operating points.

The methods based on the magnetic anisotropy of the machine, which are applied in the lower speed range and at zero speed, have been improved. Here, it is worth highlighting the flux derivative estimator quasi-direct (FDE-QD) presented in [4–6], which shows the dynamics of the closed-loop self-sensing control (SSC) of a PMSM coupled to an encoder-based control. The method utilises the advantages of current oversampling (sampling of 100 measured values during a PWM period) [7, 8] and requires a square-wave injection (SWI) along the estimated $\hat{d}$-axis of the PMSM [9]. This motivates the development of a method based on a similar estimator structure for the middle and high speed ranges that does not require a test signal and still offers the highly responsive dynamics and accuracy of rotor position estimation. A test signal is no longer necessary in the higher speed range as the EMF becomes greater and can thus be evaluated. Therefore the proposed method is named the back-electromotive force estimator quasi-direct (EMF-QD).

Estimation methods based on a current or flux observer require all machine parameters and are limited in their dynamics [10], whereas EMF-QD has fewer parameter dependencies and offers good dynamics. By evaluating the voltage equation during a passive switching state (PS) of the converter, the estimation becomes independent from the value of the output voltages of the inverter. By simplifying the optimisation algorithm of the EMF-QD, independence from the value of the permanent magnet flux linkage can be achieved.

The introduction of the EMF-QD brings the possibility of self-sensing control in highly saturated machines with highly active control dynamics, which offers a major advantage over conventional EMF-based approaches.

For the FDE-QD, the use of current oversampling and the evaluation of individual switching states of the inverter makes a vast difference to the dynamics achieved. In order to use these advantages in higher speed ranges, the EMF-QD also utilizes the current oversampling. Therefore the discrete converter voltages are evaluated and the position estimation can be examined individually in one PWM period. A similar method based on current oversampling was presented in [11]. An essential difference of [11] is the linearisation of the voltage equation by assuming the inductance values to be saturation independent. Another significant difference of [11] is the evaluation of the voltage equation, as described below, which gives the estimate a high dynamic range with good noise behaviour while all parameters can be set analytically.

The voltage differential equation of the PMSM, used as the basis for the EMF-QD, results in a voltage prediction error which can be transferred into a rotation of the dq-coordinate reference frame. Since this voltage error is minimised within one calculation period (quasi-directly), the method is, in general, able to provide a significantly better dynamic response in the rotor position estimation than a method which minimises the estimation error over several calculation steps. The calculation of the rotor position and speed according to [5] provides a structure that can be parameterised analytically and does not need to be adjusted by experiment. This provides a major advantage over methods that use a phase-locked loop (PLL) for estimation and have to be tuned by experiment [10–12].

The first section of this paper describes the mathematical principles behind minimising the estimation error and derives the basic error equation for the estimation method on the basis of the voltage differential equation of the PMSM. Next, the implemented estimation structure is described, and in the third section, the estimation method is examined in simulations and experiments.

Non-linear Model of the PMSM

The voltage equation of the PMSM can be expressed as the sum of an ohmic part and a voltage that is induced by the derivative of the flux. The voltage equation in the stator-fixed $\alpha\beta$-reference frame is:

$$\vec{u}_{\alpha\beta} = R\vec{i}_{\alpha\beta} + \frac{d\vec{\Psi}_{\alpha\beta}}{dt}. \tag{1}$$

The stator resistance R is considered to be isotropic. The vectors $\vec{i}_{\alpha\beta}$ and $\vec{\Psi}_{\alpha\beta}$ are the vectors of the stator currents and flux linkage in the $\alpha\beta$-reference frame, respectively. The flux linkage can be expressed in rotor-fixed dq-coordinates with the help of the electrical rotor position γ_{el}:

$$
\begin{aligned}
\vec{u}_{\alpha\beta} &= R\vec{i}_{\alpha\beta} + \frac{d\mathbf{T}\vec{\Psi}_{dq}}{dt} \\
&= R\vec{i}_{\alpha\beta} + \mathbf{T}\frac{d\vec{\Psi}_{dq}}{dt} + \frac{d\mathbf{T}}{dt}\vec{\Psi}_{dq} \\
&= R\vec{i}_{\alpha\beta} + \mathbf{T}\frac{d\vec{\Psi}_{dq}}{d\vec{i}_{dq}}\frac{d\vec{i}_{dq}}{dt} + \mathbf{J}\omega_{el}\mathbf{T}\vec{\Psi}_{dq}
\end{aligned}
\tag{2}
$$

with

$$\mathbf{T} = \begin{pmatrix} \cos(\gamma_{el}) & -\sin(\gamma_{el}) \\ \sin(\gamma_{el}) & \cos(\gamma_{el}) \end{pmatrix}, \quad \omega_{el} = \frac{d\gamma_{el}}{dt}, \quad \mathbf{J} = \begin{pmatrix} 0 & -1 \\ 1 & 0 \end{pmatrix} \tag{3}$$

The deviation of the flux linkage caused by the currents is introduced as the differential inductance matrix $\mathbf{L}'_{dq}$ which is described in Equation (4). The flux linkage can be separated into one part that is dependent on the permanent magnet and another part dependent on the stator current (6). The current-dependent part is defined as $\mathbf{L}_{dq}\vec{i}_{dq}$, where $\mathbf{L}_{dq}$ represents the secant inductance matrix as defined in (5).

$$\mathbf{L}_{dq}^{'} := \frac{d\vec{\Psi}_{dq}}{d\vec{i}_{dq}} = \begin{pmatrix} L_d^{'}(\vec{i}_{dq}) & L_{dq}^{'}(\vec{i}_{dq}) \\ L_{dq}^{'}(\vec{i}_{dq}) & L_q^{'}(\vec{i}_{dq}) \end{pmatrix} \quad (4) \qquad \mathbf{L}_{dq} := \begin{pmatrix} L_d(\vec{i}_{dq}) & L_{dq}(\vec{i}_{dq}) \\ L_{dq}(\vec{i}_{dq}) & L_q(\vec{i}_{dq}) \end{pmatrix} \quad (5)$$

$$\vec{\Psi}_{dq} = \vec{\Psi}_{L,dq} + \vec{\Psi}_{PM,dq} = \mathbf{L}_{dq}\vec{i}_{dq} + \Psi_{PM}\begin{pmatrix} 1 \\ 0 \end{pmatrix} \tag{6}$$

The voltage equation in the $\alpha\beta$-reference frame can be derived from equations (4), (6) and (5) is given in (7).

$$\vec{u}_{\alpha\beta} = R\vec{i}_{\alpha\beta} + \mathbf{T}\mathbf{L}_{dq}^{'}\frac{d\vec{i}_{dq}}{dt} + \mathbf{J}\omega_{el}\mathbf{T}\mathbf{L}_{dq}\vec{i}_{dq} + \mathbf{J}\omega_{el}\mathbf{T}\vec{\Psi}_{PM,dq} \tag{7}$$

Transforming all parameters into the $\alpha\beta$-reference frame using the product rule for the derivation of $\mathbf{T}^{-1}\vec{i}_{\alpha\beta}$ and the transformation of the inductances with $\mathbf{L}_{\alpha\beta}^{'} = \mathbf{T}\mathbf{L}_{dq}^{'}\mathbf{T}^{-1}$ and $\mathbf{L}_{\alpha\beta} = \mathbf{T}\mathbf{L}_{dq}\mathbf{T}^{-1}$, the non-linear differential equation yields:

$$\vec{u}_{\alpha\beta} = R\vec{i}_{\alpha\beta} + \mathbf{L}_{\alpha\beta}^{'}\frac{d\vec{i}_{\alpha\beta}}{dt} + \left(\mathbf{J}\mathbf{L}_{\alpha\beta} - \mathbf{L}_{\alpha\beta}^{'}\mathbf{J}\right)\omega_{el}\vec{i}_{\alpha\beta} + \mathbf{J}\omega_{el}\vec{\Psi}_{PM,\alpha\beta}. \tag{8}$$

By using Equation (8), the saturation dependence of the inductances is taken into account and can be used in the following to derive the estimation structure.

Derivation of Rotor Position Estimation Method

In this section, the optimisation method is presented and the error equation used for rotor position estimation is derived.

gradient descent method

The proposed estimator uses an online optimisation method based on the gradient descent method (GDM) and analogous to FDE-QD [4] to minimise a cost function. The variable of the GDM is the estimated rotor position estimation error $\hat{\gamma}_{err} = \gamma_{el} - \hat{\gamma}_{el}$, with $\hat{\gamma}_{el}$ being the estimated rotor position. The cost function (9) is defined and minimised by proceeding iteratively in the direction of the negative gradient of this function (10) [13].

$$E = \frac{1}{2}\vec{e}(\hat{\gamma}_{err})^T\vec{e}(\hat{\gamma}_{err}) \quad (9) \qquad g = \vec{e}^T(\hat{\gamma}_{err})\frac{\partial\vec{e}(\hat{\gamma}_{err})}{\partial\hat{\gamma}_{err}} \quad (10)$$

The result of one iterative step can be expressed as:

$$\hat{\gamma}_{err}(k+1) = \hat{\gamma}_{err}(k) - \eta \cdot \text{sign}(g(k)). \tag{11}$$

The step-width of the GDM is η. The error equation $\vec{e}_{\text{EMF-QD},\hat{d}\hat{q}}(\hat{\gamma}_{err})$ is derived below.

Error Equation EMF-QD

The use of the current oversampling offers the possibility of measuring the current slopes of each switching state in stator-fixed coordinates. However, when these measured slopes are transformed into the rotating reference frame using the transformation matrix $\mathbf{T}$, the resulting values do not represent the real current slopes in the rotating dq-reference frame. This is due to the change in rotor position during each individual switching state. This deviation in the rotor position can be neglected at standstill and in low speed operation. At higher speed this would introduce an additional error into the estimation and therefore needs to be considered. The solution to consider this error is the interpretation of the current derivative $\frac{d}{dt}\vec{i}_{\alpha\beta}$ as a measured value $\vec{\xi}_{\alpha\beta}$.

$$\frac{d\vec{i}_{\alpha\beta}}{dt} = \vec{\xi}_{\alpha\beta} \tag{12}$$

Inserting this relationship into Equation (8) and transforming it into the rotating dq-reference frame results in:

$$\vec{u}_{dq} = \mathbf{T}^{-1}R\vec{i}_{\alpha\beta} + \mathbf{T}^{-1}\mathbf{L}_{\alpha\beta}^{'}\vec{\xi}_{\alpha\beta} + \mathbf{T}^{-1}\left(\mathbf{J}\mathbf{L}_{\alpha\beta} - \mathbf{L}_{\alpha\beta}^{'}\mathbf{J}\right)\omega_{el}\vec{i}_{\alpha\beta} + \mathbf{T}^{-1}\mathbf{J}\omega_{el}\vec{\Psi}_{PM,\alpha\beta}. \tag{13}$$

If all values are transformed from the $\alpha\beta$-reference frame into the dq-reference frame, the equation (14) results when considering the current gradients as measured values. This equation takes into account the non-linearity of the inductance matrix $\mathbf{L}'_{dq}$.

$$\vec{u}_{dq} = R\vec{i}_{dq} + \mathbf{L}'_{dq}\vec{\xi}_{dq} + \left(\mathbf{JL}_{dq} - \mathbf{L}'_{dq}\mathbf{J}\right)\omega_{el}\vec{i}_{dq} + \mathbf{J}\omega_{el}\vec{\Psi}_{PM,dq} \tag{14}$$

The estimation of the rotor position works in the rotating reference frame with the measurements in the estimated reference frame as input. To transform the variables from the $\alpha\beta$-reference frame into the estimated $\hat{d}\hat{q}$-reference frame, the transformation matrix $\hat{\mathbf{T}}$ is used. After inserting this into Equation (13), the voltage equation becomes (15).

$$
\begin{aligned}
\vec{u}_{dq} &= \mathbf{T}^{-1}R\hat{\mathbf{T}}\vec{i}_{\hat{d}\hat{q}} + \mathbf{T}^{-1}\mathbf{L}'_{\alpha\beta}\hat{\mathbf{T}}\vec{\xi}_{\hat{d}\hat{q}} + \mathbf{T}^{-1}\left(\mathbf{JL}_{\alpha\beta} - \mathbf{L}'_{\alpha\beta}\mathbf{J}\right)\omega_{el}\hat{\mathbf{T}}\vec{i}_{\hat{d}\hat{q}} + \mathbf{T}^{-1}\mathbf{J}\omega_{el}\mathbf{T}\vec{\Psi}_{PM,dq} \\
&= \mathbf{T}^{-1}R\hat{\mathbf{T}}\vec{i}_{\hat{d}\hat{q}} + \mathbf{T}^{-1}\mathbf{T}\mathbf{L}'_{dq}\mathbf{T}^{-1}\hat{\mathbf{T}}\vec{\xi}_{\hat{d}\hat{q}} + \mathbf{T}^{-1}\mathbf{T}\left(\mathbf{JL}_{dq} - \mathbf{L}'_{dq}\mathbf{J}\right)\mathbf{T}^{-1}\omega_{el}\hat{\mathbf{T}}\vec{i}_{\hat{d}\hat{q}} + \mathbf{T}^{-1}\mathbf{J}\omega_{el}\mathbf{T}\vec{\Psi}_{PM,dq} \\
&= R\tilde{\mathbf{T}}\vec{i}_{\hat{d}\hat{q}} + \mathbf{L}'_{dq}\tilde{\mathbf{T}}\vec{\xi}_{\hat{d}\hat{q}} + \left(\mathbf{JL}_{dq} - \mathbf{L}'_{dq}\mathbf{J}\right)\omega_{el}\tilde{\mathbf{T}}\vec{i}_{\hat{d}\hat{q}} + \mathbf{J}\omega_{el}\vec{\Psi}_{PM,dq}
\end{aligned}
\tag{15}
$$

with

$$\hat{\mathbf{T}} = \begin{pmatrix} \cos(\hat{\gamma}_{el}) & -\sin(\hat{\gamma}_{el}) \\ \sin(\hat{\gamma}_{el}) & \cos(\hat{\gamma}_{el}) \end{pmatrix} \qquad \tilde{\mathbf{T}} = \mathbf{T}^{-1}\hat{\mathbf{T}} = \begin{pmatrix} \cos(\gamma_{err}) & \sin(\gamma_{err}) \\ -\sin(\gamma_{err}) & \cos(\gamma_{err}) \end{pmatrix} \tag{16}$$

Figure 1 shows an example progression of voltage and current along the estimated $\hat{d}$-axis during a PWM period.

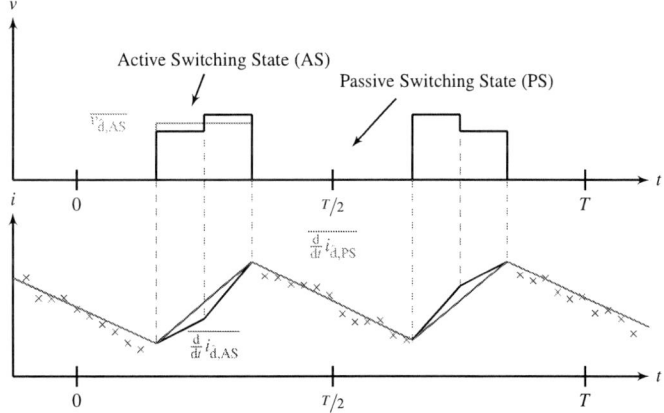

Fig. 1: Example voltage and current progression during a PWM period

By using current oversampling, many current measured values are available during one PWM period. By applying a least mean squares (LMS) algorithm during the PS, the slopes and mean values during active switching state (AS) and PS of the converter can be determined. To decouple the estimation algorithm from mapping errors of the converter voltage, Equation (15) is evaluated during the PS ($\vec{u}_{dq,PS} = 0$). The error equation during a PS is thus:

$$\vec{e}_{EMF-QD,\hat{d}\hat{q}} = R\tilde{\mathbf{T}}\overline{\vec{i}_{\hat{d}\hat{q},PS}} + \mathbf{L}'_{dq}\tilde{\mathbf{T}}\overline{\vec{\xi}_{\hat{d}\hat{q},PS}} + \left(\mathbf{JL}_{dq} - \mathbf{L}'_{dq}\mathbf{J}\right)\omega_{el}\tilde{\mathbf{T}}\overline{\vec{i}_{\hat{d}\hat{q},PS}} + \mathbf{J}\omega_{el}\vec{\Psi}_{PM,dq} \tag{17}$$

Implementation

To find the error equation of the GDM, the $\hat{d}$-axis of the voltage equation is considered. It is assumed that the method is in steady state and therefore $\omega_{el} \approx \hat{\omega}_{el}$ applies. By linearisation, the error equation can be derived:

$$
\begin{aligned}
e_{EMF-QD,\hat{d}} &= \breve{R}\left(\overline{i_{\hat{d},PS}} + \hat{\gamma}_{err}\overline{i_{\hat{q},PS}}\right) + \left(\breve{L}'_d - \breve{L}'_{dq}\hat{\gamma}_{err}\right)\overline{\xi_{\hat{d},PS}} + \left(\breve{L}'_d\hat{\gamma}_{err} + \breve{L}'_{dq}\right)\overline{\xi_{\hat{q},PS}} \\
&\quad + \left[\left(\breve{L}'_d - \breve{L}_q\right)\left(\overline{i_{\hat{q},PS}} - \overline{i_{\hat{d},PS}}\hat{\gamma}_{err}\right) - \left(\breve{L}_{dq} + \breve{L}'_{dq}\right)\left(\overline{i_{\hat{d},PS}} + \overline{i_{\hat{q},PS}}\hat{\gamma}_{err}\right)\right]\hat{\omega}_{el}.
\end{aligned}
\tag{18}
$$

With regard to the implementation, the parameters of the PMSM are expressed by measured values, as they may contain parameter errors. These are indicated by a superscripted ˅. Furthermore, it should be noted that γ_{err} is replaced by $\hat{\gamma}_{err}$. $\hat{\gamma}_{err}$ denotes the estimated disorientation of the $\hat{d}\hat{q}$-reference frame, which corresponds to γ_{err} in ideal function. With the help of (19), the gradient (20) can be formulated.

$$\frac{\partial e_{\text{EMF-QD},\hat{d}}}{\partial \hat{\gamma}_{err}} = \breve{R}\overline{i_{\hat{q},\text{PS}}} + \breve{L}'_d\overline{\breve{\xi}_{\hat{q},\text{PS}}} - \left(\left(\breve{L}'_d - \breve{L}_q \right)\overline{i_{\hat{d},\text{PS}}} + \left(\breve{L}_{dq} + \breve{L}'_{dq} \right)\overline{i_{\hat{q},\text{PS}}} \right)\hat{\omega}_{el} \tag{19}$$

$$g_{\text{EMF-QD}} = e_{\text{EMF-QD},\hat{d}}\frac{\partial e_{\text{EMF-QD},\hat{d}}}{\partial \hat{\gamma}_{err}} \tag{20}$$

Equation (19) is examined below for the dependence of the terms. For this purpose, the value $\overline{\xi_{\hat{q},\text{PS}}}$ is substituted by ideal values using Equation (13) during the PS:

$$\overline{\xi_{\hat{q},\text{PS}}} \overset{\hat{\gamma}_{err}\to 0}{=} \overline{\xi_{q,\text{PS}}} = -\frac{1}{L'_q}\Big[R\overline{i_{q,\text{PS}}} + \overline{\xi_{d,\text{PS}}}L'_{dq} \tag{21}$$
$$+ \left(\left(L_d - L'_q \right)\overline{i_{d,\text{PS}}} - \left(L_{dq} + L'_{dq} \right)\overline{i_{q,\text{PS}}} \right)\omega_{el} + \omega_{el}\Psi_{\text{PM}} \Big].$$

For the dependence on parameters and with regard to a lower speed limit $\omega_{el,min}$, (19) results in

$$\left(\frac{\partial e_{\text{EMF-QD},\hat{d}}}{\partial \hat{\gamma}_{err}} \right)_{\text{para}} = R\overline{i_{q,\text{PS}}}\left(1 - \frac{L'_d}{L'_q} \right) - \frac{L'_d}{L'_q}\overline{\xi_{d,\text{PS}}}L'_{dq} - \frac{L'_d}{L'_q}(L'_d - L_q)\overline{i_{d,\text{PS}}}\omega_{el,min} \tag{22}$$
$$+ \frac{L'_d}{L'_q}(L_{dq} + L'_{dq})\overline{i_{q,\text{PS}}}\omega_{el,min} - \frac{L'_d}{L'_q}\Psi_{\text{PM}}\omega_{el,min} - (L'_d - L_q)\overline{i_{d,\text{PS}}}\omega_{el,min}$$
$$- (L_d - L'_q)\overline{i_{d,\text{PS}}}\omega_{el,min}.$$

Next, the operating range in which (22) depends on the value of the induced voltage is determined in a simplified way. To figure out this operating range, $\overline{i_{d,\text{PS}}} = 0$ is set as the worst-case estimate. This results in the minimum speed at which the term of the induced voltage dominates:

$$\left| \omega_{el,min} \right| = \left| \frac{\overline{\xi_{d,\text{PS}}}L'_{dq} - R\overline{i_{q,\text{PS}}}(\frac{L'_q}{L'_d} - 1)}{(L_{dq} + L'_{dq})\overline{i_{q,\text{PS}}} - \Psi_{\text{PM}}} \right|. \tag{23}$$

If the rated operating point of the machine is taken as a reference (Table I) and the mutual inductance is neglected ($L_{dq} = L'_{dq} = 0$), the minimum speed above which the simplified gradient applies is: $n_{\text{mech,min}} \approx 128\,1/\text{min}$. Above this speed, the gradient is defined by

$$g_{\text{EMF-QD,simplified}} = e_{\text{EMF-QD},\hat{d}}\left(\frac{\partial e_{\text{EMF-QD},\hat{d}}}{\partial \hat{\gamma}_{err}} \right)_{\text{simplified}} = e_{\text{EMF-QD},\hat{d}}\left(-\frac{\breve{L}'_d}{\breve{L}'_q}\hat{\omega}_{el}\breve{\Psi}_{\text{PM}} \right). \tag{24}$$

The lower speed limit of the entire SSC using EMF-QD is identified in the experimental investigation section. Since only the sign of the gradient $g_{\text{EMF-QD,simplified}}$ is evaluated, the result of the GDM is independent of the absolute value of Ψ_{PM}. In Figure 2, a structure for speed and position estimation analogous to that presented in [4] is shown. The termination criterion for the GDM is a fixed number of iterations. In steady-state operation, the estimation error $\hat{\gamma}_{err}$ becomes 0 while the integrator part provides the estimated angular frequency $\hat{\omega}_{el}$.

Simulational and Experimental Results

In order to investigate the stability and dynamics of EMF-QD in the range of constant flux, a machine model based on the non-linear voltage equation of the PMSM is simulated. The rated data of the machine under test and related test bench data are provided in Table I. Figure 3 shows the structure of the SSC using the EMF-QD as

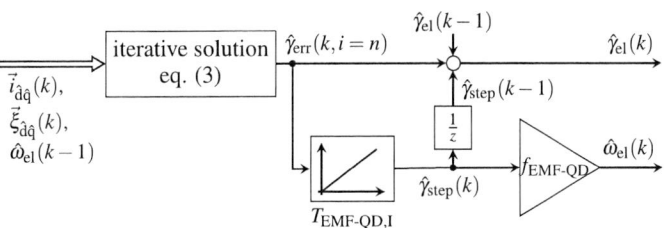

Fig. 2: Implemented structure for rotor position and speed estimation

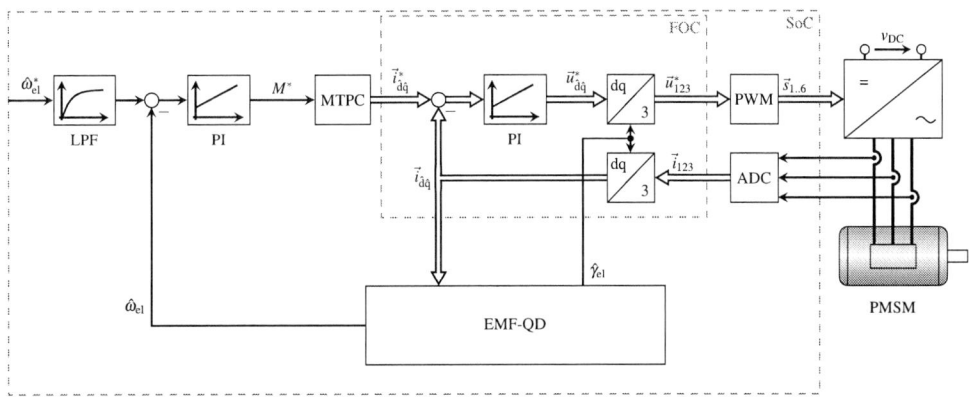

Fig. 3: Block diagram of the SSC used with PMSM and converter (encoder for benchmarking is not shown)

rotor position and speed estimator. For the investigations, closed-loop speed control is used. The speed setpoint $\hat{\omega}^*_{el}$ is filtered by a first-order low-pass filter with the same time constant as the speed pi controller. The current setpoints for the field-oriented control (FOC) are derived using a maximum torque per current (MTPC) algorithm. The slopes of the measured currents are determined with an LMS algorithm during the PS using a sampling rate of 1 MSPS and are fed to the estimator. The controller and estimator parameters are given in Table II and apply to both the simulated and experimental investigations.

<table>
<tr><td colspan="3">TABLE I
PMSM RATED VALUES AND TEST BENCH DATA</td><td colspan="3">TABLE II
CONTROL PARAMETERS</td></tr>
<tr><td>Rated current (RMS)</td><td>I_r</td><td>6.7 A</td><td>Initial step width</td><td>$\eta_{EMF\text{-}QD,init}$</td><td>$5.2 \cdot 10^{-3}$ rad</td></tr>
<tr><td>Rated torque</td><td>M_r</td><td>0.5 Nm</td><td>Time constant EMF-QD</td><td>$T_{EMF\text{-}QD,N}$</td><td>3 ms</td></tr>
<tr><td>Rated rotational speed</td><td>n_r</td><td>5850 min^{-1}</td><td>P-gain current control</td><td>$K_{EMF\text{-}QD,i,P}$</td><td>1.5 V/A</td></tr>
<tr><td>d-axis inductance (idle)</td><td>L_d</td><td>198 µH</td><td>Time const. current control</td><td>$T_{EMF\text{-}QD,i,N}$</td><td>0.5 ms</td></tr>
<tr><td>q-axis inductance (idle)</td><td>L_q</td><td>255 µH</td><td>P-gain speed control</td><td>$K_{EMF\text{-}QD,\omega,P}$</td><td>0.056 A · min</td></tr>
<tr><td>Phase resistance</td><td>R</td><td>0.39 Ω</td><td>Time const. speed control</td><td>$T_{EMF\text{-}QD,\omega,N}$</td><td>19 ms</td></tr>
<tr><td>Flux linkage</td><td>Ψ_{PM}</td><td>8.05 mVs</td><td></td><td></td><td></td></tr>
<tr><td>Number of pole pairs</td><td>p</td><td>4</td><td></td><td></td><td></td></tr>
<tr><td>DC-link voltage</td><td>V_{DC}</td><td>48 V</td><td></td><td></td><td></td></tr>
<tr><td>PWM frequency</td><td>f_{PWM}</td><td>10 kHz</td><td></td><td></td><td></td></tr>
</table>

Two test scenarios are defined to evaluate the quality of the estimated rotor position and speed. In Figure 4, the load step response of SSC is depicted. The speed control operates at a setpoint of 1500 rpm. The load torque is set, reversed and reduced to zero. The upper plot shows the currents $i_{\hat{d}}$ and $i_{\hat{q}}$ in estimated coordinates followed by the estimated speed $\hat{n}_{mech}$ as well as the speed setpoint $\hat{n}^*_{mech}$. The speed deviation from the measured speed is defined as $n_{err} = n_{mech} - \hat{n}_{mech}$ and depicted in the third row. The bottom line shows the rotor position estimation error $\gamma_{err} = \gamma_{el} - \hat{\gamma}_{el}$. Figure 5 shows two setpoint steps from 1000 1/min to 2000 1/min and in the opposite direction. The two figures show that the estimation error of the EMF-QD is very small during both the dynamic process and in steady-state operation under load.

To demonstrate the stability of the method in the range of constant flux, Figure 6 shows the estimation error as a function of the load torque and speed of the PMSM. Figure 6(a) shows the root-mean-square of the estimation error

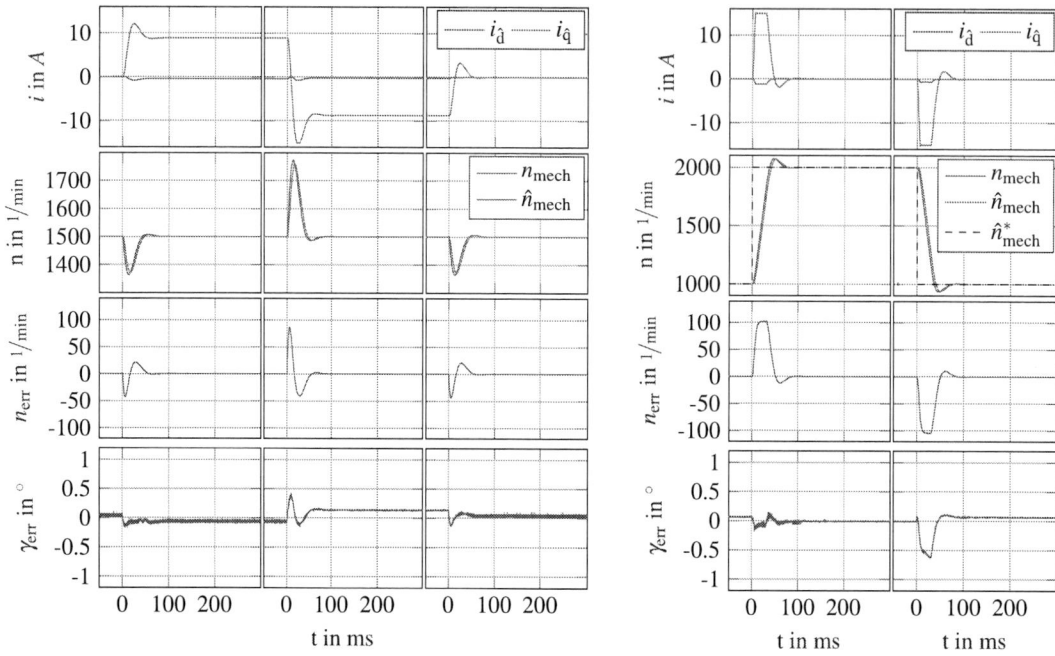

Fig. 4: Simulations of closed-loop SSC using EMF-QD with $1500 \, ^1/_{min}$ speed setpoint. Rated torque is applied (left), reversed (centre) and reduced to zero (right)

Fig. 5: Simulations of closed-loop SSC using EMF-QD. Speed setpoint is increased to $2000 \, ^1/_{min}$ and reduced to $1000 \, ^1/_{min}$

during steady-state operation, and Figure 6(b) the maximum value of the estimation error in dynamic operation. For this purpose, a step from $0 \, \mathrm{Nm}$ to the specified load torque is performed. In the range of high inverter output levels, the error increases due to short duration of the PS and therefore leads to a reduction in the signal-to-noise ratio (SNR). In the very low speed range, the simplified gradient (10) is not valid and the SNR of the error equation is reduced.

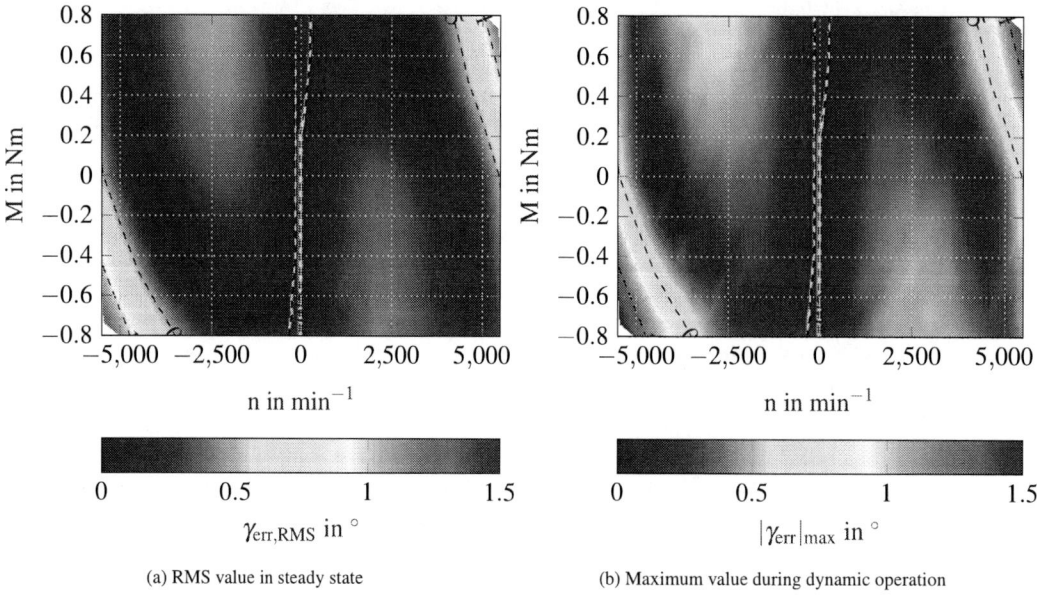

(a) RMS value in steady state

(b) Maximum value during dynamic operation

Fig. 6: Simulations of rotor position estimation error in the range of constant flux

Experimental Results

The experimental test setup uses the rapid prototyping system IAL ControlCube, which contains a Xilinx Zynq 7000 (SoC) combining an ARM Dual Core Cortex A9 processor and an Artix-7 FPGA as programmable logic (PL). The system uses shunt-based current measurement with an A/D-converter resolution of 14 bits at a conversion frequency of 1 MSPS. The MOSFET-based inverter takes advantage of a predictive compensation method for the non-linear output voltages [14]. The test scenarios defined in simulations are repeated with the experimental test setup. The results are shown in Figure 7 and Figure 8. The measurements show a fluctuation in the estimated values due to the SNR of the current measurement and possible parameter errors. Nevertheless, the measurements show a similar shape to the simulated results and thus prove the stability and dynamic effects of the SSC on the physical system. The rotor position estimation error γ_{err} remains smaller than 9 degrees electrical across the entire range.

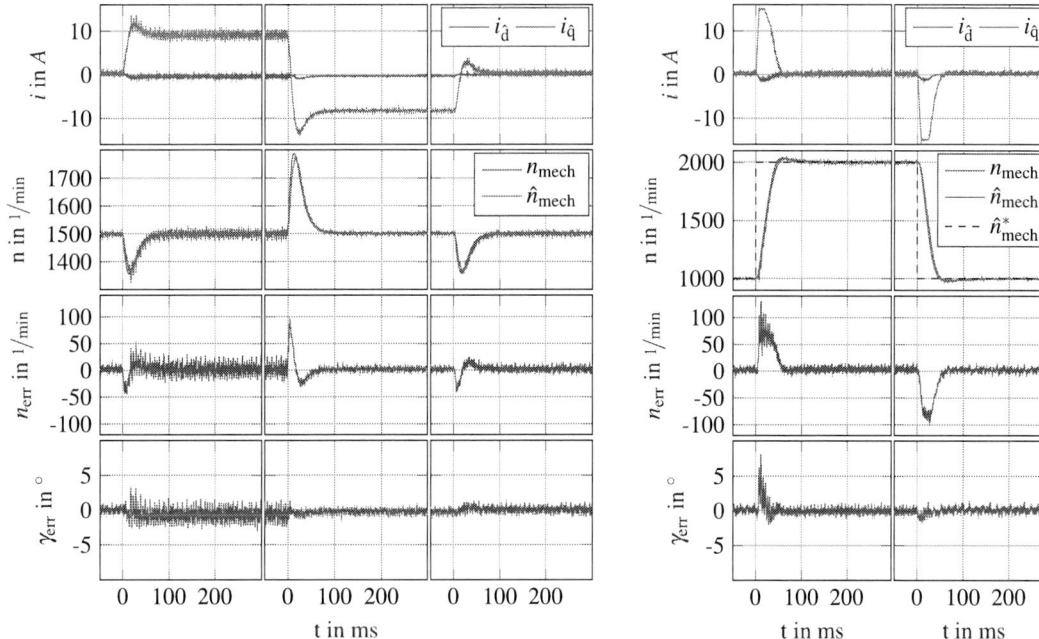

Fig. 7: Experimental results of closed-loop SSC using EMF-QD with 1500 1/min speed setpoint. Rated torque is applied (left), reversed (centre) and reduced to zero (right)

Fig. 8: Experimental results of closed-loop SSC using EMF-QD. Speed setpoint $\hat{n}^*_{mech}$ is increased to 2000 1/min and reduced to 1000 1/min

For the investigation of the lower speed limit of the EMF-QD, a speed setpoint step down to the calculated $n_{mech,min}$ limit and back up to $\hat{n}^*_{mech} = 400 \, 1/min$ is shown in Fig 10. It should be noted that the dynamic response of SSC are reduced for the experiment in order to achieve lower noise in the speed estimate. The following parameters are used: $T_{EMF\text{-}QD,N} = 20\,ms$, $\eta_{EMF\text{-}QD,init} = 31 \cdot 10^{-3}\,rad$, $T_{EMF\text{-}QD,\omega,N} = 99.7\,ms$, $K_{EMF\text{-}QD,\omega,P} = 0.011\,A \cdot min$. It can be seen that due to the overshoot in the estimated and measured speeds, a minimum speed of about 60 1/min is reached. This value could also be achieved in a steady state without load. The fact that the lower speed limit of the EMF-QD is below $n_{mech,min}$ is reasonable because $n_{mech,min}$ was calculated using a worst-case scenario at rated torque. In order to determine the disturbance behaviour at lower speeds, a load torque step to half the rated torque is shown in Fig. 10. The speed setpoint is $\hat{n}^*_{mech} = 400 \, 1/min$ and the dynamic response is on the higher level (cf. Table II). Although the system remains stable in steady state at full load torque, a step up to this value leads to very low speeds, so that the system is not stable for a short time, meaning that this test can't be shown here.
The lower speed limit is more than sufficient for stable operation of the SSC, as the anisotropy-based method is used below 10 % of the rated speed.

Conclusions

This paper presents a novel structure for rotor position estimation in PMSMs at higher speeds. The method based on the voltage equation evaluates the current slopes within a PWM period and minimises the rotor position estimation error within a control cycle. High dynamics and stability were demonstrated over the range of constant flux of the machine and validated by experimental results.

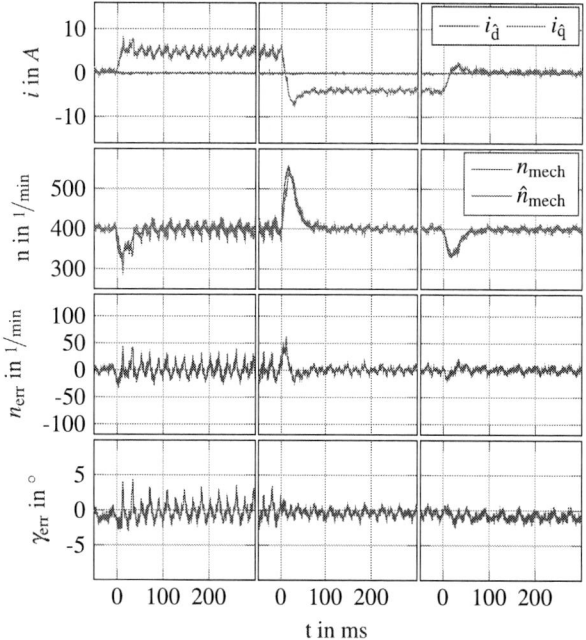 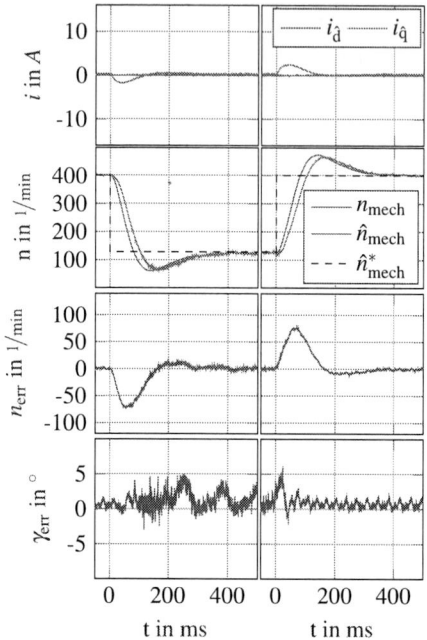

Fig. 9: Investigation of the low speed limit at 400 1/min speed setpoint. 0.5 of rated torque is applied (left), reversed (centre) and reduced to zero (right)

Fig. 10: Investigation of the low speed limit at idle with reduced dynamic response. Speed setpoint $\hat{n}^*_{mech}$ is reduced to 128 1/min and increased to 400 1/min

References

[1] S.-K. Sul and S. Kim, "Sensorless Control of IPMSM: Past, Present, and Future," *IEEJ Journal of Industry Applications*, vol. 1, no. 1, pp. 15–23, 2012.

[2] D. Paulus, J.-F. Stumper, P. Landsmann, and R. Kennel, "Robust encoderless speed control of a synchronous machine by direct evaluation of the back-EMF angle without observer," in *2010 First Symposium on Sensorless Control for Electrical Drives*, Jul. 2010, pp. 8–13.

[3] Z. Chen, M. Tomita, S. Doki, and S. Okuma, "An extended electromotive force model for sensorless control of interior permanent-magnet synchronous motors," *IEEE Transactions on Industrial Electronics*, vol. 50, no. 2, pp. 288–295, Apr. 2003.

[4] N. Himker, G. Lindemann, K. Wiedmann, B. Weber, and A. Mertens, "A Family of Adaptive Position Estimators for PMSM Using the Gradient Descent Method," *IEEE Journal of Emerging and Selected Topics in Power Electronics*, pp. 1–1, 2021.

[5] N. Himker, G. Lindemann, and A. Mertens, "Iterative Tracker for Anisotropy-Based Self-Sensing Control of PMSM," in *2019 IEEE 10th International Symposium on Sensorless Control for Electrical Drives (SLED)*, Sep. 2019.

[6] G. Lindemann, N. Himker, and A. Mertens, "Enhanced Observer with Adaptive Reference Frame for Self-Sensing Control of PMSM," in *2019 IEEE 10th International Symposium on Sensorless Control for Electrical Drives (SLED)*, Sep. 2019.

[7] Y. Duan and M. Sumner, "A novel current derivative measurement using recursive least square algorithms for sensorless control of permanent magnet synchronous machine," in *Proceedings of The 7th International Power Electronics and Motion Control Conference*, vol. 2, Jun. 2012, pp. 1193–1200.

[8] P. Landsmann, J. Jung, M. Kramkowski, P. Stolze, D. Paulus, and R. Kennel, "Lowering injection amplitude in sensorless control by means of current oversampling," in *Sensorless Control for Electrical Drives (SLED), 2012 IEEE Symposium On*, Sep. 2012, pp. 1–6.

[9] S. Kim, J. I. Ha, and S. K. Sul, "PWM Switching Frequency Signal Injection Sensorless Method in IPMSM," *IEEE Transactions on Industry Applications*, vol. 48, no. 5, pp. 1576–1587, Sep. 2012.

[10] K. Wiedmann and A. Mertens, "Novel MRAS approach for online identification of key parameters for self-sensing control of PM synchronous machines," in *2012 15th International Power Electronics and Motion Control Conference (EPE/PEMC)*. Novi Sad, Serbia: IEEE, Sep. 2012, pp. LS4b–1.2–1–LS4b–1.2–8.

[11] P. Landsmann, "Sensorless Control of Synchronous Machines by Linear Approximation of Oversampled Current," Ph.D. dissertation, Technische Universität München, 2014.

[12] B. Weber, G. Lindemann, and A. Mertens, "Reduced observer for anisotropy-based position estimation of PM synchronous machines using current oversampling," in *Proc. IEEE Int. Symp. Sensorless Control for Electrical Drives (SLED)*, Sep. 2017, pp. 121–126.

[13] S. P. Boyd and L. Vandenberghe, *Convex Optimization*. Cambridge, UK ; New York: Cambridge University Press, 2004.

[14] B. Weber, T. Brandt, and A. Mertens, "Compensation of switching dead-time effects in voltage-fed PWM inverters using FPGA-based current oversampling," in *2016 IEEE Applied Power Electronics Conference and Exposition (APEC)*, Mar. 2016, pp. 3172–3179.

Design Considerations for Fast On-State Voltage Measurement Circuits

Mathias C. J. Weiser, Manuel Rueß and Ingmar Kallfass
University of Stuttgart, Institute of Robust Power Semiconductor Systems
Pfaffenwaldring 47
Stuttgart, Germany
Phone: +49 (0) 711-685-61704
Fax: +49 (0) 711-685-58700
Email: mathias.weiser@ilh.uni-stuttgart.de
URL: https://www.ilh.uni-stuttgart.de

Acknowledgments

This work has received funding by the German Science Foundation in the frame of the project 440549658, grant number KA 3062/18-1.

Keywords

≪Component for measurements≫, ≪Device characterization≫, ≪Voltage sensor≫, ≪Wide bandgap devices≫, ≪Design optimization≫

Abstract

The recovery time of on-state voltage measurement circuits (OVMCs) is a critical, yet often overlooked aspect of the proper switching characterization of power semiconductor devices. In this work, we aim to provide a deeper understanding of the problem and propose ways to enhance the dynamic performance.

Introduction

Gallium nitride high electron mobility transistors (GaN-HEMTs) are promising devices for power systems with extremely large power density due to their ability to enable power handling at switching frequencies above 1 MHz [1]. Unfortunately, these devices feature parasitic effects like the 'dynamic on-state resistance' or 'current collapse', which describes a momentary increase of the drain-to-source resistance (R_{DS}) in on-state after a high-voltage blocking state [2, 3, 4]. The characterization of this effect is challenging: while the off-state V_{DS} can reach more than 500 V for high voltage devices, the on-state V_{DS} can drop significantly below 1 V, depending on the device and the operating point. While measuring V_{DS} with an oscilloscope, the measurement accuracy is determined by the vertical resolution. E.g., for a measurement with an oscilloscope operating at a ± 400 V measurement range and an effective 10-bit resolution (ENOB), the least significant bit represents a voltage of $V_{lsb} = \frac{800\,\text{V}}{2^{10}} \approx 0.78\,\text{V}$.

As this measurement resolution is insufficient to accurately characterize the switching behavior, OVMCs are employed. The aim of these circuits is to limit the maximum output voltage to confine the limited vertical resolution of the oscilloscope to the range of interest. While operating below the clipping voltage V_{clip}, which denotes the maximum observable voltage at the OVMC output, the OVMC should allow an accurate measurement of V_{DS} while limiting the maximum output voltage V_{out} for higher voltages. Furthermore, the transition between both states should be as quick as possible.

Fig. 1 shows a simplified schematic of a commonly used OVMC operating principle [5, 6, 7, 8]. This kind of OVMC consists of a silicon carbide (SiC) Schottky diode, a clipping diode and a current source. For a large V_{in}, the SiC diode D_{in} blocks the incoming voltage and the current flows through the clamping

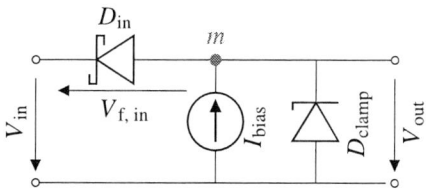

Fig. 1: Basic schematic of a diode-based OVMC.

diode D_{clamp}. V_{out} is therefore limited to the voltage drop over the clipping branch. For low input voltages, i.e., when the device is in on-state, the current flows through the power diode into the DUT drain terminal. With the assumption that the forward voltage of the power diode $V_{f,in}$ stays constant throughout the measurement, V_{DS} can be reconstructed as $V_{DS,on} = V_{in} = V_{out} - V_{f,in}$.

Diode-based OVMCs are frequently used for the characterization of the dynamic on-state resistance of GaN-HEMTs, as SiC Schottky diodes features smaller capacitances than alternative technologies, which use, e.g., HV-capable FETs instead of the diode to accomplish the HV-lockout [11, 12]. The limitations of the circuit shown in Fig. 1 are the measurement offset caused by the Schotty diode forward drop, and the temperature dependency of this parameter, which aggravates the characterization at varying temperatures. Both of these drawbacks have been addressed with the usage of subtraction networks made from operational amplifiers (OpAmps) [9, 10].

The measurement node m highlighted in Fig. 1 is the most critical node in this setup. The quicker it can follow the input voltage, the faster the acquisition of $V_{DS,on}$. Even though SiC Diodes feature capacitances significantly smaller than 100 pF, at several hundreds of V of reverse voltage, a charge of several nC can form on the device. During the turn-on transition of the DUT, this charge has to be fully neutralized to change the diode operation from blocking state to forward bias, plus the time needed to shift the measurement node to the potential which corresponds to the correct V_{DS}, before any accurate measurements of V_{DS} can take place. In this context, we define the recovery time of the OVMC as the time needed to neutralize the charge on the SiC diode and allow the internal voltage node to return to zero-bias conditions. After this, in order to correctly measure V_{DS}, V_{out} has to increase further until steady state has been reached.

Measurement Setup

Measurements were carried out using a half bridge, which consists of two 650 V, 15 A e-mode GaN-HEMTs and an isolated driver circuit. The half bridge can be used as-is, or a load inductor can be added between the supply voltage and the switch node to form a double pulse test (DPT) setup. During the design optimization phase, a 100 μH wirewound air-core inductor was used for this purpose. The coil current was measured at the wire exiting the coil using a 30 A, 100 MHz current clamp. The rise- and fall-times of the half bridge are set to approx. 10 ns to allow for smooth transitions between the states and avoid switching noise. The switching behavior of the DPT setup is shown in Fig. 2. The OVMC is connected to the switch node and the output side ground, measuring the low-side device.

The proposed OVMC is based on the approach shown in Fig. 1. The bias current is generated via a PNP-type current mirror. With a single resistor, the bias current can be altered. The circuit is setup to support variations of recovery acceleration branches and clipping branches to explore the impact of several design parameters on the reaction time.

Both circuits are shown in Fig. 3. For all double-pulse experiments, the evaluation of the proposed OVMC takes place during the second pulse. The static transfer behavior of the OVMC was captured with a parametric 3 kV/50 A curve tracer. It is shown in Fig. 4.

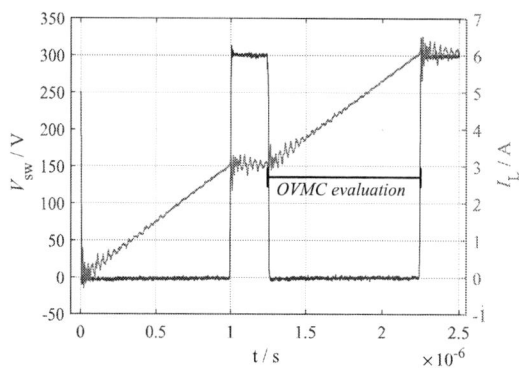

Fig. 2: Current- and voltage waveforms of the DPT.

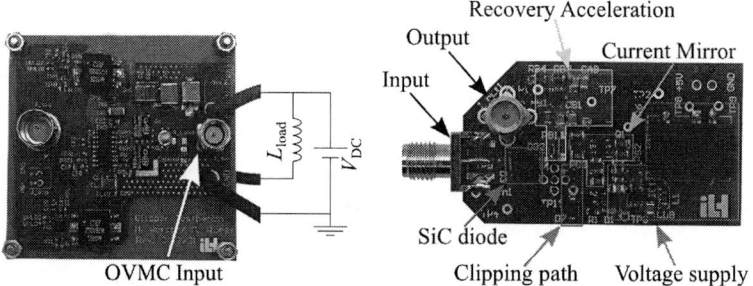

Fig. 3: Physical layout of the test PCB (left) and the proposed OVMC (right).

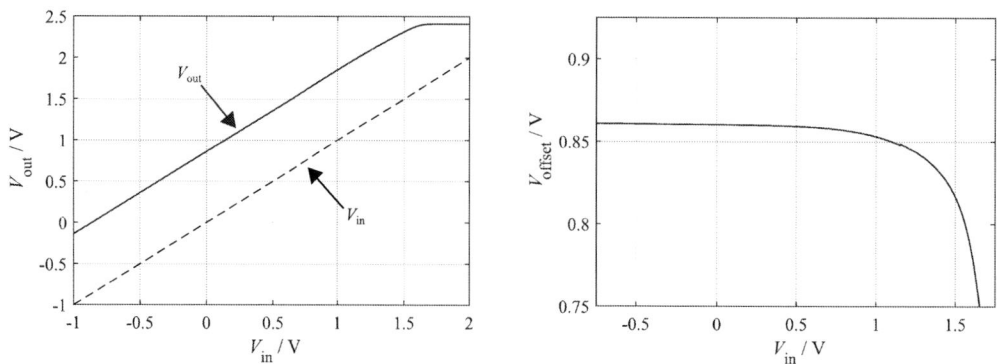

Fig. 4: Static transfer behavior (left) and output offset (right) of the proposed OVMC.

Characteristics of the Stored Charge

To highlight the implication of the stored charge, the OVMC is used as shown, with no additional way to neutralize the stored charge apart from the source current I_{bias}. The half bridge is used without load and the DC voltage is set to 300 V. The bias resistors of the OVMC are set to 3.3 kΩ, 620 Ω and 120 Ω, which corresponds to $I_{bias,1} = 1\,mA$, $I_{bias,2} = 5\,mA$ and $I_{bias,3} = 25\,mA$, respectively. At this voltage, the data sheet value of the stored charge is listed as $Q_j \approx 2.5\,nC$. For a first approximation, the expected recovery time t_{rec} can be calculated as Eq. 1.

$$t_{rec,\,n} = \frac{\Delta Q_j}{I_{bias,\,n}} \quad \Rightarrow \quad t_{rec,1} = 2.5\,\mu s; \quad t_{rec,2} = 500\,ns; \quad t_{rec,3} = 100\,ns \tag{1}$$

The output voltage transients of the OVMC after the switching event are shown in Fig. 5.

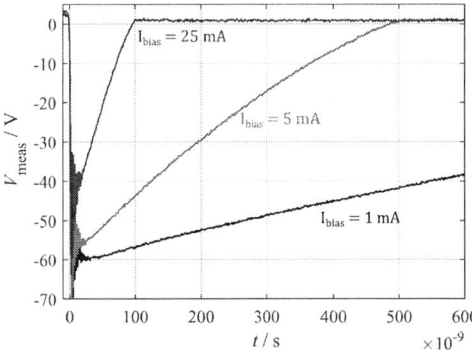

Fig. 5: Recovery time of the OVMC for different bias currents.

Immediately after the switching event, the charge on the parasitic diode capacitance ΔQ_j pulls the internal measurement node to almost -70 V, after which the stored charge is depleted by I_{bias}. Higher currents deplete the charge faster, but the negative voltage spike is almost identical in all three cases. Even with a current of 25 mA, the recovery time of the circuit is approx. 100 ns. However, it has to be noted that larger currents also mean an increasing influence of the OVMC on the measurement setup, as I_{bias} represents an additional current that flows through the DUT, which might not be desired and can alter the calculated resistance of the DUT if not taken into account. Therefore, I_{bias} cannot be increased indefinitely. It also has to be noted that the observed recovery times are in very good agreement with the estimated ones, which suggests that the observed reaction time of the OVMC is exclusively due to the stored charge.

Variation of the Proposed OVMC

To optimize the reaction time of the circuit, parts of the circuit have been altered and the effect of the changes was evaluated at the second pulse of the DPT. The variations made in this work comprise the clipping branch and a recovery acceleration branch, which are connected between the measurement node and ground. The bias current in the following experiments was set to $I_{bias,2} = 5$ mA.

Recovery Acceleration

The most straightforward way to accelerate the charge neutralization is by using fast-switching Schottky diodes, which create a conducting path to ground once the voltage at the measurement node decreases below 0 V. Fundamentally, this equals a negative voltage limitation of the output voltage to $V_{out,min} = -V_{f,Schottky}$. Schottky diodes of varying sizes and blocking voltages are widely available. As for any other semiconductor device, larger maximum currents inevitably lead to larger device dimensions, and thus, larger device capacitances, which in turn slow the device down. As the goal is to supply as much charge in as short of a time as possible, these two properties need to be balanced in order to reach maximum performance. For this reason, the impact of four Schottky diodes as recovery acceleration devices is compared. During this experiment, the fastest variant of the clamping branch has been used to focus on the impact of the recovery diodes. The key parameters of the diodes are listed in Tab. I. The results of the four diodes to the reaction time of the circuit are shown in Fig. 6.

D_1 and D_4 lead to slow reaction behavior, probably due to the low current carrying capability of D_1 and the large capacitance of D_4. D_2 and D_3, however, lead to similarly favorable results, reducing the reaction time to roughly 80 ns. This shows that the choice of the diode has a significant impact on the reaction time and a medium-sized diode offers the best trade-off between capacitance contribution and current carrying capability.

In order to reduce the reaction time even further, the minimum measured voltage can be confined even more to the desired use-case by connecting the anode of the recovery diode to a set potential instead

Table I: Data sheet values of the used diodes at $T = 25\,^{\circ}\mathrm{C}$.

	C_D at $V_D = 0\,\mathrm{V}$ (pF)	I_D at $V_D = 0.5\,\mathrm{V}$ (mA)	$V_{\mathrm{rm,max}}$ (V)
D_1	3.0	1.6	70
D_2	18	15	40
D_3	37	200	40
D_4	250	3000	30

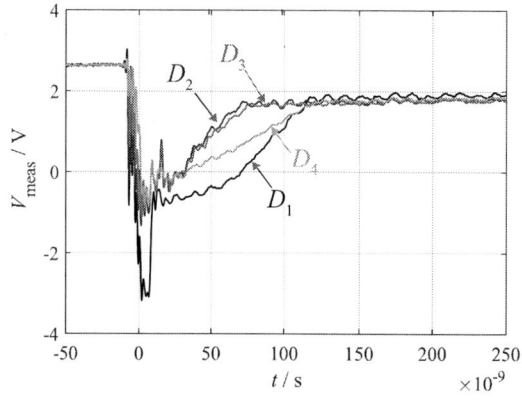

Fig. 6: Reaction behavior of the OVMC with the four diodes.

of ground. A voltage follower can then be used to create the desired offset, and high-frequency buffer capacitors can be used to store the recovery charge. The resulting schematic is shown in Fig. 7. D_2 is used in both cases as diode. In Fig. 8, the results on the reaction behavior are shown.

Clipping Branch Considerations

In order to achieve a voltage limitation at the output, a voltage clipping branch has to be established. In this branch, a highly nonlinear device like a diode is used, which shorts the voltage at the measurement node if above a certain voltage level. Z-diodes are often used for this purpose, as the clipping voltage can be accurately set by selecting a Z-diode with the desired breakdown voltage V_Z. However, Z-diodes usually feature comparably large capacitances, making them rather unsuitable picks for fast-switching circuits. One approach, which has been used in the past to reduce the overall capacitance, is to connect a fast-switching Schottky diode in series to minimize the total capacitance contribution.

A third approach for voltage clipping is to use a separate, independently powered branch. The Z-diode is constantly biased in reverse conduction operation. The measurement node is connected to the anode of the Z-diode by a Schottky diode, which redirects I_{bias} if the measurement voltage exceeds V_Z. Therefore, the capacitance contribution of the clipping branch is minimized: Schottky diodes feature significantly smaller capacitances and the Z-diode is constantly operated in reverse operation, minimizing its capac-

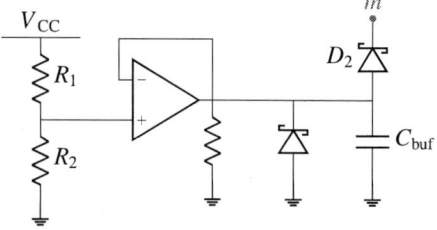

Fig. 7: Topology of the voltage follower recovery acceleration circuit.

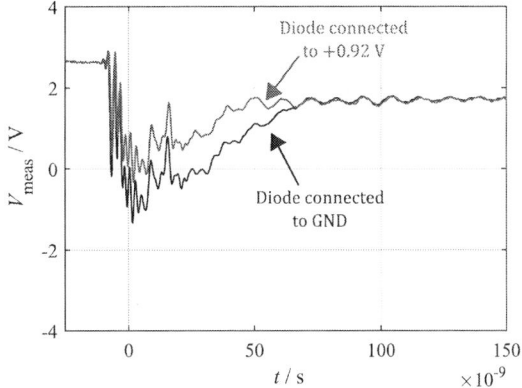

Fig. 8: Reaction behavior of the voltage follower recovery acceleration circuit vs. the GND-connected diode approach.

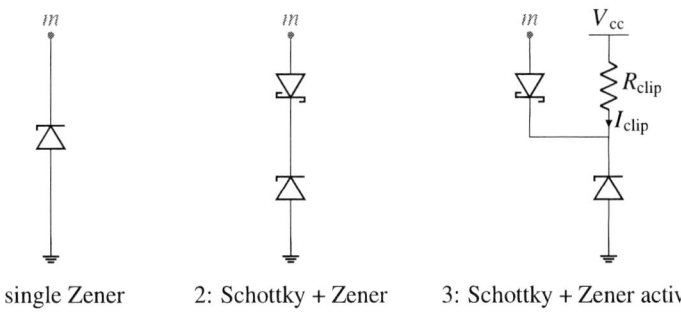

1: single Zener   2: Schottky + Zener   3: Schottky + Zener active

Fig. 9: Topologies of the three clipping branches.

itance. Schematics of the three approaches are shown in Fig. 9. The result on the measured voltage is shown in Fig. 10. During this experiment, D_2 was used as recovery acceleration diode and connected between the measurement node and ground. D_1 was used for the combined Schottky+Zener approaches.

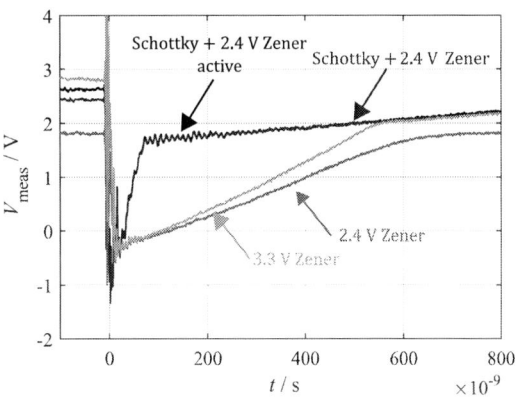

Fig. 10: Recovery behavior for the different clipping branch topologies.

Both Z-diodes feature very long reaction times of approx. 600 ns and more. The 2.4 V Zener diode leads to the slowest transient and the OVMC output voltage saturates before the correct $V_{measure}$ can be reached. Both combined approaches lead to similarly good results, both featuring reaction times of approx. 80 ns. The main drawback of the constantly reverse-biased approach is the obligatory power supply, making the combined passive approach an ideal solution, especially for passive probes.

Practical Reaction Time Estimation

The different parts of the circuit that yielded the best results in the experiments before are now combined to form the optimized version of the OVMC. The bias resistor was chosen as $R_{\text{bias}} = 300\,\Omega$, resulting in a bias current of $I_{\text{bias},4} = 10\,\text{mA}$. To reduce the switching noise contribution in the measurement, the air-core coil was removed from the setup and a $24\,\mu\text{H}$ SMD coil was instead added on the PCB, yielding much better confined current return. The drawback is that this setup does not allow for a measurement of the current anymore, and due to the different electrical parameters, the current profile in this setup will be different to the one obtained before. However, since the low-side transistor showed near-perfect ohmic behavior for the on-state currents in the experiments conducted before, the pulse lengths have been adjusted to 25% of their original value, which matches the inductance ratio between the SMD coil and the air-core coil. Therefore, similar currents as observed the experiments conducted before can be expected. The output waveform of the proposed OVMC is shown in Fig. 12. The altered measurement setup is shown in Fig. 11. The switching behavior is again evaluated during the second pulse.

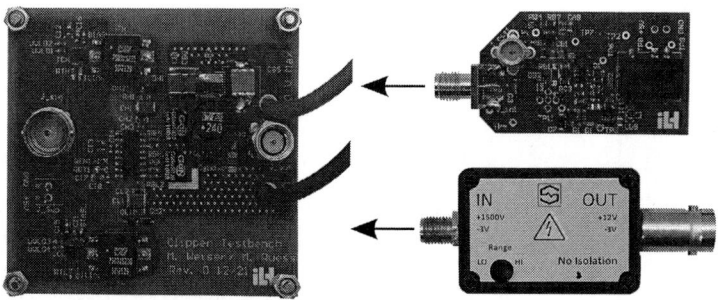

Fig. 11: Altered setup for the reaction time estimation with the two compared OVMCs.

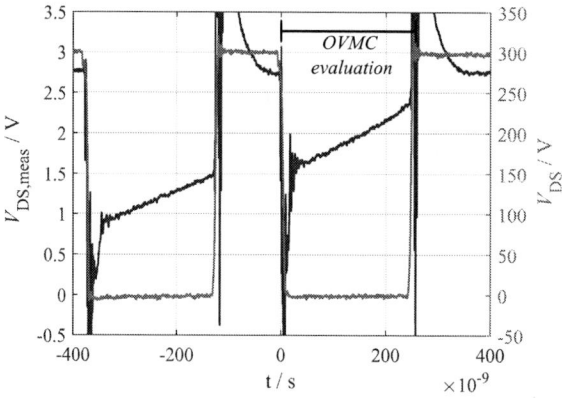

Fig. 12: Voltage waveforms of the altered DPT setup.

To put the performance into perspective, the proposed OVMC is compared to a commercial clipper probe, which is shown in Fig. 11. To compare both output voltages, the voltage offset of the proposed OVMC was evaluated in idle condition with $V_{\text{in}} = 0\,\text{V}$ as $V_{\text{offset}} = 938.9\,\text{mV}$. The reaction time was approximated by fitting a linear function to the reconstructed V_{in} of the proposed OVMC in the linear region in between $t_1 = 100\,\text{ns}$ and $t_2 = 200\,\text{ns}$ after the beginning of the second pulse, and comparing the measured output with the fitline, which is shown on the left hand side of Fig. 13. From this data, the maximum reaction time of the proposed OVMC can be approximated by calculating the difference between the measured curves and the fitted line and calculating the point at which the difference between both curves stays less than the maximum measurement error of the measurement. The measurement deviation for both probes and the calculated reaction times are shown in the right hand side of Fig. 13.

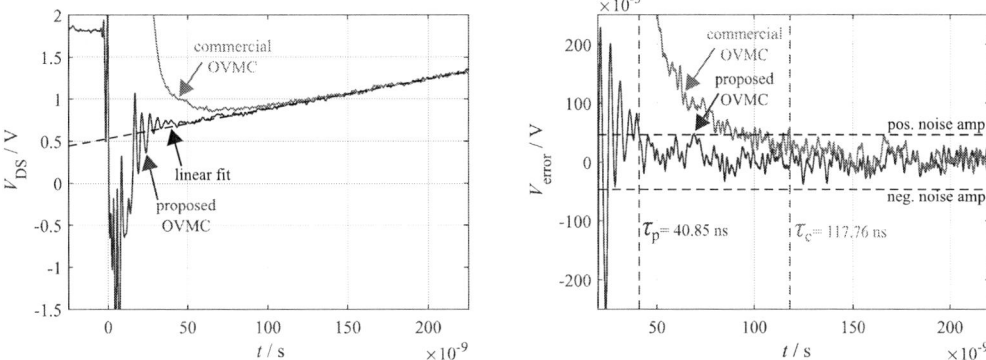

Fig. 13: Linear fit of the drain-source voltage evolution (left) and the resulting reaction time (right).

With this metric, the proposed OVMC is able to reach a reaction time of 40.85 ns, while the reaction time of the commercial clipper circuit is 117.76 ns. It has to be noted that this estimate is based on the time difference between the start of the falling edge of V_{DS} and the time when the switching noise has fully decayed. It is possible that the switching noise, which is present in the measured output voltage from 20 ns until 40 ns, originates from the measurement circuit itself. In this case, the reaction time of the proposed OVMC could potentially be even lower, correctly measuring the switching noise. The reported reaction time therefore represents a conservative approximation.

Conclusion

This work aims to give an overview over some of the design aspects that can be utilized to speed up the reaction time of OVMCs to facilitate their planning and construction. One commonly used OVMC topology has been examined, and the impact of several of its parts with regards to practical performance has been evaluated. A few general conclusions regarding the design of these circuits can be drawn.

- The stored charge on the voltage blocking device is critical to the OVMC reaction time.
- The way this charge is depleted is one of the most impactful performance determinators.
- A simple diode to supply charges from ground is one way to improve the circuit recovery by a lot. However, this leads to a limitation of the minimum measurable voltage.
- By using a voltage follower, the minimum output voltage can be confined more accurately, further improving reaction time. This way, the reaction time can be tailored to the personal needs.
- In general, the capacitance on the measurement node should be kept as small as possible. Devices with a large capacitance contribution should not be connected directly to this node. Instead, they can be connected in series with low capacitance devices to form a series connection, thus reducing the total capacitance contribution.

By combining all these factors, a reaction time of less than 50 ns can be readily reached, enabling the quick acquisition of the dynamic on-state resistance.

References

[1] Gamand F.: A 10-MHz GaN HEMT DC/DC Boost Converter for Power Amplifier Applications, IEEE Transactions on Circuits and Systems II: Express Briefs, Vol. 59, Issue: 11, 2012, pp. 776 - 779

[2] Cai Y.: Impact of GaN HEMT Dynamic On-State Resistance on Converter Performance, 2017 IEEE Applied Power Electronics Conference and Exposition (APEC)

[3] Meneghesso G.: Current Collapse and High-Electric-Field Reliability of Unpassivated GaN/AlGaN/GaN HEMTs, IEEE Transactions on Electron Devices, Vol. 53, No. 12, 2006, pp. 2932 - 2941

[4] Del Alamo J.: GaN HEMT Reliability, Microelectronics Reliability, Vol. 49, No. 9, 2009, pp. 1200-1206

[5] Li R.: Dynamic ON-State Resistance Test and Evaluation of GaN Power Devices Under Hard- and Soft-Switching Conditions by Double and Multiple Pulses, IEEE Transactions on Power Electronics, Vol. 34, No. 2, 2019

[6] Yang F.: Design of a Fast Dynamic On-resistance Measurement Circuit for GaN Power HEMTs, 2018 IEEE Transportation Electrification Conference and Expo (ITEC)

[7] Badawi N.: A new Method for Dynamic Ron Extraction of GaN Power HEMTs, PCIM Europe 2015

[8] Foulkes T.: Developing a Standardized Method for Measuring and Quantifying Dynamic On-State Resistance via a Survey of Low Voltage GaN HEMTs, 2018 IEEE Applied Power Electronics Conference and Exposition (APEC)

[9] Guacci M.: On-State Voltage Measurement of Fast Switching Power Semiconductors, CPSS Transactions on Power Electronics and Applications, Vol. 3, No. 2, 2018

[10] Rossetto L.: A Fast ON-State Voltage Measurement Circuit for Power Devices Characterization, IEEE Transactions on Power Electronics, Vol. 37, No. 5, 2022

[11] Li K.: Modelling GaN-HEMT Dynamic ON-state Resistance in High Frequency Power Converter, 2020 IEEE Applied Power Electronics Conference and Exposition (APEC)

[12] Weiser M. C. J.: A Novel Approach for the Modeling of the Dynamic ON-State Resistance of GaN-HEMTs, IEEE Transactions on Electron Devices, Vol. 68, No. 9, 2021

Analytical, FEM and Experimental Study of the Influence of the Airgap Size in Different Types of Ferrite Cores

Asier Arruti[1], Francisco Jose Perez-Cebolla[2], Jon Anzola[1], Iosu Aizpuru[1], Mikel Mazuela[1]
[1]Mondragon Unibertsitatea, Mondragon, Spain
[2]University of Zaragoza, Zaragoza, Spain
E-Mail: aarruti@mondragon.edu, fperez@unizar.es

Acknowledgements

This research has been supported by the Department of Education of the Basque Government under the Non Doctoral Research Staff Training Program through grant PRE_2020_1_0229.

Keywords

«Flux model», «Magnetic device», «Passive component», «Permeability»

Abstract

This work reviews and compares different airgap reluctance calculation approaches with experimental results, focusing on the Schwarz-Christoffel transformation. The modelling of the airgap reluctance in two dimensions is tested against FEM simulations for typical airgap geometries. Then, an approach to obtain the three-dimensional reluctance is shown, and the limited experimental data shown in previous works is expanded to validate the different airgap calculation methods in EE cores. For the case of pot cores, a geometrical transformation is proposed and validated, allowing the application of the Schwarz-Christoffel methodology to other common core geometries.

Introduction

The analysis and design of inductors is a critical task in power electronics applications. Reluctance models based on equivalent magnetic circuits have long been used for this; they are analogue to electric circuits, making them familiar to electronic engineers, and they allow for quick and easy calculations of the magnetic flux density and inductance.

When using highly permeable magnetic cores, an airgap is added to the magnetic path in the cores. This airgap has a high reluctance, so that it can be used to adjust the effective reluctance of the core, also contributing to alleviate the temperature dependency of the material and helps to avoid saturation of the magnetic core. Unfortunately, the airgap reluctance is not only the most impactful element in the reluctance model, it is also the most difficult to calculate due to the magnetic field distribution.

Various techniques have been proposed to calculate this value [1]–[8]. The simplest approach to estimate the airgap reluctance from Fig. 1 is to assume that the same magnetic field distribution of the core applies to the airgap [1]. Here, w_x and w_y are the widths of the core in the x and y directions, while l is the length of the airgap in the z direction, so that the reluctance is then (1).

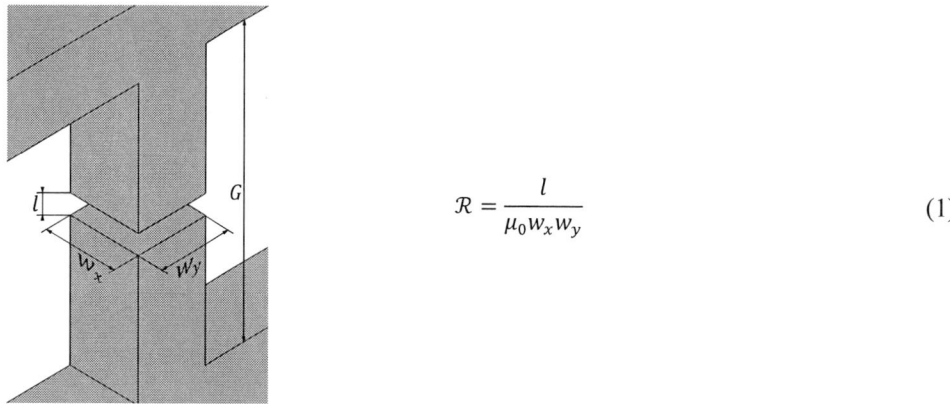

$$\mathcal{R} = \frac{l}{\mu_0 w_x w_y} \tag{1}$$

Fig. 1: Visual representation of the airgap added to a magnetic material path.

Inside the core, due to the high relative permeability, practically all of the magnetic flux is confined in the magnetic area, but this is not true in the airgap, and a fringing flux appears. The existence of the fringing field increases the effective area and reduces the airgap reluctance, and thus, (1) will not give accurate results for the reluctance.

Some other approaches to estimate the airgap reluctance have been proposed. The most commonly known technique is based on increasing the cross-section as a function of the airgap [3]–[5]. In general, this is done by adding the airgap length to the widths of the cross-section, so that the airgap is then (2). To our knowledge, there is no mathematical proof of (2), and it appears to be developed based on experimental data. Similarly, the use of the correction factor F (3) is proposed in [6]–[8], which is developed based on experimental data. In this case a new variable G is introduced, defined as the winding height (Fig. 1). The correction factor F is then used to calculate the airgap reluctance as shown in (4).

$$\mathcal{R} = \frac{l}{\mu_0 (w_x + l)(w_y + l)} \tag{2}$$

$$F = 1 + \frac{l}{\sqrt{w_x w_y}} \ln \frac{2G}{l} \tag{3}$$

$$\mathcal{R} = \frac{l}{\mu_0 w_x w_y} / F = \frac{1}{\frac{\mu_0 w_x w_y}{l} + \mu_0 \sqrt{w_x w_y} \ln \frac{2G}{l}} \tag{4}$$

Computer based numerical solutions such as finite element methods (FEM) can be used to accurately model the fringing field and resolve the airgap reluctance [9]–[11], but these approaches are much more computationally expensive. This makes then prohibitively slow to use in inductor design and optimization tools, were thousands of different cases have to be quickly analysed and compared.

Airgap calculation using the Schwarz-Christoffel transformation

Another technique for the calculation of the airgap is proposed in [12]. Unlike the previous methods, this approach uses the Schwarz-Christoffel transformation, which makes use of the analogy between the electric field and capacitance, and the magnetic field and reluctance. Note that the technicalities of the Schwarz-Christoffel transformation will not be discussed in this publication, and it will focus on the application of the base reluctance defined in [12].

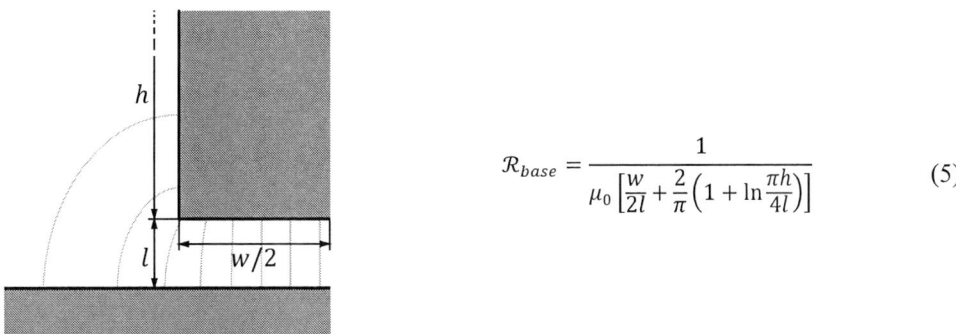

$$\mathcal{R}_{base} = \frac{1}{\mu_0 \left[\frac{w}{2l} + \frac{2}{\pi} \left(1 + \ln \frac{\pi h}{4l} \right) \right]} \tag{5}$$

Fig. 2: Base reluctance based on the Schwarz-Christoffel transformation.

For the two-dimensional geometry of Fig. 2, the two-dimensional base reluctance can be defined as a function of the geometry (5). This base reluctance serves as the building block to solve more complex two-dimensional cases. Although the approach from [12] is confined to two-dimensional problems, later works [13]–[15] propose a methodology to apply it in three-dimensional cases.

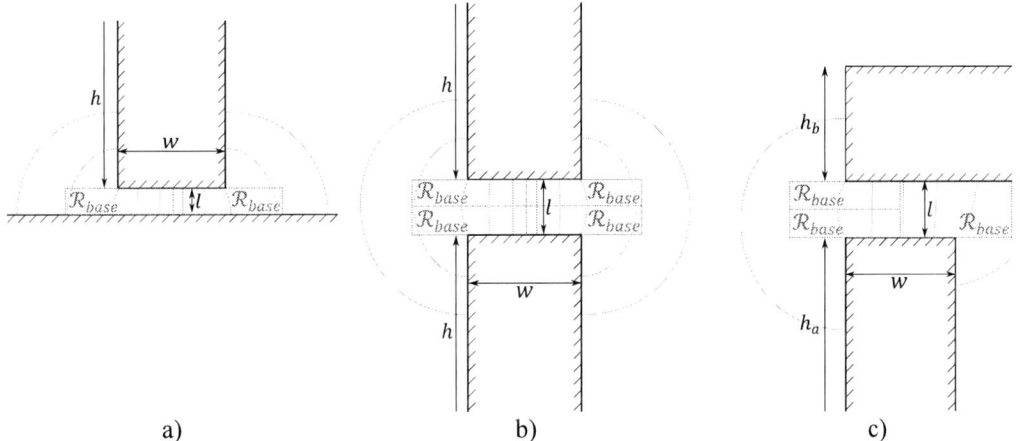

Fig. 3: a) Post-plate airgap made from two base reluctances in parallel, b) post-post airgap made from two post-plate airgaps in series (composed of four base reluctances), and c) edge airgap made from combining post-post and post-plate airgaps (composed of three base reluctances).

Using this base reluctance, both [12] and [13] identify different airgap types based on typical magnetic core assemblies. The nomenclature employed in this work follows the naming used in [12]. The common cases shown in Fig. 3 are analysed, were the equivalent reluctance expressions are based in the following geometrical assumptions:

- The post-plate airgap from Fig. 3a (type 3 according to [13]) is formed by two base reluctances in parallel, so that the two-dimensional reluctance can be directly obtained from (6). In [13] the two base reluctances are solved individually and added in parallel. Both approaches achieve the same result.

$$\mathcal{R}_{post-plate} = \mathcal{R}_{base} || \mathcal{R}_{base} = \frac{1}{\mu_0 \left[\frac{w}{l} + \frac{4}{\pi} \left(1 + \ln \frac{\pi h}{4l} \right) \right]} \tag{6}$$

- The post-post airgap from Fig. 3b (type 1 according to [13]) is made from two post-plate airgaps in series, so that the two-dimensional reluctance is simplified to (7). In [13] each reluctance is solved individually and then combined accordingly. The mathematical expressions obtained by both approaches are the same.

$$\mathcal{R}_{post-post} = \mathcal{R}_{post-plate} + \mathcal{R}_{post-plate} = \frac{1}{\mu_0 \left[\frac{w}{l} + \frac{2}{\pi}\left(1 + \ln\frac{\pi h}{2l}\right)\right]} \tag{7}$$

- The edge airgap from Fig. 3c (type 2 according to [13]) is made from half a post-post airgap in parallel with half a post-plate airgap, obtaining the expression (8). According to [12] the post-post reluctance can be solved by substituting h with the smallest post height between h_a and h_b. The technique proposed by [13] is slightly different, and the post-post reluctance is divided into three base reluctances, using the corresponding post height for each case instead of selecting the smallest value between h_a and h_b. Due to this, the edge airgap reluctances obtained from [12] and [13] differ slightly.

$$\mathcal{R}_{edge} = 2\big(\mathcal{R}_{post-post}||\mathcal{R}_{post-plate}\big) = \frac{1}{\mu_0 \left[\frac{w}{l} + \frac{1}{\pi}\left(1 + \ln\frac{\pi \min(h_a, h_b)}{2l}\right) + \frac{2}{\pi}\left(1 + \ln\frac{\pi h_a}{4l}\right)\right]} \tag{8}$$

The validity of these equations has been tested using a two-dimensional FEM simulations using FEMM [16], and mostly good agreements between analytical and FEM results have been obtained. Examples of various simulations are shown in Fig. 4, where a fine mesh area is defined around the airgap to obtain accurate results (shown as a rectangle). Furthermore, a comparison between the simulations and the analytical results is shown in Fig. 4. The post-plate and post-post airgaps maintain an error below 1.5% until $l > w/10$, while the edge airgap error for $l > w/10$ remains below 5%. For the edge airgap, the different approaches from [12] and [13] have been tested. The accuracy achieved by [12] is slightly worse than that obtained by applying [13], although the difference between these is less than 1% till $l > w/10$.

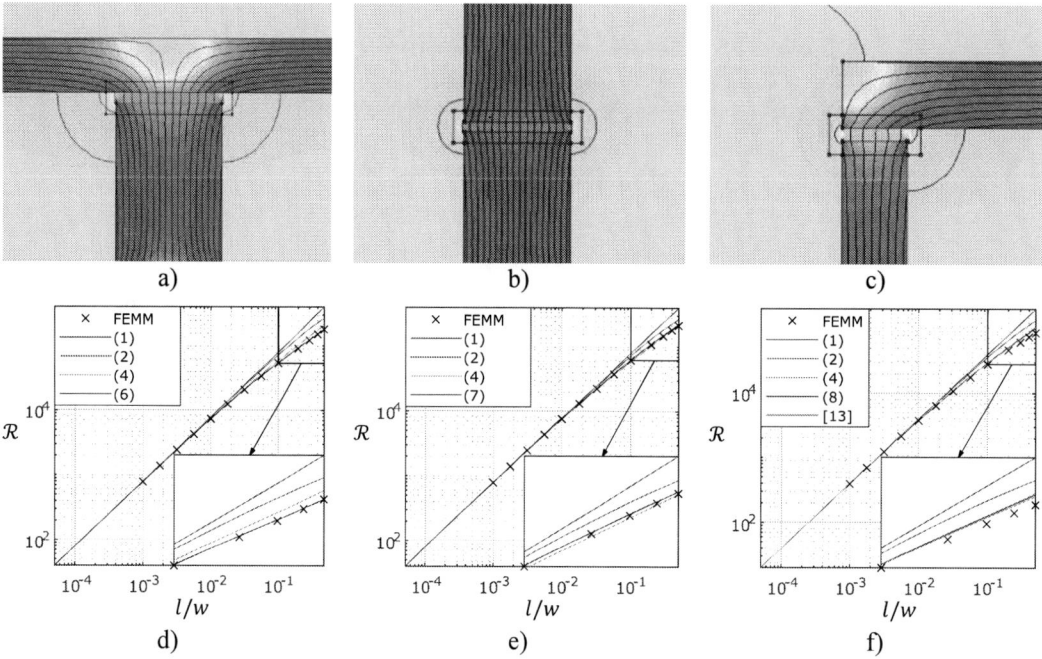

Fig. 4: FEM simulations of the magnetic flux density in a) post-plate, b) post-post and c) edge airgaps with the corresponding dimensions of an EE8020/EI8020 core with an airgap of 2 mm. The two dimensional reluctances obtained from the simulations and analytical approaches for various airgap lengths are shown in d), e) and f) respectively.

According to [13], it is possible to combine the two-dimensional reluctances to obtain the equivalent three-dimensional reluctance. To do so, it is necessary to obtain the fringing factors (σ) in the xz (9) and yz (10) planes shown in Fig. 5, which define how the two-dimensional airgap reluctances change due to the fringing fields. Then, by applying both fringing factors to the three-dimensional reluctance without fringing field (1), the reluctance of the airgap with fringing field (11) is obtained.

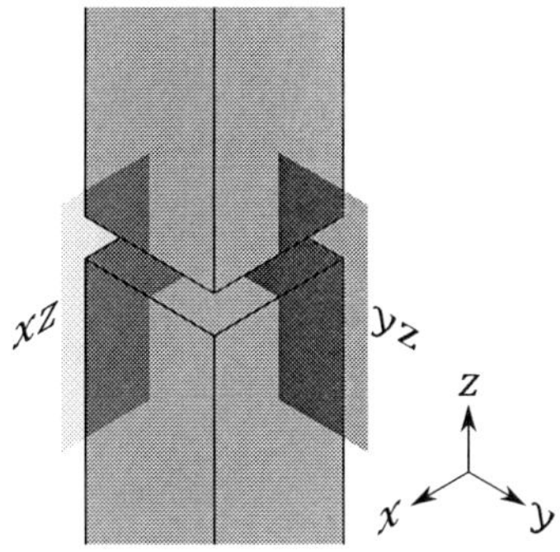

$$\sigma_{xz} = \frac{\mathcal{R}_{xz}}{\frac{l}{\mu_0 w_x}} \tag{9}$$

$$\sigma_{yz} = \frac{\mathcal{R}_{yz}}{\frac{l}{\mu_0 w_y}} \tag{10}$$

$$\mathcal{R} = \sigma_{xz}\sigma_{yz} \frac{l}{\mu_0 w_x w_y} \tag{11}$$

Fig. 5: Representation of the xz and yz planes where the two-dimensional airgap reluctances are defined.

Comparisons with experimental results using EPCOS ferrite N27 EE55/28/21 cores and 1.0, 1.5 and 2.0 mm airgaps are presented in [13]. Due to the limited amount of experimental data (3 points), it is deemed necessary to perform more tests to verify the validity of the Schwarz-Christoffel transformation. Cosmo Ferrites CF139 EE5521, EE6527, and EE8020 cores have been tested [17]. The inductance has been measured using a BK Precision 891 LCR Meter at 10 kHz. The airgap reluctance is obtained by subtracting the effective reluctance of the core without airgap to the effective reluctance of the core with airgap. The results for these are shown in Tab. 1.

Tab. 1: Measured and calculated values of the airgap reluctance, as well as the difference between them, for three different EE cores and airgap lengths between 0.5 and 2 mm.

Core	l (mm)	Equation (1) (μH^{-1})	Equation (2) (μH^{-1})	Equation (4) (μH^{-1})	Approach [13] (μH^{-1})	Measured (μH^{-1})
EE5521	0.5	2.23 (+14.7%)	2.09 (+7.28%)	**1.92 (−1.15%)**	1.80 (−7.19%)	1.94
	1.0	4.46 (+37.0%)	3.91 (+20.2%)	3.50 (+7.43%)	**3.14 (−3.61%)**	3.26
	1.5	6.69 (+54.6%)	5.52 (+27.6%)	4.87 (+12.6%)	**4.21 (−2.79%)**	4.33
	2.0	8.91 (+66.2%)	6.94 (+29.3%)	6.11 (+13.8%)	**5.10 (−4.96%)**	5.37
	2.5	11.1 (+83.1%)	8.19 (+34.5%)	7.24 (+18.9%)	**5.87 (−3.63%)**	6.09
EE6527	0.5	1.47 (+10.5%)	1.40 (+4.61%)	**1.30 (−2.57%)**	1.23 (−7.78%)	1.33
	1.0	2.95 (+29.4%)	2.65 (+16.1%)	2.39 (+4.98%)	**2.18 (−4.50%)**	2.28
	1.5	4.42 (+44.8%)	3.77 (+23.5%)	3.36 (+10.0%)	**2.95 (−3.24%)**	3.05
	2.0	5.89 (+59.1%)	4.78 (+29.1%)	4.23 (+14.3%)	**3.62 (−2.40%)**	3.70
	2.5	7.37 (+70.9%)	5.69 (+32.1%)	5.04 (+17.0%)	**4.19 (−2.74%)**	4.31
EE8020	0.5	1.99 (+18.4%)	1.87 (+11.4%)	**1.71 (+1.84%)**	1.61 (−3.97%)	1.68
	1.0	3.98 (+40.0%)	3.53 (+24.1%)	3.10 (+9.13%)	**2.80 (−1.54%)**	2.84
	1.5	5.97 (+59.7%)	5.00 (+33.7%)	4.30 (+15.1%)	**3.74 (+0.01%)**	3.74
	2.0	7.96 (+74.9%)	6.30 (+38.5%)	5.37 (+18.1%)	**5.51 (−0.87%)**	4.55
	2.5	9.95 (+88.5%)	7.47 (+41.5%)	6.35 (+20.3%)	**5.17 (−2.93%)**	5.28

According to the new data presented, the Schwarz-Christoffel transformation based approach appears to be more accurate than the other techniques as the airgap length is increased. Note that only tests with EE cores are presented, thus only post-post reluctances (7) must be solved. Since the post-post reluctance is composed of two post-plate reluctances in series, there is reason to believe that (6) will achieve a similar accuracy, although the same cannot be said about (8). Curiously enough, for an airgap of 0.5 mm, the results from (4) are slightly more accurate than the Schwarz-Christoffel approach, although this changes as the airgap is increased. As the airgap length approaches zero, the different methods quickly converge, so similar accuracies can be expected for very low airgaps.

Application to other geometries

The previous data shows good agreements between experimental and analytical calculations based on the Schwarz-Christoffel transformation, but it has only been validated for EE cores. To ensure that the Schwarz-Christoffel transformation is valid to all common magnetic geometries, it is important to test other commonly employed core geometries, such as pot cores (Fig. 6). The cylindrical shape of these cores achieves a lower wire mean turn length, which reduces the copper losses. At the same time, the magnetic material surrounds the windings longer than in EE cores, which helps reduce EMI radiations.

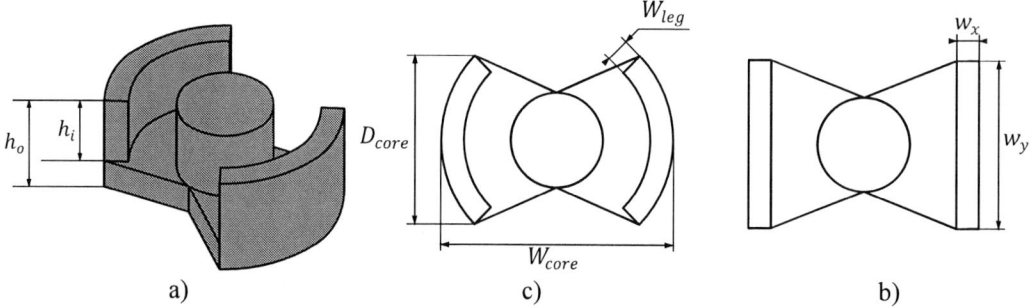

a)   c)   b)

Fig. 6: a) Example of a PM pot core, b) the cross-section and b) geometrical transformation of the outer legs into rectangular sections.

Due to the cross-section shape of the pot cores, some geometrical changes are necessary to apply the Schwarz-Christoffel transformation. For the Cosmo Ferrites CF139 PM5039, PM6249, PM7459, and PM8770 cores tested [17], in the outer legs the value of w_x is equal to the width of the leg (W_{leg}), but the value of w_y must be derived from the pot core geometry. In this work, the transformation of the outer leg to a rectangular section is proposed, so that w_y is then (12). The heights of the outer legs are different in the inside and outside of the core, which are defined as h_i and h_o. In that case, the two dimensional reluctances of the outer legs are (13) and (14), and the approach from [13] can be used to obtain the three dimensional reluctance. The centre leg does not require any geometrical transformation, since as demonstrated in [13], the σ_{xz} and σ_{yz} factors are the same and represent the radial fringing factor (σ_r), so that the airgap reluctance is then (15).

$$w_y = \left(W_{core} - W_{leg}\right) \cos^{-1}\left(\frac{W_{core}}{D_{core}}\right) \tag{12}$$

$$\mathcal{R}_{xz} = \frac{1}{\mu_0 \left[\frac{w_x}{l} + \frac{2}{\pi}\left(1 + \ln\frac{\pi h_o}{2l}\right)\right]} \tag{13}$$

$$\mathcal{R}_{yz} = \frac{1}{\mu_0 \left[\frac{w_y}{l} + \frac{2}{\pi}\left(1 + \ln\frac{\pi\sqrt{h_i h_o}}{2l}\right)\right]} \tag{14}$$

$$\mathcal{R} = \sigma_r^2 \frac{l}{\mu_0 w_x^2 \, \pi/4} \tag{15}$$

The results of the experimental tests for the four PM cores can be observed in Fig. 7. The data shows that the Schwarz-Christoffel transformation is the only approach capable of accurately estimating the impact of the airgap reluctance for long airgap lengths. It appears that the Schwarz-Christoffel transformation losses its accuracy at low airgap lengths, but there is not enough data to strongly validate this conclusion. Due to the tolerance of the gap spacers and the sensibility of the reluctance, it is not possible to test airgaps lower than 0.5 mm accurately.

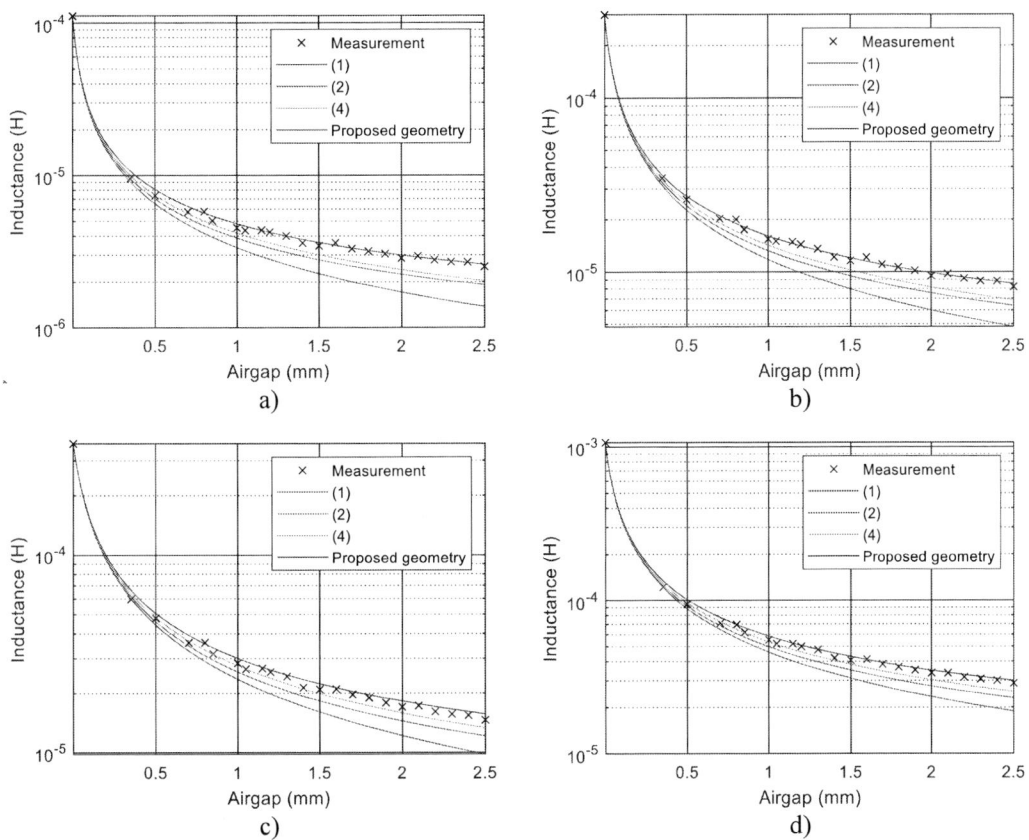

Fig. 7: Comparison of analytical and experimental results for a) PM5039, b) PM6249, c) PM7459 and d) PM8770 cores.

Conclusion

In this paper, various approaches to estimate the airgap reluctance of a magnetic core are studied, focusing on the effect of the fringing field. The Schwarz-Christoffel transformation based approach previously shown in [12], [13] is analysed and experimentally tested to solve the most common airgap geometries that appear in practical cases. The limited amount of experimental results presented in the literature is expanded for a larger airgap length range and more EE type cores. The analytical results are experimentally validated and reveal that the Schwarz-Christoffel approach is the only one capable of correctly assessing the reluctance for large airgap lengths.

The validity of the Schwarz-Christoffel transformation has also been tested in four PM pot cores, with airgap lengths ranging from 0.35 to 2.5 mm. A geometrical transformation to apply the Schwarz-Christoffel transformation to the cylindrical shape of these cores is proposed. According to the results, the Schwarz-Christoffel approach appears to be the most accurate at large airgap lengths, although its accuracy at low airgap lengths seems to be worse than in EE cores. Due to the tolerances of the gap spacers used, it has not been possible to analyse in detail the accuracy of these ranges, leaving this part of the study for future work.

References

[1]   G.F. Partridge B.Sc. F.Inst.P. (1936) LII. The inductance of iron-cored coils having an air gap , The London, Edinburgh, and Dublin Philosophical Magazine and Journal of Science, 22:148, 665-678, DOI: 10.1080/14786443608561717.

[2] R. Lee, L. Wilson, and C. E. Carter, *Electronic Transformers and Circuits*. John Wiley & Sons Inc, 1988.

[3] N. Mohan, T. M. Undeland, and W. P. Robbins, *Power Electronics: Converters, Applications, and Design*. John Wiley & Sons, Inc, 2002.

[4] W. G. Hurley and W. H. Wölfle, *Transformers and Inductors for Power Electronics: Theory, Design and Applications*. John Wiley & Sons, Inc, 2013.

[5] A. Ayachit and M. Kazimierczuk, "Sensitivity of Effective Relative Permeability for Gapped Magnetic Cores with Fringing Effect," *IET Circuits, Devices & Systems*, vol. 11, no. 3, pp. 209–215, 2017.

[6] W. T. McLyman, *Transformer and Inductor Design Handbook*. Marcel Dekker, Inc, 2004.

[7] A. Van den Bossche and V. C. Valchev, *Inductors and Transformers for Power Electronics*. CRC Press, Taylor & Francis Group, 2005.

[8] M. K. Kazimierczuk, *High-frequency magnetic components*. John Wiley & Sons, Inc, 2009.

[9] R. Salas and J. Pleite, "Nonlinear Modeling of E-Type Ferrite Inductors Using Finite Element Analysis in 2D," *Materials*, vol. 7, no. 8, pp. 5454–5469, 2014.

[10] M. Beraki, J. P. F. Trovao, and P. Marina, "Characterization of Variable Inductors using Finite Element Analysis," *Simulation Modelling Practice and Theory*, vol. 97, 2019.

[11] S. Saeed, J. Garcia, M. S. Perdigão, V. S. Costa, B. Baptista, and A. M. S. Mendes, "Improved Inductance Calculation in Variable Power Inductors by Adjustment of the Reluctance Model Through Magnetic Path Analysis," *IEEE Transactions on Industry Applications*, vol. 57, no. 2, pp. 1572–1587, 2021.

[12] A. Balakrishnan, W. T. Joines, and T. G. Wilson, "Air-Gap Reluctance and Inductance Calculations for Magnetic Circuits Using a Schwarz–Christoffel Transformation," *IEEE Transactions on Power Electronics*, vol. 12, no. 4, pp. 654–663, 1997.

[13] J. Mühlethaler, J. W. Kolar, and A. Ecklebe, "A Novel Approach for 3D Air Gap Reluctance Calculations," 2011.

[14] Z. Yang, H. Suryanarayana, and F. Wang, "An Improved Design Method for Gapped Inductors Considering Fringing Effect," 2019.

[15] E. L. Barrio, A. Urtasun, A. Ursúa, L. Marroyo, and P. Sanchis, "Optimal DC gapped inductor design including high-frequency effects," 2015.

[16] D. C. Meeker, *Finite Element Method Magnetics, Version 4.2 (21Apr2019 Build), https://www.femm.info*.

[17] "COSMO FERRITES LIMITED." https://www.cosmoferrites.com/product-store/soft-ferrites (accessed Mar. 30, 2022).

Design Method of a High Frequency GaN-based Half-Bridge with Bottom-Side Cooled Transistors Using Multi-PCB Assembly

Loris PACE
AMPERE/ECOLE CENTRALE DE
LYON
36 av. Guy de Collongue
Ecully, France
E-Mail: loris.pace@ec-lyon.fr

Florian CHEVALIER,
Thierry DUQUESNE and Nadir IDIR
L2EP/UNIVERSITE DE LILLE
Bat. Esprit, Av. Paul Langevin
Villeneuve d'Ascq, France
E-Mail: nadir.idir@univ-lille.fr

Keywords

Half Bridge, Power Density Optimization, Gallium Nitride (GaN), Parasitic inductance, Thermal Design

Abstract

This paper proposes a Multi-PCB (M-PCB) design of a half-bridge using bottom-side cooled GaN transistors with optimal layout and thermal management. The fabrication process only requires limited technological capabilities and is well-suited for industrial applications. Electrical and thermal performances are evaluated by measurements and simulation and compared with previous works.

Introduction

The constant growth in the demand for embedded static converters with high power density and high efficiency requires the development of new design approaches. Operating at higher frequencies (over 1 MHz) while reducing power losses are the key solutions to improve efficiency and miniaturize power converters. The advantages of GaN power transistors in terms of electron mobility and intrinsic parasitics make them the ideal candidates for High Frequency (HF) power conversion [1]. Recent works have highlighted the benefits brought by GaN devices in designing efficient and compact power converters for various applications in domains such as automotive, aeronautics, servers... [2]-[5].

Although GaN transistors have attractive characteristics for high frequency power conversion, their switching times in the range of nanosecond causes high overvoltage and ringing due to the parasitic elements of the commutation cells. In order to avoid a dual impact on efficiency and cooling systems of the power devices, it is necessary to reduce the commutation loop parasitic inductances. To achieve this goal, Printed Circuit Board (PCB) embedded power converters based on Electronic Design Automation (EDA) techniques have emerged these last years [6].

When designing power converters on PCB using bottom-side cooled GaN transistors, their thermal management can be easily improved by designing thermal vias through the FR4 substrate and optimizing their layout [7]. This method has been widely used for the PCB integration of unpackaged power chips [8], [9]. The recent improvements in GaN transistors packaging allow to consider applying equivalent design methods for the integration of packaged power devices in PCB assemblies. Also, it has been demonstrated that vertical commutation loops can consequently reduce the total parasitic inductances by limiting the loop area and canceling a part of the magnetic flux [10], [11]. These results lead to lower power losses and lower Electromagnetic Interferences (EMI). However, authors in [10] have highlighted that the design of a vertical loop and thermal vias were not compatible on a single PCB.

In this context, this work proposes the Multi-PCB (M-PCB) design for GaN-based half-bridge using a three-layer PCB assembly. The proposed solution allows the combination of an optimized commutation loop design with an optimized thermal management for bottom-side cooled GaN transistors. The second section of this paper details the fabrication process while Section III and IV give the Electromagnetic (EM) and thermal analysis of the proposed structure. In section V, the switching waveforms of the DC-DC converter are analyzed while operating at 1 MHz, 300 V input voltage and 6 A output current. The experimental results are compared with EM/circuit simulations. The advantages of the proposed solution are then highlighted by testing the power converter in continuous mode and comparing the obtained thermal results with previous work using the same devices in equivalent conditions. Finally, a discussion about future improvements is proposed in the conclusion.

Design of the M-PCB Half-Bridge

The electrical scheme of the considered GaN-based half-bridge is presented in Fig. 1(a). The commutation cell is constituted of two bottom-side cooled p-GaN HEMT (GS66502B: 650 V - 8 A @ 25 °C) and DC-link ceramic capacitors. Finally, two isolated single gate drivers SI8271 are used to control the power transistors. This enables the optimization of the gate loop design for the GaN devices. It should be noted that the following design is also compatible with a half-bridge gate driver but not proposed in this study.

In order to optimize the commutation loop design while offering the best thermal performances for these devices, it is proposed to separate electrical and thermal paths by achieving the three-layer PCB assembly as shown in Fig. 1(b). A vertical commutation loop is realized between the DC-link capacitor located on the top layer and the GaN transistors located on the intermediate layer. The heat transfer is performed from the bottom side of the GaN transistors packaging (source) to the bottom layer by means of thermal vias. An Aluminum Nitride (AlN) substrate is inserted between the bottom layer of the assembly and the heat sink to ensure dielectric insulation while ensuring the best thermal conductivity and spreading the heat on the heat sink area.

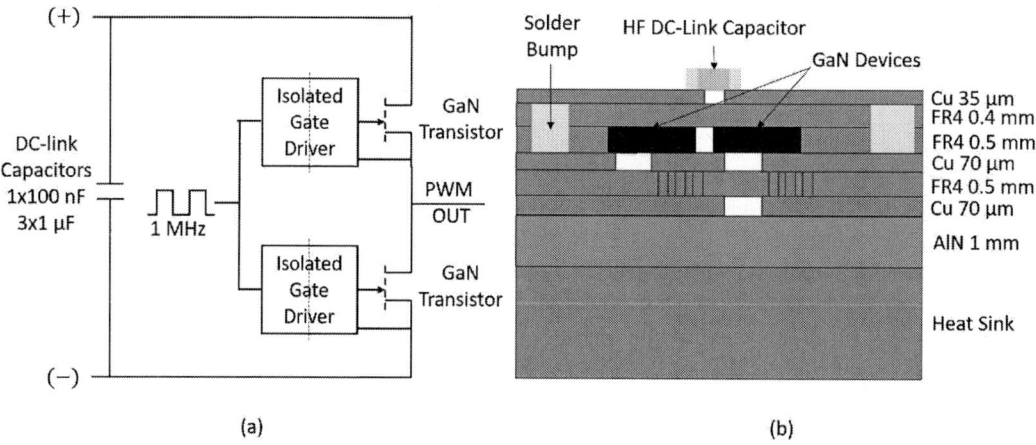

(a)   (b)

Fig. 1: Presentation of the proposed M-PCB half-bridge (a) simplified electrical scheme (b) 3-layer design (cross-sectional view)

The manufacturing process of the structure described in Fig. 1(b) is divided into four steps as shown in Fig. 2:

1. Fabrication of the circuit for the GaN transistors on a double-sided PCB (FR4: 0.5 mm, Cu: 70 µm) with metalized thermal vias (Fig. 2(a)).
2. Deposit of a tin layer on gate, drain, source connections of each devices and soldering of the transistors (Fig. 2(b)).
3. Fabrication of a 0.5 mm FR4 windowed layer and fixation on the previous PCB to reach the height of the transistors (Fig. 2(c)).
4. Assembly of the upper single-sided PCB (FR4: 0.4 mm, Cu: 35 µm) including the half-bridge circuit on the structure. Soldering of gate, drain and source connections of each transistor on the top side circuit (Fig. 2(d)).

Fig. 3 shows the fabricated parts of the M-PCB half-bridge and its final assembly. In the next sections the electrical and thermal performances of the proposed structure are analyzed through simulations and measurements.

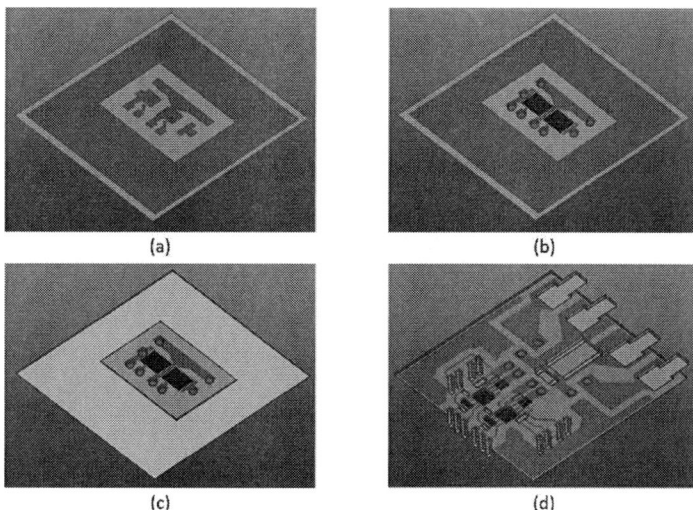

Fig. 2: M-PCB half-bridge fabrication process

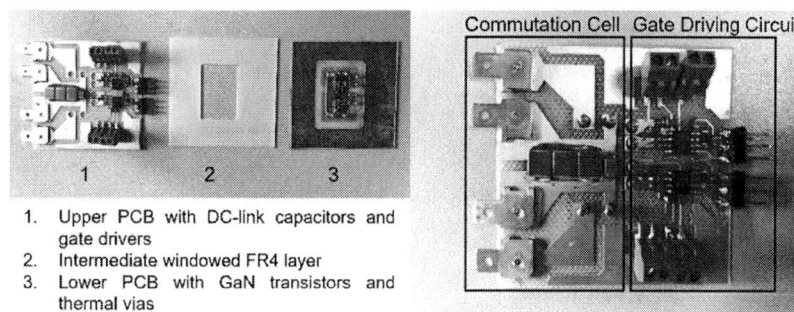

1. Upper PCB with DC-link capacitors and gate drivers
2. Intermediate windowed FR4 layer
3. Lower PCB with GaN transistors and thermal vias

Fig. 3: Presentation of the fabricated parts and the assembly of the proposed M-PCB half-bridge

Loop Inductances Characterization

The design optimization of the considered converter is carried out by simulation. Thus, an EM modeling of the M-PCB half-bridge is performed using Advance Design System (ADS®) software in order to extract the parasitic inductance values of the commutation and gate loops. The design of the PCB implemented in the software is presented in Fig. 4(a). As shown in Fig. 4(b), the tin vias connecting transistors terminals to the rest of the circuit are modeled by 2 mm diameter cylinders and the thermal vias are modeled by 0.4 mm hollow cylinders with 20 μm copper filing. In order to get the best accuracy for the determination of the parasitic inductances, the EM model of the transistors packaging proposed in [11] is added to the simulation. The EM simulation is performed from 1 MHz to 100 MHz with logarithmic frequency sweep and from 1 Hz to 2 GHz with an adaptive frequency sweep.

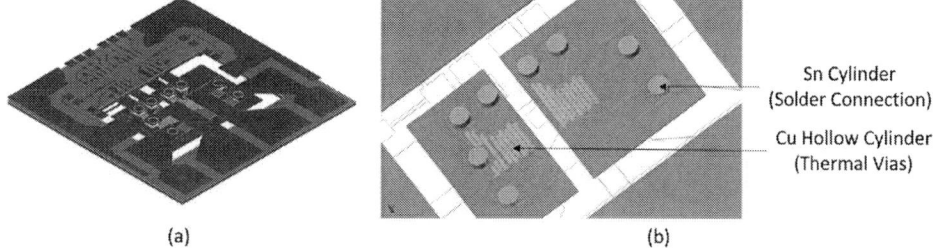

Fig. 4: PCB implementation in ADS software (a) 3D global view (b) geometries for the different vias

1-Port S-parameter EM/circuit co-simulations are then performed to characterize the different converter loops in the circuit. For each simulation, the loop inductance L_{loop} is obtained from the simulated S_{11} parameter using (1). Fig. 5 gives the commutation loop and the two gate loop inductances evolutions over frequency from 1 MHz to 100 MHz.

$$L_{loop} = \frac{Im\left(Z_0 \left(\frac{1+S_{11}}{1-S_{11}}\right)\right)}{\omega}$$

(1)

Where, ω is the pulsation in rad/s and Z_0 is the 50 Ω characteristic impedance.

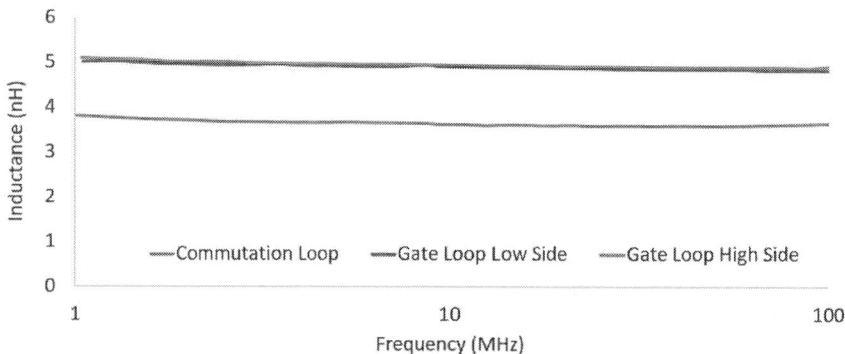

Fig. 5: Simulation results of the converter loop parasitic inductances versus frequency

For the half-bridge configuration, Table I compares the extracted commutation loop inductance at 1 MHz with the obtained values in lateral and vertical design given in [11]. The proposed optimized design of the commutation loop offers drastically lower inductance value than the lateral and vertical design in [11]. It should be noted that the encapsulation inductance of a single GaN transistor has been measured to 2.5 nH using S-parameters [11].

TABLE I
SIMULATED COMMUTATION LOOP INDUCTANCE AT 1 MHz

Lateral Layout (nH) [11]	Vertical Layout (nH) [11]	Optimized Vertical Layout (nH, This Work)
13	4.4	3.8

Thermal Analysis of the M-PCB Half-Bridge

The geometry of the M-PCB half-bridge is implemented in COMSOL® software to evaluate the temperature of the GaN transistors for different dissipated power levels. Fig. 6(a) shows the temperature distribution obtained for a dissipated power of 5 W in each GaN transistor. The obtained power-temperature relation is also experimentally analyzed using a Fluke TI-32 thermal imaging camera as shown in Fig. 6(b).

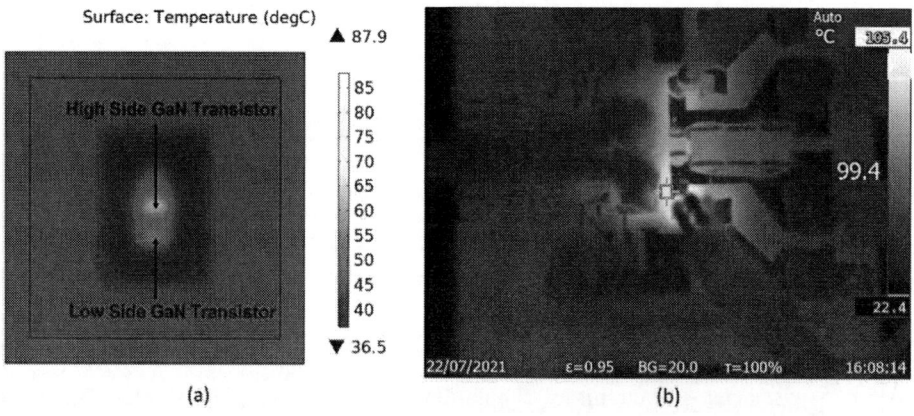

Fig. 6: Analysis of the GaN transistors temperature (a) COMSOL simulation (b) measurement

For measurements, the V_{GS} of the transistors are set to 6 V using two isolated power supply, the DC-link voltage is adjusted to inject a controlled DC current and reach different dissipated power levels in the devices. The obtained simulation results are compared to experimental data for different dissipated power in the GaN transistors (Fig. 7).

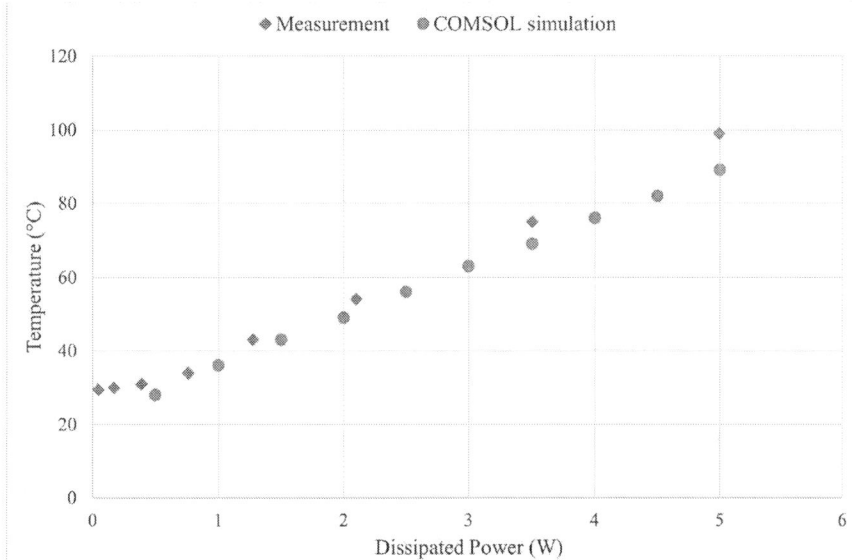

Fig. 7: Measured and simulated temperature of the low-side GaN transistor for different dissipated power levels

These results show slight differences between measurement and simulation when increasing the dissipated power. This could be explained by a non-homogenous convective environment in the experimental setup. In the simulation, a convective heat flux using external natural convection by air with spherical geometry is considered. In the experiment, the prototype is placed on a table. From these results, the thermal resistance from case to ambient is 12 °C/W by simulation and 15 °C/W by measurement.

Converter Performances Evaluation in DC-DC Operation

The DC-DC converter is first tested in burst mode at 1 MHz. Fig. 8 gives the obtained switching waveforms for both low and high side GaN transistors after reaching the electrical steady state (10 periods). The DC-link voltage is 300 V and the switched current is 6 A. The duty cycle is set to 50 % and the dead times are 50 ns (10 % of the switching period in total). Gate driving voltages are -3/+6 V and gate resistances are 27 Ω for turn-on and 2.2 Ω for turn-off. V_{GS} and V_{DS} voltages are measured using optically isolated voltage probes IsoVu®. The currents in the transistors were not measured because the current probe introduction would drastically modify the parasitic inductance of the commutation loop. The output load consists of an 80 μH inductor in series with a 25 Ω resistor.

Fig. 9 and Fig. 10 show the comparison between measured and simulated waveforms at turn-on and turn-off of the low side and high side transistor respectively. The simulations are performed in ADS® by coupling circuit models of the components and an EM model of the PCB as detailed by authors in [11], [12]. The PCB fabrication files are imported in the software and an EM simulation (Momentum in ADS software) is performed taking into account interconnects parasitic elements as well as inductive and capacitive couplings between layers. The GaN transistor model for device GS66502B obtained using S-parameter and static I(V) characterization as described in [12] is used in this work. Passive components such as DC link capacitors and RL load have been modeled based on impedance characterizations up to 110 MHz using the impedance analyzer 4294A. The gate driver model considers internal resistances as well as rise and fall times according to the datasheet.

Design Method of a High Frequency GaN-based Half-Bridge with Bottom-Side
Cooled Transistors Using Multi-PCB Assembly

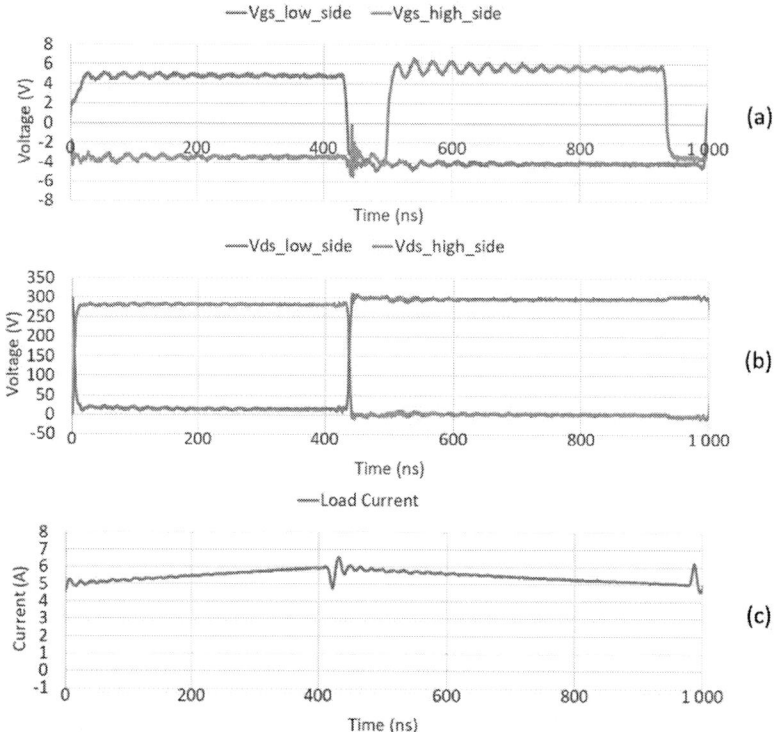

Fig. 8: Switching waveforms at 1 kW / 1 MHz (a) low and high side V_{GS} (b) low and high side V_{DS} (c) load current

Good agreement is observed between experimental and simulation results (Fig. 9 and Fig. 10). One can observe that rise and fall times of V_{GS} and V_{DS} of both transistors are in good accordance between measurement and simulation except at turn-on of the high-side GaN transistor. Authors assumed that the slight different observed could be attributed to a specific behavior of the gate driver in these test conditions.

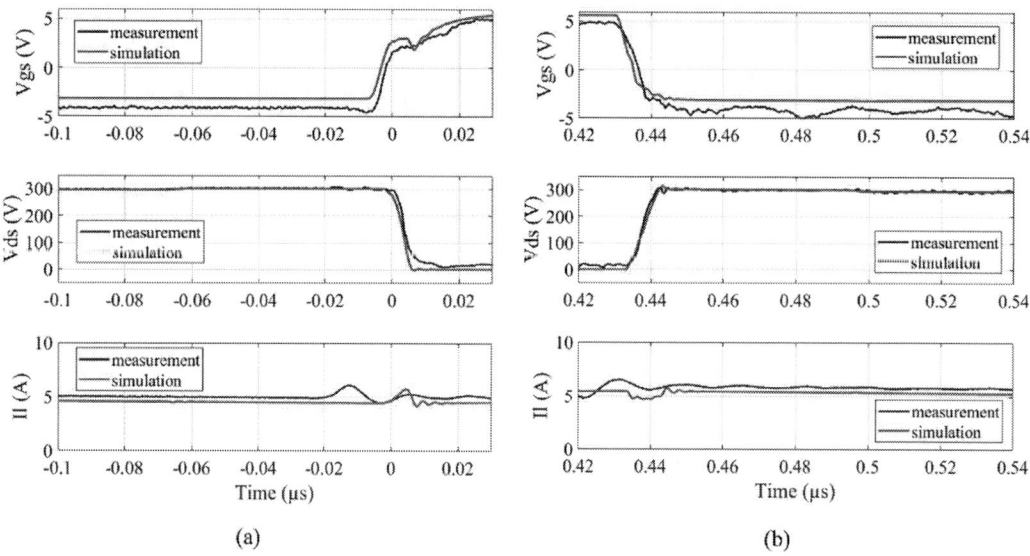

Fig. 9: Waveforms of the low-side GaN transistor and load current at (a) turn-on and (b) turn-off of the device

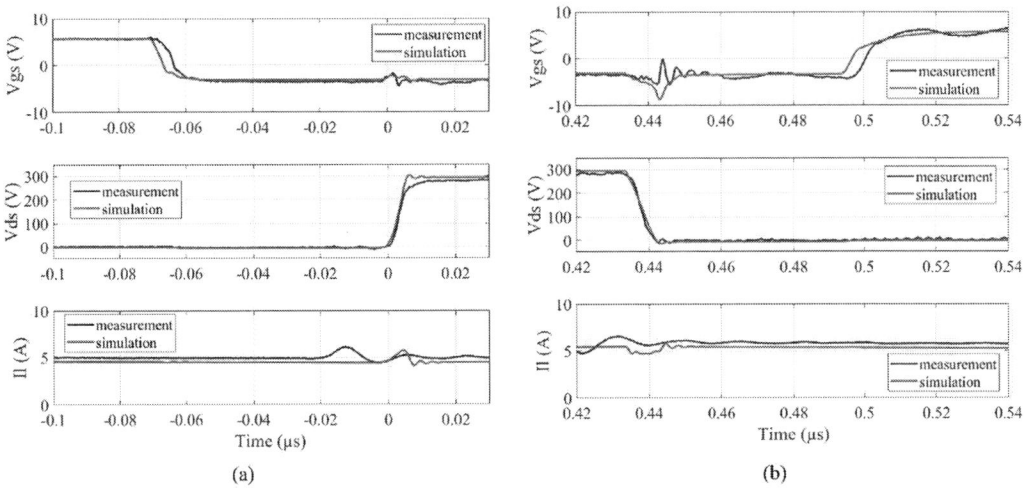

(a)   (b)

Fig. 10: Waveforms of the high-side GaN transistor and load current at (a) turn-off and (b) turn-on of the device

The switching waveforms of the GaN transistors are obtained with a maximal V_{DS} overshoot of 4 % and a maximal V_{GS} overshoot of 6 %. This confirms the advantage of the M-PCB converter design method in the optimization of the commutation loop. The V_{DS} voltage slew rate is approximatively 50 V/ns for an output current of 6 A.

The converter is then tested in continuous mode with an operating frequency of 1 MHz. The DC-link voltage is 200 V and the output current is 2 A. The duty cycle is set to 50 % and the dead times are 50 ns. The temperature steady state is reached after 15 minutes. The temperature of the GaN transistors is measured at 70 °C in steady state as shown in Fig. 11. Compared with results in [12], where the same GaN device is used in the same test conditions (power, frequency), the device temperature is 20 °C lower in this work. This enhancement is mainly due to lower switching losses and a more efficient thermal design in the proposed M-PCB converter. However, it should be noted that in [12] the GaN power transistor is tested in association with a SiC diode and not in Half-Bridge conditions.

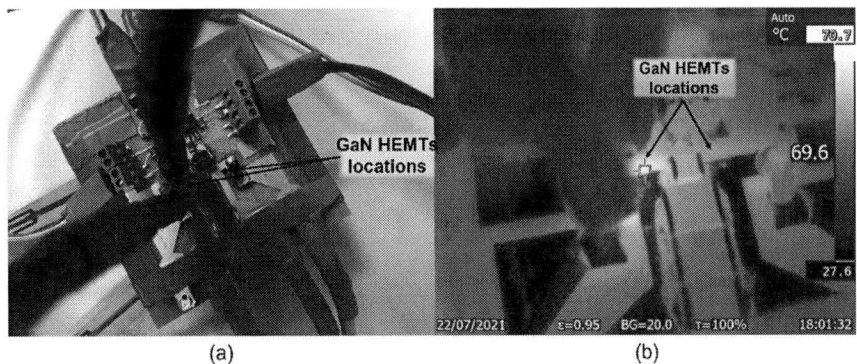

(a)   (b)

Fig. 11: Steady state GaN devices temperature at 200 W / 1 MHz (a) test setup (b) thermal measurements

Conclusion

In this paper a M-PCB design method of a half-bridge based on bottom-side cooled GaN-HEMT transistors is proposed for high frequency power conversion. The presented multi-layer PCB assembly has allowed to reduce the parasitic inductance of the commutation loop while offering an efficient cooling by thermal vias. The power converter has been electrically and thermally characterized using EM/circuit and 3D thermal simulations. A good accordance has been found between simulation and experimental results showing the possibility to optimize the design of power converters by simulation for power electronics designers. Finally, the tests of the converter that operating at 1 MHz have highlighted the benefits of the proposed solution compared with previous ones in terms of switching losses and temperature of the GaN transistors. Future work will focus on designing the GaN devices circuit (lower PCB) on an AlN substrate in order to overtake the thermal performances of thermal vias by combining lower thermal resistance and dielectric insulation.

References

[1] I. Omura, W. Saito, T. Domon and K. Tsuda, "Gallium Nitride power HEMT for high switching frequency power electronics," in 2007 International Workshop on Physics of Semiconductor Devices, Mumbai, India, 16-20 Dec. 2007 .

[2] X. Zhou, B. Sheng, W. Liu, Y. Chen, L. Wang, Y.-F. Liu and P. C. Sen, "A High-Efficiency High-Power-Density On-Board Low-Voltage DC–DC Converter for Electric Vehicles Application," IEEE Transactions on Power Electronics, vol. 36, no. 11, pp. 12781 - 12794, Nov. 2021.

[3] M. Schiestl, F. Marcolini, M. Incurvati, F. G. Capponi, R. Stärz, F. Caricchi, A. S. Rodríguez and L. Wild, "Development of a High Power Density Drive System for Unmanned Aerial Vehicles," IEEE Transactions on Power Electronics, vol. 36, no. 3, pp. 3159 - 3171, March 2021.

[4] M. H. Ahmed, M. A. d. Rooij and J. Wang, "High-Power Density, 900-W LLC Converters for Servers Using GaN FETs: Toward Greater Efficiency and Power Density in 48 V to 6V/12 V Converters," IEEE Power Electronics Magazine, vol. 6, no. 1, pp. 40 - 47, March 2019.

[5] F. Salomez, S. Vienot, B. Zaidi, A. Videt, T. Duquesne, H. Pichon, E. Semail and N. Idir, "Design of an integrated GaN inverter into a multiphase PMSM," in 2020 IEEE Vehicle Power and Propulsion Conference (VPPC), Gijon, Spain, 18 Nov.-16 Dec. 2020.

[6] C. Buttay, C. Martin, F. Morel, R. Caillaud, J. L. Leslé, R. Mrad, N. Degrenne and S. Mollov, "Application of the PCB-Embedding Technology in Power Electronics – State of the Art and Proposed Development," in 2018 Second International Symposium on 3D Power Electronics Integration and Manufacturing (3D-PEIM), College Park, MD, USA, 25-27 June 2018.

[7] S. Zhang, E. Laboure, D. Labrousse and S. Lefebvre, "Thermal management for GaN power devices mounted on PCB substrates," in 2017 IEEE International Workshop On Integrated Power Packaging (IWIPP), Delft, Netherlands, 5-7 April 2017.

[8] S. Moench, R. Reiner, P. Waltereit and a. al., "PCB-Embedded GaN-on-Si Half-Bridge and Driver ICs With On-Package Gate and DC-Link Capacitors," IEEE Transactions on Power Electronics, vol. 36, no. 1, pp. 83 - 86, Jan. 2021.

[9] B. Weiss, R. Reiner, P. Waltereit, R. Quay and O. Ambacher, "Operation of PCB-embedded, high-voltage multilevel-converter GaN-IC," in 2017 IEEE 5th Workshop on Wide Bandgap Power Devices and Applications (WiPDA), Albuquerque, NM, USA, 30 Oct.-1 Nov. 2017.

[10] B. Sun, K. L. Jørgensen, Z. Zhang and M. A. Andersen, "Research of Power Loop Layout and Parasitic Inductance in GaN Transistor Implementation," IEEE Transactions on Industry Applications, vol. 57, no. 2, pp. 1677 - 1687, March-April 2021.

[11] L. Pace, N. Idir, T. Duquesne and J.-C. D. Jaeger, "Parasitic loop inductances reduction in the PCB layout in GaN-based power converters using S-parameters and EM simulations," Energies, vol. 14, no. 5, p. 1495, March 2021.

[12] L. Pace, F. Chevalier, A. Videt, N. Defrance, N. Idir and J.-C. D. Jaeger, "Electrothermal Modeling of GaN Power Transistor for High Frequency Power Converter Design," in 2020 22nd European Conference on Power Electronics and Applications (EPE'20 ECCE Europe), Lyon, France, 7-11 Sept. 2020.

A 30 kW Dynamic Wireless Inductive Charging System for EVs

Zariff Meira Gomes
VEDECOM
Versailles, France
zariff.meira@vedecom.fr

José Renes Pinheiro
Federal University of Bahia
Federal University of Santa Maria
Salvador, Brazil
jrenes@gepoc.ufsm.br

Gilney Damm
University Gustave Eiffel
Marne-la-Vallee, France
gilney.damm@univ-eiffel.fr

Karim Kadem
VEDECOM
Versailles, France
kadem.karim@vedecom.fr

Hassan Moussa
VEDECOM
Versailles, France
hassan.moussa@vedecom.fr

Acknowledgments

This research was supported by VEDECOM. We thank our colleagues who provided insight and expertise that greatly assisted the research, although they may not agree with all of the interpretations and conclusions of this paper.

Keywords

≪Charging infrastructure for EV´s≫ ≪Wireless power transmission≫, ≪Contactless Energy Transfer≫, ≪Resonant converter≫, ≪Converter control≫.

Abstract

This paper presents a solution for real-time charging of electric vehicles along the road while moving. This is called the Dynamic Wireless Inductive Charging System (DWICS) and can mitigate the cost of the charging infrastructure, the size/weight/cost of batteries, and hence improve the cost and range of electric vehicles. This solution has the advantage of creating a common road infrastructure shared by cars, buses and trucks and presenting a high energy efficiency. In this work, a $30kW$ DWICS for electric vehicles is presented with an efficiency up to 90%. It is developed the overall scheme, its mathematical model, control scheme, simulations, and experiments, showing this solution is viable for future electric roads.

1 Introduction

Nowadays, energy transition is a very important point to consider and discuss in different political, public and techniques-scientific forums. Industry and transport sectors are changing towards increasingly electrification, energy efficiency, safety, cleanliness and economic competitiveness.

More efficient and less polluting, electric vehicles become a popular alternative to fossil-based transport. However, their autonomy, range, price, and acceptability are the main stumbling block to their promotion.

The charging infrastructure is a weak point in most countries and is a major concern and problem in countries that do not have rules and standardization of charging stations yet [1, 2].

The electrification of heavy-duty vehicles is facing important battery problems, concerning size, weight, and cost. Recycling and raw materials consumption are yet others challenges for such vehicles. With the deployment of adequate charging infrastructure, the size of the battery can be reduced or even be modular.

A fast-charging station network outside the roads is the obvious solution to face these problems, however, static charging stations are not the ideal solution. They implicate in long recharging time, parking lot congestion issues, they do not obtain a large reduction of battery size, they need heavy charging cables, high power local substation, open the path for vandalism and safety, among other problems.

Another solution, named Dynamic Wireless Charging, is one of the three technologies considered in a study made for the French Ecology Transition Ministry regrouping many industry stakeholders, universities, and research institutes to define the French strategy for the Electric Road Systems (ERS). The final report presents and concludes that the wireless charging solution is the best solution in terms of interoperability, carbon consumption, and security, but is the less deployed for high power energy transfer. Other carbon-free technologies were considered, like hydrogen, but battery-based electrification has been the final choice.

This work presents a Dynamic Wireless Charging System (DWCS), that allows any kind of vehicle to be charged while driving on an adapted road. The presented system will be implemented in a real environment for an urban scenario for INCIT-EV project, in Paris in 2022. The total power system installed will be up to 120kW. Two light vehicles and a light-duty vehicle will be recharged up to 30kW and 90kW respectively in dynamic, semi-dynamic, and static situations. The technology of the ground system has no limitation on the size of vehicle or its speed. It can be further installed in heavy-duty size vehicles without any change on the ground infrastructure.

2 System description

A Dynamic Wireless Charging System is composed of a static primary subsystem under the ground and a mobile secondary subsystem inside the vehicle as described in figure 1.

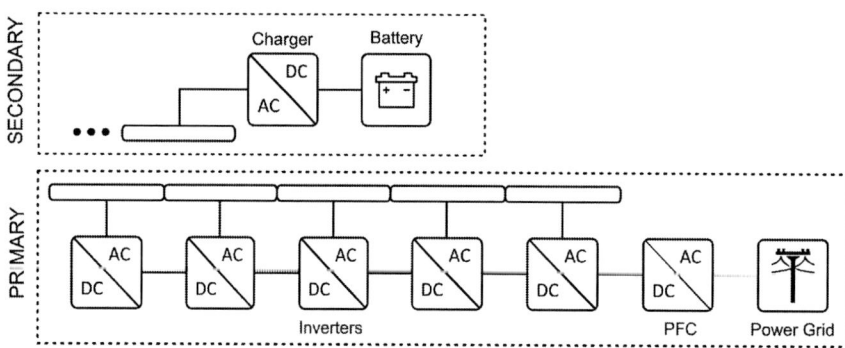

Fig. 1: Dynamic wireless charging system components.

The primary side is composed of side-by-side coils powered by DC/AC resonant converters working up to 100kHz. These converters are powered by a DC bus provided by an AC/DC grid converter with power factor correction.

The secondary system is composed of one or many coils connected to an AC/DC converter. This converter is a composition of an AC/DC rectifier and a DC/DC converter to adapt the current and voltage to the battery state of charge. The DC/DC is also responsible for the power control transfer defined by the vehicle.

The coils on the ground and inside the vehicles are the same for the presented system. The symmetry is an advantage in terms of system modeling and costs. Their dimensions were chosen to keep them completely covered by the vehicle during the charging section and avoid electromagnetic field emissions outside the vehicle limits. These coils will be completely integrated into the ground by an industrial road construction procedure for the Paris demonstration.

The inverters will be integrated into the pavement at the roadside composing a trench all along the charging lane, invisible to the public. Each coil has its own converter associated. This lane architecture was chosen for many advantages in an urban scenario despite the first impression that it can be expensive.

A first advantage is a low distance between the inverter and the coils, this is an important aspect to avoid electromagnetic emissions related to long cables in an urban scenario.

Another advantage is the performance, in this case, each inverter can optimize the power transfer for any relative position between the coils. In this technology, as it will be shown next, the nominal power is kept the same even when the alignment between coils is at 50% (between two ground coils). In addition, this lane architecture allows performing a multi-coil system where two or more coils are activated at the same time to transfer twice or more the rated power without any oversized electronics. For that, the vehicle needs to have a number of coils related to its power needs. Therefore, the same charging infrastructure can be used by any kind of vehicle. In the context of growing vehicle electrification, Wireless Charging has been investigated worldwide as a promising disruptive technology [9, 10, 11]. Indeed it stands as an attractive alternative with aesthetic, safety, convenience, modular, clean and fully automated potential advantages. Different international standardization bodies (SAE, ISO, IEC, GB) are currently addressing the recommended practices and requirements for further wireless charging technology deployments [12].

Extension from stationary to dynamic wireless charging is another challenge that started in 1976 at the Lawrence Berkley National Laboratory where the technical feasibility of a first dynamic system was evaluated. In 1992, Partners for Advanced Transit and Highway (PATH) demonstrated $60kW$ power transfer with 60% efficiency across a $7.6cm$ air gap [8]. Since then different technologies for dynamic wireless charging have been demonstrated, and are summarized in [7].

3 Power electronics topology and modulation

The topologies of the converters were chosen to be completely bidirectional and based on the same hardware converter and SiC switches (ROHM BSM180D12P2C101). The bidirectional functionality will not be tested and will be treated at another time. The system presented don't have the DC/DC converter on the secondary but it will be considered in future works (see figure 2).

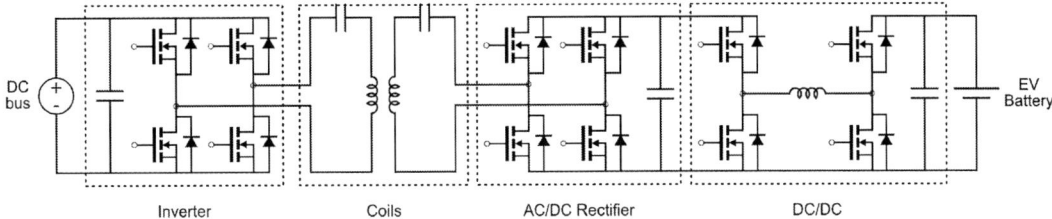

Fig. 2: Wireless power transfer system converters.

The inverter is a single-phase full bridge converter commanded by the classic bipolar modulation. Its efficiency can reach up to 99% in a standard charging section. The AC/DC rectifier is an active converter to reduce losses in the charging operation, reaching approximately the same efficiency level as the inverter.

The current-controlled DC/DC converter is bidirectional, such as to enable the charge of the intermediate capacitor between the rectifier and the DC/DC. The pre-charge of this capacitor is fundamental to avoid over-current at the system switch-on. In a charging stage, the DC/DC is commanded as a buck converter to adapt the nominal voltage (normally a little higher than the battery voltage), current, and desired power of the charging system.

4 System model

A series-series resonance circuit was chosen for this system. This topology allows us to reduce the current supported by the power converters despite the high voltage on the passive components. Figure 3 shows the simplified circuit considering only one primary coil and one secondary coil.

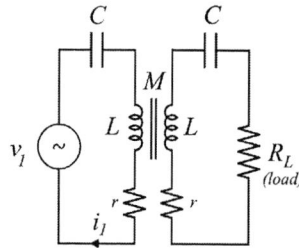

Fig. 3: Simplified two coil circuit.

The primary and secondary coils are considered equal in terms of inductance and capacitance. Therefore, from the basic circuit equations, the relation between the input voltage and primary current can be described as equation 1:

$$\frac{v_1}{i_1} = r + \frac{\omega^2 M^2 (r + R_L)}{(r + R_L)^2 + \left(\omega L - \frac{1}{\omega C}\right)^2} + j \left(\omega L - \frac{1}{\omega C} - \frac{\omega^2 M^2 \left(\omega L - \frac{1}{\omega C}\right)}{(r + R_L)^2 + \left(\omega L - \frac{1}{\omega C}\right)^2} \right) \tag{1}$$

Where:

v_1 is the primary output voltage from the inverter

i_1 is the primary output current from the inverter

ω is the frequency of the voltage source

M is mutual inductance between the primary and secondary coil

r is the self resistance of the coil

L is the self inductance of the coil

C is the resonance capacitor

R_L is the load resistance representing the secondary power demand

In the resonance condition, the imaginary part of this equation is zero, which means that the primary current and voltage are in phase. For this, the frequency to be applied to the voltage source is given by the equation 2:

$$\omega = \sqrt{\frac{-(C^2 R^2 - 2LC) + \sqrt{C^4 R^4 - 4C^3 R^2 L + 4k^2 L^2 C^2}}{2(L^2 C^2 - k^2 L^2 C^2)}} \tag{2}$$

Where $R = r + R_L$. From the differential equations of the circuit, the space state model is deduced for simulation in Equation (3).

$$\begin{bmatrix} L & M & 0 & 0 \\ M & L & 0 & 0 \\ 0 & 0 & 1 & 0 \\ 0 & 0 & 0 & 1 \end{bmatrix} \begin{bmatrix} \frac{d}{dt}i_1 \\ \frac{d}{dt}i_{21} \\ \frac{d}{dt}v_{C1} \\ \frac{d}{dt}v_{C2} \end{bmatrix} = \begin{bmatrix} -r & 0 & -1 & 0 \\ 0 & -r & 0 & -1 \\ \frac{1}{C} & 0 & 0 & 0 \\ 0 & \frac{1}{C} & 0 & 0 \end{bmatrix} \begin{bmatrix} i_1 \\ i_{21} \\ v_{C1} \\ v_{C2} \end{bmatrix} + \begin{bmatrix} 1 & 0 \\ 0 & -1 \\ 0 & 0 \\ 0 & 0 \end{bmatrix} \begin{bmatrix} v_1 \\ v_{rec} \end{bmatrix} \tag{3}$$

Table I shows the nominal parameters measured in the real system prototype and used for the simulations and tests.

Table I: System parameters

Nominal load power	30kW
Nominal voltage	435V
Nominal current	80A
Resonant capacitor	33nF
Coil inductance	$135\mu H$
Coil resistance	$25m\Omega$
Operating frequency	78kHz to 92kHz
Load resistance	5.3Ω
Ground clearance	17cm

5 Control and operating mode

The resonance condition is the main target of the control strategy on the primary side. The resonance ensures zero voltage switching, reducing power losses on the inverter. In addition, under this condition, the power transfer rate is the highest possible but it still depends on the converters architecture, transistors technology, modulation and mission profile.

In a dynamic inductive charging system, the relative position between the primary coil and the secondary is constantly changing while the vehicle is moving. It means that mutual inductance and even self-inductance are changing all the time and depends on the relative position between primary and secondary coil, the speed of the vehicle and the ground clearance. Figure 4 shows the typical variation of the mutual inductance between a primary and a secondary coil versus the relative position between them. The position $0mm$ is when the coils are 100% aligned. To compensate for this variation and keep the resonance condition in every relative positions, the frequency of the primary voltage needs to be regulated.

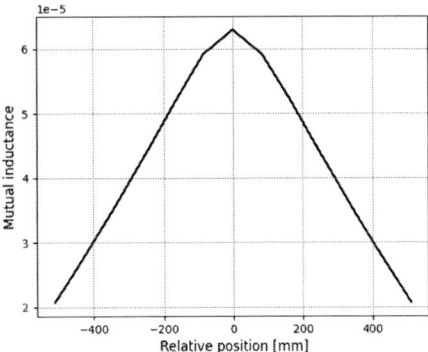

Fig. 4: Mutual inductance variation.

Different from other technologies, with the frequency regulation, the system is able to send at least the nominal power at all alignment conditions. The frequency is regulated based on the phase between the primary voltage and current. The phase error is the input for the controller that defines a reference frequency to the power inverter as described in Figure 5.

The maximum permissible latency of the phase angle detection is an important and limiting factor to define the maximum vehicle speed allowed, the system losses and the control system performance. The relation between the variation of the mutual inductance and its phase angle variation result is used as the key parameter. It means that, considering a hardware circuit phase detection, with a fixed and low phase latency measurement, the maximum latency will be limited by the allowed maximum losses on the converter related to the hard switching modulation condition. Other system specification will affect or be affected by the phase detection latency like: the pulse with modulation (PWM) resolution, the interruption frequency, the function algorithm and the maximum admissible phase error to the controller and the vehicle speed.

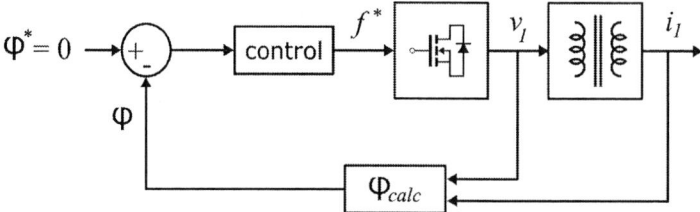

Fig. 5: Mutual inductance variation.

In the described system, only one coil on the ground is powered at each time. The coil with the best alignment is activated and it passes to the next one after a 50% misalignment. This detection is made by establishing a short-circuit on the next coil and measuring the induced current. After that, this induced current is compared with the current on the active coil to define when the system changes the state of the coils using a criteria of higher peak value.

6 Results

The validation of the system model was performed using PSIM software in a static simulation. The result is presented in Figure 6 with a $150V$ voltage on the primary side and a peak primary current at about $40A$ with a zero phase between then.

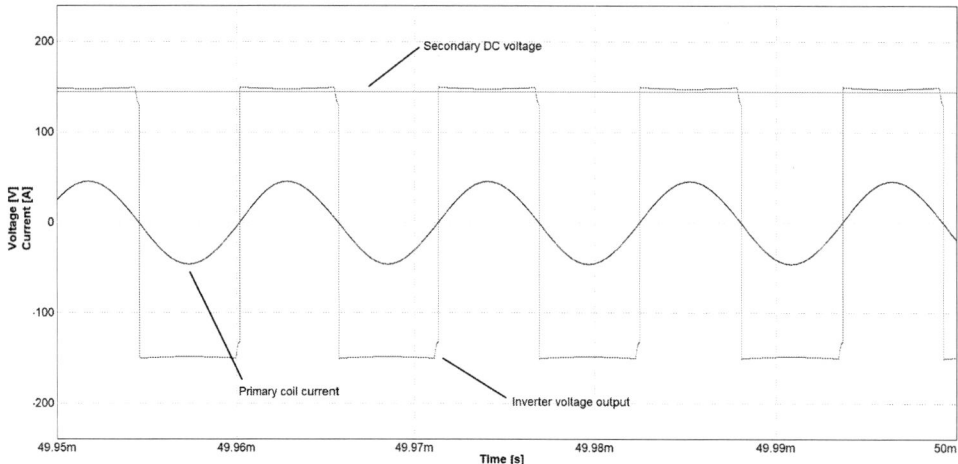

Fig. 6: Simulation result of the system model with real parameters.

This result is compared to measurements on the prototype presented in Figure 7 where the same voltage was applied and the same current is provided to the primary coil.

After the static tests made on a low voltage and power step, the dynamic behavior of the system was evaluated. The switching process, the resonant condition, and the total efficiency of the system were measured in simulation and real experiments. In Figure 8, simulation results are presented. At the beginning of the simulation, only the inverter N is active, sending power to the secondary coil. The inverter $N+1$ is short-circuited to measure the induced current. When the coils are 50% misaligned, the inverter N stops transmitting power, and inverter $N+1$ starts the energy transfer, which can be seen around $18ms$ of simulation time.

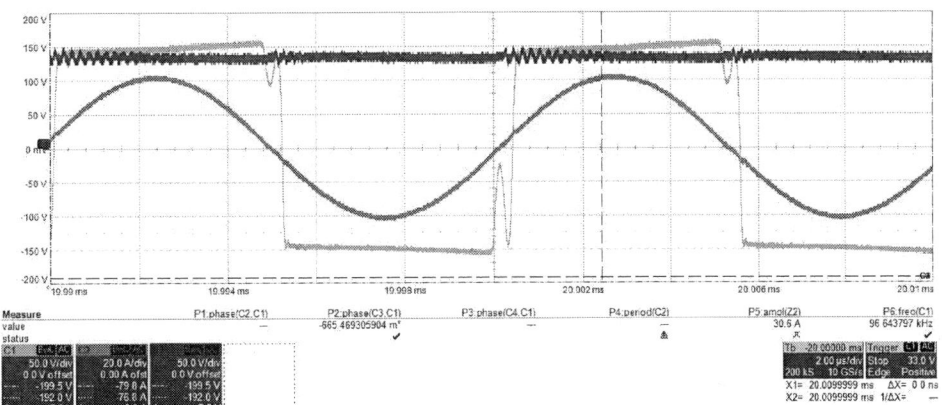

Fig. 7: Measurement of primary voltage (yellow), primary current (rose) and secondary DC voltage (blue).

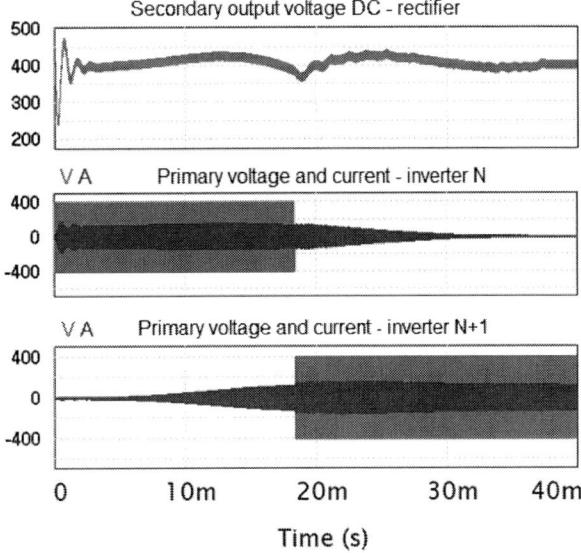

Fig. 8: Dynamic simulation results.

The experiments were performed using a wireless inductive charging systems testing bench composed of a robotic arm, power analyzers, electromagnetic field probes, and oscilloscopes, all synchronized by a supervision system (see Figure 9).

Fig. 9: Wireless inductive charging systems testing bench.

The primary coils are placed on the table forming the charging lane. While the secondary coil is attached to the robotic arm that is able to simulate the movement of a vehicle at a maximum speed of $13km/h$.

The result is presented in Figure 10 for a $30kW$ power target on the output of the system. The efficiency of the system varies between 82% and 90%, and a minimum of $30kW$ is assured at most part of the time. The instants where the $30kW$ target power is not respected, are related to a delay between the activation and deactivation of an inverter (active coil switching). The efficiency can be improved by using an active rectifier on the secondary system and improving shielding and coil design.

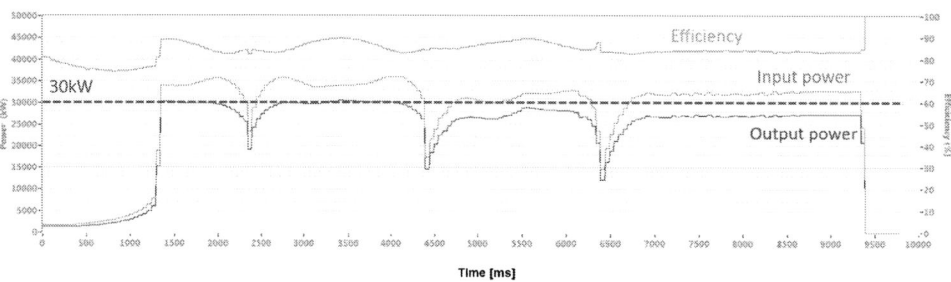

Fig. 10: Wireless inductive charging systems testing bench.

7 Conclusion

A Dynamic Wireless Charging System for electric vehicles was introduced in this paper. This system represents a good solution for reducing battery size, weight, and cost, while dramatically extending the range for vehicles, either cars or trucks (heavy-duty vehicles). This solution allows an effective energy transition from fossil fuels to electricity, even in long-range applications.

In the present work, it is proposed a simplified mathematical model, a control scheme for regulating the primary side frequency, simulations, and finally real experiments. A rather good correlation between simulation and measurement results is found for the control strategy.

This control scheme allows to keep the system in resonance, despite the movement of the vehicle, which create time-varying alignment and coupling, and as a consequence, time-varying parameters. In addition, it assures a power transfer of 30 kW with an transfer power rate up to 90% in the experiments.

Some improvements can still be made in the system to increase the performance, like using an active rectifier and optimizing the commuting process of an activated and deactivated inverter. Such steps will be carried out in future works, as well as new experiments in a real track.

References

[1] A. Ahmad, M. S. Alam, and R. Chabaan : A comprehensive review of wireless charging technologies for electric vehicles. IEEE Transactions on Transportation Electriffication, 4(1):38-63, March 2018.

[2] E. Aydin, Y. Kosesoy, E. Yildiriz, and M. Timur Aydemir: Comparison of hexagonal and square coils for use in wireless charging of electric vehicle battery. In 2018 International Symposium on Electronics and Telecommunications (ISETC), pp. 1-4, Nov 2018.

[3] J. Zhang, H. Zhou, H. Li, H. Liu, B. Li, C. Liu, and J. Yan: Multi-objective planning of charging stations considering vehicle arrival hot map. In 2017 IEEE Conference on Energy Internet and Energy System Integration (EI2), pages 1-6, Nov 2017.

[4] S. Jeong, Y. J. Jang, and D. Kum: Economic analysis of the dynamic charging electric vehicle. IEEE Transactions on Power Electronics, 30(11):6368-6377, Nov 2015.

[5] J. Shin, B. Song, S. Lee, S. Shin, Y. Kim, G. Jung, and S. Jeon: Contactless power transfer systems for on-line electric vehicle (olev). In 2012 IEEE International Electric Vehicle Conference, pages 1-4, March 2012.

[6] Antoine Caillierez: Etude et mise en oeuvre du transfert de l'énergie électrique par induction: application à la route électrique pour véhicules en mouvement: PhD thesis, 2016.

[7] D. Patil, M. K. McDonough, J. M. Miller, B. Fahimi, and P. T. Balsara, "Wireless power transfer, for vehicular applications: overview and challenges," IEEE Trans Transp Electrif, vol. 4, no. 1, pp. 3–37, doi: 10.1109/TTE.2017.2780627.

[8] "Roadway Powered Electric Vehicle Project Track Construction and Testing Program Phase 3D." [Online]. Available: https://escholarship.org/uc/item/1jr98590. [Accessed: 20-Aug-2020].

[9] J. M. Miller, O. C. Onar, and M. Chinthavali: "Primary-Side Power Flow Control of Wireless Power Transfer for Electric Vehicle Charging," IEEE J. Emerg. Sel. Top. Power Electron., vol. 3, no. 1, pp. 147–162, 2015, doi: 10.1109/jestpe.2014.2382569.

[10] S. Y. Choi et al. : "Advances in Wireless Power Transfer Systems for Roadway-Powered Electric Vehicles," vol. 3, no. 1, pp. 18–36, 2015, doi: 10.1109/JESTPE.2014.2343674.

[11] "Fabric project." : [Online]. Available: http://www.fabric-project.eu/www.fabric-project.eu/index.html,. [Accessed: 20-Aug-2020].

[12] Y. J. Jang: "Survey of the operation and system study on wireless charging electric vehicle systems," Transp. Res. Part C Emerg. Technol., vol. 95, pp. 844–866, 2018, doi: 10.1016/j.trc.2018.04.006.

Dynamic Control of the Switching Behavior of SiC MOSFETs in Converter Operation

Jochen Henn, Laurids Schmitz and Rik W. De Doncker
RWTH Aachen University, Institute for Power Electronics and Electrical Drives (ISEA)
Jaegerstrasse 17-19
Aachen, Germany
Phone: +49 241 80 96920
Fax: +49 241 80 92203
Email: post@isea.rwth-aachen.de
URL: https://www.isea.rwth-aachen.de

Acknowledgments

The project on which this report is based was funded by the Federal Ministry of Education and Research under the funding code 16EMO0375. The author is responsible for the content of this publication.

Keywords

≪Intelligent gate driver≫, ≪EMI/EMC≫, ≪Smart Gate Drivers≫, ≪Wide Bandgap≫.

Abstract

This paper presents a control approach to regulate voltage slope and oscillation amplitude in a silicon carbide inverter during operation. Based on sensor measurements, the control adapts the switching characteristics using an adaptive gate driver and thus adjusts the level of voltage slopes, oscillation amplitude and consequently, electromagnetic emissions to a target level.

1 Introduction

In recent years the advantages of wide bandgap (WBG) semiconductor devices became accessible and led to their widespread adaption [1]. Future power electronic converters using WBG semiconductors can use advanced gate driving technologies to mitigate the drawbacks of faster switching devices such as electromagnetic interference (EMI) [2].

Adaptive gate drivers are the basis for dynamic control of the switching behavior. All of the following examples either use a fixed voltage source and adapt the gate resistance or apply a controlled gate current. An array of segmented current sources is fabricated into an application-specific integrated circuit (ASIC) to form an adaptive gate driver in [3]. Instead of current sources, an array of transistors with different $R_{DS,on}$ can be used as presented in [4] and [5]. For this paper a discrete adaptive current driver as introduced in [6] is used. The results in the literature [7–9] regarding the optimization of the next switching event promise an enhanced operating behavior under the usage of adaptive gate drivers, because they extend the safe operating area (SOA) at equal or less losses.

When adaptive gate drivers are equipped with feedback circuits, they can adapt their gate driving characteristics in order to optimize the operating behavior. Closing the feedback loop while the switching event is still in progress is a challenge for WBG converters and has so far only been achieved for IG-BTs [10–12]. In addition to the relatively low bandwidth feedback needed for IGBT based converters, research has shown highly dynamic feedback loops based on current mirrors [13], RC based voltage sensors [14] and load current based adaption [3].

Similar to the goal of this paper, the following publications use feedback loops and adaptive gate drivers to optimize the operating behavior in regard to switching characteristics. In [15] the authors use a dedicated ASIC to drive gallium nitride (GaN) devices at 650 V and adapt a single parameter in between switching events. With a focus on crosstalk suppression, the authors of [9] and [16] use a direct feedback of the active switching event on the gate voltage. These approaches limit the overall flexibility of the gate driver but could be combined with an adaptive driver. Finally, [5] and [17] optimize the switching behavior during converter operation but either lack interpretation of the overshoot amplitude or use external equipment to measure the switching characteristics.

In contrast to the aforementioned work, this paper presents an enhanced PI based control algorithm which allows a dynamic control of different switching characteristics by using discrete on-board sensors and an adaptive gate driver. Thus, first, the setup of the dynamic control loop is presented by giving a description of the utilized sensors and summarizing the applied algorithm. After this the hardware used to validate the dynamic control is introduced. The validation of the algorithm is conducted based on the measurement results presented in the second to last section. Finally, the paper is concluded by summarizing the contributions to dynamic control algorithms.

2 Dynamic Control Loop

To adjust the switching characteristics and therefore the electromagnetic emissions (EME) to a predefined level, the feedback loop depicted in Fig. 1 is introduced.

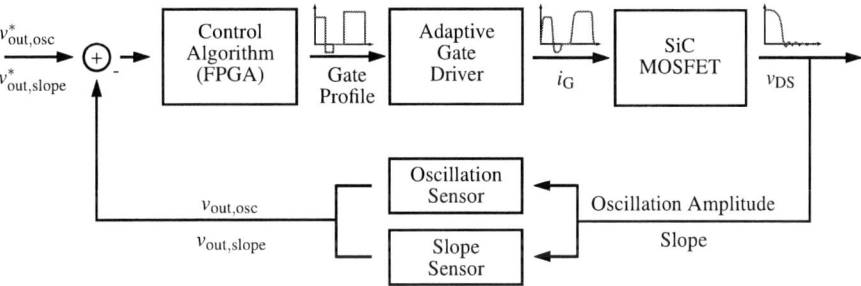

Fig. 1: Structure of the proposed dynamic feedback loop.

It consists of two sensors which measure the slope and the oscillation amplitude of each v_{DS} transition following a switching event. Additionally, the control algorithm runs on the field-programmable gate array (FPGA), which determines the adaptions to the gate profiles based on the sensor measurements. Finally, this results in a change of the switching characteristics of the silicon carbide (SiC) MOSFETs which in turn will be measured by the sensors and used as feedback for the next switching cycle.

The v_{DS} measurement after the activation of the control algorithm, depicted in Fig. 2, illustrates the effect of the control presented in this paper. The algorithm is activated after the fourth falling edge at $t = 150\,\mu s$. Consequently, the negative overshoot drops from 121 V to 51 V and the voltage slope decreases from 33.7 V/ns to 28 V/ns according to the preset slope target value. In Fig. 2 the reduction of the negative overshoot is a result of the adapted voltage slope. The effect of the control on the voltage slope is visible in Fig. 6. As a result, the electromagnetic emissions are adapted as well.

Evaluation and Control Algorithm

After each switching event, an adaption of the gate profile is determined by the control algorithm. The adaption of gate profiles is possible in several ways. A gate profile itself is divided into n different states. They are comprised of a reference gate current amplitude a_i and a state duration d_i, where i marks the position in the gate profile, which is depicted in Fig. 3a. The reference gate current amplitude can be positive or negative, depending on the sign of the targeted current value. An amplitude between -31 and +31, which is set by two 5 bit R-2R-ladders within the adaptive gate driver, is related to the desired

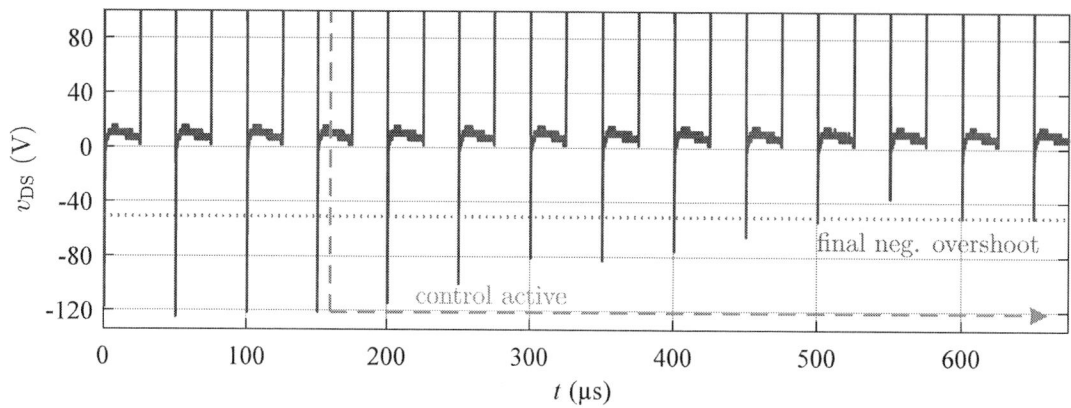

Fig. 2: Negative overshoot of v_{DS} for the first iterations of the control.

maximum turn-off and turn-on gate currents. The resolution of the duration is limited by the length of an FPGA clock cycle of 2.5 ns.

In the standard gate profile in Fig. 3a, the first state with a positive amplitude increases v_{GS} above the threshold voltage. The short second state with a negative amplitude and the longer third state with an amplitude of 0 A are used to maintain a constant level of v_{GS} after the initial charging. Finally, the fourth state is used to guarantee the increase of v_{GS} to its maximum of 15 V at the end of the transition to reduce conduction losses.

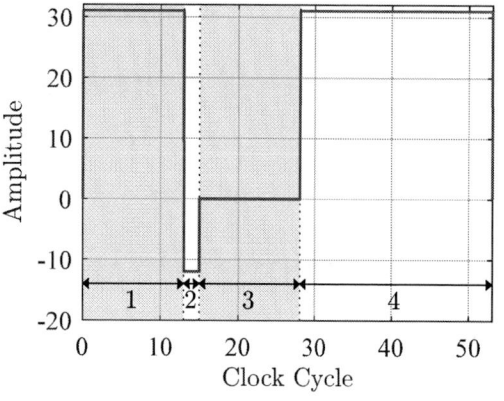 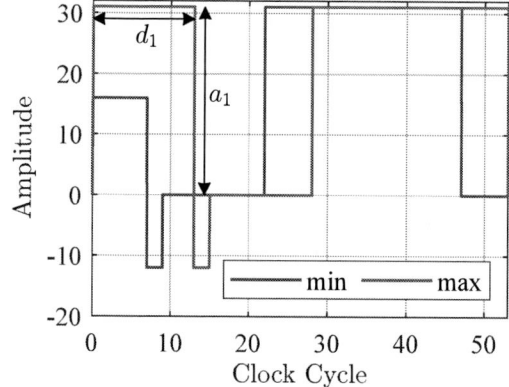

(a) Standard turn-on reference current profile used as the default profile when the control is not active.

(b) Minimum and maximum allowed reference current profiles that still ensure proper switching behavior.

Fig. 3: Profile set used for the measurements presented.

Within the set of gate profiles shown in Fig. 3, several parameters are controlled consecutively. For the control presented in this paper, the duration and the amplitude of state 1, d_1 and a_1, are selected as control parameters. These two parameters allow for fast adaptions of slope and oscillation amplitude of v_{DS}. Adaptions to d_1 lead to quick and coarse changes, while changes of a_1 allow finer adjustments and are used as the secondary parameter. All parameters are limited to a defined range, which is visualized by the minimum and maximum profile in Fig. 3b. The reason for these limitations is that parameter values outside of a certain range may disturb the v_{DS} transition too much or stop it entirely. Furthermore, the synergy between the two consecutively controlled parameters can only be guaranteed within a certain range. The standard reference profile depicted in Fig. 3a is used as the starting point of the algorithm.

The control algorithm on the FPGA compares the sensor analog digital converter (ADC) readout with a provided target value that represents the desired emission level after each v_{DS} transition. The resulting error is processed by a PI element with the resulting PI output value being the main input for the control algorithm. It is compared to thresholds as visualized in Fig. 4 and as a result, the direction as well as the step size of the adaption are determined.

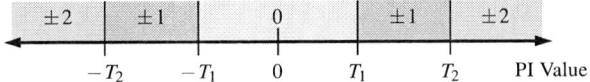

Fig. 4: Control threshold levels used to determine the adaption step size and sign.

If the absolute PI output value is higher than T_1, the currently controlled parameter is adapted by a step of 1 while it is adapted by 2 if the absolute value is higher than the second threshold level, T_2. The sign of the adaption step depends on the sign of the PI output value and the configuration for the control parameter.

A parameter is adjusted until one of the following two conditions is met: a) the valid range defined by minimum and maximum profile would be violated by the calculated adaption step, b) the previous adaption step led to an increased error, indicating the local optimum for this parameter is reached. In those cases, the control algorithm continues with adaptions to the following parameter.

Using the proportional and integral gain of the PI element, the characteristics of the control can be adapted to various requirements allowing for a trade-off between speed and accuracy. If there is a need for the adaption of additional control parameters or control thresholds and thus higher adaption step sizes, they can be added to the control algorithm.

Sensor Implementation

Two sensors are designed to measure the slope of the voltage transitions and as the magnitude of the oscillations on v_{DS} to be able to develop an algorithm that can control and limit both emission causes. Both sensors comprise a probe, a filter and a sample & hold element (S&H element) as introduced in [14]. The probe derives a high pass filtered version of v_{DS}, the subsequent filter selects the frequencies needed for the determination of slope and oscillation amplitude, respectively, while the concluding S&H element allows for analog-to-digital conversions by the FPGA. For the slope sensor, a low pass filter with a bandwidth of 65 MHz is selected while the cut-off frequencies of the oscillations sensor's band pass filter are 60 MHz and 100 MHz. These filter bands are designed to offer optimal separation for the used power module at a dc-link voltage of 800 V while allowing detection of all expected slope and oscillation amplitude levels. Nevertheless, due to the proximity of slope- and oscillation-induced spectral components in the frequency domain, a full separation cannot be accomplished. For a dc-link voltage of 400 V used for the measurement presented in this paper, the separation between slope and oscillation amplitude is impaired due to a decrease in the resonant frequency of the switching cell. This results in an increased cross correlation between the two measurement values. Therefore, both sensor measurements can be used interchangeably in this case.

3 Hardware Setup

The hardware used to develop and evaluate the presented control algorithm is depicted in Fig. 5. In the upper right corner in Fig. 5, which depicts the top side of the driver module, the discrete slope sensor elements are located. The oscillation sensor is placed on the bottom side of the gate driver module. Next to the sensor, the current gate driver as introduced in [6] is positioned. The turn-on part of the gate driver is on the top side, while the turn-off part is on the bottom side. The FPGA that evaluates the sensor readouts and controls the adaptive gate driver is situated in the center. As basis for the control algorithm the ADCs integrated into the FPGA are used. They can be sampled at a resolution of 12 bit and a sample rate of 1 MHz. With a base clock frequency of 100 MHz, the FPGA can complete the feedback evaluation and adaption of the profiles in roughly 500 ns. Thus, the evaluation and control algorithm needs at least 1.5 μs between two switching events.

Six of these driver modules are connected between a low-voltage controller Board and a high-voltage

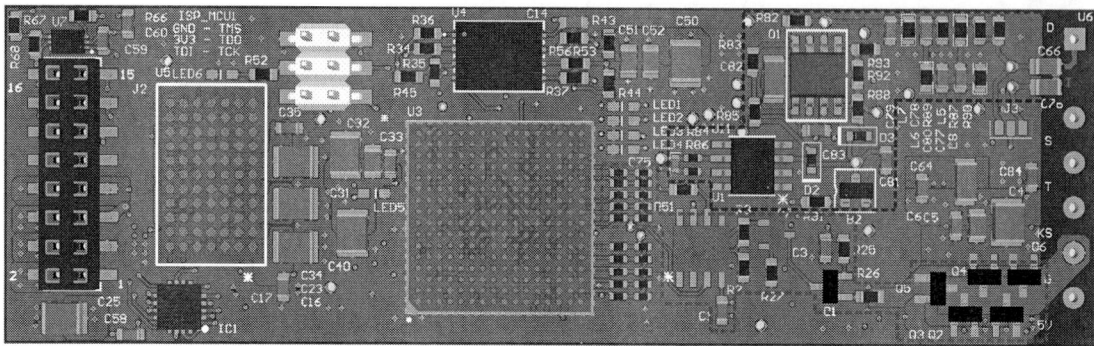

Fig. 5: Top view of a driver module printed circuit board (PCB). It contains an FPGA (marked with light blue), an adaptive current gate driver (marked with purple) and the slope sensor above the driver (blue).

PCB containing the switching cell and the dc-link. Together they operate as a three phase inverter using the 21 mΩ variant of the 1200 V SiC MOSFETs manufactured by Wolfspeed (C3M0021120K).

4 Measurement Results

Measurements are taken at 400 V without any load connected to the inverter. In this synthetic operating point, the switching slopes are fast and consistent over time, which allows better analysis of the effects of the control algorithm on the switching behavior. The goal for the control algorithm in this scenario was to reach and hold a target value of 90 for the slope sensor's feedback. The 8 MSB of the sensor ADC readouts are evaluated resulting in a range from 0 to 255.

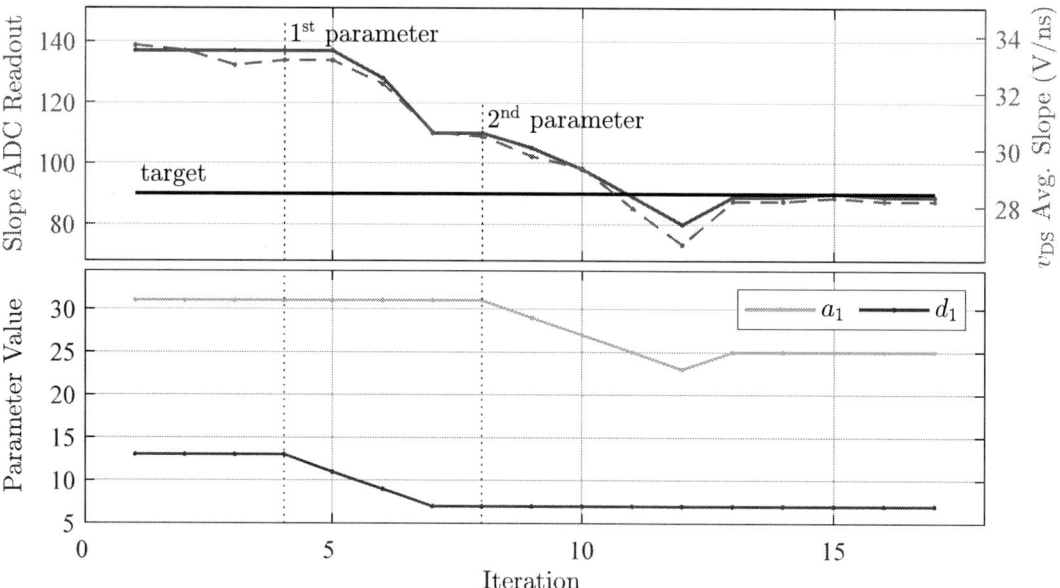

Fig. 6: Control sequence to reduce the voltage slope using two parameters.

Figure 6 visualizes the slope sensor's ADC readout and the change in the control parameter values as a reaction of the control algorithm. Additionally, the average slope values as they can be calculated from an oscilloscope measurement are plotted into the top graph. They show a very similar behavior and were exclusively used to validate the sensors. The control algorithm relies only on the sensor values gathered on the PCB itself. The control is initiated at iteration 4 and the first parameter is decreased because the

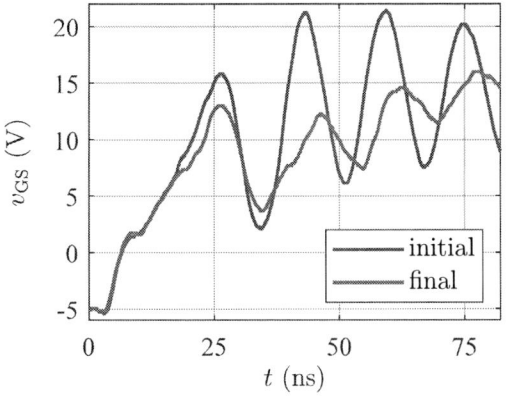
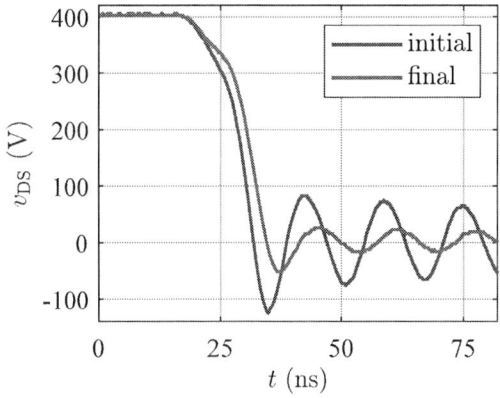

(a) v_{GS} for initial and final iteration.

(b) v_{DS} for initial and final iteration.

Fig. 7: Comparison of v_{GS} and v_{DS} waveforms of initial and final control iteration.

target value is lower than the input from the ADC. After three iterations the target value is not reached, but the duration parameter hit its lower limit and the secondary parameter is chosen to be adapted next. At iteration 12 the control causes a negative overshoot below the target value, which is detected and reversed in the next iteration. One way of reducing this negative overshoot in the control would be to decrease the integral gain at the cost of a slower approach to the target value.

The effect of using coarse changes on the primary parameter is visible in Fig. 6. According to the internal thresholds and Fig. 1 parameter d_1 is decreased by two counts at iterations five and six, because the sensor feedback is still far from the target. For these two steps the error is reduced by 57 %. Decreasing a_1 as the second parameter results in finer adjustments, because the amplitude's influence on the switching event is not as severe. The three adaptions at iteration eight to ten reduce the error by the remaining 43 % relative to the initial error.

Figure 7 depicts the initial and final voltage waveforms for v_{GS} and v_{DS}. The v_{DS} measurements for the initial and final iteration are also visible in Fig. 2 at $t = 150\,\mu s$ and $t = 650\,\mu s$. At $t = 20\,ns$ in Fig. 7a the difference in the gate profiles is visible. After reaching the threshold voltage the gate charges slower, which results in a decreased slope and overshoot on v_{DS}.

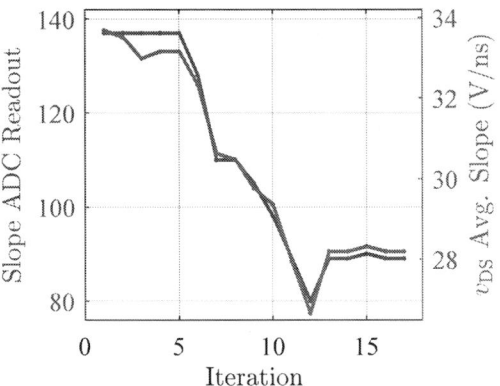
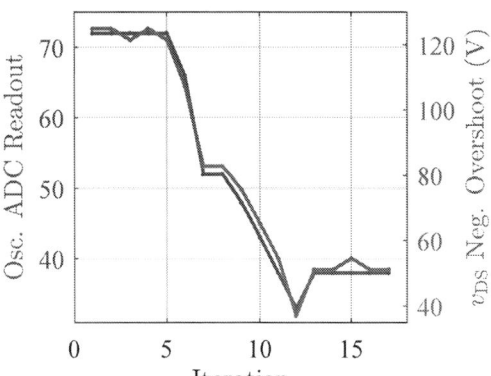

(a) Slope ADC readouts in comparison to average 90 % to 10 % slope during control.

(b) Oscillation ADC readouts in comparison to maximum negative overshoot during control.

Fig. 8: ADC readouts in direct comparison with oscilloscope measurements.

The direct comparison between the sensor readings and the data gathered from oscilloscope measurements in Fig. 8 indicates very similar behavior for both sensors but also shows differences, which will be addressed here. The biggest difference between the sensors readings and measurements is the relative

Table I: Comparison of relative changes in slope and overshoot characteristics between sensor readings and oscilloscope measurements.

	Slope	Overshoot Amplitude
Change in Oscilloscope Measurement	17 %	57 %
Change in Sensor Reading	35 %	47 %

scaling of the results. Table I contains the relative changes between the initial and final switching characteristics. Although there is a substantial relative difference between the slope sensor and the external measurement, they show a linear dependency as shown by the matching waveforms in Fig. 8a. Additionally, the oscilloscope measurements of both characteristics, the slope and the overshoot amplitude, show more noise during the first four switching iterations. This is caused by resolution limitations in the oscilloscope and the influence of disturbances onto the measurement probe. As mentioned above, during these first iterations the control is inactive and the gate driver applies the standard gate profile. The constant sensor readout during these switching instances underlines the advantage of placing the sensors and their ADCs on the same reference potential and close to the semiconductor devices. Furthermore, the control algorithm benefits from the robust and clean input.

5 Conclusion

In this paper, a feedback loop that allows the control of switching characteristics in a SiC inverter has been successfully developed. The control is based on two sensors, a slope and an oscillation sensor, that allow the independent evaluation of slope and oscillation amplitude of the drain-source voltage in an inverter. In order to adjust the level of the electromagnetic emissions, the control algorithm that runs on an FPGA is able to incrementally adapt different parameters of the gate profile of the SiC-MOSFETs based on the feedback of the sensors. An enhanced PI-based control algorithm allows closing the feedback loop and controlling the emission level during the operation of the inverter solely based on the feedback from the onboard sensors. Hence, it can operate autonomously and does not need additional information. Moreover, this approach offers the advantage that it can counteract external error sources such as temperature changes and aging effects, which cannot be covered by a feed-forward control. Measurement results at 400 V validate the algorithm's ability to iteratively adjust parameters within the gate profiles to decrease both causes of emissions, slope and oscillation amplitude, from switching event to switching event.

References

[1] A. Elasser and T. Chow, "Silicon carbide benefits and advantages for power electronics circuits and systems," *Proceedings of the IEEE*, vol. 90, pp. 969–986, 2002.

[2] J. Henn, C. Lüdecke, M. Laumen, S. Beushausen, S. Kalker, C. H. van der Broeck, G. Engelmann, and R. W. de Doncker, "Intelligent gate drivers for future power converters," *IEEE Transactions on Power Electronics*, vol. 37, no. 3, pp. 3484–3503, Mar. 2022.

[3] S. Kawai, T. Ueno, H. Ishihara, S. Takaya, K. Miyazaki, and K. Onizuka, "A 1ns-resolution load adaptive digital gate driver ic with integrated 500ksps adc for drive pattern selection and functional safety targeting dependable sic application." Vancouver, BC, Canada: IEEE, 2021, pp. 5417–5421.

[4] Y. S. Cheng, D. Yamaguchi, T. Mannen, K. Wada, T. Sai, K. Miyazaki, M. Takamiya, and T. Sakurai, "Digital active gate drive with optimal switching patterns to adapt to sinusoidal output current in a full bridge inverter circuit," in *Proc. IECON 2019 - 45th Annual Conf. of the IEEE Industrial Electronics Society*, vol. 1, Oct. 2019, pp. 1684–1689.

[5] D. Yamaguchi, Y. S. Cheng, T. Mannen, H. Obara, K. Wada, T. Sai, M. Takamiya, and T. Sakurai, "An optimization method of a digital active gate driver under continuous switching operation being

capable of suppressing surge voltage and power loss in pwm inverters," *IEEE Transactions on Industry Applications*, vol. PP, pp. 1–1, 2021.

[6] J. Henn, L. Heine, and R. W. De Doncker, "A high bandwidth active sic gate driver for dynamic adjustment of electromagnetic emissions in electric vehicles," in *PCIM Europe digital days 2020; International Exhibition and Conference for Power Electronics, Intelligent Motion, Renewable Energy and Energy Management*. VDE, 2020, pp. 1–7.

[7] Z. Li, R. W. Maier, and M.-M. Bakran, "Mitigating drain source voltage oscillation with low switching losses for SiC power MOSFETs using FPGA-controlled active gate driver," in *2020 22nd European Conference on Power Electronics and Applications (EPE'20 ECCE Europe)*. IEEE, sep 2020.

[8] E. Raviola and F. Fiori, "Experimental investigations on the tuning of active gate drivers under load current variations," in *2021 International Conference on Applied Electronics (AE)*, 2021, pp. 1–4.

[9] T. Shao, T. Q. Zheng, H. Li, J. Liu, Z. Li, B. Huang, and Z. Qiu, "The active gate drive based on negative feedback mechanism for fast switching and crosstalk suppression of sic devices," *IEEE Transactions on Power Electronics*, vol. PP, pp. 1–1, 2021.

[10] Y. Lobsiger and J. W. Kolar, "Closed-loop di/dt and dv/dt igbt gate driver," *IEEE Transactions on Power Electronics*, vol. 30, no. 6, pp. 3402–3417, 2014.

[11] L. Chen and F. Z. Peng, "Closed-loop gate drive for high power igbts," in *2009 Twenty-Fourth Annual IEEE Applied Power Electronics Conference and Exposition*. IEEE, 2009, pp. 1331–1337.

[12] S. Beushausen, F. Herzog, and R. W. De Doncker, "Gan-based active gate-drive unit with closed-loop du/dt-control for igbts in medium-voltage applications," in *PCIM Europe digital days 2020; International Exhibition and Conference for Power Electronics, Intelligent Motion, Renewable Energy and Energy Management*, July 2020, pp. 1–8.

[13] P. Bau, M. Cousineaul, B. Cougo, F. Richardeau, and N. Rouger, "Modeling and design of high bandwidth feedback loop for dv/dt control in cmos agd for gan," in *2020 32nd International Symposium on Power Semiconductor Devices and ICs (ISPSD)*, Sep. 2020, pp. 106–109.

[14] J. Henn, C. Fronczek, and R. W. De Doncker, "Design of sensors for real-time active electromagnetic-emission control in sic traction inverters," in *International Exhibition and Conference for Power Electronics, Intelligent Motion, Renewable Energy and Energy Management (PCIM Europe)*. VDE, 5 2021.

[15] P. Bau, M. Cousineau, B. Cougo, F. Richardeau, and N. Rouger, "Cmos active gate driver for closed-loop dv/dt control of gan transistors," *IEEE Transactions on Power Electronics*, vol. 35, no. 12, pp. 13 322–13 332, Dec 2020.

[16] H. Li, Z. Qiu, T. Shao, Y. Zeng, H. Du, and C. Yin, "A low level-clamped active gate driver for crosstalk suppression of sic mosfet based on dv/dt detection." Vancouver, BC, Canada: IEEE, 2021, pp. 5348–5353.

[17] Z. Wu, H. Jiang, Z. Zheng, X. Qi, H. Mao, L. Liu, and L. Ran, "Dynamic dv/dt control strategy of sic mosfet for switching loss reduction in the operational power range," *IEEE Transactions on Power Electronics*, vol. PP, pp. 1–1, 2021.

A Series Resonant Balancing Converter for Bipolar DC Grids on Ships

Sachin Yadav, Zian Qin, Pavol Bauer
Delft University of Technology
Mekelweg 4
Delft, The Netherlands
Phone: +31 (0) 15 27 84399
Email: s.yadav-1@tudelft.nl, z.qin-2@tudelft.nl, p.bauer@tudelft.nl
URL: https://www.tudelft.nl/

Acknowledgments

This work was supported by Nederlandse Organisatie voor Wetenschappelijk Onderzoek (NWO), grant 17628.

Keywords

≪Resonant converter≫, ≪DC-DC≫, ≪Bi-directional converters≫, ≪Bipolar DC≫.

Abstract

Balancing converters are an integral part of a bipolar dc grid. In this paper, we propose a balancing converter based on a series resonant converter topology. The converter operates in the capacitive region with phase shift between the upper and lower H-bridges. The converter operation is analyzed and verified with LTSpice simulation. A prototype is developed to verify the operation.

Introduction

Climate change is one of the main issues that concerns the existence of human beings across the globe. Greenhouse gas (GHG) emissions are considered the main culprit for this. The ship industry emits around 3% of all the GHG in the world [1]. To reduce the emissions, the International Maritime Organisation (IMO) has come out with regulations mandating ship manufacturers to increase the efficiency of future ships [2]. One of the ways to improve the efficiency of ships is electrification of the power train. DC grids on ships are the major contender for this [3]. These grids have several advantages such as enhanced dynamic performance, peak shaving, zero emission operation and strategic loading [3].

The dc grids can be unipolar or bipolar. Bipolar dc grids have several advantages when compared with unipolar dc grids [4] [5]. They have improved reliability because of multiple conductors/poles; if one pole is faulty then the other can be used to supply half of the power. They have increased flexibility because of more than one voltage level. Hence, bipolar dc grids are being considered for future dc grids on ships.

Balancing converters are an integral part of bipolar dc grids. Many topologies can be found in literature for these converters [6] [7] [8] [9] [10] [11]. One of the ways to form a bipolar dc grid on ships is using a three level converter to interface the synchronous generator. The balancing function can be achieved with the 3-phase multilevel converter topologies like Neutral Point Clamped and T-Type converter [6] [7]. These converters can be controlled in such a manner that voltage at the dc side remains balanced and no current flows in the neutral [12]. But, an unbalanced dc operation leads to increased total harmonic distortion on the ac side thus leading to higher losses in the machine interfaced with the converter [13]. Also, the size of the dc-link capacitors are needed to be increased to attenuate the effect of the unbalances [14]. It is also difficult to have a distributed architecture when these converters are used.

Other topologies of balancing converters include buck-boost, Cuk, and SEPIC converters [5] [10] [11]. These topologies are called inverting topologies which can be directly utilized as balancing converters in a bipolar dc grid [8]. Another interesting topology is a series-resonant converter topology. This converter is a non-inverting type topology. Hence, it can not be used as a balancing converter without some modifications. In this work, we designed a series-resonant balancing converter to show its utility as a balancing converter in bipolar dc grids.

The structure of this paper is as follows. The need for the balancing converter is detailed in the next section. Thereafter, the operating principle for the series resonant balancing converter is discussed. In the next section, the converter prototype along with the specifications in brief are shown. Subsequently, the results of the simulation and laboratory setup for the operation of the balancing converter are shown and discussed. Finally, the paper is concluded in the last section.

Need for balancing converter

Balancing converters are essential for bipolar dc grids when unequal loads are connected between +n and n- poles. Unequal power loading has an effect of shifting of voltage at the neutral of the system. In a bipolar grid with three conductor, there are multiple ways that the loads can be connected to the distribution lines. The loads can either be connected between the positive (+) and negative (-) poles or between a pole (+ or -) and neutral. In case of an isolated system (IT system), if there are any unbalanced connection of loads present in the system then the neutral voltage can shift from zero. It should be noted that there is no balancing converter present in the system. Also, the droop controlled converter only controls voltage between the two poles and not of the neutral. A 3 node 2 line 3 conductor system was simulated with unbalanced loads connected to grid using the method in [15] and [16]. The test case is given in Table I and the resulting node voltages are given in Fig. 1.

Table I: Case for simulating unbalanced power flow in a three node 2 line bipolar dc grid.

Time (ms)	n_{2+-}	n_{2+n}	n_{2n-}	n_{3+-}	n_{3+n}	n_{3n-}
0	0	0	0	0	0	0
10	-25kW	0	0	0	-2.5kW	0
20	-25kW	0	-1kW	-23.5kW	-2.5kW	0
30	-25kW	0	0	0	0	0
40	0	0	0	0	0	0

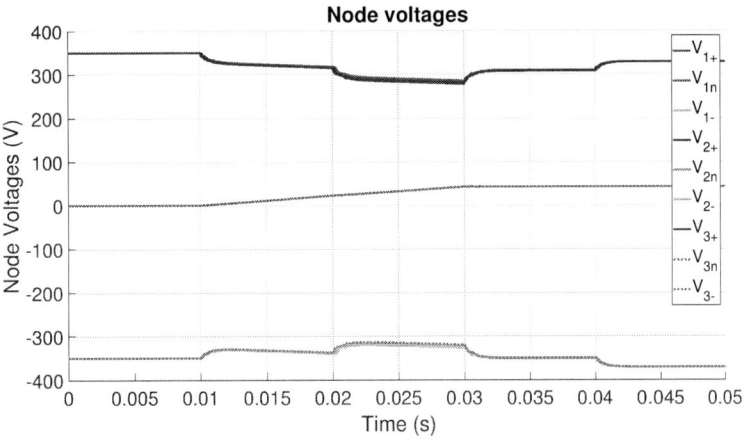

Fig. 1: Node voltages of a 3 node 2 line system with the effect of unbalancing on the neutral voltage.

It can be observed in Fig. 1, when the loads between the neutral and the poles is turned on then the voltage at the neutral shifts from zero. The shift in the neutral voltage can lead to various issues in the system. If the neutral voltage shift a lot then the drives can suddenly shut down in under-voltage or over-current fault. On a ship, this can lead to perilous condition for the people and equipment. Apart from that, the system efficiency decreases when current flows in the neutral conductor during the unbalance [4]. Hence, balancing converters are needed to balance the power flowing through both the poles.

Converter operating principle

The converter is based on a series-resonant converter topology as shown in Fig. 2. This topology is conventionally used as non-inverting. However, in our application, this topology is used as an inverting topology.

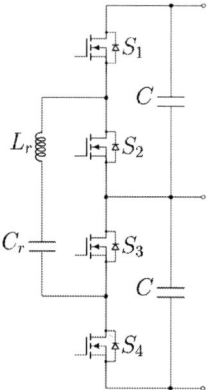

Fig. 2: Balancing converter based on a series resonant converter topology.

The operating principle of the converter is given below. The dead time between the switching has been neglected for this analysis.

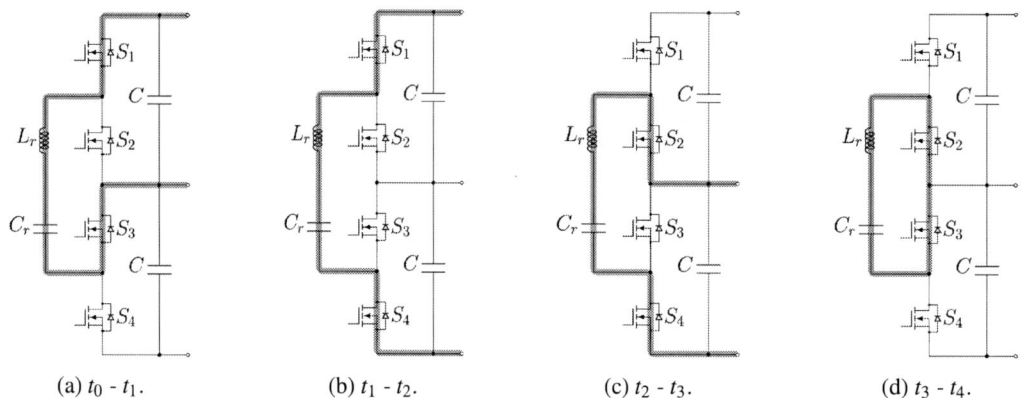

(a) t_0 - t_1.   (b) t_1 - t_2.   (c) t_2 - t_3.   (d) t_3 - t_4.

Fig. 3: Operation of the resonant converter in different time periods.

- **t_0 - t_1:** In this state, switches S1 and S3 are conducting. The resonant capacitor C_r is charging and current is limited by the resonant inductor L_r.
- **t_1 - t_2:** In this state, switch S3 is turned off and switches S1 and S4 are conducting. Hence, there is a higher voltage present across the resonant tank. Due to this stage, the charge is pumped into the resonant capacitor. This extra charge stored in the capacitor is used to boost the voltage of the capacitor.
- **t_2 - t_3:** In this state, switch S1 is turned off and S2 is turned on. The resonant tank is connected to the neutral and negative pole through switches S2 and S4. The inductor current reverses direction and the capacitor is also discharged as the energy is transferred to the load.

- t_3 - t_4: In this state, switch S4 is turned off and S3 is turned on. Hence, the resonant tank current freewheels through the switch diodes.

It should be noted that the stages t_1 - t_2 and t_3 - t_4 can be generated by a phase shift between the upper half bridge (consisting switches S1 & S2) and lower half bridge (consisting switches S3 & S4).

The power flow through the converter can be found out by solving the second order differential equations for the resonant tank. The power flow as a function of the operating frequency is given by (1). All the quantities except switching frequency are kept fixed.

$$P(f_{sw}) = 4C_r f_{sw} V_d \left[\frac{V_d}{2} - \frac{\sqrt{4I_{l1}^2 Z_0^2 sec^2(\frac{\omega_0}{4f_{sw}}) + V_d^2}}{2} - I_{l1} Z_0 tan\left(\frac{\omega_0}{4f_{sw}} \right) \right] \quad (W) \tag{1}$$

Also, the phase shift needed to achieve the change in current during the interval t_1-t_2 is given by (2):

$$\phi(f_{sw}) = \frac{2\pi f_{sw}}{\omega_0} \left[\pi - 2tan^{-1}\left(\frac{V_d + \sqrt{4I_{l1}^2 Z_0^2 sec^2(\frac{\omega_0}{4f_{sw}}) + V_d^2} + 2I_{l1} Z_0 tan(\frac{\omega_0}{4f_{sw}})}{2I_{l1} Z_0} \right) \right] \quad (rad) \tag{2}$$

where, V_d is the input voltage between one pole and neutral, I_{l1} is the current through the inductor at instant t_2, Z_0 is the characteristic impedance of the resonant tank, and ω_0 is the resonant angular frequency of the tank. The power flow when the current I_{l1} is 3.5A is shown in Fig. 4. It should be noted that the phase shift also changes with frequency according to (2).

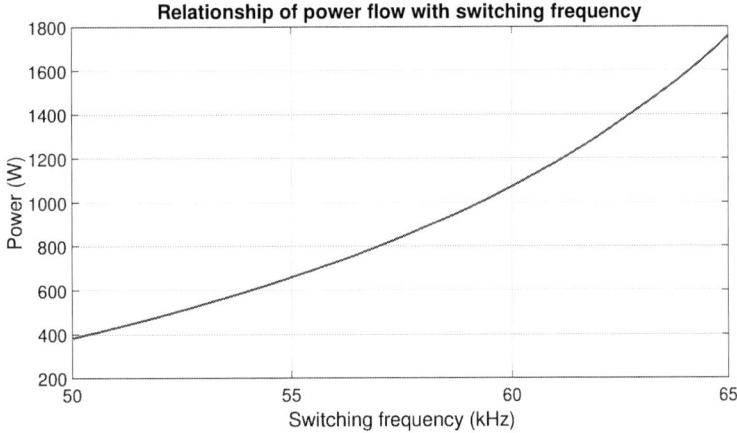

Fig. 4: Power flow dependence on frequency of the converter.

Converter prototype

A balancing converter was designed for a bipolar dc grid with voltage of ±350 VDC. The main specifications of the converter are given in Table II. The resonant frequency of the tank was found to be approximately 78 kHz.

Table II: Specifications of the designed converter.

Parameter	Value
Rated power	2 kW
Resonant inductance	7 μH
Resonant capacitance	594 nF

The prototype of the converter is shown in Fig. 5. The resonant converter are put on the top of the PCB; the resonant inductor, semiconductor half bridge modules and dc link capacitors are installed on the both side of the PCB.

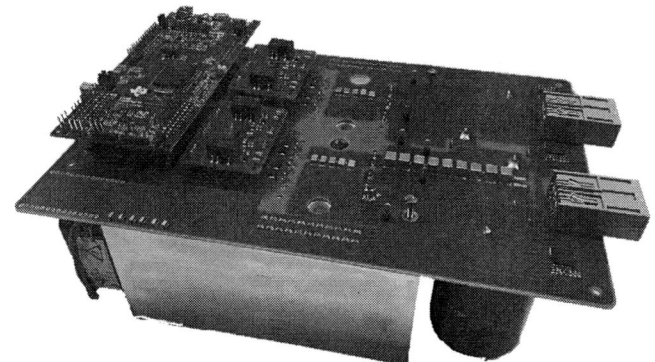

Fig. 5: Prototype of the converter.

Results

The schematic for testing the balancing converter is shown in Fig. 6. The main components are the source converters, load converter, balancing converter and the lumped line elements. The source converter are used to create a bipolar dc supply. The source converter in our setup is Agilent Technologies N5772A. The load converter is connected only between the neutral and negative pole to simulate a total imbalance of power flow. When the system is powered on without the balancing converter then the power is supplied only through the source converter connected between the neutral and negative pole. The balancing converter is used to balance the power flowing from both the source converters. When the balancing converter is turned on, the current flowing through the two source converter should be equal. This will be illustrated through the test results shown in subsequent figures. The load converter in our study is Delta Elektronika SM1500-CP-30. The lumped line elements are π-model for a 100m long two conductor cable.

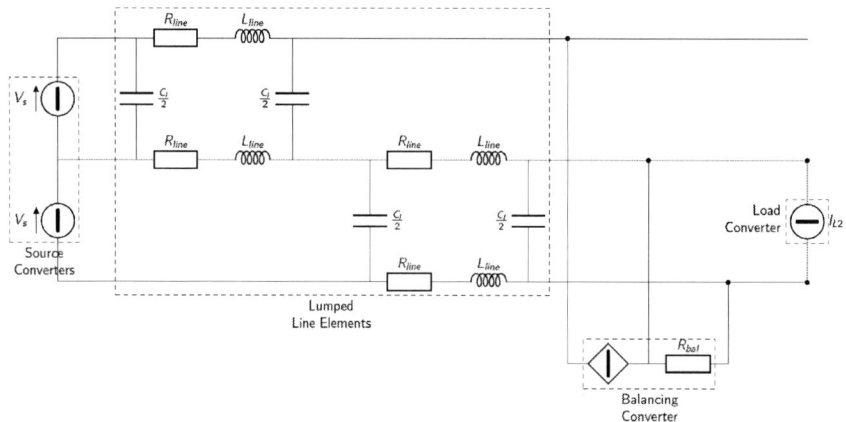

Fig. 6: Test bench schematic diagram.

The test setup in the laboratory is shown in Fig. 7. It consists of the source and load converters on the right; the balancing converter and lumped lime elements are on the left. In this setup, two lumped line elements are connected in series. Due to this, the impedance in the neutral line increases as two resistors and inductors are connected in series. This assumption is correct as the inductance and resistance increases as the cable cross section area decreases [17].

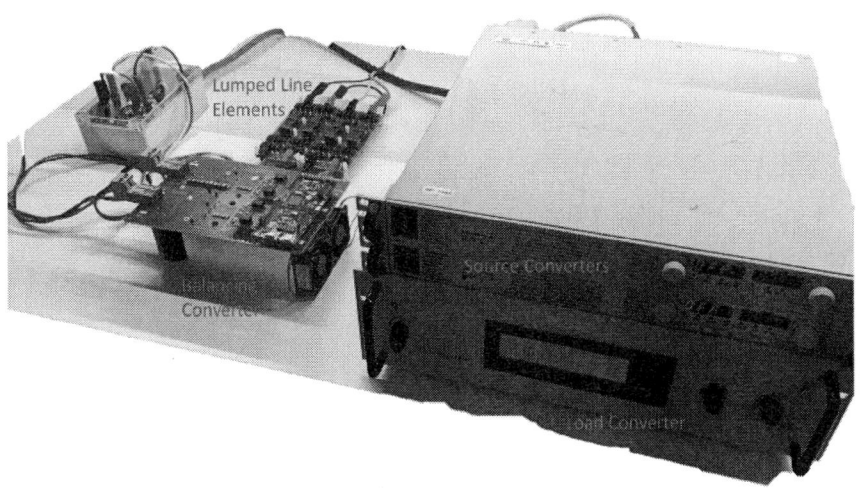

Fig. 7: Lab test setup.

The model of converter along with the other components of setup were simulated in LTSpice. The resulting resonant capacitor voltage and resonant inductor current are shown in Fig. 8. In this simulation, the load is connected between the neutral and negative poles. The load is a current sink with a value of 5A. The source converter are programmed to supply a voltage of ±350 VDC. The switches operate at 65 kHz with 50% duty cycle. A dead time of 100ns is inserted between the switch pairs S_1, S_2 and S_3, S_4. The phase shift between the upper half bridge and lower half bridge is set at 3.6 degrees. The frequency and phase shift required for a power flow of 1.75 kW were calculated using (1) and (2).

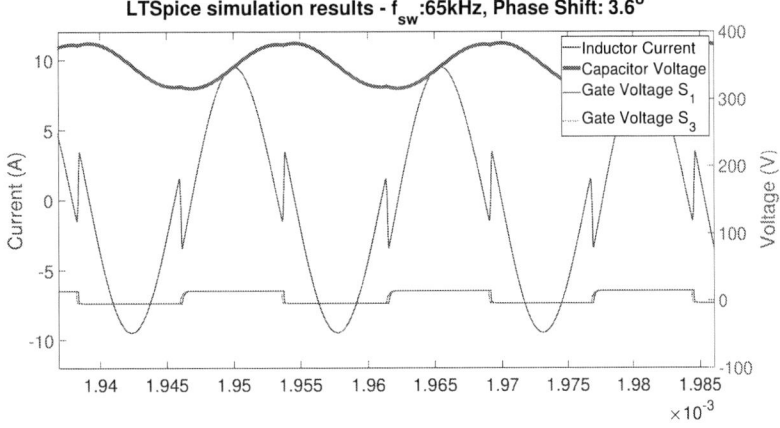

Fig. 8: Results of the LTSpice simulation

The results of the tests done in the laboratory are shown in Fig. 9 and Fig. 10. In this work, the converter is tested only at a single operating point. A closed loop control is beyond the scope of this work. The source converter supply ±350 VDC and the load converter is set to sink a current of 5 A as shown in Fig. 10. The oscilloscope signals consists of the inductor current, capacitor voltage and the gate signal for switches S1 and S3. The inductor current is measured using a Rogowski coil having a sensitivity of 50mV/A. The balancing converter operates with a phase shift of 3.6 degrees between the switch S1 and S3 and switching frequency of 65 kHz. The current output from the source converter is balanced with a minor difference of 0.06 A between the upper and lower source converters.

It can be seen from the oscilloscope output that the real life results match closely with the simulation results but not perfectly. One of the reasons of the imperfections can be the varying parameters of inductance, capacitance and resistance during the tests. Also, the microcontroller has a certain resolution of setting the frequency and phase shift which differ slightly from the simulation. A phase shift of 3.6 deg at a frequency of 65kHz corresponds to approximately 154 ns. Precisely achieving such low phase shift was found out to be difficult during the testing of the converter. Hence, there are deviations between the actual frequency and phase shift with those in the simulation.

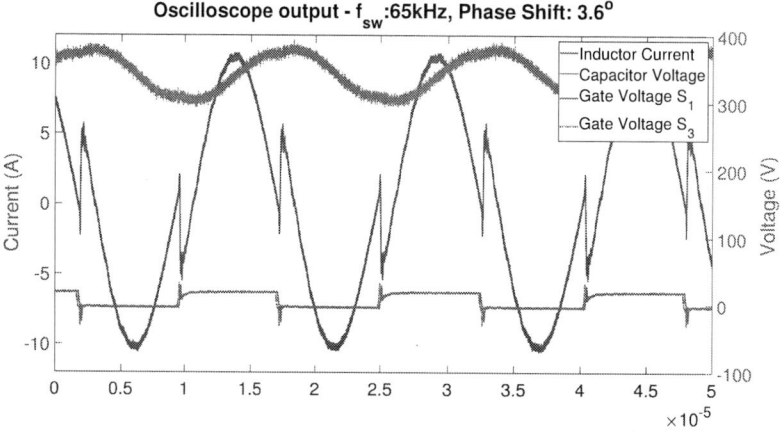

Fig. 9: Experimental results - oscilloscope output of the tests done with the converter.

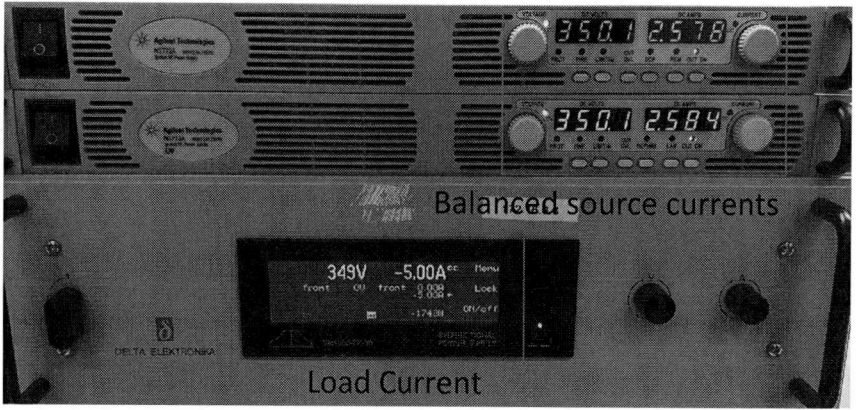

Fig. 10: Experimental results - voltage and current of the source and load converters.

Conclusion

In this paper, the use of a series resonant converter for balancing a bipolar dc grid is shown. The proposed converter is operated in the capacitive region. By changing the switching frequency of the converter and phase shift between the upper and lower half bridges of the converter, the power flow can be changed. The simulation results for balancing a load of 1.75kW connected between the neutral and negative pole is shown. A prototype of the balancing converter was build to experimentally verify the simulation results. The experimental results match closely with the simulations however not perfectly. The reason for the mismatch are also discussed. In the future, a closed loop control shall be developed for the converter.

References

[1] T. Smith, J. Jalkanen, B. Anderson, J. Corbett, J. Faber, S. Hanayama, E. O'keeffe, S. Parker, L. Johanasson, L. Aldous et al., "Third IMO GHG Study," 2014.

[2] S. Karim, "Reduction of emissions of greenhouse gas (ghg) from ships," in Prevention of Pollution of the Marine Environment from Vessels. Springer, 2015, pp. 107–126.

[3] A. J. Sorensen, R. Skjetne, T. Bo, M. R. Miyazaki, T. A. Johansen, I. B. Utne, and E. Pedersen, "Toward Safer, Smarter, and Greener Ships: Using Hybrid Marine Power Plants," IEEE Electrification Magazine, vol. 5, no. 3, pp. 68–73, Sep. 2017. [Online]. Available: http://ieeexplore.ieee.org/document/8025702/

[4] B. S. H. Chew, Y. Xu, and Q. Wu, "Voltage Balancing for Bipolar DC Distribution Grids: A Power Flow Based Binary Integer Multi-Objective Optimization Approach," IEEE Transactions on Power Systems, vol. 34, no. 1, pp. 28–39, Jan. 2019.

[5] S. Yadav, Z. Qin, and P. Bauer. "Bipolar DC grids on ships: possibilities and challenges," e & i Elektrotechnik und Informationstechnik (2022): 1-10.

[6] J. Lago, J. Moia, and M. L. Heldwein, "Evaluation of power converters to implement bipolar DC active distribution networks – DC-DC converters," in 2011 IEEE Energy Conversion Congress and Exposition. Phoenix, AZ, USA: IEEE, Sep. 2011, pp. 985–990.

[7] J. Moia, J. Lago, A. J. Perin, and M. L. Heldwein, "Comparison of three-phase PWM rectifiers to interface Ac grids and bipolar DC active distribution networks," in 2012 3rd IEEE International Symposium on Power Electronics for Distributed Generation Systems (PEDG). Aalborg: IEEE, Jun. 2012, pp. 221–228.

[8] F. Wang, Z. Lei, X. Xu, and X. Shu, "Topology Deduction and Analysis of Voltage Balancers for DC Microgrid," IEEE Journal of Emerging and Selected Topics in Power Electronics, vol. 5, no. 2, pp. 672–680, Jun. 2017.

[9] G. Van den Broeck, J. Beerten, M. Dalla Vecchia, S. Ravyts, and J. Driesen, "Operation of the full-bridge three-level DC–DC converter in unbalanced bipolar DC microgrids," IET Power Electronics, vol. 12, no. 9, pp. 2256–2265, Aug. 2019.

[10] P. Najafi, A. Houshmand Viki, and M. Shahparasti, "Evaluation of Feasible Interlinking Converters in a Bipolar Hybrid Microgrid," Journal of Modern Power Systems and Clean Energy, vol. 8, no. 2, pp. 305–314, 2020.

[11] S. Rivera, R. Lizana F., S. Kouro, T. Dragicevic, and B. Wu, "Bipolar DC Power Conversion: State-of-the-Art and Emerging Technologies," IEEE Journal of Emerging and Selected Topics in Power Electronics, vol. 9, no. 2, pp. 1192–1204, Apr. 2021.

[12] S. Rivera, B. Wu, S. Kouro, V. Yaramasu, and J. Wang, "Electric Vehicle Charging Station Using a Neutral Point Clamped Converter With Bipolar DC Bus," IEEE Transactions on Industrial Electronics, vol. 62, no. 4, pp. 1999–2009, Apr. 2015.

[13] B. Wu and M. Narimani, High-power converters and AC drives. John Wiley & Sons, 2017.

[14] J. Pou, R. Pindado, D. Boroyevich, and P. Rodr´ıguez, "Evaluation of the low-frequency neutral-point voltage oscillations in the three-level inverter," IEEE transactions on industrial electronics, vol. 52, no. 6, pp. 1582–1588, 2005.

[15] N. H. van der Blij, L. M. Ramirez-Elizondo, M. T. Spaan, and P. Bauer, "A state-space approach to modelling dc distribution systems," IEEE Transactions on Power Systems, vol. 33, no. 1, pp. 943–950, 2017.

[16] S. Yadav, N. H. Van Der Blij, and P. Bauer, "Modeling and stability analysis of radial and zonal architectures of a bipolar dc ferry ship," in 2021 IEEE Electric Ship Technologies Symposium (ESTS). IEEE, 2021, pp. 1–8.

[17] William A. Thue, ed. Electrical Power Cable Engineering. CRC Press, 2017.

A V2G-enabled Seven-level Buck PFC Rectifier for EV Charging Application

Anekant Jain, Ritika Agarwal, Krishna Kumar Gupta, Sanjay K. Jain
Thapar Institute of Engineering and Technology,
Patiala, India
E-Mail: ajain_phd19@thapar.edu, ritikaagarwal290@gmail.com,
krishna.gupta@thapar.edu, skjain@thapar.edu

Keywords

« AC-DC converter », «Charging infrastructure for EV's », « Electric vehicle », « Multi-level converters », « Power factor correction ».

Abstract

This article presents a novel bidirectional multilevel buck rectifier with power factor correction for the charging systems of currently available commercialized electric vehicles. As it synthesizes seven voltage levels, the proposed rectifier entails low harmonics. This rectifier enables grid-to-vehicle (G2V) operation in buck mode and vehicle-to-grid (V2G) operation in boost mode. The proposed topology utilizes the switched capacitors principle to achieve a self-voltage balancing of the capacitors. Experimental results are presented to validate the proposed rectifier.

Introduction

The most significant innovation in the automobile industry is the electric vehicle (EV). Electric vehicles are gaining popularity these days at an exponential rate. As a result of this expansion, more energy-efficient charging infrastructure is needed[1]. EVs provide several advantages: environmental preservation, lower running costs, zero tailpipe emission, no noise pollution and more convenient. EVs may act as a load i.e., grid-to-vehicle (G2V) and as a generator i.e., vehicle-to-grid (V2G) modes [2], [3]. As the power of a typical EV is double that of the normal residential load and fast charging will impose pressure on the grid network. If the EV charging does not employ state-of-the-art conversion, grid disruptions occur such as undesirable peak loads, harmonics, and poor power factor may occur [4]. To overcome this problem, a V2G technology refers to the process of feeding the electricity contained in an electric car's batteries back into the electrical grid. As a result, the V2G technology makes energy injection back into the grid easier. This technology forms part of a smart grid, an electrical network system that uses information technology to manage energy consumption. In general, the power converter is unidirectional for G2V mode, which includes both onboard and off-board charging systems. To enable V2G mode the grid-connected bidirectional AC-DC converter is required, which can provide sinusoidal input current with a unity power factor (UPF)[5]. The main research directions for bidirectional AC-DC converters are increasing power density, minimizing input and output current ripple, maintaining UPF and offering power adjustment capability.

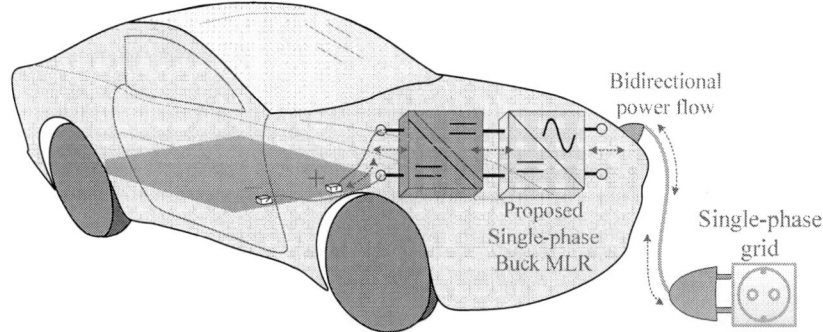

Fig. 1: Schematic diagram of the proposed charging system.

This article is focused on V2G enabled bidirectional single-phase on-board EV charging system. The design and development of this work can charge the vehicle battery as well as offer active power support to the utility grid as described in this article[6]. Fig. 1 depicts a schematic diagram of the proposed charging system, which is divided into two stages; first one is the bidirectional PFC converter and second is bidirectional DC-DC converter.

A conventional AC-DC converter converts the incoming AC grid voltage into a regulated boost DC link voltage[7]. A high voltage DC-DC buck converter is necessary to reduce this voltage to the nominal voltage acceptable for EV battery charging[8]. In this setup, the total standing voltage (TSV) between the rectifier and DC-DC converter switches is equal to the DC-link voltage i.e., high. Moreover, it required a high-voltage DC-bus capacitor [9]. The use of buck PFC rectification can fix these difficulties [10]. However, this results in discontinuous conduction mode (DCM), requiring larger inductive and capacitive filters on both DC and AC sides[11],[12]. Furthermore, DCM topologies' high-frequency operation considerably increases the switching losses. A diode bridge and a DC-DC buck converter are other ways to generate a buck DC voltage. Such two-stage systems have poorer efficiency, increased power losses, and higher production costs for medium and high-power applications[13].

A multilevel rectifier (MLR) is a revolutionary technology that can allow bidirectional power transfer. Compared to two-level converters, these converters employ lower-rated power switches, resulting in a higher-quality voltage waveform, less dv/dt stress across the switches, and lower THD line current[14]. As a result, of high-power applications, the power losses and the filter size decrease. In the AC-DC conversion of the EV charging system, conventional MLRs topologies, including neutral point clamped, flying capacitor, and T-type, have been used to enhance output voltage conversion [14]–[16]. These MLRs, on the other hand, have the following flaws:

- This functioned as a boost mode.
- The boost DC output voltage equalizes the blocking voltage on power switches.
- As the voltage level rises, the number of capacitors also rises, requiring more complex control techniques to balance the voltages of the capacitors.
- A high DC-bus capacitor is required.

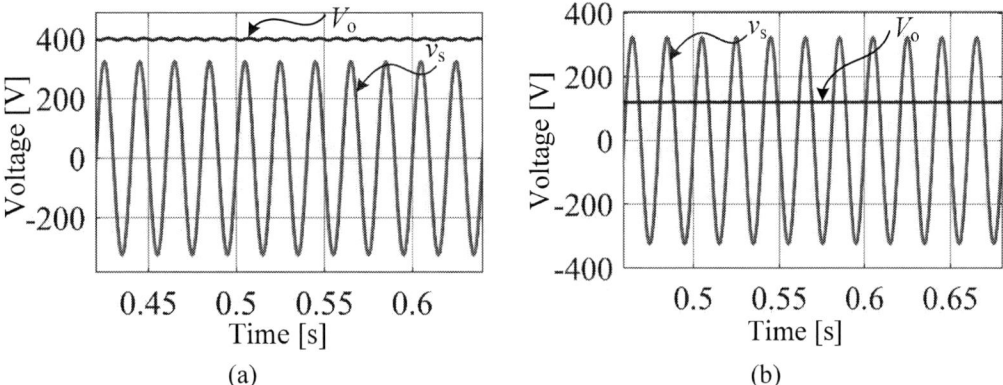

Fig. 2: The basic difference in terms of input and output voltage between (a) existing MLRs topologies and (b) proposed MLR topology.

To overcome the aforementioned constraints of conventional MLRs, a novel switched capacitors (SCs) based MLR topology is developed. The difference between the present and proposed work with input AC and output DC voltage is depicted in Fig. 2.

The power electronics interfaces for EV charging systems can support batteries ranging from 48V (e-bikes) to 400V (PHEV) [17], with the ability to charge the battery in both constant current and constant voltage modes, depending on the battery's state-of-charge (SOC). A bidirectional conventional multilevel PFC converter operates in boost mode for charging (i.e., rectifier) and buck mode for discharging (i.e., inverter). For G2V application power need to flow from high AC grid voltage to low DC battery voltage. So, in the first stage of EV charging, there is no need to use a boost AC-DC rectifier. Moreover, for V2G, low EV battery voltage connects with high AC grid voltage. In that instant also, there is no need to use a buck inverter. This use is responsible for the requirement of a high voltage step-up and step-down DC-DC converter and high voltage DC-bus capacitor. A power flow of V2G enabled EV charging system with conventional and proposed PFC rectifier shown in Fig.3.

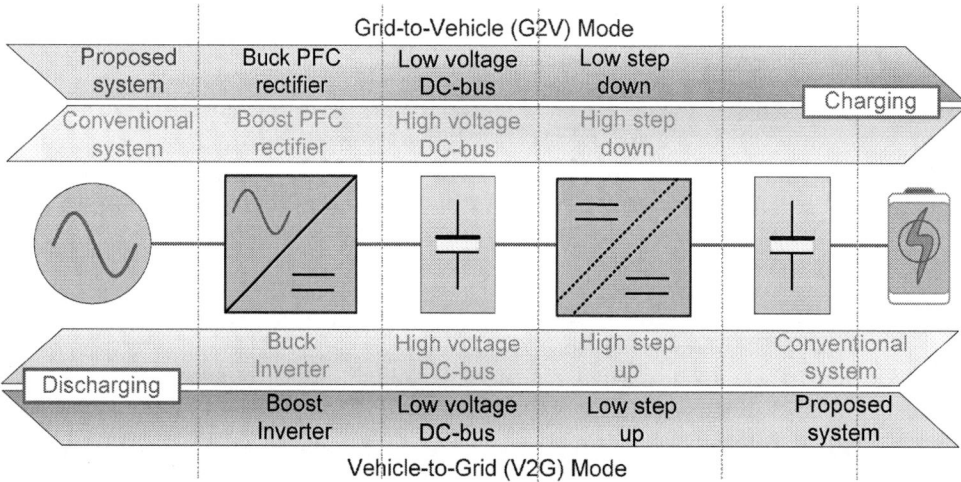

Fig. 3: Basic power flow in V2G enabled charging system and comparative effects on conventional and proposed AC-DC rectifier.

This article proposed a novel topology with the following properties to address these concerns for PFC rectifiers:

- It achieves a unity power factor and works as a V2G and G2V application.
- On the input side, it synthesizes seven levels, greatly increasing the harmonic profile.
- Ten switches, two switched capacitors, and one DC link capacitor is required.
- There is a need to balance only one DC-link capacitor voltage, the others are balanced by themselves without the use of complicated controlling methods.
- It operates in buck mode, giving it a wide output range.
- It works in continuous conduction mode (CCM), which means no huge filters are required.
- Six of ten switches have low peak inverse voltages that are the same as the DC-link voltage, and four switches have peak inverse voltages that are three times the DC-link voltage.

The proposed topology, voltage and current controller, and level-shifted pulse-width modulation (LSPWM) approach for producing gate pulses are all described in great depth. To verify the suggested work, experimental testing is carried out under steady-state and transient conditions.

Proposed buck rectifier topology

The seven-level bidirectional buck rectifier, as illustrated in Fig. 4, is made up of two H-bridges with two additional switches that link to two capacitors. By switching, connecting the two capacitors C_1 and C_2 in parallel and/or series with the DC bus capacitor represents the charging and discharging states. The operations '1' and '0' denote the ON and OFF states of the relevant switch, respectively; 'C' and 'D' represent the charging and discharging states of the capacitors are shown in Table I.

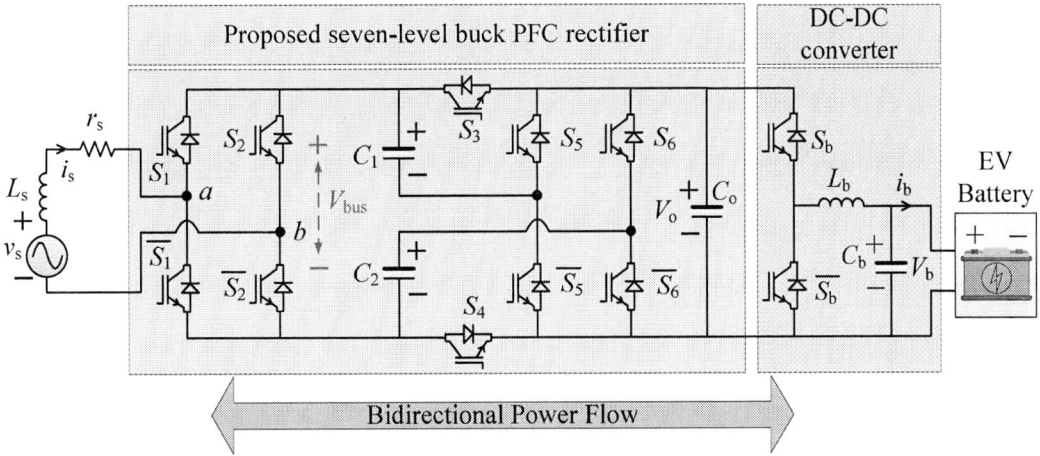

Fig. 4: Seven-level buck PFC topology for bidirectional power flow

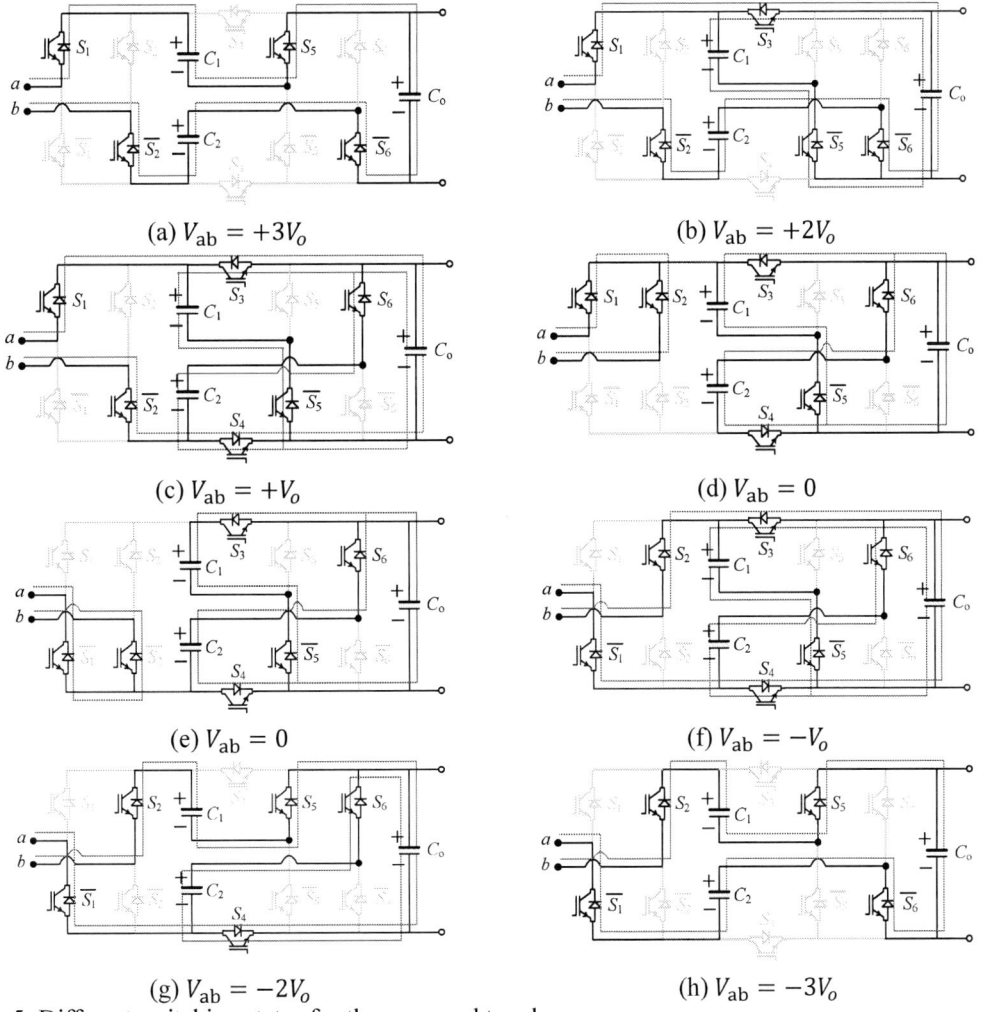

(a) $V_{ab} = +3V_o$

(b) $V_{ab} = +2V_o$

(c) $V_{ab} = +V_o$

(d) $V_{ab} = 0$

(e) $V_{ab} = 0$

(f) $V_{ab} = -V_o$

(g) $V_{ab} = -2V_o$

(h) $V_{ab} = -3V_o$

Fig. 5: Different switching states for the proposed topology

Table I: Switching states for the proposed topology

State	i_s	V_{ab}	Switches						Capacitors		
			S_1	S_2	S_3	S_4	S_5	S_6	C_1	C_2	C_o
T_1	$i_s > 0$	$+3V_o$	1	0	0	0	1	0	C	C	C
T_2	$i_s > 0$	$+2V_o$	1	0	1	0	0	0	D	C	C
T_3	$i_s > 0$	$+V_o$	1	0	1	1	0	1	D	D	C
T_4	$i_s \geq 0$	0	1	1	1	1	0	1	D	D	D
T_5	$i_s \leq 0$	0	0	0	1	1	0	1	D	D	D
T_6	$i_s < 0$	$-V_o$	0	1	1	1	0	1	D	D	C
T_7	$i_s < 0$	$-2V_o$	0	1	0	1	1	1	C	D	C
T_8	$i_s < 0$	$-3V_o$	0	1	0	0	1	0	C	C	C

The front-side H-bridge comprises four transistors S_1, $\overline{S_1}$, S_2 and $\overline{S_2}$ are responsible for converting AC to inverted multilevel voltage. Because the bus voltage V_{bus} can be three distinct DC levels of $+V_o$, $+2V_o$, and $+3V_o$, the front-side H-bridge can create seven different voltage levels at the terminals "a" and "b" that represents as V_{ab}, namely $0, \pm V_o, \pm 2V_o$, and $\pm 3V_o$. This H-bridge has voltage stress of $3V_o$. The load-side of the seven-level rectifier has two essential features. One advantage is that all components can bear the same low voltage stress V_o, which is advantageous for high-frequency operation. Another difference is that the two capacitors C_1 and C_2 perform the same function to create distinct output levels and balance the same voltage $+V_o$ (i.e., $V_{C_1} = V_{C_2} = V_o$).

Various switching states for the proposed rectifier are described herewith:

1. **State T_1 and T_8 ($V_{ab} = \pm 3V_o$):** During this state, in the positive half cycle, the switches S_1, $\overline{S_2}$, S_5 and $\overline{S_6}$ are turned ON. In the path shown in red, it can be seen that the capacitors C_1, C_2 and C_o are in series with the AC source, such that the voltage $V_{ab} = (+V_{C_1} + V_{C_2} + V_o) = +3V_o$. Moreover, during the negative half cycle, the switches $\overline{S_1}$, S_2, S_5 and $\overline{S_6}$ are turned ON, such that $V_{ab} = (-V_{C_1} - V_{C_2} - V_o) = -3V_o$. In both states, all capacitors are charged as shown in Fig. 5(a) and 5(h).

2. **State T_2 and T_7 ($V_{ab} = \pm 2V_o$):** During this state, in the positive half cycle, the switches S_1, $\overline{S_2}$, S_3, $\overline{S_5}$ and $\overline{S_6}$ are turned ON. In the path shown in red, it can be seen that the capacitors C_2 and C_o are in series with the AC source, such that the voltage $V_{ab} = (+V_{C_2} + V_o) = +2V_o$. Moreover, during the negative half cycle, the switches $\overline{S_1}$, S_2, S_5 and S_6 are turned ON, such that $V_{ab} = (-V_{C_1} + V_o) = -2V_o$. In-state T_2 capacitors C_2 and C_o are charged and C_1 is discharged to the load. Furthermore, in-state T_7, C_1 and C_o are charged and C_2 is discharged. Discharging path is shown in blue as shown in Fig. 5(b) and 5(g).

3. **State T_3 and T_6 ($V_{ab} = \pm V_o$):** During this state, in the positive half cycle, the switches S_1, $\overline{S_2}$, S_3, S_4, $\overline{S_5}$ and S_6 are turned ON. In the path shown with red, it can be seen that the capacitors C_o is in series with the AC source, such that the voltage $V_{ab} = +V_o$. Moreover, during the negative half cycle, the switches $\overline{S_1}$, S_2, S_3, S_4, $\overline{S_5}$ and S_6 are turned ON, such that $V_{ab} = -V_o$. In states T_2 and T_7 capacitors C_o charged and C_1, and C_2 are discharged to the load as shown in Fig. 5(c) and 5(f).

4. **State T_4 and T_5 ($V_{ab} = 0$):** During this state, in the positive half cycle, the switches S_1, S_2, S_3, S_4, $\overline{S_5}$ and S_6 are turned ON. In the path shown in red, it can be seen that not any capacitors are in series with the AC source, such that the voltage $V_{ab} = 0$. Moreover, during the negative half cycle, the switches $\overline{S_1}$, $\overline{S_2}$, S_3, S_4, $\overline{S_5}$ and S_6 are turned ON, such that $V_{ab} = 0$. In states T_4 and T_5 all capacitors are self-balanced and discharged to the load as depicted in Fig. 5(d) and 5(e).

The modulation approach for generating the gate pulses for the switches and the reference signal created by a suitable controller is detailed in the next section.

Modulation scheme and suitable controller

Several modulation approaches, such as multicarrier PWM and space vector modulation, can be utilized to control the output voltage of the proposed MLR. The proposed MLR is demonstrated in this part using the level shifted-PWM (LSPWM) approach, as seen in Fig. 6(a). Four high-frequency level-shifted carrier signals and reference signals produced by the controller were employed in the single-phase [18].

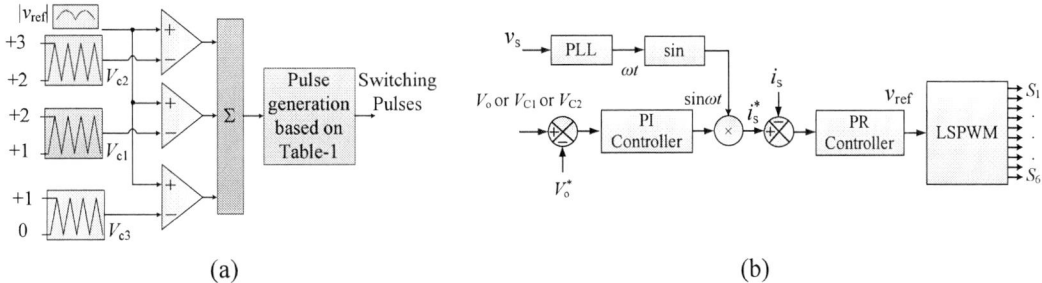

(a)   (b)

Fig. 6: Modulation and control strategy (a) level-shifted pulse width modulation for generating switching pulses (b) output voltage and PFC controller

In the proposed system cascaded PI and PR controllers are respectively used to control the DC voltage output (V_o) and the grid current (i_s), thereby providing a regulated buck DC output voltage and unity power factor at the input side. The key benefit of this proposed SC-based rectifier is that control of any one of the capacitor voltages automatically balances the voltages of the remaining capacitors. The block diagram of the implemented controller is shown in Fig. 6(b). A phase-lock loop (PLL) extracts the voltage angle and generates the synchronized current reference. The outer loop of the cascaded controller is used to regulate the voltage whose output goes to the current controller (inner loop) as the reference signal. Using the PI controller, DC voltages across the capacitors are regulated to DC voltage reference (V_o^*) [16], [19]. The current controller can be a simple gain proportional controller or a proportional-integral (PI) controller, with the inner loop having a faster dynamic than the outer loop. As a result, when using a PI controller to manage a sinusoidal input current signal, the PI's integral gain should be small enough not to affect the inner loop's speed. However, when a PI compensator is applied to a sinusoidal signal, it causes steady-state errors that appear as a DC component in the current harmonic spectrum. One approach for such a situation is to use a proportional resonant (PR) controller with an infinite gain at the fundamental grid frequency.

Experimental results

A laboratory setup was established using discrete power switches MOSFETs and a suitable gate driver IC to assess the proposed buck rectifier and closed-loop control. The output voltage and input current are sensed using a Hall Effect-based voltage sensor and current sensor. The MOSFET gate pulses are generated by the OPAL-RT OP4510, which interfaces to the hardware through MATLAB/ Simulink on the host computer. A 10-microsecond sample time is used to construct the controller and switching mechanism. A single-phase 230V RMS AC input is employed in buck mode, with 120V as a DC output. Table II summarizes the experimental verification parameters. Experimental results are taken for the resistive load and battery. Both steady-state and dynamic situations are used to evaluate the system's performance. A sudden shift in resistive DC load, reference voltage and grid voltage are examples of operating condition variations. Moreover, sudden change in battery performance from charging (i.e., G2V) to discharging mode (i.e., V2G).

The experimental results are based on two scenarios: one with an EV battery and another with resistive load. The steady-state results are produced when the rectifier converts 325V single-phase peak AC to 120V DC (in buck mode) and is further connected to the DC-DC buck converter to the EV battery. Fig.7 shows the steady-state condition when the battery is in charging condition. Whereas, a seven-level voltage (V_{ab}) is generated at the point of the input terminal of the rectifier, the bus voltage (V_{bus}), and all other capacitors voltages.

Table II: Parameters for experimental verifications

Parameters	Value	Unit
Single-phase input voltage	230	V (RMS)
Input grid frequency	50	Hz
Filter inductor	4	mH
Capacitors	1600	μF
Switching frequency	10	kHz
DC voltage	120	V
Battery	48V, 30AH battery	

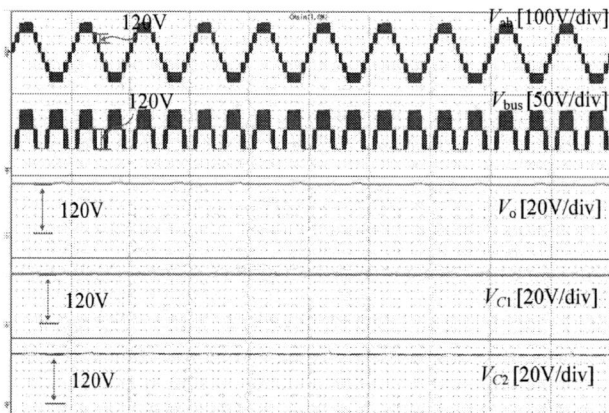

Fig. 7: Steady-state condition with level voltage (V_{ab}), bus voltage (V_{bus}) and all capacitors' voltages

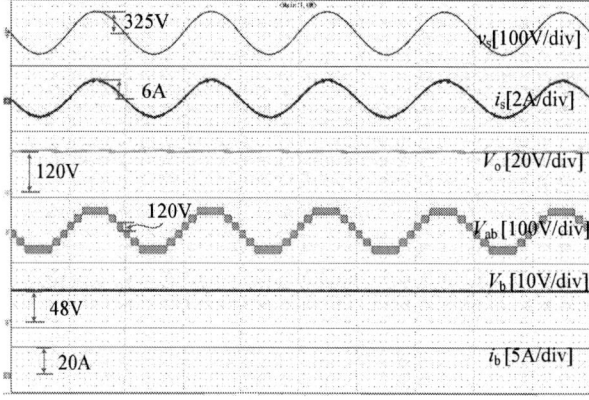

Fig. 8: Steady-state response in battery charging with unity power factor, battery voltage and current

To demonstrate applicability in single-phase charging, a power electronics interface consisting of the proposed PFC rectifier and a conventional bidirectional buck-boost DC-DC converter is built. The battery's voltage and current are 48V and 20A, respectively. Fig. 8 shows the waveforms of the battery charging, which show that the grid voltage and current are in phase. The output voltage of the rectifier is set to 120V DC and then regulated to 48V using a DC-DC converter. The V2G mode allows battery energy to be injected back into the grid. EVs can use grid-to-vehicle (G2V) charging and vehicle-to-grid (V2G) discharging modes. To use V2G mode, the interface must be capable of bidirectional power flow. In its single-phase variations, the proposed topology facilitates both charging and discharging. Experiments in both modes (G2V and V2G) of operation are shown in Fig 9. The battery current seems to be reversed when a sudden shift in the flow of the battery current is directed. The grid current is 180° degrees out of phase with the grid voltage in this scenario.

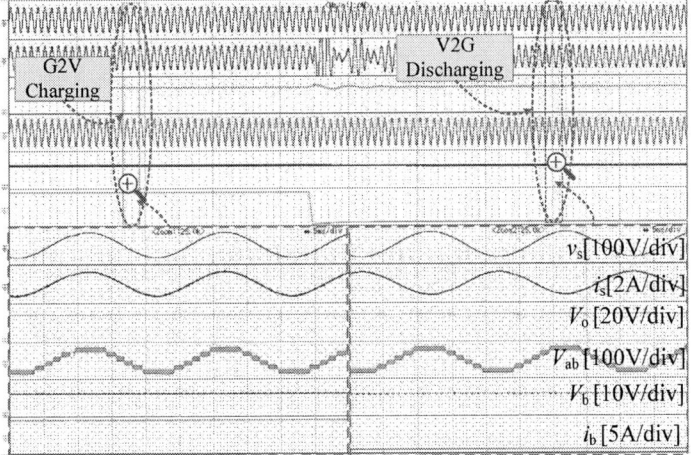

Fig. 9: Sudden change in G2V charging to V2G discharging mode

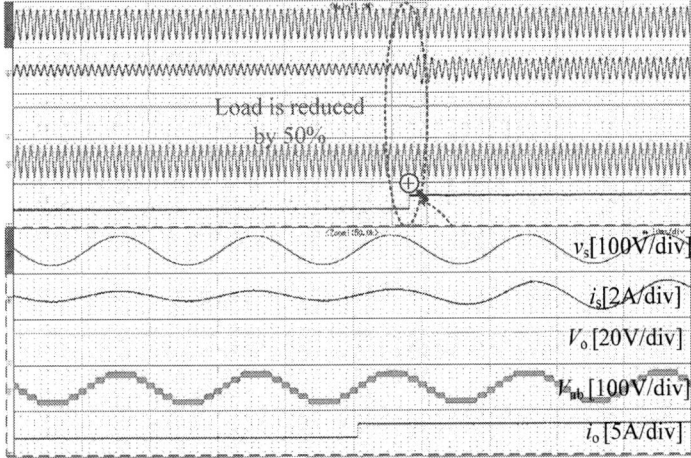

Fig. 10: Transients response with resistive load; Suddenly reduces the load by 50%

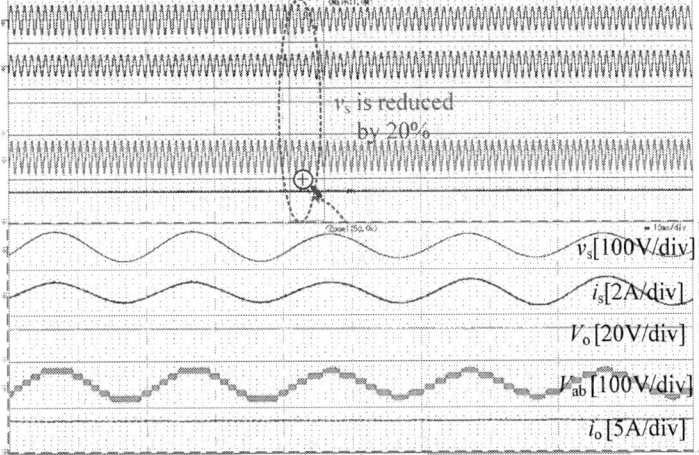

Fig. 11: Transients response with resistive load; Suddenly reduces the grid voltage by 20%

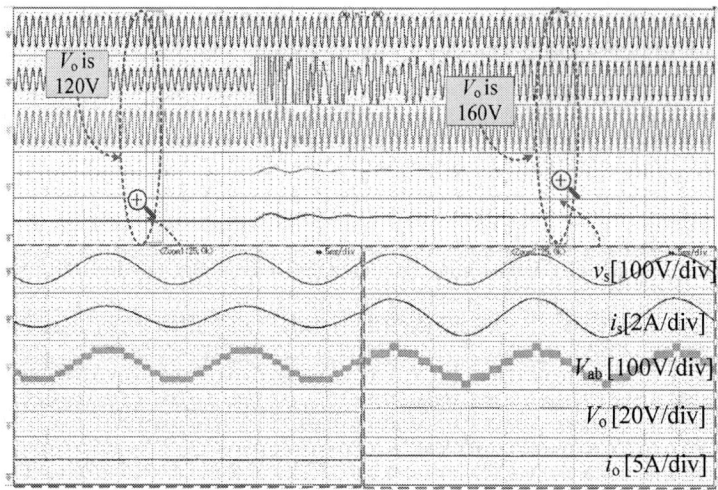

Fig. 12: Transients response with resistive load; Suddenly increases the reference voltage 120V to 160V

The proposed rectifier's dynamic performance is validated by resistive load experimentation. A 50% load rapidly decreases, as result doubling the load current and instantaneously stabilizing the load voltage at 120V. Furthermore, as seen in Fig. 10, the rectifier maintains a power factor of unity. In another case grid voltage suddenly reduces by 20% as shown in Fig. 11, as results all other voltages and currents are stable instantaneously and achieve a unity power factor. In a different case, depicted in Fig.12, if the output DC reference is increased by 120 to 160, V_o and i_o will also vary, and the load voltage will settle at 120V to 160V as expected. These outcomes validate the controller's tracking ability. The converter maintains a unity power factor and seven voltage levels at the rectifier's input even when the output voltage is raised.

Conclusion

This article presents a novel single-phase multilevel buck PFC rectifier that is most suitable for EV charging and V2G operation. LSPWM and a voltage and current controller are used to balance the voltage and maintain the unity power factor. A proto-model implementation was used to evaluate the performance on a single-phase AC input of 230 V RMS and a DC output of 120 V for various dynamic circumstances. The following conclusions were reached:

- It synthesises input into seven levels, improving the harmonic profile of the waveform.
- It offers a wide range of output voltages, making it suited for many applications.
- Because of its buck mode of operation, the proposed rectifier is appropriate for EV battery charging.
- It works in continuous conduction mode (CCM), which means no huge filters are required.
- It includes a built-in self-voltage balancing capability that eliminates the need for additional circuitry.
- It can transfer power in both directions.

References

[1] C. C. Chan, "The state of the art of electric and hybrid vehicles," *Proceedings of the IEEE*, vol. 90, no. 2, pp. 247–275, 2002, doi: 10.1109/5.989873.

[2] M. C. Kisacikoglu, M. Kesler, and L. M. Tolbert, "Single-phase on-board bidirectional PEV charger for V2G reactive power operation," *IEEE Transactions on Smart Grid*, vol. 6, no. 2, pp. 767–775, Mar. 2015, doi: 10.1109/TSG.2014.2360685.

[3] V. Kumar and K. Yi, "Single-Phase, Bidirectional, 7.7 kW Totem Pole On-Board Charging/Discharging Infrastructure," *Applied Sciences*, vol. 12, no. 4, p. 2236, Feb. 2022, doi: 10.3390/app12042236.

[4]   M. C. Kisacikoglu, B. Ozpineci, and L. M. Tolbert, "EV/PHEV bidirectional charger assessment for V2G reactive power operation," *IEEE Transactions on Power Electronics*, vol. 28, no. 12, pp. 5717–5727, 2013, doi: 10.1109/TPEL.2013.2251007.

[5]   U. K. Madawala and D. J. Thrimawithana, "A bidirectional inductive power interface for electric vehicles in V2G systems," *IEEE Transactions on Industrial Electronics*, vol. 58, no. 10, pp. 4789–4796, Oct. 2011, doi: 10.1109/TIE.2011.2114312.

[6]   A. Tariq, "A Modified Battery Charger with Power Factor Correction for Plug-In Electrical Vehicles," in *The 1st International Conference on Energy, Power and Environment*, Mar. 2022, p. 103. doi: 10.3390/engproc2021012103.

[7]   D. Rothmund, T. Guillod, D. Bortis, and J. W. Kolar, "99.1% Efficient 10 kV SiC-Based Medium-Voltage ZVS Bidirectional Single-Phase PFC AC/DC Stage," *IEEE Journal of Emerging and Selected Topics in Power Electronics*, vol. 7, no. 2, pp. 779–797, Jun. 2019, doi: 10.1109/JESTPE.2018.2886140.

[8]   S. Haller, M. F. Alam, and K. Bertilsson, "Reconfigurable Battery for Charging 48 V EVs in High-Voltage Infrastructure," *Electronics (Switzerland)*, vol. 11, no. 3, Feb. 2022, doi: 10.3390/electronics11030353.

[9]   J. S. Lee, U. M. Choi, and K. B. Lee, "Comparison of tolerance controls for open-switch fault in a grid-connected T-type rectifier," *IEEE Transactions on Power Electronics*, vol. 30, no. 10, pp. 5810–5820, Oct. 2015, doi: 10.1109/TPEL.2014.2369414.

[10]   T. B. Soeiro, T. Friedli, and J. W. Kolar, "Swiss rectifier - A novel three-phase buck-type PFC topology for Electric Vehicle battery charging," in *Conference Proceedings - IEEE Applied Power Electronics Conference and Exposition - APEC*, 2012, pp. 2617–2624. doi: 10.1109/APEC.2012.6166192.

[11]   X. Xie, C. Zhao, L. Zheng, and S. Liu, "An improved buck PFC converter with high power factor," *IEEE Transactions on Power Electronics*, vol. 28, no. 5, pp. 2277–2284, 2013, doi: 10.1109/TPEL.2012.2214060.

[12]   H. Choi, "Interleaved boundary conduction mode (BCM) buck power factor correction (PFC) converter," *IEEE Transactions on Power Electronics*, vol. 28, no. 6, pp. 2629–2634, 2013, doi: 10.1109/TPEL.2012.2222930.

[13]   X. Wu, J. Yang, J. Zhang, and Z. Qian, "Variable on-time (VOT)-controlled critical conduction mode buck PFC converter for high-input AC/DC HB-LED lighting applications," *IEEE Transactions on Power Electronics*, vol. 27, no. 11, pp. 4530–4539, 2012, doi: 10.1109/TPEL.2011.2169812.

[14]   C. A. Teixeira, D. G. Holmes, and B. P. McGrath, "Single-phase semi-bridge five-level flying-capacitor rectifier," in *IEEE Transactions on Industry Applications*, 2013, vol. 49, no. 5, pp. 2158–2166. doi: 10.1109/TIA.2013.2258877.

[15]   Y. Xu, Y. Zou, C. Wang, W. Chen, and B. Liu, "A Single-Phase High-Power-Factor Neutral-pointer Clamped Multilevel Rectifier," 2007.

[16]   H. Vahedi, P. A. Labbe, and K. Al-Haddad, "Single-Phase Single-Switch Vienna Rectifier as Electric Vehicle PFC Battery Charger," Dec. 2015. doi: 10.1109/VPPC.2015.7353019.

[17]   H. Ramakrishnan and J. Rangaraju, "Power Topology Considerations for Electric Vehicle Charging Stations," 2020. [Online]. Available: www.ti.com

[18]   M. P. Kazmierkowski and L. Malesani, "Current control techniques for three-phase voltage-source pwm converters: A survey," *IEEE Transactions on Industrial Electronics*, vol. 45, no. 5, pp. 691–703, 1998, doi: 10.1109/41.720325.

[19]   D. F. Cortez and I. Barbi, "A Three-Phase Multilevel Hybrid Switched-Capacitor PWM PFC Rectifier for High-Voltage-Gain Applications," *IEEE Transactions on Power Electronics*, vol. 31, no. 5, pp. 3495–3505, May 2016, doi: 10.1109/TPEL.2015.2467210.

Experimental Demonstration of a 2.2kW Active-Clamp Converter for High-Current Wide-Voltage-Transfer Ratio Applications

Philipp Rehlaender, Bastian Korthauer, Frank Schafmeister, Joachim Böcker
Power Electronics and Electrical Drives, Paderborn University
Warburger Str. 100
Paderborn, Germany
Phone: +49 (0) 5251 60 2159
Fax: +49 (0) 5251 60 3443
Email: rehlaender@lea.upb.de
URL: http://lea.upb.de

Acknowledgments

The authors would like to thank Delta Energy Systems (Germany) GmbH for funding this research.

Keywords

≪High frequency power converter≫, ≪Switched-mode power supply≫, ≪DC-DC converter≫, ≪Battery charger≫, ≪Cooling≫.

Abstract

Active-clamp forward converters are typically not applied for converters with power ratings larger than 500 W as power as the power transfer between primary and secondary is discontinuous leading to large magnetic components. This paper, however, proves that silicon carbide (SiC) devices enable this topology as a well-suitable topology for a wide voltage-transfer ratio due to the low switching losses allowing high switching frequencies reducing the size of the magnetic components. A laboratory prototype is designed using a very precise model. It employs SiC MOSFETs of 900 V for the primary and Si synchronous rectifiers of 100 V. The semiconductors are cooled with cost-efficient copper inlays, which effectiveness is demonstrated through a thermal FEM simulation. The developed prototype achieves a maximum efficiency of 95% while maintaining an efficiency above 92% for almost the entire operating region. The high efficiency and the power density of approximately 2 kW/l confirm the proposed concept. Finally, the converter is benchmarked to two LLC resonant converters (Si, SiC) and the prototype is used to verify a highly accurate steady-state model showing errors below 1 %.

1 Introduction

Automotive on-board DC-DC converters are the connecting link between the traction battery with a nominal voltage of around 400 V and the auxiliary battery with a nominal voltage of 12 V. The actual voltages, however, largely depend on the state-of-charge of the batteries. The voltages may vary by a factor of two or even more (200-420 V, 8-16 V) such that the connecting converter needs to cover a very wide voltage-transfer ratio. Earlier publications showed that traditional topologies (LLC, phase-shifted full bridge) suffer from the wide-voltage transfer ratio and pointed out that the active-clamp forward converter (ACFC) and the LLC with operating mode variations are suitable topologies for this application [1, 2]. In contrast to the LLC resonant converter, the ACFC only employs two primary switches making it a cost-effective solution. Typically, however, the ACFC is not used for such large output powers as energy is only transferred once per period to the secondary side making it a single-pulse topology. This results in a larger transformer and a large output inductor L_g compared to other topologies.

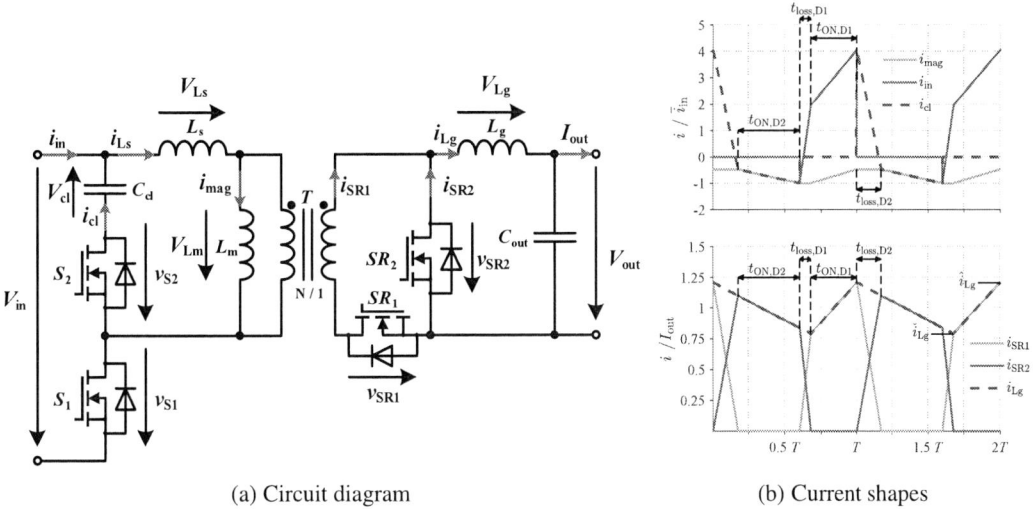

(a) Circuit diagram　　　　　　　　(b) Current shapes

Fig. 1: Circuit diagram and current shapes of the active-clamp forward converter (ACFC)

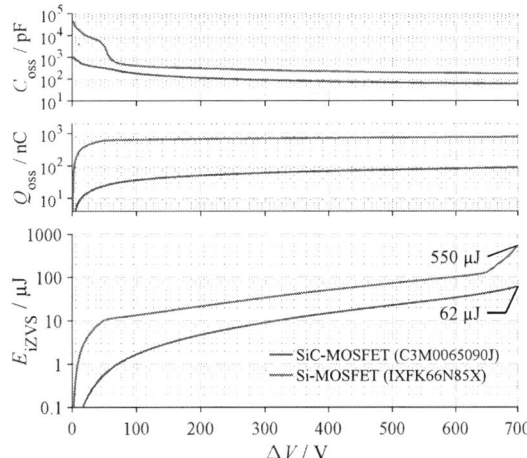

Fig. 2: Advantage of SiC semiconductors over Si semiconductors. The output capacitance C_{oss}, output charge Q_{oss} and incomplete ZVS lossses E_{iZVS} are significantly smaller.

The topology of the ACFC offers the potential of zero-voltage switching (ZVS) for both switches as the turn-off current of both switches is positive (cf. Figure 1b). The large transferred current results in a large turn-off current for the main switch S_1 easily enabling ZVS for the auxiliary switch S_2. The lower turn-off current of the auxiliary switch can be utilized to achieve ZVS for main switch S_1. Due to the significantly smaller current, only a small energy is available to achieve ZVS. However, due to the outstanding performance of wide-bandgap semiconductors, only a small energy is necessary to achieve ZVS and the losses associated with the turn-on at a residual voltage [3] are much smaller (cf. Figure 2). In preceding work [2], the ACFC was analytically benchmarked vs. the LLC resonant converter and phase-shifted full bridge showing that the topology is an attractive alternative as it yields in low transformer currents and low primary switch currents. This paper, consequently, presents the design of an active-clamp forward converter as an on-board DC-DC converter. The analysis and parameter selection of the converter is presented in Section II, the evaluation is described in Section III before the paper is concluded.

2 Converter Design

When designing the ACFC, the system is conventionally modeled without the series inductance L_s and with large output and magnetizing inductances (L_g and L_m respectively). This yields block-shaped currents and a neglection of the duty-cycle loss intervals $t_{loss,D1}$ and $t_{loss,D2}$ [4–7]. This overestimates the length of the demagnetizing interval $t_{ON,D2}$. With this assumption, the clamp capacitor voltage $V_{cl,com}$ is commonly calculated as

$$V_{cl,com} = V_{in} \frac{D}{1-D} \tag{1}$$

where $D = \frac{NV_{out}}{V_{in}}$. However, as the demagnetizing interval $t_{ON,D2}$ is shortened by the duty-cycle loss intervals $t_{loss,D1}$ and $t_{loss,D2}$ the clamp capacitor voltage V_{cl} becomes much larger resulting in errors of more than 60% [1, 8]. Consequently, [1, 8] described the detailed and accurate modeling procedure of an ACFC operating in continuous conduction mode (CCM). While this model was only verified with simulation results, this work will also demonstrate its accuracy by verifying it with experimental results. Consequently, this model will be used during the following design procedure. However, since this model only described the analytical derivation yielding a complex analytical expression, the next section shortly describes the analytical derivation and expression of a simplified calculation of the clamp capacitor voltage, which still gives good results.

2.1 A simplified model of the clamp-capacitor voltage V_{cl}

The preceding simplified modeling methods [4–7] assumed an ideal transformer such that the transformed input voltage appears at the secondary side of the transformer during the energy-transfer interval $t_{ON,D1}$ (cf. Figure 1b). However, this is not the case since a significant portion of the input voltage may be applied to the series inductance L_s. Thus, the output inductor voltage is reduced and $V_{Lg,D1}$ must be accurately calculated by solving the following four equations [1, 8].

$$
\begin{aligned}
V_{in} &= V_{Ls,D1} + V_{Lm,D1} \\
V_{Lm,D1} &= V'_{Lg,D1} + V'_{out} + V'_{D} \\
\Delta i_{Ls} &= \Delta i_{Lm} + \Delta i'_{Lg} \\
\frac{V_{Ls,D1}}{L_s} &= \frac{V_{Lm,D1}}{L_m} + \frac{V'_{Lg,D1}}{L'_g}
\end{aligned}
\tag{2}
$$

For $V_{Lg,D1}$ and $\widetilde{V}_{out} = V_{out} + V_D = V_{out} + V_d + R_d I_{out}$ (V_d is the forward voltage of the rectifier, R_d is the resistance of the rectifier), this yields

$$V_{Lg,D1} = -\frac{L_g N \left(-L_m V_{in} + N\widetilde{V}_{out}(L_s + L_m) \right)}{N^2 L_g (L_m + L_s) + L_m L_s} \tag{3}$$

Thus, the duty cycle $\widetilde{D}$ can be calculated as $\widetilde{D} = \frac{\widetilde{V}_{out}}{V_{Lg,D1} + \widetilde{V}_{out}}$. The duty-cycle loss intervals $t_{loss,D1}$ and $t_{loss,D2}$ (cf. Figure 1b) can be estimated as $t_{loss,D1} = \frac{I_{out}L_s}{NV_{in}}$ and $t_{loss,D2} = \frac{I_{out}L_s}{NV_{cl}}$. The voltage-time area of the voltage applied to the magnetizing inductance L_m during $t_{ON,D1}$ and $t_{ON,D2}$ must be equal:

$$V_{cl}(T - t_{loss,D1} - t_{loss,D2} - \widetilde{D}T) = V_{in}\widetilde{D}T. \tag{4}$$

In this equation, the voltage V_{in} is used for the magnetizing voltage V_{Lm} during $t_{ON,D1}$, even though the real voltage $V_{Lm,D1}$ can be accurately calculated by solving (2). This was done as it yields a slightly higher approximation of the clamp capacitor voltage to simplify the lengthy accurate calculation of the clamp capacitor voltage as it was done in the preceding work of [1, 8]. Equation (4) can be solved for the clamp capacitor voltage V_{cl} yielding

$$V_{cl} = -\frac{V_{in}(\sigma + NTV_{in}V_{out})}{\sigma - NTV_{Lg,D1}V_{in}} \tag{5}$$

with (3) and

$$\sigma = I_{out}L_s \left(V_{Lg,D1} + V_{out} \right) \tag{6}$$

Table I: Operating points with the maximum component stress

Abbreviation	OP (V_{in}, V_{out}, I_{out})	component
$OP_{Gmax,FL}$	240 V, 14 V, 160 A	SR_1, S_1
$OP_{Gmax,FL}\|\hat{V}_{in}$	420 V, 16 V, 140 A	T, L_g, V_{rev}, V_{SR2}
$OP_{Gmin,FL}$	420 V, 8 V, 160 A	SR_2, S_2
$OP_{Gmax,DL}$	200 V, 16 V, 100 A	C_{cl}, V_{SR1}
OP_{Gnom}	345 V, 14 V, 64 A	SR_1, S_1

This calculation is much simpler than the calculation developed in [1]. The simplified model and the complex model of [1] are benchmarked in 3.3.

2.2 Definition of characteristic design operating points

The circuit is designed to run with an input voltage $V_{in} \in [200\,V, 420\,V]$ and an output voltage of $V_{out} \in [8\,V, 16\,V]$ at a maximum power of $P_{out} = 2.24\,kW$ and a maximum output current of $I_{out} = 160\,A$. For an input voltage smaller than $V_{in} = 240\,V$, the maximum output current is proportionally reduced from its maximum current at the operating output voltage to $I_{out} = 100\,A$ at $V_{in} = 200\,V$. This is the derated operating region. The maximum current stress of the primary and secondary semiconductors is fairly obvious. The forward synchronous rectifier SR_1 is most stressed at the operating point of $V_{in} = 240\,V$, $V_{out} = 14\,V$ as this is the operating point with the maximum current and duty cycle. The same also applies for S_1. For output voltages larger 14 V, the current is reduced due to the maximum power rating. For freewheeling synchronous rectifier SR_2 and the auxiliary switch, the operating point with the maximum stress is $V_{in} = 420\,V$, $V_{out} = 8\,V$ as this is the operating point with the smallest duty cycle. For the maximum reverse voltage of the primary semiconductors ($\hat{v}_{rev} = \hat{v}_{cl} + V_{in}$), two operating points need to be considered as depicted in Figure 3a. While the clamp capacitor voltage increases with increasing duty cycle (cf. Figure 3b) such that $OP_{Gmax,DL}$ is the operating point with the largest clamp capacitor voltage $\hat{v}_{cl}$, the increasing input voltage provokes a larger blocking voltages.

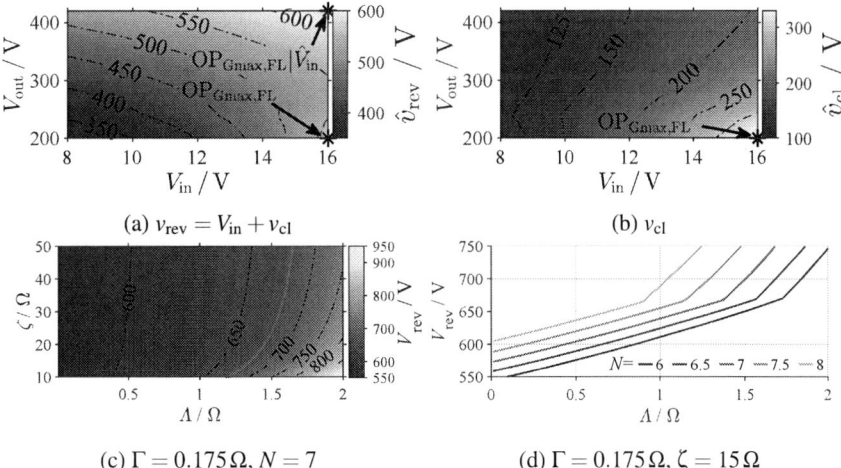

(a) $v_{rev} = V_{in} + v_{cl}$

(b) v_{cl}

(c) $\Gamma = 0.175\,\Omega$, $N = 7$

(d) $\Gamma = 0.175\,\Omega$, $\zeta = 15\,\Omega$

Fig. 3: Operating point dependent reverse voltage v_{rev} (a) and clamp capacitor voltage v_{cl} (b) for a sample set of parameters. Parameter-dependent reverse voltage v_{rev} for the dependency on Λ and ζ (c) and Λ and n (d). Depicted is the maximum for $OP_{Gmax,FL}\|\hat{V}_{in}$ and $OP_{Gmax,DL}$.

2.3 Circuit parameter selection

The definition of a suitable switching frequency is one of the most crucial steps in the design of a power electronic circuit. However, when comparing a set of designs, the calculated stress values only apply for a single switching frequency since the inductances scale anti-proportionally with the switching frequency [2, 9]. Thus, the following analysis normalizes the system parameters with the switching frequency such

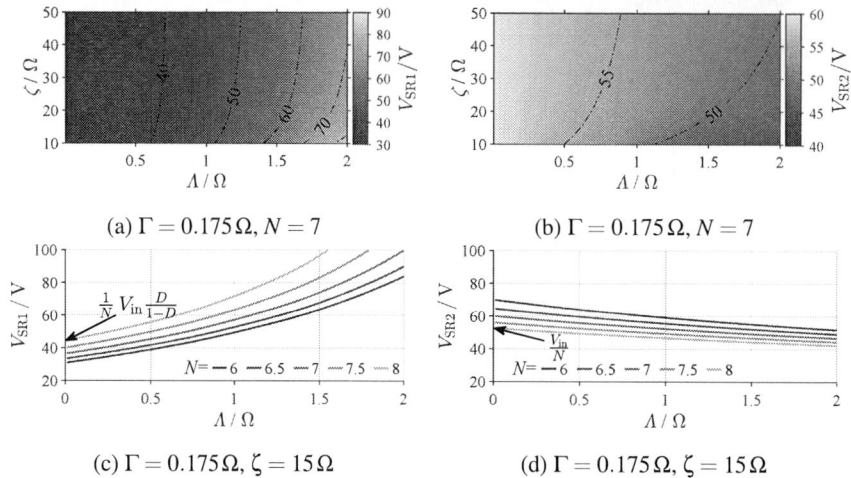

(a) $\Gamma = 0.175\,\Omega, N = 7$ (b) $\Gamma = 0.175\,\Omega, N = 7$

(c) $\Gamma = 0.175\,\Omega, \zeta = 15\,\Omega$ (d) $\Gamma = 0.175\,\Omega, \zeta = 15\,\Omega$

Fig. 4: Parameter-dependent synchronous rectifier reverse voltage v_{SR1}, v_{SR2} for the dependency on Λ and ζ (a,b) and Λ and n (c,d). For (a,c), the operating point is Depiction as the maximum for $OP_{Gmax,DL}$ and $OP_{Gmax,DL}$; for (b,d), the operating point is $OP_{Gmax,FL}|\hat{V}_{in}$.

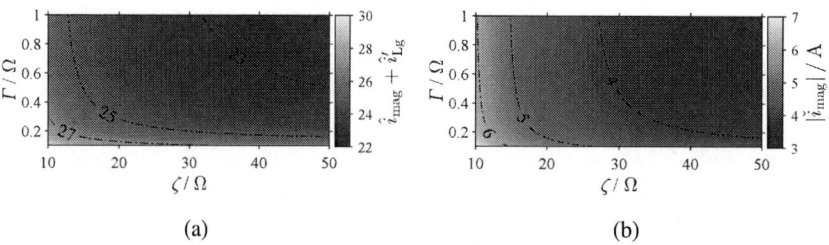

(a)   (b)

Fig. 5: Parameter dependent turn-off current for the main switch S_1 (a) and auxiliary switch S_2 (b). For (a), the operating point is $OP_{Gmax,FL}$; for (b), it is $OP_{Gmin,FL}$.

that the frequency can be selected in a later step. To define switching frequency-independent system parameters, the following three variables are defined:

$$\zeta = L_m f_{sw}$$
$$\Lambda = L_s f_{sw} \tag{7}$$
$$\Gamma = L_g f_{sw}.$$

where ζ is a measure for the magnetizing inductance L_m, Λ for the series inductance L_s and Γ for the output inductance L_g. Through this normalization, current and voltage values can directly be analyzed. Figure 3c shows an analysis of the maximum reverse voltage in relation to the normalized system parameters. For system parameters in the upper left of the red line, the operating point with the maximum reverse voltage is $OP_{Gmax,FL}|\hat{V}_{in,max}$, for system parameters in the lower right, it is $OP_{Gmax,DL}$. The reverse voltage increases significantly with a larger series inductance L_s (Λ). The maximum reverse voltage can be reduced by a definition of a suitable transformation ratio N (cf. Figure 3d). To limit the reverse voltage of the primary semiconductors, N and L_s must be chosen carefully. As the clamp capacitor voltage also influences the blocking voltage of the secondary semiconductors, this choice also influences the design of the secondary semiconductors. Figure 4a and Figure 4c show the blocking voltage of the synchronous rectifier SR_1. For $\Lambda \to 0$, the blocking voltage can be calculated through (1). However, similar to the calculation of the clamp capacitor voltage, the series inductance has a severe impact on the blocking voltage and may lead to a significant increase whereas an increased series inductance, in turn, results in a reduced blocking voltage for the freewheeling synchronous rectifier SR_2 (cf. Figure 4a, Figure 4c).

To accurately design the transformer, the influence of the magnetizing current must be considered. For this analysis, it is considered that the series inductance L_s is integrated in the transformer such that the flux density can be calculated as [10]

$$b(t) = \frac{(L_m + L_s)i_{mag}}{N_{prim}A_{eff}}.$$

(8)

To keep the analysis frequency-independent, the stress value $\mathcal{B}_T$ is introduced, which is of the unit *volts*.

$$b(t) = \frac{\overbrace{(\zeta + \Lambda)i_{mag}}^{\mathcal{B}_T(t)}}{f_{sw}N_{prim}A_{eff}}.$$

(9)

The stress variable $\mathcal{B}_T$ is analyzed to investigate the flux ripple of the transformer. Of special interest is the minimum transformer flux $\check{b}_{min}$ to avoid saturation, which is influenced by the system parameters through the minimum magnetizing current. To investigate the minimum flux, the normalized stress value $\check{\mathcal{B}}_T$ is displayed in Figure 6a whereas the flux ripple is investigated through $\frac{1}{2}\Delta\mathcal{B}_T$ in Figure 6b. The analysis shows that while the flux ripple is barely influenced through the design of the inductances, the minimum magnetizing current may significantly influence the minimum transformer flux. For a large magnetizing inductance L_m (corresponding with a large ζ), and a large series inductance L_s (corresponding with a large Λ), the minimum magnetizing current and thus the minimum transformer flux becomes very large. Eventually it is about four times half the magnetizing current or flux ripple ($\check{\mathcal{B}}_T \approx 4 \cdot \frac{1}{2}\Delta\mathcal{B}_T$) such that for those designs, the transformer cross section A_{eff} must be designed for significantly larger surface areas.

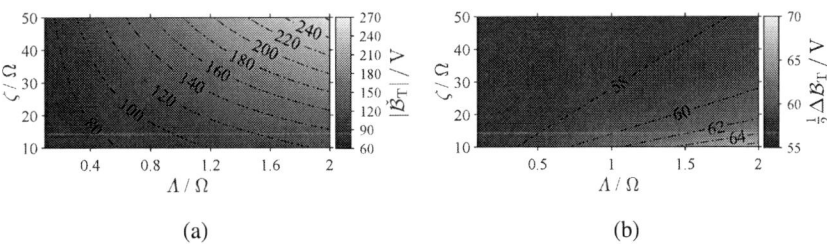

(a)   (b)

Fig. 6: Parameter dependent minimum magnetizing flux $\check{\mathcal{B}}_T$ (a) and flux ripple $\frac{1}{2}\Delta\mathcal{B}_T$ (b) with its dependency on Λ and ζ. The operating point is $\text{OP}_{Gmax,FL}|\hat{V}_{in}$.

2.4 ZVS consideration

The conventional ACFC operation can be used to achieve ZVS for both MOSFETs. The main switch S_1 is turned off at a large positive transformer current value i_{Ls}. During the turn-on of the auxiliary switch S_2, the current reverses (cf. Figure 1b) such that at the turn-off of this semiconductor, the transformer current is negative. With a proper design of the magnetizing inductance and the series inductance, both switches can be switched at zero voltage (ZVS). For reasons of brevity, only the switching transition of the main switch S_1 will be considered since the absolute value of the turn-off current of the auxiliary switch $|\check{i}_{mag}|$ is much lower compared to the turn-off value of the main switch $|\hat{i}_{mag} + \hat{i}_{Lg}|$. Furthermore, it is assumed that $\check{i}_{mag} + \frac{\check{i}_{Lg}}{N} \geq 0$. If this condition is not fulfilled, i_{Ls} is negative at the end of $t_{loss,D1}$ (cf. Figure 1b) such that ZVS can be achieved more easily.

When the auxiliary switch is turned-off, the output capacitance of S_1 is discharged while the one of S_2 is charged. The magnetizing inductance L_m does not participate in the transition as it is clamped to the transformed secondary-side blocking voltage v'_{SR1}. At the beginning of the transition, the energy available for ZVS is $E_{Ls} = \frac{1}{2}L_s\check{i}^2_{mag}$. The energy can, therefore, be influenced by the minimum magnetizing current value through the design of L_m and the series inductance L_s. While for other topologies, the ZVS condition can directly be calculated [3], this is not possible for the ACFC since three semiconductors (S_1, S_2 and SR_1) are involved during this condition. While preceding work [1] neglected the influence of the

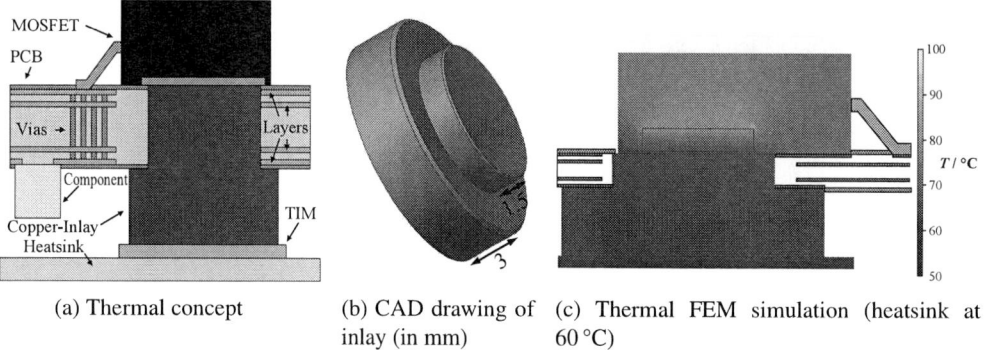

(a) Thermal concept

(b) CAD drawing of inlay (in mm)

(c) Thermal FEM simulation (heatsink at 60 °C)

Fig. 7: Employed cooling concept (a) for the primary and secondary semiconductors. A copper inlay is soldered into the PCB. Construction drawing of the inlay (b) and FEM thermal simulation of the inlay (c).

output capacitance of the synchronous rectifier, this results in large errors. Thus, computer simulations that considered the nonlinear output capacitance of the primary and secondary semiconductors were used to design the circuit for ZVS in this work.

2.5 Definition of the system parameters

From the above analysis, the inductance values were defined as follows: $\Lambda = 0.75\,\Omega$, $\Gamma = 0.175\,\Omega$ and $\zeta = 11.25\,\Omega$. While this does result in increased turn-off currents, the influence on the design of the transformer and reverse voltage are kept low. For the main switch S_1, the semiconductor $C3M0065090J$ (Cree 900 V, 65 mΩ) was selected, for the auxiliary MOSFET, the switch $C3M0120090J$ (Cree 900 V, 120 mΩ) was chosen. Both are bottom-cooled SMD components. The cooling concept will be discussed in Section 2.6. The secondary semiconductors are two parallel MOSFETs of type $IAUT300N10S5N015$ (Infineon Optimos 100 V, 1.5 mΩ) for SR_1 and SR_2. The switching frequency was set to 250 kHz to achieve a balance of small magnetic components and switching losses. With the definition of the switching frequency, the transformer was build up with a 7:1 turns ratio consisting of seven primary litz wire windings, three parallel copper sheet windings of 500 μm thickness and the transformer core ER54/38/20.

2.6 Cooling concept

To dissipate the heat of SMD semiconductors, a number of different methods has been used in the past [11]. By introducing thermal vias, the heat can be dissipated through the PCB vias connecting the top and bottom layer [12–15]. However, the possible heat dissipation is low resulting in a large thermal resistance. A better heat dissipation can be achieved through pedestals, where a piece of copper is placed between the top and bottom layer [11, 16, 17]. Similarly, it is also possible to insert a piece of ceramic (AlN) between the top and bottom layer for a reduced thermal resistance [15]. While these methods allow the application of a multi-layer PCB, the insertion of copper or a piece of ceramic is costly. Another possibility is the use of an insulated metal substrate [18–21] where the heat dissipation is very good but the layer number is limited to two. However, the heat dissipation is very good. In this work, the heat dissipation is achieved through a thermal inlay, which can be placed as an SMD component. Figure 7a shows the setup. The inlay is placed into the PCB and connected by solder to the sides. The component is placed on the other side on the top layer. A thermal interface material (TIM) is placed between the copper inlay and the heatsink to achieve the required isolation. A construction drawing of the inlay is depicted in Figure 7b. A FEM simulation is depicted in Figure 7c for a power dissipation of 20 W and a heatsink temperature of 60 °C.

3 Evaluation

The developed dual-rail prototype is depicted in Figure 8a. The dimensions of the prototype for two parallel ACFCs are (width×length×height) 193 mm × 28 mm × 210 mm without the aluminum chassis

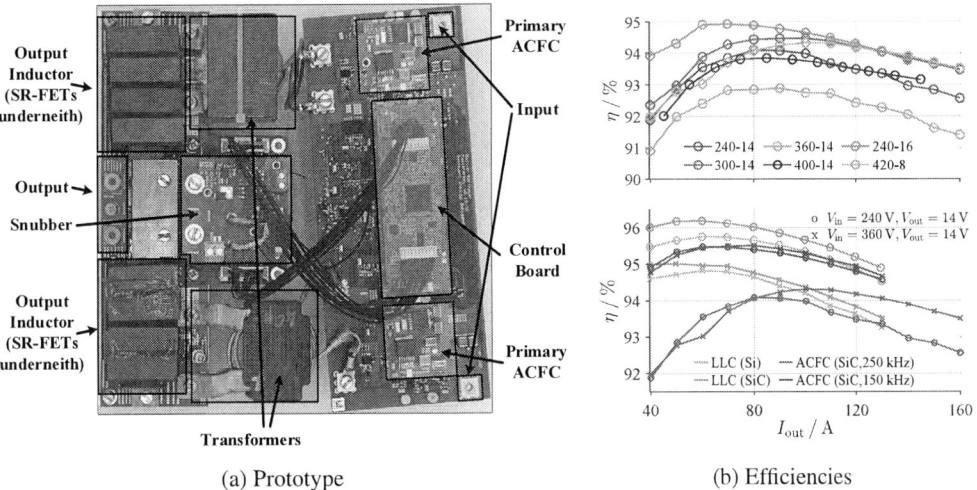

(a) Prototype   (b) Efficiencies

Fig. 8: (a) Developed dual rail 4.48 kW (2×2.24 kW) ACFC prototype and (b) measured efficiencies.

and the transformer as the largest component. With the designed chassis the dimensions are $220\,\text{mm} \times 43\,\text{mm} \times 235\,\text{mm}$ yielding in a power density of about $2\,\text{kW/l}$ considering the maximum transferred power of $2\times160\,\text{A}\cdot14\,\text{V} = 4.48\,\text{kW}$.

The right side of the prototype shows the primary PCB with the two times two primary semiconductors and the control board in the center. The transformers are depicted in the center with the $7:1$ transformer on the bottom and the $15:2$ transformer on the top. Between the transformers is the snubber circuit, which will be addressed in section 3.1. The left side shows the secondary PCB with the output inductor, which is assembled as a ferrite around a busbar. The synchronous rectifiers are placed below this ferrite.

3.1 Secondary semiconductor overshoot reduction

While the primary SiC MOSFETs are characterized by their small die area with a small output capacitance, the secondary rectifier MOSFETs are made of silicon, which results in a large output capacitance. In operation, this capacitance resonates with the series inductance of the transformer. Since the output capacitance is highly non-linear, the resulting overshoot may surpass twice the blocking voltage [22]. An exemplary shape of the secondary reverse voltages is depicted in Figure 9a. In preceding works, it was shown that the high energy of the oscillation prevents the application of a traditional $RC(D)$ snubber and that a regenerative snubber must be employed. The analysis and design of this snubber is covered in [22]. The employed solution uses a buck converter that controls the clamping voltage of the snubber, which is connected to the output. An exemplary operation with the employed snubber is depicted in Figure 9b. The overvoltages are clamped by the snubber voltage, which is controlled to about $60\,\text{V}$. A nominal operation could not be achieved without the developed snubber. Only by employing the clamping snubber, an operation at higher input voltages and output loads is possible.

3.2 Stationary results

The developed converter achieves a top efficiency of about $95\,\%$. Figure 8b (top) shows the measured efficiencies for different operating points over the output current. Since the high core losses are almost load independent, the top efficiency is limited. However, for increased output currents, the efficiency drop is quite limited and the converter achieves similar efficiencies for different operating points. To address the high core losses, the converter was modified to a 15:1 transformer (depicted in Figure 8a, top) with a magnetizing inductance of $105\,\mu\text{H}$ and an increased output inductance of $1.5\,\mu\text{H}$, which was operated at $150\,\text{kHz}$ (the converter is operated up to an output current of $130\,\text{A}$). The efficiencies are compared with the previous described ACFC and an LLC converter (with Si and SiC MOSFETs) for the same application in Figure 8b. The design of the resonant tank of the LLC converter with Si semiconductors has previously been covered in [23, 24]; the LLC converter with SiC semiconductors

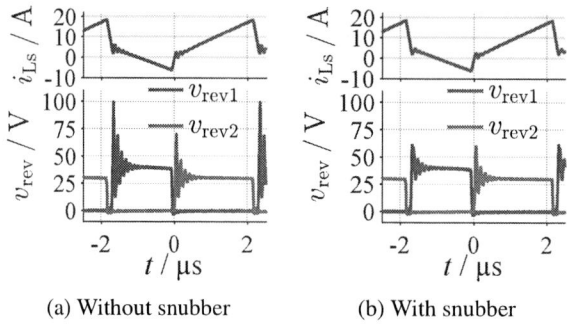

(a) Without snubber   (b) With snubber

Fig. 9: Synchronous rectifier reverse voltages; (a) operation without snubber, (b) operation with employed snubber (snubber voltage at 60 V).

C3M0060065J (Cree 650 V, 60 mΩ) employs the same resonant tank compared to the LLC converter with Si MOSFETs *IPW65R080CFDA* (Infineon super junction 650 V, 80 mΩ). The development of the SiC converter was described in [25]. Both LLC resonant converters are operated in a number of different operating modes to cover the wide voltage-transfer ratio, whereas the operating concept of the ACFC is much simpler. By considering the efficiency comparison of Figure 8b (bottom), it is evident that the efficiency of the LLC resonant converters reduces significantly for smaller voltage-transfer ratios (360 V to 14 V) whereas for an operation near resonance (240 V to 14 V), the efficiency is very high. Compared to the LLC resonant converter, the ACFC offers relatively stable efficiencies, independently from the operating point, which are also relatively stable over the output load.

It has also been tested whether the converter achieves ZVS for the nominal operating point at ($V_{in} = 345$ V, $V_{out} = 14$ V, $I_{out} = 64$ A). Figure 10a shows the nominal operation with the drain-source v_{S1} and gate voltage $v_{G,S1}$ of the main switch S_1 and the primary transformer current i_{Ls}. Figure 10b shows a zoom to the switching transient. It is evident that ZVS was achieved and that the transformer current is still negative when the drain-source voltage v_{S1} reaches zero.

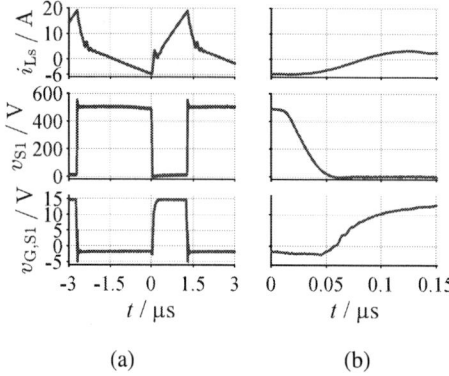

(a)   (b)

Fig. 10: ZVS analysis for the nominal operating point. (a) Measured operation and (b) zoom to the switching transient.

3.3 Model evaluation

The values of the transformer were measured with a Wayne Kerr 6500B to $L_{11} = 43\,\mu$H, $M_{12} = 6.32\,\mu$H and $L_{22} = 902\,$nH. If the connectors and the PCB tracks are estimated with a stray inductance of $L_{con} = 45\,$nH, the equivalent inductances are $L_s = 3.27\,\mu$H, $L_m = 37.8\,\mu$H and $n = 6.48$. These values were used for the calculation of the clamp-capacitor voltage in Figure 11a and the depiction of the calculated and measured primary transformer currents in Figure 11b and Figure 11c. The prediction of the detailed model fits very well. The maximum error is below 1 %. The simplified model, which was developed in

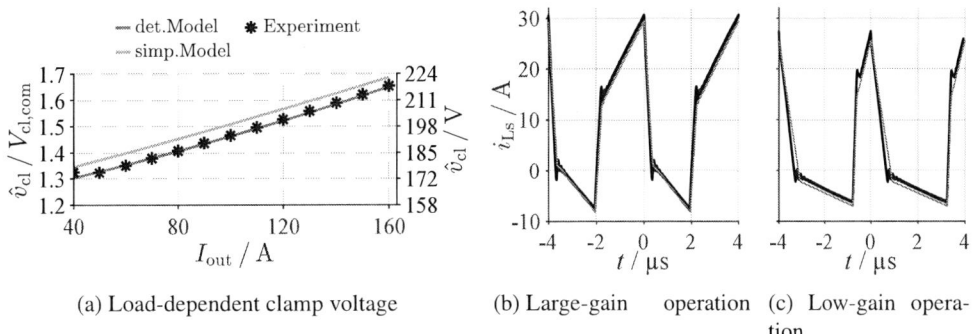

(a) Load-dependent clamp voltage   (b) Large-gain operation   (c) Low-gain operation

Fig. 11: (a) Validation of the clamp capacitor voltage model (detailed and simplified for the operating point $V_{\text{in}} = 290\,\text{V}$, $V_{\text{out}} = 14\,\text{V}$, $I_{\text{out}} = 130\,\text{A}$ (a). Modeled and measured primary transformer currents for the operating point (b) $V_{\text{in}} = 240\,\text{V}$, $V_{\text{out}} = 14\,\text{V}$, $I_{\text{out}} = 130\,\text{A}$ and (c) $V_{\text{in}} = 420\,\text{V}$, $V_{\text{out}} = 8\,\text{V}$, $I_{\text{out}} = 130\,\text{A}$.

2.1 also shows accurate results with errors below 3 %. The voltage is, however, slightly overestimated. Considering the measured and calculated current shape displayed in Figure 11b and Figure 11c, it is evident that the current slope of the model in the $t_{\text{loss,D1}}$ interval fits the measured current slope. Hence, this assumption of a connection stray inductance of $L_{\text{con}} = 45\,\text{nH}$ can be considered valid.

4 Conclusion

Silicon-carbide semiconductors enable the active-clamp forward converter as an attractive topology for applications with a wide voltage-transfer ratio. This paper presented the design and the experimental evaluation of an active-clamp forward converter applied as a single-stage onboard DC-DC converter of 2.24 kW. An experimental prototype revealed a maximum efficiency of 95 % while maintaining an efficiency larger than 90 % over almost the entire operating region. The achieved power density is 2 kW/l. Copper inlays that can be assembled as SMD components proved to be an effective cooling method achieving a maximum thermal resistance from junction to heatsink of $R_{\text{th}} = 2\,\text{K/W}$. Moreover, a developed high-accuracy steady-state model was validated on the prototype showing a maximum deviation of below 1 % compared to the conventional modeling approach that shows an error of about 70 %. An additional presented simplified model can be used for easy and accurate prediction of the clamp-capacitor voltage.

References

[1] P. Rehlaender, T. Grote, F. Schafmeister, and J. Böcker. "Analytical Modeling and Design of an Active Clamp Forward Converter Applied as a Single-Stage On-Board DC-DC Converter for EVs". In: *PCIM Europe*. VDE Verlag GmbH and IEEE, 2019.

[2] P. Rehlaender, F. Schafmeister, J. Böcker, and T. Grote. "Analytical Topology Comparison for a Single Stage On-Board EV-Battery Converter". In: *2019 IEEE 28th International Symposium on Industrial Electronics (ISIE)*. IEEE, 2019, pp. 2477–2482.

[3] M. Kasper, R. Burkat, F. Deboy, and J. Kolar. "ZVS of Power MOSFETs Revisited". In: *IEEE Transactions on Power Electronics* (2016), pp. 8063–8067.

[4] Ö. Bulut and M. T. Aydemir. "Design and Loss Analysis of a 200-W GaN Based Active Clamp Forward Converter". In: *International Conference on Electrical and Electronics Engineering* 5 (2018), pp. 97–100.

[5] A. Dheeraj and V. Rajini. "Comparison of active clamping circuits for isolated forward converter". In: *6th International Conference on Renewable Energy Research and Applications (ICRERA)*. IEEE, 2017, pp. 839–841.

[6] B.-R. Lin, F.-Y. Hsieh, D. Wang, and K. Huang. "Analysis, design and implementation of active clamp zero voltage switching converter with output ripple cancellation". In: *IEEE Proceedings - Electric Power Applications* 153 (2006), pp. 653–663.

[7] B.-R. Lin, C.-S. Yang, S.-C. Tsay, and D. Wang. "Analysis and Implementation of an Active Clamp ZVS Forward Converter". In: *2005 IEEE International Conference on Industrial Technology*. IEEE, 2005, pp. 1427–1432.

[8] P. Rehlaender, T. Grote, F. Schafmeister, and J. Böcker. "Interleaved Active Clamp Forward Converters as Single Stage On-Board DC-DC Converters for EVs – an Accurate Model and Design Considerations". In: *PCIM Europe*. VDE Verlag GmbH and IEEE, 2019.

[9] R. M. Burkart and J. W. Kolar. "Comparative eta-rho-sigma Pareto Optimization of Si and SiC Multilevel Dual-Active-Bridge Topologies With Wide Input Voltage Range". In: *IEEE Transactions on Power Electronics* 32.7 (2017), pp. 5258–5270.

[10] M. Albach. *Induktivitäten der Leistungselektronik: Spulen, Trafos und ihre parasitären Eigenschaften.* Wiesbaden: Springer Vieweg, 2017.

[11] B. Strothmann, T. Piepenbrock, F. Schafmeister, and J. Böcker. "Heat dissipation strategies for silicon carbide power SMDs and their use in different applications". In: *PCIM Europe digital days*. Ed. by Mesago Messe Frankfurt GmbH. Stuttgart, 2020.

[12] C. Negrea and P. Svasta. "Modeling of thermal via heat transfer performance for power electronics cooling". In: *2011 IEEE 17th International Symposium for Design and Technology in Electronic Packaging (SIITME)*. IEEE, 2011, pp. 107–110.

[13] D. S. Gautam, F. Musavi, D. Wager, and M. Edington. "A comparison of thermal vias patterns used for thermal management in power converter". In: *IEEE Energy Conversion Congress and Exposition*. IEEE, 2013, pp. 2214–2218.

[14] B. S. McCoy and M. A. Zimmermann. "Performance evaluation and reliability of thermal vias". In: *Nineteenth Annual IEEE Applied Power Electronics Conference and Exposition*. IEEE, 2004, pp. 1250–1256.

[15] J. Shao, F. Wei, X. Zhao, and J. Solovey. "Thermal Solutions for Surface Mount Power Devices". In: *PCIM Europe digital days 2020*. Ed. by Mesago Messe Frankfurt GmbH. Stuttgart, 2020.

[16] C.-C. Lee and W.-Y. Chen. "Coin insertion technology for PCB thermal solution". In: *5th International Microsystems Packaging Assembly and Circuits Technology Conference*. IEEE, 2010, pp. 1–4.

[17] C. Wei, J. Shao, B. Agrawal, D. Zhu, and H. Xie. "New Surface Mount SiC MOSFETs Enable High Efficiency High Power Density Bi-directional On-Board Charger with Flexible DC-link Voltage". In: *Applied Power Electronics Conference and Exposition (APEC)*. IEEE, 0.2019, pp. 1904–1909.

[18] X. Shao, Y. Cai, H. Li, and M. Wu. "Research of heat dissipation of RGB-LED backlighting system on LCD". In: *7th IEEE International Conference 2009*, pp. 807–812.

[19] F. Ludwig, T. Heidrich, and A. Mockel. "Integrated high-speed PMSM drive with IMS PCB-technology for mobile applications". In: *11th International Conference on Power Electronics and Drive Systems*. IEEE, 2015, pp. 1070–1073.

[20] X. Jorda, X. Perpina, M. Vellvehi, J. Millan, and A. Ferriz. "Thermal characterization of Insulated Metal Substrates with a power test chip". In: *21st International Symposium on Power Semiconductor Devices & IC's*. IEEE, 2009, pp. 172–175.

[21] A. Jafari, M. Samizadeh Nikoo, R. van Erp, and E. Matioli. "Optimized Kilowatt-Range Boost Converter Based on Impulse Rectification With 52 kW/l and 98.6% Efficiency". In: *IEEE Transactions on Power Electronics* 36.7 (2021), pp. 7389–7394.

[22] B. Korthauer, P. Rehlaender, F. Schafmeister, and J. Böcker. "Design and Analysis of a Regenerative Snubber for a 2.2 kW Active-Clamp Forward Converter with Low-Voltage Output". In: *APEC '21*. Piscataway, NJ: IEEE Service Center, 2021.

[23] P. Rehlaender, T. Grote, S. Tikhonov, M. Schröder, F. Schafmeister, and J. Böcker. "A 3,6 kW Single-Stage LLC Converter Operating in Half-Bridge, Full-Bridge and Phase-Shift Mode for Automotive Onboard DC-DC Conversion". In: *PCIM Europe 2020*. Ed. by Mesago Messe Frankfurt GmbH. Stuttgart, 2020.

[24] P. Rehlaender, F. Schafmeister, and J. Bocker. "Interleaved Single-Stage LLC Converter Design Utilizing Half- and Full-Bridge Configurations for Wide Voltage Transfer Ratio Applications". In: *IEEE Transactions on Power Electronics* 36.9 (2021), pp. 10065–10080.

[25] L. Hankeln. "Entwicklung eines LLC Konverters mit Siliziumkarbid-Halbleitern für den Einsatz als 400 V zu 12 V Bordnetztwandler". master thesis. Paderborn: Paderborn University, 2021.

A Simplified Model for the Battery Ageing Potential Under Highly Rippled Load

Tomáš Kacetl, Jan Kacetl, Nima Tashakor, Stefan Goetz

Technische Universität Kaiserslautern

Kaiserslautern, Germany

E-Mail: tomas.kacetl@porsche-engineering.de, tashakor@eit.uni-kl.de, jan.kacetl@porsche-engineering.de

Acknowledgements

The authors acknowledge the financial support by the Federal Ministry of Education and Research of Germany in the project "Open6GHub" (grant number: 16KISK004).

Keywords

«Aging», «Batteries», «Modelling», «Impedance Analysis», «Impedance Analysis»

Abstract

Influence of rippled load on lithium batteries is receiving increased attention. According to recent studies, accelerated ageing strongly depends on the frequency control of the ripple. We use electrochemical models to derive a highly simplified regression model that catches the asymptotic behavior and allows parameter identification and calibration to specific cells.

Introduction

High power and energy density stimulates the use of lithium-ion batteries in a wide range of applications, such as portable devices, battery energy storage systems (BESS), and electric vehicles [1-3]. During its lifetime, materials of the cell components undergo constant degradation associated with chemical and structural changes, which results either in loss of lithium inventory, loss of active material, or increase of the cell impedance [4, 5]. In addition to ambient temperature and state of charge (SoC), which plays an exclusive role in calendar ageing, character of the load shows to have dominant influence during cycling [6].

Demands on the battery load progressively increase with advances in semiconductor technology and scaling of power converters [7-9]. Especially the automotive industry pushes for peak performance and high-power charging, where batteries experience loads of multiples of the C-rate (ampere–hour capacity) [10, 11]. The discharge rate and related increase of the cell temperature during high current load follow Arrhenius law [12, 13] per

$$y_k = A \exp\left(-\frac{E_a}{RT} + \alpha(k)(I_k - I_{1C})\right) k^z. \tag{1}$$

In addition, the conversion of energy in a traction inverter further introduces a wide range of harmonics into the battery load [14]. Although the battery side of the inverter is typically stabilized by a DC-link capacitor, cost and space constraints drive a trend to reduce them so that a large share of the ripple propagates to the battery [15, 16]. Furthermore, alternative topologies such as battery-integrated modular multilevel converters and dynamically reconfigurable batteries generate even higher load ripple [17, 18]. Nevertheless, the ripple is commonly not considered in ageing models, and even cell manufacturers test and specify maximum current ratings for continuous load down to periods in second range, which is ripple-wise far from actual load conditions in automotive applications [17]. Similarly, also battery models tend to approximate battery behavior for a direct current (DC) load and neglect dynamics of the battery in millisecond range [19].

Nevertheless, studies experimenting with rippled load are recently appearing [20]. Despite yet some inconsistent conclusions of individual studies regarding the influence of the load ripple on battery ageing, the frequency of the ripple proves to have unambiguous impact. The studies almost uniformly report vanishing additional ageing at higher frequencies, which is unanimously assigned to the double–layer and electrode capacitance [21]. Shunting behavior of the battery cells at higher frequencies leads to lower activation of faradaic processes at the electrode–electrolyte interface, and the high frequency ripple is supplied from charge accumulated at the interface[22]. Despite molecule-level modeling approaches based on a pseudo 2D model of the cell as presented in [23], there is not yet any Arrhenius-like model which would approximate the effect in a simple form and might even run online on a vehicle's battery management system.

We combine features of the 2D model with equivalent circuit modeling, which yields and verifies the simplified model approximating the ageing behavior and which allows identification and calibration to specific cell. The simplified model can be used to benchmark modulation techniques, selection of switching frequency and necessity of filter installation.

In Section II, we review battery modeling approach and electro-chemical processes in the battery cell. Section III presents a combined modeling approach, which forms a simple relation between frequency and ageing. Section IV as the experimental part of the manuscript reveals shunting behavior of the battery cell concluded in Section V.

Battery cell modeling

Electrochemical model

Charging and discharging of battery cells is necessarily connected to (electro-)chemical activity and reactions, so-called faradaic processes. Through

$$C_6 + Li^+ + e^- \leftrightarrow LiC_6, \tag{2}$$

the lithium molecules are de-intercalated from one electrode, dissolved in the electrolyte, transported, and consequently intercalated into the counter electrode [24].

Intercalation on the electrode–electrolyte interface is governed by the Butler-Volmer (BV) equation per

$$j_{\text{int}}(\eta) = A\, j_{0,\text{int}} \left(\exp\left(\frac{\alpha_{\text{int}} F}{RT} \left(\eta - U_{\text{eq}}^{\text{int}} \right) \right) - \exp\left(-\frac{(1-\alpha_{\text{int}})F}{RT} \left(\eta - U_{\text{eq}}^{\text{int}} \right) \right) \right), \tag{3}$$

where

j_{int} intercalation current in A;
A surface of the porous electrode in m^2;
$j_{0,int}$ exchange current density in A m^{-2};
α_{int} charge transfer coefficient;
η overpotential in V;
T temperature in K;
F Faraday constant, $F = 96\,485$ C mol^{-1};
R gas constant, $R = 8.3145$ J mol^{-1} K^{-1};
U_{eq}^{int} equilibrium potential in V;

and the overpotential η represents difference of the electrochemical potential between solid electrode and liquid electrolyte phase. For small overpotentials ($\eta < 10$ mV), the equation can further be approximated and expressed as charge transfer resistance R_{CT} per

$$j_{\text{int}}(\eta) = A\, j_{0,\text{int}} \frac{nF}{RT} \eta \tag{4}$$

$$R_{\text{CT}} \approx \frac{\eta}{j_{\text{int}}} = \frac{RT}{nF i_{0,int}}, \tag{5}$$

where n is the number of electrons participating in the reaction [25].

Except for the faradaic (intercalation) current, the electrode–electrolyte interface can also contribute to the load by accumulated charge at the interface. In contrast to faradaic processes, this source of charge immediately responds to any disturbance in load and progressively builds up local overpotentials necessary for the activation of the intercalation processes. The effect is modelled as a parallel plate capacitor C_{dl}, and this non-faradaic current can described as [19]

$$
\begin{aligned}
j_{dl} &= A\, C_{dl}\, \frac{\partial \eta}{\partial t}, \\
j_{load} &= j_{int} + j_{dl}.
\end{aligned}
\tag{6}
$$

Further processes, such as ionic flux in the electrolyte is governed by Stefan–Maxwell diffusion in the electrolyte, Fick's second law diffusion in the solid electrode, etc. More details can be found in literature [24].

Solid-electrolyte interface layer and lithium plating

Aside from intercalation of the lithium in the electrode, especially the anode exhibits further chemical activity. Surface graphite electrodes react with the electrolyte and passivate the surface during initial cycles. The solid electrolyte interface layer (SEI) allows transport of lithium ions, but prevents further reaction with the electrolyte [26]. Nevertheless, growth of the SEI layer continues at a certain rate during lifetime of the battery cell. Equations (6) and (7) describe side reactions of electrolyte components—ethylene carbonate (EC) and dimethyl carbonate (DMC)—and production of key inorganic components of the SEI multilayer [27] per

$$
2Li^+ + 2(C_3H_4O_3)(EC) + 2e^- \leftrightarrow (CH_2OCO_2Li)_2 + C_2H_4,
\tag{7}
$$
$$
2Li^+ + C_3H_6O_3(DMC) + 2e^- \leftrightarrow Li_2CO_3 + C_2H_6.
\tag{8}
$$

Both reactions happen at the same interface as the main reaction (3). Kinetics of the reactions and SEI growth is governed by cathodic BV per

$$
j_{SEI}^{EC} = -F k_0^{EC} c_{EC}^s \left[\exp\left(-\frac{\alpha_{c,EC} F}{RT} \left(\eta - U_{SEI}^{EC} \right) \right) \right],
\tag{9}
$$
$$
j_{SEI}^{DMC} = -F k_0^{DMC} c_{DMC}^s \left[\exp\left(-\frac{\alpha_{c,DMC} F}{RT} \left(\eta - U_{SEI}^{DMC} \right) \right) \right].
\tag{10}
$$

At high overpotentials, the cell can additionally deposit lithium in metallic form directly onto surface of the graphite anode rather than intercalate into the lattice. Deposition of the metallic lithium, also called as lithium plating, is governed by

$$
Li^+ + e^- \leftrightarrow Li_{(s)},
\tag{11}
$$
$$
j_{pl} = -i_0^{pl} \left[\exp\left(-\frac{\alpha_{c,pl} F}{RT} \left(\eta - U_{eq}^{pl} \right) \right) \right].
\tag{12}
$$

The side reactions are irreversible and assumed to be a major sources of ageing of the cell [28]. Lithium plating and SEI growth are typically irreversible reactions, which contribute to the loss of lithium inventory and cause capacity fade [29]. In addition, growth of the SEI layer hinders mass transport, increases electrical impedance of the battery cell, and lowers performance. Metallic lithium grows dendrites, which may in catastrophic scenarios puncture the separator and cause internal electric short circuits as well as thermal runaway of the battery cell [30].

Randles' equivalent circuit

In contrast to complex electrochemical models, equivalent circuits of battery cells are widely used for their simplicity. Components of the cell, their properties, and internal chemical processes are approximated with combination of few electrical elements [31]. A reduced number of parameters further allows parameter identification with experimental data [32]. Figure 1 shows Randles' equivalent circuit, where both electrodes and intercalation processes are represented by a single charge-transfer resistance R_{CT}. Diffusion processes are generalized by the Warburg impedance Z_W, which can be further approximated by a string of R–C elements (see Figure 1) [33]. Charge in the capacitor C_{dl} alternatively contributes to non-faradaic current and corresponds to electrode capacitance in Equation 6. Mass transportation limitations through the SEI layer are reflected in another R–C component, where resistance R_{SEI} progressively increases as the cell ages. Resistance R_0 sums resistances of various components (electrolyte, contacts etc.), and inductance L_s represents parasitic inductance of cells as well as electrical interconnection. Finally, the potential of electrodes is simulated by internal voltage source dependent on the state of charge (SoC).

While electrochemical models offer more insight in the chemical processes, parameter identification of the models is despite several attempts not feasible with conventional electrical measurements [34-36]. Nevertheless, description of the internal processes and their dependencies can be conveniently combined with easily identifiable equivalent circuit model to get a simple degradation model per our further suggestions.

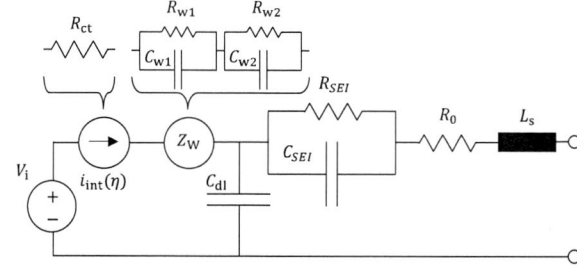

Figure 1. Randles equivalent circuit modeling approach with charge transfer and diffusion approximation.

Combined modeling approach

Finding a simple form of frequency dependency of the cell ageing requires shrinking and generalization of the electrochemical model combined with properties of the equivalent circuit. Equations (9)–(12) describe loss of lithium inventory as a function of the over-potential, which simultaneously drives kinetics of the intercalation in the Butler–Volmer equation (3) and can be found in the equivalent circuit as charge transfer resistance R_{CT} (see Figure 1).

Inspecting Equations 9–12, we assume an exponential relation between kinetics of the side reactions and overpotential. Since all corresponding reactions result in loss of lithium inventory, we sum them up as

$$j_{ageing} = j_{SEI}^{EC} + j_{SEI}^{DMC} + j_{pl} \tag{13}$$

and further rewrite the individual equations in the form of

$$
\begin{aligned}
j_{SEI}^{EC} &= -F k_0^{EC} c_{EC}^s \exp\left(U_{SEI}^{EC}\right)\left[\exp\left(-\frac{\alpha_{c,EC} F}{RT}\eta\right)\right], \\
j_{SEI}^{EC} &= -F k_0^{DMC} c_{DMC}^s \exp\left(U_{SEI}^{DMC}\right)\left[\exp\left(-\frac{\alpha_{c,DMC} F}{RT}\eta\right)\right], \\
j_{pl} &= -i_0^{pl} \exp\left(\frac{\alpha_{c,pl} F}{RT}U_{eq}^{pl}\right)\left[\exp\left(-\frac{\alpha_{c,pl} F}{RT}\eta\right)\right],
\end{aligned}
\tag{14}
$$

The reaction rate of the considered side-reactions is an essential indicator of battery ageing processes, and we therefore introduce an ageing potential term, which benchmarks the reaction rates under specific load conditions. Per the suggestion in [28, 37], the charge coefficients of the side reactions are equal and we can substitute them with a common coefficient $\alpha_{c,ag}$ in Eqs. (13)–(14) as

$$
\begin{aligned}
&\left(\alpha_{c,EC} = \alpha_{c,DMC} = \alpha_{c,pl}\right) = \alpha_{c,ag}, \\
&j_{ageing} = \left(k_{EC} + k_{DMC} + k_{pl}\right)\exp\left(-\frac{\alpha_{c,ag} F}{RT}\eta\right).
\end{aligned}
\tag{15}
$$

As remarked earlier, we can find the electrochemical overpotential directly in the charge–transfer resistance of the equivalent circuit. Using Eqs. (5) and (15), the ageing rate can be written as a function of the intercalation current j_{int} per

$$j_{\text{ageing}} = k_{\text{ag}} \exp\left(-\frac{\alpha_{\text{c,ag}}F}{RT} R_{\text{ct}} i_{\text{int}}\right). \tag{16}$$

Considering DC load, the intercalation current i_{int} is after all transients equal to the load current, which directly yields an exponential relation between ageing and load current, which can be cross validated to the Arrhenius law (1).

Nevertheless, during transients the electrode capacitance contributes to the load current according to Eq. (6) until the interfacial overpotential reaches high enough values to run intercalation reactions (ergo (dis-)charging current) at a sufficient rate. Similarly, the alternating current or any current ripple is partly supplied from the electrode capacitance. In principle, the battery cell offers two alternatives: a faradaic current related to chemical activity and a non-faradaic current related to charging of the electrode capacity.

The equivalent circuit (Figure 1) and impedance of the cell (Figure 2) represent this phe-

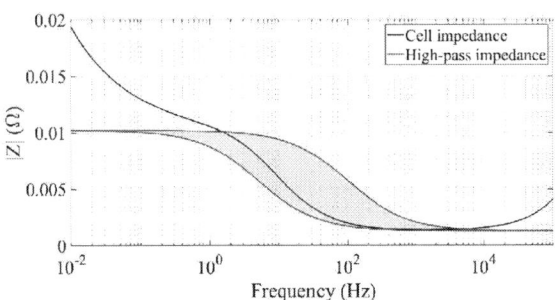

Figure 2. Cell impedance compared to high-pass characteristic.

nomenon under variable frequency. Electrode capacitance C_{dl} acts as shunting path, which allows the flow of high–frequency components around the electrochemical reactions at minimum impedance and adds high–pass characteristic to the cell as Figure 2 outlines. The high impedance of diffusion processes limiting the electrochemistry at low frequencies is progressively bypassed by shunting of the load through the electrode capacitance, where the impedance reaches its minimum (given by electrolyte resistance and metallic contacting of electrodes). At higher frequencies, the impedance rises as the inductance of the cells dominates. Nevertheless, for a substantial part of the frequency range, the cell impedance can be approximated as

$$Z_{\text{cell}} \sim \frac{1}{\sqrt{f_c^2 + f^2}}, \tag{17}$$

where f_c corresponds to the cut-off frequency of the high pass characteristic. The magnitude of the AC component of the intercalation current is then proportional to the load current as follows

$$\hat{I}_{\text{int}} = \frac{1}{\sqrt{f_c^2 + f^2}} \hat{I}_{\text{load}}. \tag{18}$$

The ageing potential (15) consist of two load components: a DC component I_{DC} equal to the average of the load current, and an AC component $\hat{I}_{\text{AC}}$ corresponding to the effective value of the AC component of the load current

$$j_{\text{ag}}^{AC} = \exp\left(-\frac{\alpha_{\text{c,ag}}F}{RT} R_{\text{ct}} \left(I_{\text{DC}} + \frac{1}{\sqrt{f_c^2 + f^2}} \hat{I}_{\text{AC}}\right)\right). \tag{19}$$

To separately extract the influence of the frequency on ageing, we introduce a so-called ageing potential as a proportion between the effects of pure DC load (where $\widehat{I}_{AC} = 0$), and load with an AC component per

$$AP = \frac{j_{ag}^{AC}}{j_{ag}^{DC}} = \frac{k_{ag}^{AC}}{k_{ag}^{DC}} \exp\left(-\frac{\alpha_{c,ag}F}{RT} R_{ct} \frac{1}{\sqrt{f_c + f^2}} \widehat{I}_{AC}\right),$$

$$AP = A \exp\left(\frac{B}{\sqrt{C + f^2}}\right). \tag{20}$$

The ageing potential model (20) factorizes the contribution of an AC component of certain frequency to the reference DC load. The suggested model is further validated through measurement of specific cell properties and presented in following section.

Experiment

Test setup

In the previous sections, we derived a simple ageing-potential model using combined cell modelling, which assumes high-pass characteristic of the cell impedance. We will further perform a battery measurement, fit the battery models presented in Section II, evaluate presumed attributes of the battery cell and estimate the cell ageing. Extracted data will be used to validate applicability of the suggested ageing-potential model (20).

Figure 3. Experimental setup with custom ripple generator.

We perform the measurement on a 6s module containing Sony/Murata US18650VTC5A 2600 mAh lithium-ion cells (Table I). The cells use a lithium nickel manganese cobalt oxide (NMC) cathode and a graphite anode. The data are measured at 80% SoC under an ambient temperature of 25°C.

The cell is loaded with a profile (see Fig. 4) which sweeps over frequencies ranging from 10 Hz up to 50 kHz. Rectangle pulses of the load profile are generated by custom electronics based on switched-inductor operation (see Fig. 3), where the output stage controls the magnitude and ripple of the inductor current. The test bench can reproduce load conditions of traction inverters. The load profile consists of initial 10 seconds of a DC current of 5 A to remove relaxation processes, which would affect further measurement (the DC character of the load is disturbed every 20 ms by switching the low side transistor for 30 μs to bootstrap the gate driver of the high side switches of the electronics). This initial phase is followed by increasing the amplitude up to 10 A while reducing the duty cycle down to 50 % (average current is kept at 5 A during the whole measurement). We measure the terminal voltage and the battery current at the battery terminals by 4-wire method. The current is sensed by a Hall-effect probe, where both current and voltage probes have a measurement bandwidth > 100 kHz.

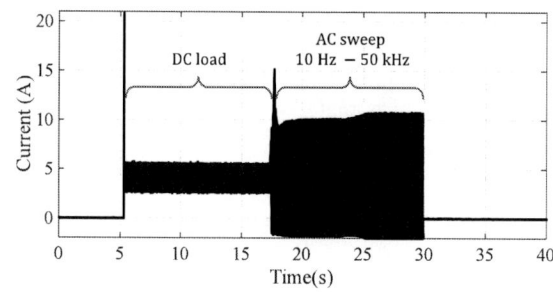

Figure 4. Measured load profile with initial DC load and a frequency sweep.

TABLE I
BATTERY SPECIFICATION

Load parameter	Value
Cell type	US18650VTC5A (Murata/Sony)
Nominal cell capacity	2 600 mAh
Internal cell impedance	10 mΩ
Cell max discharge current	30 A
Battery configuration	6s 1p

Battery modeling and parameter estimation

The measurement data are used to identify parameters of the battery model of Section II, where the Warburg impedance is approximated with two R–C pairs. Rather than a charge transfer resistance only, the intercalation processes are represented by the full Butler–Volmer (Equation 3) to avoid limitations of the approximation. The fidelity of the model can be reviewed in Fig. 5, which compares measured and simulated voltage responses at various frequencies.

Identified parameters are displayed in Table II, and the model is further used to simulate overpotentials at loads of various frequencies. The identified battery model is experimentally loaded with a set of sinusoidal load profiles (Fig. 5), and the overpotential at the electrode–electrolyte interface is used to identify kinetics of the side reactions described in Section II.b. Loss of the lithium inventory following from Eqs. (9)–(12) is compared to a reference experiment with purely DC component. The ageing potential is simulated for frequencies ranging from 1 Hz up to 100 kHz and further used to vali-

TABLE II
BATTERY MODEL PARAMETERS

Parameter	Value
Open circuit voltage, V_{OCV}	22 V
Exchange current, $i_{0,int}$	0.44 A
Charge transfer coefficient	0.5
Battery resistance, R_0	77.5 mΩ
Battery inductance, L_0	533 nH
SEI resistance, R_{SEI}	67 mΩ
SEI capacitance, C_{SEI}	23 mF
Electrode capacitance, C_{dl}	2.6 mF
Diffusion resistance, R_{W1}	0.6 mΩ
Diffusion capacitance, C_{W1}	3.5 mF
Diffusion resistance, R_{W2}	30 mΩ
Diffusion capacitance, C_{W2}	258 F
Temperature, T	298.15 K
Gas constant, R	8.314462 J K^{-1} mol^{-1}
Faraday constant, F	96 485.33 C mol^{-1}

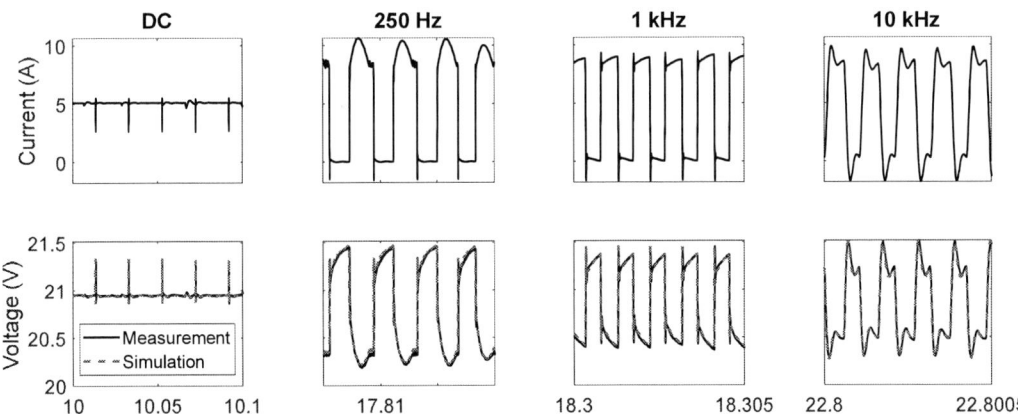

Figure 5. Synopsis of the regression model and measurement

date our proposed ageing potential model.

Results

In addition to the ageing potential, the model also offers insights into the frequency influence on the battery loss and the source impedance of the cell. The AC impedance of the cell is lower than pure DC until the inductance starts limiting thanks to the shunting capability of the electrode capacitance (see Figure 6). Nevertheless, the impedance rapidly rises at higher frequencies as the cell inductance starts to dominate. Further, the cell exhibits increased losses related to an increased root mean square (RMS) current with superimposed AC component. Interestingly, the losses are progressively reduced at higher frequencies due to the decreased impedance of the cell. Similarly, the model explains the known drop of battery ageing potential at higher frequencies [23, 38-42]. The reduced ageing can be widely attributed to the shunting effect of the electrode capacitance, which is translated to decreased values of the overpotential. At higher frequencies, values of the overpotential are effectively filtered down to average values, which comparable to the DC load and so is the ageing at these frequencies.

We further use the data of the relative ageing potential to evaluate the feasibility of our simplified model (20). The parameters of the model are identified using least-square regression. The results of our simplified approximation are compared to the extracted values of ageing potential in Figure 7. The resultant ageing potential function can be expressed as

$$AP = 1.93 \exp\left(\frac{5.16\cdot10^6}{\sqrt{6.79\cdot10^5+f^2}}\right), \qquad (21)$$

where the R-squared coefficient of our further simplified ageing-potential approximation function (21) is $R^2 = 0.995$. Our simplified model fits well both low– and high–frequency regions and features a slightly sharper transition than the more detailed analytical model.

Conclusion

We introduced an ageing-potential model for the evaluation of ageing under rippled load. The model is derived from low-level chemistry models, from which we extracted the core of our ageing-potential model. The model was further conveniently combined with equivalent circuit modeling, which brought about the possibility to identify the model and calibrate it to a cell.

We further forked off a highly simplified regression model that catches the asymptotic behavior and allows very simple regression. The complexity of the model is on a similar

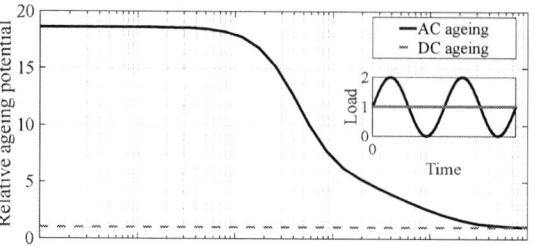
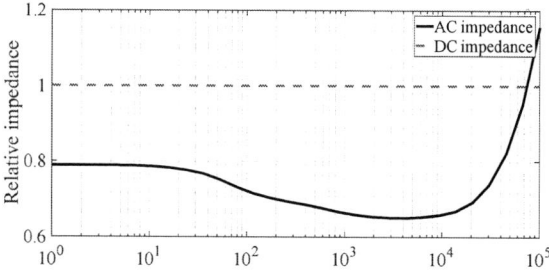
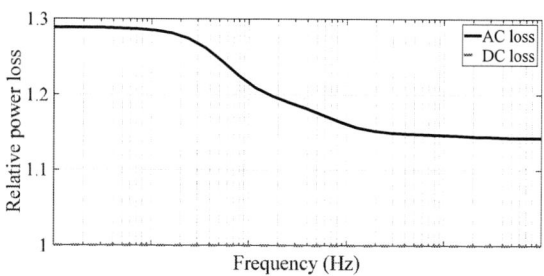

Figure 6. Results of the combined modeling approach. The model exploits ageing potential under rippled condition together with power loss and source impedance of examined cell.

level as the conventional Arrhenius model and offers practical usability. Despite all simplifications, the suggested model features high fidelity with R-squared coefficient > 0.99 with respect to data extracted from cell measurements and chemical modeling.

The suggested model may serve for benchmarking various battery cells in applications with inherently high load ripple (e.g., battery-based modular multilevel converters, electric vehicle drive trains with reduced dc–link capacitors), appropriate selection of switching frequency, and control strategies, as well as proper design of DC-link capacitors or input filter for battery interfaced inverters.

References

[1] N. Tashakor, B. Arabsalmanabadi, F. Naseri, and S. Goetz, "Low-Cost Parameter Estimation Approach for Modular Converters and Reconfigurable Battery Systems Using Dual Kalman Filter," *IEEE Transactions on Power Electronics,* vol. 37, no. 6, pp. 6323-6334, 2022, doi: 10.1109/TPEL.2021.3137879.

[2] I. Aghabali, J. Bauman, P. J. Kollmeyer, Y. Wang, B. Bilgin, and A. Emadi, "800-V Electric Vehicle Powertrains: Review and Analysis of Benefits, Challenges, and Future Trends," *IEEE Transactions on Transportation Electrification,* vol. 7, no. 3, pp. 927-948, 2021, doi: 10.1109/TTE.2020.3044938.

[3] B. Yang *et al.*, "Classification, summarization and perspectives on state-of-charge estimation of lithium-ion batteries used in electric vehicles: A critical comprehensive survey," *Journal of Energy Storage,* vol. 39, p. 102572, 2021/07/01/ 2021, doi: https://doi.org/10.1016/j.est.2021.102572.

[4] J. Vetter *et al.*, "Ageing mechanisms in lithium-ion batteries," *Journal of Power Sources,* vol. 147, no. 1, pp. 269-281, 2005/09/09/ 2005, doi: https://doi.org/10.1016/j.jpowsour.2005.01.006.

[5] B. Arabsalmanabadi, N. Tashakor, S. Goetz, and K. Al-Haddad, "Li-ion Battery Models and A Simplified Online Technique to Identify Parameters of Electric Equivalent Circuit Model for EV Applications," presented at the IECON 2020 The 46th Annual Conference of the IEEE Industrial Electronics Society, 18-21 Oct. 2020, 2020.

[6] E. Redondo-Iglesias, P. Venet, and S. Pelissier, "Modelling Lithium-Ion Battery Ageing in Electric Vehicle Applications—Calendar and Cycling Ageing Combination Effects," vol. 6, no. 1, p. 14, 2020. [Online]. Available: https://www.mdpi.com/2313-0105/6/1/14.

[7] M. A. Rahman, K. d. Craemer, J. Büscher, J. Driesen, P. Coenen, and C. Mol, "Comparative Analysis of Reconfiguration Assisted Management of Battery Storage Systems," presented at the IECON 2019 - 45th Annual Conference of the IEEE Industrial Electronics Society, 14-17 Oct. 2019, 2019.

[8] Y. Zhu, W. Zhang, J. Cheng, and Y. Li, "A novel design of reconfigurable multicell for large-scale battery packs," presented at the 2018 International Conference on Power System Technology (POWERCON), 6-8 Nov. 2018, 2018.

[9] S. Ci, N. Lin, and D. Wu, "Reconfigurable Battery Techniques and Systems: A Survey," *IEEE Access,* vol. 4, pp. 1175-1189, 2016, doi: 10.1109/ACCESS.2016.2545338.

[10] S. Rivera, S. Kouro, S. Vazquez, S. M. Goetz, R. Lizana, and E. Romero-Cadaval, "Electric Vehicle Charging Infrastructure: From Grid to Battery," *IEEE Industrial Electronics Magazine,* vol. 15, no. 2, pp. 37-51, 2021, doi: 10.1109/MIE.2020.3039039.

[11] N. Wassiliadis *et al.*, "Review of fast charging strategies for lithium-ion battery systems and their applicability for battery electric vehicles," vol. 44, p. 103306, 2021.

[12] X. Han, M. Ouyang, L. Lu, and J. J. J. o. P. S. Li, "A comparative study of commercial lithium ion battery cycle life in electric vehicle: Capacity loss estimation," vol. 268, pp. 658-669, 2014.

[13] C. Liu, Y. Wang, and Z. J. E. Chen, "Degradation model and cycle life prediction for lithium-ion battery used in hybrid energy storage system," vol. 166, pp. 796-806, 2019.

[14] B. P. McGrath and D. G. Holmes, "A general analytical method for calculating inverter DC-link current harmonics," vol. 45, no. 5, pp. 1851-1859, 2009.

[15] G. Bon-Gwan and N. Kwanghee, "A DC-link capacitor minimization method through direct capacitor current control," *IEEE Transactions on Industry Applications,* vol. 42, no. 2, pp. 573-581, 2006, doi: 10.1109/TIA.2006.870036.

[16] N. Tashakor, J. Kacetl, J. Fang, Z. Li, and S. Goetz, "Dual-Port Dynamically Reconfigurable Battery with Semi-Controlled and Fully-Controlled Outputs," *arXiv preprint arXiv:2206.01435,* 2022.

[17] Z. Li, R. Lizana, S. M. Lukic, A. V. Peterchev, and S. M. Goetz, "Current Injection Methods for Ripple-Current Suppression in Delta-Configured Split-Battery Energy Storage," *IEEE Transactions on Power Electronics,* vol. 34, no. 8, pp. 7411-7421, 2019, doi: 10.1109/TPEL.2018.2879613.

[18] Z. Li, R. Lizana, A. V. Peterchev, and S. M. Goetz, "Ripple current suppression methods for star-configured modular multilevel converters," in *IECON 2017 - 43rd Annual Conference of the IEEE Industrial Electronics Society,* 29 Oct.-1 Nov. 2017 2017, pp. 1505-1510, doi: 10.1109/IECON.2017.8216256.

[19] N. Legrand, S. Raël, B. Knosp, M. Hinaje, P. Desprez, and F. Lapicque, "Including double-layer capacitance in lithium-ion battery mathematical models," *Journal of Power Sources,* vol. 251, pp. 370-378, 2014/04/01/ 2014, doi: https://doi.org/10.1016/j.jpowsour.2013.11.044.

[20] W. Vermeer, M. Stecca, G. R. C. Mouli, and P. Bauer, "A Critical Review on The Effects of Pulse Charging of Li-ion Batteries," in *2021 IEEE 19th International Power Electronics and Motion Control Conference (PEMC),* 25-29 April 2021 2021, pp. 217-224, doi: 10.1109/PEMC48073.2021.9432555.

[21] T. Kacetl, J. Kacetl, J. Fang, M. Jaensch, and S. Goetz, "Degradation-Reducing Control for Dynamically Reconfigurable Batteries," *arXiv preprint,* 2022.

[22] A. Jossen, "Fundamentals of battery dynamics," *Journal of Power Sources,* vol. 154, no. 2, pp. 530-538, 2006/03/21/ 2006, doi: https://doi.org/10.1016/j.jpowsour.2005.10.041.

[23] A. Bessman, R. Soares, O. Wallmark, P. Svens, and G. Lindbergh, "Aging effects of AC harmonics on lithium-ion cells," *Journal of Energy Storage,* vol. 21, pp. 741-749, 2019/02/01/ 2019, doi: https://doi.org/10.1016/j.est.2018.12.016.

[24] M. Doyle, T. F. Fuller, and J. Newman, "Modeling of Galvanostatic Charge and Discharge of the Lithium/Polymer/Insertion Cell," *Journal of The Electrochemical Society,* vol. 140, no. 6, pp. 1526-1533, 1993/06/01 1993, doi: 10.1149/1.2221597.

[25] A. Bessman *et al.*, "Challenging Sinusoidal Ripple-Current Charging of Lithium-Ion Batteries," *IEEE Transactions on Industrial Electronics,* vol. 65, no. 6, pp. 4750-4757, 2018.

[26] A. Wang, S. Kadam, H. Li, S. Shi, and Y. J. n. C. M. Qi, "Review on modeling of the anode solid electrolyte interphase (SEI) for lithium-ion batteries," vol. 4, no. 1, pp. 1-26, 2018.

[27] S. Atalay, M. Sheikh, A. Mariani, Y. Merla, E. Bower, and W. D. J. J. o. P. S. Widanage, "Theory of battery ageing in a lithium-ion battery: Capacity fade, nonlinear ageing and lifetime prediction," vol. 478, p. 229026, 2020.

[28] X.-G. Yang, Y. Leng, G. Zhang, S. Ge, and C.-Y. J. J. o. P. S. Wang, "Modeling of lithium plating induced aging of lithium-ion batteries: Transition from linear to nonlinear aging," vol. 360, pp. 28-40, 2017.

[29] P. Arora, R. E. White, and M. Doyle, "Capacity Fade Mechanisms and Side Reactions in Lithium‐Ion Batteries," *Journal of The Electrochemical Society,* vol. 145, no. 10, pp. 3647-3667, 1998/10/01 1998, doi: 10.1149/1.1838857.

[30] W. Cai *et al.*, "The Boundary of Lithium Plating in Graphite Electrode for Safe Lithium‐Ion Batteries," vol. 133, no. 23, pp. 13117-13122, 2021.

[31] C. Brivio, V. Musolino, M. Merlo, and C. Ballif, "A Physically-Based Electrical Model for Lithium-Ion Cells," *IEEE Transactions on Energy Conversion,* vol. 34, no. 2, pp. 594-603, 2019, doi: 10.1109/TEC.2018.2869272.

[32] S. M. M. Alavi, A. Mahdi, S. J. Payne, and D. A. Howey, "Identifiability of Generalized Randles Circuit Models," *IEEE Transactions on Control Systems Technology,* vol. 25, no. 6, pp. 2112-2120, 2017, doi: 10.1109/TCST.2016.2635582.

[33] A. G. Li, K. Mayilvahanan, A. C. West, and M. J. J. o. P. S. Preindl, "Discrete-time modeling of Li-ion batteries with electrochemical overpotentials including diffusion," vol. 500, p. 229991, 2021.

[34] L. Xu, X. Lin, Y. Xie, and X. J. E. S. M. Hu, "Enabling high-fidelity electrochemical P2D modeling of lithium-ion batteries via fast and non-destructive parameter identification," vol. 45, pp. 952-968, 2022.

[35] B. Rajabloo, A. Jokar, M. Désilets, and M. J. J. o. T. E. S. Lacroix, "An inverse method for estimating the electrochemical parameters of lithium-ion batteries," vol. 164, no. 2, p. A99, 2016.

[36] R. Masoudi, T. Uchida, and J. J. J. o. P. S. McPhee, "Parameter estimation of an electrochemistry-based lithium-ion battery model," vol. 291, pp. 215-224, 2015.

[37] M. Safari, M. Morcrette, A. Teyssot, and C. J. J. o. T. E. S. Delacourt, "Multimodal physics-based aging model for life prediction of Li-ion batteries," vol. 156, no. 3, p. A145, 2008.

[38] L. R. Chen, J. J. Chen, C. M. Ho, S. L. Wu, and D. T. Shieh, "Improvement of Li-ion Battery Discharging Performance by Pulse and Sinusoidal Current Strategies," *IEEE Transactions on Industrial Electronics,* vol. 60, no. 12, pp. 5620-5628, 2013, doi: 10.1109/TIE.2012.2230599.

[39] M. Uno and K. Tanaka, "Influence of High-Frequency Charge–Discharge Cycling Induced by Cell Voltage Equalizers on the Life Performance of Lithium-Ion Cells," *IEEE Transactions on Vehicular Technology,* vol. 60, no. 4, pp. 1505-1515, 2011, doi: 10.1109/TVT.2011.2127500.

[40] M. J. Brand, M. H. Hofmann, S. S. Schuster, P. Keil, and A. Jossen, "The Influence of Current Ripples on the Lifetime of Lithium-Ion Batteries," *IEEE Transactions on Vehicular Technology,* vol. 67, no. 11, pp. 10438-10445, 2018.

[41] L. Chen, S. Wu, D. Shieh, and T. Chen, "Sinusoidal-Ripple-Current Charging Strategy and Optimal Charging Frequency Study for Li-Ion Batteries," *IEEE Transactions on Industrial Electronics,* vol. 60, no. 1, pp. 88-97, 2013, doi: 10.1109/TIE.2012.2186106.

[42] Y. Lee and S. Park, "Electrochemical State-Based Sinusoidal Ripple Current Charging Control," *IEEE Transactions on Power Electronics,* vol. 30, no. 8, pp. 4232-4243, 2015, doi: 10.1109/TPEL.2014.2354013.

System Modeling and Design of a Hybrid Renewable Energy System for a Cable Network Head-End Station in Rural Area

Tobias Schillinger, Thomas Schuhmann and Martin Eckart
University of Applied Sciences Dresden
Friedrich-List-Platz 1
01069 Dresden, Germany
Phone: +49 (0) 351-4623068
Email: tobias.schillinger@htw-dresden.de
URL: https://www.htw-dresden.de/ema

Acknowledgements

This project is co-financed with tax funds on the basis of the budget passed by the Saxon State parliament.

Keywords

≪Renewable energy systems≫, ≪Simulation≫, ≪Energy system management≫, ≪Wind energy≫, ≪PV active generator≫

Abstract

This paper focuses on system modeling of a small scale renewable hybrid energy system for a cable network head-end station in rural areas located in central European lower mountain and lowland regions. Based on one year measured energy demand and local weather data the entire system model allows a location dependent energetic simulation and optimization for individual configurations of the photovoltaic, wind energy and battery storage systems. Using selected examples, different system configurations at varying locations are simulated and compared to each other with regard to the number of photovoltaic modules, size of the battery and power of the small wind turbine.

Introduction

The expansion of digital infrastructure in rural areas is accompanied by increasing energy demand of the signal processing systems and requires a high availability of the grid connection. The consequence of the loosely interconnected grid in rural areas is a high vulnerability with respect to power failures. Furthermore this development stands in contrast with the aim of reducing the reliance on fossil fuels. For these reasons it is necessary to crossover to decentralized sustainable energy systems with a high degree of self sufficiency, consisting of small wind energy, photovoltaic and battery storage systems. A further aspect is the relating reduction of the energy procurement costs. For this system the usage of second life lithium-ion traction batteries with a minimum residual capacity of approx. 80 % comes into question [1]. This paper describes a two stage energetic simulation of a hybrid renewable grid-connected energy system with a solar energy conversion system (SECS), small wind energy conversion system (WECS) and a battery energy storage system (BESS) in MATLAB/SIMULINK using the example of a cable network head-end station. Typically such stations convert incoming fiber optic signals and provide local broadband networks for small towns [2]. The structure of the proposed system is depicted in Fig. 1. The three subsystems, SECS, WECS & BESS, are modeled separately. Considering the losses of the power converters, every system is simulated with the MATLAB Parallel Computing Toolbox to obtain a subsystem model with reduced complexity based on characteristic diagrams. These lookup table (LUT) based subsystems are connected with an energy management system (EMS) to an entire system model which is

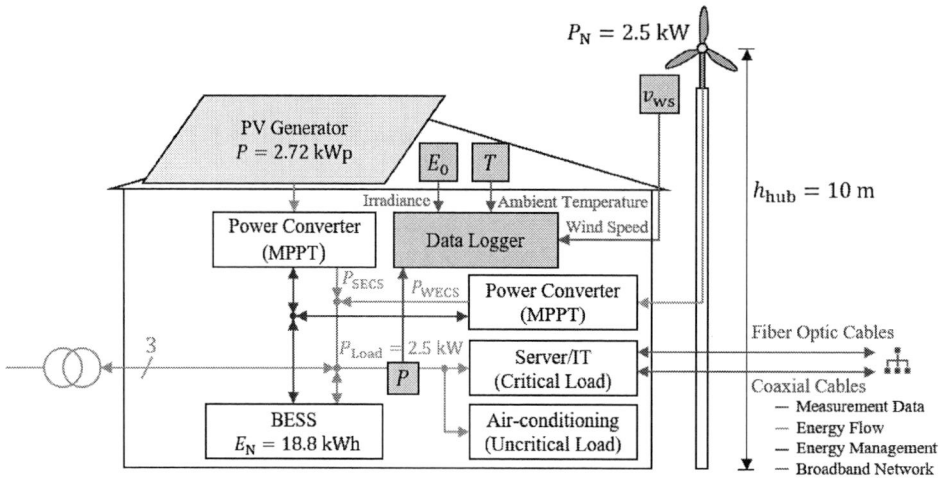

Fig. 1: Structure of the grid connected cable network head-end station with renewable energy and measurement system.

used for processig the measured annual data. These data were logged over a full year and include the energy demand of the three phase load, solar irradiance, ambient temperature and the wind speed. The load is divided into critical and not critical loads towards grid power failures.

This paper is organized as follows: The modeling aspects of the detailed subsystems and the entire system are given in the first section. The subsequent section presents the results of the simulation study based on four different exemplary locations as shown in Table V. In the conclusion, based on the simulation study, criteria for an optimal design of hybrid small-scale regenerative energy systems are discussed.

Modeling

System overview

The technical structure of the entire system model is shown in Fig. 2. The model is divided in three separate SIMULINK-models, SECS, WECS and BESS. A maximum power point tracking algorithm (MPPT) evaluates voltage and current from the PV array respectively the uncontrolled rectifier (B6U) of the wind generator system and controls the boost converter in the SECS and WECS. The two-level inverters are controlled by an inner current control and an outer DC voltage control loop. A phase locked loop calculates the phase angle for the Park transformation. The buck-boost converter controls the power flow for charging or discharing the battery. All power electronic components are modeled with Simscape insulated-gate-bipolar-transistors (IGBT). The operation point dependent losses of each IGBT are calculated based on [3] in combination with a thermal network of the heat sink. Typical IGBT characteristics are chosen from *Semikron skm50gb123D*.

Detailed modeling of system components

Two level inverter control

The single phase equivalent circuit of the three phase grid connected to the voltage source converter (VSC) is shown in Fig. 3. $\underline{v}_{VSC}$ is the voltage of the VSC, R_f the equivalent grid resistance, L_f the inductance, $\underline{v}_f$ and $\underline{i}_f$ are the grid voltage respectively current with the anguular frequency ω_f. The following equation (1) describes the space vector model of this grid transformed to a coordinate system rotating with ω_f, the dq-system.

$$\vec{v}_{VSC}^{dq} = v_{VSC,d} + j v_{VSC,q} = R_f \vec{i}_f^{dq} + L_f \frac{d\vec{i}_f^{dq}}{dt} + j\omega_f L_f \vec{i}_f^{dq} + \vec{v}_f^{dq} \tag{1}$$

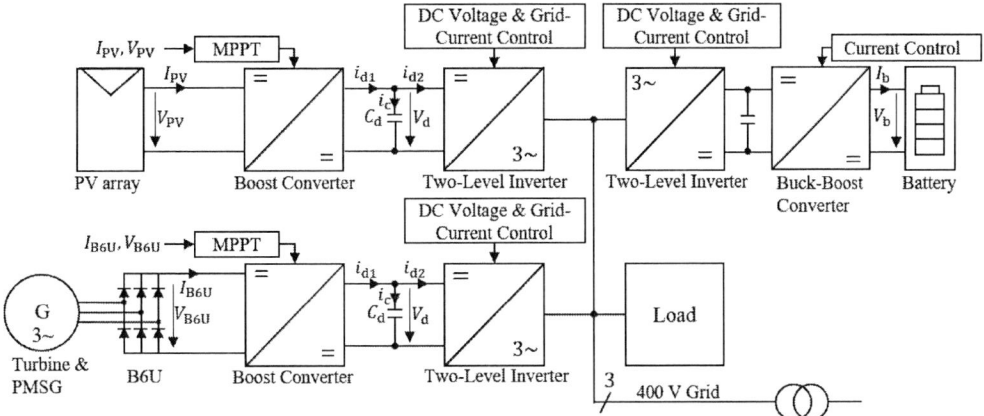

Fig. 2: Overview of the grid connected renewable energy system model.

After splitting up the $\mathcal{L}$-transfrom of (1) to its components and adding the following equations (2) for feed forward control, the transfer functions of the current control system are resulting to (3).

$$v_{\mathrm{ff,d}}(s) = -\omega_{\mathrm{f}} L_{\mathrm{f}} i_{\mathrm{f,q}}(s) + v_{\mathrm{f,d}}(s) \qquad v_{\mathrm{ff,q}}(s) = \omega_{\mathrm{f}} L_{\mathrm{f}} i_{\mathrm{f,d}}(s) + v_{\mathrm{f,q}}(s) \tag{2}$$

$$G_{\mathrm{f,d}}(s) = \frac{i_{\mathrm{f,d}}(s)}{v_{\mathrm{VSC,d}}(s)} = \frac{1/R_{\mathrm{f}}}{1 + s\frac{L_{\mathrm{f}}}{R_{\mathrm{f}}}} \qquad G_{\mathrm{f,q}}(s) = \frac{i_{\mathrm{f,q}}(s)}{v_{\mathrm{VSC,q}}(s)} = \frac{1/R_{\mathrm{f}}}{1 + s\frac{L_{\mathrm{f}}}{R_{\mathrm{f}}}} \tag{3}$$

The PI current controllers are adjusted with the Magnitude Optimum. The transfer function $G_{\mathrm{PI,i}}$, integral time $T_{\mathrm{I,i}}$, proportional gain $K_{\mathrm{P,i}}$ and the equivalent time constant T_{eq} considering the pulse frequency f_{p} are given in (4).

$$G_{\mathrm{PI,i}}(s) = K_{\mathrm{P,i}} \frac{1 + sT_{\mathrm{I,i}}}{sT_{\mathrm{I,i}}} \qquad T_{\mathrm{I,i}} = \frac{L_{\mathrm{f}}}{R_{\mathrm{f}}} \qquad K_{\mathrm{P,i}} = \frac{L_{\mathrm{f}}}{2\,T_{\mathrm{eq,i}}} \qquad T_{\mathrm{eq,i}} = \frac{1}{2\,f_{\mathrm{p}}} \tag{4}$$

For active power control the set point $i_{\mathrm{f,d,ref}}$ is calculated with (5) while $i_{\mathrm{f,q,ref}} = 0$.

$$P_{\mathrm{f}} = \frac{3}{2}\mathrm{Re}\left\{ \vec{v}_{\mathrm{f}}^{\,\mathrm{dq}} \vec{i}_{\mathrm{f}}^{\,\mathrm{dq}*} \right\} = \frac{3}{2} v_{\mathrm{f,d}}\, i_{\mathrm{f,d}} \tag{5}$$

The current i_{d1} that comes from the preconnected system, e. g. PV array or uncontrolled rectifier, partially loads the DC link capacitance C_{d}, see (6). It results the input current i_{d2} of the VSC. Excluding the losses of the VSC, with the power balance (7) the integral time $T_{\mathrm{I,v}}$ and proportional gain $K_{\mathrm{P,v}}$ of the DC voltage PI controller $G_{\mathrm{PI,v}}$ (8) are calculated with the Symmetrical Optimum, as described in [4].

$$i_{\mathrm{c}} = i_{\mathrm{d1}} - i_{\mathrm{d2}} = C_{\mathrm{d}} \frac{\mathrm{d}v_{\mathrm{d}}}{\mathrm{d}t} \tag{6}$$

$$P_{\mathrm{d}} = P_{\mathrm{f}} = i_{\mathrm{d2}}\, v_{\mathrm{d}} \tag{7}$$

$$G_{\mathrm{PI,v}}(s) = K_{\mathrm{P,v}} \frac{1 + sT_{\mathrm{I,v}}}{sT_{\mathrm{I,v}}} \qquad T_{\mathrm{I,v}} = 32\,T_{\mathrm{eq,i}} \qquad K_{\mathrm{P,v}} = \frac{C_{\mathrm{d}}}{32\,T_{\mathrm{eq,i}} \frac{3}{2} \frac{v_{\mathrm{f,d}}}{V_{\mathrm{d,ref}}}} \tag{8}$$

Furthermore the switching signals for the IGBTs are generated with space vector modulation. The controlled two level inverter as a part of every subsystem is parametrized as follows:

$$R_{\mathrm{f}} = 1\,\Omega \quad L_{\mathrm{f}} = 1\,\mathrm{mH} \quad \omega_{\mathrm{f}} = 2\pi\,50\,\mathrm{s}^{-1} \quad \hat{v}_{\mathrm{f}} = 400\sqrt{2/3}\,\mathrm{V} \quad f_{\mathrm{p}} = 10\,\mathrm{kHz} \quad C_{\mathrm{d}} = 10\,\mathrm{mF}$$

$$V_{\mathrm{d,ref}} = 800\,\mathrm{V} \quad T_{\mathrm{I,i}} = 1\,\mathrm{ms} \quad K_{\mathrm{P,i}} = 10\,\mathrm{V/A} \quad T_{\mathrm{I,v}} = 1.6\,\mathrm{ms} \quad K_{\mathrm{P,v}} = 10.26\,\mathrm{A/V}$$

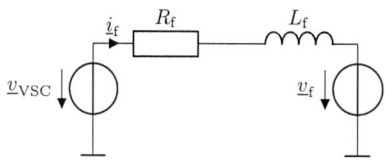

Fig. 3: Single phase equivalent circuit model of three phase grid with voltage source converter (VSC).

Solar energy conversion system

The Simscape photovoltaic (PV) array model is based on the dual resistance equivalent circuit [5], parametrized with data for the module Yingli YL170 (23)P, given in Table Ia. For MPPT a perturb-and-observe algorithm (P&O) is used, as described in [6]. The boost converter [7] inductance can be calculated by (10), the duty cycle with (9). Using the example of eight PV modules connected in series the resulting parameters are listed in Table Ib.

$$D_{V_{MPP}} = 1 - \frac{V_{MPP}}{V_{d,ref}} \qquad (9) \qquad L = \frac{V_{MPP} D_{V_{MPP}}}{\Delta i_L f_p} \qquad (10)$$

Table I: Parameterization of Simscape PV array with module data & boost converter.

(a) Parameterization of Simscape PV module.

Maximum Power	P_{max} / W	170
Cells per module	N_{cell}	48
Open circuit volage	V_{ocv} / V	29
Short circuit current	I_{sc} / A	8.1
Voltage at MPP	V_{MPP} / V	23
Current at MPP	I_{MPP} / A	7.39

(b) Parameterization of boost converter.

Switching frequncy	f_p / kHz	10
Modules in series	n_{PV}	8
Voltage at MPP	V_{MPP} / V	184
Duty cycle	$D_{V_{MPP}}$	0.717
Current ripple	$\Delta i_L / A$	$0.2\,I_{MPP}$
Inductance	L / mH	8.93

Wind energy conversion system

The system model of the WECS is simulated as described in [8]. The P&O-algorithm varies the duty cycle of the boost converter. Using the Simscape permanent magnet synchronous generator (PMSG) model, the simulation is carried out using the configuration given in Table IV.

Table II: Parameter PMSG and Turbine for three different power ratings.

Nominal power	P_N / kW	2.5	5	7.5
Stator resistance	R_s / Ω	7.017	1.93	1.4
d-axis inductance	L_d / mH	41.27	13.41	7.82
q-axis inductance	L_q / mH	125	50	7.82
Permanent magnet flux linkage	ψ_{PM} / Vs	0.79	2.45	3.18
Number of pole pairs	p		5	5	6
Turbine radius	r / m	1.5	2.2	2.65

Battery energy storage system

A simplified single resistance model with a variable open circuit voltage (OCV) $V_{b,0}$ is used, shown in Fig. 4. As an example the battery capacity $Q_{b,N}$ is chosen from the built in lithium-ion traction battery of the BMWi3, see [9]. Experimental measurements on this battery have shown an inner resistance about $R_b = 0.1\,\Omega$. The Joule losses $P_{b,loss}$ over R_b are calculated with (11). The state of charge (SOC) (12) results from the integration of the battery current I_b.

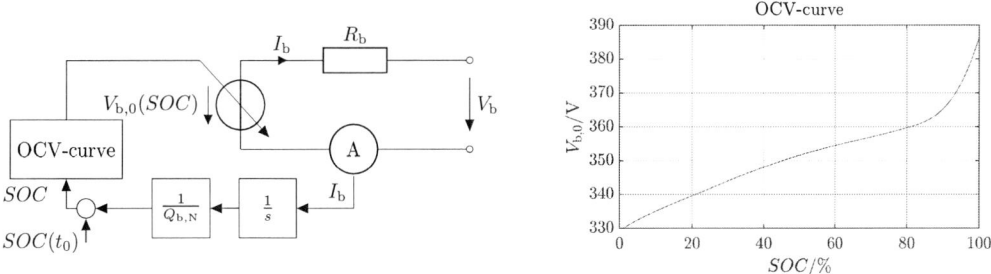

Fig. 4: Single resistance equivalent circuit & simulated OCV-curve of chosen lithium-ion battery.

$$P_{b,loss} = I_b^2 R_b \tag{11}$$

$$SOC(t) = SOC(t_0) + \frac{1}{Q_{b,N}} \int I_b \, dt \quad \text{with} \quad SOC_{min} \leq SOC \leq SOC_{max} \tag{12}$$

$$SOC_{min} = 0.05 \quad SOC_{max} = 0.95$$

Two stage approach for system modeling

The two stage model approach is used to reduce the simulation effort. For this the detailed models of the SECS, WECS and BESS are simulated with predefined test vectors, see Table III, using the MATLAB Parallel Computing Toolbox. The fixed step size for simulation is $1\,\mu s$.

Table III: Test vectors for detailed loss modeling of SECS, WECS and BESS.

	SECS $E_0 / (\text{W/m}^2)$	SECS $T / {}^\circ\text{C}$	WECS $v_w / (\text{m/s})$	BESS $P_{\text{BESS,ref}} / \text{W}$
Minimum	100	-20	1	-10000
Increment	100	5	1	1000
Maximum	1400	40	15	10000

Using the example of the SECS, Fig. 5 shows the power flow of the system, considering the thermal losses from the switching components and cable connections. For this system all input combinations of E_0 and T are simulated until the steady state end values from thermal losses are reached. The simulation output is the resulting active power that is fed into the grid. Every output value is assigned to its input vector configuration and is saved to a LUT, shown in Fig. 6. The overall efficiency η for the SECS is calulated with the output power of the PV modules and the measured power after the VSC.

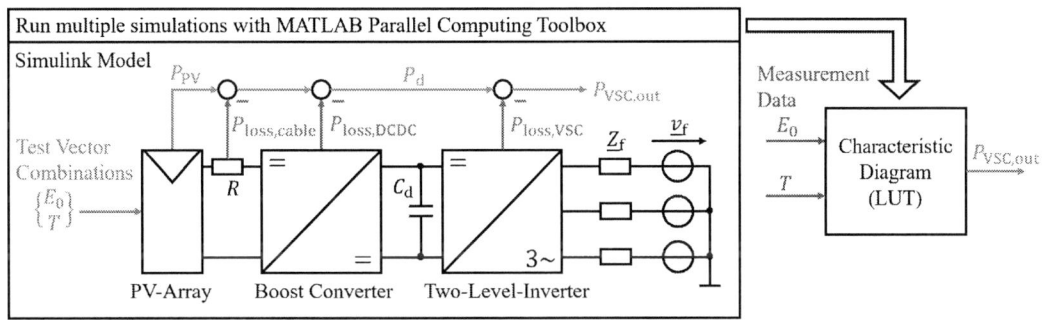

Fig. 5: Two model approach using the example of the solar energy conversion system.

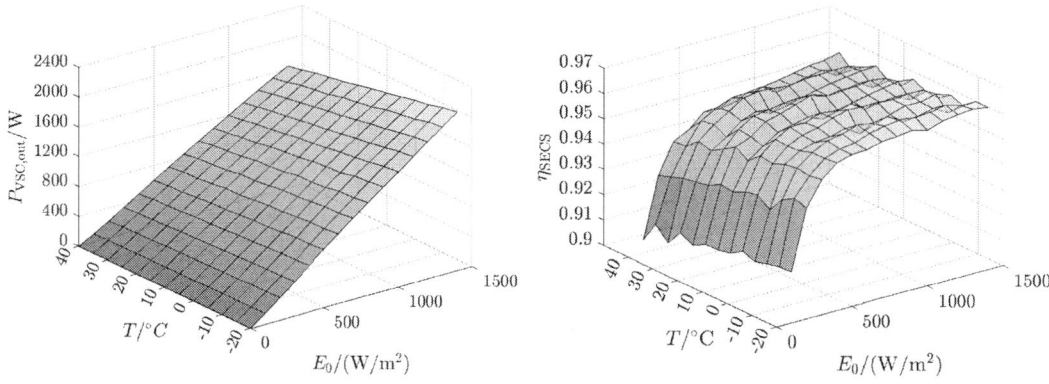

Fig. 6: SECS characteristic- and efficiency diagramm.

Entire system model

The entire system model contains the LUT based SECS, WECS and BESS, shown in Fig. 7. The EMS is realized with SIMULINK-Stateflow and designed for a high grade of energy self consumption of the head-end station. For this the sign of the power difference P_{Diff} is considered. With a positive sign the battery charges until SOC reaches the upper limit SOC_{max}. From then on power is fed into the grid. When P_{Diff} changes the sign the battery discharges. If SOC reaches a threshold at 50 % the uncritical load gets disconnected. When reaching the lower limit SOC_{min} the BESS is turned off and the power is supplied by the grid. If the BESS maximum charging resp. discharging power is exceeded the energy is fed into or supplied by the grid. A conversion of the global irradiance E_0 to South- and East-West alignments of the photoltaic modules is included. The time resolution of the measured data is 10 minutes. The timestamp allows the calculation of the sun position, as described in [10]. The measured wind speed at sensor height $v_{\text{w},\text{h}_{\text{ws}}}$ is converted to the wind speed at hub height $v_{\text{w},\text{h}_{\text{hub}}}$ (13) including the wind turbine radius r and roughness length z_0, given in Table II, Table IV and [11]. In addition to the energetic simulation with the measured data from Hilmersdorf further simulations with full-year datasets of the same time resolution for additional locations [12] are executed, see Table V. Using the MATLAB Parallel Computing Toolbox all combinations of the input configuration parameters, given by Table IV, are simulated. Every simulation iteration results the cumulated energies, calculated with (14). Furthermore the degree of self sufficiency α (15) and self consuption ε (16) are defined as follows.

$$v_{\text{w},\text{h}_{\text{hub}}} = v_{\text{w},\text{h}_{\text{ws}}} \frac{\ln\left(\frac{h_{\text{hub}}-2r}{z_0}\right)}{\ln\left(\frac{h_{\text{ws}}-2r}{z_0}\right)} \tag{13}$$

$$E = \int P\,dt \tag{14}$$

$$\alpha = \frac{E_{\text{SECS}} + E_{\text{WECS}} - E_{\text{feed-in}} - E_{\text{charge}} + E_{\text{discharge}}}{E_{\text{SECS}} + E_{\text{WECS}} - E_{\text{feed-in}} - E_{\text{charge}} + E_{\text{discharge}} + E_{\text{demand}}} \tag{15}$$

$$\varepsilon = \frac{E_{\text{SECS}} + E_{\text{WECS}} - E_{\text{feed-in}}}{E_{\text{SECS}} + E_{\text{WECS}}} \tag{16}$$

Simulation results

One simulation of a detailed system with test input vectors takes about one hour to generate its own characteristic diagram. However the LUT based entire system model needs about 15 minutes for simulating 648 system configurations, given in Table IV. The measured energy demand of the head-end station is about $E_{\text{Load}} = 16.5\,\text{MWh}$ per year. Using the example of a single configuration Table VI shows the annual self sufficiency of the four different locations.

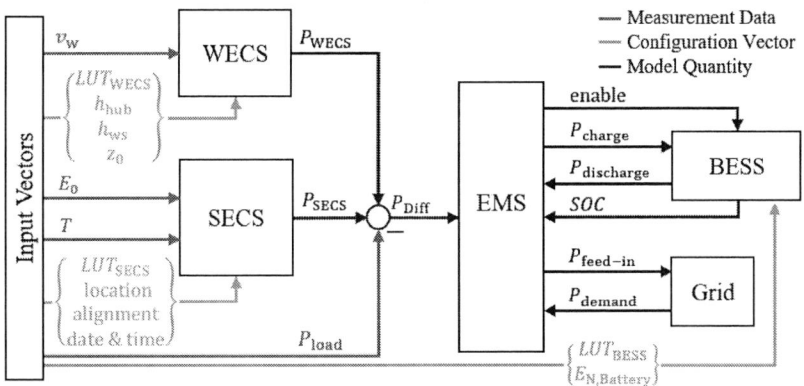

Fig. 7: Structure of entire renewable energy system model.

Table IV: Parameter space for simulating the entire system model.

Parameter	Quantity	Symbol	Parameter space
Location	4	-	{Fichtelberg; Görlitz; Hilmersdorf; Leipzig}
WECS Nominal power	3	$P_{\text{WECS,N}}$	$\{2.5; 5; 7.5\}\,\text{kW}$
Hub height	3	h_{hub}	$\{10; 15; 20\}\,\text{m}$
Number of PV modules	3	n_{PV}	$\{8; 16; 24\}$
Nominal battery capacity	3	$E_{\text{b,N}}$	$\{18.8; 37.6; 56.4\}\,\text{kWh}$ with $V_{\text{b,N}} = 360\,\text{V}$
Alignment of PV Modules	2	α	$\{180° \text{ South}; 90° \text{ East}, 270° \text{ West}\}$

$\Rightarrow$ 648 Simulations

Table V: Location with associated lat. φ, long. λ, metres above mean sea level (MAMSL), height of wind sensor h_{ws} and roughness length z_0.

Location	Characteristic	MAMSL	$\varphi/°$	$\lambda/°$	h_{ws}/m	z_0/m
Fichtelberg	mountain top	1213	50.4283	12.9536	29	0.2
Görlitz	lowland, near city	239	51.1621	14.9506	12	0.03
Hilmersdorf	highland, industrial park	604	50.6761	13.1175	10	0.3
Leipzig	lowland, near airport	131	51.4347	12.2396	10	0.02

The maximum self sufficiency is reached at the Hilmersdorf and Fichtelberg site. For the mentioned configurations the monthly normalized energies on the energy demand for the SECS and WECS at the four locations are shown in Fig. 8. It can be seen that for this configuration the monthly WECS and SECS energies are fairly evenly distributed at the Hilmersdorf site, so this system configuration can supply a base load over the year. Furthermore a higher number of PV modules at the Görlitz and Leipzig site are neccesary to increase the energy availability during the summer months. Fig. 9 shows the self sufficiency for variable SECS, WECS and BESS configurations with a defined hub height and alignment of the PV modules. With an increasing system size the self sufficiency increases significantly.

This simulation tool requires a one year load, irradiance, temperature and wind speed profile at a specified location. In the first step these data are prepared with regard to uniform time resolution and time stamp. In addition the parameters of the prefered PV module type, wind turbine and battery are important for parameterization of the detailed system models. With predefined input vector combinations the characteristic diagrams of each detailed system are calculated. In the next step all combinations of the system configuration vectors consisting of the characteristic diagrams and parameter space, defined in e. g. Table IV, are simulated with the entire system model. The simulation results are a data basis for system optimization.

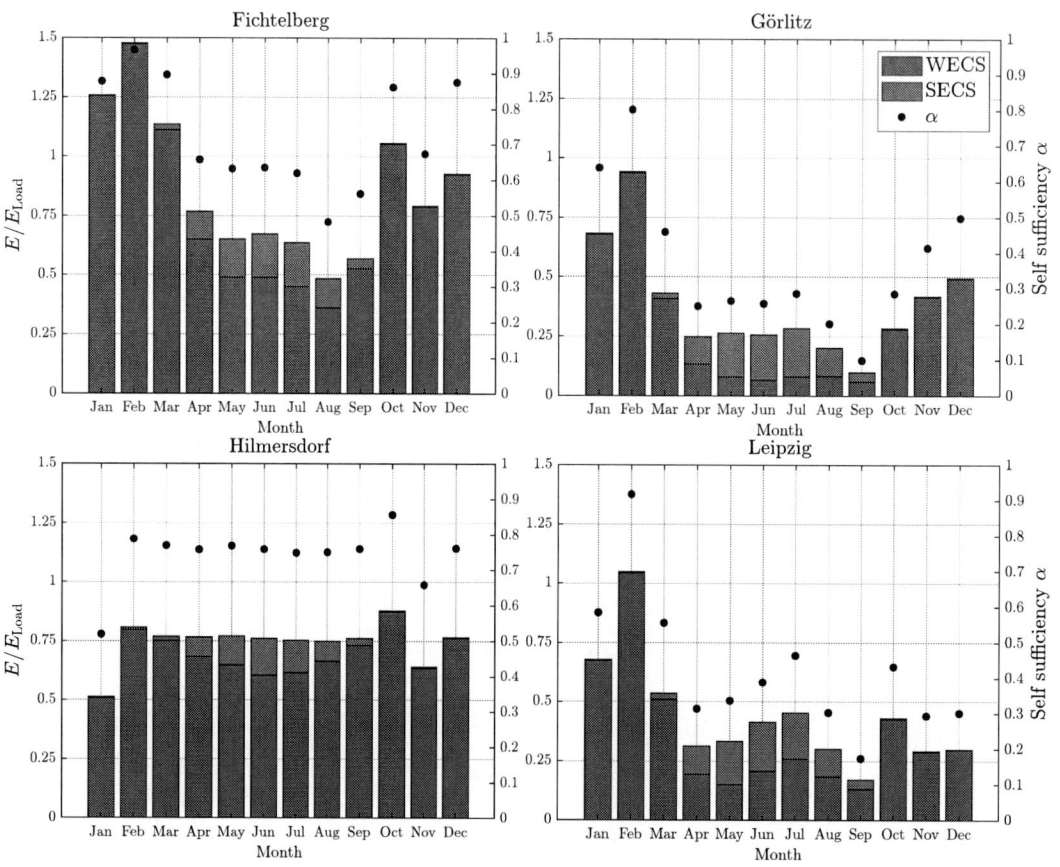

Fig. 8: Monthly normalized energies for SECS & WECS, system configuration mentioned in Table VI.

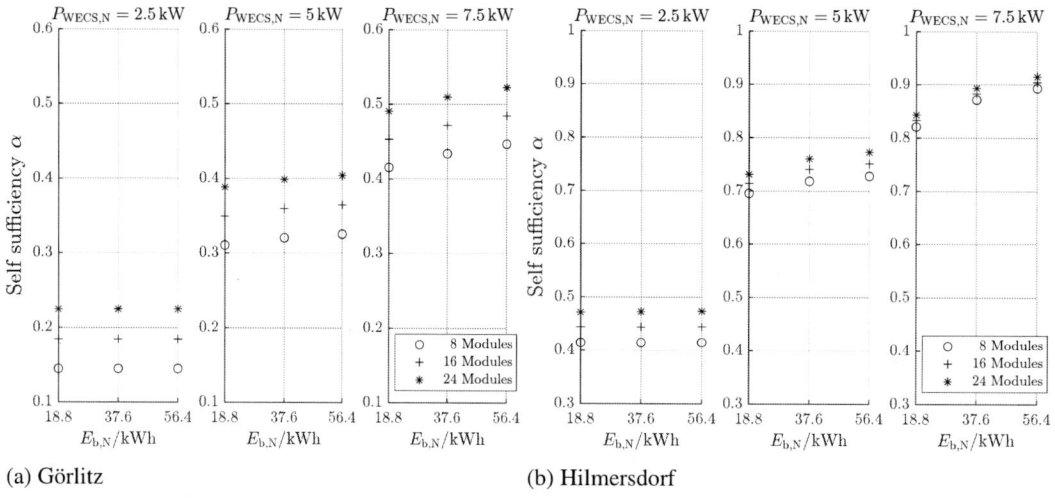

(a) Görlitz

(b) Hilmersdorf

Fig. 9: Self sufficiency for variation of SECS, WECS and BESS configuration for $h_{hub} = 15\,m$ and $\alpha = 180°$ South.

Table VI: Location dependent annual self sufficiency for a single system configuration.

Configuration	$P_{\text{WECS,N}} = 5\,\text{kW}$	$h_{\text{hub}} = 15\,\text{m}$	$E_{\text{b,N}} = 37.6\,\text{kWh}$	$n_{\text{PV}} = 16$	$\alpha = 180°$ South
Location	Fichtelberg	Görlitz	Hilmersdorf	Leipzig	
Self sufficiency α	0.73	0.36	0.74	0.41	

Conclusion

This paper presents an energetic simulation with an entire system model of a hybrid renewable energy system consisting of a solar energy conversion system (SECS), wind energy conversion system (WECS) and a battery energy storage system (BESS) combined with an energy management system.

For reducing the simulation effort, in the first step, the detailed subsystem models, SECS, WECS and BESS are modeled separately in MATLAB/SIMULINK considering system losses e. g. power converter losses or joule losses of connections. Subsequent each subsystem is simulated with test input vectors to obtain a subsystem model with reduced complexity based on characteristic diagrams. In the second step these lookup table based models are brought together in combination with an energy management system to an entire system model. Thus, the entire system model allows an energetic simulation with full year periods of irradiance, temperature, wind speed and energy demand with regard to variable system configurations. Furthermore the entire system model can be used to optimize the detailed system specification regarding to site conditions and energy management requirements. Based on these calculations, a system design is currently already being developed and will be realized in 2022.

References

[1] Fischhaber S., Regett A., Schuster S. F., Hesse H.: Begleit- und Wirkungsforschung Schaufenster Elektromobilität (BuW): Ergebnispapier Nr. 18, Second- Life-Konzepte für Lithium-Ionen-Batterien aus Elektrofahrzeugen, 2016

[2] Stopka, U. et al.: Breitbandstudie Sachsen 2030. Studie im Auftrag des SMWA, TU Dresden, 2013

[3] Giroux P.: Loss Calculation in a BuckConverter Using SimPowerSystems and Simscape, https://www.mathworks.com/matlabcentral/fileexchange/35980-loss-calculation-in-a-buck-converter-using-simpowersystems-and-simscape, MATLAB Central File Exchange, Retrieved March 23, 2022

[4] Winkelnkemper M.: Reduzierung von Zwischenkreiskapazitäten in Frequenzumrichtern für Niederspannungsantriebe, TU Berlin, 2005

[5] Nguyen B. N., Nguyen V. T., Duong M. Q. , Le K. H., Nguyen H. H., Doan A. T.: Propose a MPPT Algorithm Based on Thevenin Equivalent Circuit for Improving Photovoltaic System Operation. Front. Energy Res., 18 February 2020

[6] Singh S., Manna S., Mansoori M. I. H., Akella A.K.: Implementation of Perturb & Observe MPPT Technique using Boost converter in PV System, CISPSSE-2020 India

[7] Hasaneen B. M., Mohammed A.: Design and Simulation of DC/DC Boost Converter, 12th International Middle-East Power System Conference, 2008, pp. 335-340

[8] Zammit D., Staines C. S., A Micallef A., Apap M.: Optimal Power Control for a PMSG Small Wind Turbine in a Grid- Connected DC Microgrid, CoDIT'18, Greece, 2018

[9] BMW Group, PressClub Global, https://www.press.bmwgroup.com/deutschland/article/attachment/T0189822DE/276981, Retrieved March 30, 2022

[10] Quaschning V.: Understanding Renewable Energy Systems, 2nd ed., London, 2016

[11] Namyslo J., Koßmann M.: Bestimmung effektiver Rauigkeitslängen an Windmessstationen aus topographischen Karten (TK-Verfahren), Deutscher Wetterdienst, www.dwd.de/DE/leistungen/gutachtenqpr/z0_aus_topo_karten.html, Retrieved March 23, 2022

[12] Deutscher Wetterdienst, Open Data Server: opendata.dwd.de, Retrieved March 23, 2022

Comparison of System-Level Availability in Industrial Grids

G. Emmers, J. Driesen
KU LEUVEN, Department of Electrical Engineering, ELECTA - ENERGYVILLE
Kasteelpark Arenberg 10
3001 Leuven, Belgium
Phone: +32 (0) 16-379165
Email: glenn.emmers@kuleuven.be

Acknowledgments

This work is performed as part of the MultiDC-ICON project, HBC.2019.0084. Funded by the Institute for the Promotion of Innovation through Science and Technology in Flanders (VLAIO) and Flanders Make, the strategic research center for the manufacturing industry in Belgium.

Keywords

≪LVDC≫, ≪Reliability≫, ≪Industrial Application≫, ≪Uninterruptible Power Supply (UPS)≫

Abstract

This paper evaluates the availability of traditional AC based and state-of-the-art LVDC based industrial grids. The main focus is on industrial grids with a high penetration of power electronics, used to power production lines with many drives. Firstly, the Monte Carlo approach is presented to analyze the availability of several grid topologies relevant to industry. Secondly, this method is used to validate a Multi-State Markov approach. The Multi-State Markov approach is extended with a battery dependency model, to incorporate battery life in the analysis. Finally, after validation, the Multi-State Markov approach is used to determine the influence of the AC grid availability and the battery backup time on the availability of the presented use cases. LVDC systems prove to be more available at the load connection points and, depending on converters chosen, also have the potential to outperform UPS systems at distribution board level.

Introduction

In recent years, the interest in DC has been growing because, compared to a traditional AC exploitation, there is a potentially increased efficiency, an increased transmission capacity and an increase in compatibility with modern loads and storage [1]. Especially in industry, the dependency on variable speed drives has grown significantly, with the main drivers being the increased efficiency and the decreased CAPEX due to less conversion steps [2]. These variable speed drives introduce a DC step in between the two AC systems. Oftentimes multiple back-end DC/AC converters are connected to a single front-end AC/DC converter creating a larger DC bus to facilitate regenerative braking. When building further upon this concept, the logical next step is the implementation of a battery on the DC bus. Adding a battery to an industrial grid offers plenty of advantages, such as distributed UPS properties for the entire grid to increase the system availability and peak shaving when the power system demand is characterized by short but high power peaks [3]. These peaks are crucial for the design, exploitation and cost of the electrical network [4]. Not only is the electrical network designed to supply the peak power, in some cases industrial customers are also charged by their maximum power demand. The adaptation of LVDC technology allows for a simple and efficient implementation of battery storage on the DC bus. This storage can then be used to flatten the load curve, where energy is stored during periods of low demand and released during periods of higher demand [5].

Besides the peak shaving aspect, a battery significantly increases the availability of an LVDC network, making it act as a UPS system. The battery will isolate the DC grid from anything happening on the AC side, creating fault ride-through capabilities. Many industrial systems are plagued with very short-term interruptions (<1s), which are statistically not recorded but occur significantly more frequently than unplanned supply interruptions of more than three minutes [6]. Additionally, these LVDC power systems can be significantly larger in size and power than a UPS system, covering more loads [7].

New in this paper is the comparison of a traditional AC based industrial grids with new and unconventional LVDC based industrial grids, with a special focus on the effects of battery backup time and AC grid availability in the availability analysis of all systems considered. Conclusions can be drawn based on how the amount of power electronic conversion steps influence the system availability and how redundancy can be used to increase the system availability, yet using less components than traditional solutions.

This paper first describes the mathematical framework used to evaluate the systems under consideration. Next follows a description of the four system architectures under consideration and the analyses performed. Finally, the paper is ended with a conclusion.

Mathematical framework

This section starts with discussing the assumptions made regarding the reliability modeling, followed by providing the two methodologies used to analyze the availability of the selected use cases, to finish with a definition of system failure.

Reliability modeling

The analysis presented in this paper includes several assumptions to keep the model concise and tractable. These assumptions are the following:

- All components are independent from each other and can be modeled as a two state Markov process.
- Given that the components can be modeled as a two state Markov process, it is assumed that each component i has a constant failure rate λ_i (failures/hr) and a constant repair rate μ_i (repairs/hr), which means they have an exponential distribution. This is true under the assumption that all equipment is operational in their useful life, rather than the wear-out phase or the phase of infant mortality.
- Batteries with finite charge cannot be modeled as a two state Markov process, therefore another assumption regarding the incorporation of battery life in the availability modeling is required.
- System unavailability U_s can be calculated as the probability where down-time in the battery power supply exceeds battery reserve time. The battery power supply consists of the main power supply, distribution transformer, a front-end rectifier and in some cases a DC/DC converter. In [10] this is presented as follows:

$$U_s = \lambda_A \int_T^\infty f_A(t)dt \frac{\int_T^\infty (t-T)f_A(t)dt}{\int_T^\infty f_A(t)dt}$$

  With λ_A the failure rate of the battery power supply, $f_A(t)$ the probability density function of the failure duration time and T the battery reserve time. If the probability density function is then dissolved in terms of each element that is part of the battery power supply, this leads to:

$$U_s = \sum_{k=1}^n U_k e^{-\mu_k T}$$

  With n the amount of components part of the battery power supply and U_k the unavailability of this component. In this paper T is initially considered to be five hours, after validation of the model T is varied to analyze its influence.
- One and only one transition can occur at a given time instant t.

- After the repair of one out of two components that were in a faulty state together, the other component is assumed to be repaired before the firstly repaired component fails again. This assumption is made to allow the battery to at least partially recharge once it has run out of charge.
- Availability as used in this paper is considered to be the steady state availability, which is defined to be the limit of the instantaneous availability for time going to infinity [11]. This availability can be expressed as $A = \frac{MTBF}{MTBF+MTTR}$, with MTBF the Mean Time Between Failures ($\frac{1}{\lambda}$) and MTTR the Mean Time To Repair ($\frac{1}{\mu}$). Consequently, the availability can also be expressed in terms of the failure rate and the repair rate as: $A = \frac{\mu_i}{\mu_i+\lambda_i}$. Typically, steady state availability is specified in nines notation, where the number of nines represents the amount of nines in the fraction of time that the system is available [12]. Table I gives an overview of the availability specified in nines notation and its corresponding downtime.

Table I: Availability and downtimes [13]

Availability (number of 9s)	Downtime per year	Downtime per day
1	36.5 days	2.4 hours
2	3.65 days	14.4 minutes
3	8.76 hours	1.44 minutes
4	52.56 minutes	8.66 seconds
5	5.26 minutes	864.3 milliseconds

Methodologies

Many options exist when it comes to performing a reliability analysis of electrical installations with Markovian properties. The most common ones are the reliability block diagram or the fault tree analysis, especially when the systems under consideration are relatively small and calculations are simple to perform. As systems get larger, these analytical methods become tedious to solve. For that reason the Universal Generating Operator (UGO) approach is used, which is a method for solving a Multi-State problem. The assumptions made above limit us to the use of Markov-models, including a battery in this analysis would ideally require a semi-Markov approach, which is a significantly more difficult problem to set up. Alternatively the method described above is used to estimate the system unavailability due to an empty battery. Because the results based on this battery unavailability model as described above are an estimate, the results found with the UGO are validated with a Monte Carlo Simulation (MCS). In contrary to the UGO, the MCS is a time-based solution to the availability analysis rather than a stochastic one. This heavily simplifies the implementation of time-dependent problems, such as the battery problem at hand. The disadvantage of the MCS approach, compared to the UGO approach, is the time it takes to solve the problem. Solving a UGO can be done within a time span of seconds to minutes, whereas an MCS approach can take up to several minutes to hours, depending on the size of the system to solve and the accuracy required.

Universal Generator Operator

The UGO technique is used to calculate a Multi State System performance distribution based on the stochastic performance of the elements forming that system. What makes the UGO technique so unique is the fact that this approach uses simple recursive procedures and makes complicated combinatorial algorithms obsolete by providing a systematic method for enumerating the systems states [14].

The performance of any element e in the system is represented by a Universal Generating Function:

$$\omega_e(z) = \sum_{s \in S} p_s \cdot z^{v_s}$$

With S the state-space including its unique performances v_s and p_s the associated probabilities[15]. In the case of this analysis, the unique performances of the elements are taken to be the minimum and the maximum power level. These power levels then have their respective probability associated to them. The probabilities for each state are calculated for each component separately in its designated Markov model. In this analysis every component has two states, each determined by a constant failure rate and a constant repair rate, which in turn determines the final states probabilities.

To find the system performance at a specific point in the network, the Universal Generating Operator Ω can be constructed, based on the methodologies presented in [14]:

$$\Omega_u([\omega_e(z)]) = \sum_{s_1}^{S_{e_1}} \ldots \sum_{s_n}^{S_{e_n}} (p_{s_1} \cdot \ldots \cdot p_{s_n}) \cdot z^{f^{str}(v_{s_1,\ldots,s_n})}$$

With f^{str} the structure function to express the performance towards that specific point in the system[15]. To implement the battery backup time into the UGO, the method as discussed in Section "Reliability modeling" is used. This leads to an additional unavailability of the system, which is equal to an additional probability for the system to be in either an unavailable or an available state in the steady state solution. The new probabilities following from the calculated unavailability, are treated as the solved state transition diagram of an extra component, which can be placed at the position of the distribution board in the grids described in the next section. All UGO analyses are performed using MULTISTATESYSTEMS.JL[1], a JULIA package developed to solve multi-state systems.

Monte Carlo Simulation

The validation of the results generated with the UGO are performed with a Monte Carlo Simulation in MATLAB as presented in [16]. The chosen implementation is the Direct Simulation Method, which samples the transition times of all elements individually. The element with the shortest transition time is then selected to be the one to make the transition at that moment t. This particular implementation of the MCS allows for simple time-dependent adjustments, such as the time dependency of the battery state. The algorithm was extended to change the battery state from as soon as the state transition time went beyond the battery backup time, in this case five hours. To perform the MCS several minimal cut sets have to be defined, which will be elaborated upon in the next section.

Case study

Systems in comparison

In this section four different topologies are presented, the first two of which are traditional AC implementations, followed by two state-of-the-art LVDC implementations.

Conventional AC distribution system

The first system under consideration is the conventional AC distribution system as shown in Fig. 1. This system is connected to the main power supply through a distribution transformer, followed by a distribution board towards the different machines, which contain variable speed drives (VSD's). Typically these VSD's contain a two-stage power electronic conversion, with a front-end AC/DC converter connected to the AC distribution to create an internal DC bus, which in turn feeds a back-end DC/AC inverter to supply the motors of the machines. In case machines contain multiple drives, it is common practice to connect these back-end converters to a common active front-end. This benefits overall system efficiency as recovered braking energy is directly used by another drive.

Conventional AC UPS distribution system

The second system under consideration is the conventional AC distribution system with an additional UPS as shown in Fig. 2. In this case the system consists of the traditional AC distribution, but critical machines are being connected through a UPS installation. This increases the resilience against grid-side disturbances, but is detrimental to the system efficiency. Many types of UPS-setups exist, but the most common one is a static double-conversion type. This type of UPS consists of an AC/DC converter feeding a DC bus, to which a battery and a DC/AC inverter are connected [8].

---

[1] Available at https://github.com/timmyfaraday/MultiStateSystems.jl

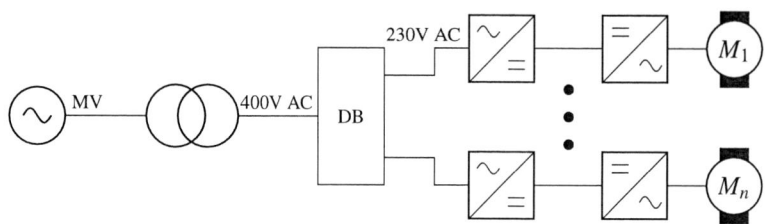

Fig. 1: Conventional AC distribution system.

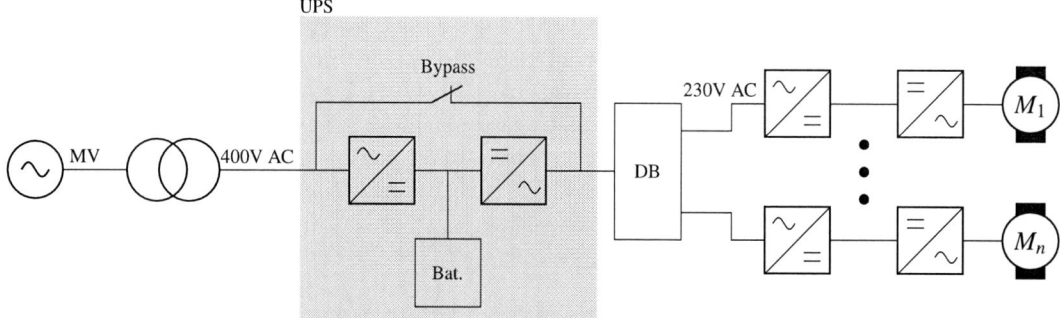

Fig. 2: Conventional AC distribution system with UPS.

LVDC distribution system

The third system under consideration is the LVDC distribution system as shown in Fig. 3. This system architecture extends the benefits of a common DC bus as seen for the conventional AC distribution system, towards the entire plant. Not only is energy transfer possible within a machine, energy transfer between machines is facilitated as well. Fewer conversion steps are required and common components such as filters and capacitors are eliminated. Furthermore, there is an ever decreasing cost of electrical storage, which allows to incorporate storage either centrally or per machine [7],[9]. This allows to add UPS properties to the entire system, comparable to the AC UPS case, but with a significantly lower amount of converters required. Notice that the converters in a DC system can be smaller in power level than for AC systems, because power transfer between several components is facilitated and the battery can be used for peak shaving.

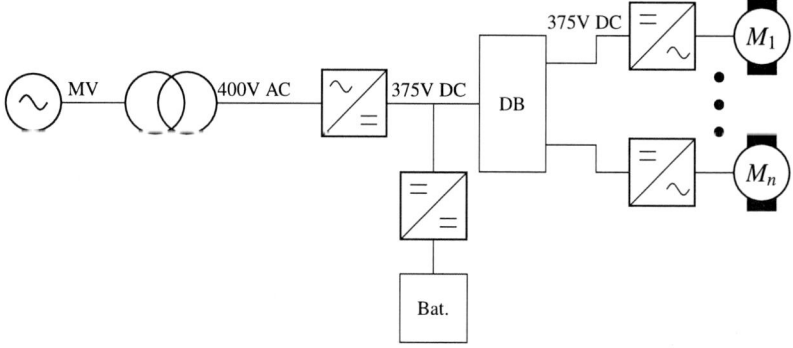

Fig. 3: State-of-the-art LVDC distribution system.

Redundant LVDC distribution system

Finally, the previously described LVDC distribution system is extended with a redundant front-end AC/DC converter, which is shown in Fig. 4. LVDC systems are highly modular, which makes it easy to add an additional AC/DC converter for extra redundancy. By doing so, the front-end as single point of failure is eliminated and the degrees of redundancy of the AC UPS case and the LVDC case are leveled out, which makes for a fair comparison. Whereas the AC UPS case has a path through the bypass switch, the front end and the battery, the redundant LVDC case has a path through two front-ends and the battery. By adding an extra converter to the system, the influence of redundancy before the distribution board is investigated.

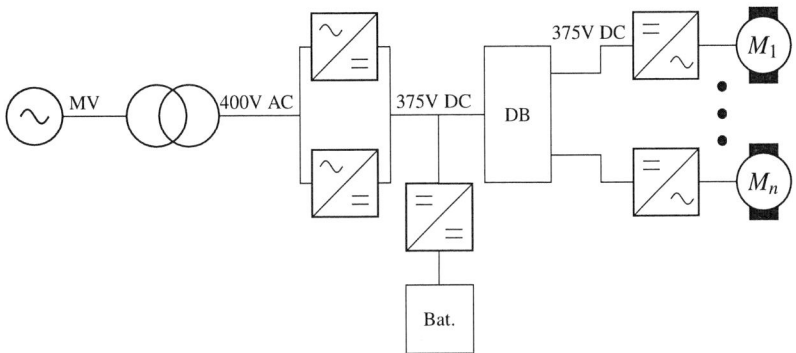

Fig. 4: State-of-the-art redundant LVDC distribution system.

Availability Analysis

System Failure Definition

Two points of interest are investigated in the further analysis of the systems under investigation. The first point of interest is the availability of the distribution system at the distribution board, indicated by DB in the figures above. It is important to take into account that the characteristics of the power available at the distribution board are different for AC systems and DC systems. As a consequence, it is impossible to connect DC loads directly to an AC distribution board and vice versa. The second point under investigation is the load itself, which is the main concern of the system exploiter. The system is said to be available as long as it is in a state to perform as required, meaning the distribution board or the load gets power of a sufficient quality. Table II shows the failure rates and repair rates of the components under consideration in this analysis, expressed in failures per hour and repairs per hour respectively.

Table II: Reliability Data [8]

Component	i	λ_i (f/hr)	μ_i (r/hr)
Main Power Supply	MPS	0.0003142	0.8058
Bypass Switch	SW	4.2166700e-6	0.4269
Battery	BAT	2.0722222e-6	0.0658
Distribution Transformer	DT	8.5777778e-6	0.9552
Front-end Rectifier	FE	2.9666667e-6	0.1400
Inverter/Rectifier	INV	2.3305555e-6	0.0936
DC/DC Converter	DC	8.0580000e-6	0.1250

Universal Generating Operator

The availability analysis with the UGO requires the additional unavailability caused by an empty battery to be calculated manually. An illustrative example applied to the AC UPS case follows below. The battery is the single power source in case one of the following events occur:

- Grid failure
- Transformer failure
- Bypass switch failure & front-end failure

This leads to the following calculations, as presented in section "Reliability Modeling":

$$U_1 = U_{MPS}e^{-\mu_{MPS}*5}$$

$$U_2 = U_{DT} e^{-\mu_{DT}*5}$$

$$U_3 = U_{FE} U_{SW} e^{-\mu_{SW}*5}$$

$$U_s = \sum_{k=1}^{3} U_k$$

A first thing to note is the choice to only incorporate the repair-rate of the bypass switch in the equation for the unavailability (U_3) caused by the simultaneous bypass switch and front-end failure. The reasoning behind this is that the bypass switch has the highest repair-rate and that the system will be operational once this component is repaired. Secondly, it is also assumed that the front-end is repaired before the switch fails again. Next, the DC/DC converter is not included in these calculations for cases that have a DC/DC converter as part of the battery power supply. This is justified because the state of charge of the battery will not change in case the DC/DC converter fails, consequently the battery cannot act as backup power supply. Finally, the calculated unavailability is then added in series of the distribution board unavailability within the UGO. As mentioned before, the UGO analysis is performed with JULIA package MULTISTATESYSTEMS.JL[2].

Monte Carlo Simulation

The implementation of the MCS as described above, requires minimal cut sets to be defined for the cases under investigation. Table III gives an overview of the minimal cut sets defined for each case, which is a combination of elements that cause the system to become unavailable when being in a failed state. The abbreviations used in Table III can be found in Table II. The failure of the battery includes both failure states; broken battery or out of charge. The table-elements marked in gray indicate the additional cut sets used to investigate the availability at the load.

Table III: Minimal Cut Sets for the MCS implementation

AC cut sets	UPS cut sets	LVDC cut sets	LVDC Red. cut sets
$\{MPS\}$	$\{MPS, BAT\}$	$\{MPS, BAT\}$	$\{MPS, BAT\}$
$\{DT\}$	$\{MPS, INV_{UPS}\}$	$\{MPS, DC\}$	$\{MPS, DC\}$
$\{INV_{ac/dc}\}$	$\{DT, BAT\}$	$\{DT, BAT\}$	$\{DT, BAT\}$
$\{INV_{dc/ac}\}$	$\{DT, INV_{UPS}\}$	$\{DT, DC\}$	$\{DT, DC\}$
	$\{SW, BAT\}$	$\{FE, BAT\}$	$\{FE1, FE2, BAT\}$
	$\{SW, INV_{UPS}\}$	$\{FE, DC\}$	$\{FE1, FE2, DC\}$
	$\{INV_{ac/dc}\}$	$\{INV_{dc/ac}\}$	$\{INV_{dc/ac}\}$
	$\{INV_{dc/ac}\}$		

Results and discussion

Table IV and Table V show the results of the UGO method and MCS method respectively, both with a battery backup time of five hours and an AC mains grid availability of three nines. The results of both methods are not the exact same, but they are definitely close enough to consider them equal. Consequently, the MCS method not only validates the UGO method, it also proves that the implementation of the battery life as such, is an appropriate method to consider battery life without using more complex semi-Markov methods.

The results clearly show that the AC implementation has the lowest availability, both at the distribution board and at the load, although the discrepancy at the load decreases relative to other setups. This is because the AC grid availability is more dominant in this analysis than the converter reliabilities, therefore contributing less to the overall availabilities calculated. Next, the new LVDC implementation has the lowest availability at the distribution board out of the final three topologies, leading to more system-wide outages. This is explained by the level of redundancy, which is lower than for the AC UPS

---

[2] Available at https://github.com/timmyfaraday/MultiStateSystems.jl

Table IV: Availability Data Resulting From UGO

System (UGO)	At Distribution Board		At Load	
	Availability	Unavailability	Availability	Unavailability
Standard AC	3 nines	39.876e-5	3 nines	44.853e-5
AC UPS	5 nines	0.7032e-5	4 nines	5.6822e-5
LVDC	4 nines	1.7497e-5	4 nines	4.2336e-5
Redundant LVDC	5 nines	0.6972e-5	4 nines	3.1867e-5

Table V: Availability Data Resulting From MCS

System (MCS)	At Distribution Board		At Load	
	Availability	Unavailability	Availability	Unavailability
Standard AC	3 nines	39.889e-5	3 nines	44.905e-5
AC UPS	5 nines	0.7068e-5	4 nines	5.6638e-5
LVDC	4 nines	1.7441e-5	4 nines	4.2474e-5
Redundant LVDC	5 nines	0.6949e-5	4 nines	3.2139e-5

case and the redundant LVDC case. Yet, the unavailability of the load is lower for the LVDC case than for the AC UPS case. The reduction in converters pays off towards the end of the distribution chain. Finally, the redundant LVDC case has the lowest unavailability by a very small margin. Both the AC UPS and the redundant LVDC system have a similar level of redundancy, which translates into similar levels of availability at the distribution board level. The difference is larger at the load level, once more because of the reduced amount of converters in the DC case. Beware that the system integration and exploitation cost for both LVDC system is significantly lower than for the AC UPS case, driven by a lower amount of installed conversion steps.

Battery life dependency

The results so far were calculated using a constant battery life and a constant AC grid availability. In this section, the results show the dependency on battery backup time. Because the MCS method requires significantly more time to solve than the UGO method, this analysis is performed using the validated UGO method exclusively. Fig. 5 shows the availability at (a) distribution board and (b) load level on a logarithmic scale and a linear scale respectively, for a battery backup time ranging from 1 hour to 50 hours. Fig. 5 (a) shows that the battery backup time has an influence on the distribution board availability until a battery backup time of about 15 hours for both the AC UPS case and redundant LVDC case. The battery life has a much larger influence on the availability of the standard LVDC case, which is caused by the front-end as single point of failure. The small difference in availability between the AC UPS case and the redundant LVDC case beyond 15 hours of battery life is explained by a difference in reliability of the DC/AC converter on the one hand and the DC/DC on the other hand. Fig. 5 (b) shows that loads connected to LVDC systems are more available at any battery size, which can be explained by the additional converter between the distribution board and the loads for AC cases.

Variable AC grid availability

This section analyzes the effect of the AC grid availability on the availability of both the distribution board and the load. Fig. 6 (a) shows the unavailabilities of the distribution board for the different cases, including an AC Grid case that reflects the AC grid availability on the DB availability. In essence this mimics a facility where the system operator already made investments to improve the availability of the AC distribution. Interesting about Fig. 6 (a), are the crossover points between the different cases and the AC Grid, indicated where it might or might not be worthwhile to make extra investments on another distribution topology, for a given AC distribution grid availability. The LVDC case has a crossover at around 5 nines, the AC UPS case has a crossover at 7 nines and finally the Redundant LVDC case has a crossover at 13 nines. Each case is more available than the AC grid before the crossover point and less available behind the crossover point. To put this into perspective, even TSO busbar availabilities rarely

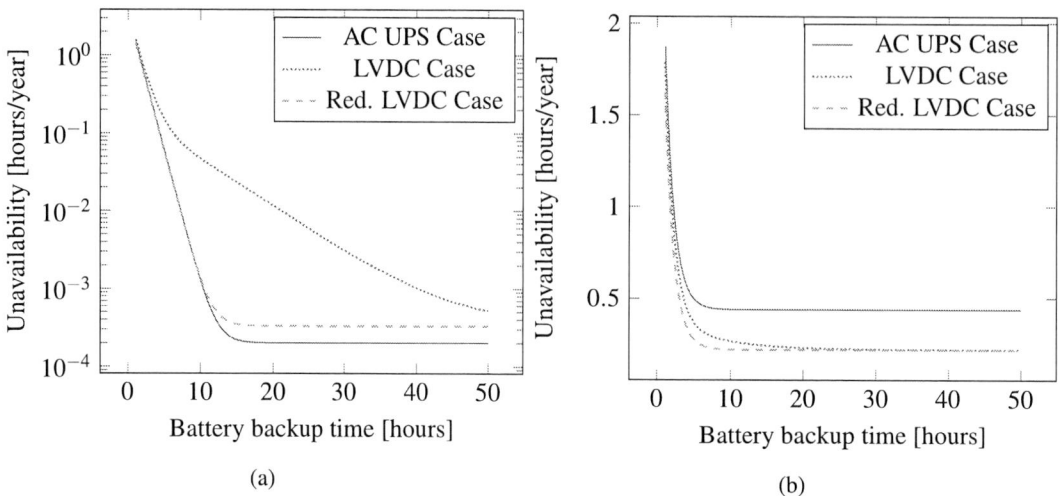

Fig. 5: Unavailability in hours per year in function of the battery backup time for (a) the distribution board and (b) the loads

go beyond 5 nines. Finally, Fig. 6 (b) shows that for any given AC grid availability, the loads connected to both LVDC cases outperform the loads for the AC cases availability-wise.

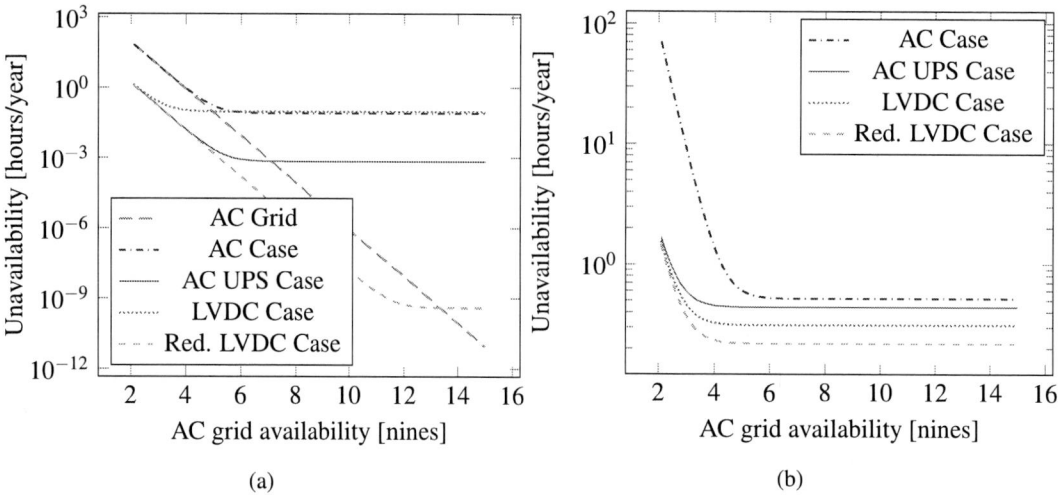

Fig. 6: Unavailability in hours per year in function of the AC grid availability for (a) the distribution board and (b) the loads

Conclusion

This paper first describes the mathematical framework to perform an availability analysis on industrial distribution grids. In this analysis a method to include battery backup time in a Universal Generating Operator methodology based on Markov methods is validated using a Monte Carlo Simulation. The results of both methods show promising results regarding the availability of loads in LVDC grids, compared to existing AC implementations. Finally, the analysis performed with the UGO method exclusively shows the influence of both the battery backup time and AC grid availability on both the distribution board availability and the load availability. These results show that LVDC implementations, with a significantly smaller component count, have the potential to outperform AC solutions at distribution board level and most definitely outperform them at load level.

References

[1] J. J. Justo, F. Mwasilu, J. Lee, and J. W. Jung, "AC-microgrids versus DC-microgrids with distributed energy resources: A review," Renewable and Sustainable Energy Reviews, vol. 24, pp. 387–405, 2013. [Online]. Available: http://dx.doi.org/10.1016/j.rser.2013.03.067

[2] R. Saidur, S. Mekhilef, M. B. Ali, A. Safari, and H. A. Mohammed, "Applications of variable speed drive (VSD) in electrical motors energy savings," Renew. Sustain. Energy Rev., vol. 16, no. 1, pp. 543–550, 2012.

[3] C. Rahmann, B. Mac-Clure, V. Vittal, and F. Valencia, "Break-even points of battery energy storage systems for peak shaving applications," Energies, vol. 10, no. 7, 2017.

[4] Z. Wang and S. Wang, "Grid power peak shaving and valley filling using vehicle-to-grid systems," IEEE Trans. Power Deliv., vol. 28, no. 3, pp. 1822–1829, 2013.

[5] R. Martins, H. C. Hesse, J. Jungbauer, T. Vorbuchner, and P. Musilek, "Optimal component sizing for peak shaving in battery energy storage system for industrial applications," Energies, vol. 11, no. 8, 2018.

[6] Industrie- und Handelskammer in Bayern, "BIHK-Studie: Energiewende im Strommarkt – Versorgungsqualität" 2017.

[7] B. K. Johnson, "An Industrial Power Distribution System Featuring Ups Properties," Ieee, 1993.

[8] V. Sithimolada and P. W. Sauer, "Facility-level DC vs. typical AC distribution for data centers: A comparative reliability study," IEEE Reg. 10 Annu. Int. Conf. Proceedings/TENCON, pp. 2102–2107, 2010.

[9] ZVEI - Zentralverband Elektrotechnik- und Elektronikindustrie e.V., "Gleichspannungsnetze in der industriellen Produktion," p. 12, 2017.

[10] K. Yotsumoto, S. Muroyama, S. Matsumura, and H. Watanabe, "Design for a Highly Efficient Distributed Power Supply System Based on Reliability Analysis," in 10th int. Telecommunications Energy Conf. 1988, pp. 545-550.

[11] "Steady State Availability," 2022. [Online]. Available: http://www.electropedia.org/iev/iev.nsf/display?openform&ievref=192-08-07

[12] A. Kwasinski, "Quantitative evaluation of DC microgrids availability: Effects of system architecture and converter topology design choices," IEEE Trans. Power Electron., vol. 26, no. 3, pp. 835–851, 2011.

[13] B. R. Shrestha, T. M. Hansen, and R. Tonkoski, "Reliability analysis of 380V DC distribution in data centers," 2016 IEEE Power Energy Soc. Innov. Smart Grid Technol. Conf. ISGT 2016, pp. 0–4, 2016.

[14] A. Lisnianski, I. Frenkel, and Y. Ding, Multi-State System Reliability Analysis and Optimization for Engineers and Industrial Managers. London: Springer, 2010.

[15] G. Abeynayake, T. Van Acker, D. Van Hertem, and J. Liang, "Analytical Model for Availability Assessment of Large-Scale Offshore Wind Farms including Their Collector System," IEEE Trans. Sustain. Energy, vol. 12, no. 4, pp. 1974–1983, 2021.

[16] E. Zio, The Monte Carlo Simulation Method for System Reliability and Risk Analysis. Springer, 2005.

Ageing Mitigation and Loss Control in Reconfigurable Batteries in Series-Level Setups

Tomáš Kacetl, Jan Kacetl, Nima Tashakor, Stefan Goetz

Technische Universität Kaiserslautern

Kaiserslautern, Germany

E-Mail: tomas.kacetl@porsche-engineering.de, tashakor@eit.uni-kl.de, jan.kacetl@porsche-engineering.de

Acknowledgements

The authors acknowledge the financial support by the Federal Ministry of Education and Research of Germany in the project "Open6GHub" (grant number: 16KISK004).

Keywords

«Modular Multilevel Converters (MMC) », «Converter control», «Current observer», «Batteries»

Abstract

This paper presents a novel control method that reduces the low-frequency ripple of dynamically reconfigurable battery technology to reduce cell ageing and loss. It furthermore shifts the residual ripple to higher frequencies where the lower impedance reduces heating and the dielectric capacitance of electrodes and electrolyte shunt the current around the electrochemical reactions.

Introduction

Reconfigurable batteries

Electromobility and grid storage are rapidly developing applications of power electronics and batteries. They use battery packs as an energy tank and a semiconductor inverter to generate the ac output for the motor or grid. Conventionally, cells are hard-wired in a battery pack with certain fixed parallel and serial configuration [1]. In combination with an inverter, the ac side of the inverter supplies the grid or an electric motor with ac current, whereas the dc link of the inverter loads the battery with a current resulting from the operation of the inverter [2].

Alternatively, modular circuit structures such as modular multilevel converters (MMC) or cascaded H bridges (CHB) with batteries offer interesting advantages and can form battery systems with immediate multiphase ac output [3, 4]. In contrast to hard-wired batteries, the distributed power electronics can dynamically reconfigure the module interconnection and control the power of individual modules. Thus, reconfigurable MMC–battery systems offer excellent balancing of the state of charge [5, 6] and state of health [7-10], introduce fault tolerance by bypassing defective modules or even semiconductors [11-14], and increase the effectively available capacitance of battery systems [5, 15, 16]. Several companies are developing or already market commercial systems based on battery-integrated CHB/MMC [17-19]. Further advantages over and comparison of

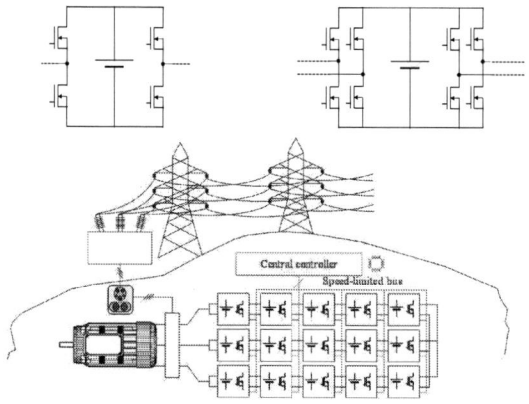

Figure 1. Top: Diagram of the system topology: battery module with CHB (left) and CHB2 (right) switch topology. Bottom: Overall system topology in drive trains or with grid connection as storage or charging vehicle.

reconfigurable with hard-wired batteries can be found in the literature [20, 21].

Module load-ripple

In all CHB circuits, the load currents of the individual module batteries depend on the macro-level topology. In case of a star configuration and ac output, the modules are divided into phase strings, where each of the phase strings supplies one output phase, for instance feeding a motor or the grid. As a result, the module load is rippled, where the spectrum of the module load contains a strong 2nd harmonic of the output ac frequency [22-24]. CHBs without parallel module connectivity alternate between a series module state, where modules run on phase load, and bypass state, where modules have zero load [25]. Alternating these states introduces components of variable frequency in the module load spectrum. Thus, they depend on the specific control strategy. Such ripple load on the battery cells occurs additionally to the load current and does not contribute to the active output power but constitutes reactive power fluctuations. As the reactive ripple current loads the equivalent resistances of module components and connections, it generates unnecessary additional loss as well as heating. Such extra heating can easily cause derating in thermally limited automotive batteries, and may further degrade components.

Among the various CHB topologies, those with parallel connectivity (e.g., CHB[2] in Figure 1, sometimes also denoted as modular multilevel series parallel converters, MMSPC) [25-28] offer better load distribution among battery modules, lower effective source impedance, and lower ripple load, which is a major advantage particularly for battery applications [29]. In CHB[2] or MMSPC, the parallel states may substitute the inactive bypass state. Paralleling modules eliminates no-load states and lowers the load of the active modules to bring both closer to the mean [30, 31].

In addition to the averaging effect of two parallel modules within a phase string instead of bypassing one, MMSPC topologies allow a reduction of the module load ripple through internal compensation currents and voltages as well as a double neutral point (see Figure 1 at the right end), which allows module paralleling across the previously widely independent phase module strings [22, 29-35]. Dynamical module paralleling across the module strings through this double neutral point can exchange power and improve the load distribution.

High-frequency-shunting dielectric capacitance

During operation, the elements and materials of the battery cells in the reconfigurable battery undergo degradation processes so that the cells gradually lose their capacity and performance [36-41]. The degradation is a result of many physical and particularly electrochemical processes, also called faradaic processes. Rippled load, however, does not necessarily lead to faradaic processes. Electrically charged electrodes and the adjacent electrolyte form a charge double layer [42-45]. Due to the small spacing of relatively large charges, the capacitance of the double layer can be immense. Furthermore, the electrolytes of modern batteries are strongly polar for high lithium-ion solubility, entailing a large dielectric constant as a side product [46]. Thus, the resulting merely dielectric capacitance can absorb enough charge for short current pulses without further chemical reactions. In contrast to the electrochemical capacitance, the dielectric capacitance offers lower impedance and cell heating.

Electrochemical reactions have limited kinetics, often constrained by diffusion, and electrically appear as low-pass system. To initiate sufficient chemical reactions, a high current has to be maintained for an extended period of time in the range of milliseconds [47]. Electrochemical models show that short pulses are almost completely buffered by the dielectric capacitance of the electrodes [48].

The actual contribution of the rippled load to ageing is lately receiving more attention, and experiments agree with the degradation-neutral dielectric shunting at higher frequencies [49]. Measurements suggest the existence of a corner frequency or transition band above which the dielectric charge absorption capability of electrodes dominates and leads to a decrease in ripple-related battery ageing [50]. Operation in this transition band is therefore accompanied by a decrease in the cell impedance for the battery ripple.

This paper for the first time solves a major problem of MMCs with batteries and other reconfigurable battery systems, specifically their large low-frequency ripple load, which particularly arises in realistic real-world setups that use bandwidth-limited sensors, control busses, and/or control hardware that introduces latency. The control approach aims to exploit the filtering effect of the dielectric electrode capacitance. The method introduces a battery ripple modulation loop in the scheduling algorithm, which shifts the load ripple toward higher frequencies. Considering practical limitations of the monitoring and communication system speed, the control method uses state observers to sufficiently increase the bandwidth of the battery ripple modulation loop. According to aforementioned studies, load content in the high frequency range is absorbed by the electrode capacitance, which significantly lowers the ageing potential driven by electrochemical reactions. Utilization of the electrode capacitance is furthermore associated with lower impedance and consequently losses [51].

Control in battery integrated CHB

CHB topologies, including CHB², incorporate low-voltage semiconductor switches into each module. The high number of individually governable active components provides the degrees of freedom to control additional objectives, such as active balancing of the module charge [52-57]. While online-optimizing model-predictive approaches can trade off all degrees of freedom but suffer from the high computational load, phase-shifted carrier control, for instance, is a rather simple way to manage the complexity and can further use the parallel mode to maximize utilization of the modules [27, 31, 33, 58-61]. The majority of the control methods introduces and considers only certain useful switch states on the module level to reduce the complexity, such as parallel P, series plus S+, bypass plus B+, series minus S–, and bypass minus B– [26]. The complexity can be further reduced by introducing feasible series–parallel configurations of the whole string of modules, so called string states S.

A recently presented approach for controlling those expands above multi-objective optimization approaches of module states and includes criteria such as SoC, temperature T, phase current demand i*, measured phase current im, and the previous string state S^{-1} for the selection of the next string state to comply with the voltage demand v^*, discretized in the voltage modulator to v_d^* [62]. However, this method as other solutions with already product-ready topologies with a bus system to distribute the commands, decouple the battery control objectives from the output control to deal with the limited bus capacity. Instead a fast, strictly real-time loop of the controller selects optimal module-string states S_o from an optimized *state list* provided by a slower loop with more time (see Figure 2). Update times on the second level are typical, but introduces persistent and regular switching patterns that get translated to low-frequency and even sub-harmonic ripple content in the module load. These low-frequency patterns are a result of the interaction of various control objectives and the reality of limited control bandwidth and feedback speed.

Increasing the update rate is limited by relying on slow acquisition of module information, which is often collected through data communication busses with considerable latency [63-67]. The communication bus latency in addition to signal processing can readily reach $10 - 100$ ms, while the switching rate of the phase voltage period is in the microsecond range. The other existing solutions struggle (and fail) in view of one fundamental trade-off: the need for lab-grade low-latency sensors, fast direct, bus-less connection of all sensors as well as gate sig-

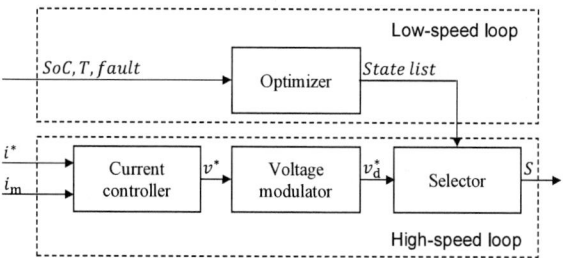

Figure 2. Parallel asynchronous optimization for MMC control according to the state of the art, where a low-speed loop optimizes module states or state transitions and stores them in a list or look-up table, which a high-speed loop of the actual controller, modulator, and scheduler uses to actuate the transistors in the modules. Parallel asynchronous optimization methods substantially reduced the computational burden for online optimization.

nals, and high-performance embedded control for rapid scheduling as described, for instance, in Li et al. [59] clashes with the conditions in more realistic larger systems as used in commercial setups. In commercial systems, a larger number of modules is typically connected to a more economic off-the-shelf controller via a communication bus, more affordable industry-grade sensors provide slower and lower-bandwidth data, and off-the-shelf economic processing power introduces bottle necks as described in Specht et al. [62], Rietmann et al. [68], and Hao et al. [64].

Proposed control approach

To achieve optimal battery treatment, the load distribution and the ripple modulation become the major, constrained by the voltage demand of the modulator. We use two major components, a strictly real-time compliant state selection as well as the high-bandwidth but asynchronous and not strictly real-time state optimization. The latter writes information into look-up table, which the former reads, enabling concurrent operation. The state optimizer comprises a set of ripple modulators, one assigned to each module, an optimization routine for the selection of optimal module-string states, and a battery-ripple observer, which closes the control loop. The battery-ripple observer instead of waiting for slow measurement data is the core element that enables the high bandwidth. Figure 3 outlines the complete control loop.

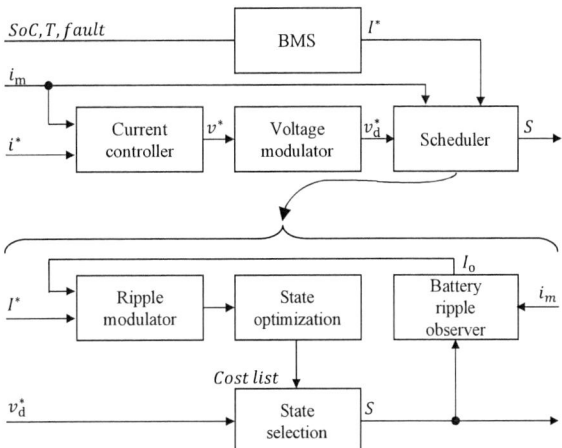

Figure 3. Block diagram of the proposed MMC control algorithm with fast quasi-inline optimization inside the scheduler block, which is enabled through a fast module current observer and further detailed at the bottom.

Our algorithm aims for a maximization of the bandwidth, which reduces artefactual patterns in switching and reduces loss as well as battery ageing potential associated with low-frequency load ripple. Efficient suppression of switching patterns follows from matching the module control loop and the phase control loop in speed. Therefore, in contrast to previous suggestions in the literature, all blocks of the ripple modulator preferably run within a switching period of the phase control loop. The ripple modulator loop still has a fundamentally asynchronous design and is not strictly time-critical so that any delay in execution does not halt control but might only introduces a short artefactual switching pattern and small ripple with length respective to the delay.

Battery-ripple modulator

The battery-ripple modulator guarantees discharging and charging of the modules at the demanded rate I^* and provides an interface for any higher-level entity that balances the SoC and for any BMS functionality. The use of proportional units in the demanded discharge rate reference I^* distributing the load between the modules ignores one degree of freedom, i.e., scaling, and solves eventual contradictions with the phase-current demand i^*.

The modulator needs to implement a controller with highly integrational character, also known as reset controller, which brings a controlled variable close to the demanded value. The requirement follows from the distinct distribution J of the phase load i_m among modules in each string state S, which does not necessarily allow the demanded current distribution in each step. The integrated value of the battery-ripple is passed to the optimization routine, which modulates an appropriate sequence of string states and provides the demand on average while controlling the battery ripple.

Battery-ripple observer

The ripple-modulator loop requires considerably fast feedback to run at maximum speed. An acquisition of the module current and transmission of the measured value to the controller represents either unacceptable propagation delay and/or heavy load of a data bus [62, 64, 67-69]. The typical period of the data acquisition is on the order of milliseconds, which is comparable to the load frequency and may not be sufficient for proper control of the module load frequency. Our modulator architecture solves the slacking feedback using an observation technique. The observation technique needs to sufficiently approximate the module load but primarily provide minimal delay. We further derive a simplified observation technique with afore-mentioned qualities.

The actual current load i_{bi} of module i is a result of the phase string state (series–parallel configuration of modules), current load of the phase I_L, and voltage V_B as well as impedance ratios of modules R_B vs. their interconnection paths R_S (R_{LS} designates resistance of the low-side and R_{HS} designates resistance of the high-side interconnection). In principle, the string configuration comprises a set of parallel groups, where each parallel group forms one level of the output voltage and is loaded by the phase current. Equation set

$$\begin{bmatrix} R_{D1} & R_{U2} & 0 & 0 & 0 \\ R_{L1} & R_{D2} & R_{U3} & 0 & 0 \\ R_{L2} & R_{L2} & R_{D3} & R_{U4} & 0 \\ R_{L3} & R_{L3} & R_{L3} & R_{D4} & R_{U5} \\ 1 & 1 & 1 & 1 & 1 \end{bmatrix} \cdot \begin{bmatrix} i_{b1} \\ i_{b2} \\ i_{b3} \\ i_{b4} \\ i_{b5} \end{bmatrix} = \begin{bmatrix} B_1 \\ B_2 \\ B_3 \\ B_4 \\ I_L \end{bmatrix},$$

where

$$R_{Dj} = -(R_{Bj} + R_{LSj} + R_{HSj})$$
$$R_{Lj} = -(R_{HSj} + R_{LSj})$$
$$R_{Uj} = R_{Bj}$$

$$B_j = V_{Bj+1} - V_{Bj} - R_{LSj} \cdot I_L \tag{1}$$

governs further distribution of the phase current among paralleled modules, where the dimension of the problem is equal to the number of parallel modules, and subscript j designates the index of the module in the phase string [70].

Considering constant values of all resistances allows pre-calculation of the impedance matrix and significantly simplifies the algorithm complexity solving (1). The current distribution is then reduced to a function of module voltage differences, which are kept minimal and change only slowly. Under the assumption of relatively constant module voltages within the relevant periods, the whole problem can be further simplified to a look-up table where the observer pre-estimates the expected module current distribution of each feasible string state in the look-up table.

State optimization

The optimization routine also deals with the special setting in case of additional parallel connectivity. To equally distribute the phase load, the majority of modules preferably stay in the active state and rather control their contribution to the phase current by appropriate clustering in parallel groups. Similar to the battery-ripple observer, all string states in the optimization block are represented by the current distribution (in a look-up table). The optimization routine selects output state S_o with optimal current distribution $J_{i,m}$ respecting the demand of the module current controllers J_i^*. Our algorithm uses a typical least-square criterion of optimality to evaluate each state. The least-squares criterion guarantees sufficient effort to meet the regulator demands and simultaneously reduces conduction losses by preventing states with far outlying load distribution (e.g., excessive use of bypassing). Results of the evaluation of each feasible state are stored in a look-up table, which interfaces and decouples the ripple modulator loop and the strictly real-time phase-current control loop as outlined above (see Figure 4).

To achieve a sufficient speed of the optimization routine, we constrain transitions between consecutive string configurations by limiting the number of switches that can toggle in each step. This rule also limits the number of commutating switches and consequently reduces switching losses

Experimental ripple suppression measurements

We built a single-phase laboratory test setup for experimental evaluation and prepared two control methods: the presented one and as a reference from the state of the art the method of [62]. The aim of the demonstration is to illustrate the effect of the fast feedback and control loop in preventing fixed switching patterns for long intervals.

The test setup includes five CHB[2] modules with silicon field-effect transistors (IAUT300N10S5N015, Infineon). The dc bus of each module contains a six-cell LiFePO$_4$ battery (22.5 V, 6s, 6.2 Ah). The system controller uses a Mars ZX3 module (Enclustra) with Xilinx's Zynq-7020 system-on-chip. The

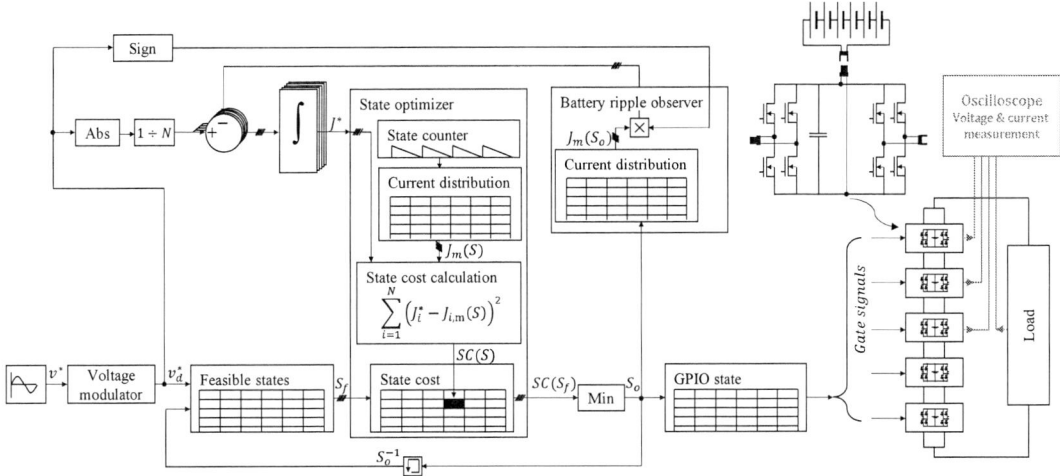

Figure 4. Block diagram of the control algorithm adapted for sensorless operation and open-loop control. The ripple-modulator loop works directly with current distribution J_m without the need for scaling it with the phase-current magnitude. The demand of the ripple modulator considers an equal distribution of load and is calculated from the discrete phase-voltage demand divided by the total number of modules N. The ripple modulator uses an integrator to cumulate any disturbances from the demanded load distribution. The cumulated value is used in the optimizer, which evaluates all states according to given criteria and assigns their cost SC in the state cost table. The phase-voltage loop independently selects feasible states according to the demanded voltage and finds the state with minimal cost function, which most effectively compensates cumulated load disturbances. The load distribution of the state is observed, fed back, and cumulated in the ripple modulator.

control algorithm runs fully on the FPGA part. The setup implements a sigma–delta modulator running at 20 kHz.

The implementation of the proposed control method follows the structure given in Figure 4. The method of Specht et al. (see Figure 2) serves as a reference. It delivers measured module load data at an update rate of ~10 Hz with an execution cycle time at 100 ms for the measurement loop, which in turn leads to slacking in the feedback and which we implemented accordingly.

Figure 5 displays the output of the system, while Figure 6 presents the module current comparison between the proposed method with high bandwidth and the reference method from the literature. The slacking feedback of the reference method is noticeable in the module current as it results in artifactual load patterns (see Figure 6a). These artifacts obviously form intervals that correspond to the cycle time of the feedback loop around 100 ms. The output states (and consequently the load distribution) do hardly vary within such an interval, and their repetition generates a regular pattern in the module current

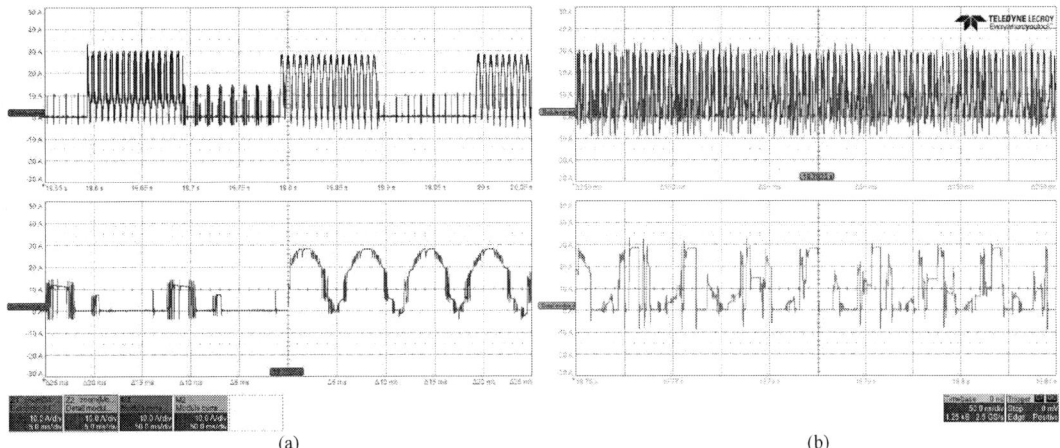

(a)   (b)

Figure 6. Measurement of the current load of an individual module over time of (a) the reference method (with characteristic long-lasting switching patterns) and (b) the proposed method (with distributed switching) measured under phase load specified in Table II. The lower panel of each subplot represents a magnified section out of the full trace.

provided by the batteries. The patterns cover the entire range from full phase current to practically no battery current in a module at all.

In stark contrast to this established method, the proposed control allows full utilization of the switching rate and evenly distributes the switching events in time. As intended, the distribution of the switching prevents extensive utilization of individual modules and naturally reduces the low frequency content. Yet, the module load can contain some minor residual switching patterns, which we found to be mostly emerging in case of long propagation delay. As long as the period length of the pattern stays negligible compared to the phase frequency of 50 Hz, however, the control method efficiently suppresses low-frequency content through distributed switching (see Figure 6b).

The impact of the proposed control method on the module current is more obvious in the frequency domain. The spectrum of the reference method's module current exhibits dominant peaks at twice the phase load frequency ($f = 2{\times}50$ Hz $= 100$ Hz). Futher peaks can be observer at 5 Hz, which repeats at 10 Hz, 15 Hz, etc. These peaks follow from the update period of 100 ms and the patterns in the module load. The reference method displays relatively low content at frequencies above 100 Hz, which rises again around the switching frequency.

In contrast to the prior art, the frequency content of our proposed method is practically negligible for low frequencies, just starts at 100 Hz, and features increased content up to a fraction of the switching frequency. The reduction of lower frequencies and the partial shift to higher ones is a result of appropriate switching distribution and prevention of pattern formation. Depending on the modulation index, the ripple is shifted to a region corresponding to the individual module switching frequency.

Effective impedance reduction through nonfaradaic shunting

Low-frequency ripple contains larger charge quantities, generates losses, and can age batteries either through the associated heating stress or potentially also electrochemical degradation [48]. Above a certain frequency, however, the impedance of battery cells drops steeper (closer to a dielectrically capacitive f^{-1}) than the diffusion-limited $f^{-1/2}$ behavior of faradaic reactions until at very high frequencies the inductance sets a minimum [71]. At such high frequencies, not only the losses decrease but also faradaic processes cease as they cannot follow those charge oscillations anymore; the detected reduction in ageing potential at higher frequencies concurs with this effect [49]. Impedance spectroscopy indicates the transition from faradaic processes, i.e., electrochemical charge-transfer reactions, as the key source of the currents to the dielectric electrode capacitance for most cells above 100 Hz − 1 kHz.

This behavior can be represented with a small-signal approximation of the widely used Randles' equivalent circuit for the electrochemical interface (see inset of Figure 7) [72]. With increasing frequency, the dielectric electrode capacitance bypasses the slow diffusion-limited charge-transfer processes, reducing loss. Consequently, the proposed control method should lead to lower losses.

To quantify the losses, we performed load tests of 30 seconds each and extracted the value of the internal voltage V_i shortly after the load test with a delay of 15 seconds for voltage settling to evaluate the internal power loss. We further pool results from multiple measurements and modules to incorporate the entire range of potential conditions.

As intended, the power loss of the proposed scheduling method is throughout all modulation indices substantially lower than the conventional method from the literature as Figure 8 indicates. The solid lines represent the average values, whereas the lighter range indicates the observed variety of values across all measurements. The spread is mainly from module to module, while repetitive measurements within the same module at the same SoC did not vary by more than 0.1 % of the power loss. Our method reduces the power loss by up to 12 W, which equals

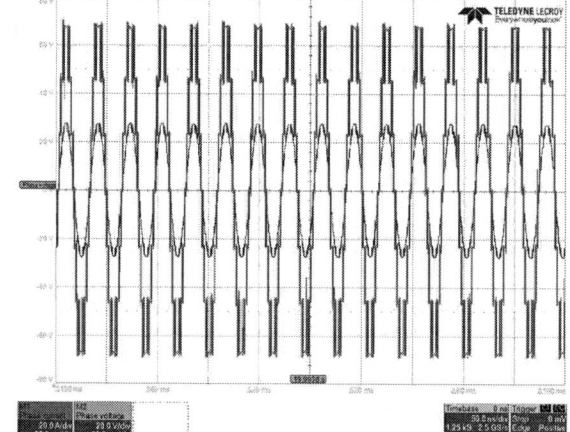

Figure 5. Experimental load of the eleven-level MMSPC phase string measured at m = 0.7, Ipk = 25 A.

a reduction in battery loss by ~20 % compared to the reference method. The loss improvement gradually decreases close to 100 % modulation index, where high utilization of modules reduces the degrees of freedom in circuit reconfiguration for optimization of the ripple.

Conclusion

This paper presents a novel control method for modular multilevel converters with integrated batteries, which suffer from substantial ripple load on the batteries and associated degradation and additional heating. The method aims for battery applications and their specific need for better battery treatment. Based on previous observations that the impedance of the battery cells as well as ageing potential tend to decrease with frequency, since higher frequencies are absorbed by the dielectric electrode capacitance.

In conventional methods, the limited speed and nonnegligible latency of sensor data collection from the individual modules in addition to often slow update rates of scheduling tables typically generate regular patterns in the module load. Accordingly, also module-balancing loops of the state of the art have to adjust their bandwidth to these conditions to avoid instability and driving oscillations. To solve this issue, we developed a battery-ripple observer, which allows us to create a fast feedback loop despite unavoidable latencies so that our control algorithm can actively modulate the module ripple and operate with comparably high bandwidth. We designed a control algorithm to minimize particularly low-frequency content of the module load spectrum.

We evaluated our novel control technique experimentally and compared it to the state of the art. Due to above-mentioned limitations, the conventional method from the literature indeed produced fluctuating module load below 100 Hz and even below 10 Hz,

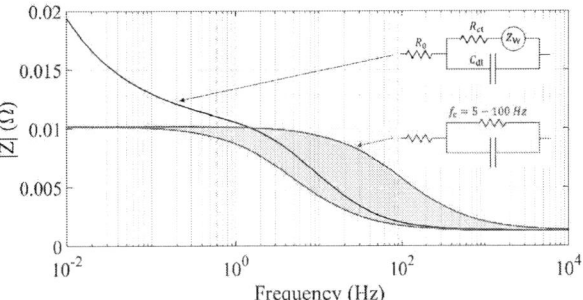

Figure 7. Impedance profile over frequency of Randles' model compared to a reduced first-order high-pass filter impedance to represent the dielectric shunting around the diffusion limitation of the faradaic component. Inset: Randles' equivalent circuit with the Warburg impedance ZW representing the diffusion limitation of the charge-transfer reactions, the charge-transfer equivalent resistance R_{ct} of ion travel near the electrode interface, the double-layer capacitance C_{dl}, and the electrolyte/separator equivalent resistance; reduced small-signal filter representation with range of cut-off frequencies f_c.

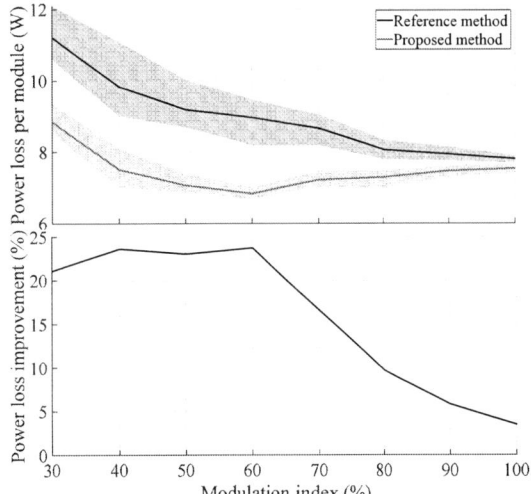

Figure 8. Top: Battery power loss averaged across multiple modules and experiments for various modulation indices. The suppression of low-frequency ripple currents and shift of some of those to higher frequencies, where the batteries can absorb the current dielectrically and accordingly show lower effective impedance, clearly reduce the loss and heating in the batteries. Bottom: Power loss improvement of the proposed method with observer relative to reference. The loss improvement gradually decreases on the way to maximum modulation index, where high utilization of modules reduces the available degrees of freedom in circuit reconfiguration for optimization of the ripple

which corresponds to the module data acquisition period, latencies, and bandwidth limits of the feedback, whereas our proposed method could prevent the formation of harmonics in this frequency range and shift fluctuations to the band around 5 kHz. Consequently, our control approach does effectively exploit lowered impedance at higher frequencies and reduces losses in the battery modules by up to 20 %.

References

[1] L. Zhang, J. Qin, Q. Duan, and W. Sheng, "Component Sizing and Voltage Balancing of MMC-based Solid-State Transformers Under Various AC-Link Excitation Voltage Waveforms," in *2019 IEEE Applied Power Electronics Conference and Exposition (APEC)*, 2019, pp. 371-375.

[2] O. Satilmis and E. Mese, "Investigating DC link current ripple and PWM modulation methods in Electric Vehicles," in *2013 3rd International Conference on Electric Power and Energy Conversion Systems*, 2013, pp. 1-6.

[3] I. Trintis, S. Munk-Nielsen, and R. Teodorescu, "A new modular multilevel converter with integrated energy storage," in *IECON 2011 - 37th Annual Conference of the IEEE Industrial Electronics Society*, 2011, pp. 1075-1080.

[4] N. Tashakor, B. Arabsalmanabadi, F. Naseri, and S. Goetz, "Low-Cost Parameter Estimation Approach for Modular Converters and Reconfigurable Battery Systems Using Dual Kalman Filter," *IEEE Transactions on Power Electronics,* vol. 37, no. 6, pp. 6323-6334, 2022.

[5] M. Vasiladiotis, N. Cherix, and A. Rufer, "Impact of Grid Asymmetries on the Operation and Capacitive Energy Storage Design of Modular Multilevel Converters," *IEEE Transactions on Industrial Electronics,* vol. 62, no. 11, pp. 6697-6707, 2015.

[6] F. Gao, L. Zhang, Q. Zhou, M. Chen, T. Xu, and S. Hu, "State-of-charge balancing control strategy of battery energy storage system based on modular multilevel converter," in *2014 IEEE Energy Conversion Congress and Exposition (ECCE)*, 2014, pp. 2567-2574.

[7] S. Chowdhury, M. N. B. Shaheed, and Y. Sozer, "An Integrated State of Health (SOH) Balancing Method for Lithium-Ion Battery Cells," in *2019 IEEE Energy Conversion Congress and Exposition (ECCE)*, 2019, pp. 5759-5763.

[8] Q. Zhang, F. Gao, L. Zhang, and N. Li, "Multiple time scale optimal operation of MMC battery energy storage system," in *2015 IEEE Energy Conversion Congress and Exposition (ECCE)*, 2015, pp. 1-7.

[9] N. Li, F. Gao, T. Hao, Z. Ma, and C. Zhang, "SOH Balancing Control Method for the MMC Battery Energy Storage System," *IEEE Transactions on Industrial Electronics,* vol. 65, no. 8, pp. 6581-6591, 2018.

[10] Z. Ma, T. Hao, F. Gao, N. Li, and X. Gu, "Enhanced SOH balancing method of MMC battery energy storage system with cell equalization capability," in *2018 IEEE Applied Power Electronics Conference and Exposition (APEC)*, 2018, pp. 3591-3597.

[11] F. Gao, X. Gu, Z. Ma, and C. Zhang, "Redistributed Pulsewidth Modulation of MMC Battery Energy Storage System Under Submodule Fault Condition," *IEEE Transactions on Power Electronics,* vol. 35, no. 3, pp. 2284-2294, 2020.

[12] S. M. Goetz, C. Wang, Z. Li, D. L. K. Murphy, and A. V. Peterchev, "Concept of a distributed photovoltaic multilevel inverter with cascaded double H-bridge topology," *International Journal of Electrical Power & Energy Systems,* vol. 110, pp. 667-678, 2019/09/01/ 2019.

[13] C. Wang, F. R. Lizana, Z. Li, A. V. Peterchev, and S. M. Goetz, "Submodule short-circuit fault diagnosis based on wavelet transform and support vector machines for modular multilevel converter with series and parallel connectivity," in *IECON 2017 - 43rd Annual Conference of the IEEE Industrial Electronics Society*, 2017, pp. 3239-3244.

[14] C. Wang, A. V. Peterchev, and S. M. Goetz, "Online Switch Open-Circuit Fault Diagnosis Using Reconfigurable Scheduler for Modular Multilevel Converter with Parallel Connectivity," in *2019 21st European Conference on Power Electronics and Applications (EPE '19 ECCE Europe)*, 2019, pp. P.1-P.10.

[15] M. Vasiladiotis and A. Rufer, "Analysis and Control of Modular Multilevel Converters With Integrated Battery Energy Storage," *IEEE Transactions on Power Electronics,* vol. 30, no. 1, pp. 163-175, 2015.

[16] Z. Li, A. Yang, G. Chen, Z. Zeng, A. V. Peterchev, and S. M. Goetz, "A High-Frequency Pulsating DC-Link for Electric Vehicle Drives with Reduced Losses," in *IECON 2021 – 47th Annual Conference of the IEEE Industrial Electronics Society*, 2021, pp. 1-6.

[17] J. V. Muenzel and D. Crowley, "Battery system including circuit module for selectively connecting a plural of battery cell units," Feb. 25, 2020.

[18] D. L. Miller, "Battery charge balancing system having parallel switched energy storage elements," May 16, 2000.

[19] S. Götz, J. Kacetl, T. Kacetl, and M. Jaensch, "Electric energy storage system," DE 10 2018 109 921, Aug. 8, 2019.

[20] S. Ci, N. Lin, and D. Wu, "Reconfigurable Battery Techniques and Systems: A Survey," *IEEE Access,* vol. 4, pp. 1175-1189, 2016.

[21] M. A. Rahman, K. d. Craemer, J. Büscher, J. Driesen, P. Coenen, and C. Mol, "Comparative Analysis of Reconfiguration Assisted Management of Battery Storage Systems," in *IECON 2019 - 45th Annual Conference of the IEEE Industrial Electronics Society*, 2019, vol. 1, pp. 5921-5926.

[22] Z. Li, R. Lizana, A. V. Peterchev, and S. M. Goetz, "Ripple current suppression methods for star-configured modular multilevel converters," in *IECON 2017 - 43rd Annual Conference of the IEEE Industrial Electronics Society*, 2017, pp. 1505-1510.

[23] Z. Ma, F. Gao, C. Zhang, W. Li, and D. C. Niu, "Variable DC-Link Voltage Regulation of Single-Phase MMC Battery Energy Storage System for Reducing Additional Charge Throughput," *IEEE Transactions on Power Electronics,* pp. 1-1, 2021.

[24] S. Qiu and B. Shi, "An Enhanced Battery Interface of MMC-BESS," in *2019 IEEE 10th International Symposium on Power Electronics for Distributed Generation Systems (PEDG)*, 2019, pp. 434-439.

[25] J. Fang, F. Blaabjerg, S. Liu, and S. M. Goetz, "A Review of Multilevel Converters With Parallel Connectivity," *IEEE Transactions on Power Electronics,* vol. 36, no. 11, pp. 12468-12489, 2021.

[26] S. M. Goetz, Z. Li, X. Liang, C. Zhang, S. M. Lukic, and A. V. Peterchev, "Control of Modular Multilevel Converter With Parallel Connectivity—Application to Battery Systems," *IEEE Transactions on Power Electronics,* vol. 32, no. 11, pp. 8381-8392, 2017.

[27] S. M. Goetz, A. V. Peterchev, and T. Weyh, "Modular Multilevel Converter With Series and Parallel Module Connectivity: Topology and Control," *Power Electronics, IEEE Transactions on,* vol. 30, no. 1, pp. 203-215, 2015.

[28] H. Bahamonde, S. Rivera, Z. Li, S. Goetz, A. Peterchev, and R. Lizana, "Different parallel connections generated by the Modular Multilevel Series/Parallel Converter: an overview," in *IECON 2019 - 45th Annual Conference of the IEEE Industrial Electronics Society,* 2019, vol. 1, pp. 6114-6119.

[29] C. Korte, E. Specht, M. Hiller, and S. Goetz, "Efficiency evaluation of MMSPC/CHB topologies for automotive applications," in *2017 IEEE 12th International Conference on Power Electronics and Drive Systems (PEDS),* 2017, pp. 324-330.

[30] Z. Li, R. Lizana, S. M. Lukic, A. V. Peterchev, and S. M. Goetz, "Current Injection Methods for Ripple-Current Suppression in Delta-Configured Split-Battery Energy Storage," *IEEE Transactions on Power Electronics,* vol. 34, no. 8, pp. 7411-7421, 2019.

[31] Z. Li, R. Lizana, Z. Yu, S. Sha, A. V. Peterchev, and S. M. Goetz, "Modulation and Control of Series/Parallel Module for Ripple-Current Reduction in Star-Configured Split-Battery Applications," *IEEE Transactions on Power Electronics,* vol. 35, no. 12, pp. 12977-12987, 2020.

[32] Z. Li, R. Lizana, A. V. Peterchev, and S. M. Goetz, "Ripple Current Suppression Methods for Star-Configured Modular Multilevel Converters," *Proc. IEEE Energy Conversion Congress and Exposition,* vol. 43, pp. 1-6, 2017.

[33] Z. Li, R. Lizana, A. V. Peterchev, and S. M. Goetz, "Predictive control of modular multilevel series/parallel converter for battery systems," in *2017 IEEE Energy Conversion Congress and Exposition (ECCE),* 2017, pp. 5685-5691.

[34] Z. Li, R. Lizana, S. Sha, Z. Yu, A. V. Peterchev, and S. Goetz, "Module Implementation and Modulation Strategy for Sensorless Balancing in Modular Multilevel Converters," *IEEE Transactions on Power Electronics,* vol. 34, no. 9, pp. 8405-8416, 2018.

[35] C. Korte, E. Specht, S. M. Goetz, and M. Hiller, "A Control Scheme to Reduce the Current Load of Integrated Batteries in Cascaded Multilevel Converters," in *2019 10th International Conference on Power Electronics and ECCE Asia (ICPE 2019 - ECCE Asia),* 2019, pp. 1-8.

[36] J. Vetter *et al.*, "Ageing mechanisms in lithium-ion batteries," *Journal of Power Sources,* vol. 147, no. 1, pp. 269-281, 2005/09/09/ 2005.

[37] C. R. Birkl, M. R. Roberts, E. McTurk, P. G. Bruce, and D. A. Howey, "Degradation diagnostics for lithium ion cells," *Journal of Power Sources,* vol. 341, pp. 373-386, 2017/02/15/ 2017.

[38] P. Novák *et al.*, "The complex electrochemistry of graphite electrodes in lithium-ion batteries," *Journal of Power Sources,* vol. 97-98, pp. 39-46, 2001/07/01/ 2001.

[39] V. Agubra and J. Fergus, "Lithium Ion Battery Anode Aging Mechanisms," *Materials,* vol. 6, pp. 1310-1325, 03/01 2013.

[40] P. Verma, P. Maire, and P. Novák, "A review of the features and analyses of the solid electrolyte interphase in Li-ion batteries," *Electrochimica Acta,* vol. 55, no. 22, pp. 6332-6341, 2010/09/01/ 2010.

[41] G.-A. Nazri and G. Pistoia, *Lithium Batteries: Science and Technology.* 2009.

[42] H. Helmholtz, "Ueber einige Gesetze der Vertheilung elektrischer Ströme in körperlichen Leitern mit Anwendung auf die thierisch-elektrischen Versuche," *Annalen der Physik,* https://doi.org/10.1002/andp.18531650603 vol. 165, no. 6, pp. 211-233, 1853/01/01 1853.

[43] O. Stern, "Zur Theorie der elektrolythischen Doppelschicht," *Zeitschrift für Elektrochemie und angewandte physikalische Chemie,* https://doi.org/10.1002/bbpc.192400182 vol. 30, no. 21-22, pp. 508-516, 1924/11/01 1924.

[44] M. Gouy, "Sur la constitution de la charge électrique à la surface d'un électrolyte," (in French), vol. 9, no. 1, pp. 457-468, 1910 1910.

[45] D. L. Chapman, "LI. A contribution to the theory of electrocapillarity," *The London, Edinburgh, and Dublin Philosophical Magazine and Journal of Science,* vol. 25, no. 148, pp. 475-481, 1913/04/01 1913.

[46] K. Xu, "Electrolytes and Interphases in Li-Ion Batteries and Beyond," *Chemical Reviews,* vol. 114, no. 23, pp. 11503-11618, 2014/12/10 2014.

[47] P. E. de Jongh and P. H. L. Notten, "Effect of current pulses on lithium intercalation batteries," *Solid State Ionics,* vol. 148, no. 3, pp. 259-268, 2002/06/02/ 2002.

[48] N. Legrand, S. Raël, B. Knosp, M. Hinaje, P. Desprez, and F. Lapicque, "Including double-layer capacitance in lithium-ion battery mathematical models," *Journal of Power Sources,* vol. 251, pp. 370-378, 2014/04/01/ 2014.

[49] A. Bessman, R. Soares, O. Wallmark, P. Svens, and G. Lindbergh, "Aging effects of AC harmonics on lithium-ion cells," *Journal of Energy Storage,* vol. 21, pp. 741-749, 2019/02/01/ 2019.

[50] M. Uno and K. Tanaka, "Influence of High-Frequency Charge–Discharge Cycling Induced by Cell Voltage Equalizers on the Life Performance of Lithium-Ion Cells," *IEEE Transactions on Vehicular Technology,* vol. 60, no. 4, pp. 1505-1515, 2011.

[51] L. Chen, "A Design of an Optimal Battery Pulse Charge System by Frequency-Varied Technique," *IEEE Transactions on Industrial Electronics,* vol. 54, no. 1, pp. 398-405, 2007.

[52] S. Debnath, J. Qin, B. Bahrani, M. Saeedifard, and P. Barbosa, "Operation, Control, and Applications of the Modular Multilevel Converter: A Review," *IEEE Transactions on Power Electronics,* vol. 30, no. 1, pp. 37-53, 2015.

[53] J. D. Mike Dommaschk, Ingo Euler, Jörg Lang, Quoc-Buu Tu, Klaus Würflinger, "Driving of a phase module branch of a multilevel converter," U. S., 2010.

[54] F. Deng and Z. Chen, "A Control Method for Voltage Balancing in Modular Multilevel Converters," *IEEE Transactions on Power Electronics,* vol. 29, no. 1, pp. 66-76, 2014.

[55] M. Hagiwara and H. Akagi, "Control and Experiment of Pulsewidth-Modulated Modular Multilevel Converters," *IEEE Transactions on Power Electronics,* vol. 24, no. 7, pp. 1737-1746, 2009.

[56] M. Guan, Z. Xu, and C. Hairong, "Control and modulation strategies for modular multilevel converter based HVDC system," in *IECON 2011 - 37th Annual Conference of the IEEE Industrial Electronics Society,* 2011, pp. 849-854.

[57] K. Ilves, L. Harnefors, S. Norrga, and H. Nee, "Predictive Sorting Algorithm for Modular Multilevel Converters Minimizing the Spread in the Submodule Capacitor Voltages," *IEEE Transactions on Power Electronics,* vol. 30, no. 1, pp. 440-449, 2015.

[58] Z. Li, R. Lizana, A. V. Peterchev, and S. M. Goetz, "Distributed balancing control for modular multilevel series/parallel converter with capability of sensorless operation," in *2017 IEEE Energy Conversion Congress and Exposition (ECCE),* 2017, pp. 1787-1793.

[59] S. M. Goetz, Z. Li, A. V. Peterchev, X. Liang, C. Zhang, and S. M. Lukic, "Sensorless scheduling of the modular multilevel series-parallel converter: enabling a flexible, efficient, modular battery," in *2016 IEEE Applied Power Electronics Conference and Exposition (APEC),* 2016, pp. 2349-2354.

[60] Z. Li, R. L. F, S. Sha, Z. Yu, A. V. Peterchev, and S. M. Goetz, "Module Implementation and Modulation Strategy for Sensorless Balancing in Modular Multilevel Converters," *IEEE Transactions on Power Electronics,* vol. 34, no. 9, pp. 8405-8416, 2019.

[61] N. Tashakor, M. Kilictas, E. Bagheri, and S. Goetz, "Modular Multilevel Converter With Sensorless Diode-Clamped Balancing Through Level-Adjusted Phase-Shifted Modulation," *IEEE Transactions on Power Electronics,* vol. 36, no. 7, pp. 7725-7735, 2021.

[62] E. Specht, C. Korte, and M. Hiller, "Reducing Computation Effort by Parallel Optimization for Modular Multilevel Converters," in *IECON 2018 - 44th Annual Conference of the IEEE Industrial Electronics Society,* 2018, pp. 3991-3996.

[63] H. Tu and S. Lukic, "A hybrid communication topology for modular multilevel converter," in *2018 IEEE Applied Power Electronics Conference and Exposition (APEC),* 2018, pp. 3051-3056.

[64] H. Tu and S. Lukic, "Comparative study of PES Net and SyCCo bus: Communication protocols for modular multilevel converter," in *2017 IEEE Energy Conversion Congress and Exposition (ECCE),* 2017, pp. 1487-1492.

[65] Y. Park, J. Yoo, and S. Lee, "Practical Implementation of PWM Synchronization and Phase-Shift Method for Cascaded H-Bridge Multilevel Inverters Based on a Standard Serial Communication Protocol," *IEEE Transactions on Industry Applications,* vol. 44, no. 2, pp. 634-643, 2008.

[66] C. L. Toh and L. E. Norum, "A high speed control network synchronization jitter evaluation for embedded monitoring and control in modular multilevel converter," in *2013 IEEE Grenoble Conference,* 2013, pp. 1-6.

[67] G. S. Tomas Kacetl, Kacelt Jan, Simon Daniel, Jaensch Malte, Specht Eduard, Dibos Hermann Helmut, Weyland Axel, "Verfahren und System zu einer Vorauswahl von Schaltzuständen für einen Multilevelkonverter," DE102020117264, Feb. 06, 2021.

[68] S. Rietmann, S. Fuchs, A. Hillers, and J. Biela, "Field Bus for Data Exchange and Control of Modular Power Electronic Systems with High Synchronisation Accuracy," in *2018 International Power Electronics Conference (IPEC-Niigata 2018 -ECCE Asia),* 2018, pp. 2301-2308.

[69] M. Slepchenkov and R. Naderi, "Module-based energy systems capable of cascaded and interconnected configurations, and methods related thereto " USA Patent US 11,135,923 B2, 2019.

[70] G. Gunlu, "Dynamically Reconfigurable Independent Cellular Switching Circuits for Managing Battery Modules," *IEEE Transactions on Energy Conversion,* vol. 32, no. 1, pp. 194-201, 2017.

[71] B. V. Ratnakumar, M. C. Smart, and S. Surampudi, "Electrochemical impedance spectroscopy and its applications to lithium ion cells," in *Seventeenth Annual Battery Conference on Applications and Advances. Proceedings of Conference (Cat. No.02TH8576),* 2002, pp. 273-277.

[72] J. E. B. Randles, "Kinetics of rapid electrode reactions," *Discussions of the Faraday Society,* 10.1039/DF9470100011 vol. 1, no. 0, pp. 11-19, 1947.

Characterization of Conventional and Advanced Current Measurement Techniques Suitable for WBG Semiconductor Devices

Severin Klever, André Thönnessen, Rik W. De Doncker
Institute for Power Electronics and Electrical Drives
RWTH Aachen University
Jägerstr. 17-19, 52066 Aachen, Germany
Phone: +49 241 80 96920, Fax: +49 241 80 92203
Email: post@isea.rwth-aachen.de
URL: https://www.isea.rwth-aachen.de

Keywords

≪Current Sensor≫, ≪Wide Bandgap Devices≫, ≪Double Pulse Test≫, ≪Parasitic Inductance≫, ≪Impedance Measurement≫, ≪Frequency-Domain Analysis≫, ≪Time-Domain Analysis≫

Abstract

With wide bandgap semiconductor devices, conventional current sensors like Rogowski coils and coaxial shunts reach their limits in bandwidth or parasitic inductance. Therefore, novel techniques are continuously being developed, which promise improved characteristics. In order to evaluate the performance of current sensors, it is necessary to compare them in a uniform environment. In this work, different sensors for currents up to 100 A and frequencies of several 100 MHz are characterized in the frequency- and time-domain. The characteristics bandwidth, parasitic inductance and delay are determined by measurement and compared to each other. Among the advanced measurement techniques is a design consisting of coaxially arranged discrete resistors. A prototype is built based on this idea and tested against the other sensors. The prototype outperforms the conventional sensors and is validated in a double pulse test with Gallium nitride devices.

1 Introduction

wide bandgap (WBG) semiconductors like Gallium nitride (GaN) or Silicon carbide (SiC) are present in modern power electronics. Their main advantage are the possibility of very high switching frequencies, which enable high power densities. However, high switching frequencies and fast switching transitions also bring challenges, such as the increase of switching losses or EMC problems as well as the accelerated aging of insulation materials [1]. Therefore, it is essential to have a precise knowledge of the switching behavior of the semiconductor devices used and to carry out exact measurements during the design process of power electronic components. A common method is the characterization via the double pulse test (DPT).

One key requirement for measuring the drain-source currents of WBG devices is a transfer function with a sufficiently high measurement bandwidth f_{BW}, which can be approximated from the rise time t_{rise} using equation (1) [2].

$$f_{\mathrm{BW}} = \frac{0.35}{t_{\mathrm{rise}}} \tag{1}$$

In addition, the switching behavior is affected significantly by parasitic elements such as capacitances and inductances inserted by the measurement probe. Therefore, another requirement for current measurements is that it should affect the switching behavior as little as possible. For these reasons, current sensors must be compared in a consistent environment, as not all data sheets provide complete insight into the device or the data is simply not available for self-assembled sensors. Different current sensors for DPTs are compared, which can be divided into two classes: resistive measurement devices and inductive measurement devices. Rogowski coils from the manufacturer *PEM UK* and coaxial shunts from the manufacturer *T&M Research* are examined. In addition to these conventional measuring devices, two advanced sensors are investigated: a special rogowski coil that was developed at *Fraunhofer IZM* [3] and a newly developed coaxial shunt according to an idea from [4] based on resistors in the surface-mounted device (SMD)-format. Alternative measurement methods like magneto-resistive or hall effect sensors are not considered in this work because their bandwidths are not sufficient to characterize WBG semiconductors with rise times in the nanosecond range [5]. Combinations of different sensors are also excluded, although there are promising approaches with tunnel magneto-resistance sensors combined with Rogowski coils [6]. Apart from the characteristics analyzed in this work, there are further requirements such as a high noise immunity, galvanic isolation, a low temperature drift and cost [7].

2 Inductive Current Measurement

One simple technique for current measurement is contactless measurement by evaluating the magnetic field generated by the current. For so-called Rogowski coils, the law of induction is applied and describes the current to be measured via the integral of the induced voltage v_{ind}. Equation (2) shows the corresponding formula of an ideal Rogowski coil with the coil cross-section A and the number of turns N [8].

$$i_{\mathrm{sense}}(t) = \int v_{\mathrm{ind}}(t) \frac{N}{A\mu_0} \mathrm{d}t \tag{2}$$

The main advantages are a wide current measurement range without magnetic saturation, flexible mounting around conductors, a non-invasive and fully isolated measurement, as well as a high bandwidth in the range from 0.1 Hz to 1 GHz depending on the construction. The main disadvantage is the lower frequency boundary, i.e. no DC current can be measured. Furthermore, it is not practical to measure the current on a trace of a printed circuit board (PCB) with multiple layers. The same is the case for closed core current transformers, for example from *Pearson Electronics* [9]. These achieve a high bandwidth and a high accuracy, but are too bulky to be used for drain source current measurements and also have the problem of magnetic saturation. In the following, two measuring devices are presented that are based on the principle of induction.

Conventional Rogowski Coils

The manufacturer *PEM UK* offers a diversified range of Rogowski coils [10]. In this work the models *CWT MiniHF50 3* and *CWT UltraMini 3* are analyzed. These products have bandwidths of 50 MHz and 30 MHz and maximum currents of 600 A and 60 A, respectively. Although these Rogowski coils are very flexible, they have a limited field of application and cannot be easily used to measure the drain-source current i_{ds} on a PCB. Fig. 1a shows an exemplary model that includes the coil itself (yellow) and an unit for integration and amplification of the signal.

Specialized Rogowski Coils

Another way to measure currents is to use a Rogowski coil that is inserted into an Ω-shaped tube, shown in Fig. 1c [11]. The current to be measured (i_{sense}) flows over the surface of the tube and generates a magnetic flux $\vec{B}$ in the axial direction. Sensors using this technique promise very low insertion inductances and high bandwidths. The present model from *IPM Design* [12] can be used in a frequency range from 20 kHz to 260 MHz and allows to measure currents up to 1.55 kA according to the datasheet (see fig. 1b). Depending on the application, customized Rogowski coils of this type can also be used without analog integration. In this case, the current waveform can be calculated in a postprocessing step via the measured voltage v_{ind} [13]. In both cases, the disadvantage of the Ω-tube method is a pure AC measurement with a lower frequency boundary of several kHz, so additional effort is needed for calibration.

EPE'22 ECCE Europe

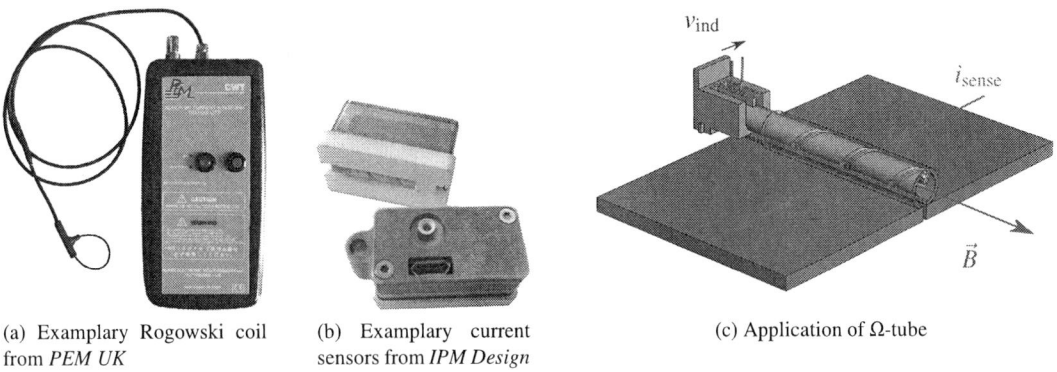

(a) Examplary Rogowski coil from *PEM UK*

(b) Examplary current sensors from *IPM Design*

(c) Application of Ω-tube

Fig. 1: Inductive current measurement devices

3 Resistive Current Measurement

Another method is resistive current measurement via a current viewing resistor (CVR). A defined and constant resistor R is inserted into the current path and the voltage drop across this resistor is measured via an oscilloscope. The measured voltage is proportional to the current according to Ohm's law. The advantage of this method is a simple and accurate measurement, suitable for low and high frequencies. The disadvantage is the impact on the circuit to be measured by inserting an additional parasitic inductance. This, together with the unavoidable frequency dependency of the resistor due to the skin effect, limits the bandwidth of current measurements with CVRs.

Conventional Coaxial Shunts

The manufacturer *T&M Research* offers a diversified range of coaxial shunts (Fig. 2a) [14]. In this work, the models *SDN-10* (100 mΩ) and *SDN-414-05* (50 mΩ) are analyzed. In the data sheets, the manufacturer specifies bandpass frequencies of 2 GHz, which must not be confused with the measurement bandwidth f_{BW}. In [15], it was shown that the actual bandwidth strongly varies between different samples, with samples ranging from 49.2 MHz to 147.6 MHz. Furthermore, the manufacturer does not provide any information about the parasitic inductances of the devices. It is therefore necessary to analyze the available coaxial shunts in more detail and, if necessary, to look for alternatives.

SMD Coaxial Shunt Prototype

In [4], a novel approach to build a CVR is introduced. The current sensor presented is based on vertically placed SMD resistors, which form a low-cost, low-inductance coaxial structure. Depending on the configuration, the inductance is claimed to be in the range from 0.12 nH to 0.22 nH and the bandwidth in the range from 528 MHz to 1630 MHz. In this work, a custom CVR is built based on this idea. The number of resistors is increased to 30 and standard thick film chip resistors in a 0402-Package with 1.58 Ω each are used (*Vishay CRCW04021R58FKED*). Fig. 2b and 2c show the cross sections of the sensor. The resistors are placed inside the disc-shaped PCB and are connected in parallel, resulting in a total resistance of $R_{CVR} = \frac{1.58\,\Omega}{30} = 52.7\,\text{m}\Omega$. First, the current flows upwards at the metallized outer edge of the PCB-disc and back inwards via the resistors. Fig. 2d shows the solder footprint for attaching the sensor to the circuit under test with the top layer traces in red and the solder pads in purple. The voltage across the resistors is measured via a BNC connector and the current is calculated according to $I_{\text{sense}} = \frac{V_{\text{res}}}{52.7\,\text{m}\Omega} = 19.0 V_{\text{res}}$. Each resistor has a rated current of 1.5 A, so a total current of 45 A is thermally allowed assuming that the current is distributed evenly over all resistors. The maximum allowed pulse current is above this value. Inductance, bandwidth and further characteristics are analyzed in the following and compared to the other measuring devices.

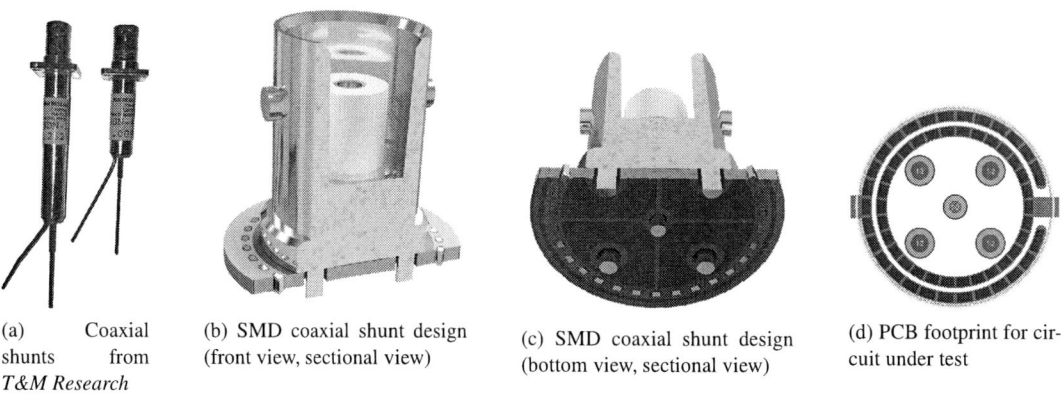

(a) Coaxial shunts from *T&M Research*

(b) SMD coaxial shunt design (front view, sectional view)

(c) SMD coaxial shunt design (bottom view, sectional view)

(d) PCB footprint for circuit under test

Fig. 2: Current shunt resistors

4 Characterization of Current Sensors

In this work, the following six sensors are characterized

1. *CWT UltraMini 3* Rogowski coil from *PEM UK*
2. *CWT MiniHF50 3* Rogowski coil from *PEM UK*
3. Ω-tube current sensor from *IPM Design*
4. *SDN-10* coaxial shunt from *T&M Research*
5. *SDN-414-05* coaxial shunt from *T&M Research*
6. SMD coaxial shunt prototype

For each sensor a PCB is designed (Fig. 3). In order to characterize the sensors by different measurements, each board can be connected to an impedance analyzer, a vector network analyzer (VNA) or a current pulse generator. This allows the characterization of each sensor in both the time and frequency domains. The several characterization methods are described in detail in the following sections.

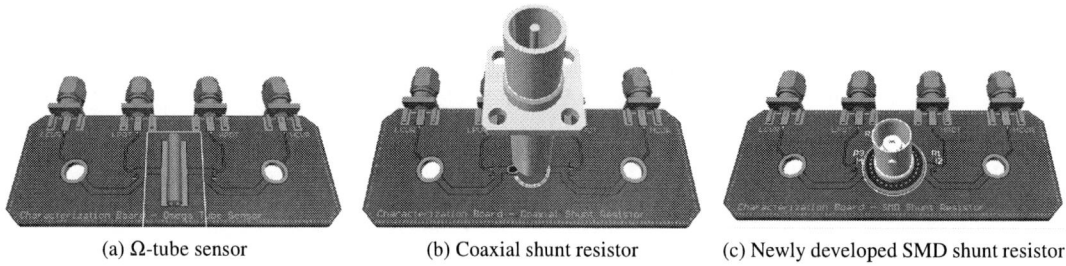

(a) Ω-tube sensor

(b) Coaxial shunt resistor

(c) Newly developed SMD shunt resistor

Fig. 3: 3D-views of sensor PCBs

Characterization of the Input impedance

One key characteristic of each measuring device is its impedance. In this work, the input impedance is defined as the impedance inserted into the circuit when the sensor is used and is defined by the voltage drop along the sensor: $Z_{\text{in}} = \frac{V_{\text{sensor}}}{I_{\text{sensor}}}$. To measure this impedance, the *4294A* impedance analyzer from *Agilent/Keysight* is used, which allows measurements in the range from 40 Hz to 110 MHz. The assembled prototyping boards allow a 4-point measurement via SMA connectors. Fig. 4a shows the absolute values of the measured impedances of each sensor under test. The two regular Rogowski coils are excluded from this method, since no measurable change in impedance can be detected when placing the coil around the PCB trace. The curves show a minor irregularity at 15 MHz, which is due to the calibration of the impedance analyzer.

EPE'22 ECCE Europe

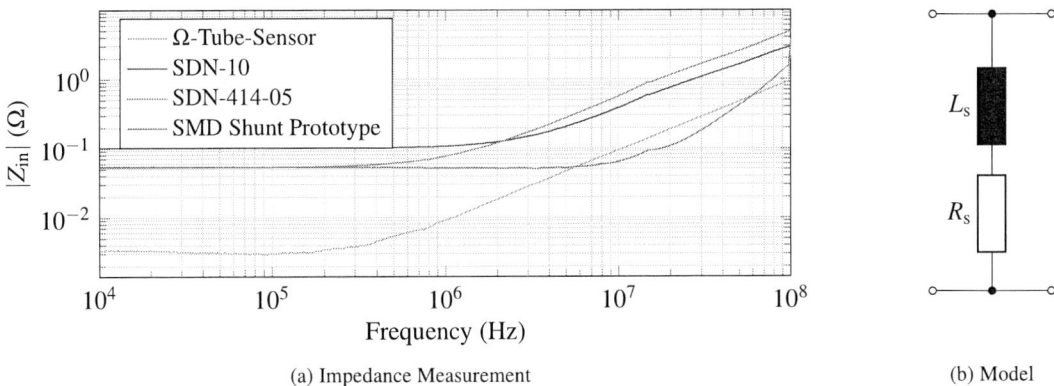

(a) Impedance Measurement

(b) Model

Fig. 4: Insertion impedances of different sensors

The impedance analyzer allows the measured impedances to be fitted directly to an electrical equivalent circuit (fig. 4b). For the measurement, a simple first-order model was chosen, consisting of the series resistance R_s and an the series inductance L_s. Table I shows the related parameters for each sensor measured. No significant capacitive component could be measured in the measuring range, thus all sensors behave inductive and do not form any resonance that would limit the field of application. It is observed that the impedance of the SMD shunt increases sharply between $10\,\text{MHz}$ and $100\,\text{MHz}$. The reason for this might be the skin effect, which superimposes the increase due to inductance. Equation (3) shows the formular for the skin depth δ rearranged to the frequency [16]. For ρ the specific resistance of copper is used and for μ the value of vacuum permeability. If the skin depth δ is set to half the thickness of the $35\,\mu\text{m}$ PCB trace, the frequency at which the skin effect begins to have an impact can be calculated. Above a frequency of $23.3\,\text{MHz}$, the resistance will increase. This approximation is consistent with the measurement.

$$\delta = \frac{\rho}{\sqrt{\pi \cdot f \cdot \mu}} \Leftrightarrow f = \frac{\rho^2}{\pi \cdot \delta^2 \cdot \mu} = \frac{(1.68 \times 10^{-8}\,\Omega\text{m})^2}{\pi \cdot (0.5 \cdot 35\,\mu\text{m})^2 \cdot 12.57 \times 10^{-8}\,\frac{\text{H}}{\text{m}}} = 23.3\,\text{MHz} \tag{3}$$

The actual bandwidths of the sensors cannot be derived from the input impedance, since the impedance analyzer does not measure the voltage at the BNC connector. To determine the bandwidth it is necessary to use the transfer function introduced in the following section.

Characterization of the Frequency Response

The bandwidth of the current sensors can be determined from their S-parameters. For this purpose, the 2-port S-parameters of each sensor are measured with a *Tektronix TTR506A* VNA in a frequency range from $100\,\text{kHz}$ to $6\,\text{GHz}$ [17]. The desired transfer function $Z_{\text{tr}} = \frac{V_{\text{out}}}{I_{\text{in}}}$ equals the Z-parameter Z_{21}, which can be obtained using equation (4) with a reference impedance of $Z_0 = 50\,\Omega$ [18]. The influence of the PCB traces can be eliminated by short-, open-, and load-calibration for which additional PCBs have been designed. The bandwidth of each device is defined via the $3\,\text{dB}$ cutoff frequency, at which the gain deviates by more than factor $\sqrt{2}$ from the nominal gain G_N (see equation (5)). The cutoff gain G_c is sometimes above and sometimes below the nominal gain, depending on if the transfer function shows either an amplifying or an attenuating behavior.

$$Z_{21} = \left.\frac{V_2}{I_1}\right|_{I_2=0} = \frac{2S_{12}}{(1-S_{11})(1-S_{22}) - S_{12}S_{21}} Z_0 \tag{4}$$

$$G_c = 10^{\frac{\pm 3\text{dB}}{20}} G_N = \begin{cases} 1.41 \cdot G_N & \text{upper limit} \\ 0.71 \cdot G_N & \text{lower limit} \end{cases} \tag{5}$$

Fig. 5 shows the measured transfer functions and the corresponding 3 dB-points, which are summarized in table I. The coaxial shunts show a low pass behavior with an attenuation of the frequency components above the 3 dB limit. This may provide the opportunity to extend the frequency range by compensation of the transfer function with postprocessing [19]. Above a frequency of approx. 100 MHz, the *CWT* Rogowski coils behave unpredictably and therefore are not suitable for measurements in this range. The lower frequency limit of the Rogowski coils cannot be determined with this measurement because it is below the measurement range of the VNA.

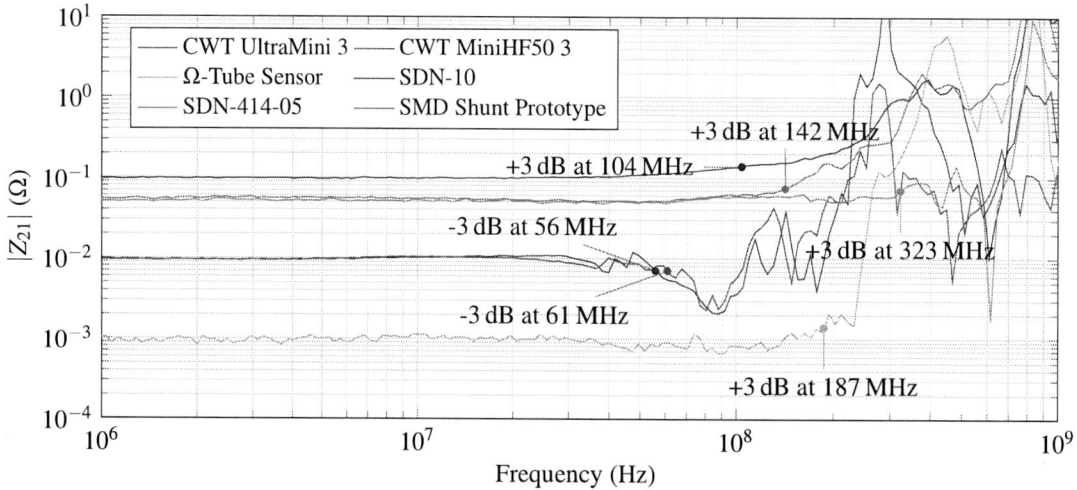

Fig. 5: Measured transfer functions

Characterization of the Step Response

Next, the step response in the time domain is investigated. This allows to check the suitability for high currents, if the gain is accurate and allows to determine the delay. For this purpose, the presented sensors are connected to the *KSZ 100D* current clamp calibrator from *Nagel Electronic* [20], which generates rectangular current pulses with amplitudes of 20 A, 50 A and 100 A at a slew rate of approx. $1\,\frac{A}{ns}$. The PCBs with the sensors under test can be directly connected via 4 mm laboratory jacks, so no cables are needed. The voltage across the shunt resistors is measured via an *IsoVu TIVM1* differential probe to ensure an insulated measurement with a bandwidth of 1 GHz and a defined delay of $t_{probe} = 18.3$ ns [17]. The use of a differential probe is necessary because the output of the current pulse generator is at a potential of 200 V. To measure the delay of the sensors, a trigger signal is available that drops from 16 V to zero. The trigger signal is measured with a *LMR-240* coaxial cable that has a length of 1 m and a defined propagation delay of $t_{cable} = 3.97$ ns. The actual delay t_{delay} of the respective current sensor including the sensor connection cable can be calculated using equation (6). The event $t_{current}$ is defined by the time the current reaches 10 % of its amplitude and the event $t_{trigger}$ by the time the trigger signal drops below 90 %.

$$t_{delay} = t_{current} - t_{trigger} - t_{cable} \tag{6}$$

Fig. 6 shows the step response of each sensor under test. All sensors exhibit a low-pass behavior and are suitable for currents up to 100 A. The signals are scaled with the previously determined gains. Especially the Rogowski coils are affected by oscillations. This is partly due to the fact that they are designed for much higher currents and have a low gain. The rise time of the current pulse is not sufficient to determine the differences in bandwidth. The waveforms are shifted by t_{cable} so that the actual delays of the sensor feedback are visible and marked with dashed lines. The delay of the differential probe t_{probe} is included for the shunt resistors. Table I summarizes the results.

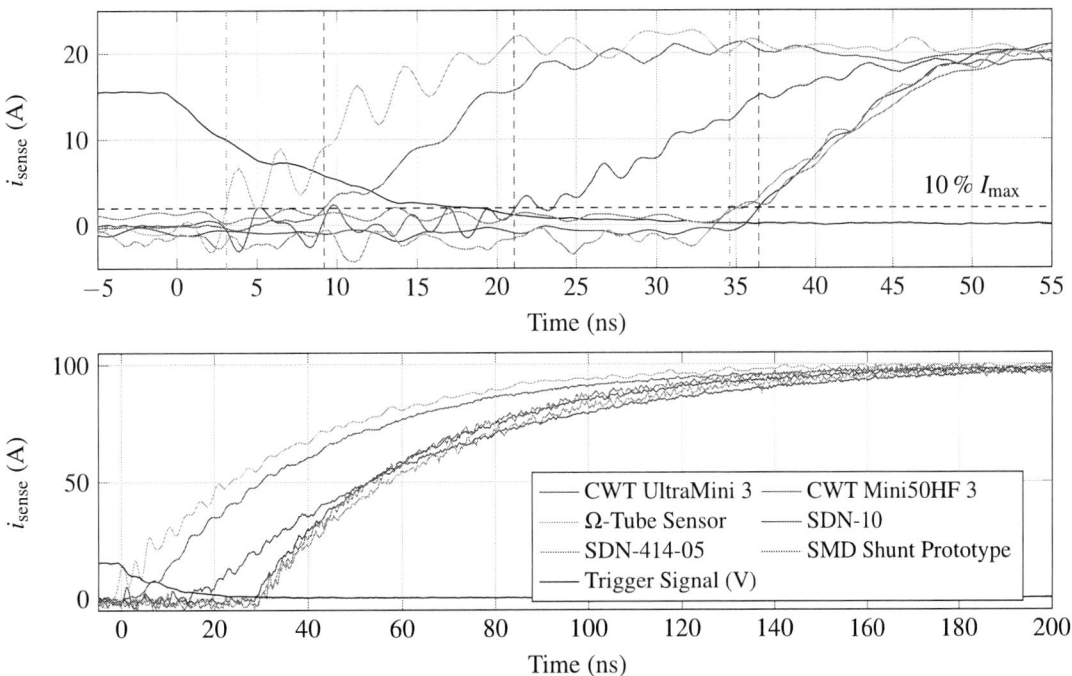

Fig. 6: Step responses for 20 A and 100 A pulses

5 Summary of Results and Suitability for WBG-Applications

Table I summarizes the results of the previous measurements. The theoretical values for the minimum rise time $t_{\mathrm{rise,min}}$ are calculated from the measured bandwidth $f_{\mathrm{BW,meas}}$ using equation (1). The nominal bandwidths $f_{\mathrm{BW,N}}$ are extracted from the device data sheets.

Table I: Comparison of examined current sensors

Device	Gain ($\frac{\mathrm{mV}}{\mathrm{A}}$)	$f_{\mathrm{BW,N}}$	$f_{\mathrm{BW,meas}}$	$t_{\mathrm{rise,min}}$	t_{delay}	L_{s}	R_{s}
CWT UltraMini 3	10	30 MHz	56 MHz	6.25 ns	21.1 ns	-	-
CWT MiniHF50 3	10	50 MHz	61 MHz	5.74 ns	9.2 ns	-	-
Ω-tube sensor	0.1	260 MHz	187 MHz	1.87 ns	3.1 ns	1.45 nH	3.07 mΩ
SDN-10	100	2 GHz	104 MHz	3.37 ns	18.2 ns	5.54 nH	104 mΩ
SDN-414-05	50	2 GHz	323 MHz	1.08 ns	16.3 ns	8.58 nH	54.3 mΩ
SMD shunt prototype	53	-	142 MHz	2.46 ns	5 ns	978 pH	53.0 mΩ

Bandwidth

When it comes to bandwidth, the conventional Rogowski coils are not sufficient for WBG applications with rise times in the single-digit nanosecond range. The conventional coaxial shunts do not achieve the values from the data sheets. However, the *SDN-414-05* still offers the highest bandwidth. The Ω-tube sensor and the SMD shunt prototype can compete with this and allow to measure rise times of approx. 2 ns. Nevertheless, this is only a theoretical value and for a precise measurement the minimum rise time of the measuring device should be three to five times lower than the rise time of the signal to be measured [21].

Parasitic Inductance

Even a slight increase of the loop inductance will affect the switching behaviour of WBG semiconductors and therefore introduce a measurement error [22]. In [23], a GaN switching cell achieves loop inductance of 1.19 nH, which is significantly lower than the inductance of the coaxial shunts. It is therefore necessary to consider whether the additional inductance is acceptable before each use of a current sensor. Unfortunately, the use of non-invasive Rogowski coils is usually not possible, since there are no PCB traces that can be enclosed by the coil. In terms of this aspect, the Ω-tube sensor and the SMD shunt prototype are the most suitable devices. The decision between both is application specific and depends on the overall design of the circuit.

Delay & Gain

The different delays must be taken into account when double pulse tests are performed [24]. This process is called de-skewing and can be done either by subtracting the previously determined delays in the time domain or by algorithmically aligning characteristic points of the switching waveforms [25]. If this aspect is not taken into account, even a small misalignment of current and voltage will lead to very large measurement errors in the switching loss calculation. With regard to the gain, a particularly high gain is desirable so that the accuracy in amplitude of the oscilloscope can be maximized and a sufficient signal-to-noise ratio is achieved. The present Ω-tube sensor has a very small gain, which is due to its high measuring range up to 1.55 kA. To overcome this disadvantage, it is possible to adjust the winding configuration of the inserted coil [12]. A higher number of turns increases the induced voltage but also leads to an increased parasitic capacitance between the windings.

6 Validation of the SMD Shunt with Double Pulse Tests

Finally, the performance of the developed SMD shunt is validated with double pulse tests to measure the commutation of the drain-source current i_{ds}. The device under test (DUT) is a halfbridge consisting of two 650 V GaN transistors from *GaN Systems* (*GS66516T* [26]). The gate resistors are 18 Ω when switching on and 1 Ω when switching off. Since the setup is floating, the ground potential of the BNC connector can be freely chosen and is connected to the source contact of the low side transistor S_{LS} according to fig. 7a. This allows the use of a regular passive probes for the simultaneous measurement of the drain-source voltage v_{ds} and the gate-source voltage v_{gs}. However, this leads to an inverted signal for i_{ds}, which must be taken into account in the configuration of the oscilloscope.

The double pulse test was performed at a load current of $i_L = 45$ A and a DC voltage of $v_{DC} = 400$ V. The resulting waveforms shown in fig. 7b deviate from the classical double pulse test, since the topology from [27] was used to avoid an increase of the load current during the second pulse. Fig. 7c shows the hardware with the SMD shunt on the PCB. The transistors are located on the bottom layer. Fig. 7d and fig. 7e show the zoomed waveforms for turn off and turn on. The current is turned off with a fall time of $t_{fall} = 5.6$ ns with a maximum slope of $\frac{di}{dt}_{max,off} = -45.5 \frac{A}{ns}$. At turn on the rise time is $t_{rise} = 12.4$ ns and the maximum slope is $\frac{di}{dt}_{max,on} = 6 \frac{A}{ns}$.

7 Conclusion

In this work, six current sensors have been compared. Conventional current measuring devices were considered as well as advanced techniques such as the inductive measurement via an Ω-tube and a newly developed coaxial shunt based on SMD resistors. After the measurement principles were introduced, several measurements were carried out in the time and frequency domain. The characteristics bandwidth, insertion impedance and delay could be determined for each sensor. Based on the characteristics, it was shown that not all devices are suitable for characterizing WBG semiconductor devices. The bandwidth is often insufficient or the insertion inductance is too high. In contrast to the widely used conventional coaxial shunts, the newly derived SMD coaxial shunt meets the requirements completely. The prototype achieved a bandwidth of 142 MHz and a insertion inductance of 978 pF and could be tested successfully in a double pulse test with GaN devices. Therefore, this approach will be pursued further in future work. There may be potential for improvement in terms of the arrangement, number and type of resistors.

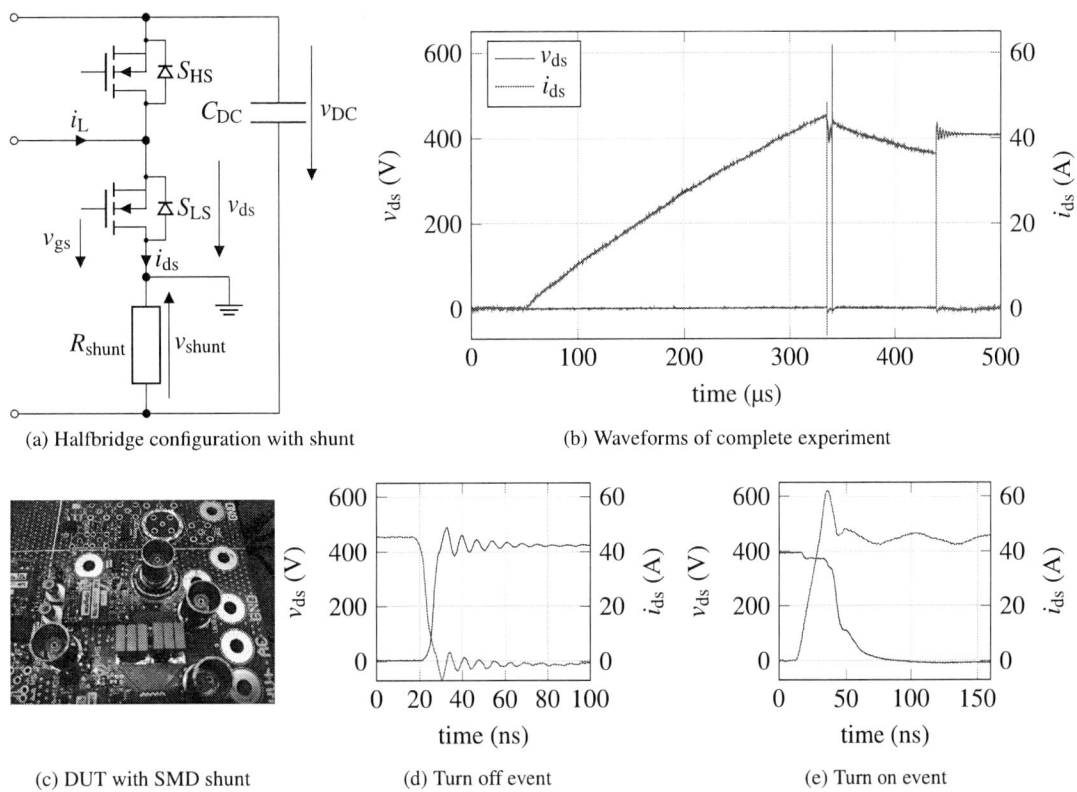

(a) Halfbridge configuration with shunt　　　　(b) Waveforms of complete experiment

(c) DUT with SMD shunt　　　(d) Turn off event　　　(e) Turn on event

Fig. 7: Double pulse test

References

[1] V. C. Grau, "Development of a test bench to investigate the impact of steep voltage slopes on the lifetime of insulation systems for coil windings," Dissertation, Rheinisch-Westfälische Technische Hochschule Aachen, Aachen, 2021.

[2] C. Mittermayer and A. Steininger, "On the determination of dynamic errors for rise time measurement with an oscilloscope," *IEEE Transactions on Instrumentation and Measurement*, vol. 48, no. 6, pp. 1103–1107, 1999.

[3] K. Klein and E. Hoene, "High speed current measurements for ultra low inductance power modules," in *ECPE Workshop "Current Measurement for Power Electronics Applications and in Lab Scale"*, Fraunhofer IZM, Oct. 2017.

[4] W. Zhang, Z. Zhang, F. Wang, *et al.*, "High-bandwidth low-inductance current shunt for wide-bandgap devices dynamic characterization," *IEEE Transactions on Power Electronics*, vol. 36, no. 4, pp. 4522–4531, Apr. 2021.

[5] S. Sprunck, M. Muench, and P. Zacharias, "Transient current sensors for wide band gap semiconductor switching loss measurements," in *Proceedings of the PCIM Europe, Nuremberg, Germany*, May 2019, pp. 1–8.

[6] P. Ziegler, N. Tröster, D. Schmidt, *et al.*, "Wide bandwidth current sensor for commutation current measurement in fast switching power electronics," in *22nd European Conference on Power Electronics and Applications (EPE'20 ECCE Europe)*, Sep. 2020, P.1–P.9.

[7] Z. Xin, H. Li, Q. Liu, *et al.*, "A review of megahertz current sensors for megahertz power converters," *IEEE Transactions on Power Electronics*, vol. 37, no. 6, pp. 6720–6738, 2022.

[8] M. H. Samimi, A. Mahari, M. A. Farahnakian, *et al.*, "The rogowski coil principles and applications: A review," *IEEE Sensors Journal*, vol. 15, no. 2, pp. 651–658, Feb. 2015.

[9] Pearson Electronics, *https://www.pearsonelectronics.com*, online, May 2022.

[10] PEM UK, *http://www.pemuk.com*, online, Mar. 2022.

[11] E. Hoene, A. Ostmann, B. Lai, *et al.*, "Ultra-low-inductance power module for fast switching semiconductors," *Proceedings of the PCIM Europe, Nuremberg, Germany*, pp. 198–205, Jan. 2013.

[12] IPM Design, *https://www.ipmdesign.de*, online, Mar. 2022.

[13] A. Stippich, "Exploiting the full potential of silicon carbide devices via optimized highly integrated power modules," Dissertation, Rheinisch-Westfälische Technische Hochschule Aachen, Aachen, 2021.

[14] T&M Research Products, Inc., *https://www.tandmresearch.com*, online, Mar. 2022.

[15] W. Zhang, Z. Zhang, and F. Wang, "Review and bandwidth measurement of coaxial shunt resistors for wide-bandgap devices dynamic characterization," in *IEEE Energy Conversion Congress and Exposition (ECCE)*, Sep. 2019, pp. 3259–3264.

[16] J. D. Jackson, *Classical electrodynamics*. American Association of Physics Teachers, 1999.

[17] Tektronix, *https://www.tek.com*, online, Mar. 2022.

[18] S. Ramo, J. R. Whinnery, and T. Van Duzer, *Fields and waves in communication electronics*. John Wiley & Sons, 1994.

[19] W. Tun Latt, U.-X. Tan, C. N. Riviere, *et al.*, "Transfer function compensation in gyroscope-free inertial measurement units for accurate angular motion sensing," *IEEE Sensors Journal*, vol. 12, no. 5, pp. 1207–1208, 2012.

[20] Nagel Electronic, *https://www.nagel-electronic.de*, online, Mar. 2022.

[21] Tektronix, Inc., *Abcs of probes*, online, 2018.

[22] P. Weiler, B. Vermulst, M. Roes, *et al.*, "Design limitations of heat spreaders for gallium nitride power modules," in *PCIM Europe digital days 2020*, Jul. 2020, pp. 1–7.

[23] S.-S. Yang, J.-H. Soh, and R.-Y. Kim, "Parasitic inductance reduction design method of vertical lattice loop structure for stable driving of gan hemt," in *IEEE 4th International Future Energy Electronics Conference (IFEEC)*, 2019, pp. 1–8.

[24] J. Schweickhardt, K. Hermanns, and M. Herdin, *Tips & tricks on double pulse testing*, Rohde & Schwarz Application Note, Mar. 2021.

[25] S. Yin, Y. Liu, Y. Gu, *et al.*, "Automatic v - i alignment for switching characterization of wide band gap power devices," in *2018 1st Workshop on Wide Bandgap Power Devices and Applications in Asia (WiPDA Asia)*, May 2018, pp. 75–78.

[26] GaN Systems, *https://gansystems.com*, online, May 2022.

[27] J. Gottschlich, M. Kaymak, M. Christoph, *et al.*, "A flexible test bench for power semiconductor switching loss measurements," in *2015 IEEE 11th International Conference on Power Electronics and Drive Systems*, Jun. 2015, pp. 442–448.

Zero-Sequence Voltage Reduces DC-Link Capacitor Demand in Cascaded H-Bridge Converters for Large-Scale Electrolyzers by 40%

Roland Unruh, Frank Schafmeister, Joachim Böcker
Paderborn University
Warburger Str. 100
Paderborn, Germany
Tel.: +49/ (05251) 60-3492
Fax.: +49/ (05251) 60-3443
E-Mail: unruh@lea.uni-paderborn.de
URL: http://lea.upb.de

Keywords

≪Cascaded H-Bridge≫, ≪Solid-State Transformer≫, ≪Zero sequence voltage≫, ≪Third harmonic injection≫, ≪Capacitor voltage balancing≫

Acknowledgments

The authors would like to thank the German Research Foundation (DFG) for the funding the research under the project number 456097802.

Abstract

Cascaded H-bridge Converters (CHBs) are a promising solution in converting power from a three-phase medium voltage of $6.6\,\text{kV}...30\,\text{kV}$ to a lower DC-voltage in the range of $100\,\text{V}...1\,\text{kV}$ to provide pure DC power to applications such as electrolyzers for hydrogen generation, data centers with a DC power distribution and DC microgrids. CHBs can be interpreted as modular multilevel converters with an isolated DC-DC output stage per module, require a large DC-link capacitor for each module to handle the second harmonic voltage ripple caused by the fluctuating input power within a fundamental grid period. Without a zero-sequence voltage injection, star-connected CHBs are operated with approximately sinusoidal arm voltages and currents. The floating star point potential enables to utilize different zero-sequence voltage injection techniques such as a third-harmonic injection with $\frac{1}{6}$ of the grid voltage amplitude or a Min-Max voltage injection. Both well-known methods have the advantage to reduce the peak arm voltage and thereby the number of required modules by $13.4\,\%$ (to $\frac{\sqrt{3}}{2}$). This paper proves analytically that the third-harmonic injection with $\frac{1}{6}$ of the grid voltage amplitude reduces the second harmonic voltage ripple by only $15.1\,\%$ compared to no-voltage injection for unity power factor operation and balanced grid voltages. Then it is shown, that the Min-Max injection has the often overlooked advantage of reducing the second harmonic voltage ripple by even $18.8\,\%$. By applying the here proposed zero-sequence voltage injection in saturation modulation, the second harmonic voltage ripple of the DC-link capacitors is reduced by even $24.3\,\%$, while still requiring the same number of modules as the Min-Max injection. For a realistic number of reserve modules, the overall energy ripple in the DC-link capacitors is reduced by $40\,\%$.

I. Introduction

High-power DC loads such as industrial water electrolyzers [1–3] require high DC currents up to $5\,\text{kA}$ in the low-voltage range of $100\,\text{V}...1\,\text{kV}$ [4, 5]. The same applies for data centers with a $400\,\text{V}$ and $800\,\text{V}$ DC power distribution [6, 7] as well as for ultra-fast electric vehicle charging [8]. The state of the art is a centralized line frequency medium voltage transformer to convert the three-phase medium voltage of $6.6\,\text{kV}...30\,\text{kV}$ to a low AC voltage that is rectified for the load [5, 9, 10]. The transformer has usually secondary windings in delta and star configuration to allow a 12-pulse rectifier to generate the required DC load voltage. However, a 12-pulse thyristor rectifier increases the specific energy consumption of an

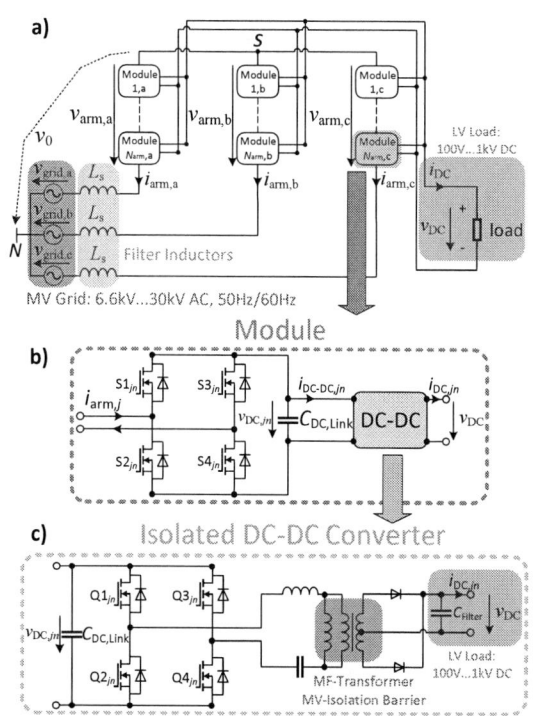

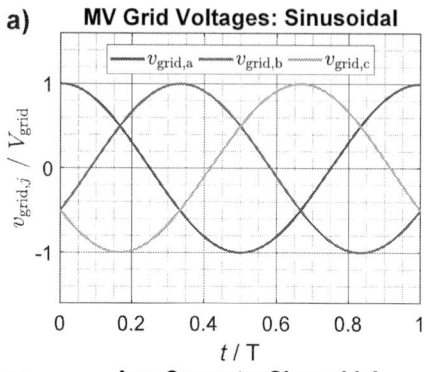

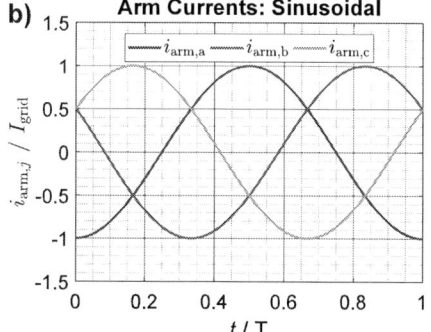

Fig. 1: The three-phase cascaded H-bridge converter a) consists of identical modules b). Each DC-DC converter c) transfers power to the load [20].

Fig. 2: The line-to-neutral grid voltages $v_{\text{grid},j}$ in a) and arm currents $i_{\text{arm},j}$ in b) are sinusoidal. Harmonic current injections are not allowed [14].

electrolyzer stack by around 9 % [11] due to the DC current ripple. Additionally, thyristor rectifiers show a low power factor for high firing angles at low load [12]. A 12-pulse diode rectifier [5] with subsequent buck converters improves the power factor and the specific energy consumption [11]. Still, it requires additional active and/or passive filters to achieve low grid-side current harmonics [13].

Therefore, a system is required that achieves nearly sinusoidal grid currents [14] and a controllable power factor [15–17] for inductive, capacitive and unity power factor operation. Additionally, it should not rely on a single component such as a central line- or medium-frequency transformer [18]. This can be achieved by using the three-phase cascaded H-bridge Converter (CHB) [19] with star configuration shown in Fig. 1 a). In this cascaded structure, an isolated full-bridge DC-DC converter [20] is connected to each module capacitor $C_{\text{DC,Link}}$, which converts the module voltage to a lower output voltage v_{DC} shown e.g. in Fig. 1 c). The transferred power of each DC-DC converter is controlled on each module locally by adjusting the switching frequency and phase-shift [21, 22]. The bulky line-frequency transformer (LFT) is replaced by multiple smaller medium-frequency transformers (MFTs) within the DC-DC converters providing the required galvanic isolation. The low DC-voltage v_{DC} can be used directly for DC applications such as electrolysis and data center distribution buses or be converted via DC-AC converters into a three-phase voltage in order to replace a low-frequency distribution transformer [23].

For the purposes of this paper, the load will be considered as a large-scale electrolyzer with unidirectional power flow. However, the developed zero-sequence injection can be transferred to bidirectional CHBs or to large-scale photovoltaic (PV) CHBs because in both cases the power factor is $\cos(\phi) = 1$, if there is no significantly unbalanced PV generation. However, the results can not be transfered to cascaded multilevel static compensators (STATCOMs) directly because they have a power factor of $\cos(\phi) \approx 0$.

The main focus of this article, is to minimize the required capacitance of the module capacitors $C_{\text{DC,Link}}$ by reducing the second harmonic voltage ripple by shifting the star points potential s in Fig. 1 a). Altering the output power of the isolated DC-DC converters is also possible [24], but it increases the losses and is disregarded in this article. Still, the presented techniques are expected to be beneficial for open-loop operated DC-DC converters due to the reduced peak arm input power.

This paper is organized as follows: the second harmonic voltage ripple is calculated with normalized arm currents and voltages for no zero-sequence injection in Section II. In Section III, the benefits of the commonly used third-harmonic voltage injection with $\frac{1}{6}$ of the grid-voltage amplitude are briefly shown. In Section IV, it is proven numerically, that the also commonly used Min-Max voltage injection requires the same arm voltages as the third-harmonic injection, but allow to further reduce $C_{\text{DC,Link}}$ by 4.4% without increasing the second harmonic voltage ripple. Going further, a new method is presented in Section V, which usually operates one of the three arms at its voltage limit and reduces the second harmonic voltage ripple by between 24% and 40% without the need for additional hardware. Finally, Section VI presents the conclusion.

II. Nominal Operation: No Zero-Sequence Voltage Injection

For the purposes of the paper, the three-phase grid voltages are balanced and can be expressed as follows:

$$
\begin{aligned}
v_{\text{grid,a}}(t) &= V_{\text{grid}} \cos(\omega t) \\
v_{\text{grid,b}}(t) &= V_{\text{grid}} \cos(\omega t - 2\pi/3) \\
v_{\text{grid,c}}(t) &= V_{\text{grid}} \cos(\omega t + 2\pi/3)
\end{aligned}
\tag{1}
$$

Ideally, the arm currents are controlled to be sinusoidal and in phase with the grid voltages [25]. The negative sign results from the fact, that power is transferred from the grid to the load. The resulting currents are expressed as:

$$
\begin{aligned}
i_{\text{arm,a}}(t) &= -I_{\text{grid}} \cos(\omega t) \\
i_{\text{arm,b}}(t) &= -I_{\text{grid}} \cos(\omega t - 2\pi/3) \\
i_{\text{arm,c}}(t) &= -I_{\text{grid}} \cos(\omega t + 2\pi/3)
\end{aligned}
\tag{2}
$$

The grid voltages and arm currents are shown in Fig. 2. The arm voltages show a high-frequency component due to the modulation, which can be neglected for sufficiently high switching frequencies [26] and DC-link voltage balancing [27].

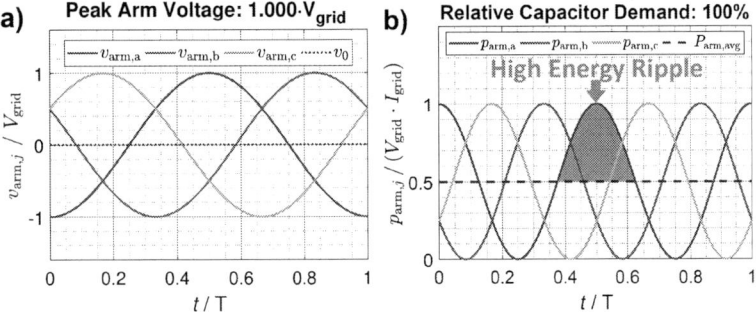

Fig. 3: Resulting arm voltages a) and input power b) for no zero-sequence voltage injection. The required arm voltage and DC-link capacitance are high and normalized to 100%.

Furthermore, the voltage drop across the filter inductors L_{s} is approximately 1% of the grid voltage. Therefore, the filter inductors are neglected in the following analytical calculations. The resulting arm voltages may contain a zero-sequence $v_0(t)$ and are:

$$
\begin{aligned}
v_{\text{arm,a}}(t) &= -v_{\text{grid,a}}(t) + v_0(t) = -V_{\text{grid}} \cos(\omega t) + v_0(t) \\
v_{\text{arm,b}}(t) &= -v_{\text{grid,b}}(t) + v_0(t) = -V_{\text{grid}} \cos(\omega t - 2\pi/3) + v_0(t) \\
v_{\text{arm,c}}(t) &= -v_{\text{grid,c}}(t) + v_0(t) = -V_{\text{grid}} \cos(\omega t + 2\pi/3) + v_0(t)
\end{aligned}
\tag{3}
$$

The resulting input arm powers are the product of the arm currents and voltages:

$$p_{\text{arm,a}}(t) = v_{\text{arm,a}}(t)i_{\text{arm,a}}(t) = V_{\text{grid}}I_{\text{grid}}\cos^2(\omega t) - v_0(t)I_{\text{grid}}\cos(\omega t)$$

$$p_{\text{arm,b}}(t) = v_{\text{arm,b}}(t)i_{\text{arm,b}}(t) = V_{\text{grid}}I_{\text{grid}}\cos^2(\omega t - 2\pi/3) - v_0(t)I_{\text{grid}}\cos(\omega t - 2\pi/3) \quad (4)$$

$$p_{\text{arm,c}}(t) = v_{\text{arm,c}}(t)i_{\text{arm,c}}(t) = V_{\text{grid}}I_{\text{grid}}\cos^2(\omega t + 2\pi/3) - v_0(t)I_{\text{grid}}\cos(\omega t + 2\pi/3)$$

It should be noted, that $v_0(t)$ is the zero-sequence voltage, i.e. the voltage difference of the star point to the grid neutral N. It is often used to achieve power balance for an unbalanced grid or faulty modules [28–31]. The resulting arm voltages and input arm powers are shown in Fig. 3 for the nominal operation of $v_0(t) = 0$. In order to achieve steady-state, the isolated DC-DC converters in each arm have to transfer the average power of $P_{\text{arm,avg}} = \frac{1}{T}\int_0^T V_{\text{grid}}I_{\text{grid}}\cos^2(\frac{2\pi}{T}t)dt = \frac{1}{2}V_{\text{grid}}I_{\text{grid}}$ to the load. However, the input arm power is not constant, which results in a second harmonic voltage ripple of the DC-link voltages.

Now, the net input power of the DC-link capacitors in arm 'a' is calculated for nominal operation and it is identical for the other two arms:

$$p_{\text{DC,Link,Nominal,a}}(t) = v_{\text{arm,a}}(t)i_{\text{arm,a}}(t) - P_{\text{arm,avg}} = V_{\text{grid}}I_{\text{grid}}\cos^2(\omega t) - \frac{1}{2}V_{\text{grid}}I_{\text{grid}}$$

$$= \frac{1}{2}V_{\text{grid}}I_{\text{grid}}\cos(2\omega t) = P_{\text{arm,avg}}\cos(2\omega t) \quad (5)$$

Between $t_1 = \frac{3}{8}T = \frac{3}{8}\cdot\frac{2\pi}{\omega}$ and $t_2 = \frac{5}{8}T = \frac{5}{8}\cdot\frac{2\pi}{\omega}$, the input power $p_{\text{DC,Link,Nominal,a}}(t)$ of the DC-link capacitors in arm 'a' is positive, which causes the second harmonic voltage ripple. The energy, which has to be stored in the DC-link capacitors is highlighted in green in Fig. 3 and can be calculated as:

$$\Delta E_{\text{Ripple,Nominal,a}} = \int_{t_1}^{t_2} p_{\text{DC,Link,Nominal,a}}(t)dt = \int_{\frac{3}{8}\cdot\frac{2\pi}{\omega}}^{\frac{5}{8}\cdot\frac{2\pi}{\omega}} P_{\text{arm,avg}}\cos(2\omega t)dt = P_{\text{arm,avg}}\cdot\frac{1}{\omega} \quad (6)$$

For a transferred power of $P_{\text{arm,avg}} = 1\,\text{MW}$ per arm and a grid frequency of $50\,\text{Hz}$, the DC-link capacitors have to buffer the energy of $\Delta E_{\text{Ripple,Nominal,a}} = 3.2\,\text{kJ}$ for each arm.

III. Third-Harmonic Zero-Sequence Voltage Injection

As shown in the previous Section II, the nominal stored energy in the DC-link capacitors is significant and increases the costs and volume of the overall system. Consequently, a commonly utilized technique is a third-harmonic voltage injection with $\frac{1}{6}$ of the grid voltage amplitude [32–35]. The resulting zero-sequence voltage injection can be expressed as:

$$v_{0,\text{3rd}}(t) = \frac{1}{6}V_{\text{grid}}\cos(3\omega t) \quad (7)$$

The arm voltages and input arm powers are calculated with Eq. (3) and Eq. (4) respectively and are shown in Fig. 4.

They show two major advantages compared to no voltage injection shown in Fig. 3. Firstly, the required arm voltages are reduced to $\frac{\sqrt{3}}{2}V_{\text{grid}}$, which allows to reduce the number of modules in each arm by around 13.4 %. Secondly, the peak input power of each arm is reduced by $\frac{1}{6}$, which reduces the enclosed area between an arm's input power and the DC-DC converter power $P_{\text{arm,avg}}$ by 15.1 %.

Now, these advantages of the third-harmonic injection are analytically derived with the same procedure as for the case with no injection. Again, only arm 'a' is considered and modulation artifacts shall be out of

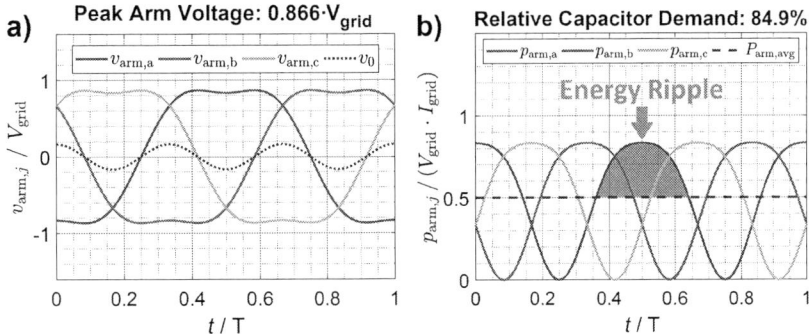

Fig. 4: Resulting arm voltages a) and input power b) for the third harmonic zero-sequence voltage injection. The required arm voltage is reduced to $\frac{\sqrt{3}}{2}V_{\text{grid}} = 0.866\,V_{\text{grid}}$ and the energy ripple to $84.9\,\%$.

the scope of this analysis. The net input power of the DC-link capacitors in arm 'a' can be expressed as:

$$
\begin{aligned}
p_{\text{DC,Link,3rd,a}}(t) &= v_{\text{arm,a}}(t)i_{\text{arm,a}}(t) - P_{\text{arm,avg}} \\
&= V_{\text{grid}}I_{\text{grid}}\cos^2(\omega t) - v_{0,\text{3rd}}(t)I_{\text{grid}}\cos(\omega t) - P_{\text{arm,avg}} \\
&= V_{\text{grid}}I_{\text{grid}}\cos^2(\omega t) - \tfrac{1}{6}V_{\text{grid}}I_{\text{grid}}\cos(3\omega t)\cos(\omega t) - P_{\text{arm,avg}} \\
&= 2P_{\text{arm,avg}}\cos^2(\omega t) - \tfrac{1}{3}P_{\text{arm,avg}}\cos(3\omega t)\cos(\omega t) - P_{\text{arm,avg}} \\
&= \tfrac{1}{6}P_{\text{arm,avg}}(5\cos(2\omega t) - \cos(4\omega t))
\end{aligned}
\tag{8}
$$

The power $p_{\text{DC,Link,3rd,a}}(t)$ is positive between $t_{1,\text{3rd}} = 0.3601 \cdot \text{T} = 0.3601 \cdot \frac{2\pi}{\omega}$ and $t_{2,\text{3rd}} = 0.6399 \cdot \text{T} = 0.6399 \cdot \frac{2\pi}{\omega}$ (numerically calculated), which is only a slight deviation compared to the nominal operation. Consequently, the DC-link capacitors in arm 'a' have to buffer the energy of:

$$
\begin{aligned}
\Delta E_{\text{Ripple,3rd,a}} &= \int_{t_{1,\text{3rd}}}^{t_{2,\text{3rd}}} p_{\text{DC,Link,3rd,a}}(t)\,dt \\
&= \int_{0.3601\cdot\frac{2\pi}{\omega}}^{0.6399\cdot\frac{2\pi}{\omega}} \tfrac{1}{6}P_{\text{arm,avg}}(5\cos(2\omega t) - \cos(4\omega t))\,dt = P_{\text{arm,avg}} \cdot \tfrac{1}{\omega} \cdot \mathbf{0.849} \\
&= \Delta E_{\text{Ripple,Nominal,a}} \cdot \mathbf{0.849}
\end{aligned}
\tag{9}
$$

IV. Min-Max Zero-Sequence Voltage Injection

Another commonly used technique is the Min-Max zero-sequence voltage injection and it shares many similarities with the third-harmonic injection shown in Section III. The offset voltage $v_0(t)$ is calculated with a well-known equation to minimize the instantaneous value of arm voltages [36–38]:

$$
v_{0,\text{MinMax}}(t) = \frac{\max(v_{\text{grid,a}}(t), v_{\text{grid,b}}(t), v_{\text{grid,c}}(t)) + \min(v_{\text{grid,a}}(t), v_{\text{grid,b}}(t), v_{\text{grid,c}}(t))}{2}
\tag{10}
$$

The resulting Min-Max voltage injection is shown in Fig. 5. Exactly like by the 3rd harmonic injection, the peak arm voltage is reduced by 13.4% (to $\frac{\sqrt{3}}{2}$). However, the Min-Max injection has a benefit regarding the second harmonic voltage ripple. Now the DC-link capacitors in arm 'a' have to buffer the energy:

$$
\Delta E_{\text{Ripple,MinMax,a}} = 0.812\,\Delta E_{\text{Ripple,Nominal,a}} = 0.956\Delta E_{\text{Ripple,3rd,a}}
\tag{11}
$$

In Fig. 6 the three discussed voltage injection techniques are shown for one arm to explain why the Min-Max voltage injection performs better than the 3rd harmonic injection. For $t = 0.5\,\text{T}$, the arm current $i_{\text{arm,a}}(t)$ reaches its peak value as shown in Fig. 2b). Without a zero-sequence injection the arm voltage $v_{\text{arm,a}}(t)$ also reaches its peak value for $t = 0.5\,\text{T}$. This has the consequence, that for $t = 0.5\,\text{T}$ the module capacitors in arm 'a' receive 2/3 of the total input power while accounting for only 1/3 of all modules.

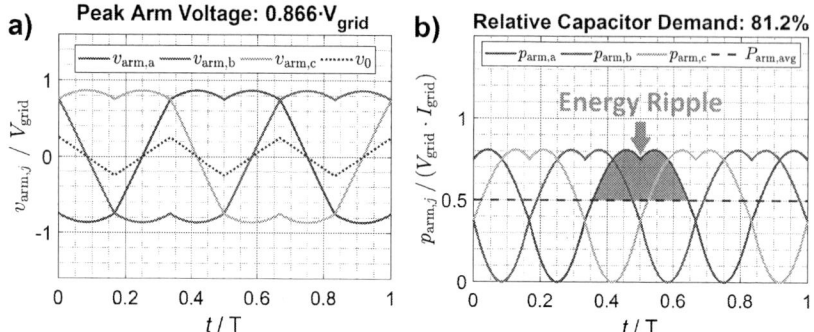

Fig. 5: Resulting arm voltages a) and input power b) for the Min-Max zero-sequence voltage injection. It requires a 4.4 % smaller DC-link capacitance $C_{DC,Link}$ compared to the third-harmonic injection.

With the Min-Max zero-sequence voltage injection, the arm voltage of arm 'a' is reduced by $1/4$ compared to the nominal case for $t = 0.5\,T$. As the input arm power is the product of the arm voltage and the (still same) arm current, the reduction of the input power of arm 'a' at $t = 0.5\,T$ is clearly visible. As a result, the enclosed area between p_{MinMax} and $P_{arm,avg}$ is reduced by 4.4 % compared to the third-harmonic voltage injection p_{3rd} and even by 18.8 % compared to no zero-sequence voltage injection p_{Nom}.

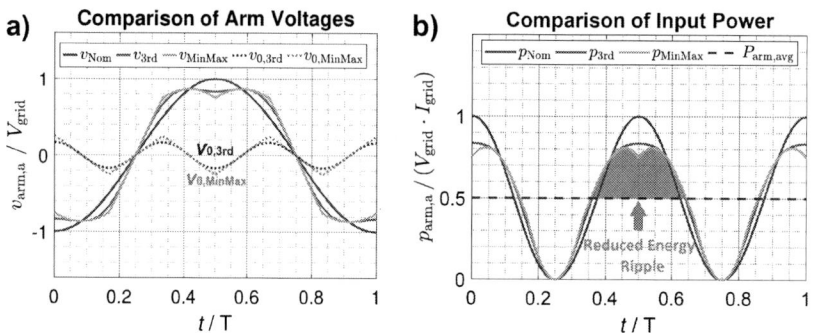

Fig. 6: Comparison of 3rd-harmonic voltage injection to Min-Max and no voltage injection for arm 'a'.

V. Novel Saturation Zero-Sequence Voltage Injection

The third-harmonic and especially the Min-Max zero-sequence voltage injections show a significant reduction of the energy which has to be buffered by the DC-link capacitors $C_{DC,Link}$ in each fundamental cycle. The zero-sequence voltage injection $v_0(t)$ can be selected freely as long as the resulting arm voltages are not larger than the maximum possible arm voltages $\hat{v}_{arm,j}(t) = v_{DC,jn}(t) \cdot N_{arm,j}$ for $j = \{a, b, c\}$. The maximum possible arm voltage $\hat{v}_{arm,j}(t)$ of each arm depends on the nominal DC-link voltage and the number of modules connected in series. One of the three arms is operating at its positive voltage limit, if the voltage injection of $v_{0,Max}(t)$ is selected [31]:

$$v_{0,Max}(t) = \min(\hat{v}_{arm,a}(t) + v_{grid,a}(t), \hat{v}_{arm,b}(t) + v_{grid,b}(t), \hat{v}_{arm,c}(t) + v_{grid,c}(t)) \tag{12}$$

The following voltage injection allows to operate one arm at its negative voltage limit:

$$v_{0,Min}(t) = -\min(\hat{v}_{arm,a}(t) - v_{grid,a}(t), \hat{v}_{arm,b}(t) - v_{grid,b}(t), \hat{v}_{arm,c}(t) - v_{grid,c}(t)) \tag{13}$$

In practice, the controller must account for the instantaneous voltage ripple, so the maximum available arm voltage $\hat{v}_{arm,j}(t)$ for each arm $j = \{a, b, c\}$ is time, load and capacitor size dependent. For analytical purposes, these factors are neglected and constant maximum available arm voltages $\hat{v}_{arm,a}(t) = \hat{v}_{arm,b}(t) =$

$\hat{v}_{\text{arm,c}}(t) = \hat{v}_{\text{arm},j}$ are assumed. Eq. (12) and Eq. (13) are simplified to:

$$v_{0,\text{Max}}(t) = \hat{v}_{\text{arm},j} + \min\left(v_{\text{grid,a}}(t), v_{\text{grid,b}}(t), v_{\text{grid,c}}(t)\right)$$

$$v_{0,\text{Min}}(t) = -\hat{v}_{\text{arm},j} - \min\left(-v_{\text{grid,a}}(t), -v_{\text{grid,b}}(t), -v_{\text{grid,c}}(t)\right)$$

(14)

Any selected voltage injection must fulfill the condition $v_{0,\text{Min}}(t) \le v_0(t) \le v_{0,\text{Max}}(t)$. A third harmonic zero-sequence voltage injection with the grid voltage amplitude halves the DC-link capacitor demand [34], but requires over 77 % additional modules per arm compared to the Min-Max injection. Therefore, the new proposed method uses the third-harmonic injection to calculate the offset voltage. In case, that the offset voltage violates the criterion $v_{0,\text{Min}}(t) \le v_0(t) \le v_{0,\text{Max}}(t)$, it is set to $v_{0,\text{Min}}(t)$ or $v_{0,\text{Max}}(t)$:

$$v_{0,\text{Saturation}}(t) = \max\left(\min\left(V_{\text{grid}}\cos(3\omega t), v_{0,\text{Max}}(t)\right), v_{0,\text{Min}}(t)\right)$$

(15)

The resulting arm voltages for the proposed method are plot in Fig. 7. The arm voltage limit of $\hat{v}_{\text{arm},j} = \frac{\sqrt{3}}{2}V_{\text{grid}}$ is selected to have the same peak arm voltages as the Min-Max injection. It extends the advantage of the Min-Max Injection by further reducing the arm voltage at the time instant of maximum arm current. The resulting energy ripple is numerically calculated and finally decreased to 75.7 %. This means, that the new method requires less than 24 % DC-link capacitance when compared to the standard method.

In order to verify, that the trajectory of the zero-sequence is optimal, a genetic algorithm similar to [39] is implemented. It approximates the shape of the zero-sequence with $v_{0,\text{Genetic}}(t) = \hat{v}_{0,3}\cos(3\omega t) + \hat{v}_{0,9}\cos(9\omega t) + \hat{v}_{0,15}\cos(15\omega t)$ and optimizes to coefficients $\hat{v}_{0,3}$, $\hat{v}_{0,9}$ and $\hat{v}_{0,15}$ to minimize the energy ripple and ensure $\hat{v}_{\text{arm},j} = \frac{\sqrt{3}}{2}V_{\text{grid}}$. The resulting arm voltages shown in Fig. 7c) are indeed very similar to Fig. 7a) and show no advantage in respect to the energy ripple reduction. The optimal coefficients depend on the available arm voltage, as it can be seen in Fig. 8c). Again, the optimal zero-sequence calculated by the genetic algorithm is achieved with the much simpler proposed equation Eq. (15).

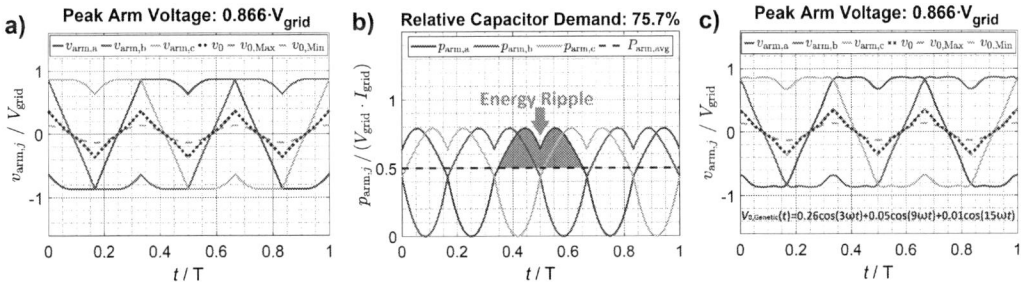

Fig. 7: Resulting arm voltages a) and input power b) for the proposed saturation zero-sequence voltage injection. The required arm voltage is reduced to $\frac{\sqrt{3}}{2}V_{\text{grid}} = 0.866\,V_{\text{grid}}$ and the energy ripple to 75.7 %.

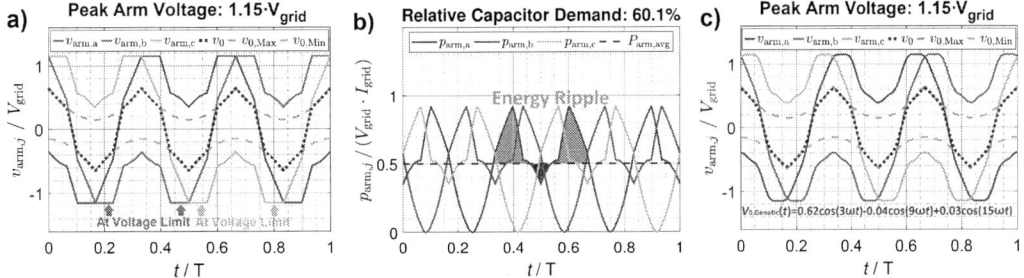

Fig. 8: Resulting arm voltages and input power for the proposed saturation zero-sequence voltage injection. An available arm voltage of $1.15\,V_{\text{grid}}$ reduces the energy ripple to 60.1 %. So, $C_{\text{DC,Link}}$ can be reduced by 40 %.

Usually, the arm voltage is not strictly limited to $\frac{\sqrt{3}}{2}V_{\text{grid}}$ because several reserve modules are available to compensate faulty modules and grid-side overvoltages. Additionally, the cell voltages of the PEM electrolyzers can reach 2.2 V at full load at $3\frac{A}{\text{cm}^2}$ due to the ohmic resistance of the cell membrane, gas pressure, water temperature as well as other factors [2], while it is only 1.4 V at light load. In large-scale electrolyzers, several hundred cells are connected in series, but the resulting load voltage v_{DC} still has the same relative variation as a single cell.

In order to operate the isolated DC-DC converter (LLC resonant converter) at or at least close to the efficient resonance condition [20] at light load, the DC-link voltage of each module has to be lower than at full load. This additional available arm voltage (at full load) can be used to decrease the second harmonic voltage ripple. The resulting arm voltages of the proposed method are shown in Fig. 8a), if the peak available arm voltage is 15 % higher than the peak grid voltage. This results in a further decrease of the second harmonic voltage ripple and therefore of the capacitor size requirement. It should be noted, that usually one arm is operating at its positive or negative voltage limit. This is the case, if the offset voltage $v_{0,\text{Saturation}}(t)$ is set to $v_{0,\text{Max}}(t)$ or $v_{0,\text{Min}}(t)$. As a positive side effect, this also reduces the number of switching actions in the AC-DC stage by nearly 33 % because usually one arm is clamped.

The following table summarizes the required arm voltages and resulting energy ripples for standard methods as well as the newly proposed method. The proposed saturation injection outperforms the commonly used 3rd harmonic and Min-Max injection methods even if only a small arm voltage of $\frac{\sqrt{3}}{2}V_{\text{grid}} = 0.866 V_{\text{grid}}$ is available. In addition, the energy ripple decreases, if more reserve modules are available. For unrealistically high available arm voltages of 54 % above grid voltage ($\hat{v}_{\text{arm},j} \geq \frac{16}{9}\frac{\sqrt{3}}{2}V_{\text{grid}} = 1.54 V_{\text{grid}}$), the proposed voltage injection is equivalent to $v_{0,\text{Saturation}}(t) = V_{\text{grid}}\cos(3\omega t)$. The 2nd harmonic voltage ripple is converted to a 4th harmonic voltage ripple, which has twice the frequency and therefore half the energy ripple.

Zero-Sequence Voltage Injection	Normalized Peak Arm Voltage	Normalized Energy Ripple
Nominal	1.000	1.000
3rd Harmonic	0.866	0.849
Min-Max	0.866	0.812
Saturation (Proposed)	0.866	0.757
Saturation (Proposed)	1.15	0.601
Saturation (Proposed)	1.54	0.500

Tab. 1: Required arm voltage and resulting energy ripple for investigated methods. The proposed saturation modulation performs best and requires the least amount of DC-link capacitance.

VI. Conclusion

A new saturation-based zero-sequence voltage injection for modular cascaded H-bridge (CHB) with star configuration has been proposed and analytically verified. First, the commonly used techniques of the 3rd harmonic and Min-Max zero-sequence voltage injection were evaluated. As a result, the Min-Max injection reduces the second harmonic voltage ripple by 18.8 % while the third-harmonic injection with $\frac{1}{6}$ of the grid voltage amplitude reduces the voltage ripple by only 15.1 %. This is achieved by distributing the input power more evenly across the three arm capacitors by preventing too high instantaneous arm voltages in combination with high arm currents.

The proposed zero-sequence injection technique usually operates the arm with the second largest absolute current at its voltage limit to minimize the arm voltage of the arm with the maximum current. This has the advance of reducing the energy to be buffered by the module DC-link capacitors between 24.3 % and 40 % depending on the available arm voltage. Commonly, several reserve modules are available to compensate for grid-overvoltages and modules faults. However, most of the time this is not required and a large arm voltage reserve is available, which is utilized by the proposed voltage injection technique. Finally, the costs and volume of the DC-link capacitors are reduced by 40 % compared to the nominal operation of CHBs operating as AC-DC converters for large-scale water electrolyzers.

References

[1] M. Kuprat, M. Bending, and K. Pfeiffer. "Possible role of power-to-heat and power-to-gas as flexible loads in German medium voltage networks". In: *Frontiers in Energy* 11 (2017), pp. 135–145.

[2] Geert Hauke Tjarks. "PEM-Electrolysis-Systems for the Integration in Power-to-Gas Applications". In: *Doctoral dissertation, Ph. D. thesis, RWTH Aachen University* (2017).

[3] Thomas Grube, Larissa Doré, André Hoffrichter, Laura Elisabeth Hombach, Stephan Raths, et al. "An option for stranded renewables: electrolytic-hydrogen in future energy systems". In: *Sustainable Energy Fuels* 2 (2018), pp. 1500–1515.

[4] J. Solanki. "High Power Factor High-Current Variable-Voltage Rectifiers". In: *Dissertation, Paderborn University, Germany* (2015).

[5] Mengxing Chen, Shih-Feng Chou, Frede Blaabjerg, and Pooya Davari. "Overview of Power Electronic Converter Topologies Enabling Large-Scale Hydrogen Production via Water Electrolysis". In: *Applied Sciences* 12.4 (2022).

[6] A. Pratt, P. Kumar, and T. V. Aldridge. "Evaluation of 400V DC Distribution in Telco and Data Centers to Improve Energy Efficiency". In: *INTELEC 07 - 29th Int. Telecommunications Energy Conf.* 2007, pp. 32–39.

[7] Jonas Huber, Peter Wallmeier, Ralf Pieper, Frank Schafmeister, and Johann W Kolar. "Comparative Evaluation of MVAC-LVDC SST and Hybrid Transformer Concepts for Future Datacenters". In: (2022).

[8] Hao Tu, Hao Feng, Srdjan Srdic, and Srdjan Lukic. "Extreme Fast Charging of Electric Vehicles: A Technology Overview". In: *IEEE Transactions on Transportation Electrification* 5.4 (2019), pp. 861–878.

[9] Burin Yodwong, Damien Guilbert, Matheepot Phattanasak, Wattana Kaewmanee, Melika Hinaje, and Gianpaolo Vitale. "AC-DC Converters for Electrolyzer Applications: State of the Art and Future Challenges". In: *Electronics* 9.6 (2020).

[10] J. R. Rodriguez, J. Pontt, C. Silva, E. P. Wiechmann, P. W. Hammond, et al. "Large current rectifiers: State of the art and future trends". In: *IEEE Transactions on Industrial Electronics* 52.3 (2005), pp. 738–746.

[11] Joonas Koponen, Vesa Ruuskanen, Antti Kosonen, Markku Niemelä, and Jero Ahola. "Effect of Converter Topology on the Specific Energy Consumption of Alkaline Water Electrolyzers". In: *IEEE Transactions on Power Electronics* 34.7 (2019), pp. 6171–6182.

[12] Vesa Ruuskanen, Joonas Koponen, Antti Kosonen, Markku Niemelä, Jero Ahola, and Aki Hämäläinen. "Power quality and reactive power of water electrolyzers supplied with thyristor converters". In: *Journal of Power Sources* 459 (2020).

[13] Sanzhong Bai and Srdjan M. Lukic. "New Method to Achieve AC Harmonic Elimination and Energy Storage Integration for 12-Pulse Diode Rectifiers". In: *IEEE Transactions on Industrial Electronics* 60.7 (2013), pp. 2547–2554.

[14] K. D. McBee and M. G. Simões. "Evaluating the Long-Term Impact of a Continuously Increasing Harmonic Demand on Feeder-Level Voltage Distortion". In: *IEEE Trans. Ind. App.* 50.3 (2014), pp. 2142–2149.

[15] H. Akagi, S. Inoue, and T. Yoshii. "Control and Performance of a Transformerless Cascade PWM STATCOM With Star Configuration". In: *IEEE Transactions on Industry Applications* 43.4 (2007), pp. 1041–1049.

[16] Q. Song and W. Liu. "Control of a Cascade STATCOM With Star Configuration Under Unbalanced Conditions". In: *IEEE Transactions on Power Electronics* 24.1 (2009), pp. 45–58.

[17] T. Tanaka, K. Ma, H. Wang, and F. Blaabjerg. "Asymmetrical Reactive Power Capability of Modular Multilevel Cascade Converter Based STATCOMs for Offshore Wind Farm". In: *IEEE Transactions on Power Electronics* 34.6 (2019), pp. 5147–5164.

[18] J. Solanki, N. Fröhleke, J. Böcker, and P. Wallmeier. "A modular multilevel converter based high-power high-current power supply". In: *IEEE Int. Conference on Industrial Technology (ICIT)*. 2013, pp. 444–450.

[19] Roland Unruh, Frank Schafmeister, and Joachim Böcker. "Evaluation of MMCs for High-Power Low-Voltage DC-Applications in Combination with the Module LLC-Design". In: *2020 22nd European Conference on Power Electronics and Applications (EPE'20 ECCE Europe)*. 2020.

[20] Roland Unruh, Frank Schafmeister, and Joachim Böcker. "11kW, 70kHz LLC Converter Design with Adaptive Input Voltage for 98% Efficiency in an MMC". In: *2020 IEEE 21st Workshop on Control and Modeling for Power Electronics (COMPEL)*. 2020, pp. 1–8.

[21] J. Kim, C. Kim, J. Kim, J. Lee, and G. Moon. "Analysis on Load-Adaptive Phase-Shift Control for High Efficiency Full-Bridge LLC Resonant Converter Under Light-Load Conditions". In: *IEEE Transactions on Power Electronics* 31 (2016), pp. 4942–4955.

[22] P. Rehlaender, R. Unruh, F. Schafmeister, and J. Böcker. "Alternating Asymmetrical Phase-Shift Modulation for Full-Bridge Converters with Balanced Switching Losses to Reduce Thermal Imbalances". In: *IEEE Applied Power Electronics Conference and Exposition (APEC)*. 2021.

[23] J. E. Huber and J. W. Kolar. "Volume/weight/cost comparison of a 1MVA 10 kV/400 V solid-state against a conventional low-frequency distribution transformer". In: *ECCE*. 2014, pp. 4545–4552.

[24] Zheqing Li, Yi-Hsun Hsieh, Qiang Li, Fred C. Lee, and Chunyang Zhao. "Evaluation of Double-Line-Frequency Power Flow in Solid-State Transformers". In: *2021 IEEE Fourth International Conference on DC Microgrids (ICDCM)*. 2021.

[25] T. Zhao, G. Wang, S. Bhattacharya, and A. Q. Huang. "Voltage and Power Balance Control for a Cascaded H-Bridge Converter-Based Solid-State Transformer". In: *IEEE Transactions on Power Electronics* 28.4 (2013), pp. 1523–1532.

[26] H. Bärnklau, A. Gensior, and S. Bernet. "Derivation of an equivalent submodule per arm for modular multilevel converters". In: *15th Int. Power Electronics and Motion Control Conf. (EPE/PEMC)*. 2012.

[27] Youngjong Ko, Anatolii Tcai, and Marco Liserre. "DC-Link Voltage Balancing Modulation for Cascaded H-Bridge Converters". In: *IEEE Access* 9 (2021), pp. 103524–103532.

[28] L. Wang, D. Zhang, Y. Wang, B. Wu, and H. S. Athab. "Power and Voltage Balance Control of a Novel Three-Phase Solid-State Transformer Using Multilevel Cascaded H-Bridge Inverters for Microgrid Applications". In: *IEEE Transactions on Power Electronics* 31.4 (2016), pp. 3289–3301.

[29] F. V. Amaral, T. M. Parreiras, G. C. Lobato, A. A. P. Machado, I. A. Pires, and B. de Jesus Cardoso Filho. "Operation of a Grid-Tied Cascaded Multilevel Converter Based on a Forward Solid-State Transformer Under Unbalanced PV Power Generation". In: *IEEE Trans. on Industry App.* 54.5 (2018), pp. 5493–5503.

[30] T. Zhao, X. Zhang, M. Wang, W. Mao, F. Li, et al. "Module Power Balance Control and Redundancy Design Analysis of Cascaded PV Solid State Transformer under Fault Conditions". In: *IEEE Journal of Emerging and Selected Topics in Power Electronics* (2020).

[31] R. Unruh, J. Lange, F. Schafmeister, and J. Böcker. "Adaptive Zero-Sequence Voltage Injection for Modular Solid-State Transformer to Compensate for Asymmetrical Fault Conditions". In: *2021 23rd European Conference on Power Electronics and Applications (EPE'21 ECCE Europe)*. 2021.

[32] Yifan Yu, Georgios Konstantinou, Branislav Hredzak, and Vassilios G. Agelidis. "On extending the energy balancing limit of multilevel cascaded H-bridge converters for large-scale photovoltaic farms". In: *2013 Australasian Universities Power Engineering Conference (AUPEC)*. 2013, pp. 1–6.

[33] Y. Yu, G. Konstantinou, B. Hredzak, and V. G. Agelidis. "Power Balance of Cascaded H-Bridge Multilevel Converters for Large-Scale Photovoltaic Integration". In: *IEEE Transactions on Power Electronics* 31.1 (2016), pp. 292–303.

[34] Y. Hu, X. Zhang, W. Mao, T. Zhao, F. Wang, and Z. Dai. "An Optimized Third Harmonic Injection Method for Reducing DC-Link Voltage Fluctuation and Alleviating Power Imbalance of Three-Phase Cascaded H-Bridge Photovoltaic Inverter". In: *IEEE Trans. on Industrial Electronics* 67.4 (2020), pp. 2488–2498.

[35] Y. Hu, Z. Li, H. Zhang, C. Zhao, F. Gao, et al. "High-Frequency-Link Current Stress Optimization of Cascaded H-Bridge-Based Solid-State Transformer With Third-Order Harmonic Voltage Injection". In: *IEEE Journal of Emerging and Selected Topics in Power Electronics* 9.1 (2021), pp. 1027–1038.

[36] S. Rivera, B. Wu, S. Kouro, H. Wang, and D. Zhang. "Cascaded H-bridge multilevel converter topology and three-phase balance control for large scale photovoltaic systems". In: *2012 3rd IEEE International Symposium on Power Electronics for Distributed Generation Systems (PEDG)*. 2012, pp. 690–697.

[37] B. P. McGrath, D. G. Holmes, and T. Lipo. "Optimized space vector switching sequences for multilevel inverters". In: *IEEE Transactions on Power Electronics* 18.6 (2003), pp. 1293–1301.

[38] S. Haghbin and T. Thiringer. "DC bus current harmonics of a three-phase PWM inverter with the zero sequence injection". In: *2014 IEEE Transportation Electrification Conf. and Expo (ITEC)*. 2014, pp. 1–6.

[39] Simon Fuchs, Min Jeong, and Jürgen Biela. "Reducing the Energy Storage Requirements of Modular Multilevel Converters with Optimal Capacitor Voltage Trajectory Shaping". In: *2020 22nd European Conference on Power Electronics and Applications (EPE'20 ECCE Europe)*. 2020.

Thermal behavior impact on the electric motor shape multi-objective optimization

Aissam Riad MEDDOUR[*1], Anthony BABIN[1], Nassim RIZOUG[1], Christopher VAGG[2], Richard BURKE[2], Laid DEGAA[1]

[1]École supérieure des Techniques Aéronautiques et de Construction Automobile (ESTACA)
Parc universitaire Laval-Changé, Rue Georges Charpak, 53000
Laval, France
[2]University of Bath
Newlands Ln, Emersons Green, Bristol BS16 7PT
Bath, United Kingdom
*Email: aissam.meddour@estaca.fr

Keywords

≪Electric vehicle ≫,≪Multi-objective optimization≫, ≪Permanent magnet motor ≫, ≪Modelling≫, ≪Energy storage≫

Abstract

The paper investigates the impact of thermal behaviour on the electric car motor's shape optimization for a realistic driving cycle. The motor model is created by integrating electromagnetic-thermal modelling on Ansys electromagnetic and Motor-CAD, which is then linked to Matlab's genetic optimization algorithm evaluating the motor's price and performance.

Introduction

Transport is one of the primary sectors of involvement in government-launched climate plans; in France, for example, a policy targeted at eliminating the sale of diesel and gasoline automobiles by 2040 was implemented in July 2017[1] [2].
The car sector is now a part of these new government policies. As a result, they need to address the electrical transition of its products while maintaining consumer appeal by delivering efficient and reasonably affordable solutions compared to the thermal vehicle that has already infiltrated the automotive industry. The electric powertrain is one of the most critical components in transition, accounting for 8 to 20 % of the total cost of the vehicle, of which 33 to 43 % is the price of the electric motor alone [3]. As a result, it is critical to consider the many factors that influence this component's performance, cost, and lifetime. Because of their performance and longevity, permanent magnet synchronous motors have grown popular [4]. However, the price of this motor is mainly determined by the magnet usage rate, which is regarded as rare and expensive on the earth [5]. Furthermore, lowering the magnet usage rate necessitates a bigger motor to maintain the same performance (not to mention the oversizing of the power electronics)[5]. However, reducing the magnet size will inevitably increase the current flowing through the winding affecting the thermal performance and longevity of the motor. This demonstrates the need to consider both thermal and electromagnetic variables when assessing the performance/cost of an electric motor [6]. Previous studies on motor design and optimization methodologies concentrated mostly on electromagnetic performance, such as torque ripple, torque density, and efficiency [7] [8] [9]. The temperature of the motor was used to check the loss reduction only [10]. Although overall temperature can be reduced by decreasing the motor's loss when optimizing the motor, the geometric shapes, size, and airflow between the stator and rotor will still affect the heat dissipation of the motor. Furthermore, minimizing the loss may not result in a reduction in the temperature of specific motor components [11].

Several other studies have addressed the subject of optimizing the geometry of an electric motor by considering the thermal behaviour of the latter; we can cite, for example, the study of [12] in which is studied the variation of the temperature of a motor with variable reluctance, the temperature is evolved with the help of an electrothermal model and the study carried out for several operating speeds.

We can also mention [13] in, which was investigated the impact of the temperature of materials on the optimization of the price and efficiency of the analytical motor model optimal. [14] proposed to study the impact of the thermal aspect on a DC motor used for a short-time duty with a constant rotational speed using an analytical actuator model. In this article, the PMSM of the Toyota Prius [15] is regarded as the study's topic. Instead of testing it simply over fixed working points or non-realistic conditions, its optimization is carried out considering The Worldwide harmonized Light vehicles Test Cycles (WLTC) using the optimization toolbox in Matlab. For more precession, a finite element(FE) electromagnetic model created using ANSYS electronics will examine the EM performances; this model will be linked to an analytical evaluation to boost computation speed.

The EM data will be used in the thermal lumped circuit under MotorCAD software to assess temperature at various motor components; those temperatures will then be used to re-evaluate the electric motor losses and repeat the process until the temperature stabilizes.

The EM and thermal evaluations will be included in the optimization algorithm, which will calculate the cost function and then suggest a new motor geometry to be revalued until a stop condition is satisfied.

The rest of this paper is structured as follows:

Section 1 presents the driving cycle and the vehicle dynamic model, followed by the electromagnetic and thermal model developed for the electric motor. In addition to the cost function construction, Section 3 describes the multiobjective optimization method and its settings. Furthermore, in section 4, the optimization results are reviewed, and a thorough comparison is made between an optimization that considers the thermal component and another that does not.

1 Mission specifications & vehicle dynamic model

The optimisation is performed by taking into account the wltc driving cycle [16], which is considered as the standard in-vehicle testing after replacing the outdated NEDC [17]. The amount of mechanical energy "consumed" by a vehicle when driving a pre-specified driving pattern is primarily determined by three factors [18]

- Losses due to aerodynamic friction;

- Rolling friction losses;

- The brakes dispersed the energy.

The driving cycle characteristics are integrated into the vehicle dynamic model to predict the output power and torque requirement while considering the various forces that may occur on the moving vehicle. The technique is depicted in the diagram in figure 1.

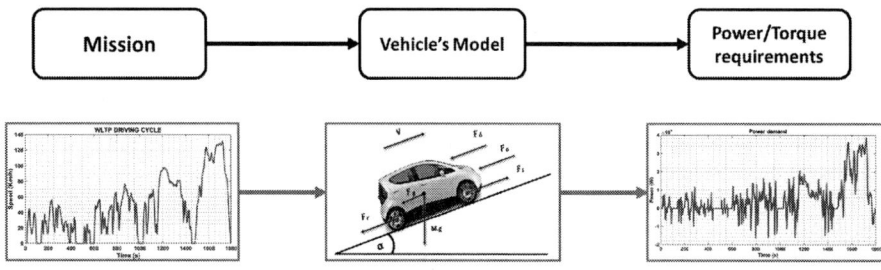

Fig. 1: Power demands estimation

2 Motor Model

Electromagnetic model

The finite element modelling method provides high precision in motor performance evaluation [19] using the Ansys electronics software.

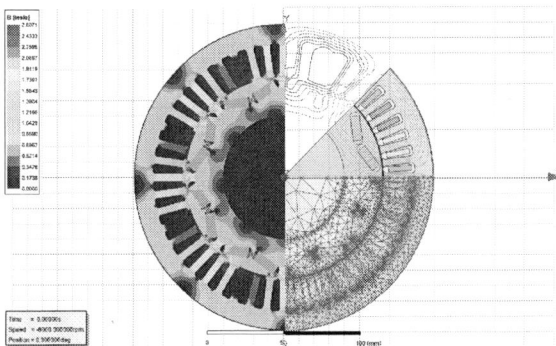

Fig. 2: Electromagnetic motor model

Using finite element analysis in an optimisation loop might result in extremely high computing costs, making this type of solution impractical.

This prompted us to use a semi-analytical approach for mapping the engine's electrical magnitudes over a few finite element simulations at a single speed and then extrapolating the results analytically over the complete operational nodes of the motor, spanning the driving cycle's operating points.

The analytical part's method is based on a d-q-axis flux-linkage model [20], coupled with a finite element method to predict the d- and q-axis inductances and flux.

Thermal model

There are two schools of thought in thermal modelling: finite element modelling, which is highly accurate but takes a lot of computational time, and thermal network modelling, which produces decent results when properly designed but does not require a lot of processing power [21].

The thermal network method (TNM) operates on the idea of dividing the engine into fundamental thermal parts that represent a mix of heat transfer actions via conduction, convection, and radiation [22].

The second modelling approach is selected; it interweaves the thermal model with the electromagnetic finite element model built on Ansys Electronics.

The geometrical parameters used to construct the electromagnetic model will also be used to build the thermal model, as illustrated in the figure 3. The temperature will be estimated in the different compartments of the motor under operating conditions corresponding to the WLTC driving cycle.

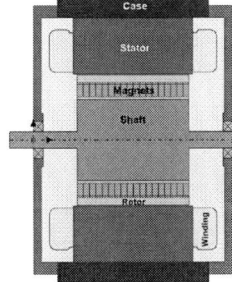

Fig. 3: Thermal motor model

This approach was used to perform the thermal modelling of the inside permanent magnet motor, using MOTOR-CAD software. The built thermal circuit of the motor is shown in figure 4.

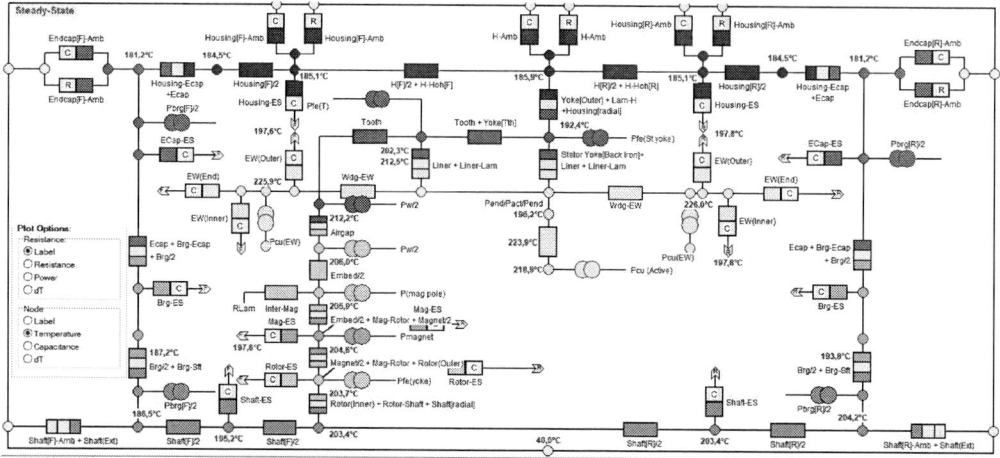

Fig. 4: Lumped thermal circuit

3 Optimization algorithm

The flowchart in figure 5 is the oganigram explaining the used optimization method.

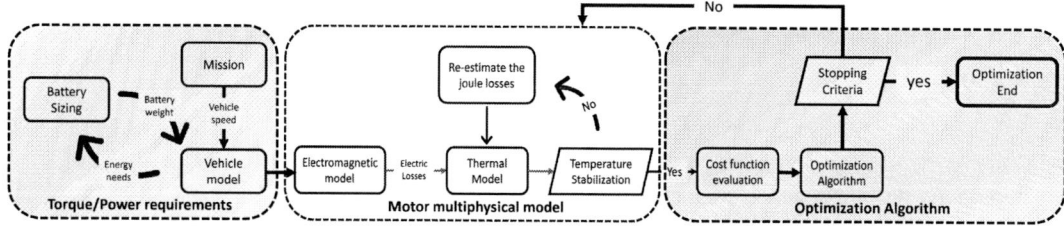

Fig. 5: Optimization methodology flowchart

Torque/Power requirements

As stated in Section 1, the WLTP considered mission will be introduced into the vehicle dynamic model to estimate the energy and power needs of the driving cycle.

These specifications will be entered into the storage system sizing algorithm to size a battery that can handle the mission while taking its weight into account [23].

The battery's weight will be re-introduced back into the vehicle dynamic model to recalculate the new power and torque needs, which will include the weight of the storage system.

Motor multiphysical model

The geometrical parameters of the engine given by the optimization method will be used to build the engine's electromagnetic and thermal models.

The engine's EF/Analytical model using Matlab and Ansys electronics will compute its performance and electrical losses at the drive cycle's representative operation points.

The electrical losses will be transmitted to the equivalent thermal circuit model on MotorCad to estimate the temperature of the different electric motor parts. The temperature of the windings will be reintroduced in an analytical model to reestimate the joule losses expressed with the equation 1, which make up the majority of the electrical losses. These new calculated losses will be reintroduced on MotorCAD to evaluate the latest temperature, and so on, until the temperature (n+1 - n) does not represent any difference.

$$R = R_{ref}\left[1 + \alpha(T - T_{ref})\right] \tag{1}$$

With R_{ref} the reference resistance calculated at the temeprature Reference T_{ref} 20 degree, and Alpha the temeprature coefficient for the conductive material.

Optimization algorithm

The cost of the motor's construction materials are stated by the equation 2 [24] .The formula is weighted in relation to the cost of iron weight.

$$AMC = 24 m_{PM} + 7 m_{Cu} + m_{Iron} \tag{2}$$

This equation , in addition to the average sum of losses , will be assessed in the form of an objective function expressed by the equation 3,

$$F = AMC + Average Sum of losses \tag{3}$$

The genetic optimization algorithm will propose new geometrical parameters to create the new motor model. The objective function to be reduced will be examined, and if one of the stopping conditions is met, the optimization process will be halted under the guise of finding the best model. The optimization is restarted for a fresh iteration in the opposite situation.

The table I summarizes the optimization algorithm's settings.

Table I: Optimisation algorithm configuration

Parameters	Value	Parameters	Value
Number of variables	6	Generation number	20
Migration fraction	0.3	Migration interval	4
Elite count	5	Population size	85

Based on the literature [24] [25] [26], six motor geometrical parameters are optimized by being adjusted until the optimum solution meeting our target function is identified. While avoiding a perfect local solution, the genetic algorithm seeks a vast solution surface. It might, however, result in the construction of an infeasible motor. Consequently, upper and lower parameter limitations in the table II are applied.

Table II: Optimisation algorithm boundary limits

Parameters	Upper(mm)	Lower(mm)	Parameters	Upper(mm)	Lower(mm)
Rotor inner diameter	75	192	Stator outer diameter	220	310
Magnet's width	25	39	Magnet's thickness	4.3	12
Slot opening	0.9	4.6	Motor length	70	160

4   Results discussion

This part summarizes the optimization results comparaison.

In order to confirm the importance of considering the thermal aspect into the designing procedure,two optimization processes have been initiated: the first will include the motor's electromagnetic model in addition to the battery sizing process, and the second case takes into account the thermal aspect in the optimisation loop to re-evaluate the losses from the proposed motor geometry.

The simulation results for the two cases' results are presented in the figure 6.

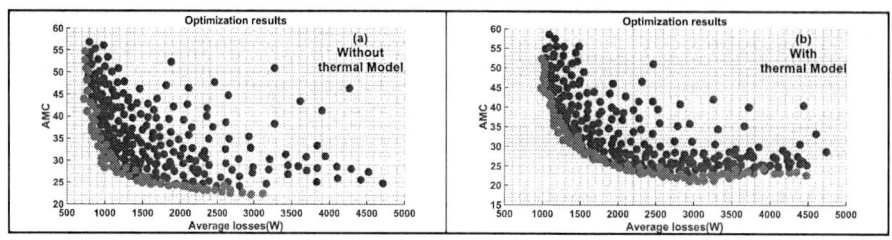

Fig. 6: Optimization results, (a):Without thermal model (b):With thermal model

In both situations, the genetic algorithm converged to an ideal population, shown in red, is called the Pareto front, where the best balance between material cost and average electrical losses was achieved. However, the optimal choice can't be made without a tradeoff between the optimization objectives. According to the manufacturer's specification notes, the final optimal pick among the optimal population could only be made. We will focus on the population near the Pareto front's elbow for the comparison.

The results obtained in scenario (a) demonstrate that the optimization method converged toward a motor model with an indexed cost between 22 and 30 and an average electrical losses between 750 and 3200W over the working points of the driving cycle utilized as calculation benchmarks.

On the other hand, the Pareto front of the second scenario (b), where the thermal model was included, has the same shape as the previous one. However, we can notice a slight increase in the average amount of the electrical losses due to the stator resistance variation impacted by the temperature increase at the winding.

At the same time, it is noticeable that the Pareto front of the second scenario presents a considerable rise in terms of indexed cost compared to the first case. To analyse this perception in detail, we display the two optimal motor models at the centre of the Pareto front bend in the figure 7 to compare their geometry.

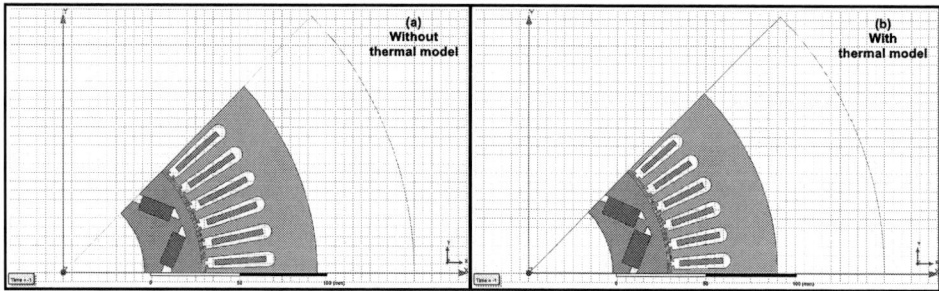

Fig. 7: Motor optimal geometry, (a):Without thermal model (b):With thermal model

As illustrated in the figure 7, when the thermal aspect is considered in the optimisation loop, the slot surface is reduced. The permanent magnet size is more important to ensure the desired performance; increased size of the permanent magnets guarantees the desired performance while reducing the current amplitude flowing through the winding and thus reducing the joule losses.

The increase in the use of permanent magnets justifies the rise in the price of the active construction materials of the motor compared to the first case where the thermal aspect was not taken into account.

Conclusion

This paper proposes a co-modelling optimization of the Toyota Prius synchronous permanent magnet motor. To achieve this, an electromagnetic model is implemented in Ansys electronics, connected to a thermal model built under motor-CAD software. The co-modelling allowed the optimization to consider thermal behaviour to re-evaluate the actual performances and joules losses.

The optimization is performed for the WLTC driving cycle while considering the battery sizing impact on the power/torque requirements.

The final optimization results are compared with previous results obtained for a simple optimization achieved using a simple electromagnetic model. The comparison confirmed the impact direction of the thermal aspect consideration and its importance, leading to a costly motor due to increased permanent magnet usage rate pushed by the lowering joule losses objective. In the future, experimental testing will be carried out to provide a more realistic view of the overall efficiency of the suggested system.

References

[1] M. R. Bernard, "Update on electric vehicle uptake in European cities," p. 18.

[2] G. Broadbent, D. Drozdzewski, and G. Metternicht, "Electric vehicle adoption: An analysis of best practice and pitfalls for policy making from experiences of Europe and the US," *Geography Compass*, vol. 12, p. e12358, Dec. 2017.

[3] A. König, L. Nicoletti, D. Schröder, S. Wolff, A. Waclaw, and M. Lienkamp, "An Overview of Parameter and Cost for Battery Electric Vehicles," *World Electric Vehicle Journal*, vol. 12, p. 21, Feb. 2021.

[4] Y. Chin and J. Soulard, *A permanent magnet synchronous motor for traction applications of electric vehicles*, Jul. 2003, pages: 1041 vol.2.

[5] J. Widmer, R. Martin, and M. Kimiabeigi, "Electric vehicle traction motors without rare earth magnets," *Sustainable Materials and Technologies*, vol. 29, Mar. 2015.

[6] E. Gundabattini, R. Kuppan, D. G. Solomon, A. Kalam, D. P. Kothari, and R. Abu Bakar, "A review on methods of finding losses and cooling methods to increase efficiency of electric machines," *Ain Shams Engineering Journal*, vol. 12, no. 1, pp. 497–505, Mar. 2021. [Online]. Available: https://www.sciencedirect.com/science/article/pii/S2090447920301854

[7] X. Sun, Z. Shi, Y. Cai, G. Lei, Y. Guo, and J. Zhu, "Driving-Cycle-Oriented Design Optimization of a Permanent Magnet Hub Motor Drive System for a Four-Wheel-Drive Electric Vehicle," *IEEE Transactions on Transportation Electrification*, vol. 6, no. 3, pp. 1115–1125, Sep. 2020. [Online]. Available: https://ieeexplore.ieee.org/document/9141333/

[8] S. Huang, J. Zhang, J. Gao, and K. Huang, "Optimization the Electromagnetic Torque Ripple of Permanent Magnet Synchronous Motor," *Electrical and Control Engineering, International Conference on*, vol. 0, pp. 3969–3972, Jun. 2010.

[9] O. R. Solomon, "Efficiency Optimization and Control of Permanent Magnet Synchronous Brushless Motors in Three-Phase Pulse Width Modulated Voltage Source Inverter Drives," PhD, West Virginia University Libraries, Jan. 2008. [Online]. Available: https://researchrepository.wvu.edu/etd/4423

[10] X. Sun, Y. Shen, S. Wang, G. Lei, Z. Yang, and S. Han, "Core Losses Analysis of a Novel 16/10 Segmented Rotor Switched Reluctance BSG Motor for HEVs Using Nonlinear Lumped Parameter Equivalent Circuit Model," *IEEE/ASME Transactions on Mechatronics*, vol. 23, no. 2, pp. 747–757, Apr. 2018, conference Name: IEEE/ASME Transactions on Mechatronics.

[11] X. Sun, B. Wan, G. Lei, X. Tian, Y. Guo, and J. Zhu, "Multiobjective and Multiphysics Design Optimization of a Switched Reluctance Motor for Electric Vehicle Applications," *IEEE Transactions on Energy Conversion*, vol. 36, no. 4, pp. 3294–3304, Dec. 2021. [Online]. Available: https://ieeexplore.ieee.org/document/9427101/

[12] H. Rouhani, J. Faiz, and C. Lucas, "Lumped thermal model for switched reluctance motor applied to mechanical design optimization," *Mathematical and Computer Modelling*, vol. 45, no. 5-6, pp. 625–638, Mar. 2007. [Online]. Available: https://linkinghub.elsevier.com/retrieve/pii/S0895717706002871

[13] G. Bramerdorfer, A. Cavagnino, and S. Vaschetto, "Importance of thermal modeling for design optimization scenarios of induction motors," in *2017 IEEE Energy Conversion Congress and Exposition (ECCE)*. Cincinnati, OH: IEEE, Oct. 2017, pp. 4666–4672. [Online]. Available: http://ieeexplore.ieee.org/document/8096796/

[14] Q. Li, M. Dou, B. Tan, H. Zhang, and D. Zhao, "Electromagnetic-Thermal Integrated Design Optimization for Hypersonic Vehicle Short-Time Duty PM Brushless DC Motor," *International Journal of Aerospace Engineering*, vol. 2016, pp. 1–9, 2016. [Online]. Available: https://www.hindawi.com/journals/ijae/2016/9725416/

[15] T. A. Burress, S. L. Campbell, C. Coomer, C. W. Ayers, A. A. Wereszczak, J. P. Cunningham, L. D. Marlino, L. E. Seiber, and H.-T. Lin, "Evaluation of the 2010 Toyota Prius Hybrid Synergy Drive System," Tech. Rep. ORNL/TM-2010/253, 1007833, Mar. 2011. [Online]. Available: http://www.osti.gov/servlets/purl/1007833-qNciEv/

[16] T. J. Barlow, S. Latham, I. S. Mccrae, and P. G. Boulter, "A reference book of driving cycles for use in the measurement of road vehicle emissions," *TRL Published Project Report*, 2009, iSBN: 9781846088162. [Online]. Available: https://trid.trb.org/view/909274

[17] J. Pavlovic, B. Ciuffo, G. Fontaras, V. Valverde, and A. Marotta, "How much difference in type-approval CO2 emissions from passenger cars in Europe can be expected from changing to the new test procedure (NEDC vs. WLTP)?" *Transportation Research Part A Policy and Practice*, vol. 111, Mar. 2018.

[18] L. Guzzella and A. Sciarretta, "Electric and Hybrid-Electric Propulsion Systems," in *Vehicle Propulsion Systems: Introduction to Modeling and Optimization*, L. Guzzella and A. Sciarretta, Eds. Berlin, Heidelberg: Springer, 2013, pp. 67–162. [Online]. Available: https://doi.org/10.1007/978-3-642-35913-2_4

[19] J. Santiago Ochoa, "FEM Analysis Applied to Electric Machines for Electric Vehicles." [Online]. Available: http://uu.diva-portal.org/smash/record.jsf?pid=diva2%3A436792&dswid=-3483

[20] G. Qi, J. T. Chen, Z. Q. Zhu, D. Howe, L. B. Zhou, and C. L. Gu, "Influence of Skew and Cross-Coupling on Flux-Weakening Performance of Permanent-Magnet Brushless AC Machines," *IEEE Transactions on Magnetics*, vol. 45, no. 5, pp. 2110–2117, May 2009, conference Name: IEEE Transactions on Magnetics.

[21] A. Zeaiter, "Thermal Modeling and Cooling of Electric Motors: Application to the Propulsion of Hybrid Aircraft," p. 240.

[22] Y. Chin, E. Nordlund, and A. Staton, "Thermal analysis - Lumped-circuit model and finite element analysis," *undefined*, 2003.

[23] R. Sadoun, "Intérêt d'une Source d'Energie Electrique Hybride pour véhicule électrique urbain – dimensionnement et tests de cyclage," p. 139.

[24] A. Fatemi, "Design Optimization of Permanent Magnet Machines Over a Target Operating Cycle Using Computationally Efficient Techniques," *undefined*, 2016. [Online]. Available: https://www.semanticscholar.org/paper/Design-Optimization-of-Permanent-Magnet-Machines-a-Fatemi/4013fede76ec22f25100860c13be25c1265bab56

[25] "Parametric Sensitivity Analysis and Design Optimization of an Interior Permanent Magnet Synchronous Motor | IEEE Journals & Magazine | IEEE Xplore."

[26] S. Ahmadi, T. Lubin, A. Vahedi, and N. Taghavi, "Sensitivity-Based Optimization of Interior Permanent Magnet Synchronous Motor for Torque Characteristic Enhancement," *Energies*, vol. 14, no. 8, p. 2240, Jan. 2021, number: 8 Publisher: Multidisciplinary Digital Publishing Institute. [Online]. Available: https://www.mdpi.com/1996-1073/14/8/2240

Modelling Approaches of Power Systems Considering Grid-Connected Converters and Renewable Generation Dynamics

Jaume Girona-Badia
Vinícius Albernaz Lacerda
Eduardo Prieto-Araujo
Oriol Gomis-Bellmunt
Centre d'Innovacio Tecnològica en Convertidors
Estatics i Accionaments (CITCEA-UPC)
Barcelona, Spain

Stephan Kusche
Florian Pöschke
Horst Schulte
Department of Engineering I, Control Engineering
HTW Berlin - University of Applied Sciences
Berlin, Germany

Abstract—**This paper presents a comparative analysis of several modelling approaches of key elements used in simulations of power systems with renewable energy sources. Different models of synchronous generators, transmission lines, converters, wind generators and PV power plants are compared to assess the most suitable models for grid-connection studies. It also analyses how the dynamics of PV power plants and the mechanical dynamics of wind generators affect the electrical variables on the grid side. The models were compared in terms of precision, computational time and ease of use through simulations of load connection, short-circuits, disconnection of generators and lines in a benchmark system modelled in Simulink.**

Index Terms—**EMT simulation, mechanical dynamics, phasor simulation, power systems modelling, renewable generation dynamics.**

I. INTRODUCTION

The electrical power system is experiencing a deep penetration of renewable energy sources (RES) worldwide. Several countries have defined targets to increase the integration of RES, such as wind, solar, geothermal, hydro, ocean and biomass [1], [2], using different solutions such as DC grids, microgrids and Virtual Power Plants [3].

In order to assess how present and future power systems will perform with high penetration of RES, researchers and industry need to use proper power systems models considering a variety of technologies. However, various modelling approaches have been proposed depending on the type of study, and there is not a single choice on how to model transmission lines, synchronous generators (SGs), converters and RES.

While important recommendations and guidelines were recently available [4], [5], those are often high-level and based on the researchers' experience, and important questions still need to be addressed, such as the influence of PV power plants dynamics and mechanical dynamics of wind generators (WGs)

This project has received funding from the European Union's Horizon 2020 research and innovation programme under grant agreement No 883985 (POSYTYF project).

on electrical variables and how these dynamics interact with other elements in the system.

Therefore, this paper presents a comparative analysis among several modelling approaches, considering different levels of detail of the main components of power systems with RES. It also considers the dynamics of PV power plants and mechanical dynamics of wind generators (WGs) and analyses how these dynamics affect the electrical variables on the grid side.

The remainder of this paper is organized as follows. Section II briefly introduces the models of SGs, transmission lines, converters and RES used in this study. Section III presents the methodology of the comparative analysis, including the simulated system and the tests performed. The result are shown in Section IV followed by discussions. Finally, the conclusions are drawn in Section V.

II. MODELLING APPROACHES

Synchronous generators, transmission lines, loads and converters are conventional elements used when simulating power systems with RES. Several models of converters were proposed, varying from detailed models, in the semiconductors domain, to high-level phasor models [6]. Multiple choices can also be found for synchronous generators, transmission lines and wind generators. The proper choice for each element will depend on the level of detail, the phenomena being analysed and the time available for simulation. While other studies have analysed these modelling approaches deeply focusing on specific components, it is also important to assess the interaction between them. If a very detailed transmission line model is used with an approximated generator model, the overall simulation might not be detailed enough for some specific scenarios. This interaction is covered in this paper, where groups of models are analysed together.

A. Wind turbine modelling

The wind energy system in this work is modeled as a variable-speed turbine equipped with a synchronous generator fed by a back-to-back converter that also establishes the connection to the electrical grid. The associated aeromechanical energy conversion is nonlinear and depends on the current wind speed and the turbine states such as rotor speed or pitch angle. Different control loops govern the operation of the wind energy system with the power control of the wind energy conversion system being dominant. Depending on the current operating point and thus wind speed, the power control either maximizes the power output in partial-load region, limits the power output to rated in full-load region or generates the desired power output following a setpoint signal [7]. For the multi-MW class, the resulting closed-loop dynamics have timescales in the range of seconds [8], [9], and thus may represent relevant dynamics for the interaction with other participating units in the electrical grid.

To study the effects of including these dynamics within the power system simulation, two different models are used to display wind power. The first comprises algebraic power relations that statically portray the produced power depending on the current wind speed as only input to the model. Therein, the power produced by the wind turbine P_w is calculated using the wind speed v and a power coefficient look-up table for $c_P(\frac{\omega R}{v}, \beta)$ depending on the rotor speed ω and the pitch angle β, such that the power is given by $P_w = \frac{1}{2} c_P(\frac{\omega R}{v}, \beta) \rho \pi R^2 v^3$ [10], where R denotes the rotor radius of the wind turbine. Whenever the power output setpoint is below the extractable power of the wind, the power output of the model is immediately set to the desired value. This modeling implies a perfect control and following of the turbine states. Thus, this modeling approach neglects the dynamics when transitioning between operating points that are governed by the control loops and system characteristics such as rotor inertia or pitch dynamics.

The second model is capable of displaying the interaction of the mechanical turbine states and the control loops to form a dynamical description. It relies on the Takagi-Sugeno modeling framework that uses a convex description of linear models to describe nonlinear dynamics. To derive the model, first a linearization analysis of NREL's 5 MW reference turbine [11] using FAST [12] is conducted at several operating point within the relevant operating space. Subsequently, an observer-based controller in the Takagi-Sugeno framework is designed with respect to the aforementioned power control problem using linear matrix inequalities to derive the necessary controller gains. Further, the pitch and generator torque dynamics are modeled as first-order transfer functions to account for limited pitch rates and the generator dynamics. Details about the applied modeling and control structure can be found in [7], [9]. The model (both, open- and closed-loop) is validated using FAST and provides a proper description of the aerodynamic conversion dynamics governed by the control loops. As a result, the model-based control design directly yields a closed-loop system description that can be implemented in power system simulations capable of portraying the nonlinear dynamics inherited in the wind energy conversion process.

B. Photovoltaic power station modelling

In general, the photovoltaic (PV) power station consists of multiple arrays of PV cells connected to the grid via an optional DCDC converter and an inverter. Though the DCDC converter is not present in all facilities, it is used in this work because it facilitates power tracking control.

The model of one PV cell is realised as explicit single diode model, i.e. a current source (photon current i_{ph}) in parallel with a diode (diode current i_d) in parallel with a resistor (R_h) to accommodate losses [13], such that t (Kirchhoff's first law)

$$i_{pv} = i_{ph} - \underbrace{i_s \left[\exp\left(\frac{q_e v_{pv}}{A_n k_B T_c} \right) - 1 \right]}_{i_d} - \frac{v_{pv}}{R_h}. \qquad (1)$$

In this form, the diode I-V characteristics is described by the theory of Shockley [14], using the Boltzmann constant k_B, elementary charge q_e and using the tunable parameters ideality factor of the diode A_n, photon current i_{ph} and saturation current i_s. The tunable parameters are fitted on the I-V and P-V characteristics of the PV cell obtained under standard test conditions (STC). The cell temperature T_c is only subject to slow changes in time and therefore assumed to be constant at the STC value of 25°C. On the other hand, the irradiation S may change rapidly, and affects the photon current following (α_T being a PV cell dependent parameter):

$$\frac{i_{ph}}{i_{ph}^{STC}} = \frac{S}{S^{STC}} \left[1 + \alpha_T (T_c - T_c^{STC}) \right]. \qquad (2)$$

Agglomeration of multiple PV cells in series and parallel raises the output voltage and current. In total the PV voltage at the maximum power point (MPP) is 6.28 kV, converted to 11 kV DC voltage and power production of 90 MW peak.

Power transformation is done in a first step via a boost converter [15] and in a second step via an inverter. The boost converter is controlled via the duty cycle D, which determines the conversion ratio between PV voltage v_{pv} and DC link voltage v_{dc}. Within the continuous conduction mode, $v_{pv} = (1 - D)v_{dc}$ holds [16]. A higher level controller is used to track either the MPP or demand power point (DPP) using a perturb and observe (P&O) method, e.g. [17], [18], [19]. The basic idea of this method is to perturb the PV voltage and observe the change in the power output. The direction of the voltage steps is kept if the power increases and reversed otherwise. Usually the feedback information is the PV power: $P_{pv} = v_{pv} \times i_{pv}$ (MPP). If we aim for a demanded power P_{dem}, feedback is replaced by $P_{pv}^* = -|P_{pv} - P_{dem}|$ (DPP).

C. Synchronous machines modelling

1) Simplified model: The simplified SG model consists of a voltage-behind-impedance model with variable frequency, governed by the swing equation. The model diagram is depicted in Fig. 1, where mechanical speed ω_m and the internal voltage E_s are calculated by

$$2H \frac{d\omega_m}{dt} = P_m - P_e \qquad (3)$$

$$E_s = \frac{1}{1 + \tau_f s} V_f \qquad (4)$$

where H is the SG inertia constant, P_m is the mechanical power, P_e is the electrical power, τ_f is the field circuit time constant in p.u., and V_f is the field voltage in p.u.

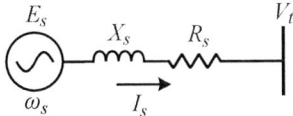

Fig. 1. Simplified SG model single-line diagram.

The first-order transfer function with time constant τ_f links the SG internal voltage to the field voltage. Adding τ_f to the simplified model allows using the same excitation system used in detailed SG models. τ_f can be calculated from the SG's field resistance and inductance or numerically by fitting a step response in the field circuit of the complete model.

2) IEEE Model 2.2: The IEEE Model 2.2 [20] is a precise yet simple electrical model in the dq axis. This model takes into account the dynamics of the stator, field, and damper windings. One of the benefits is that standard data supplied by manufacturers is usually based on the inherited parameters [20]. The equivalent circuit is represented in the rotor reference frame, depicted in Fig. 2, where the voltages are calculated as

$$v_d = \frac{\mathrm{d}\psi_d}{\mathrm{d}t} - i_d R_s - \omega_s \psi_q \qquad (5a)$$

$$v_q = \frac{\mathrm{d}\psi_q}{\mathrm{d}t} - i_q R_s - \omega_s \psi_d \qquad (5b)$$

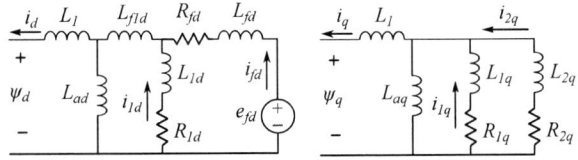

Fig. 2. IEEE Model 2.2 single-line diagram.

3) IEEE 2.2 Model with saturation: According to [20], the SG saturation significantly affects both rotor angle and excitation currents and thus should be considered in transient stability studies. In this study, the saturation was modelled as factor k added to the excitation, following the description and parameters of [21], [22].

D. Transmission lines modelling

Transmission lines play an important role in power system simulation, either by influencing steady-state power flow or by adding additional dynamics to the system. Three models were implemented in this study, described next.

1) PI model: The PI model is widely used in several power system studies due to its simplicity and suitability to model power flows and electromechanical transients. However, the PI model is only precise for a limited frequency range [23].

2) Bergeron model: The Bergeron model is also calculated for a single frequency, but it is more accurate than PI model as it uses distributed parameters and considers the travelling waves trough the lines. In this model there is no direct connection between the two line terminals. Voltages and currents at one end affect indirectly the other end after a time delay due to the travelling time.

3) Frequency-dependent model: The frequency-dependent model precisely represents the transmission line or cable throughout the whole frequency range of interest. The line resistances, inductances, capacitances and conductances are calculated from the physical line geometry and are frequency dependent. The travelling waves are accurately represented in this model and the travelling wave speed is also frequency-dependent, making it suitable to model electromagnetic transients [23].

E. Converter modelling

Several models have been proposed for the conventional two-level VSC, varying from very detailed models to high-level RMS models [6], [24]. Two models were compared in this study, described next.

1) EMT average model: One widely-used EMT model of VSCs is the average (AVG) model. The AVG model neglects the converter's high-frequency switching and considers that the VSC electrical model on the AC side is simply a controlled voltage source, defined as an amplification of the modulation index [25].

2) Phasor model: Phasor models aim to capture only the slow dynamics of the power grids, such as the electromechanical transients, with time constants generally bigger than 100 ms. In this model, the grid differential equations are substituted by algebraic equations. As the phasors are assumed to be rotating at nominal angular speed, voltages and currents have their dynamics around 0 Hz instead of 50 or 60 Hz. This allows to dramatically increase the simulation time step and consequentially the simulation speed.

In phasor simulation, the VSCs are represented as current sources with magnitude and angle defined by the control system. As phasor simulation aims mainly to speed up and simplify simulations, several approximations can be performed in the VSC control system to allow bigger simulation time steps. In this study, the reference output current given by the outer loop is directly sent to the current source, thus the output current-loop dynamics are neglected.

III. METHODOLOGY

To assess the influence of the aforementioned models, several studies were performed using an adapted Cigre European HV transmission network benchmark system [26], modelled in Simulink. The system is composed of four synchronous generators, eight transmission lines and one VSC, and represents a generic equivalent transmission system.

Table I summarizes the models simulated in each test.

1) Setpoint tracking: In this test, active and reactive power setpoints of the RES were set to 100 MW and 30 Mvar at $t = 1$ s and $t = 1$ s, respectively.

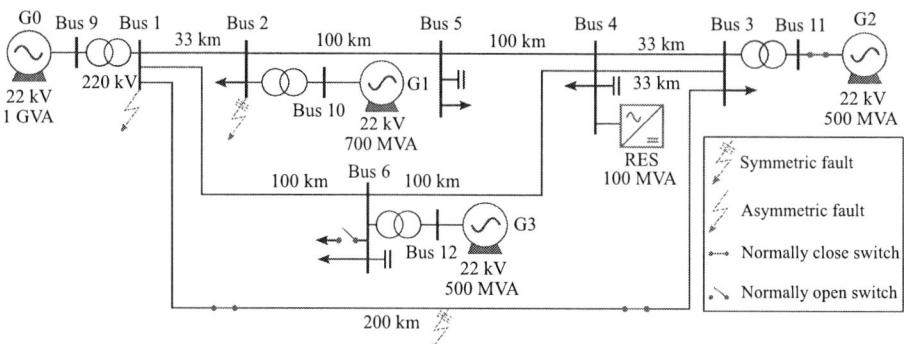

Fig. 3. Simulated system single-line diagram. Modified from [26].

2) Load connection: In this test, a 100 MW, 20 Mvar load was connected to the bus 6 at $t = 5\,s$, dropping the system frequency and voltage.

3) Symmetrical faults: In this test, a 5 Ω three-phase fault was applied to bus 2 at $t = 5\,s$, lasting for 200 ms.

4) Asymmetrical faults: In this test, a 10 Ω mono-phase fault at phase B was applied to bus 1 at $t = 5\,s$, lasting for 500 ms.

5) Loss of generation: In this test, the generator G2 is disconnected from the system at $t = 5\,s$, producing a slow but large transient in the system.

6) Line outage: In this test, a permanent 1 Ω three-phase fault was applied to the line connecting bus 1 to bus 3. The line was isolated by two ideal circuit breakers 100 ms after the fault.

All tests were simulated using a fixed time step and the Euler method (*ode1* in Simulink). The single-line diagram of the simulated system is depicted in Fig. 3.

TABLE I
SIMULATED MODELS

Component	Model
Synchronous generator	Model 2.2 [20] with saturation **model 2.2 [20] without saturation** inertia-only model
Transmission line	Frequency-dependent (FD) model Bergeron model calculated at 50 Hz, **nominal PI model calculated at 50 Hz**
Converter	**EMT average model**, Phasor model
Renewable generation	Dynamic PV power plant, dynamic wind turbine, algebraic power laws, **ideal DC voltage source**

The tests were performed as follows. First, one model of each element (SG, transmission lines, converter, RES) was chosen to form a base group, which are indicated in bold in Table I. Afterwards, each test was performed varying one element per time in relation to the base group. This allowed to identify the influence of each model in the overall system behaviour.

Three key aspects were analysed for each model: Precision, execution time and simplicity. Finally, the key aspects were combined to express the suitability of each model for the evaluated scenarios.

IV. RESULTS AND DISCUSSION

This section presents the simulation results, followed by discussions. Due to the extensive number of tests, only a few cases and variables are presented, which are representative of each test.

A. Influence of synchronous generator model

Figures 4-9 present active power and speed for tests 1 and 6, respectively. From Figs. 4-9 it can be observed that SGs' model 2.2 with and without saturation show a similar response. However, the system was simulated with light load, which can present optimistic results. When heavy load is simulated with contingencies, saturation should be considered. The inertia-only model, although requiring less parameters to simulate, presented more severe voltage oscillations due to the lack of damper windings. This effect is more noticeable during and post fault, as shown in Fig. 7.

The SGs' electric model slightly influenced the system's total simulation time. Model 2.2 without saturation was 12 % faster than the inertia-only model and 18.4% faster than the model 2.2 with saturation.

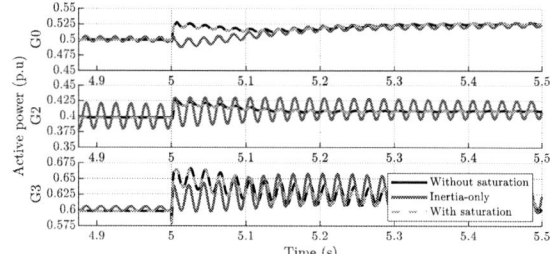

Fig. 4. SGs active power during a load connection for each SG model.

Fig. 5. SGs speed during a load connection for each SG model.

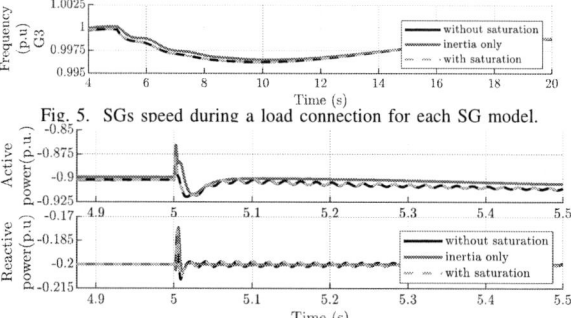

Fig. 6. RES power during a load connection for each SG model.

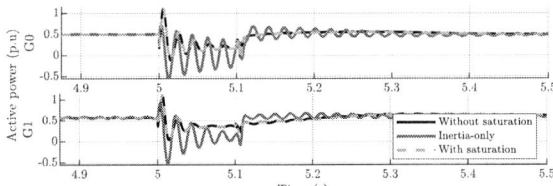

Fig. 7. SGs active power during a line outage for each SG model.

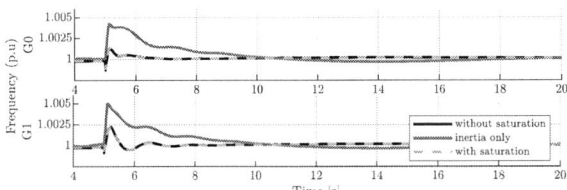

Fig. 8. SGs speed during a line outage for each SG model.

B. Influence of the transmission lines model

Figures 10-12 and 13-15 show the results for tests 2 and 3 (load connection and symmetric fault), respectively.

In can be observed in both tests that the FD model represents higher frequencies due to wave propagation and resonance frequencies. These high-frequency components are also present though less represented in the Bergeron model and almost absent in PI nominal. However, the same low-frequency tendency is well represented in the three models. The fastest simulation was performed with the PI model, followed by the Bergeron model 16% slower on average than the the simulation with PI model. The FD model resulted in simulations 2113.6% slower than the PI model on average.

Therefore, the PI model would be adequate if essentially electromechanical transients are simulated. Using high-order models would add minor precision at the cost of a larger simulation time. Nevertheless, if electromagnetic transients are simulated or if the converter switching or control dynamics are a concern, the PI model might under-represent important high-frequency components, yielding optimistic results. The Bergeron model might be adequate between both extremes, where the nominal frequency dominates the response but the simulated lines' length are too large to neglect travelling wave effects. These guidelines might change for underground cables as their frequency-dependent behaviour is more significant.

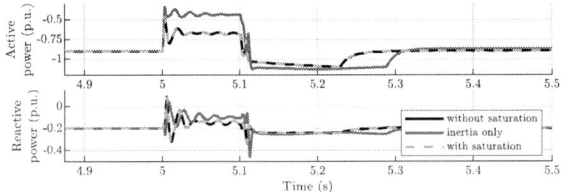

Fig. 9. RES power during a line outage for each SG model.

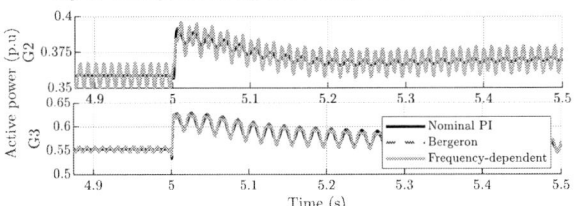

Fig. 10. SGs active power during a load connection for each line model.

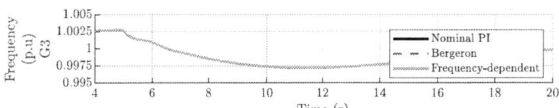

Fig. 11. SGs speed during a load connection for each line model.

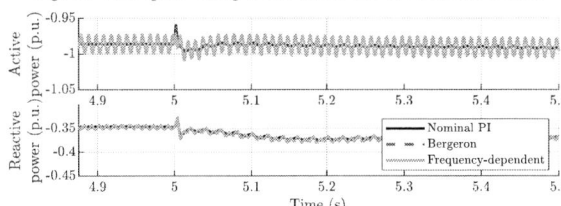

Fig. 12. RES power during a load connection for each line model.

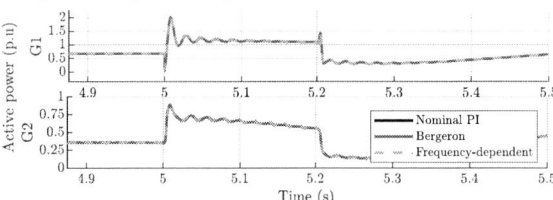

Fig. 13. SGs active power during a symmetric fault for each line model.

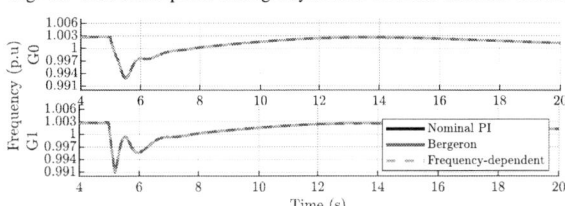

Fig. 14. SGs speed during a symmetric fault for each line model.

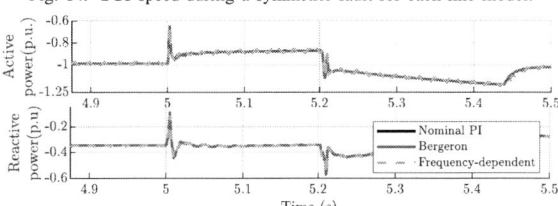

Fig. 15. RES power during a symmetric fault for each line model.

C. Influence of the converter model

Figures 16-22 show the results for tests 1, 4 and 5 (setpoint tracking, asymmetric fault and loss of generation), respectively.

In the setpoint tracking test, the VSC output power was nearly the same in both EMT and Phasor models.

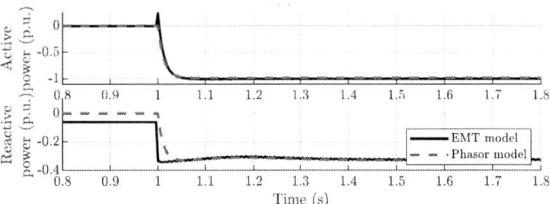

Fig. 16. RES power during a setpoint tracking for each VSC model.

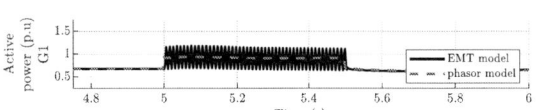

Fig. 17. SG active power during an asymmetric fault for each VSC model.

During the asymmetric fault, SGs in EMT model showed an oscillatory torque produced by the transient, where only the average value is captured by the phasor model (Fig. 17). However, it should be highlighted that this effect is due to the simulation type (EMT or Phasor), and not due to the VSC model. Similar behaviour can be observed in the RES output power (Fig. 19).

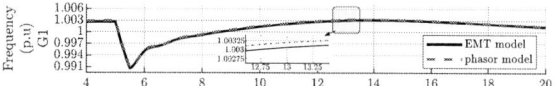

Fig. 18. SG speed during an asymmetric fault for each VSC model.

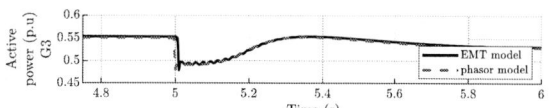

Fig. 19. RES power during an asymmetric fault for each VSC model.

In the generator disconnection test, both EMT and phasor models presented similar responses, despite for an offset in the SGs speed (Fig. 20-22). This slight deviation in the phasor model might be due to model approximations as the stator fluxes derivatives are neglected.

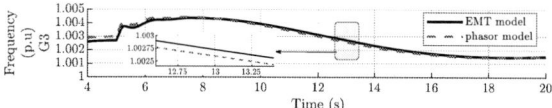

Fig. 20. SG active power during a generator disconnection for each VSC model.

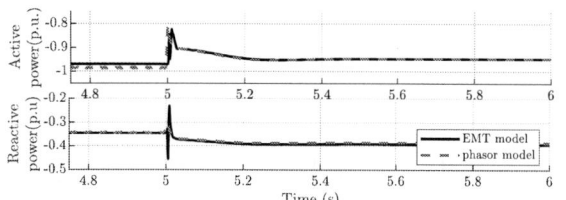

Fig. 21. SG speed during a generator disconnection for each VSC model.

The proper choice on using EMT or Phasor models will depend on the type of study and the system's size. Studies on small systems, focused on fast dynamics such as modulation and control studies, dynamic analysis of the PLL or detailed short circuit studies might require EMT simulation. On the other hand, studies on large systems, where electromechanical variables are being analysed, can be precisely simulated using phasor models in a fraction of the EMT simulation time. In the tests performed, the simulation with phasor models was 1277% faster in average when compared to simulations with EMT models.

Fig. 22. RES power during a generator disconnection for each VSC model.

D. Influence of RES

Figures 23-28 show the results for tests 3 and 4 (symmetric and asymmetric faults), respectively. By analysing the results, it can be observed that the RES model had minor influence on the SGs variables during short-circuit tests. This was observed because the RES dynamics are slower and consequently do not produce significant deviations during the short period of the fault.

In the VSC output power, a noticeable difference between the RES models could be observed.

But in the VSC power, the difference is noticeable, especially post-fault.

The fastest simulation was performed using the ideal DC source model, followed by static PV, which was 72,5% slower, followed by the static wind, which was 85,39 % slower than the ideal DC source. Finally, the slowest models were the dynamic PV (106.1% slower than the ideal DC source) and the dynamic wind (164.82% slower than the ideal DC source).

Therefore for studies in systems with low-share of RES and dominated by SGs, the ideal DC source would provide adequate accuracy. Conversely, for studies in systems with high-share of RES, static and dynamic models should be used. Static models are good options as they are faster to simulate than dynamic models and provide a similar response to those in several events.

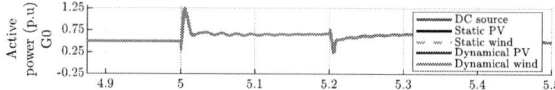

Fig. 23. SG active power during a symmetric fault for each RES model.

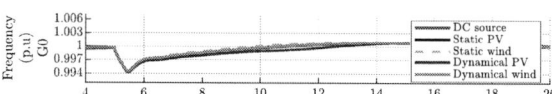

Fig. 24. SG speed during a symmetric fault each RES model.

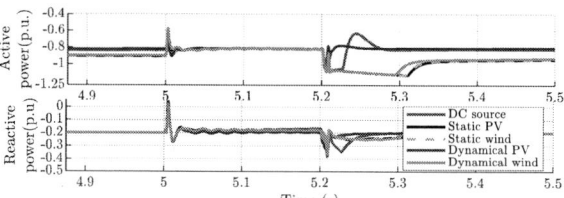

Fig. 25. RES power during a symmetric fault for each RES model.

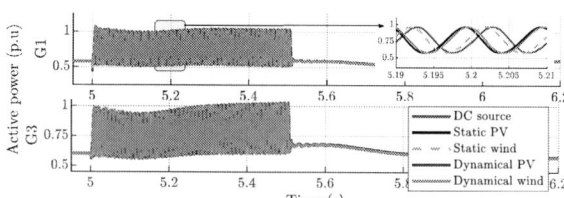

Fig. 26. SGs active power during an asymmetric fault for each RES model.

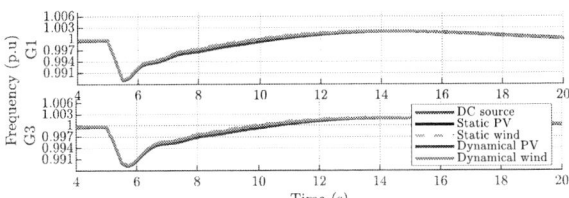

Fig. 27. SGs speed during an asymmetric fault for each RES model.

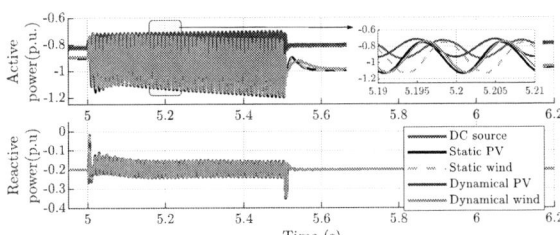

Fig. 28. RES power during an asymmetric fault for each RES model.

V. Conclusion

This paper presented a comparative analysis amongst several models currently used to simulate power systems with RES in transient stability studies. Several models of SGs, transmission lines, converters and RES were compared in terms of precision and computational time. The simulated benchmark system was generic and could well represent typical transmission systems.

From the tests performed, it could be observed that the most suitable models for the study of an electromechanical events in the system tested were: the Model 2.2 [20] without saturation, nominal PI model, phasor model and ideal DC voltage source. These models were able to track the fundamental components of the variables been simulated and at the same time have less computational burden. However, if the system or the study purpose changes, the best models will change. For example, a system with high load might require modelling SG saturation to achieve a precise response. Moreover, If the converter switching is modelled, a frequency-dependent transmission line model would be needed to represent precisely the high-frequency components associated with the switching. The EMT model represents fast oscillations needed for fast dynamic studies. In this study, the SGs dominate the system presented. However, if the share of RES increases, the RES model dynamics will have a more significant influence on the electromechanical events and thus dynamic RES models should be used.

References

[1] European Comission, "Energy roadmap 2050," Luxembourg: Publications Office of the European Union, Tech. Rep., 2012.

[2] European Council, "Directive 2009/28/EC of the European Parliament and of the Council of 23 April 2009 on the promotion of the use of energy from renewable sources," Brussels, Belgium, Tech. Rep., 2009.

[3] B. Marinescu, O. Gomis-Bellmunt, F. Dörfler, H. Schulte, and L. Sigrist, "Dynamic virtual power plant: A new concept for grid integration of renewable energy sources," *arXiv:2108.00153*, 2021.

[4] G. D. Carne, M. Liserre, M. Langwasser, M. Ndreko, R. Bachmann, R. W. D. Doncker, R. Dimitrovski, B. J. Mortimer, A. Neufeld, and F. Rojas, "Which deepness class is suited for modeling power electronics?: A guide for choosing the right model for grid-integration studies," *IEEE Industrial Electronics Magazine*, vol. 13, no. 2, pp. 41–55, jun 2019.

[5] M. Paolone, T. Gaunt, X. Guillaud, M. Liserre, S. Meliopoulos, A. Monti, T. V. Cutsem, V. Vittal, and C. Vournas, "Fundamentals of power systems modelling in the presence of converter-interfaced generation," *Electric Power Systems Research*, vol. 189, p. 106811, dec 2020.

[6] CIGRE Working Group B4-57, *Technical Brochure 604: Guide for the Development of Models for HVDC Converters in a HVDC Grid.* CIGRE, 2014.

[7] F. Pöschke, E. Gauterin, M. Kühn, J. Fortmann, and H. Schulte, "Load mitigation and power tracking capability for wind turbines using linear matrix inequality-based control design," *Wind Energy*, 2020, dOI: 10.1002/we.2516.

[8] J. Bjork, D. V. Pombo, and K. H. Johansson, "Variable-speed wind turbine control designed for coordinated fast frequency reserves," *IEEE Transactions on Power Systems*, p. 1–1, 2021. [Online]. Available: http://dx.doi.org/10.1109/TPWRS.2021.3104905

[9] F. Pöschke and H. Schulte, "Evaluation of different apc operating strategies considering turbine loading and power dynamics for grid support," *Wind Energy Science Discussions*, vol. 2021, pp. 1–14, 2021. [Online]. Available: https://wes.copernicus.org/preprints/wes-2021-80/

[10] M. Hansen, *Aerodynamics of Wind Turbines.* Routledge, 2015.

[11] J. M. Jonkman, S. Butterfield, W. Musial, and G. Scott, "Definition of a 5-MW Reference Wind Turbine for Offshore System Development," National Renewable Energy Laboratory, Tech. Rep., 2009.

[12] J. M. Jonkman and M. L. Buhl, "FAST Users Guide," National Renewable Energy Laboratory, Tech. Rep., 2005.

[13] W. Xiao, *Photovoltaic Power System: Modeling, Design, and Control.* Wiley, 2017.

[14] W. Shockley, "The theory of p-n junctions in semiconductors and p-n junction transistors," *The Bell System Technical Journal*, vol. 28, no. 3, pp. 435–489, 1949.

[15] R. W. Erickson, "DC–DC Power Converters," in *Wiley Encyclopedia of Electrical and Electronics Engineering.* American Cancer Society, 2007.

[16] R. W. Erickson and D. Maksimovic, *Fundamentals of Power Electronics*, 2nd ed. Springer US, 2001.

[17] W. Teulings, J. Marpinard, A. Capel, and D. O'Sullivan, "A new maximum power point tracking system," in *Proceedings of IEEE Power Electronics Specialist Conference - PESC '93*, Jun. 1993, pp. 833–838.

[18] Y. Kim, H. Jo, and D. Kim, "A new peak power tracker for cost-effective photovoltaic power system," in *IECEC 96. Proceedings of the 31st Intersociety Energy Conversion Engineering Conference*, vol. 3, Aug. 1996, pp. 1673–1678 vol.3, iSSN: 1089-3547.

[19] N. Femia, G. Petrone, G. Spagnuolo, and M. Vitelli, "Optimization of perturb and observe maximum power point tracking method," *IEEE Transactions on Power Electronics*, vol. 20, no. 4, pp. 963–973, Jul. 2005.

[20] IEEE, *Std 1110-2019 - IEEE Guide for Synchronous Generator Modeling Practices and Parameter Verification with Applications in Power System Stability Analyses.*

[21] T. V. Cutsem and L. Papangelis, "Description, Modeling and Simulation Results of a Test System for Voltage Stability Analysis," University of Liège, Belgium, Tech. Rep., 2013.

[22] T. Van Cutsem, M. Glavic, W. Rosehart, C. Canizares, M. Kanatas, L. Lima, F. Milano, L. Papangelis, R. A. Ramos, J. A. d. Santos, B. Tamimi, G. Taranto, and C. Vournas, "Test systems for voltage stability studies," *IEEE Transactions on Power Systems*, vol. 35, no. 5, pp. 4078–4087, 2020.

[23] N. Watson and J. Arrillaga, *Power Systems Electromagnetic Transients Simulation, 2nd ed.* Institution of Engineering and Technology, 2018.

[24] S. Khan and E. Tedeschi, "Modeling of MMC for fast and accurate simulation of electromagnetic transients: A review," *Energies*, vol. 10, no. 8, p. 1161, aug 2017.

[25] W. Lu and B.-T. Ooi, "Optimal acquisition and aggregation of offshore wind power by multiterminal voltage-source HVDC," *IEEE Transactions on Power Delivery*, vol. 18, no. 1, pp. 201–206, jan 2003.

[26] CIGRE Task Force C6-04, *Technical Brochure 575: Benchmark Systems for Network Integration of Renewable and Distributed Energy Resources.* CIGRE, 2014.

Efficiency and Lifetime Analysis of Several Airborne Wind Energy Electrical Drive Concepts

Bakr Bagaber, Daniel Heide, Bernd Ponick and Axel Mertens
LEIBNIZ UNIVERSITY HANNOVER
Institute for Drive Systems and Power Electronics
Hannover, Germany
Phone: +49 (0) 511-762-3766
Email: bakr.bagaber@ial.uni-hannover.de
URL: www.ial.uni-hannover.de

Acknowledgments

This work was supported by the German Ministry of Economics and Technology (BMWi) – 0324217D.

Keywords

≪Wind-generator systems≫, ≪Voltage Source Converter (VSC)≫, ≪Parallel operation≫, ≪Thermal cycling≫, ≪Lifetime≫, ≪Permanent Magnet Synchronous Generator≫

Abstract

Several electrical drive concepts and control strategies for airborne wind energy systems are compared in this work. The results suggest that the proposed modified kite trajectory control and the parallel drive concept can indeed reduce the required silicon chip area and prolong the converter's lifetime without significantly impacting the overall system efficiency. The compromise is however, a higher system complexity and larger torque ripples which could impact the noise profile of the electrical machine. The investigation also reveals that the power converter size is influenced by the installation location and the associated wind class. Where installations around land agricultural areas have the biggest impact on the drive train because of the high wind turbulence.

Introduction

Airborne wind energy systems (AWES) are a new class of wind generators promising to harness wind at high altitudes above 200 meters in a cost-effective way [2]. Several concepts of AWES are under research, among which the pumping cycle (PC) type has already reached a market commercialization stage [3]. The system consists of a flying soft kite connected to a ground-based electrical machine by means of a strong tether; a control pod attached to the kite is used for steering. The principle of operation can be understood with the help of Fig. 1a. The cycle is initialized by positioning the kite at a suitable altitude against the wind direction in phase one. In phase two, reel-out (generation) starts by maneuvering the kite in a crosswind direction at a certain wind window angle (ϑ). The crosswind principle exploits the lift-to-drag ratio (E) to induce high apparent wind speed -and lift force- which maximizes the extraction of power [4]. Once the tether is almost entirely winched out, the third phase starts (transition phase). The kite is steered towards the zenith position ($\vartheta = 90$). During this transfer phase, the machine decelerates rapidly towards zero, while the torque remains high. Eventually, the fourth phase (reel-in) begins. The electric machine accelerates as a motor to pull the kite into lower elevations. Reel-in is usually accomplished under maximum speed at very wide angles ($\vartheta > 90$) to reduce the tether force.

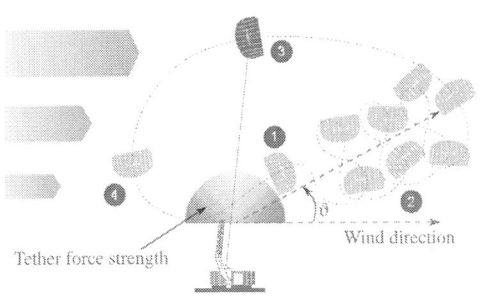

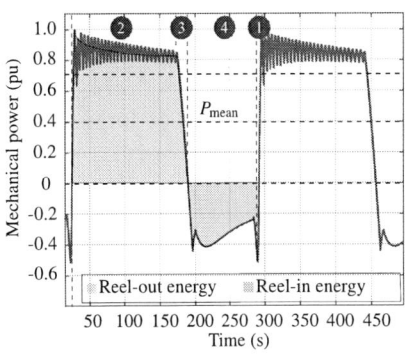

(a) Complete pumping cycle.

(b) Typical power profile.

Fig. 1: principle of operation of a typical PC-AWES: (1) Start/restart, (2) Reel-out (generation), (3) Transfer (generation), (4) Reel-in (consumption). Modified from [1].

The resulting variable power profile is shown in Fig. 1b. The impact of this unusual load-cycle on the dimensioning and lifetime of a standard machine drive comprising a three-phase voltage source converter (VSC) driving a permanent magnet synchronous machine (PMSM) was investigated in [5]. It was concluded that the thermal cycles because of the reel-out/reel-in duality, as well as the thermal cycles due to speed reversal (low frequency at maximum torque), produce large stress on the free-wheeling diodes baseplate to case solder joints. Therefore, the converter needs to be significantly over-scaled to reach the target lifetime of 20 years. This would increase the capital cost and the switching losses of the converter.

To mitigate this problem, an alternative drive concept based on the parallel connection of a VSC and a passive diode rectifier as depicted in Fig. 2 was proposed in [6]. This concept is capable of decoupling the influence of the reel-out/reel-in thermal cycles, which should allow for a cost-effective and efficient over dimensioning of the passive diode rectifier. It was also found that the size of the VSC can be further reduced through an adjusted kite control during the transfer phase to allow for speed reversal at lower torques. Reducing the switching frequency at lower operational speeds can further contribute to a reduction of the required VSC size [6].

The goal of this work is to first extend the comparison between the two drive concepts in terms of their lifetime and efficiency, including the DC-DC converter under ideal and turbulent wind conditions. Sec-

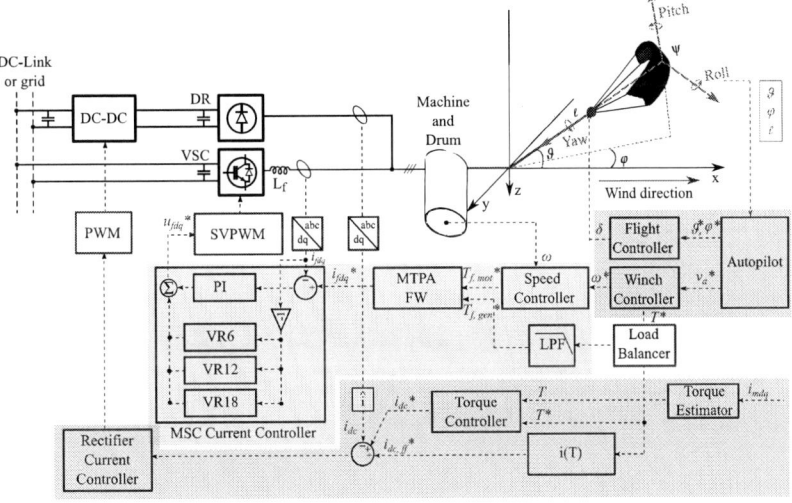

Fig. 2: Overview of the system model and control.

ond, to evaluate the impact of the parallel concept on the electrical machine. The investigation is based on the simulation model of Fig. 2, which is used for estimating the thermal performance of the power electronic converters. A 2D FEM model is used for designing the PMSM and analyzing its mechanical, electrical, and thermal performance. Finally, analytical models are used for sizing the converters to meet the lifetime goal of 20 years.

AWES Drive Concepts

In this work, two drive concepts are compared. The first is a standard machine drive consisting of a PMSM and a VSC. The VSC is controlled using maximum torque per ampere (MTPA) and field-weakening (FW). The second concept depicted in Fig. 2 was first introduced in [6]. It comprises a PMSM driven by a VSC paralleled with a diode rectifier (DR) and a DC-DC converter.

The control principle of the parallel drive concept can be explained as follows, during reel-out, the system operates under torque control mode, in which the kite controller adjusts the winching speed. The diode rectifier controls the machine torque using a unidirectional DC-DC converter, and the VSC acts as a current sink through which a small part of the machine current deviates according to the command of the load balancer. This control concept allows the cheaper, more robust diode rectifier to carry most of the reel-out power using a near unity power factor. Control of the VSC current is only possible through the use of the series decoupling inductor L_f, which allows the converter to operate as a controlled current source/sink. It also facilitates the selective damping of the odd non-triple harmonics (5th, 7th, 11th, 13th, ...) generated by the diode rectifier in the VSC branch using several vector-resonant (VR) controllers. A depiction of the system currents with and without the VR controllers is availabe in Fig. 3. During reel-in, the VSC operates in speed control mode using standard MTPA and FW techniques; the rectifier branch is turned off, and the kite controller adjusts the system torque [6].

System modeling

Pumping Cycle AWES Model

The dynamic model of the PC-AWES can be explained with the help of Fig. 2. The kite position can be described by means of three state variables in the polar coordinate system, the wind window angle ϑ, the

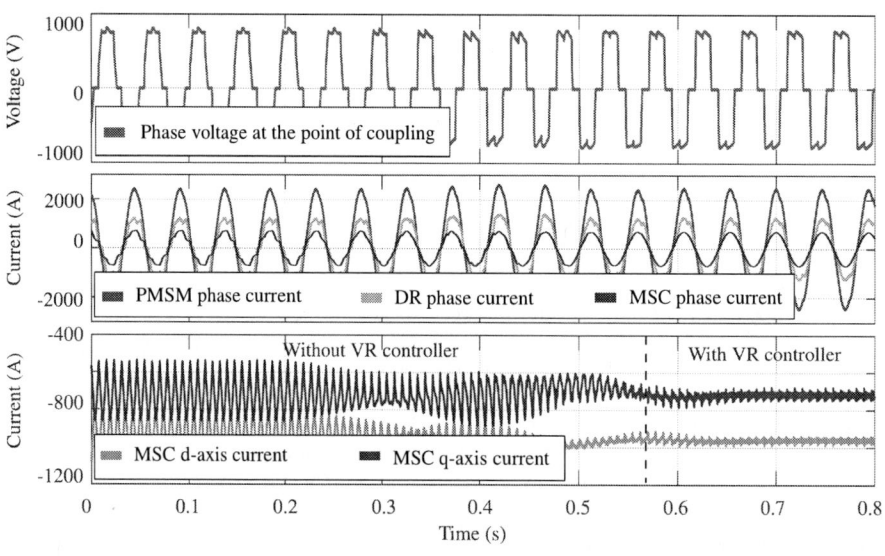

Fig. 3: Phase voltage and currents of the parallel drive PMSM, VSC and diode rectifier.

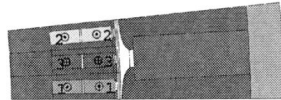

Fig. 4: Cross-section of one of the machine poles.

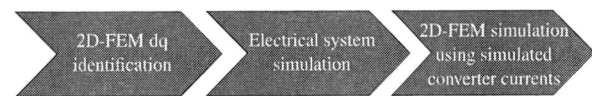

Fig. 5: Machine simulation process.

azimuth angle φ, and the tether length l [7, 4], whereas the orientation of the kite around the roll-axis is defined by a fourth state variable ψ.

Assuming a mass-less kite and the tether in a state of aerodynamic equilibrium with a homogeneous wind field along the x-axis, the dynamic and kinematic equations of motion can be simplified into a first-order system of equations

$$\dot\psi = g_k v_a \delta + \dot\varphi \cos\vartheta \tag{1}$$

$$\dot\vartheta = \frac{v_w}{l}(E\cos\vartheta\cos\psi - \sin\vartheta) - \frac{\dot l}{l}E\cos\psi \tag{2}$$

$$\dot\varphi = -\frac{v_w E\cos\vartheta - \dot l E}{l\sin\vartheta}\sin\psi \tag{3}$$

$$\dot l = v_{winch}, \tag{4}$$

where v_w is the wind speed, l is tether lengths, v_a is the apparent wind speed of the kite and E is the glide ratio.

The control system of the AWES comprises a central autopilot, which with the help of a flight controller defines the optimal flight trajectory of the kite. The steering dynamics are simplified in Equation (1) by the turn-rate law (TRL) $g_k v_a \delta$, where g_k is an empirical system parameter that quantifies the maneuvering response of the kite, and δ is a non-dimensional control input to the pod controller. A winch controller is used to calculate the optimal winching speed for maximum energy yield according to the method described in [4]. An overview of the kite system parameters is available in Table I.

PMSM Model

Two models for the PMSM are employed. The first is an analytical voltage behind reactance model used in the simulation environment of Fig. 2. The second is a 2D FEM model used for designing the machine and estimating its losses. The chosen electrical machine is a low-speed, high-torque 64-pole ferrite permanent magnet synchronous generator with an external rotor. The generator is rated for 690 V line-to-line RMS voltage and has a three-phase stator winding. The parameters of the electrical machine are listed in Table I, the cross-section of one pole of the machine 2D FEM model is depicted in Fig. 4.

Analysis and parametrization of the electrical machine can be understood with the help of Fig. 5. Starting with the FEM identification of the dq-model parameters with ideal current supply, the machine parameters can be identified. This data is then fed into the simulation model of Fig. 2 where the electrical current harmonics are calculated for each converter topology. These currents form the basis for the last step, where another FEM simulation is run to accurately determine the flux distribution, forces, and iron losses for each drive concept. The simulations are based on angular step-wise magneto-static FEM calculations with 192 angular steps per electrical period using FEMAG FEM software. Reference temperatures for all simulations are 120°C for the winding and 40°C for the magnets, respectively.

The losses are calculated and considered in a post-processing routine. For this purpose, the extended Jordan model with the loss density for each frequency component of the flux distribution is applied

Table I: System Parameters

Component	Parameter	Symbol	Value	Unit
Kite	Reel-out power	P_{mech}	1.8	MW
	Projected area	A	300	m^2
	Aerodynamic coef.	C_R	1	-
	Glide ratio	E	5	-
PMSM	Rated power	P_N	1.75	MW
	Rated volrage	U_N	690	V
	Rated torque	T_N	220	kNm
	Maximum speed	N_{max}	200	min^{-1}
	Rated power factor	$cos(\Phi_N)$	0.9	-
	Pair of poles	p	32	-
	d-axis inductance	L_d	0.74	mH
	q-axis inductane	L_q	0.95	mH
	Outer diameter	D_o	2240	mm
	Core length	L_{FE}	1200	mm
VSC/Boost converter	DC-link voltage	U_{DC}	1.2	kV
	Switching frequency	f_{sw}	2	kHz
	Decoupling inductor	L_f	800	μH
	Boost inductor	L_{boost}	277	μH
Reference IGBT/Diode Module	Reference module	-	[8]	-
	Reference heatsink	-	[9]	-
	LPF (IGBT)	f_{cr}	0.8, 8.7, 81.6, 304.2	Hz
	LPF (Diode)	f_{cr}	0.8, 8.4, 55.2, 130.1	Hz

according to

$$p_{fe} = \sum_{v} (c_\mathrm{h} \cdot f_v^{c_{f,h}} + c_\mathrm{w} \cdot f_v^{c_{f,w}}) \cdot B_v^{c_B}, \tag{5}$$

where c_h and $c_{f,h}$ are the loss coefficients for hysteresis losses, c_w and $c_{f,w}$ for eddy-current losses in the core material M600-50A. The eddy current losses in the magnets were calculated based on [10], but were found to be negligible because of the very low ferrite magnet electrical conductivity and hence are not considered further. Finally, the additional winding losses due to the current harmonics are calculated with the help of

$$p_{v,w} = m \cdot R_{1,DC} \cdot \sum_{\mu} k_{r,\mu}(f_\mu) \cdot I_\mu(f_\mu)^2, \tag{6}$$

where $k_{r,\mu}$ is the frequency dependent AC resistance rise factor caused by the current displacement due to the skin and proximity effects.

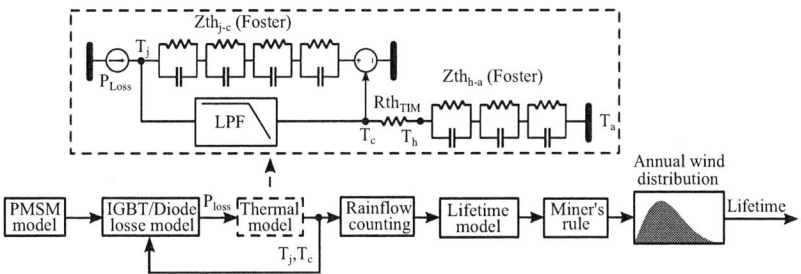

Fig. 6: Overview of the thermal and lifetime model.

Power Electronic Converter Thermal and Lifetime Models

Estimation of the converter lifetime is implemented according to Fig. 6. The converter losses are calculated from

$$P_{cond}(SF) = \left\{ D, 1-D \right\} \cdot U_{on}\left(\frac{I}{SF}, T_j\right) \cdot I \tag{7}$$

$$P_{sw}(SF) = f_{sw} \cdot SF \cdot E_{sw}\left(U_{DC}, \frac{I}{SF}, T_j\right) \tag{8}$$

as a function of the discrete instantaneous duty cycle (D), device current (I), the junction temperature (T_j), and the silicon chip area scaling factor (SF) defined according to

$$SF = \frac{Chip\ Area^{new}}{Chip\ Area^{old}}. \tag{9}$$

The scaling assumes a linear relationship between the chip area and the power losses as explained in [11]. The thermal model of each discrete device (IGBT or diode) is modeled in the frequency domain according to the method described in [12], the calculated power module critical frequencies f_{cr} used in the low-pass filter (LPF) are listed in Table I. This method allows for better estimation of the power module case temperature when compared to the standard Foster model. The thermal resistance (R_{th}) is scaled according to

$$R_{th}^{scaled}(SF) = \frac{R_{th}^{base}}{SF}, \tag{10}$$

whereas the thermal time constant is assumed to remain constant regardless of the SF.

Finally, the number of thermal cycles for different wind speeds is calculated using the Rainflow counting algorithm. The empirical model in [13] is used to calculate the lifetime of the devices, while Miner's rule and a Weibull wind distribution at 300 m [14] are used to extend the calculation for one year. More details about the investigation algorithm can be found in [5].

In this work, the investigations in [5, 6] are further extended by examining the impact of wind turbulence on the lifetime of electrical system. Three different wind classes corresponding to offshore (c_0 case), open land (c_1 case), and agricultural land (c_2 case) [15] are considered.

Selection of Power Converter Topology

The parallel drive requires a DC-DC stage to serve the purpose of controlling the current of the passive diode rectifier, as well as to boost the unregulated rectifier voltage (U_{DR}) to the DC-Link voltage level ($U_{DC-Link}$), especially at lower operational speeds [16]. Two different DC-DC converter topologies are compared in this work. The first is the standard boost converter shown in Fig. 7a, which, when combined

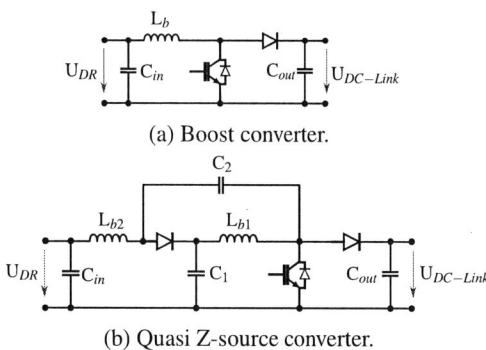

(a) Boost converter.

(b) Quasi Z-source converter.

Fig. 7: DC-DC converter topologies.

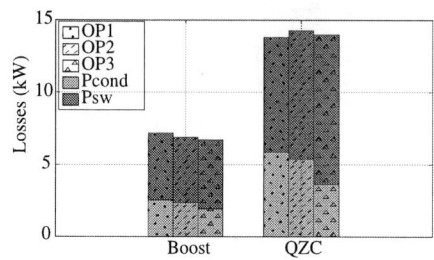

Fig. 8: DC-DC converter losses for different operating points. The machine speed in OP1, OP2 and OP3 is 0.1 pu, 0.5 pu, and 1 pu of the rated value respectively; the torque in all OPs is equal to the rated torque.

with the DR, provides a reliable and low cost topology that has been successfully commercialized in classical wind applications ranging from few kilowatts to several Megawatts [16].

The second considered topology is the quasi Z-source converter (QZC) shown in Fig. 7b. It is a member of the impedance network converter family which are characterized by their excellent voltage boost capability, reduced-size energy link, and higher reliability. The standard Z-source topology has been investigated for wind applications in [17]. The quasi variation is an improved version of the topology that exhibits lower stress on the components, a smaller sized energy link, and eliminated inrush current problem [18].

The topologies are compared in terms of the size and losses of the active components for three operating points. The dimesnioning results suggest that, in order to keep the device temperatures within the rated range, the QZC uses 50% more silicon chip area than the boost converter. A comparison of the losses is also depicted in Fig. 8. The results indicate that, regardless of the operating point, the boost converter produces roughly 50% fewer losses than the QZC while requiring 50% less silicon chip area. Therefore, the boost topology is considered a more viable option and has been used in later investigations.

Power Converter Lifetime Sizing and Efficiency

The dimensioning of the power electronic converters for both systems is conducted based on the target lifetime of 20 years, according to Fig. 6. The dimensioning is carried out for different control algorithms and wind classes. The impact of those variations can be understood with the help of Fig. 9 where speed, torque, and converter diode temperatures are plotted for a variable wind speed condition.

Two distinct kite controls are adopted, the first being a standard kite control (SKC) in which the machine reverses rotation under maximum torque. This causes large temperature spikes during this low-frequency operation as represented by the blue curve of Fig. 9d. The second kite control strategy was first introduced in [6]. This modified kite control strategy (MKC) allows for a slower transfer phase during which the torque drops by 20% to 60% of the rated value depending on the wind condition. The impact on the converter is also depicted by the yellow curve of Fig. 9d where the spikes in the diode junction temperature $T_{j,FWD}$ are clearly reduced. Another improvement to the control system considers the reduction of the switching frequency by 50% during this low-frequency operation. The impact on the VSC diode temperatures is depicted by the red and violet curves of Fig. 9d for the standard power converter system.

However, there remain temperature cycles because of the reel-out/reel-in duality of the system, which, according to [5] has the highest impact on the converter lifetime. The parallel drive concept tries to mitigate this problem by maintaining a relatively equal loading of the VSC during both phases, which reduces the temperature variation ΔT and prolongs the lifetime. This can be clearly seen by the blue curve of Fig. 9e. On the contrary, the rectifier and the boost converter would have to withstand much larger temperature variation ΔT as shown by the red and the yellow curves of Fig. 9e. The VSC power is

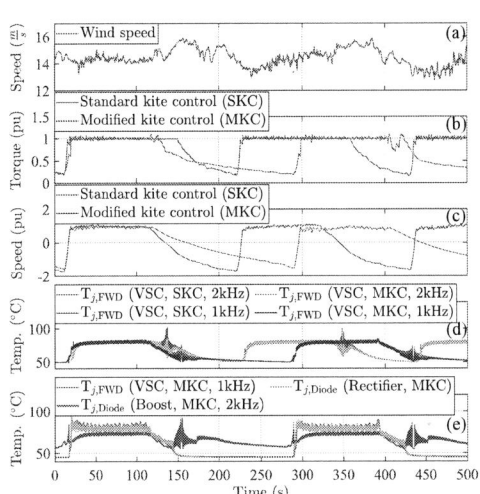

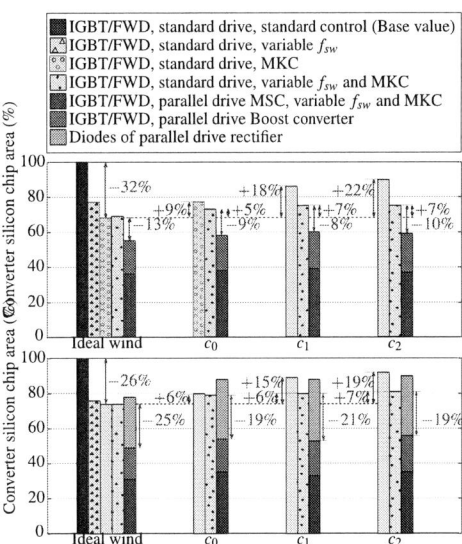

Fig. 9: FWD temperatures for the different control strategies and drive concepts.

Fig. 10: Sizing results of the power converters for a target lifetime of 20 years.

also filtered with a frequency of 0.2 Hz. Therefore, temperature cycles due to wind turbulence are largely decoupled from the VSC as depicted in the same figure.

The impact of these different control strategies on the power converters sizing for a 20 years lifetime goal is shown in Fig. 10. The standard drive with standard control is taken as the base value (shown in black) to which all the other cases are compared. The results are depicted for four different wind conditions. They suggest that, under ideal wind conditions, the standard drive VSC size (shown in yellow) can be reduced by the largest margin when adopting a MKC strategy. This is not the case under turbulent wind conditions, where adopting a combined MKC and a lower switching frequency reduces the converter size the most. Also to be noticed is, depending on the wind class, the standard drive VSC would have to be over-scaled by up to 22% for the IGBT and 19% for the FWD compared to the ideal wind case. This is especially true for land installations in agricultural areas, whereas offshore installations require only an over-scalling of 9% and 6% for the IGBT and FWD respectively.

The impact of the parallel drive on the total converter size can be understood from the red variation bars of Fig. 10. Depending on the wind speed, the IGBT chip area can be further reduced by 8% to 13% compared to the standard drive with optimized control. The combined chip area of the VSC and the boost converter FWD's can be reduced by 19% to 25% respectively. There is, however, an additional diode chip area required for the DR. These silicon chips are usually much cheaper than the fast switching FWDs which should result in an overall cost reduction. This reduction is however compromised by the higher complexity of the system.

A breakdown of the converter losses for the rated operating point is depicted in Fig. 11. The results suggest that the VSC of the parallel drive generates fewer losses; the savings are compromised by the extra losses in the boost converter and the diode rectifier; The difference is however negligible. A better understanding can be deducted from Fig. 12 where the absolute difference in the efficiency of both drive concepts during the generation phase is shown. The difference is negligible for most operating points except for the very low-speed region around the speed reversal point.

Analysis of PMSM Performance and Efficiency

The PMSM performance variation between both drives is investigated for the generator operation only, as the motor operation mode is identical in both converters and thus not considered furthermore. While the current angle can be selected for the standard converter in such a way that optimal efficiencies are

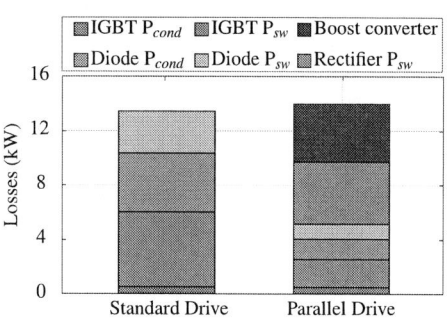

Fig. 11: Nominal losses of the converters.

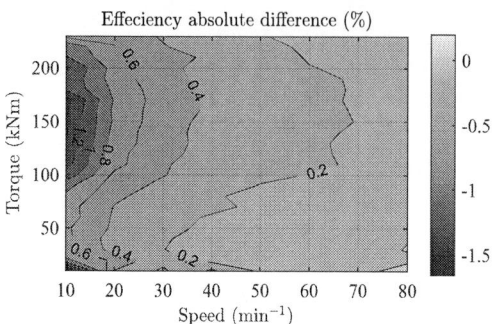

Fig. 12: Power converters efficiency difference. Positive values indicate the standard drive is better.

achieved in each operating point (efficiency optimization), this is not possible for the parallel converter because of the diode rectifier. The parallel converter current angles are therefore larger than in case of the standard converter as depicted in Fig. 13, which increases the field-weakening d-axis current and the resulting phase current for a given mechanical load. On the one hand, this results in higher winding losses as shown in Fig. 14. On the other hand, it reduces the main magnetic flux, which reduces the iron losses. Furthermore, additional winding and iron losses are caused by the current harmonics created by the DR.

An indicator of the current harmonics loss contribution is the total harmonic distortion (THD) depicted in Fig. 15, which describes the relation of the amplitudes of the harmonics to the fundamental amplitude. It should be noted that these harmonics flow only through the DR and not the VSC. This is accomplished by using several vector-resonant controllers [6] as shown in Fig. 2. However, the THD remains below 20% for all generator operating points. The additional winding losses due to the current harmonics are roughly proportional to the square of the THD if the AC resistance rise factor $k_{r(f)}$ caused by the current displacement is neglected. Otherwise, the current displacement increases the winding resistance with growing frequencies.

Because of the large reluctance in the magnetic circuit (main flux saturation and large paths with low permeability through the air gap and magnets), a significant amount of current is needed to build a magnetic flux capable of increasing the machine iron losses. Since the THD of the current is on average around 10% and field weakening current is higher for the parallel converter, the current harmonics almost only lead to an increase in winding losses, while iron losses remain comparable for both drives. Fig. 16 demonstrates that the current harmonics also lead to a considerable increase in the torque oscillations.

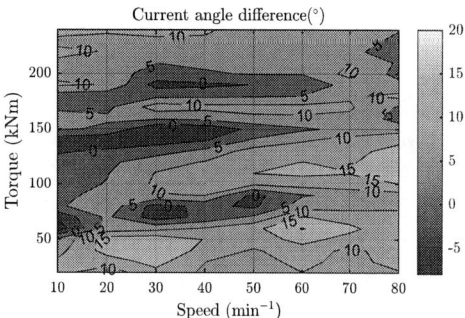

Fig. 13: PMSM current angle difference between standard and parallel drives. Positive values indicate the standard drive is better.

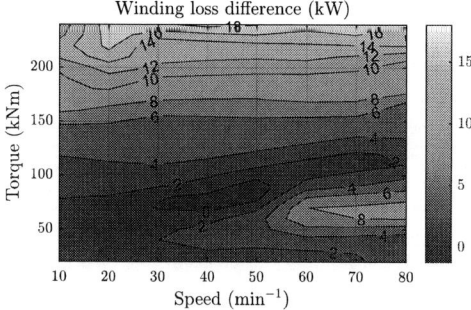

Fig. 14: Winding loss difference between the standard and parallel drives. Positive values indicate the standard drive is better.

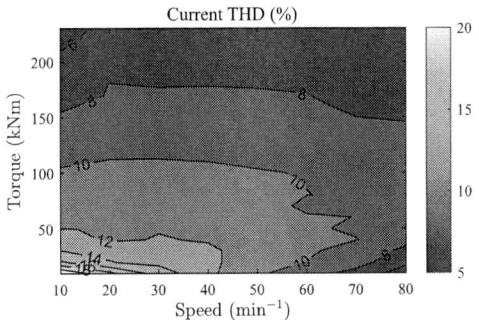

Fig. 15: Machine current THD for the parallel drive.

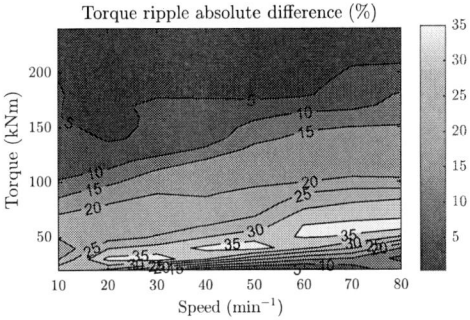

Fig. 16: Torque ripple difference. Positive values indicate the standard drive is better.

The magnetic noise emission of the machine is therefore likely to increase.

The PMSM efficiency for the standard drive stands out to be highly efficient at higher generative speeds and less efficient for low-speed high-torque operation points, as needed in the transfer phase. This effect is even worse when the machine is fed by the DR. In Fig. 17, the efficiency of the parallel converter fed generator is compared with the efficiency of the standard converter fed generator. On average, the efficiency of the parallel converter fed generator is only around 1% lower. At low speeds with high loads, the decrease in efficiency is higher. However, the modified kite control strategy reduces the impact of the low transfer phase efficiency by flying the kite at lower torques, thus avoiding the low-efficiency region of the machine.

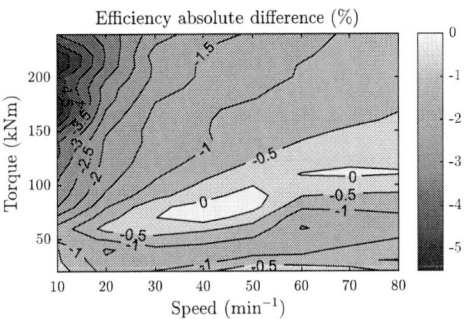

Fig. 17: Generator efficiency. Negative values indicate the standard drive is better.

Conclusion

In this work, two power drive systems and two control strategies were analyzed and compared in terms of their impact on the power electronic converter size, the machine performance, and the overall system efficiency. We concluded that adjusting the transfer phase kite control to allow for a lower torque during speed reversal could reduce the required converter size by around one-third.

Using a novel parallel power converter concept comprising a voltage source converter paralleled with a diode rectifier was investigated for decoupling the effect of reel-out/reel-in thermal cycles. It was found that the IGBT chip area can be reduced by an additional 13% compared to the standard drive concept with adjusted control, while the required fast switching diode chip area is expected to decrease by roughly 25%. However, an additional 30% of slow cheap diodes are required for the passive diode rectifier. The total cost should however remain lower since the significant saving of the expensive IGBTs and FWD's should outweight the slight increase in the cheap rectifier cost. The impact of the parallel drive on the power converter efficiency was found to be negligible.

The impact of the installation location for different wind classes was also looked at. It was concluded that onshore installations could increase the required converter size anywhere between 5% to 22% depending on the control strategy and the roughness of the surrounding terrains. On the contrary, offshore installations would require only a marginal over-sizing of the power converters.

The impact of the parallel converter concept on the PMSM was also analyzed. The results indicate that the machine efficiency decreases by less than 1% around the rated operating range. This value increases to 5% at very-slow high-torque regions required during the transfer phase. This operating range can however be avoided by opting for the modified kite control strategy. The impact of the diode rectifier current harmonics on the machine was also investigated. It was concluded that the PMSM current has a THD of 10% on average, which gives rise to torque ripples and could increase the machine acoustic noise emissions.

References

[1] S. P. GmbH, "SkySails Power GmbH Image Brochure." https://skysails-group.com/downloads/.

[2] M. Diehl, U. Ahren, and R. Schmehl, *Airborne Wind Energy*. Springer, Berlin, Heidelberg, 2013.

[3] S. P. GmbH, "Kite Power For Mauritius." https://skysails-power.com/kite-power-for-mauritius/.

[4] M. Erhard and H. Strauch, "Flight control of tethered kites in autonomous pumping cycles for airborne wind energy," *Control Engineering Practice*, vol. 40, pp. 13–26, July 2015.

[5] B. Bagaber, P. Junge, and A. Mertens, "Lifetime Estimation and Dimensioning of the Machine-Side Converter for Pumping-Cycle Airborne Wind Energy System," in *2020 22nd European Conference on Power Electronics and Applications (EPE'20 ECCE Europe)*, Sept. 2020.

[6] B. Bagaber and A. Mertens, "A Parallel Voltage Source Converter and Diode Rectifier PMSM Drive Concept for Decoupling the Thermal Cycles in the Machine-Side Converter of an Airborne Wind Energy Generator," in *2021 23rd European Conference on Power Electronics and Applications (EPE'21 ECCE Europe)*, Sept. 2021.

[7] M. Erhard and H. Strauch, "Control of Towing Kites for Seagoing Vessels," *IEEE Transactions on Control Systems Technology*, vol. 21, pp. 1629–1640, Sept. 2013.

[8] ABB, "Data Sheet, Doc. No. 5SYA 1461-01 10-2020: 5SNA 2400N170300 HiPak IGBT Module."

[9] SEMIKRON, "SEMIKRON_DataSheet_SKiiP_2414_GB17E4_4DUW_V2_20603236."

[10] D. Zhang, A. Ebrahimi, C. Wohlers, J. Redlich, and B. Ponick, "On the analytical calculation of eddy-current losses in permanent magnets of electrical machines," in *IECON 2020 The 46th Annual Conference of the IEEE Industrial Electronics Society*, pp. 1052–1056, Oct. 2020.

[11] A. Merkert, T. Krone, and A. Mertens, "Characterization and Scalable Modeling of Power Semiconductors for Optimized Design of Traction Inverters with Si- and SiC-Devices," *IEEE Transactions on Power Electronics*, vol. 29, no. 5, pp. 2238–2245, 2014.

[12] K. Ma, M. Xu, and B. Liu, "Modeling and Characterization of Frequency-Domain Thermal Impedance for IGBT Module Through Heat Flow Information," *IEEE Transactions on Power Electronics*, vol. 36, pp. 1330–1340, Feb. 2021.

[13] ABB, "Application Note 5SYA 2043-04 : Load-cycling capability of HiPak modules."

[14] K. D. Centre, "Dutch Offshore Wind Atlas." https://data.knmi.nl.

[15] T. Haas and J. Meyers, "AWESCO Wind Field Datasets [Data set]," *Zenodo*, 2019.

[16] V. Yaramasu, B. Wu, P. C. Sen, S. Kouro, and M. Narimani, "High-power wind energy conversion systems: State-of-the-art and emerging technologies," *Proceedings of the IEEE*, vol. 103, pp. 740–788, May 2015.

[17] U. Supatti and F. Z. Peng, "Z-source inverter based wind power generation system," in *2008 IEEE International Conference on Sustainable Energy Technologies*, pp. 634–638, Nov. 2008.

[18] O. Ellabban and H. Abu-Rub, "Z-Source Inverter: Topology Improvements Review," *IEEE Industrial Electronics Magazine*, vol. 10, pp. 6–24, Mar. 2016.

Design and Performance Analysis of Single-phase Axial Flux Permanent Magnet Motor for Coaxial Cascade

Chu Wang, Xiaowei Hu, Xiaoya Wang, Weiwei Geng, Qiang Li and Jingning Hou
Nanjing University of Science and Technology
Nanjing 210094, Jiangsu Province, China
Tel.: +86 – 15850724704
Fax: +(025)84315468-7085
E-Mail: gww@njust.edu.cn
URL: http://www.njust.edu.cn/

Keywords

« Axial machines », « All Electric Aircraft », « Electrical machine », « Finite-element analysis », « Permanent magnet motor », « Synchronous motor ».

Abstract

This paper is about the modeling, design and verification of single-phase axial-flux permanent magnet (AFPM) motor for coaxial cascade. The topology principle with yokeless and segmented armature (YASA) is proposed and discussed. The single-phase AFPM motor is designed and coaxially cascaded for forming a multiphase AFPM motor to compare with three-phase AFPM motor. The comparative results show that the three-phase AFPM motors formed with coaxial has higher power density and fault-tolerant ability.

Introduction

In recent years, with the rapid development of electric aircraft, the electric propulsion system of many electric aircraft needs higher power density, torque density and efficiency due to the limitation of installation space and weight [1]-[3]. High power / torque density motor has always been the key basic component of electric propulsion system. Because of its compact axial construction and high torque density, AFPM motors are widely concerned in electric vehicles, aerospace and other fields [4]-[6].

However, there are still significant challenges in the application of three-phase AFPM motor in the case of rotorcraft or UAV. For the traditional three-phase AFPM motor, it is difficult to meet the general space requirements of rotor wing integration due to the high length-diameter. Furthermore, the electro-magnetic coupling among three-phase windings and winding short circuit. The polyphase AFPM motor system composed of single-phase multistage coaxial motor is suitable for applications with long axial length and can improve the fault-tolerance [7]-[12].

This paper presents a new topology of three-phase AFPM motor formed by single-phase AFPM motor coaxial cascade. Finite element method is used to compare the electromagnetic performance, torque capacity and fault-tolerance features between single-phase AFPM coaxial motor and traditional three-phase coaxial motor.

Description of AFPM motor with multiple single-phase coaxial cascade

The single-phase motor usually refers to the asynchronous AC motor. The single-phase AFPM motor proposed in this paper is used to connect three independent single-phase motors in series on the same shaft.

Fig. 1 shows the topology structure and equivalent magnetic circuit of single-phase AFPM motor. The solid line with arrow is the closed loop of the magnetic flux path generated by a pair of N and S poles. The specific magnetic flux path is described as follows. The magnetic flux starts from the N-pole permanent magnet, passes through the air-gap, stator core and air-gap to the N-pole permanent magnet of another rotor mountain, then passes through the rotor core to form a closed magnetic circuit.

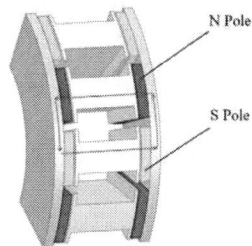

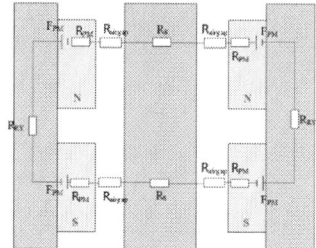

(a) Topological structure   (b) Principle of magnetic flux path

Fig. 1: Topology principle of single-phase AFPM motor

Three coaxially cascaded motors are independently working at the same time to form an equivalent three-phase rotating magnetic field which is staggered in coaxial direction, as shown in Fig. 2. Consequently, there is no electromagnetic coupling and mutual inductance among phase windings. The required torque is generated with coaxial superposition of three independent single-phase motors. The explosive view and structure of three-phase motor for coaxial cascade are shown in Fig. 3.

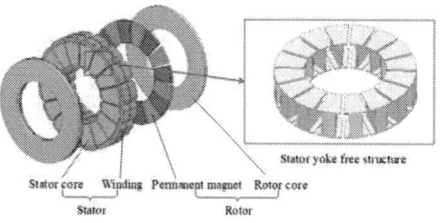

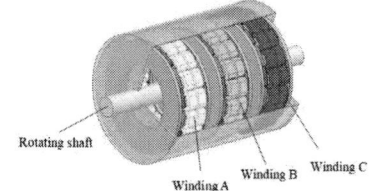

(a) Explosive view of single-phase AFPM motor   (b) Structure of single-phase coaxial motor

Fig. 2: Schematic diagram of the single-phase motor for coaxial cascade

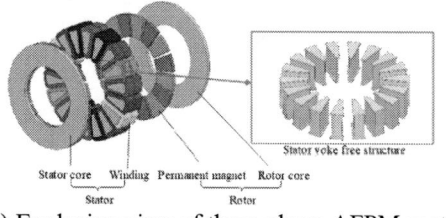

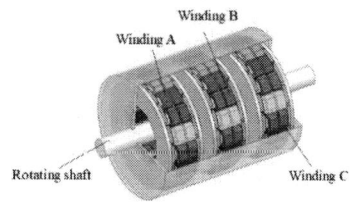

(a) Explosive view of three-phase AFPM motor   (b) Structure of three-phase coaxial motor

Fig. 3: Schematic diagram of the three-phase motor for coaxial cascade

Design and modeling of AFPM motors

Each rotor axis of the rotor UAV is driven by an electric motor. The properly designed motor can achieve high mobility, has the advantages of small size and light quality, and can meet the basic requirements of UAV drive quickly and accurately. Table. I. shows the design requirements for a driving power matching for some kind of rotor aircraft. To better analyze the basic properties and advantages of single-phase AFPM motors, a 16-slot 16-pole AFPM motor was designed and cascaded to form a three-phase AFPM motor, compared to the conventional 16-slot 18-pole three-phase AFPM motor. ANSYS Maxwell software is used for finite element analysis and simulation model is established. In order to target theoretical calculation and simulation analysis, variables must be controlled so that the two AFPM motors have the same size, current density, slot filling factor, etc. Size parameters and winding information of the designed motor are shown in Table. II. It can be seen that compared with the traditional three-phase AFPM motor, we have the same poles and different grooves, the materials of two motors and other structural parameters are both the same.

Table I: Motor design requirements

Parameter	Value
Continuous Output (kW)	5
Peak Power (kW)	10
Continuous Output Torque (N·M)	25
Peak Output Torque (N·M)	50
Rated Speed (rpm)	2000
Peak Speed (rpm)	3000
Effective Material Weight (kg)	<3
The Efficiency	>90%
Working Voltage (V)	36

Table II: The motor structure parameters

Parameter	Three-phase AFPM Motor	Single-phase Coaxial Motor
Slot Number	18	16
Pole Number	16	16
Outer Diameter of Motor (mm)	90	90
Inner Diameter of Motor (mm)	52	52
Air-Gap Length (mm)	0.5	0.5
Axial Length (mm)	33×3	33×3
Thickness of Permanent Magnet (mm)	3	3
Number of Turns	16	16
Number of Parallel Branches	1	1
Slot Filling Factor	0.6	0.6
Core Material	1J22	1J22
Permanent Magnet Materials	N52	N52

All core parts of single-phase coaxial motor and three-phase coaxial motor are made of the same material. In particular, it is pointed out that the stator core and rotor core are made of 1J22 material, and the permanent magnet is N52 rare earth permanent magnet with residual magnetism of 1.45 T. 1J22 is a high saturation magnetic induction strength iron diamond vanadium soft magnetic alloy. Due to the high saturation magnetic induction strength, the volume can be greatly reduced when making motors with the same power. Fig. 4 shows the magnetization curve of 1J22 material.

Comparative analysis of electromagnetic performance

a. Performance comparison at no load

Fig. 5 shows the waveforms of no-load back EMF of two motors, with RMS voltage of 18.65 V and 18.69 V respectively. It can be seen that the waveform of single-phase coaxial motor is more sinusoidal. Fig. 6 shows the phase back EMF at no load of the two motors. Among them, the phase voltage RMS values of single-phase motors and three-phase motors are 10.9V and 10.2V respectively. The waveforms of the two motors at load Fourier decomposition are shown in Fig. 7 .

Fig. 8 shows the air-gap flux density waveforms of single-phase AFPM motor and three-phase AFPM motor. It can be seen that the magnetic dense waveform of the single-phase motor is smoother. This is because the number of pole slots for a single-phase generator corresponds to each tooth. However, the tooth groove of the three-phase motor makes the pole cannot correspond one to one, resulting in the magnetic leakage of the pole and the local magnetic density protrusion.

Fig. 9 shows the Fourier decomposition of air gap-flux density waveform. It can be seen that the fundamental amplitude BS of the air-gap magnetic density of single-phase coaxial motor and three-phase coaxial motor are both 1.2 T. The amplitude of the third and fifth harmonics of the three-phase AFPM motor are significantly higher than that of the single-phase coaxial motor.

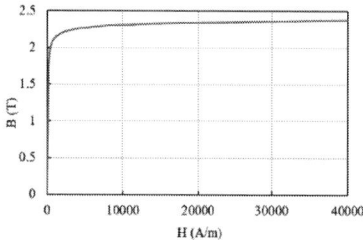

Fig. 4: 1J22 material magnetization curve.

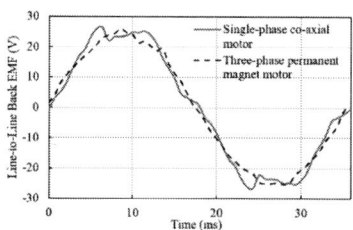

Fig. 5: Line-line back EMF at no load

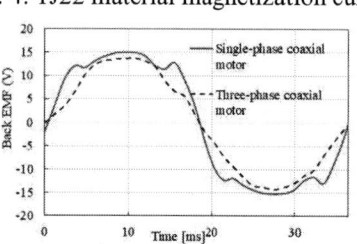

Fig. 6: Phase back EMF at no load

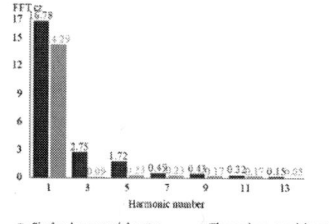

Fig. 7: Harmonic analysis of phase back

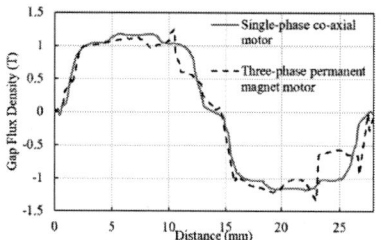

Fig. 8: Air-gap magnetic density

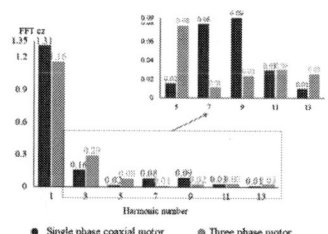

Fig. 9: Harmonic analysis of air-gap

Fig. 10 shows the magnetic field distribution of two kinds of motors under no-load state. The rotor core is partially saturated at the tangential magnetized permanent magnet and the stator is partially saturated at the tooth tip, but it has little impact on the electromagnetic torque production.

Fig. 11 (a) shows the waveforms of cogging torque of two kinds of motors. It can be seen that the peak-peak values of cogging torque of two kinds of motors are 2.6 N·m and 1.7 N·m respectively The cogging torque of three-phase coaxial permanent magnet motor is smaller.

b. The load performance contrast

In order to compare the torque output capacity of the two motors, two AFPM motors initially operate at the same current density and back EMF peak value at no load. In contrast, the single-phase coaxial motor has larger electromagnetic torque with less fluctuation, which means that when the two motors have the same volume, the single-phase coaxial motor has greater torque density and power density.

Fig. 11(b) shows that the output average torque of single-phase coaxial motor is about 28 N·m, and the output average torque of three-phase coaxial motor is relatively smaller, which is about 21 N·m. It can be seen that the output torque of single-phase coaxial motor is larger. Due to the influence of electromagnetic coupling, part of the magnetic saturation state is offset, three-phase coaxial motor has smaller torque than single-phase coaxial motor.

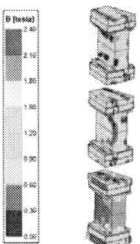 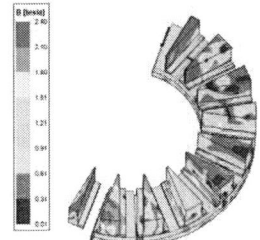

(a) Magnetic field of single-phase coaxial motor at no load

(b) Magnetic field of three-phase coaxial motor at no load

Fig. 10: Magnetic field distribution at no load

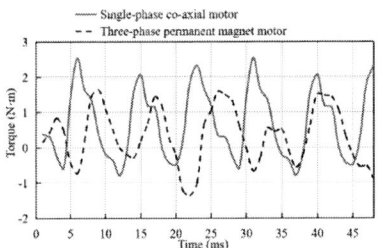 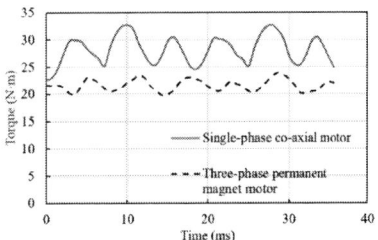

(a) No load cogging torque waveform.

(b) Electromagnetic torque waveforms.

Fig. 11 Motor torque curve

Fig. 12 shows that the stator teeth and stator yoke of the two motors which present have partial magnetic saturation. The result shows that the stator teeth and stator yoke of the two motors are partially magnetically saturated and the magnetic density saturation of the single-axis motors is relatively higher.

The average rated output power of three-phase coaxial motor and single-phase coaxial motor is 4.54 kW and 5.92 kW respectively. Through the average output power and effective material weight of the motor, we can obtain that the rated power density of the motor with two different topologies are 1.59 kW/kg and 2.00 kW/kg respectively. The result shows that single-phase coaxial motor has higher power density and higher material utilization than three-phase coaxial motor.

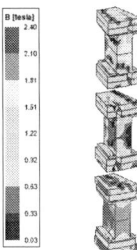 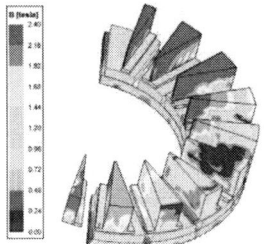

(a) Magnetic field of single-phase coaxial motor at load.

(b) Magnetic field of three-phase coaxial motor at load.

Fig. 12: Magnetic field distribution at load

In Fig. 13, the torque current ratio curve of single-phase coaxial motor and traditional three-phase coaxial motor (under id = 0 control) is compared. It can be seen from the figure that single-phase motor for coaxial cascade can produce higher torque under the same current.

Fig. 14 shows the power/torque speed curves of two different topologies of single-phase coaxial motor and three-phase coaxial motor. By observing the reverse electric momentum curve of the motor, changing the current and current angle of the motor, the torque curve at different speeds is obtained, and the power curve at different speeds is obtained through the calculation formula of torque and power. It shows that the topology of single-phase multistage series coaxial motor has better torque output capacity than the traditional three-phase series coaxial motor.

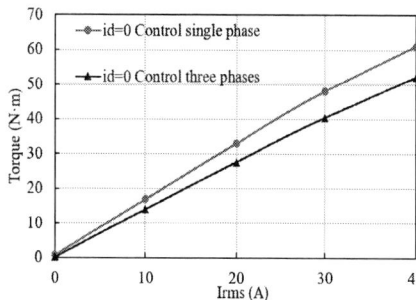 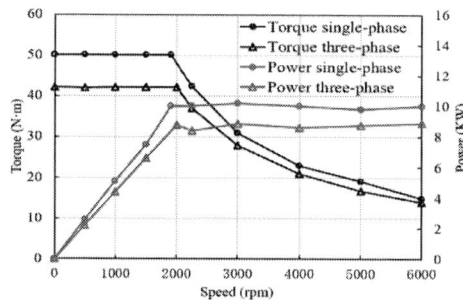

Fig. 13: Torque vs phase current curve (id = 0 control) Fig. 14: Power / torque vs speed curves

c. Efficiency calculation of single-phase coaxial motor

The loss of the motor determines the efficiency of the motor, among which the core loss, permanent magnet eddy current loss and copper loss are the main parts of the total loss of the motor. Table IV shows the comparison of electro-magnetic performance parameters between single-phase coaxial motor and three-phase coaxial motor.

Table IV: The comparison of two motors

Parameter	Three-phase Coaxial Motor	Single-phase Coaxial Motor
No Load Back EMF	26.37 V	26.43 V
Rated Output Power	4.5 kW	5.9 kW
Power Density	1.59 kW/kg	2.00 kW/kg
Total Weight	2.848 kg	2.935 kg
Core Loss	2.7 W	2.5 W
Winding Loss	219 W	223 W
Eddy Current Loss	90 mW	120 mW
Total Loss	221.79 W	225.62 W
Motor Efficiency (η)	95.0 %	96.2 %

Conclusion

This paper compares traditional three-phase AFPM motor and single-phase AFPM motor for coaxial combination with the same material and size from the aspects of motor topology and electromagnetic performance. It is concluded that compared with three-phase AFPM motor, single-phase AFPM motor for coaxial combination has the higher output torque, the better no-load waveform. The power density of single-phase coaxial motor for coaxial cascading to form three-phase AFPM motor is increased by 25.8%. The efficiency of single-phase coaxial motor is 96.2%, which is higher than that of the traditional three-phase coaxial motor.

References

[1] B. B. Choi, "Propulsion Powertrain Simulator: Future turboelectric distributed-propulsion aircraft," IEEE Electrification Magazine, vol.2, no.4, pp. 23-34, Dec. 2014: 10.1109/MELE.2014.2364901.

[2] K. P. Duffy and R. H. Jansen, "Turboelectric and Hybrid Electric Aircraft Drive Key Performance Parameters,". 2018 AIAA/IEEE Electric Aircraft Technologies Symposium (EATS), 2018, pp. 1-19.

[3] S. Agrawal, A. Banerjee and R. F. Beach, "Brushless Doubly-Fed Reluctance Machine Drive for Turbo-Electric Distributed Propulsion Systems,". 2018 AIAA/IEEE Electric Aircraft Technologies Symposium (EATS), 2018, pp. 1-17.

[4] C. Ye, Y. Du, J. Yang, X. Liang, F. Xiong and W. Xu,"Research of an Axial Flux Stator Partition Hybrid Excitation Brushless Synchronous Generator," in IEEE Transactions on Magnetics Vol. 54, no. 11, pp. 1-4, Nov. 2018.

[5] W. Geng, J. Hou and Q. Li: "Electromagnetic Analysis and Efficiency Improvement of Axial-Flux Permanent Magnet Motor with Yokeless Stator by Using Grain-oriented Silicon Steel," IEEE Transactions on Magnetics

[6] A.M.El-Refaie, "Fractional-slot concentrated-windings synchronous permanent magnet machines: Opportunities and challenges," IEEE Trans .Ind. Electron., vol. 57, no. 1, pp. 107–121, Jan. 2010.

[7] H. S. Che, M. J. Duran, E. Levi, M. Jones, W. P. Hew and N. Abd Rahim, "Postfault Operation of an Asymmetrical Six-Phase Induction Machine with Single and Two Isolated Neutral Points," IEEE Trans. on Power Electronics, vol.29, no.10, pp.5406-5416, Oct. 2014.

[8] M.Ruba and D. Fodorean, "Analysis of Fault-Tolerant Multiphase Power Converter for a Nine-Phase Permanent Magnet Synchronous Machine," IEEE Trans. on Industry Appl., vol.48, no.6, pp.2092-2101,Nov.-Dec. 2012.

[9] M. Barcaro, N. Bianchi and F. Magnussen, "Six-Phase Supply Feasibility Using a PM Fractional-Slot Dual Winding Machine," IEEE Trans. on Ind. Appl., vol.47, no.5, pp.2042-2050, Sept./Oct. 2011.

[10] Xuefeng Jiang; Wenxin Huang; Ruiwu Cao; Zhen yang Hao; Jie Li; Wen Jiang, "Analysis of a Dual-Winding Fault-Tolerant Permanent Magnet Machine Drive for Aerospace Applications," IEEE Trans. On Magnetics, vol.51, no.11, pp.1-4, Nov. 2015.

[11] J. W. Bennett, G. J. Atkinson, B. C. Mecrow, D. J. Atkinson, "Fault-Tolerant Design Considerations and Control Strategies for Aerospace Drives," IEEE Trans. on Industrial Electronics, vol.59, no.5, pp.2049-2058, May 2012.

[12] Vansompel, H.; Sergeant, P.; Dupre, L.; Bossche, A. "A Combined Wye-Delta Connection to Increase the Performance of Axial-Flux PM Machines with Concentrated Windings." IEEE Trans. Energy Convers. 2012, 27, 403–410.

Comparison of Pulse Current Capability of Different Switches for Modular Multilevel Converter-based Arbitrary Wave shape Generator used for Dielectric Testing of High Voltage Grid Assets

Dhanashree Ashok Ganeshpure[1], Ajeeth Phrassanna Soundararajan[1], Thiago Batista Soeiro[3], Mohamad Ghaffarian Niasar[1], Peter Vaessen[1,2] and Pavol Bauer[1]

[1]DELFT UNIVERSITY OF TECHNOLOGY	[2]KEMA LABORATORIES	[3]EUROPEAN SPACE AGENCY
Delft, The Netherlands	Arnehm, The Netherlands	Noordwijk, The Netherlands
D.A.Ganeshpure@tudelft.nl		
Phone: +31(0)152788819		

Acknowledgments

The authors would like to thank KEMA Laboratories, Arnhem for this research financial support. Also, the authors are grateful for the practical help from the staff of the ESP lab of TU Delft.

Keywords

≪Pulse Current Capability≫, ≪Si and SiC Device Technologies≫, ≪Modular Multilevel Converter≫, ≪Dielectric Testing of Grid Assets≫

Abstract

This article compares the pulse current capability of various Semiconductor (SM) device technologies for Modular Multilevel Converter (MMC)-based High Voltage (HV) Arbitrary Waveform Generator (AWG) for dielectric testing of grid assets to find the most suitable SM device technology which can perform well in generating lightning impulse that demands a high peak current for a relatively short time. For the typical HV loads of the AWG, Lightning Impulse (LI) test may require a pulse current to rise to 1.7 kA in 0.2 µs. It is essential to highlight that most other dielectric tests performed with an HV AWG demand a relatively low current such as less than 10 A. Therefore, TO-packaged semiconductors would be well-suited for a large number of tests other than short impulses. To optimize the size and cost of the HV AWG, this paper evaluates the pulse current capabilities of TO-packaged semiconductors for the above-mentioned current requirement to generate LI waveform. The first comparison is made among Non-Punch Through (NPT) Si IGBT, Field Stop (FS) Si IGBT, Si MOSFET, and SiC MOSFETs with roughly the same current rating of 40 A. It is found that the Si MOSFET gives the fastest rise time of 0.42 µs and the NPT IGBT gives the highest current amplification factor of almost 12 times greater than its own rated current. However, 3rd Generation SiC MOSFET combines Si MOSFET and NPT IGBT capabilities to generate a fast rise time and high peak pulse current. Additionally, the FS IGBT is compared with the SiC MOSFET. The SiC MOSFET performs better in peak current capability and the obtained rise time. All in all, the research results and the stringent HV AWG requirements for LI show that the application requires a relatively complex switch implementation with far superior current capability than in normal operation. Therefore, a parallel connection of several TO-packaged devices is necessary to generate LI from MMC-based HV AWG.

Introduction

Medium Voltage (MV) and High Voltage (HV) equipment such as transformers, switchgear, and cables in the electrical power system are experiencing new and more severe electrical stresses due to the rise of Distribution Generation (DG) systems and large-scale renewable energy integration by power electronic converters [1][2][3]. For the reliable operation of the current and future electrical power system, MV and HV equipment need to be tested with electric stresses with complex wave shape such as the superposition of impulse waveforms with AC and DC, as shown in Fig. 1. Currently, these waveforms are generated using a superimposed circuit as shown in Fig. 2 [3]. On right side of the figure, the impulse generator or Marx generator is shown which can generate impulse waveforms with varied rise time ranging from $1\,\mu s$ to $250\,\mu s$ and tail time ranging from $50\,\mu s$ to $2500\,\mu s$. The left side of the Fig. 2 can be either the transformer or rectifier circuit to generate the AC or DC part from the voltage waveforms as shown in Fig. 1(a) and (b) [2][3]. This arrangement needs to be customized and tuned for one particular test. Hence, generating such complex waveforms is a time-consuming process. Additionally, the third waveform depicted in Fig. 1(c) consists of AC, DC, and impulse [1][4], requiring three different test sources to generate the desired waveform, increasing the complexity even further. Therefore, a Modular Multilevel Converter (MMC)-based Arbitrary Wave shape Generator (AWG) is proposed in [4] which can generate all exemplary testing waveforms by itself and its schematic is shown in Fig. 3(a). It has been found that the programmability required to generate waveforms mentioned above is present in MMC-based AWG. However, the challenge lies in generating short impulses such as Lightning Impulse (LI) present in the waveforms from Fig. 1 as compared with the low frequency part from these waveforms. The LI waveform, as shown in Fig. 3, has a rise time of $1.2\,\mu s$ and tail time of $50\,\mu s$ [5] and the same waveform is studied in detail to understand the challenges to generate the LI waveform and other complex waveforms from MMC-based AWG.

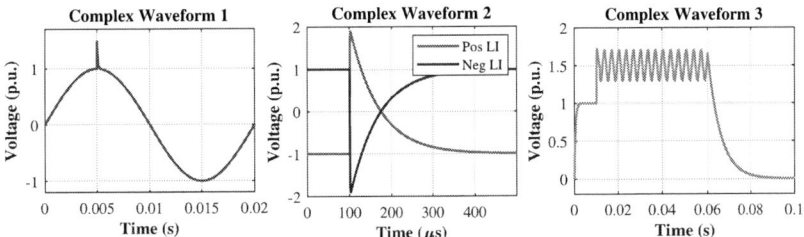

Fig. 1: Typical complex voltage waveforms required for the dielectric tests of grid assets [2][3][4]

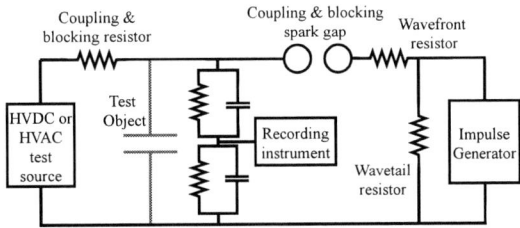

Fig. 2: Test Circuit for Generating Superimposed Waveforms shown in Fig. 1 [3]

There are several challenges for generating LI from an MMC, and they can be summarized as large bandwidth requirement, compensation of the jitter in several series-connected switches, and above all, the switches need to withstand a pulse current with high amplitude. The latter occurs because of the stringent requirement of applying steep dv/dt electric stress across the dielectric insulation of the grid assets that can be modelled as an electric capacitance [6]. The typical value of dielectric capacitance found in MV and HV grid equipment ranges from $100\,pF$ to $10\,nF$ [7]. Fig. 3(b) shows the equivalent circuit of a MMC-based AWG to calculate the exact current required to generate LI. Based on this equivalent circuit, the submodule current required to generate LI waveform across a MV distribution transformer of $36\,kV$ rating which has a $10\,nF$ equivalent dielectric capacitance [7] is calculated to be $1.7\,kA$ in $0.2\,\mu s$, as shown in Fig. 3(c). The rise of $0.2\,\mu s$ is calculated as the difference of time instants when the current magnitude is 10 % and 90 % of the peak current. Note that although the equipment is rated for $36\,kV$, it is tested at a much higher voltage amplitude such as $250\,kV$ during a LI test [5]. Table I summarizes the pulse current required for generating LI across the MV distribution transformer.

Table I: Application requirement on pulse current

Pulse Current (A)	Rise time (µs)	Slew rate (A/µs)
1685	0.2	8425

Table II: IEC Standards for Performing Dielectric Tests on Most Common HV Grid Assets

Instrument	General Dielectric Standard	Switchgear	Distribution Transformer	AC Cable	Instrument Transformer
IEC Standard used	IEC 60060-1 IEC 60060-2	IEC 62271-1	IEC 60076-1 IEC 60073-3	IEC 60502	IEC 61869-1

If the high current rated IGBT modules are chosen to be incorporated into the MMC considering the generation of short impulses, the MMC-based HV test source will be bulky and costly. Moreover, other low-frequency waveform tests do not need a high current. Therefore, this paper aims to verify the pulse current capability of commercially available TO-packaged discrete SM devices instead of SM modules, which could be suited for the LI tests of grid assets.

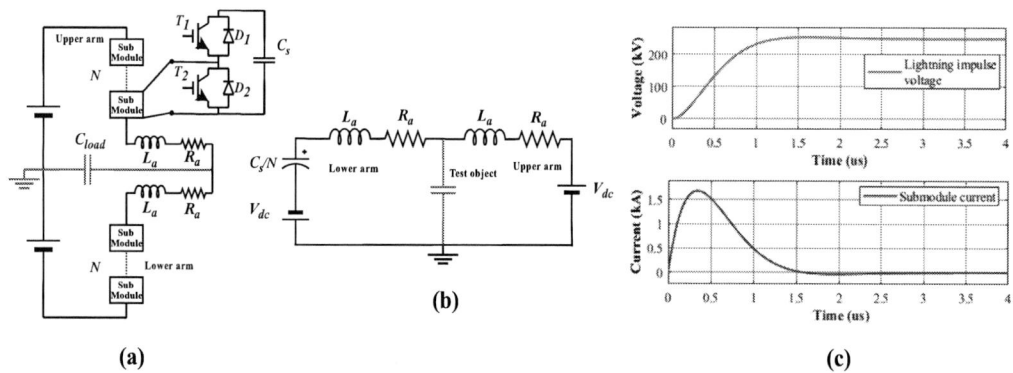

Fig. 3: (a) Schematic of the MMC-based HV AWG (b) Equivalent circuit of MMC for LI operation (c) Response of the MMC when the LI is generated

Please note that the MMC-based HV AWG does not generate continuous short impulses with a fast repetition rate. HV test sources are generally not used continuously like the HVDC converters [4]. They are used in shifts with definite off-time between two impulses and as per the testing requirement at the testing facility. Furthermore, the requirement of generating short impulses such as LI is different for different HV equipment. The testing requirements for each HV equipment are defined in their respective IEC standard, as shown in Table II. For example, LI is only applied seven times for MV transformers [8]. Additionally, it is possible to allow sufficient time between two impulses to prevent cumulative stress for the switches. In power electronic converters, SM devices are required to sustain the higher current for a short time due to various switching instants in the grid [9]. The switching instances in the grid are lower in the number compared to the impulse waveform generation from HV AWG. Hence, after studying the pulse current capability of switches, it is essential to test the short circuit reliability of SM devices to know how many pulses a device can withstand without undergoing severe degradation and how much off-time is necessary between two pulses to ensure the switches are not cumulatively overloaded.

Apart from the HV dielectric testing application, such a pulsed power operation of SM devices is extensively used in plasma applications [10]. Mostly, these plasma applications have a resistive load at a much lower voltage level. Even when it is capacitive load, the capacitance is in the range of 150 pF to 600 pF [11]. Additionally, the MMC-based AWG is a multi-purpose HV test source where the bandwidth of the desired waveform can vary from a very low value (such as DC) to a very large value (short impulses such as LI). This poses a unique requirement for the SM devices, where most tests need small current magnitudes, and only LI and other impulse tests require pulse currents with a high amplitude. All manufacturers of SM devices mention pulse current capability with other electrical properties in their datasheets. However, the single pulse current amplitude mentioned in these data sheets is rather limited to 2 to 3 times the rated continuous current at 25 °C junction temperature. Additionally, the short circuit test results on different semiconductor devices indicate that these devices can withstand much higher pulse current for short duration [9] [12] [13] [14]. However, these studies have analysed only the pulse current capability of a single switch technology, and benchmarking among other technologies is missing. In [14], pulse current capability of Silicon Carbide (SiC) MOSFET is compared with Field Stop (FS) IGBT.

This paper presents a comparison of 8 switches with different switch technologies and manufacturers to obtain the highest peak current with a fast rise time. The scientific contribution of this paper in relation to the pulse current capability of an MMC submodule implemented with several discrete SM technologies is summarized as follows:

- Comparison among Non-Punch Through (NPT) IGBT technology, Si Cool MOSFET technology, and SiC MOSFET technology

- Comparison between 2nd and 3rd generation SiC MOSFET technology

- Comparison between FS IGBT technology and 3rd SiC MOSFET technology

- Investigation of the NPT IGBTs with a different blocking voltage rating

Theoretical Background and Adopted Methodology

The current capability of a particular SM device is limited by various factors such as its on-state resistance, thermal stability limit, and package limit [15]. Fig. 4(b) shows these factors for a CREE-Wolfspeed device named C2M0160120D. Fig. 4(a) s hows the thermal impedance of the same switch for a single pulse with variable duty (0.01 to 0.5). As it is visible, the thermal impedance drops significantly as the pulse length is shortened, and thus this property enables the switch to safely conduct pulse currents of magnitudes much greater than its own rated current. As highlighted in Fig. 4(b), this article aims to find a safe operating point outside the typical SOA given in the SM datasheets for a single pulse application without damaging the device. The idea is to perform the pulse current tests in an MMC submodule prototype especially designed for the AWG application where TO-packaged SM devices can be implemented. Therein, various switch technologies are tested, and later their performances are benchmarked to find the most suitable SM technologies for the MMC-based HV AWG.

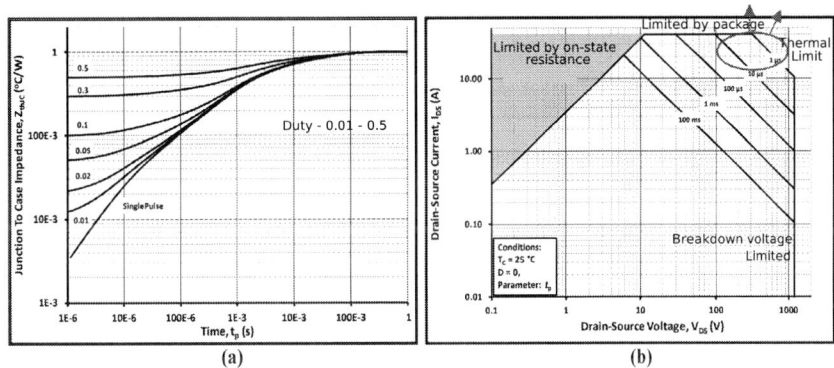

Fig. 4: C2M0160120D (a) Transient Thermal Impedances (Junction - Case) (b) Safe Operating Area

Since the requirements of the HV AWG application demands high pulse currents similar to the short circuit tests, SM devices are tested by discharging the charge stored in the capacitor into the Device Under Test (DUT) via an auxiliary loop that has minimal stray inductance and resistance to obtain the highest peak pulse current with the fastest rise time. Various external electrical factors determine the magnitude and rise time of the peak pulse current. The factors that affect magnitude of peak pulse current are Gate to Source/Emitter voltage (V_{GS}/V_{GE}) and Drain/Collector to Source/Emitter voltage (V_{DS}/V_{CE}). V_{GS}/V_{GE} determines the thickness of the conduction channel and hence higher values of V_{GS}/V_{GE} results in a thicker conduction channel. Hence more electrons can pass through the channel. V_{DS}/V_{CE} determines the potential difference applied across the bulk of the SM device. Higher values of V_{DS}/V_{CE} results in higher potential difference applied across the bulk during the linear region of conduction and hence higher drain current. The gate resistance (R_G) determines the rise time directly as the time constant of the conduction channel is directly proportional to R_G [12]. Lower value of gate resistance results in a faster turn-on of the DUT. Thus, the lowest value of R_G is preferred. At high values of V_{DS}/V_{CE}, all DUTs undergo saturation because of pinch-off, limiting the peak pulse current magnitude to its final value [16]. Additionally, such a high magnitude of current heats up the DUT, resulting in increased bulk resistance of the DUT, reducing the peak magnitude of the current even further [17]. Therefore, first, the effect of the above-mentioned electrical parameters is studied experimentally using single SM device technology. Later, the electrical behavior of different SM device technologies with the same aforementioned electrical parameters are compared to one another to find out the most suitable technology that could satisfy the high current demand of the HV AWG application.

To achieve the goal of this article, multiple SM devices are chosen, and their details are summarized in Fig. 5. First, Cree-Wolfspeed C2M0160120D device is selected to test the experimental setup for constant V_{GS} while applying variable V_{DS} and vice-versa. Additionally, the same tests are repeated on three different devices to understand their behavior statistically. The three boxes in Fig. 5 represent different groups for comparing the pulse current behaviors of various switch technologies. IGBTs are classified into NPT, Punch Through (PT), and Trench Gate Field Stop (FS) based on the presence of an n+ buffer layer. Among these, NPT IGBT (SGW25N120) and FS IGBT (IHW40N120R5) from Infineon are selected to investigate their pulse current capabilities. Additionally, more NPT IGBTs from IXYS with higher blocking voltage capability, such as 1.7 kV (IXGH24N170) and 3.0 kV (IXBH20N300), are added to the list. MOSFETs are inherently famous for their faster switching characteristics over current carrying capacity. Therefore, Cool Si MOSFET (IPW90R120C3) from Infineon is selected to obtain a faster rise time. Since SiC MOSFET combines higher current carrying capability and fast switching characteristics, multiple devices from CREE-Wolfspeed with different current capabilities are selected to compare with other technologies.

No.	Device name	Manufacturer	V_{BR} (V)	I_D/I_C 25 °C (A)	I_{pulse} (A)	Op range for V_{GS} (V)
1.	C2M0160120D SiC MOSFET	CREE-Wolfspeed	1200	18	40	-5 /+20
2.	C3M0075120D SiC MOSFET	CREE-Wolfspeed	1200	32	80	-5 /+15
3.	C2M0080120D SiC MOSFET	CREE-Wolfspeed	1200	36	80	-5 /+20
4.	IPW90R120C3 Cool Si MOSFET	Infineon	900	36	96	± 20
5.	SGW25N120 NPT IGBT	Infineon	1200	46	84	± 20
6.	IXGH24N170 NPT IGBT	IXYS	1700	50	150	± 20
7.	IXBH20N300 (NPT IGBT)	IXYS	3000	50	150	± 20
8.	IHW40N120R5 FS IGBT	Infineon	1200	80	200	± 20
9.	C3M0016120D SiC MOSFET	CREE-Wolfspeed	1200	81	200	-5 /+15

Fig. 5: Details of the tested switches

Experimental Setup and Results

Fig. 6(a) shows the schematic of the test circuit to characterize the selected SM devices with an installed capacitance of 198 µF. In the actual test setup, the circuit is realized using a half-bridge submodule of an MMC where the high side switch is shorted using a wire for current measurement, and the low side switch acts as a DUT. Indeed, even if the submodule PCB design is optimized, some stray inductance and stray resistance are added to the auxiliary loop. Their respective values are measured from the setup when the DUT is not inserted in the DUT holder. Hence, the stray inductances and resistances are measured with an impedance analyzer (Agilent 4294 A) in two steps. Firstly, these values are measured from the positive DC voltage to the Drain of the DUT. Secondly, these values are measured from the Source of the DUT to the negative DC voltage. By adding these values, the loop inductance and resistance can be calculated as 120 nH and 160 mΩ. This means that the different switching technologies from different manufacturers can add different inductance and resistance to the auxiliary loop. The packaging of these discrete SM devices can add significant inductance to the loop, as illustrated in [18][19]. That is why it is crucial to investigate the pulse current capability of different SM devices experimentally. Furthermore, the test circuit is studied in the LTspice with a SiC MOSFET from Cree-Wolfspeed (C2M0160120D) with its actual spice model, and the simulation results are compared with the experimental setup in Fig. 6(b). If the stray inductance and resistance are reduced from 120 nH and 160 mΩ to 30 nH and 40 mΩ in the spice model, the peak pulse current and the rise time of current waveform do not change. That means the DUT is saturating and limiting the pulse current over the stray inductance and resistance. The drain/collector current flowing through the DUT is measured using a Panasonic current sensor model 411. The V_{DS}/V_{CE} and V_{GS}/V_{GE} are measured using a differential probe from Keysight N2791A.

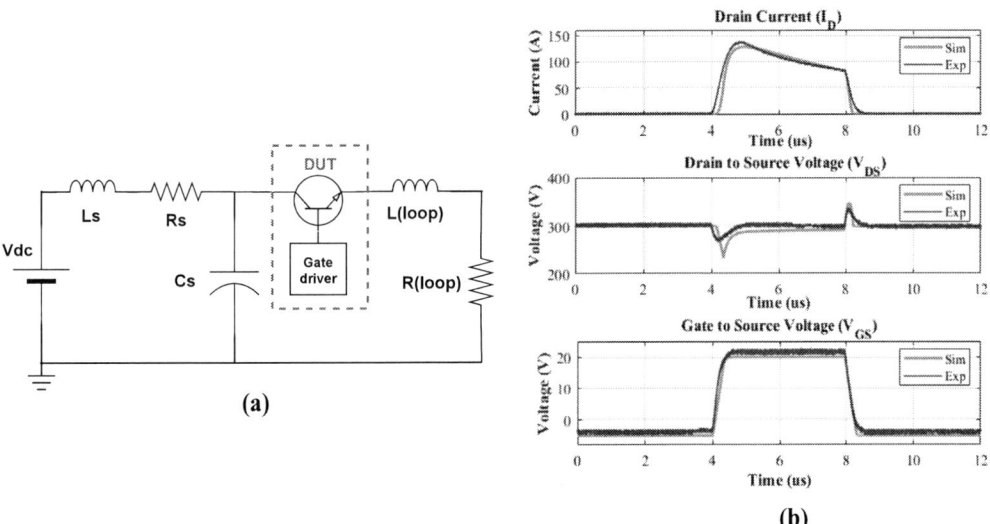

Fig. 6: (a) Schematic of the test setup (b) Comparison of LT-spice simulation and experimental results of C2M0160120D

The experimental setup is shown in Fig. 7. As discussed in the previous section, several circuit parameters affect the pulse current characteristic of the DUT, such as V_{GS}/V_{GE}, V_{DS}/V_{CE}, and the turn-on gate resistance ($R_{g,on}$). To obtain the fastest rise time from the DUT, the external turn-on gate resistance is set to $0\,\Omega$ for testing limits of all DUTs. Please note that the discrete SM devices have an internal gate resistance that limits the current rise time. However, the turn-off transient of the switch is limited by a much higher turn-off gate resistance ($R_{g,off}$), and the selected value is $10\,\Omega$ for all DUTs. The effect of V_{GS}/V_{GE} and V_{DS}/V_{CE} are studied with a SiC MOSFET (C2M0160120D), as shown in Fig. 8. With higher values of V_{GS}/V_{GE} and V_{DS}/V_{CE}, higher peak pulse currents are obtained with a faster rise time. However, the peak pulse current is more sensitive to V_{GS}/V_{GE} since it directly affects the thickness of the conduction channel that carries the current. Moreover, for the same V_{GS}/V_{GE}, the peak current capability is saturated for higher values of V_{DS}/V_{CE}, as it is visible in Fig. 8(b). These findings complement the theoretical understanding of these external electrical factors affecting the pulse current magnitude. Additionally, these tests are performed on multiple devices, their peak pulse currents are almost the same, and a maximum difference of 1.2 % is found. Hence, for other DUTs, statistical analyses are not performed.

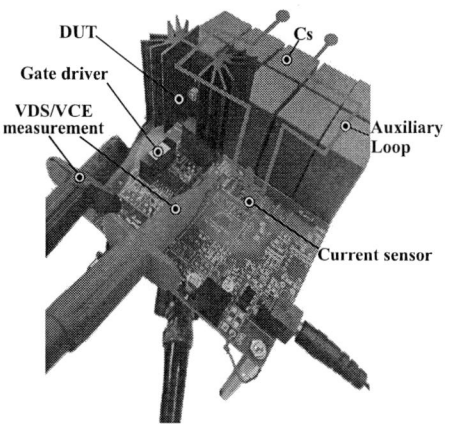

Fig. 7: Experimental Setup

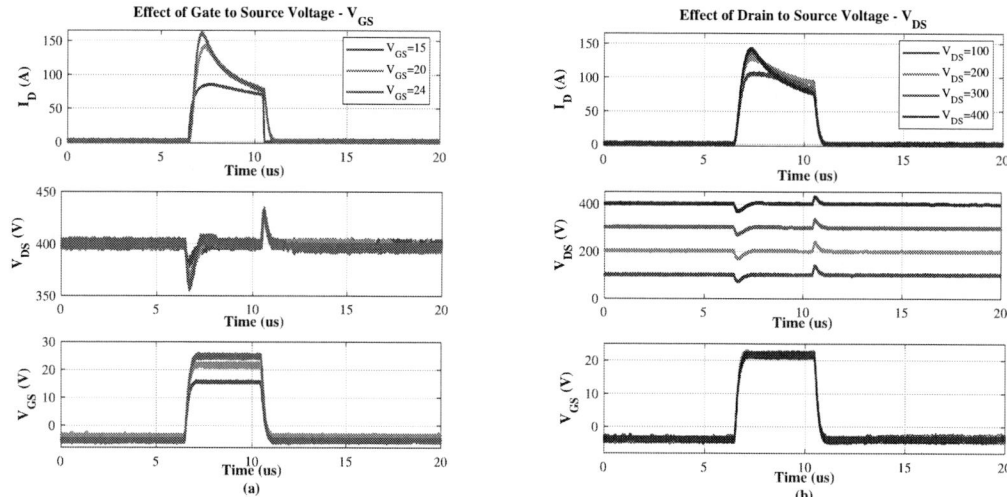

Fig. 8: C2M0160120D (a) Effect of the turn-on-gate-to-source voltage for constant $V_{DS} = 400$ V (b) Effect of applied drain-to-source voltage $V_{GS} = 20$ V

The first box from the list of tested switches consists of the NPT IGBT, Si Cool MOSFET, and SiC MOSFETs from 2nd and 3rd Generation with roughly the same rated current capability. The pulse current obtained from these four different technologies are compared at Fig. 9(a) for $V_{GS} = 20$ V. Since V_{GE}/V_{GS} and V_{CE}/V_{DS} waveforms follow a similar profile as shown in Fig. 8, they are excluded in Fig. 9 and further figures. It is visible that the NPT IGBT conducts a much higher magnitude of pulse current than all three technologies. However, Table III presented the rise time and obtained a slew rate, where NPT IGBT has the slowest rise time. Whereas the tested Si Cool MOSFET delivers the fastest rise time of $0.450\,\mu s$ considering its superior switching performance. However, Si Cool MOSFET has an almost flat current profile compared to SiC MOSFET and Si IGBT, where higher V_{GS} values does not increase the peak current but only reduce the rise time. In the case of SiC MOSFETs, the current profile is significantly different at the peak and at the end of $4\,\mu s$. The 3rd generation SiC MOSFET performs better than the 2nd generation for $V_{GS} = 20$ V both in terms of the peak current and the rise time. However, this behaviour has changed at $V_{GS} = 24$ V in terms of the peak current obtained, as shown in Fig. 9(b). This change can be attributed to the difference in the operating range of V_{GS} of 2nd and 3rd Generation SiC

MOSFETs.

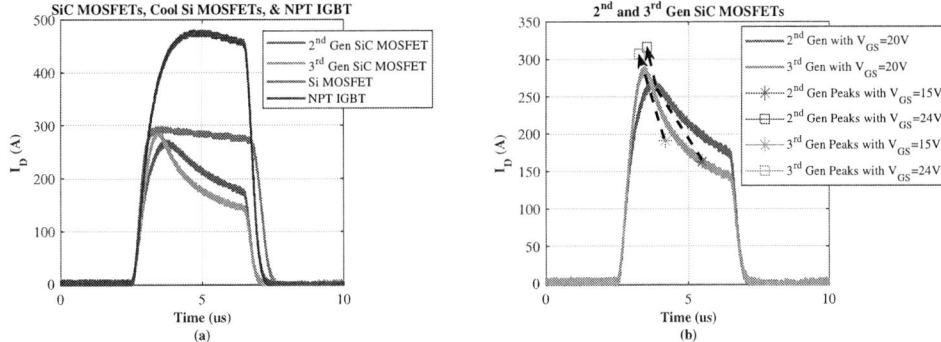

Fig. 9: V_{GS} = 20 V and V_{DS} = 300 V: Comparison of (a) SiC MOSFETs, Si MOSFET, and NPT IGBT (b) 2nd and 3rd Generation SiC MOSFETs

Table III: Comparison among 2nd and 3rd Generation SiC MOSFETs, Si MOSFET, and NPT IGBT

	C3M0075120D 3rd Gen SiC MOS I_c=32 A, I_p=80 A			C2M0080120D 2nd Gen SiC MOS I_c=36 A, I_p=80 A			IPW90R120C3 Cool Si MOS I_c=36 A, I_p=96 A			SGW25N120 NPT IGBT I_c=46 A, I_p=84 A		
VGS (V)	15	20	24	15	20	24	15	20	24	15	20	24
Pulse current obtained (A)	191	286	307	163	271	314	297	296	295	281	476	570
Times the rated current	6	8.9	9.6	4.5	7.53	8.7	8.2	8.2	8.2	6.1	10.3	12.4
Rise time (μs)	1.01	0.57	0.46	1.43	0.73	0.55	0.53	0.42	0.44	1.06	1.04	1.04
Slew rate (A/μs)	189	502	667	114	371	634	560	705	670	265	458	548

Fig. 10(a) compares the pulse current capability of NPT IGBTs with different breakdown voltage ratings. The peak current obtained from 1.7 kV rated switch is the highest. However, the rise time to reach the peak current increases as the blocking voltage rating of the devices increases since their switching performances are poorer with a higher blocking voltage rating. Fig. 10(b) compares the performance of the 3rd Generation SiC MOSFET and FS IGBT at V_{GS} = 24 V. SiC MOSFET performs much better with respect to both the peak current capability and the rise time, and it is summerized in Table IV, which was also observed in [14].

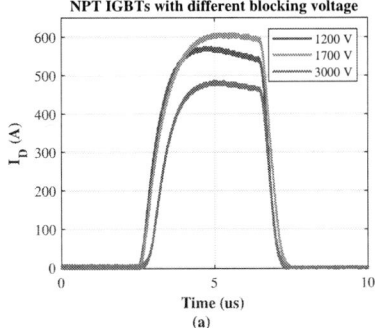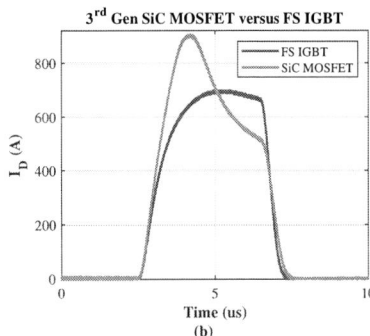

Fig. 10: V_{GS} = 24 V and V_{DS} = 300 V: (a) Effect of blocking voltage for NPT IGBTs (b) Comparison of SiC MOSFETs and FS IGBT

Table IV: Comparison between 3rd Generation SiC MOSFET and FS IGBT

	C3M0016120D 3rd Gen SiC MOSFET I_c=81 A, I_p=200 A			IHW40N120R5 FS IGBT I_c=80 A, I_p=200 A		
VGS (V)	15	20	24	15	20	24
Pulse current obtained (A)	579	809	905	287	554	697
Times the rated current	7.2	10	11.2	4.5	6.9	8.7
Rise time (μs)	2.47	1.24	1.06	1.27	1.32	1.36
Slew rate (A/μs)	234	652	854	226	420	521

Discussion

The quantitative comparison from the last section shows that the 3rd generation SiC MOSFET with 30 A current rating has a comparable rise time as the Si Cool MOSFET (0.040 μs faster). Additionally, the amplification factor of the peak current with respect to the rated current for the same SiC MOSFET is 9.6, which is 9.3 % higher than the Cool MOSFET but 30 % lower than the NPT IGBT. However, the slew rate obtained from SiC MOSFET is higher than the NPT IGBT by 17.8 %. In Table III, Cool Si MOS delivered the highest slew rate of 705 A/μs, which is 5.6 % higher than the SiC MOSFET. When the 30 A rated SiC MOSFET is compared with the 81 A rated SiC MOSFET, the latter delivers a much higher peak current, however, with a slower rise time. Additionally, the 81 A rated SiC MOSFET has achieved the highest slew rate among all tested switches, which is 17.5 % higher than the Si Cool MOSFET.

Apart from the quantitative comparison, it is important to understand qualitatively why different SM device technologies behave differently in Fig. 9 and Fig. 10. The Si MOSFET has the flattest current profile, whereas the drain current of SiC MOSFETs rises to a peak and reduces to a lower value at the end of 4 μs. Compared to these two technologies, Si IGBTs exhibit similar behavior of flat current profile with the Si MOSFET, with a much slower rise time. In all three device technologies, the DUT is saturated, and the conduction channel is pinched off considering the high value of the applied V_{DS}/V_{CE} [16]. The pinched-off state of the conduction channel adds significantly higher resistance for the drain/collector current to flow, and hence the peak value of the drain/collector current is limited. Additionally, when such a high current flows in the device, the temperature inside the substrate increases, and hence the overall bulk resistance increases. This relationship is linear in Si MOSFET and Si IGBT [20]. Hence, they have a flat pulse current profile in Fig. 9 and Fig. 10. However, SiC MOSFETs exhibit unique behaviour. When the temperature of the SiC device rises, the conduction channel exhibits a negative temperature coefficient. That is, the conduction channel resistance drops significantly, and charge carrier mobility increases in the conduction channel leading to the conduction of higher current magnitudes at the beginning of the current profile. But later, the resistance of the drift region dominates the overall resistance of the device as it starts to increase as the temperature increases leading to low values of drain current at the end of the current profile [12][20]. Due to these two effects, the current in SiC MOSFETs rises to a higher value and starts reducing to a lower value.

Apart from these dynamic factors, the obtained pulse current from various SM devices generally depends on the thickness of the device since the on-state resistance or bulk resistance directly depends on it. Among the different types of IGBTs, FS IGBT has the smallest thickness compared to NPT and PT IGBTs [21]. Therefore, it gives a comparable performance to a SiC MOSFET. With superior SiC material properties, the thickness of SiC MOSFET is significantly lowered [22]. Combining this property with the positive coefficient of channel mobility with temperature, SiC MOSFETs have superior performance. Among the Si MOSFET and IGBT which have flat profiles, the IGBT allows a higher peak pulse current considering its lower on-state resistance [16]. Nevertheless, this highest slew rate obtained from the 81 A SiC MOSFET is ten times smaller than the requirement of the application. Additionally, the obtained peak current and rise time are two times lower and five times slower than the requirement. Hence, it can be concluded that the SiC MOSFET with a lower current rating and lower voltage rating will have superior switching performance, which may give the fast rise time of $0.2\,\mu s$. However, multiple lower current rated SiC MOSFETs need to be connected in parallel to meet the requirement of the peak current, similar to discussed in [23]. This makes the switch implementation complex to generate such a severe dv/dt stress across the capacitive load.

Conclusion

This article investigates the pulse current capability for various discrete SM devices for the MMC-based HV AWG for dielectric testing of grid assets. This application, particularly the lightning impulse tests, poses a unique operating condition for the switches with a very high peak pulse current and a very fast rise time. Among SiC MOSFETs, Si Cool MOSFET, and Si NPT IGBT with roughly 40 A current rating, NPT IGBT has the highest amplification factor for the peak current capability, and Si Cool MOSFET has the fastest rise time leading to the highest slew rate as well. 3rd generation SiC MOSFET has a comparable rise time to Si Cool MOSFET, but gives a higher amplification factor. The 81 A rated SiC MOSFET gives the highest amplification factor and slew rate among all tested devices; however, it has a rise time in the range of μs. Therefore, parallel operation of several low current rated SiC MOSFETs can be a solution to satisfy all criteria for this application.

References

[1] Mukherjee, S.: Cable overvoltage for MMC based VSC HVDC system: Interaction with converters, CIGRE India Journal 7.2 2018, pp 18-23

[2] Jiayang, W. U.: Effects of Transients on High Voltage Cable Insulation, Delft University of Technology, 2020

[3] IEC 62895, High voltage direct current (HVDC) power transmission

[4] D. A. Ganeshpure, T. Batista Soeiro, M. G. Niasar, P. Vaessen and P. Bauer, "Design Trade-offs of Modular Multilevel Converter-based Arbitrary Wave Shape Generator for Conventional and Unconventional High Voltage Testing," in IEEE Open Journal of the Industrial Electronics Society, doi: 10.1109/OJIES.2021.3125747

[5] IEC 60060-1, High-voltage test techniques - Part 1: General definitions and test requirements

[6] Kreuger, F. H. Industrial High Voltage. [vol. II], 4. Coordinating. 5. Testing. 6. Measuring, Delft University Press, 1992

[7] "Internal test reports, KEMA Laboratories, Arnhem, Netherlands," Confidential

[8] IEC 60076-3, Power transformers – Part 3: Insulation levels, dielectric tests and external clearances in air

[9] T. B. Soeiro, E. Mengotti, E. Bianda and G. Ortiz, "Performance Evaluation of the Body-Diode of SiC Mosfets under Repetitive Surge Current Operation," IECON 2019 - 45th Annual Conference of the IEEE Industrial Electronics Society, 2019, pp. 5154-5159.

[10] M. Hochberg et al., "A Fast Modular Semiconductor-Based Marx Generator for Driving Dynamic Loads," in IEEE Transactions on Plasma Science, vol. 47, no. 1, pp. 627-634, Jan. 2019, doi: 10.1109/TPS.2018.2876503.

[11] L. M. Redondo, A. Kandratsyeu, M. J. Barnes, S. Calatroni and W. Wuensch, "Solid-state Marx generator for the compact linear collider breakdown studies," 2016 IEEE International Power Modulator and High Voltage Conference (IPMHVC), 2016, pp. 187-192, doi: 10.1109/IPMHVC.2016.8012824.

[12] C. Ionita, M. Nawaz, K. Ilves and F. Iannuzzo, "Short-circuit ruggedness assessment of a 1.2 kV/180 A SiC MOSFET power module," 2017 IEEE Energy Conversion Congress and Exposition (ECCE), 2017, pp. 1982-1987

[13] T. Basler, J. Lutz, R. Jakob and T. Brückner, "Surge current capability of IGBTs," International Multi-Conference on Systems, Signals and Devices, 2012, pp. 1-6, doi: 10.1109/SSD.2012.6198072.

[14] S. Yin, Y. Gu, S. Deng, X. Xin and G. Dai, "Comparative Investigation of Surge Current Capabilities of Si IGBT and SiC MOSFET for Pulsed Power Application," in IEEE Transactions on Plasma Science, vol. 46, no. 8, pp. 2979-2984, Aug. 2018

[15] J. Schoiswohl, "Linear Mode Operation and Safe Operating Diagram of Power-MOSFET", Application Note by Infineon, May 2017

[16] N. Mohan, T. M. Undeland, W. P. Robbins, "Power electronics. InConverters, Applications, and Design", John Whiley & Sons, 1995

[17] B. J. Baliga, "Fundamentals of power semiconductor devices", Springer Science & Business Media, 2010.

[18] L. Zhang, S. Guo, X. Li, Y. Lei, W. Yu and A. Q. Huang, "Integrated SiC MOSFET module with ultra low parasitic inductance for noise free ultra high speed switching," 2015 IEEE 3rd Workshop on Wide Bandgap Power Devices and Applications (WiPDA), 2015, pp. 224-229

[19] F. Denk, K. Haehre, S. E. Cabrera, C. Simon, M. Heidinger, R. Kling, and W. Heering. "R DS (on) vs. inductance: comparison of SiC MOSFETs in 7pin D2Pak and 4pin TO-247 and their benefits for high-power MHz inverters." IET Power Electronics 12, no. 6 (2019): 1349-1356

[20] Z. Chen, D. Boroyevich and J. Li, "Behavioral comparison of Si and SiC power MOSFETs for high-frequency applications," 2013 Twenty-Eighth Annual IEEE Applied Power Electronics Conference and Exposition (APEC), 2013, pp. 2453-2460

[21] "Process Enhancements Increase IGBT Efficiency for Motor Drive Applications", Application Note by Digi-Key, February 2015

[22] "Comparison of SiC MOSFET and Si IGBT", Application Note by Toshiba, August 2020

[23] Weihua Jiang et al., "Compact solid-State switched pulsed power and its applications," in Proceedings of the IEEE, vol. 92, no. 7, pp. 1180-1196, July 2004, doi: 10.1109/JPROC.2004.829003

Accurate Modeling of IGBT-Based Converters in PLECS

Anne von Hoegen[1], Philipp Tillmann[1], Tetsuya Kojima[2], Rik W. De Doncker[1]

[1] Institute for Power Electronics and Electrical Drives (ISEA)
RWTH Aachen University, Germany
Email: post@isea.rwth-aachen.de

[2] Advanced Technology R&D Center, Mitsubishi Electric Corporation, Amagasaki, Japan

Keywords

≪Dead-Time≫, ≪Device Modeling≫, ≪IGBT≫, ≪Voltage Source Inverter (VSI)≫

Abstract

Control of modern electrical drive systems commonly requires a precise flux-linkage estimation. An accurate model of the commutation cell with non-ideal switches is a first step to improve the estimation of the flux. This paper analyzes und evaluates different modeling approaches for non-ideal IGBT-based converters and recommends their implementation with PLECS.

In addition to modeling the dead time of the controller, compensation methods for further major contributors to volt-second errors of the switching cell are presented: turn-on and turn-off delay of the semiconductor devices, their on-stage voltage drops and the non-linear behavior during commutation time. Furthermore, this work presents how to parameterize the model using datasheet values of an IGBT-based power module.

Introduction

Position sensorless control based on the estimated flux-linkage is widely used in modern electrical drive systems [1], [2]. However, this type of control often suffers from performance problems in low speed region where the flux observer becomes inaccurate. At the same time, at very high speeds where optimized pulse patterns are applied as in [3] and [4], an accurate flux observer as well as an exact knowledge of the flux-linkage is crucial. A precise model of IGBT-based converter yields more realistic phase-voltage trajectories, which are especially relevant for the low speed range. Therefore, an accurate model of the commutation cell of a converter is a first step to improve the estimation of volt-seconds and thus the flux observer.

IGBT-diode combinations are often modeled as ideal switches although they impose volt-second errors. Most typical causes for volt-second errors are dead time, device on-stage voltage drop, and switching delays, which are not considered in these ideal models. Especially in the low speed range, the relative volt-second error due to these deviations is no longer negligible. First, various IGBT modeling approaches for the major contributors of non-ideal behavior are presented to compensate the volt-second error. Next, the modeling approaches are evaluated with regard to implementation in PLECS as this software tool for system-level simulations of electrical circuits provides fast simulation of power electronics. PLECS also provides direct embedding of these models in MATLAB/Simulink which allows the subsequent design of control algorithms. Appropriate approaches of modeling an IGBT switching cell are first compared using a half-bridge circuit. The most suitable approach regarding accuracy, simulation time, and analytical description is then investigated for the application of a three-phase converter. The influence of various model-parameters variation on the phase-voltage trajectories is presented.

Accurate Modeling of Volt-Second Errors

Several approaches to accurately model the commutation process in an IGBT-based converter are analyzed regarding their applicability in PLECS. All feasible methods are implemented and compared regarding their accuracy and simluation time using the example of a half-bridge converter. The approach with the best performance are furtherly investigated using a three-phase bridge converter.

IGBT Modeling Approaches

The first modeling approach for an IGBT is presented in [5] in 1988 and it is known as Hefner model. Hefner developed the first analytical model that consistently describes both the IGBT steady-state characteristics and the switching transients for all loading conditions. By using devices with different parameter conditions, he experimentally verified his model to be very accurate. However, along with the preciseness of his model, it is also rather time-consuming. A more practical but less accurate model than the Hefner model is introduced in [6] and [7] more than ten years later. The authors suggest an extraction from datasheets and measurement setups for experiments for the parameters of the analytical model. To further improve this model, [8] proposes an optimization of the IGBT parameters through an algorithm for faster simulation in SPICE-based environments. Nevertheless, it is a detailed mathematical model that demonstrates how to improve the parametrization of a non-ideal IGBT model and therefore is not suitable for utilization in PLECS.

In [9], an accurate modeling approach for IGBT and diode converter voltage distortion for high-performance motor drives is introduced. Here, the major contributors to volt-second errors of a voltage-fed PWM-based IGBT inverter are addressed with

- dead time T_d of controller,
- turn-on and turn-off delay $T_{on/off}$ of the devices,
- power devices on-stage voltage drops U_{IGBT} and U_{diode},
- as well as the limitied voltage slope during commutation time.

In a first step, a simplified model that neglects the non-linear behavior of the IGBT is presented. The influence of the above listed effects on an actual phase voltage u_{ph} with respect to its reference voltage u_{ph}^* depending on the sign of the phase current I_{ph} are shown in Fig. 1. For both cases, a positive phase current I_{ph} (Fig. 1a) and a negative phase current I_{ph} (Fig. 1b), the same voltage pulse u_{ph}^* is assumed. In a second step, an additional capacitance parallel to the switching cell represents the charging and discharging effect and thus the last major contributor to volt-second errors during the commutation process (Fig. 2). Accordingly, the simulated instantaneous phase voltage u_{ph} has a limited slope as the capacitance of the upper switching cell has to be discharged while the capacitance of the lower switching cell has to be charged during the commutation process. This modeling approach of converter voltage distortion is suitable for representation in PLECS and is therefore used for further investigation.

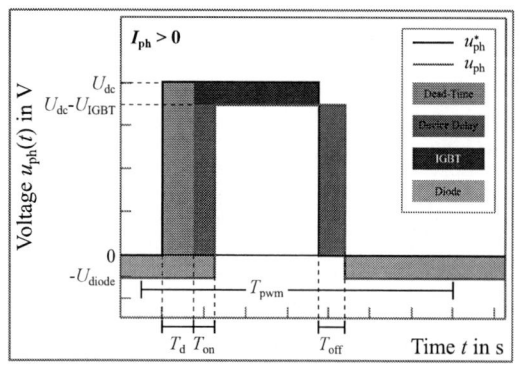
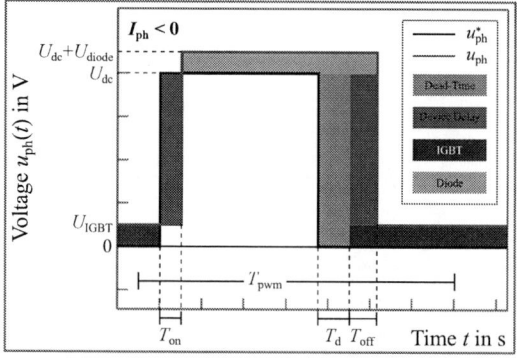

(a) Volt-second error for positive phase current I_{ph} (b) Volt-second error for negative phase current I_{ph}

Fig. 1: Volt-second error during one PWM period

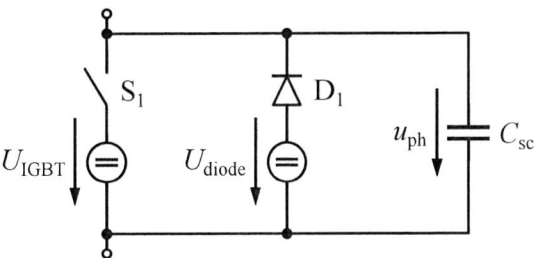

Fig. 2: Switching cell with additional capacitance C_{sc}

IGBT-Converter Simulations in PLECS

In 1999, PLECS is introduced as a toolbox for fast simulation of power electronic circuits under MAT-LAB/Simulink [10]. PLECS enables fast simulation times by using a state-space formulation for linear components. To guarantee these fast simulation times for non-linear components such as IGBTs and diodes, too, these models are consequently transferred into piece-wise linear models. Nowadays, PLECS is not only available as blockset within MATLAB/Simulink but also as standalone. Besides the disciplines of modeling and simulation of power converters, PLECS offers a comprehensive component library, which covers the electrical, as well as the magnetic, thermal, and mechanical aspects of power conversion systems and their controls [11].

As an application for the utilization of PLECS in combination with MATLAB/Simulink, [12] presents a PV inverter simulation including its control. The simulation results of the PLECS-based IGBT-inverter are compared with a common Simulink transfer function approach. It is shown that the simulation results achieved with the PLECS toolbox are as accurate as the simulation results generated with the transfer functions in Simulink but the simulation time of the PLECS model is 75 % less. A similar study comparing simulation results generated with Simulink, Psim and PLECS is conducted in [13] but with a thyristor-controlled induction generator application. The simulation performed with Simulink is as fast as the PLECS and Simulink co-simulation. To implement an alternative closed-loop control strategy, another application of this modeling method, a two-level PWM-fed voltage source converter, is published in [14]. All presented approaches have in common that they assume ideal IGBT models for their converter topologies.

In contrast to the aforementioned publications, in [15] an empirical validation of PLECS is investigated with a comparison between ideal and non-ideal m-level IGBT-inverters. Each switching cell of the inverter is configured as a non-ideal IGBT-diode combination. Since the PLECS library provides this non-ideal IGBT model, it will be compared with the approach from [9] in subsequent proceedings of this research.

Investigated Half-Bridge Module

The previously outlined quantities of dead time, delays, on-stage voltage drops and non-linearities of an IGBT are used in the following to model volt-second errors using the example of a high voltage insulated gate bipolar transistor (HVIGBT) module: the CM600DE-66X produced by Mitsubishi Electric Corporation [16]. This module is an IGBT-based halfbridge with a collector current $I_c = 600\,\mathrm{A}$ and a collector-emitter voltage $U_{ces} = 3.3\,\mathrm{kV}$. An equivalent circuit model of this inverter leg with an ohmic-inductive load is shown in Fig. 3.

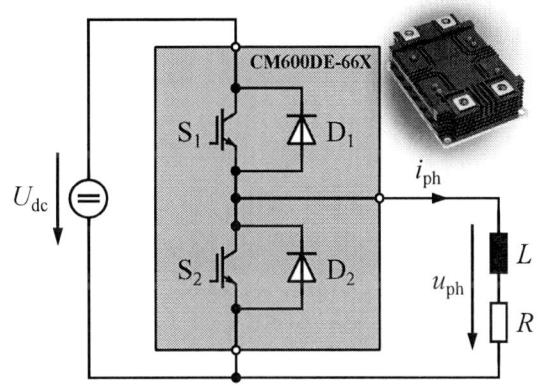

Fig. 3: Half-bridge circuit with HVIGBT module

Modeling Volt-Second Errors in PLECS

First analyses are conducted with the simplified model of [9] which exclude the non-linearities (c.f. Fig. 1). The IGBT-based half-bridge model is parameterized with module specific parameters from the datasheet introduced in the previous subsection. First, the controller of an one-phase inverter is built. Sine-triangle modulation is chosen to generate the gate signals for the PWM-based voltage signal (Fig. 4a). As there is no additional parameter to model the dead time T_d of the controller, an additional turn-on delay element Td_Si is used for each gate signal g_$Si, i \in [1, 2]$. These gate signals are passed forward to the half-bridge model which is shown in Fig. 4b. Neither the ideal IGBT model nor the non-ideal IGBT model from the PLECS library include separately declared turn-on and turn-off delays. Therefore, a further turn-on/off delay $T_{on/off}$ is implemented before the IGBT is switched on or off, respectivly. The forward voltage drops U_{IGBT} and U_{diode} of both semiconductor devices are the last contributors to the simplified model and are adjustable in the corresponding mask of the semiconductor device model.

The combination of the low speed (corresponding to a low fundamental frequency f_{sin}) and the low load (corresponding to a low reference voltage $\hat{U}_{ref}$) represents a critical operating point. The corresponding trajectory is plotted in purple in Fig. 5 with the simulation parameters given in Tab. I. The small pulse width of the reference phase voltage u^*_{ph} clearly highlights the dead time and delay effects. The deviations resulting from the device forward voltage drops are less significant since they are less than 0.2% of the dc-link voltage.

The next step is to model the non-linearities of the IGBT that are caused by charging and discharging the Miller-capacitance of the IGBT which corresponds to capacitance C_{gc} shown in the equivalent circuit diagram of a switching cell in Fig. 6a. Unfortunately, this well proven model to simulate the non-linearities of an IGBT-diode switching cell during the commutation process cannot be implemented in the PLECS environment due to the different signal types that are used. In Fig. 6b it can be seen that the collector-emitter signal is an analog signal (black colored line) while the gate signal is represented through a digital signal (green colored line). For this reason, the the two previously presented suitable modeling approaches of an IGBT-diode combination – the non-ideal IGBT of the PLECS library and the switching cell with additional capacitance – will be investigated and presented in the following.

The first model is based on the approach presented in [9]; each switching cell is extended by a parallel-connected capacitance C_{sc} that limits the voltage slopes $\frac{du_{ph}(t)}{dt}$ during commutation. The value of the parallel-connected capacitance C_{sc} presented in this paper is equivalent to the output capacitance C_{oes} taken from the datasheet. In addition, a resistor R_{sc} as equivalent series resistance (ESR) is connected to the capacitor to resolve state-source dependencies and to model a more realistic first order time lag behavior for the current within the commutation cells. The resistance R_{sc} is based on empirical factors. The second method utilizes the provided IGBT model from the PLECS library named "IGBT with Limited di/dt" that includes dynamic behavior by finite current slopes $\frac{di_{ph}(t)}{dt}$ during the turn-on and turn-off process [17].

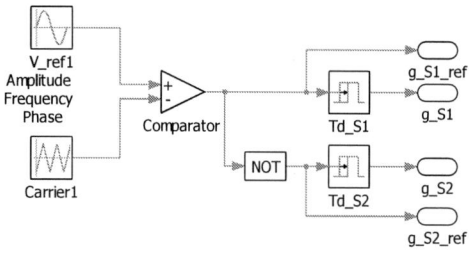

(a) Driver with dead-time T_d

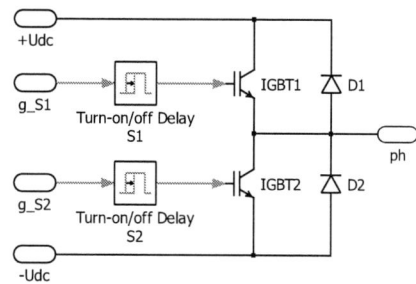

(b) Half-bridge with device delay and on-stage voltage drop

Fig. 4: Modeling of driver and half-bridge in PLECS

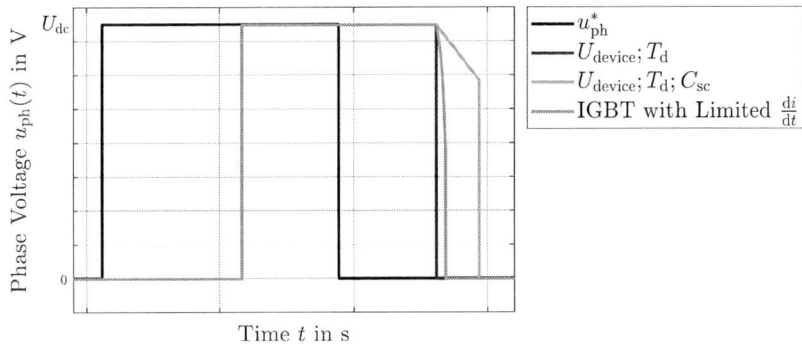

Fig. 5: Comparison of IGBT model approaches

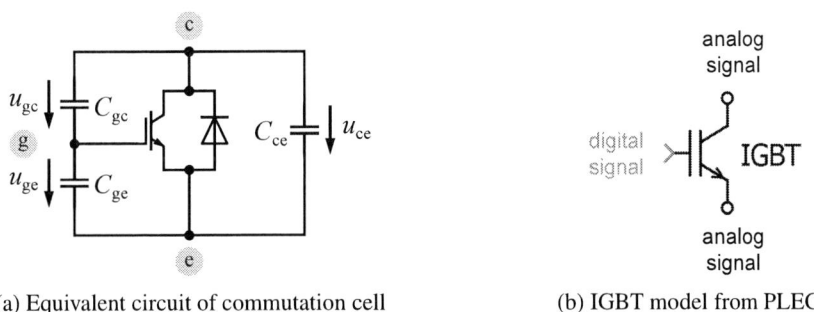

(a) Equivalent circuit of commutation cell

(b) IGBT model from PLECS

Fig. 6: Equivalent circuit of a commutation cell and data types of IGBT model in PLECS

The simulation results of the phase-voltage trajectories (Fig. 5) that represent the extended approaches including the additional non-ideal behavior of the IGBT show that the non-ideal IGBT model of the PLECS library results into a steeper voltage slope (blueish trajectory) than the approach with the additional capacitor (greenish trajectory). The simulation times of the two extended models are compared with the simulation time t_{sim} of the simplified model (purple trajectory). Regardless of their trajectories, the model extended by a capacitance C_{sc} is nearly two times faster than the PLECS library IGBT model (Tab. II). Moreover, the model with the added capacitance C_{sc} offers the advantage that the voltage trajectory can be described mathematically which is not possible with the model from the PLECS library. Both benefits – the simulation time and the availability of a mathematical description – lead to the decision to apply the IGBT model with the additional capacitor C_{sc} in the following. Accordingly, the approach using the PLECS-self-developed IGBT model is not pursued further in this work.

Table I: Simulation parameters for single-phase converter

Parameter	Symbol	Value	Parameter	Symbol	Value
Reference voltage	$\hat{U}_{\text{ref}}$	80 V	IGBT turn-on delay	$T_{\text{d,on}}$	1.25 µs
Sinus frequency	f_{sin}	10 Hz	IGBT turn-off delay	$T_{\text{d,off}}$	3.65 µs
Switching frequency	$f_{\text{sw}} = T_{\text{pwm}}^{-1}$	1 kHz	IGBT forward voltage	$U_{\text{f,IGBT}}$	3 V
Dc-link voltage	U_{dc}	1.5 kV	Diode forward voltage	$U_{\text{f,diode}}$	2.3 V
Load resistance	R_{load}	1 Ω	Capacitance	C_{sc}	3.8 nF
Load inductance	L_{load}	50 mH	ESR of capacitance	R_{sc}	10 Ω
Dead time of controller	T_{d}	10 µs			

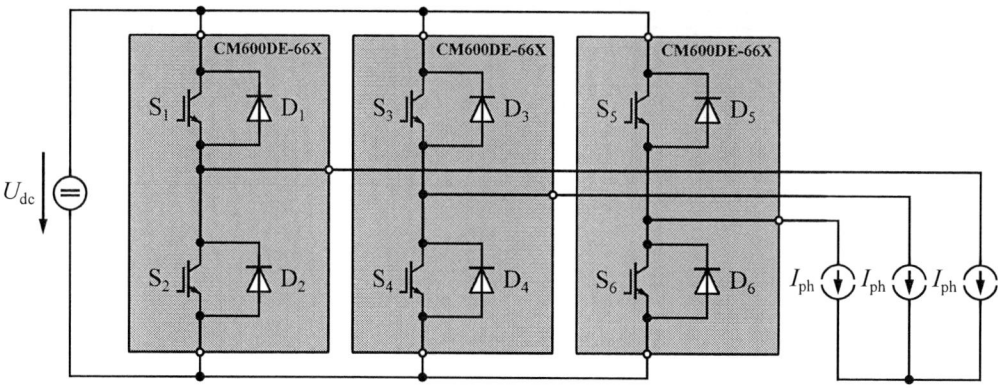

Fig. 7: Three-phase bridge converter

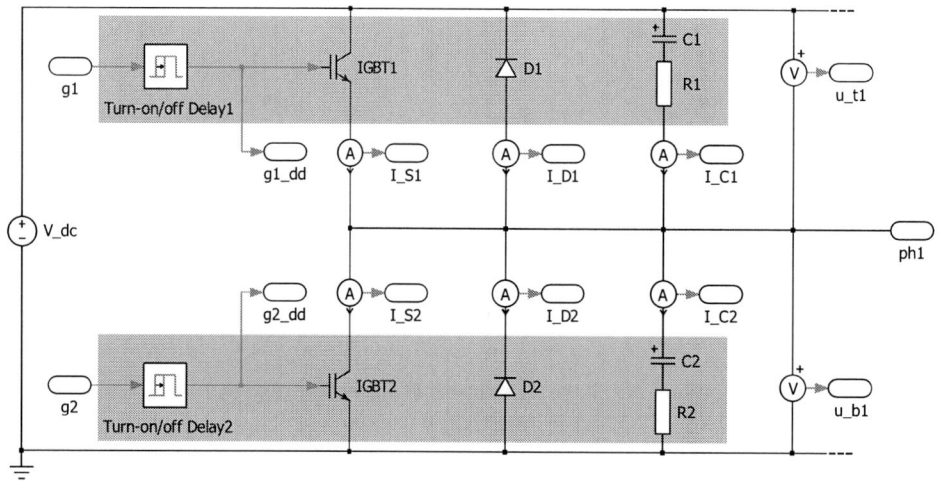

Fig. 8: Three-phase converter with IGBT turn-on/off delay and additional capacitance in PLECS

Non-Ideal Three-Phase IGBT-based Inverter

The three-phase IGBT-based inverter is modeled as shown in Fig. 7. Each switching cell consists of an IGBT-diode-capacitor combination. At first, for a proper investigation of the actual phase voltage u_{ph}, it is assumed that the load inductance $L_{load} \to \infty$. Thus, the load can be approximated by a constant current source I_{ph}. The implementation of the three-phase inverter model in PLECS is shown in Fig. 8. The modeling of each commutation cell is highlighted through a blueish background. It consists of an ideal IGBTi with the external device delay Turn $-$ on/off Delay$i$ connected to its gate signal. The ideal diode Di and the capacitance Ci with its ESR Ri is connected in parallel to limit the voltage slopes ($i \in [1,2]$).

In addition to the gate signals, the phase voltage $u_{ph} = u_{b1}$ and the currents through all three parallel-connected components within the switching cell are measured. These quantities are shown as an example w. l. o. g. for the first inverter leg.

Table II: Simulation times for single-phase converter

Modeling Approach	Simulation Time
$U_{f,IGBT/diode}$, T_d, $T_{on/off}$	t_{sim}
$U_{f,IGBT/diode}$, T_d, $T_{on/off}$, C_{sc}	$1.7 \cdot t_{sim}$
"IGBT with Limited di/dt" (PLECS library)	$3.3 \cdot t_{sim}$

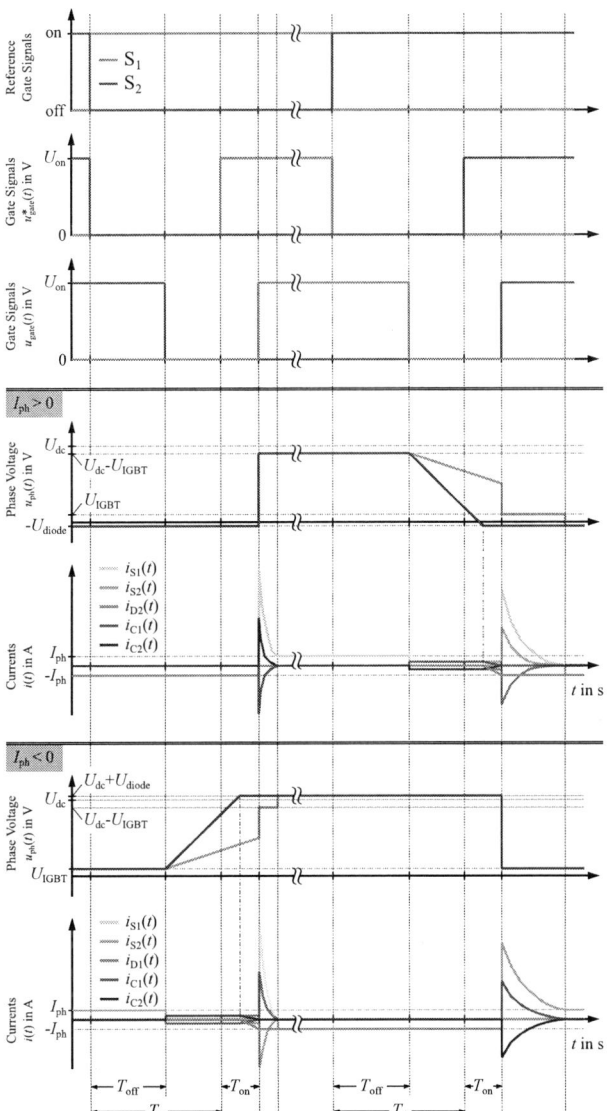

Fig. 9: Voltage and current trajectories of first inverter leg

Figure 9 shows the simulation results and is divided into three parts. The first part shows the reference gate signals for the upper and lower IGBT. Both signals are delayed by the dead time of the controller, i.e. each IGBT has a time delay before it is switched on. The third graph of the first block shows the additional turn-on and turn-off delay of the IGBT. When this signal changes to high, the semiconductor device is turned on. At this point in time, the phase voltage $u_{ph} = u_{b1}$ of the first phase drops at the load. From here, two cases must be distinguished depending on the sign of the load current I_{ph} which can be seen in the second and third part of Fig. 9. For both instances, the modeling approach for the phase voltage u_{ph} according to [9] is confirmed for an utilization in PLECS. Depending on the magnitude of the load current value I_{ph}, the voltage slope shows either an ideal ramp (dark red colored trajectory) or there is a step within the phase voltage signal (bright red colored trajectory). This voltage step is also reflected in the high currents within the switching cell during simulation.

The limiting current $I_{ph,lim}$ for the transition between the different phase-voltage characteristics can be determined analytically or from simulations (Fig. 10). For a current constant in time, the electric charge is the product of current and time (1), left. At the same time, the relationship between the electric charge

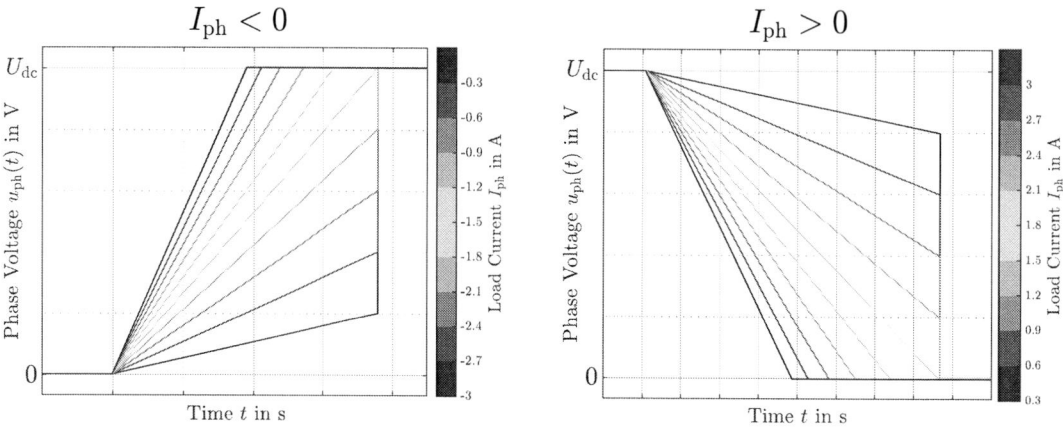

Fig. 10: Simulated phase-voltage trajectories u_{ph} for various constant phase-currents I_{ph}

and the voltage applies to the capacitors C_{cs} of the two switching cells of each inverter leg connected in parallel (1), right.

$$Q = I_{\mathrm{ph}} \cdot t \quad \text{and} \quad Q = 2C_{\mathrm{cs}} \cdot U_{\mathrm{dc}} \tag{1}$$

The resulting limiting current $I_{\mathrm{ph,lim}}$ is as follows in (2).

$$\left| I_{\mathrm{ph,lim}} \right| = \frac{2C_{\mathrm{cs}} \cdot (U_{\mathrm{dc}} \pm U_{\mathrm{f,IGBT}} \mp U_{\mathrm{f,diode}})}{(T_{\mathrm{d}} - T_{\mathrm{off}}) + T_{\mathrm{on}}} \tag{2}$$

The comparison of the simulative and analytical approach proves that the calculated current limits $I_{\mathrm{ph,lim}}$ of both approaches are nearly identical. For the investigated HVIGBT module, the positive and negative current limits are $\pm 1.5\,\mathrm{A}$. The similarity stems from the similar forward-voltage drops of IGBT and diode. This approach is also chosen in [18] but there is no method yet on how to dimension the additional capacitance C_{cs}.

As Fig. 11 shows, for a positive fixed load current (here $I_{\mathrm{ph}} = I_{\mathrm{ph,lim}}$), varying the capacitance C_{cs} also affects the slope of the phase voltage u_{ph} during commutation. In this case, the limit case is at the output capacitance $C_{\mathrm{cs}} = C_{\mathrm{oes}} = 3.8\,\mathrm{nF}$ (bright orange trajectory) taken from the datasheet. Smaller values of the capacitance $C_{\mathrm{cs}} < C_{\mathrm{oes}}$ lead to steeper phase-voltage slopes due to a faster discharging process.

Accordingly, the capacitance value C_{cs} allows to adjust the trajectory of the phase voltage. In this work, the output capacitance

$$C_{\mathrm{oes}} = C_{\mathrm{cs}} \tag{3}$$

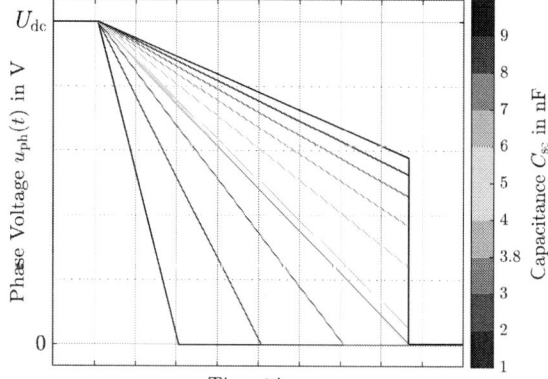

Fig. 11: Simulated phase-voltage trajectories u_{ph} for various capacitance C_{sc}

from the corresponding datasheet is adopted as in [19]. It describes that the output capacitance limits the gradient of the voltage at switching transitions. By measuring the output capacitance across the collector and emitter connections, the specific parameter C_{oes} is determined for the datasheet. Thus, the voltage drop acorss this capacitance relates to the phase voltage of the bottom switching cell. Consequently, in a following step, a characterization of the corresponding module by measuring the collector-emitter voltage must be conducted to validate the model.

Conclusion and Outlook

Controller dead time, semiconductor device delay, its forward voltage drop, and the non-linear behavior of voltage-fed PWM-based IGBT inverters are the major contributors to volt-second errors that could lead to miscalculation of flux-linkage estimation at very low speed. The investigated approach for accurate modeling of IGBT-based converters in PLECS is easy to implement, results to fast simulation times, and is therefore suitable for an accurate modeling of non-ideal IGBT-based converters. It is demonstrated how to implement the dead time of the controller, the delay of the semiconductor devices as well as their forward voltage drop and how to choose these parameters from the corresponding datasheet.

Additionally, two different approaches to model the limited gradient of the phase voltage during commutation are presented: Firstly, the IGBT model with limited current slope that is offered by the PLECS library and secondly, an extension of the switching cell by a capacitance with its ESR to directly limit the voltage slope. The latter shows a faster simulation time than the current-limited IGBT model from PLECS and it is further possible to derive an analytical description for the additional capacitance added to each switching cell. Finally, the influence of various parameters, like the phase current and the additional capacitance on the phase-voltage trajectory, is presented.

In a next step, a characterization of the presented HVIGBT module has to be performed by measurements on a double-pulse testbench to validate the half-bridge model. Validation of the quantity of the additional capacitance is of particular interest so that it is extractable directly from the corresponding datasheet for modeling purposes.

References

[1] J. S. Lee, C. Choi, J. Seok and R. D. Lorenz, "Deadbeat-Direct Torque and Flux Control of Interior Permanent Magnet Synchronous Machines With Discrete Time Stator Current and Stator Flux Linkage Observer," in IEEE Transactions on Industry Applications, vol. 47, no. 4, pp. 1749-1758, July-Aug. 2011, doi: 10.1109/TIA.2011.2154293.

[2] M. Schubert, D. Scharfenstein and R. W. De Doncker, "On the torque accuracy of stator flux observer based induction machine control," 2016 IEEE Symposium on Sensorless Control for Electrical Drives (SLED), 2016, pp. 1-8, doi: 10.1109/SLED.2016.7518791.

[3] J. Holtz and N. Oikonomou, "Synchronous Optimal Pulsewidth Modulation and Stator Flux Trajectory Control for Medium-Voltage Drives," in IEEE Transactions on Industry Applications, vol. 43, no. 2, pp. 600-608, March-april 2007, doi: 10.1109/TIA.2006.889893.

[4] N. Hartgenbusch, I. Ralev, R. W. De Doncker and T. Kojima, "Stator Flux Trajectory Control combined with Optimized Pulse Patterns for Interior Permanent Magnet Machines," 2019 22nd International Conference on Electrical Machines and Systems (ICEMS), 2019, pp. 1-6, doi: 10.1109/ICEMS.2019.8921717.

[5] A. R. Hefner and D. L. Blackburn, "An Analytical Model for the Steady-State and Transient Characteristics of the Power Insulated Gate Bipolar Transistor," Solid State Electronics, vol. 31, p. 1513 1532, 10, 1988, doi: 10.1016/0038-1101(88)90025-1.

[6] P. O. Lauritzen, G. K. Andersen and M. Helsper, "A Basic IGBT Model with Easy Parameter Extraction," 2001 IEEE 32nd Annual Power Electronics Specialists Conference (IEEE Cat. No.01CH37230), 2001, pp. 2160-2165 vol. 4, doi: 10.1109/PESC.2001.954440.

[7] R. Withanage, N. Shammas, S. Tennakoon, C. Oates and W. Crookes, "IGBT Parameter Extraction for the Hefner IGBT Model," Proceedings of the 41st International Universities Power Engineering Conference, 2006, pp. 613-617, doi: 10.1109/UPEC.2006.367551.

[8] Y. Gao, N. Li, S. Guo, and H. Liu, "The Modeling and Parameters Identification for IGBT Based on Optimization and Simulation," Bio-Inspired Computational Intelligence and Applications, 2007, pp. 628-638, doi: 10.1007/978-3-540-74769-7_67.

[9] N. Bedetti, S. Calligaro and R. Petrella, "Accurate Modeling, Compensation and Self-Commissioning of Inverter Voltage Distortion for High Performance Motor Drives," 2014 IEEE Applied Power Electronics Conference and Exposition, 2014, pp. 1550-1557, doi: 10.1109/APEC.2014.6803513.

[10] J. H. Alimeling and W. P. Hammer, "PLECS-Piece-Wise Linear Electrical Circuit Simulation for Simulink," Proceedings of the IEEE 1999 International Conference on Power Electronics and Drive Systems (Cat. No.99TH8475), 1999, pp. 355-360 vol.1, doi: 10.1109/PEDS.1999.794588.

[11] PLECS, The Simulation Platform for Power Electronic Systems. https://www.plexim.com/products/plecs, Plexim GmbH, 2022.

[12] M. Ciobotaru, T. Kerekes, R. Teodorescu and A. Bouscayrol, "PV Inverter Simulation Using MATLAB/Simulink Graphical Environment and PLECS Blockset," 32nd Annual Conference on IEEE Industrial Electronics, 2006, pp. 5313-5318, doi: 10.1109/IECON.2006.347663.

[13] D. Baimel, R. Rabinovici and S. Ben-Yakov, "Simulation of Thyristor Operated Induction Generator by Simulink, Psim and Plecs," 18th International Conference on Electrical Machines, 2008, pp. 1-6, doi: 10.1109/ICELMACH.2008.4799931.

[14] M. Biweta and M. Mamo, "Closed Loop Control Strategy of Back to Back PWM Converter Fed by PMSG Using PLECS Toolbox on Matlab/Simulink for Wind Energy Application," IEEE AFRICON, 2017, pp. 1313-1318, doi: 10.1109/AFRCON.2017.8095672.

[15] M. S. Sharma and A. P. R. Taylor, "Comparative PLECS Modelling of Ideal and Non-Ideal m-Level IGBT Inverters at 11kV," Australasian Universities Power Engineering Conference, 2009, pp. 1-6.

[16] High Voltage Insulated Gate Bipolar Transistor, 5th- CM600DE-66X datasheet. https://www.mitsubishielectric.com/semiconductors/content/product/powermodule/hvigbt_ipm/x_series/cm600de-66x_e.pdf, Mitsubishi Electric Corporation, 2022.

[17] PLECS documentation, "IGBT with Limited di/dt," Plexim GmbH, 2022.

[18] M. Schubert and R. W. D. Doncker, "Self-Sensing Torque Control of an IM Drive with Instantaneous Phase-Voltage Sensing," 2019 International Conference on Industrial Engineering, Applications and Manufacturing (ICIEAM), 2019, pp. 1-6, doi: 10.1109/ICIEAM.2019.8742913.

[19] AN2011-05 Industrial IGBT Modules - Explanation of Technical Information, Application Note AN 2011-05, V1.2 November 2015, IFAG IPC APS, Infineon Technologies AG, 2015.

Novel Analytical Method for Estimating the Junction-to-Top Thermal Resistance of Power MOSFETs

José Miguel Sanz-Alcaine[1], Francisco Jose Perez-Cebolla[1], Carlos Bernal-Ruiz[1], Asier Arruti[2], Iosu Aizpuru[2]

[1] University of Zaragoza, Zaragoza, Spain
[2] Mondragon Unibertsitatea, Mondragon, Spain
Email: jm.sanz@unizar.es, fperez@unizar.es

Acknowledgments

The work has been supported by the VEGAN project (KK-2021/00044) of the Elkartek programme of the Basque Country.

Keywords

≪Thermal model≫,≪Device modelling≫, ≪Packaging≫, ≪Power semiconductor device≫, ≪Double-side cooling (DSC)≫.

Abstract

This papers proposes a new methodology for estimating the thermal resistance from the junction-to-top capsule surface. By placing the transistor in a vertical position, without being soldered to any PCB, and sensing the dissipated power and the temperatures of the device, it is possible to characterize the internal thermal resistance.

1 Introduction

Modern power semiconductors such as Wide bandgap (WBG) devices are well known by their system benefits such as smaller die size, lower $R_{DS(ON)}$ or good switching behaviour [1]. However, these benefits bring to new design challenges that must me solved. Some of them are a more complex loss characterization, critical gate drive design or higher thermal resistance which brings the need to new thermal design approaches.

Classically, the only heat propagation path considered by semiconductor manufacturers is the one from the junction to the bottom capsule of the device [2]. This is because the path from the junction to the top capsule is made of epoxy, which has a low thermal conductivity [3]. However, this new thermal design challenge has led power electronics designers to take the most out of their transistors packages by placing heatsinks at the top side [4]. To correctly assess the thermal behaviour of the transistor, a proper method for measuring the thermal resistance from the junction-to-top capsule (R_{JC_T}) is needed.

A few modern devices now include this parameter into their datasheets as in [5]. However, the methodology to determine this thermal resistance is still not clear and under discussion [6, 7]. In fact, the JEDEC standard describes the proper procedure for measuring the thermal resistance from the junction to-bottom capsule (R_{JC}) and the thermal resistance from the junction-to-ambient (R_{JA}) in [8], but not for the (R_{JC_T}). An example of the different parts of a SMD power transistor is illustrated in Fig. 1.

Commonly the thermal data provided on MOSFET datasheets is very limited. Transistors manufacturers provide typical and sometimes maximum values of R_{JC} and R_{JA}, assuming for its determination a model with a single path of heat propagation, as shown in Fig. 2 but do not report R_{JC_T} values.

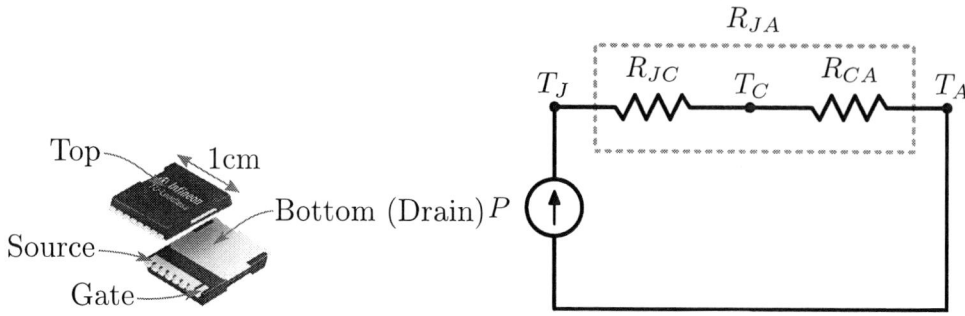

Fig. 1: Parts and dimensions of a SMD power transistor.

Fig. 2: Compact thermal model considering a single path of heat transfer to the environment.

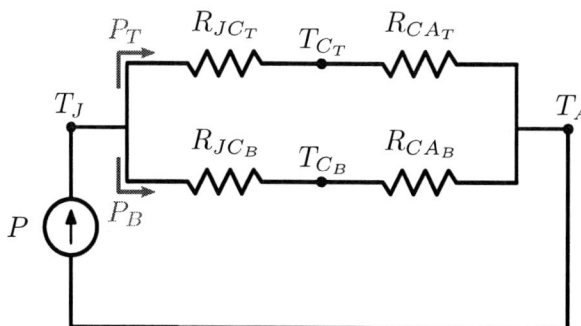

Fig. 3: Compact thermal model considering two paths of heat transfer to the environment.

In order to mitigate the above problem, a calculation procedure is proposed. It is based on a propagation compact model of heat considering two heat propagation paths as in Fig. 3.

R_{JC_T}, R_{JC_B}, R_{JA_T} and R_{CA_B} are respectively the thermal resistances between the junction and the top capsule surface, the thermal resistance between the junction and the bottom capsule surface, the thermal resistance between the top capsule surface and the environment and the thermal resistance between the bottom capsule surface and the environment. On the other hand, T_J and T_A are respectively the junction temperature and the ambient temperature. Finally, P, P_T and P_B specify respectively the total power dissipated by the device, the fraction of power circulating through the top path and the fraction of power circulating through the bottom path.

2 Methodology

With the transistor located inside a natural convection still air chamber, in an upright position and without being soldered to any PCB as in Fig. 4, it is reasonable to assume a compact thermal model in which the heat generated in the junction is transmitted to the outside following two paths. One from the junction to the bottom capsule surface and another path from the junction to the top capsule surface. Moreover, in a thermal steady state and in the absence of external excitation, it is appropriate to assume that the temperature of the top and bottom surfaces are equal to the temperature of the junction, being in turn equal to the ambient temperature. Under the above assumptions, the thermal behavior of the transistor can be modeled in a first approximation as an ideal lineal system, analogous to the electrical circuit shown in Fig. 3.

For a proper and repeatable temperature measurement, different sensors can be used taking into account their limitations. If a thermocouple is chosen, as shown in Fig. 4, the position in which it is placed on the package surface, as well as the adhesive used to fix it on this, can be critical for a good measurement as described in [9] and [10]. On the other hand, if an infrared camera is chosen, the emisivity factor and the focal distance must be taken into account. Due to the low emissivity coefficient of the bottom

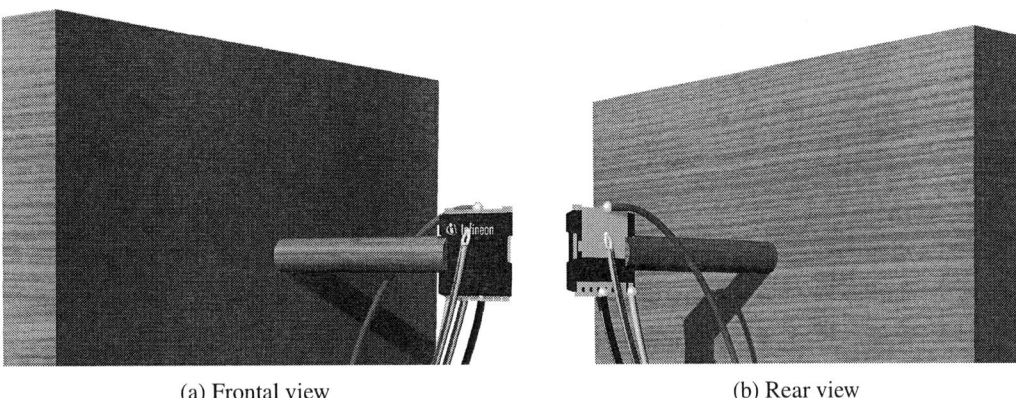

(a) Frontal view   (b) Rear view

Fig. 4: 3D Model of the transistor to air test-bench.

metal capsule, the reflected temperature has a strong influence on the temperature measurement, thus it would be beneficial to increase its emisivity factor. Electrical tape or specific paints that withstand high temperatures can be applied for increasing this parameter.

From the analysis of Fig. 3, it is deduced that the temperature difference between the junction and both package surfaces is equal to

$$T_J - T_{C_T} = P_T \cdot R_{JC_T} = (1-D) \cdot P \cdot R_{JC_T} \tag{1}$$

and

$$T_J - T_{C_B} = P_B \cdot R_{JC_B} = D \cdot P \cdot R_{JC_B} \tag{2}$$

where D is the fraction of total power, P, flowing through the bottom path of the circuit shown in Fig. 3. Subtracting the above expressions, it is possible to find the relationship between both surfaces

$$T_{C_B} - T_{C_T} = [R_{JC_T} - D(R_{JC_T} + R_{JC_B})] \cdot P \tag{3}$$

which yields to

$$P_B = D \cdot P = \frac{R_{JC_T} + R_{CA_T}}{R_{JC_T} + R_{CA_T} + R_{JC_B} + R_{CA_B}} \cdot P. \tag{4}$$

Therefore,

$$D = \frac{R_{JC_T} + R_{CA_T}}{R_{JC_T} + R_{CA_T} + R_{JC_B} + R_{CA_B}} = \frac{R_{JA_T}}{R_{JA_T} + R_{JA_B}} \tag{5}$$

and the power flowing through the top path is

$$P_T = (1-D) \cdot P = \frac{R_{JC_B} + R_{CA_B}}{R_{JC_T} + R_{CA_T} + R_{JC_B} + R_{CA_B}} \cdot P. \tag{6}$$

In consequence, the fraction $1 - D$ is equal to

$$1 - D = \frac{R_{JC_B} + R_{CA_B}}{R_{JC_T} + R_{CA_T} + R_{JC_B} + R_{CA_B}} = \frac{R_{JA_B}}{R_{JA_T} + R_{JA_B}}. \tag{7}$$

Same way

$$T_{C_B} - T_A = P_B \cdot R_{CA_B} = D \cdot P \cdot R_{CA_B} \tag{8}$$

and

$$T_{C_T} - T_A = P_T \cdot R_{CA_T} = (1 - D) \cdot P \cdot R_{CA_T} \tag{9}$$

If Eq. 8 and Eq. 9 get combined, the relationship between power and temperature difference between top and bottom surfaces is expressed as

$$T_{C_B} - T_{C_T} = P \cdot \frac{R_{JC_T} R_{CA_B} - R_{JC_B} R_{CA_T}}{R_{JA_T} + R_{JA_B}} = P \cdot R_{EQ} \tag{10}$$

where

$$R_{EQ} = \frac{R_{JC_T} R_{CA_B} - R_{JC_B} R_{CA_T}}{R_{JA_T} + R_{JA_B}} \tag{11}$$

On the other hand, the heat transfer in natural convection is described by the Newton's Law of Cooling

$$\dot{Q} = hA(T - T_f)^b \tag{12}$$

where $\dot{Q}$ is the heat transferred per unit time, A is the area of the object, h is the heat transfer coefficient, T is the object's surface temperature, T_f is the fluid temperature, and b is a scaling exponent. The convective coefficient h depends on the absolute temperature, the material property of the fluid, the flow rate of the fluid, the dimensions of the considered surface, the orientation of the considered surface, and the surface texture.

In this case, the influence of the surface texture is quasi-negligible [11]. Therefore, since the area of both surfaces is approximately equal, as well as their lengths, with the transistor placed in a vertical position the thermal resistances R_{CA_T} and R_{CA_B} should also be approximately the same.

For illustration purposes, Fig. 5 shows how ambient temperatures, and therefore the resistances (R_{CA_T} and R_{CA_B}), are equal in both sides of the transistor with this placed in a vertical position. This can be corroborated by observing the result from Finite Element Method (FEM) simulations [12].

Consequently, assuming R_{CA_T} and R_{CA_B} equal to R_{CA}, R_{EQ} results in

$$R_{EQ} = \frac{R_{CA}(R_{JC_T} - R_{JC_B})}{R_{JC_T} + R_{JC_B} + 2R_{CA}}. \tag{13}$$

Also, if Eq. 8 is divided by Eq. 9, it is possible to determine the value of D as

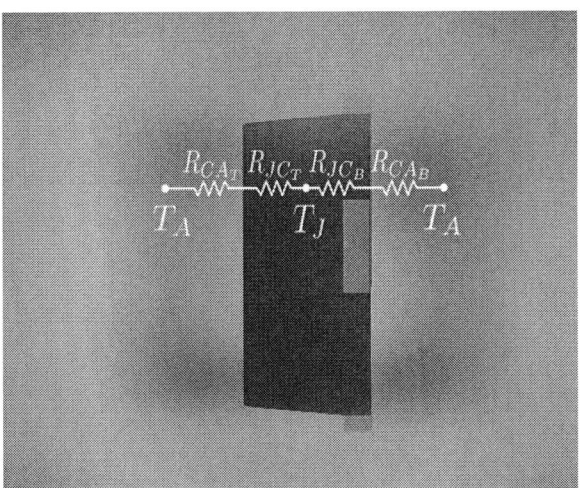

Fig. 5: Illustration of how heat spreads symmetrically when the transistor is not attached to any PCB and it is placed vertically.

$$\frac{T_{C_B} - T_A}{T_{C_T} - T_A} = \frac{D}{1-D} \rightarrow D = \frac{T_{C_B} - T_A}{T_{C_B} + T_{C_T} - 2T_A} \tag{14}$$

Making use of Eq. 4 and Eq. 6 it is also possible to determinate the power along each path. The dependence of D with the total power, and therefore with the temperature, is assumed negligible, as will be shown in the next section.

Regarding R_{CA_B} and R_{CA_T}, both resistances can be determined in a similar way by replacing the value of D in Eq. 8 and Eq. 9 respectively.

On the other hand, considering the single-path model used by the manufacturer and the typical value of thermal resistance capsule-junction R_{JC} provided by it, it is possible to estimate the temperature of the junction:

$$T_J = P \cdot D \cdot R_{JC_{Datasheet}} + T_{C_B} \tag{15}$$

Knowning T_J and D it is possible to determinate R_{JC_T}. This is:

$$R_{JC_T} = \frac{T_J - T_{C_T}}{P \cdot (1-D)} \tag{16}$$

3   Experimental Results

For validating the methodology, the proposed test bench shown in Fig. 4 has been built. The Device Under Test (DUT) is the Infineon 100V OptiMOS™ 5 power MOSFET IPT015N10N5 [13].

Temperatures in both bottom and top surfaces have been recorded for different power values. It is observed that, as no heatsink is used, high temperatures are reached for low power values as depicted in Fig. 7. These temperature values given by the thermocouples have been validated by an infrared camera.

By applying Eq. 14, the negligible dependence of D with power is corroborated as shown in Fig. 8. This allows to estimate the junction temperature (T_J) of the DUT through Eq. 15 and therefore calculate R_{JC_T} by using Eq. 16. This thermal resistance value is shown for different powers values in Fig. 9.

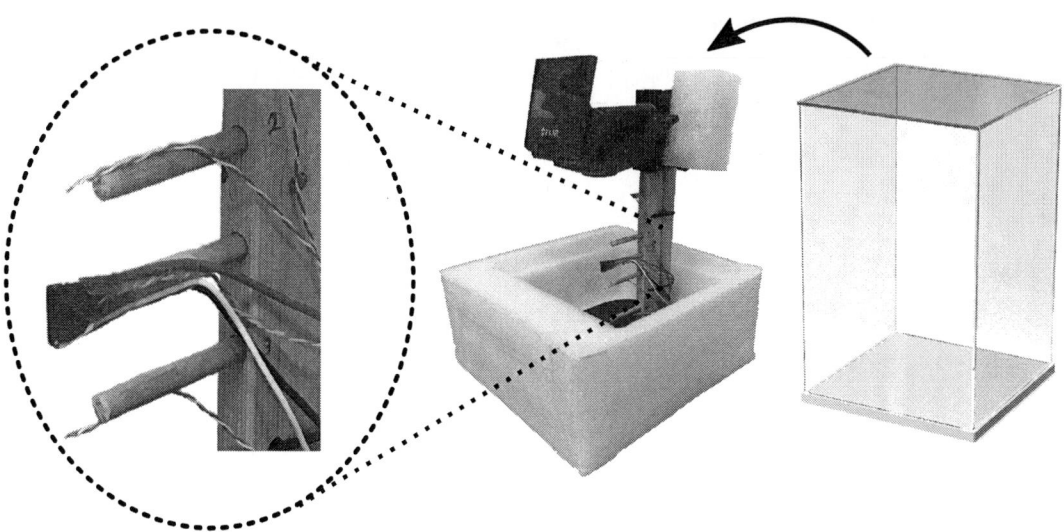

Fig. 6: Images of the built test bench to characterize the transistor inside a still air chamber, in vertical position and without being soldered to any PCB.

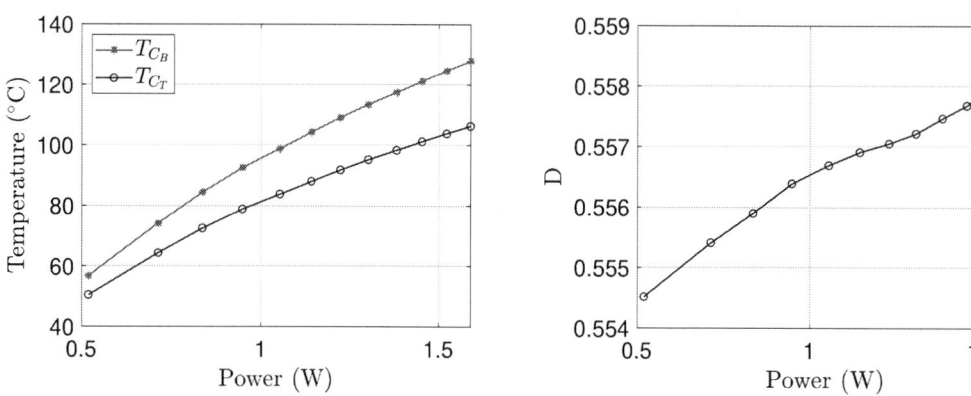

Fig. 7: Experimental relationship between the capsules temperatures (top and bottom) and the total power dissipated in the MOSFET.

Fig. 8: Experimental relationship between the fraction of the power in the lower path and the total power dissipated in the MOSFET.

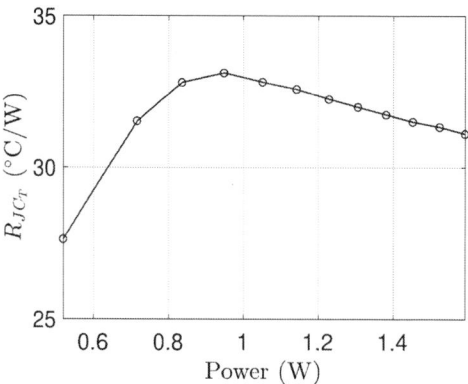

Fig. 9: Experimental values of the thermal resistance from the junction to the top capsule *vs* the power in the transistor.

4 Conclusion

This paper proposes a novel methodology for estimating the thermal resistance from junction-to-top capsule of power semiconductor devices. This characterization procedure takes into account the power from the junction to the top surface. The carried out research is based on the behaviour of a transistor located inside a natural convection still air chamber, in vertical position and without being soldered to any PCB. Under these operating conditions, and as is common in many types of transistor packages, with roughly equal dimensions and geometry on the top and bottom surfaces of the capsule, it is possible to determine the value of thermal resistances in a two-path equivalent thermal circuit, as well as the fraction of power associated with each path. This methodology is applied to a real transistor and results are shown.

References

[1] Giuseppe Iannaccone, Christian Sbrana, Iacopo Morelli, and Sebastiano Strangio. Power electronics based on wide-bandgap semiconductors: Opportunities and challenges. *IEEE Access*, 9:139446–139456, 2021.

[2] Thermal Resistance Theory and Practice. Application note, Infineon Technologies, January 2000.

[3] Gongyue Tang, Tai chong Chai, and Xiaowu Zhang. Thermal optimization and characterization of sic-based high power electronics packages with advanced thermal design. *IEEE Transactions on Components, Packaging and Manufacturing Technology*, 9(5):854–863, 2019.

[4] TO-Leaded Top Side Cooling. Application note, Infineon Technologies, May 2021.

[5] ISC230N10NM6 -OptiMOS™ 6 Power-Transistor, 100V. Datasheet, Infineon Technologies, July 2021.

[6] Heng Yun Zhang, Xiao Wu Zhang, B. L. Lau, Sharon Lim, Liang Ding, and M. B. Yu. Thermal characterization of both bare die and overmolded 2.5-d packages on through silicon interposers. *IEEE Transactions on Components, Packaging and Manufacturing Technology*, 4(5):807–816, 2014.

[7] Y. Tal and A. Nabi. A simple analytic method for converting standardized ic-package thermal resistances (θ_{ja}, θ_{jc}) into a two-resistor model (θ_{jb}, θ_{jt}). In *Seventeenth Annual IEEE Semiconductor Thermal Measurement and Management Symposium (Cat. No.01CH37189)*, pages 134–144, 2001.

[8] JESD51-14 - Methodology for the Thermal Measurement of Component Packages (Singular Semiconductor Device). Standard, JEDEC-Solid State Technology Association, January 2010.

[9] Qinghong He, Shane Smith, and Guohua Xiong. Thermocouple attachment using epoxy in electronic system thermal measurements — a numerical experiment. In *2011 27th Annual IEEE Semiconductor Thermal Measurement and Management Symposium*, pages 280–291, 2011.

[10] Precautions When Measuring the Rear of the Package with a Thermocouple. Application note, ROHM Semiconductor, October 2020.

[11] Papa Momar Souare, Mamadou Kabirou Toure, Stephanie Allard, Benoit Foisy, Bijan Borzou, Eric Duchesne, and Julien Sylvestre. High precision numerical and experimental thermal studies of microelectronic packages in still air chamber tests. In *2018 7th Electronic System-Integration Technology Conference (ESTC)*, pages 1–8, 2018.

[12] Pavel P. Khramtsov Oleg G. Martynenko. *Free-Convective Heat Transfer: With Many Photographs of Flows and Heat Exchange*. Springer Berlin, Heidelberg, 2005.

[13] IPT015N10N5 - 100V OptiMOS™ 5 power MOSFET in TOLL. Datasheet, Infineon Technologies, October 2016.

DC-side Impedance for Handling Interoperability of Multi-vendor Multi-terminal HVDC Systems

Ashkan Nami[*], Adil Abdalrahman, Ying-Jiang Häfner, Malaya Kumar Sahu,
and Khirod Kumar Nayak
Hitachi Energy - HVDC, Ludvika, Sweden
[*]Email: ashkan.nami1@hitachienergy.com

Keywords

≪Interoperability≫, ≪DC stability≫, ≪HVDC≫, ≪DC impedance≫, ≪Impedance scanning≫

Abstract

Using equivalent impedance to address possible interoperability issues is a well-established practice in electric rail systems and it is also a feasible approach for investigating DC-side control interactions, thereby avoiding or mitigating possible interoperability issues in multi-vendor multi-terminal high voltage direct current (HVDC) systems. This paper presents different aspects related to DC-side equivalent impedances which may be necessary to be considered in practice.

Introduction

In addressing the challenges of shifting from a fossil fuel-based to renewable energy-based powers, the application of high voltage direct current (HVDC) transmissions is continuously increased. In particular, considering the intermittency and spatial disparity of renewable energy generation, interconnecting the renewable powers from a wide area on the DC side and developing multi-terminal DC (MTDC) networks have recently received an increasing interest [1]-[3] in order to achieve operational flexibility and power supply sustainability. In MTDC, there will be at least one power converter at each terminal or node. The dynamic behaviors and steady-state characters of power converters are highly dependent on control design. The converter control is the core intellectual property (IP) of power converter manufactures from all types of industries including HVDC. Due to black-box character of HVDC control, the interoperability issue draws attention to multi-vendor multi-terminal HVDC system developers as well as researchers. It should be mentioned that the interoperability issue related to power converters is not new and established practices as well as standards/guidelines are available since decades [4]. It has been shown in railway application that using equivalent impedance to characterize the active element such as power converter and to assess the stability is very effective. Therefore, using equivalent impedance for HVDC converter can be a good approach to address the interoperability issue related to multi-vendor MTDC systems.

Since the finding of DC oscillation in a MTDC with multi-vendors [5], a number of studies related to using impedance to assess DC network stability have been published in the literature. In [6]-[8], a stability analysis based on an analytical analysis of models of HVDC converters or entire MTDC systems is proposed in which detailed knowledge of the converter control systems is needed. However, analytical expression of the equivalent converter DC-side impedance is very complex. Modular multilevel converter (MMC) based HVDC stations may be far more complicated to be described by simple analytical forms due to physical limits in the main circuit equipment (voltage, current rating, temperature, etc.). Furthermore, the dynamics of each converter cell may be difficult to be represented in simple analytical forms [9]. Although the use of an average model could avoid the complexity, it may miss some significant dynamic properties. Moreover, significant details of the converter and its corresponding control are needed in those methods which is not suitable for multi-vendor applications [10].

Therefore, methods which allow protecting the IP of vendors (while giving more realistic representation) can facilitate impedance-based stability analysis in multi-vendor MTDC systems. For this purpose, methods based on the detailed impedance measurements utilizing a frequency sweep approach [11]-[14] can be used. Reference [15] presents an effective equivalent impedance calculation method for control stability assessment in HVDC grids with the main contribution of impedance combination of different elements. This leads to highly-accurate impedances of radial MTDC networks being achieved in a very short time. However, the considered MTDC is based on symmetrical monopole meaning each terminal has only one converter. References [16] and [17] introduce a PSCAD-compatible software toolbox that enables the accurate impedance measurement of MMC while in [17], a bipolar HVDC system with metallic return is considered. However, none of those references address (a) the details of considered impedance scanning setups, (b) HVDC relevant scanning frequency range, and (c) impact of AC network conditions on the scanned impedances.

The aim of this paper is to address the aforementioned issues and as a result, to share some relevant study experiences as a vendor with a lot of real project experiences. In the following section, the study experiences is presented which is split into five different subsections: 1) the scanning setup in both monopole and bipolar, 2) the scanning frequency range, 3) the impact of the AC network on the equivalent DC-side impedance, 4) the impact of load level of converter, and 5) the impact of DC system configuration. Finally, a summary of the studies is given.

DC-side Impedance Scanning – Study Experiences

DC-side Impedance Scanning Setup

To enable the active power flow over the HVDC link from one station to other station, one station is operated in DC voltage control (U_{dc}-control) mode while other station is operated in active power control (P-control) mode. There are other control modes as well such as islanding (frequency-voltage) control mode of operation where converter regulates neither DC voltage nor active power. Accordingly, a converter under test sees other station as a current source (when it is operated in U_{dc}-control mode) or a voltage source (when it is operated in power or islanding control mode) at its DC-side. Thus, depending on the control mode, the impedance scan circuit on DC-side varies. It is also worthwhile to investigate the impact of one pole converter on another pole converter in a bipolar system. Therefore, the test circuits are discussed for both monopole and bipolar systems.

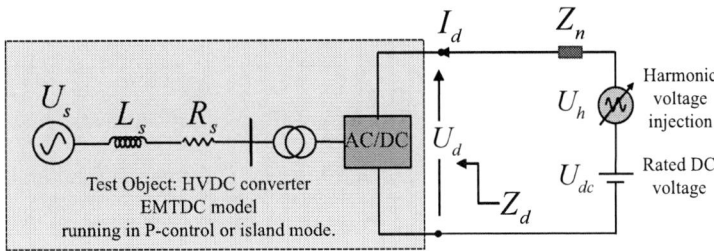

Fig. 1: Test circuit for measuring impedance seen from the DC-side of a converter operating in P-control or islanding control mode.

Fig. 1 shows the test circuit for scanning the DC-side impedance of a P-controlling or islanding control station in a monopole configuration. Note that a detailed switching model of converter provided by the manufacturer is used in the test circuits proposed throughout this paper. Moreover, any passive element connected in series and shunt to ground on the DC-side can be excluded from the test object leading to only active element being seen from the DC-port. Therefore, those passive elements can be considered as a part of DC network which are not shown in Fig. 1 for the sake of clarity. U_{dc}-controlling station can be represented by an equivalent DC voltage source behind a series impedance (Z_n) or by an ideal DC voltage source whose voltage is set equal to the rated DC voltage of HVDC link. As the objective is to measure the impedance inserted by the converter which is an active element, a variable AC voltage source is also added in series with the DC voltage source in order to perform an impedance scanning at various frequencies. Therefore, the impedance scanning can be performed by driving sufficient harmonic currents to the test object and measuring the voltage drops across it. It is worth mentioning that the AC voltage source magnitude (U_h) needs to be kept below a certain level in order to ensure that any protection is not activated while injecting harmonic voltages. The frequency of the AC voltage source can be varied over a wide range of interest. Finally, the impedance seen from the DC-side (Z_d) is measured as given by (1).

$$Z_d = \frac{U_d}{I_d} \tag{1}$$

where U_d is the voltage across the test object and I_d is the current driven to the test object.

Fig. 2 shows DC-side impedance scanning results for a part of frequency range of interest from two different setups: the test circuit shown in Fig. 1 and a point-to-point HVDC system. From these results, it can be concluded that either setup can be used as there is very little difference between the results of the two setups in both real and imaginary parts of the DC-side impedance (Z_{re} and Z_{im}, respectively).

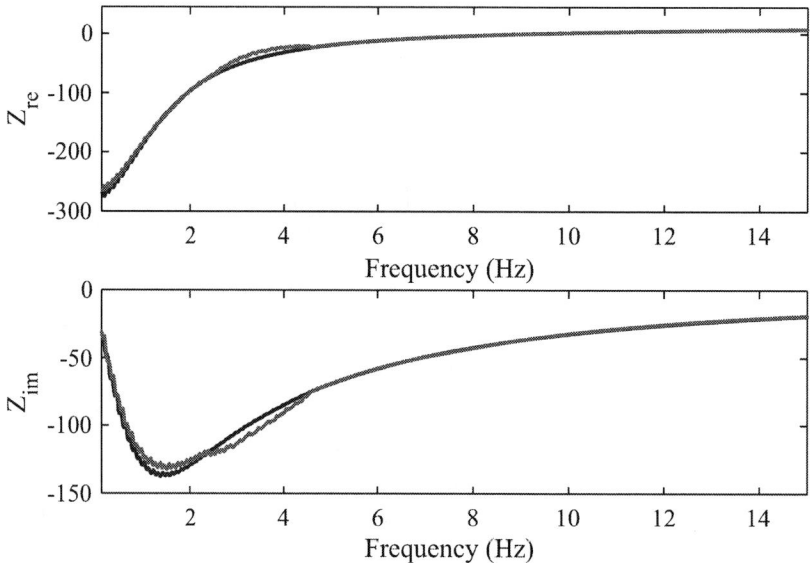

Fig. 2: DC-side impedance scanning results for P-controlling station from the proposed test circuit (blue color) and a point-to-point HVDC system (red color).

Moreover, Fig. 3 shows the test circuit for scanning the DC-side impedance of a bipolar converter station operated in P-control or islanding control mode. Note that the bipolar configuration has a neutral pole which can be a ground electrode via ground impedance, or a metallic return to other station or can be floating point (rigid-bipolar). To investigate the impact of one pole converter dynamics on the other pole converter dynamics, the DC-side impedance can be measured for individual pole converters while any neutral impedance connected to ground is kept as a part of outer circuit (see Fig. 3). The DC-side impedances of both positive and negative pole converters (Z_{dp} and Z_{dn}, respectively) can be represented by following equations:

$$Z_{dp} = \frac{U_{dp}}{I_{dp}} \tag{2}$$

$$Z_{dn} = \frac{U_{dn}}{I_{dn}} \qquad (3)$$

where U_{dp} and U_{dn} are the voltages across the positive and negative test objects, respectively. I_{dp} and I_{dn} are the currents driven to the positive and negative test objects, respectively.

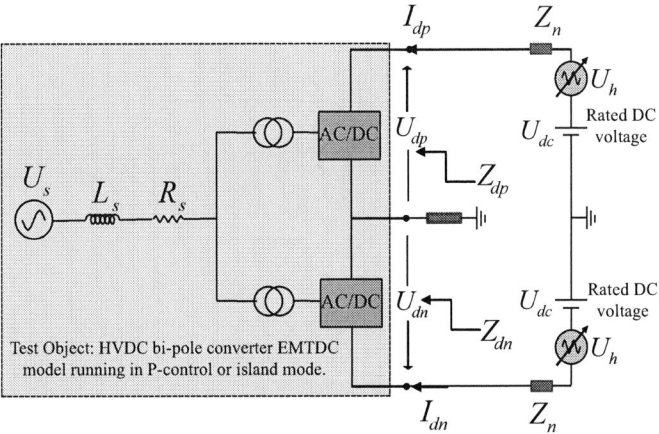

Fig. 3: Test circuit for measuring impedances seen from the DC-side in a bipolar configuration while operating in P-control or islanding control mode.

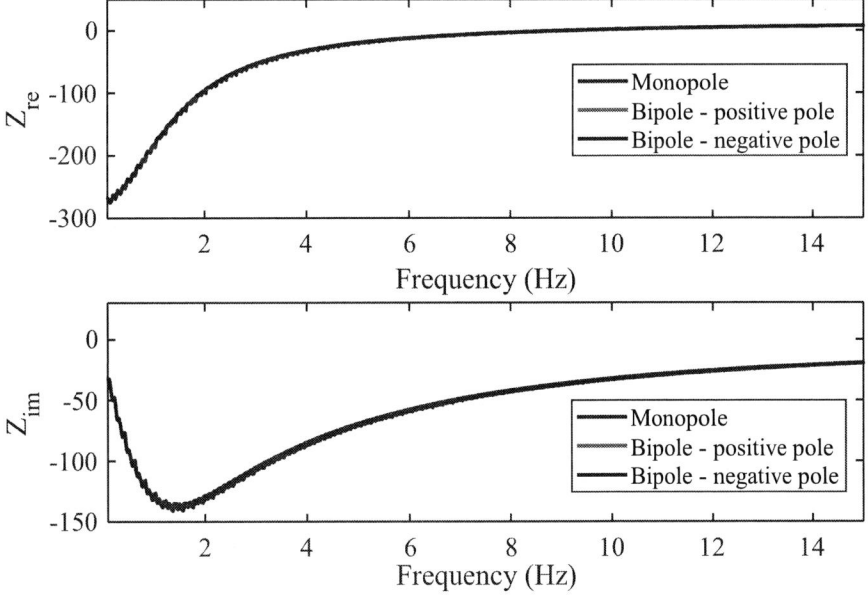

Fig. 4: DC-side impedance seen for monopole (blue color), for positive pole converter (red color) and for negative pole converter (black color) in bipolar with ground return configuration while operating in P-control mode.

It turns out that there is no cross-coupling impedance when both active and passive elements in bipolar system are symmetrical, and the impedances expressed in (1), (2) and (3) are almost equal under all scanned frequencies for the station with direct or indirect grounding. Fig. 4 shows a comparison of scanned DC-side impedances for a part of frequency range of interest from the test circuits depicted in Fig. 1 and Fig. 3. According to this comparison, the pole self-impedance in bipolar is similar to the

DC-side impedance in monopole. This means that it is possible to consider only the monopole system for stability analyzing which, in generic, could significantly simplify the stability analysis.

Finally, Fig. 5 shows the test circuit for measuring DC-side impedance of U_{dc}-controlling station in monopole configuration. Similarly, any passive element connected at DC pole can be kept as a part of DC network While P-controlling station can be represented by a DC current source behind a parallel admittance (Y_n) or an ideal DC current source whose current is set equal to the rated DC current. In order to scan the impedance seen from the converter DC terminal at different frequencies, an AC current source is connected in parallel with the DC current source (see Fig. 5). It is also worth mentioning that the AC current source magnitude (I_h) needs to be kept very low to ensure that any protection is not activated while injecting harmonic currents.

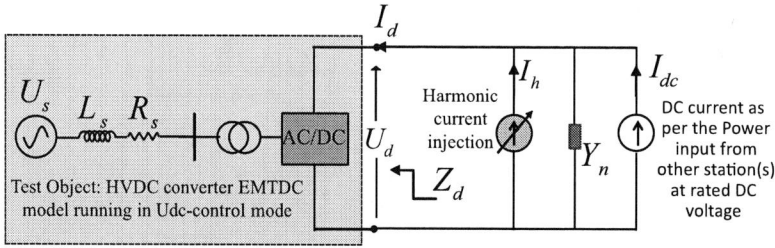

Fig. 5: Test circuit for measuring impedance seen from the DC-side of a U_{dc}-control station.

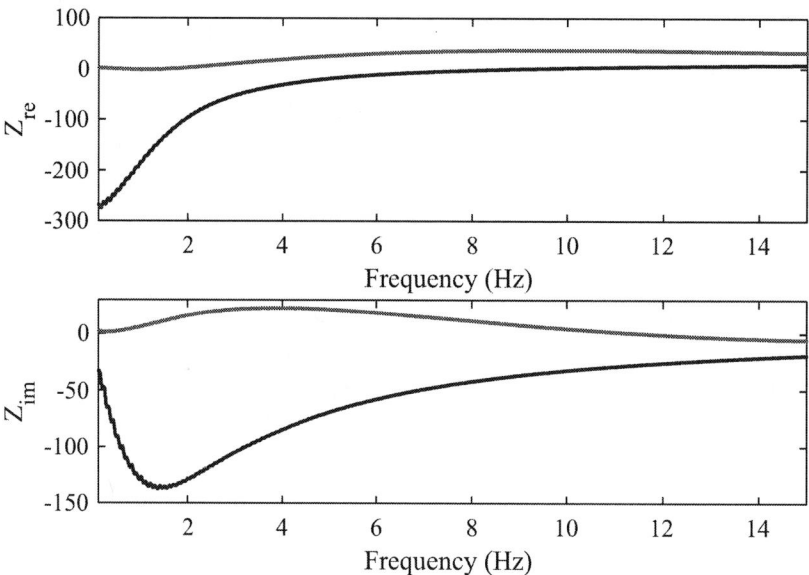

Fig. 6: Example of DC-side impedance of MMC in low frequency for both P-controlling station (blue color) and U_{dc}-controlling station (red color).

Scanning Frequency Range

From different publications, it is observed that the considered scanning frequency range is 1 to 1000 Hz. However, it may not be required to scan the frequency all the way up to 1000 Hz as the converter dynamics has an effect on the DC-side impedance in much lower frequency range. This is because the individual cells of an MMC have a quite high time constant even though the MMC converter has much lower time constant, i.e. a factor of $\frac{1}{100}$ lower. Furthermore, the bandwidths of outer control loops (which mostly impact on the DC-side impedance) are basically much lower than 1000 Hz. Besides, there could be a risk of low frequency instability (normally occurs at frequency below the bandwidths of outer control loops) which is caused by non-linearity in the control system [4] (due to voltage and current limitations).

Fig. 6 shows an example of the scanned DC-side impedances of U_{dc}-control and P-control stations for low frequencies. As it can be seen, there is a clear difference between U_{dc}-control and P-control modes in both Z_{re} and Z_{im} for frequencies below 1 Hz. On the other hand, the possible resonance frequency may be much lower than 1000 Hz in the case of a large line current limiting inductance on the DC-side (e.g. 100 mH). Thus, the possible range of scanning frequency depends on a number of factors, but in general, a range of 0.1 to 350 Hz may cover majority of applications.

Impact of the AC Network on the Equivalent DC-side Impedance

The AC network strength could also impact on the scanned DC-side impedance. First, injecting a harmonic current ,as shown in Fig. 5, results in a cell voltage ripple (having the same frequency as the injected current) which in turn generates the corresponding side band AC voltages. Then, the generated AC voltages drive a distorted AC current through the equivalent impedances of the AC network and converter resulting in a distorted power flow into the converter. Finally, this distortion can be seen as an error in the control system which tries to compensate it. Therefore, the impact of the AC network strength on DC-side impedance could be more pronounced in some control modes. Fig. 7 shows an example of scanned DC-side impedances of MMC for both AC network short circuit ratios (SCRs) of 40 and 3 under the same control mode as well as same power level. As it can be observed, there is a clear impact of AC network SCR on both Z_{re} and Z_{im}. Therefore, for stability analyzing, it may be necessary to scan the DC-side impedance for two extreme AC network conditions (i.e. very strong and very weak).

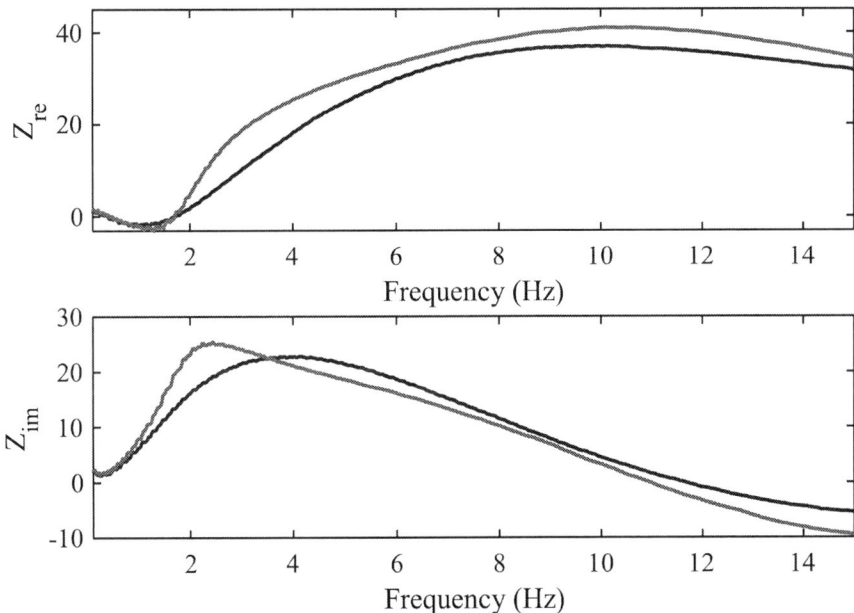

Fig. 7: Example of DC-side impedance of MMC showing the impact of AC network SCR while operating in U_{dc}-control mode. blue color: SCR=40 and red color: SCR=3.

Impact of Load Level of Converter

In order to investigate the impact of the converter load, an example of scanned DC-side impedance of MMC for both inverter and rectifier modes of P-control station is shown in Fig. 8. As it can be seen, for inverter mode operation, the converter provides negative resistance as frequency approaches to zero; while for rectifier mode operation, it is opposite way. The similar test can be done for the U_{dc}-control station, and it is observed that there is no similar effect on the resistance when it changes from positive load to negative load. This is an expected behavior for P-controlling station and U_{dc}-controlling station.

Impact of DC System Configuration

It is worth mentioning that although it may not be realistic to design rigid-bipolar in a DC grid, it should be possible to operate with partial rigid-bipolar in the case of damage to one section of metallic return line. Thus, it is also worthwhile to investigate the impact of a change to the operating configuration on the DC-side impedance. It may be noted that the pole balancing is achieved via balancing current when the station is grounded (direct or indirect via metallic return), and the pole balancing is achieved via balancing voltage when the station is not grounded as in rigid-bipolar. Fig. 9 shows a comparison of the positive pole impedances from the ground return setup and the rigid-bipolar setup for both U_{dc}-control and P-control stations. From these results, it can be concluded that the configuration change has no impact on the DC-side impedance as far as the balancing control works as expected meaning steady-state balance is maintained.

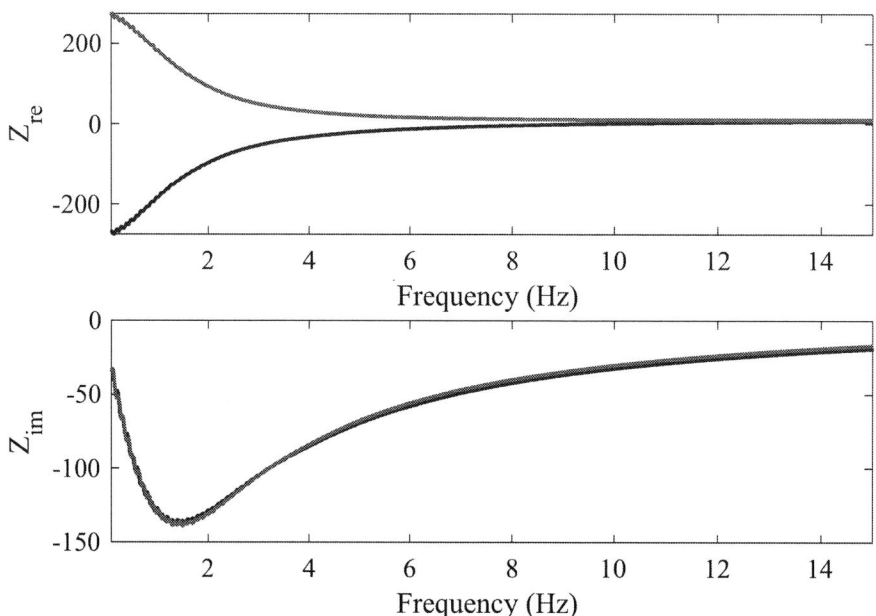

Fig. 8: Example of DC-side impedance of MMC showing the impact of load level of converter while operating in P-control mode. blue color: inverter mode and red color: rectifier mode.

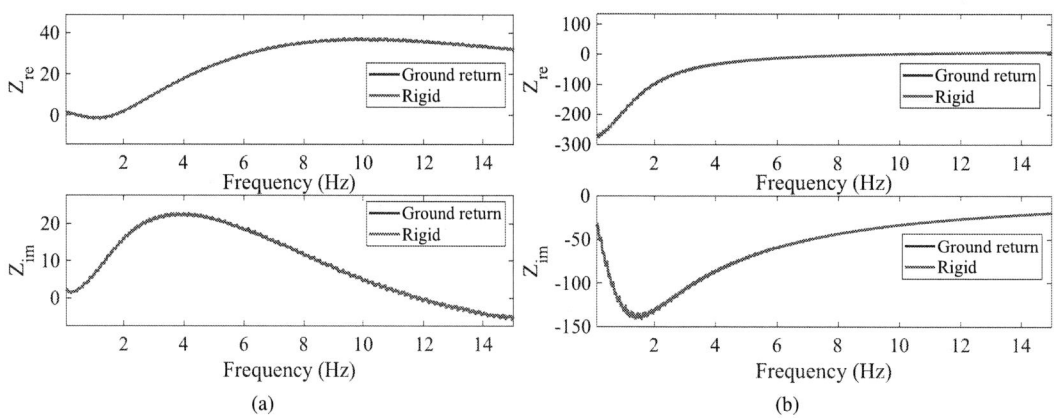

Fig. 9: Comparison of impedance of the positive pole converter seen from the DC-side in a bipolar with ground return (blue color) and rigid-bipolar (red color) configurations. (**a**) U_{dc}-control; (**b**) P-control.

Conclusion

The DC-side impedance scanning as well as its corresponding circuit setups have been discussed in this paper. Some typical simulation results of the scanned DC side impedances have been presented. The study experience and results have given the following conclusions and recommendations:

1. The DC-side impedance seen for the P-control station is different from the one for the DC control station justifying that impedance seen from the DC-side depends on various control modes.

2. The rectifier and inverter operations give significantly different resistance at very low frequency (close to zero frequency) in P-control mode. However, in U_{dc}-control mode, the impedance is not affected by the power flowing direction.

3. It is advisable to consider the AC network strength when scanning DC-side impedance, and it may be sufficient to consider two boundary conditions meaning the network conditions with minimum and maximum SCRs.

4. The DC-side impedance obtained in the monopole configuration is the same as the impedance when the same converter operates together with the other pole converter, i.e, balanced poles operation in the bipolar configuration. Thus, it may be adequate to perform the stability analysis in a monopole setup which could make the analysis significantly simplified.

Appendix

The simulation throughout this paper has been implemented in PSCAD/EMTDC using the system parameters detailed in Table I.

Table I: System parameters

Parameter		Value
Point-to-point HVDC Link	$U_{ac,rated}$	400 kV
	P_{rated}	1000 MW (per pole)
	$U_{dc,rated}$	525 kV
	f_0	50 Hz

References

[1] PROMOTioN. Deliverable 12.4: Final Deployment Plan; Technical Report, PROMOTioN Project; PROMO-TioN–Progress on Meshed HVDC Offshore Transmission Networks: Arnhem, The Netherlands, 2020.

[2] Rodriguez, Pedro, and Kumars Rouzbehi. "Multi-terminal DC grids: challenges and prospects." Journal of Modern Power Systems and Clean Energy 5.4 (2017): 515-523.

[3] Nghiem, A., and I. Pineda. "Wind Energy in Europe: Scenarios for 2030; WindEurope: Brussels, Belgium, 2017." https://windeurope.org/wp-content/uploads/files/about-wind/reports/Wind-energy-in-Europe-Scenarios-for-20 30: 32.

[4] prEN 50388-2:2017: "Railway Applications – Fixed installations and rolling stock – Technical criteria for the coordination between traction power supply and rolling stock to achieve interoperability – Part 2: stability and harmonics".

[5] BestPaths, "Deliverable D9.3: BEST PATHS DEMO#2 Final Recommendations for Interoperability of Multivendor HVDC Systems".

[6] Lyu, Jing, Xu Cai, and Marta Molinas. "Impedance modeling of modular multilevel converters." IECON 2015-41st Annual Conference of the IEEE Industrial Electronics Society. IEEE, 2015.

[7] Jamshidifar, Aliakbar, and Dragan Jovcic. "Small-signal dynamic DQ model of modular multilevel converter for system studies." IEEE Transactions on Power Delivery 31.1 (2015): 191-199.

[8] Li, Zhenyu, et al. "Accurate impedance modeling and control strategy for improving the stability of DC system in multiterminal MMC-based DC grid." IEEE Transactions on Power Electronics 35.10 (2020): 10026-10049.

[9] A. Antonopoulos, et al. "On Dynamics and Voltage Control of the Modular Multilevel Converter" 2009 13th European Conference on Power Electronics and Applications.

[10] Saad, Hani, Albane Schwob, and Yannick Vernay. "Study of resonance issues between HVDC link and power system components using EMT simulations." 2018 Power Systems Computation Conference (PSCC). IEEE, 2018.

[11] Sun, Jian. "Impedance-based stability criterion for grid-connected inverters." IEEE transactions on power electronics 26.11 (2011): 3075-3078.

[12] Quester, Matthias, et al. "Online impedance measurement of a modular multilevel converter." 2019 IEEE PES Innovative Smart Grid Technologies Europe (ISGT-Europe). IEEE, 2019.

[13] Quester, Matthias, et al. "Frequency behavior of an MMC test bench system." 2020 6th IEEE International Energy Conference (ENERGYCon). IEEE, 2020.

[14] Liao, Yicheng, et al. "Stability and Sensitivity Analysis of Multi-Vendor, Multi-Terminal HVDC Systems." arXiv preprint arXiv:2111.12013 (2021).

[15] F. Loku et al. "Equivalent Impedance Calculation Method for Control Stability Assessment in HVDC Grids," MDPI Energies 2021, 14, 6899. Available at: https://www.mdpi.com/1996-1073/14/21/6899/pdf.

[16] Yang, Dongsheng, et al. "Automation of impedance measurement for harmonic stability assessment of mmc-hvdc systems." 18th Wind Integration Workshop. 2019.

[17] H. Wu et al. "Development of an AC/DC Impedance Matrix Measurement Toolbox for MTDC System," 20th Wind Integration Workshop, November 2021.

Utilizing the Electroluminescence of SiC MOSFETs as Degradation Sensitive Optical Parameter

Lukas A. Ruppert, Michael Laumen, Rik W. De Doncker
Institute for Power Electronics and Electrical Drives
RWTH Aachen University
Jaegerstrasse 17-19, Aachen, Germany
Phone: +49 241 80 99552
Email: post@isea.rwth-aachen.de
URL: www.isea.rwth-aacchen.de

Acknowledgments

This research was funded by the German Research Foundation (DFG) within the project ElluSense (BR 6266/2-1).

Keywords

≪Condition Monitoring≫, ≪Reliability≫, ≪Diagnostics≫, ≪Degradation≫, ≪Aging≫, ≪Silicon Carbide (SiC)≫, ≪MOSFET≫, ≪Threshold voltage instability≫, ≪Threshold voltage shift≫

Abstract

Bias Temperature Instability (BTI) is a major reliability challenge of SiC MOSFETs due to the high defect density of the gate oxide compared to Si MOSFETs. Charge trapping at these defects causes gradual threshold voltage drift and degradation of the on-state resistance, which leads to higher device losses and temperatures. To avoid critical failures, the health state of the gate oxide can be monitored by measuring the BTI-related degradation. This work demonstrates that the electroluminescence (EL) emitted by the body diode of SiC MOSFETs can be utilized as such an aging precursor. As trapped charges at the gate oxide influence the current proportions between the channel and the light-emitting body diode during third quadrant operation, the EL exhibits a sensitivity to BTI that can be sensed as part of online condition monitoring. This paper investigates the EL sensitivity to BTI by means of accelerated aging tests and EL measurements on an industry-standard SiC power MOSFET. The measurement results show a strong increase in the EL intensity for a positive threshold voltage drift, thereby validating the proposed sensing approach.

Introduction

Toward developing highly-integrated power electronic converters and pushing their power semiconductors against electrical and thermal limits, SiC power MOSFETs have widely been adopted in power electronic applications due to their superior physical properties [2]. Moreover, SiC MOSFETs play an increasingly important role in safety-critical applications, such as autonomous vehicles [3] or electrical aircrafts [4], which demands high reliability of the semiconductor components. However, SiC MOSFETs are affected by gate oxide degradation due to the comparatively high defect density of the gate oxide interface compared to conventional Si devices [5]–[7], which may hinder long-term reliability of power electronic converters. One major reliability challenge related to gate oxide degradation is charge trapping at the gate oxide induced by BTI that causes threshold voltage instability [6], [8]. Gradual threshold voltage shifts can increase the on-state resistance of the device and thereby lead to higher losses and temperatures [6], as well as higher thermomechanical stress.

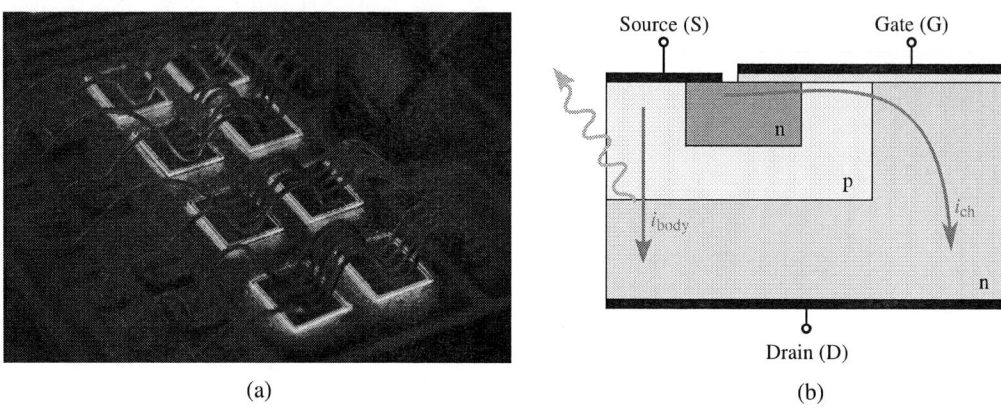

(a)   (b)

Fig. 1: Electroluminescence of a SiC power module (a) and simplified cross section of a power MOSFET (VDMOS) [1] (b)

To monitor the state-of-health of the gate oxide and prevent critical failures, literature proposed online condition monitoring techniques that measure aging precursors during converter operation: Gate charge time [9], gate-plateau voltage [10], [11], body diode voltage [12], [13], and input capacitance [14]. Most of these degradation sensitive *electrical* parameters had already been studied in detail in the context of junction temperature sensing [15]–[17] and have been adopted for degradation sensing. A key challenge in measuring these electrical parameters is their sensitivity to electromagnetic emission (EME) that is particularly challenging in fast switching wide bandgap converters and demands high requirements for electromagnetic compatibility (EMC) of the sensing circuitry [15].

The novel condition monitoring technique that is proposed in this work overcomes this limitation because it is based on a degradation sensitive *optical* parameter, the EL of SiC MOSFETs. As an optical parameter, the EL can be extracted via optical fibers and spatially transmitted to an external sensing circuitry, thus providing low sensitivity to EME due to galvanic isolation. Additionally, the intrinsic galvanic isolation of the EL simplifies condition monitoring in high-voltage converters, where isolating electrical signals requires great effort.

The EL occurs during conduction of the MOSFET body diode [18], as illustrated in Fig. 1. Since the spectrum of the emitted light is dependent on the device current and temperature [19], various monitoring solutions have been proposed in literature that measure the emitted light with photodiode sensors to extract current and temperature information [1], [19]–[24]. While these publications have proven the suitability of EL sensing for high-bandwidth current and temperature sensing, this work shows that EL-based condition monitoring can additionally measure BTI-related threshold voltage drifts. The presented aging precursor is based on the sensitivity of the EL to BTI since charge trapping at the gate oxide influences the proportions of the light-emitting body diode current and channel current during third quadrant operation.

This paper is organized as follows: The first section gives a brief overview of the EL in SiC MOSFETs and investigates key sensitivities of the EL spectrum. Subsequently, the second section analyzes the influence of BTI-induced threshold voltage drifts on the EL and shows how the EL can be used as an aging precursor in terms of condition monitoring. Finally, the sensing approach is validated with an accelerated aging test on a SiC power MOSFET and EL measurements.

Sensitivities of Electroluminescence

The EL in MOSFETs occurs when the internal p-n structure, i.e. the body diode, is forward biased [18]. This is usually the case when the MOSFET is turned off during third quadrant operation ($i_{SD} > 0$) such that minority carriers are injected in the body diode. Due to radiative recombinations of minority carriers, the body diode acts like a parasitic light-emitting diode whose intensity increases with an increasing

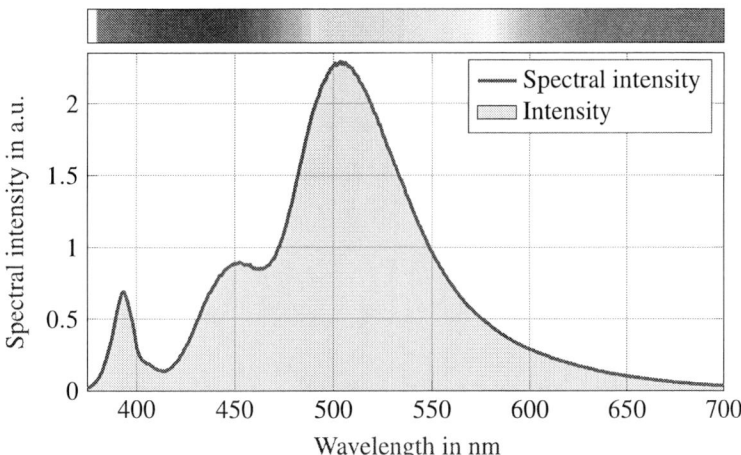

Fig. 2: EL spectra of a 4H-SiC MOSFET body diode

number of minority carriers [25]. The body diode of SiC MOSFETs emits light with sufficient intensity to be used as part of condition monitoring techniques despite the indirect band gap of SiC and the resulting low light emission efficiency.

The EL spectrum of 4H-SiC MOSFETs comprises two major peaks in the visible spectrum [26]–[28] as shown in Fig. 2. While the first peak is located around 390 nm in the ultraviolet region, the second peak is in the blue-green region around 500 nm. The spectral peaks exhibit characteristic current and temperature dependencies which have been analyzed in [1], [20], [24], [29]. Based on these dependencies, different monitoring techniques have been suggested in literature that measure the intensity of the light emission with photodiode sensors to extract temperature and/or current information. For instance, [29] and [22] proposed a current measurement based on the principle that a change in the device current influences body diode current and thus the EL intensity. Besides the device current and temperature dependency of the EL spectrum, the light emission is also sensitive to the gate bias voltage [1].

It is important to understand the influence of the gate bias on the EL because it is based on the same mechanism as the influence of the proposed aging precursor. As illustrated in Fig. 1b, the device current of a MOSFET comprises two components during third quadrant operation: the channel current i_{ch} and the body diode current i_{body}. Only the latter leads to light emission of the body diode. The proportion of these two components strongly depends on the applied gate bias voltage V_G. For instance, the lower the gate bias voltage, the higher the share of the body diode current and EL intensity. This can be observed in Fig. 3a showing the spectral intensity of the EL for different gate bias voltages V_G: the spectral intensity of the entire spectrum decreases with rising V_G. For a more detailed analysis, the spectral intensity is integrated to calculate the total intensity of the light emission A_{EL} as function of V_G (compare Fig. 3b). At low and high gate bias voltages, the EL intensity is saturated as the device current flows completely through the body diode or channel, respectively. Thus, the EL exhibits only low sensitivity on V_G for these bias conditions because the derivative $\frac{dA_{EL}}{dV_G}$ is low. However, the sensitivity increases when the gate bias causes significant current flows through both the channel and the body diode, e.g., between $V_G = -3\,\text{V}$ and $V_G = 0\,\text{V}$ for the given operating point in Fig. 3.

Electroluminescence as an Aging Precursor for Bias Temperature Instability

BTI is a major reliability challenge of SiC power MOSFETs due to the comparatively high defect density of the gate oxide, especially the gate oxide interface, compared to Si MOSFETs [5]–[7]. Charge trapping at these defects induces positive or negative threshold voltage drifts when a positive or negative gate bias

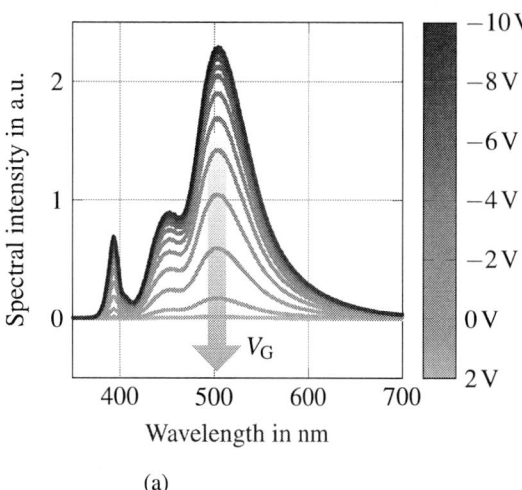
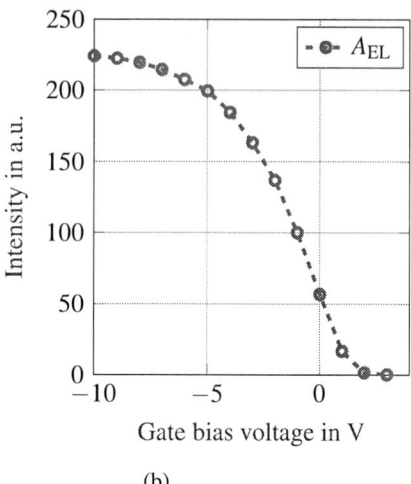

(a)   (b)

Fig. 3: Spectral intensity (a) and intensity (b) for different gate bias voltages V_G measured at $i_{SD} = 10\,A$ and $T_j = 95\,°C$

voltage is applied [6], [30]:

$$V_{G,th} = V_{G,th,0} + \frac{Q_{it}}{C_{ox}} \quad, \tag{1}$$

where $V_{G,th,0}$ is the initial threshold voltage, Q_{it} is the interface trapped charge and C_{ox} is the gate oxide capacitance [6]. BTI-induced threshold voltage variations can be divided into fully reversible drifts that can recover with short recovery time constants and more permanent drifts that are considered as long-term effects and thereby may gradually degrade the electrical performance of the power MOSFET. For instance, a gradual positive drift of $V_{G,th}$ reduces the voltage overdrive $(V_G - V_{G,th})$ during on-state of the MOSFET and thus increases the channel resistance

$$R_{ch} = \frac{L}{W \mu_n C_{ox} (V_G - V_{G,th})} \quad. \tag{2}$$

In (2) μ_n is the electron mobility whereas L and W are the channel length and width, respectively. Such a degradation of the on-state resistance causes higher losses and thermal stress that may ultimately lead to fatigue of the device. By monitoring BTI-sensitive parameters of the MOSFET during operation, abnormal threshold voltage drifts and thus premature aging can be detected.

The EL of SiC MOSFETs is a promising candidate for such an aging precursor. As BTI-induced charge trapping changes the threshold voltage $V_{G,th}$ according to (1), BTI influences the proportions of the body diode and channel current in a similar way as a varying V_G [1]. For a positive drift of the threshold voltage, the EL intensity increases due to lower channel and higher body diode current (and vice versa). Consequently, EL is sensitive to BTI-induced threshold voltage drifts and thereby can be utilized as an aging precursor by monitoring the light intensity during operation.

For this purpose, a gate bias voltage must be applied that provides a high sensitivity of the EL to BTI. As shown in the previous section, the sensitivity is high for bias conditions that allow current conduction through both the channel and the body diode while the EL exhibits almost no sensitivity to threshold voltage variations when the channel is completely closed or opened. Thus, a gate bias voltage with high $\frac{dA_{EL}}{dV_G}$ should be applied for the proposed condition monitoring approach. By repetitively measuring the EL at such a gate bias voltage and a defined operating point for the junction temperature and device current, variations in the light intensity will indicate BTI-induced gate oxide degradation.

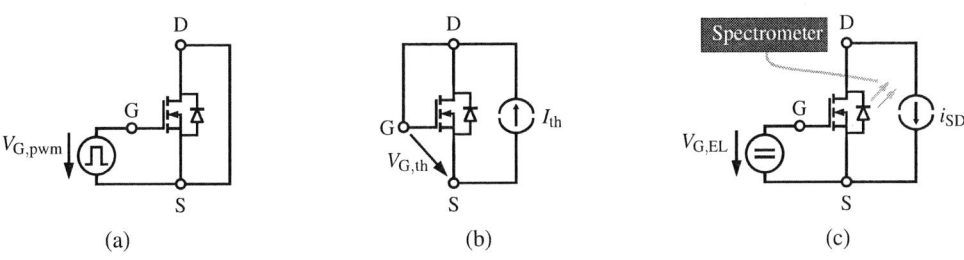

Fig. 4: Equivalent circuit of the accelerated aging test setup (a). After every $18 \cdot 10^8$ switching cycles, the threshold voltage (b) and EL (c) are measured.

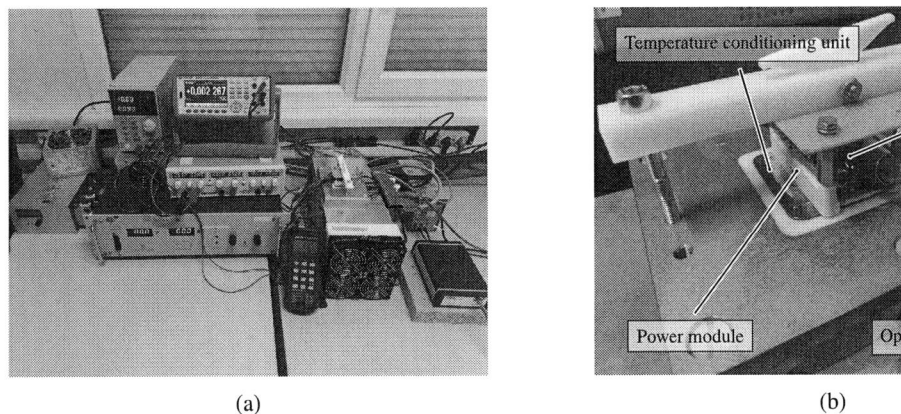

Fig. 5: Measurement setup (a) of the accelerated aging and characterization test as well as an enlarged view (b) of the SiC power module, optical fiber and temperature conditioning unit

Accelerated Aging Test

To verify the feasibility of the proposed degradation sensing method, an accelerated AC aging test, whose technique was proposed in [31], was performed on a trench SiC power MOSFET within an industry-standard power module. For this purpose, the MOSFET was subjected to bipolar gate signals at maximum allowed data sheet voltage levels of 20 V and -10 V at room temperature and a highly elevated switching frequency of 500 kHz while the drain and source terminals were shortened ($V_{DS} = 0$ V). The equivalent circuit of the aging test setup is depicted in Fig. 4a.

The accelerated aging test was interrupted every hour, which corresponds to $\Delta N_{sw} = 18 \cdot 10^8$ switching cycles, to measure the BTI-induced threshold voltage drift and the EL intensity (compare Fig. 4b and Fig. 4c). The threshold voltage was recorded by a Keysight 34461A digital multimeter that measures the gate-source voltage V_G for an injected drain-source current of $I_{th} = 2$ mA while shortening the gate and drain terminals. For reproducible conditions for the threshold voltage measurements, the temperature of the power module was controlled at 45 °C by using type-K thermocouples that were glued to the die surface and a temperature condition unit presented in [32].

After each $V_{G,th}$ measurement, a fiber-coupled spectrometer (AvaSpec-ULS2048CL) measured the EL spectrum for an injected device current i_{SD} of 10 A and a junction temperature T_j of 115 °C. These EL measurements were performed for two gate bias voltages: -10 V and 0 V. As discussed in the previous sections, a gate bias of 0 V is associated with a strong sensitivity of the EL to threshold voltage shifts for the given operating point while almost no sensitivity is expected at -10 V due to the completely closed channel. Thus, the 0 V measurements are meant to validate the proposed aging precursor while the purpose of the -10 V measurements is to proof consistent measurement conditions since the EL intensity is insusceptible to BTI at this gate bias condition. The entire measurement setup is shown in Fig. 5.

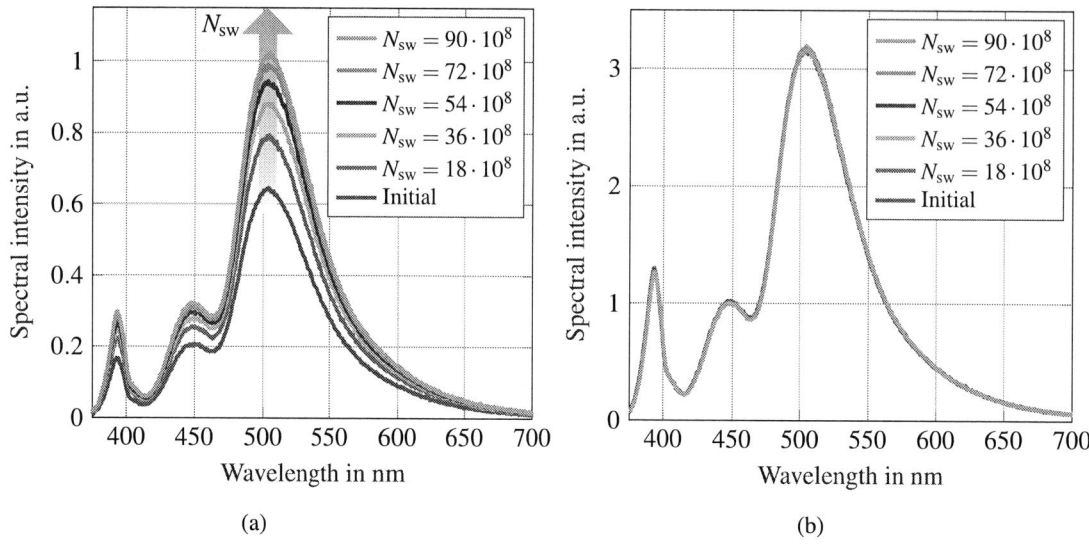

(a)   (b)

Fig. 6: EL spectra of the SiC MOSFET body diode recorded after each stress cycle at $i_{SD} = 10\,\text{A}$ for gate bias voltages of $V_G = 0\,\text{V}$ (a) and $V_G = -10\,\text{V}$ (b)

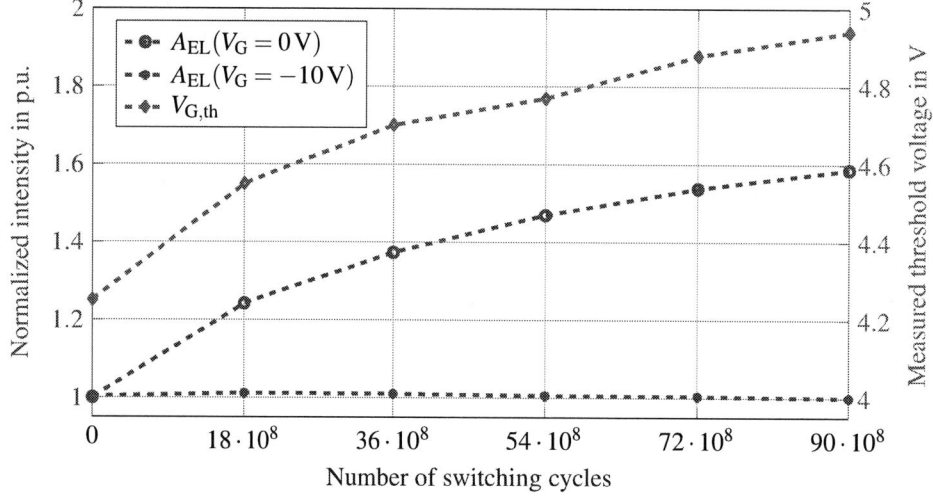

Fig. 7: EL intensity (blue), calculated by integrating the spectral intensity from Fig. 6 and normalized to initial value, and measured threshold voltage (orange) as a function of the number of stress switching cycles (marker: measurement data; lines: linear regression)

The results of the EL measurements are depicted in Fig. 6a and Fig. 6b for $V_G = 0\,\text{V}$ and $V_G = -10\,\text{V}$, respectively. The spectral intensity of the EL spectra that are recorded at $V_G = 0\,\text{V}$ show a significant increase with rising stress cycles N_{sw} compared to the spectral intensity of the initial EL spectrum whereas the EL spectra show almost no change for $V_G = -10\,\text{V}$.

To verify that the change of the EL is associated with a threshold voltage drift, Fig. 7 compares the recorded threshold voltage with the total EL intensity A_{EL}. Again, A_{EL} is calculated by integrating the data from Fig. 6 over the entire spectrum and normalizing to the initial value to highlight the relative change caused by the accelerated aging test. The results show a gradual positive threshold voltage shift from $V_{G,th} = 4.25\,\text{V}$ to $V_{G,th} = 4.94\,\text{V}$ which is clearly correlated with the increase of A_{EL} for $V_G = 0\,\text{V}$. This can be explained by trapped negative charges in the gate oxide or at the gate oxide interface that are induced by the applied AC bias stress and cause a rise of the threshold voltage [30]. Since this positive threshold voltage shift increases the proportion of the body diode current and decreases the proportion of the channel current, the EL intensity rises with increasing stress cycles N_{sw}. Consequently, the measurement results verify the sensitivity of the EL to BTI-induced threshold voltage variation and thereby the suitability of the EL as a degradation sensitive optical parameter.

The measurement setup that is used for this work is limited to a laboratory environment and is not suitable for highly dynamic degradation detection in real application. However, previous publications, e.g., [20], have already proven the feasibility of EL-based online condition monitoring by utilizing photodiode sensors and highly dynamic extraction circuits that can easily be adapted for the proposed sensing method. Another challenge is that the presented condition monitoring technique relies on EL measurements at defined junction temperature and device current whose information might not be available in converters. Additionally, the gate bias voltage V_G is usually limited to the gate supply voltages. Those depend on the power semiconductor as well as the specific application and thus cannot be chosen arbitrarily as suggested in this work. To cope with these challenges, the proposed degradation sensing method could be integrated in existing EL-based temperature and current sensing techniques by utilizing gate drivers with multiple voltage levels. This would provide junction temperature and current information while one gate driver voltage level that is designated for degradation sensing would allow EL measurements with high sensitivity to BTI-induced threshold drifts.

Conclusion

This work presented a new aging precursor for BTI-related gate oxide degradation utilizing the EL of SiC MOSFETs. Due to intrinsic galvanic isolation of EL sensing, e.g. via optical fibers, the proposed aging precursor is particularly suitable for applications with high EME and high voltages. By conducting accelerated AC aging tests and EL measurements on a SiC power MOSFET, it has been demonstrated that the EL intensity is strongly sensitive to BTI-induced threshold voltage variations and thus can detect critical V_G shifts. To maximize the sensitivity of the EL to these threshold voltage drifts, gate bias voltages are applied that neither completely close nor open the MOSFET channel, but allow a current flow through both the channel and body diode. Future work will investigate how this condition monitoring technique can be integrated in existing EL-based current and temperature sensing solutions.

References

[1] L. A. Ruppert, S. Kalker, C. H. van der Broeck, and R. W. De Doncker, "Analyzing spectral electroluminescence sensitivities of SiC MOSFETs and their impact on power device monitoring," in *International Exhibition and Conference for Power Electronics, Intelligent Motion, Renewable Energy and Energy Management (PCIM)*, VDE, May 2021.

[2] A. Stippich, C. H. van der Broeck, A. Sewergin, A. H. Wienhausen, M. Neubert, *et al.*, "Key components of modular propulsion systems for next generation electric vehicles," *CPSS Transactions on Power Electronics and Applications*, vol. 2, no. 4, pp. 249–258, Dec. 2017.

[3] F. Blaabjerg, H. Wang, I. Vernica, B. Liu, and P. Davari, "Reliability of power electronic systems for ev/hev applications," *Proceedings of the IEEE*, vol. 109, no. 6, pp. 1060–1076, 2021.

[4] L. Dorn-Gomba, J. Ramoul, J. Reimers, and A. Emadi, "Power electronic converters in electric aircraft: Current status, challenges, and emerging technologies," *IEEE Transactions on Transportation Electrification*, vol. 6, no. 4, pp. 1648–1664, 2020.

[5] R. Singh, "Reliability and performance limitations in SiC power devices," *Microelectronics Reliability*, vol. 46, no. 5, pp. 713–730, May 2006.

[6] T. Aichinger, G. Rescher, and G. Pobegen, "Threshold voltage peculiarities and bias temperature instabilities of SiC MOSFETs," *Microelectronics Reliability*, vol. 80, pp. 68–78, Jan. 2018.

[7] J. Wang and X. Jiang, "Review and analysis of SiC MOSFETs' ruggedness and reliability," *IET Power Electronics*, vol. 13, no. 3, pp. 445–455, 2020.

[8] K. Puschkarsky, T. Grasser, T. Aichinger, W. Gustin, and H. Reisinger, "Review on SiC MOSFETs High-Voltage Device Reliability Focusing on Threshold Voltage Instability," *IEEE Transactions on Electron Devices*, vol. 66, no. 11, pp. 4604–4616, Nov. 2019.

[9] M. Xie, P. Sun, K. Wang, Q. Luo, and X. Du, "Online Gate-Oxide Degradation Monitoring of Planar SiC MOSFETs Based on Gate Charge Time," *IEEE Transactions on Power Electronics*, vol. 37, no. 6, pp. 7333–7343, Jun. 2022.

[10] Z. Ni, Y. Li, X. Lyu, O. P. Yadav, and D. Cao, "Miller plateau as an indicator of SiC MOSFET gate oxide degradation," in *2018 IEEE Applied Power Electronics Conference and Exposition (APEC)*, ISSN: 2470-6647, Mar. 2018, pp. 1280–1287.

[11] U. Karki and F. Z. Peng, "Precursors of Gate-Oxide Degradation in Silicon Carbide MOSFETs," in *2018 IEEE Energy Conversion Congress and Exposition (ECCE)*, ISSN: 2329-3748, Sep. 2018, pp. 857–861.

[12] J. Xin, M. Du, Z. Ouyang, and K. Wei, "Online Monitoring for Threshold Voltage of SiC MOSFET Considering the Coupling Impact on BTI and Junction Temperature," *IEEE Transactions on Electron Devices*, vol. 68, no. 4, pp. 1772–1777, Apr. 2021.

[13] E. Ugur, C. Xu, F. Yang, S. Pu, and B. Akin, "A New Complete Condition Monitoring Method for SiC Power MOSFETs," *IEEE Transactions on Industrial Electronics*, vol. 68, no. 2, pp. 1654–1664, Feb. 2021.

[14] M. Farhadi, F. Yang, S. Pu, B. T. Vankayalapati, and B. Akin, "Temperature-Independent Gate-Oxide Degradation Monitoring of SiC MOSFETs Based on Junction Capacitances," *IEEE Transactions on Power Electronics*, vol. 36, no. 7, pp. 8308–8324, Jul. 2021.

[15] S. Kalker, L. A. Ruppert, C. H. van der Broeck, J. Kuprat, M. Andresen, *et al.*, "Reviewing thermal monitoring techniques for smart power modules," *IEEE Journal of Emerging and Selected Topics in Power Electronics*, pp. 1–1, 2021.

[16] C. H. van der Broeck, A. Gospodinov, and R. W. De Doncker, "IGBT junction temperature estimation via gate voltage plateau sensing," *IEEE Transactions on Industry Applications*, vol. 54, no. 5, pp. 4752–4763, 2018.

[17] N. Fritz, M. Friedel, T. A. Polom, and R. W. de Doncker, "Online junction temperature monitoring of power semiconductor devices based on a wheatstone bridge," in *IEEE Energy Conversion Congress and Exposition (ECCE)*, Oct. 2021, pp. 2740–2746.

[18] J. Winkler, J. Homoth, and I. Kallfass, "Utilization of Parasitic Luminescence from Power Semiconductor Devices for Current Sensing," in *PCIM Europe 2018; International Exhibition and Conference for Power Electronics, Intelligent Motion, Renewable Energy and Energy Management*, Jun. 2018, pp. 1–8.

[19] S. Kalker, C. H. van der Broeck, and R. W. De Doncker, "Utilizing Electroluminescence of SiC MOSFETs for Unified Junction-Temperature and Current Sensing," in *2020 IEEE Applied Power Electronics Conference and Exposition (APEC)*, ISSN: 2470-6647, Mar. 2020, pp. 1098–1105.

[20] J. Winkler, J. Homoth, and I. Kallfass, "Electroluminescence-Based Junction Temperature Measurement Approach for SiC Power MOSFETs," *IEEE Transactions on Power Electronics*, vol. 35, no. 3, pp. 2990–2998, Mar. 2020.

[21] G. Susinni, S. A. Rizzo, F. Iannuzzo, and A. Raciti, "A non-invasive SiC MOSFET Junction temperature estimation method based on the transient light Emission from the intrinsic body diode," *Microelectronics Reliability*, 31st European Symposium on Reliability of Electron Devices, Failure Physics and Analysis, ESREF 2020, vol. 114, p. 113 845, Nov. 2020.

[22] S. Kalker, C. H. van der Broeck, L. A. Ruppert, and R. W. De Doncker, "Next Generation Monitoring of SiC MOSFETs via Spectral Electroluminescence Sensing," *IEEE Transactions on Industry Applications*, pp. 1–1, 2021.

[23] L. A. Ruppert, S. Kalker, and R. W. De Doncker, "Junction-Temperature Sensing of Paralleled SiC MOSFETs Utilizing Temperature Sensitive Optical Parameters," in *2021 IEEE Energy Conversion Congress and Exposition (ECCE)*, ISSN: 2329-3748, Oct. 2021, pp. 5597–5604.

[24] H. Luo, J. Mao, C. Li, F. Iannuzzo, W. Li, and X. He, "Online Junction Temperature and Current Simultaneous Extraction for SiC MOSFETs With Electroluminescence Effect," *IEEE Transactions on Power Electronics*, vol. 37, no. 1, pp. 21–25, Jan. 2022.

[25] J. Winkler, J. Homoth, and I. Kallfass, "Electroluminescence in power electronic applications: Utilization of p-n junctions in power semiconductors as unintentional light emitting diodes for current and temperature sensing," Oct. 2018.

[26] M. Anikin, A. Lebedev, N. Poletaev, A. chuk, A. Syrkin, and V. Chelnokov, "Deep centers and blue-green electroluminescence in 4H-SiC," Oct. 1993.

[27] A. M. Strel'chuk, E. V. Kalinina, and A. A. Lebedev, "Temperature dependence of the band-edge injection electroluminescence of 4h-sic pn structure," in *Silicon Carbide and Related Materials 2012*, ser. Materials Science Forum, vol. 740, Trans Tech Publications Ltd, Mar. 2013, pp. 569–572.

[28] S. G. Sridhara, L. L. Clemen, R. P. Devaty, W. J. Choyke, D. J. Larkin, *et al.*, "Photoluminescence and transport studies of boron in 4h sic," *Journal of Applied Physics*, vol. 83, no. 12, pp. 7909–7919, 1998.

[29] J. Winkler, J. Homoth, H. Bartolf, and I. Kallfass, "Study on Transient Light Emission of SiC Power MOSFETs Regarding the Sensing of Source-Drain Currents in Hard-Switched Power Electronic Applications," in *PCIM Europe 2019; International Exhibition and Conference for Power Electronics, Intelligent Motion, Renewable Energy and Energy Management*, May 2019, pp. 1–8.

[30] K. Puschkarsky, H. Reisinger, T. Aichinger, W. Gustin, and T. Grasser, "Understanding BTI in SiC MOSFETs and Its Impact on Circuit Operation," *IEEE Transactions on Device and Materials Reliability*, vol. 18, no. 2, pp. 144–153, Jun. 2018.

[31] P. Salmen, M. W. Feil, K. Waschneck, H. Reisinger, G. Rescher, and T. Aichinger, "A new test procedure to realistically estimate end-of-life electrical parameter stability of SiC MOSFETs in switching operation," in *2021 IEEE International Reliability Physics Symposium (IRPS)*, ISSN: 1938-1891, Mar. 2021, pp. 1–7.

[32] G. Engelmann, M. Laumen, J. Gottschlich, K. Oberdieck, and R. W. De Doncker, "Temperature-Controlled Power Semiconductor Characterization Using Thermoelectric Coolers," *IEEE Transactions on Industry Applications*, vol. 54, no. 3, pp. 2598–2605, May 2018.

Characterization of GaN-on-AlN/SiC transistors towards monolithic integrability

Nick Wieczorek[1], Xiaomeng Geng[1], Carsten Kuring[1], Oliver Hilt[2], Frank Brunner[2],
Mihaela Wolf[2], Joachim Würfl[2] and Sibylle Dieckerhoff[1]

[1] TECHNISCHE UNIVERSITÄT BERLIN
Einsteinufer 19
Berlin, Germany
[2] FERDINAND-BRAUN-INSTITUT
LEIBNIZ-INSTITUT FÜR HÖCHSTFREQUENZTECHNIK
Gustav-Kirchhoff-Str.4
Berlin, Germany
Tel.: +49 / (0)30 314-70024
Fax: +49 / (0)30 314-25526
E-Mail: n.wieczorek@tu-berlin.de
[1] URL: https://www.pe.tu-berlin.de
[2] URL: https://www.fbh-berlin.de

Keywords

«Monolithic power integration», «Gallium Nitride (GaN)», «Device characterisation», «HEMT», «Double pulse test»

Abstract

A GaN-on-AlN/SiC technology is proposed for monolithic GaN power switch integration. As opposed to conventional GaN-on-Si devices, the insulating SiC substrate results in immunity to back-gating effects and enables monolithic integration without degradation of switching characteristics resulting from the shared substrate. This is validated for a discrete half-bridge, with the substrate of both transistors shorted together, as well as for a monolithic half-bridge. Hard switching transients up to 300 V in a double-pulse test with both half-bridges reveal faster switching transients and reduced switching losses for the monolithically integrated half-bridge.

1 Introduction

GaN-on-Si high electron mobility transistors (HEMTs) have achieved a mature technology state during recent years and are increasingly employed in commercial power electronic applications. Low on-resistance combined with small intrinsic parasitic capacitances allow for efficient high frequency operation in hard- and soft- switched converters. Due to increased switching speed in comparison to conventional silicon (Si) power transistors, monolithic integration of GaN HEMTs is a suitable approach to minimize parasitic stray inductances of the switching cell in order to achieve stable and clean switching transitions. Monolithic integration of two (or more) conventional GaN-on-Si HEMTs has been successfully demonstrated for Monolithic Half-Bridges (MHB) [1, 2] as well as monolithic bidirectional GaN HEMTs [3, 4]. However, substrate coupling effects can result in oscillations, reduced switching speed and increased on-resistance. Advanced substrate termination schemes can be employed to overcome these limitations but inevitably increase the circuit complexity [5, 6, 7, 4]. On-chip isolation by means of metal oxide layers is a further approach to suppress the substrate coupling effect in GaN-on-Si HEMTs but increase the device fabrication complexity [8]. Instead, the lateral GaN HEMTs presented in this paper are fabricated on a semi-isolating SiC substrate using an AlN buffer layer enabling reduced substrate coupling effects as well as reduced thermal resistance from the GaN channel towards the substrate. The immunity against back-gating has been previously validated for discrete GaN-on-AlN/SiC HEMTs in static and dynamic characterizations [9]. In this paper, dynamic characterizations are executed in a half-bridge topology and hard-switched double-pulse test using the same discrete GaN-on-AlN/SiC HEMTs with a gate width of $w_G = 92$ mm. In difference to [9] the substrate nodes of both discrete power transistors forming the half-bridge are electrically shorted in order to share a common bulk-potential similar to an MHB. Furthermore, hard-switched double-pulse tests are performed with an MHB fabricated on the same wafer and based on the same GaN-on-AlN/SiC technology with identical gate widths as the discrete devices. Reduced switching losses for the MHB could be realized because of the monolithic integration and associated changes in PCB layout. Additional device characterizations cover the temperature-dependent on-resistance as well as voltage-dependent parasitic device capacitances. All static characterizations of the proposed GaN-on-AlN/SiC HEMT are compared to a conventional normally-off p-GaN gate GaN HEMT based on a mature GaN-on-Si technology platform [10].

2 Device fabrication and technology

An unintentionally doped AlN layer as buffer material is used for the GaN-on-AlN/SiC transistor instead of a compensation doped GaN buffer as it is used for most conventional power-electronic GaN transistors. The AlN buffer acts as efficient back-barrier giving a good transistor-channel confinement and thus high device breakdown strength [11]. The AlGaN/GaN/AlN hetero structure is MOCVD-grown on top of 4" semi-insulating 4H-SiC substrates. While SiC substrates are more expensive than Si substrates, this is offset by the increased complexity of nitrite growth on Si. Unless sophisticated strain management methods are used, nitrite layers on Si will crack during cool down from MOCVD growth temperatures [12].

Normally-on AlGaN/GaN/AlN HEMTs with Ir-based Schottky-type gates are processed on the 4" wafers. All structures are defined by *i*-line optical stepper lithography. Ohmic source and drain contacts consist of a Ti/Al//Ni/Au based metallization, alloyed with RTA at 870°C in N_2 ambient. The devices were isolated by nitrogen implantation. The hetero structure is further passivated with 180 nm PECVD-deposited SiN_x. An Ir/Ti/Au gate metal was electron-beam evaporated on top of the 0.7 μm gate trench inside the SiN_x passivation layer. Source-connected Field Plates (SFP) are positioned on top of the SiNx layer 2 μm in front of the gate metal towards drain, see Fig. 1a. A benzocyclobutene (BCB) encapsulation was used to prevent any surface arcing during high voltage operation. Fabricated power-switching devices with 92 mm gate width, 0.7 μm gate length, 1 μm source-gate separation and 18 μm gate-drain separation (Fig. 1b) showed a static on-state resistance of approx. 100 mΩ and a maximum pulse current of 70 A at 1.0 V gate bias. MHB structures consisting of two corresponding 92 mm power-switches (Fig. 1c) were fabricated on the same wafer as well.

For comparison, similar GaN-on-Si transistors (Fig. 1e), having the same gate-source and gate-drain separation are studied [10]. Instead of an AlN buffer on SiC substrate, a carbon-doped GaN buffer structure was grown on a conductive Si substrate, see Fig. 1d. In contrast to the GaN-on-AlN/SiC transistor (Fig. 1a), a p-GaN based gate module is used for the device's normally-off characteristic and the SFP is enclosing the gate from the top (Fig. 1d).

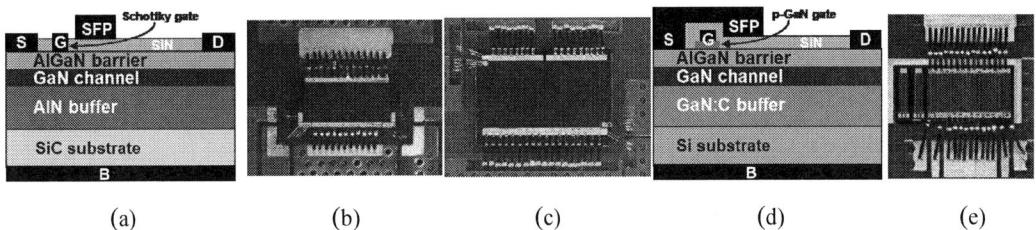

| (a) | (b) | (c) | (d) | (e) |

Fig. 1: Schematic cross-sections and mounted power devices on custom submount-PCBs: (a) Structure, (b) 92 mm transistor chips and (c) 2×92 mm MHB chip of the GaN-on-AlN/SiC transistor structure with Schottky-type gate. (d) Structure and (e) 92 mm transistor chip of the AlGaN/GaN-on-Si transistor with a p-GaN gate module for normally-off operation

3 Static characteristics

3.1 Static on-resistance R_{DSon}

The static on-resistance of GaN HEMTs is known to be strongly temperature-dependent due to reduced electron mobility in the two-dimensional electron gas (2DEG) at increased device temperature. The temperature-dependent on-state resistance is measured inside a temperature cabinet while a time-constant drain current I_D=100 mA is sourced into the device under test (DUT). The fabricated 92 mm GaN-on-AlN/SiC achieve a 29.7 % lower static on-resistance $R_{DSon,GaN-on-AlN/SiC}$ =101.4 mΩ compared to their GaN-on-Si counterparts ($R_{DSon,GaN-on-Si}$ =144.2 mΩ) at an ambient temperature T=30°C. At elevated temperature T=150°C the static on-resistance is increased up to 198.9 mΩ and 287.5 mΩ in the GaN-on-AlN/SiC and GaN-on-Si HEMT, respectively. Due to a similar transistor design concerning the AlGaN/GaN hetero-junction where the 2DEG is formed, the relative static R_{on}-increase of +96.15 % (GaN-on-AlN/SiC) and +99.37 % (GaN-on-Si) from 30°C to 150°C is similar for both GaN HEMTs.

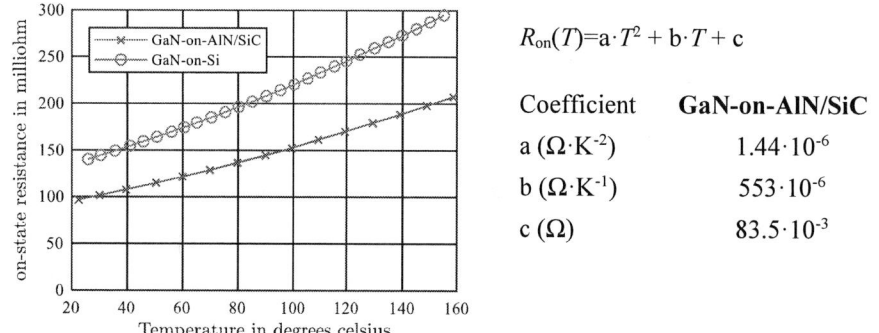

$$R_{on}(T)=a \cdot T^2 + b \cdot T + c \qquad (1)$$

Coefficient	GaN-on-AlN/SiC	GaN-on-Si
a ($\Omega \cdot K^{-2}$)	$1.44 \cdot 10^{-6}$	$2.20 \cdot 10^{-6}$
b ($\Omega \cdot K^{-1}$)	$553 \cdot 10^{-6}$	$799 \cdot 10^{-6}$
c (Ω)	$83.5 \cdot 10^{-3}$	$118.2 \cdot 10^{-3}$

Fig. 2: Comparison of the temperature-dependent static on-resistance of 92 mm GaN-on-AlN/SiC and GaN-on-Si HEMTs

3.2 Static I-V characteristics

The static I-V characteristics of the GaN-on-AlN/SiC power transistor are acquired using the test bench introduced in [13] and compared with the I-V characteristics of a 210 mm GaN-on-Si transistor variant, see Fig. 3. The I-V curves are measured at gate-source voltage V_{GS}=−3 V to 1 V, drain-source voltage V_{DS}=−5 V to 20 V and different bulk-source bias V_{BS}=−400 V to 400 V for the GaN-on-AlN/SiC transistor and V_{GS}=−3 V to 6 V, V_{DS}=−5 V to 20 V and different bulk-source bias V_{BS}=−450 V to 100 V for the GaN-on-Si transistor. The pulse length was 5 µs to avoid device self-heating during I-V characterization and the data is normalized by the transistors' gate width.

The presented GaN-on-AlN/SiC transistor (Fig. 3a) shows a maximum saturation current of approximately 0.76 A/mm (70 A for the 92 mm transistor) with the bulk-terminal connected to source. The maximum current density of the GaN-on-Si transistor (0.6 A/mm) is less because of its designed normally-off characteristic (Fig. 3b). The bulk-source bias V_{BS} has strong impact on the maximum drain current for the GaN-on-Si transistor which decreases from 0.6 A/mm to 0.2 A/mm with increasing negative V_{BS}, see Fig. 4b. The transistor channel resistivity increases in parallel. Root cause is the capacitive coupling of the conductive Si substrate to the 2DEG transistor channel via the GaN buffer and channel layers of approx. 5 µm thickness. The 2DEG electron density reduces with increasing negative bulk potential. In contrast, V_{BS} has negligible influence on the output characteristics of the GaN-on-AlN/SiC transistor and the I-V curves remain almost identical with varied V_{BS} in the range of −400 V to 400 V and a fixed V_{GS} of 1 V (Fig. 4a). The electrically insulating SiC substrate with 340 µm thickness strongly reduces the capacitance between the bulk (backside of the SiC substrate) and the transistor channel. As a consequence, GaN-on-AlN/SiC transistors are immune to back-gating effects in their static characteristics.

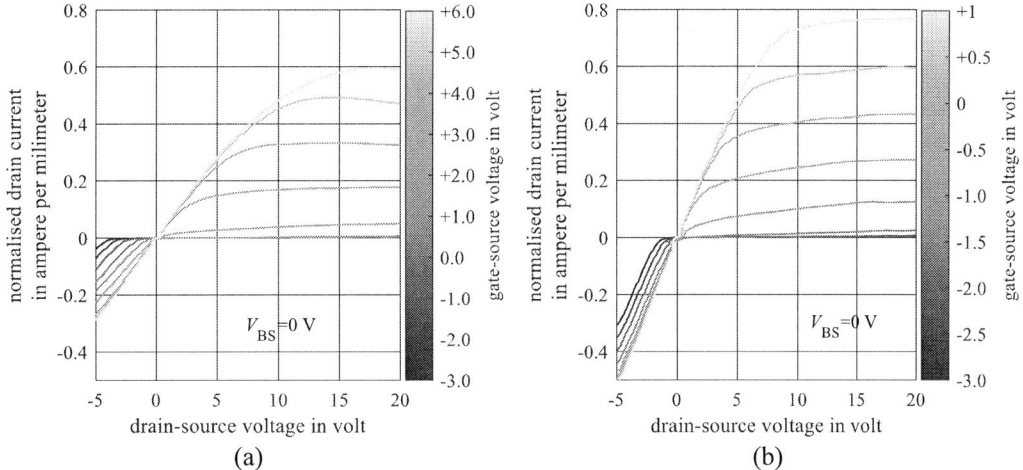

Fig. 3: Output characteristic of (a) GaN-on-Si and (b) GaN-on-AlN/SiC HEMT transistor with bulk-terminal connected to source-terminal for different gate-source voltages.

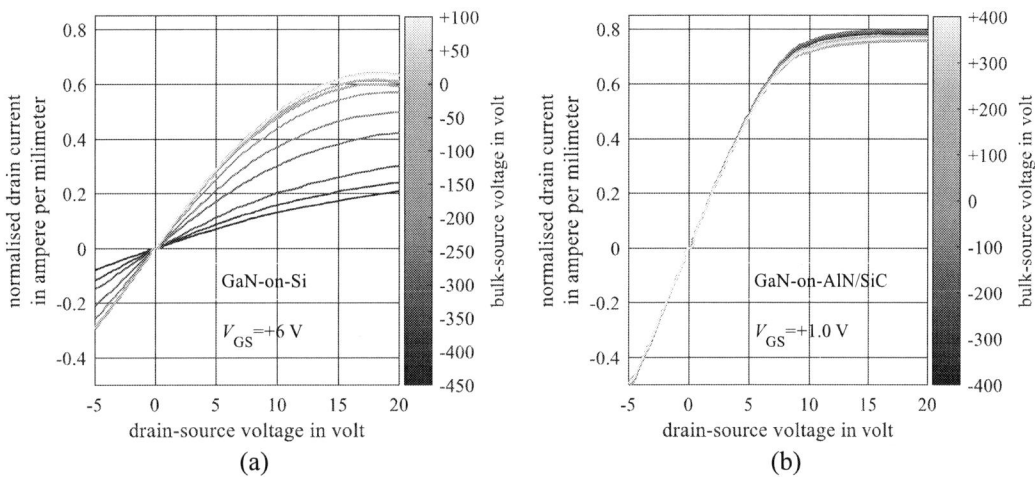

Fig. 4: Output characteristic of (a) GaN-on-Si and (b) GaN-on-AlN/SiC HEMT transistor at a fixed gate-source voltage for different bulk-source bias voltages.

3.3 Static C-V characteristics

GaN transistors show much lower parasitic capacitances due to their lateral structure compared with Si-counterparts, enabling fast switching [14]. The input capacitances C_{ISS}, output capacitance C_{OSS} and reverse transfer capacitance C_{RSS} are acquired via small signal measurements. The measurements are carried out with a Keithley parameter analyzer (Table I and Fig. 5) with an RMS voltage of 100 mV and a frequency of 200 kHz, while the bulk-terminal of the GaN-on-AlN/SiC transistor is shorted to the source-terminal. The drain-source voltage V_{DS} is swept from 0 V to 600 V, the gate-source voltage V_{GS} is kept at a constant value of -6 V. The transistor capacitances show a strong dependence on drain-source voltage V_{DS}. C_{ISS}, C_{OSS} and C_{RSS} have a value of 107 pF, 124 pF and 60 pF, respectively, at V_{DS}=0 V, which decreases with increasing V_{DS}, and reaches a value of 54 pF, 11 pF and 1.6 pF at V_{DS}=600 V (Fig. 6) The low capacitances and the calculated correspondingly low total output charge Q_{OSS} of 9.8 nC and stored energy E_{OSS} of 1.2 μJ at V_{DS}=400V, indicate that the GaN-on-AlN/SiC transistors are promising for fast switching and high-frequency applications. In comparison to the 92 mm GaN-on-Si device, the GaN-on-AlN/SiC transistor has substantially lower input and output capacitances but an increased transfer capacitance at a high drain-source voltage (>120 V). This is beneficial to reduce of switching losses but increases susceptibility to cross conduction. The coupling of source, gate and drain via the substrate is reduced for the GaN-on-AlN/SiC device because of the insulating nature of the

SiC substrate. As opposed to the GaN-on-Si device, the source-connected field plate of the GaN-on-AlN/SiC transistor is not overlapping the gate. That's why the GaN-on-AlN/SiC transistor has a higher C_{RSS}.

Table I: CV profiling setup

CVU	Keithley 4215-CVU
Drain SMU	Keithley 2657A
Drain bias tee	Keithley 2650-RBT-3K
Gate SMU	Keithley 2635B
Gate bias tee	Keithley 2600-RBT-200
Source bias tee	Keithley 2600-RBT-200

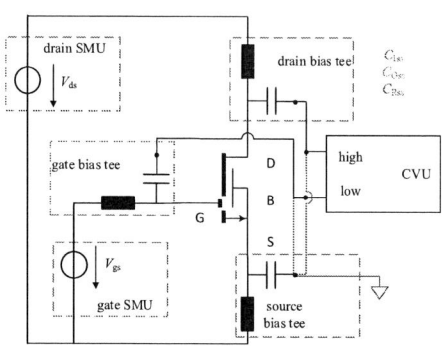

Fig. 5: Schematic of C-V measurements

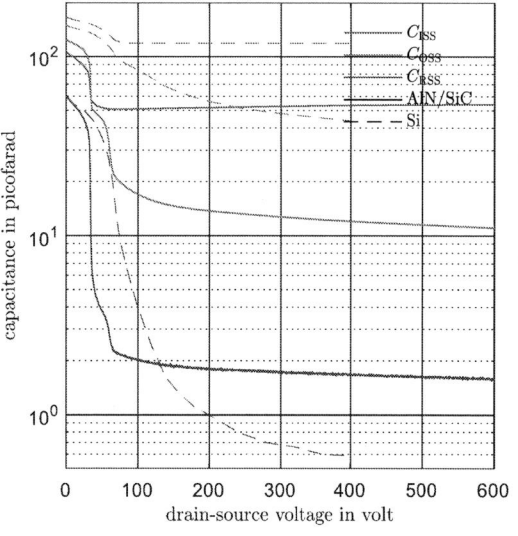

Fig. 6: Capacitance voltage profile of the GaN-on-AlN/SiC transistor with $V_{GS}=-6$ V, measured with a 200 kHz and 100 mV RMS small signal measurement, compared to capacitance voltage measurements of the GaN-on-Si transistor, recorded using a HP 4280A C-V Plotter at 1 MHz small signal frequency.

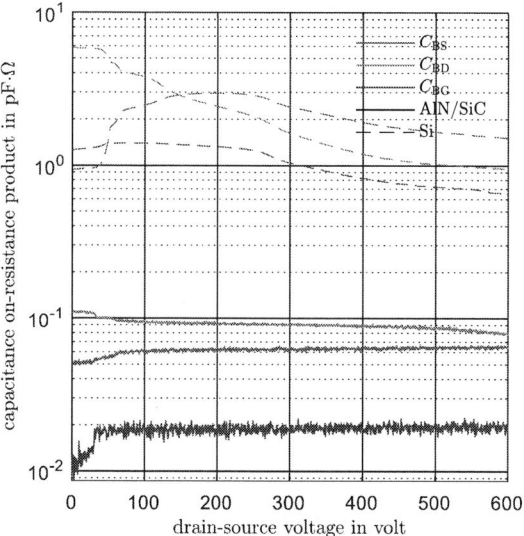

Fig. 7: Bulk capacitances of the GaN-on-AlN/SiC transistor with $V_{GS}=-6$ V compared to a commercial GaN-on-Si transistor (GS66508P) with $V_{GS}=0$ V, measured with a 200 kHz and 300 mV RMS small signal.

Additionally, the bulk-source, bulk-gate and bulk-drain capacitances of the GaN-on-AlN/SiC transistor and a commercially available GaN-on-Si transistor with exposed bulk node (GS66508P) are compared in Fig. 7. Because the transistor layout of the commercial device is unknown, the data is normalized to the nominal on-state resistance, 100 mΩ and 50 mΩ, respectively. The GaN-on-AlN/SiC transistor shows significant lower normalized bulk capacitances (between one and two orders of magnitude) than the GaN-on-Si transistor. The reduced coupling from source, gate and drain to the substrate is in favor of fast switching transients.

3.4 Gate-source characteristics

The GaN-on-AlN/SiC transistor Schottky-diode which is formed between the Ir gate metal on top of the AlGaN barrier and the 2DEG beneath features a barrier height of ~0.9 eV. This leads to a gate current as forward diode current, when the transistor is in a typical on-state condition with $V_{GS} = 1$ V and $V_{DS} < 1$ V. The characteristics of the gate diode (Fig. 8) are measured with bulk shorted to source while drain is left open. The measurements are performed using a Keithley 2635B SMU. At a typical off-state gate-source voltage of −6 V the gate diode has a leakage current of −4.3 µA. The diode conducts and current flows through the gate in the on-state. This non-insulating gate characteristic, which is similar to GaN Gate Injection Transistors (GIT) with p-GaN gate [15], brings new challenges to driving the transistor, since a large gate current pulse is required during switching transitions to achieve fast switching, while a mA-range continuous gate current is desired during on-state to keep the device in on-state and minimize the conduction losses simultaneously [15, 16].

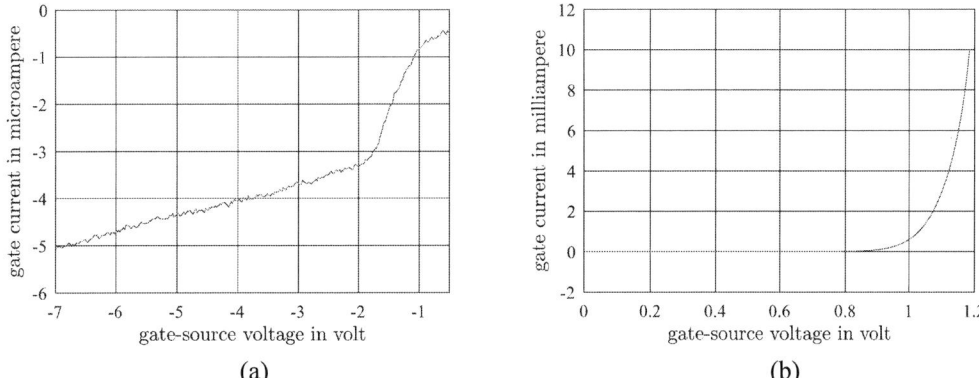

Fig. 8: (a) Gate reverse characteristic and (b) forward characteristic of gate-source diode, measured with bulk connected to source and open drain.

4 Dynamic characteristics

4.1 Measurement setup

Current collapse is a critical issue for monolithic GaN power transistor integration since the transistors share a common substrate potential (Fig. 9a) and cannot be connected to their respective source-terminal separately to suppress the back-gating effects (see Fig. 4b for the case of GaN-on-Si).

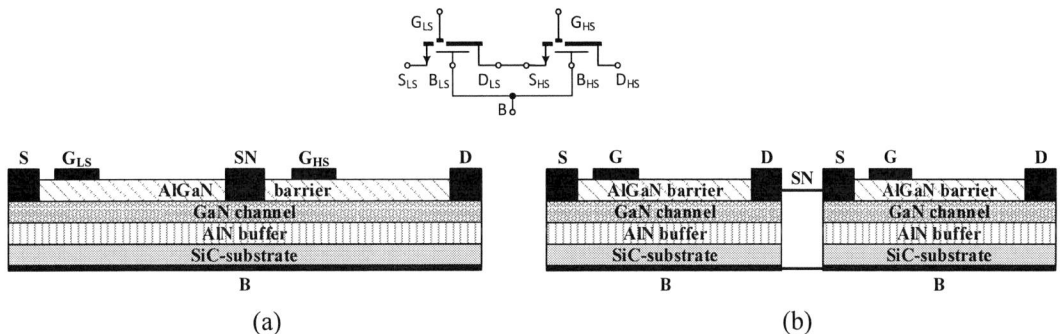

Fig. 9: (a) monolithic half-bridge chip and (b) half-bridge built via two discrete chips.

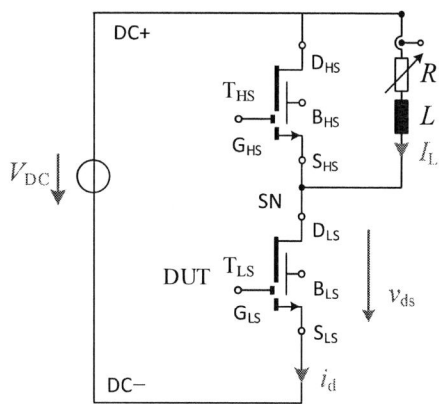

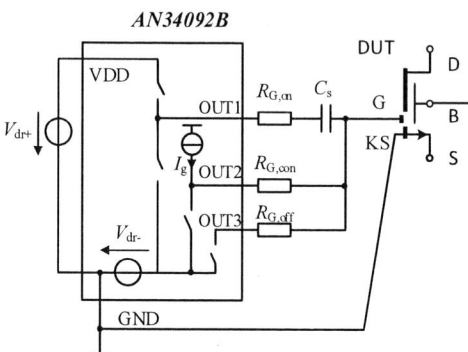

Fig. 10: Schematic of half-bridge used for double-pulse tests.

Fig. 11: Schematic of gate-loop circuit.

To study the benefits of the presented GaN-on-AlN/SiC HEMT technology for monolithic integration, a half-bridge using discrete transistors is built and evaluated with the different bulk terminations listed in Table II. Only the bulk-potential is varied while all other conditions are kept completely identical. Conf. 1 represents the typical bulk termination used for discrete transistors, avoiding back-gating effects and enabling fast switching [14]. As this configuration is not feasible for MHBs with a shared bulk potential, configurations with the common bulk potential either floating (conf. 2) or connected to low-side source, switch-node and high-side drain for conf. 3 to 5, respectively are studied. Further, the measurements for conf. 2 to 5 are repeated for the real MHB chip (Fig. 1c) to verify the validity of the results for real monolithic applications and to study the benefits from monolithic integration compared with discrete devices.

Table II: Configuration of different bulk termination

Conf. 1	Conf. 2	Conf. 3	Conf. 4	Conf. 5
Bulk of each transistor connected to their respective source-terminal	Bulk of both transistors are connected and floating	Bulk of both transistors are connected to S_{LS}	Bulk of both transistors are connected to switching note	Bulk of both transistors are connected to D_{HS}

The double pulse tests are performed in a half-bridge configuration with an inductive load at different load currents I_L and blocking voltages V_{dc} (Fig. 10). Due to the non-insulating gate characteristics of the GaN-on-AlN/SiC HEMT, the Panasonic driver *AN34092B* [17] with an integrated current source is used (Fig. 11). The turn-on gate resistor $R_{G,on}$ is connected in series with a high-pass capacitor C_s to achieve fast turn-on (OUT1) and the integrated current source (OUT2) offers a constant current in mA-range in on-state. Both gate resistors $R_{G,off}$ and $R_{G,con}$ have an influence on the turn-off speed. The gate configuration is listed in Table III. The drain current of the low-side transistor (DUT) is measured using a high-bandwidth, low inductance SMD shunt [18] and the drain-source voltage is measured via a TPP0850 850 MHz bandwidth voltage probe.

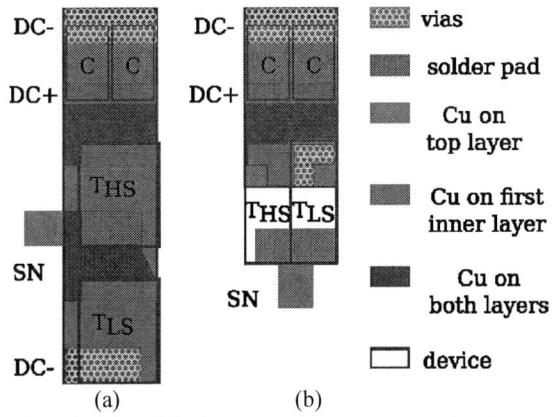

vias
solder pad
Cu on top layer
Cu on first inner layer
Cu on both layers
device

(a)   (b)

Fig. 12: Simplified layout of the commutation loop of the discrete (a) and monolithic (b) half-bridge. Both layouts are to scale.

Table III: Gate circuit configuration

Parameters	Symbol	Value
Turn-on gate resistor	$R_{G,on}$	5.6 Ω
Turn-off gate resistor	$R_{G,off}$	2 Ω
Continuous gate resistor	$R_{G,con}$	3.9 Ω
Boost capacitor	C_s	330 pF
Continuous gate current turn-on	$I_{G,on}$	5 mA
Turn-on gate voltage	$V_{dr,on}$	5 V
Turn-off gate voltage	$V_{dr,off}$	−5 V

The PCBs for the discrete half-bridge and the MHB where designed almost identical to ensure comparability between the different transistors. The gate drivers, DC-link capacitors and shunts use the same circuitry and layout. However, the commutation loop is changed. While the discrete half-bridge uses a completely vertical commutation loop, the commutation loop of the MHB is vertical at the shunt and DC-link capacitors and horizontal on the MHB as shown in Fig. 12. For the discrete half-bridge, the switching node's copper overlaps the copper on the inner layer, which is connected to the low-side transistor's source. In contrast, there is no overlapping copper at the switching node with the MHB leading to a reduction in parasitic capacitances of the switching node. However, horizontal commutation loops tend to have an increased parasitic inductance, compared to vertical loops [19], which is mitigated by the monolithic integration.

4.2 Measurement results

With conventional bulk termination (conf.1) the GaN-on-AlN/SiC HEMT shows good high-speed switching performance with a turn-on voltage slew rate of −63 V/ns and a turn-off slew rate of 63 V/ns at a blocking voltage of 250 V and an inductor current of 7 A (Fig. 13) over an interval from 20 % to 80 %.

The switching transitions and the switching losses for the different studied bulk terminations are shown in Fig. 14 and Fig. 15(a) respectively. The bulk terminations show only negligible influence on the discrete half-bridge's dynamic characteristics. While GaN-on-Si devices show a strong impact of the bulk termination on the dynamic R_{on} [4], no such effect is observed here. Furthermore, both the current and voltage transients are nearly unchanged for conf. 1 through 4 (Fig. 14). In configuration 5 the drain current peak during turn-on is slightly elevated, leading to increased switching losses (Fig. 15a). Although the capacitive coupling between the top and bottom side of GaN-on-AlN/SiC HEMT is significantly reduced by the larger separation of the conductive layers as compared to GaN-on-Si, the metallized backside of the SiC substrate may still couple to the devices top. Different bulk terminations could change the device's capacitances and could lead to increased cross conduction and correspondingly higher losses. However, since these measurements were conducted consecutively in the order of increasing configuration number, device degradation cannot be ruled out as well.

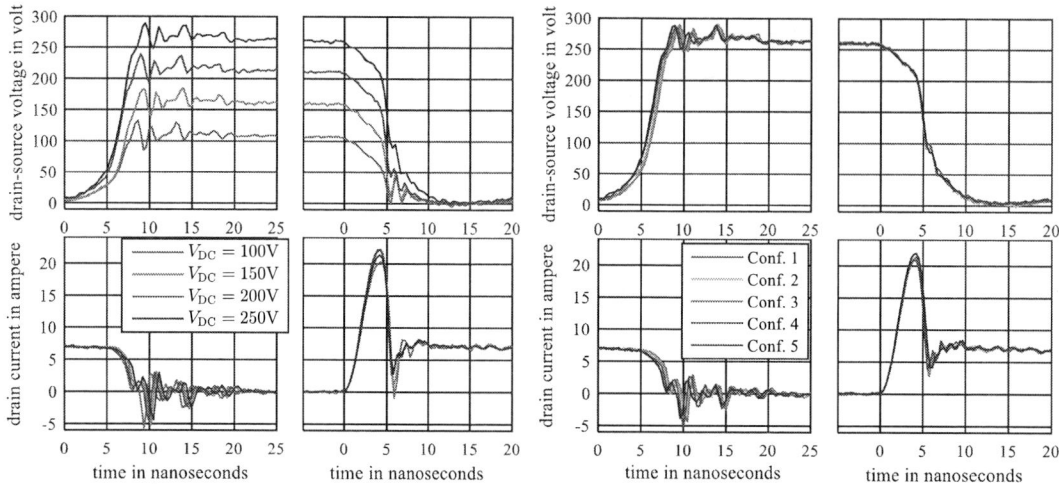

Fig. 13: Turn-off (left) and turn-on (right) transients for conf. 1 at an inductor current of 7 A and 100 V to 250 V blocking voltage, for the emulated MHB.

Fig. 14: Turn-off (left) and turn-on (right) transients at an inductor current of 7 A and blocking voltage of 250 V in different configurations, for the emulated MHB.

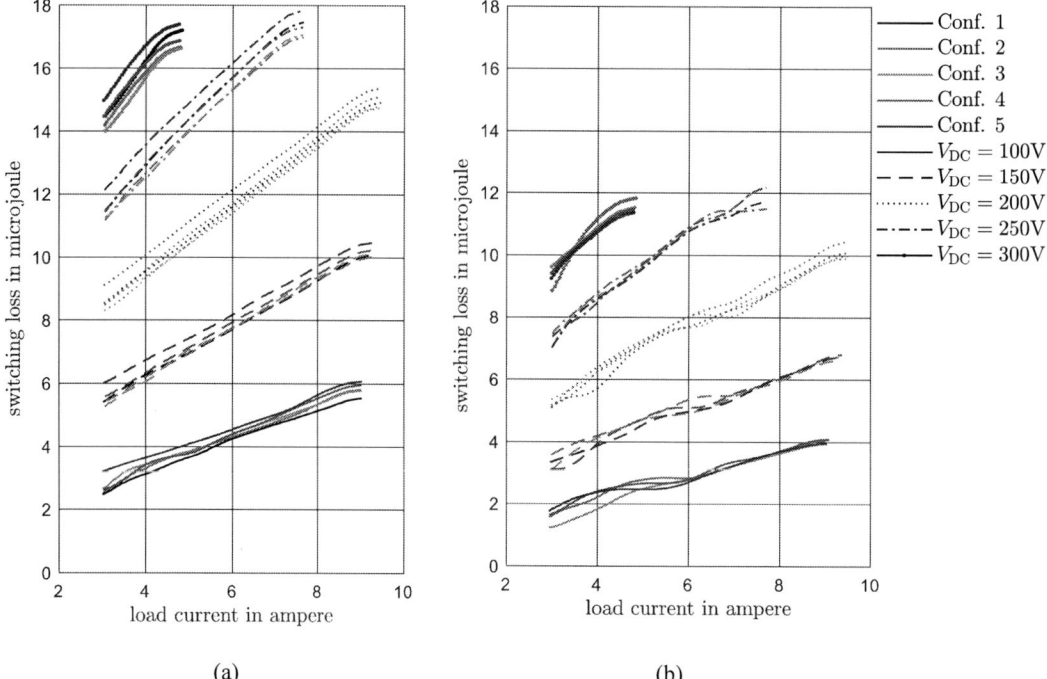

(a)   (b)

Fig. 15: Switching losses for the emulated MHB (a) and the actual MHB (b).

The real MHB, with bulk connected to the lower side source, achieves slew rates of −68 V/ns at turn-on and 133 V/ns at turn-off at a blocking voltage of 250 V and an inductor current of 7 A (Fig. 16) over an interval from 20 % to 80 %. The monolithic half-bridge shows a much higher turn-off slew rate and much lower current peak at turn-on (15 A vs. 20 A) as compared to the discrete half-bridge setup (Fig. 18), reducing the switching losses by 30 % (Fig. 15). This demonstrates the benefit of the monolithic device integration in comparison with discrete devices. The improved switching performance of the MHB is a result of the lower parasitic inductance of the monolithic chip itself, as well as of the reduced parasitic capacitances of the PCB's commutation loop. However, it is currently not possible to separate both contributions. But we may state that, independent of the studied PCB-based commutation loop, an MHB generally allows for smaller commutation loop designs. The switching transients in Fig. 17 clearly show that the four different bulk terminations have no influence on the switching transients and performance of the MHB, which verifies a sufficient immunity to back-gating effects of the presented technology also for the real monolithic half-bridge.

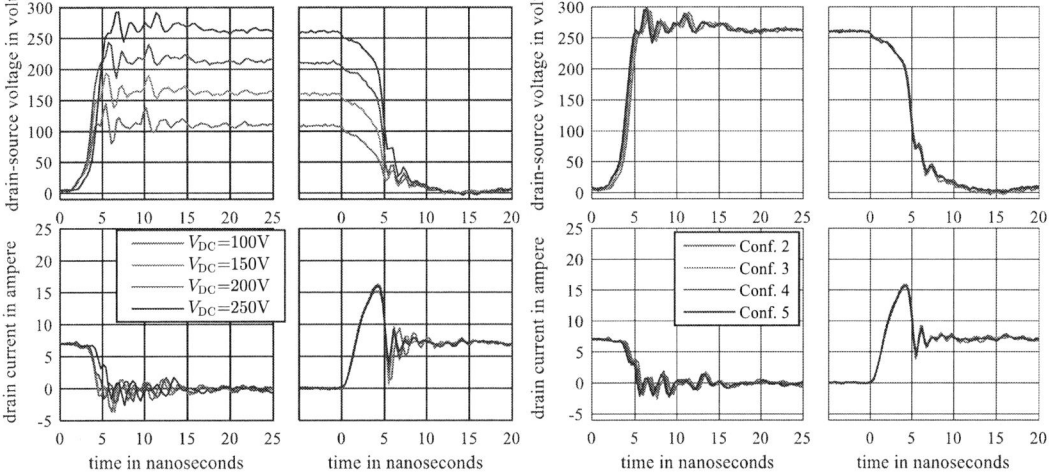

Fig. 16:Turn-off (left) and turn-on (right) transients for conf. 1 at an inductor current of 7 A and 100 V to 250 V blocking voltage, for the MHB.

Fig. 17: Turn-off (left) and turn-on (right) transients at an inductor current of 7 A and blocking voltage of 250 V in different configurations, for the MHB.

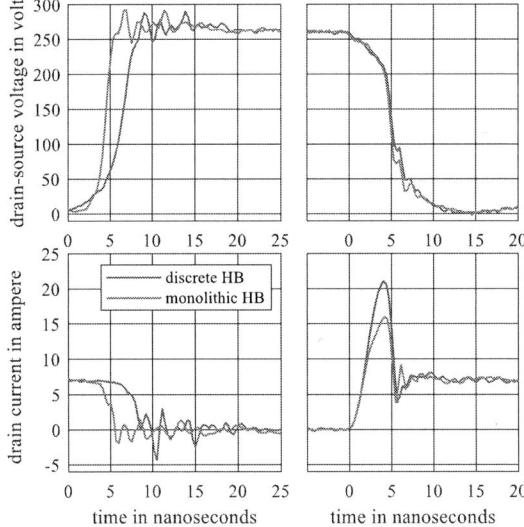

Fig. 18: Turn-off (left) and turn-on (right) transients for conf. 2 at an inductor current of 7 A and 250 V blocking voltage, for the discrete and monolithic HB.

5 Conclusion

Discrete transistors and monolithically integrated half-bridges based on a GaN-on-AlN/SiC technology are characterized to demonstrate efficient power switching with GaN monolithic integration up to 300 V. The immunity of the transistor to back-gating effects, in both its static and dynamic characteristics, is shown through I-V-measurements and double-pulse tests. The reduced coupling between the transistor's top and bottom side leads to low output and input capacitances, resulting in fast switching. The hard-switching tests reveal faster switching transients and reduced switching losses for the monolithically integrated half-bridge as compared to the half-bridge with discrete transistors.

References

[1] S. Moench, B. Weiss, R. Reiner, P. Waltereit, R. Quay, O. Ambacher and I. Kallfass, "Instabilities by Parasitic Substrate-Loop of GaN-on-Si HEMTs in Half-Bridges," in *2018 IEEE 6th Workshop on Wide Bandgap Power Devices and Applications (WiPDA)*, 2018.

[2] B. Weiss, R. Reiner, V. Polyakov, P. Waltereit, R. Quay, O. Ambacher and D. Maksimović, "Substrate biasing effects in a high-voltage, monolithically-integrated half-bridge GaN-Chip," in *2017 IEEE 5th Workshop on Wide Bandgap Power Devices and Applications (WiPDA)*, 2017.

[3] C. Kuring, O. Hilt, J. Böcker, M. Wolf, S. Dieckerhoff and J. Würfl, "Novel monolithically integrated bidirectional GaN HEMT," in *2018 IEEE Energy Conversion Congress and Exposition (ECCE)*, 2018.

[4] C. Kuring, N. Wieczorek, O. Hilt, M. Wolf, J. Böcker, J. Würfl and S. Dieckerhoff, "Impact of Substrate Termination on Dynamic On-State Characteristics of a Normally-off Monolithically Integrated Bidirectional GaN HEMT," in *2019 IEEE Energy Conversion Congress and Exposition (ECCE)*, 2019.

[5] S. Bahl, M. Senesky, N. Tipirneni, D. Anderson and S. Pendharkar, "Bi-directional gallium nitride switch with self-managed substrate bias". Patent US20140374766A1 (20.06.2013).

[6] M. Imam, H. Kim, K. Leong, B. Pandya and G. Prechtl, "Semiconductor device having a bidirectional switch and discharge circuit". US / EU Patent US20190326280A1 (23.04.2018) / EP3562040A1 (15.04.2019).

[7] K. Leong, "Bidirectional switch with passive electrical network for substrate potential stabilization". Patent US10224924B1 (22.08.2017).

[8] A. Lidow, "Gallium Nitride Integrationand the End of Discretes," in *2021 IEEE 8rd Workshop on Wide Bandgap Power Devices and Applications (WiPDA)*, 2021.

[9] S. Heucke, O. Hilt, X. Geng, C. Kuring, J. Würfl and S. Dieckerhoff, "Substrate Bias Effects up to 400 V of Normally-On GaN-on-AlN/SiCHEMTs in Static and Dynamic Tests," in *CIPS 2022; 12th International Conference on Integrated Power Electronics Systems*, 2022.

[10] O. Hilt, R. Zhytnytska, J. Böcker, E. Bahat-Treidel, F. Brunner, A. Knauer, S. Dieckerhoff and J. Würfl, "70 mΩ/600 V normally-off GaN transistors on SiC and Si substrates," in *2015 IEEE 27th International Symposium on Power Semiconductor Devices IC's (ISPSD)*, 2015.

[11] O. Hilt, F. Brunner, E. B. Treidel, M. Wolf and J. Würfl, "GaN-channel HEMTs with AlN buffer for high-voltage switching," in *2021 Device Research Conference (DRC)*, 2021.

[12] A. Dadgar, "Sixteen years GaN on Si," *physica status solidi (b)*, vol. 252, p. 1063–1068, February 2015.

[13] J. Böcker, H. Just, O. Hilt, N. Badawi, J. Würfl and S. Dieckerhoff, "Experimental analysis and modeling of GaN normally-off HFETs with trapping effects," in *2015 17th European Conference on Power Electronics and Applications (EPE'15 ECCE-Europe)*, 2015.

[14] F. Yang, C. Xu and B. Akin, "Experimental Evaluation and Analysis of Switching Transient's Effect on Dynamic on-Resistance in GaN HEMTs," *IEEE Transactions on Power Electronics,* vol. 34, pp. 10121-10135, 2019.

[15] B. Zojer, "Driving 600 V CoolGaN™ high electron mobility transistors," *Infineon Technologies AG, Application Note,* vol. 2, 2018.

[16] X. Geng, C. Kuring, O. Hilt, M. Wolf, J. Würfl and S. Dieckerhoff, "Design and Optimization of the Driver Circuit for Non-Insulating Gate GaN-Transistors Enabling Fast Switching and High-Frequency Opera-tion," in *CIPS 2022; 12th International Conference on Integrated Power Electronics Systems*, 2022.

[17] Panasonic, "AN34092B-Single-Channel GaN-Tr High-Speed Gate Driver," *Application Note,* 2017.

[18] J. Böcker, S. Schoos and S. Dieckerhoff, "Experimental Comparison and 3D FEM Based Optimization of Current Measurement Methods for GaN Switching Characterization," in *2018 20th European Conference on Power Electronics and Applications (EPE'18 ECCE Europe)*, 2018.

[19] C. Kuring, M. Wolf, X. Geng, O. Hilt, J. Bocker, J. Würfl and S. Dieckerhoff, "GaN-based multi-chip half-bridge power module integrated on high-voltage AlN ceramic substrate," *IEEE Transactions on Power Electronics,* 2022.

Optimal frequency for Dynamic Wireless Power Transfer

Mincui LIANG, Khalil EL KHAMLICHI DRISSI, Christophe PASQUIER
Université Clermont Auvergne, Clermont Auvergne INP, CNRS, Institute Pascal
Campus Universitaire des Cézeaux, 4 Avenue Blaise Pascal
Aubière France
Email: mincui.liang@uca.fr; khalil.drissi@uca.fr; christophe.pasquier@uca.fr;
URL: http://www.institutpascal.uca.fr/index.php/en/

Acknowledgments

This work was sponsored by a public grant overseen by the French National Research Agency as part of the "Investissements d'Avenir" through the IMobS3 Laboratory of Excellence (ANR-10-LABX-0016) and the IDEX-ISITE initiative CAP 20-25 (ANR-16-IDEX-0001).

Keywords

≪Electric vehicle≫, ≪Wireless power transmission≫, ≪DC-AC converter≫, ≪Resonant converter≫, ≪Time domain analysis≫

Abstract

To control the system for maximum power transfer, it is necessary to understand its behaviour for large range of frequencies. Based on the time domain theoretical analysis of a Series-Series compensated WPT, the optimal frequencies to reach maximum power are proposed, considering the misalignment between the primary and secondary coils. It is further validated by a simulation case study. The theoretical and simulation results match well.

Introduction

Transportation plays a critical role in achieving the net-zero emission goal as it is responsible for 17% of greenhouse gas emissions. Therefore, The COP26 suggested that countries and states should accelerate the transition to zero-emission vehicles in 10 years. Consequently, the sales of global zero-emission vehicles are projected to increase approximately by 70% in 2040 [1]. Electric Vehicles (EVs) are the most mature zero-emission vehicle technology. As the demand for EVs increases, the need for transport infrastructures and supplementary facilities will increase correspondingly, especially power transfer systems for charging.

Power transfer systems can be classified into wired power transfer and Wireless Power Transfer (WPT). Compared to wired power transfer, WPT is safer and can seamlessly integrate with autonomous connection in the future. Practically, WPT has two main applications for EVs, static charging and dynamic charging. Dynamic Wireless Power Transfer (DWPT) not only inherits the advantages of WPT but also can solve the well-known battery range limitations by reducing the size of the battery and charging it while driving. Although DWPT is not yet mature for the commercial use, many DWPT system prototypes using magnetic resonance technology exist [2]. This research focus on magnetic resonant coupling DWPT technology. However, there are a few challenges of designing the controller for a dynamic magnetic resonance DWPT system. On the one hand, adding a resonant tank as a compensation network can increase the quality factor and thus amplify the power transfer. On the other hand, it also makes the system sensitive to misalignment, which causes a significant reduction in efficiency with respect to the transmission distance. Precisely, the main reason why DWPT is more difficult to control than the static

WPT is that sensitive parameters, like coupling coefficient k, are not fixed since the lateral and vertical displacements of EVs vary. Thus, it is essential to investigate the potential influences of those sensitive parameters on the behaviour of the DWPT systems so as to control the systems for maximum power transfer.

This paper aims to provide a holistic overview of the current progress and trend in the WPT and propose a novel method to identify the optimal operating frequencies for the DWPT. Specifically, it is organized in a logical order as follows: 1) to review the Series-Series compensated WPT (SS-WPT) which topology is considered sufficient for modelling the DWPT at the first stage, 2) to review operating frequencies for the WPT, 3) to construct a theoretical model for finding those optimal frequencies to reach maximum power transfer with respect to different coupling coefficients of misalignment, and 4) to carry out a case study using the developed theoretical model and validate the system performances using dynamic simulation.

Series-series compensated WPT

There are four mono-resonant converter topologies of compensation circuits for WPT applications, i.e. Series-Series (SS), Series-Parallel (SP), Parallel-Parallel (PP) and Parallel-Series (PS). Other topologies use different combinations of multi-inductor and multi-capacitor components designed for better performance and for larger power transmission. SS topology is the most straightforward and favourable solution to achieve Maximum Wireless Power Transfer (Max-WPT) and guarantee high efficiency. Misalignment and frequency bifurcation phenomenon are two main challenges of the DWPT systems [3]. From the circuit theory perspective, the SS topology is adequate to address the challenges. SS topology design is load-impedance-independent and mutual-inductance-independent for various misalignment situations [4]. Besides, SS topology can reduce the voltage rating of the power supply to reduce losses [5]. Furthermore, the current source-type primary series–secondary series (I-SS) topology has a k-independent feature in terms of maximum power transfer, high power efficiency and load independent output characteristics [2]. Therefore, SS topology is sufficient to test the proposed controller at the first stage. A simple SS-WPT compensation network, consisting of the primary side with a DC/AC converter, the wireless link and the secondary side with a resistive load, is considered during this study, as shown in Fig. 1. The switching strategy applied to the DC/AC converter corresponds to 1) SW1 and SW4 are on and SW2 and SW3 are off for the first half period, $0 < t < \frac{T}{2}$, and 2) SW2 and SW3 are on and SW1 and SW4 are off for the second half period, $\frac{T}{2} < t < T$. At this stage, Power Factor Corrector (PFC) and DC/DC on-board charger are not considered.

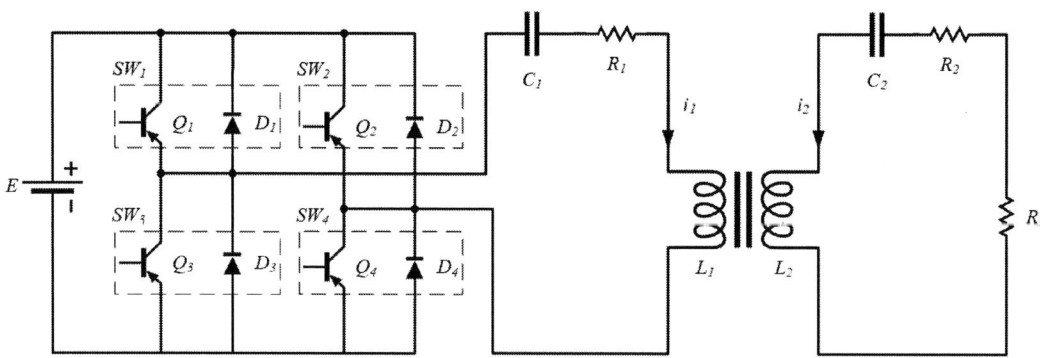

Fig. 1: Circuit diagram of a series-series WPT compensation network

Switching frequencies for DWPT

Practically, switching frequencies for DWPT vary depending on the control aims of the system. It is preferable to maintain the system operation when input voltage and input current are in phase to reduce the electrical stress on semiconductor devices and thus reduce switching losses, known as soft switching. The optimal frequencies of the Zero Voltage Switching (ZVS) [6] control is for high power transfer.

In [7], the optimal frequency of the proposed Maximum Energy Efficiency Tracking method is slightly lower than the resonance frequency. In this paper, the optimal frequency is the frequency corresponding to the maximum power point of input or its corresponding output power. We named it as Maximum Power Point Frequency (MPPF). There are several MPPFs in the SS-WPT system for different coupling coefficients. According to the IEC 61980-3 standard, the switching frequency of the WPT systems for EVs is between 79 kHz to 90 kHz [8]. It is considered as a high control frequency for power electronics and a narrow operation frequency range. From our point of view, following the IEC 61980-3 standard, the best choice is the MPPF below the primary resonance frequency, reducing the switching losses and maximizing the transferred power.

Theoretical modeling of the SS-WPT

In this section, a theoretical model of the SS-WPT is thoroughly presented in order to understand how different parameters correlate and affect the system performances. Fig. 2 (a) gives an equivalent circuit of the SS-WPT compensation network presented in Fig. 1. Two current-controlled voltage sources are introduced to represent the mutual inductance transformer, with two corresponding currents i_1 and i_2.

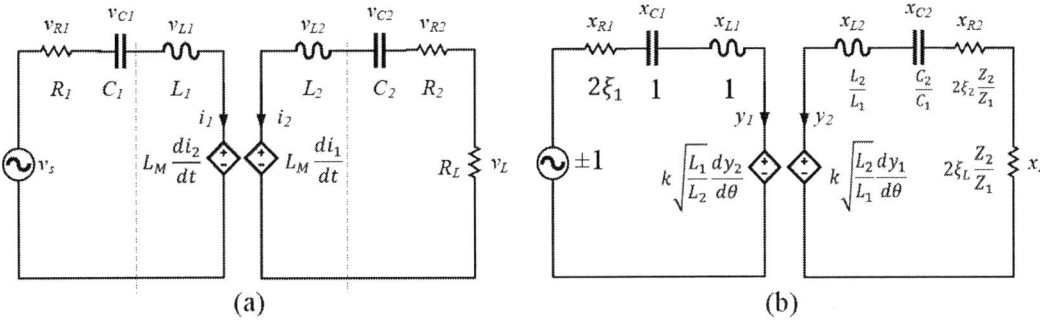

(a)   (b)

Fig. 2: (a) SS-WPT equivalent circuit with current-controlled voltage source; (b) Corresponding circuit with dimensionless elements

The mathematical relationship in time domain is given below in (1). The equivalent main voltage source $v_s = E$ when $0 < t < \frac{T}{2}$ and $v_s = -E$ when $\frac{T}{2} < t < T$. T is the switching period and t is the operating time.

$$\begin{cases} v_s = v_{C1} + R_1 i_1 + v_{L1} + L_M \frac{di_2}{dt} \\ 0 = v_{C2} + (R_2 + R_L) i_2 + v_{L2} + L_M \frac{di_1}{dt} \\ i_1 = C_1 \frac{dv_{C1}}{dt} \\ i_2 = C_2 \frac{dv_{C2}}{dt} \end{cases} \tag{1}$$

By applying theoretical modelling, the system can reach almost 100% efficiency, assuming losses are negligible and all semiconductor devices are ideal. The theoretical model does not include the model of the switching losses. However, the switching losses need to be considered since the finding of the optimal operational frequency is not based on Zero Current Switching (ZCS) control strategy. Therefore, an equivalent resistor is introduced to include the switching losses in the model, which is equal to 5 times of the coil losses on each side. The resistor of each designed coil using litz wire is 0.03 Ω in our setup, and hence we put two equivalent resistors (0.15 Ω each) on both sides to represent the switching, capacitor and coil losses. This value will be adapted according to our setup. The efficiency is then lower than one and close to the real situation.

SS-WPT model with dimensionless elements

For the ease of analysis and representation in phase plane, the model uses dimensionless elements, which are scalable and can be adjusted for any input source, coil parameters, compensation capacitors and re-

Table I: Dimensionless and corresponding not-dimensionless elements of the model

Elements	Voltage	Current	Resistor	Time	Inductor	Capacitor
Not-dimensionless	$v_{C_1}, v_{C_2}, v_s, v_L$	i_1, i_2	R_1, R_2, R_L	t	$L_{1,2}$	$C_{1,2}$
Dimensionless	x_1, x_2, u, x_L	y_1, y_2	$2\xi_1, 2\xi_2, 2\xi_L$	$\frac{\theta}{\omega_{r1}}$	1	1

sistive loads. The equivalent dimensionless circuit diagram is given in Fig. 2 (b). Parameters of dimensionless and not-dimensionless elements are defined in Table I. Additionally, the switching frequency is $f = \frac{1}{T} = \frac{\omega}{2\pi}$. The normalized or dimensionless angular frequency is $\omega_n = \frac{\omega}{\omega_{r1}}$, where $\omega_{r1} = 2\pi f_{r1}$ and $\omega_{r2} = 2\pi f_{r2}$. More specifically, the angular frequencies of the primary and secondary circuits are $\omega_{r1} = \frac{1}{\sqrt{L_1 C_1}}$ and $\omega_{r2} = \frac{1}{\sqrt{L_2 C_2}}$ respectively. The characteristic impedance in the primary and secondary circuits are $Z_1 = \sqrt{\frac{L_1}{C_1}}$ and $Z_2 = \sqrt{\frac{L_2}{C_2}}$ respectively. The mathematical model with dimensionless elements is given below in (2). $\theta = \omega_{r_1} t$, when $0 < \theta < \frac{\pi}{\omega_n}$, $u = 1$; when $\frac{\pi}{\omega_n} < \theta < \frac{2\pi}{\omega_n}$, $u = -1$.

$$
\begin{cases}
u = x_1 + 2\xi_1 y_1 + \frac{dy_1}{d\theta} + kn\frac{dy_2}{d\theta} \\
0 = \frac{x_2}{n^2} + a(2\xi)y_2 + \frac{dy_2}{d\theta} + \frac{k}{n}\frac{dy_1}{d\theta} \\
y_1 = \frac{dx_1}{d\theta} \\
y_2 = \frac{1}{(an)^2}\frac{dx_2}{d\theta}
\end{cases}
\tag{2}
$$

where, n is the ratio of the turns of two inductors, $n = \sqrt{\frac{L_2}{L_1}}$; $(an)^2$ is the ratio of two capacitors, $(an)^2 = \frac{C_1}{C_2}$; a is the ratio of two resonance frequency, $a = \frac{\omega_{r2}}{\omega_{r1}}$; an^2 is the ratios of primary and secondary impedance, $an^2 = \frac{Z_2}{Z_1}$; k is the coupling coefficient; ξ is the sum of dimensionless resistors on secondary side, $\xi = \xi_2 + \xi_L$; $x_{1,2}$ is dimensionless voltage of C_1 and C_2, $x_{1,2} = \frac{v_{C_{1,2}}}{E}$; $y_{1,2}$ is dimensionless current of L_1 and L_2, $y_{1,2} = \frac{Z_1}{E}i_{1,2}$; P is dimensionless power, p is real power, $P_0 = \frac{E^2}{Z_1}$, $P = \frac{p}{P_0}$.

To transform the model (2) into a state space equation in (3), we get, when $0 < \theta < \frac{\pi}{\omega_n}$,

$$
\begin{bmatrix} \dot{y}_1 \\ \dot{y}_2 \\ \dot{x}_1 \\ \dot{x}_2 \end{bmatrix} + \begin{bmatrix} \frac{2\xi_1}{1-k^2} & \frac{-2\xi kna}{1-k^2} & \frac{1}{1-k^2} & \frac{-k}{n(1-k^2)} \\ \frac{-2\xi_1 k}{n(1-k^2)} & \frac{2\xi a}{(1-k^2)} & \frac{-k}{n(1-k^2)} & \frac{1}{n^2(1-k^2)} \\ -1 & 0 & 0 & 0 \\ 0 & -(an)^2 & 0 & 0 \end{bmatrix} \begin{bmatrix} y_1 \\ y_2 \\ x_1 \\ x_2 \end{bmatrix} = \begin{bmatrix} \frac{1}{1-k^2} \\ \frac{-k}{n(1-k^2)} \\ 0 \\ 0 \end{bmatrix}
\tag{3}
$$

which is equivalent to $\dot{X} + AX = B$, where

$$
A = \begin{bmatrix} \frac{2\xi_1}{1-k^2} & \frac{-2\xi kna}{1-k^2} & \frac{1}{1-k^2} & \frac{-k}{n(1-k^2)} \\ \frac{-2\xi_1 k}{n(1-k^2)} & \frac{2\xi a}{(1-k^2)} & \frac{-k}{n(1-k^2)} & \frac{1}{n^2(1-k^2)} \\ -1 & 0 & 0 & 0 \\ 0 & -(an)^2 & 0 & 0 \end{bmatrix} ; B = \begin{bmatrix} \frac{1}{1-k^2} \\ \frac{-k}{n(1-k^2)} \\ 0 \\ 0 \end{bmatrix}
$$

and X is state vector, A is system matrix, and B is control matrix. By solving the state space equation (3), one derives the state vector, X. It contains the dimensionless variation of the primary inductor current, y_1, the secondary inductor current, y_2, the primary capacitor voltage, x_1 and the secondary capacitor voltage, x_2. Those variables are important for the state plane analysis and for the energy tank and power evaluation.

Power representation

The input power is the total power supplied by the DC source and expressed as the mean value of the input current multiplying by the DC voltage. The dimensionless input power is equal to the mean value of y_1 for a half period multiplying by the dimensionless DC source, 1. Since there is a relationship between two half periods when reaching steady state for all variables, $X(t) + X(t + \frac{T}{2}) = 0$, it thus only needs to evaluate the primary dimensionless coil current and dimensionless input voltage.

The definition of the dimensionless input power is given in (4).

$$P_{input} = 1 * \bar{y_1} = \frac{\omega_n}{\pi} \int_0^{\frac{\pi}{\omega_n}} y_1(\theta)d\theta \tag{4}$$

The output power is the power consumed by the resistive load and expressed as the mean value of the square of the output current multiplying by the value of the resistive load. For the dimensionless output power, it expresses as the mean value of the square of y_2, which is $\bar{y_2^2}$. The half period is sufficient to determine this mean value. Multiplying this mean value by the dimensionless load resistor, $2\xi_L$, we get the dimensionless output power,

$$P_{output} = 2\xi_L * \bar{y_2^2} = 2\xi_L * \frac{\omega_n}{\pi} \int_0^{\frac{\pi}{\omega_n}} y_2^2(\theta)d\theta \tag{5}$$

The definition of the efficiency of the SS-WPT system is given in (6).

$$\eta = \frac{P_{output}}{P_{input}} \tag{6}$$

Case study

A 7.4 kW SS-WPT compensation network

In this section, a 7.4 kW SS-WPT compensation network is represented using the developed theoretical modelling method and validated against the dynamic simulation results. The values of the system variables used in this model for theoretical modeling and simulation are given in Table II. The different misalignment situations with different coupling coefficients ($k = 0.1$ to 0.9) are investigated so as to reach optimal frequencies for each k and to ensure a Maximum Wireless Power Transfer.

Table II: Parameter specification of system components

$C_1, C_2 (nF)$	$L_1, L_1 (\mu H)$	Power (kW)	Q	f_r (kHz)	Z_c (Ω)	R_1, R_2(Ω)	R_L (Ω)	E(V)
33	82.5	7.4	5	96.5	50	0.3	10	302

The theoretical results show that the input power and output power largely depend on the normalized switching frequency, ω_n, and the coupling coefficient, k. Fig. 3 presents the relationships between the input and output power and these two parameters. When k is below 0.3, a significant maximum power point is reached. Specifically, when the normalized frequency is equal to 1 and k is at 0.1, the input

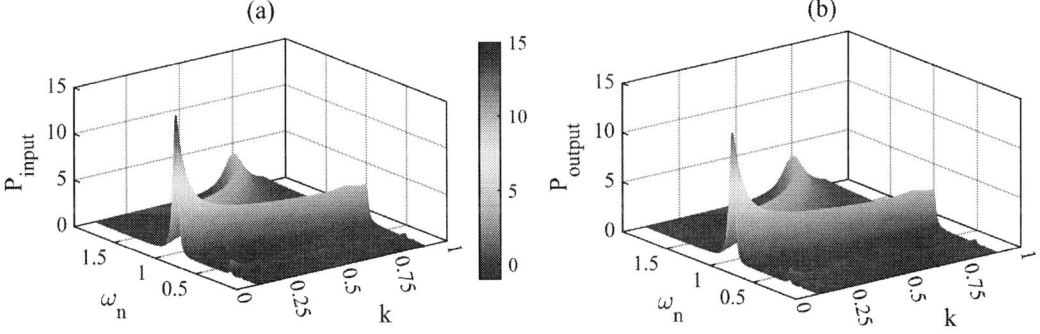

Fig. 3: Dimensionless power for normalized frequencies and corresponding k: (a) P_{input} ; (b) P_{output}

power and output power reach the global maximum. When k is higher than 0.3, the peak power point is almost constant.

Fig. 4 gives a closer look at how the input power and output power interact with the normalized switching frequencies ($0 < \omega_n < 2$) and the coupling coefficients ($k = 0.1$ to 0.9). The envelope of the power curves indicates the maximum power points corresponding to their switching frequencies, MPPFs, and coupling coefficients, with highlighted optimal power peaks.

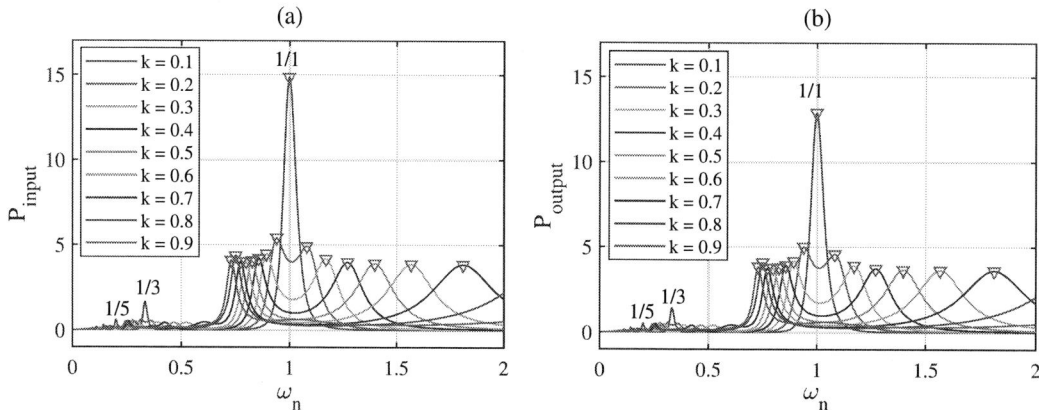

Fig. 4: Dimensionless power against ω_n for different k: (a) P_{input}; (b) P_{output}

Fig. 5 uses phase planes to represent the highlighted peak power points in Fig. 4 with respect to their corresponding MPPFs and coupling coefficients. Specifically, Fig. 5(a) - (c) correspond to the peak input power cases in Fig. 4 (a). Fig. 5(a) shows the corresponding MPPF and coupling coefficient for each peak input power point, which are highlighted with respect to the MPPFs in blue ($\omega_n < 1$) and red ($\omega_n > 1$). Fig. 5 (b) and (c) draw phase planes to show the peak input powers for the MPPFs lower or higher than 1, respectively. The cases with respect to the peak output power points, corresponding to Fig. 4 (b), are shown similarly in Fig. 5(d) - (f). When the resonance system is symmetric, meaning L_1 is equal to L_2 and C_1 is equal to C_2, the MPPFs of the maximum input power and maximum corresponding output power are very close. As seen in Fig. 5, the energy tank levels, operating at normalized frequency lower than 1, are generally higher than operating at normalized frequency higher than 1. It is thus straightforward to use Fig. 5 to select the switching frequency according to the physical limitation of the energy tank.

Fig. 6 shows the maximum input power and maximum output power differences and their transferring efficiencies, with respect to their MPPFs and corresponding coupling coefficients. For the peak power point, it can choose to operate at normalized frequency lower than 1 or higher than 1. Specifically, when k is moving from 0.1 to 0.3, the peak power input and output drops dramatically, indicating significant changes in power. When k is larger than 0.3, the power starts to reach a similar power level for both normalized frequency ranges. It means that operating at normalized frequency lower than 1 has the similar performance as higher than 1. It is thus preferable to operate the systems with a normalized MPPF lower than 1 since lower frequency has less demand on the component selection than higher frequencies. The two green line represent the efficiency of the system at switching frequency lower or higher than 1. From 0.1 to 0.2, the system can deliver maximum input and output power but with the lowest efficiency. When k is increasing, the input and output power and operating efficiency are similar for both MPPF ranges. However, the system performs slightly better in efficiency when the switching frequency is higher than 1, and a little better in maximum input and output power when the frequency is smaller than 1. When losses are lower, the differences get smaller, corresponding to higher efficiency. The maximum output power depends on the resistive load without considering losses. The definition of the maximum output power is given in (7).

$$P_{max} = \frac{8}{\pi^2} \frac{(nE)^2}{R_L} \tag{7}$$

Optimal frequency for Dynamic Wireless Power Transfer   LIANG Mincui

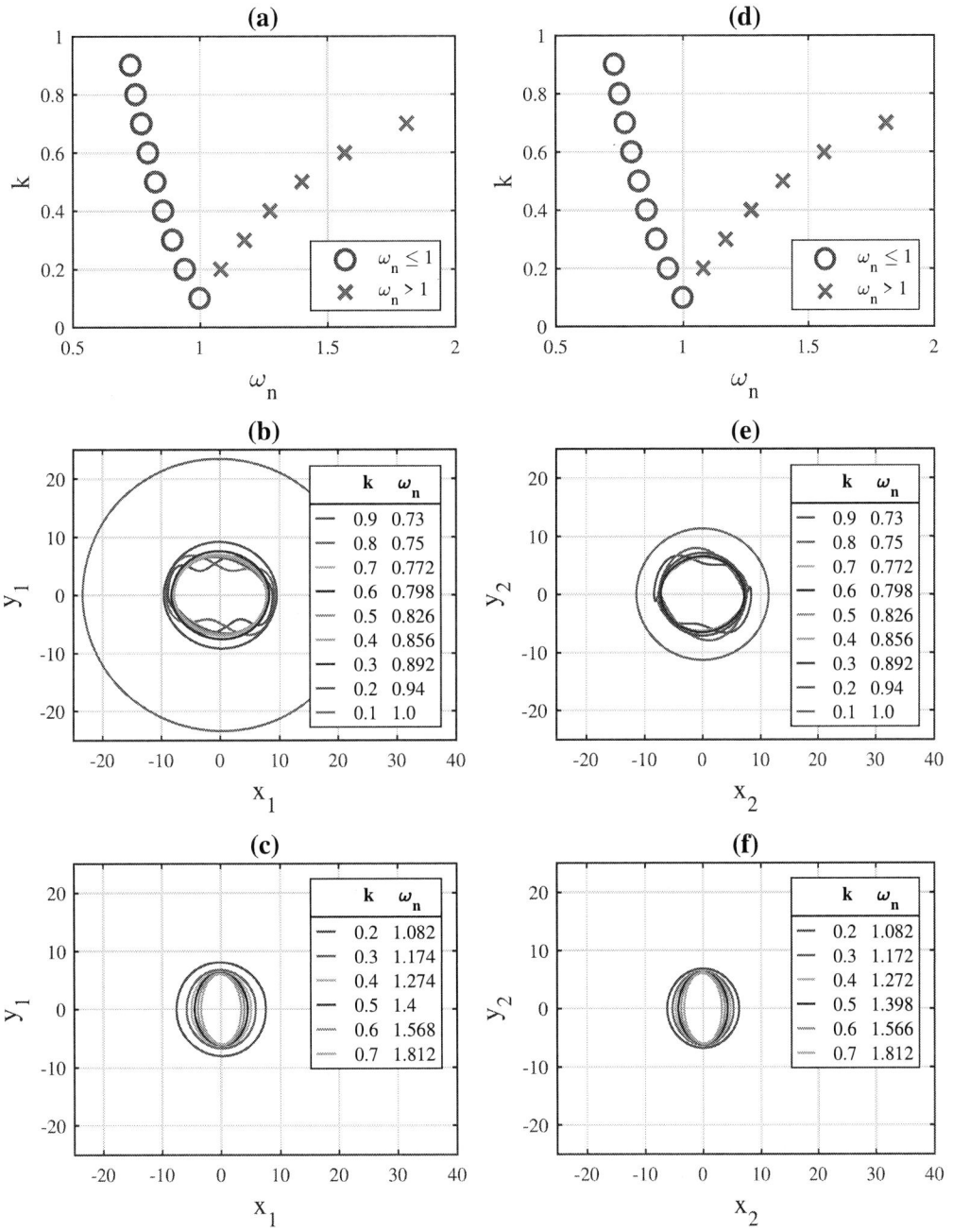

Fig. 5: MPPF and corresponding *k* and their phase plane : (a) ω_n and *k* for maximum P_{input} ; (b) Corresponding phase plane ($\omega_n <= 1$) for (a) ; (c) Corresponding phase plane ($\omega_n > 1$) for (a) ; (d) ω_n and *k* for maximum P_{output} ; (e) Corresponding phase plane ($\omega_n <= 1$) for (d) ; (f) Corresponding phase plane ($\omega_n > 1$) for (d)

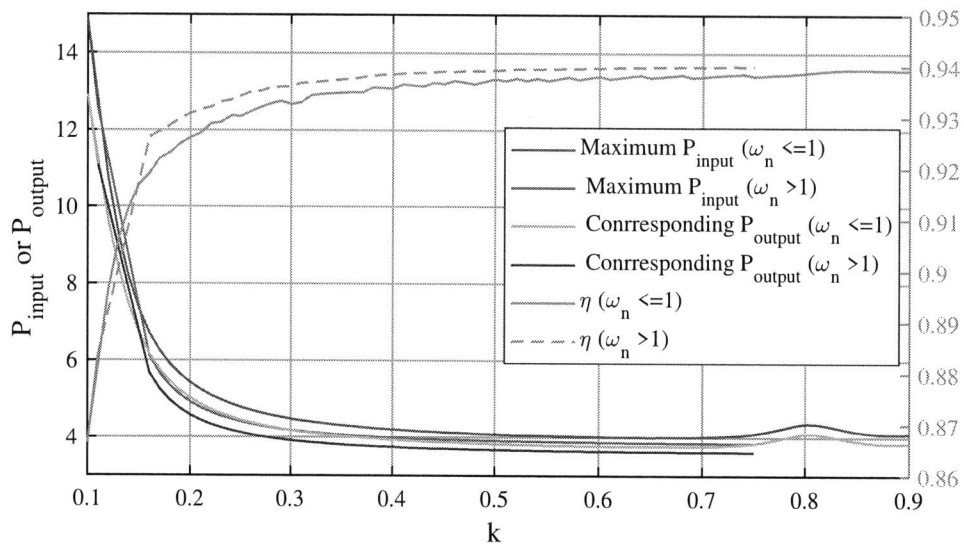

Fig. 6: Maximum power and efficiency for dimensionless P_{input} and P_{output}

MPPFs corresponding to the maximum input power and maximum corresponding output power are load dependent. However, differences between their MPPFs are minor and are getting smaller when k is increasing, as shown in Fig. 7. Table III further specifies the Normalised Root Mean Square Error (NRMSE) for the dataset of Fig. 7 (a). When k is equal to 0.1, the difference is the biggest and the value is less than 1%. Similar conclusions can be drawn for Fig. 7 (b).

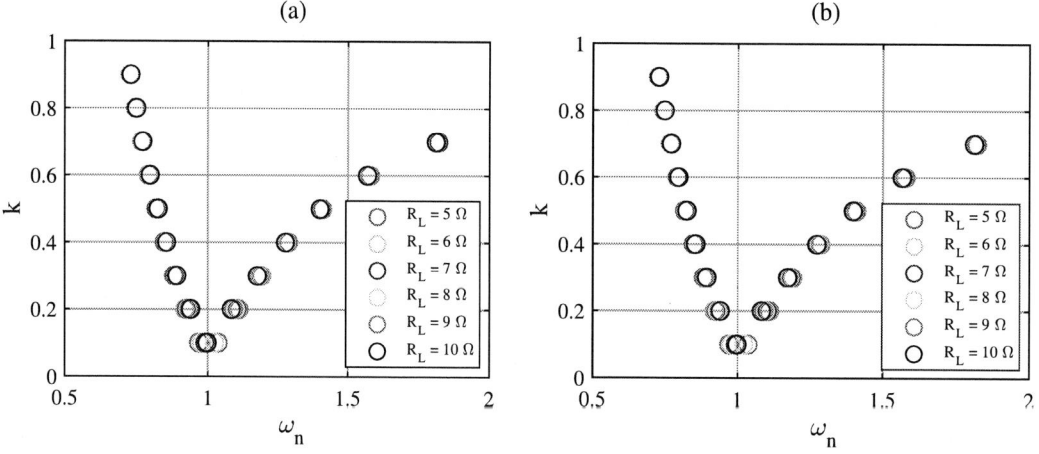

Fig. 7: MPPFs for different resistive loads : (a) MPPFs for maximum input power; (b) MPPFs for maximum output power

Table III: NRMSE of MPPFs for different resistive loads for maximum input power

k	0.1	0.2	0.3	0.4	0.5	0.6	0.7	0.8	0.9
NRMSE (%)	0.985	0.745	0.441	0.299	0.264	0.210	0.174	0.170	0.175

Note: NRMSE - Normalised Root Mean Square Error, $NRMSE = \frac{RMSE}{\bar{O}}$, where $\bar{O}$ is the mean value of the dataset

Simulation and validation

A Simulink model is developed to simulate and validate the performances of the SS-WPT in Fig. 1, with respect to the normalized frequency from 0.5 to 1.6 and the coupling coefficients from 0.1 to 0.6, as the difference is negligible, when coupling coefficient is higher than 0.6. In Fig. 8, the red stair line represents normalized switching frequency scenarios used in the simulation. The color of each power line corresponds to different coupling coefficient from 0.1 to 0.6. It can be seen that, at primary resonance frequency ($\omega_n = 1$ and $k = 0.1$), it has the global maximum input power. When k increases, the frequency bifurcation phenomenon becomes obvious especially for the situations where primary and secondary resonance tank elements are the same. It is clear that the simulation results and the theoretical results in Fig. 4 agreed very well, which can be further speculated in Fig. 8.

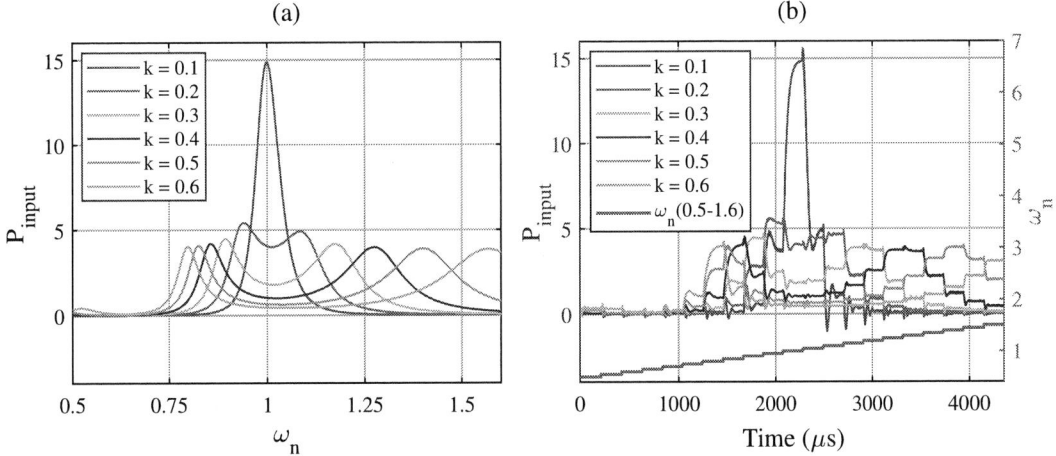

Fig. 8: Result comparison: (a) Theoretical results; (b) Simulation results

In order to evaluate optimal frequencies for each coupling coefficient, the simulation model in time domain was run repeatedly for each k from 0.1 to 0.6 and a range of frequencies from 0.5 to 1.6 including the optimal frequencies representing the dynamic behavior for each given frequency. The simulation model includes 21 normalized angular frequencies (ω_n) between 0.5 to 1.6, plotted as a red stair line in Fig. 8 (b). The simulation runs at each given frequency for 207 μs and reaches the steady state. The results in Fig. 8 give the input power from 0 to 207 μs for each given frequency. The maximum input power at steady state can be found at the corresponding MPPF for each k.

The results presented in Table IV further prove an excellent agreement between the theoretical and sim-

Table IV: Simulation and theoretical results for maximum input power

k	0.6	0.5	0.4	0.3	0.2	0.1	0.2	0.3	0.4	0.5	0.6
MPPF	0.80	0.83	0.86	0.89	0.94	1.0	1.08	1.17	1.27	1.40	1.57
Theor. P_{in}	4.02	4.08	4.20	4.48	5.42	14.87	4.92	4.18	3.99	3.91	3.87
Sim. P_{in}	3.90	3.90	4.03	4.40	5.19	14.29	4.84	4.04	3.88	3.8	3.70
NRMSE (%)	1.51	2.23	2.10	0.87	2.16	2.00	0.83	1.71	1.33	1.41	2.26
Theor. P_{out}	3.77	3.82	3.93	4.18	5.02	12.89	4.58	3.91	3.74	3.67	3.64
Sim. P_{out}	3.70	3.70	3.70	4.10	4.84	11.70	4.36	3.69	3.53	3.45	3.38
NRMSE (%)	0.94	1.64	3.04	0.95	1.85	4.82	2.42	2.91	2.95	3.10	3.67
Theor. η	0.94	0.94	0.94	0.93	0.93	0.87	0.93	0.94	0.94	0.94	0.94
Sim. η	0.95	0.95	0.92	0.93	0.93	0.82	0.9	0.91	0.91	0.91	0.91
NRMSE (%)	0.64	0.67	0.84	0.08	0.22	2.76	1.68	1.19	1.53	1.57	1.60

Note: P_{in} - Input power; P_{out} - Output power ; Theor. - Theoretical; Sim. - Simulation

ulation results. As the theoretical modeling results in higher efficiency, larger equivalent resistor losses on both sides were introduced to better present and compare the results. Furthermore, if the system can avoid operating at the global maximum power point, the normalized output power is almost constant and the resistive loads have little effect on the MPPFs. Without considering the global maximum point when k is equal to 0.1, the theoretical normalized average output power is 4.03 which is equivalent to real output power of 7.34 kW. For our case study with 7.4 kW output power, the SS-WPT system achieve excellent performances when operating at the selected optimal frequencies.

Conclusion

Series-Series compensation topology has the potential to reach maximum wireless power transfer and high efficiency under different misalignment conditions. A time-domain theoretical study of an SS-WPT compensation network is conducted so as to find Maximum Power Point Frequencies (MPPFs) for maximum power transfer of a DWPT system with different coupling coefficients. An excellent agreement is noticed between the theoretical and simulation results during validation. Based on the phase plane analysis, it is easy to normalize constraints of various components in terms of maximum voltages and currents necessary for the right choice of the passive and active components of the SS-WPT. Under steady-state conditions, the method facilitates qualitative understanding of system operation including identification of the operating modes. In the future research, different control strategies are to be developed to reach the MPPFs and allows the system to operate at maximum power points for any misalignment conditions.

References

[1] United Nations, "Transport," UN Climate Change Conference (COP26) at the SEC – Glasgow 2021, 16-Sep-2021. [Online]. Available: https://ukcop26.org/transport/. (Accessed: 15-Nov-2021).

[2] Chun T. Rim; Chris Mi, "Introduction to Dynamic Charging," in Wireless Power Transfer for Electric Vehicles and Mobile Devices, IEEE, 2017, pp.155-160, doi: 10.1002/9781119329084.ch8.

[3] A. Triviño-Cabrera and J. Sánchez, "A Review on the Fundamentals and Practical Implementation Details of Strongly Coupled Magnetic Resonant Technology for Wireless Power Transfer," Energies, vol. 11, no. 10, p. 2844, Oct. 2018.

[4] Triviño-Cabrera, A., González-González, J.M., Aguado, J.A. (2020). Wireless Chargers for Electric Vehicles. In: Wireless Power Transfer for Electric Vehicles: Foundations and Design Approach. Power Systems. Springer, Cham. https://doi.org/10.1007/978-3-030-26706-32

[5] K. Aditya and S.S. Williamson, "Comparative study of series–series and series–parallel compensation topologies for electric vehicle charging," in 2014 ISIE Conf., pp. 426–430.

[6] Y. Jiang, L. Wang, Y. Wang, J. Liu, M. Wu and G. Ning, "Analysis, Design, and Implementation of WPT System for EV's Battery Charging Based on Optimal Operation Frequency Range," in IEEE Transactions on Power Electronics, vol. 34, no. 7, pp. 6890-6905, July 2019, doi: 10.1109/TPEL.2018.2873222.

[7] W. X. Zhong and S. Y. R. Hui, "Maximum Energy Efficiency Tracking for Wireless Power Transfer Systems," in IEEE Transactions on Power Electronics, vol. 30, no. 7, pp. 4025-4034, July 2015, doi: 10.1109/TPEL.2014.2351496.

[8] Electric vehicle wireless power transfer (WPT) systems - Part 3: Specific requirements for the magnetic field wireless power transfer systems, IEC 61980-3 ED1, 2020

A Wide-Input-Voltage-Range 50W Series-Capacitor Buck Converter with Ancillary Voltage Bus for Fast Transient Response in 48V PoL Applications

Nameer Khan[1], James Xu[1], Gerard Villar Piqué[2], John Pigott[2], Henk Jan Bergveld[2],
Alaa El Sherif[2], and Olivier Trescases[1]

[1]Edward S. Rogers Sr. Department of Electrical & Computer Engineering, University of Toronto
[1]10 King's College Road, Toronto, ON, Canada, M5S 3G4
[2]NXP Semiconductors Inc.
Email: nameer.khan@mail.utoronto.ca

Acknowledgments

This research was supported by NXP Semiconductors Inc. and the Natural Science and Engineering Research Council of Canada.

Keywords

≪Automotive applications≫, ≪DC-DC converter≫, ≪High frequency power converter≫, ≪Voltage Regulator Modules (VRM)≫, ≪Power converters for EV≫, ≪Multi-Level converters≫

Abstract

This paper presents an auxiliary-assisted 5:1 Series-Capacitor (SC) buck converter for fast transient response in 48V Point-of-Load (PoL) automotive applications. The 5:1 SC buck converter, which is implemented as the main hybrid stage, regulates the auxiliary buffer capacitor voltage, V_{AUX}, and consequently, delivers the average load power. By regulating V_{AUX}, the main hybrid stage facilitates the creation of an ancillary voltage bus for on-board electronics. The high-frequency auxiliary stage precisely regulates the output voltage, V_{OUT} with a Ripple-Based Constant On-Time (RB-COT) control scheme, by utilizing the energy stored in the buffer capacitor, C_{AUX}. With no stringent transient requirements on V_{AUX}, the main stage is optimized to deliver the average load power, while the auxiliary stage is designed for high-frequency operation. A 50W, 48V-to-0.8V experimental prototype was built to validate the proposed architecture. Load transients from 5 A to 29 A were applied to the prototype, resulting in an output voltage deviation of 60 mV. The system achieves a peak efficiency of 89.5% with an output capacitance of only 720 μF and an auxiliary capacitance of 120 μF.

Introduction

The proliferation of on-board electronics in Electric Vehicles (EVs) has constrained power delivery and led to the adoption of 48V voltage distribution networks [1]. Furthermore, the stringent transient regulation tolerances of automotive processors are currently achieved using large, costly decoupling capacitors. These challenges necessitate an efficient Point-of-Load (PoL) power architecture capable of delivering the increasing power demands of automotive processors from a variable high-voltage bus (24 V - 70 V) to the sub-1V processor core voltage (0.675 V - 0.8 V). Hybrid switched-capacitor converters [2, 3] are promising candidates for efficient 48V PoL converters due to their large native conversion ratio and higher energy density compared to magnetic-based converters. Despite these benefits, fast transient response in 48V PoL converters remains a challenge and has not been suitably addressed in past literature.

To improve the transient response in dc-dc converters, low-cost auxiliary converters have been employed to assist the main stage during load transients [4, 5]. By using a smaller inductor than in the main-stage, auxiliary converters alleviate the physical limit imposed by the main-stage LC-filter on the system transient response. 48V distribution networks present challenges for auxiliary converters due to the small duty-cycle requirements, with on-times that are often not practical with state-of-the-art power stages. A separate, lower-voltage bus can supply the auxiliary converter, at the expense of additional cost and volume which remain as concerns for space-critical automotive applications. Given these challenges, a power-stage architecture that minimizes the cost of the auxiliary converter by using only low-voltage devices while also providing a loosely regulated ancillary bus for on-board electronics is desirable.

The auxiliary converter presented in [6] provides charge to the output during load transients from a buffer capacitor that is decoupled from the input-voltage bus. By pre-charging the buffer capacitor based on the largest expected load transient, the auxiliary converter provides the necessary charge to satisfy the core voltage transient tolerances. While [6] replenishes the charge in the buffer capacitor, sporadic regulation of the auxiliary capacitor voltage prevents its use as a secondary voltage bus. An active electronic capacitor is proposed in [7] that behaves as a bi-directional current source. The active capacitor is implemented as a Gyrator Resonant Switched-Capacitor Converter, which constrains the conversion ratio for optimal efficiency and requires five additional switches. While both works propose a bi-directional auxiliary converter for fast transient response, the regulation of the auxiliary voltage, V_{AUX}, for ancillary on-board electronics has not been covered by past literature.

In this work, an auxiliary-assisted power architecture is proposed to regulate the auxiliary capacitor voltage, V_{AUX}, using the main hybrid dc-dc stage, while output voltage regulation is provided by the auxiliary stage, as shown in Fig. 1. With the proposed approach, the main hybrid dc-dc stage provides the steady-state load current, I_{LOAD}, the current drawn by the auxiliary-stage losses, and the auxiliary load current, $I_{LOAD,AUX}$. With no stringent transient requirements on V_{AUX}, the main stage can be optimized for DC power delivery, while the auxiliary stage can be designed for high-frequency operation. In addition, creating an ancillary bus from V_{OUT} provides a more efficient conversion stage compared to a voltage bus created by a V_{IN}-tapped power stage.

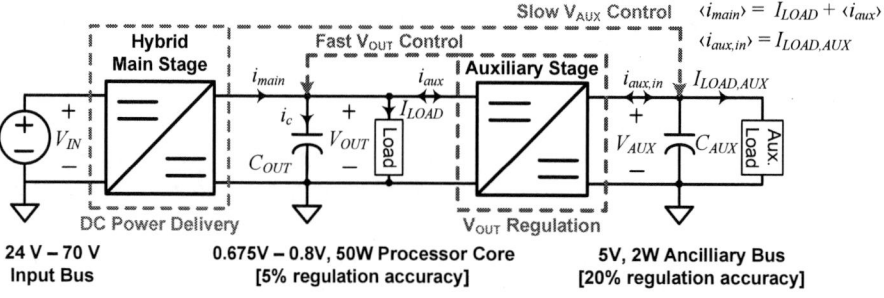

Fig. 1: Proposed auxiliary-assisted power architecture.

This paper is organized as follows. The system architecture, theory of operation, and Power-Delivery-Network (PDN) design are first presented. The proposed system is validated using closed-loop simulation results. An experimental prototype is then presented to characterize the dynamic performance and efficiency of the proposed system. Finally, conclusions are presented at the end of the paper.

System Architecture

The key system parameters and power-stage architecture are presented in Table I and Fig. 2(a), respectively. For the main stage, a 5:1 Series-Capacitor (SC) buck converter was selected for its high native conversion ratio, inherent current sharing, and high output-current capability [8, 9], with its ideal steady-state operating waveforms presented in Fig. 2(b). For the auxiliary converter, a bi-directional two-level converter was selected to minimize cost and size. Typically, auxiliary converters only operate during load transients to provide output charge, whereas in this work, the auxiliary converter operates continuously to regulate V_{OUT}. For fast transient response, the auxiliary-stage switching frequency, $f_{sw,aux}$, was maximized to accommodate a small auxiliary inductor, L_{AUX}, and consequently, increase the inductor current slew rate, $m_{f,aux}$. A Ripple-Based Constant On-Time (RB-COT) control scheme, similar to [10], was chosen for V_{OUT} regulation due to its simplicity, minimal control delay, and V_{OUT}-based sensing. The inherent current sharing of the SC-buck converter enables the use of a fixed-frequency voltage-mode PWM control scheme for V_{AUX} regulation.

Table I: System Parameters

V_{in} [V]	V_{out} [V]	I_{load} [A]	P_{load} [W]	C_{out} [μF]	$f_{sw,main}$ [kHz]	L_{1-5} [nH]	C_{1-4} [μF]	$f_{sw,aux}$ [MHz]	L_{aux} [nH]	V_{aux} [V]	C_{aux} [μF]
24 - 70	0.675 - 0.8	62.5	50	720	160	470	10	1.6	36	5	120

As the transient requirements on V_{AUX} are less stringent than at V_{OUT}, the SC-buck switching frequency, $f_{sw,main}$, can be relaxed to optimize the main stage for efficient DC power delivery by selecting reliable Silicon (Si) devices with low $R_{ds,on}$ for M_{1-10}. Likewise, the low steady-state current of the auxiliary stage enables the use of higher-$R_{ds,on}$, and therefore low-Q_g, devices for $M_{11,12}$ to minimize switching losses at a higher switching frequency. Typically, auxiliary converters provide charge from the high input voltage bus resulting in a requirement for high-FOM devices. With V_{AUX} being regulated by the main stage, a low V_{AUX} can be set for high $f_{sw,aux}$ and the use of low-FOM silicon devices for $M_{11,12}$.

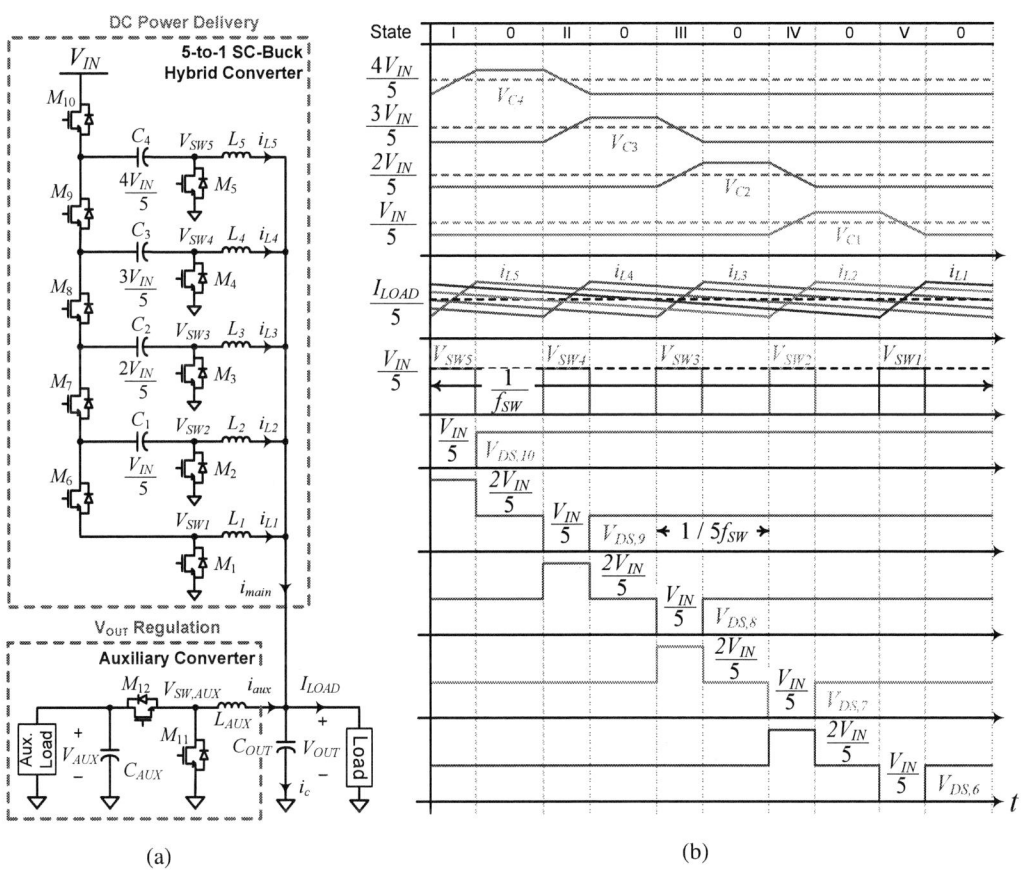

Fig. 2: (a) Auxiliary-assisted power architecture. (b) Ideal steady-state operating waveforms of 5:1 SC-buck converter.

Theory of Operation

The proposed system is dependent on the seamless interdependent operation of the V_{OUT} and V_{AUX} control loops, which are presented in Fig. 3(a). As such, the control bandwidth of each loop must be carefully selected to ensure system stability. The closed-loop operating waveforms of the proposed system during load transients are shown in Fig. 3(b). When a load step-up transient occurs, V_{OUT} decreases, which is detected by the RB-COT control scheme. The auxiliary converter increases its output current, I_{AUX}, to prevent further charge removal from C_{OUT}, which restores V_{OUT} to $V_{OUT,ref}$. The auxiliary converter continues to provide the transient current, $I_2 - I_1$, from C_{AUX}, which consequently, creates a deviation on V_{AUX}. The auxiliary regulation loop detect the change in V_{AUX} and adjust i_{main} until $V_{AUX} = V_{AUX,ref}$.

With the auxiliary converter providing V_{OUT} regulation, the main stage regulates V_{AUX}, which creates a regulated ancillary voltage bus. It is important to note that the main stage is not directly connected to C_{AUX} and is dependent on the high control bandwidth of the auxiliary stage to transfer charge to C_{AUX}. Due to the dependence of the auxiliary voltage loop on the COT-based control loop, stability analysis is required to fully optimize the auxiliary voltage control loop, which is beyond the scope of this paper.

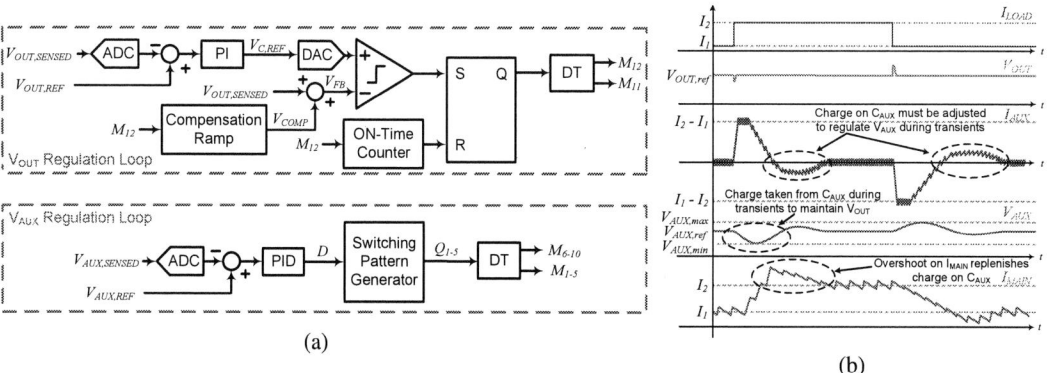

(a)

(b)

Fig. 3: (a) Control architecture of the proposed auxiliary-assisted dc-dc converter. (b) Representative closed-loop operating waveforms of the proposed system.

Design of Power Delivery Network (PDN)

While the output capacitance, C_{OUT}, is selected to satisfy the stringent transient requirements during load transients, the high-frequency impedance of the Power Delivery Network (PDN), must also be carefully designed. A poorly designed PDN causes large voltage deviations, ΔV_{ESL}, to occur during steady-state operation due to the inductive divider between the Equivalent Series Inductance (ESL) of C_{OUT} and the power-stage inductors, L_{MAIN} and L_{AUX}. The ΔV_{ESL} due to the main and auxiliary stages can be calculated as follows:

$$\Delta V_{ESL,main} = \frac{V_{IN}/N}{1 + \frac{L_{MAIN}}{ESL}}, \; \Delta V_{ESL,aux} = \frac{V_{AUX}}{1 + \frac{L_{AUX}}{ESL}}, \tag{1}$$

where N is the native conversion ratio of the hybrid dc-dc converter. Based on (1), ΔV_{ESL}, is determined by the ratio of L_{MAIN} and L_{AUX} to the ESL, V_{IN}, and V_{AUX}. To analyze the PDN of a typical processor, a commercial autonomous-driving development platform, shown in Fig. 4(a), which houses an automotive-grade processor, is selected as a reference design, with its PDN impedance plot shown in Fig. 4(b). Due to a large L_{MAIN}, the reference design is less sensitive to the ESL of C_{OUT}. Conversely, the large power-stage inductor limits the system transient response, which large costly decoupling capacitors to reduce the PDN impedance at low frequencies.

In the proposed architecture, a high-frequency two-level converter with a small auxiliary inductor, L_{AUX}, of 36 nH are used to improve the system transient response. As a result, the proposed system reduces the required amount of large decoupling capacitors but the resultant ratio of L_{AUX} to ESL is much higher than the ratio of L_{MAIN} to ESL. Thus, high-frequency decoupling capacitors must be included in the proposed system to reduce the ESL of C_{OUT}, as shown in Fig. 4(b). There is a tradeoff between low-frequency, large bulky electrolytic capacitors for load-transient-based voltage deviation and high-frequency, small ceramic capacitors for ESL-based voltage deviation. Given that processors already require high-frequency ceramic capacitors to compensate for the socket inductance amongst other parasitic inductive effects, it is more desirable to reduce costly and bulky low-frequency electrolytic capacitors.

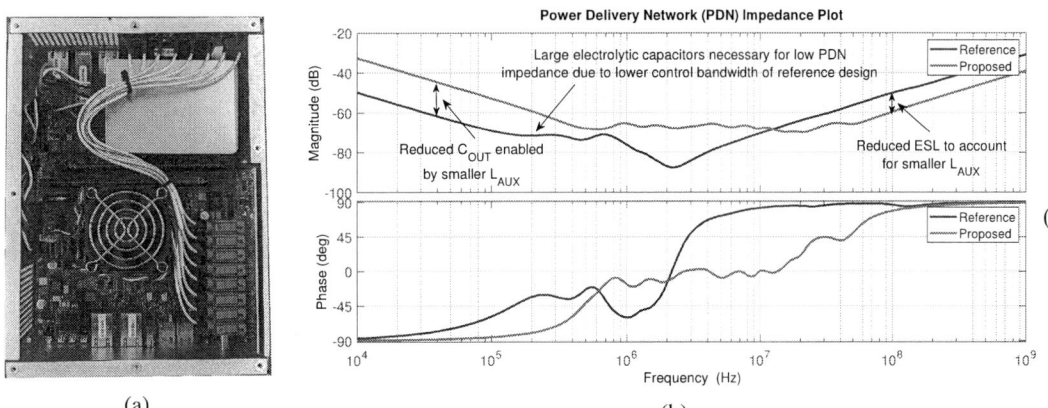

(a)   (b)

Fig. 4: (a) Reference autonomous-driving development platform with a 45W automotive-grade processor. (b) Calculated PDN impedance plots of reference design and proposed system.

Simulation Results

The system was first verified using a PLECS closed-loop simulation. The steady-state waveforms are shown in Fig. 5(a). The main-stage current ripple, ΔI_{MAIN}, creates a disturbance in V_{OUT} at $5 \times f_{sw,main}$ which the auxiliary converter attempts to correct by adjusting I_{AUX}. Load steps from 30 A to 60 A were applied to demonstrate the fast transient response and voltage regulation of the ancillary bus, as shown in Fig. 5(b) and (c). During load step-up transients, with $C_{AUX} = 120\,\mu F$, an output voltage deviation of 30 mV is observed while V_{AUX} decreases by 0.6 V before the main SC-buck stage replenishes the charge on C_{AUX}. During load step-down transients, an output voltage deviation of 35 mV is observed, while V_{AUX} increases by 0.42 V before the SC-buck main stage responds. Using the Time-Optimal Control (TOC) comparison presented in [11], adding the auxiliary converter allows the output capacitance to be reduced by 62% compared to operating only a TOC-based main stage.

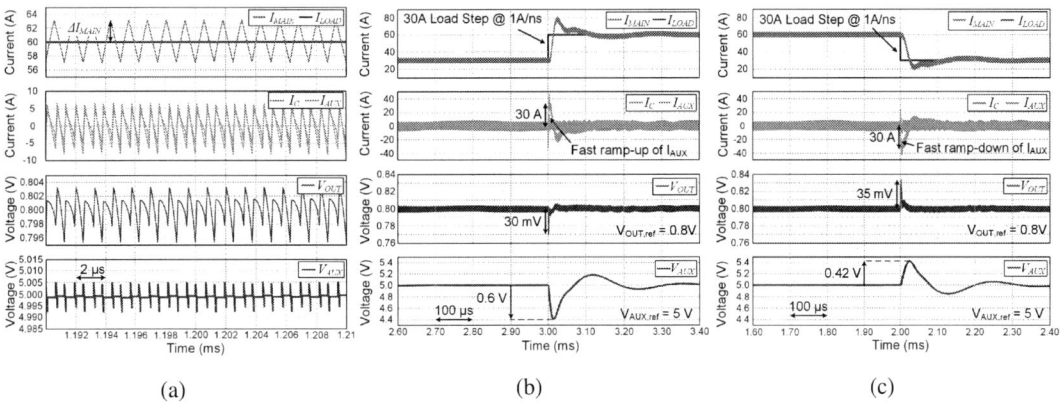

(a)   (b)   (c)

Fig. 5: Simulated closed-loop operating waveforms of the proposed system at 48V-to-0.8V conversion ratio during: (a) steady-state, (b) load step-up transient from 30 A to 60 A and (c) load step-down transient from 60 A to 30 A.

Experimental Results

The power consumption of the automotive-grade processor from the autonomous-driving development platform, shown in Fig. 4(a), was measured to determine its typical profile, as shown in Fig. 6. The load profile was determined by measuring the main-stage inductor current based on the DCR current sensing capacitor. To remove the ESL voltage ripple and observe the low-frequency load profile, the measured capacitor voltage was filtered using a Savitzky-Golay filter. The 45W automotive-grade processor continuously performs computationally intensive tasks and generates dynamic load profiles.

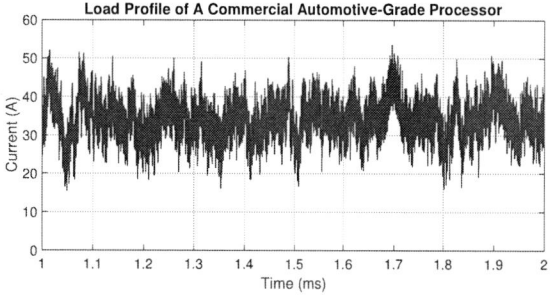

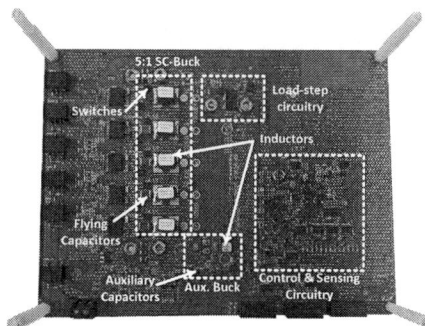

Fig. 6: Measured load profile of the automotive-grade processor from the commercial autonomous-driving development platform.

Fig. 7: Experimental prototype of the proposed system.

A prototype composed of the 5:1 SC-buck stage, auxiliary stage, and control circuitry was built, as shown in Fig. 7, to characterize the dynamic response and efficiency of the proposed system. A high di/dt load-step circuit was implemented on the PCB. For M_{6-10}, the switches are implemented using 60V, 3.3mΩ BSZ040N06LS5 devices, while 25V, 1.3mΩ NTTFS1D8N02P1E devices are used for M_{1-5}. For the auxiliary converter, a 16V integrated silicon power module, CSD95377Q4M, was selected for its low-inductance package, making it suitable for a high-frequency operation of 1.6 MHz. The COT-based control circuitry was placed on the right side of the board, as shown in Fig. 7, while the digital PID controller for V_{AUX} regulation was implemented on the FPGA on the PCB backside.

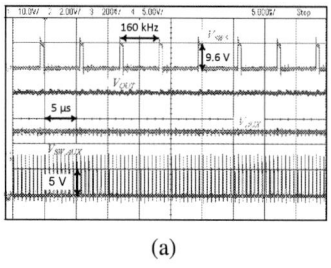
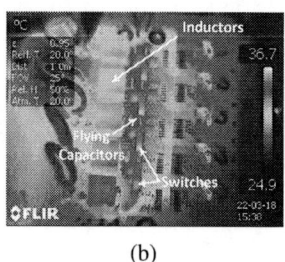
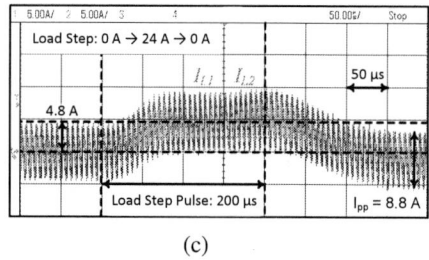

(a)   (b)   (c)

Fig. 8: (a) Measured steady-state waveforms of the proposed system at 48V-to-0.8V conversion ratio at 30 A. (b) Thermal image of the 5:1 SC-buck converter at 50 W and 48V-to-0.8V conversion ratio. (c) Measured current waveforms of the 5:1 SC-buck converter during load transient from 0 A to 24 A.

The measured steady-state waveforms of the proposed system are shown in Fig. 8(a). The prototype was operated at 50 W with forced air-cooling until the system reached thermal steady-state, as shown in Fig. 8(b). The peak temperature in the system was 36.7°C and was observed on the high-side SC-buck switches. To highlight the inherent current-sharing of the SC-buck converter, the inductor phase currents during load transients are shown in Fig. 8(c). To demonstrate the voltage regulation of V_{AUX}, the high-speed load-step circuitry, shown in Fig. 7, is used to apply load steps from 5 A to 29 A for 200 µs. The resulting step response on V_{AUX} is shown in Fig. 9(a). The main stage detects the disturbance on V_{AUX} and adjusts i_{main} to correct the auxiliary capacitor voltage with V_{AUX} settling within 0.8 ms. A deviation of -0.6V and +0.9V is observed during transients, which is within 20% of the steady-state value of 5 V. Magnified versions of Fig 9(a) are presented as Fig. 9(b) and 9(c) to demonstrate the fast transient response of the high-speed COT-based auxiliary converter. A maximum output voltage deviation of 60 mV is observed during either load-step transition. Furthermore, due to the high bandwidth of the COT control loop, V_{AUX} is regulated without creating significant disturbances on V_{OUT}.

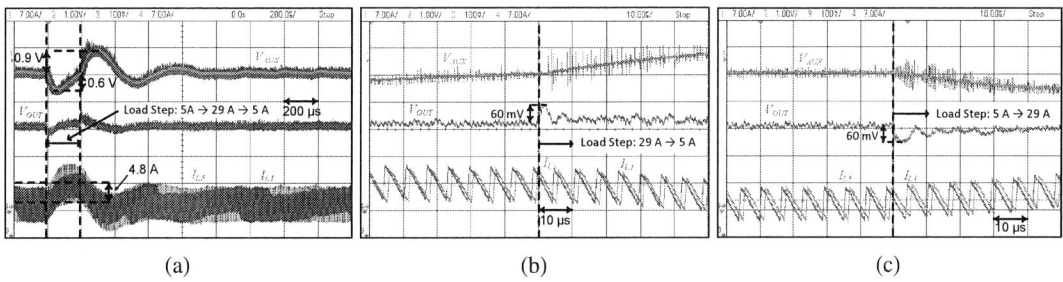

Fig. 9: (a) Operating waveforms of the proposed system during load-step transients from 5 A to 29 A. Magnified (b) step-down and (c) step-up load transient waveforms from 5 A to 29 A.

The measured system efficiency versus load power for varying input voltage, V_{IN}, is shown in Fig. 10(a). The system achieves a peak efficiency of 89.5% and an efficiency of 86.8% at the rated output power and nominal conversion ratio of 48V-to-0.8V. As expected, the efficiency improves with lower V_{IN} due to reduced switching and core losses. The measured system efficiency for varying V_{OUT} is also shown in Fig. 10(b). At V_{OUT} = 0.675 V, a peak efficiency of 87.8% is achieved at P_{LOAD} = 17 W while an efficiency of 86.5% is achieved at P_{LOAD} = 42 W.

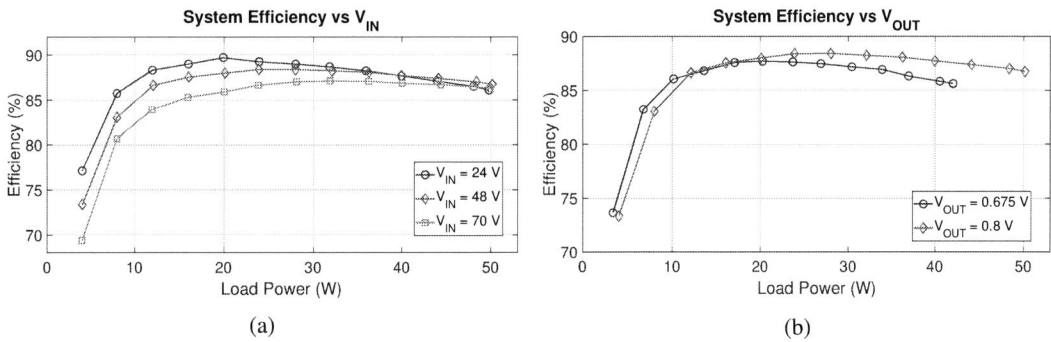

Fig. 10: Measured system efficiency for varying: (a) V_{IN} at V_{OUT} = 0.8 V and (b) V_{OUT} at V_{IN} = 48 V.

The proposed system is compared to state-of-the-art auxiliary-assisted converters in Table II. With the highest conversion ratio and highest current rating, this work leverages an auxiliary inductor current slew rate of 22.2 A/µs to minimize the output capacitance, while achieving high efficiency in the main stage.

Table II: State-of-the-Art Auxiliary-Assisted Converters

Year	V_{in} [V]	V_{out} [V]	I_{load} [A]	C_{out} [µF]	L_{aux} [nH]	$m_{f,aux}$ [A/µs]	ΔI_{load} [A]	ΔV_{out} [mV]	V_{aux} [V]	C_{aux} [µF]
This Work	48	0.8	62.5	720	36	22.2	24	60	5	120
2021 [4]	15	3.3	15	220	500	6.6	11	62	15	-
2013 [6]	12	5	10	47	420	11.9	9	388	8.5 - 10	40
2015 [7]	12	1.5	20	200	20	-	15	100	3	30
2015 [12]	5	1.5	15	140	100	15	8.2	45	5	-

Conclusion

An auxiliary-assisted control scheme for a wide-input-voltage-range 50W 5:1 SC-Buck converter was presented for 48V-to-0.8V PoL applications. The auxiliary converter provides V_{OUT} regulation for fast transient response while the 5:1 SC-buck stage efficiently delivers the dc load power and regulates V_{AUX} to create an ancillary voltage bus for on-board electronics. The system achieves a peak efficiency of

89.5% with only $720\,\mu F$ of output capacitance and $120\,\mu F$ of auxiliary capacitance. During load steps of 24 A, a maximum ΔV_{OUT} of 60 mV and a ΔV_{AUX} of 0.9V is observed.

References

[1] C. Jürgen Bilo et al, "48-Volt Electrical Systems, A Key Technology Paving the Road to Electric Mobility." ZVEI-German Electrical and Electronic Manufacturers Association, 2016.

[2] G.-S. Seo, R. Das, and H.-P. Le, "A 95%-Efficient 48V-to-1V/10A VRM Hybrid Converter Using Interleaved Dual Inductors," in *2018 IEEE Energy Conversion Congress and Exposition (ECCE)*, 2018, pp. 3825–3830.

[3] C. Chen, J. Liu, and H. Lee, "A 92.7%-Efficiency 30A 48V-to-1V Dual-Path Hybrid Dickson Converter for PoL Applications," in *2021 IEEE Energy Conversion Congress and Exposition (ECCE)*, 2021, pp. 1989–1994.

[4] D. Kim, M. Hong, J. Baek, J. Lee, J. Shin, and J.-W. Shin, "Soft-Switching Auxiliary Current Control for Improving Load Transient Response of Buck Converter," *IEEE Transactions on Power Electronics*, vol. 36, no. 3, pp. 2488–2494, 2021.

[5] W. Lu, Z. Zhao, Y. Ruan, S. Li, and H. H.-C. Iu, "Two-Period Frame Transient Switching Control for Buck Converter Using Coupled-Inductor Auxiliary Circuit," *IEEE Transactions on Industrial Electronics*, vol. 66, no. 10, pp. 8040–8050, 2019.

[6] Z. Shan, S.-C. Tan, and C. K. Tse, "Transient Mitigation of DC-DC Converters for High Output Current Slew Rate Applications," *IEEE Transactions on Power Electronics*, vol. 28, no. 5, pp. 2377–2388, 2013.

[7] O. Kirshenboim, A. Cervera, and M. M. Peretz, "Improving loading and unloading transient response of a voltage regulator module using a load-side auxiliary gyrator circuit," in *2015 IEEE Applied Power Electronics Conference and Exposition (APEC)*, 2015, pp. 913–920.

[8] Y. Chen, D. M. Giuliano, and M. Chen, "Two-Stage 48V-1V Hybrid Switched-Capacitor Point-of-Load Converter with 24V Intermediate Bus," in *2020 IEEE 21st Workshop on Control and Modeling for Power Electronics (COMPEL)*, 2020, pp. 1–8.

[9] Y. Chen, H. Cheng, D. M. Giuliano, and M. Chen, "A 93.7% Efficient 400A 48V-1V Merged-Two-Stage Hybrid Switched-Capacitor Converter with 24V Virtual Intermediate Bus and Coupled Inductors," in *2021 IEEE Applied Power Electronics Conference and Exposition (APEC)*, 2021, pp. 1308–1315.

[10] P. S. Shenoy, O. Lazaro, R. Ramani, M. Amaro, W. Wiktor, J. Khayat, and B. Lynch, "A 5 MHz, 12 V, 10 A, monolithically integrated two-phase series capacitor buck converter," in *2016 IEEE Applied Power Electronics Conference and Exposition (APEC)*, 2016, pp. 66–72.

[11] V. R. Namburi, M. Ashourloo, and O. Trescases, "Fast Transient Response of GaN-Based Hybrid Dickson Converter using Quasi-Fixed-Frequency Control for 48-V-to-1-V Direct Conversion in Automotive Applications," in *2020 IEEE Applied Power Electronics Conference and Exposition (APEC)*, 2020, pp. 33–40.

[12] V. Šviković, J. J. Cortés, P. Alou, J. A. Oliver, O. García, and J. A. Cobos, "Multiphase Current-Controlled Buck Converter With Energy Recycling Output Impedance Correction Circuit (OICC)," *IEEE Transactions on Power Electronics*, vol. 30, no. 9, pp. 5207–5222, 2015.

Four-Level Boost Inverter Based on ANPC Topology with Switched-Capacitor Branch

Robert Stala[1], Adam Penczek[1], Stanisław Piróg[1], Aleksander Skała[1], Andrzej Mondzik[1], Zbigniew Waradzyn[1], Krishna Kumar Gupta[2], Pallavee Bhatnagar[3], Sanjay K. Jain[2], Kasinath Jena[4]

[1]AGH UNIVERSITY OF SCIENCE AND TECHNOLOGY, Krakow, Poland
[2]THAPAR INSTITUTE OF ENGINEERING AND TECHNOLOGY, Patiala, India
[3]IES COLLEGE OF TECHNOLOGY, Bhopal, India
[4]SCHOOL OF ELECTRICAL ENGINEERING, KIIT Deemed to be University, Bhubaneswar, India

[1]al. A. Mickiewicza 30, Krakow, Poland
Tel.: +48 / (12) – 617 40 56
Fax: +48 / (12) – 633 22 84
E-mail: stala@agh.edu.pl
URL: http://www.agh.edu.pl; http://www.pelab.agh.edu.pl

Acknowledgements

This paper has been prepared within the bilateral exchange of scientists between the Republic of Poland and the Republic of India: Polish National Agency for Academic Exchange – NAWA (Poland) / Department of Science & Technology, Government of India (India). Grant numbers: PPN/BIN/2019/1/00053/U/00001 and DST/INT/POL/P-39/2020.

Keywords

«Multi-level inverters», «DC-AC converter», «Voltage Source Inverter (VSI)», «Boost», «Switched capacitor».

Abstract

This paper presents a novel single-stage four-level inverter with voltage-boosting ability. In comparison to the cascaded topology, the proposed converter allows for the elimination of a power choke on the DC side. The inverter topology and its special switching algorithm are demonstrated and verified by simulations and experiments.

Introduction

Multilevel inverters (MLIs) have been widely investigated for applications at low, medium, and high voltages [1–2]. The special features of MLIs are low voltage stress of switches, low total harmonic distortion, low du/dt load stress, low filtering requirements, and high modularity [3]. Traditional MLIs use three topologies: cascaded H-bridge (CHB), flying capacitor (FC), and neutral point clamped (NPC) inverters. However, these topologies do not ensure any voltage gain, and, therefore, applications that involve low-voltage DC input (such as photovoltaic systems and electric vehicle drives) require a DC-DC boost converter at the input or a transformer at the output. This can cause additional losses and costs [4]. In addition to traditional MLIs, hybrid multilevel topologies, such as active neutral point clamped (ANPC) inverters, have been continuously researched. An example is a five-level NPC inverter (5L-ANPC) [5] that combines the features of a 3L-NPC and a 3L-FC to generate a five-level output voltage waveform. Its advantage is a good trade-off between the reduced number of components and single DC-link operation, but it has no voltage-boosting capability.

Recent research has focused on switched-capacitor-based MLIs (SCMLIs), which can overcome the drawbacks of traditional MLIs. The switched-capacitor principle involves a parallel connection of the capacitors to the DC source for charging and a series connection for discharging. Thus, the SCs add several levels to the output waveform and simultaneously offer a voltage gain [6–17]. The topologies presented in [6–7] are based on the concept of SC and possess modularity characteristics. However, their drawback is a high voltage stress of semiconductor switches, making them unsuitable for high-voltage applications. In [8], the authors demonstrate how a single unit of the basic cell may generate a seven-level output voltage waveform. As the design includes a symmetrical voltage source, the requirement for numerous sources and polarity generation via an H-bridge makes the architecture more expensive and less desirable. The topologies described in [12–14] use a larger number of switching components while achieving a low gain. Moreover, another seven-level structure [16] achieves a gain of less than unity and employs additional switching components. The SCMLI in [17] synthesizes seven levels with a voltage boosting capability three times the input voltage, but has the disadvantage of causing excessive voltage stress of switches. The topologies described in [9–11], [15] offer a limited voltage gain. Further examples of inverters, suitable for single- and three-phase high-power systems, achieved on the basis of NPC or ANPC multilevel inverters, are demonstrated in [18–24].

This paper presents a novel single-stage four-level inverter with voltage-boosting ability (Fig. 1). In comparison to the cascaded topology, the proposed converter allows for the elimination of a power choke on the DC side. The inverter topology and its special switching algorithm are demonstrated and verified by simulations and experiments.

Principle of operation

The proposed inverter is based on the ANPC topology with an auxiliary switched capacitor (SC) resonant branch. The switched capacitor (C_S) transfers energy between the DC-link capacitors (C_1 and C_2) and the third DC capacitor C_3. The SC branch can operate during the states that generate medium voltages on the output; therefore, this circuit can be used in all cases of modulation. Figs. 2 and 3 present the idea of inverter operation with marked current paths.

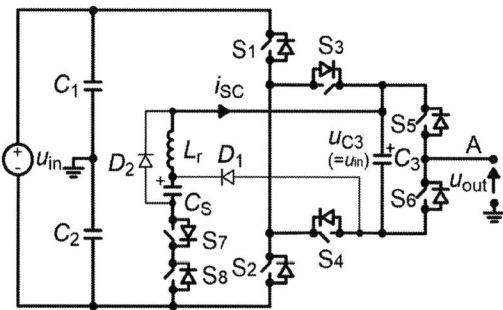

Fig. 1: The proposed four-level switched-capacitor boost inverter in a single-phase implementation

The diagrams presented in Fig. 2 show that the states where the low positive voltage is generated (SP1a and SP1b) are redundant. In state SP1a, the switched capacitor C_S can be recharged in the circuit with the DC-link capacitors and in state SP1b - with capacitor C_3. This allows to maintain the voltage on the DC-link parts equal during the operation with both active or reactive power:

$$U_{C3} = U_{in} \tag{1}$$

The switching pattern for the modulation on the highest positive levels should contain the switching states SP1a and SP1b alternately and is the following:

$$SPp = \{SP1a, SP2, SP1b, SP2, SP1a, \ldots\} \tag{2}$$

According to the principle of symmetry, the switching pattern for the modulation on the lowest negative output voltage levels is the following:

$$SPn = \{SN1a, SN2, SN1b, SN2, SN1a, \ldots\} \tag{3}$$

The states SN2, SN1a, and SN1b are presented in Fig. 3.

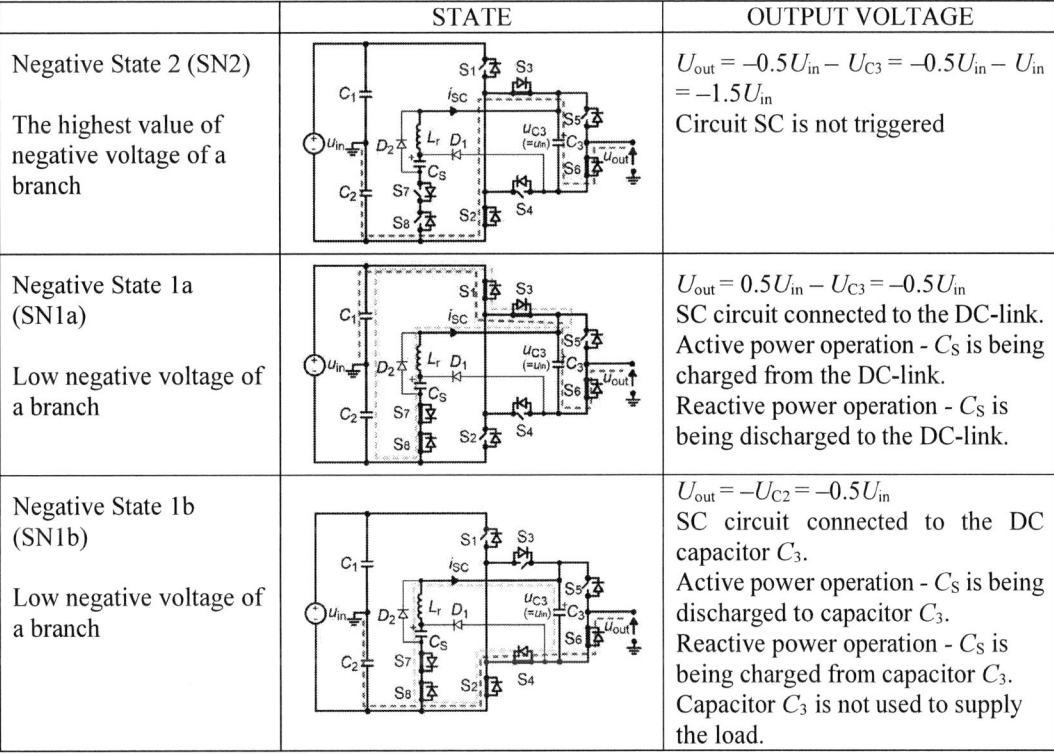

	STATE	OUTPUT VOLTAGE
Positive State 2 (SP2) The highest positive voltage of a branch		$U_{out} = 0.5U_{in}+U_{C3} = 0.5U_{in}+U_{in} = 1.5U_{in}$ SC circuit is not triggered.
Positive State 1a (SP1a) Low positive voltage of a branch		$U_{out} = U_{C1} = 0.5U_{in}$ SC circuit connected to the DC-link. Active power operation - C_S is being charged from the DC-link. Reactive power operation - C_S is being discharged to the DC-link. Capacitor C_3 is not used to supply the load.
Positive State 1b (SP1b) Low positive voltage of a branch		$U_{out} = -U_{C2}+U_{C3} = 0.5U_{in}$ SC circuit connected to the DC capacitor C_3. Active power operation - C_S is being discharged to capacitor C_3. Reactive power operation - C_S is being charged from capacitor C_3.

Fig. 2: Single phase states with the highest positive voltage ($U_{out} = 1.5U_{in}$) and two redundant states with low positive voltage ($U_{out} = 0.5U_{in}$)

	STATE	OUTPUT VOLTAGE
Negative State 2 (SN2) The highest value of negative voltage of a branch		$U_{out} = -0.5U_{in} - U_{C3} = -0.5U_{in} - U_{in} = -1.5U_{in}$ Circuit SC is not triggered
Negative State 1a (SN1a) Low negative voltage of a branch		$U_{out} = 0.5U_{in} - U_{C3} = -0.5U_{in}$ SC circuit connected to the DC-link. Active power operation - C_S is being charged from the DC-link. Reactive power operation - C_S is being discharged to the DC-link.
Negative State 1b (SN1b) Low negative voltage of a branch		$U_{out} = -U_{C2} = -0.5U_{in}$ SC circuit connected to the DC capacitor C_3. Active power operation - C_S is being discharged to capacitor C_3. Reactive power operation - C_S is being charged from capacitor C_3. Capacitor C_3 is not used to supply the load.

Fig. 3: Single-phase states with the highest value of negative voltage ($U_{out} = -1.5U_{in}$) and two redundant states with low negative voltage ($U_{out} = -0.5U_{in}$)

For modulation on medium levels, the capacitor C_3 may be not affected or may operate as a flying capacitor when the following switching patterns are used:

$$SPpn(1) = \{SP1a, SN1b, \ldots\} \tag{4}$$

In this switching pattern, capacitor C_3 is not affected. Transistors S_7 and S_8 may be turned off during this part of the fundamental frequency period (marked in Fig. 4).

$$SPpn(2) = \{SP1b, SN1a, \ldots\} \tag{5}$$

In this case, capacitor C_3 is charged and discharged as a flying capacitor.

Figure 4 presents the switching pattern for a single-phase inverter in the proposed topology. Four-level modulation is accomplished together with energy exchange between capacitor C_3 and DC-link capacitors C_1 and C_2. The energy exchange is supported by the switched capacitor C_S. The branch composed of the capacitor C_S and the resonant inductor L_r is connected in parallel to the DC link or the capacitor C_3. The current of the L_rC_S branch is oscillatory. Since the duty cycle of switching signals varies, the current oscillation in the L_rC_S circuit is terminated after various time intervals. When the converter operates in non-ZCS mode, diodes D_1 and D_2 allow the current of the resonant choke to circulate, which is presented in Fig. 5.

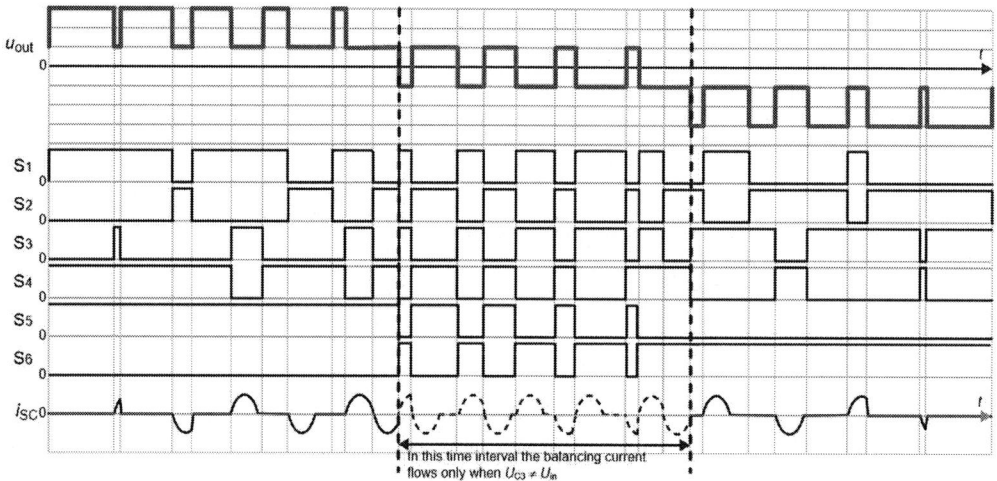

Fig. 4: Operation of a single-phase inverter: output voltage, modulation pattern, and idealized waveform of current in the resonant inductor L_r ($i_{SC} = i_{Lr}$)

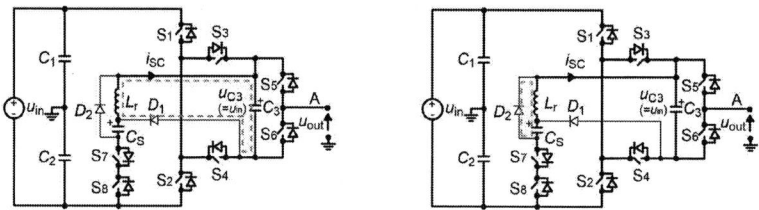

Fig. 5: Operation of auxiliary diodes in non-ZCS mode during the dead time

The SC circuit converts only a fraction of the total converter power and utilizes a very low-volume resonant choke. In comparison to the cascaded topology composed of a boost and an inverter, the proposed converter allows the removal of a power choke on the DC side.

Simulation results

Simulation tests of the proposed circuit have been carried out in MATLAB/Simulink environment for the following parameters: $U_{in} = 300$ V, $L_r = 2$ µH, $C_S = 4.7$ µF, $C_3 = 470$ µF where U_{AC_rms} is 230 V and switching frequency f_{sw} is 50 kHz. As can be seen in Fig. 6, an additional LC filter with the following parameters was used at the output: $L_F = 300$ µH, $C_F = 4.7$ µF (resonant frequency is $f_g = 4.23$ kHz).

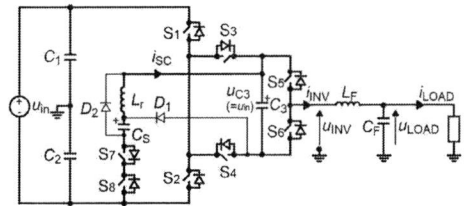

Fig. 6: Schematic of the simulated circuit

Figure 7 presents the switching signals together with the output current and voltage waveforms in a time period of 20 ms, obtained for $P_{out} = 1$ kW. The current and voltage waveforms of the components are presented in Figs. 8a) and 8b), respectively.

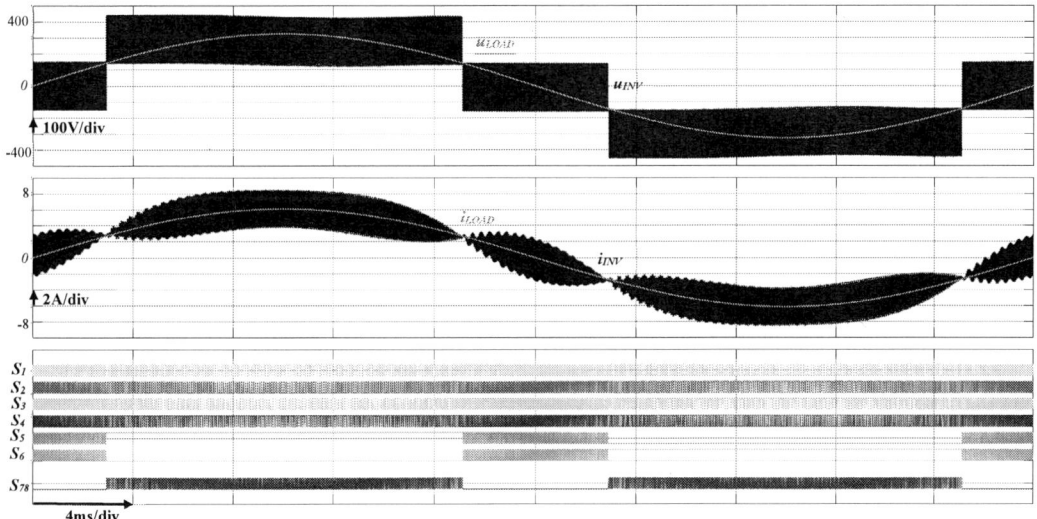

Fig. 7: PWM switching signals, output current, and voltage waveforms in a time period of 20 ms (for $P_{out} = 1$ kW).

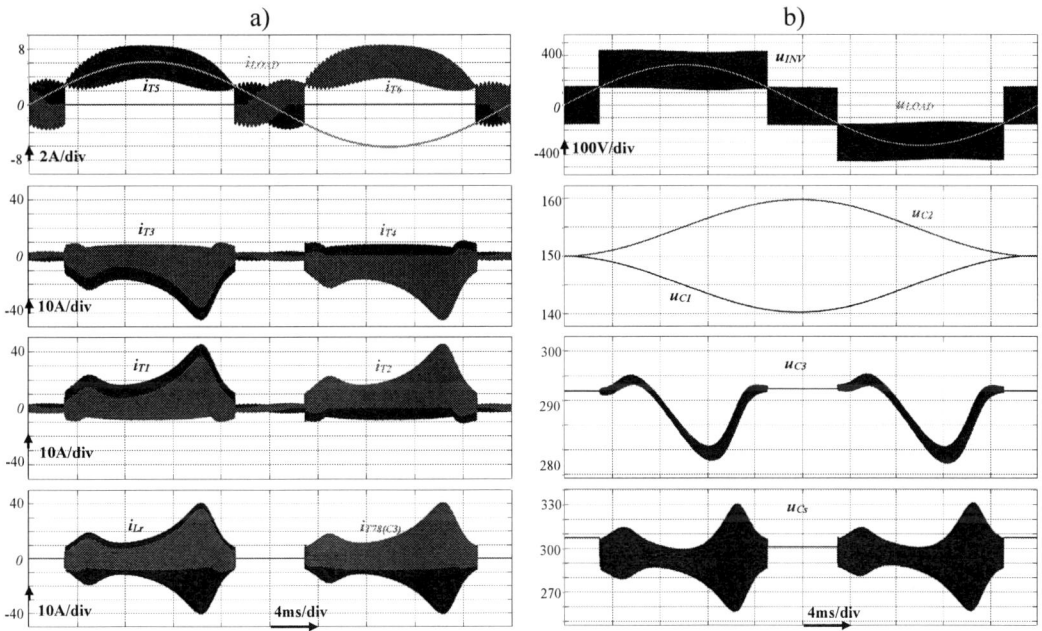

Fig. 8: Current (a) and voltage (b) waveforms of the inverter components

From the simulation results presented in Figs. 7 and 8 it is seen that the four-level modulation is used, the voltage ripple of C_3 is low, the DC-link voltages are self-balanced, and the balancing circuit does not operate in the whole switching period.

The inverter can produce the maximum positive voltage $U_{INVmax} = 450$ V and the minimum voltage $U_{INVmin} = -450$ V from the output to the neutral point of the DC-link divider. The equivalent DC voltage U_{eq2-L} of the two-level inverter in this output voltage range is:

$$U_{eq2-L} = U_{INVmax} + U_{INVmin} = 3U_{in} = 900 \text{ V} \tag{6}$$

In Table I, the estimated voltage and current stresses of the components are listed. The voltage stress across the transistors in the proposed inverter is 3 times lower than that in the equivalent two-level inverter.

Table I. Current and voltage stresses of components at $P_{load} = 1$ kW ($i_{OUT\_rms} = 4.34$ A)

Component	I_{rms} [A]	I_{rms}/I_{OUT_rms}	U_{max}/U_{in}	U_{max}/U_{eq2L}
S_1 (S_2)	7.84	1.8	1	1/3
S_3 (S_4)	7.84	1.8	1	1/3
S_5 (S_6)	3.18	0.7	1	1/3
S_7 (S_8)	10.13	2.3	1	1/3
D_1	0.48	0.1	2	2/3
D_2	0.48	0.1	2	2/3
L_r	10.15	2.3	1	1/3
C_S	10.13	2.3	1	1/3

Experimental results

Figure 9 presents a photograph of the experimental setup and its parameters are assembled in Table II.

Fig. 9: The experimental inverter

Table II. Parameters of the experimental setup

Parameter	Value
Input DC voltage	100 V
Output voltage frequency	50 Hz
Switching frequency	50 kHz
Modulation index	0.75
Capacitors C_1, C_2	150 µF + 4.7 µF
Capacitor C_3	150 µF + 4.7 µF
Capacitor C_S	4.7 µF
Inductor L_r	2 µH

Figure 10a) presents oscillograms of waveforms obtained in the experimental inverter, showing the output voltage u_{INV} of the inverter, its output current i_{LOAD}, the voltage u_{C3} across the capacitor C_3 and the current i_{Lr} of the resonant inductor L_r. Furthermore, Figs. 10b)–10d) show waveforms in zoomed time intervals, presenting the modulation method used according to the switching pattern for a single-phase inverter presented in Fig. 4. They concern the time intervals marked in Fig. 10a).

All of the presented waveforms are in good correlation with those obtained by simulations depicted in Fig. 7 (u_{INV}, i_{LOAD}) and Fig. 8 (i_{Lr}).

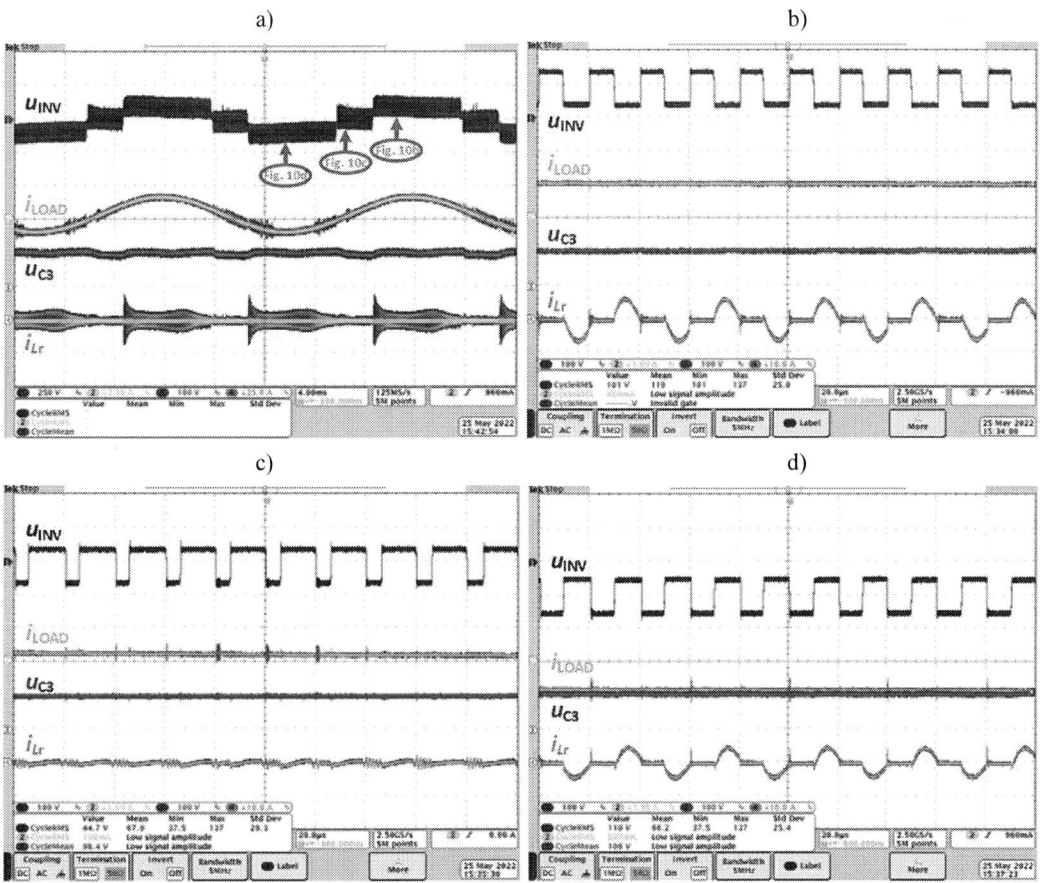

Fig. 10: Waveforms obtained in the experimental inverter setup: output voltage u_{INV}, output current i_{LOAD}, voltage u_{C3} of capacitor C_3, and current i_{Lr} of resonant inductor

Conclusion

In this paper, a novel concept of an inverter topology has been presented. The inverter allows to operate with a voltage gain higher than 1. DC-AC systems with low DC input voltage can operate without a DC-DC boost converter. The proposed inverter utilizes a switched-capacitor branch with a resonant inductor of very low volume. The proposed inverter has positive characteristics compared to the counterpart classic solution with a front-end DC-DC boost converter. The voltage stress of the transistors in the proposed inverter is 3 times lower than that in an equivalent two-level inverter. The entire DC-AC boost/inverter can be designed for MOSFET transistor use. In the boost converter, a larger volume of the inductor is required, and higher voltage stresses of the semiconductor devices occur when operating with the voltage gain $G = 3$. The inverter presented contains a relatively large number of devices, and its efficiency may be deteriorated by losses associated with the operation of the balancing circuit. However, in a classic DC-AC system, the input DC-DC converter operates with high voltage stress of switches and converts 100% of energy. In the proposed solution, the voltage

stress of the switches is lower than the DC-link voltage of an equivalent classic converter. Furthermore, the balancing circuit converts only part of the total energy of the inverter.
The presented topology can be extended to a three-phase converter supplied from a single DC source.

References

[1] Rodriguez J., Lai J. S. and Peng F. Z.: Multilevel inverters: A survey of topologies, control, and applications. IEEE Transactions on Industrial Electronics, 2002, Vol. 49 no 4, pp. 724–738

[2] Abu-Rub H., Holtz J., Rodriguez J. and Baoming G.: Medium-voltage multilevel converters—state of the art, challenges, and requirements in industrial applications. IEEE Transactions on Industrial Electronics, 2010, Vol. 57 no 8, pp. 2581–2596

[3] Gupta K. K., Ranjan A., Bhatnagar P., Sahu L. K. and Jain S.: Multilevel inverter topologies with reduced device count: A review, IEEE Transactions on Power Electronics, 2016, Vol. 31 no 1, pp. 135–151

[4] Kerekes T., Séra D. and Máthé L.: Three-phase photovoltaic systems: structures, topologies, and control, Electric Power Components and Systems, 2015, Vol. 43 no 12, pp. 1364–1375

[5] Barbosa P., Steimer P., Steinke J, Winkelnkemper M. Celanovic N.: Active-neutral-point-clamped (ANPC) multilevel converter technology, 2005 European Conference on Power Electronics and Applications, 2005, Dresden, 11-14 September.

[6] Hinago Y. and Koizumi H.: A switched-capacitor inverter using series/parallel conversion with an inductive load, IEEE Transactions on Industrial Electronics, 2012, Vol. 59 no 2, pp. 878–887

[7] Ye Y., Cheng K. W. E., Liu J. and Ding K.: A step-up switched capacitor multilevel inverter with self-voltage balancing, IEEE Transactions on Industrial Electronics, 2014, Vol. 61 no12, pp. 6672–6680

[8] Raman S. R., Cheng K. W. E. and Ye Y.: Multi-input switched-capacitor multilevel inverter for high-frequency AC power distribution, IEEE Transactions on Power Electronics, 2018, Vol. 33 no 7, pp. 5937–5948

[9] Salem A., Ahmed M., Orabi E. M. and Ahmed M.: New Three-Phase Symmetrical Multilevel Voltage Source Inverter, IEEE Journal on Emerging and Selected Topics in Circuits and Systems, 2015, Vol. 5 no 3, pp. 430–442

[10] Raushan R., Mahato B. and Jana K. C.: Comprehensive analysis of a novel three-phase multilevel inverter with the minimum number of switches, IET Power Electronics, 2016, Vol. 9 no 8, pp. 1600–1607

[11] Belkamel H., Mekhilef S., Masaoud A. and Naeim M. A.: Novel three-phase asymmetrical cascaded multilevel voltage source inverter, IET Power Electronics, 2013, Vol. 6 no 8, pp. 1696–1706

[12] Lee S. S., Bak Y., Kim S. M., Joseph A. and Lee K. B.: New Family of Boost Switched-Capacitor Seven-Level Inverters (BSC7LI), IEEE Transactions on Power Electronics, 2019, Vol. 34 no 11, pp. 10471–10479

[13] Zeng J., Lin W. and Liu J.: Switched-Capacitor-Based Active-Neutral-Point-Clamped Seven-Level Inverter With Natural Balance and Boost Ability, IEEE Access, 2019, 7, pp. 126889–126896

[14] Sathic M. J., Sandeep N. and Blaabjerg F.: High Gain Active Neutral Point Clamped Seven-Level Self-Voltage Balancing Inverter, IEEE Transactions on Circuits and Systems II: Express Briefs, 2020, Vol. 67 no 11, pp. 2567-2571

[15] Siwakoti Y. P., Mahajan A., Rogers D. J. and Blaabjerg F.: A Novel Seven-Level Active Neutral-Point-Clamped Converter With Reduced Active Switching Devices and DC-Link Voltage, IEEE Transactions on Power Electronics, 2019, Vol. 34 no 11, pp. 10492-10508

[16] Abhilash T., Annamalai K. and Tirumala S. V.: A Seven-Level VSI With a Front-End Cascaded Three-Level Inverter and Flying Capacitor Fed H-Bridge, IEEE Transactions on Industry Applications, 2019, Vol. 55 no 6, pp. 6073-6088

[17] Roy T., Sadhu P. K., Dasgupta A. and Aarzoo N.: A novel three-phase multilevel inverter structure using switched capacitor basic unit for renewable energy conversion systems, International Journal of Power Electronics, 2019, Vol. 10 no 1/2, pp.133–154

[18] Pineda C. W. A. and Rech C.: Modified Five-Level ANPC Inverter with Output Voltage Boosting Capability, Proceedings of IECON 2019 - 45th Annual Conference of the IEEE Industrial Electronics Society, Lisbon, Portugal, 14-17

[19] Siwakoti Y. P.: A new six-switch five-level boost-active neutral point clamped (5L-Boost-ANPC) inverter, 2018 IEEE Applied Power Electronics Conference and Exposition (APEC), San Antonio, 4-8 March

[20] Zeng J., Lin W. and Liu J.: Switched-Capacitor-Based Active-Neutral-Point-Clamped Seven-Level Inverter With Natural Balance and Boost Ability, IEEE Access, 2019, 7, pp. 126889-126896

[21] Lee S. S. and Lee K.: Dual-T-Type Seven-Level Boost Active-Neutral-Point-Clamped Inverter, IEEE Transactions on Power Electronics, 2019, vol. 34, no. 7, pp. 6031-6035

[22] Ye Y., Chen S., Sun R., Wang X. and Yi Y.: Three-Phase Step-Up Multilevel Inverter With Self-Balanced Switched-Capacitor, IEEE Transactions on Power Electronics, 2021, vol. 36, no 7, pp. 7652-7664

[23] Akagi H.: Multilevel converters: Fundamental circuits and systems, Proc. IEEE, 2017, vol. 105, no. 11, pp. 2048–2065

[24] Kumari M., Siddique M. D., Sarwar A., Tariq M., Mekhilef S., Iqbal A.: Recent trends and review on switched-capacitor-based single-stage boost multilevel inverter, International Transactions on Electrical Energy Systems, 2021; Vol. 31 no 3, e12730

Comparative Evaluation of Partially-Rated Energy Storage Integration Topologies for High Voltage Modular Multilevel Converters

Zoe Blatsi[1], Sebastian Neira[1,2], Stephen Finney[1] and Michael M.C. Merlin[1]

[1] The University of Edinburgh, Edinburgh, United Kingdom

[2] Pontificia Universidad Católica de Chile, Santiago, Chile

Email: z.blatsi@ed.ac.uk

Keywords

≪Ancillary Services≫, ≪Energy Storage≫, ≪HVDC transmission≫, ≪Modular Multilevel Converter≫.

Abstract

This paper compares three partially-rated MMC topologies (Partially Rated Storage – PRS, Stack Parallel Branch – SPB, Inductor Parallel Branch – IPB) which integrate energy storage solutions for HVDC-scale Modular Multilevel Converters to provide with extra degrees of flexibility in the grid. The paper compares (i) the ES power that can be contributed from each topology under a given converter design and (ii) the trade-offs in terms of losses, and control adaptation required on top of a standard half-bridge MMC design in order to provide ES power. The results indicate that MMC stacks with full-bridge submodules have consistently higher ES power capability than their half-bridge counterparts – this comes at the expense of higher losses and extra devices, which for certain applications could be avoided.

Introduction

The HVDC technology is the key-enabling technology complimenting the long-existing HVAC backbone of modern power transmission systems. HVDC is employed in the connection of offshore wind farms to the grid [1], bulk power transfer from generation to consumption centres [2], stability enhancement of the existing AC grid [3], and the interconnection of asynchronous systems [4]. In this context, the Modular Multilevel Converter (MMC) is the state-of-the-art topology for HVDC applications, featuring high efficiency and power quality while at the same time achieving low insulation stress and high redundancy [5–7].

Utility-Scale Energy Storage Systems (ESS) can mitigate the consequences of the displacement of synchronous generators (SGs) due to the increased adoption of renewables, such as the reduction in inertia available for frequency support [8]. Additionally, ESSs can provide marketed services such as energy shifting and blackstart. Both these services can reduce curtailment of renewables and decrease the downtime of the grid [9]. The commercially available ESSs are installed in the medium voltage grid, where they usually consist of a single- or double-stage of ac-dc conversion stage, or alternatively, they are embedded in Cascaded H-Bridge submodules (SMs) [10].

In the HVDC context, the energy stored in the SM capacitors of an MMC is not sufficient for substantial inertia provision or primary frequency control [11]. However, integration of partial-voltage, partial-power rated ESSs within the MMC can assist with ac and dc power decoupling. In the following, the existing topologies for integration of partial power rating within an HVDC-scale MMC are presented and compared with regards to their P_{es}/P_{conv} power capability, losses and complexity of control implementation. The analysis is validated with simulation results obtained from MATLAB/Simulink.

Partially–Rated Energy Storage Topologies in MMC stacks

MMCs purposed for HVDC transmission typically consist of Half-Bridge (HB) SMs, which utilise the minimum number of devices and are therefore hard to beat in terms of converter efficiency. Within the MMC structure, the connection of the ES units to the converter can be implemented:

- on the *SM* level, where the ES units are connected to the SM capacitor via a dc-dc converter such as a bidirectional dc-chopper or an interleaved full-bridge converter [12, 13]. This arrangement allows the ES units to be subdivided in shorter series strings, and increases the reliability and availability of the ES provision, in case of fault. The ES-SMs can be equally distributed among all arms [14], or concentrated in a stack or a phase [15, 16]. In the latter case, circulating currents are required to maintain the energy balance among all arms. The symmetrical distribution of ES-SMs is studied here, resulting in the Partially-Rated Storage (PRS) MMC shown in Fig. 1a.
- on the *stack* level, where the ESSs are integrated within a separate cascaded string of SMs, external to the converter and connected in parallel to a portion of each MMC stack, as displayed in Fig. 1b. A symmetrical distribution of stack-parallel ESS branches results in the Stack-Parallel Branch MMC. This topology could be useful in the context of industrial practice for keeping the ESS –which have high energy density and are a liability for electrical fires– in a separate hall or fire-resistant container.
- on the *phase* level, where the separate ESS branch is connected in parallel to the ac inductors, known as the Inductor-Parallel Branch MMC (Fig. 1c).

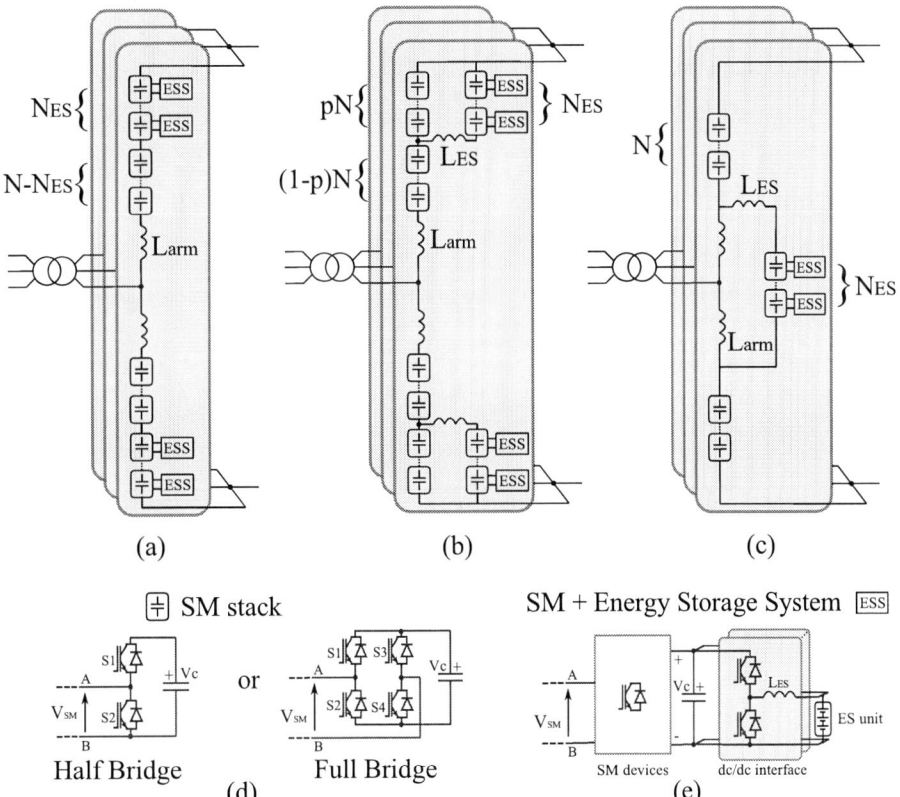

Fig. 1: Integration topologies: on SM level (a), on stack level (b), and on phase level (c). (d) Half-Bridge and Full-Bridge SM topologies. (e) ES-SM showing the connection of the SMs to the ES units via the dc/dc interface.

In the following, the topologies are tested upon an HB-MMC rated to operate within a certain P/Q range and ac voltage variation, as required by national grid codes [8]. The main converter parameters are presented in Table I.

Table I: Converter parameters

MMC	DC Voltage V_{dc}	± 375 kV	AC Voltage (PCC) V_{LL}	430 kV
	Rated power P_{rated}	1000 MW	Equivalent capacitive energy E	$40 \frac{kJ}{MVA}$
	Arm inductor L_{arm}	0.1 pu	Transformer leakage L_{xfo}	0.14 pu
	Nominal SM voltage V_{sm}	2.7 kV	SM capacitance C_{sm}	6.6 mF
	Controller frequency f_c	10 kHz	Number of SMs per arm N	278
PRS	ES-SM Voltage Rating (arm)	$0.1V_{dc}$	Number of ES-SMs	28
SPB	ES-Branch Voltage Rating (arm)	$0.1V_{dc}$	Number of ES-parallel HB-SMs	28
IPB	ES-Branch Voltage Rating (phase)	$0.2V_{dc}$		

Partially-Rated Storage MMC

This topology has been extensively studied in [14] for grid applications. A part of the MMC stack capacitors is continually charged by their respective ESS (Fig. 1e). These SMs are inserted preferentially so that power exchange occurs with the ac or dc grid. For the right amount of power to be delivered to/from the ESS regardless of the current direction whilst at the same time maintaining the ES-SM voltages around nominal, negative voltage capability is required by the ES-SMs.

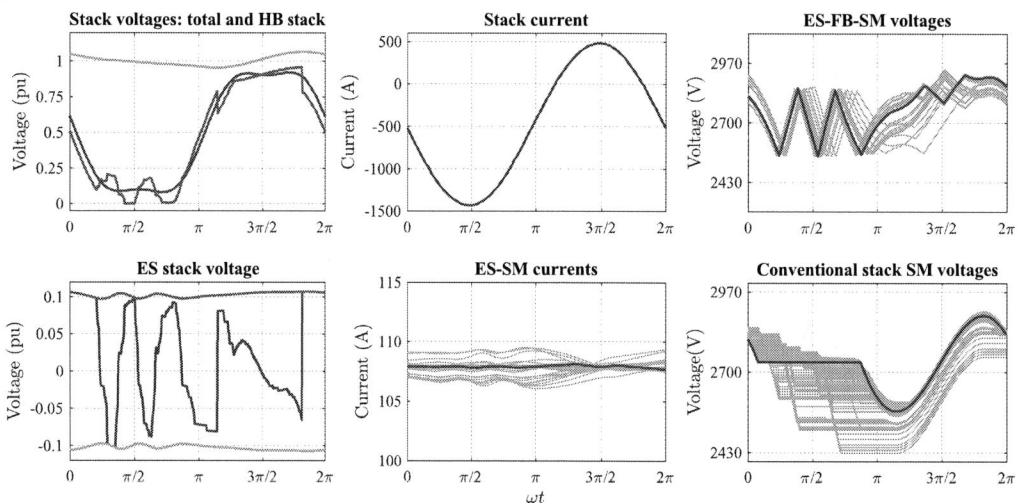

Fig. 2: Steady-state waveforms for an SPB-MMC that operates at $P_{ac} = -1$pu, $P_{es} = 0.05$pu (injecting) and $P_{dc} = P_{ac} - P_{es} = -1.05$pu.

Fig. 2 shows indicative results for a power decoupling of 5% between ac and dc. In the top left figure, the total inserted stack voltage (blue) and the HB stack (red) are shown. The yellow waveform denotes the available voltage of the entire stack. In the bottom left, the inserted voltage of the ES-SM portion of the stack is displayed, bounded by its available ES-SM voltage. The individual ESS currents contributed to each ES-SM capacitor are displayed in the bottom centre, with ES-SM 1 highlighted. In the right side, the ES-SM and HB-SM capacitor voltages are shown, with SM 1 of each stack highlighted. The HB-SMs follow a rotational sorting algorithm (bottom right), whereas the ES-SMs are confined within their hysteresis band (top right).

Stack-Parallel Branch MMC

This topology has been already studied in [17] as the parallel combination of a three-phase Cascaded H-Bridge Converter to a portion of the MMC stacks. The adopted approach required high intra-stack balancing currents and voltages within the stacks to redistribute the energy deviation built up because the two parts of the stack process unequal dc and ac powers. A generalisation of this concept is presented in the following. The stack is subdivided in two parts, of which one is connected in parallel to an external ESS. The ES-parallel part of the stack is denoted by p, and the rest of the stack by $1 - p$. For a stack with throughput powers P_{ac}^* and P_{dc}^*:

$$P_{ac} = 3V_{ac}I_{ac}cos(\varphi) \tag{1}$$

$$P_{dc} = V_{dc}I_{dc} \tag{2}$$

$$V_{stack}^* = V_p^* + V_{1-p}^*, \tag{3}$$

The idea proposed in the following is to keep the ac and dc power of $1 - p$ substack equal by assigning a voltage with dc and ac components that will result in a zero end-of-cycle net energy deviation. That way the stack energy remains balanced without the requirement for extra balancing currents and voltages shifting energy from one part of the stack to the other, as shown in [17]. The voltages of the substacks are:

$$V_{1-p}^* = (1-p)(V_{dc}^* + V_{ac}^* \frac{P_{dc}}{P_{ac}}) \tag{4}$$

$$V_p^* = V_{stack}^* - V_{1-p}^* = pV_{dc}^* + V_{ac}^*(1 - (1-p)\frac{P_{dc}}{P_{ac}}) \tag{5}$$

Given that the voltage assigned to $1 - p$ stack results in ac-dc power equilibrium, it is the p substack parallel to ESS that processes the difference between ac and dc power. The energy keeps around nominal with the aid of a separate control loop that acts on the energy by compensating current.

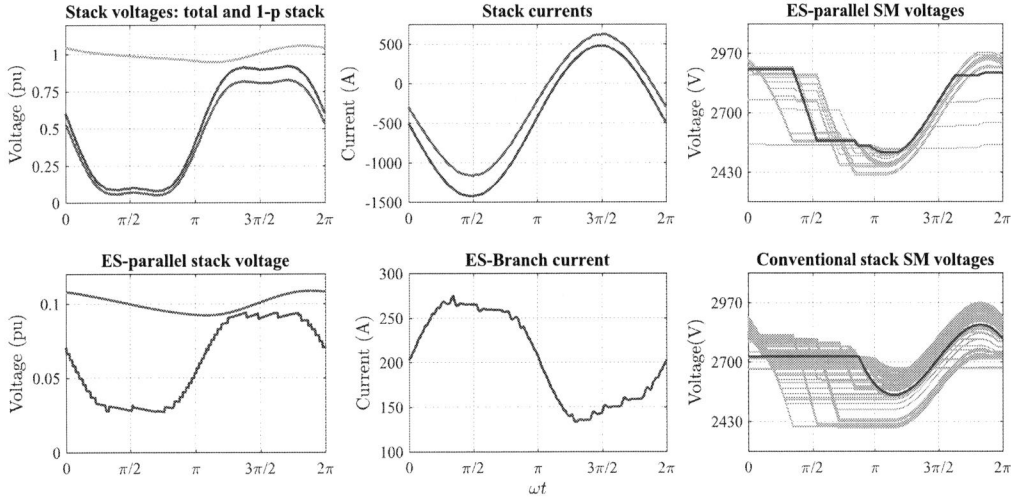

Fig. 3: Steady-state waveforms for an SPB-MMC that operates at $P_{ac} = -1$pu, $P_{es} = 0.05$pu (injecting) and $P_{dc} = P_{ac} - P_{es} = -1.05$pu.

Fig. 3 displays 5% of extra power being injected to the dc grid. Top left shows the total stack voltage V_{stack} (blue), the inserted voltage V_{1-p} (red), and the total available voltage (yellow). At the bottom left the ES-parallel part of the voltage V_p is shown in blue, with its available voltage. At the centre (top), the stack currents I_{1-p} (blue), I_p (red) are shown, and below them the ES branch current, which is their difference. In the right, the SM voltage waveforms follow a standard rotational algorithm pattern. By examining closely Eq. (5), it is easily understood that in the cases where the ac voltage component is higher than the dc one, then the substack voltage would need to overmodulate. This can be overcome by upgrading the p substack HB-SMs with FB-SMs.

Inductor-Parallel Branch MMC

This topology, first introduced in [18], uses an ESS branch connected in parallel to the stack inductors of the MMC, keeping the topology of the main power path unmodified. Power is exchanged by using a new circulating current operating at harmonic frequency h:

$$i^*_{circ} = \frac{I^*_{dc}}{3} + I^*_{ES}\sin\left(h(\omega t + \theta) + \varphi_z\right) \tag{6}$$

The amplitude I^*_{ES} and phase angle φ_z are selected offline to maximise power capability whilst maintaining a moderate increase in losses. h is the harmonic order of the current, which is chosen to be even –in this case $h = 8$. The ES branch operates generating a limited magnitude ac voltage at the selected harmonic frequency.

$$v^*_{ES} = 2h\omega L_s I^*_{ES}\cos\left(h(\omega t + \theta) + \varphi_z\right) \tag{7}$$

Fig. 4 illustrates the operation of the topology for a power decoupling of 5% between the dc and ac sides.

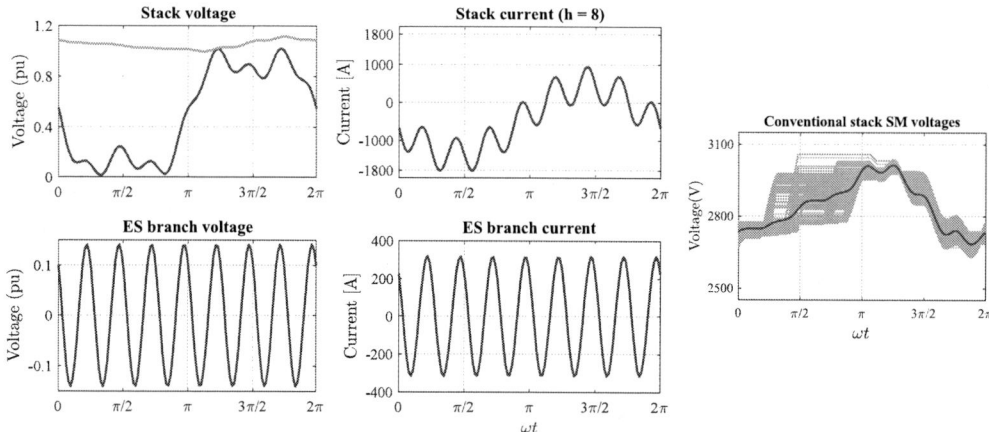

Fig. 4: Steady-state waveforms for an IPB-MMC that operates at $P_{ac} = -1$pu, $P_{es} = 0.05$pu (injecting) and $P_{dc} = P_{ac} - P_{es} = -1.05$pu. Exchange of power between the ES branch and the stack occurs through voltage and current of 8th harmonic. On the right, the SM voltages of a stack.

The harmonic components superimposed on the stack voltage and current (top waveforms) exchange power with the ES branch (bottom waveforms).

Comparative evaluation

P_{es}/P_{conv} power capability

The power capability of a given topology is defined by four operational constraints:
1. the current rating of the components in the power path (semiconductor devices and capacitors),
2. the maximum available stack voltage,
3. the minimum attainable stack voltage,
4. the maximum SM capacitor voltage.

Fig. 5–Fig. 7 are generated by sweeping across P_{ac} and P_{es} – the area within all contours illustrates the range of operation that each topology can reach without requiring component upgrades.

For the PRS-MMC, P_{es}/P_{conv} capability ranges from +0.07 to -0.13pu across the ac power range [-1.3, 1.25] pu, as displayed in Fig. 5. The power capability is consistent across the entire ac power range, owing to the negative voltage insertion capability of the ES-SMs, and the second harmonic circulating current, it is therefore suitable for ESS integration within MMC-based interconnectors, where it can be used for energy shifting.The capability is limited by the current rating in high load conditions, and the maximum allowable SM capacitor voltage. The solid red line of Fig. 5 indicates the SM voltage limit for the HB-SMs only, whereas the green dashed line indicates the voltage limit for all the SMs, including the FB-ES-SMs. For $P_{es} > 0$ the voltage limit is defined by the FB-ES-SM capacitors, since I_{es} from each ES element is *added* to respective capacitor. Higher voltage rating for these capacitors might allow higher capability. For $P_{es} < 0$, the ES current for the FB-ES-SM stacks is *subtracted* from the stack current, causing smaller voltage excursion to these submodules than the HB-SMs.

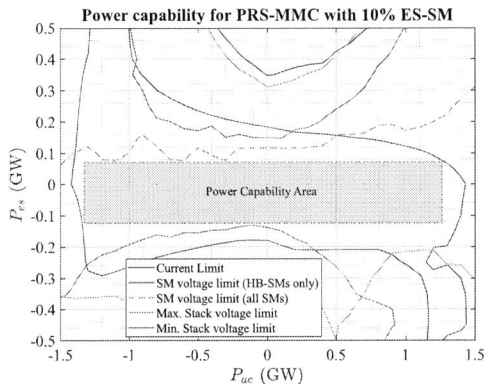

Fig. 5: Capability plot for the PRS-MMC with 10% of FB-ES-SM within the MMC stack.

Fig. 6: SPB for the SPB-MMC with ESS branch parallel to 10% of the MMC stack.

The SPB-MMC P_{es}/P_{conv} capability is shown within the contour lines of Fig. 6, where it is shown that ES power integration can only occur when $|P_{dc}| > |P_{ac}|$. The fact that the ES capability area is asymmetrical, makes this variant unsuitable for grid services where the ESS is required to cycle energy to and from the grid depending on the condition of the grid. However, it is a good candidate for MMC applications where the power flows unidirectionally most of the time, for example offshore wind turbine HVDC connections, where the offshore (rectifying) converter can integrate energy from secondary renewable unidirectional energy sources, such as photovoltaic (PV) panels or fuel cells [19–21]. Conversely, at the receiving converter there is the flexibility of supplying loads which are installed in parallel to the p stack.

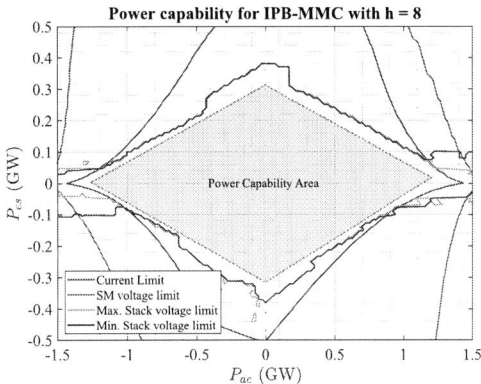

Fig. 7: Capability plot for the IPB-MMC using the eighth harmonic.

The P_{es}/P_{conv} capability of the IPB-MMC starts from a value of ± 0.07pu at 1pu of P_{ac} power, and increases almost linearly to more than 0.3pu for zero power. This is due to the fact that in low-load conditions, there is current headroom in the stacks to exchange power with the ES branches – the limiting factor is the maximum and minimum stack voltage limits. The IPB-MMC presents a higher feasible area for partial-load operation, which makes it favourable for energy storage integration in renewable generation collector converter. Suitable applications would be offshore and onshore windfarm converters, wave and tidal energy converters which operate frequently at low-load.

Losses

After establishing the capability of each topology, it is necessary to assess the losses. The study considers the use of a 4.5kV 2000A IGBT from ABB Semiconductors (ABB 5SNA2000K450300) [22]. The losses are examined for two cases: a full-load ($P_{ac} = \pm 1$pu) and a partial load ($P_{ac} = \pm 0.3$pu) operational scenario with $\pm 5\%$ ES decoupling. The estimation includes the losses in the main MMC stack devices – the ES branch dc/ac or dc/dc converters are not included.

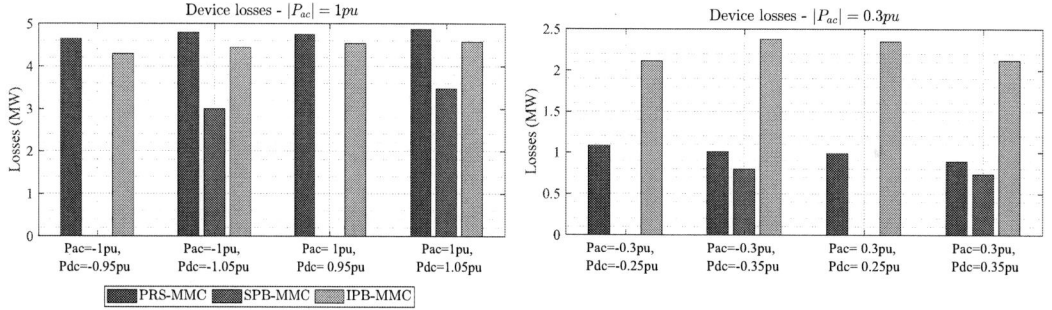

Fig. 8: Loss estimation for full ac power $P_{ac} = \pm 1$pu and $P_{dc} = P_{ac} - P_{es}$.

Fig. 9: Loss estimation for partial ac power $P_{ac} = \pm 0.3$pu and $P_{dc} = P_{ac} - P_{es}$.

Fig. 8 displays the estimated losses in the MMC stack for full-power setpoints PRS-MMC has expectedly the highest losses due to the extra IGBTs in the FB-ES-SMs. Similarly, IPB-MMC suffers from high losses due to its high current required for power transfer, and the slightly higher switching frequency caused by the harmonic currents and voltages in the stacks. Interestingly, the SPB has significantly lower losses than the other two variants owing to its low number devices and lower current in the p substack. However this comes at the expense of lack of capability for $|P_{dc} < |P_{ac}|$. The results for partial load are shown in Fig. 9. The losses are overall smaller, however the high harmonic current of IPB-MMC causes its losses to be at least twice as high as PRS-MMC. However, IPB covers a much greater capability plot area in low load.

Control implementation

The presented variants feature significant differences in their control scheme, summarised in Fig. 10. A qualitative difference is observed between IPB and the other two topologies; the fact that ESSs are placed *outside* the MMC stacks means that a conventional MMC controller can effectively operate as is, with the only modification being the superposition of harmonic current on the references to the current controller.

On the contrary, the other two topologies that interact with a part of the MMC stack capacitors, require modifications in the low-level control – to prioritise the part of the stack that serves as an ESS energy buffer, and to ensure voltage balancing across all parts of the stack. Furthermore, higher level functions such as the stack reference generation requires the addition of a harmonic current in the case of PRS, and the energy management feature has to exclude the ES-SMs submodules from the energy calculation, since ESS current keeps the submodule voltages around nominal.

Conclusion

This paper evaluates three different MMC variants: the PRS-, SPB- and IPB- MMC topologies with regards to their suitability for HVDC-scale partial ES storage provision. The results show that ES integration within an MMC structure is a trade-off between the ES capability, the losses, and the control

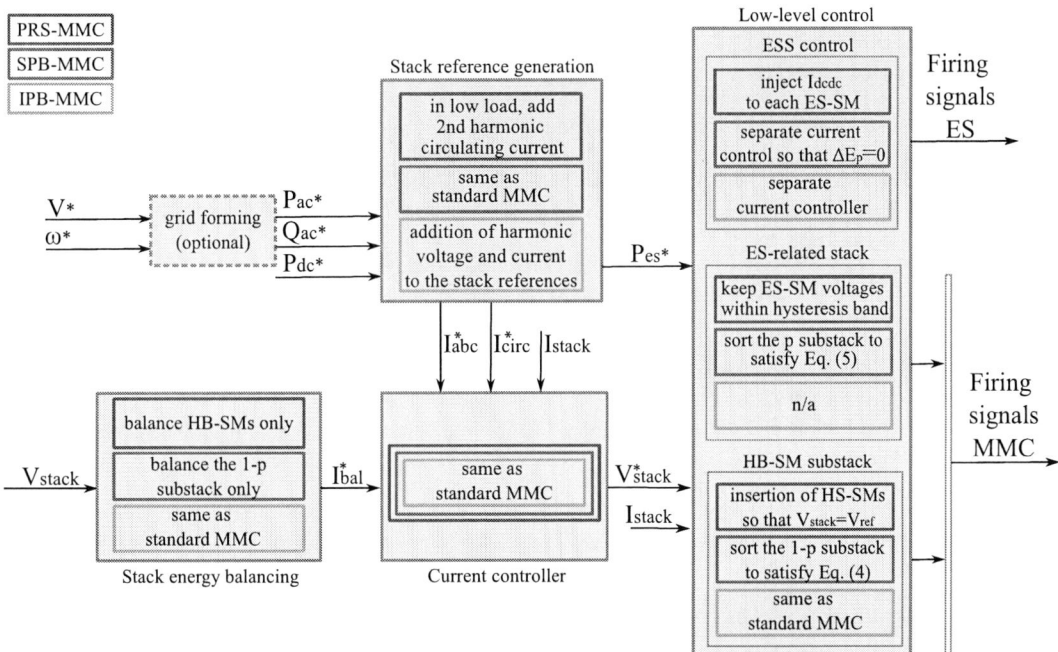

Fig. 10: Control adaptations required for each topology.

adaptation. The PRS-MMC with FB-ES-SMs is a flexible variant that can provide ES margins sufficient for grid service provision, at the expense of extra device losses in full-load conditions. On the other hand, the SPB-MMC with HB SMs has significantly lower MMC stack device losses but the lack of negative voltage margin does not allow the converter to overmodulate, which is necessary in low-load conditions. By embedding ESS parallel to the inductors, the IPB-MMC avoids a complex low-level control implementation. The transfer of power through a set of harmonic voltages and currents results in higher ES capability for low-load conditions, where there is more current headroom –however, current causes a significant loss penalty. For an HVDC-scale MMC rated for dc-ac transmission, the voltage ratings are quite strict, not allowing significant margin to perform other operations on top of dc-ac transmission without equipment upgrade.

References

[1] BorWin 1 — Hitachi ABB. https://search.abb.com/library/Download.aspx?DocumentID=HVDC0068 &LanguageCode=en&DocumentPartId=&Action=Launch (accessed Dec. 06, 2021).

[2] H. Pang and X. Wei, "Research on Key Technology and Equipment for Zhangbei 500kV DC Grid," 2018 International Power Electronics Conference (IPEC-Niigata 2018 -ECCE Asia), 2018, pp. 2343-2351, doi: 10.23919/IPEC.2018.8507575.

[3] D. V. Hertem and M. Delimar, "High voltage direct current (HVDC) electric power transmission systems," 2013.

[4] P. L. Francos, S. S. Verdugo, H. F. Álvarez, S. Guyomarch and J. Loncle, "INELFE — Europe's first integrated onshore HVDC interconnection," 2012 IEEE Power and Energy Society General Meeting, 2012, pp. 1-8, doi: 10.1109/PESGM.2012.6344799.

[5] A. Lesnicar and R. Marquardt, "An innovative modular multilevel converter topology suitable for a wide power range," in 2003 IEEE Bologna Power Tech Conf. Proceedings, Jun. 2003, vol. 3, p. 6 pp. Vol.3-. doi: 10.1109/PTC.2003.1304403.

[6] A. Antonopoulos, L. Angquist, and H.-P. Nee, "On dynamics and voltage control of the Modular Multilevel Converter," in 2009 13th European Conf. Power Electron. and Appl., Sep. 2009, pp. 1–10.

[7] M. M. C. Merlin and T. C. Green, "Cell capacitor sizing in multilevel converters: cases of the modular multilevel converter and alternate arm converter," IET Power Electron., vol. 8, no. 3, pp. 350–360, 2015, doi: 10.1049/iet-pel.2014.0328.

[8] "Grid Code (GC) — National Grid ESO." https://www.nationalgrideso.com/industry-information/codes/grid-code (accessed Dec. 06, 2021).

[9] "Inertia Report — Continental Europe — Inertia Report — ENTSO-E Docs." https://docs.entsoe.eu/dataset/inertia-report-continental-europe/resource/ 148d6b97-4fb9-4eaa-8977 -124e581ddd1c (accessed Jun. 15, 2022).

[10] E. Spahic, C. P. Susai Sakkanna Reddy, M. Pieschel and R. Alvarez, "Multilevel STATCOM with power intensive energy storage for dynamic grid stability - frequency and voltage support," 2015 IEEE Electrical Power and Energy Conference (EPEC), 2015, pp. 73-80, doi: 10.1109/EPEC.2015.7379930.

[11] J. Porst, M. Richter, M. Biller and M. Luther, "Potential of Modular Multilevel Converter Technology providing Synthetic Inertia and AC Frequency Support for AC Grids," NEIS 2020; Conf. Sust. En. Supply and En. Storage Syst., 2020, pp. 1-7.

[12] L. Baruschka and A. Mertens, "Comparison of Cascaded H-Bridge and Modular Multilevel Converters for BESS application," in 2011 IEEE Energy Conversion Congress and Exposition, Phoenix, AZ, USA, Sep. 2011, pp. 909–916. doi: 10.1109/ECCE.2011.6063868.

[13] F. Errigo, F. Morel, C. Mathieu de Vienne, L. Chedot, A. Sari, and P. Venet, "A Submodule with Integrated Supercapacitors for HVDC-MMC providing Fast Frequency Response," IEEE Trans. Power Del., pp. 1–1, 2021, doi: 10.1109/TPWRD.2021.3086864.

[14] P. D. Judge and T. C. Green, "Modular Multilevel Converter With Partially Rated Integrated Energy Storage Suitable for Frequency Support and Ancillary Service Provision," in IEEE Trans. Power Del., vol. 34, no. 1, pp. 208-219, Feb. 2019, doi: 10.1109/TPWRD.2018.2874209.

[15] F. Errigo, L. Chédot, P. Venet, A. Sari, P. Dworakowski, and F. Morel, "Assessment of the Impact of Split Storage within Modular Multilevel Converter," in IECON 2019 - 45th Annual Conf. of the IEEE Industrial Electron. Society, Oct. 2019, vol. 1, pp. 4785–4792. doi: 10.1109/IECON.2019.8927698.

[16] G. Henke and M.-M. Bakran, "Balancing of modular multilevel converters with unbalanced integration of energy storage devices," in 2016 18th Eur. Conf. Power Electron. and Appl. (EPE'16 ECCE Europe), Sep. 2016, pp. 1–10. doi: 10.1109/EPE.2016.7695265.

[17] Z. Blatsi, S. Neira, P. Judge, M. Merlin, and S. Finney, "Modular Multilevel Converter with Stack-Parallel Cascaded H-Bridge Energy Storage Branch," in 2021 IEEE 21th Workshop Control and Modeling for Power Electron., IEEE COMPEL 2021, Nov. 2021, doi: 10.1109/ECCE-Asia49820.2021.9479410.

[18] S. Neira, Z. Blatsi, P. Judge, M. Merlin, and J. Pereda, "Modular Multilevel Converter with Inductor Parallel Branch Providing Integrated Partially Rated Energy Storage," in 2021 IEEE 12th Energy Conversion Congress Exposition - Asia (ECCE-Asia), May 2021, pp. 676–681. doi: 10.1109/ECCE-Asia49820.2021.9479410.

[19] S. Sau, A. C. Nair and B. G. Fernandes, "Theoretical Analysis and Comparison of Capacitor Requirement in Modular Converters for Grid Integration of High Power Solar PV," 2019 IEEE Energy Conversion Congress and Exposition (ECCE), 2019, pp. 5514-5521, doi: 10.1109/ECCE.2019.8912549.

[20] A. I. Elsanabary, G. Konstantinou, S. Mekhilef, C. D. Townsend, M. Seyedmahmoudian and A. Stojcevski, "Medium Voltage Large-Scale Grid-Connected Photovoltaic Systems Using Cascaded H-Bridge and Modular Multilevel Converters: A Review," in IEEE Access, vol. 8, pp. 223686-223699, 2020, doi: 10.1109/ACCESS.2020.3044882.

[21] A. Abdelhakim and F. Blaabjerg, "Current-fed Modular Multilevel Converter (CMMC) for Fuel Cell and Photovoltaic Integration," 2020 IEEE 21st Workshop on Control and Modeling for Power Electronics (COMPEL), 2020, pp. 1-6, doi: 10.1109/COMPEL49091.2020.9265695.

[22] 5SNA2000K450300 Data Sheet, Doc. No. 5SYA 1431-02 01-2018. [Online]. Available: https://new.abb.com/products/5SNA2000K450300

Influence of Current Collapse due to V_{ds} Bias Effect on GaN-HEMTs I_d-V_{ds} Characteristics in Saturation Region

Xuyang Lu[1,2], Arnaud Videt[1], Ke Li[2], Soroush Faramehr[2], Petar Igic[2], Nadir Idir[1]

[1] Univ. Lille, Arts et Metiers Institute of Technology, Centrale Lille, Junia
ULR 2697 - L2EP Lille F-59000, France
[2] Coventry University, Centre for Clean Growth and Future Mobility (CGFM)
Coventry CV1 2TL, UK
Email: xuyang.lu.etu@univ-lille.fr, arnaud.videt@univ-lille.fr
ke.li@coventry.ac.uk, soroush.faramehr@coventry.ac.uk
petar.igic@coventry.ac.uk, nadir.idir@univ-lille.fr

Keywords

≪Gallium Nitride (GaN)≫, ≪Double pulse test≫, ≪Device characterisation≫, ≪Switching losses≫, ≪Threshold voltage shift≫

Abstract

A new method is proposed in this paper to investigate the influence of current collapse effect on the I_d-V_{ds} characteristics of GaN-HEMTs in high voltage region based on a modified H-bridge circuit. The measured I_d-V_{ds} characteristics with and without the V_{ds} bias are compared, which shows the effect of charge trapping due to the V_{ds} bias on device I_d-V_{ds} characteristics in saturation region. These data will be used for a device model including the current collapse effect in full I_d-V_{ds} region.

Introduction

Gallium Nitride High-Electron-Mobility Transistors (GaN-HEMTs) are strong candidates for high-power-density and high-efficiency power converters due to the fast switching speed and low power losses. To accurately model the switching transients of high voltage GaN-HEMTs, an appropriate drain-current versus drain-source voltage (I_d-V_{ds}) characteristics covering the entire switching trajectory is indispensable. However, the I_d-V_{ds} characteristics provided by the datasheet are only in low V_{ds} voltage region of few volts which cannot predict the high voltage switching trajectory. The reason is that most of the I_d-V_{ds} characteristics are measured by curve tracer, which is not suitable for high voltage measurement due to the equipment power limitation. Moreover, GaN transistors suffer from the current collapse effect due to the trapped electrons in device structures induced by the off-state V_{ds} bias, on-state V_{gs} bias and hot electrons, which degrades device characteristics resulting in the increased dynamic on-state resistance (R_{dson}) and the threshold voltage (V_{th}) shift [1], [2]. Many research work focused on the device dynamic R_{dson} degradation in Ohmic region, which increases device conduction losses [3–5]. But there is not much reported data for this characteristics degradation under saturation region, which though is highly relevant to the device switching waveforms and switching losses. The impact of current collapse on the device turn-on losses are reported in [6], which may be attributed to a positive V_{th} shift, but there is no quantitative study on characteristics shift. This characteristics shift will have an impact on device switching transients, which may increases switching losses, trigger sustained oscillation etc [10]. Furthermore, the main advantage of GaN-HEMTs is the high frequency application so the influence of current collapse effect on device switching transients deserve much attention. Consequently, it is necessary to have a method to construct the I_d-V_{ds} characteristics of GaN-HEMTs in high voltage region and investigate the current collapse degradation in saturation region.

This work proposes a new method to measure the I_d-V_{ds} characteristics of GaN-HEMTs in high voltage region, which is based on a modified H-bridge to control the V_{ds} initial voltage bias. The I_d-V_{ds} characteristics are extracted from the turn-on switching waveforms during the whole Miller plateau. In order to reduce the influence of circuit parasitic inductance (L_{para}) and device output capacitance (C_{oss}) on measurement results, a large turn-on gate resistance (Rg_{on}) is used to slow down the switching-on speed. The I_d-V_{ds} characteristics are compared between a fresh device, referred to as unbiased (short: ub) and a device after V_{ds} voltage bias (short: b), which are recorded respectively as I_d-V_{ds}^{ub} and I_d-V_{ds}^{b} characteristics to demonstrate the V_{ds} bias effect. The paper is constituted by following parts. At first, the proposed experimental setup to measure the high voltage I_d-V_{ds} characteristics are presented. Afterward, the influence of circuit L_{para} and device C_{oss} on measurement results as well as device self-heating are analysed. In the third section, the measured I_d-V_{ds}^{ub} and I_d-V_{ds}^{b} characteristics are compared, in which a positive shifted V_{th} and the decreased current in high-voltage saturation region are observed. The paper is concluded at last with a discussion on future work.

Experimental setup

To investigate the influence of current collapse of GaN-HEMTs on saturation region, it is essential to have an experiment that can not only measure the high voltage I_d-V_{ds} characteristics but also control the V_{ds} initial bias, which is highly related to current collapse effect aforementioned. Several research works have focused on constructing the I_d-V_{ds} characteristics of SiC MOSFETs in high voltage region based on the double-pulse test [7], [8]. In [7], the high voltage I_d-V_{ds} characteristics are constructed from measured I_d and V_{ds} points, where V_{gs} equals to the Miller plateau voltage (V_{pl}). The drawback is that only one data point can be acquired per test and each test requires different settings to obtain different V_{ds} and I_d values. More importantly, this method do not control initial V_{ds} bias of the DUT, which is not suitable for GaN-HEMTs characterization because the initial V_{ds} bias can affect device characteristics as mentioned before. In [9], a circuit to control the initial V_{ds} bias induced trapping effect of GaN-HEMTs in conventional DPT is proposed, but the impact on full I_d-V_{ds} characteristics of the device is not mentioned. Overall, two advantages of the proposed method are that the initial V_{ds} voltage bias can be controlled to avoid the V_{ds} trapping effect and less tests are required as only I_d needs to be adjusted during each test.

The experiment board consists of a main DPT board and an auxiliary board as shown in Fig. 1(a), which is proposed in [10] at first. Point A and B are not hard-connected as the conventional DPT board, which can be separated by the T_H to avoid the initial drain bias of DUT. Hence, this experiment board supports two test modes as detailed below.

Conventional DPT mode

Under this mode points A and B in Fig. 1(a) are connected, therefore, the experiment board will work in the conventional DPT mode. The top-side device T_1 works as a freewheeling diode by shorting the gate and source terminal. The voltage drop and current flowing through the DUT are V_{ds} and I_d respectively.

The DUT is controlled by a double-pulse signal V_g as shown in Fig. 1(b). From t_0 to t_1, the DUT is in turn-off state undertaking DC source voltage, called the initial drain bias. At t_2, the inductive load is charged and I_d is the same as load current I_L, which can be calculated by equation (1):

$$I_d = I_L = V_{DC} \times \frac{t_{on}}{L} \tag{1}$$

In eq(1), L is the load inductance and t_{on} equals to $t_2 - t_1$. The time interval between t_2 and t_3 is short, and the load current is freewheeling through transistor T_1 during $t_2 - t_3$. Thus, the turn-off and turn-on switching transient of the DUT at same current and voltage can be observed at t_2 and t_3. Moreover, the commutation speed at turn-on and turn-off of the DUT can be adjusted by using different gate resistor Rg_{on} and Rg_{off}. However, the DUT under this mode undertakes V_{ds} voltage bias before the test is

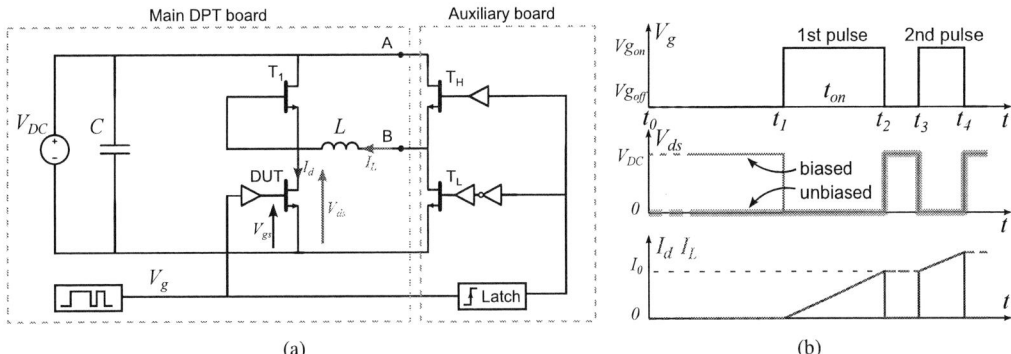

Fig. 1: Experimental setup. (a) Schematic of experimental setup. (b) Typical waveforms of experimental setup.

started at t_1, therefore, the obtained DUT turn-on and turn-off switching waveforms include the V_{ds} bias effect [11]. To compare with the switching waveforms without V_{ds} bias effect, an unbiased mode of the modified DPT will be introduced below.

Unbiased DPT mode

When A and B in Fig. 1(a) are separated by auxiliary board, the experiment board will work in unbiased DPT mode. The top-side device T_H in auxiliary board is synchronously controlled with DUT by V_g while the low-side device T_L is complementary to the T_H by adding a NOT gate in the driver side. Before test starting (during t_0 - t_1), DUT and T_H are in off-state and T_L is turned on, so the DUT is free of the initial drain bias since T_H blocks V_{DC} for the DUT. Afterward, during the rising edge of the first pulse at t_1, the transistors in auxiliary board are latched (T_H and T_L are in on and off state respectively) and the modified DPT board will resume to conventional DPT operation as shown in Fig. 1(b).

The proposed method to obtain high voltage I_d-V_{ds} characteristics is established on the Miller plateau of turn-on switching waveforms. To clearly show the Miller plateau and eliminate the impact of parasitic circuit elements, a large Rg_{on} = 1 kΩ is chosen to slow down the turn-on switching speed. All power transistors in the H-bridge are GaN devices (GS66502B 650 V/7.5 A). The V_{gs} of DUT are measured by a 500 MHz passive probe (N2873A) and the V_{ds} is measured by a high voltage differential probe (N2790). A current probe (N2783B) is used to measure the I_d. The DC source voltage is set at 200 V, which determines the V_{ds} range in measured I_d-V_{ds} characteristics. And I_d can be controlled by adjusting the duration of the first pulse t_{on} based on equation (1). Therefore, full current and voltage of DUT can be obtained. These setup will be implemented in both conventional DPT mode and unbiased DPT mode to get two sets of data, to investigate the influence of drain bias on device I_d-V_{ds} characteristics in saturation region.

Measurement results and error analysis

Measurement results

Turn-on switching waveforms of the DUT measured in the unbiased DPT mode are shown in Fig. 2(a), where curve A, B, C and D respectively represent different switching waveforms (V_{gs}, I_d and V_{ds}) with incremental t_{on} from 500 ns to 2000 ns. The I_d and V_{gs} keep nearly constant while V_{ds} decrease during the Miller plateau so that different I_d and V_{ds} data with the same V_{gs} can be obtained in this region. For example, when $V_{gs,ref}$ = 1.8 V, it has an intersection with curve A (t_{on} = 500 ns) and the corresponding I_d and V_{ds} point can be obtained, which is one measured point. As $V_{gs,ref}$ has other intersections with different V_{gs} from curve B, C, and D, a set of measured points can be obtained, which consists an I_d-V_{ds} curve with V_{gs} = 1.8 V as shown in Fig. 2(b). By repeating this over a range of $V_{gs,ref}$ values, the high voltage I_d-V_{ds}^{ub} characteristics with different V_{gs} can be obtained. Correspondingly, if the switching waveforms are from conventional DPT, a high voltage I_d-V_{ds}^{b} characteristics can be obtained.

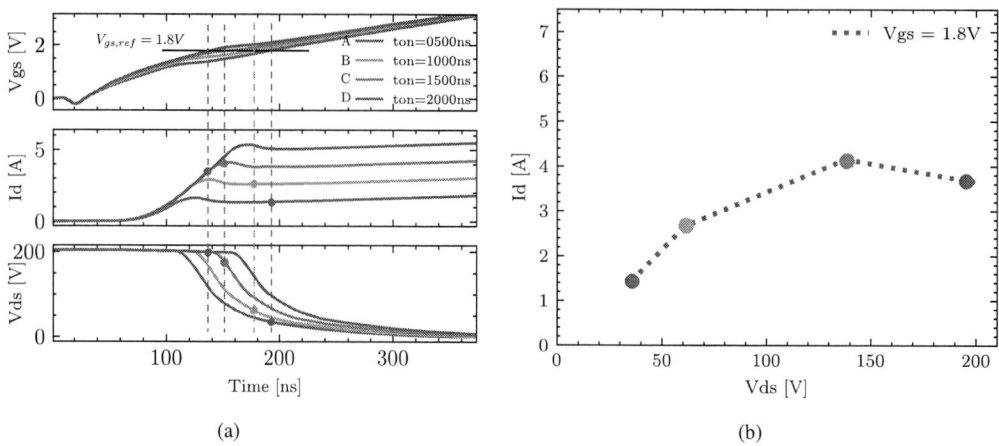

(a)   (b)

Fig. 2: High voltage I_d-V_{ds} measurement principle. (a) Turn-on switching waveforms for different t_{on} from 500 ns to 2000 ns. (b) I_d-V_{ds}^{ub} characteristics at $V_{gs} = 1.8$ V.

The I_d-V_{ds} characteristics measurement method using curve tracer is steady-state measurement, where the channel of DUT is completely formed before drain current flowing through it, which may not correspond to the real switching process of device and involves significant self-heating. While the proposed method is based on device turn-on switching transient, which follows a real dynamic V_{ds}, I_d and V_{gs} trajectory of the DUT in hard-switching, with low self-heating. However, even with slow commutations the GaN device switching transients remain sensitive to parasitic circuit elements, and the junction temperature might still noticeably increase and influence the I_d-V_{ds} data for high voltage and current values. Thus, it is necessary to discuss and quantify these impacts.

Error analysis and compensation

There are several factors (e.g., measurement noise, parasitic circuit elements, device output capacitance and junction temperature) that can affect measured results, which will be discussed in the following parts.

Measurement noise

Accurate intersections of a fixed V_{gs} value and measured V_{gs} waveforms in Fig. 2(a) are significant in this method. However, measured V_{gs} waveforms are affected by measurement noise as shown in Fig. 3, which may interfere the location of intersections, leading to inaccurate V_{gs}, V_{ds} and I_d values. The noise frequency is mainly above 60 MHz according to the signal spectrum analysis. A Butterworth filter with 60 MHz cut-off frequency is used to filter the high-frequency measurement noise based on the zero-phase filtering method. The filtering result is shown in Fig. 3(b). Note that all of the measured V_{gs}, V_{ds} and I_d are filtered using this method to improve the accuracy of high voltage I_d-V_{ds} characteristics.

Influence of parasitics parameters

The I_d-V_{ds} characteristics represent the relation between the device channel current I_{ch} and the voltage drop on gate-source capacitance V_{Cgs}. However, the I_{ch} and V_{Cgs} cannot be directly measured in DPT due to the parasitic elements in device packaging as shown in Fig. 4(a). Therefore, the impact of these parasitic parameters on the difference between measured I_d, V_{gs} and device internal I_{ch}, V_{Cgs} should be discussed.

During the turn-on transient, the $C_{oss} = C_{gd} + C_{ds}$ of the DUT will discharge and the corresponding I_{Coss} current cannot be measured by current probes. Hence, the I_{Coss} should be calculated to compensate measured I_d as equation (2) shown.

$$I_{ch} = I_d + I_{Coss} = I_d + C_{oss}\frac{dV_{ds}}{dt} \tag{2}$$

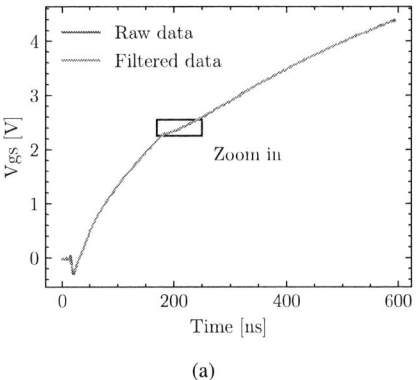
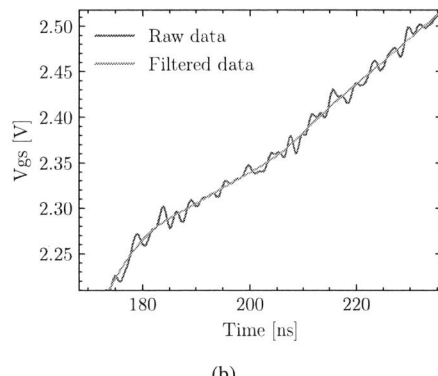

(a)   (b)

Fig. 3: V_{gs} waveform with and without filtering. (a) V_{gs} waveform in turning-on. (b) Zoomed-in V_{gs}.

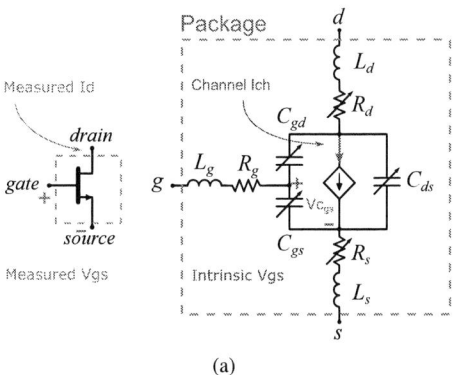
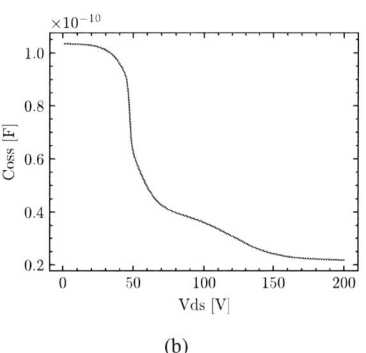

(a)   (b)

Fig. 4: Parasitic elements inside the device package. (a) Symbol and device parasitic elements of GaN-HEMTs. (b) C_{oss} - V_{ds} of GS66502B.

C_{oss} is a non-linear capacitance, which is V_{ds} voltage dependent. An accurate C_{oss} model of GS66502B is presented in [12] as shown in Fig. 4(b), which is used to calculate $I_{C_{oss}}$.

As shown in Fig. 4(a), the measured V_{gs} is the potential between gate and source terminals including the voltage drop on the L_g, L_s and R_g, R_s. Moreover, the mutual inductance between power loop and gate loop M_p will also exert additional voltage drop on measured V_{gs}. These voltage drops should be compensated to get the V_{Cgs}. The impact of parasitic inductance L_g, L_s and mutual inductance M_p can be neglected because the low $\frac{dI_d}{dt}$ during Miller plateau and the value of L_g and L_s are at most several hundreds pico Henry [13]. As for the internal gate resistance R_g, the voltage drop is determined by the gate current I_g that can be calculated by equation (3).

$$I_g = \frac{V_g - V_{Miller}}{R_{gon}} \tag{3}$$

Where the output voltage of gate driver V_g is 6 V and the minimal V_{Miller} is about 1.5 V (based on measurement) and R_{gon} is 1 kΩ. So the maximum I_g is around 4.5 mA. According to the device model from manufacturer, the R_g is 225 mΩ so the voltage drop is around 1 mV, which can also be neglected. Hence, the difference between measured V_{gs} and intrinsic V_{Cgs} is the voltage drop on the R_s, which can be compensated based on equation (4).

$$V_{Cgs} = V_{gs} - (I_d + I_g)R_s \tag{4}$$

R_s is considered to be constant (14.3 mΩ) because of the extreme low temperature sensitivity (0.1 mΩ/K) based on the result in [13]. Consequently, the error between measured I_d, V_{gs} and device internal I_{ch}, V_{Cgs} can be compensated by the above method.

The comparison of the measured I_d-V_{ds} characteristics with and without the error compensation are shown in Fig. 5. The influence of the measurement noise and parasitics parameters on the high voltage I_d-V_{ds} characteristics are obvious in high current region.

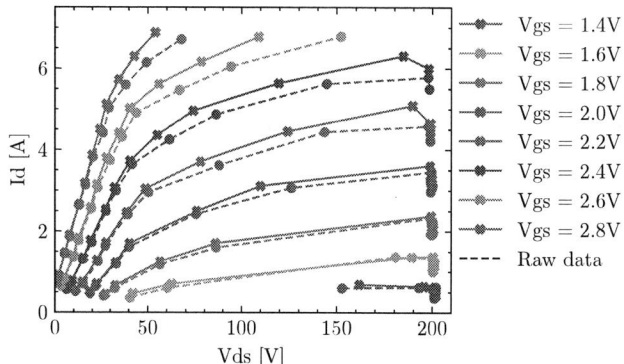

Fig. 5: I_d-V_{ds}^b characteristics with and without error compensation from conventional DPT.

Self-heating of the DUT

During the measurement, the DUT will turn on and off twice as shown in Fig. 1(b). The turn-off switching speed is fast as a small Rg_{off} is used, which will not cause much switching losses. However, a large Rg_{on} is used to slow down the turn-on speed, which may cause noticeable self-heating.

The RC thermal SPICE model of GS66502B provided by the datasheet is adopted to estimate the junction temperature T_j during the measurement, which is shown in Fig. 6. Where the R_n and C_n represent the thermal resistance and capacitance of different layers in the device package, and the R_{CA} represents the thermal resistance between case and ambient. The T_j and T_c are the junction and case temperature respectively. The ambient temperature is set to 25 °C. The T_j can be calculated when the device power losses are imported into this model as the power source.

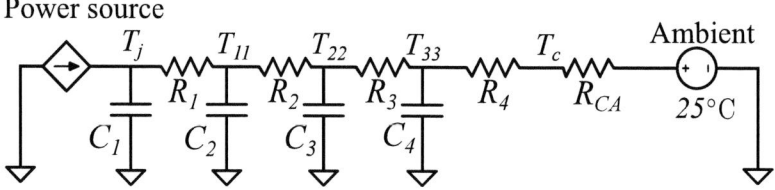

Fig. 6: RC thermal model simulation.

As shown in Fig. 7(a), only the second turn-on transient causes obvious switching losses and increased temperature. And the first turn-on loss can be neglected in both unbiased and conventional DPT mode because the first turn-on transient is zero current switching (ZCS). Hence, the increased T_j is mainly caused by the second turn-on transient, which is similar in unbiased model and conventional DPT mode. Note that the initial values of T_j at the second turn-on transient are different as a result of the heat accumulation from previous switching losses. This is due to the large thermal propagation time constant R_2C_2, which slows down the temperature propagation speed from junction to case. Anyhow, the maximum change of T_j is less than 7 °C, which validates that R_s can be considered as a constant resistance during the test.

Based on the processed V_{gs}, I_d, V_{ds} and T_j, a high voltage I_d-V_{ds}^b characteristics with temperature distribution can be obtained in Fig. 7(b), where the temperature variation is small even in high voltage region, so the measured I_d-V_{ds} characteristics can be seen as in constant temperature.

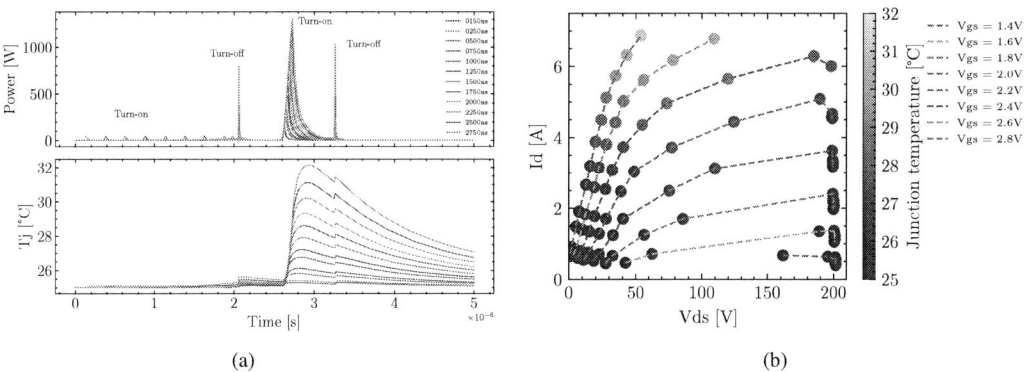

(a)   (b)

Fig. 7: Junction temperature and high voltage I_d-V_{ds}^b characteristics from conventional DPT. (a) Power and T_j versus time. (b) Temperature distribution in I_d-V_{ds}^b characteristics.

Influence of drain bias on measured I_d-V_{ds} characteristics in saturation region

The high voltage I_d-V_{ds}^{ub} characteristics constructed from the unbiased DPT mode are expected closer to the original device characteristics because there is no initial V_{ds} bias. Thus, the I_d-V_{ds}^{ub} characteristics are compared to the simulation result from the manufacturer model, which are shown in Fig. 8(a). Note that the I_d-V_{ds} characteristics of the manufacturer model is the same as that in the datasheet. The measured I_d-V_{ds}^{ub} characteristics has a noticeable difference with it in the manufacturer model, which mainly represents as higher I_d in saturation region. Since the switching trajectory is mainly located on the saturation region and it is highly related to switching losses, it is reasonable to assume that in the real working condition the device switching losses are different with that predicted by simulation using the manufacturer model.

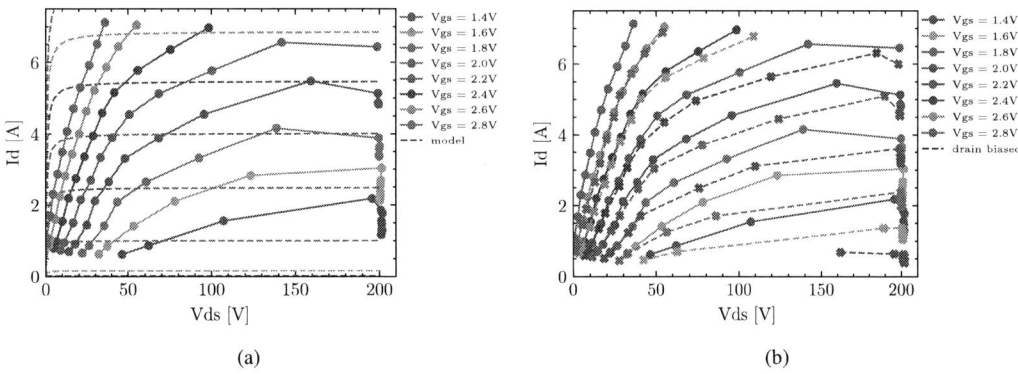

(a)   (b)

Fig. 8: I_d-V_{ds} characteristics comparison. (a) Measured I_d-V_{ds}^{ub} characteristics versus manufacturer model. (b) Comparison of I_d-V_{ds}^{ub} and I_d-V_{ds}^b.

Measured I_d-V_{ds}^b and I_d-V_{ds}^{ub} characteristics are compared in Fig. 8(b). A distinct decreased current is observed in the drain biased condition, which reveals the drain bias trapping effect on saturation region. This comparison shows a clear perspective that the drain bias trapping effect can affect device switching losses by changing the device switching trajectory.

Moreover, the device transfer characteristics from different measurement modes and manufacturer model are compared in Fig. 9. As shown in Fig. 9(a), the measured transfer characteristics has an negative shift compared with the manufacturer model (when V_{ds} is 10 V). On the other hand, the drain biased transfer characteristics at $V_{ds} = 200$ V have a positive shift compared with the unbiased transfer characteristics. To quantify these shifts, the I_d is represented on the logarithmic axis as shown in Fig. 9(b). The V_{th} of device is determined when I_d is above 20 mA. A 0.4 V positive V_{th} shift is observed between the unbiased and

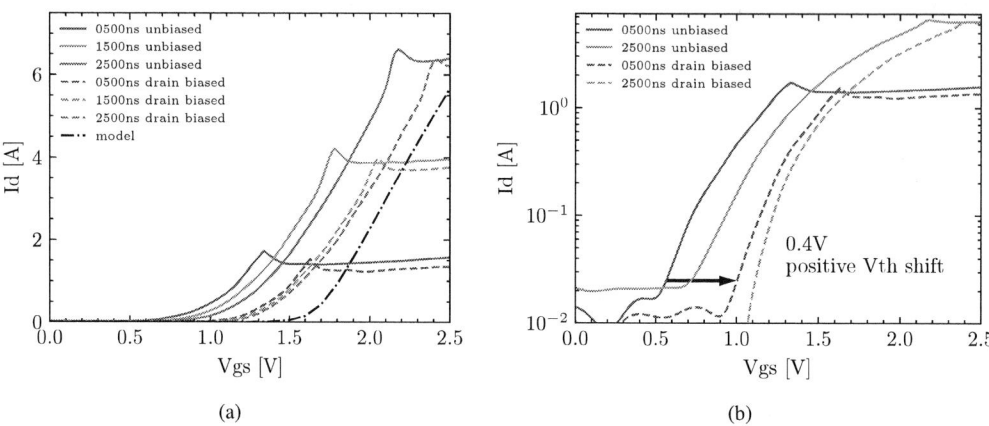

(a)   (b)

Fig. 9: Transfer characteristics comparison when $V_{ds} = 200V$. (a) Measured transfer characteristics versus manufacturer model. (b) Transfer characteristics with and without drain bias in logarithm scale.

biased transfer characteristics when the first turn-on pulse time $t_{on} = 500$ ns. In addition, a slight positive V_{th} shifts are observed when t_{on} increases from 500 ns to 2500 ns in both biased and unbiased transfer characteristics. The positive V_{gs} bias induced trapping effects may contribute to this V_{th} shift since the t_{on} is positively correlated to the positive V_{gs} bias time during the first turn-on pulse as shown in Fig. 1(b). Additionally, the positive V_{th} shift of GaN-HEMTs under positive V_{gs} bias was reported in [14], [15].

It should be noted that the turn-on speed of the DUT is very slow for a GaN device due to the large turn-on gate resistance, hence, the overlapping of the V_{ds} and I_d waveforms is larger than that in the normal turn-on transient, which provides more opportunities for hot electrons trapping to occur in device structure. However, the hot electrons trapping and voltage biased trapping are coupled together in this measurement setup. Consequently, it is necessary to have the quantitative characterization for the hot electrons and V_{gs} bias trapping effect separately. The constructed I_d-V_{ds} characteristics and the trapping effect will be considered in device modelling in our future work.

Conclusion

This paper proposed a new method to evaluate the influence of current collapse effect on the I_d-V_{ds} characteristics of GaN-HEMTs in high voltage region. The influence of measurement noise, parasitic circuit elements and junction temperature are considered and they are extracted to obtain the I_d-V_{ds} characteristics representing the relation between channel current and gate source voltage. The I_d-V_{ds}^b and I_d-V_{ds}^{ub} characteristics are compared and the decreased current in high voltage region and positive V_{th} shift in transfer characteristics are observed. The future work will focus on device modelling based on the measured data, which will consider both of the voltage bias and hot electron trapping effect in the full I_d V_{ds} range to enhance the accuracy of simulated switching waveforms of GaN-HEMTs.

References

[1] S. Yang, S. Han, K. Sheng, and K. J. Chen, "Dynamic on-resistance in GaN power devices: Mechanisms, characterizations, and modeling," *IEEE Journal of Emerging and Selected Topics in Power Electronics*, vol. 7, no. 3, pp. 1425–1439, 2019.

[2] G. Zulauf, M. Guacci, and J. W. Kolar, "Dynamic on-resistance in GaN-on-Si HEMTs: Origins, dependencies, and future characterization frameworks," *IEEE Transactions on Power Electronics*, vol. 35, no. 6, pp. 5581–5588, 2020.

[3] K. Li, A. Videt, N. Idir, P. L. Evans, and C. M. Johnson, "Accurate measurement of dynamic on-state resistances of GaN devices under reverse and forward conduction in high frequency power converter," *IEEE Transactions on Power Electronics*, vol. 35, no. 9, pp. 9650–9660, 2020.

[4] F. Yang, C. Xu, and B. Akin, "Experimental evaluation and analysis of switching transient's effect on dynamic on-resistance in GaN-HEMTs," *IEEE Transactions on Power Electronics*, vol. 34, no. 10, pp. 10121–10135, 2019.

[5] R. Li, X. Wu, S. Yang, and K. Sheng, "Dynamic on-state resistance test and evaluation of GaN power devices under hard- and soft-switching conditions by double and multiple pulses," *IEEE Transactions on Power Electronics*, vol. 34, no. 2, pp. 1044–1053, 2019.

[6] K. Li, A. Videt, N. Idir, P. Evans, and M. Johnson, "Experimental investigation of GaN transistor current collapse on power converter efficiency for electrical vehicles," in *2019 IEEE Vehicle Power and Propulsion Conference (VPPC)*, pp. 1–6, 2019.

[7] H. Sakairi, T. Yanagi, H. Otake, N. Kuroda, and H. Tanigawa, "Measurement methodology for accurate modeling of SiC MOSFET switching behavior over wide voltage and current ranges," *IEEE Transactions on Power Electronics*, vol. 33, no. 9, pp. 7314–7325, 2018.

[8] M. Pulvirenti, L. Salvo, G. Scelba, A. G. Sciacca, M. Nania, G. Scarcella, and M. Cacciato, "Characterization and modeling of SiC MOSFETs turn on in a half bridge converter," in *2019 IEEE Energy Conversion Congress and Exposition (ECCE)*, pp. 1960–1967, 2019.

[9] R. Hou, Y. Shen, H. Zhao, H. Hu, J. Lu, T. Long, "Power loss characterization and modeling for GaN-based hard-switching half-bridges considering dynamic on-state resistance," *IEEE Transactions on Transportation Electrification*, vol. 6, no. 2, pp. 540–553, 2020.

[10] A. Videt, K. Li, N. Idir, P. Evans, and M. Johnson, "Analysis of GaN converter circuit stability influenced by current collapse effect," in *2020 IEEE Applied Power Electronics Conference and Exposition (APEC)*, pp. 2570–2576, 2020.

[11] J. M. Tirado, J. L. Sanchez-Rojas, and J. I. Izpura, "Trapping effects in the transient response of AlGaN/GaN HEMT devices," *IEEE Transactions on Electron Devices*, vol. 54, no. 3, pp. 410–417, 2007.

[12] K. Li, P. L. Evans, C. M. Johnson, A. Videt, and N. Idir, "A GaN-HEMT compact model including dynamic Rdson effect for power electronics converters," *Energies*, vol. 14, no. 8, 2021.

[13] L. Pace, N. Defrance, A. Videt, N. Idir, J.-C. De Jaeger, and V. Avramovic, "Extraction of packaged GaN power transistors parasitics using S-parameters," *IEEE Transactions on Electron Devices*, vol. 66, no. 6, pp. 2583–2588, 2019.

[14] L. Sayadi, G. Iannaccone, S. Sicre, O. Häberlen, and G. Curatola, "Threshold voltage instability in p-GaN gate AlGaN/GaN HFETs," *IEEE Transactions on Electron Devices*, vol. 65, no. 6, pp. 2454–2460, 2018.

[15] J. O. Gonzalez, B. Etoz, and O. Alatise, "Characterizing threshold voltage shifts and recovery in Schottky gate and ohmic gate GaN-HEMTs," in *2020 IEEE Energy Conversion Congress and Exposition (ECCE)*, pp. 217–224, 2020.

Deep-Learning fault detection and classification on a UAV propulsion system

Pierre-Yves BRULIN[1,2], Fouad KHENFRI[1], Nassim RIZOUG[1]
[1]ESTACA'Lab, S2ET
Parc universitaire Laval-Changé, Rue Georges Charpak, 53061 Laval, France
Email: first-name.last-name@estaca.fr
URL: http://www.estaca.fr
[2]HEXADRONE
ZA La Sagne, 99 Chem. de la Borie, 43330 Saint-Ferréol-d'Auroure, France
Email: pierre-yves.brulin@hexadrone.fr
URL: http://www.hexadrone.fr

Keywords

≪Fast Fault Detection≫, ≪Deep Learning≫, ≪Machine Learning≫, ≪Permanent magnet motor≫, ≪Condition Monitoring≫

Abstract

A fault detection and identification method using a Deep-Learning classification method is used to identify several faults that may occur on a UAV propulsion system. Training is performed from a dataset acquired from a simplified multiphysics simulation of the system which allows for the generation of large datasets of modular, interconnected and scalable components of various sizes and performances. We aim to provide a model able to identify faults occurring on a propulsion system using a reduced set of input signals.

Introduction

Drones or swarm of drones are now gaining interest in several industrial applications such as package delivery, environmental monitoring, aerial photography, 3D mapping, industrial and infrastructure inspections, and much more. The growth in the use of drones involves many complex independent systems to monitor. On a large fleet, human interventions on each machine are necessary to maintain its performances and integrity. On the other hand, every levels of maintenance cannot be performed by all the operators of the same machine. Fig. 1 shows the five levels of maintenance as described in [1]. The 1st and 2nd maintenance level performed respectively by the end-user and by an authorized technician remains limited in the actions that can be performed on the machine and does not involve any action on its internal components. Beyond the 3rd maintenance level, the diagnosis of failures is considered but implies specialized knowledge of the system and allow a repair limited to the exchange of components. Beyond the 4th level, the system is entrusted to a dedicated intervention team, or sent back to the manufacturer, for a partial or total reconstruction.

An internal system fault can only be identified after an examination by a specialized operator at the 3rd level. At this level, the cost and time required for maintenance are very high and can be subject to errors in the identification of defective components. In order to overcome this problem, different methods are developed to detect and isolate faults (FDI) that occur on complex and critical systems (i.e. airplanes, cars, drones, ...). In [2], fault detection (FD) and fault isolation (FI) methods are classified into 3 categories: hardware redundancy, signal processing and analytical redundancy.

The hardware redundancy method is not applicable to the propulsion system due to the limitations of UAV in terms of their carrying capacity which drastically limits the number of subsystems that can be

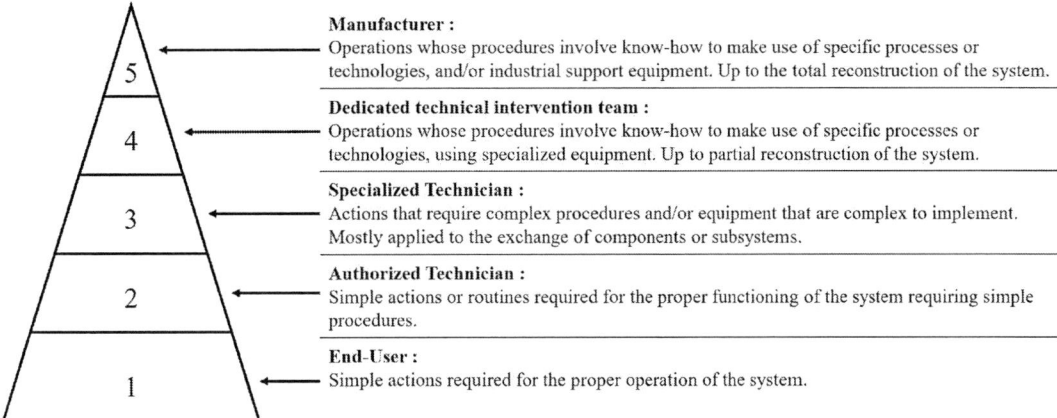

Fig. 1: Hierarchy of the five levels of maintenance according to [1]

installed on the same machine. Thus this method is mainly limited to the sensors that are critical to the attitude control of the machine for which it demonstrates a strong capacity to isolate faults on these components.

The signal processing method and the model-based analytical redundancy method are numerous as shown in [3], but are mainly restricted to a predefined system and limit the scalability of the considered FDI system. [4] and [5] propose the detection of faults on a PMSM by the real-time comparison between a simulated model and the measurements of a motor on a test bench. The presented procedure consists in the prior identification of the propulsion system parameters in order to obtain a healthy representation close to reality. The detection of faults is then carried out by comparing the measured and simulated signals to detect any anomaly on the system operation. For [5], the study focuses on the speed and the current flowing through the motor input. In [4] the measured currents from the battery and the stator phases of the motor, as well as their frequency in operation are studied. This frequency study of the currents is similar to [6] which studies the stator phase currents for the monitoring of open phase faults, using only current sensors for the detection. This work relies mainly on the use of flowcharts designed to identify a behavior that would not follow the expectation of a healthy system. Finally, [7] proposes the determination of faults related to the eccentricity of the motor and the propeller, based on the frequency analysis of the system. The signal is then compared to trend curves to determine if the system is balanced or not.

In this paper, we rely on the data-driven analytical redundancy method to maximize the fault detection and classification processing so that our solution is not restricted to a single specific system configuration. Several recent and relevant research works focus on the data-driven analytical method [8-11]. In [8], the authors use fuzzy logic to detect the state of the motor and its propeller by combining a frequency study of the stator phase currents for faults that may occur on the motor, and the accelerations at the end of the motor block for faults on the propeller. Authors of [9] proposes a method for identifying faults that may occur on the propellers of a multirotor UAV based on the frequency study of vibrations on the UAV chassis using the accelerometer sensors already used by the flight controller. In [10] an embedded deep-learning processing is used to detect faults that could occur to the whole UAV system and whose natures are very varied, since the behaviors to be identified also include the risks of cyber-attacks. The presented two-step logic works initially by detecting if there is a faulty behavior ongoing by using a CNN-BiLSTM model in an encoder-decoder logic for unsupervised learning, and then if a fault is detected to classify it. This processing uses several temporal data from the UAV's onboard sensors (IMU, accelerometers, gyroscope, etc.) to identify specific patterns and temporal dependencies that would demonstrate a fault impacting the UAV. Finally [11] presents the CNN-BiLSTM architecture model as a preferred detection method for the analysis of temporal signals from complex industrial machines.

We seek to ensure reliable detection and identification of faults that may occur on propulsion system by minimizing the sensors required to operate our fault detection in the face of cost and weight constraints on the UAV.

This paper is organized as follows: we will first present the generation of our dataset from a simplified multiphysics model to collect the faulty behaviors to be detected. Then, we will present the neural network architecture able to answer the problem.

Generation of the training dataset

We focus on the study of a Permanent Magnet Synchronous Motor (PMSM) equipped with a propeller for propulsion and its Electronic Speed Controller (ESC) driver, allowing us not to restrict this application to multirotor UAVs only. Nonetheless the case of multirotor UAVs allows a wide application setting, since we will find several instances of this independent system according to the multirotor configuration. The use of this type of motor equipped with a propeller can also be shared with other propulsion systems as the ones of fixed-wing UAVs, or even surface vehicles (USV) such as hovercrafts.

As fault data is scarce on this type of system and the cost of each components to allow the acquisition of this type of data is high, the use of real components to collect fault states, which will result in the destruction of the component, is not viable. In this perspective, we propose the use of a simplified simulated model of this propulsion system in which we can virtually inject faults of different natures and generate as many scenarios as necessary.

The model of the propulsion system is composed of the following components:
1. The battery for powering the components;
2. The Electronic Speed Controller (ESC) which controls the motor;
3. The Permanent Magnet Synchronous Motor (PMSM);
4. The propeller allowing the conversion of the rotation speed into propulsion force.

All these components are integrated into a multiphysics model using simplifying assumptions to simulate both the electrical and thermal behavior of each components. We use the observed performance of the system to generate a dataset of temporal signals for the fault-induced responses.

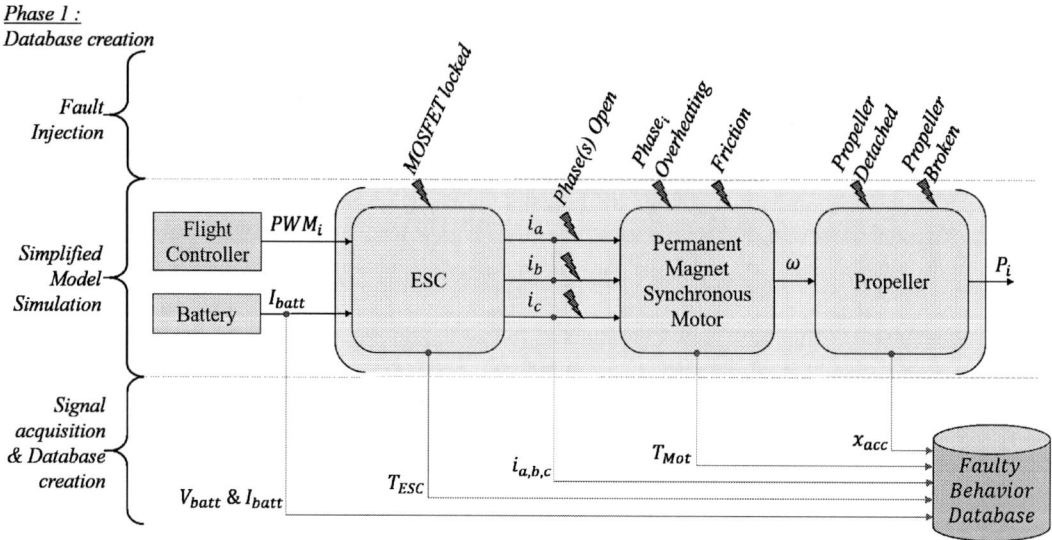

Fig. 2: Creation of the dataset of faulty behaviors from the simulated propulsion system

Fig. 2 shows the initial generation of our dataset of fault-prone behaviors. The authors of [4] and [5] present a list of faults that can occur on a propulsion system. We choose to inject the following faults one at a time in our model to create our dataset:
- On the Electronic Speed Controller:

– One of the MOSFETs cannot commutate;
- On the Permanent Magnet Synchronous Motor:
  – One of the stator phases is open;
  – Excessive overheating of one of the motor phases (causing the internal resistance of the phase to increase);
  – The rotor is subject to abnormal friction (e.g. damaged bearings or friction);
- On the propeller:
  – The propeller is detached (the motor is running idle);
  – The propeller has a break, inducing an eccentricity on the rotor.

On our simplified model, we simulate the initial causes of each fault considered and we observe all the signals selected on the system following the introduction of this fault. This allows us to make correlations between the injected faults and the system behavior. We also make the initial assumption that only one fault can occur at a time in our model.

In order to perform the detection and identification of each of these faults, we acquire and store in our dataset the following signals at different input throttle commands:
- The voltage and current drawn at the ESC input;
- The currents of each phase of the motor;
- The accelerations perceived at the rotor;
- The motor temperature;
- The temperature of the ESC, whose power dissipation comes mainly from the MOSFETs that allow the current flow in each phase of the stator (conduction losses due to their resistance, losses induced by the switching associated with the control frequency of the FET and a set of losses resulting from the recirculation of currents during switching [12]);

Due to the nature of the components and sensors assumed for signal collection, we also introduce signals that vary at random frequencies and amplitudes, in order to simulate spurious and background noise that will be superimposed on the signal.

We present in Table I the characteristics of one of the propulsion systems whose parameters we apply to our simulated model for the creation of our dataset.

Table I: Example of a combination of characteristics of components used in the propulsion system

Component	Characteristic	Value
ESC: T-Motor Alpha 60A	Input Voltage	25.2 V
	Peak Current	60 A
Motor: T-Motor U8II 190KV	Configuration	36N42P
	Rs	$48 \pm 3m\Omega$
	Ψ_m	$1.484e^{-3}Wb$
	Ipeak	43.7 A
	Ppeak	1048.8 W
Propeller: T-Motor MF2815	Diameter	$28.4" / 721.8e^{-3}m$
	Max Thrust / RPM	15 kg / 5000 RPM
	C_T / C_Q	2e-5 / 75e-8

Training and Fault Detection

Fig. 3 shows the neural network training and detection phase applied on a real system components whose characteristics were used as parameters in the simulation for the acquisition. The training is done entirely from the timeseries recorded in the dataset. For training our neural network, we reserve one third of our acquired dataset for validation while the rest will be used to train the model.

The input of our model consists of windows of 250 elements per the number of signals. Since faults are manually injected into our simulated model, we can also use the transient phases of our fault-prone

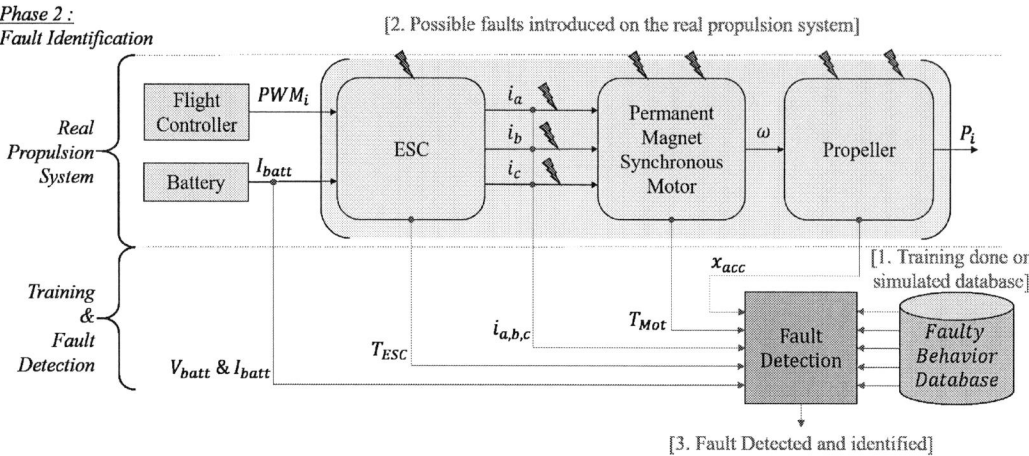

Fig. 3: Training of the fault detection system via the recorded dataset and identification of the faults occurring on a real system

system for training. In order to label each window for the classification training, we consider that a window of signals is showing a fault if one has been injected during more than 75% of the window acquisition.

Since the signals measured from our model are of different natures, we must adapt the data in order to standardize the inputs of our model. We normalize our data by window and not on the whole dataset. This is necessary, because some faults may cause some of the captured signals to diverge. Since our model aims to fit different components configurations, the data cannot be normalized between specific minimum/maximum values for every configurations. Thus, we normalize our sample data between the maximum values of each signal window.

$$X_{i\text{window}}^{*} = \frac{X_{i\text{window}}}{\max\left(|X_{i\text{window}}|\right)} \qquad (1)$$

In healthy operation, we expect to observe similar or repeatable patterns on the signals, while defective behavior should cause deviations identifiable by the layers of our neural network.

Neural Network Architecture

Fault detection relies on the use of a deep-learning multiheaded model based on one-dimensional CNN (Convolutional Neural Network) and LSTM (Long-Short Term Memory) layers as shown in Fig. 4.

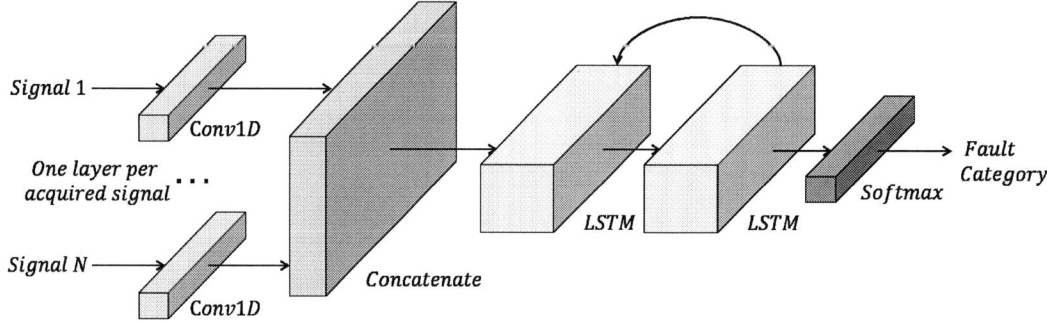

Fig. 4: Neural network architecture for fault detection and identification

The structure of our neural network based on a multiheaded model allows our model to receive input data of diverse nature from different sensors, where each head is able to extract the features of each signal independently. The input of our model can be considered as multivariate timeseries, therefore one-dimensional CNN layers are used as input layers at each one of the model heads. A one-dimensional Convolutional layer operates by drawing smaller sequences of data from long one-dimensional sequences ready to be interpreted by downstream layers.

After concatenation, the final Bidirectionnal-LSTM layers can determine the long-term dependencies of the model. Recurrent Neural Networks (RNN) are subject to the long-term problem of gradient dissipation during model training. This means that traditional RNNs cannot natively capture long-term dependencies for weight update. LSTMs were designed as to preserve both short and long-term network memories, to prevent backpropagation errors from disappearing or exploding. Capable of capturing dynamic timeseries states of systems, LSTMs have been successfully applied in various applications such as speech recognition, handwriting recognition and natural language processing.

The training is performed using the Adam optimizer and the categorical crossentropy as the loss function. Input windows are sequentially fed into our neural network using the sliding window approach, which is based upon using the previous system states to reinforce the temporal characteristics and dependencies of our input signals caught by the underlying layers.

Performance evaluation

The classification allows us to identify faults that may occur in the propulsion system. In Fig. 5, we present the confusion matrix reflecting the challenges our classifier faces in differentiating between healthy and faulty system states. Our neural network performs well on a balanced validation dataset which contains more than 1500 sample windows for each state.

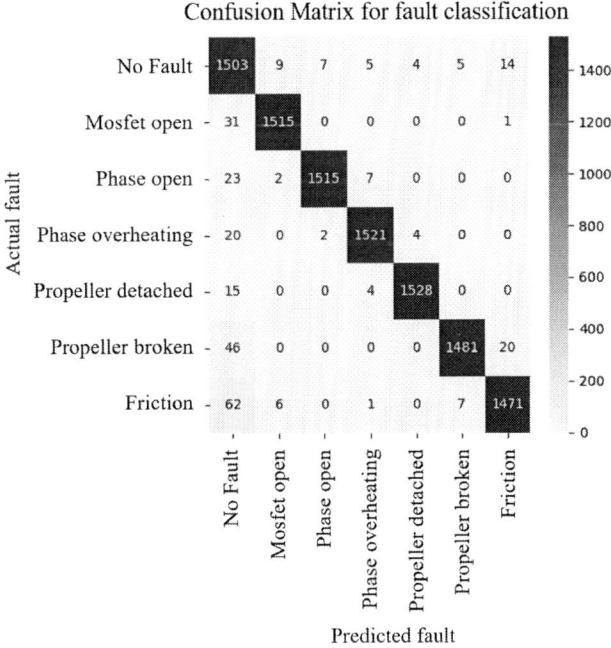

Fig. 5: Confusion matrix for the fault classifier on the validation dataset

We observe that the case where the model performs the worst corresponds to the identification of the friction fault on the motor as opposed to the healthy state. This is due to the fact that we have introduced different levels of possible friction on our simulated model, sometimes at low levels that are difficult for our neural network classifier to perceive. Furthermore, the relative confusion between the healthy and faulty states by our classifier is mainly explained by the addition of the transient phases of the faulty system for the training.

Conclusion

We present a method for detecting and identifying faults in the components of a UAV propulsion system. The data acquired from a multiphysics model using simplifying assumptions allows us to model the electrical and thermal behavior of the system when it is affected by a fault. This dataset generation method allows us to generate a balanced set of samples for training a deep learning neural network classifier to classify the different degraded behaviors of the system.

The multiheaded CNN-BiLSTM architecture allows us to perform a robust classification of the faulty behaviors from a reduced set of signals and sensors on the system, minimizing the cost and the load constraint on an actual machine.

We plan to apply this method on real components on a test bench in order to validate the detection of real faults introduced on the propulsion system.

References

[1] "NF X60-000," *Afnor EDITIONS.* https://www.boutique.afnor.org/fr-fr/norme/nf-x60000/maintenance-industrielle-fonction-maintenance/fa063074/1561 (accessed Mar. 21, 2022).

[2] G. K. Fourlas and G. C. Karras, "A Survey on Fault Diagnosis and Fault-Tolerant Control Methods for Unmanned Aerial Vehicles," *Machines*, vol. 9, no. 9, p. 197, Sep. 2021, doi: 10.3390/machines9090197.

[3] S. Nandi, H. A. Toliyat, and X. Li, "Condition Monitoring and Fault Diagnosis of Electrical Motors—A Review," *IEEE Trans. Energy Convers.*, vol. 20, no. 4, pp. 719–729, Dec. 2005, doi: 10.1109/TEC.2005.847955.

[4] G. Jouhet, L. E. González-Jiménez, M. A. Meza-Aguilar, W. A. Mayorga-Macías, and L. F. Luque-Vega, "Model-Based Fault Detection of Permanent Magnet Synchronous Motors of Drones Using Current Sensors," in *New Trends in Robot Control*, J. Ghommam, N. Derbel, and Q. Zhu, Eds. Singapore: Springer, 2020, pp. 301–318. doi: 10.1007/978-981-15-1819-5_15.

[5] J. Lee, W. Lee, S. Ko, and H. Oh, "Fault Classification and Diagnosis of UAV motor Based on Estimated Nonlinear Parameter of Steady-State Model," *IJMERR*, pp. 22–31, 2020, doi: 10.18178/ijmerr.10.1.22-31.

[6] A. Khlaief, M. Boussak, and M. Gossa, "Open phase faults detection in PMSM drives based on current signature analysis," in *The XIX International Conference on Electrical Machines - ICEM 2010*, Rome, Italy, Sep. 2010, pp. 1–6. doi: 10.1109/ICELMACH.2010.5607977.

[7] F. C. Veras, T. L. V. Lima, J. S. Souza, J. G. G. S. Ramos, A. C. Lima Filho, and A. V. Brito, "Eccentricity Failure Detection of Brushless DC Motors From Sound Signals Based on Density of Maxima," *IEEE Access Pract. Innov. Open Solut.*, vol. 7, pp. 150318–150326, 2019, doi: 10.1109/ACCESS.2019.2946502.

[8] F. Pourpanah, B. Zhang, R. Ma, and Q. Hao, "Anomaly Detection and Condition Monitoring of UAV Motors and Propellers," in *2018 IEEE SENSORS*, Oct. 2018, pp. 1–4. doi: 10.1109/ICSENS.2018.8589572.

[9] X. Zhang, Z. Zhao, Z. Wang, and X. Wang, "Fault Detection and Identification Method for Quadcopter Based on Airframe Vibration Signals," *Sensors*, vol. 21, no. 2, p. 581, Jan. 2021, doi: 10.3390/s21020581.

[10] V. Sadhu, S. Zonouz, and D. Pompili, "On-board Deep-Learning-Based Unmanned Aerial Vehicle Fault Cause Detection and Identification," May 06, 2020. http://arxiv.org/abs/2005.00336 (accessed Apr. 12, 2021).

[11] M. Canizo, I. Triguero, A. Conde, and E. Onieva, "Multi-head CNN–RNN for multi-time series anomaly detection: An industrial case study," *Neurocomputing*, vol. 363, pp. 246–260, Oct. 2019, doi: 10.1016/j.neucom.2019.07.034.

[12] A. Gopalan and A. Lawrence, "Calculating Power Dissipation for a H-Bridge or Half Bridge Driver." 2012.

A Compact Solid State Transformer for Replacing Conventional Medium Power Transformer in Weight-Critical Applications

Leon Fauth, Felix Willer, and Jens Friebe
Institute for Drive Systems and Power Electronics - Leibniz University Hannover
Welfengarten 1
Hannover, Germany
Phone: +49 (0) 511-762-14571
Email: Leon.Fauth@ial.uni-hannover.de
URL: https://www.ial.uni-hannover.de/en/

Keywords

≪Solid-State Transformer≫, ≪Resonant converter≫, ≪Gallium Nitride (GaN)≫

Abstract

In certain weight-sensitive applications requiring galvanic isolation, conventional transformers can be of disadvantage due to their high weight and volume. Solid state transformers allow to overcome this issue. The volume can be further reduced by eliminating large dc storage capacitors, resulting in a straight AC/AC converter system based on a resonance converter.

Introduction

The operation of electrical equipment often requires galvanic isolation to the grid voltage for safety reasons, like in industrial grids or aircraft and marine applications. Conventionally, transformers are used, because of their well-proven design, robustness, and simplicity. However, in specific applications, tight requirements on volume, weight, or other parameters can exist, like for example in aircraft applications. Because of the usual grid frequency of 50 or 60 Hz, conventional transformers need a large active core cross sectional area and a high number of turns to avoid saturation. This leads to an overall high volume and weight. Also, conventional transformers allow for no controllability or intelligence. As an alternative, so-called solid state transformers (SST) can fulfil the same task. A SST will consist of a rectification block, a resonance converter, and an output inverter. If the operation frequency of the resonance tank is chosen sufficiently high, the volume of the magnetic components can be drastically reduced [1]. For a well-executed design, the system efficiency can be higher compared to conventional transformers. Also, the application of power electronics allows for a controllability during operation, for example to stabilize the output voltage amplitude. As design challenges, the higher component count and complexity has to be stated. If these issues can be solved during the design process, solid state transformers can be a good alternative for special applications. This paper shows the design of an ultra-compact SST. Also, the possibility to eliminate large DC storage capacitors is highlighted and the resulting waveforms and consequences on the control are shown [2]. The SST is designed as a direct replacement for a currently available off-the-shelf transformer. The reference off-the-shelf design has an overall volume of 150 x 143 x 113 mm^3 = 2.42 L and a weight of 7.9 kg. The rated power is 630 V A, the maximum efficiency is given as 93%. Isolation between primary and secondary side is rated as 5 kV. It allows the transformation of various input voltage levels to a fixed output voltage by different winding taps. As an extension of the current state of research, the SST in this paper is designed for input voltages up to $U_{in} = 400$ V, which results from a direct connection between two lines in a three-phase system, if no neutral line is available. Regardless of the input voltage, it will be converted to a stable output voltage of 230 V by adapting the

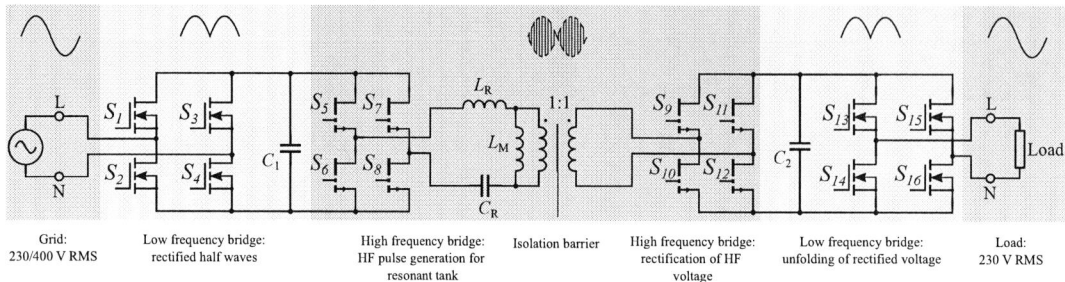

Fig. 1: Circuit diagram of the proposed system. All power electronics stages are used symmetrically on the primary and secondary side.

switching frequency of the resonance converter. For this operation, the resonance tank needs to be designed to allow for a sufficient minimum power level. Also, the compact realization of the hardware is highlighted with an integrated resonance inductor and transformer for the resonance converter.

Description of Topology

The chosen topology for the replacement solution is shown in Fig. 1. It consists of a fullbridge for input voltage rectification, the resonance converter, an output rectification and a fullbridge for unfolding the rectified voltage to a sine wave. The input fullbridge will rectify the grid voltage. To allow for reactive power capability, the switches S_1 to S_4 need to be active switches, for active power conversion, diodes can be used, but will typically result in higher losses. As the switching frequency is the same as the grid frequency, switching losses can be neglected, which allows the selection of power semiconductors with low conduction losses at lower cost. In this example, super junction MOSFETs optimized for a low channel resistance are selected. As stated before, for a pure AC to AC application, the capacitor C_1 can have a very low capacitance. If for example $C_1 = 1\,\mu\text{F}$ is chosen, the capacitor voltage consists of sinusoidal half waves following the grid voltage. The capacitor is only used as a high frequency buffer for the following resonant converter. The resonant stage is designed as a series resonance converter (SRC) or a LLC-converter depending on the ratio of leakage and magnetizing inductance. A fullbride (switches S_5 to S_8) is operated at a high switching frequency with 50% duty cycle. By that, a pulsed voltage with a sinusoidal envelope is generated over the resonance tank. The resonance frequency will be in the range of several hundreds of kilohertz, to allow for small magnetic components. As the resonance converter shall be operated at or above resonance to allow soft on-switching of the power semiconductors, gallium nitride switches with low switching energy are chosen. The transformer has to be designed for the high frequency components. Also, the required isolation level needs to be provided. In this case, 5 kV are selected based on the reference design. The high-frequency fullbridge consisting of the switches S_9 to S_{12} is used to rectify the output voltage of the resonant converter. A low capacitance value is used for C_2. The resulting voltage is again consisting of sinusoidal half waves. Similar to the input fullbridge, diodes can be used if only active power needs to be transferred, but will result in higher conduction losses. For reactive power conversion, active switches are used, which need to be synchronized to the switching of the resonance converter fullbridge. Finally, the sinusoidal output voltage is generated by unfolding by the switches S_{13} to S_{16}. In summary, this leads to a fully symmetrical layout as shown in Fig. 1. Furthermore, under ideal conditions the switching signals from the primary side can also be used for the high- and low-frequency fullbridge on the secondary side. During the experimental tests it needs to be verified that even under real conditions, like non-ideal behaviour of gate drivers and propagation delay, a direct use of the primary side switching signals is possible.

Design of the Resonance Tank

The design strategy for the AC to AC resonance converter is shown in Fig. 2 and will be explained in the following. The design of the resonance tank, especially the magnetic components, holds several challenges. While designing the resonance elements L_R and C_R, the operation point for reducing the

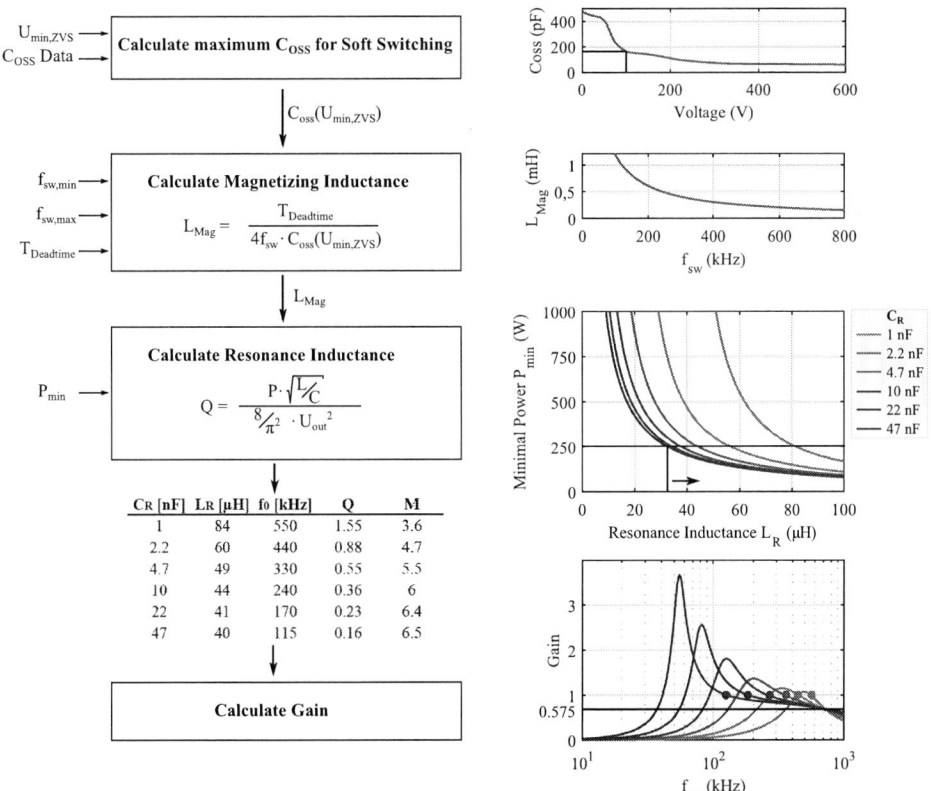

CR [nF]	LR [µH]	f0 [kHz]	Q	M
1	84	550	1.55	3.6
2.2	60	440	0.88	4.7
4.7	49	330	0.55	5.5
10	44	240	0.36	6
22	41	170	0.23	6.4
47	40	115	0.16	6.5

Fig. 2: Calculation of the resonance tank elements. From top to bottom: 1) Fixing of voltage threshold for soft switching for sinusoidal input voltages. 2) Dependency of maximum magnetizing inductance on the switching frequency. 3) Selection of resonance tank elements based on a minimal power for a gain of 0.575. 4) Resulting gain curves for different feasible combinations.

input voltage of $U_{in} = 400\,\mathrm{V}$ to an output voltage of $U_{out} = 230\,\mathrm{V}$ can mean restrictions on the minimum power. Therefore, careful consideration of chosen resonance frequency, inductance and capacitance value is necessary. Other challenges are due to the sinusoidal input voltage of the resonance converter, which influences the design of the magnetizing inductance of the transformer, affecting the operation area for soft on-switching.

For the conventional DC to DC operation of a resonant converter, the magnetizing inductance L_M can be designed to guarantee soft on-switching of the power semiconductors. For the here proposed AC to AC structure, certain trade-offs need to be made, as the input voltage of the resonance converter and therefore the blocking voltage of the power semiconductors are time-dependent. For low voltages, a very small magnetizing inductance value would be necessary to provide a sufficiently high magnetizing current for soft on-switching. This could result in a unreasonable high transformer core cross sectional area. On the other hand, switching losses due to hard switching would be comparably low, as the blocking voltage is still small. As a compromise, L_M can be chosen large enough to allow soft on-switching starting from a previously determined voltage threshold. This can be done based on the function of the output capacitance C_{oss} over the voltage as shown in Fig. 2, top image. It can be noticed that the output capacitance is increased strongly for voltages less then $100\,\mathrm{V}$. As a result, in this design the magnetizing inductance is designed to allow soft switching only when in the input voltage is larger than $100\,\mathrm{V}$, which occurs for 80% of the time for a sinusoidal input voltage with a amplitude of $325\,\mathrm{V}$. From that, the highest allowable magnetizing inductance can be calculated, if a certain maximum switching frequency is fixed (see Fig. 2, second image from top). The overall peak switching frequency for this design is set to $f_{sw,max} = 800\,\mathrm{kHz}$. This is the highest switching frequency which can be used for attenuation of a higher input voltage to the desired output voltage of 230 V. It has to be stated that the design of the magnetizing inductance is only relevant for the operation point with unity gain, because the resonant converter will then operate at or slightly above resonance frequency. For the operation at higher input voltages, the resonance tank will behave inductive allowing for soft switching either way [3]. This means that the critical peak frequency for determining the magnetizing inductance will be lower than the overall peak frequency of $f_{\mathrm{sw,max}} = 800\,\mathrm{kHz}$ and needs to be calculated by iterating the design loop shown in Fig. 2.

To allow connection to different grid configurations as described before, the resonance converter has to be able to provide a gain of close to one for an input voltage of $U_{\mathrm{in}} = 230\,\mathrm{V}$ and also a gain of close to $A = 0.575$ for an input voltage of $U_{\mathrm{in}} = 400\,\mathrm{V}$. Both operation points have to be available without reconfiguration of the resonance elements, meaning by only adjusting the switching frequency. For this task, generally a design with a high quality factor Q is beneficial, because only smaller frequency variations are required for large changes in the attenuation. Also, it needs to be considered that the quality factor is linearly dependent on the output power. As for the applications aimed for in this paper the load can vary widely, this also can mean an influence on the resonance tank design. As a result, the resonance tank elements L_{R} and C_{R} need to be chosen carefully, as shown in Fig. 2, third image from top. For an exemplary minimum power of $P_{\mathrm{min}} = 250\,\mathrm{W}$, the feasible combinations of L_{R} and C_{R} as well as the resulting parameters are shown in the table in Fig. 2. The resulting gain plots are shown below. It can be seen that high values of C_{R} will lead to a lower minimal power requirement and have the possibility of designing a low L_{R}. On the other hand, low values of C_{R} can also lead to a sufficiently low minimal power, but will result in overall higher switching frequencies for the operation at unity gain, increasing the losses.

For typical applications, the resonance tank will mostly operate in the area of unity gain. To ensure zero voltage switching for all load conditions, a low magnetizing inductance is necessary. This results in a lower ratio between leakage and magnetizing inductance, consequently, the resonance converter is designed as a LLC converter. This can also lead to a smaller distance between the switching frequency for unity gain and attenuation by $A = 0.575$, allowing for a tighter design in terms of switching losses.

Hardware Prototype

The potential in volume reduction is presented in Fig. 3, were a conventional transformer, a SST with DC storage and the approach shown in this paper are compared. The here proposed solution of a SST without DC storage shows the overall lowest volume which is only 17% of the conventional transformer. The weight is 760 g, which is less than 10 % of the reference design. This shows the suitability for weight-sensitive applications like aviation.

The prototype is shown in Fig. 4. All components are mounted on a u-profile heatsink, allowing for a optimal utilization of available space. The PCBs holding the power semiconductors are mounted on the outside of the heatsink arrangement. Additionally, the high frequency stages are mounted on a metal substrate circuit board, which reduces the thermal resistance. This allows the usage of bottom cooled devices, which greatly simplifies the construction. The magnetic components can be mounted in the inside of the heatsink arrangement. The resonance capacitors are directly mounted at the PCB close to the switching node of the half bridges. By minimizing the area, a large parasitic capacitance of the resonance tank against earth can be avoided. This helps in reducing the losses.

As a LLC design is preferred as mentioned before, both the resonance inductor and the transformer can be build up as a single component. The typical ratio of leakage and magnetizing inductance for a LLC lead to a coupling factor of $k = 0.67$ to 0.8. Such low values can not be reached with conventional winding arrangements, were both windings are wound on the middle core of the ferrite core, even if primary and secondary windings are separated by a special two-chamber coilformer. Instead, a design as shown in Fig. 5, left, is chosen. The primary and secondary windings are located on the outer legs of the core. The middle core is used to control the leakage inductance by adjusting the middle air gap $d_{\mathrm{AG,leak}}$. The air gap d_{AG} in the outer legs of the core influences both the leakage and the coupling inductance and is used to realize the desired value of the magnetizing inductance. Consequently, the leakage inductance can

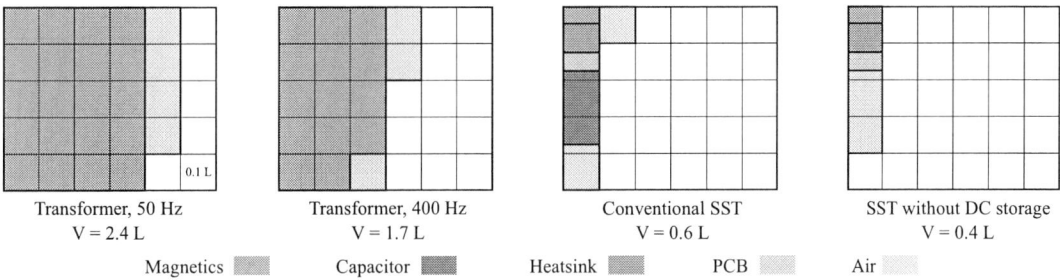

Fig. 3: Volume comparison of conventional transformer for 50 and 400 Hz applications, conventional SST design and SST design without DC storage.

Fig. 4: Design of the SST. All PCBs are mounted on a u-shape heatsink, the magnetic elements are mounted inside. Maximum dimensions are 100 x 58 x 78 mm. Overall weight is 760 g.

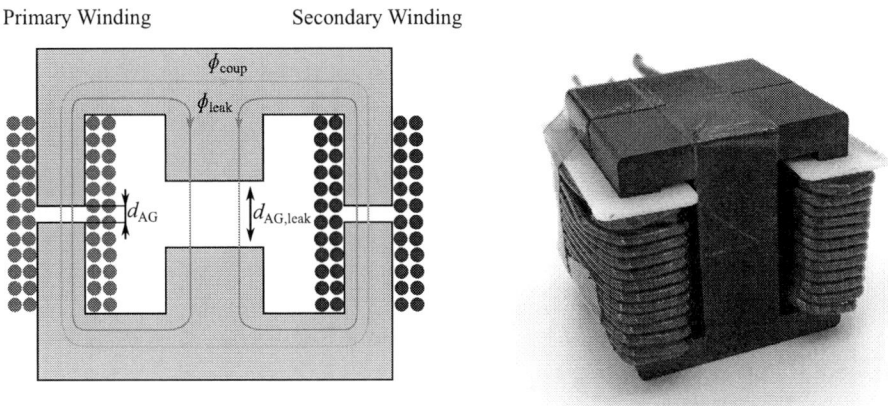

Fig. 5: Design of the integrated resonance inductor and transformer (left), realisation of the design (right).

be designed largely independently from the coupling inductance. Because of that, discrete components of resonance inductor and transformer are not necessary, reducing the overall volume. To accommodate for the reduced core cross sectional area of the outer legs of the ferrite core, two E-cores are stacked in parallel. The resulting dimensions are 43 x 42 x 42 mm^3. As both windings are fully separated from each other, the required isolation can be realised. The prototype is shown in Fig. 5, right. The coupling factor was realized as $k = 0.67$.

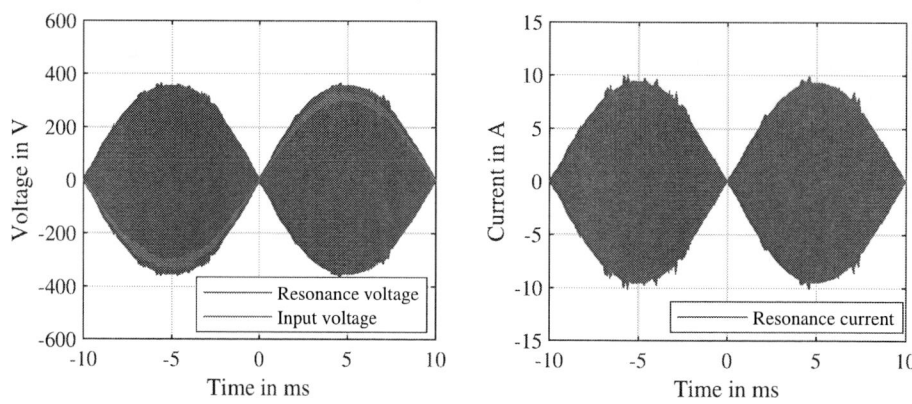

Fig. 6: Waveforms of input voltage and resonance voltage (left) and resonance current (right). Both resonance voltage and current are following the sinusoidal envelope of the rectified input voltage.

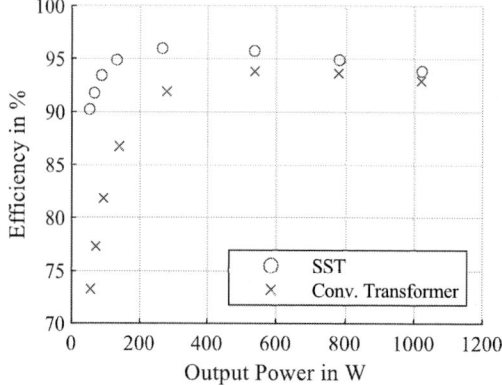

Fig. 7: Comparison of the efficiency of the Solid State Transformer and the reference conventional transformer. The peak efficiency of the SST is 96 %, the efficiency at maximum power is 94 %. The maximum semiconductor temperature at full load was 69.6 °C at the resonance converter fullbridge. The ambient temperature was 26 °C.

To simplify the control, off-the-shelf ICs and analogue control blocks are used. The low frequency fullbridges (S_1 to S_4 and S_{13} to S_{16}) are controlled by a zero-voltage-detection IC, which measures the input voltage. The signal is transferred to the secondary side with a digital isolator. The resonance converter is controlled by an integrated LLC controller. As a feedback to control the output voltage, the voltage at C_2 is rectified and measured. This signal is fed back to the controller by a standard optocoupler circuit. Because no microcontroller is needed, the overall design and PCB layout is simplified.

Measurement Results

For verification of the design, several measurements are conducted. In Fig. 6, the input voltage as well as resonance tank voltage and current are shown. As described before, the input voltage is rectified to sine half waves. This is visible as a sine wave envelope in the resonance voltage and current. The waveforms of input and output voltage and the resulting frequency spectrum obtained by oscilloscope measurement are shown in Fig. 8. As expected, clearly visible peaks at multiples of the switching frequency of the resonance converter are generated. Due to the bipolar modulation, with two half bridges switching at opposing times, the only common mode noise is generated due to slight mismatches of the half bridges at switching [5]. Additionally, the common mode noise is directly coupled back to the noise source via y-capacitors placed close to the switching cells. As a result, only differential mode disturbance contributes

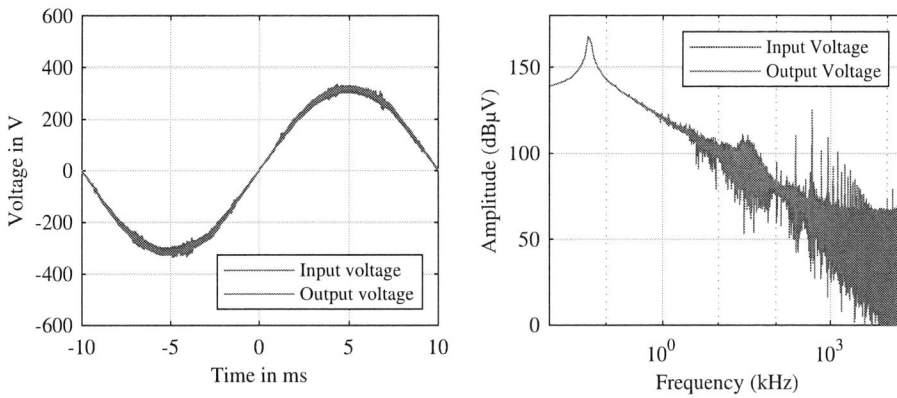

Fig. 8: Waveforms of input and output voltage for operation at unity gain (left) and corresponding frequency spectra (right) obtained by FFT of the oscilloscope measurement data (peak measurement).

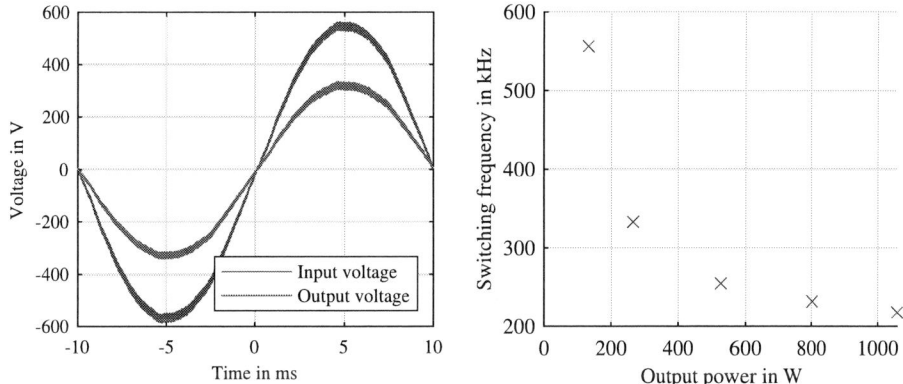

Fig. 9: Stepdown operation from $U_{in} = 400\,V$ to $U_{out} = 230\,V$. Resulting waveforms for input and output voltages (left) and switching frequency over output power (right). The stepdown operation is possible down to a minimum power of 132 W, which requires a switching frequency of 556 kHz.

to the noise at the input and output of the SST, which can be reduced with little effort by adding small x-capacitors. For the practical application, additional EMI measurements are necessary.

The prototype was operated up to a power of 1 kW. The efficiency in dependency of the output power is shown in Fig. 7. For comparison, also the efficiency of the reference conventional transformer is shown. For all operation points, the efficiency of the SST is higher, with a peak efficiency of 96%. Especially at low power, the losses of the SST are drastically lower, making the application beneficial for applications with varying load and long phases of partial load operation.

Another benefit is the possibility of controlling the output voltage by adapting the gain of the resonance tank. This can be used to stabilize the output voltage during voltage fluctuations of the grid. Also it is possible to provide an output voltage of 230 V for the connection of the SST directly between two phases of the grid, if no neutral point is available. This is shown in Fig. 9, left plot. As stated before, based on the properties of the resonance tank, a certain minimal load is required. With the current design, the step down operation is possible starting from $P_{out} = 132\,W$ up to 1 kW, see Fig. 9, right plot. For all loads, switching frequency was below the maximum limit of $f_{sw,max} = 800\,kHz$.

Conclusion

Solid state transformer can be a valid alternative to conventional transformers in certain applications. They are beneficial in terms of volume, weight, efficiency and controllability. If the DC storage is strongly reduced, the input voltage of the resonant converter is comprised of sinusoidal half waves, which can be converted to the secondary side. By that, a pure AC-to-AC converter is gained. While this allows for the reduction of the overall volume, certain points during the design process need to be carefully evaluated. Based on this, a prototype SST with a LLC converter is designed. The magnetic elements are realized as a single component, further reducing the volume. This is achieved by using the outer legs of an E-core for the coupling and modulating the leakage inductance by adapting the air gap in the centre leg. The design is then proven by measurements. An overall peak efficiency of 96% can be realized, which is higher than the efficiency of the conventional design while only having 10% of the weight. Especially at partial load operation, the efficiency is drastically higher compared to the conventional reference transformer. Also, a gain reduction to allow for different operation points was shown. For future research, the proposed solution has to be examined in greater detail in terms of electromagnetic interference. Also, the operation at reactive load needs to be covered in detail.

References

[1] M. D. Seeman, S. R. Bahl, D. I. Anderson, and G. A. Shah, "Advantages of GaN in a high-voltage resonant LLC converter," in 2014 IEEE Applied Power Electronics Conference and Exposition - APEC 2014, Fort Worth, TX, USA, Mar. 2014 - Mar. 2014, pp. 476–483.

[2] A. N. Rahman, S.-K. Chen, and H.-J. Chiu, "Single Phase AC-AC Solid State Transformer based on Single Conversion Stage," in 2019 IEEE Workshop on Wide Bandgap Power Devices and Applications in Asia (WiPDA Asia), Taipei, Taiwan, May. 2019 - May. 2019, pp. 1–5.

[3] H. Wen, J. Gong, C.-S. Yeh, Y. Han, and J. Lai, "An Investigation on Fully Zero-Voltage-Switching Condition for High-Frequency GaN Based LLC Converter in Solid-State-Transformer Application," in 2019 IEEE Applied Power Electronics Conference and Exposition (APEC), Anaheim, CA, USA, Mar. 2019 - Mar. 2019, pp. 797–801.

[4] K. Tan, R. Yu, S. Guo, and A. Q. Huang, "Optimal design methodology of bidirectional LLC resonant DC/DC converter for solid state transformer application," in IECON 2014 - 40th Annual Conference of the IEEE Industrial Electronics Society, Dallas, TX, USA, Oct. 2014 - Nov. 2014, pp. 1657–1664.

[5] M. H. Hedayati and V. John, "Filter Configuration and PWM Method For Single-Phase Inverters With Reduced Conducted EMI Noise," IEEE Trans. on Ind. Applicat., vol. 51, no. 4, pp. 3236–3243, 2015, doi: 10.1109/TIA.2014.2387483.

Comparative Study of Single-phase and Three-phase DAB for EV Charging Application

Nicola Blasuttigh[1], Hamzeh Beiranvand[2,3], Thiago Pereira[2], Marco Liserre[2,3]

[1]Department of Engineering and Architecture, University of Trieste, Trieste, Italy
[2]Chair of Power Electronics, Kiel University, Kiel, Germany
[3]Kiel Nano, Surface and Interface Science KiNSIS, Kiel University, Kiel, Germany
Email: nicola.blasuttigh@phd.units.it

Acknowledgment

Funded by the European Union - European Regional Development Fund (EFRE), the German Federal Government and the State of Schleswig-Holstein.

Keywords

≪Dual Active Bridge (DAB)≫, ≪Three-phase system≫, ≪Analytical losses computation≫ ≪Power converters for EV≫, ≪Battery≫.

Abstract

Bidirectional converters enable vehicle-to-grid (V2G) operations in electric vehicle (EV) charging stations. In this context, dual-active bridge (DAB) DC-DC converter is a preferable solution due to galvanic isolation and reduced volumes compared to other systems. Single-phase DAB (1ph-DAB) and three-phase DAB (3ph-DAB) topologies are usually compared in terms of efficiency and performances with the same rated power. Conversely, this paper focus on a comparison concerning device losses and stresses, medium-frequency transformer (MFT) design and capacitor filter sizing for the same power per switch, considering DABs and batteries coupled in a V2G application. Thereby, the impact of the battery state-of-charge (SoC) variation relative to the grid-side DC voltage is studied. Theoretical analysis and simulations results reveal that, in some respects, 1ph-DAB performance is superior to that of the 3ph-DAB with the proposed comparison approach. While, the main advantage of 3ph-DAB over 1ph-DAB is the reduced size of filter capacitors.

Introduction

Gas emissions restrictions and increasing air pollution suggest that other transportation technologies should be exploited leading to the future loss of oil's role [1, 2]. Nevertheless, conventional vehicles account for most of the global mobility fleet although electrification of transportation is leading to change future outlooks. EVs integration can bring several advantages related to energy sustainability such as

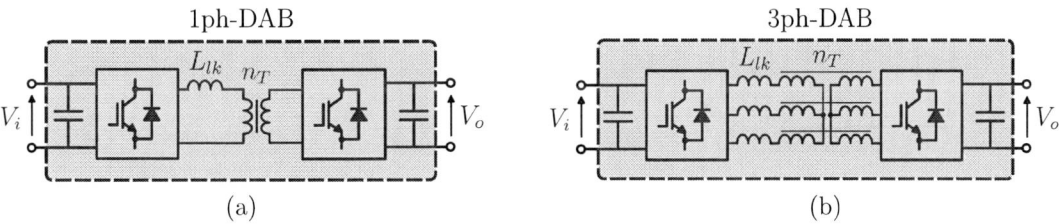

Fig. 1: Schemes for (a) 1ph-DAB and (b) 3ph-DAB

decarbonization and renewable energy integration. A relevant aspect is related to charging stations distribution and charging time, which greatly influences the EV penetration. When EVs are connected to the grid, battery charging has always been the main target. However, new opportunities are investigated for grid support, called V2G technologies [3]. In this context, it is necessary to design highly efficient systems that provide high power levels as powerful DC solutions [4–6].

DAB converter is one of the most popular converter for bidirectional power flow functionality given its high-power density, high efficiency and galvanic isolation [7, 8]. Two topologies are mainly used in the development of a bidirectional DC-DC bridge converter: single-phase and three-phase DABs. However, other several topologies and multi-phase structures are investigated with particular interest [9, 10].

In the existing literature, 1ph-DAB and 3ph-DAB performances are usually unfairly compared from the system utilisation perspectives (i.e. same rated power), leading to the 3ph-DAB outperforming the 1ph-DAB in terms of lower device losses, lower current harmonic content and capacitor size [11, 12]. Nevertheless, a more proper and comprehensive study is shown in [13] where the two topologies are compared with the same silicon area for low-voltage high-current applications. However, the paper focuses on an on-board charger that connects the LV battery to the internal HV-DC bus while the proposed study aims to analyse an off-board charging station solution for direct charging of the existing higher voltage batteries considering V2G applications. Furthermore, the cited efficiency analysis does not accurately consider the MFT losses nor its design. Analytical calculations on filter capacitor sizing are not shown and the gate driver losses are not included in the study. For this reason, considering the higher capability of 3ph-DAB compared to 1ph-DAB, the proposed comparison approach involves analysing the two topologies for a nominal power level related to the actual difference in the switching legs of the converters. Doing this, the 3ph-DAB is not favored. The study includes devices and gate driver losses, MFT design and capacitor sizing where all these aspects are considered with different battery voltages in order to estimate DAB's losses as a function of different EV state-of-charge. This, supported by the ever-increasing studies on DAB and batteries integration as well as the impact of output voltage variations on converter losses [14].

Single-Phase (1ph) and Three-Phase (3ph) DAB Comparison

The detailed structures for 1ph-DAB and 3ph-DAB are shown in Fig.1. Both converters consist of two back-to-back bridges connected together by a transformer and a series inductance. Its operation is based on the exchange of active and reactive power between two alternating voltage sources, which is regulated through the relative phase shift between them. As it is well known, as the active power is mainly defined by phase shift, the reactive power is exchanged in relation to voltage source amplitudes. This means that, especially under unbalanced voltage conditions such as EV batteries charging and discharging process, current stresses and losses must be carefully assessed. Several modulation techniques with different

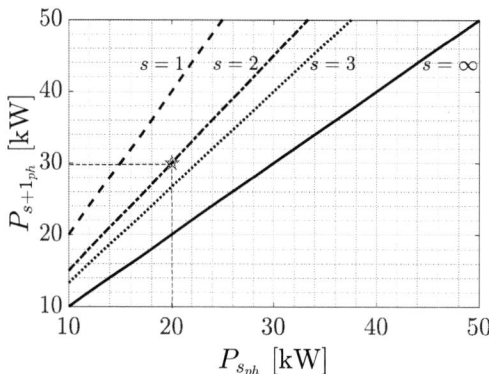

Fig. 2: Characteristic of (2) for different number of phases s which provides the power of the $s+1$-phases converter ($P_{s+1_{ph}}$) in order to have the same power per switch of the s-phase converter with power $P_{s_{ph}}$.

Table I: DABs parameter specifications

Topology	P_N [kW]	V_i [V]	V_o [V]	n_T [-]	L_{lk} [μH]	φ_{max} [rad]	f_{sw} [kHz]
1ph-DAB	20	800	600-800	1:1	61.1	$\pi/12$	20
3ph-DAB	30	800	600-800	1:1	27.7	$\pi/12$	20

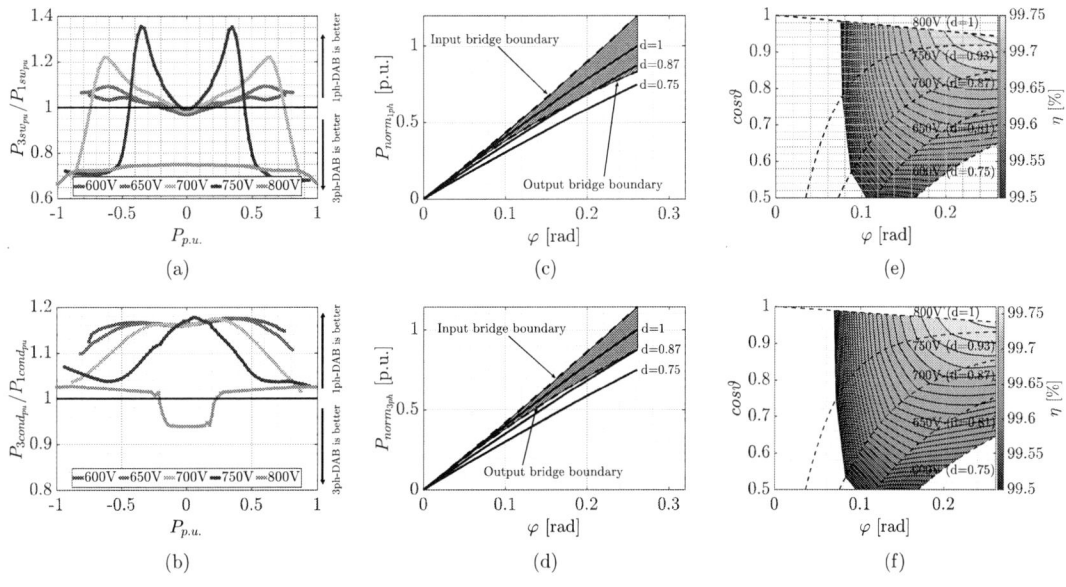

Fig. 3: Different comparison analysis between the two topologies in terms of device switching (a) and conduction (b) losses, ZVS range (c,d) and MFT losses (e,f) for 1ph-DAB and 3ph-DAB, respectively.

purposes are developed in literature although only single phase-shift (SPS) modulation is considered in this paper. More details on DAB operations can be found in [7] for both topologies.

Since the aim is to compare the converters for the same power per switch, the design rated power must be chosen with appropriate values. More in general, given an s-phase converter with a power $P_{s_{ph}}$ and a d-phase converter with $d > s$, the latter is able to provide a power $P_{d_{ph}}$, as described in (1):

$$P_{d_{ph}} = \prod_{M=s}^{d-1} \frac{M+1}{M} P_{s_{ph}}, \quad \forall \quad 1 \leq s < d \tag{1}$$

where s and d are the number of phases of the first and the second converter, respectively. A particular form of (1) can be defined for consecutive number of phases ($d = s + 1$) as shown in (2):

$$P_{s+1_{ph}} = \left(\frac{s+1}{s} \right) \cdot P_{s_{ph}} \tag{2}$$

Figure 2 shows the characteristic of (2) for different number of phases s which, given the power of the s-phases converter $P_{s_{ph}}$, defines the the power of the $s + 1$-phases converter $P_{s+1_{ph}}$ in order to have the same power per switch. The two nominal powers of the proposed DABs are highlighted by the red mark and the converters specifications are listed in Table I.

Specifically, the comparison approach focuses only on the DC-DC converters installed and operated in an existing DC grid. Thus, no assessment is made on AC grid-side rectifiers. Also, regarding a possible voltage and current imbalances in the 3ph-DAB transformer, it is assumed that the design is carefully performed to avoid them.

Power Losses on the Semiconductor Devices

Conduction and switching losses are the main losses for a switching device. These losses are estimated through simulations by varying the normalised power from -1 to 1 [p.u.], with different battery voltage to evaluate the impact of the SoC. Figures 3 (a) and (b) show the ratio between 3ph-DAB and 1ph-DAB switching and conduction losses in p.u., respectively, with different normalised power and output voltages. Each switching device consists of a SiC-MOSFET and its anti-parallel body-diode, both described by the manufacturer thermal model. The loss increment in 3ph-DAB is noticeable in both cases for most conditions. More specifically, 3ph-DAB switching losses are up to 35% higher for specific power conditions. Conversely, for all power conditions where no voltage mismatch occur switching losses are up to 30% lower with respect to 1ph-DAB. Also regarding conduction losses, 3ph-DAB performs worse causing losses to increase up to 18%. These considerations lead the 1ph-DAB to be more suitable in case of variable voltage applications such as V2G. Although gate driver losses are usually omitted due to their feeble contribution, in this work they are taken into account based on the model shown in [15]. Gate driver losses contribution has been also taken into account, resulting in 50% more losses in 3ph-DAB than in 1ph-DAB due to the larger number of devices.

ZVS Range Comparison

The total efficiency of DAB is highly dependent on the conditions of use. Particularly, the switching losses can worsen during hard-switching. Input and output ZVS boundary conditions can be found by specifying particular constraints to the inductance current as shown in [7] and [16] for 1ph-DAB and 3ph-DAB, respectively. Zero-voltage switching (ZVS) range is compared between the two systems to evaluate the operating spectrum. Figures 3 (c) and (d) show the normalized power relative to the rated power (solid lines) and the ZVS operating range (dashed lines) for the 1ph-DAB and 3ph-DAB, respectively, where $d = n_T V_o / V_i$ is the dc conversion ratio and φ the phase-shift angle. As shown, the ZVS operating range for 1ph-DAB is wider [13], allowing it to enter soft-switching mode earlier than in the three-phase case (see $d = 0.87$ line as reference example).

MFT Design and Losses Evaluation

MFTs were designed based on DABs maximum voltage mismatch and rated power. The purpose of the transformer design is to evaluate its losses at all operating points of the DAB in order to calculate a more accurate overall losses than that obtained with simulations for switching devices only. The transformer apparent power is calculated as in [7]. Considering the worst case scenario (i.e. $d = 0.75$ and $\varphi_{max} = \pi/12$), the apparent power equations are obtained in (3).

$$S_{T_{1ph}} = \frac{1}{2} V_i I_{1_{RMS}} (1+d), \qquad S_{T_{3ph}} = \frac{1}{\sqrt{2}} V_i I_{3_{RMS}} (1+d) \qquad (3)$$

where $I_{1_{RMS}}$ and $I_{3_{RMS}}$ are the RMS primary current for 1ph-DAB and 3ph-DAB, respectively. For both topologies, 3C90 Ferroxcube E100/60/28 e-core has been used, changing the number of parallel cores to meet the design specifications. Both MFTs have a fully interleaved structure and a single layer winding. In order not to unnecessarily increase the transformer losses, the leakage inductance is made as small as possible preferring to include an external inductance to obtain the design value shown in Table I. Moreover, shell-type core magnetic structure has been chosen for both MFT.

Winding and core losses are the two main contribution to the total losses in a transformer. Since the MFT's frequency is much more higher than a line-frequency transformer, skin and proximity effects begin a crucial aspect in the correct evaluation of the AC resistance. Usually, Dowell's equations are used to determined the so-called resistance factor, which relates the DC resistance of the winding with its AC value. Following the calculations in [17], where the effect of interleaving structure on leakage inductance and winding losses is studied, it is possible to obtain copper losses as a function of d and φ for both MFTs.

Core losses are due mainly to two effects: eddy currents, which are induced in the core by the time-changing magnetic field and hysteresis losses. Steinmetz's equation provides a simple and easy way to calculate core losses for sinusoidal voltages. Nevertheless, this equation is not accurate anymore with all those power electronics applications where the voltage excitation is far from sinusoidal. For all these cases, the Improved Generalized Steinmetz Equation (iGSE) [18] was introduced and it is used within this design analysis. The main variables to consider for specific core losses evaluation are the peak-to-peak value and the time derivative of the flux density $B(t)$ as well as the magnetic material parameters. Using a T-model for the MFT as in [19], it is possible to calculate the magnetic flux density integrating the voltage across the magnetizing inductance v_m as:

$$B(t) = \frac{1}{NA_c} \int_0^t v_m(t)dt \quad \text{where} \quad v_m(t) = \frac{v_p(t) + v_s'(t)}{2} \tag{4}$$

where N is the primary winding turns number, A_c is the cross section area of the core column and $v_p(t)$ and $v_s'(t)$ are the primary and secondary reflected MFT voltages, respectively. In general, the magnetic flux density peak changes mainly with the primary and secondary voltage values (and thus d). Many authors have studied the flux density reduction due to blanking times [20,21]. However, only the primary voltage is considered while the total voltage across L_m should be used to calculated the right reduction of $B(t)$ during load operations. In this way it is possible to evaluate the flux density reduction starting from the no-load condition as a function of d and φ allowing to add this losses to the winding losses and finally calculate the total losses of the MFT.

By substituting $v_m(t)$ in (4) and integrating, flux density time-varying equation and its peak value are found. At this point, iGSE equation can be evaluated including the flux derivative over time for the entire period T. Finally, the specific magnetic losses for 1ph-MFT are shown in (5). The same consideration can be done for the 3ph-MFT.

$$P_{s_{1ph}} = \frac{k_i}{\pi}(2B_m)^{\beta - \alpha}\left(\frac{V_i}{2NA_c}\right)^\alpha \cdot \left[\varphi|d - 1|^\alpha + (\pi - \varphi)(d + 1)^\alpha\right] \tag{5}$$

Figures 3 (e) and (f) show the MFT total efficiency for both topologies as a function of phase-shift and power factor $\cos\vartheta$. Dashed lines highlight the MFT operation points for different battery voltages. As can be seen, the efficiency difference between the transformers is negligible and strongly depends on the construction topology of the windings and the ferromagnetic core. It is worth noting that for $d = 1$ (no voltage mismatch), there is a share of reactive power that increases as the phase shift increases for both cases. However, it can be shown analytically that the slope of this curve for the 3ph-DAB is $-3/(4\pi)$ versus $1/\pi$ for the 1ph-DAB, leading 3ph-DAB to process less reactive power for the same φ.

Current Efforts

An important aspect of converters design is the evaluation of the thermal limits of the switching devices and the MFT stresses. In particular, attention must be paid to the peak and RMS value of the full load current with mismatched voltages. For this reason, the current on L_{lk} is analytically calculated and compared performing a sweep analysis of d and φ. Figure 4 (a) shows the percentage difference between 1ph-DAB and 3ph-DAB where it can be seen that 1ph-DAB peak currents is up to 30% lower at high d values. As d decreases, this difference varies up to 10% in favor of 3ph-DAB. It is noticeable to see how the use of the three-phase topology is advantageous with regard to the peak current stress aspect for conditions where high power demands are paired with low DC conversion ratios. On the contrary, during all operations with low voltage mismatch, the 1ph-DAB shows a remarkable peak current reduction.
RMS current values for the same working conditions are analytically calculated and the RMS reduction of 1ph-DAB with respect 3ph-DAB is shown in Fig.4 (b). The RMS current difference varies between 4% and almost 9% for all the conditions, obtaining higher gap during low power. Despite the small difference, this results in the conduction loss ratio values shown in Fig.3 (b), being proportional to the square of the percentage difference.

From the single MOSFET perspective, since each device conducts only for half period, the RMS current is calculated as $1/\sqrt{2}$ times the RMS value of the inductor current whereas the peak current is the same of the inductor current, for both converters. For this reason, the ratios between 1ph-DAB and 3ph-DAB single device peak and RMS currents are exactly the same as shown in Fig.4 (a) and (b), respectively.

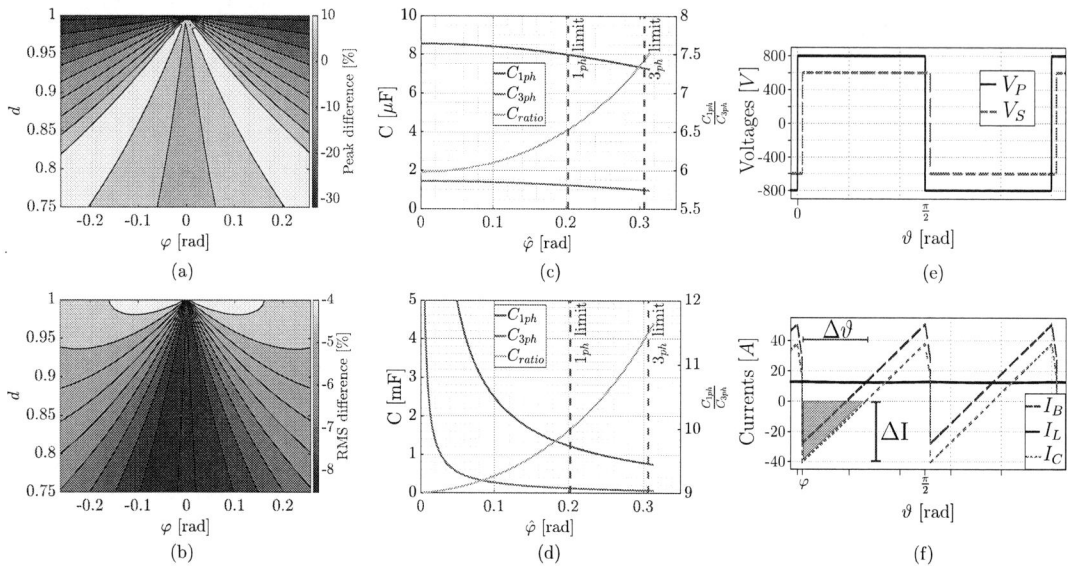

Fig. 4: Different comparison analysis between the two topologies in terms of peak (a) and RMS (b) current differences, filter capacitor values for voltage (c) and current (d) ripple constraints and DAB's voltages (e) and currents (f) waveforms for capacitor sizing calculation.

Filter Capacitor

In order to perform normal operations, DC/DC converters need input and output capacitors which are designed too meet voltage ripple specifications by using the best capacitor technology for each application. The input and output properties of power converters, as well as the allowed maximum voltage ripple, vary depending on the converter application. Low ESR and ESL values and high capacitance densities are particularly needed for filter capacitor solutions due to the tendency towards high switching frequencies and the need for small converter volumes [22]. Electrolytic capacitors are most commonly used when high capacitance values are needed. However, when high voltages are required, it is more appropriate to use film capacitors due to their lower internal resistance and higher current capability.

The purpose of this section is to analyse the worst scenario from the point of view of battery-side filter capacitor sizing. In fact, in order to design the filter appropriately and to meet the ripple voltage and current constraints, the worst case (i.e. the one with the minimum value of d and the minimum phase-shift angle) must be considered to evaluate the maximum amount of charge variation ΔQ.

Considering the voltage and current waveform depicted in Fig.4 (e) and (f), it is possible to calculate the electrical charge ΔQ accumulated during the capacitor charging process (blue shadowed area) as in (6), where I_B is the output bridge current, I_C is the capacitor current and I_L is the battery current.

$$\Delta Q = \Delta V \cdot C = \frac{\Delta\vartheta}{2} \cdot \Delta I = \frac{\Delta\vartheta}{2} \cdot I_C(\varphi), \quad I_C(\varphi) = I_B(\varphi) - I_L \tag{6}$$

In order to calculate the angle variation $\Delta\vartheta$, the output current equation of the bridge from φ to $\pi/2$ has to be considered and set equal to I_L, as in (7), whereas ϑ is the angle variable.

$$I_B(\vartheta) = I_B(\varphi) + \frac{V_i - nV_o}{L_{1ph}}\vartheta = \frac{V_i\left(2\frac{\varphi}{\pi}-1\right)+nV_o}{4fL_{1ph}} + \frac{V_i - nV_o}{L_{1ph}}\vartheta = I_L \quad \forall \quad \vartheta \mid \varphi \le \vartheta \le \frac{\pi}{2} \tag{7}$$

Solving (7) for ϑ, substituting in (6) and solved for C, the minimum capacity needed to comply the voltage ripple constraint is obtained in (8) and with similar considerations for 3ph-DAB in (9):

$$C_{1ph} = \frac{\Delta Q}{\hat{V}_o \delta_{V\%}} 100 = \frac{V_i \left[\pi^2 (1 - \hat{d}) + 2\hat{\varphi}[n(\pi - \hat{\varphi}) - \pi] \right]^2 \cdot 100}{8\omega^2 \pi^2 L_{1ph} (1 - \hat{d}) \hat{V}_o \delta_{V\%}} \tag{8}$$

$$C_{3ph} = \frac{\Delta Q}{\hat{V}_o \delta_{V\%}} 100 = \frac{V_i \left[2\pi^2 (1 - \hat{d}) + 3\hat{\varphi}[n(4\pi - 3\hat{\varphi}) - 4\pi] \right]^2 \cdot 100}{432\omega^2 \pi^2 L_{3ph} (1 - \hat{d}) \hat{V}_o \delta_{V\%}} \tag{9}$$

where $\delta_{V\%}$ is the ripple voltage define as $\delta_{V\%} = \Delta V / \hat{V}_o \cdot 100$ and ΔV is the peak-peak DC voltage. In the above equations, $\hat{\varphi}$, $\hat{d}$ and $\hat{V}_o$ are the minimum value of φ, d and V_o, respectively, which leads to a worst condition.

As can be seen in (6), ΔQ is only valid for a specific φ and d ranges because, by increasing them, the blue shadowed shape changes no longer being a simple triangle. In mathematical terms, it is possible to state that (6) holds only when $-I_B(\varphi) - I_L \geq 0$ is true. Solving $-I_B(\varphi) - I_L \geq 0$, two solutions are found. Whereas one is trivial ($\varphi = 0$ and $d \leq 1$), the second one gives the maximum angle for which the capacitor equation is true for a given d. Now, consider the two topologies and doing same considerations for the 3ph-DAB, (10) shows the φ range for which (6) holds.

$$1\text{ph-limit: } 0 \leq \varphi \leq \frac{1}{2} \left(2\pi - \sqrt{2}\pi\sqrt{d+1} \right), \quad 3\text{ph-limit: } 0 \leq \varphi \leq \frac{1}{3} \left(\pi - \pi\sqrt{2d-1} \right) \tag{10}$$

Fig.4 (c) illustrates (8) and (9) for different values of $\hat{\varphi}$ considering a $\delta_{V\%}$=5%. The capacitors ratio ($C_{ratio} = C_{1ph}/C_{3ph}$) is also calculated just to emphasize the capacitance difference which show that C_{1ph} requires more or less 6 times the C_{3ph} size. The limits of (10) are depicted by the dashed vertical lines for $\hat{d} = 0.75$. However, the previous considerations are not useful to limit the current ripple in case of a equivalent battery model, i.e. an impedance in series with a DC voltage source. In this case, the constraint is defined on the current ripple whereas the voltage ripple will be a consequence. The relationship between $\Delta I(s)$ and $\Delta Q(s)$ in the Laplace's domain is represented in (11).

$$\Delta I(s) = \frac{\Delta V(s)}{R_{bat} + sL_{bat}} = \frac{\Delta Q(s)}{C(R_{bat} + sL_{bat})} \tag{11}$$

where R_{bat} and L_{bat} are the total output resistance and inductance of the battery, respectively. As expected, the capacitor size is dependent on the output equivalent impedance and on the allowed current ripple. Since the battery impedance is quite small (few $m\Omega$), the capacitance tends to increase to very high values. A possible solution is to introduce a LC filter to limit the current variations. Fig.4 (d) shows the capacitors value for the two topologies with a 5% ripple current constraint with a pure resistive battery impedance ($R_{bat} = 0.2\Omega$). With these constraints, C_{3ph} appears to be up to 10 times smaller than C_{1ph} for the considered parameters, resulting in a less expensive and less bulky capacitors to be used.

Conclusions and Future Works

In case of variable conditions typical of battery charging process, this paper aims to fairly evaluate 1ph-DAB and 3ph-DAB. Device and MFT losses, ZVS ranges, peak and RMS current stresses as well as capacitor filter sizing have been compared between the two topologies, leading to the following findings and considerations. (1) For most conditions, 3ph-DAB switching and, more clearly, conduction losses are higher compared with 1ph-DAB up to 30% and 18%, respectively. (2) ZVS boundaries for 1ph-DAB result in wider soft-switching operations and thus, for a greater range of normalised transferred power than in the 3ph-DAB case, the 1ph-DAB working points lie within the soft-switching region. (3) Particular operating conditions can reduce peak current stress by up to 30% for the 1ph-DAB while 3ph-DAB RMS current is up to 9% higher despite its lower harmonic content. (4) 3ph-DAB requires much smaller capacitors due to reduced voltage and current ripple resulting in smaller volumes and weights.

References

[1] R. Cherif, F. Hasanov, and A. Pande, "Riding the Energy Transition: Oil beyond 2040," *Asian Economic Policy Review*, vol. 16, no. 1, pp. 117–137, 2021. [Online]. Available: https://onlinelibrary.wiley.com/doi/abs/10.1111/aepr.12317

[2] J. A. Sanguesa, V. Torres-Sanz, P. Garrido, F. J. Martinez, and J. M. Marquez-Barja, "A Review on Electric Vehicles: Technologies and Challenges," *Smart Cities*, vol. 4, no. 1, pp. 372–404, Mar. 2021. [Online]. Available: https://www.mdpi.com/2624-6511/4/1/22

[3] C. Liu, K. Chau, D. Wu, and S. Gao, "Opportunities and challenges of vehicle-to-home, vehicle-to-vehicle, and vehicle-to-grid technologies," *Proceedings of the IEEE*, vol. 101, no. 11, pp. 2409–2427, 2013, publisher: IEEE.

[4] H. van Hoek, M. Neubert, and R. W. De Doncker, "Enhanced modulation strategy for a three-phase dual active bridge—Boosting efficiency of an electric vehicle converter," *IEEE Transactions on Power Electronics*, vol. 28, no. 12, pp. 5499–5507, 2013, publisher: IEEE.

[5] A. Sharma and S. Sharma, "Review of power electronics in vehicle-to-grid systems," *Journal of Energy Storage*, vol. 21, pp. 337–361, 2019, publisher: Elsevier.

[6] M. Di Benedetto, A. Lidozzi, L. Solero, F. Crescimbini, and S. Bifaretti, "Hardware design of SiC-based Four-Port DAB Converter for Fast Charging Station," in *2020 IEEE Energy Conversion Congress and Exposition (ECCE)*. IEEE, 2020, pp. 1231–1238.

[7] R. W. De Doncker, D. M. Divan, and M. H. Kheraluwala, "A three-phase soft-switched high-power-density DC/DC converter for high-power applications," *IEEE transactions on industry applications*, vol. 27, no. 1, pp. 63–73, 1991, publisher: IEEE.

[8] L. M. Cúnico, Z. M. Alves, and A. L. Kirsten, "Efficiency-Optimized Modulation Scheme for Three-Phase Dual-Active-Bridge DC–DC Converter," *IEEE Transactions on Industrial Electronics*, vol. 68, no. 7, pp. 5955–5965, 2020, publisher: IEEE.

[9] A. Garcia-Bediaga, I. Villar, A. Rujas, I. Etxeberria-Otadui, and A. Rufer, "Analytical models of multiphase isolated medium-frequency DC–DC converters," *IEEE Transactions on Power Electronics*, vol. 32, no. 4, pp. 2508–2520, 2016, publisher: IEEE.

[10] F. Krismer, "Modeling and optimization of bidirectional dual active bridge DC-DC converter topologies," Ph.D. dissertation, ETH Zurich, 2010.

[11] D. Segaran, D. G. Holmes, and B. P. Mcgrath, "Comparative analysis of single and three-phase dual active bridge bidirectional DC-DC converters," in *2008 Australasian Universities Power Engineering Conference*. IEEE, 2008, pp. 1–6.

[12] T. Jimichi, M. Kaymak, and R. W. De Doncker, "Comparison of single-phase and three-phase dual-active bridge DC-DC converters with various semiconductor devices for offshore wind turbines," in *2017 IEEE 3rd International Future Energy Electronics Conference and ECCE Asia (IFEEC 2017 - ECCE Asia)*, Jun. 2017, pp. 591–596.

[13] H. van Hoek, M. Neubert, A. Kroeber, and R. W. De Doncker, "Comparison of a single-phase and a three-phase dual active bridge with low-voltage, high-current output," in *2012 International Conference on Renewable Energy Research and Applications (ICRERA)*, Nov. 2012, pp. 1–6.

[14] R. Haneda and H. Akagi, "Design and Performance of the 850-V 100-kW 16-kHz Bidirectional Isolated DC–DC Converter Using SiC-MOSFET/SBD H-Bridge Modules," *IEEE Transactions on Power Electronics*, vol. 35, no. 10, pp. 10013–10025, Oct. 2020, conference Name: IEEE Transactions on Power Electronics.

[15] H. Beiranvand, E. Rorok, and M. Liserre, "Theoretical Evaluation of Semiconductor Loss Components Behavior in ISOP-DAB Converters," in *2019 IEEE 13th International Conference on Compatibility, Power Electronics and Power Engineering (CPE-POWERENG)*, Apr. 2019, pp. 1–7.

[16] H. A. B. Siddique, "The three-phase dual-active bridge converter family : modeling, analysis, optimization and comparison of two-level and three-level converter variants; 1. Auflage," Ph.D. dissertation, RWTH Aachen University, 2019.

[17] B. Chen, "Analysis of Effect of Winding Interleaving on Leakage Inductance and Winding Loss of High Frequency Transformers," *Journal of Electrical Engineering & Technology*, vol. 14, no. 3, pp. 1211–1221, May 2019. [Online]. Available: https://doi.org/10.1007/s42835-019-00129-6

[18] K. Venkatachalam, C. Sullivan, T. Abdallah, and H. Tacca, "Accurate prediction of ferrite core loss with nonsinusoidal waveforms using only Steinmetz parameters," in *2002 IEEE Workshop on Computers in Power Electronics, 2002. Proceedings.*, Jul. 2002, pp. 36–41.

[19] N. Fritz, M. Rashed, S. Bozhko, F. Cuomo, and P. Wheeler, "Analytical modelling and power density optimisation of a single phase dual active bridge for aircraft application," *The Journal of Engineering*, pp. 3671–3676, 2019.

[20] I. Villar, A. Garcia-Bediaga, U. Viscarret, I. Etxeberria-Otadui, and A. Rufer, "Proposal and validation of medium-frequency power transformer design methodology," in *2011 IEEE Energy Conversion Congress and Exposition*, Sep. 2011, pp. 3792–3799.

[21] I. Villar, U. Viscarret, I. Etxeberria-Otadui, and A. Rufer, "Global Loss Evaluation Methods for Nonsinusoidally Fed Medium-Frequency Power Transformers," *IEEE Transactions on Industrial Electronics*, vol. 56, no. 10, pp. 4132–4140, Oct. 2009.

[22] H. Van Hoek, "Design and operation considerations of three-phase dual activa bridge converters for low-power applications with wide voltage ranges," Ph.D. dissertation, RWTH Aachen University, 2017.

Dynamic Load Emulation for Automotive Power IC Robustness Validation

Alexander Ulbing[1,2], Daniel Kostynski[1], Markus Sievers[1,2]

[1]KAI Kompetenzzentrum fuer Industrie- und Automobilelektronik
[2]Technical University of Graz - Institute of Electronics
[1]Europastr. 8, [2]Infeldgasse 12
[1]9524 Villach, Austria, [2]8010 Graz, Austria
Phone: +43 (51777) 19968
Email: alexander.ulbing@k-ai.at
URL: [1]http://www.k-ai.at, [2]https://www.tugraz.at/institutes/ife/home

Acknowledgement

This work was funded by the Austrian Research Promotion Agency (FFG, Project No. 884573).

Keywords

≪Automotive Application≫, ≪Machine Emulation≫, ≪Test Bench≫, ≪Integrated Circuit (IC)≫, ≪Power Hardware-in-the-loop≫.

Abstract

This paper addresses the gap in application related stress testing for automotive power IC qualifications as well as for development-related testing to make ICs more robust. To make test systems more efficient and reduce the cost per test slot the concept of dynamic load emulation has been evaluated. An approach taken by industrial power converters has been adapted to fit the needs of the low power automotive drive domain. To prove the concept and show how application relevant stress could be applied to the device under test simulations are done. In a further step a dedicated hardware test bench has been created and the applicability of the concept within the automotive domain was verified. Several measurements are shown to demonstrate the functionality as well as possible improvements and next steps are discussed.

Introduction

Since the early beginnings of the automotive industry electrical engineers have been in the pursuit of making cars more convenient, performant, efficient and more environmentally friendly. Above all though, the need for reliability has been a define factor in this global industry, especially with the increasing electrification. The semiconductor industry is a key enabler for many reasons and in many areas. While the focus in the last years has been on the replacement of fuses and relays, nowadays, motor control topics become more and more important. For example, driven by the above mentioned credo to make cars more reliable, brushed Direct Current (DC) motors are increasingly replaced by Brushless Direct Current (BLDC) motors. In addition to the reduced mechanical wear and tear of the missing brushes welcome side effects such as electromagnetic interference normally caused by the brushes are reduced as well. Other advantages are higher efficiency (15 % to 20 %), higher torque and a higher dynamic response. Additional diagnostic features and Field Oriented Control (FOC) make BLDC motors the best fit for the use in Advanced Driver Assistance Systems (ADAS), Heating, Ventilation, Air Conditioning (HVAC) air flaps, x-by-wire applications and various other automation topics within cars [1, 2]. These applications are more and more controlled by dedicated automotive power Integrated Circuits (ICs) with a high level of integration. For example, a power IC can encompass not only the power stage but also the driver and sometimes even the controller itself. This trend will certainly continue such that power ICs will incorporate more and more functionality as well as complex power electronic circuitry.

Throughout the operative lifetime of automotive power ICs the loads connected to them and the resulting load profiles have a tremendous impact on the IC's lifetime. Therefore, it is necessary to start robustness validations and reliability investigations according to application related load profiles at an early stage of chip development. Due to the high amount of test throughput there is a need for a concept that emulates such load profiles without having to use the many specific loads. To handle this issue it is necessary to address the topic of *load emulation*. One of the strategies that can be used is hardware-in-the-loop (HIL) testing. While this takes already place for converter topologies described by Zade *et al.* in [3] and Kadam *et al.* in [4], module testing as described by Ibrahim *et al.* in [5] or for high power electric motor applications as described by Oliveira *et al.* in [6] has not been applied to the characterization and qualification of automotive power ICs.

Before any Automotive Power IC, that drive and control the aforementioned electronic applications, can go to the market rigorous stress tests have to be conducted. Besides the usual standards such as the *AEC-Q100* it is necessary to validate the parts' robustness and reliability under application relevant stress condition with a suitable test bench. Economic and ecological aspects are some of the main consideration points when designing such a test setup. First of all, it is necessary to reduce the complexity of the system development through modularization. Steinwender *et al.* have introduced such a modular test system in [7]. This brings about a significant cost reduction during the design phase of the system in addition to the re-usability of many system components. Since a change of the Device Under Test (DUT) requires only the directly associated printed circuit board (PCB) to be re-designed and changed. The remaining system can stay untouched.

The system presented in this work represents a feasibility study to investigate the aforementioned impact of load profiles with a rapid controller platform. For the design of a test system that needs to stress many devices in parallel the mentioned modular system approach will be used. The following sections show how a setup has been developed to first simulate relevant but simple load profiles and then validate them with a hardware setup. The final sections will discuss the results and offer an outlook on the next steps.

Simulation Model Derivation

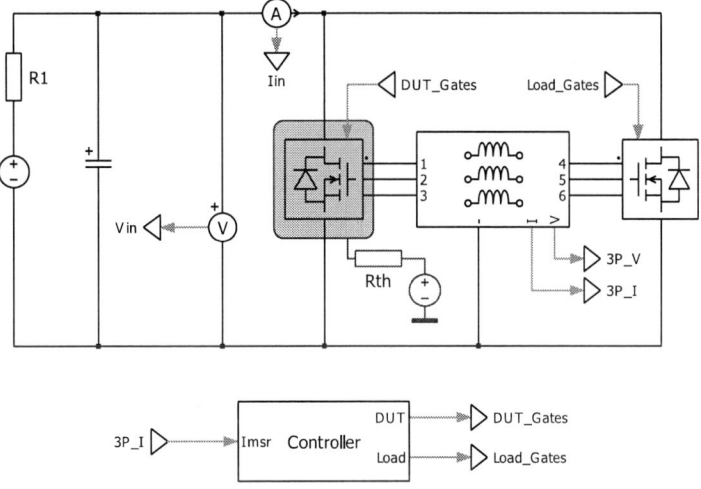

Fig. 1: Overview of the simulation

As described in the previous section the aim is to develop a modular platform that allows various application relevant stress scenarios to be applied to integrated power semiconductor products at various stages of their development cycles. Such integrated power semiconductor products can embody many different topologies. Nevertheless, a B6 configuration is a commonly implemented solution and will be the focus of the herein presented investigation. In order to investigate a variety of test scenarios a commercially available Rapid Control Prototyping (RCP) platform (here PLECS®) is used to first create a model of the system and then apply the operating points of interest to the DUTs.

Similar to the approach described by Choi *et al.* in [8] two B6 bridges are connected to the same DC link at their DC+ and DC- terminals. The phase terminals are interconnected through inductors. Thereby, one of the converters will act as the DUT and the other as the controllable load that provides the application relevant stress pulses. An overview of this setup is depicted in Figure 1.

The setup has the unique advantage that the load current is circulated within the test system, such that the power supply will only have to supply the energy that is dissipated within the test system. Since [8, 9] already provide a thorough discussion on controller concepts, especially with respect to the simulation of motors characteristic the focus herein is to demonstrate the ability to generate and apply a variety of stress pulses. Since these pulses are intended to investigate the ability of the developed system some of them may be uncharacteristic to motor application. This approach is of special interest when the final application profile of the investigated product is not yet set in stone or the robustness and reliability of the device needs to be investigated during the development process. To accomplish such stress pulses one of the requirements is that each leg of the load module must have a current measurement.

Next to the electrical performance of the system the thermal behavior of the DUT module and load module is of interest. The thermal characteristics of the power semiconductor are taken into account by the use of a thermal model of the power semiconductor provided by the manufacturer, which will require additional tuning with the implemented demonstrator hardware. As shown in Figure 1 additional components, such as the heat sink, are modeled with the appropriate thermal simulation library elements.

Each controller uses a reference current signal (I_{ref}), which is located within the controller block to streamline the code generation for the RCP system. As shown in Figure 2 the input and output blocks of the controller are special functional blocks to implement the hardware functionality of the RCP hardware. During the simulation stage these blocks enable the later used hardware to be simulated. Therefore, besides the reference current the measured inductor currents (I_{msr}) are fed to the controller. Based on these inputs, the controller block generates the PWM control signals for all twelve power semiconductors. While the DUT module is operated with a fixed duty cycle, D, of 50 % for most scenarios the load module uses individual current controllers to generate the desired load profiles for each phase.

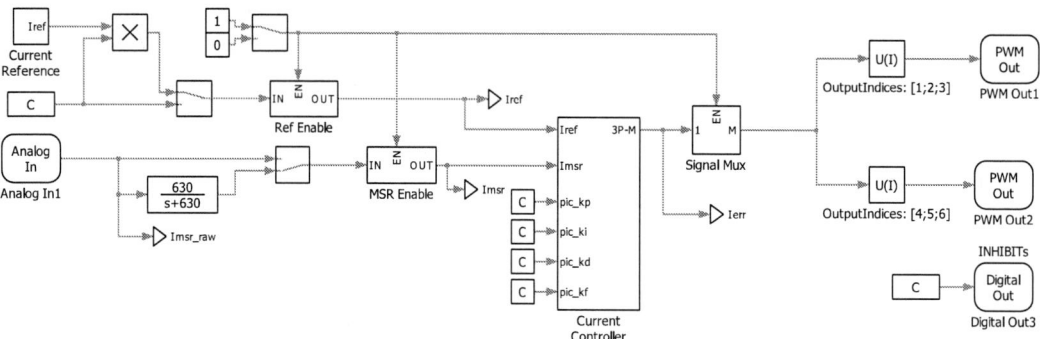

Fig. 2: Overview of the controller implementation

At the heart of the current controller is a PID controller that controls the duty cycle of each half-bridge in the load module. The PID controller uses commonly known anti-windup and saturation features to improve the controller response. Besides it being a well known controller concept the simulation results presented in the next section show that the dynamic requirements of the targeted application profiles can be covered with such a PID controller. Therefore, no further discussion on the controller implementation is provided at this point.

Simulation Results

To verify the concept within the simulation environment, two scenarios are used. The goal of the first scenario is to simulate the stress exerted on the DUT during the start of a motor application. This start-up behavior is simulated by applying a constant value for I_{ref}. Besides simulating the abrupt load change this scenario also helps to analyze the controller response and stability after a step change at the input. The second current profile presented below is a repetitive load change the DUTs may experience throughout its lifetime.

Start-up of the System

As a first step a static operating point is applied to the system at rest. The controller is then forced to perform a step response to reach the static working point. This commonly used approach shows the theoretic functionality of the implemented controller. In Figure 3 the controller behavior for each Half-Bridge is shown.

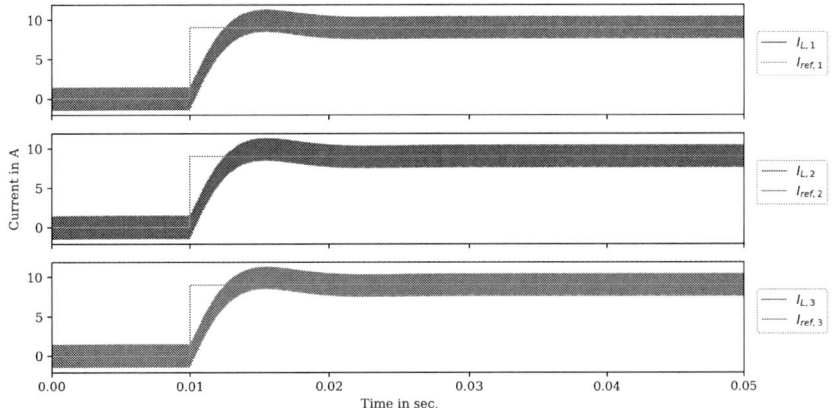

Fig. 3: Overview of the main system parameters at startup

The static working point is defined by a unit step of the output current from 0 A to 9 A as it is shown in Figure 3. Furthermore, the curves demonstrate the controller behavior and the dynamic of the system very well. After 10 ms and a small overshoot, the input parameters have already settled and the power losses in the system remain constant.

Repetitive load change

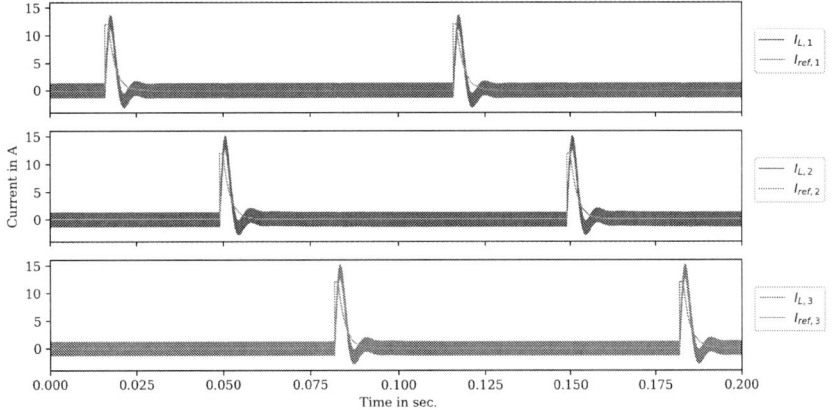

Fig. 4: Overview of the main system parameters during repeated load change

After verifying the basic functionality, an arbitrary reference is added. The aim is to check the performance of the controller and the system when repetitively changing the reference current. The arbitrary reference current is applied alternatingly to all three phases. With this kind of input the half bridge ICs undergo a repetitive stress which can be clearly seen by the currents depicted in Figure 4.

Hardware Prototype

To apply the proposed concept to a real DUT, a hardware (HW) prototype has been developed and is shown in Figure 5a. On the prototype board the DUT is a fully integrated Half-Bridge IC with integrated current sense structures and driver stages. Thereby it is possible to connect to the devices directly with the RCP system. The selected IC is placed three times on the prototype board in a B6 bridge configuration. Due to the fact that the HW setup is intended verify the concept at this stage it serves as a first experimental platform to investigate necessary adaptations and performance limitations. Therefore, the same type of board has been used here for the DUT side and the load side. As a result two of the shown prototype boards in Figure 5a have been connected via inductors at the midpoint of each half-bridge share the same input power supply. This HW setup is shown in Figure 5b.

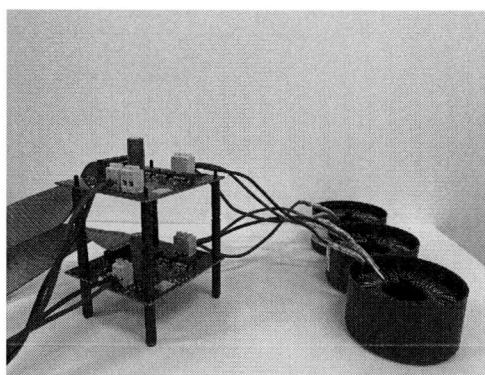

(a) CAD model of the prototype board

(b) Lab Setup with stacked prototype boards and loads

Fig. 5: Overview of the used hardware

Measurement Results

In a first step it was intended to show that the setup is able to reach a stable steady state after a load change has been applied. Therefore, a static reference current of 9 A was applied to the controller after it had been running with approximately 0 A. Since the DUT board was operated with a fixed duty cycle there is already a current ripple visible before the applied step input. As shown in Figure 6 the controller performs well when applying a load step. Similar to the simulation, the controller shows a small overshoot at the beginning but the steady state is reached fairly quickly. Due to the very low switching frequency of 10 kHz and interferences on the measurement signal, that can be mainly attributed to ground bouncing of the system, small control deviations are noticeable.

As a next step it was necessary to show that the system is able to handle repetitive load pulses. Figure 7 shows the measured phase currents of such an operating point. The figure also includes the simulation results showing that for this operating point the performance is very similar. Since the selected DUTs have a much higher rating that the pulses that are of interest for future DUTs the case temperature was measured but revealed only a negligible increase. Therefore, a more detailed investigation of the thermal behavior will be addressed once the DUTs have been determined.

Although the here presented results show a very stable operation the investigations conducted up to this point show that the signal integrity is strongly affected by the fact that everything is referenced to the

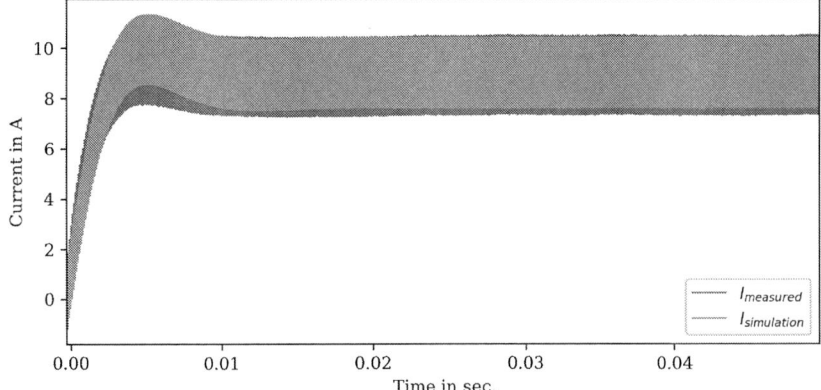

Fig. 6: Recorded oscilloscope measurement results compared to the simulated step response of the system for a single phase.

same ground potential in the automotive power ICs. While the reference to the same ground potential simplifies the HW setup and the connections to the RCP system the controllability of the system is strongly affected by the noise introduced by the switching of the power transistors. Higher ground bouncing due to higher switching currents also contribute to higher interferences on the measurement signals which require additional filtering. The gained insights will serve for a revision of the HW in the near future with the intention to establish a platform suitable for investigation throughout the development process of new power semiconductor technologies and products.

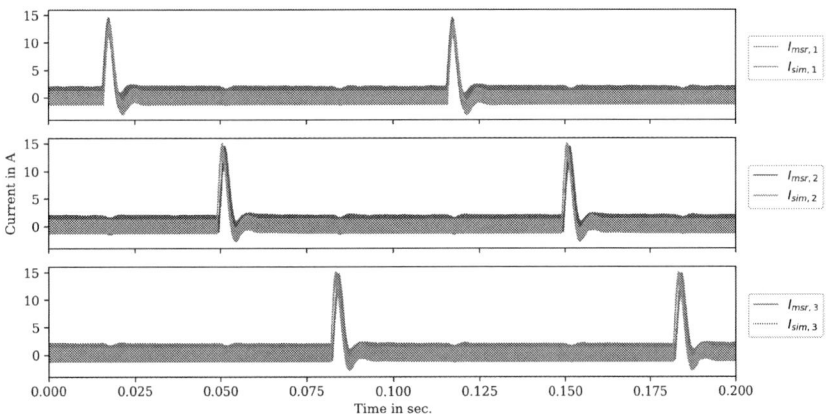

Fig. 7: Recorded oscilloscope measurement results compared to the simulated profiles of a repetitive load pulses.

Conclusion

In conclusion the concept of having a modular platform to perform application related stress to automotive power ICs has been presented and verified in this paper. While this concept is known in the industrial sector it is necessary to start a discussion for automotive reliability and robustness investigations that go beyond the tests required by the known standards. Within the simulation chapter, the approach of running two B6 bridges back-to-back was verified using discrete power semiconductor models since no suitable models were readily available of the later on used automotive power ICs. Using the developed modular HW platform connected to the RCP software the functionality of the controller has been verified for two scenarios. With the help of the RCP system and the HW platform two operating points have been presented to demonstrate the functionality of the system. The first operating point presented was a single load step to demonstrate the system's response. The second presented operating point showed the dynamic performance and stability of the system under repetitive pulses. Overall it is shown that the developed HW platform is ready to use for further investigations, especially taken various load mission profiles under consideration. Future work will focus on improving the setup with respect to signal integrity and space as well as investigations and test runs on new control concepts. Additional investigations will apply different load profiles and stress levels to the DUTs and monitor the resulting stress at the power ICs. Last but not least, additional measurements will be implemented to acquire more diagnostic data of each stressed device to verify which factors impacts the most.

References

[1] P. Yedamale, "Brushless DC (BLDC) motor fundamentals", *Microchip Technology Inc*, vol. 20, no. 1, pp. 3–15, 2003.

[2] B. Lequesne, "Automotive electrification: The nonhybrid story", *IEEE Transactions on Transportation Electrification*, vol. 1, no. 1, pp. 40–53, 2015. DOI: 10.1109/TTE.2015.2426573.

[3] A. Zade, D. Venkatramanan, and V. John, "Power converter based impedance emulation of passive loads for anti-islanding tests", in *2018 IEEE International Conference on Power Electronics, Drives and Energy Systems (PEDES)*, 2018, pp. 1–6. DOI: 10.1109/PEDES.2018.8707594.

[4] A. H. Kadam and S. S. Williamson, "A common dc-bus-configured traction motor emulator using a virtually isolated three-phase ac-dc bidirectional converter", *IEEE Access*, vol. 9, pp. 80 621–80 631, 2021. DOI: 10.1109/ACCESS.2021.3085029.

[5] A. Ibrahim, R. Lallemand, Z. Khatir, M. Berkani, and D. Ingrosso, "Condition monitoring and evaluation of ron degradation during power cycling in switching mode of sic-mosfets power modules", in *International Conference on Integrated Power Electronics Systems (CIPS)*, 2022, pp. 150–155, ISBN: 978-3-8007-5757-2.

[6] C. M. R. de Oliveira, M. L. de Aguiar, A. G. de Castro, P. R. U. Guazzelli, W. C. d. A. Pereira, and J. R. B. d. A. Monteiro, "High-accuracy dynamic load emulation method for electrical drives", *IEEE Transactions on Industrial Electronics*, vol. 67, no. 9, pp. 7239–7249, 2020. DOI: 10.1109/TIE.2019.2942566.

[7] B. Steinwender, S. Einspieler, M. Glavanovics, and W. Elmenreich, "Distributed power semiconductor stress test amp; measurement architecture", in *2013 11th IEEE International Conference on Industrial Informatics (INDIN)*, 2013, pp. 129–134. DOI: 10.1109/INDIN.2013.6622870.

[8] U.-M. Choi, S. Jørgensen, and F. Blaabjerg, "Advanced accelerated power cycling test for reliability investigation of power device modules", *IEEE Transactions on Power Electronics*, vol. 31, no. 12, pp. 8371–8386, Dec. 2016, ISSN: 1941-0107. DOI: 10.1109/TPEL.2016.2521899.

[9] I. Vernica, F. Blaabjerg, and K. Ma, "Mission profile emulator for the power electronics systems of motor drive applications", in *2017 19th European Conference on Power Electronics and Applications (EPE'17 ECCE Europe)*, 2017, P.1–P.10. DOI: 10.23919/EPE17ECCEEurope.2017.8099241.

DAB frequency decoupling control with current minimization

Simon UICICH [1,2], Jean-Yves GAUTHIER [1], Xuefang LIN-SHI [1],
Bruno ALLARD [1], Arnaud PLAT [2]

[1] Univ Lyon, INSA Lyon, Universite Claude Bernard Lyon 1, Ecole Centrale de Lyon, CNRS,
Ampere, UMR5005,
69621 Villeurbanne, France
E-Mail :jean-yves.gauthier@insa-lyon.fr

[2] AIRBUS OPERATIONS,
26 Chem. de l'Espeissière,
31300 Toulouse, France
E-Mail :arnaud.plat@airbus.com

Keywords

«Dual Active Bridge (DAB)», «Efficiency » «Optimization», «Converter control», «High frequency power converter », «Non-linear control»

Abstract

This paper applies a new Dual-Active-Bridge (DAB) triple-phase-shift (TPS) model modulation to simplify converter analysis, to simplify control and improve efficiency, and reduce transient current stress in the power stage throughout a wide operating space. Achieved current stress is comparable to state-of-the-art. The approach is validated through Simulink-Simscape.

Introduction

The Dual Active Bridge (DAB) is an interesting converter due to the amount of ways in which the phase shifts between its half bridge legs can be controlled to minimize component stress at a given operating condition. Typical modulation approaches achieve this by defining analytical relationships between each phase shift in order to shape inductor current, i.e. triangular current modulation, trapezoidal current modulation (Fig. 1), etc. The goal of this normally achieving low loss (i.e.: minimizing inductor RMS value at given load conditions and achieving certain current values at semiconductor switching instants to guarantee zero voltage switching -ZVS-). In this sense triple-phase-shift (TPS) is the one with the highest potential for optimization since it allows independent control of the phase shift between all 4 half bridges. A typical modulation design approach then uses precise power component loss models for global modulation optimization. Operating space regions are thus matched to modulations, and controllers for each one are developed [1][2]. Straightforward as this is, it has several disadvantages that have been addressed by subsequent research.

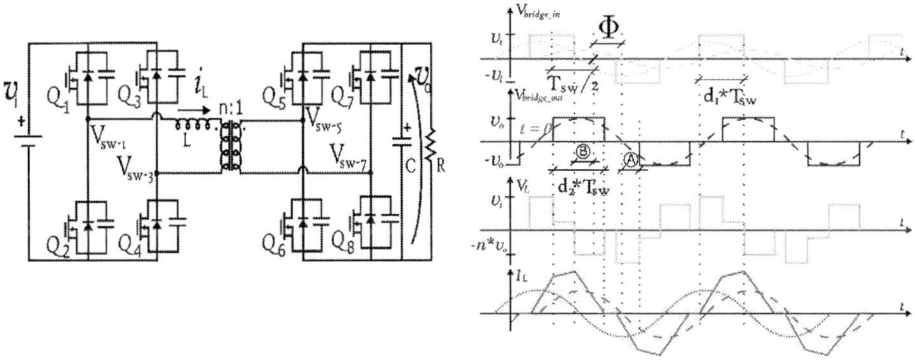

Fig. 1: DAB Schematic and relevant waveforms associated to Triple Phase Shift modulation operation. Relevant wave-forms are $V_{bridge\_in} = V_{sw-} - V_{sw-3}$, $V_{bridge\_out} = V_{sw-5} - V_{sw-7}$, inductor voltage $V_L = V_{bridge\_in} - V_{bridge\_out}$ and current I_L shown in yellow, blue, green and purple respectively. Note: $A = T_{sw}0.5(1 - d_2)$ and $B = T_{sw}0.5(1 - d_1)$

This approach however suffers from design complexity [3][4], design sensitivity to parasitic [5] [6], control variable continuity and output dynamics [1] amongst other issues. Some authors have addressed this. e.g.: with a loss minimization P&O approach to simplify design [7]; optimal approaches for conduction loss min. [8][9] for reduced performance sensitivity and with non-linear approaches limited to low frequency switching operation [10][11] to solve dynamic issues.

Approaching converter control with a generalized multi-frequency averaging approach (MFA) [12][13] avoids these issues. It evaluates the dynamics of frequency components of state space variables through pseudo-Fourier Series coefficients. [14] develops the model for the case of TPS as in [7][9][15] and verifies resemblance to actual circuit behavior. The approach is also found as fundamental component analysis or FCA model in literature[9]. [9]'s approach is similar, but it is restricted to conduction loss reduction ignoring switching loss reduction, and as [16] has shown at low load even this is ineffective due to the FCA model's limited precision. Additionally, neither [17] nor [9] take advantage of the current overshoot prevention, operation with transformer turns ratios bigger than 1 and the possibility of constraining input harmonics which the model can also assess. The stability of the controller is also not guaranteed under large signal state dynamics (small signal modelling). All these disadvantages are the object of the current article's proposed approach.

The paper is organized as follows: Section **FCA model** recalls the DAB fundamental component analysis already developed in the literature. Section **Control Approach** transforms large signal equations to a two-time scale system to facilitate closed-loop control design. Section **Optimisation Strategy** then develops an algorithm based on the model and controller structure aimed to minimize converter RMS current. Section **Validation and Simulation**, verifies the proposed systems performance in Matlab/Simulink, comparing the result to SPS and the state of the art in MFA control. Section **Conclusion** takes stock at the advantages and disadvantages of the presented approach.

FCA model

As underlined previously, the generalized multi-frequency averaging approach can be used to obtain the dynamics of frequency components of state-space variables for the case of TPS modulation. The state-space model obtained is different according to the choice of the time reference and the states. In [9], the reference $t = 0$ is chosen as depicted in Fig. 1. The states used in the model are output voltage average value and transformer current's at the switching frequency, f_{sw}. Then, the FCA model can be expressed as in Eqs. (1) when using in-phase, α, and quadrature, β, fundamental components represented in dashed and dotted lines respectively as in Fig. 1.

$$\frac{dv_0}{dt} = -\frac{v_0}{C \cdot R} + \frac{n}{2C} \cdot d_{out,1\alpha} \cdot i_{L,1\alpha} \tag{1.a}$$

$$\frac{di_{L,1\alpha}}{dt} = \frac{v_{in}d_{in,1\alpha}}{L} - \frac{n \cdot v_0 d_{out,1\alpha}}{L} - \frac{R_p \cdot i_{L,1\alpha}}{L} + \omega_0 \cdot i_{L,1\beta} \tag{1.b}$$

$$\frac{di_{L,1\beta}}{dt} = \frac{v_{in}d_{in,1\beta}}{L} - \frac{R_p \cdot i_{L,1\beta}}{L} - \omega_0 \cdot i_{L,1\alpha} \tag{1.c}$$

Where R_p is conduction loss related parasitic resistance, ω_0 given as 2π times f_{sw}, $d_{in,1\alpha}$, $d_{in,1\beta}$, and $d_{out,1\alpha}$, are the fundamental components of the Fourier Series expansion of each full-bridge stages' associated switching function as described by Eqs. (2). In turn, they depend on the actual full order circuit phase shifts d_1, d_2, ϕ as depicted in Fig. 1.

$$d_{in,1\alpha} = \frac{4 \cdot sin(\pi \cdot d_1) \cdot cos(2 \cdot \pi \phi)}{\pi} \tag{2.a}$$

$$d_{in,1\beta} = \frac{4 \cdot sin(\pi \cdot d_1) \cdot sin(2\pi \cdot \varphi)}{\pi} \tag{2.b}$$

$$d_{out,1\alpha} = \frac{4 \cdot sin(\pi d_2)}{\pi} \tag{2.c}$$

The FCA model provides information on the response of inductor current, the physical variable used to reduce loss and which impacts switching loss and input harmonics. Thus, it is attractive to develop a control scheme to regulate it, independently of optimization.

Control Approach

The FCA model is a very nonlinear one which can be written as Eqs. (3):

$$\frac{dx}{dt} = f(x, u) \tag{3.a}$$

$$x = [v_0 \quad i_{L,1\alpha} \quad i_{L,1\beta}]^t \quad u = [u_1 \quad u_2 \quad u_3]^t = [d_{out,1\alpha} \quad d_{in,1\alpha} \quad d_{in,1\beta}]^t \tag{3.b}$$

Having a single controller for this model will contribute to system simplification. As [14] shows, model state variable response to a step in one of the control inputs can be divided into two dynamics. A faster one observed in the response of inductor current, and a slower one related to the output capacitor and load response. Thus, the dynamical system of Eq. (1) can be seen as a two-time scale system. A mathematical formalism has been developed for two time-scale systems (corresponding to the so-called "frequency separation" in the linear system framework). The idea is to consider that the dynamic of the slow subsystem is very slow compared to the dynamic of the fast one, i.e. it can be neglected for the fast subsystem control loop design.

The fast time-scale subsystem is composed of the currents $i_{L,1\alpha}$ and $i_{L,1\beta}$.

$$\frac{di_{L,1\alpha}}{dt} = \frac{v_{in}}{L} u_2 - \frac{R_p \cdot i_{L,1\alpha}}{L} + \omega_0 \cdot i_{L,1\beta} - f_1 \tag{4.a}$$

$$\frac{di_{L,1\beta}}{dt} = \frac{v_{in}}{L} u_3 - \frac{R_p \cdot i_{L,1\beta}}{L} - \omega_0 \cdot i_{L,1\alpha} \tag{4.b}$$

Where $f_1 = \frac{n \cdot v_0 d_{out,1\alpha}}{L}$ can be considered as a known disturbance for the current control design.

The slow time-scale system is composed of the output voltage v_0.

$$\frac{dv_0}{dt} = -\frac{v_0}{C \cdot R} + \frac{n}{2C} \cdot i_{L,1\alpha} . u_1 \tag{5}$$

Where the dynamic of $i_{L,1\alpha}$ can be neglected to design the voltage control.

This implies two decoupled Linear Time-Invariant (LTI)systems:
- $i_{L,1\alpha}$ and $i_{L,1\beta}$ will be controlled by u_2 and u_3 through Eqs. (4) .
- v_0 can be controlled by u_1 from the 1^{st}order system Eq. (5).

Eqs. (4) lead to expressions for decoupled control while a PI control for Eq. (5) will close the loop for output voltage regulation and create the reference value of $i_{L,1\alpha}$ thus needed. An optimization strategy developed in the next section determines the reference value of $i_{L,1\beta}$ and $u_1(d_{out,1\alpha})$. The next step is then to match each "$u_{1,2,3}$" of the FCA model control space to state variables. It is important to match well to avoid control effort saturating the actual control variables " d_1, d_2, ϕ" – dynamic range between $(-0.5; 0.5)$.

Fig. 2 shows the resulting system diagram. For a given output voltage, two degrees of freedom arise: the reference for $i_{L,1\beta}$ and $d_{out,1\alpha}$. Preliminary simulations showed that the current controller implementation needed to provide transient prevention of saturation in "$\boldsymbol{D}$", for extreme cases. The final part of the control system implementation requires calculating the actual control variables d_1, d_2, ϕ from $d_{in,1\alpha}$, $d_{in,1\beta}$, and $d_{out,1\alpha}$, and using them in a TPS modulator to control Q_1 to Q_8. This is represented by the d_1, d_2, ϕ calculation block from Fig. 2.

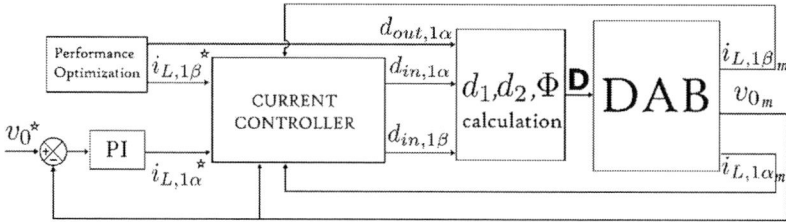

Fig. 2. Block diagram of the proposed control system. Measured values are subscripted with an "m" whereas reference values are marked with a star.

In addition to controlled peak values on power component currents, closed-loop control of $i_{L,1\alpha}$ and $i_{L,1\beta}$ has another key advantage. Namely, if the operating space were divided into regions paired to different optimization approaches – as is normally the case–, approaches which are normally based on current amplitudes, variation of L would not impact the selection of the optimal solution since the controller/optimizer would have this information.

Optimization Strategy

To attract interest, frequency decoupling control should be compatible with classic performance optimization techniques. The classic RMS current minimization problem from [9] is thus first approached. There, it is shown that, for $M = \frac{n * v_o}{v_{in}} < 1$, minimum RMS current is obtained with the control variables from Eqs. (6):

$$d_{out,1\alpha} = d_{out,1\alpha\,\mathrm{MAX}} = \frac{4}{\pi}, \tag{6.a}$$

$$d_{in,1\beta} = \frac{2\,\omega_0\,L\,P}{d_{out,1\alpha}\,v_{in}\,v_o n} = k_2\frac{\pi}{4}, \tag{6.b}$$

$$d_{in,1\alpha}\big|_{k_2 \le \frac{16}{\pi^2}\sqrt{1-M^2}} = \frac{4}{\pi}M \text{ , and} \tag{6.c}$$

$$d_{in,1\alpha}\big|_{k_2 > \frac{16}{\pi^2}\sqrt{1-M^2}} = d_{in,1\alpha\,\mathrm{MAX}} = \frac{16}{\pi^2} - d_{in,1\beta}^2 \tag{6.d}$$

Note that the condition on k_2 equates to a limitation on output power "P", with Eq. (6.c) corresponding to low power and Eq. (6.d) corresponding to high power levels. When using the reference from Eq. (6.c) $i_{L,1\alpha}$ is minimized and $i_{L,1\beta} = 0$. In [9], the optimization result showed that minimum RMS current is always obtained, independently of the power level, at $d_{out,1\alpha} = d_{out,1\alpha\,\mathrm{MAX}}$, for $M < 1$. However, it is evident to show that analysis is not valid for $M > 1$. What does still hold from the analysis in [9] for the case of up-conversion though, is that $\min(I_{RMS})$ can be obtained when $d_{in,1\alpha}^2 + d_{in,1\beta}^2 = \frac{16}{\pi^2}$. Considering Eqs. (7) representing the steady state behavior of Eq. (1):

$$d_{out,1\alpha}i_{L,1\alpha} * n = \frac{2.v_o}{R} = \frac{2P}{v_o} \tag{7.a}$$

$$d_{in,1\beta} = \frac{L\omega_0}{v_{in}}i_{L,1\alpha} = \frac{L\omega_0 2P}{v_o n\,v_{in}\,d_{out,1\alpha}} \tag{7.b}$$

$$i_{L,1\beta} = \frac{1}{L\omega_0}\big(d_{out,1\alpha}nv_o - v_{in}d_{in,1\alpha}\big) \tag{7.c}$$

If $d_1 = d_{1-\mathrm{MAX}} = 0.5$, Eq. (8) holds. With Eq. (8) and Eq. (7.b) and Eq. (7.a), the steady state value of $i_{L,1\beta}$ can be obtained as in Eq. (9).

$$d_{in,1\alpha}^2 + d_{in,1\beta}^2 = \frac{16}{\pi^2} \tag{8}$$

$$i_{L,1\beta} = \frac{1}{L\omega_0}\left(d_{out,1\alpha}nv_o - v_{in}\sqrt{\left(\frac{4}{\pi}\right)^2 - \left(\frac{L\omega_0 2P}{v_o n\,v_{in}\,d_{out,1\alpha}}\right)^2}\right) \tag{9}$$

Thus, minimizing RMS current equates to Eq. (10).

$$\min\left(i_{L,1\alpha}^2 + i_{L,1\beta}^2\right) = \min\left[\left(\frac{2P}{d_{out,1\alpha}\,v_0\,n}\right)^2 + \left(\frac{1}{L\omega_0}\left(nv_o\,d_{out,1\alpha} - v_{in}\sqrt{\left(\frac{4}{\pi}\right)^2 - \left(\frac{L\omega_0 2P}{v_0 n\, v_{in}\, d_{out,1\alpha}}\right)^2}\right)\right)^2\right]$$

(10)

With the result for both the FCA model input and $d_{out,1\alpha}$ and state variable $i_{L,1\beta}$ detailed in Eq. (11).

$$d_{out,1\alpha_{opt}} = \sqrt{\left(\frac{\pi P L \omega_0}{2 v_{in} v_o n}\right)^2 + \left(\frac{4 v_{in}}{\pi v_o n}\right)^2}$$

(11.a)

$$i_{L,1\beta_{opt}} = \frac{1}{L\omega}\left(v_o n\, d_{out,1\alpha_{opt}} - v_{in}\sqrt{\left(\frac{4}{\pi}\right)^2 - \left(\frac{L\,\omega_0\,2P}{v_o n\, v_{in}\, d_{out,1\alpha_{opt}}}\right)^2}\right)$$

(11.b)

The result for other control variables can be immediately derived from Eqs. (7). Here, the fact that $d_{out,1\alpha} \le d_{out,1\alpha\,MAX} = \frac{4}{\pi}$, implies a limitation on power like the one found for the case of $M < 1$. This can be used to transform the result from Eq. (11.a) and express the constraint as in Eq. (12). If Eq. (12) does not hold, the value of all three d variables is as for the high power case when $M < 1$, i.e.: Eqs. (6.a), (6.b) and (6.d).

$$P \le \frac{8\, v_{in} v_o n}{\pi^2 L \omega_0}\sqrt{1 - \frac{1}{M^2}}$$

(12)

In Fig. 3 both resulting minimum RMS 1st harmonic components are plotted as a function of conversion gain and load. It represents the optimization results here shown, Eqs. (11), and current components obtained with the pre-existing solution for M<1 ([9]). Since the optimization algorithm also needs to generate the $d_{out,1\alpha_{opt}}$ reference, it is also included in Fig. 3. It is interesting to note that minimum currents are obtained for low conversion ratios, contrary to [9].

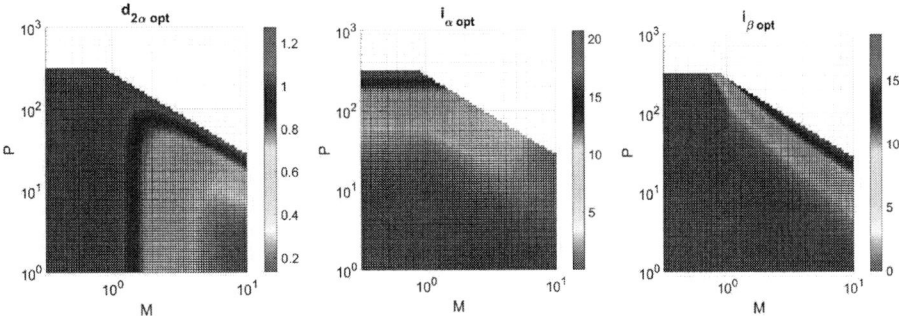

Fig. 3. Result of the optimization analysis for the two current components and $d_{2\alpha}$ variable plotted as a function of conversion ratio and output power. Power is in watts, currents in amperes. Note how higher conversion ratios lead to higher currents at full load.

Simulation Validation

The proposed approach was applied to validate the performance of a 150W, 28V to 12V isolated power supply designed for avionics applications operating at $f_{sw} = 1MHz$ with turns ratio $n = 2$, $L = 0.26uH$ and $C = 3mF$ anr with approximately $R_p = 40m\Omega$ parasitic resistance. It is especially interesting for this type of applications because they involve high v_{in} (between 18V to 46V) and P variation, i.e.: a wide operating range, where typical DAB design struggles to achieve high performances. The simulation was performed on the Simscape full-order converter model of Fig. 4, with a discrete control frequency, $f_{CONTROL} = f_{sw} = 1MHz$, and 1ns phase shift quantization, a value readily achievable many commercial FPGAs and micro-controllers

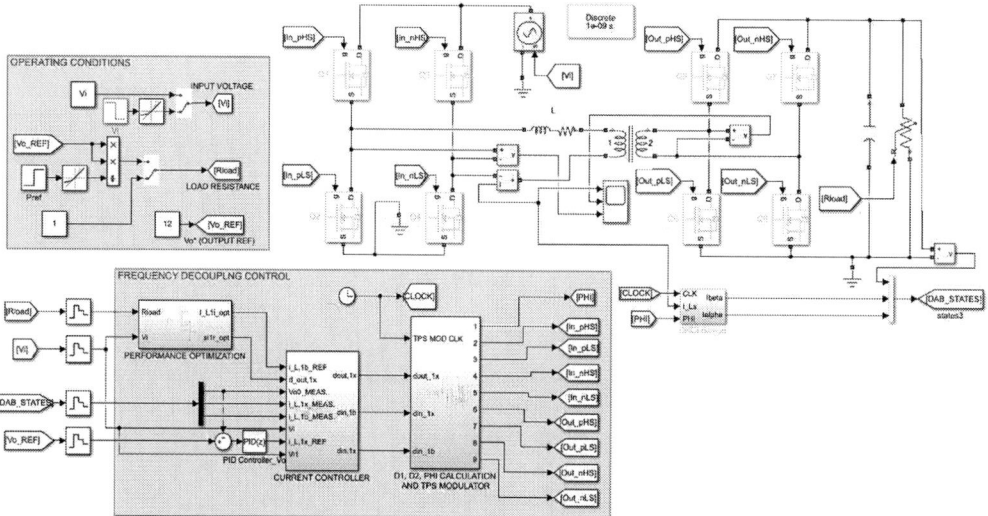

Fig. 4. Schematic of the Simscape model used to validate the proposed control approach. The diagram from Fig. 2 is implemented in the sub-system labelled "FREQUENCY DECOUPLING CONTROL" at the bottom of the image.

From a dynamic performance perspective, with the application covering a wide operating range, one first test that should be performed on the proposed controller structure is to ensure the tracking of state references under operating point variations. Fig. 5 shows control system proper state tracking under a step input voltage variation from minimum to maximum v_i. Note how voltage undershoot as the controller adapts is limited to $100mV$, lasting less than $2ms$.

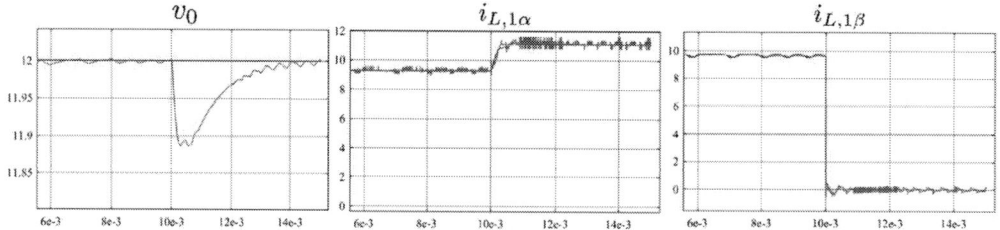

Fig. 5. System's states transient behavior under an input voltage step from 18V to 46V at nominal load. Controller references depicted in blue, state depicted in red. Currents are in amperes, voltages in volts, time is in seconds. The voltage step is applied at $t = 10ms$.

On the other hand, Fig. 6 depicts the proposed approach's inductor current response to a fast load variation. There, a step from 10W to 150W was applied at maximum v_{in}, i.e.: 46V. To highlight the fact that controlled current transitions are guaranteed even if implementing multiple optimization schemes, the proposed optimization scheme was implemented only for power levels above 20% nominal power, i.e. 30W. For power levels below that limit, $i_{L,1\beta}$ and $d_{out,1\alpha}$ references were defined such that steady state current would resemble that of a triangular current modulation scheme as in [2]. Thus, the transition to full power required switching optimization schemes. A smooth current transition can be appreciated, time resolution impeding discerning individual switching periods. The slight overshoot during the beginning of the transient is a DC offset, linked to the way in which phase shifts are updated and is readily avoidable as in [18].

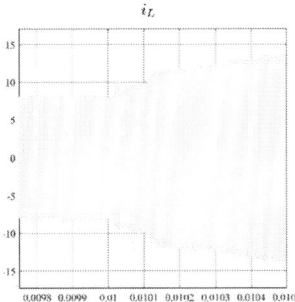

Fig. 6. Transformer current transient behavior under an output load step from 10W to 150W at 46V input voltage. Current is in amperes, time is in seconds. The load step is applied at $t = 10ms$.

Regarding steady state performance, the optimization approach should be able to reduce inductor RMS current in comparison to classical modulation approaches. Fig. 7 shows RMS current values throughout the operating space for both the proposed approach (referenced "FDM", as in "fundamental duty modulation" [9]) and for simple phase shift, noted "SPS". The improvement regarding SPS is considerable, especially in the extremes of the operating region (high/low input voltage), worst case low load and nominal load currents are reduced by more than 60% and 20% respectively.

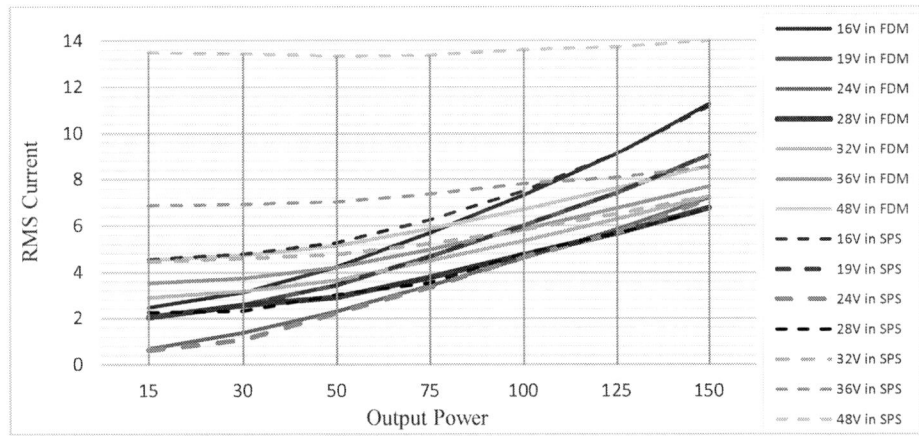

Fig. 7. RMS current throughout the operating range for both the proposed current minimization approach (full lines) vs for SPS (dashed lines). Current represented in amperes, power in watts.

To better evaluate and compare the performance of the current minimization scheme in steady state, the approach from [8] was employed to control the previously described converter under TPS. [8] is a current minimization scheme based on traditional operating mode analysis. RMS currents obtained throughout the operating range are shown in Fig. 8, labelled "TPS". RMS currents obtained when operating under other popular modulation approaches are also represented in Fig. 8. Label "TRM" is for classic triangular current mode [2], where the inductance is charged by v_{in} and then discharged by v_o in a manner similar to a buckboost converter. Currents labelled "Buck" and "Boost" correspond to classic down and up-conversion TRM modulations from [1]. The difference with the one labelled "TRM" being that, for these two, during one of the inductance charge/ discharge phases, the inductance is simultaneously excited by $v_{in} \; and \; v_o$. Analyzing Figs. 7 and 8, it can be concluded that all TRM modulations generate higher RMS currents than the proposed approach throughout the input voltage range at power levels above 20% nominal power, i.e. 30W. On the other hand, the proposed approach generates comparable current levels to the TPS modulation from [8] (less than 10% difference) throughout the input voltage range for power levels above 50% nominal power, i.e. 75W.

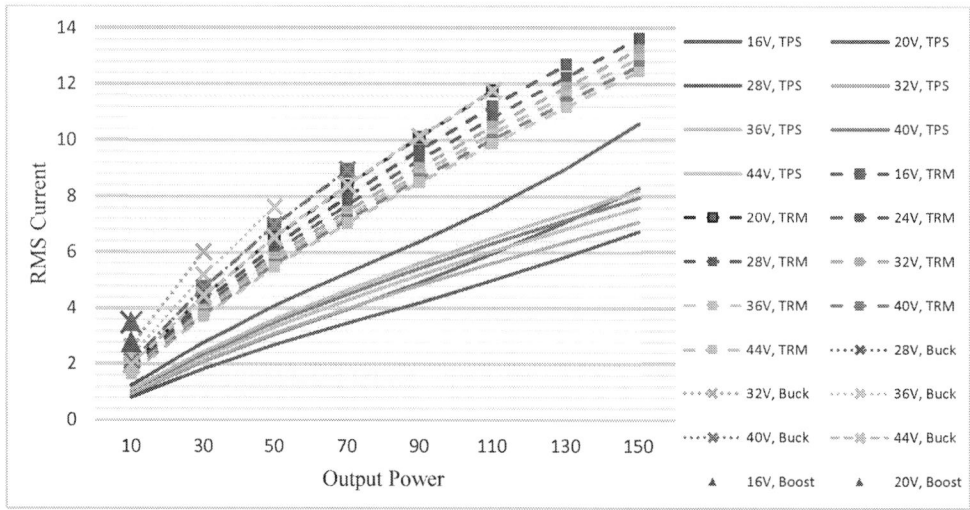

Fig. 8. RMS current throughout the operating range for state of the art ("TPS") and popular TRM modulations ("TRM", "Buck", "Boost") throughout the operating range. Current represented in amperes, power in watts.

Conclusion

A generic control system compatible with triple phase shift modulation was developed. It was shown to provide a simple mechanism to control a DAB converter without the need to analyze operating modes. The current minimization analysis was expanded from [9] to cover M>1 (up conversion). At moderate to high power levels, it was shown to result in RMS currents comparable to state-of-the-art TPS modulations based on operating mode analysis. Transient simulations showed proper tracking of controller references under varying operating conditions and optimization schemes. This shows the approach effectively decoupling optimization from dynamic control of a DAB. Follow up research will demonstrate using the model to reduce input harmonics, further improve converter dynamics and maintain ZVS throughout the load range, even for light load as in [17] and apply the proposed algorithms on a testbench.

References

[1] Krismer F.: Modeling and optimization of bidirectional dual active bridge DC-DC converter topologies, ETH Zurich PhD Thesis 2010.

[2] Yade O.: Commande Prédictive d'un Convertisseur Dual Active Bridge, École Supériéreur Polytéchnique de Dakar, mémoire de fin d'étude, 2015.

[3] Li X.: An Optimized Phase-Shift Modulation For Fast Transient Response in a Dual-Active-Bridge Converter, IEEE Transactions on Power Electronics 29(6):2661-2665

[4] Gu Q.: Current Stress Minimization of Dual-Active-Bridge DC–DC Converter Within the Whole Operating Range, IEEE Journal of Emerging and Selected Topics in Power Electronics, Volume: 7, Issue: 1, pp 129-142

[5] Riedel J.: ZVS Soft Switching Boundaries for Dual Active Bridge DC–DC Converters Using Frequency Domain Analysis, IEEE Transactions on Power Electronics, Vol.: 32, Issue: 4, pp 3166 – 3179

[6] Demumieux P.: Design of a Low-Capacitance Planar Transformer for a 4 kW/500 kHz DAB Converter, 2019 IEEE Applied Power Electronics Conference and Exposition (APEC),

[7] Hebala O.: Generic Closed-Loop Controller for Power Regulation in Dual Active Bridge DC–DC Converter With Current Stress Minimization, IEEE Transactions on Industrial Electronics, Vol.: 66, Issue: 6, pp 4468-4478

[8] Das D.: Optimal Design of a Dual-Active-Bridge DC–DC Converter, IEEE Transactions on Industrial Electronics, Vol.: 68, Issue: 12, pp 12034-12045

[9] Choi W.: Fundamental Duty Modulation of Dual-Active-Bridge Converter for Wide-Range Operation, IEEE Transactions on Power Electronic, Vol.: 31, Issue: 6, pp 4048 – 4064

[10] Oggier G.: Fast transient boundary control of the Dual Active Bridge Converter using the Natural Switching Surface, 2012 IEEE Energy Conversion Congress and Exposition (ECCE)

[11] Takagi K.: Dynamic control and dead-time compensation method of an isolated dual-active-bridge DC-DC converter, 2015 17th European Conference on Power Electronics and Applications (EPE'15 ECCE-Europe)

[12] Caliskan V.A.: Multifrequency averaging of DC/DC converters, IEEE Transactions on Power Electronics, Vol.: 14, Issue: 1

[13] Qin H.: Generalized Average Modeling of Dual Active Bridge DC–DC Converter, IEEE Transactions on Power Electronics, Vol.: 27, Issue: 4

[14] Uicich S.: General DAB 1st Harmonic TPS State Space Model, IECON 2021 – 47th Annual Conference of the IEEE Industrial Electronics Society

[15] Mueller J.A.: An Improved Generalized Average Model of DC–DC Dual Active Bridge Converters, IEEE Transactions on Power Electronics, Vol.: 33, Issue: 11, pp 9975 – 9988

[16] Mou D.: Hybrid Duty Modulation for Dual Active Bridge Converter to Minimize RMS Current and Extend Soft-Switching Range Using the Frequency Domain Analysis, IEEE Transactions on Power Electronics, Vol.: 36, Issue: 4

[17] Mou D.: Five-Degree-of-Freedom Modulation Scheme for Dual Active Bridge DC–DC Converter, IEEE Transactions on Power Electronics, Vol.: 36, Issue: 9

[18] Bu Q.: A Comparative Review of High Frequency Transient DC Bias Current Mitigation Strategies in Dual Active-Bridge DC-DC Converters Under Phase-Shift Modulations, IEEE Trans. Ind. Appl., vol. 58, pp. 2166-2182, December 2021.

Design and Performance Analysis of a Modified Proportional Multi-Resonant (PMR) Controller for Three-Phase Voltage-Source Inverters

Ahmad Ali Nazeri, Mahmoud Saeidi, and Peter Zacharias
Centre of Competence for Distributed Electric Power Technology
Faculty of Electrical Engineering / Computer Science
University of Kassel, Kassel, Germany
Email: ahmad.nazeri@student.uni-kassel.de,
{mahmoud.saeidi, peter.zacharias}@uni-kassel.de

August 15, 2022

Acknowledgments

This work was financially supported by the German Academic Exchange Service (DAAD) Germany and providing fully funded scholarship to Ahmad Ali Nazeri to support his doctoral studies.

Keywords

≪Current control≫, ≪Discretization≫, ≪Harmonics≫, ≪Non-linear load≫, ≪Resonant control≫

Abstract

The three-phase voltage source inverter (VSI) can be operated in grid-connected and/or stand-alone mode where the VSI is connected to the grid and/or critical loads at the point of common coupling (PCC). Proper voltage control is needed for the output voltage regulation in stand-alone operation and current control is needed for the grid current control. This paper presents a step-by-step design procedure, an extensive system stability analysis, and methods of discretization for the current control of three-phase power converters in the synchronous (dq) and stationary reference frame (SRF). A proportional-integral multi-resonant (PI-MR) controller in the synchronous reference frame (SynRF) is implemented for the regulation of the inner current loop. Moreover, the inverter inductor current controller in a stationary frame is proposed to provide active damping, and improve transients, and steady-state performance. The traditional PI-MR controller is compared for different load conditions with the modified practical proportional multi-resonant (PMR) controller in parallel with the harmonic compensators of orders 5th, 7th, 11th, and 13th to reduce low-order load current harmonics. The PMR controller shows superior performance with lower total harmonic distortion (THD) than the conventional PI and PI-MR controllers for highly nonlinear load conditions. Moreover, the modified PMR controller has almost zero steady-state error, improved tracking of the reference signal, and better disturbance rejection compared to the conventional PI-MR control. A comprehensive design guideline of the proposed controller with a wider range of system stability margin is analyzed with harmonic damping of the three-phase VSI. Proper discretization methods for each controller have been outlined. The system is simulated in MATLAB/Simulink environment and experimentally implemented on a TMS320F28335 floating-point digital signal processor (DSP) for a 7.5 kW inverter to validate the performance of the controllers.

Introduction

The distributed generation (DG) systems such as photovoltaic (PV), wind energy, and fuel cell are extensively integrated with the power electronic converters [1].

The voltage source inverters (VSIs) are widely interfaced in different power conversion applications such as island mode (microgrids), distributed generation, shunt active filters, and uninterruptible power supplies (UPS) [1]. Normally, the pulse-width-modulation (PWM) inverters can be operated in stand-alone and/or grid-connected mode [2, 3]. A constant voltage constant frequency (CVCF) PWM inverter should be used to regulate the output voltage with low total harmonic distortion (THD) and fast transient performance [4]. In, grid-connected mode, a fast current controller must be able to inject a nearly sinusoidal current into the grid with low THD according to the IEEE standards [5]. The three-phase VSI can be connected to the grid and/or the load at the point of common coupling (PCC). A voltage and current controller is needed for the output voltage and grid current regulation when the three-phase VSI is connected in stand-alone feeding loads or grid-connected injecting the grid current [6]. The basic working principle of PWM VSI is to convert the dc voltage to a sinusoidal ac output connected with an LC filter. The performance of the VSI is evaluated by the content of THD, the transient response, and overall efficiency [6]. The rectifier loads connected at the point of PCC affect the output current and voltage and introduce distortions into the output of the PWM VSI, which degrades the power quality [2]. The objective of a control system for the three-phase VSI system is to keep the output current nearly sinusoidal with minimum zero steady-state error and low THD under rectifier load [7].

To address this problem, there is a need for a current controller, which improves the dynamic and transient performance of the system, offers a wide range of stability margins, and provides almost zero steady-state error [8]. Several control strategies in the literature have been reported such as conventional proportional-integral (PI) control [9], proportional-resonant (PR) control [10], repetitive control (RC) [11], deadbeat model and predictive control [12]. The RC based on the internal model principle [11], perfectly tracks the periodic reference, eliminating the periodic errors from the rectifier loads. The RC is normally combined with another feedback controller or a deadbeat controller to improve the system stability, which increases the complexity and tuning of the controller [13]. Moreover, the RC uses large memory and has a slow response to non-periodic disturbance [13]. The traditional PI control is simple to implement but it has a steady-state error with poor tracking capability and poor disturbance rejection. On the other hand, the PR control eliminates the zero steady-state error, better tracking of the reference signal, and has better disturbance rejection [7]. The PI control in SynRF (dq) is identical to the ideal PR control in SRF ($\alpha\beta$) [10]. The PI control in the SynRF combined with the multiple resonant control (MR) in the SRF namely PI-MR can be used in current control to reduce the harmonics introduced by the rectifier loads [14].

This paper presents a modified practical proportional multiple-resonant (PMR) current control in parallel with the harmonic compensators (HCs) of orders 5th, 7th, 11th, and 13th used to mitigate the low-order load current harmonics. Moreover, a step-by-step design procedure of the PMR current control with the analysis of system stability is presented. A wide range of stability margins for the current control is derived based on the design methodology. The dynamic and transient performance of the modified PMR current control is compared with the traditional PI-MR for highly nonlinear loads. Also, the proper discretization methods for the PI-MR and PMR controllers are derived for the practical implementation of the controller on TMS320F28335 32-bit floating-point digital signal processor (DSP). The modified PMR controller can also be implemented in the voltage control in stand-alone operation and/or grid-current control for grid-connected mode. The proposed methodology is simulated in MATLAB/Simulink environment and experimentally implemented on a 7.5 kW three-phase VSI, which effectively tracks the reference signal with nearly zero steady-state error and reduces the output current THD.

Harmonic Damping Scheme for Three-Phase VSI System

Fig. 1 shows the block diagram of a three-phase VSI with an LC filter. The dc-ink voltage V_{dc} is assumed to be constant. The inverter is connected to an LC filter L_f, C_f, and the load Z_L. R_f is the equivalent series resistance (ESR) of the filter inductance L_f. i_L, i_c, and i_o are the inverter, capacitor and load currents respectively where v_o is the output voltage. The reference rectifier load is connected with the load resistance R_d as depicted in Fig. 1. The system parameters are given in Table I. The traditional PI controller in SynRF (dq) can be used to regulate the output current of the inverter [14]. The three-phase output currents are transformed to the dq frame and given to the PI controller as shown in Fig. 2.

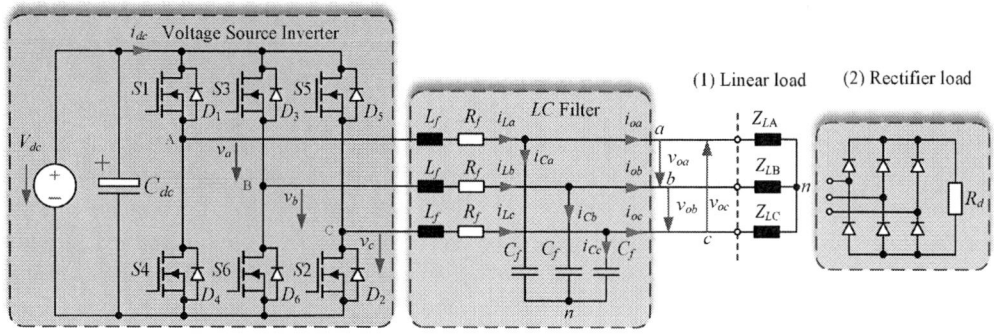

Fig. 1: Diagram of a three-phase VSI with an *LC* filter

The standard PI controller is given in (1a) forces the error in the *dq* axis to zero and decouples the *d* and *q* components. This control scheme is similar to the ideal PR controller in αβ frame [16] given in (1b). The ideal PR controller has an infinite gain at the nominal grid frequency ω_o [8], has no phase shift and gain at other frequencies [16], and has a better tracking capability of AC reference signals compared to the PI control [10]. A multi-resonant (MR) controller in αβ frame given in (1c) can be used in parallel to the standard PI controller in *dq* frame to compensate for the harmonics in the load current when a highly non-linear load is connected as shown in Fig. 2. The traditional PI with the MR control scheme is referred as PI-MR controller as depicted in Fig. 2. The transfer functions of each controller is expressed as

$$G_{\text{PI},ci}(s) = K_{pi} + \frac{K_{i1}}{s} \tag{1a}$$

$$G^s_{\text{PR},ci}(s) = K_{pi} + \frac{K_{i1}s}{s^2 + \omega_o^2} \tag{1b}$$

$$G^s_{\text{MR},ci}(s) = \sum_{n=1}^{k} \frac{K_{in}s}{s^2 + (n\omega_o)^2} \tag{1c}$$

$$G^s_{\text{PMR},ci}(s) = K_{pi} + \frac{K_{i1}\omega_d s}{s^2 + 2\omega_d s + \omega_o^2} + \sum_{n=1}^{k} \frac{K_{in}\omega_d s}{s^2 + 2\omega_d s + (n\omega_o)^2}. \tag{1d}$$

Table I: System parameters

Parameter	Symbol	Value	Unit
Nominal active power	P_{max}	7.5	[kW]
Base voltage	V_R	325	[V]
Base current	I_B	15.3	[A]
Base impedance	Z_B	21.3	[Ω]
Base capacitance	C_B	150	[μF]
DC input voltage	V_{dc}	400	[V]
Nominal frequency	f_o	50	[Hz]
Switching frequency	f_{sw}	20	[kHz]
Sampling time	T_s	50	[μs]
PWM Time delay	T_d	75	[μs]
Filter inductance	L_f	2.6	[mH]
Filter winding resistance	R_f	0.1	[Ω]
Filter capacitance	C_f	6	[μF]
Connected load	R_d	12.74	[Ω]

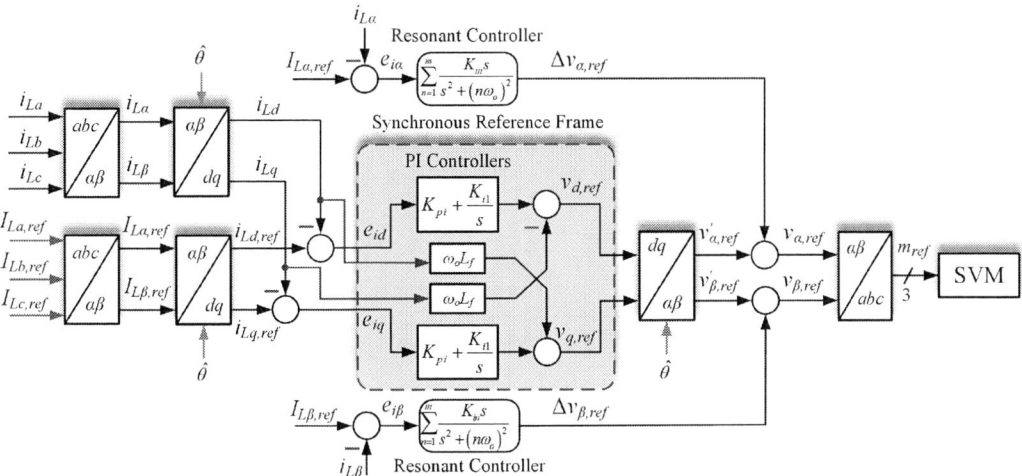

Fig. 2: Control structure of the standard PI with multi-resonant (MR) current control

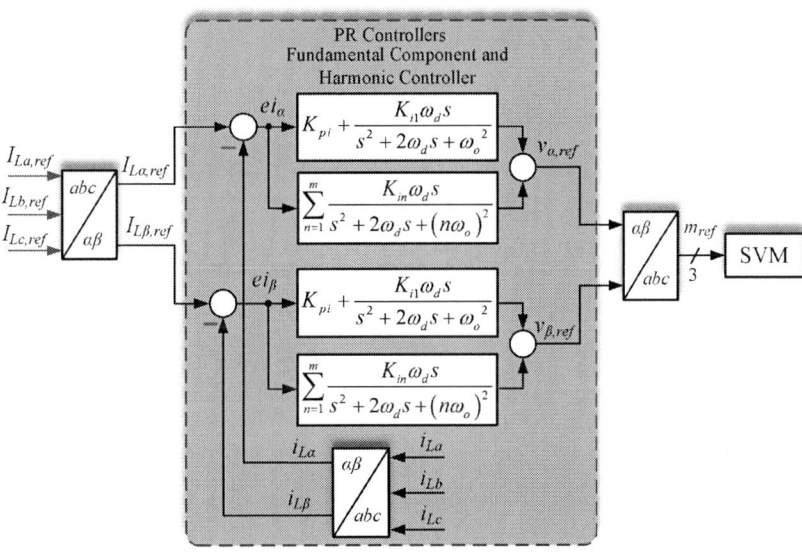

Fig. 3: Control block diagram of the practical proportional multi-resonant (PMR) current control

where 's' denotes the $\alpha\beta$-frame. K_{pi} is the proportional gain, K_{i1} and K_{in} are the fundamental and multiple harmonic integral gains. n denotes the harmonic order, ω_o is the fundamental grid frequency, and ω_d is the cut-off frequency of the PMR controller. The modified PMR current control in $\alpha\beta$ frame given in (1d), which has a finite gain at the target frequency, has better transient performance, and is capable to regulate the output current close to the reference signal compared to the traditional PI and PI-MR controllers [15]. However, there is a trade-off between the damping ratio ω_d for stable a operation and the controller gains at the selected frequency to force the zero steady-state error [13]. Choosing smaller ω_d makes the filter more sensitive to the frequency variations, leads to slow transient response [13], and makes it difficult for the practical implementation of the DSP due to its deteriorating performance and limited resolution [17]. The modified PMR controller has half of the resonance gain proposed in [10, 16] and harmonic compensating terms are normally added in parallel to the fundamental controller for the excellent harmonic damping capability with fast transient performance as shown in Fig. 3.

Controllers Design

Design Procedure of the PI Current Control

The standard PI current control given in (1a) is taken with the transport and sampling delay e^{-sT_d} in the forward path and the plant illustrated in Fig. 4, which gives the open-loop transfer function of the current control as

$$G_{\text{OL},ci}(s) = \underbrace{K_{pi}\left(1 + \frac{K_{i1}}{sK_{pi}}\right)}_{\text{PI current controller}} \underbrace{\left(e^{-sT_d}\right)}_{\text{PWM delay}} \underbrace{\left(\frac{K_f}{sT_f + 1}\right)}_{\text{Inverter}}. \tag{2}$$

where T_d is the combined sampling and transport delay times, $K_f = V_{dc}/R_f$, and $T_f = L_f/R_f$ are the plant transfer functions. Substituting $s \to j\omega_{ci}$ in (2) and applying the Nyquist stability criterion such as when the phase response reaches $-\pi$, the magnitude response must be less than 1 to ensure no encirclement of the -1. Deriving the open-loop magnitude and phase response of the forward current loop at the cross-over frequency ω_{ci} [18]

$$\begin{aligned} \angle G_{\text{OL},ci}(j\omega_{ci}) &= \angle\left\{\left[K_{pi}\left(1 + \frac{K_{i1}}{j\omega_{ci}K_{pi}}\right)\right] \times e^{-j\omega_{ci}T_d} \times \left(\frac{K_f}{j\omega_{ci}T_f + 1}\right)\right\} \\ &= (-\pi + \phi_m) \\ &\approx \tan^{-1}\left(\frac{\omega_{ci}K_{pi}}{K_{i1}}\right) - \frac{\pi}{2} - \omega_{ci}T_d - \tan^{-1}(\omega_{ci}T_f). \end{aligned} \tag{3}$$

At ω_{ci}, the phase contribution of $\tan^{-1}(\omega_{ci}K_{pi}/K_{i1}) \approx \pi/2$ and $\tan^{-1}(\omega_{ci}T_f) \approx \pi/2$. Therefore, the maximum crossover frequency $\omega_{ci,\max}$ can be written as [17]

$$\omega_{ci,\max} = \frac{\pi/2 - \phi_m}{T_d}. \tag{4}$$

The maximum magnitude of K_{pi} can be found by setting the (3) as $|G_{\text{OL},ci}(j\omega_{ci,\max})| = 1$ which gives,

$$K_{pi} = \left(\frac{\omega_{ci,\max}K_{pi}}{K_{i1}K_f}\right)\frac{\sqrt{1 + \omega_{ci,\max}^2 T_f^2}}{\sqrt{1 + \frac{\omega_{ci,\max}^2 K_{pi}^2}{K_{i1}^2}}}. \tag{5}$$

Considering $\omega_{ci,\max}K_{pi}/K_{i1} \gg 1$ and $\omega_{ci,\max}T_f \gg 1$. Substituting K_f and T_f in (5) with simplification, which results in the proportional gain K_{pi} as [18],

$$K_{pi} = \frac{\omega_{ci,\max}T_f}{K_f} = \frac{\omega_{ci,\max}L_f}{V_{dc}}. \tag{6}$$

The integral gain K_{i1} can be determined by setting $\tan^{-1}(\omega_{ci,\max}K_{pi}/K_{i1}) \approx \pi/2$,

$$K_{i1} = \frac{\omega_{ci,\max}K_{pi}}{\tan 89.1°} = \frac{\omega_{ci,\max}K_{pi}}{30}. \tag{7}$$

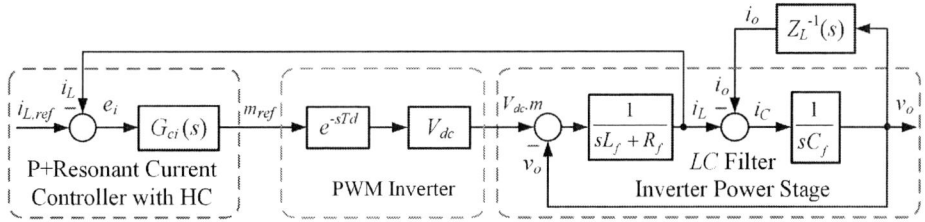

Fig. 4: Closed loop control structure of the system in the stationary reference frame

Design Procedure of the PMR Current Control

The open-loop transfer function of the PMR current control in $\alpha\beta$ frame shown in Fig. 4 and is expressed as

$$G_{\text{OL},ci}^{s}(s) = \underbrace{\left(K_{pi} + \frac{K_{i1}\omega_d s}{s^2 + 2\omega_d s + \omega_o^2} \right)}_{\text{PMR current controller}} \underbrace{\left(e^{-sT_d} \right)}_{\text{PWM delay}} \underbrace{\left(\frac{K_f}{sT_f + 1} \right)}_{\text{Inverter}}. \tag{8}$$

where ω_d limits the forward gain at ω_o to $G_{\text{PMR},ci}^{s}(s) = K_{pi} + K_{i1}/\omega_{ci}$ with the approximation that the cross-over frequency ω_{ci} is much greater than the ω_o, the phase angle of the $G_{\text{OL},ci}(s)$ in (8) can be approximated as [18]

$$\begin{aligned}
\angle G_{\text{OL},ci}(j\omega_{ci}) &= \angle \left\{ K_{pi} \left[1 + \frac{K_{i1}}{K_{pi}} \times \frac{j\omega_{ci}}{(\omega_o^2 - \omega_{ci}^2) + j\omega_{ci}\omega_d} \right] \times e^{-j\omega_{ci}T_d} \times \left(\frac{K_f}{j\omega_{ci}T_f + 1} \right) \right\} \\
&= (-\pi + \phi_m) \\
&\approx \tan^{-1}\left(\frac{\omega_{ci}K_{pi}}{K_{i1}} \right) - \frac{\pi}{2} - \omega_{ci}T_d - \tan^{-1}(\omega_{ci}T_f).
\end{aligned} \tag{9}$$

which gives the phase margin as

$$\omega_{ci} = \frac{\tan^{-1}(\omega_{ci}K_{pi}/K_{i1}) - \phi_m}{T_d}. \tag{10}$$

From (10), the maximum cross-over frequency $\omega_{ci,\max}$ for a given phase margin ϕ_m with the approximation of $\tan^{-1}(\omega_{ci,\max}K_{pi}/K_{i1}) \approx \pi/2$, which results in similar results given in (4). The K_{pi} and K_{i1} parameters can be found in similar manner as described in the section of standard PI control design with the approximations. Therefore, the fundamental component parameters are of K_{pi} and K_{i1} given in (6) and (7) can be used respectively. The integral gain of the MR control is designed based on the K_{i1} as $K_{in} = K_{i1}/n$. The current controller is designed in the $\alpha\beta$ frame based on (1) used in the open-loop transfer function $G_{\text{OL},ci}^{s}(s)$ which is written as

$$G_{\text{OL},ci}^{s}(s) = G_{ci}^{s}(s) \times e^{-sT_d} \times \frac{K_f}{sT_f + 1}. \tag{11}$$

Fig. 5a depicts the comparison of the frequency response of the $G_{\text{OL},ci}^{s}(s)$ of PI+MR controller (see 1a and 1c) with the PMR+MR controller (see 1d) in the $\alpha\beta$ frame extracted from (11). The combined

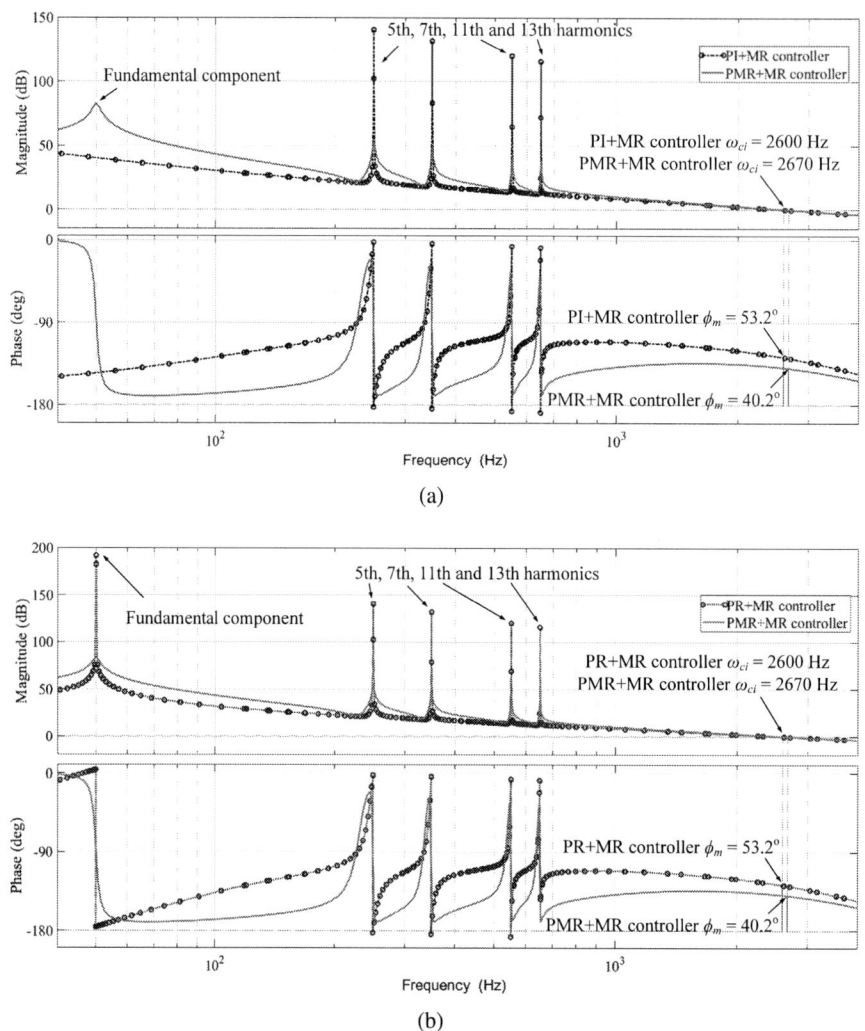

(a)

(b)

Fig. 5: Frequency response of the current control loop: (a) open-loop transfer function in synchronous and stationary reference frames of PI+MR and the PMR+MR controllers; (b) open-loop transfer function in stationary reference frame of PR+MR and PMR+MR controllers

sampling and transport delay time is $T_d = 75\mu s$. The set phase margin $\phi_m = 55°$, and the resulted cross-over frequency $\omega_{ci,max} = 2 \times \pi \times 2593$ rad/s. However, the final phase margin of the PI+MR controller is slightly reduced to 53.2° due to the resonance at the output frequency with $\omega_{ci,max} = 2 \times \pi \times 2600$ rad/s. Fig. 5b presents the comparison of the open-loop frequency response of the PR+MR controller (see 1b and 1c) and PMR+MR controller extracted from (11) where the phase margin $\phi_m = 40.2°$ is reduced due to presence of the higher-order harmonic controllers with $\omega_{ci,max} = 2 \times \pi \times 2670$ rad/s [13].

Methods of Discretization

Most of the voltage and current controllers are implemented digitally on a DSP, so the suitable discretization method for the resonant controller should not be ignored [17]. Due to the narrow band and infinite gain of the resonant controller, it is sensitive to the discretization process because the slight displacement in the resonant poles leads to the high loss of controller performance. Moreover, discretization has also an effect on the zeros, as it modifies its displacement with its continuous domain transfer function [17]. The computational delay e^{-sT_d} influences the performance of the system and can lead to system instability. Hence, the sampling and transport delay has been implemented in the design of the controller. Therefore, great care should be taken while implementing the resonant controllers on DSP [16]. The ideal resonant controller in (1c) is discretized using the forward and backward Euler's method and the PMR controller in (1d) is discretized using the Tustin method.

Forward Euler and Backward Euler Implementation

The results of a MR controller in (1c), when the direct integrator is discretized using the forward Euler $s = \frac{1-z^{-1}}{z^{-1}T_s}$ and the feedback integrator is discretized using the backward Euler $s = \frac{1-z^{-1}}{T_s}$, which results in

$$
\begin{aligned}
G_{\mathrm{MR},ci}(z) &= \frac{K_{in}\left(\frac{z^{-1}T_s}{1-z^{-1}}\right)}{1+n^2\omega_o^2\left(\frac{T_s}{1-z^{-1}}\right)\left(\frac{z^{-1}T_s}{1-z^{-1}}\right)} = \frac{\left(K_{in}T_s z^{-1}\right)/\left(1-z^{-1}\right)}{\left[\left(1-z^{-1}\right)^2 + n^2\omega_o^2 T_s^2 z^{-1}\right]/\left(1-z^{-1}\right)^2} \\
&= \frac{K_{in}T_s z^{-1}\left(1-z^{-1}\right)}{\left(1-z^{-1}\right)^2 + n^2\omega_o^2 T_s^2 z^{-1}} = \frac{K_{in}T_s\left(z^{-1}-z^{-2}\right)}{1+\left(n^2\omega_o^2 T_s^2 - 2\right)z^{-1}+z^{-2}}.
\end{aligned}
\tag{12}
$$

Comparing (12) to the standard z-domain discrete transfer function in (13) to obtain the coefficients of the MR control and digitally implemented in the direct form II transpose as shown in Fig. 6b and given

$$
\frac{Y(z)}{e(z)} = \frac{b_0 + b_1 z^{-1} + b_2 z^{-1}}{1 + a_1 z^{-1} + a_2 z^{-2}}.
\tag{13}
$$

where $b_0 = 0$, $\qquad b_1 = K_{in}T_s$, $\qquad b_2 = -b_1$, $\qquad a_1 = n^2\omega_o^2 T_s^2 - 2$, $\qquad a_2 = 1$.

Tustin (trapezoid) Implementation

The results of a PMR controller in (1d) is discretized using Tustin $s = \frac{2}{T_s}\frac{1-z^{-1}}{1+z^{-1}}$ as shown in Fig. 6a and given as [16]

$$
\begin{aligned}
G_{\mathrm{PMR},ci}(z) &= \frac{K_{in}\omega_d\left(\frac{2}{T_s}\frac{1-z^{-1}}{1+z^{-1}}\right)}{\left(\frac{2}{T_s}\frac{1-z^{-1}}{1+z^{-1}}\right)^2 + 2\omega_d\left(\frac{2}{T_s}\frac{1-z^{-1}}{1+z^{-1}}\right) + \left(n\omega_o\right)^2} \\
&= \frac{(2/T_s)K_{in}\omega_d\left(1-z^{-2}\right)}{(4/T_s^2)\left(1-z^{-1}\right)^2 + (4\omega_d/T_s)\left(1-z^{-2}\right) + n^2\omega_o^2\left(1+z^{-1}\right)^2} \\
&= \frac{2K_{in}\omega_d T_s\left(1-z^{-2}\right)}{4\left(1-2z^{-1}+z^{-2}\right) + 4\omega_d T_s\left(1-z^{-2}\right) + n^2\omega_o^2 T_s^2\left(1+2z^{-1}+z^{-2}\right)} \\
&= \frac{2K_{in}\omega_d T_s\left(1-z^{-2}\right)}{\left(4+4\omega_d T_s + n^2\omega_o^2 T_s^2\right) + \left(2n^2\omega_o^2 T_s^2 - 8\right)z^{-1} + \left(4-4\omega_d T_s + n^2\omega_o^2 T_s^2\right)z^{-2}}.
\end{aligned}
\tag{14}
$$

To obtain the z-domain discrete transfer function given (13), we divide the nominator and denominator of (14) by temp $= \left(4+4\omega_d T_s + n^2\omega_o^2 T_s^2\right)$, we get

$$
G_{\mathrm{PMR},ci}(z) = \frac{\left(2K_{in}\omega_d T_s/\mathrm{temp}\right)\left(1-z^{-2}\right)}{1 + \left[\left(2n^2\omega_o^2 T_s^2 - 8\right)/\mathrm{temp}\right]z^{-1} + \left[\left(4-4\omega_d T_s + n^2\omega_o^2 T_s^2\right)/\mathrm{temp}\right]z^{-2}}.
\tag{15}
$$

where $b_0 = 2K_{in}\omega_d T_s/\mathrm{temp}$, $\qquad b_1 = 0$, $\qquad b_2 = -b_0$,
$a_1 = \left(2n^2\omega_o^2 T_s^2 - 8\right)/\mathrm{temp}$, $\qquad a_2 = \left(4-4\omega_d T_s + n^2\omega_o^2 T_s^2\right)/\mathrm{temp}$.

Fig. 6b depicts the discrete implementation of the modified PMR controller in the direct form II transpose as given in (13) and (15). The difference equation needed for the DSP implementation of the PR controller is given as [16]

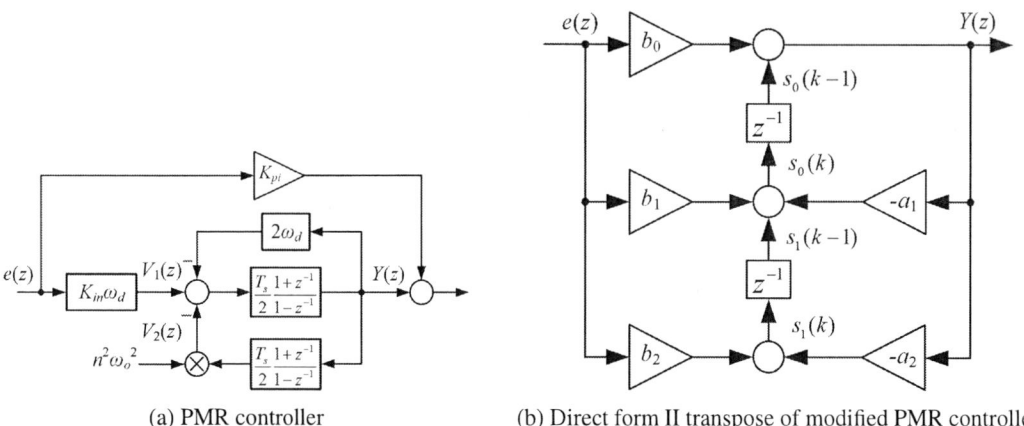

(a) PMR controller   (b) Direct form II transpose of modified PMR controller

Fig. 6: Discrete-time implementation of the PMR controller

Table II: Parameters of the current controller

Control schemes	K_{pi}	K_{i1}	$K_{in(5,7,11,13)}$	ω_d [rad/s]	ω_{ci} [rad/s]	ϕ_m
PI-MR	1.37	933.61	186, 133, 84, 71	-	16336	53.2°
PMR-MR	1.78	276.63	55, 39, 25, 21	5	16776	40.2°

$$Y(z) = b_0 e(z) + b_1 e(z) z^{-1} - a_1 Y(z) z^{-1} + b_2 e(z) z^{-2} - a_2 Y(z) z^{-2}. \tag{16}$$

Fig. 7: Experimental prototype of the system

Simulation and Experimental Validations

The three-phase VSI illustrated in Fig. 1 with the control schemes shown in Fig. 2, and 3 were implemented with the space vector modulation (SVM) in MATLAB/Simulink. Fig. 7 shows the experimental setup at the laboratory. The experimental setup includes a constant programmable DC power source

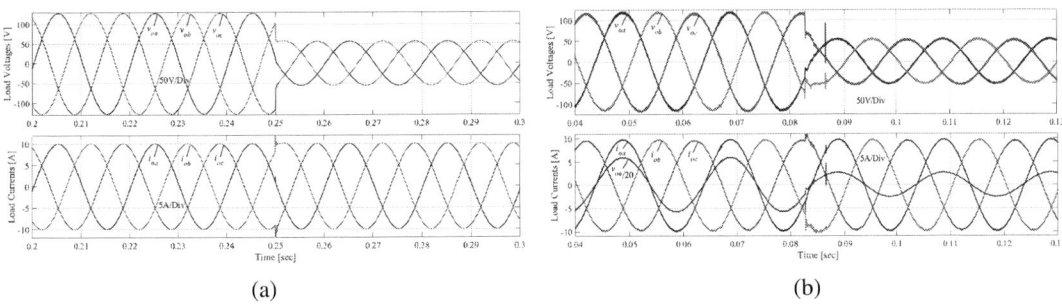

(a)   (b)

Fig. 8: Dynamic performance of the three-phase VSI with the step-change in the load employing standard PI control; (a) simulation, (b) experimental

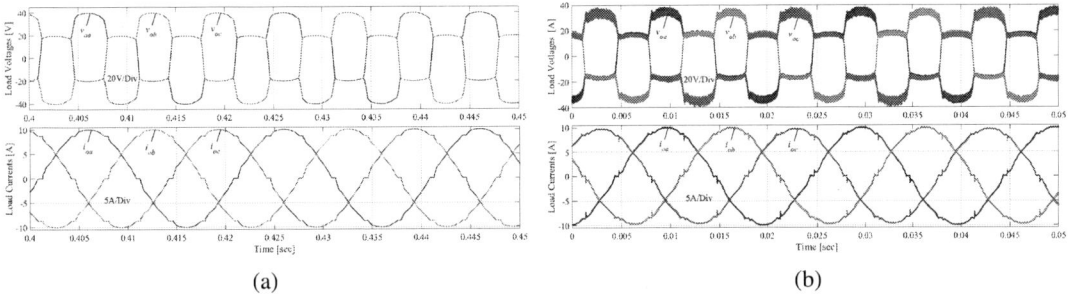

(a)   (b)

Fig. 9: The performance of the PI-MR control under highly rectifier load; (a) simulation, (b) experimental

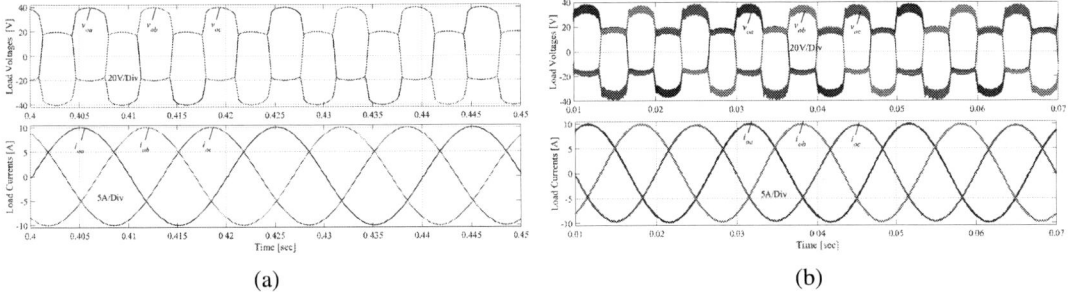

(a)   (b)

Fig. 10: The performance of the proposed modified PMR-MR control; (a) simulation, (b) experimental

from Regatron as V_{dc} connected to a three-phase CoolSiC MOSFET of 7.5kW inverter, a three-phase LC filter, a three-phase voltage and current sensors, and the load as illustrated in Fig. 7. An 8-channel MSO68B oscilloscope was used to display the results. The current and voltage THDs were extracted from the precision power analyzer ZES-Zimmer. The proposed control schemes were implemented on a 32-bit floating-point TMS320F28335 DSP from Texas Instruments. The sampling was executed in the middle of every sampling period T_s. The base voltage V_B, base current I_B, base impedance Z_B, and the base capacitance C_B given in Table I were used to calculate the control parameters in the per-unit system, and the measured voltages and currents were scaled into the per-unit system with the control parameters given in Table II were used for the simulation and experiments. The discrete-time PI controller (see 1a) in the dq frame was implemented using the backward difference approximation whereas the MR controller (see 1c) was discretized and implemented using the forward and backward Euler method. The PMR controller (see 1d) was carefully discretized using the Tustin approximation [16]. The reference signals $V_{\alpha,ref}$ and $V_{\beta,ref}$ are the input signals to the SVM to generate the control signals. The 12-bit digital to analog converter (DACs) was used to show the internal signals on the oscilloscope. The inverter was tested under resistive load with the step-change in the load and highly rectifier load as shown in Fig. 7. Fig. 8a shows the simulated dynamic response of the three-phase inverter with an LC filter employing

Table III: Experimental data of the inverter

Control schemes	V_o [V]	THD$_v$ [%]	I_o [A]	THD$_i$ [%]
under the rectifier load				
PI	40	28	10	16
PI-MR	40	28	10	2.54
PR-MR	40	28	10	2.40
PMR-MR	40	28	10	0.55
under the RC load				
PI	130	0.78	10	1.44
PI-MR	130	0.47	10	0.74
PR-MR	130	0.40	10	0.71
PMR-MR	130	0.24	10	0.38
under the resistive load				
PI	153	0.67	12	0.67
PI-MR	153	0.48	12	0.47
PR-MR	153	0.47	12	0.45
PMR-MR	153	0.36	12	0.33

standard PI controller in a dq frame where Fig. 8b depicts the experimental results. A 10Ω resistive load is connected in parallel to R_d where a step change in the load happens from full load to half load at 0.25s and 0.085s in the simulation and experiment respectively, where the smooth transition happens in the base currents of 10A as shown in Fig. 8. The reference currents in per-unit are 0.7A of the base current I_B. Fig. 9 presents the simulated and experimental implementation of the PI-MR controller under highly nonlinear load. Fig. 10 depicts the discrete implementation of the modified PMR controller where a highly nonlinear load is connected to the inverter. The modified PMR controller in the $\alpha\beta$ frame shown in Fig. 10a (simulation) and Fig. 10b (experimental) has superior transient performance compared to the standard PI and PI-MR control in the dq frame. It is evident that the modified PMR controller discretized with Tustin has better harmonic damping capability compared to PI-MR controller. Table III outlines the RMS and the THD values of the output voltage and current for different control schemes. Table III illustrates the effective performance of the PMR-MR controller over the traditional PI, PI-MR, and PR-MR control schemes for resistive, RC, and rectifier loads.

Conclusion

This paper explores the harmonic damping schemes of the three-phase inverter for off-grid renewable energy and grid-connected systems. The comparison of the standard current control employing PI and PI-MR control is compared with the proposed modified PMR control supplying linear and highly nonlinear loads. A step-by-step design procedure of the current control loop for the PI-MR and PMR control is presented. The ideal PR control in the $\alpha\beta$ frame is equivalent to the PI control in the dq frame. A multi-resonant (MR) controller in the $\alpha\beta$ frame with the harmonic components of 5th, 7th, 11th, and 13th, are added in parallel to the fundamental PI controller in the dq frame. The proposed modified PMR controller in $\alpha\beta$ frame with the harmonic components of 5th, 7th, 11th, and 13th added to the fundamental controller for the mitigation of the harmonics in the load currents. A proper design methodology of the controllers, system stability analysis, and dynamic and transient performance with the proper methods of discretization are the key points of this paper. The proposed controller offers better dynamic performance, a higher stability margin, ease of digital implementation, and is capable to suppress the load current harmonics with zero-steady state error. The output currents show a lower total harmonic distortion of 0.55 % with the load voltage total harmonic distortion of 28 %. It is found that the proposed modified PMR controller offers fast dynamic and transient performance, good tracking accuracy, and better disturbance rejection under linear and nonlinear loads.

References

[1] F. Blaabjerg, Z. Chen and S. B. Kjaer, "Power electronics as efficient interface in dispersed power generation systems," IEEE Transactions on Power Electronics, vol. 19, no. 5, pp. 1184-1194, 2004.

[2] A. A. Nazeri, P. Zacharias, F. M. Ibanez and I. Idrisov, "Paralleled Modified Droop-Based Voltage Source Inverter for 100% Inverter-Based Microgrids," 2021 IEEE Industry Applications Society Annual Meeting (IAS), 2021, pp. 1-8, 2021.

[3] A. Ali, P. Shanmugham and S. Somkun, "Single-phase grid-connected voltage source converter for LCL filter with grid-current feedback," 2017 International Electrical Engineering Congress (iEECON), 2017, pp. 1-6, 2017.

[4] S. Yang, P. Wang, Y. Tang and L. Zhang, "Explicit Phase Lead Filter Design in Repetitive Control for Voltage Harmonic Mitigation of VSI-Based Islanded Microgrids," IEEE Transactions on Industrial Electronics, vol. 64, no. 1, pp. 817-826, 2017.

[5] D. Chen, J. Zhang and Z. Qian, "An Improved Repetitive Control Scheme for Grid-Connected Inverter With Frequency-Adaptive Capability," IEEE Transactions on Industrial Electronics, vol. 60, no. 2, pp. 814-823, 2013.

[6] R. Wai, C. Lin, Y. Huang and Y. Chang, "Design of High-Performance Stand-Alone and Grid-Connected Inverter for Distributed Generation Applications," IEEE Transactions on Industrial Electronics, vol. 60, no. 4, pp. 1542-1555, 2013.

[7] S. Somkun, "High performance current control of single-phase grid-connected converter with harmonic mitigation, power extraction and frequency adaptation capabilities," IET Power Electronics, vol. 14, no. 2, pp. 352–372, 2021.

[8] A. A. Nazeri, P. Zacharias, F. M. Ibanez and S. Somkun, "Design of Proportional-Resonant Controller with Zero Steady-State Error for a Single-Phase Grid-Connected Voltage Source Inverter with an LCL Output Filter," 2019 IEEE Milan PowerTech, pp. 1-6, 2019.

[9] N. M. Abdel-Rahim and J. E. Quaicoe, "Analysis and design of a multiple feedback loop control strategy for single-phase voltage-source UPS inverters," IEEE Transactions on Power Electronics, vol. 11, no. 4, pp. 532-541, 1996.

[10] D. N. Zmood and D. G. Holmes, "Stationary frame current regulation of PWM inverters with zero steady-state error," IEEE Transactions on Power Electronics, vol. 18, no. 3, pp. 814-822, 2003.

[11] Francis, Bruce A., and William M. Wonham. "The internal model principle for linear multivariable regulators, Applied mathematics and optimization," vol. 2, no. 2, pp. 170-194, 1975.

[12] S. Kouro, P. Cortes, R. Vargas, U. Ammann and J. Rodriguez, "Model Predictive Control—A Simple and Powerful Method to Control Power Converters," IEEE Transactions on Industrial Electronics, vol. 56, no. 6, pp. 1826-1838, 2009.

[13] Somkun, Sakda. "Unbalanced synchronous reference frame control of singe-phase stand-alone inverter." International Journal of Electrical Power & Energy Systems, vol. 107, pp. 332-343, 2019.

[14] M. Liserre, R. Teodorescu and F. Blaabjerg, "Multiple harmonics control for three-phase grid converter systems with the use of PI-RES current controller in a rotating frame," IEEE Transactions on Power Electronics, vol. 21, no. 3, pp. 836-841, 2006.

[15] A. Roshan, R. Burgos, A. C. Baisden, F. Wang and D. Boroyevich, "A D-Q Frame Controller for a Full-Bridge Single Phase Inverter Used in Small Distributed Power Generation Systems," APEC 07 - Twenty-Second Annual IEEE Applied Power Electronics Conference and Exposition, pp. 641-647, 2007.

[16] Teodorescu, Remus, Frede Blaabjerg, Marco Liserre, and P. Chiang Loh. "Proportional-resonant controllers and filters for grid-connected voltage-source converters." IEE Proceedings-Electric Power Applications, vol. 153, no. 5, 750-762, 2006.

[17] A. G. Yepes, F. D. Freijedo, J. Doval-Gandoy, Ó. López, J. Malvar and P. Fernandez-Comesaña, "Effects of Discretization Methods on the Performance of Resonant Controllers," IEEE Transactions on Power Electronics, vol. 25, no. 7, pp. 1692-1712, 2010

[18] D. G. Holmes, T. A. Lipo, B. P. McGrath and W. Y. Kong, "Optimized Design of Stationary Frame Three Phase AC Current Regulators," IEEE Transactions on Power Electronics, vol. 24, no. 11, pp. 2417-2426, 2009.

Proposition and Comparison of Several Solutions for High Induced Voltage across Inactive Transmitting coils in a Series-Series Compensation DIPT System

Wassim KABBARA[1,2], Tanguy PHULPIN[1], Mohamed BENSETTI[1], Antoine CAILLIEREZ[2], Serge LOUDOT[2] and Daniel SADARNAC[1]

[1.] LABORATOIRE DE GENIE ELECTRIQUE ET ELECTRONIQUE DE PARIS, CNRS, CENTRALESUPELEC, SORBONNE UNIVERSITE, UNIVERSITE DE PARIS SACLAY
3 rue Joliot Curie, 91192 Gif-sur-Yvette, France

[2.] RENAULT
1 Avenue du Golf, 78084 Guyancourt, France

E-Mail: wassim.kabbara@centralesupelec.fr
URL: https://www.geeps.centralesupelec.fr

Keywords

«Wireless power transmission», «Robustness», «Resonant converter», «System Integration»

Abstract

Dynamic inductive power transfer technology (DIPT) has recently seen significant development. It is proposed as an alternative solution for increasing the range of electrical vehicles (EV) while decreasing the battery size. Transmitting coils located under the road transfer the power to a receiving coil integrated within the moving EV by inductive coupling. In a series-series DIPT system with multiple transmitting coils, a high induced voltage can occur on the adjacent inactive transmitting coils, thus creating numerous risks. This high induced voltage risks reinjecting power to the grid and thus significantly decreases the system's performance and efficiency. In this article, we present and compare several solutions for the high-induced voltage problem. A four-quadrant switch is designed, modeled, and realized based on two technologies: Saturable reactor and IGBT transistors. Then the 4Q-switch solution is compared with the short-circuiting method. Experimental validation is performed using a series-series DIPT platform using a 300 W resistive load with variable frequency control.

Introduction

Electrical vehicles (EVs) are witnessing significant advancement on many levels [1]. One central research area is how to increase an EV's autonomy without increasing its battery's size. Dynamic Inductive Power Transfer (DIPT) has been a technology in development for several years now and is proposed to increase the range of EVs without using bigger batteries [2]–[4]. It consists of multiple transmitting coils embedded in the road, sending energy by magnetic induction to an embedded receiver coil inside the EV, thus charging wirelessly while in motion. According to a case study in Lisbon [5], dynamic wireless charging would enable more drivers to adopt EVs by 13 to 18%. In addition, the study showed that the maximum change experienced by the state of charge of the EV would be reduced by 2/3, confirming that using this technology significantly reduces battery usage.

In DIPT, compensation circuits are often added to the system due to high leakage flux [6]. Symmetrical Series-Series (S-S) compensation topology is often used in EV charging applications due to its high tolerance for coupling variations and higher peak efficiency [7], [8]. Fig. 1-a presents a typical series-series DIPT system with three transmitting coils on the road and one receiving coil in the EV with an equivalent load (R_{Load}).

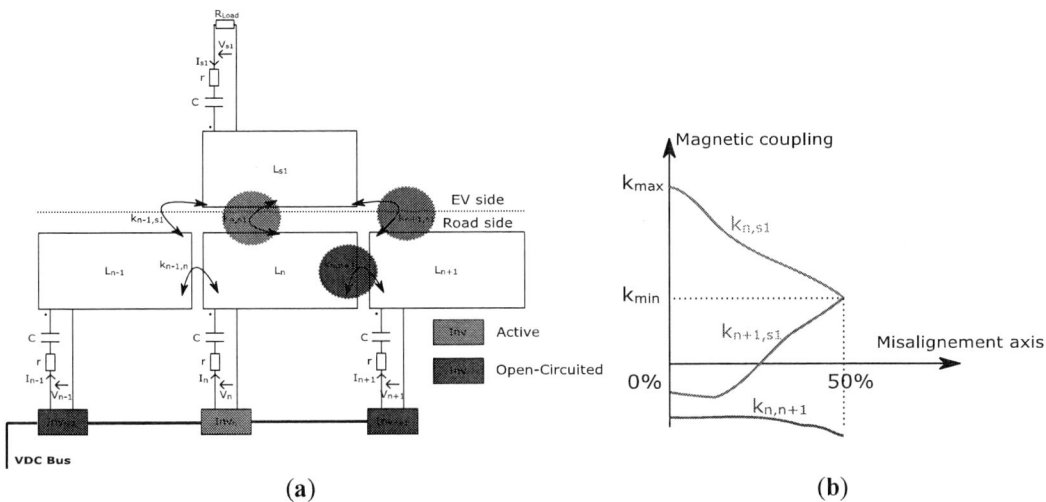

(a)   (b)

Fig. 1: (**a**) Model of a series-series DIPT system with a resistive load RLoad; (**b**) Typical magnetic coupling between transmitting and receiving coils of identical sizes

The electromagnetic leakage fields generated by the active transmitter and the receiver coils can create a high induced voltage across the neighboring inactive transmitting coils due to magnetic coupling between these coils, as illustrated in Fig. 1-b. This induced voltage increases with the misalignment of the receiver coil. It could become higher than the V_{DC_bus} used on the transmitter side, thus reinjecting power to the grid and degrading the system's performance and efficiency.

A possible solution for this problem would be spacing the transmitting coils enough to lower the magnetic couplings over the inactive neighboring transmitting coils (ex: $k_{n,n+1}$ & $k_{n+1,s1}$). This would solve the problem but lead to zero power transfer zones. Using one long transmitter coil as proposed in [9] would eliminate this problem; however, it would be difficult to respect the maximum radiated magnetic field around the vehicle as declared by the INCIRP [10]. [11], [12] propose activating and synchronizing multiple transmitting coils. However, the perfect symmetry of the system's magnetic and electrical parameters should be respected. Otherwise, the power transfer would not be stable due to magnetic coupling between transmitters. In [11], [12], the overvoltage problem across the inactive transmitters was not mentioned since they have activated all transmitter coils simultaneously. In [13], a similar synchronization method was adopted, and a solution for the induced voltage across neighboring transmitters was proposed by limiting the reinjected current (80% DC to DC efficiency at 100 W load, 6.78 MHz). In [14], a coil geometry that significantly reduces the magnetic coupling between two adjacent transmitters was proposed. However, this solution does not eliminate the coupling between the receiver coil and the next inactive transmitter ($k_{n+1,s1}$).

This paper proposes using a 4 Quadrant switch in series with the transmitting coils to solve the high induced voltage problem across inactive transmitting coils in a symmetrical series-series compensation network. A 4Q-switch is proposed, and realized based on two technologies: Saturable reactors [15], [16] and IGBT transistors [17], [18]. Then the 4Q-switch solution is compared with the resonant short-circuiting method [19]–[21]. Comparative experiments were performed while transferring 300 watts on a symmetrical series-series DIPT platform using a resistive load and variable frequency inverters around 85 kHz to achieve zero phase angle control. Moreover, analyses of the results are given, and conclusions and perspectives are presented at the end.

Choice and design of the 4Q-Switches

The positioning of the 4Q-switches is presented in Fig. 2. When the switch is in series with an active inverter, it should act as a short circuit and support positive and negative currents. On the other hand, when the switch is in series with an inactive inverter, it should act as an open circuit and support positive and negative induced voltages. Therefore, the choice of a 4Q-switch was made. There are several possible techniques to create a 4Q-switch. This article investigates the use of two technologies: saturable reactors and IGBT transistors. Using an adequate control circuit, both technologies could achieve the required modes: Low Impedance (LI) mode & High Impedance (HI) mode.

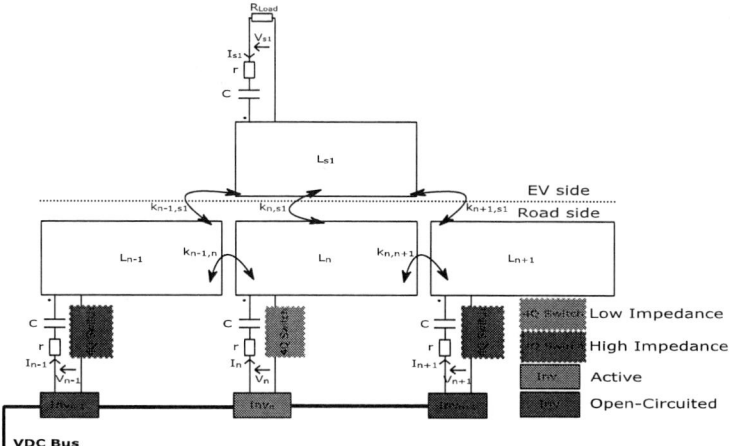

Fig. 2: Using a 4Q-switch with a series-series DIPT system and a resistive load RLoad

The saturable reactor, presented in Fig. 3-a, is a reliable device with low maintenance costs and long life. Its fundamental model is a transformer with two windings: control and power windings. The design used in this article is based on the design presented in [16]. It consists of 2 toroid cores, a control winding, and two power windings in series. This structure doubles the power winding's inductance and protects the control winding from the high induced voltage. A prototype (Fig. 4-a) of the designed saturable inductor is realized based on the specifications presented in Table I.

Table I: Specifications of the realized saturable reactor

Symbol	Definition	Value	Unit
n	Number of turns of the AC winding	8	turns
N	Number of turns of the control winding	210	turns
B_m	Operating peak flux density	1.25	T
A_e	Effective area of the magnetic path	$1.125*10^{-4}$	m^2
l_e	Effective length of the magnetic path	0.102	m
μ_r	Relative permeability of the magnetic core	$3.8*10^4$	/
L_{HI}	Inductance of saturable reactor in High Impedance mode	1.57	mH
L_{LI}	Inductance of saturable reactor in Low Impedance mode	1.01	µH
R_{CW}	DC Resistance of the control winding	4.2	Ω

The chosen topology of the 4Q-switch using the IGBT technology studied in this article is the reverse-blocking IGBT (RB-IGBT) described in [17]. The chosen reverse-blocking, presented in Fig. 3-b, is based on the model developed in [18]. The FGH60N60SMD IGBT module is used with a manual switch to choose between HI-mode or LI-mode (Fig. 4-b). The chosen module has an embedded anti-parallel diode. It has a forward breakdown voltage of 600 V in HI-mode and supports up to 60 A at 100°C in LI-mode, which is sufficient for the intended DIPT application in the experimental phase.

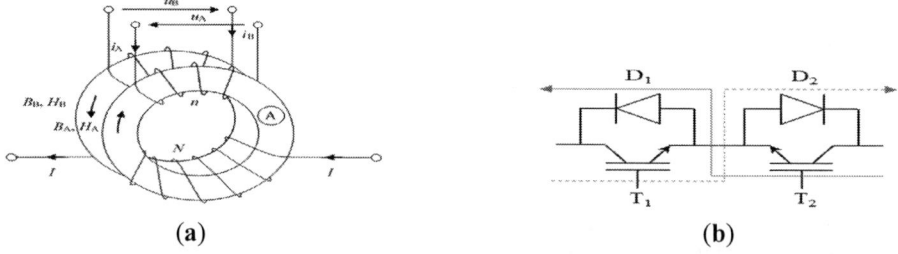

(a) (b)

Fig. 3: (a) Structure of the saturable reactor; (b) Structure of the reverse-blocking IGBT 4Q-switch

(a) Saturable Reactor

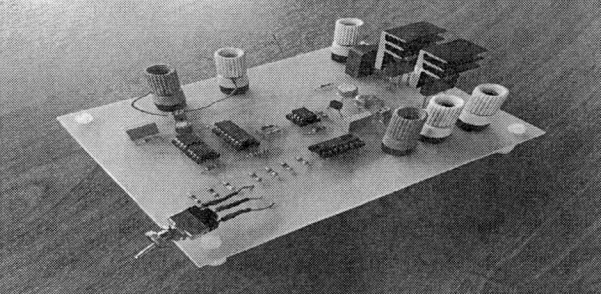
(b) IGBT-RB based on FGH60N60SMD

Fig. 4: Realization of 2 types of saturable reactors

Resonant short circuit mode

Fig. 5 shows the implementation of the Resonant Short Circuit (RSC) mode, proposed in [19], studied in [20], and applied in [21]. The RSC-mode is to short circuit the inverter's output connected to the neighboring inactive coils of the active coil. RSC mode activates the lower two transistors in the H-Bridge and opens the upper ones. Therefore the neighboring inactive coils enter into a series RLC resonant state where the current circulating depends on the induced voltage across the coils and the operating frequency. The advantage of this method is that it does not use any additional components to solve the induced voltage problem across inactive transmitters. However, special care should be taken to limit the resonant current circulating. The active coil cannot operate close to the resonant frequency of the transmitters set to RSC-mode. Otherwise, the current will diverge due to low equivalent impedance. The main added losses to the system using the RSC-mode are due to the dissipation within two turned on transistors (used to achieve the short circuit), the total equivalent series resistance of the coil with the series capacitor, and finally, the magnetic and shielding losses.

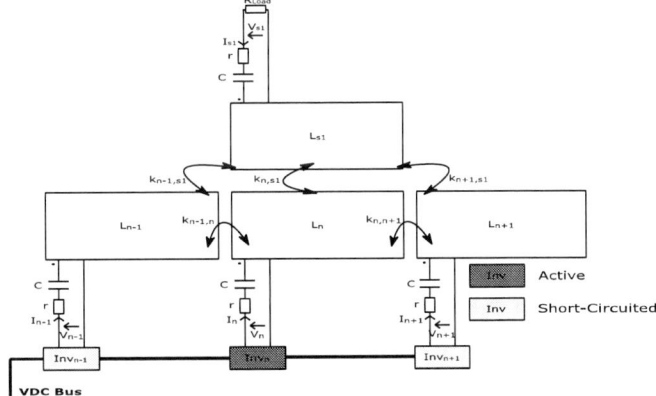

Fig. 5: Using the short-circuiting solution with a series-series DIPT system and a resistive load R_{Load}

Experimental Results

The three presented methods are compared via a DIPT test bench. The block schematic of the used system is given in Fig. 6-a, and a photo of the test bench is presented in Fig. 6-b. The 3 inverters share an identical topology (full H-bridge) with the same control card containing 1 DSP and 1 FPGA for control and monitoring. IPB017N10N5LF MOSFET modules are used in the H-bridge. The selected DSP (TMS320F28335) manages the system's state machine, the control loop regulation, the PWM control signals, and the communication. The selected FPGA (AGLN250V2-xxGxx) executes the voltage/current phase measurements and manages the PWM control signals sent from the DSP. The DSP is programmed using the compiled code from a Matlab Simulink model using the Texas Instrument C2000 package in Matlab coupled with Code Composer Studio. All tests are done under identical output power (300 W) consumed by a resistive load. Moreover, the active inverter operates at zero current switching – ZCS (inverter's output voltage and current are in phase). The measurements are done for two positions of the receiver coil, at 0% misalignment (coil$_n$ and coil$_{s1}$ face each other) and 50% misalignment (coil$_{s1}$ is centered between coil$_n$ and coil$_{n+1}$). No lateral misalignment is considered in this study. The values of the system's parameters are given in Table II. Fig. 7 shows the induced voltage across the inactive coil$_{n+1}$ when coil$_n$ is activated. The induced peak voltage at 50 % misalignment reaches 80 V, which is considerably higher than the DC bus connected to the input of the inverter. Therefore, it is necessary to adopt one of the three presented solutions to avoid reinjecting power to the DC bus via the anti-parallel diode of the inverter connected to the inactive coil$_{n+1}$.

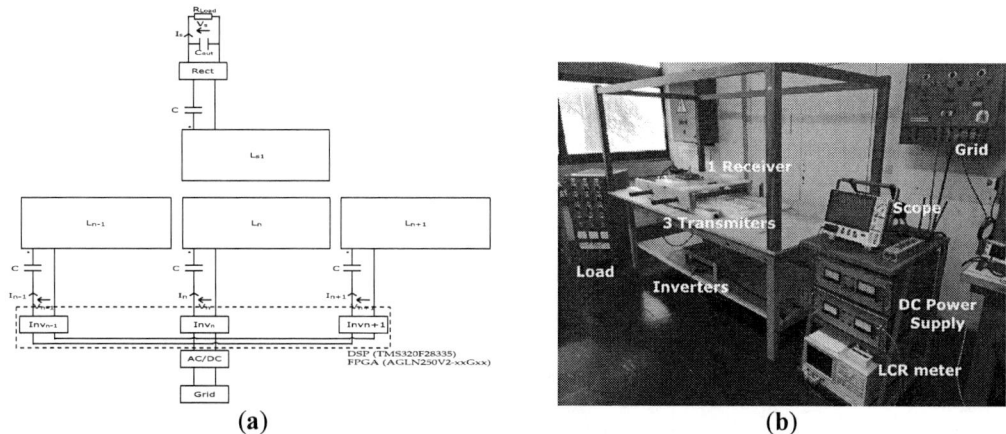

Fig. 6: (a) Block schematics of the test bench; (b) Photo of the test bench

Table II: DIPT Test Bench Parameters

Symbol	Definition	Value	Unit
P_{Load}	Output power consumed by the load	300	W
R_{Load}	Resistive load	3.2 \ 3.7	Ω
V_{DC}	DC bus feeding the inverters	39.6 \ 45.23	V
r	Total equivalent series resistance of the coil & series capacitor	90	mΩ
C	Series compensation capacitor	66	nF
L	Transmitter/Receiver coil's inductance	65	μH
C_{out}	Output capacitor bank	581	μF
f	Operating frequency	83 \ 89	kHz
$k_{n-1,n}$	Coupling between coil$_n$ & coil$_{n-1}$	0.2	%
$k_{n,n+1}$	Coupling between coil$_n$ & coil$_{n+1}$	0.2	%
$k_{n-1,s1}$	Coupling between coil$_{n-1}$ & coil$_{s1}$	4 \ 0	%
$k_{n,s1}$	Coupling between coil$_n$ & coil$_{s1}$	27 \ 17	%
$k_{n+1,s1}$	Coupling between coil$_{n+1}$ & coil$_{s1}$	4 \ 14.6	%

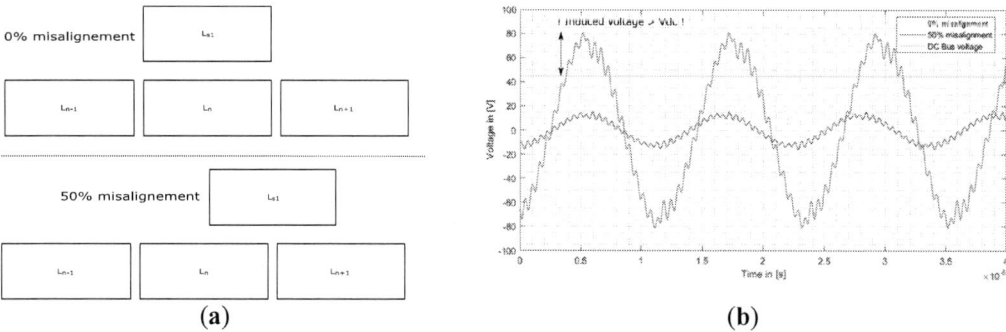

Fig. 7: (a) 0% vs 50% misalignment; (b) Induced voltage across the inactive coil$_{n+1}$ when coil$_n$ is activated

Two saturable reactors have been introduced in the first setup (Fig. 8-a). The first saturable reactor is added in series with the active $coil_n$. At the same time, the second saturable reactor is added in series with the inactive $coil_{n+1}$. In the second setup (Fig. 8-b), we replace the two saturable reactors with RB-IGBT switches and repeat the same measurements under exact conditions. The 4Q-switch, connected in series with the active transmitter, is set to LI-mode. The second switch, connected in series with the inactive transmitter, is set to HI-mode.

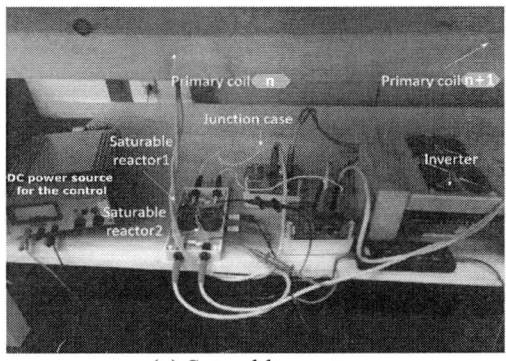
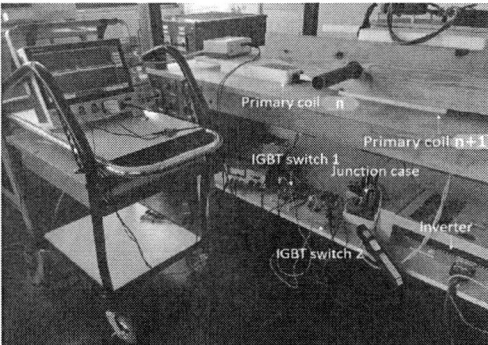

(a) Saturable reactor   (b) RB-IGBT

Fig. 8: 4Q-switches implementation in the platform

In Fig. 9, the voltage and current across the saturable reactor switch in LI-mode are shown in blue. The instantaneous and the average power consumed by the saturable reactor switch are presented in red. The voltage peaks at 20 V, and the current peaks at 13 A for both considered positions (0% and 50% misalignment). We note that voltage leads the current by almost 90° due to the inductive nature of the switch. The total dissipated power by the core and AC winding is the measured average power dissipated over one period: 3.7 W at 0% and 3 W at 50% misalignment. The copper losses in the DC control winding consume 4.2 W. On the other hand, no current was detected circulating across the $coil_{n+1}$ for both considered positions due to the switch in HI-mode. Thus the total losses within the added 2 saturable reactor switches are only due to the one in LI-mode. In Fig. 10, the voltage and current across the RB-IGBT switch in LI-mode are shown in blue. The instantaneous and the average power consumed by the IGBT AC switch are presented in red. The voltage oscillates peaks at 5 V peak and current peaks at 13 A for both considered positions (0% and 50% misalignment). The total dissipated power by LI-mode switch is the measured average power dissipated over one period: 18.7 W at position 0% and 21.8 W at 50% misalignment. The losses in the control circuit are negligible. In addition, no current was detected circulating across the $coil_{n+1}$ for both considered positions due to the switch in HI-mode. Thus the total losses within the added 2 RB-IGBT switches are only due to the one in LI-mode.

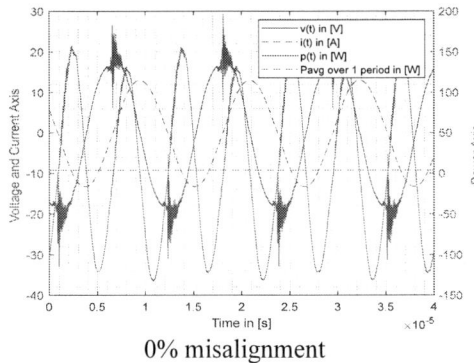
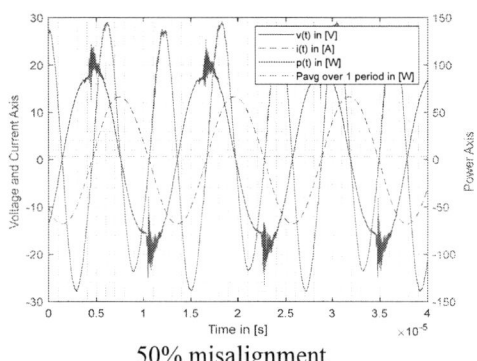

0% misalignment   50% misalignment

Fig. 9: Current, voltage, and power loss across the saturable reactor in LI mode

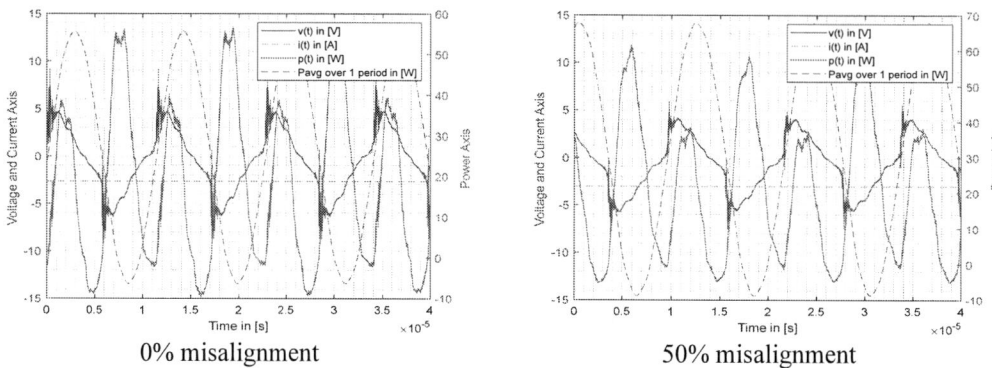

0% misalignment   50% misalignment

Fig. 10: Current, voltage, and power loss across the IGBT-RB Switch in LI mode

The third setup does not need additional 4Q-switches to prevent reinjecting power to the transmitter's DC bus. Instead, the inactive transmitter $coil_{n+1}$ is set to RSC-mode by closing the two lower transistors of the H-bridge $inverter_{n+1}$. In the presence of an induced voltage, the $transmitter_{n+1}$ enters into a series RLC resonant mode. Losses are divided between the inverter in a short circuit (2 transistors conducting at any given time) and the passive RLC components of the transmitter. The total measured equivalent series resistance of the coil and series capacitor at the operating frequency is around 90 mΩ. In Fig. 11, the voltage and current across the $transmitter_{n+1}$ in RSC-mode are presented in blue. The instantaneous and the average power consumed by the $inverter_{n+1}$ are presented in red. At 0% misalignment, the induced voltage across $transmitter_{n+1}$ is very low due to the low magnetic couplings $k_{n,n+1}$, and $k_{n+1,s1}$ at this position. Therefore the current is limited to 1 A peak only. However, at 50% misalignment, the induced voltage across $transmitter_{n+1}$ is much higher due to higher magnetic couplings $k_{n+1,s1}$ at this position. Therefore the current circulating increased to 8.9 A peak. It is noted that the voltage drop across the $inverter_{n+1}$ is relatively low due to the use of transistors with low diode forward voltage and low $RDS_{(ON)}$ (1.7 mΩ). The total dissipated power by the $inverter_{n+1}$ in short-circuit mode is the measured average power dissipated over one period: 0.06 W at position 0 % and 1.18 W at 50 % misalignment. For the total losses in $transmitter_{n+1}$, we add the losses in the RLC equivalent circuit (90 mΩ): 0.06 W at position 0 % and 3.65 W at 50 % misalignment.

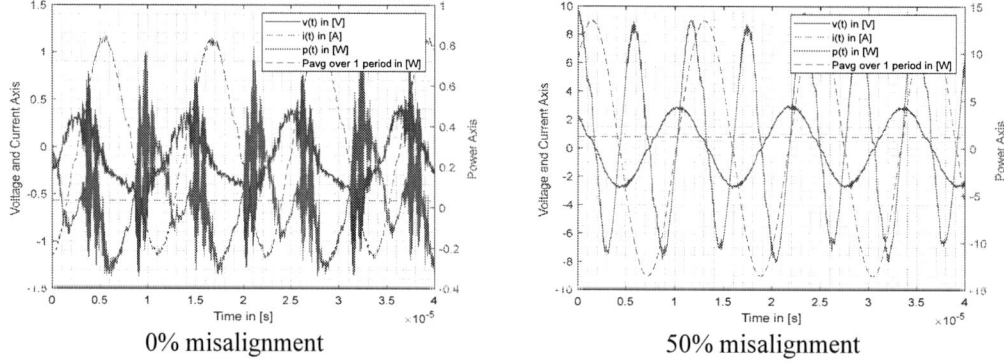

0% misalignment   50% misalignment

Fig. 11: Current, voltage, and power loss across the $inverter_{n+1}$ in resonant short circuit mode

Comparison and Analysis

The loss comparison of the three tested solutions is shown in Table III for two positions of the receiver coil. Concerning the two 4Q-switch methods, the presented losses are dissipated by the 4Q-switch in LI-mode located in series with the active $transmitter_n$. On the other hand, the presented losses of the RSC-mode are dissipated within the $transmitter_{n+1}$. All measurements are done with identical power consumed by the load at 300 W and in ZCS mode for the active inverter.

Table III: Losses Comparison

Misalignment	Saturable Reactor 4Q-Switch	IGBT reverse Blocking 4Q-switch	Resonant Short Circuit
0 %	7.9 W	18.7 W	0.12 W
50 %	7.2 W	21.8 W	4.83 W

Concerning the saturable reactor in LI-mode, a high-frequency current flows through the AC winding, and a constant current flows through the control winding, so there are copper losses in both windings besides the magnetic losses in the core. However, the core operates in the saturation region. Thus, the incremental permeability is small, and the peak variation flux density ΔB_m is also minimal. Therefore, the hysteresis losses are minimal. Furthermore, considering the high resistivity of the chosen Fe-based soft magnetic material, we obtain low losses by eddy currents, leading to low overall core losses. The experimental measurements confirm this remark. There is no control current in high impedance mode, and the high impedance blocks the high-frequency current. Hence, the copper losses in both windings are almost null. Moreover, core losses are also null due to the absence of current circulating in the AC winding.

Under this scenario, the saturable reactor presents fewer losses than the 4Q-switch based on IGBT. However, it is to be noted that the saturable reactor is designed to withstand a 15 A peak current while the chosen IGBT module can withstand 60 A nominal. Moreover, the maximum permissible induced voltage the saturable reactor can withstand is defined by the maximum allowable reinjected current to the DC bus. On the other hand, saturable reactors can support high-temperature variation and are more reliable than semiconductors due to their simplicity and robustness. Heat dissipation is not an issue when the vehicle moves fast across the transmitter coils. However, an effective thermal dissipation in static applications with higher power transfer (example: during traffic jams). Furthermore, the harsh external environment can also lead to failure. Additional improvement on the saturable reactor could be envisioned by increasing the control's winding number of turns to use a lower control current. This will lower the overall losses of the saturable reactor.

Comparing the RSC mode results with both 4Q-switch methods, we observe significantly fewer losses and lower costs since no additional elements are required. The lower losses are mainly due to the geometry of the coils used. The used geometry in the DIPT platform has a very low magnetic coupling between transmitter coils. Therefore, the induced voltage on the neighboring inactive transmitter coil was relatively limited. Thus the current circulating in the transmitter in RSC mode was also limited, explaining the lower losses. For this reason, the losses of the RSC mode solution are highly dependent on the choice of the geometry for the transmitter and receiver coils, unlike the 4Q-switch methods with fixed losses proportional to the transmitted power.

Conclusions and Perspectives

This article compares several solutions for high induced voltage across inactive transmitting coils in a series-series compensation DIPT system. A 4Q-switch is presented and realized based on two technologies: Saturable reactors and IGBT transistors. The 4Q-switch solution is compared with the short-circuiting method. Experimental validation is performed on a series-series DIPT platform by transferring a 300 W to a resistive load at a variable frequency control between 83 kHz and 89 kHz. Results show that the saturable reactor presents fewer losses than the 4Q-switch based on IGBT. Moreover, the 4Q-switch offers better reliability and overvoltage withstand. On the other hand, tests also showed that the RSC-mode dissipated less power with no added costs. However, the losses in RSC-mode depend on the coils' chosen geometry. In further research, a comparison between the three solutions will be performed at higher power transfer (several kW) using several coil geometries. Moreover, increasing the number of turns of the control windings for the saturable reactor, thus using lower control current, results in lower losses.

References

[1] F. Un-Noor, S. Padmanaban, L. Mihet-Popa, M. Mollah, and E. Hossain, "A Comprehensive Study of Key Electric Vehicle (EV) Components, Technologies, Challenges, Impacts, and Future Direction of Development," *Energies*, vol. 10, no. 8, p. 1217, Aug. 2017, doi: 10.3390/en10081217.

[2] G. A. Covic and J. T. Boys, "Modern Trends in Inductive Power Transfer for Transportation Applications," *IEEE J. Emerg. Sel. Top. Power Electron.*, vol. 1, no. 1, pp. 28–41, Mar. 2013, doi: 10.1109/JESTPE.2013.2264473.

[3] R. Bosshard and J. W. Kolar, "Inductive power transfer for electric vehicle charging: Technical challenges and tradeoffs," *IEEE Power Electron. Mag.*, vol. 3, no. 3, pp. 22–30, Sep. 2016, doi: 10.1109/MPEL.2016.2583839.

[4] L. Hutchinson, B. Waterson, B. Anvari, and D. Naberezhnykh, "Potential of wireless power transfer for dynamic charging of electric vehicles," *IET Intell. Transp. Syst.*, vol. 13, no. 1, pp. 3–12, Jan. 2019, doi: 10.1049/iet-its.2018.5221.

[5] G. Duarte, A. Silva, and P. Baptista, "Assessment of wireless charging impacts based on real-world driving patterns: Case study in Lisbon, Portugal," *Sustain. Cities Soc.*, vol. 71, p. 102952, Aug. 2021, doi: 10.1016/j.scs.2021.102952.

[6] N. Liu and T. G. Habetler, "Design of a Universal Inductive Charger for Multiple Electric Vehicle Models," *IEEE Trans. Power Electron.*, vol. 30, no. 11, pp. 6378–6390, Nov. 2015, doi: 10.1109/TPEL.2015.2394734.

[7] K. Aditya and S. S. Williamson, "Comparative study of series-series and series-parallel topology for long track EV charging application," in *2014 IEEE Transportation Electrification Conference and Expo (ITEC)*, Dearborn, MI, Jun. 2014, pp. 1–5. doi: 10.1109/ITEC.2014.6861793.

[8] Y. Chen, H. Zhang, C.-S. Shin, K.-H. Seo, S.-J. Park, and D.-H. Kim, "A Comparative Study of S-S and LCC-S Compensation Topology of Inductive Power Transfer Systems for EV Chargers," in *2019 IEEE 10th International Symposium on Power Electronics for Distributed Generation Systems (PEDG)*, Xi'an, Jun. 2019, pp. 99–104. doi: 10.1109/PEDG.2019.8807684.

[9] J. L. Villa, J. Sallán, A. Llombart, and J. F. Sanz, "Design of a high frequency Inductively Coupled Power Transfer system for electric vehicle battery charge," *Appl. Energy*, vol. 86, no. 3, pp. 355–363, Mar. 2009, doi: 10.1016/j.apenergy.2008.05.009.

[10] "ICNIRP GUIDELINES FOR LIMITING EXPOSURE TO TIME-VARYING ELECTRIC AND MAGNETIC FIELDS (1 Hz TO 100 kHz)," *Health Phys.*, vol. 99, no. 6, pp. 818–836, Dec. 2010, doi: 10.1097/HP.0b013e3181f06c86.

[11] K. Kim and J.-W. Choi, "Influences of Magnetic Couplings in Transmitter Array of MIMO Wireless Power Transfer System," in *2019 IEEE Wireless Power Transfer Conference (WPTC)*, London, United Kingdom, Jun. 2019, pp. 531–535. doi: 10.1109/WPTC45513.2019.9055594.

[12] F. Lu, H. Zhang, H. Hofmann, and C. C. Mi, "A Dynamic Charging System With Reduced Output Power Pulsation for Electric Vehicles," *IEEE Trans. Ind. Electron.*, vol. 63, no. 10, pp. 6580–6590, Oct. 2016, doi: 10.1109/TIE.2016.2563380.

[13] A. Pacini, A. Costanzo, S. Aldhaher, and P. D. Mitcheson, "Load- and Position-Independent Moving MHz WPT System Based on GaN-Distributed Current Sources," *IEEE Trans. Microw. Theory Tech.*, vol. 65, no. 12, pp. 5367–5376, Dec. 2017, doi: 10.1109/TMTT.2017.2768031.

[14] G. A. Covic, M. L. G. Kissin, D. Kacprzak, N. Clausen, and H. Hao, "A bipolar primary pad topology for EV stationary charging and highway power by inductive coupling," in *2011 IEEE Energy Conversion Congress and Exposition*, Phoenix, AZ, USA, Sep. 2011, pp. 1832–1838. doi: 10.1109/ECCE.2011.6064008.

[15] P. Mali, *Magnetic amplifiers: principles and applications*. New York: J.F. Rider, 1960.

[16] M. S. Perdigao, M. F. Menke, A. R. Seidel, R. A. Pinto, and J. M. Alonso, "A Review on Variable Inductors and Variable Transformers: Applications to Lighting Drivers," *IEEE Trans. Ind. Appl.*, vol. 52, no. 1, pp. 531–547, Jan. 2016, doi: 10.1109/TIA.2015.2483580.

[17] C. Benboujema, A. Schellmanns, N. Batut, J. B. Quoirin, and L. Ventura, "Low losses bidirectional switch for AC mains," p. 10.

[18] A. Trentin, L. de Lillo, L. Empringham, P. Wheeler, and J. Clare, "Experimental Comparison of a Direct Matrix Converter Using Si IGBT and SiC MOSFETs," *IEEE J. Emerg. Sel. Top. Power Electron.*, vol. 3, no. 2, pp. 542–554, Jun. 2015, doi: 10.1109/JESTPE.2014.2381001.

[19] A. Caillierez, D. Sadarnac, A. Jaafari, and S. Loudot, "Dynamic inductive charging for electric vehicle: modelling and experimental results," p. 1.7.01-1.7.01, Jan. 2014, doi: 10.1049/cp.2014.0423.

[20] P.-A. Gori, D. Sadarnac, A. Caillierez, and S. Loudot, "Sensorless inductive power transfer system for electric vehicles: Strategy and control for automatic dynamic operation," in *2017 19th European Conference on Power Electronics and Applications (EPE'17 ECCE Europe)*, Warsaw, Sep. 2017, p. P.1-P.10. doi: 10.23919/EPE17ECCEEurope.2017.8099233.

[21] W. Kabbara, M. Bensetti, T. Phulpin, A. Caillierez, S. Loudot, and D. Sadarnac, "A Control Strategy to Avoid Drop and Inrush Currents during Transient Phases in a Multi-Transmitters DIPT System," *Energies*, vol. 15, no. 8, Art. no. 8, Jan. 2022, doi: 10.3390/en15082911.

Modeling and Measuring the Bearing Capacitance of Radially Loaded Bearings

Stefan Quabeck, Daniel C. Rodriguez and Rik W. De Doncker
Institute for Power Electronics and Electrical Drives, RWTH Aachen University
Jaegerstraße 17-19
Aachen, Germany
Phone: +49 (0) 241-80 96920
Fax: +49 (0) 241-80 92203
Email: post@isea.rwth-aachen.de
URL: http://www.isea.rwth-aachen.de

Keywords

≪Bearing currents≫, ≪Parasitic elements≫, ≪Impedance measurements≫, ≪Reliability≫

Abstract

Bearing currents can severely reduce the lifetime of electrical machines. The parasitic bearing capacitance causing these currents depends on speed, lubricant temperature, and bearing load. In radially loaded bearings, the capacitance of each bearing ball varies over its angular position, exhibiting high capacitance in the load zone and lower capacitance outside of it. This work proposes a method for measuring the bearing capacitance over the angular position and provides a model that incorporates this effect.

Introduction

Bearing faults are one of the most prevalent fault types in electrical machines [1]. They can be caused by bearing currents that damage bearing balls and raceways, and thus reduce bearing lifetime [2], [3]. Modeling and understanding the electrical behavior of rolling element bearings is essential for improving their lifetime and reducing current in the bearings.

Many publications investigate the electrical behavior of bearings. In [4], the author presents a high-frequency model of an electrical machine that includes the bearing as a lumped model. Lumped models are also used in many other publications as they provide reasonable accuracy at low computational cost, allowing to predict bearing currents [5] or to test bearing current countermeasures quickly in a simulation environment [6]. A method for measuring the parasitic capacitance of a bearing at varying speeds and loads is given in [7]. The authors apply a rectangular voltage waveform to the bearing via a resistor and use the rise time of the bearing voltage to calculate the bearing capacitance. The same measurement method is used in [8] to estimate the bearing capacitance in order to design a capacitive shunt with lower impedance, which guides most of the current around the bearing. The influence of vibrations on the bearing capacitance is investigated in [9].

Many publications estimate the bearing capacitance by means of the Hertzian contact area and other parallel parasitic capacitances. In those publications, the bearing is modeled as one lumped, time-invariant capacitance. This work investigates the parasitic capacitance of a single bearing ball in more detail. The influence of a varying radial load depending on the angular position of the bearing ball is measured. It is shown that the sum of the parasitic bearing capacitances is not as constant over one rotation as other publications suggest.

Fundamentals

In this section, the theoretical background required for modeling the electrical behavior of the bearing is illustrated.

Bearing Capacitance

The parasitic capacitance of rolling element bearings consists of three parts: the capacitance between the inner raceway and outer raceway of the bearing, the capacitance between the inner raceway and the bearing balls, and the capacitance between the outer raceway and the bearing balls. The capacitance between the inner and outer raceways of the bearing is determined by the average dielectric properties of the materials between the races, i.e. the amount of air, lubricant, and steel. This parasitic capacitance C_{IO} can be approximated as a cylinder capacitor with a length W, an inner diameter d_I, and an outer diameter d_O as

$$C_{IO} = \frac{2\pi\varepsilon W}{\ln \frac{d_I}{d_O}}. \tag{1}$$

Depending on the amount of lubricant in the bearing and its dielectric properties, ε can strongly vary. The capacitances between the bearing balls and the inner and outer raceways mostly depend on the oil film thickness, the dielectric properties of the lubricant, and the Hertzian contact area. The oil film thickness h can be calculated as presented in [10] as

$$h = 2.69 \cdot (\alpha E)^{0.49} \cdot U^{0.68} \cdot W^{-0.067} \cdot (1 - 0.61 e^{-0.73\chi}), \tag{2}$$

where W is the dimensionless load, U is the dimensionless velocity, E is the equivalent modulus of elasticity and $\alpha = 2.3 \cdot 10^{-8} m^2/\text{N}$. Fig. 2b illustrates the film thickness for different lubricant temperatures and rotational speeds assuming a constant load. When the film thickness is low, i.e. for low rotational speeds and high temperatures, the parasitic capacitance is high but the breakdown voltage is low, leading to a high number of breakdowns. When the film thickness increases, the capacitance decreases but the breakdown voltage increases, reducing the number of breakdowns.

In radially loaded bearings, the load is distributed non-uniformly between the bearing balls. We assume that the main load on the bearings is the gravitional force acting on the rotor. Thus, the point with the highest load is expected to be at the bottom of the bearing. The rolling velocity of a bearing ball also changes depending on its angular position. In the load zone, the load gradually becomes higher and the slip of the rolling element decreases. Thus, the oil film is thinnest at an angle of 180° as illustrated in Fig. 6a. Afterward, the load decreases and the slip of the ball increases, increasing the oil film thickness. As the parasitic capacitance of each individual ball is approximately inversely proportional to the oil film thickness, the capacitance reaches its peak value at an angle of 180° as illustrated in Fig. 1. The influence of temperature and speed on the lubricant film thickness in a rolling element bearing is investigated in [5] and [7].

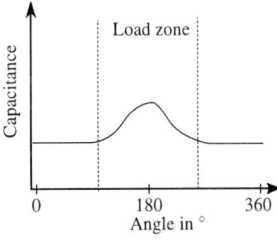

Fig. 1: Expected capacitance

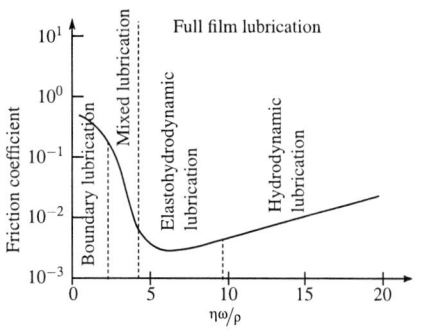
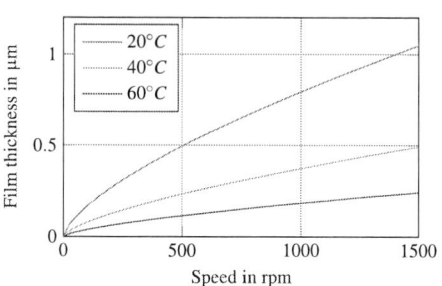

(a) Friction coefficient as a function of the bearing number as per [11]

(b) Lubricant film thickness over speed at different temperatures as per [5]

Fig. 2: Bearing properties at different speeds and temperatures

Lubrication Regimes

The oil film thickness strongly depends on the rotational speed of the machine, the contact pressure, i.e., the radial load, and the viscosity of the lubricant, which in turn changes with temperature. Fig. 2a shows the friction coefficient of a rolling element bearing over the bearing number, which consists of the lubricant viscosity η, the speed ω and the contact pressure p. In boundary lubrication, more than 90 % of the radial load rests on surface peaks of the bearing raceways and the balls. At low speeds, low viscosity, i.e., high temperature, and high contact pressure, the electrical behavior of the bearing is therefore assumed to be ohmic. Assuming that the radial load in a typical electrical machine application is constant, with increasing speed, a greater part of the radial load is carried by the lubricant, but there is still direct contact between the raceways and the bearing balls. This lubrication regime is called mixed lubrication. When the bearing is operated in the (elasto)hydrodynamic lubrication regime, i.e., higher speeds and low radial load or low temperature, friction is minimal and there is no direct contact between the raceways and the bearing balls. The electrical behavior is now capacitive and mostly determined by the properties of the lubricant film. Only when the bearing is in (elasto)hydrodynamic lubrication, a constant oil film that can withstand a certain voltage is formed, and therefore, a shaft voltage greater than zero can occur.

Breakdown

There are mainly two events that can lead to a rapid discharge of the bearing capacitance. First, when the bearing leaves hydrodynamic lubrication, due to either a decrease in speed or an increase in radial load or temperature, direct contact between the bearing raceways and balls can occur. The energy stored between the raceways and the bearing balls, i.e., the electrodes of the bearing capacitance, is then discharged through these contact points. Second, when the shaft voltage is higher than the breakdown voltage of the lubricant film, an arc discharge occurs within the bearing. To avoid the first effect, the shaft voltage has to be zero at all times. To avoid the second effect, it is sufficient to keep the shaft voltage below the breakdown voltage. However, the breakdown voltage changes with temperature, speed and radial load. Since the main part of the lifetime of a well-designed bearing is spent in (elasto)hydrodynamic lubrication, electrical discharges are the more relevant breakdown phenomenon.

Model

As already discussed in the previous section, the parasitic capacitance of a rolling element bearing consists of the capacitance between the inner and outer races and the parasitic capacitances between the bearing balls and the bearing raceways. Traditional modeling approaches use a lumped RC-circuit with an additional switch for discharge modeling. In [10], [12], a distributed bearing model is presented. Fig. 3 shows the distributed and lumped equivalent circuit models for a rolling element bearing. Note that two versions of the lumped equivalent circuit are depicted. Fig. 3a is used for most purposes, Fig. 3b

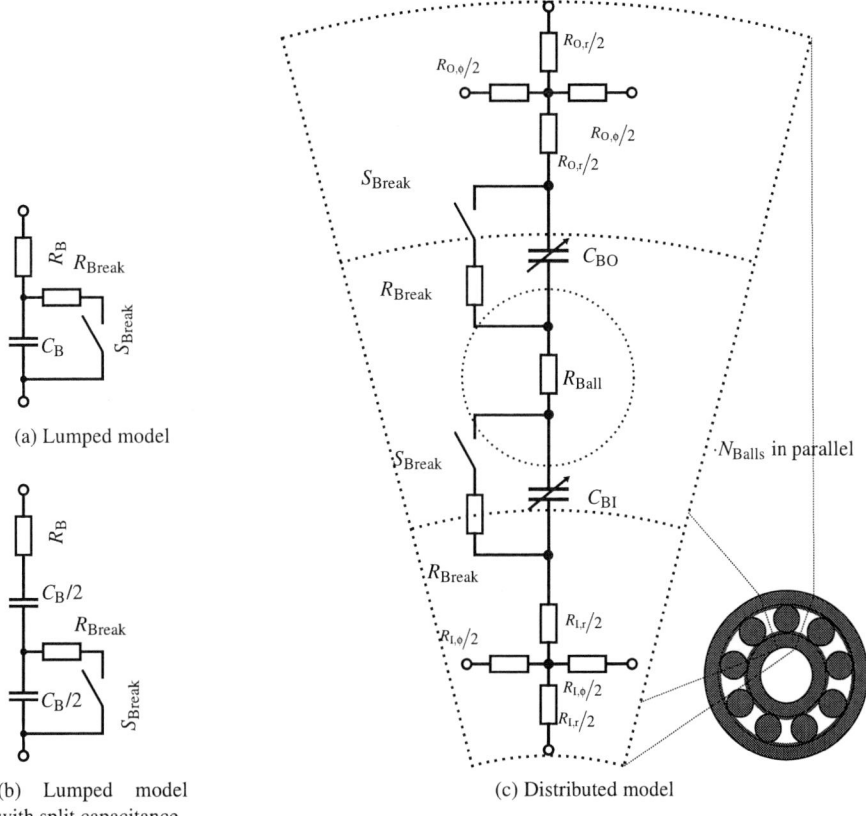

(a) Lumped model

(b) Lumped model with split capacitance

(c) Distributed model

Fig. 3: Lumped and distributed bearing models

is only needed to model partial bearing voltage breakdowns.

While the lumped models exhibit the same electrical behavior from the outside for full breakdowns, and are thus preferable for most modeling purposes, the distributed model shown in Fig. 3c provides further insight into the internal electrical behavior of a rolling element bearing. It allows to individually model the parasitic capacitance of every single bearing ball to the raceways. This enables the discernment of three effects.

First, when a constant voltage is applied to the bearing and the bearing rotates, there will be currents flowing within the bearing, redistributing the charge due to the changes in capacitance. The amplitude of these currents strongly depends on the rotational speed and on the gradient of the parasitic capacitance when a ball enters and exits the load zone.

Second, when a breakdown occurs, the model suggests that the breakdown has two stages. Initially, the parasitic capacitance of the bearing ball where the breakdown occurs is discharged almost immediately as the impedance is low. Then, due to the resulting voltage drop, current from the remaining bearing capacitances, i.e., the parasitic capacitances between the bearing raceways and the parasitic capacitances of the remaining bearing balls, also flows into the breakdown arc until the voltage is low enough and the arc is extinguished. Furthermore, current from outside the bearing can also feed the arc, depending on the source impedance. In an electrical machine, there is only capacitive coupling between the rotor on the inner race and the windings, which act as a source in this case, meaning that the external current flow will decrease quickly.

Third, when only a partial breakdown occurs, which is caused by direct contact between the bearing ball and one of the raceways (e.g., in mixed lubrication or due to irregularities in the raceway surface), the voltage will not be reduced to zero. This effect takes place because only one of the ball-to-raceway capacitances in Fig. 3c is discharged.

Fig. 4 illustrates these effects. Fig. 4a compares the behavior of the lumped and distributed models during a full breakdown. Since the models have been parametrized in a way that their impedances match over a wide frequency range, there is no difference in the breakdown behavior from the outside. However, when looking at the internal charge redistribution currents between the bearing balls in Fig. 5 in the last section of this paper, it can be seen that the internal current amplitudes are much higher than the current seen from the outside of the bearing. This is because most of the energy that is dissipated in a breakdown comes from the parasitic capacitances of the bearing itself and is thus not measurable from the outside.

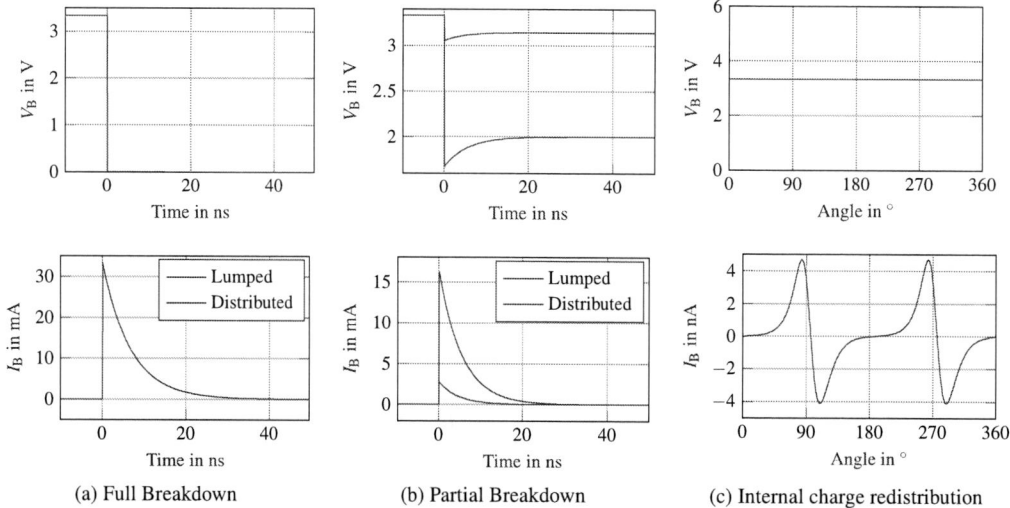

(a) Full Breakdown   (b) Partial Breakdown   (c) Internal charge redistribution

Fig. 4: Bearing currents and voltages under various breakdown conditions

The partial breakdown phenomenon is illustrated in Fig. 4b. Note that the lumped model has been split into two parts to allow for a fairer comparison, as depicted in Fig. 3b. The voltage drops to approximately 50 % in the lumped model, which was expected from the distribution of the capacitances. In the distributed model, however, the voltage only drops to approximately 93 %. This is because only a small part of the bearing energy is actually discharged, in this case $1/22$ of it. The rest of the energy is redistributed within the bearing. Therefore, the distributed model exhibits a more realistic behavior in case of a partial breakdown.

Finally, Fig. 4c shows the internal charge redistribution within the bearing when no breakdown occurs. A constant voltage of 3.33 V is applied to the bearing. At a speed of 3.000 rpm the current resulting from the change in capacitance amounts to approximately 4.2 nA and can thus be considered irrelevant for the bearing lifetime, especially when compared to the currents occurring during switching events or breakdowns. The unbalanced positive and negative current peaks are due to the calculation method used

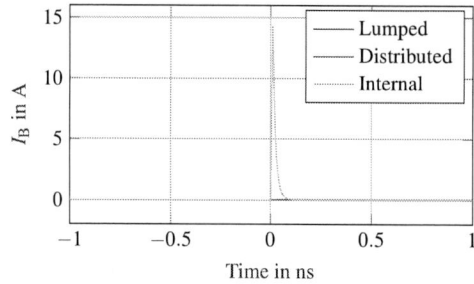

Fig. 5: Breakdown current flowing within the bearing

for the derivation of the bearing capacitance. The charge redistribution current can be calculated as

$$I_\text{B} = V_\text{B} \frac{\mathrm{d}C_\text{B}}{\mathrm{d}t} \rightarrow i_\text{B}(\theta) = V_\text{B} \frac{\mathrm{d}C_\text{B}}{\mathrm{d}\theta} \cdot \omega_\text{Cage}. \tag{3}$$

Even at 16.000 rpm, which is the maximum allowed speed for the bearing under test, the charge redistribution current is expected to be well below the critical thresholds for bearing damage.

Test Bench

In order to confirm the capacitive behavior of a single ball in a ball bearing, which was derived analytically in the previous sections, bearing capacitance measurements are conducted on a radial load test bench.

Radial Load Test Bench

The radial load test bench that is used for these measurements consists of two test bearings and one load bearing in the middle. It is shown in Fig. 6c. The load bearing is connected to a spring to set the radial load. The test bench is driven by an induction machine via a belt drive, which controls the speed of the test bench. Oil is pumped through the bearing in axial direction and heating elements are used to set the temperature and thereby the bearing temperature via a closed-loop control with a temperature sensor at the bearing.

To minimize the effect of additional parasitic capacitances that disturb the measurements, various steps are taken. The load bearing and the rear test bearing are replaced with ceramic bearings. Only the capacitive behavior of the front test bearing is investigated. A specially manufactured 6006 bearing with steel raceways, ten ceramic balls and one steel ball is used as front test bearing. Additionally, the front plate of the test bench, where the front test bearing is mounted, is replaced with polyoxymethylene (POM).

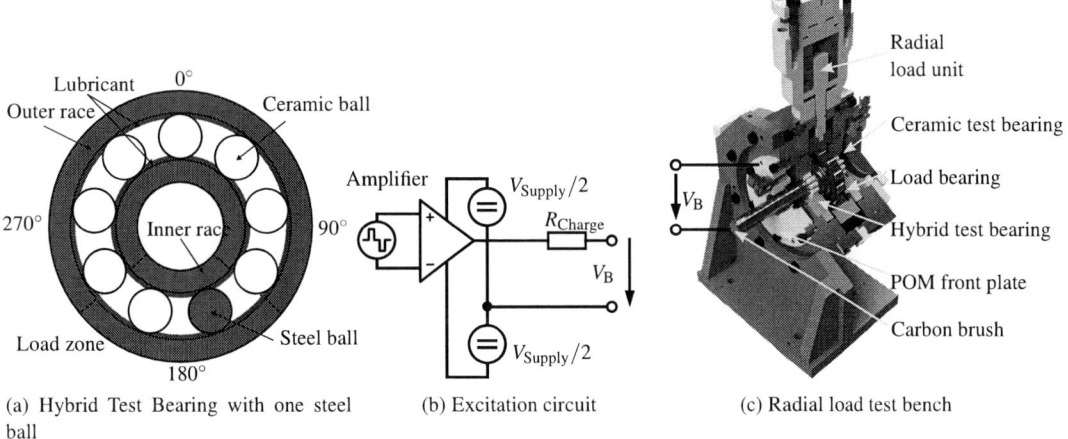

(a) Hybrid Test Bearing with one steel ball

(b) Excitation circuit

(c) Radial load test bench

Fig. 6: Measurement setup

Electrical Excitation

Measuring the bearing capacitance with a reasonable angular resolution at high speeds requires a fast measurement method. Since impedance analyzers or LCR-meters usually have long sweep or settling times, they cannot be used here. Instead, a rectangular voltage waveform is applied to the bearing via a large resistor. The amplitude is set to $\pm 5\,\text{V}$. The excitation circuit is shown in Fig. 6b.

The resistor R_Charge is set to $10\,\text{k}\Omega$ to ensure a charging process that is fast enough to allow the highest required excitation frequency of $20\,\text{kHz}$, but that is still slow enough so that changes in its rise time are measurable with the temporal resolution of the oscilloscope of $80\,\text{MSa/s}$. All data is directly prepro-cessed in the oscilloscope and only the rise time, fall time, pulse width, amplitude, and duty cycle of the measured pulse voltage are recorded. The rise and fall times depend on the charging resistance and on the bearing capacitance. Since the charging resistance R_{Charge} is set to a constant value of $10\,\text{k}\Omega$, the rise and fall times $\tau_{\text{rise/fall}}$ are directly proportional to the bearing capacitance:

$$C_{\text{B}} = \frac{\tau_{\text{rise/fall}}}{R_{\text{Charge}}}. \tag{4}$$

The circuit is evaluated and calibrated by conducting capacitance measurements on several different capacitors in the range of $10\,\text{pF}$ to $1\,\text{nF}$ prior to applying it to the bearing and shows good measurement accuracy.

Pulse width, duty cycle, and voltage amplitude are constant during the measurement and are not influ-enced by the bearing capacitance. However, when a breakdown occurs, these parameters deviate from the set values and thus allow to check the data for validity. If the bearing capacitance is not fully charged before the excitation voltage changes of polarity or if it is discharged prematurely, a breakdown occured and the bearing did not behave purely capacitive in that period. Under these circumstances, the rise and fall times should not be used for capacitance estimation and those specific measurements are thus omitted. In the measurement results, these breakdowns manifest themselves as large capacitance spikes, which often exceed the actual capacitance value by more than 100%.

Operating Conditions

The bearing capacitance measurements are conducted at a constant radial load of $100\,\text{Nm}$. The bearing capacitance is measured at four different speeds and four different temperatures: $300\,\text{rpm}$, $1.000\,\text{rpm}$, $3.000\,\text{rpm}$ and $6.000\,\text{rpm}$ and at $20\,^\circ\text{C}$, $30\,^\circ\text{C}$, $50\,^\circ\text{C}$ and $80\,^\circ\text{C}$. Automatic transmission fluid is used as lubricant.

Measurement Results

Since the position of the steel ball within the bearing is unknown, the bearing capacitance can only be given over the relative angular position. It can be assumed, however, that the peak of the bearing capacitance corresponds to an angular position of 180° as shown in Fig. 1. The capacitance waveforms are therefore shifted, so that 180° coincides with the peak capacitance, in order to be able to compare them more easily.

Bearing Capacitance

The results of the single steel-ball bearing capacitance measurements are discussed in this section. Fig. 7 shows the measured bearing capacitance over the rotational angle for different speeds and temperatures. Note that the large spikes in the bearing capacitance values indicate breakdowns of the lubricant film where the capacitance could not be measured. At $300\,\text{rpm}$, the bearing capacitance cannot be reasonably extracted from the conducted measurements. Only the capacitance values for $20\,^\circ\text{C}$ allow for a rough estimation. The high number of breakdowns suggests either that the bearing is still in mixed lubrication or that the excitation voltage was too high for the still very thin lubricant film.

At $1.000\,\text{rpm}$, the bearing capacitance can be extracted for lower temperatures. The curve at $20\,^\circ\text{C}$ indicates a maximum capacitance of $40\,\text{pF}$. At $30\,^\circ\text{C}$ the capacitance reaches approximately $50\,\text{pF}$. The increasing number of breakdowns suggests that the excitation voltage should be set to a lower value.

At $3.000\,\text{rpm}$, the capacitance can be extracted for temperatures up to $50\,^\circ\text{C}$, above which the number of breakdowns is too high again. Still, the increase in peak capacitance as temperature increases matches with the expectations from Fig. 2b.

At $6.000\,\text{rpm}$, the capacitance can finally be extracted for all temperatures. Only at $80\,^\circ\text{C}$ breakdowns are noticable. The increase of the peak capacitance value with increasing temperature is in accordance with expectations from Fig. 2b.

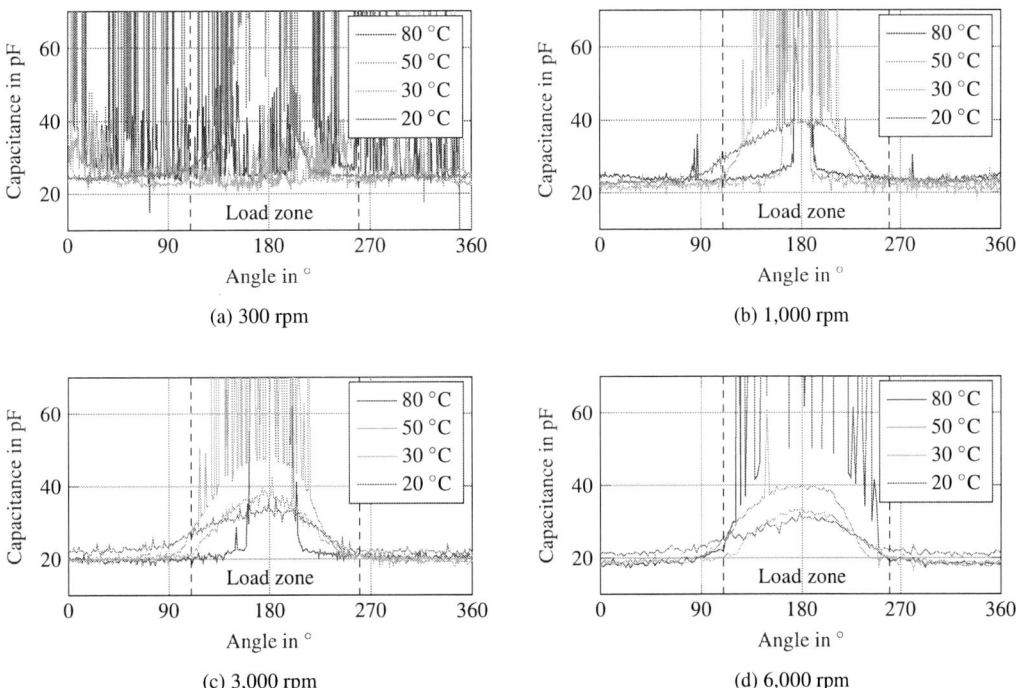

(a) 300 rpm

(b) 1,000 rpm

(c) 3,000 rpm

(d) 6,000 rpm

Fig. 7: Measured parasitic capacitance of a single bearing ball

The capacitance outside the load zone is nearly constant, with only slight deviations over temperature and speed. This indicates that the reduction of viscosity, due to the increase in temperature, and the higher radial forces, due to the increased speed, have nearly no influence on the lubricant film thickness outside the load zone. The capacitance decreases from approximately 25 pF to 20 pF over the whole speed and temperature range. This reduction in capacitance can be attributed to the reduction in raceway-to-raceway capacitance due to the reduced amount of lubricant in the bearing at a higher rotational speed. In addition, at lower temperatures, the higher viscosity of the lubricant leads to higher amounts of lubricant in the bearing and thus increases the raceway-to-raceway capacitance. It was shown that the capacitance peak is strongly influenced by both the rotational speed and the lubricant temperature. The dielectric properties of the lubricant are assumed to be constant over temperature [7].

When overlaying the capacitance measurements with a phase shift of $360°/N_{\text{Balls}}$, a nearly constant capacitance value with a small ripple results, as shown in Fig. 8. However, since this ripple of approximately

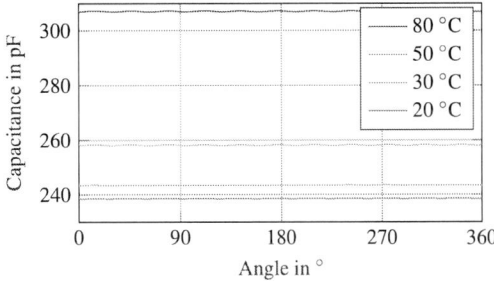

Fig. 8: Parasitic capacitance of the bearing calculated from filtering and overlaying measurements

0.1 pF to 0.3 pF is below 0.1 % of the total bearing capacitance, this is not confirmed with measurements. For further improvement of the measurements, the excitation voltage should be reduced and the measurement frequency should be increased, so that a moving-average filter can be used to reduce the amount of noise on the capacitance measurement.

Conclusions and Outlook

In this work, a distributed parameter model for rolling element bearings has been presented. To confirm the assumption that the parasitic capacitance of individual bearing balls changes over the angular position, a hybrid bearing with one steel ball and ten ceramic balls is investigated. By charging and discharging the bearing with a high-frequency rectangular voltage waveform via a large resistor, the value of the parasitic capacitance is extracted from the rise and fall time. It is shown that at low speeds or high temperatures, the bearing is either in mixed lubrication or the lubricant film is too small to withstand the charging voltage of $\pm 5\,\mathrm{V}$. However, at higher speeds and lower temperatures, bearing capacitance values can be computed. The measurement results show that the parasitic capacitance of an individual ball increases by up to 150 % when the ball enters the load zone. When overlaying the measured values to model the parasitic capacitance of a full steel bearing, a small capacitance ripple occurs, which was neglected in previous publications.

In future work, the influence of the excitation voltage should be investigated in more detail. With a lower excitation voltage and an increased excitation frequency in combination with additional filters, the capacitance can be determined more precisely and for a wider range of speeds and temperatures.

Additionally, it is shown analytically that due to the changes in capacitance with rotation, even when a constant voltage is applied to the bearing, internal charge rebalancing currents will flow in the bearing. These currents are in the range of $4\,\mathrm{nA}$ for the investigated operating point and are thus assumed as not harmful to the bearing. However, for very high speeds and large bearing capacitances, e.g., in high-speed cylinder bearings, these currents will increase and should be considered in the design phase.

Finally, the model has shown that the internal currents flowing within the bearing during a breakdown are much higher than the current that can be measured from the outside. Precisely modeling the internal currents allows a more accurate estimation of the energy dissipated in the bearing during a breakdown and of the resulting damage. This effect still needs to be confirmed with measurements.

References

[1] J. Kammermann, I. Bolvashenkov, S. Schwimmbeck, and H. Herzog, "Reliability of induction machines - statistics, tendencies, and perspectives," in *2017 IEEE 26th International Symposium on Industrial Electronics (ISIE)*, Jun. 2017, pp. 1843–1847.

[2] M. Kriese, E. Wittek, S. Gattermann, H. Tischmacher, G. Poll, and B. Ponick, "Influence of bearing currents on the bearing lifetime for converter driven machines," in *International Conference on Electrical Machines*, 2012, pp. 1735–1739.

[3] H. Tischmacher and S. Gattermann, "Bearing currents in converter operation," in *The XIX International Conference on Electrical Machines - ICEM*, 2010.

[4] A. Muetze, "Bearing currents in inverter-fed ac-motors," Ph.D. dissertation, Technische Universitaet Darmstadt, 2004.

[5] M. Kriese, E. Wittek, S. Gattermann, H. Tischmacher, G. Poll, and B. Ponick, "Prediction of motor bearing currents for converter operation," in *The XIX International Conference on Electrical Machines - ICEM 2010*, 2010, pp. 1–6.

[6] S. Quabeck, V. Grau, and R. W. De Doncker, "Modeling and mitigation of bearing currents in electrical traction drives," in *2020 23rd International Conference on Electrical Machines and Systems (ICEMS)*, Nov. 2020, pp. 1101–1106.

[7] E. Wittek, M. Kriese, H. Tischmacher, S. Gattermann, B. Ponick, and G. Poll, "Capacitances and lubricant film thicknesses of motor bearings under different operating conditions," in *The XIX International Conference on Electrical Machines - ICEM 2010*, 2010, pp. 1–6.

[8] D. C. Ludois and J. K. Reed, "Brushless mitigation of bearing currents in electric machines via capacitively coupled shunting," *IEEE Transactions on Industry Applications*, vol. 51, no. 5, pp. 3783–3790, 2015.

[9] E. Wittek, M. Kriese, H. Tischmacher, S. Gattermann, B. Ponick, and G. Poll, "Capacitance of bearings for electric motors at variable mechanical loads," in *2012 XXth International Conference on Electrical Machines*, 2012, pp. 1602–1607.

[10] P. Han, G. Heins, D. Patterson, M. Thiele, and D. M. Ionel, "Modeling of bearing voltage in electric machines based on electromagnetic fea and measured bearing capacitance," *IEEE Transactions on Industry Applications*, vol. 57, no. 5, pp. 4765–4775, 2021.

[11] L. Gao, "16 - lubrication modelling of hip joint implants," in *Computational Modelling of Biomechanics and Biotribology in the Musculoskeletal System (Second Edition)*, ser. Woodhead Publishing Series in Biomaterials, Z. Jin, J. Li, and Z. Chen, Eds., Second Edition, Woodhead Publishing, 2021, pp. 415–436.

[12] J. Erdman, R. Kerkman, D. Schlegel, and G. Skibinski, "Effect of pwm inverters on ac motor bearing currents and shaft voltages," *IEEE Transactions on Industry Applications*, vol. 32, no. 2, pp. 250–259, 1996.

Comprehensive Control of Matrix Converters in On-Board Electric Drive Applications

Galina Mirzaeva
THE UNIVERSITY OF NEWCASTLE
University Drive
Callaghan, NSW2308, Australia
Phone: +61 (2) 4921-6083
Fax: +61 (2) 4921-6993
Email: Galina.Mirzaeva@newcastle.edu.au

Acknowledgments

This research was partially supported by the Australian Government through the Australian Research Council's Discovery Projects funding scheme (project DP2201039287).

Keywords

≪AC-AC Converter≫, ≪Matrix Converter≫, ≪Converter control≫, ≪Power converters for EV≫, ≪State and disturbance observers≫.

Abstract

Matrix Converters are an attractive alternative to the traditional AC/DC/AC converters, particularly, in electric transportation. Their advantages include low weight and volume, high reliability and efficiency. However, variable frequency and variable power operation, typical for electric vehicles, poses control challenges. This paper addresses these challenges and proposes a robust control solution for Matrix Converters with a number of novel features. The paper compares the proposed and the conventional control schemes for Matrix Converters, and shows improvement with respect to power factor control, input resonance suppression, reference tracking and harmonics performance. Findings of the paper are supported by simulations results.

Introduction

A Direct Matrix Converter (DMC) is illustrated in in Fig.1a. Unlike the conventional two-stage AC/DC/AC conversion option, Matrix Converter (MC) does not require an intermediate DC link energy storage. This leads to a more compact, power efficient and reliable design [1]. Only a small RLC filter is needed on the MC input side, to prevent harmonic pollution of the supply.

Over the past decade, the MC technology has seen many applications including AC drives, power quality compensators, renewable energy integration and grid interface [2]. Currently, MC are being introduced to electric vehicles [3]. This has posed new challenges to the MC control. A larger number of control objectives need to be satisfied simultaneously, while the available degrees of freedom in MC are very limited. As with any electric drive application, the load side objectives include accurate reference tracking and low harmonic distortion. Additionally, at the input side, active damping of the input filter resonance, low harmonic distortion and power factor correction at the supply are desired.

Conventional MC control schemes include closed-loop control of the output (load) side current and have similarities to Field Oriented Control (FOC) used with standard AC drives [4]. The control action, in the form of a switched output voltage vector $\bar{v}_o = v_o \angle \alpha_o$, is typically implemented by Space Vector

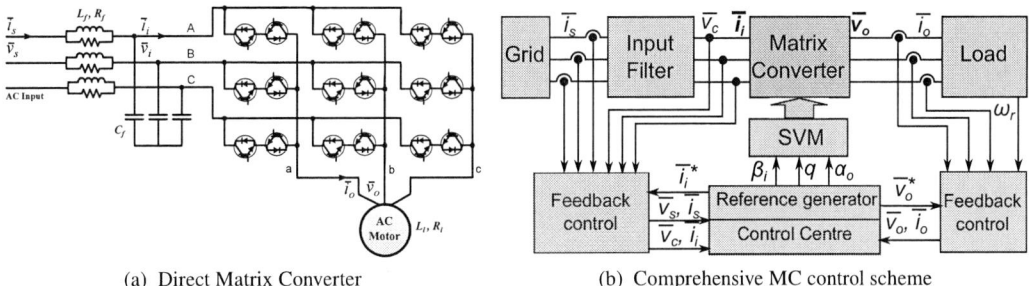

(a) Direct Matrix Converter   (b) Comprehensive MC control scheme

Fig. 1: Direct Matrix Converter and the proposed Comprehensive control.

Modulation (SVM). SVM for MC, discussed in detail in [5], has room for additional control of the angle β_i of the input current vector ($\bar{i}_i = i_i \angle \beta_i$). This is typically utilised to achieve unity power factor at the supply side [6]. If active damping of the input filter resonance is also desired, some form of trade-off between the input and the output control is applied.

For example, control of the MC output current can be performed indirectly, by closed-loop control of the MC input current [7]. This approach allows for including extra loops for power factor control and active damping. An alternative, MPC, approach is explored in [8, 9]. It provides indirect active damping by penalising the input current error together with the load current error in a cost function. To achieve harmonic performance comparable to SVM, modulated MPC can be used [10].

The authors of the current paper propose a new Comprehensive control method, called so because it seamlessly matches the competing control objectives on both MC sides. The Comprehensive control scheme is illustrated in Fig.1b. Standard measurements performed at the supply and load sides of MC are used by "Reference generator" to produce references that are compatible with each other. State feedback control is then used to regulate both the load- and the supply-side current around the corresponding references. The use of observers guarantees zero steady-state errors at both sides.

The proposed control scheme provides excellent tracking performance of the load-side current, as well as unity power factor, active damping and harmonic suppression at the supply side. These features, combined with weight and volume advantages of an MC, make it a very attractive solution for on-board drives in air, sea or road electric transportation. This paper compares the proposed control scheme with the conventional MC control, with respect to the above performance indicators. It demonstrates, by simulation, the advantages of the proposed scheme.

Comprehensive control for Matrix Converter

Supply side model

An RLC filter (with parameters R_f, L_f, C_f) typically connects the supply with the input side of the MC. The associated voltages and currents in each phase $\phi \in \{\alpha, \beta\}$, can be modelled in continuous time by the following second order dynamic system:

$$\frac{d(\mathbf{x}_i^\phi)}{dt} = \mathbf{A}_i \mathbf{x}_i^\phi + \mathbf{B}_i \begin{bmatrix} v_s^\phi \\ i_i^\phi \end{bmatrix}; \quad \mathbf{x}_i^\phi \doteq \begin{bmatrix} i_s^\phi \\ v_i^\phi \end{bmatrix}; \quad \mathbf{A}_i = \begin{bmatrix} -\frac{R_f}{L_f} & -\frac{1}{L_f} \\ \frac{1}{C_f} & 0 \end{bmatrix}; \quad \mathbf{B}_i = \begin{bmatrix} \frac{1}{L_f} & 0 \\ 0 & -\frac{1}{C_f} \end{bmatrix} \tag{1}$$

where v_s^ϕ and i_s^ϕ are the supply voltage and current in phase $\phi \in \{\alpha, \beta\}$, and v_i^ϕ, i_i^ϕ are the voltage and current at the input of the MC in the same phase.

According to the Internal Model Principle [11], rejection of a disturbance requires its inclusion in the system model. Hence it is proposed to include in the supply side model a disturbance model of the form:

$$\frac{d(\mathbf{x}_i^\phi)}{dt} = \mathbf{A}_i \mathbf{x}_i^\phi + \mathbf{B}_i \begin{bmatrix} v_s^\phi \\ i_i^\phi \end{bmatrix} + \begin{bmatrix} d_1^\phi \\ d_2^\phi \end{bmatrix}; \quad \frac{d(\mathbf{d}_i)}{dt} = \mathbf{A}_{di}\mathbf{d}_i; \quad \mathbf{d}_i \doteq \begin{bmatrix} d_1^\alpha \\ d_1^\beta \\ d_2^\alpha \\ d_2^\beta \end{bmatrix}; \quad \mathbf{A}_{di} = \begin{bmatrix} 0 & -\omega_i & 0 & 0 \\ \omega_i & 0 & 0 & 0 \\ 0 & 0 & 0 & -\omega_i \\ 0 & 0 & \omega_i & 0 \end{bmatrix} \quad (2)$$

where $d_1^{\alpha,\beta}$ and $d_2^{\alpha,\beta}$ are harmonic disturbances, at the supply frequency ω_i, for $i_s^{\alpha,\beta}$ and $v_c^{\alpha,\beta}$, respectively.

Load side model

The load side model corresponds to an RL load with parameters R_l, L_l. Similarly to the supply side, it is proposed to include a disturbance at ω_o in the load side model as:

$$\frac{d(i_o^\phi)}{dt} = -a_o i_o^\phi + b_o v_o^\phi + d_o^\phi; \quad \frac{d(\mathbf{d}_o)}{dt} = \mathbf{A}_{do}\mathbf{d}_o; \quad \mathbf{d}_o \doteq \begin{bmatrix} d_o^\alpha \\ d_o^\beta \end{bmatrix}; \quad \mathbf{A}_{do} = \begin{bmatrix} 0 & -\omega_o \\ \omega_o & 0 \end{bmatrix} \quad (3)$$

where v_o^ϕ, i_o^ϕ are the load voltage and current in one of the output phases $\phi \in \{\alpha, \beta\}$; $a_o = R_l/L_l$; $b_o = 1/L_l$; and d_o^ϕ is harmonic disturbance, at the load side frequency ω_o, for i_o^ϕ in the same phase.

For the control computation, discrete-time state-space equivalents of (2) and (3) are used.

Input and output side observers

To achieve integral action at the desired fundamental frequencies, it is suggested to complement the extended models (2) and (3) by the observers, for all state and disturbances variables, as follows.

$$\frac{d(\hat{\mathbf{x}}_i^\phi)}{dt} = \mathbf{A}_i \hat{\mathbf{x}}_i^\phi + \mathbf{B}_i \begin{bmatrix} v_s^\phi \\ i_i^\phi \end{bmatrix} + \begin{bmatrix} \hat{d}_1^\phi \\ \hat{d}_2^\phi \end{bmatrix} + \mathbf{J}_i(\mathbf{x}_i^\phi - \hat{\mathbf{x}}_i^\phi); \quad \frac{d(\hat{\mathbf{d}}_i)}{dt} = \mathbf{A}_{di}\hat{\mathbf{d}}_i + \mathbf{J}_{di}(\mathbf{x}_i^\phi - \hat{\mathbf{x}}_i^\phi); \quad (4)$$

$$\frac{d(\hat{i}_o^\phi)}{dt} = a_o \hat{i}_o^\phi + b_o v_o^\phi + \hat{d}_o^\phi + \mathbf{J}_o(i_o^\phi - \hat{i}_o^\phi); \quad \frac{d(\hat{\mathbf{d}}_o)}{dt} = \mathbf{A}_{do}\hat{\mathbf{d}}_o + \mathbf{J}_{do}(i_o^\phi - \hat{i}_o^\phi). \quad (5)$$

where $\mathbf{J}_i$, $\mathbf{J}_o$, $\mathbf{J}_{di}$ and $\mathbf{J}_{do}$ are matrices of observers gains of appropriate dimensions.

Introduction of the observers has the following important effect. Irrespective of the model accuracy, if a steady state trajectory is reached, then $\hat{\mathbf{x}}_i \to \mathbf{x}_i$, $\hat{i}_o \to i_o$, at the respective disturbance frequencies. In other words, the system models may include parameter errors and unmodelled terms, such as, for instance, a missing back-emf in (3) or incorrectly defined supply voltage in (1). Regardless of that, due to the inclusion of the disturbance models in (2) and (3), "integral action" is provided by observers (4) and (5), and the estimated system states $(\hat{i}_s, \hat{v}_i, \hat{i}_o)$ will have zero errors at frequencies ω_i and ω_o, respectively.

Steady state input and output side references

The load and the supply sides are required to track sinusoidal steady state references, and quickly transition to a new steady state if the load conditions change. In Fig.1b the sinusoidal steady state references (i_s^*, v_i^* and i_o^*) and the corresponding steady state MC signals (i_i^*, v_o^*) are calculated by "Reference generator" as follows. When in steady state, the MC supply side and load side models satisfy the trajectories obtained by using standard circuit theory. Additionally, the supply and the load side references and MC signals are linked by two steady state conditions, namely, real power balance across the switches and the desired (e.g. unity) power factor at the supply. This yields the following eight steady state equations:

$$Input \begin{cases} -L_f\omega_i i_s^{*\beta} + R_f i_s^{*\alpha} = v_s^{*\alpha} - v_i^{*\alpha} + L_f \hat{d}_1^\alpha \\ L_f\omega_i i_s^{*\alpha} + R_f i_s^{*\beta} = v_s^{*\beta} - v_i^{*\beta} + L_f \hat{d}_1^\beta \\ -C_f\omega_i v_i^{*\beta} = i_s^{*\alpha} - i_i^{*\alpha} + C_f \hat{d}_2^\alpha \\ C_f\omega_i v_i^{*\alpha} = i_s^{*\beta} - i_i^{*\beta} + C_f \hat{d}_2^\beta \end{cases} \quad Output \begin{cases} -L_l\omega_o i_o^{*\beta} + R_l i_o^{*\alpha} = v_o^{*\alpha} + L_l \hat{d}_o^\alpha \\ L_l\omega_o i_o^{*\alpha} + R_l i_o^{*\beta} = v_o^{*\beta} + L_l \hat{d}_o^\beta \end{cases}$$

$$Extra \begin{cases} v_i^{*\alpha} i_i^{*\alpha} + v_i^{*\beta} i_i^{*\beta} = v_o^{*\alpha} i_o^{*\alpha} + v_o^{*\beta} i_o^{*\beta} \\ v_s^{*\alpha} i_s^{*\beta} - v_s^{*\beta} i_s^{*\alpha} = 0 \end{cases} \quad (6)$$

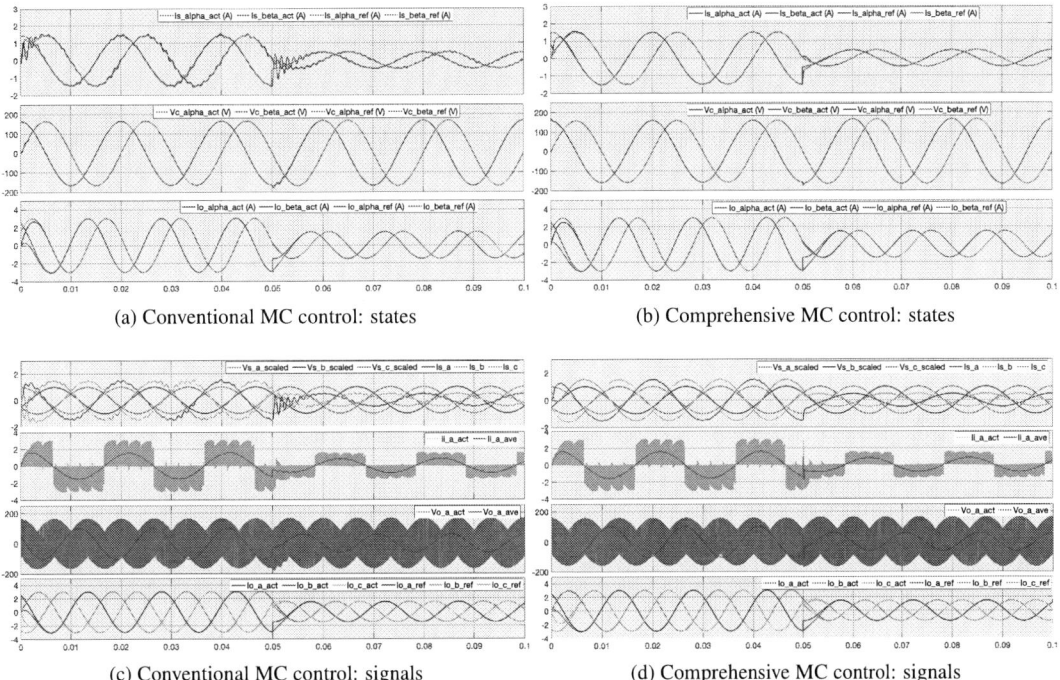

(a) Conventional MC control: states

(b) Comprehensive MC control: states

(c) Conventional MC control: signals

(d) Comprehensive MC control: signals

Fig. 2: Performance comparison of conventional (left) and Comprehensive (right) control schemes.

Note that the disturbance estimates ($\hat{d}_o$, $\hat{d}_1$ and $\hat{d}_2$) provided by the observers (4) and (5) are included in the equations (6). Eight equations (6) are solved together in the following way. The current references ($i_o^{*\alpha}$, $i_o^{*\beta}$) come from the load side (e.g. motor) control and are assumed known. Then ($v_o^{*\alpha}$, $v_o^{*\beta}$) can be determined from the two *Output* equations. This defines the right-hand side of the first *Extra* equation.

Next, state variables ($i_s^{*\alpha}$, $i_s^{*\beta}$), ($v_i^{*\alpha}$, $v_i^{*\beta}$) are expressed from the four *Input* equations in terms of the MC input ($i_i^{*\alpha}$, $i_i^{*\beta}$), known supply voltage ($v_s^{*\alpha}$, $v_s^{*\beta}$) and known disturbance estimates. Substitution into the two *Extra* equations allows one to find ($i_i^{*\alpha}$, $i_i^{*\beta}$). Other unknowns can be then found by back substitution.

Dynamic control

Dynamic control utilises state estimate feedback to regulate the state observer about the previously calculated sinusoidal steady state. The control laws for the input and output sides are given by, respectively:

$$ i_i = i_i^* - K_1(\hat{i}_s - i_s^*) - K_2(\hat{v}_i - v_i^*); \qquad v_o = v_o^* - K_o(\hat{i}_o - i_o^*); \qquad \text{for} \quad \alpha, \beta \tag{7} $$

The control inputs i_i and v_o are determined from (7) independent of each other, and may contradict to the requirement of the instantaneous real power balance across the MC switches. This is resolved by "Control centre" shown in Fig.1b, as follows.

The supply side RLC filter, due to its resonant nature, requires an accurate and "aggressive" control. Hence the control gains K_1 and K_2 are designed to achieve "fast-acting" control poles. When defining the SVM switching pattern, the value of i_i determined from (7) is given preference to v_o. The v_o value found from (7) is projected onto the power balance constraint, and the constrained v_o^{con} nearest to v_o is determined and applied. Hence the applied values i_i and v_o^{con} are perfectly matched by the power balance.

Consequently, in transient, the load-side voltage v_o is driven by the instantaneous power balance with i_i, but its trend is to match the load-side current i_o with i_o^*. When approaching steady state, the load-side tracking is error-free, due to the use of the disturbance observer. In fact, steady state errors are eliminated on both sides due to inclusion of the disturbance estimates in the steady state reference calculations (6).

Comparison of Comprehensive to the conventional control of MC

To validate the proposed solution, a detailed simulation of an MC-based variable speed drive was set up in the Matlab/Simulink environment. The input side filter included $R_f = 2.5\Omega$; $L_f = 2.5\text{mH}$; $C_f = 14.2\mu\text{F}$. The input source was $V_s = 207V$ (rms, line-to-line) at frequency 50Hz. The load side model represents an induction motor with $\sigma L_s = L_o = 52.3\text{mH}$; $R_s = R_o = 5\Omega$; and back-emf representing energy conversion from electromagnetic into mechanical.

Fig.2 compares performance of the conventional and the proposed control schemes under the following scenario: initially the load-side current reference is 3A; at time 0.05s this reference drops to 1.5A.

Figs.2a shows the state variables: $i_{s\alpha}$, $i_{s\beta}$, $v_{c\alpha}$, $v_{c\beta}$ for the MC input side and $i_{o\alpha}$, $i_{o\beta}$ for the MC output side under conventional control. The output-side states $i_{o\alpha}$, $i_{o\beta}$ are controlled by PI current control, and the input states are driven by the requirement of unity power factor. Fig.2b shows the same variables for the proposed control. The supply-side states $i_{s\alpha}$, $i_{s\beta}$ are controlled by state feedback current control, and the output-side states are driven to their steady-state references by integral action of the observers.

Fig.2c and Fig.2d compare the important MC input and output signals for the two control schemes. These signals include: the supply-side currents $i_s^{(abc)}$ plotted next to the scaled supply voltages $v_s^{(abc)}$ to illustrate the unity power factor; switched input-side current waveform $i_i^{(a)}$ and its low-frequency average; switched output-side voltage waveform $v_o^{(a)}$ and its low-frequency average; and the load-side currents $i_o^{(abc)}$ plotted next to their references to illustrate the reference tracking.

It can be observed from Fig.2 that, under the conventional control scheme, the change of the load-side current reference is accompanied by a resonant trace in the supply-side current. Additionally, even under steady state, the resonant component can be still observed in the supply-side current.

Both effects completely disappear when the proposed Comprehensive control scheme is applied. Under Comprehensive control, high-quality reference tracking and fast dynamic performance at the load side are achieved simultaneously with unity power factor and resonance suppression at the supply side.

Application of MC with Comprehensive control to on-board electric drives

Fig.3 illustrates performance of the proposed control under a simulated scenario of electric drive application. An additional problem that needed to be addressed in this case is power factor optimisation. The supply-side power factor has a complex and nonlinear dependence on the operating conditions [12]. Unity power factor cannot be always achieved [13].

The proposed control algorithm was modified so as to optimise power factor under every operating condition. Namely, if the power balance and the unity power factor requirements (labelled as *Extra* conditions in expressions (6)) are compatible, then unity power factor is achieved. Otherwise, the power balance condition is given preference, and power factor is made as close to unity as possible. This is achieved by implementing the following logic.

- The output side current control provides the reference vector $\bar{i}_o^*$. By solving the *Output* conditions in expressions (6), the $\bar{v}_o^*$ can be found. From that the output power requirement $P_o^* = v_o^{*\alpha}i_o^{*\alpha} + v_o^{*\beta}i_o^{*\beta}$ is determined;

- Two *Extra* conditions in expressions (6) conditions (power balance $P_i^* = P_o^*$ and the vector alignment $\bar{i}_s||\bar{v}_c$), are used simultaneously to determine the references vectors $\bar{i}_i^*$, $\bar{i}_s^*$ and $\bar{v}_c^*$;

- It is known from the literature [14] that the voltage transfer ratio for the MC, for a given input power factor, is limited by

$$q = \frac{V_o}{V_c \cos\varphi_i} \le \frac{\sqrt{3}}{2} \quad \text{from which} \quad V_o^{max} = \frac{\sqrt{3}}{2}V_i\cos\varphi_i \tag{8}$$

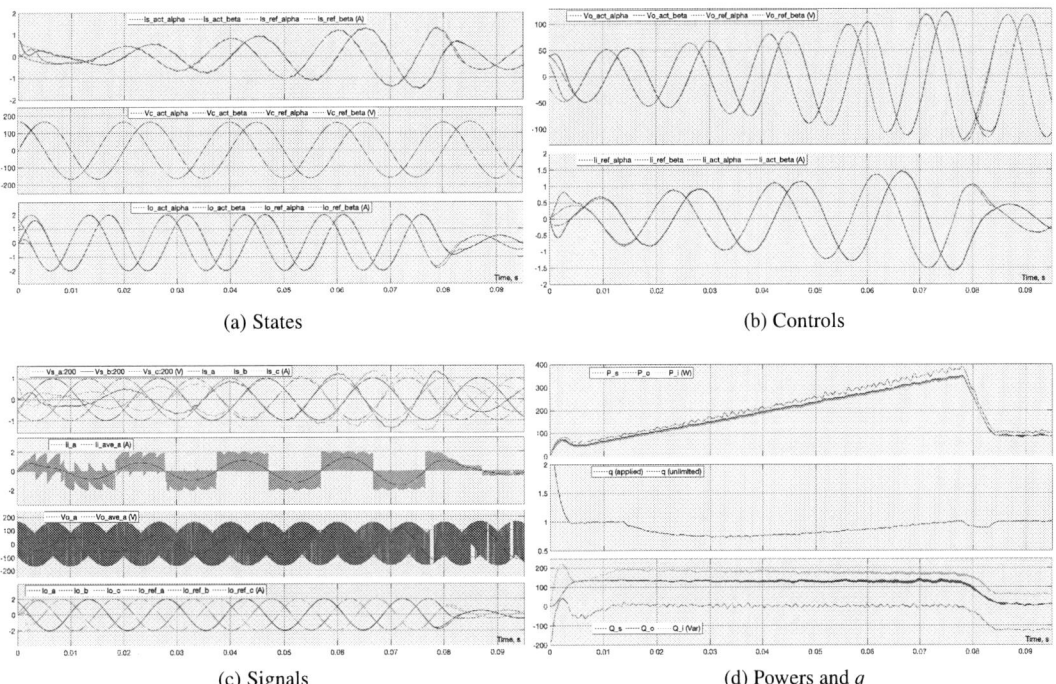

(a) States   (b) Controls

(c) Signals   (d) Powers and q

Fig. 3: Simulation plots for a MC-based variable speed drive operation.

On the other hand, from the input/output power balance, it follows that

$$\frac{V_o}{V_c \cos \varphi_i} = \frac{I_i}{I_o \cos \varphi_o} \quad \text{then} \quad I_i^{max} = \frac{\sqrt{3}}{2} I_o \cos \varphi_o \tag{9}$$

While condition (8) is more common in the literature, condition (9) is easier to check. The $\bar{i}_i^*$ magnitude is then compared to I_i^{max}. If $|\bar{i}_i^*| \leq I_i^{max}$ then the two *Extra* conditions are compatible, and unity power factor at the grid side can be achieved;

- Otherwise, the two *Extra* conditions are not compatible. Then the power balance condition is given preference, and power factor is made as close to unity as is possible for the given limitation I_i^{max} on the $\bar{i}_i^*$ magnitude. Then the second *Extra* condition in (6) is replaced by $(i_i^{*\alpha})^2 + (i_i^{*\beta})^2 = I_i^{max}$, and the two updated *Extra* conditions are solved together. The updated solution for $\bar{i}_i^*$ and the corresponding solutions for the dependent vectors $\bar{i}_s^*$ and $\bar{v}_c^*$, are determined. The best possible (but not unity) grid-side power factor is automatically achieved.

- If, in the previous step, the solution for $\bar{i}_i^*$ cannot be found, then the demanded $(\bar{i}_o^*, \bar{v}_o^*)$ combination cannot be produced under any grid-side power factor, and the MC enters an overmodulation mode.

Fig.3a shows the system state variables: $i_{s\alpha}$, $i_{s\beta}$, $v_{c\alpha}$, $v_{c\beta}$ for the MC input side and $i_{o\alpha}$, $i_{o\beta}$ for the MC output side. Fig.3b illustrates the control variables: $v_{o\alpha}$, $v_{o\beta}$ and $i_{i\alpha}$, $i_{i\beta}$. The actual look of the MC-generated signals (v_{oa} and i_{ia}) can be observed in Fig.3c, together with other phase signals of interest. Finally, Fig.3d illustrates the real powers (P_s at the supply, P_o at the MC output and P_i at the MC input), the normalised voltage transfer ratio $q = \frac{2}{\sqrt{3}} \frac{V_o}{V_c \cos \varphi_i}$ and the respective reactive powers (Q_s, Q_o and Q_i).

Initially, at the motor start and until $t = 0.0135$s, unity power factor at the MC supply side could not be achieved. As the motor speeds up and consumes more real power, the supply-side voltage and current come into alignment. This happens between $t = 0.0135$s and $t = 0.078$s. After reaching the rated speed at $t = 0.078$s, the load torque reduces and, finally, settles at a new constant level corresponding to less than $1/3$ of the rated power at $t = 0.09$s. The supply-side power factor is 0.625 (leading).

An important observation from Fig.3 is that the proposed control scheme provides excellent reference tracking at the load side, operates in a stable and smooth manner, with no resonance excitation in the supply-side current and with maximum achievable supply-side power factor for every load-side condition. This makes Comprehensive control driving an MC-based electric drive a very attractive solution for electric transport applications.

Conclusion

This paper has explored a new Comprehensive control scheme for Matrix Converters intended for on-board electric drive applications. The paper has shown performance advantages of this scheme in comparison to the existing control methods. The paper has illustrated performance of Comprehensive control under simulated electric vehicle application scenario and has demonstrated excellent dynamics and reference tracking, smooth and stable operation, optimal handling of the power factor and effective resonance suppression.

References

[1] Khosravi M., Amirbande M., Khaburi D.A., Rivera M., Riveros J. et al: Review of Model Predictive Control strategies for Matrix Converters, 2019 IET Journal on Power Electronics, Vol 12 no 12, pp. 3021-3032

[2] Zhang J., Yang H., Wang T., Li L., Dorrell D.G. et al: Field-oriented control based on hysteresis band current controller for a permanent magnet synchronous motor driven by a direct matrix converter, 2018 IET Journal on Power Electronics, Vol 11 no 7, pp. 1277-1285

[3] Mirzaeva G., Seron M., Goodwin G.C.: Matrix Converters with input resonance suppression for mobile mining vehicles, 2020 IEEE IAS Annual Meeting, pp.1-5

[4] Lee K., Blaabjerg F.: Sensorless DTC-SVM for Induction Motor Driven by a Matrix Converter Using a Parameter Estimation Strategy, 2008 IEEE Transactions on Industrial Electronics, Vol 55 no2, pp.512-521

[5] Casadei D., Serra G., Tani A., Zarri L.: Matrix converter modulation strategies: a new general approach based on space-vector representation of the switch state, 2002 IEEE Transactions on Industrial Electronics, Vol 49 no 2, pp. 370-381

[6] Nguyen H.M., Lee H.H., Chun T.W.: Input power factor compensation algorithms using a new Direct-SVM method for matrix converter, 2011 IEEE Transactions on Industrial Electronics, Vol 58 no 1, pp. 232-243

[7] Lei J., Zhou B., Bian J., Wei J., Zhu Y. et al: Feedback control strategy to eliminate the input current harmonics of matrix converter under unbalanced input voltages, 2017 IEEE Transactions on Power Electronics, Vol 32 no 1, pp. 878-888

[8] Rivera M., Rodriguez J., Wheeler P.W., Rojas C.A., Wilson A. et al: Control of a matrix converter with imposed sinusoidal source currents, 2012 IEEE Transactions on Industrial Electronics, Vol 59 no 4, pp. 1939-1949

[9] Lei J., Feng S., Wheeler P., Zhou B., Zhao J.: Steady-state error suppression and simplified implementation of direct source current control for matrix converter With Model Predictive Control, 2020 IEEE Transactions on Power Electronics, Vol 35 no 3, pp. 3183-3194

[10] Rivera M., Amirbande M., Vahedi A., Tarisciotti L., Wheeler P.: Fixed frequency model predictive control with active damping for an indirect matrix converter, CHILECON 2017, pp.1-6

[11] Francis B.A., Wonham W.M.: The internal model principle of control theory, 1976 Automatica, no 12, pp. 457-465

[12] Nguyen H.N., Nguyen M.K., Duong T.D., Tran T.T., Lim Y.C. et al: A study on input power factor compensation capability of matrix converters, 2020 Electronics, Vol 9 no1, pp.82-99

[13] Mirzaeva G., Carter D., Seron M.: Grid-side power factor optimisation for Matrix Converters in mobile mining vehicle applications, 2021 IEEE IAS Annual Meeting, pp.1-6

[14] Huber L., Borojevic D.: Space vector modulated three-phase to three-phase matrix converter with input power factor correction, 1995 IEEE Transactions on Industry Applications, Vol 31 no 6, pp. 1234-1246

Power System Simulation Tool for Quick Benchmarking of Innovative MVDC Grids in E-Mobility Applications

Daniel Siemaszko, Philippe Noisette
Hitachi Energy
Spinnereistrasse 3
Turgi, Switzerland
Tel.: +41 79 944 02 99
E-Mail: daniel.siemaszko@hitachienergy.com, philippe.noisette@hitachienergy.com
URL: https://www.hitachienergy.com/

Keywords

« Solid-State Transformer », « Simulation », « Microgrid », « Stability », « Converter control »

Abstract

The rapid development of Solid-State Transformers enables a future for microgrids that features the connection of various sources, loads, and storage elements to a common MVDC bus. The simulation of such systems present various challenges due to their growing complexity caused by the large amount of connected power converters and diversification of sources and loads. Quick benchmarking in booming areas such as E-Mobility typically require simulation models that feature superior speed with a dynamic response that fits real hardware. Within this context and objective in mind, a simulation tool has been developed with system level average models of elementary SST cells for DC grids. The implemented models are built with controlled current/voltage sources which embody dynamic behaviors of real SST systems. They may be arranged in various modular configurations, and they feature a basic failure mechanism that allows bypass of failed cells to study their impact on the full power system.

Introduction

DC microgrids are foreseen to become the backbone of interconnection between decentralized sources, loads and Energy Storage Systems [1]-[3]. It is foreseen that Solid-State Transformers (SST) featuring a DC/DC isolated conversion are to be widely used as elementary building blocks in such DC microgrids. The function of the DC/DC isolated converter can be implemented in at least two well-known ways: a so-called resonant approach with an LLC tank and the isolated Dual Active Bridge (DAB). The choice between the two may be determined based on the application requirements, namely the need for a DC/DC converter with isolation or a DC transformer. The resonant approach has been extensively studied with the introduction of the Power Electronics Transformer (PET) [4]-[8]. Mainly foreseen for traction applications, the resonant approach requires an additional front end for voltage control purposes. The Phase-shift controlled DAB approach has also been well studied and resulted in a 0.5MVA demonstrator [9]. This work showed that the DAB seems better suited for wide voltage applications with its scalability possibilities and easily realized redundancy.

MVDC in E-mobility context

In 2020 the global electric car stock increased by 43% over 2019; about 3 million new e-Cars were registered, electric bus and electric heavy-duty truck (HDT) registrations also increased in China, Europe and North America. Some efforts are underway to develop new standards for mega-chargers (MCS) (i.e Ultra ChaoJi or 3MW chargers), chargers capable of charging trucks or busses reasonably quickly [10]. Some impact to grids is inevitable given the high-power requirements of mega-chargers. Long-term planning for infrastructure is needed now to avoid negative impacts on the electrical grid. Significant investment may be needed for grid reinforcements, renewable integration [11], and energy storage support [12].

On the power system level, the recent development of modular DC structures such as Modular Multilevel Converters (MMC) or Solid-State Transformers (SST) based on Medium Frequency Transformers (MFT), allows to envision a new type of flexible grids that features energy storage and peak sharing, without increasing the available power of substations connected to the grid. The growth of a fleet of electric vehicles in existing facilities requires the installation of energy source such as rooftop solar panels, flexible BESS systems, converters upcoming H_2 fuel cells, and bidirectional chargers for grid support from vehicle batteries.

The foreseen power system would be supported by an MVDC backbone [1], with modular converters interacting the various elements allowing to run on-demand Megawatt Chargers and still rely on renewable energy sources. This is enabled by Battery Energy Storage System (BESS) that act like a buffer for PV, braking energy recovery, and allows limiting power demand from the utility grid. The MVDC grid will also connect local LVDC grids where bidirectional power flow chargers will be operated with la growing number of individual cars [13]-[15].

Considered power system configuration

In this work, a microgrid with ±10kV MVDC backbone (as illustrated on Fig. 1) is presented with the following elements:

- isolated Active Front End (AFE) connecting a 36kVAC utility grid to the ±10kV MVDC link with a so-called Synthetic Inertia that represents the ability of maintaining DC grid stability
- bidirectional SST based converters for connecting the batteries, an LVDC grid used for fast charging of electric cars, and a traction network with breaking energy recovery feature
- unidirectional SST based converters connecting Megawatt Charging loads
- unidirectional SST based converters for supplying PV and H_2 energy to the DC grid
- a so-called Energy Management System that ensures the stability of the bus by engaging the BESS, the fuel cells or the car batteries for grid support.

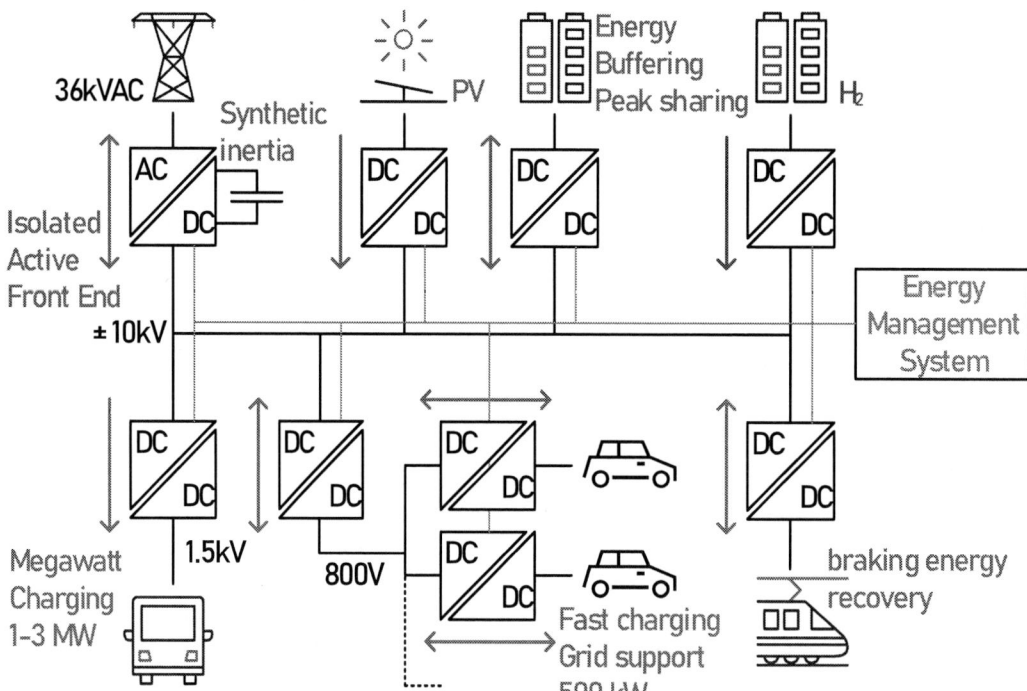

Fig. 1: MVDC network for future E-mobility context

The main elements of the considered system are isolated DC/DC converters that allows modular configurations as illustrated on Fig. 2. The left-hand side figure shows a so-called Input-Series Output-Parallel (ISOP) configuration for reaching high voltages on the one side, and high currents on the fast-charging side. The right-hand side figure shows a simple staked configuration for connecting isolated H_2 fuel cells.

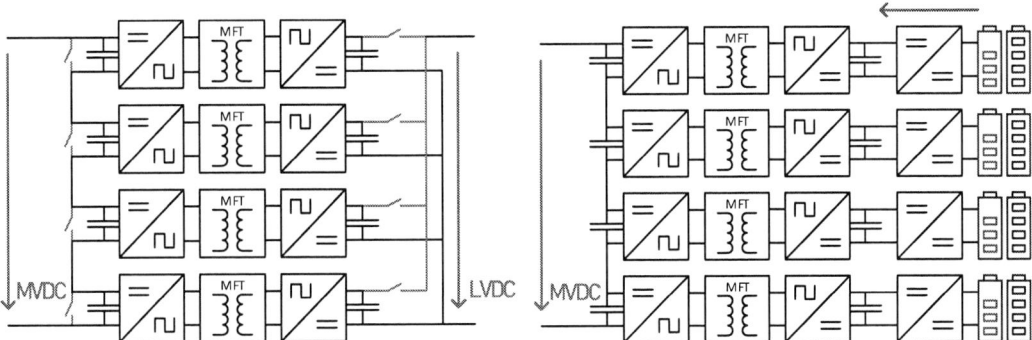

Fig. 2: Typical modular configurations for SST blocks connected

For simulation purpose, the modelling of the system illustrated in Fig. 1 presents a challenge since it is built with a high number of elementary converters, all interacting with each other. On the other hand, a high-level modelling of the system does not consider the dynamic response of the converters. A balance must be found between simulation speed and the level of accuracy one would like to reach. One would like to benchmark several types of configurations, with given scenarios to validate a power system. One would also like to study to impact of a failure of a module on a full system.

Current and voltage source modelling of SST elementary cells

Any power system can be described with their elementary Thevenin-Norton equivalencies, meaning as an arrangement equivalent controlled voltage sources and controlled current sources [16]. Those sources implement an average behaviour of power converters or power sources and can be combined with the condition that some elementary rules are followed. The most important rule is that two voltage sources can't be connected in parallel, and current sources can't be connected in series. A capacitor typically implements the behaviour of a controlled voltage source, and an inductor typically implements the behaviour of a controlled current source. The controls of those sources are typically integrators with multipliers that reflect the behaviour of the modelled elements.

The implemented SST and microgrid models with their current and voltage source equivalents could be run within any power electronics and systems simulation environment. To this specific work, the models have been run with simulation environment Simba.io by Aesim [17]. This environment is a Python based implementation and features a specific Python API which allows building and calling models from Python scripts and programs. This allows all sort of implementation, optimization, pseudo real time operation, and complex post processing with tools that are not available with other simulation environments.

Modelling of the SST elementary Cell

The SST cell is modelled with its equivalent current/voltage source model and can interface power systems that are also modelled as such. Both DC links on each side of the SST are interfaced with controlled voltage sources. The MFT windings are modelled with two controlled current sources. The functions performed by the SST cell are depicted on Fig. 3. As a convention adopted in this work, the SST is connected to a DC grid on its primary side and a current source load on its secondary side.

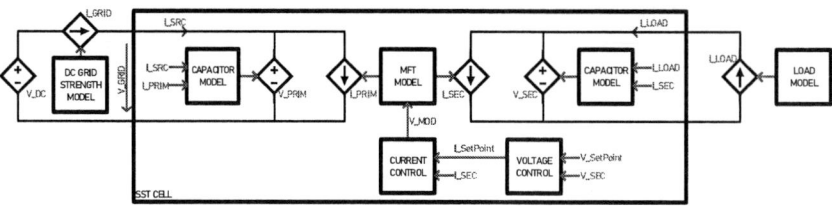

Fig. 3: System level modelling of an SST elementary cell

The dynamics of the system are reflected in the various time constants of the system elements. The behaviour of the capacitors on each side of the SST is modelled as an integrator taking the difference in currents and the capacitance value C_{DC} as described in Equ. 1 and 2. The modulator has a time constant delay defined as a function of switching frequency F_{SW}. The current in the the MFT inductance L_{MFT} is modelled as equivalent transfer function taking the applied voltage V_{MOD} input as in Equ. 3 and 4. The voltage current cascaded PI controllers described by Equ. 5 and 6 are tuned and optimized as a function of the model's physical parameters. The voltage controller maintains the secondary side voltage to a given setpoint, providing a current setpoint. The current controller maintains de MFT current to the given set points by providing the modulation voltage V_{MOD}. The parameters used for running the model are described in Table 1.

$$V_{PRIM} = \frac{1}{C_{DC}} \int (I_{SRC} - I_{PRIM}) \tag{1}$$

$$V_{SEC} = \frac{1}{C_{DC}} \int (I_{LOAD} - I_{SEC}) \tag{2}$$

$$I_{SEC} = \frac{1}{L_{MFT}F_{SW}} V_{MOD} \tag{3}$$

$$I_{PRIM} = I_{SEC}N_{MFT} \tag{4}$$

$$V_{MOD} = K_P(I_{SEC} - I_{SP}) + K_I \int (I_{SEC} - I_{SP}) \tag{5}$$

$$I_{SP} = K_P(V_{SEC} - V_{SP}) + K_I \int (V_{SEC} - V_{SP}) \tag{6}$$

Table I: List of considered SST Cell parameters

	Symbol	Value	Comment
DC nominal voltage on primary side	V_{PRIM}	1000V	Set by grid model V_{GRID}
DC nominal voltage on secondary side	V_{SEC}	2000V	Controlled by SST
nominal current on secondary side	I_{SEC}	250A	Set by load model I_{LOAD}
maximum current on secondary side	I_{SEC_MAX}	300A	Controlled by SST
MFT side SST inductance	L_{MFT}	8.5μH	Defines control's time response
DC side SST capacitor	C_{DC}	3mF	Defines control's time response
MFT switching frequency	F_{SW}	10kHz	Defines control's transfer function

The DC Grid Strength is modelled through its "equivalent impedance" represented by its resistive and inductive components R_{GRID} and L_{GRID} as in Equ. 7. The current source representing the grid impedance interfaces the DC ideal grid voltage source V_{DC} with the primary side of the SST.

$$I_{GRID} = \frac{1}{L_{GRID}} \int (V_{DC} - V_{GRID} - R_{GRID}I_{GRID}) \tag{7}$$

The model is run with a current limiter and anti wind-up on the controller's side. When external load is above SST maximum current specification, typically a burst or a short-circuit, the voltage cannot be maintained on the secondary side. Depending on the direction of the current, a short can be mitigated with or without hitting its own current limiter. On on the left-hand side plot of Fig. 4, the simulation shows the two behavioural cases of an SST running with a short burst on the load side. This allows testing the controller's dynamics in terms of voltage and current control, as well as current limitation capabilities of the model. One can see that the dynamics of the SST cell model correspond to a real SST system when compared with real hardware results from an inhouse built SST demonstrator [9]. Indeed, in both figure the voltage controller time response to a load step is about 7.5ms.

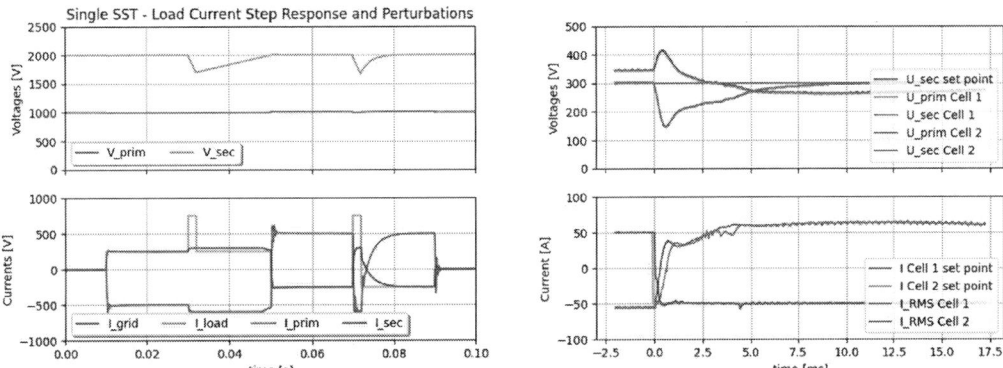

Fig. 4: Voltage and current response of the elementary SST cell with current limiter

Elementary SST connecting a Battery and an Active Front End through a PCC

The same SST Cell as described previously is run with a model of an Active Front End (AFE) on the primary side, and a model of a Battery on the secondary side. As illustrated on Fig 5, a DC bus model is implemented for connecting the two voltage sources, from the AFE and the SST primary side. The DC bus model considers the AFE side as the side providing voltage, and the Passive Front End side (PFE) as the side only drawing current.

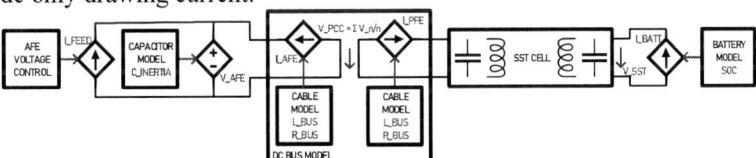

Fig. 5: Modelling of SST cell connecting a Battery to an Active Front End

The Battery is a simple controlled current source; however, it produces current only when its State of Charge (SoC) permits it. SoC should typically vary between 10% and 90% in order to maintain optimal battery lifetime, it is calculated as a function of the integral of the battery current I_{BATT} and the battery capacity Q_{BATT} as given in Equ. 8.

$$SoC = \int \frac{I_{BATT}}{Q_{BATT}} \tag{8}$$

The AFE is modelled as a controlled current source I_{FEED} connected to a controlled voltage source V_{AFE}. Equ. 9 implements the behaviour of a capacitor model C_{AFE} which value represents the inertia of the grid. The AFE current is controlled in a way to maintain the AFE voltage to the voltage set point given by V_{DC} with the use of a PI controller as given in Equ. 10. It features given dynamics, inertia and current limitation set by the system.

$$V_{AFE} = \frac{1}{C_{AFE}} \int (I_{FEED} + I_{AFE}) \tag{9}$$

$$I_{FEED} = K_P (V_{AFE} - V_{DC}) + K_I \int (V_{AFE} - V_{DC}) \tag{10}$$

The DC Bus is connecting voltage sources and must be consequently modelled as current sources. The controls of those current sources implement the dynamic response of a cable with its resistance R_{DCBUS} and line inductance L_{DCBUS}. The model features the voltage of the Point of Common Coupling (PCC), and current sharing between the connected devices. The voltage of the PCC is given by Equ. 11, the currents I_{AFE} and I_{PFE} on both sides of the PCC are given by Equ. 12 and 13.

$$V_{PCC} = \frac{V_{AFE} + V_{PFE}}{2} \tag{11}$$

$$I_{AFE} = \frac{1}{L_{DCBUS}} \int (V_{PCC} - V_{AFE} - R_{DCBUS}I_{AFE}) \tag{12}$$

$$I_{PFE} = \frac{1}{L_{DCBUS}} \int (V_{PCC} - V_{PFE} - R_{DCBUS}I_{PFE}) \tag{13}$$

Fig. 6 shows results of the model run with a load profile reflected by a current set point given to the Battery system. When SoC reaches 90%, the Battery current falls to zero even though the setpoint is set to its nominal value. The AFE is maintaining the DC bus voltage with a "Synthetic Inertia" that is defined by its capacitor model and voltage controller.

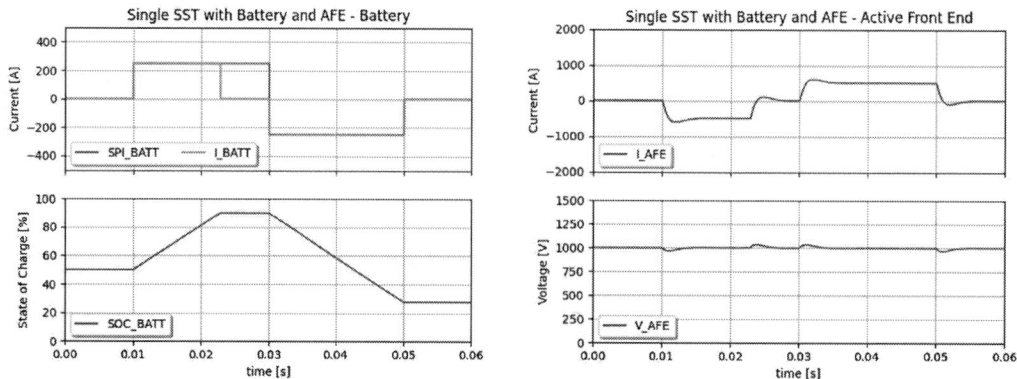

Fig. 6: Voltage and current response of SST cell connecting a Battery to an AFE

Modularity of the SST and Faults

The elementary SST cell described in the previous section can be vertically combined to generate modular configurations, such as Input-Parallel Output-Series (IPOS) and Input-Series Output-Parallel (ISOP) allowing to connect LVDC to MVDC grids [18]-[19]. Each SST cell is working independently and is implemented with bypass capabilities for simulating the effect of a cell fault on the system.

On the MVDC side, the series connection of voltage sources is not a problem. On the LVDC side however, the parallel connection of voltage sources cannot be done without specific attention on the SST local PCC side. Voltage sources must be connected through an arrangement of controlled current sources that reflect the dynamics due to cabling resistance R_{DCBUS} and coupling inductors L_{DCBUS}. This allows generating a PCC voltage that is equivalent to the mean value of all N primary side voltages of all N parallel connected SSTs as in Equ. 14, it also allows current sharing in between the cells, reflecting on what is drawn from the load side, as given in Equ.15. The final current share in between the parallel connected modules is given by Equ. 16.

$$V_{PCC} = \sum \frac{V_{PRIM\_N}}{N} \tag{14}$$

$$I_{GRID} = \sum I_{PRIM\_N} \tag{15}$$

$$I_{SST\_N} = \frac{1}{L_{DCBUS}} \int (V_{PCC} - V_{PRIM\_N} - R_{DCBUS}I_{PRIM\_N}) \tag{16}$$

IPOS configuration with faults

In the 6-cell IPOS configuration illustrated in Fig. 7, the 3-port MVDC side voltage is fully controlled by the sum of individual SST cells. This arrangement allows asymmetric loading on the MVDC side with individual voltage control of both MVDC+ and MVDC- ports. The system is run for demonstration with a fault occurring in one cell to assess voltage and current stability on MVDC side.

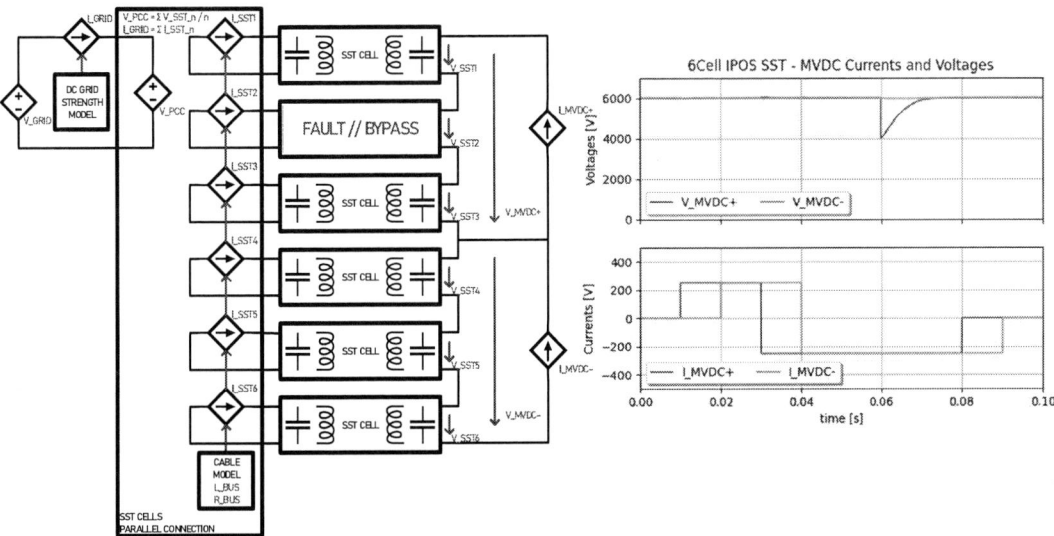

Fig. 7: IPOS configuration of 6 SST cells – Asymmetric loads and fault occurring in one module

Fig. 8 shows currents and voltages on the primary and secondary sides of the SST cells. Asymmetric currents are well distributed among the cells connected to the PCC. At some point, when the two MVDC side currents are opposite to each other, no current is taken from the LVDC grid side as current is only circulating in between the cell via the PCC. When a fault occurs, the MVDC+ voltage is maintained by two cells instead of three, their voltage setpoint is adjusted after the fault occurrence. Theses results show the accuracy of the SST local PCC model for the parallel connection of individual SST cells.

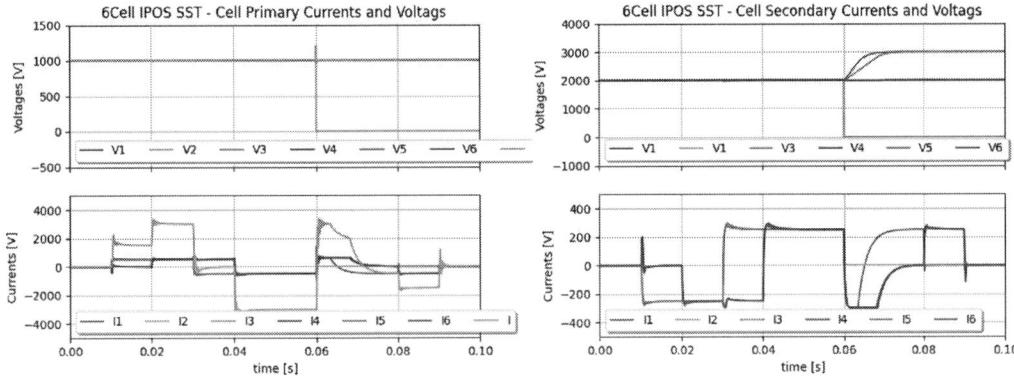

Fig. 8: Current and voltage response of the 6-cell IPOS configuration with a fault

ISOP configuration

The ISOP is slightly different from IPOS in the sense that MVDC side is given by the system and that LVDC side must be controlled. In the arrangement presented by Fig. 9, the LVDC side voltage is controlled by one SST cell, which provides a PCC for the full converter. All other cells control the neutral points voltages on the MVDC side, naturally providing voltage balancing in between the cells.

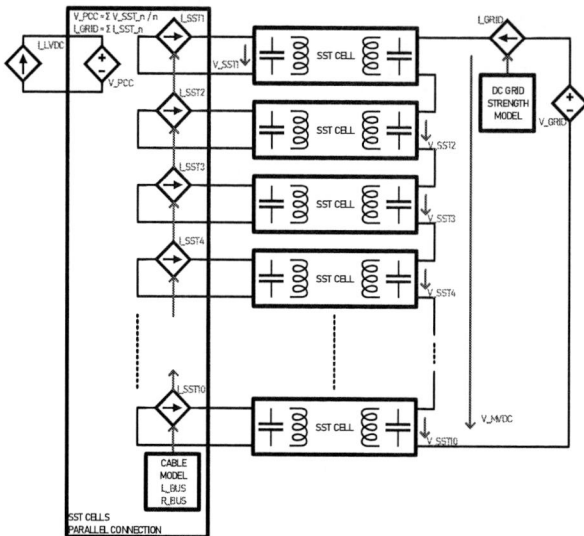

Fig. 9: ISOP configuration of 10 SST CELLs

Results depicted in Fig. 10 show a slight voltage difference between the secondary side voltage of the SST cell that is controlling the LVDC side, because it takes the voltage difference between the MVDC and the sum of all secondary side cell voltages. Otherwise, we foresee excellent voltage and current balancing for this kind of arrangement, as long as one remains within voltage and current limits of the SST cells obviously.

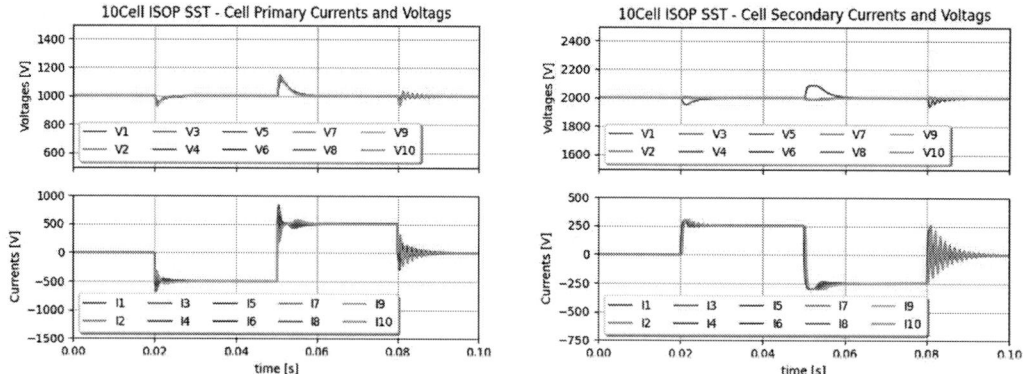

Fig. 10: Current and voltage response of the 10-cell ISOP configuration

MVDC grid model for E-mobility

The power system for the e-Mobility case presented in Fig. 1 is entirely modelled and simulated with the elementary building blocks presented so far. As illustrated on Fig. 11, an MVDC bus interfaces an AFE with several 10-cell ISOP blocks. The AFE maintains the voltage with a dynamic defined by its capacitive inertia and the voltage controller's time constant. One of the 10-cell ISOP blocks interfaces an LVDC bus where single SST blocks are connected for modelling bidirectional fast chargers for car batteries. This model features 73 SSTs working together, it has been implemented with Simba.io the model took less than a minute to simulate 1s simulation time with a time step of 1µs with a quite standard computer.

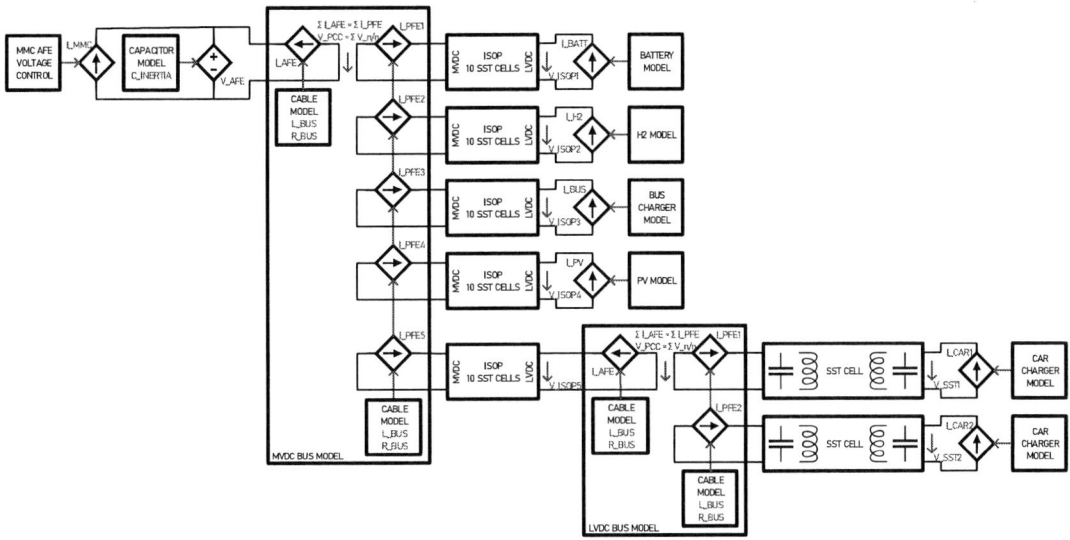

Fig. 11: Modelling of the MVDC grid for E-Mobility context

The simulated scenario involves PV energy drops, grid support from car chargers, and an MMC AFE current limiter for illustrating possible limitations of the system. Fig. 12 shows a run with PVs get covered by clouds, the battery cells fail to work and the AFE can't compensate the full power load. At that point of the scenario, the MVDC bus slightly drops before the car battery kicks in for temporarily supporting the MVDC grid and to save the day.

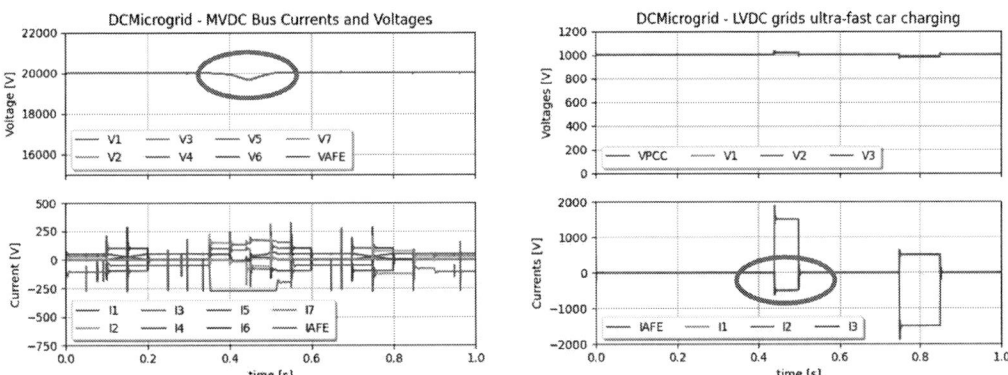

Fig. 12: Current and Voltage response of the MVDC grid for E-Mobility context

Conclusion

This paper presents successful system level modeling of power converter components with their equivalent average models and transfer functions. Their behavior and time response include dynamics of the system and the SST hardware that fits real behaviors. A complex microgrid system with independently controlled elementary SST cells has been simulated. The simulation highlighted some specific DC grid related issues such as DC grid strength and power balancing. The model allows helping the power system design, by evaluating various scenarios and configurations as well as DC protection concept. Also, when evaluating a power system upgrade, one can compare typical CAPEX vs OPEX figures together with functional response of given power systems involving battery storage, H2 fuel cells, and PV cells.

As a prospect, this type of modelling opens up an easy way to quickly simulated various complex configuration in future MVDC power grids, one foreseen application is the booming E-mobility segment that foresees a massive inclusion of renewables and progressive upgrade to DC power.

References

[1] Steinke J. K., Maibach P., Ortiz G., Canales F. and Steimer P.: MVDC Applications and Technology, PCIM Europe 2019; International Exhibition and Conference for Power Electronics, Intelligent Motion, Renewable Energy and Energy Management.

[2] Stieneker M. and De Doncker R. W.: Medium-voltage DC distribution grids in urban areas," 2016 IEEE 7th International Symposium on Power Electronics for Distributed Generation Systems (PEDG).

[3] Mura F. and De Doncker R.W. .: Design aspects of a medium-voltage direct current (MVDC) grid for a university campus, 8th International Conference on Power Electronics - ECCE Asia, 2011.

[4] C. Zhao et al., "Design, implementation and performance of a modular power electronic transformer (PET) for railway application," in Proc. 14th European Conf. on Power Electron. and Applicat. (EPE'11 ECCE Europe), Birmingham, Sep. 2011.

[5] C. Zhao et al., "Power Electronic Traction Transformer—Medium Voltage Prototype," in IEEE Trans. Ind. Electron., vol. 61, no. 7, pp. 3257-3268, July 2014.

[6] C. Zhao et al., "Power electronic transformer (PET) converter: Design of a 1.2MW demonstrator for traction applications," Int. Symposium on Power Electronics Power Electronics, Electrical Drives, Automation and Motion, 2012, pp. 855-860.

[7] D. Dujic, F. Kieferndorf, F. Canales and U. Drofenik, "Power electronic traction transformer technology," in Proc. 7th Int. Power Electronics and Motion Control Conf., 2012, pp. 636-642.

[8] D. Dujic, A. Mester, T. Chaudhuri, A. Coccia, F. Canales and J. K. Steinke, "Laboratory scale prototype of a power electronic transformer for traction applications," in Proc. 14th European Conf. on Power Electron. and Applicat. (EPE'11 ECCE Europe), Birmingham, Sep. 2011.

[9] Heinig S., Siemaszko D., Baumann R. Hubatka N., Klaeusler M., Ruiz R., Burkart R., Yuan C.: Experimental Insights into the MW Range Dual Active Bridge with Silicon Carbide Devices, International Power Electronics Conference IPEC-Himeji 2022 - ECCE ASIA.

[10] Trends and developments in electric vehicle markets, Global EV Outlook 2021, https://www.iea.org/reports/global-ev-outlook-2021/trends-and-developments-in-electric-vehicle-markets

[11] García-Triviño P., Oliveira-Assis L. D., Soares-Ramos E. P. P., Sarrias-Mena R., García-Vázquez C. A. and Fernández-Ramírez L. M.: Configuration and Control of a MVDC Hybrid Charging Station of Electric Vehicles with PV/Battery/Hydrogen System, 2021 IEEE International Conference on Environment and Electrical Engineering and 2021 IEEE Industrial and Commercial Power Systems Europe (EEEIC / I&CPS Europe).

[12] Eldeeb H. H. and Mohammed O. A., Control and Voltage Stability of A Medium Voltage DC Micro-Grid Involving Pulsed Load," 2018 IEEE International Conference on Environment and Electrical Engineering and 2018 IEEE Industrial and Commercial Power Systems Europe (EEEIC / I&CPS Europe.

[13] D. Aggeler, F. Canales, H. Zelaya-De La Parra, A. Coccia, N. Butcher and O. Apeldoorn, "Ultra-fast DC-charge infrastructures for EV-mobility and future smart grids," 2010 IEEE PES Innovative Smart Grid Technologies Conference Europe (ISGT Europe), 2010

[14] M. Dicorato, G. Forte, M. Trovato, C. B. Muñoz and G. Coppola, "An Integrated DC Microgrid Solution for Electric Vehicle Fleet Management," in IEEE Transactions on Industry Applications, vol. 55, no. 6, pp. 7347-7355, Nov.-Dec. 2019

[15] H. B. Sonder, L. Cipcigan and C. E. Ugalde-Loo, "INTEGRATING DC FAST/RAPID CHARGERS IN LOW VOLTAGE DISTRIBUTION NETWORKS," The 12th Mediterranean Conference on Power Generation, Transmission, Distribution and Energy Conversion (MEDPOWER 2020), 2020, pp. 59-65

[16] Barrade P.: Électronique de Puissance, Presses Polytechniques et Universitaires Romandes. 2006

[17] https://simba.io/

[18] J.-W. Kim, J.-S. Yon and B. H. Cho, "Modeling, control, and design of input-series-output-parallel-connected converter for high-speed-train power system," in IEEE Transactions on Industrial Electronics, vol. 48, no. 3, pp. 536-544, June 2001.

[19] L. Heinemann, "An actively cooled high power, high frequency transformer with high insulation capability," APEC. Seventeenth Annual IEEE Applied Power Electronics Conference and Exposition, 2002

An artificial intelligence pipeline for critical equipment thermal conditioning system design

Raik Orbay, Athanasios Tzanakis, Inko Marcaide, Jonas Löfgren,
Torbjörn Thiringer, Thomas Bernichon
raik.orbay@volvocars.com
Volvo Car Corporation,
405 31 Göteborg, Sweden
https://www.volvocars.com

Keywords

≪Battery Electric Vehicles≫, ≪inverters≫, ≪ machine learning≫, ≪topology optimization≫, ≪genetic algorithms≫, ≪SVM≫, ≪CFD≫, ≪FEM≫

Abstract

Efficient electric machinery often needs to be accurately thermally conditioned. Heat sinks and heating surfaces frequently used to allow for precise temperature control of the critical equipment. To tackle the thermal challenges in the art, different design methodologies, such as the parametric or the topology optimization are introduced. Compared to parametric optimization, topology optimization allows for more tailored cooling solutions on elaborate geometries related to propulsion. Being based on gradient descent algorithm from the machine learning toolbox, topology optimization may suffer from local minima. In this report, the setup is designed to alleviate the risk for local minima and instead aim for a more global optimization. Accordingly, an artificial intelligence pipeline is scripted to run several gradient-descent based topology optimization assessments under a genetic algorithm optimization loop. The resulting geometry is shown to substantially improve the cooling ability in the given packaging volume in a light duty battery electric vehicle with quantified reduction in CO_2 emissions.

Introduction

Main requirements of power electronics for propulsion are high energy efficiency, further downsizing, cradle-to-grave environmental performance and minimum costs. Accordingly, equipment like inverters for motor drive, have reached high switching speeds [1] to decrease losses [2, 3]. Despite this, finite

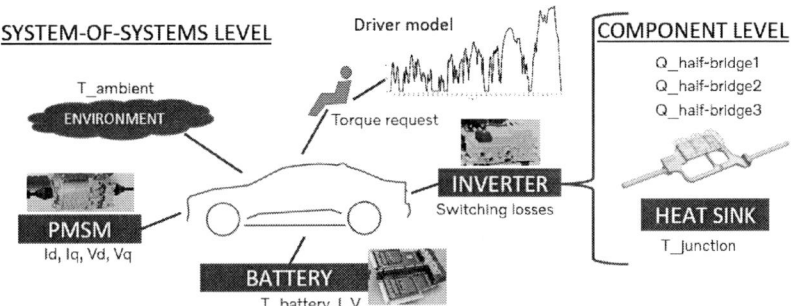

Fig. 1: A system-of-systems level tool is used to quantify inverter losses.

switching speeds fail to eliminate losses completely [4]. Efficient power electronics for propulsion often need to be accurately thermally conditioned to minimize temperature swings [5]. The critical equipment, thus, is connected to a heat sink / source via thermal bridges and kept at the design temperature regardless of the environmental conditions [6]. The coolant is conducted onto the heat exchange surface using piping systems to dissipate heat [7]. The state-of-the art for the heat sink design is by means of using automated parametric optimization [8], where a parametrized geometric entity from the solution domain for the heat sink is taken as a starting point. Several sizes of this geometric entity are populated and then quantified in their heat transfer ability in a computer aided engineering (CAE) loop to reach an optimum in the chosen objective; e.g. pressure drop, flow split, surface shear stress etc. However, as the parametric optimization locks the solution domain to one particular type of the CAD geometry as prescribed by the parametrization, it may come short of all the possible concept candidates. Differing to this in topology optimization, the parameter space is not limited to the parametrized geometric entity, but instead the whole packaging space (computation domain mesh size) is taken as the parameter set. Accordingly, given the extend of the optimization space that the topology optimization is acting in, it has substantial advantage over parametric optimization to reach the optimum. Despite this, as the topology optimization softwares generally use gradient descent algorithm to reach the optimum for the objective function, it is a well-documented behavior that the gradient-descent based approaches are prone to be captured by local optima instead of global optima. Additionally, for cooling of sensitive equipment like inverters, focusing the available cooling power to the exact position of the hot spot is crucial [7]. Usual practice of sensitive equipment cooling is achieved by serial cooling due to its packaging efficiency, compared to the parallel cooling. In this study, each of the switches are cooled in parallel by their dedicated cooling channel to minimize temperature difference among half-bridges. In a sub-optimized application, the heat dissipated by the first and second inverter switches may affect the third inverter switch junction temperatures negatively as this switch is in the downstream of the two initial switches. It is therefore crucial to balance the available cooling power among the switches. To remedy the shortcomings, a

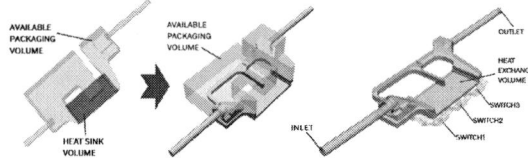

Fig. 2: Available packaging space in grey. A manually designed thermal conditioning topology is shown in yellow. The switch positions are also provided in the rightmost inset. Casing not shown.

critical part of the inverter cooling system is tuned using genetic algorithms in this report. Consequently, an artificial intelligence controlled topology optimization loop for cooling system design is shown to reach to a better optimum than which is allowed by one-step topology optimization. To reduce human intervention even further, the output of the automated topology optimization loop is fed into a machine learning algorithm to analyze concept candidate behavior and pinpoint outliers. The concept candidates populated in this study will be compared to a base cooling system concept designed by human-intense manual optimization loops, as shown in Fig 2.

Computational approach

System of systems level modeling

The initial work to quantify the boundary conditions for the inverter cooling is done using a system-of-systems tool [9] based on lumped parameter models of all the vehicle systems and components. A battery electric vehicle (BEV) is modelled in time domain using a certification cycle, as shown in Fig.1. Form this study, maximum power loss from the inverter is extracted as a continuous power, which was further fed into a 3D thermal CFD tool [10, 11, 12] for optimization, as shown on Fig.1.

Component level modeling

The topology optimization software tool allows for multiobjective topology optimization, thereupon a case with minimization of pressure drop and volume flow maximization or shear stress maximization on heat transfer surfaces was prescribed for this report. The topology optimizer comes with a new geometry with optimized key performance indices, ie., objectives, by computing volumetric sensitivities of chosen objective for changes in mesh element porosities. Two cases of topology optimization approaches are launched to compare the gains in cooling ability and decrease in pressure drop to the base case. Topology optimization of coolant path is done in a predetermined packaging volume shown in the first inset of the Fig. 2. This volume is the remaining inner region when all the geometric constraints (crash surfaces, adjacent component solidity volumes, heat protection distances, cable and plumbing routing, etc.) are united. In the first topology optimization approach, a constant packaging volume is taken as an input to the topology optimizer, as shown in Fig. 3. In a second topology optimization approach, an improvement

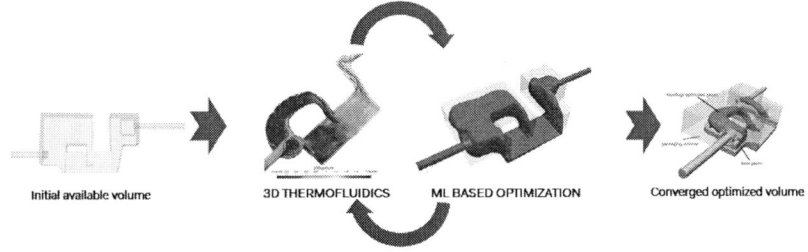

Fig. 3: Pure ML based topology optimization schematics.

in objectives are achieved by means of a pipeline using GA from the artificial intelligence toolbox, as shown in Fig. 4. In this approach, the GA will run several parallel topology optimization cases using alternative packaging limits while updating the overall dimensions of the available packaging volume at a critical region between inverter switch two and inverter switch three, as shown in Fig. 5. For the

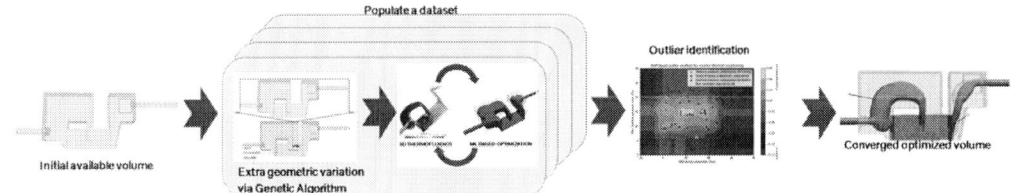

Fig. 4: Prescribed pipeline for GA to run topology optimization schematics.

purpose, the automatized pipeline shown in Fig. 4 is scripted to update the packaging volume dimensions in order to distribute the coolant power to cool the third inverter switch adequately. Another advantage of this approach is that, the topology optimization is run under a GA framework with reduced risk for local optima.

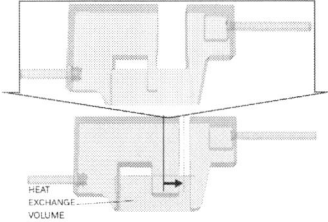

Fig. 5: Available packaging space in grey is morphed to allow for fine tuning of cooling power distribution.

Adjoint-FEM for 3D topology optimization

In this section, finite volume method (FVM), which belongs to the family Finite Element Method (FEM), computations using an adjoint-based solver Helyx [11], which is an OpenFOAM fork [10, 12] will be reported. In the adjoint-state based approaches, an initial CFD run is used as a database to evaluate sensitivities of the objectives, ie., the derivatives of the objectve functions like pressure drop, uniformity, mass flow etc. to changes in flow variables like pressure, velocity, turbulence magnitudes etc. An optimum geometry is created after regions, where the objective function(s) are negatively affected, are blocked by high porosity cells. In the whole computational domain, coupling changes in flow conditions to changes in objective functions is a prohibite task from the computational point of view. But by the introduction of Lagrangian multipliers, the costs of these computations are reduced substantially [13, 14] so that the sensitivity of objective functions to changes in computational domain are analyzed in each and every cell of the computational mesh in a duration comparable to a CFD / Conjugate Heat Transfer (CHT) run. For the current project, a multi-objective approach is chosen. The pressure loss objective is set between inlet and outlet of the heatsink tubes and minimized, as shown in Fig. 6. The second objective, the volume flow maximization objective, is prescribed in the volume in the vicinity of the heat exchange surface, shown also in Fig. 6 as the blue region. In mathematical formalism, an adjoint operation on an

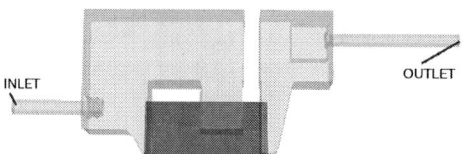

Fig. 6: Prescribed topology optimization objective functions are pressure drop between inlet and outlet and volume flow maximization in the heat exchange volume, as shown in blue in the above.

equation is conducted to be able to write its Lagrange multipliers. Modifying the approach from [13] and taking the first variation of an objective function J, which is a function of flow variables and design variables such as $J(w,\alpha)$, yields

$$\delta J = \frac{\partial J}{\partial w}\delta w + \frac{\partial J}{\partial \alpha}\delta \alpha \qquad (1)$$

The set of Navier-Stokes equations can be written in residual form as $N(w,\alpha)=0$, then the first variation of Navier-Stokes equations can be expressed similarly as

$$\delta N = \frac{\partial N}{\partial w}\delta w + \frac{\partial N}{\partial \alpha}\delta \alpha = 0 \qquad (2)$$

Introducing a Lagrange multiplier, λ

$$\lambda^T \delta N - \lambda^T \frac{\partial N}{\partial w}\delta w + \lambda^T \frac{\partial N}{\partial \alpha}\delta \alpha - 0 \qquad (3)$$

Following addition of (3) to (1),

$$\delta L \equiv \delta J = \left[\frac{\partial J}{\partial w} + \lambda^T \frac{\partial N}{\partial w}\right]\delta \alpha + \left[\frac{\partial J}{\partial \alpha} + \lambda^T \frac{\partial N}{\partial \alpha}\right]\delta \alpha \qquad (4)$$

where, by choosing the suitable λ, variations with respect to flow variables will vanish and it will be possible to define

$$\lambda^T \frac{\partial N}{\partial w}\delta w = \frac{\partial J}{\partial w}\delta w \qquad (5)$$

Finally by means of integrating by parts the left hand side of (5), a system of adjoint equations will be obtained, which will reduce the task of gradient calculations comparable to a flow field computation [14]. Subsequently, the topology engine of the adjoint solver will move the surfaces of the initial geometry to reach a final geometric solution in the allowed packaging volume.

Genetic algorithm controlled packaging volume tuning

In Darwin's theory on evolution, the individuals with the best gene pool will survive the natural selection, transmitting their genetic string. The theory comprises analogues for four biological processes in nature: reproduction, mutation, recombination, and selection. The mathematical model starts with a randomly selected population of concept designs on the parameter space (the first generation), where each of the unique design parameters form a genetic string [15]. Through an iterative process discarding the least fit concept solutions to ensure survival of the fittest, the algorithm reaches to a design point in the solution domain that minimizes the objectives of the optimization. For the reported research a two-point crossover technique to exchange genetic string values between the members of the population during the GA breeding process is chosen. The result of the breeding process is a population comprised of the 10 best parent design points (aka. the elitist strategy) plus 40 new child design points. The GA optimization process will be terminated after either certain number of iterations (generations of the GA) where changes in objectives are less than a tolerance or after a finite number of function evaluations. Total number of concept candidates for the study is about 300.

FEM for 3D thermal CFD analysis

When a topology optimization loop is finished, the best candidate was further verified using a steady-state solver for buoyant, turbulent flow of incompressible fluids [12]. For the creation of computational meshes, the hexahedral volume mesher from OpenFOAM is used. To ensure the quality of the fluid mechanical model, emphasis is put on the mesh parameters. A uniform mesh size distribution is preferred for the topology optimization, while Y^+ is acounted for. On the critical heat transfer surface average Y^+ reaches to 10.35 although its mean vaue is 1. Although this is on the higher side of the usual practice, the resulting computational mesh had minimum errors and results were verified to be conform with computations with other CFD software. Both the meshing and computations are parallelized on about 64 CPUs. Turbulence is modelled using the $k - \omega$ SST model [16], which provides a good balance between quality of results and computational time compared to simpler one-/ two-equation models. For the stability of the computations, appropriate settings as relaxation values for the solvers are chosen following tests.

Support vector machine based outlier analysis

Finally, the populated concept candidate pool is further investigated to discern interesting thermal conditioning solutions. For this purpose, a Support Vector Machine (SVM) classifier is used [17]. SVMs aim to construct a hyper-plane in a high dimensional space, to enable classification and regression using the data at hand. SVM is a supervised learning method and here it is used as an outlier detection algorithm. SVMs aim to maximize margin among classes and a good separation is achieved by the hyper-plane that has the largest distance to the nearest training data points of any class, since in general the larger the margin the lower the generalization error of the classifier [17].

Results

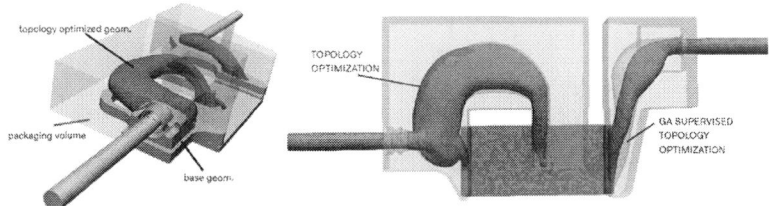

Fig. 7: The manually designed cooling system is in yellow color and the topology optimized geometry is in green in the left inset. On the right inset, the topology optimized geometry is shown in green and the GA supervised topology optimized geometry shown in beige.

In the left inset of the Fig. 7, the geometry from one-step topology optimization result is shown in green, embedded on manually optimized cooling system shown in the same inset as yellow. In the same figure right side, the minute differences among two topology optimization approaches are visualized by comparing the one-step topology optimization geometry in green with GA supervised topology optimization geometry in beige color. The morphology changes imposed by the GA supervised design are especially discernable at the outlet from the heat transfer volume. The manually designed coolant path reaches a 3942.85 W/m^{2}K. To optimize for a uniform cooling power distribution among all the swithces, the topology optimization runs had volume flow maximization in the heat exchange volume, as shown in blue in Fig. 6, additional to the pressure drop minimization objective. This process effectively distributes the cooling power in whole heat sink. In Fig.8 detail of the flow field is depicted. This case is topology optimization in the initial packaging volume. It is interesting to note that the flow is distributed into

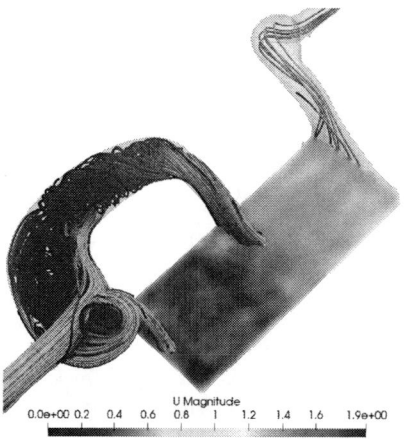

Fig. 8: Illustration of the streamlines in a gradient-descent based multiobjective topology optimization created coolant path.

two coherent structures, one entering the first switch heat exchange region and the other one to a region between the second and the third switches. This enables a heat transfer coefficient of 3950.15 W/m^{2}K, about 7 W/m^{2}K higher than the manually optimized cooling sytem, as tabulated in Table I. The topology optimization run under GA supervision reaches to 3983.75 W/m^{2}K, about 40.90 W/m^{2}K higher than the cooling system design from the human-intense manual optimization loops, with a substantially lower pressure drop. The same trend is valid for the thermal resistance of the heat sink, as the artificial intelligence supervised topology optimization result shows a decrease in thermal resistance. Additionally, the

Table I: Heat dissipation ability for studied topologies

$T_{Ref} =$ 338.15 [K]	design	$\triangle P$ [Pa]	HTC [W/m^{2}K]	themal resistance [K/W]	T_{out} [K]
	BASE	4752.42	3942.85	0.00919	346.27
	OPTIMIZED	4325.77	3950.15	0.00917	346.33
	AI PIPELINE	4199.53	3983.75	0.00909	346.38

topology optimized design assures a 0.55 kPa lower pressure drop compared to the manually optimized cooling system, which can be coupled to a 0.228 W decrease in thermal conditioning circuit pumping work. Returning back to the system-of-systems level modelling set-up described in Fig.1, it is simulated that 0.228 W decrease in thermal conditioning circuit pumping power enable a 0.0051 Wh/km reduction in consumption on a WLTP certification cycle for a representative BEV with AWD driveline [9]. For a vehicle with 400000 km / 10 years economic life, these figures can be connected to the reduction in CO_2 emissions [18]. Accordingly, a 0.0051 Wh/km consumption reduction will lead to 27.2 g CO_2 decrease in Sweden and a 470.6 g CO_2 in Europe. Finally, to further help the automation effort, a SVM based outlier analysis is also done on the one of the populated concept pool. For pedagogical reasons, only two-dimensional data is shown. In fact, the complete graph is a six dimensional hyperplane, comprising

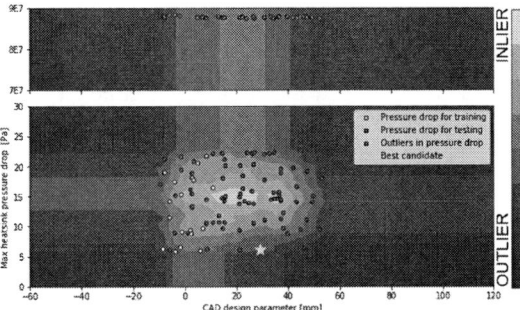

Fig. 9: Concept candidate pool investigated using machine learning. The more the lighter the background canvas color, the likely that the concept candidate is a regular candidate, ie., not an outlier.

four CAD design variables as well as key performance indices, the pressure drop and the uniformity. The candidate number 28 of the concept candidate pool is the optimum according to the GA supervised optimization loop and is well inside the expected region for the concept candidates, as depicted in Fig. 9. In the same plot, some of the divergent CFD runs are also shown in red. These are the concept candidates where the complete flowfield is occupied with high porosity cells, thus leading to extreme pressures and failed computations. All the failed concept candidates are positioned in dark (outlier) region of the canvas. This is an advantageous step in automatizing the complete pipeline as it will further reduce human interaction. In Fig. 9, two plots are shown as the outliers have extremely high pressure conditions.

Future work

The topology optimized cooling system is produced using additive manufacturing as shown in Fig. 10. The specimens are designed to be smoother than the layered design from the additive manufacturing process, thus advantageous for surfaces adjacent to fluid flows with or without heat exchange [19]. A

Fig. 10: Additively manufactured cooling system morphologies. On the left side, manually optimized design is shown; on the right side a topology optimized geometry is shown.

future publication is planned to report on findings from the experimental study.

Conclusions

A cooling system design pipeline supervised by genetic algorithm is deployed. A machine learning based outlier analysis is programmed as a filter to further help in workflow automation, which successfully captured diverged flow field computations. Finally, the GA supervised topology optimization is shown to reach to 40.90 W/m^{2}K higher heat transfer coefficient than the manually designed cooling sytem, with 0.55 kPa lower pressure drop. For a representative AWD BEV, the advantage through reduced pumping work in thermal conditioning circuit reaches to a 470.6 g CO_2 advantage in Europe for the economic life of the vehicle.

Acknowledgment

The authors would like to express their gratitude to the Swedish Energy Agency and Volvo Car Corporation for financing the project.

References

[1] A. Poorfakhraei, M. Narimani and A. Emadi, "A Review of Multilevel Inverter Topologies in Electric Vehicles: Current Status and Future Trends," in IEEE Open Journal of Power Electronics, vol. 2.

[2] A. K. Morya et al., "Wide Bandgap Devices in AC Electric Drives: Opportunities and Challenges," in IEEE Transactions on Transportation Electrification, vol. 5, no. 1,March 2019.

[3] X. Ding et al., "Analytical and Experimental Evaluation of SiC-Inverter Nonlinearities for Traction Drives Used in Electric Vehicles," in IEEE Transactions on Vehicular Technology, vol. 67, Jan. 2018.

[4] S. Jones-Jackson, R. Rodriguez, Y. Yang, L. Lopera and A. Emadi, "Overview of Current Thermal Management of Automotive Power Electronics for Traction Purposes and Future Directions," in IEEE Transactions on Transportation Electrification, vol. 8, no. 2, 2022.

[5] J. Wölfle, M. Nitzsche, J. Ruthardt and J. Roth-Stielow, "Junction Temperature Control System to Increase the Lifetime of IGBT-Power-Modules in Synchronous Motor Drives Without Affecting Torque and Speed," in IEEE Open Journal of Power Electronics, vol. 1, 2020.

[6] K. Numakura, K. Emori, Y. Yoshino, Y. Hayami and T. Hayashi, "Direct-cooled power module with a thick Cu heat spreader featuring a stress-suppressed structure for EV/HEV inverters,"IEEE Energy Conversion Congress and Exposition (ECCE), 2016.

[7] Seo, H.S. and Shin, D., "Development of Parallel and Direct Cooling System for EV/FCEV Inverter," SAE Technical Paper 2018- 01-0454, 2018.

[8] Z. Wang, M. Chinthavali, S. L. Campbell, T. Wu and B. Ozpineci, "A 50-kW Air-Cooled SiC Inverter With 3-D Printing Enabled Power Module Packaging Structure and Genetic Algorithm Optimized Heatsinks," in IEEE Transactions on Industry Applications, vol. 55, no. 6, pp. 6256-6265, Nov.-Dec. 2019.

[9] Matlab, Accessed: 2021-Jul-29. Available at https://www.mathworks.com/products/matlab.html

[10] Weller H. G., Tabor G., Jasak H., and Fureby C. A tensorial approach to computational continuum mechanics using object-oriented techniques. Computers in Physics, 12(6):620–631, 1998

[11] Engys ltd. Accessed 18-October-2018. Available at https://engys.com

[12] OpenFOAM. Accessed 2020-01-13. Available at https://openfoam.org/.

[13] Juan J. Alonso. A crash-course on the adjoint method for aerodynamic shape optimization. In 17th International Congress of Mechanical Engineering, Sao Paulo, November 2003.

[14] Eysteinn Helgason. Development of adjoint-based optimization methods for ducted flow in vehicles. Doctoral thesis, Chalmers University of Technology, Gothenburg, Sweden, 2015.

[15] Adams B.M. Bohnhof W.J. Dalbey K.R. Eddy J.P. Eldred M.S. Gay D.M. Haskell K. Hough P.D. Lefantzi S. and Swiler L.P. Dakota, Version 5.1 user's manual, 2010

[16] Florian R Menter. Two-equation eddy-viscosity turbulence models for engineering applications. AIAA journal, 32(8):1598–1605, 1994.

[17] Scikit-learn: Machine Learning in Python, Pedregosa et al., JMLR 12, pp. 2825-2830, 2011

[18] European Environmental Agency. Accessed: 2021- Oct-15. Available: https://www.eea.europa.eu/data-and-maps/daviz/co2-emission-intensity-5.

[19] inkbit. Accessed 2022-03-07. Available at https://inkbit3d.com/technology/

Aspects of stability issues of HVAC/HVDC coupled grids

Gianni BAKHOS[*], Kosei SHINODA[*], Juan-Carlos GONZALEZ-TORRES[*], Abdelkrim
BENCHAIB[*], Luigi VANFRETTI[‡], Seddik BACHA[*†]

[*]SuperGrid Institute SAS, Villeurbanne, France
[†]University of Grenoble Alpes, Grenoble INP, G2Elab, Grenoble, France
[‡]Rensselaer Polytechnic Institute, Troy, NY, USA. Co-author's email: vanfrl@rpi.edu

Corresponding author email: gianni.bakhos@supergrid-institute.com

Acknowledgments

This work was supported by a grant overseen by the French National Research Agency (ANR) as part
of the "Investissements d'Avenir" Program (ANE-ITE-002-01) of the French Government.

Keywords

«Stability analysis», «Multi-terminal HVDC», «optimization», «AC-DC».

Abstract

This paper explores MTDC control expectations and attempts to investigate the challenges preventing
usage of MTDC grid to enhance power system stability. Protection and dispatching are not in the scope
of this paper. Dynamic performance up to tens of seconds is only considered (AC voltage regulation is
therefore not studied here).

Introduction

Power system network is undergoing structural changes mainly driven by the increase of share of
renewable energies [1] implying more power electronic (PE) interfaced grids. These changes create
challenges for the TSOs to operate the system. To solve this issue, reinforcement and/or upgrading of
the grid is one solution and development and expansion of High-Voltage Direct Current (HVDC) [2]
has proven to be the most adapted upgrading solution in many situations. A more complex configuration
of power transmission through DC lines is called MTDC implicating multiple terminals of conversion
and therefore a more elaborated system compared to point-to-point HVDC. Since renewable power
production devices are replacing classical machines, their intermittence decreases reliability of primary
reserves of power system. Another problem emerges directly from the use of PEs: the inertia classically
provided to the system by AC generators is not equivalently provided directly by power converters
interfacing electrical production [3]. While PE devices allow faster and more flexible power control for
stability enhancement [4], their vulnerability against perturbations makes power system less robust.

The increasing integration of HVDC in the current grid, will lead to complex interconnected AC/DC
grids. Controllability of HVDCs could serve in stabilizing the system [5] such as in power flow
balancing and participation to frequency services to HVAC side [6]. MTDC grid can contribute to power
balancing of connected AC grid by charging or discharging DC grid capacitances. However, if no power
storage or fast production systems are connected to the system, the DC power needed power should
come from/go to another AC system, causing its imbalance. The increasing amount of HVDC, together
with the use of supplementary controllers leads to interactions – between initially decoupled AC systems
connected through DC lines or between AC and DC systems too – that affect global AC/DC power
system stability due to coupling between the modes of both grids [7]. For this reason, MTDC grid control
must be optimized: what terminal to take the power from? What amount of power to take from each
terminal? Interoperability questions emerge too (feasibility, coordination, automation of actions, etc).

This paper examines the literature of AC & DC power system stability solutions especially looking at how DC system can help enhance AC stability. The paper first investigates the DC grid stability requirements and challenges that may affect it as well as proposed solutions. It looks then for existing solutions to enhance AC stability and checks their limits. Note that AC power system controllers will not be explored but instead the possibilities offered by HVDC links will be specially studied. This paper does not attempt to survey all supplementary controllers that can be added to enhance AC & DC stability but focuses on main ones to move to expectations from these controllers in case of MTDC grid. It also studies some difficulties that could prevent MTDC direct replacement of HVDCs and proposes first steps for AC & DC power system stability enhancement through optimal usage of MTDC grid.

This paper is organized as follows. Section II summarizes offered possibilities for AC stability enhancement through HVDC supplementary controllers. Section III explores the requirements and challenges in DC grid. Section IV reviews opportunities offered by MTDC grid control to enhance AC & DC stability and the difficulties to implement the control. Section V concludes the paper.

I.   AC grid stability

a.   Requirements for AC stability

TSOs require from the classical AC system to be 'N-1 criterion'-compliant meaning that control should withstand loss of one component (line, generation unit, etc) in an initial grid operating with N components.
AC grid stability is classically divided into three main aspects [8] [9]:
- Rotor angle: transient able to split system or small disturbance creating power oscillations,
- Frequency: short term or long term,
- Voltage: large or small disturbance.

b.   HVDC control to meet requirements of AC stability

When HVDC system interfaces AC system, these stability aspects can be deteriorated or even jeopardized if no proper control is implemented and, on the opposite, enhanced in case of appropriate control. In contrast, oscillating modes may be excited and wide AC grid oscillations may occur when improper HVDC power control is applied. In [10], power reference of converters needed to be manually reduced to help damp power oscillations. Finally in the case of system split incident in January 2021 (Continental Europe Synchronous Area), one recommendation in the final report was to seek better frequency support through HVDC links [11]. Thus, the question of HVDC system control arises directly.

Since power converters used to interface AC and DC grids are high-speed components that help also integrate remote renewable energy sources, better flexibility, and controllability in dynamic hosting of AC systems can be reached. In this context of new AC/DC system, HVDC part of the system is required not only to allow for power transfer but to assist AC grid in stability conservation or enhancement [12]. The current situation is a proliferation of VSC power converters throughout the global power system, and the same is expected for future developments of the grid allowing thereby for better control of HVDC power flows. However, the impact of power electronic devices on AC system may not only be positive for stability since bad interactions may occur as it has been already observed and analyzed.

As shown in [5], some functionalities are expected from HVDC operating as either embedded or non-embedded lines in AC power systems, all of them feasible with VSCs:
- AC Voltage control
- Sub-synchronous damping
- Frequency control (Frequency Containment Reserve delivery)
- Emergency Power Control
- Power Oscillation Damping
- AC line emulation

- Synthetic Inertia

If first two controls do not influence others directly (reactive power in action), the remaining ones can even cancel themselves if not properly implemented [5] [13].
To enhance AC/DC stability, some degrees of freedom using HVDC grid through its power converters are to:

- Add controllers for power reference of converters,
- Dynamically adjust the gains of the controllers,
- Or to recalculate more convenient setpoints as converter inputs after solving for optimal power flow.

Other possibilities include modification of AC grid topology or update design requirements (line capacity limits, power converters nominal power). The focus here will be on first three approaches.
In the following, a non-exhaustive list of potential controllers will be presented as they will be reused in the last part of this paper. For each controller, many technologies exist, each having its own advantages and drawbacks despite being innovative. A zoom is made on two approaches for rotor angle stability enhancement.

o Power Oscillation Damping

Table 1: Proposed key points for POD control. These can be extended to all HVDC-based controls which will be dealt with in the following sections.

Control strategy	Observability	Data acquisition	Actuators
• Power System Stabilizer • Power Oscillation Damper (local & remote inputs) • Load shedding (load side) & bang-bang type (converter side) controls • Operating Point Adjustment • Model Predictive Control	• Signals that contain most significant information (check section IV)	• Measurements • Ttransmission system • Time delays • Processing time	• Synchronous generators • HVDC interfaced converters (LCC vs VSC)

In this section, discussion is focused on some parts of control strategy and data acquisition of Table 1. Data acquisition part may be common for all controls presented in the following sections.

Before implementing Power Oscillation Damper (POD), the PSS was a decentralized method allowing for independent power oscillation damping action for each generator. New designs such as PSS4B and adaptive tuning of PSS can play a significant role in small-signal interarea mode damping [14], [15]. However, PSS requires tuning of 6 parameters (1 gain and 5 time constants) and a centralized POD can also play a supplementary role for a given power system's optimal operation . Figure 2 in article [16] shows that remote signal used to ensure power oscillation damping can outperform local signals but delay margins limit their performance. However, only time delays' effects were studied in the article while remote signals imply using of wide-area measurement systems and availability and robustness of signal against noise were not treated in the study. Two other important points to evaluate when adding a POD are the placement of the actuator and parameter tuning.

POD controllers are widely studied today, and the emergence of machine learning and artificial intelligence allows for new ways at tuning their parameters. Figures 6.12 to 6.18 in [17] compare different machine learning-based algorithms that predict eigenvalues and classify them to the true eigenvalues of the IEEE 14-bus power system. Promising results are shown but more investigation is needed for real-time real-life application of the deep-learning-based power oscillation damping. To what extent the predictions provided are robust against system noises since a trade-off must be found between processing time and accuracy? The 'intelligent POD' (iPOD) then 'multi-band iPOD' (MiPOD) shown in [18], [19] is one of the possible improvements proposed for intelligent tuning of POD but its comparison with conventional PSSs showed that in certain circumstances the latter can have better results. It is true that, compared to conventional PSS, the 'MiPOD' needs to tune only 1 parameter per oscillation band and assures selective and adaptive damping as well, but it assumes a control dependent on wide-area measurements (availability and[20] reliability [20], transmission, processing [21]) and communication delays.

Finally, complementary action to POD control (added in the system to increase the damping of the modes at the same operating points via usage of supplementary controllers) are bang-bang type control [22], Operating Point Adjustment (OPA) [23] and Model Predictive Control that takes also the dynamics of the system into consideration instead of just performing optimal power flow. This moves the system to a new condition to have better damping of critical modes. In case of OPA for instance, POD can be the first action implemented against disturbances, while OPA can be the longer-term action to improve system's small signal stability. Moreover, OPA can be applied before the occurrence of a disturbance, providing therefore additional stability margins. Coordination between POD and OPA may be needed, too.

     o   Angle Difference Controller (ADC): AC line emulation

To enhance transient stability, one possibility is to emulate AC lines through VSC-HVDC control. Although this simple measure is normally done for better steady state of the AC system (powerflow concerns), it can also enhance dynamic behavior of the power system.
The injection of synchronizing power (not only damping power) by POD through DC lines to support AC grid in case of three-phase fault was investigated [24] in addition to the main topic of transient stability enhancement.

Although ADC controller is a simple controller, particular attention is needed on filtering time constant of ADC which may impact transient stability negatively as was learnt in the case of INELFE DC interconnection [25] [26]. Extremely quick or slow time constants of ADC are considered to enhance transient stability while other values can jeopardize system stability. Since ADCs are mostly used for steady state concerns, choosing a slow time constant is the best option to avoid transient instability.

II. DC grid stability

a.   Requirements for DC stability

In [27], DC power system stability was explored through multiple case studies at converter's level and this affects stability of whole DC system – which affects AC & DC stability. A study of DC power system stability was done in [28] for distribution systems. If stability is only considered through checking range of operation of DC components' variables at steady state in the article, it is however important to check the paths the DC voltage and powers follow to move from one operating point to the steady state equilibrium point. Should an HVDC power system be considered stable if at some point of the operation the system deviates 'significantly' from nominal values before moving to acceptable equilibrium point at steady state? Moreover, stability margins are not the same for a DC power system where converters are saturated compared to a system operating at lower power stress. The problem is more eminent when converter ratings (voltage, power, etc) are violated. Therefore, a DC power system may be considered stable if the whole trajectory including first one operating point till a steady state is reached is acceptable (within tolerated range) and allows for higher stress. Otherwise, the equilibrium point, if one is reached, may be considered critical or unstable.

As a basis for operation, the availability and reliability of HVDC systems depend on its topology and the used components. Requirements for individual components should normally be specified during planning of the HVDC grid. For DC system stability, the main variable that should be monitored in DC operation is the DC voltage. However, the current regulations available in [29] are general requirements that need to be "specified" by TSOs at each time specific network codes are needed for AC/DC interface. Therefore, to preserve DC voltage stability, some voltage profiles have been proposed as in figure 9 in [30] [31] to show to what extent DC voltage protections should stay untriggered but AC & DC interactions should be considered too to keep discrimination between AC & DC protections and avoid unvoluntary triggering of DC protections. Thus, conclusions for range of operation of DC voltage are not straightforward as coupling with AC side needs more examination and regulatory specifications are still missing.

b.   Challenges

Sudden loss of power converter is one of the main events that may jeopardize DC grid operation. This can be expressed as instant active power loss of injection or extraction and EU legislation has already taken into consideration this case [29]. Nonetheless, network code does not express direct guidelines and methodologies for calculation of the maximum allowable active power loss and it is left for TSOs to define and apply them.

While the impact on AC grid stability can be limited by this value, what about the impact on DC grid stability in case of MTDC system? Sudden loss of power converter can be much more impactful on DC side than on AC side, so what are needed controls to preserve DC stability in such case?

III. Expectations from MTDC grid for AC/DC stability enhancement

a.   From point-to-point HVDC to MTDC

- Transposition of some types of controls is not straightforward

To maintain stability of DC grids, many DC Voltage control methods exist [32]. DC Voltage control exists for point-to-point (PtP) HVDC where one converter regulates DC voltage while the other controls power flow. However, connecting an AC grid to a Multi-Terminal DC system is not as simple as connecting it to a PtP HVDC due to added complexity. Paradoxically, when a power converter is down in a PtP HVDC, all the DC link gets down, but MTDC grid should allow for maintained operability even with one converter down making it a better solution than cascaded PtP HVDCs.

Extension of DC voltage control from PtP to MTDC is possible through 'master-slave control mode' where one terminal plays the role of slack bus to control DC voltage by absorbing any power flow variation in DC grid while the remaining terminals control power flow. This strategy has however its own drawbacks since all balancing power responsibility lies on one converter and system (at least DC system) depends entirely on one component which fails to meet N-1 contingency criterion.

To cope with this issue, 'voltage-margin control mode' was invented where terminals operate in constant power control mode until certain voltage range is violated (maximal power reached for converter in master mode) where they switch to constant voltage control mode. In this strategy, it is true that system does not rely on one converter for power flow and voltage control in disturbed situation, but it relies on it for steady-state operation and large efforts of power balance still rely on one component to a certain extent. Moreover, when power control mode switches, big power disturbances may occur, and AC grid may be destabilized.

For these reasons, distributed voltage control is needed for MTDC: 'DC voltage droop control' [33] implicates more than one terminal in DC voltage and active power regulation.

The DC Voltage droop control formula that will be used later in the paper is:

$$P_{hvdc_{ref}} = K_{dc_{droop}} * (V_{dc_{meas}} - V_{dc_{base}})$$

- One control action can affect more than only targeted stability aspect

In MTDC grid, since multiple converters are involved, it is expected to benefit from the available power headroom of each one of them to add a control that helps AC system stability. However, implementing these controls must not be as straightforward as in PtP HVDC where effects are directly expected.

For instance, in 'master-slave' configuration in HVDC, frequency droop should be implemented in converter in 'slave' mode since no power reference can be changed in converter in 'master' mode responsible of DC voltage regulation. Effect of frequency droop and DC voltage controls are known in advance. When transposed to MTDC, since DC voltage droop control mode is implemented instead of 'master-slave' mode, all converters in this control mode are expected to have their power references changed according to the needed voltage control efforts for DC side. Frequency droop, in parallel, is expected to modify the power reference of given converters too according to the needed frequency control efforts for AC side. These two control efforts, normally destined to target separate stability aspects (DC voltage by DC voltage droop control & AC frequency by frequency droop control), will then interact [13] and affect each other in a way that may have better effects if the control actions are coordinated, if power injections and extractions are calculated by a higher-level controller.

Therefore, need for global stability assessment and enhancement appears evident with hybrid AC/DC systems, and it gets even more important when complexity of MTDC systems is involved to be able to benefit from the flexibility expected from MTDC.

b.   Global AC/DC system stability enhancement

i.   Necessity of assessment for prioritization of stability aspects afterwards

Before applying control to AC/DC system, assessment should be performed to have a clear vision of what stability aspect needs to be enhanced and in which priority. This means that a certain 'score' should be given to the system depending on each stability aspect state.

- How should the 'score' be established?

To do so, TSOs need indicators that reflect the power system's stability state for each aspect of stability. For a given stability aspect 'lambda', proper indicators should be defined to help evaluate margins of stability and conclude whether the system is lambda-stable or not. These indicators will help control in allocating power according to the assessed 'need' for stability enhancement. The Key Performance Indicators chosen to calculate stability score are in their turn based on measurements coming from system PMUs. To give best assessment possible, observability study may be conducted to determine which measurements can best detect stability issues in the system. For instance, observability study helps choose the measurements that theoretically contain and illustrate best among available data for oscillating mode detection [34].

ii.   Coordination of control actions

- A 'global' score of the hybrid AC/DC system

The score discussed above for each stability aspect should be combined with another stability aspect's score so that priority can be established in stability enhancement efforts. For this reason, a global stability function needs to be calculated with dynamic weighting of each stability aspect's indicators to put priority on most endangered stability aspect for enhancement.

- Should all controls be always activated? What compromise is needed between control actions?

Optimal placement of controllers can be a matter of fact due to added costs of each implemented control action. For this reason, placement of controllers or at least their activation must guarantee they maximally impact power stability for improvement, otherwise all power efforts are lost. Study may be conducted to determine what injection point of power control will theoretically have this highest impact. Since control effects are not just linear superposition of effects for each stability aspect, a compromise should be found so that control effects are the best for 'global' stability enhancement. With all degrees of maneuver offered by MTDC system (flexibility & maneuverability of power flows, additional control possibilities, effort mutualization for AC system stability enhancement, etc), control coordination becomes more prominent. Though an optimization approach requires important calculation efforts, it guarantees the system has best control actions to enhance global stability.

- 'Global stability' case study

To evaluate the 'global' stability aspect of the system, the following benchmark (Fig.1) was simulated on Modelica-based environment using works on power converters in [35]. AC grids were tested alone then tested MTDC grid was added to connect AC zones 1 & 2. Generator units are equipped with speed governors. Initial power flow (table 2) is performed to quickly reach steady state power curves before applying positive active power step at upper VSC. Before any disturbance, the power converters' references are put to 0 and no power was flowing from/to AC side through them. Following case studies are performed to show effects of frequency droop control applied through VSC3 on non-targeted stability aspects: DC voltage of MTDC system and rotor angle of AC zone 2 (small-signal stability).

Table 2: Power flow initialization of benchmark's generators and loads. Power converters deliver zero power output at initial system operation.

Component	Initial Active Power Flow (MW)	Component	Initial Active Power Flow (MW)
Generator G1	500	Load L1	1027
Generator G2	527	Load L2	300
Generator G3	500	Load L3	700
Generator G4	500		

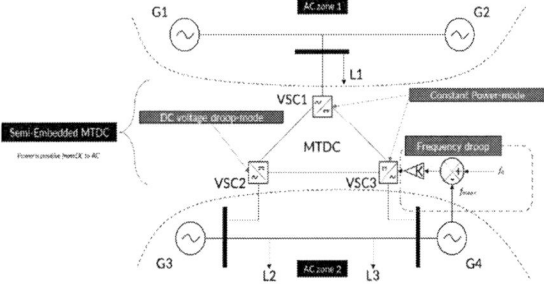

Figure 1: Used benchmark for AC & DC stability study.

    o   First case study: interaction between frequency droop and DC voltage droop controls affecting DC voltage stability. A step of P_{VSC1} asked by AC Zone 1 (could be load change, compensation of generator/line tripping, etc.)

In this case study, DC voltage drop for same DC Voltage droop gain at VSC2 (8MW/kV) was observed with different values of frequency droop gains through sweeping this parameter and running multiple simulations. The following results were observed:

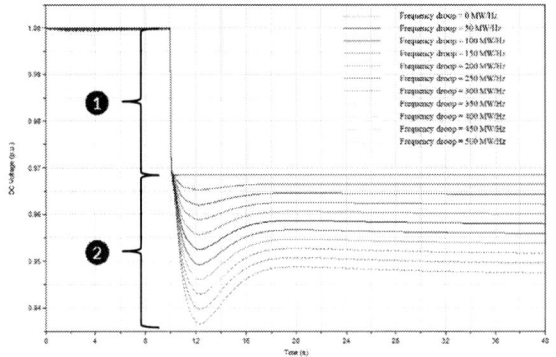

❶ DC voltage drop due to single effect of DC voltage droop control. For 100 MW step, final DC voltage value is calculated as the following:

$$V_{dc_{final}} = V_{base_{dc}} - \frac{P_{step_{VSC1}}}{K_{dc_{droop}}}$$

$$V_{dc_{final}} = 387500\ V$$

$$V_{dc_{final}} = 0.96875\ p.u.$$

Figure 2: DC voltage profiles for different frequency droop gains. Single effect of DC voltage droop control and its combined effect with frequency droop control is shown.

❷ Frequency droop control degrades DC voltage's transient and final values. Increased voltage drop comes with increased value of frequency droop controller gains.

However, as was mentioned before, global stability evaluation is needed before concluding on the usefulness of a proposed control (frequency droop in this case). Therefore, rotor angle stability was studied in a second approach.

o Second case study: effect of frequency droop on rotor angle stability. A step of P_{VSC1} asked by AC Zone 1 (load change, compensation of generator/line tripping, etc.)

In this case study, curves for different values of the step at VSC1 are compared to emphasize the effect of the single presence/absence of frequency droop control at VSC3. The following results were obtained by simulation:

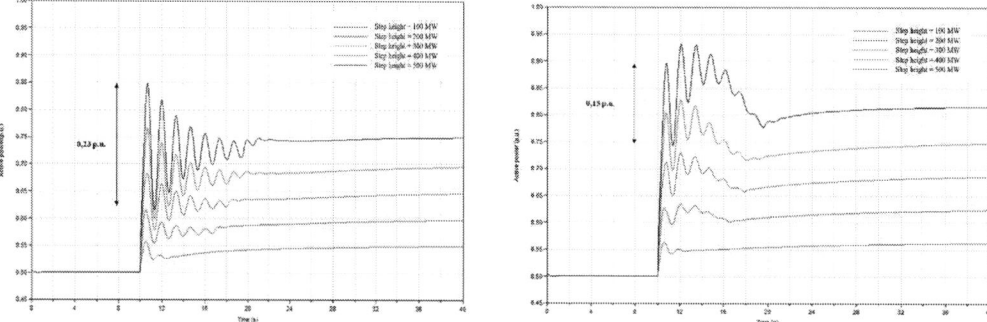

Figure 3: Power oscillations observed at AC line connecting L3 to bus connecting VSC3 and G4 for different step heights.
Left side: frequency droop is deactivated (gain = 0). Right side: frequency droop gain = 100 MW/Hz.

Frequency droop control affects positively power oscillations in this case since first swing is smaller when frequency droop gain is bigger (for highest power step at VSC1, first swing oscillation values are shown in Fig. 3). This reflects better rotor angle stability in AC system zone 2.
The equation for frequency droop controller output is:

$$P_{hvdc_{ref}} = K_{f_{droop}} * (f_{meas} - f_n).$$

As this mathematical expression suggests, frequency droop first detects difference between the nominal frequency and the measured one. Due to added power stress in AC zone 2, power starts to oscillate due to excited modes. As known, frequency acts similarly as the oscillating active power in the AC zone. Therefore, due to relatively large time reaction of generator turbine governors and the relatively small structural damping, speed of generators tends to increase with positive power oscillation and to decrease with negative oscillation. This explains measurable frequency oscillations at G4.
In this logic, as previous equation shows, frequency droop measures these oscillations in f_{meas} and tends to reduce the difference between oscillating frequency and fixed nominal frequency which leads to damping of power oscillations like what would PODs do. The damping is more eminent compared to case without frequency droop because VSC3's power injections act very fast and damping is not due to turbine governors and generator's damping in this case.

Finally, as it is clear in the above figure, the steady state of the power flow has higher values when frequency droop is implemented than when it is deactivated. This adds stress to the system and capacity limits of AC lines subject to the added stress must be respected when implementing control.
If frequency droop needs to be implemented, a compromise must be found between DC voltage stability, frequency stability and rotor angle stability while respecting all power system constraints. This approach needs to be implemented to enhance any aspect of power stability and it is currently under study.

Conclusion

MTDC grid can be involved in stability enhancement since it provides wide control opportunities and flexibility. While AC networks have already well-established network codes, proper codes for DC grids are still required nonetheless to ensure MTDCs' usage leads to better AC/DC power system operation. As was shown in the case studies, power converters' supplementary controllers have effects not only on targeted stability aspects but on other aspects as well. The need for a new and adapted approach is then evident to avoid antagonist effects of future controllers. For optimal power stability improvement, local and wide-area measurements may be needed to perform best stability assessment due to observability issues. Most important is that, as opposed to conventional enhancement approaches, 'global' stability enhancement requires coordination of control from MTDC's potential controllers as was highlighted by the shown results.

Ongoing works are on establishing rigorous assessment approaches that allow to evaluate AC & DC power system stability to enhance it in a future step.

References

[1] 'bp Statistical Review of World Energy 2021'. Statistical Review of World Energy BP, 2021. [Online]. Available: https://www.bp.com/content/dam/bp/business-sites/en/global/corporate/pdfs/energy-economics/statistical-review/bp-stats-review-2021-full-report.pdf

[2] I. Oleinikova and E. Hillberg, 'micro vs MEGA: trends influencing the development of the power system', p. 92, 2022.

[3] A. Fernández-Guillamón, E. Gómez-Lázaro, E. Muljadi, and Á. Molina-García, 'Power systems with high renewable energy sources: A review of inertia and frequency control strategies over time', *Renewable and Sustainable Energy Reviews*, vol. 115, p. 109369, Nov. 2019, doi: 10.1016/j.rser.2019.109369.

[4] W. Feng, Q. Shi, H. Cui, and F. Li, 'Optimal power allocation strategy for black start in VSC-MTDC systems considering dynamic impacts', *Electric Power Systems Research*, vol. 193, p. 107023, Apr. 2021, doi: 10.1016/j.epsr.2021.107023.

[5] 'HVDC Links in System Operations', ENTSO-E, 2019.

[6] B. Silva, C. L. Moreira, L. Seca, Y. Phulpin, and J. A. Pecas Lopes, 'Provision of Inertial and Primary Frequency Control Services Using Offshore Multiterminal HVDC Networks', *IEEE Trans. Sustain. Energy*, vol. 3, no. 4, pp. 800–808, Oct. 2012, doi: 10.1109/TSTE.2012.2199774.

[7] R. Mourouvin, K. Shinoda, J. Dai, A. Benchaib, S. Bacha, and D. Georges, 'AC/DC Dynamic Interactions of MMC-HVDC in Grid-Forming for Wind-Farm Integration in AC Systems', in *2020 22nd European Conference on Power Electronics and Applications (EPE'20 ECCE Europe)*, Sep. 2020, p. P.1-P.9. doi: 10.23919/EPE20ECCEEurope43536.2020.9215668.

[8] 'Definition and Classification of Power System Stability IEEE/CIGRE Joint Task Force on Stability Terms and Definitions', *IEEE Trans. Power Syst.*, vol. 19, no. 3, pp. 1387–1401, Aug. 2004, doi: 10.1109/TPWRS.2004.825981.

[9] N. Hatziargyriou *et al.*, 'Definition and Classification of Power System Stability – Revisited & Extended', *IEEE Trans. Power Syst.*, vol. 36, no. 4, pp. 3271–3281, Jul. 2021, doi: 10.1109/TPWRS.2020.3041774.

[10] 'Analysis of CE Inter-Area Oscillations of 1st December 2016', ENTSO-E, Boulevard Saint-Michel, 15 - 1040 Brussels - Belgium, ENTSO-E SG SPD REPORT, Jul. 2017. Accessed: Jul. 15, 2021. [Online]. Available: https://eepublicdownloads.entsoe.eu/clean-documents/SOC%20documents/Regional_Groups_Continental_Europe/2017/CE_inter-area_oscillations_Dec_1st_2016_PUBLIC_V7.pdf

[11] 'Continental Europe Synchronous Area Separation on 08 January 2021 ICS Investigation Expert Panel Final Report Main Report', ENTSO-E, Jul. 2021.

[12] B. Luscan *et al.*, 'A Vision of HVDC Key Role Toward Fault-Tolerant and Stable AC/DC Grids', *IEEE J. Emerg. Sel. Topics Power Electron.*, vol. 9, no. 6, pp. 7471–7485, Dec. 2021, doi: 10.1109/JESTPE.2020.3037016.

[13] S. Akkari, M. Petit, J. Dai, and X. Guillaud, 'Interaction between the Voltage-Droop and the Frequency-Droop Control for Multi-Terminal HVDC Systems', p. 7, Apr. 2016.

[14] Z. Assi Obaid, L. M. Cipcigan, and M. T. Muhssin, 'Power system oscillations and control: Classifications and PSSs' design methods: A review', *Renewable and Sustainable Energy Reviews*, vol. 79, pp. 839–849, Nov. 2017, doi: 10.1016/j.rser.2017.05.103.

[15] W. Peres, 'Multi-band power oscillation damping controller for power system supported by static VAR compensator', *Electr Eng*, vol. 101, no. 3, pp. 943–967, Sep. 2019, doi: 10.1007/s00202-019-00830-9.

[16] N. T. Anh, L. Vanfretti, J. Driesen, and D. Van Hertem, 'A Quantitative Method to Determine ICT Delay Requirements for Wide-Area Power System Damping Controllers', *IEEE Trans. Power Syst.*, vol. 30, no. 4, pp. 2023–2030, Jul. 2015, doi: 10.1109/TPWRS.2014.2356480.

[17] T. Bogodorova, S. A. Dorado-Rojas, and L. Vanfretti, 'DeepGrid: A Deep Learning Computing System for Resilient Grid Operations', RPI, Dec. 2020.

[18] G. N. Baltas, N. B. Lai, L. Marin, A. Tarraso, and P. Rodriguez, 'Grid-Forming Power Converters Tuned Through Artificial Intelligence to Damp Subsynchronous Interactions in Electrical Grids', *IEEE Access*, vol. 8, pp. 93369–93379, 2020, doi: 10.1109/ACCESS.2020.2995298.

[19] G. N. Baltas, N. B. Lai, A. Tarraso, L. Marin, F. Blaabjerg, and P. Rodriguez, 'AI-Based Damping of Electromechanical Oscillations by Using Grid-Connected Converter', *Front. Energy Res.*, vol. 9, p. 598436, Mar. 2021, doi: 10.3389/fenrg.2021.598436.

[20] R. Preece, J. V. Milanovic, A. M. Almutairi, and O. Marjanovic, 'Damping of inter-area oscillations in mixed AC/DC networks using WAMS based supplementary controller', *IEEE Trans. Power Syst.*, vol. 28, no. 2, pp. 1160–1169, May 2013, doi: 10.1109/TPWRS.2012.2207745.

[21] L. Vanfretti, S. Bengtsson, and J. O. Gjerde, 'Preprocessing synchronized phasor measurement data for spectral analysis of electromechanical oscillations in the Nordic Grid: PREPROCESSING SYNCHRONIZED PHASOR MEASUREMENT DATA FOR MODE ESTIMATION', *Int. Trans. Electr. Energ. Syst.*, vol. 25, no. 2, pp. 348–358, Feb. 2015, doi: 10.1002/etep.1847.

[22] K. W. V. To, A. K. David, and A. E. Hammad, 'A robust co-ordinated control scheme for HVDC transmission with parallel AC systems', *IEEE Trans. Power Delivery*, vol. 9, no. 3, pp. 1710–1716, Jul. 1994, doi: 10.1109/61.311190.

[23] O. Kotb, M. Ghandhari, R. Eriksson, R. Leelaruji, and V. K. Sood, 'Stability enhancement of an interconnected AC/DC power system through VSC-MTDC operating point adjustment', *Electric Power Systems Research*, vol. 151, pp. 308–318, Oct. 2017, doi: 10.1016/j.epsr.2017.05.026.

[24] J. C. Gonzalez-Torres, G. Damm, V. Costan, A. Benchaib, and F. Lamnabhi-Lagarrigue, 'Transient stability of power systems with embedded VSC-HVDC links: stability margins analysis and control', *IET Generation, Transmission & Distribution*, vol. 14, no. 17, pp. 3377–3388, Sep. 2020, doi: 10.1049/iet-gtd.2019.1074.

[25] P. L. Francos, S. S. Verdugo, H. F. Alvarez, S. Guyomarch, and J. Loncle, 'INELFE — Europe's first integrated onshore HVDC interconnection', in *2012 IEEE Power and Energy Society General Meeting*, San Diego, CA, Jul. 2012, pp. 1–8. doi: 10.1109/PESGM.2012.6344799.

[26] J. Renedo, L. Sigrist, L. Rouco, and A. Garcia-Cerrada, 'Impact on power system transient stability of AC-line-emulation controllers of VSC-HVDC links', in *2021 IEEE Madrid PowerTech*, Madrid, Spain, Jun. 2021, pp. 1–6. doi: 10.1109/PowerTech46648.2021.9494939.

[27] D. Carroll and P. Krause, 'Stability Analysis of a DC Power System', *IEEE Trans. on Power Apparatus and Syst.*, vol. PAS-89, no. 6, pp. 1112–1119, Jul. 1970, doi: 10.1109/TPAS.1970.292701.

[28] S. D. Sudhoff, S. F. Glover, S. D. Pekarek, E. J. Zivi, D. E. Delisle, and D. Clayton, 'Stability Analysis Methodologies for DC Power Distribution Systems', p. 10, 2003.

[29] 'COMMISSION REGULATION (EU) 2016/ 1447 - of 26 August 2016 - establishing a network code on requirements for grid connection of high voltage direct current systems and direct current-connected power park modules', p. 65.

[30] British Standards Institution, *HVDC Grid Systems and connected Converter Stations. Guideline and Parameter Lists for Functional Specifications. Part 1, Part 1,.* 2020.

[31] Conseil international des grands réseaux électriques, Ed., *Guidelines for the preparation of connection agreements or grid codes for multi-terminal DC schemes and DC grids.* Paris: Cigré, 2016.

[32] T. K. Vrana, J. Beerten, R. Belmans, and O. B. Fosso, 'A classification of DC node voltage control methods for HVDC grids', *Electric Power Systems Research*, vol. 103, pp. 137–144, Oct. 2013, doi: 10.1016/j.epsr.2013.05.001.

[33] E. Prieto-Araujo, A. Egea-Alvarez, S. Fekriasl, and O. Gomis-Bellmunt, 'DC Voltage Droop Control Design for Multiterminal HVDC Systems Considering AC and DC Grid Dynamics', *IEEE Trans. Power Delivery*, vol. 31, no. 2, pp. 575–585, Apr. 2016, doi: 10.1109/TPWRD.2015.2451531.

[34] Y. Chompoobutrgool and L. Vanfretti, 'Identification of Power System Dominant Inter-Area Oscillation Paths', *IEEE Trans. Power Syst.*, vol. 28, no. 3, pp. 2798–2807, Aug. 2013, doi: 10.1109/TPWRS.2012.2227840.

[35] A. Zama, S. Bacha, A. Benchaib, D. Frey, and S. Silvant, 'Comparison and assessment of implementation techniques for dynamics MMC type models', in *2019 21st European Conference on Power Electronics and Applications (EPE '19 ECCE Europe)*, Genova, Italy, Sep. 2019, p. P.1-P.10. doi: 10.23919/EPE.2019.8914946.

Measurement of Coss-V characteristic of the 1.7kV/900A SiC power module and estimation of the channel current

Jacek Rabkowski[1], Fernando Gonzalez-Hernando[2], Mariusz Zdanowski[1], Irma Villar[2], Uxue Larrañaga[3]

[1]Warsaw University of Technology
Koszykowa 75
00-662 Warsaw, Poland
jacek.rabkowski@pw.edu.pl

[2]Ikerlan Technology Research Center
Basque Research and Technology Alliance (BRTA),
J. Mª. Arizmendiarrieta, 2
Arrasate-Mondragón, Spain.
fgonzalez@ikerlan.es

[3]CAF Power & Automation
San Sebastián, Spain
ULarranaga@cafpower.com

Acknowledgements

This project has received funding from the Shift2Rail Joint Undertaking under the European Union's Horizon 2020 research and innovation programme under grant agreement No 101015423. The content of this publication does not reflect the official opinion of the European Union. Responsibility for the information and views expressed in the report lies entirely with the author(s).

Keywords

«Silicon Carbide (SiC)», «Zero-voltage switching», «MOSFET», «measurements», «switching losses».

Abstract

This article presents a novel method for dynamic measurements of C_{OSS}-V characteristics of SiC power modules based on the process of charging the output capacitance of the transistors. The technique has been validated for a 1.2kV/450A power module with this characteristic available in its datasheet, showing good compliance. Then, this technique has been used to determine the C_{OSS}-V characteristics of a 1.7kV/900A SiC MOSFET module, which allowed the extraction of the capacitive current while switching off the transistor. Finally, the channel current and the share of the capacitive current in the drain current were determined for various switched currents and switching speeds. According to the capacitive charge calculations for several cases, the accuracy of the method is high enough to perform switching loss estimations.

Introduction

Medium voltage SiC MOSFETs are excellent candidates for soft-switched DC-DC or Inductive Power Transfer (IPT) converters, including these operating at Zero Voltage Switching (ZVS) conditions [1]-[4]. Such systems usually operate at high switching frequencies, and a substantial part of the power losses occurring in semiconductor elements are losses generated during the switching-off process. Therefore, it is essential to correctly determine the switching losses to properly design the converter from an electro-thermal perspective. Several publications, to mention only [5]-[9], have shown that the problem of switch-off losses is quite complex. Without going into deep analysis, it can be concluded that the drain current includes the channel current, which causes conduction power losses in the transistor's channel, and the capacitive current, which is almost lossless [10]. Therefore, to correctly determine the switch-off power losses, it is necessary to distinguish these two components. This can be done by finding the capacitive current from the derivative of the drain-source voltage, for this, the C_{OSS}-V characteristic of the power device must be employed. However, this characteristic is not available for all power modules – this is the case of the 1.7kV/900A SiC power module MSM900FS17ALT discussed in this work [11] or other power modules under development.

The problem of measuring C_{OSS}-V characteristics, including for SiC power devices, was undertaken by many research teams [12]-[17]. There are classic methods using an impedance analyzer or an LRC bridge, but the main problem is the need for high voltage polarisation. Hence, various circuit concepts have arisen to allow measurements at higher voltages [12]-[13]. Another dynamic testing approach is based on running one- or two-pulse tests and observing the corresponding waveforms to determine charges and capacitances [14]-[16]. It is also possible to use a similar method in transistors' continuous operation without load [18]. Considering the specificity of the 1.7kV/900A module, i.e. medium voltage and high rated current, i.e. a large active surface and possibly a large output capacitance, the authors propose a dynamic method of measuring the C_{OSS} capacitance based on a single-pulse test. Observing the drain-source voltage and the current flowing through the power device, it is possible to determine the C_{OSS}-V characteristic precisely.

The paper is organized as follows: after the introduction, the problem of the capacitive current during the turn-off process is discussed. Then, the proposed method for measuring the C_{OSS}-V characteristic is presented. In the next section, this method is validated for a CAB425M12XM3 module, whose C_{OSS}-V characteristics can be found in its datasheet [18]. Finally, a measurement is performed for the MSM900FS17ALT module, and the capacitive currents in a system operating at a frequency of up to 25 kHz are estimated at different switching speeds. The work is concluded with a summary of the main outcomes.

Difference between the drain and channel current

Typical waveforms recorded at the turn-off process of the MSM900FS17ALT at 850V and 600A are presented in Fg.1(a), while idealized equivalents can be seen in Fig.1(b). To simplify, the impact of parasitic inductances and high-frequency oscillations are not considered. In the optimal scenario from the switching losses point of view, the gate driver can quickly discharge the gate-source capacitance C_{GS} to reach threshold voltage V_{TH} before drain-gate capacitance C_{GD} is charged and drain-source voltage V_{DS} achieves high values. The current through the channel of the power transistor i_{CH} is rapidly reduced to zero before V_{DS} is high and, as result, the power dissipated in the transistor (as the product of v_{DS} and i_{CH}) is low. These conditions can be recognized as nearly ZVS and this scenario is preferable as results in nearly zero switch-on losses and therefore leads to higher efficiency [7].

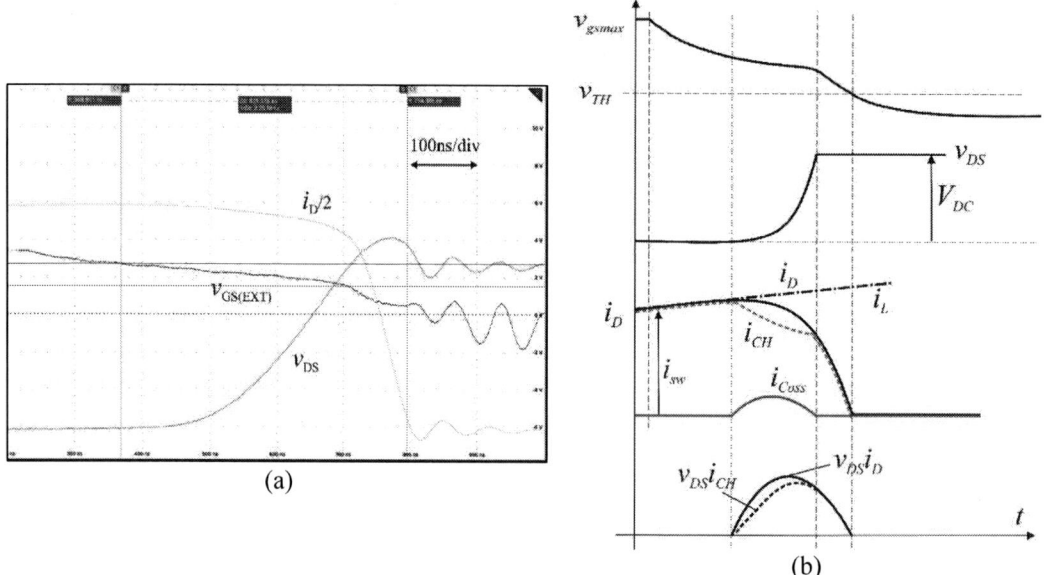

Fig. 1: Waveforms during turn-off – recorded for the MSM900FS17ALT (a) and idealized – without an influence of parasitic inductance (b)

However, this scenario is difficult to achieve in most high-current power modules. The substantial input capacitance C_{ISS} requires the large gate current to be quickly discharged but limited supply voltage of the gate driver and non-zero internal gate resistance are limiting factors. Thus, the real scenario usually observed is presented in the idealized waveforms in Fig. 1(b): during the voltage rising phase the v_{GS} is above the threshold and the channel remains open, generating power losses. Note that the resulting power loss is lower than a product of v_{DS} and i_D multiplication usually provided in the datasheets. Finally, the v_{GS} crosses v_{TH} (and the i_{CH} becomes zero) after v_{DS} has reached V_{DC}. It is worth noting that, at the end of the voltage rise phase, the observed i_D is equal to i_{CH}, since there is no displacement current in the output capacitances when the voltage has reached a constant value. In practice, it will also be increased by voltage overshoots across parasitic inductances in the switching loop, as shown in Fig. 1(a).

Coss-V test method

Similarly to [17], the proposed method uses a half-bridge power module (Fig. 2), but only a single pulse test is performed. The low-side transistor of the half-bridge is the device under test (DUT) and is permanently switched off, with its gate shorted to the source, while the upper transistor plays the role of the switch applying a positive voltage to the DUT. In consequence, the output capacitance of the lower transistor is charged via the upper transistor. There is no resistor in series to limit the current slope, but the gate resistor R_G is applied to control the switching speed and peak of the charging current.

An example of the waveforms is presented in Fig. 3, obtained for the CAB425M12XM3 power module. At the beginning of the process, v_{DS} is low, and the capacitance shows maximum values; therefore, the charging current rises fast and reaches peak value when the v_{DS} slope becomes linear. Then, while the voltage increases, the output capacitance drops and the charging current is reduced. Note that a similar current also discharges the capacitances of the upper transistors. At the end of the test, the v_{DS} slope becomes non-linear again, most likely, due to increased capacitance of the upper transistor. All in all, the single pulse takes a few microseconds, and the recorded voltage and current waveforms are employed to perform the calculations:

$$C_{OSS} = i_{COSS} \frac{dV_{DS}}{dt} \qquad (1)$$

The calculation of (1) based on the waveforms in Fig.3 was conducted in Matlab, and the obtained C_{OSS}-V characteristic is shown in Fig.4. The dashed line presents the characteristic of the power module extracted from the datasheet [18]. Despite minor differences in a few areas, the method provides accurate and realistic C_{OSS}-V characteristic and may be applied to other power modules.

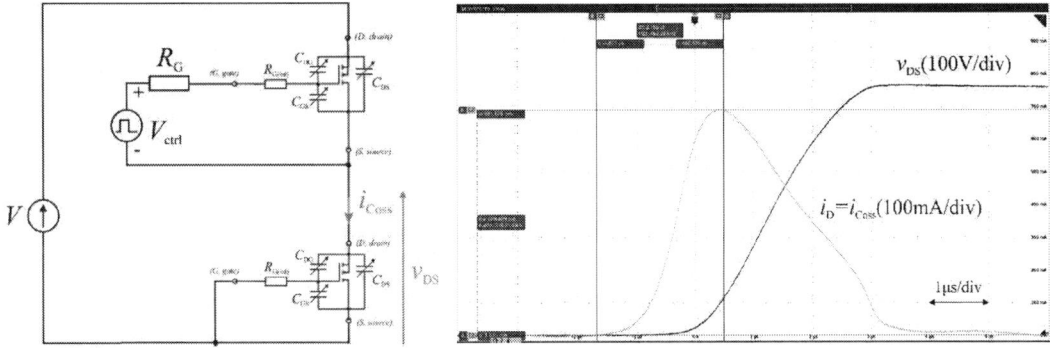

Fig. 2: Half-bridge test setup for the Coss-V characteristic measurements of the SiC power module.

Fig. 3: Waveform of the drain current and drain-source voltage during a test of CAB425M12XM3.

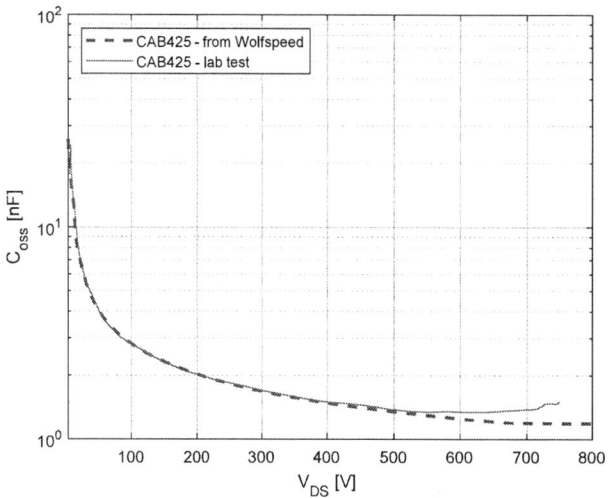

Fig. 4: C_{OSS}-V characteristics for CAB425 – datasheet and measured.

Tests of the 1.7kV/900A module

As the 1.7kV/900A SiC power module is the focus of this paper, the proposed method is employed to extract the C_{OSS}-V characteristics for this module. The test setup was the same as presented in Fig.2, but different values of the supply voltage V_{DC} and gate resistor $R_{G(ext)}$ were employed. The shape of the current and voltage waveforms in Fig.5 is similar to that observed in Fig.3, however, the C_{OSS} is higher and the peak charging current reaches almost 1.8 A. It can be also seen that the process takes more time. Again, with the described methodology, the C_{OSS}-V characteristic was determined for the tested MSM900FS17ALT, and the results are shown in see Fig. 6. Moreover, the Q_{OSS} can be integrated from the current waveform as presented in Fig.5 where the charge was approximately 11.3 µQ.

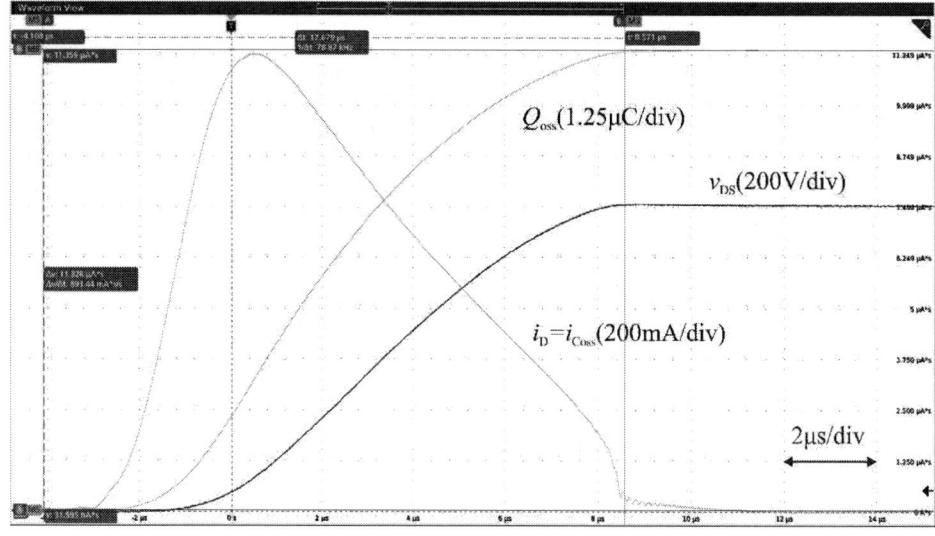

Fig. 5: Waveform of the drain current and drain-source voltage during test of MSM900FS17ALT.

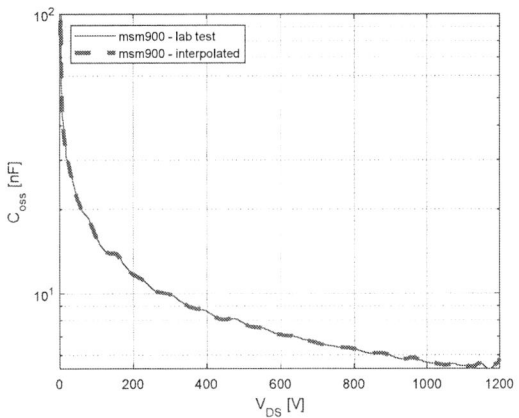

Fig. 6: Measured C-V characteristics for MSM900FS17ALT

Estimation of the i_{COSS} and i_{CH}

The MSM900FS17ALT has been arranged in a half-bridge circuit supplied from an 850V source and loaded with an inductive load (3x114uH/100A in parallel), as presented in Fig.7. Then, the transistors were controlled to generate a square voltage wave in the load with a variable switching frequency up to 25 kHz to obtain a different amplitude of the load current. These conditions lead to a triangle shape of the load current and switching conditions similar to those in soft-switched DC-DC converters. In particular, the transistors perform turn-on at zero voltage and turn-off at peak load current. Thus, most power losses appear during the turn-off event, twice per single switching period. In the circuit in Fig.7, the drain-source voltage and drain current waveforms were measured with the high-bandwidth voltage probe (P5200A) and Rogowski coil. Examples of the waveforms are presented in Fig.8 and Fig.10 for $R_{G(EXT)}$ = 3.3 Ω and without an external gate resistor, respectively, for different switched currents. The switching process is faster for the higher current and lower gate resistance; the same observation can be made for the oscillations – they become more severe for the higher current and lower gate resistance.

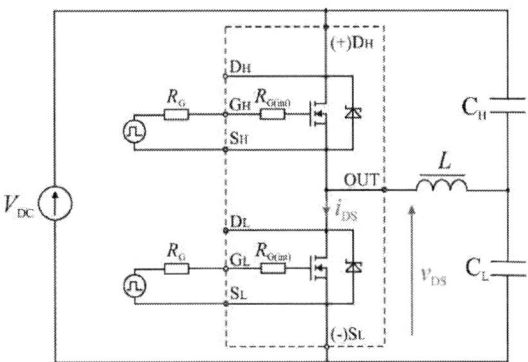

Fig. 7 Scheme of the half-bridge circuit with inductive load.

The C_{OSS}-V characteristic determined before has been employed to calculate the capacitive current i_{COSS} during the switching process as:

$$i_{COSS} = C_{OSS}(V_{DS})\frac{dV_{DS}}{dt} \tag{2}$$

also using the derivative of the drain-source voltage dV_{DS}/dt. In order to take into account the voltage drop in the stray inductance inside the power module, and therefore obtain more accurate results, V_{DS} in (2) was replaced by the internal $V_{DS}(i)$ calculated as:

$$V_{DS(i)} = V_{DS} + L_S\frac{di_D}{dt} \tag{3}$$

Where L_s is half of the internal module inductance provided by the datasheet. The obtained results are presented in Fig. 9 and Fig.11 for $R_{G(EXT)}$ = 3.3 Ω and without an external gate resistor, respectively.

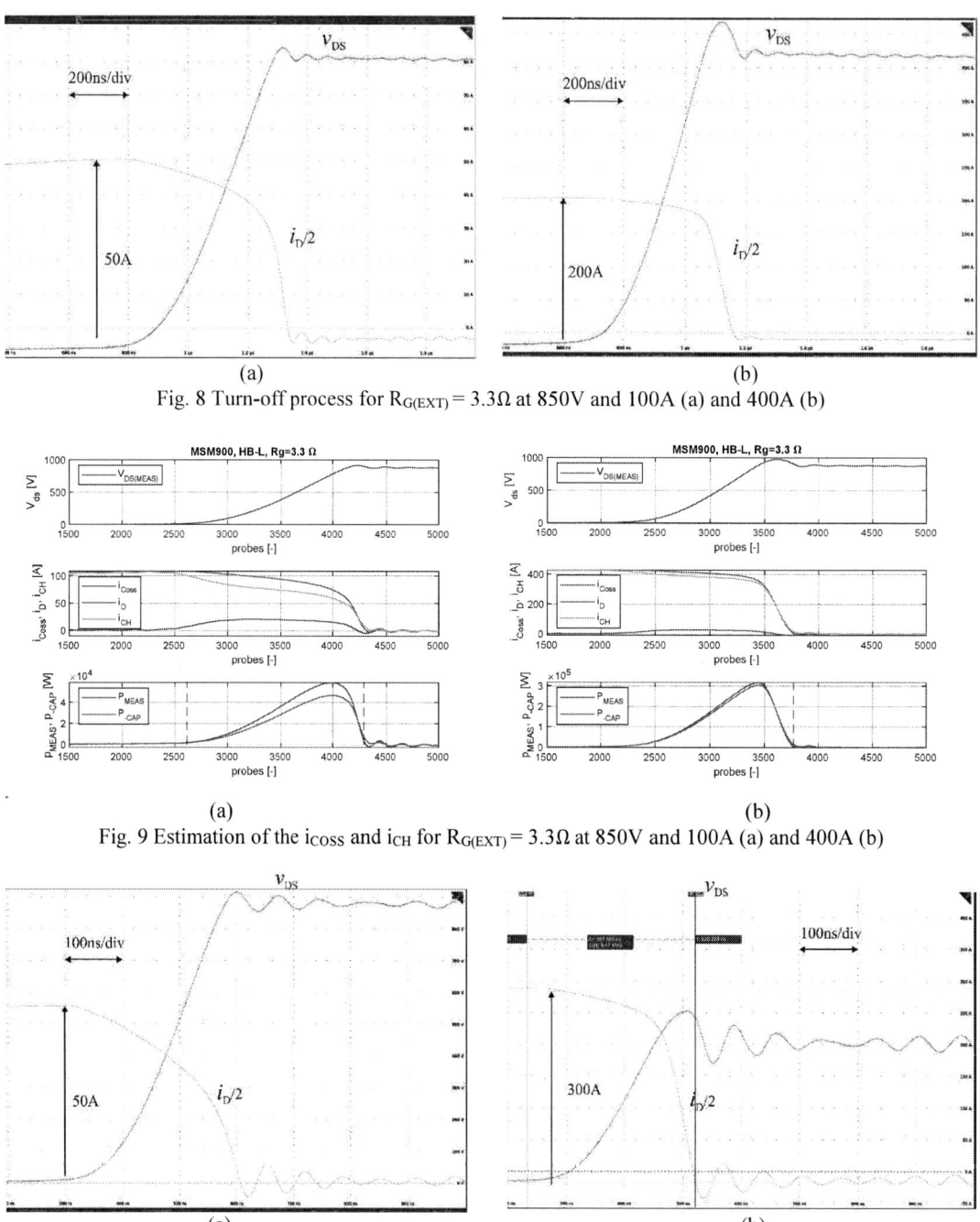

Fig. 8 Turn-off process for $R_{G(EXT)} = 3.3\Omega$ at 850V and 100A (a) and 400A (b)

(a)   (b)

Fig. 9 Estimation of the i_{COSS} and i_{CH} for $R_{G(EXT)} = 3.3\Omega$ at 850V and 100A (a) and 400A (b)

(a)   (b)

Fig. 10: Turn-off process for $R_{G(EXT)} = 0\Omega$ at 850V and 100A (a) and 600A (b)

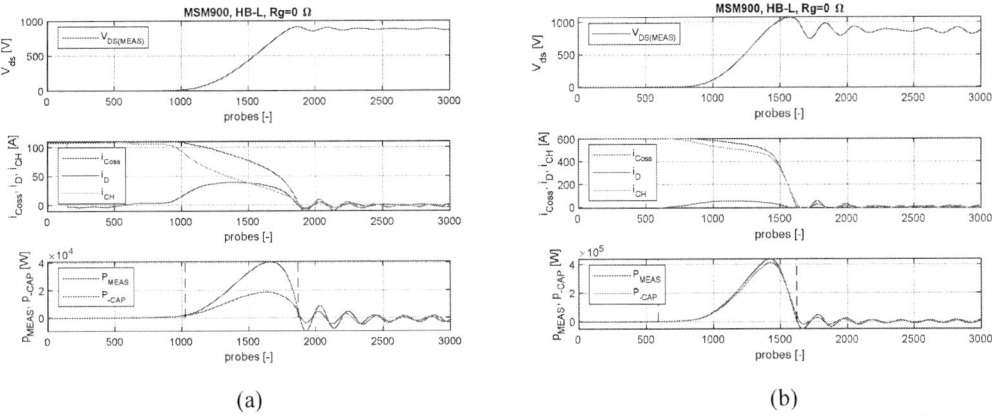

(a)   (b)

Fig. 11: Estimation of the i_{COSS} and i_{CH} for $R_{G(EXT)} = 0\Omega$ at 850V and 100A (a) and 600A (b)

Discussion

Closer analysis of the i_{COSS} waveforms presented in Fig. 9 and Fig.11 shows that they are almost independent on the switched drain current. The slight dependency is caused by a faster switching at the higher current. Major impact on the shape of the i_{COSS} has the decrease of the gate resistance resulting in much higher switching speed. In particular, the peak value of the i_{COSS} is higher for more rapid switching. Both observations agree that i_{COSS} is mainly the result of the charge displacement related to the changes in the drain potential. For the cases shown in Fig.9 and Fig.11 but also two more gate current values ($R_{G(EXT)} = 1.6\ \Omega$ and $R_{G(EXT)} = 1\ \Omega$), the waveforms of i_{COSS} were integrated to determine the charge Q_{OSS} - see results in Fig. 12. The values should be constant but vary between 8.9 and 9.7 μC (~9% of the total error), which may be considered an acceptable value resulting from the measurement mismatches. They rise with the switched current for all cases, which is most likely, a result of higher voltage overshoot (increasing with the switched current and decreasing with the gate resistance). Note that values for the same switched currents show minimal differences, thus, the error results from the measurements in the half-bridge circuit rather than from the obtained Coss-V characteristic.

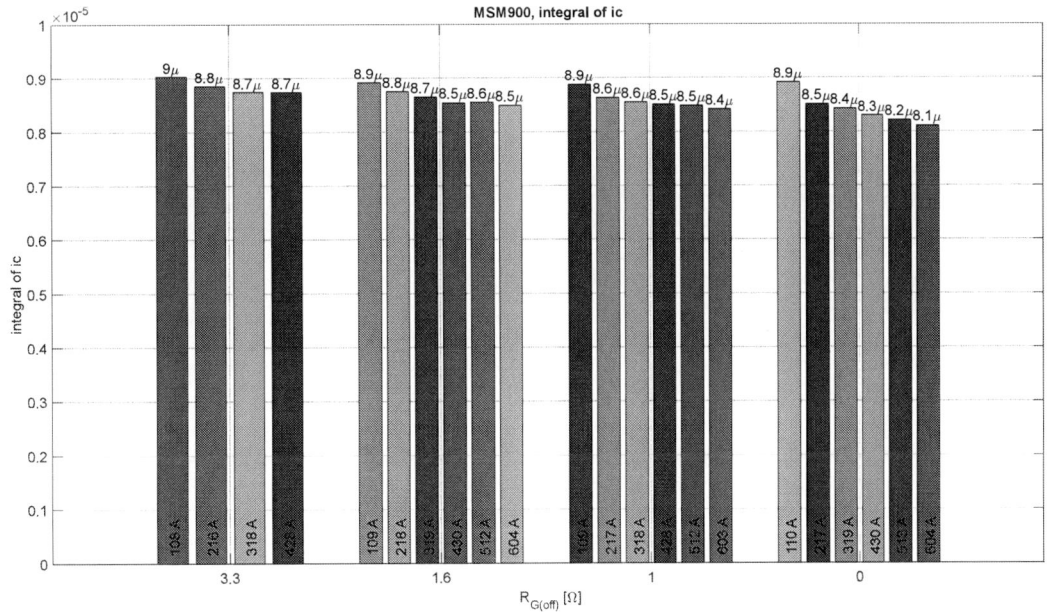

Fig. 12 Calculated Q_{OSS} for all tested cases.

Conclusion

This paper presents a novel method of C_{OSS}-V characteristics for SiC power modules. The main advantage is a simple application within the half-bridge module without additional effort – a low power high-voltage power supply and typical voltage and current probes are necessary. The setup is very similar to that required for double-pulse tests and power loss characterization of power devices. The test for the 1.2kV/425A module shown very good agreement with the C_{OSS}-V characteristic provided by the manufacturer, and the method was applied to the 1.7kV/900A module, whose datasheet is unavailable. Based on the measured characteristic, the capacitive current i_{COSS} has been calculated for this same module operating in the continuous mode at 850V at different switching speeds with currents up to 600A and switching frequencies up to 25 kHz. Then, the Q_{OSS} was determined for all cases. Under the assumption that Q_{OSS} should be constant for all tested cases, the authors have determined errors of the whole procedure on the level of +/-4.5%. It is very possible, that most of the error comes from the different voltage overshoots across the parasitic inductances. All in all, for the fast switching SiC power devices, this is an acceptable value to increase the accuracy of power losses estimation of soft-switched DC-DC converters.

References

[1] R. Chen, S. Shi, C. Zhang and W. Shi, "11 kW High-efficiency bidirectional CLLC converter with 1200 V SiC MOSFET," PCIM Asia 2021; International Exhibition and Conference for Power Electronics, Intelligent Motion, Renewable Energy and Energy Management, 2021, pp. 1-9.

[2] Z. Lu *et al.*, "Medium Voltage Soft-Switching DC/DC Converter With Series-Connected SiC MOSFETs," in *IEEE Transactions on Power Electronics*, vol. 36, no. 2, pp. 1451-1462, Feb. 2021, doi: 10.1109/TPEL.2020.3007225.

[3] I. Villar, U. Iruretagoyena, A. Rujas, A. Garcia-Bediaga and I.P. de Arenaza, "Design and implementation of a sic based contactless battery charger for electric vehicles", *IEEE Energy Conversion Congress and Exposition (ECCE)*, 2015.

[4] A. Rodriguez, I. Ayarzaguena, I. Larranzabal, J. Najera, D. Ortega, A. Pulido, M. R. Rogina, "Design and operation of a DC/DC converter to integrate energy storage into a railway traction system", *2021 23rd European Conference on Power Electronics and Applications (EPE'21 ECCE Europe)*, 2021

[5] M. Haider et al., "Analytical Calculation of the Residual ZVS Losses of TCM-Operated Single-Phase PFC Rectifiers," in *IEEE Open Journal of Power Electronics*, vol. 2, pp. 250-264, 2021

[6] B. Agrawal, M. Preindl, B. Bilgin and A. Emadi, "Estimating switching losses for SiC MOSFETs with non-flat miller plateau region," *2017 IEEE Applied Power Electronics Conference and Exposition (APEC)*, 2017, pp. 2664-2670.

[7] X. Li *et al.*, "Achieving Zero Switching Loss in Silicon Carbide MOSFET," in *IEEE Transactions on Power Electronics*, vol. 34, no. 12, pp. 12193-12199, Dec. 2019.

[8] N. Fritz, G. Engelmann, A. Stippich, C. Lüdecke, D. A. Philipps and R. W. D. Doncker, "Toward an In-Depth Understanding of the Commutation Processes in a SiC mosfet Switching Cell Including Parasitic Elements," in *IEEE Transactions on Industry Applications*, vol. 56, no. 4, pp. 4089-4101, July-Aug. 2020.

[9] Y. Xie, C. Chen, Y. Yan, Z. Huang and Y. Kang, "Investigation on Ultralow Turn-off Losses Phenomenon for SiC MOSFETs With Improved Switching Model," in *IEEE Transactions on Power Electronics*, vol. 36, no. 8, pp. 9382-9397, Aug. 2021.

[10] J. Rabkowski, M. Zdanowski, F. Gonzalez-Hernando, I. Villar, U. Larrañaga, "Experimental Study on Turn-off Process of Medium Voltage SiC MOSFET Modules", *PCIM Europe*, 2022.

[11] MSM900FS17ALT datasheet.

[12] P. Ralston, T.H. Duong, Nanying Yang, D.W. Berning, C. Hood, A.R. Hefner, et al., "High-Voltage capacitance measurement system for SiC power MOSFETs", *Energy Conversion Congress and Exposition 2009. ECCE 2009. IEEE*, pp. 1472-1479, 20–24 Sept. 2009.

[13] T. Funaki, N. Phankong, T. Kimoto and T. Hikihara, "Measuring Terminal Capacitance and Its Voltage Dependency for High-Voltage Power Devices", *Power Electronics IEEE Transactions on*, vol. 24, no. 6, pp. 1486-1493, June 2009.

[14] X. Song, A. Q. Huang, M. Lee and G. Wang, "A dynamic measurement method for parasitic capacitances of high voltage SiC MOSFETs", *2015 IEEE Energy Conversion Congress and Exposition (ECCE)*, 2015, pp. 935-941, doi: 10.1109/ECCE.2015.7309788.

[15] Cristino Salcines, Ingmar Kallfass, Hisao Kakitani, Atsushi Mikata, "Dynamic characterization of the input and reverse transfer capacitances in power MOSFETs under high current conduction", *Applied Power Electronics Conference and Exposition (APEC) 2016 IEEE*, pp. 2969-2972, 2016.

[16] M. Samizadeh Nikoo, A. Jafari, N. Perera and E. Matioli, "Measurement of Large-Signal C_{OSS} and C_{OSS} Losses of Transistors Based on Nonlinear Resonance," in *IEEE Transactions on Power Electronics*, vol. 35, no. 3, pp. 2242-2246, March 2020, doi: 10.1109/TPEL.2019.2938922.

[17] Wei Xu, Zhicheng Guo, S. Milad Tayebi, Sanjay Rajendran, Ao Sun, Ruiyang Yu, Alex Q. Huang, "Hardware Design and Demonstration of a 100kW 99% Efficiency Dual Active Half Bridge Converter Based on 1700V SiC Power MOSFET", *Applied Power Electronics Conference and Exposition (APEC) 2020 IEEE*, pp. 1367-1373, 2020.

[18] CAB425M12MX3 datasheet.

In-slot Cooling of Electrical Machines Using Traditional Techniques and Additive Manufacturing

Ahmed Hembel, Gokhan Cakal, and Bulent Sarlioglu
Wisconsin Electric Machines and Power Electronics Consortium (WEMPEC)
Department of Electrical and Computer Engineering
University of Wisconsin-Madison
Madison, WI 53706 USA
E-Mail: sarlioglu@wisc.edu

Acknowledgments

The authors would like to acknowledge the Grainger Center for Electric Machinery and Electromechanics at the University of Illinois at Urbana-Champaign and the Wisconsin Electric Machinery and Power Electronics Consortium at the University of Wisconsin – Madison for supporting this project.

Keywords

«Additive manufacturing», «Efficiency», «Electric machines», «In-slot cooling»

Abstract

This paper covers in-slot cooling methods used for the effective cooling of electric machines. Both traditional and additively manufactured cooling solutions are discussed throughout the paper. A new additively manufactured heat exchanger is also presented in the paper. It is shown that the proposed solution can boost current density to 83 A_{rms}/mm^2 using PCB as stator winding.

Introduction

Cooling of electric machines is becoming more important as researchers push the boundaries of power density for electric machines. The winding conductors are one of the critical parts that need superior cooling performance due to high conduction losses. In-slot cooling methods are getting attention for effective cooling of the winding conductors and the back iron. The cooling channels are placed inside the slots. These methods achieve superior current densities in the windings by placing the cooling channels very close to the conductors.

In the first part, this paper reviews in-slot cooling methods manufactured by traditional methods and additive manufacturing technology, which enables the manufacturing of complex geometries. Key metrics such as winding temperature, fill factor, and current density are collected from the reviewed literature. In the second part, this paper proposes in-slot cooling solutions for electric machines using a 3D printed heat exchanger with PCB winding and a flat wire winding. The performance of the presented solutions is also compared with the reviewed solutions in the literature.

Traditional in-slot cooling methods

Traditional in-slot cooling methods are categorized into three groups: liquid cooling with heat exchangers, in-slot cooling with heat pipes and heat guides, and forced air in-slot cooling. These cooling methods can be manufactured using conventional methods instead of additive manufacturing technology.

Liquid cooling with heat exchangers

Traditionally manufactured solutions utilizing heat exchangers (HX) include a variety of designs. One such design is proposed in [1], which utilizes the gap between two adjacent windings of a double-layer

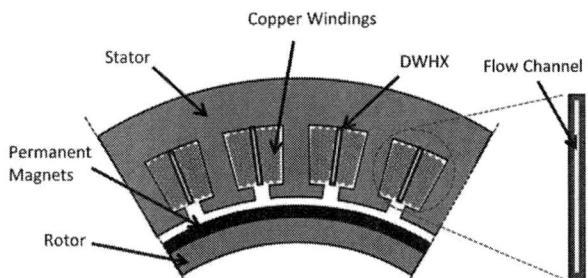

Fig. 1. Copper heat exchanger between adjacent windings [1].

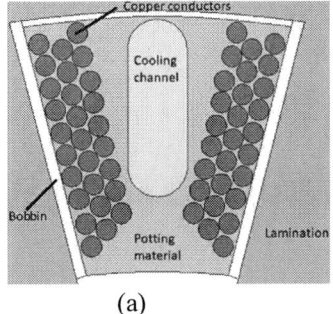

(a)

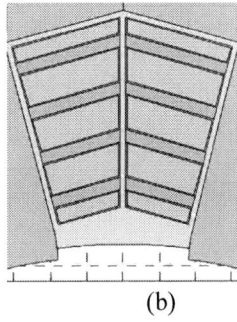

(b)

Fig. 2. a. Cooling channel formed by thermally conductive epoxy [2] b. Cooling channels interleaved between flat conductors [3].

concentrated winding. A copper HX is placed in this gap with coolant flowing inside it. This design is shown in Fig. 1. The authors report a continuous current density of 25 A_{rms}/mm^2 and a transient current density of up to 40 A_{rms}/mm^2 at a hotspot temperature of 155 °C. The flow rate is set to 5300 cc/min, and the maximum pressure drop through the system is measured as 5.14 kPa. The experimental work is carried out by removing the rotor in the machine (nonrotating machine). Therefore, the core losses are not included in the results.

Another in-slot cooling design is proposed in [2], where a cooling channel between adjacent windings is formed using a thermally conductive epoxy during global winding encapsulation, as shown in Fig. 2a. The current densities are reported to be 25 A_{rms}/mm^2 continuous and 35 A_{rms}/mm^2 peak at a hotspot temperature of 180°C and a fill factor of 0.32. Oil is used as a coolant with an inlet temperature of 60 °C and a flow rate of 6 L/min. A pressure drop of 42 kPa is reported under the operating conditions. A more aggressive approach is presented in [3], where multiple layers of flat conductors are interleaved with cooling channels, as shown in Fig. 2b. The maximum continuous current density is reported to be 50 A_{rms}/mm^2 with hotspot temperatures below 120 °C and a fill factor of 0.47. Oil is used as a coolant at a 2 L/min flow rate and 40 °C of inlet temperature. However, no manufacturing method is proposed for this highly complex design.

In-slot cooling with heat pipes and heat guides

Heat pipes are another effective in-slot cooling method with very high thermal conductivity. Heat pipes are sealed pipes made from a thermally conductive metal, mainly copper or aluminum, containing water vapor. A pressure gradient inside the pipe causes circulation of the liquid and gaseous phases, which causes a convection heat transfer inside a sealed tube. On the other hand, a heat guide is just a solid made from a thermally conductive material, which causes a conduction heat transfer.

One application of heat pipes and heat guides for in-slot cooling is shown in Fig. 3. In this study, heat is removed by employing heat pipes inside the slot bottom paired with a heat guide to cover the empty area in the middle of the slot. Two designs are evaluated: one with a cooling jacket and heat pipes and the other with just the heat pipes. The design paired with a cooling jacket achieved 24 A_{rms}/mm^2, whereas

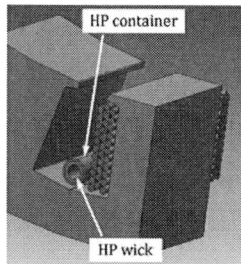
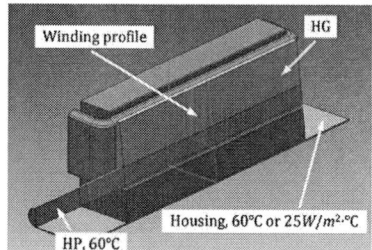

Fig. 3: In-slot heat pipe with heat guide [5]

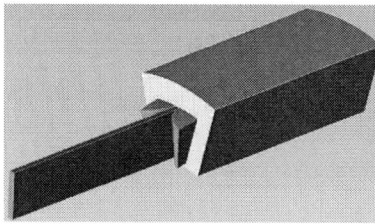

Fig. 4. a. Flat heat pipe for in-slot cooling b. Internal structure of flat heat pipe [6]

the design without the cooling jacket achieved 21 A_{rms}/mm^2. In both these cases, the winding hotspot temperature was kept at 180 °C, and a fill factor of 0.38 was achieved [4], [5]. Another heat pipe-based design is presented in [6], which uses a flat heat pipe covering the gap between adjacent windings in a double-layer concentrated wound stator, as shown in Fig. 4. The maximum current density achieved is 16.5 A_{rms}/mm^2 with a fill factor of 0.4 and a hotspot temperature of 95 °C.

A tubular linear motor is presented in [7], as shown in Fig. 5, with a double layer concentrated wound stator using a heat guide between the adjacent winding gap paired with a stator cooling jacket. This heat guide is made from aluminum, which reduces the thermal resistance between the winding hotspot and the actively cooled stator back iron. The reported current density is 35 A_{rms}/mm^2 at 180 °C hotspot temperature.

Forced air in-slot cooling

Air cooling is generally cheaper and more robust compared to any form of liquid cooling. Forcing air through slots is achieved using cast coils in a design presented in [8]. Aluminum cast coils are used with axial airflow to remove the heat combined with a stator cooling jacket. The casting process allows for slot fill factors of up to 0.9 and continuous current densities of up to 24 A_{rms}/mm^2 while keeping the hotspot temperature below 180 °C. The aluminum cast coil is shown in Fig. 6. Another forced air convection cooled aluminum winding is shown in Fig. 7 to be employed in a stator as a wave winding [9]. The peak current density is 30 A_{pk}/mm^2 with a maximum hotspot temperature of 180 °C. In this case, the windings are laminated with spaces between each layer to allow airflow, which provides convection for cooling.

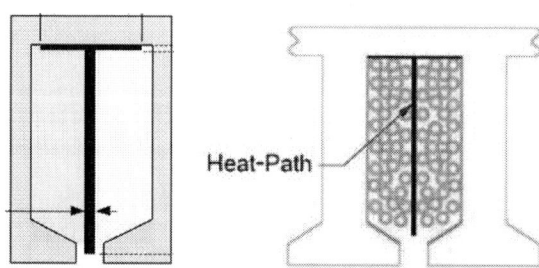

Fig. 5. Aluminum heat path between winding and back iron [7]

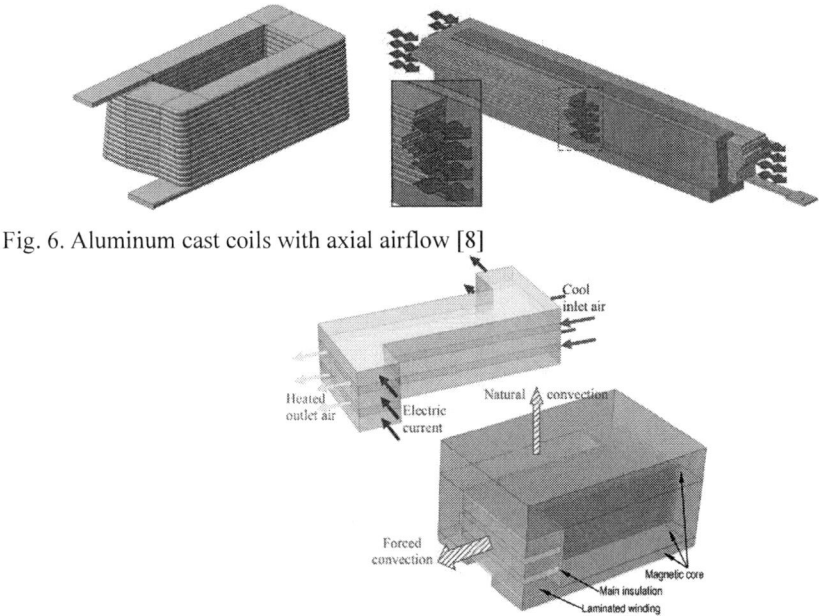

Fig. 6. Aluminum cast coils with axial airflow [8]

Fig. 7. Forced convection for AM winding [9]

Additively manufactured in-slot cooling solutions

Additively manufactured heat exchangers

AM enables design freedom for manufacturing complex geometries required for HXs. A liquid-cooled version of in-slot HX is presented in [10] for use with rectangular wire. The cooling channel is formed by 3D printing a tube insert with a water-soluble plastic (polyvinyl alcohol (PVA) in this case). This tube occupies the gap between adjacent double-layer concentrated windings in a slot, which is then encapsulated. After curing, the water-soluble insert is dissolved, and a cooling channel is formed, as shown in Fig. 8. A fill factor of 0.545 is achieved with rectangular wire. The authors report a current density of 20 A_{rms}/mm^2 with a hotspot temperature of 85 °C. The authors measured the performance of the heat exchanger on a 70 kW machine with 1.2 kW of copper losses. A 50/50 water/glycol mixture is used as a coolant with an inlet temperature of 50 °C.

Another example of in-slot cooling is presented in [11], where a 3D-printed direct winding HX (3DWHX) is made from aluminum-flake polycarbonate. The HX is inserted in the gap between the adjacent windings of a double-layer concentrated wound machine, as shown in Fig. 9. The HX can be encapsulated with the windings to provide good thermal contact. The authors report a current density of 20 A_{rms}/mm^2 tested in a motorette where the windings were not encapsulated. Theoretically, with a

Fig. 8. In-slot heat exchanger using PVA molds [10]

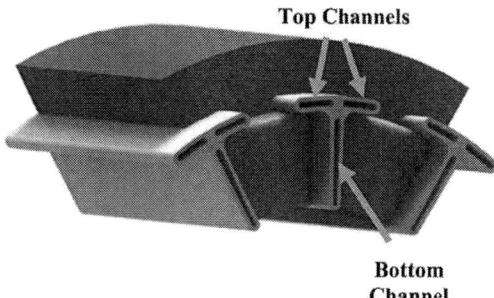

Fig. 9. 3D-printed direct winding HX for a double layer concentrated wound machine [12]

hotspot temperature of 195 °C, the current density could be increased to 33 A_{rms}/mm^2. Another similar HX is made from 3D-printed alumina, which allows a current density of 35 A_{rms}/mm^2 with a hotspot temperature of 200 °C [12]. Water is used as a coolant with a flow rate of 0.95 L/min. In both cases, the fill factor was assumed to be around 0.5 in the area occupied by windings, excluding the area occupied by the HX, and around 0.28 for the total slot area.

Additively manufactured heat guide

Like the traditionally manufactured heat guides, AM heat guides can also provide effective in-slot cooling while conforming to the complex shapes for better thermal contact and lower contact resistance. One such AM-based heat guide is presented in [13]. This heat guide, like others, occupies the gap between neighboring double-layer concentrated windings and conducts the heat produced by the windings to the stator housing, which is actively cooled. The heat guide is 3D-printed with AlSi10Mg, which can manage up to 40% additional power loss for the same temperature rise with a 180 °C hotspot temperature. This heat guide is shown in Fig. 10.

Additively manufactured hollow windings with integrated cooling

In addition to cooling channels and heat guides, 3D-printing has also been utilized to design liquid-cooled coils for electric machines. In [14], a 3D-printed AlSi10Mg coil is designed and compared to a similarly sized cast copper coil. The 3D-printed coil is shown in Fig. 11, submerged in coolant instead of a separate heat exchanger, achieving extremely high current densities. With hotspot temperature limited to 180 °C, the cast copper coil has a tested DC current density of 100 A_{rms}/mm^2, whereas the 3D-printed coil has a possible theoretical current density of 70 A_{rms}/mm^2. The fill factor is not reported for the 3D-printed coil used in this study.

A more integrated system is proposed in [15] with hollow AM coils integrated with heat pipes. The coils are printed with AlSi10Mg and are hollow from the inside to allow for the insertion of copper heat pipes, which can conduct heat at a very high rate. The system is shown in Fig. 12, where the authors report a current density of 13.9 A_{rms}/mm^2.

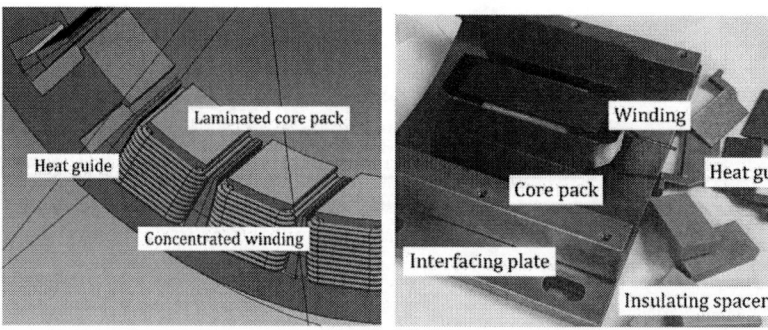

Fig. 10. AM heat guide a. 3D CAD assembly b. Prototype motorette [13]

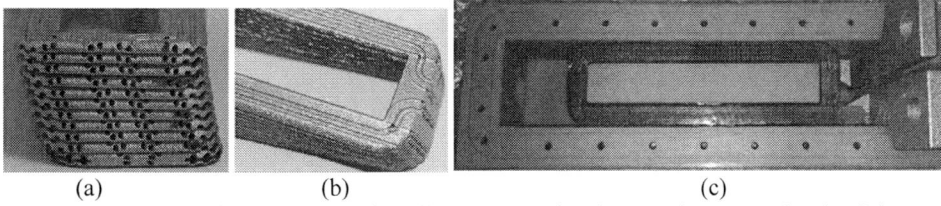

(a)   (b)   (c)

Fig. 11. 3D printed coil for submerged cooling a. Front view b. Top view c. Insulated coil in a coolant tank [14]

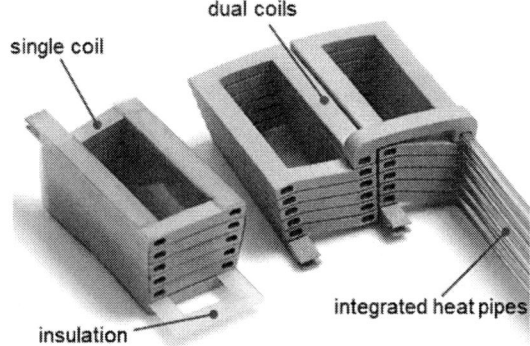

Fig. 12. AlSi10Mg coils integrated with heat pipes [15]

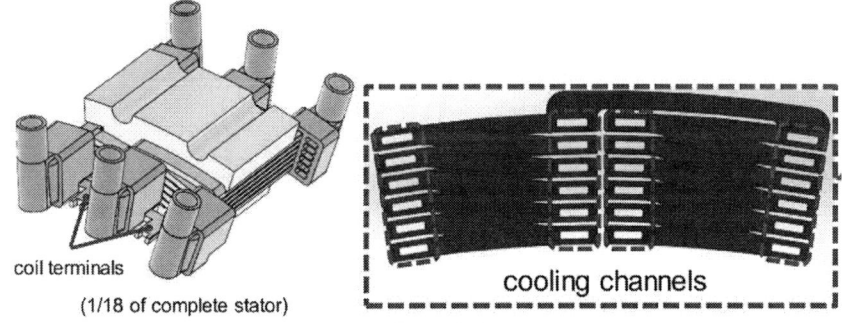

Fig. 13: Hollow AM conductors with cooling channels [16]

An even more integrated design along the same lines eliminates the heat pipes [16]. In this case, the coolant flows inside the hollow conductors, printed with AlSi10Mg, as shown in Fig. 13. The conductors are used for an aerospace SPM machine with a current density of 20 A_{rms}/mm^2 and a machine-specific power of 20.17 kW/kg.

Proposed 3D printed heat exchanger with PCB winding

The first proposed heat exchanger and winding combination utilizes PCBs and 3D printed heat exchangers to achieve a high current density. PCBs have flat conductors with large surface areas, which are extremely efficient in heat dissipation. The advantages of using a PCB-based winding are the possibility of a modular system, streamlining mass production, and reduced cost. However, the disadvantage is that the copper thickness is small compared to the substrate thickness. The PCB substrate does not play any advantageous role in electromagnetic or thermal performance. The proposed PCB winding with HX is shown in Fig. 14a. This topology is utilized for a 12-slot, 10-pole SPM machine with a double-layer concentrated winding. The slot winding area is square to make the insertion of fully assembled winding modules from the airgap side. The finned HX can be slid in from the axial side once the PCB winding is in place. Thermally conductive epoxy is used between the copper layer and the HX fin to form a good thermal contact for maximum heat transfer, as shown in green in Fig. 14b. The presented version, as shown in Fig. 14, shows 16 turns per winding.

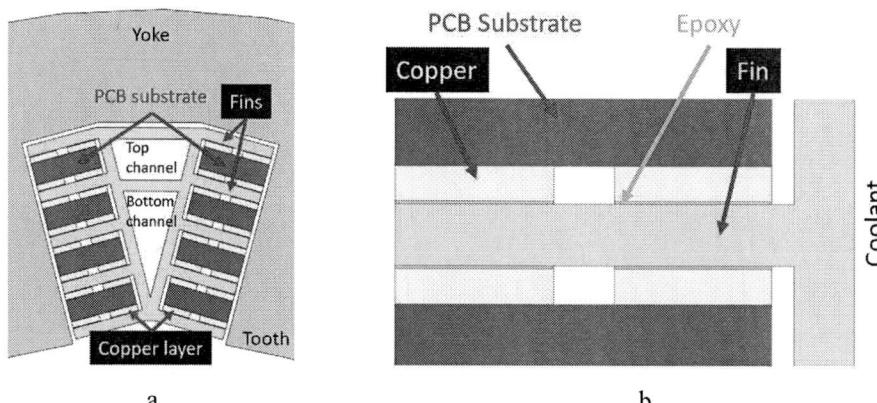

a   b

Fig. 14. a. Proposed in-slot heat exchanger b. Zoomed in PCB winding with heat exchanger

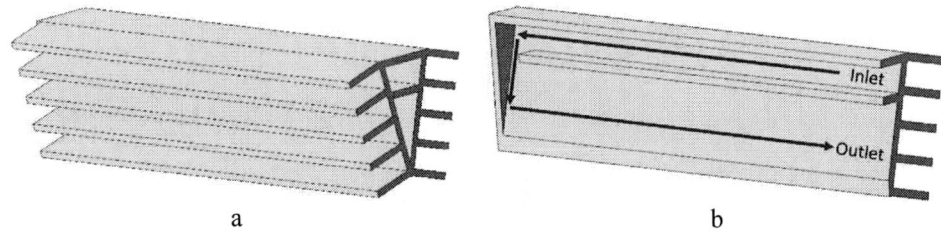

a   b

Fig. 15. a. Proposed finned HX b. Cross-section and coolant path

The HX is a triangular duct with fins extending from the walls. The duct has a water-ethylene glycol mixture flowing inside it with a return feature, as shown in Fig. 15. The return feature makes it possible to have the coolant inlet and outlet on the same side of the machine. The whole assembly (duct and fins) is designed to be 3D-printed from a thermally conductive material. Currently, two materials are under consideration. The first one is a thermoplastic polyurethane (TPU) filament with thermal conductivity of 6 W/m-K and a maximum operating temperature of 110°C, while the other is a thermally conductive nylon filament with thermal conductivity of 4 W/m-K and maximum operating temperature of 200°C. The PCB substrate has a maximum temperature limit of 170°C. The slot component parameters are given in Table I [17].

For TPU, the conductor's maximum steady-state current density is 92.9 A_{rms}/mm^2, while the slot current density is 19.8 A_{rms}/mm^2 at a hotspot temperature of 110°C. For nylon, the conductor current density is 138.5 A_{rms}/mm^2, while the slot current density is 29.5 A_{rms}/mm^2 at a hotspot temperature of 170°C, limited by the PCB substrate. The slot current densities are calculated by excluding the area occupied by the cooling channel. The higher current density for the lower thermal conductivity material is due to the higher operating temperature.

Table I. Slot and winding parameter values for PCB winding

Parameter	Value
Copper thickness	13 Oz. (0.455 mm)
PCB substrate thickness	1.6 mm
Thermal epoxy thickness	0.4 mm
Fin thickness	0.8 mm
Coolant temperature	70 °C
Heat transfer coefficient	14000 W/m²-K
Pressure drop (one channel)	150 Pa

Proposed 3D printed heat exchanger with flat wire winding

This version of the proposed heat exchanger utilizes a flat copper wire, and rectangular 3D printed heat exchangers to achieve a high current density. The advantages of this type of winding are that there is no limit on copper thickness, no substrate, and a higher fill factor with a higher number of turns can be achieved. The proposed winding with HX is shown in Fig. 16a. This topology is also utilized for a 12-slot, 10-pole SPM machine with a single-layer concentrated winding. The HX and winding are assembled outside the stator and form a module inserted in the slots from the air gap side, forcing the slots to be square. Thermally conductive epoxy is used between the copper layer and the HX, just like in the first case, to form a good thermal contact for maximum heat transfer, as shown in green in Fig. 16b. The current version, as shown in Fig. 16, shows 32 turns per winding module (in two layers).

The HX, in this case, is a rectangular duct with a water-ethylene glycol mixture flowing inside it. The return feature is formed by connecting ducts from two adjacent slots, as shown in Fig. 17, to keep the inlet and outlet on the same side of the machine. The duct is designed to be 3D-printed from the same two thermally conductive materials proposed in the first case. The slot component parameters are given in Table II.

For TPU, the conductor's maximum steady-state current density is 50.5 A_{rms}/mm^2, while the slot current density is 25.6 A_{rms}/mm^2 at a hotspot temperature of 110°C. For nylon, the conductor current density is 75.9 A_{rms}/mm^2, while the slot current density is 38.5 A_{rms}/mm^2 at a hotspot temperature of 190°C. The higher slot current densities, despite the lower conductor current densities, are attributed to higher slot fill factors achieved by using flat wire instead of PCB.

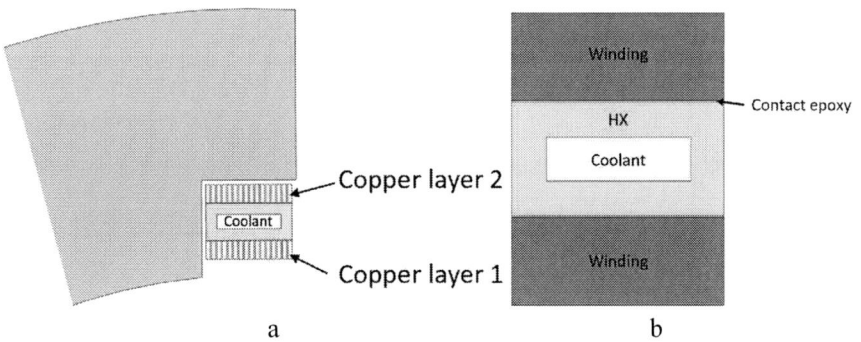

Fig. 16. a. Proposed in-slot heat exchanger with flat wire winding b. Zoomed in winding with heat exchanger

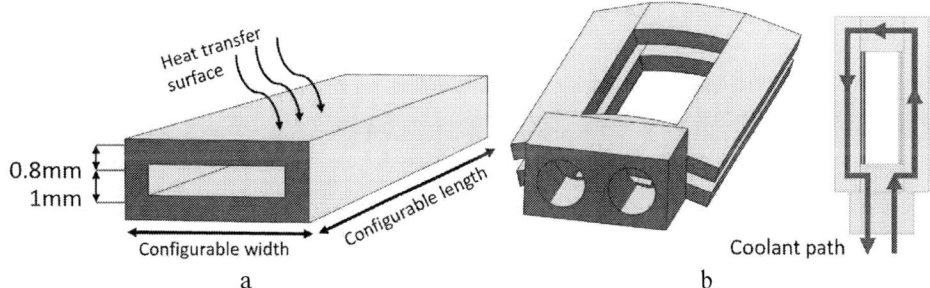

Fig. 17. a. Proposed rectangular HX, b. Coolant path

Table II. Slot and winding parameter values for flat wire winding

Parameter	Value
Copper thickness	1.4 mm
Thermal epoxy thickness	0.2 mm
HX wall thickness	0.8 mm
HX coolant channel height	1 mm
Coolant temperature	70 °C
Heat transfer coefficient	6000 W/m^2-K

Discussion

For discussion purposes, a number of papers are selected for both traditional and AM-based techniques in Table III to create a contrast between the solutions. The critical parameters such as current density, fill factor, or hotspot temperature can give an idea about the performance comparison of the selected in-slot cooling methods. It should be noted that the fill factor definition for different publications may be different.

Compared with liquid-cooled heat exchangers, the heat pipes can provide a more cost-effective solution because of less complexity and hardware requirement. However, they may not be ideal for high-power applications due to their limited cooling capacity. On the other hand, more aggressive cooling can be achieved with liquid cooling using heat exchangers. The flow rate is one of the parameters for heat exchangers that can be adjusted according to the cooling requirements of the machine. Compared to previous solutions, forced air cooling through slots provides less complexity with comparable current density ratings. However, reliability issues arise due to airflow leakage, especially at high-speed operations.

The additively manufactured solutions enable more complex cooling system designs to boost the cooling performance. Using AM technology, both metal and non-metal cooling solutions can be manufactured. The advantage of non-metal solutions, such as the epoxy design in [10], is that the induced eddy current losses on the material due to slot flux leakages can be eliminated. Additive manufacturing technology developments also allow printing materials with various properties like thermal conductivity. From the cost perspective, additive manufacturing can be more expensive compared to traditional techniques. However, with the developments in this emerging field, manufacturing costs are also decreasing.

Table III: Comparison of the in-slot cooling solutions by traditional and AM methods

	Traditional technique				Additive manufacturing				
Reference	[1]	[2]	[5]	[9]	This work	[10]	[12]	[13]	[16]
Cooling method	Heat exc.	Heat exc.	Heat pipe	Forced air	Heat exc.	Heat exc.	Heat exc.	Heat guide	Hollow cond.
Material	Cu	Epoxy	Cu	-	Nylon	Epoxy	Alum.	AlSi10 Mg	AlSi10 Mg
Coolant	Water	Oil	Liquid gas	Air	Water glycol	Water glycol	Water	-	Liquid
Current density [A$_{rms}$/mm^2]	25	25	24	21	75.9	20	35	-	20
Fill factor	0.40	0.32	0.38	-	0.51	0.55	0.28	-	-
Hotspot temp. [°C]	155	180	180	180	190	85	200	180	110
Inlet temp. [°C]	30	60	60	25	70	50	30	60	85

Conclusion

In-slot cooling methods are shown to be effective in cooling the stator winding conductors. Both traditional and additively manufacturing methods provide superior cooling performance. It is shown that the slot current density of the machine with in-slot cooling can be boosted to over 30 A_{rms}/mm^2, which allows for a higher power density for the machine. When combined with proposed in-slot heat exchanger cooling, PCB windings are shown that they can achieve superior slot current densities up to 20 A_{rms}/mm^2. Therefore, the proposed solution with PCB windings needs more attention for high power density and mass production of electric machines.

References

[1] S. A. Semidey and J. R. Mayor, "Experimentation of an electric machine technology demonstrator incorporating direct winding heat exchangers," *IEEE Transactions on Industrial Electronics*, vol. 61, no. 10, pp. 5771–5778, Oct. 2014.

[2] A. Acquaviva, S. Skoog, and T. Thiringer, "Design and verification of in-slot oil-cooled tooth coil winding PM machine for traction application," *IEEE Transactions on Industrial Electronics*, vol. 68, no. 5, pp. 3719–3727, May 2021.

[3] A. Reinap, F. J. Marquez-Fernandez, M. Alakula, R. Deodhar, and K. Mishima, "Direct conductor cooling in concentrated windings," in *XIII International Conference on Electrical Machines (ICEM)*, 2018, pp. 2654–2660.

[4] R. Wrobel and R. J. McGlen, "Opportunities and challenges of employing heat-pipes in thermal management of electrical machines," in *International Conference on Electrical Machines (ICEM)*, 2020, pp. 961–967.

[5] R. Wrobel and D. Reay, "A feasibility study of heat pipes for thermal management of electrical machines," in *IEEE Energy Conversion Congress and Exposition (ECCE)*, 2020, pp. 4230–4237.

[6] C. Dong, Y. Qian, Y. Zhang, X. Hu, and W. Zhuge, "Coupled thermal-electromagnetic parametric modeling of permanent magnet machine based on flat heat pipe cooling," in *23rd International Conference on Electrical Machines and Systems (ICEMS)*, 2020, pp. 1689–1694.

[7] M. Galea, C. Gerada, T. Raminosoa, and P. Wheeler, "A thermal improvement technique for the phase windings of electrical machines," *IEEE Transactions on Industry Applications*, vol. 48, no. 1, pp. 79–87, Jan. 2012.

[8] M. Groninger, F. Horch, A. Kock, M. Jakob, and B. Ponick, "Cast coils for electrical machines and their application in automotive and industrial drive systems," in *4th International Electric Drives Production Conference (EDPC)*, 2014, pp. 1–7.

[9] A. Reinap, F. J. Marquez-Fernandez, R. Andersson, C. Hogmark, M. Alakula, and A. Goransson, "Heat transfer analysis of a traction machine with directly cooled laminated windings," in *4th International Electric Drives Production Conference (EDPC)*, 2014, pp. 1–7.

[10] M. Schiefer and M. Doppelbauer, "Indirect slot cooling for high-power-density machines with concentrated winding," in *IEEE International Electric Machines & Drives Conference (IEMDC)*, 2015, pp. 1820–1825.

[11] W. Sixel, M. Liu, G. Nellis, and B. Sarlioglu, "Cooling of windings in electric machines via 3-D printed heat exchanger," *IEEE Transactions on Industry Applications*, vol. 56, no. 5, pp. 4718–4726, Sep. 2020.

[12] W. Sixel, M. Liu, G. Nellis, and B. Sarlioglu, "Ceramic 3-D printed direct winding heat exchangers for thermal management of concentrated winding electric machines," *IEEE Transactions on Industry Applications*, vol. 57, no. 6, pp. 5829–5840, Nov. 2021.

[13] R. Wrobel and A. Hussein, "A feasibility study of additively manufactured heat guides for enhanced heat transfer in electrical machines," *IEEE Transactions on Industry Applications*, vol. 56, no. 1, pp. 205–215, Jan. 2020.

[14] C. Wohlers, P. Juris, S. Kabelac, and B. Ponick, "Design and direct liquid cooling of tooth-coil windings," *Electrical Engineering*, vol. 100, no. 4, pp. 2299–2308, Dec. 2018.

[15] F. Wu and A. M. EL-Refaie, "Additively manufactured hollow conductors integrated with heat pipes: design tradeoffs and hardware demonstration," in *International Conference on Electrical Machines (ICEM)*, 2020, pp. 52–58.

[16] F. Wu and A. M. EL-Refaie, "Additively manufactured hollow conductors with integrated cooling for high specific power electrical machines," in *International Conference on Electrical Machines (ICEM)*, 2020, pp. 1497–1503.

[17] A. Hembel and B. Sarlioglu, "PCB winding for electric machines with integrated 3D printed heat exchanger," in *International Transportation Electrification Conference (ITEC)*, Anaheim, CA, USA, 2022.

Comparison of High-Power 2-Level and 3-Level Converters in Terms of Power Density, Costs and Performance

Ludwig Schlegel [1), 2)]
+49 (0)351 32 33 05 26
schlegel@powerelectronics.de or
ludwig.schlegel@tu-dresden.de
[1)] M&P MOTION CONTROL AND
POWER ELECTRONICS GMBH (M&P)
Dresden, Germany
www.powerelectronics.de

Prof. Dr.-Ing. Wilfried Hofmann [2)]
+49 (0)351 463-37634
wilfried.hofmann@tu-dresden.de
[2)] CHAIR OF ELECTRICAL MACHINES
AND DRIVES, ELEKTROTECHNISCHES
INSTITUT, TU DRESDEN,
Dresden, Germany
www.tu-dresden.de/ing/elektrotechnik/eti/ema

Acknowledgments

Thanks to Forschungsnetzwerk Mittelstand, Bundesministerium für Wirtschaft und Energie, Germany for the generous support.

Keywords

≪Multi-level converters≫, ≪High power density systems≫, ≪Analytical losses computation≫, ≪DC-AC converter≫, ≪AC-DC converter≫

Abstract

This paper compares high-power converters in terms of their power density, costs and performance. 2-level and 3-level NPC phase modules with 1400 A semiconductors are compared. For pulse frequencies higher than 6.75 kHz 3-level NPC converter is twice smaller and 48 % cheaper then 2-level converter, because for 2-level parallel operation of phase modules is necessary. In this case the 3-level NPC has with $0.58\,\mathrm{kVA/dm^3}$ a twice higher power density as the 2-level converter. The performance of 3-level NPC is much higher than of the 2-level, because of the twice higher possible pulse frequency of 16 kHz and therefore higher possible control frequency. Because of the higher pulse frequency and the lower voltage ripple the chokes can be quite smaller for the 3-level NPC and so the maximum current raise is a lot higher than in case of a 2-level converter.

Introduction

The 3-level neutral point clamped converter (3LNPCC) [1] has great advantages over the classical 2-level converter (2LC) in the operation of permanent magnet synchronous machines, which is shown by latest investigations [2], [3]. Depending on the operating point and pulse frequency, the use of the 3LNPCC can significantly reduce losses in the machine (20-30 %) and improve efficiency by 2-3 percentage points [2], [3]. The 3LNPCC also has advantages over the 2LC when used at the grid as an active frond end. For example, due to the lower voltage ripple, smaller AC filter capacitors and chokes can be used [4]. This reduces the volume and cost of the setup [5]. To find out how the use of the 3LNPC affects the overall drive in terms of power density, costs and performance, the losses of the 2LC and 3LNPCC are calculated here. Based on this, the junction temperature of the semiconductors and thus the maximum current carrying capacity of the converter can be determined.

The goal of this paper is to compare grid connected 2LC and 3LNPCC with over 500 kW power. In order to keep the weight of the individual components within limits (less than 40 kg because of occupational

EPE'22 ECCE Europe

safety), the power converters are constructed from individual phase modules. The aim is to use the same semiconductors in order to be able to compare the results better. For the power class, the use of 1400 A PrimePack3 semiconductor modules from Infineon (FF1400R12IP4) [6] is suitable. The Semitrans 10 modules (SKM1400GB12P4) [7] can be used as a one-to-one replacement if the Infineon modules are not available.

The 3LNPCC [1] is the most used 3- or multilevel converter [8, 9]. The mass of 21 kg from the 2-level (2L) phase module increases to 35 kg for the 3LNPC phase module. So a 5-level converter in the targeted power class with two more semiconductor modules and twice as much DC link capacitors as 3LNPCC would be too large and too heavy. In addition, the complexity of controlling so many semiconductors increases significantly.

Besides the classical 3LNPCC, there is the 3-level active NPC converter (3LANPCC) [10]. The 3LANPCC has antiparallel IGBTs to the diodes $D_{5..6}$ of 3LNPCC (compare Fig. 1b). So the connection to the neutral point can be controlled actively. The 3LANPCC is characterized by a slightly better loss distribution to the individual semiconductors. However, the effect is not very large and strongly dependent on the operating point [11]. The 3LANPCC could be constructed from PrimePack3 semiconductors, as planned. However, the 3LANPCC requires two more IGBTs per phase module than the 3LNPCC and is correspondingly more difficult in terms of control and operation. This makes the 3LANPCC much more complex than the 3LNPCC and thus unattractive for the current application.

Another alternative is the 3-level T-type NPC converter (3LTNPCC) [4]. This is built like a 2-level converter, but still has 2 bidirectionally arranged IGBTs with antiparallel diodes from the DC link center to the AC output. For small switching frequencies, the 3LTNPCC has slightly lower losses than the 3LNPCC [4]. However, 3LTNPCC cannot be constructed from PrimePack3 modules only, since bidirectionally arranged semiconductors are required. Accordingly, the 3LTNPCC cannot be considered in the current comparison. Otherwise, the control and the output voltage of 3LNPCC and 3LTNPCC are identical, as our simulations and measurements have shown.

The 3LTNPCC can very easily be configured to operate as a 2LC. To do this, simply deactivate the bidirectionally arranged IGBTs at the middle point. This effect was used in [2] to compare the effects of the 3-level converter on the load with those of the 2LC. At some operating points, it may also be beneficial to switch between the topologies in real operation [12]. However, the impact is relatively small and highly dependent on the operating point.

Besides the used converter topology, many other points have influence on the losses of the converter. For example the gate driver [6], [7], the modulation method [11] and the DC link balancing [13]. Compared to the converter topology, however, their influence is small and is therefore not considered further here.

An important aspect for comparing different converter topologies is their lifetime expectation. Felgemacher [14] clearly shows that this is very complex and it is not the subject of this paper. Just capturing and categorizing load cycles is a complex issue [15].

In recent years, many silicon carbide (SiC) semiconductors have been introduced. Also 3LNPCC with SiC semiconductors have been examined [16]. However, there are still only SiC modules with a few hundred amps available on the market. So SiC semiconductors cannot be considered in this paper.

There are already some papers comparing different power converter topologies [4, 8, 9, 10]. TEICHMANN [5] compares 2LC and 3LNPCC and shows that above 5 kHz pulse frequency the 3LNPCC achieves better efficiency than the 2LC. The paper also discusses costs of many individual components such as terminal filters, DC link capacitance, IGBT modules, gate drivers and cooling.

This paper is the first to compare the complete costs required to build the different converter topologies. This includes costs for mechanics, DC link, busbars, IGBT modules, cooling, chokes and the time of the assembly itself. In addition, the integration into a control cabinet and thus into an overall system is considered. Taking all the components is the only way to determine a realistic power density.

The introduction is followed by a general overview of the differences between 2LC and 3LNPCC. After that, the calculation of losses and junction temperatures is discussed. The paper ends with the evaluation of costs and power density, as well as a conclusion.

Differences Between 2LC and 3LNPCC

A phase module of a 2LC consists of the DC link capacitance C_1, the two transistors ($T_{1..2}$) and the two diodes ($D_{1..2}$), as shown in Fig. 1a. Via the power semiconductors it is possible to connect the positive DC link potential $DC+$ or the negative $DC-$ to the output. So there are two possible voltage levels. The 3-level NPC (3LNPC phase module [1] has two DC link halves ($C_{1/2}$), four transistors ($T_{1..4}$) and six diodes ($D_{1..6}$), as seen in Fig. 1b. With diodes D_5 and D_6 it is possible to connect the output (AC) to the DC link center N. So it is possible to connect three voltage levels to the output ($DC+$, N or $DC-$). The second investigated phase module LT1000-ML (3LNPC) consists of three IGBT modules. Which semiconductors belong to which IGBT modules is shown by the gray circles $M_{1..3}$ in Fig. 1b. In case of M_1 and M_3 one IGBT is deactivated.

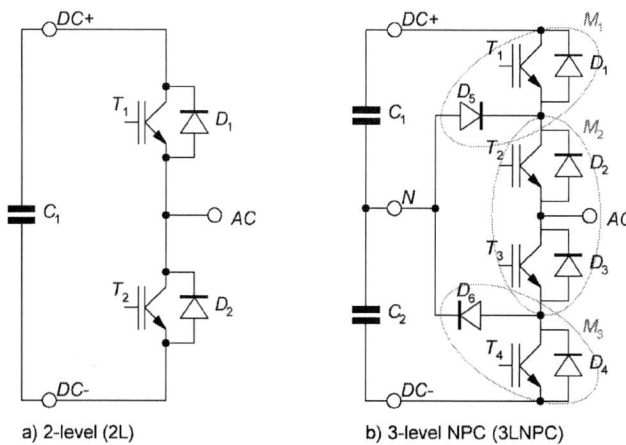

a) 2-level (2L)   b) 3-level NPC (3LNPC)

Fig. 1: 2L and 3LNPC phase modules with DC link capacitors and all there semiconductors

Three phase modules (2L or 3LNPC) can be used to assemble rectifiers for operation on three-phase mains or inverters for operation of three-phase electrical machines. If the usage of one phase module per converter phase results in too high losses and thus too high junction temperatures, two or three phase modules can also be connected in parallel. Both investigated phase modules use Infineon PrimePack3 (FF1400R12IP4) modules and are shown in Fig. 2.

The power semiconductors in the 2LC must be able to switch and block the entire DC link voltage. In the 3LNPCC, on the other hand, it is only half the DC link voltage. Accordingly, 3LNPCCs with 1200 V

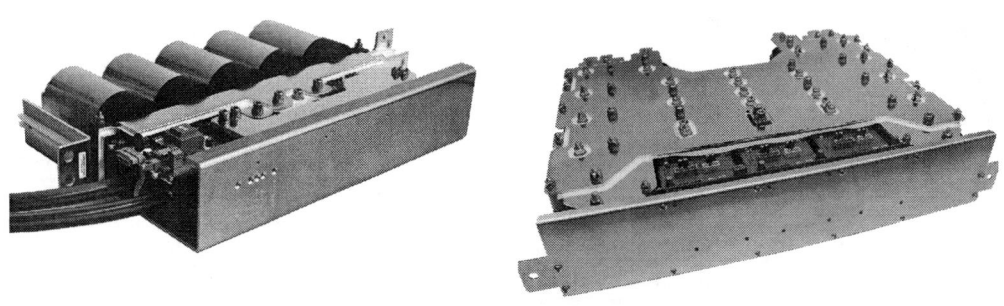

a) 2-level (LT1000)   b) 3-level (LT1000-ML)

Fig. 2: Investigated M&P phase modules LT1000 [17] and LT1000-ML [18]

semiconductors can drive twice as high voltages as 2LCs. In 2LCs usually, the switching frequency of the semiconductors is equal to the pulse frequency at the output. With 3LNPCCs, the effective switching frequency is only half the pulse frequency, since all semiconductors are never switched simultaneously. Depending on the phase angle of current and voltage, as well as the modulation level, only switches T_1 and T_3 or T_2 and T_4 are switched. Further differences between the converter topologies and the used phase modules are shown in Table I.

Table I: Comparison of 2L and 3LNPC Phase Modules Using the Example of LT1000 [17] and LT1000-ML [18]

	2-level (2L)	**3-level NPC (3LNPC)**
Semiconductors per phase module	2 IGBTs, 2 Diodes	4 IGBTs, 6 Diodes
DC link voltage with 1200 V IGBTs	750 V (typ.)	1400 V (typ.)
Price per phase module	100 %	156 %
Max. pulse frequency $f_{\mathrm{p\,max}}$	8 kHz	16 kHz
Max. switching frequency $f_{\mathrm{sw\,max}}$	8 kHz	8 kHz
Volume per phase module	23 dm³	42 dm³
Mass per phase module	21 kg	35 kg
Phase modules per control cabinet	3 (typ.)	3 (typ.)
Cooling type and cooling capacity	Water, 4 kW	Water, 6 kW

How the phase modules look like integrated in the cabinet can be seen schematically in Fig. 3. The AC filter inductance can be realized like shown as one 3-phase inductance, but also three single inductances are possible.

Loss and Junction Temperatures Calculation of High Power Converters

In general, there are three main types of losses in power converters: switching, conduction and driver losses. The driver losses can be neglected for high-power IGBT converters, since they are less than 1 % of the total losses. Losses are influenced by many factors. Some of them, such as DC link voltage, power factor, modulation depth and output current, can be well described with analytical formulas. The determination of other factors, such as driver design (gate dropping resistor) and junction temperature,

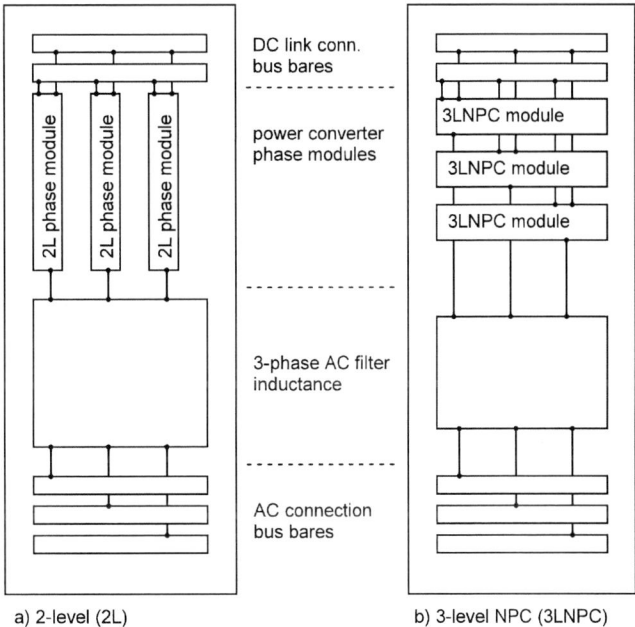

Fig. 3: 2L and 3LNPC phase modules (LT1000 [17] and LT1000-ML [18]) schematically integrated in cabinet with cables, AC filter inductance and connection bus bars

is more difficult. To investigate the influence of the drivers, measurements of turn-on and turn-off loss energy from the LT1000 were made in [19]. These measurements show that the results are comparable to those from the data sheet. The measured values are taken here accordingly. For the junction temperature, 120 °C is assumed. In this paper, 120 °C is also the maximum permissible value for all investigations.

The total losses of the phase module P_{pm} and the junction temperatures $T_{jD/T}$ of the semiconductors are calculated for a fixed operating point, various pulse frequencies f_p and different numbers of phase modules connected in parallel.

The modulation index $M = \hat{V}_{out1}/(V_{DC}/2)$ is defined using the amplitude of the fundamental harmonic of the phase voltage against the neutral point $\hat{V}_{out1}$ and the DC link voltage V_{DC}. All values that are needed for the calculations of the semiconductors are found in the datasheet [6] and are shown in Table II.

Table II: Values Needed for Calculations from the PrimePack3 (FF1400R12IP4) datasheet

Name	Value	Unit
Reference voltage V_{ref}	600	V
Collector-emitter supply voltage V_{CC} - 2-Level: $V_{CC} = V_{DC}$	750	V
Collector-emitter supply voltage V_{CC} - 3-Level: $V_{CC} = V_{DC}/2$	375	V
Reference current I_{ref}	1400	A
Thermal resistance heat sink to ambient per PrimePack3 R_{ths-a}	10	K/kW
Collector-emitter threshold voltage (IGBT) V_{CE0}	0.85	V
On-state slope resistance (IGBT) r_{CE}	0.944	mΩ
Energy dissipation during turn-on and off (IGBT) E_{on+off} [19]	420	mJ
Thermal resistance junction to case to heat sink (IGBT) $R_{thj-c-sT}$	28.8	K/kW
Forward threshold voltage (diode) V_{F0}	1.85	V
Forward slope resistance (diode) r_F	0.5	mΩ
Energy dissipation during reverse recovery (diode) E_{rr} [19]	110	mJ
Thermal resistance junction to case to heat sink (diode) $R_{thj-c-sD}$	53	K/kW

The later used formulas to calculate the losses are only valid for PWM schemes with constant carrier frequency (= pulse frequency), pure sinusoidal reference and $0 \leq \phi \leq \pi$ [10]. In case of Third Harmonic Injection Sinusoidal PWM there are a bit different formulas, but the results are nearly the same as with the given formulas [10].

The switching losses do not increase ideally linearly with the phase current and the DC link voltage. Thus, for the terms I_{out}/I_{ref} and V_{CC}/V_{ref} in the later shown formulas the additional exponents K_I and K_V are introduced [20] - [22], resulting in $(I_{out}/I_{ref})^{K_I}$ and $(V_{CC}/V_{ref})^{K_V}$. For the 3LNPCC, the switching losses for the diodes are further multiplied by an additional factor $G_I = 1.15$. For Semicron modules, $K_I = K_V = 0.6$ for the diodes and $K_I = 1$ and $K_V = 1.4$ for the IGBTs. Each of the factors results in differences of 10 ..15 %. However, the errors in current and voltage dependent switching losses respectively for IGBTs and diodes largely cancel each other out. Thus, the error related to the total losses is significantly smaller than 10 % and can therefore be left out.

Analytical Loss Calculation for the 2-level Converter

In case of a symmetrical three phase load all six IGBTs and diodes of a 2LC have the same losses. To calculate the losses modulation index M, the values from Table II and the amplitude of the fundamental harmonic $\hat{I}_{out1}$ of the phase current I_{out} are used. The conduction and switching losses of the IGBTs (P_{condT} and P_{swT}) and diodes (P_{condD} and P_{swD}) are determined with the help of the following formulas [20, 21]:

$$P_{condT} = \left(\frac{1}{2\pi} + \frac{M\cos\varphi}{8} \right) V_{CE0} \hat{I}_{out1} + \left(\frac{1}{8} + \frac{M\cos\varphi}{3\pi} \right) r_{CE} \hat{I}_{out1}^2 \tag{1}$$

$$P_{swT} = f_{sw} E_{on+off} \frac{\sqrt{2}}{\pi} \frac{I_{out}}{I_{ref}} \frac{V_{CC}}{V_{ref}} \tag{2}$$

$$P_{\text{cond}D} = \left(\frac{1}{2\pi} - \frac{M\cos\varphi}{8} \right) V_{\text{F0}} \hat{I}_{\text{out}1} + \left(\frac{1}{8} - \frac{M\cos\varphi}{3\pi} \right) r_{\text{F}} \hat{I}_{\text{out}1}^2 \tag{3}$$

$$P_{\text{sw}D} = f_{\text{sw}} E_{\text{rr}} \frac{\sqrt{2}}{\pi} \frac{I_{\text{out}}}{I_{\text{ref}}} \frac{V_{\text{CC}}}{V_{\text{ref}}} \tag{4}$$

To obtain the losses of the whole phase module $P_{\text{pm}2}$, the individual losses must be added together

$$P_{\text{pm}2} = 2 \left(P_{\text{cond}T} + P_{\text{sw}T} + P_{\text{cond}D} + P_{\text{sw}D} \right). \tag{5}$$

Analytical Loss Calculation for 3-level NPC Converter

For the loss calculation of the 3LNPCC the same input values are used as for the 2LC. In the case of the 3LNPCC the equations differ for the outer IGBTs $T_{1/4}$ and the inner IGBTs $T_{2/3}$ and also for the diodes $D_{1/4}$, $D_{2/3}$ and $D_{5/6}$ [11]. So the following ten equations are needed to get all the conduction and switching losses [11, 22]:

$$P_{\text{cond}T1/4} = \frac{M\hat{I}_{\text{out}1}}{12\pi} \left(3V_{\text{CE0}} \left[(\pi - \varphi)\cos(\varphi) + \sin(\varphi) \right] + 2r_{\text{CE}}\hat{I}_{\text{out}1} \left[1 + \cos(\varphi) \right]^2 \right) \tag{6}$$

$$P_{\text{sw}T1/4} = f_{\text{sw}} E_{\text{on+off}} \frac{1}{2\pi} (1 + \cos(\varphi)) \frac{I_{\text{out}}}{I_{\text{ref}}} \frac{V_{\text{CC}}}{V_{\text{ref}}} \tag{7}$$

$$P_{\text{cond}T2/3} = \frac{\hat{I}_{\text{out}1}}{12\pi} \left(V_{\text{CE0}} \left[12 + 3M(\varphi\cos(\varphi) - \sin(\varphi)) \right] + r_{\text{CE}}\hat{I}_{\text{out}1} \left[3\pi - 2M(1 - \cos(\varphi))^2 \right] \right) \tag{8}$$

$$P_{\text{sw}T2/3} = f_{\text{sw}} E_{\text{on+off}} \frac{1}{2\pi} (1 - \cos(\varphi)) \frac{I_{\text{out}}}{I_{\text{ref}}} \frac{V_{\text{CC}}}{V_{\text{ref}}} \tag{9}$$

$$P_{\text{cond}D1/4} = \frac{M\hat{I}_{\text{out}1}}{12\pi} \left(3V_{\text{F0}} \left[-\varphi\cos(\varphi) + \sin(\varphi) \right] + 2r_{\text{F}}\hat{I}_{\text{out}1} \left[1 - \cos(\varphi) \right]^2 \right) \tag{10}$$

$$P_{\text{sw}D1/4} = f_{\text{sw}} E_{\text{rr}} \frac{1}{2\pi} (1 - \cos(\varphi)) \frac{I_{\text{out}}}{I_{\text{ref}}} \frac{V_{\text{CC}}}{V_{\text{ref}}} \tag{11}$$

$$P_{\text{cond}D2/3} = \frac{M\hat{I}_{\text{out}1}}{12\pi} \left(3V_{\text{F0}} \left[-\varphi\cos(\varphi) + \sin(\varphi) \right] + 2r_{\text{F}}\hat{I}_{\text{out}1} \left[1 - \cos(\varphi) \right]^2 \right) \tag{12}$$

$$P_{\text{sw}D2/3} = 0 \tag{13}$$

$$P_{\text{cond}D5/6} = \frac{\hat{I}_{\text{out}1}}{12\pi} \left(V_{\text{F0}} \left[12 + 3M((2\varphi - \pi)\cos(\varphi) - 2\sin(\varphi)) \right] + \right.$$
$$\left. r_{\text{F}}\hat{I}_{\text{out}1} \left[3\pi - 4M(1 + \cos^2(\varphi)) \right] \right) \tag{14}$$

$$P_{\text{sw}D5/6} = f_{\text{sw}} E_{\text{rr}} \frac{1}{2\pi} (1 + \cos(\varphi)) \frac{I_{\text{out}}}{I_{\text{ref}}} \frac{V_{\text{CC}}}{V_{\text{ref}}} \tag{15}$$

To get the losses $P_{\text{pm}3}$ of the whole 3LNPC phase module, the individual losses must be added together:

$$P_{\text{pm}3} = 2 \left(P_{\text{cond}T1/4} + P_{\text{sw}T1/4} + P_{\text{cond}T2/3} + P_{\text{sw}T2/3} + P_{\text{cond}D1/4} + P_{\text{sw}D1/4} + P_{\text{cond}D2/3} + \right.$$
$$\left. P_{\text{sw}D2/3} + P_{\text{cond}D5/6} + P_{\text{sw}D5/6} \right) \tag{16}$$

Calculation of Junction Temperatures

The junction temperature can be calculated from the losses of the semiconductor S by $P_{\text{S}} = P_{\text{cond}S} + P_{\text{sw}S}$ via the sum of thermal resistances and the ambient (water) temperature T_{a}. One PrimePack3 consists of four semiconductors (two diodes and two IGBTs). The thermal resistance $R_{\text{ths}-\text{a}}$ per semiconductor $R_{\text{ths}-\text{aS}}$ is $R_{\text{ths}-\text{aS}} = 4R_{\text{ths}-\text{a}}$. Finally, the junction temperature of the semiconductor S results from the equation $T_{\text{jS}} = T_{\text{a}} + P_{\text{S}} \left(R_{\text{ths}-\text{aS}} + R_{\text{thj}-\text{c}-\text{sS}} \right)$ using the thermal resistance $R_{\text{thj}-\text{c}-\text{s}}$ between junction, case and sink.

Cost Evaluation and Calculation Results

The costs for the individual components depend heavily on how much raw materials (e.g. copper) and components (e.g. semiconductors and capacitors) currently cost. In order to be able to make a comparison that is as generally valid as possible, all costs are therefore given in relation to the costs for three 2L phase modules (LT1000 [17]). This base cost equals 100 % and includes the costs for semiconductors, DC link capacitance, bus bars, housing, control electronics, time for assembly, as well as current and voltage measurement. Based on experience, the cost of construction and materials (water cooling tubes, bus bars, mounting frame) is about 25 % of the base cost. The costs of the chokes (compare 3-phase AC filter inductance in Fig. 3) for the 2LC are about the same as the base cost. From experience, about 40 % of the cost of the chokes is fixed and the rest is variable cost depending on the size of the inductor. To achieve the same filtering effect, it is sufficient for the chokes of the 3LNPCC to have 70 % of the value of the 2LC [5]. Taking into account the fixed cost of 40 %, the costs is then 82 % ($= 70\% \cdot 60\% + 40\%$) of the base cost. When two 2L phase modules are connected in parallel, the current is halved and thus the choke required is also much smaller. The variable cost per choke is assumed to decrease by 50 %. Since the two control cabinets also require two chokes, the total costs for the chokes increases to 140 % ($= 2\,(60\%/2 + 40\%)$). Similarly, with three parallel 2L phase modules, the total costs increases to 180 %. The total costs of the systems is then the sum of the phase modules, the chokes, and the control cabinets. All data are summarized in Table III.

Table III: Costs and power density for the investigated configurations (p. m. - phase module)

Configurations	3LNPC, 1 p. m.	2L, 1 p. m.	2L, 2 p. m.	2L, 3 p. m.
Number of cabinets	1	1	2	3
Rel. costs for phase modules	156 %	100 %	200 %	300 %
Relative costs for cabinets	25 %	25 %	50 %	75 %
Relative costs for chokes	82 %	100 %	140 %	180 %
Relative total system costs	263 %	225 %	390 %	555 %
Costs compared to 2L, 1 p. m.	117 %	100 %	173 %	247 %
Volume of system V_{sys}	960 dm^3	960 dm^3	1920 dm^3	2880 dm^3
Power density p of system	0.58 kVA/dm^3	0.58 kVA/m^3	0.29 kVA/dm^3	0.19 kVA/dm^3

Costs for main switches, main contactors, fuses, and any necessary devices for rapid safety-related fault shutdown were not considered. This would be the same for all systems and would only equalize the system costs. Line voltage measurement, precharging and control play a subordinate role in terms of costs and were not taken into account. If a constant current ripple of e.g. 10 % of the rated current is to be achieved, different values for the chokes result depending on the pulse frequency. However, taking these into account here makes the comparison of the configurations much more difficult. Thus, values for the chokes were calculated independently of the pulse frequency.

A typical operation point for grid-connected converters (active front end) was used for the investigations:
- DC link voltage $V_{DC} = 750\,V$
- grid voltage $V_{AC} = 400\,V$
- output current $I_{out} = 800\,A$
- modulation index $M = 87.2\%$
- power factor $\cos(\varphi) = 0.93$
- water (ambient) temperature $T_a = 25\,°C$

To determine the power density $p = S/V_{sys}$ related to the volume, the apparent power $S = 554\,kVA$ and the volume of the system V_{sys} are required. The system volume results from the number of control cabinets and its volume $V_{cab} = 960\,dm^3$ ($= 20\,dm \cdot 8\,dm \cdot 6\,dm$). The different system volumes and the calculated power densities can be found in Table III.

Calculated losses per phase and maximum junction temperature can be seen in Fig. 4. For the configurations with parallel phase modules (2 m. and 3 m.) a maximum asymmetry of 10 % is considered, so the current of one module is 5 % higher and the other 5 % lower. In case of two modules the current is not

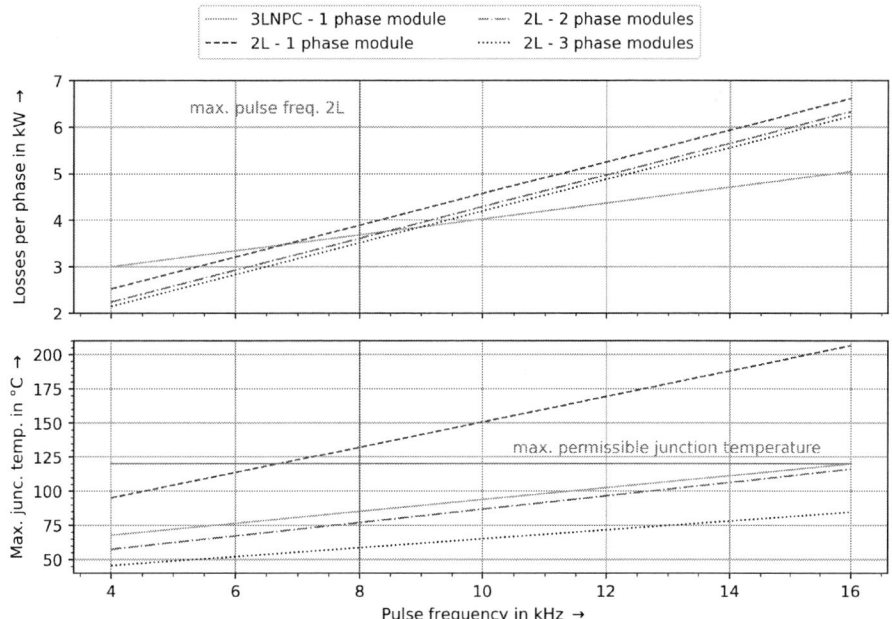

Fig. 4: Calculated results for one 3LNPC phase module and one, two or three of parallel 2L phase modules with different pulse frequencies, DC link voltage of 750 V and output current of 800 A

$I_{out}/2 = 400\,A$ but the worst case is 420 A. Up to a pulse frequency of 6.75 kHz, all configurations can be operated at the considered operating point, since the maximum junction temperature is below 120 °C.

The variant with the three parallel 2L phase modules has up to the 9 kHz pulse frequency the lowest total losses, but the costs are 2.47 times higher than the costs for 2L with one phase module and the power density is the lowest as seen in Table III.

To reach a pulse frequency of more than 6.75 kHz it is possible to use the converter with two parallel 2L phase modules. However, this solution is twice as large and 1.73 times more expensive than 2LC with one phase module. But the configurations with 2LC are limited by the gate driver to 8 kHz pulse frequency. The better option for higher switching frequencies is the 3LNPCC. At only 17 % extra cost and the same volume as the 2LC with one phase module, it offers the possibility to work with up to 16 kHz pulse frequency. If the pulse frequency equals the control frequency the 3LNPCC can react two times faster then the 2LC. Because of the smaller AC choke of the 3LNPCC a higher current rise can be achieved with the same driving voltage (DC link voltage). Due to the higher control frequency and the larger possible current rise, the performance of the 3-level is significantly better than that of the 2-level.

Conclusion

The losses and junction temperature calculations for 2-level (2L) and 3-level Neutral-Point-Clamped (3LNPC) converters with the Sinusoidal PWM, 800 A output current in a grid application are investigated. In addition, the costs for phase modules, control cabinets, chokes and their construction for various configurations were determined.

Up to a pulse frequency of 6.75 kHz it is possible to use one 2L phase module (LT1000 [17]) per converter phase for the given operating point. For higher pulse frequencies at least two 2L phase modules in parallel are necessary. But with two phase modules in parallel the system costs are 1.73 times higher. Also two cabinets instead of one cabinet are needed and the system is limited to 8 kHz pulse frequency.

For a slightly higher costs (17 %) compared to one 2L phase module, the configuration with 3LNPC phase modules (LT1000-ML [18]) provides by far the highest performance, cause of the high pulse frequency of up to 16 kHz. Furthermore the 3LNPC converter has the smallest volume and the highest power density of all investigated system configurations.

References

[1] Nabae, A.; Takahashi, I., Akagi, H..: A New Neutral-Point-Clamped PWM Inverter, IEEE Transactions on Industrial Applications, IA-17, Issue: 5, p. 518-523, 1981

[2] Schlegel, L., Knapp T., Hofmann W.: Comparison of Losses in Permanent Magnet Synchronous Machines fed with 2-level or 3-level-TNPC Converter (german: Vergleich der Verluste in permanentmagneterregten Synchronmaschinen gespeist durch 3-Level-TNPC- oder 2-Level-Stromrichter), 3. Freiberger Kolloquium Elektrische Antriebstechnik, p. 18-34, Freiberg (Germany), 2021

[3] Knapp T., Schlegel L., Hofmann W.: Comparison of Losses in Permanent Magnet Synchronous Machines fed with 2-level or 3-level-NPC Converter, PCIM, No. 94, Nuremberg (Germany), 2022

[4] Schweizer M., Friedli T., Kolar J. W.: Comparative Evaluation of Advanced Three-Phase Three-Level Inverter/Converter Topologies Against Two-Level Systems, IEEE Transactions on Industrial Electronics Vol. 60, No. 12, p. 5515-5527, 2013

[5] Teichmann R. and Bernet S.: A comparison of three-level converters versus two-level converters for low-voltage drives, traction, and utility applications, IEEE Transactions on Industrial Applications, Vol. 41, No. 3, pp. 855–865, 2005.

[6] Infineon: Technical Information FF1400R12IP4, datasheet, rev. 2.4, 2013

[7] Semikron: Technical Information SKM1400GB12P4, datasheet, rev. 5.0, 2020

[8] Franquelo, L. G.; Rodriguez, J.; Leon, J. I.; Kouro, S.; Portillo, R.; Prats, M. A.: The age of multilevel converters arrives, IEEE Industrial Electronics Magazine, Vol. 2, No. 2, p. 28-39, 2008

[9] Kashihara, Y.; Itoh, J.-i.: The performance of the multilevel converter topologies for PV inverter, 7th International Conference on Integrated Power Electronics Systems (CIPS), p. 1-6, Nuremberg (Germany), 2012

[10] Brückner, T.; Bernet, S.; Guldner, H.: The active NPC converter and its loss-balancing control, IEEE Transactions on Industrial Electronics, Vol. 52, Issue: 3, p. 855-868, 2005

[11] Brückner, T.: The active NPC converter for medium-voltage drives, thesis, TU-Dresden, 2006

[12] Kim, T.-H., Lee, W.-C.: Level Change Method for Higher Efficiency of a 3-Level T-Type Converter, 21st International Conference on Electrical Machines and Systems (ICEMS), p. 741-744, Jeju (Korea), 2018

[13] Choudhury, A.; Pillay, P.; Amar, M.; Williamson, S. S.: Reduced switching loss based DC-bus voltage balancing algorithm for three-level neutral point clamped (NPC) inverter for electric vehicle applications, IEEE Energy Conversion Congress and Exposition (ECCE), p. 3767-3773, Pittsburgh, 2014,

[14] Felgemacher, C.: Investigation of Reliability Aspects of Power Semiconductors in Photovoltaic Central Inverters for Sunbelt Regions, thesis, Kassel University, 2018

[15] Denk, M.; Bakran, M.-M.: Comparison of counting algorithms and empiric lifetime models to analyze the load-profile of an IGBT power module in a hybrid car, 3rd International Electric Drives Production Conference (EDPC), p. 1-6, Nuremberg (Germany), 2013

[16] Ahmed, M. H.; Wang, M.; Hassan, M. A. S.; Ullah, I.: Power Loss Model and Efficiency Analysis of Three-Phase Inverter Based on SiC MOSFETs for PV Applications, IEEE Access, Vol. 7, 2019

[17] M&P Motion Control and Power Electronics GmbH: Datasheet LT1000, www.powerelectronics.de/ datasheets/DS_LT1000.pdf, rev. 3, 2022

[18] M&P Motion Control and Power Electronics GmbH: Datasheet LT1000-ML, www.powerelectronics.de/ datasheets/DS_LT1000-ML.pdf, rev. 3, 2022

[19] Rosenbaum, L.: Conception of an IGBT phase module for the construction of modular power electronic devices of high power ratings (german: Konzeption eines IGBT-Phasenmoduls zum Aufbau von modularen leistungselektronischen Geräten großer Leistungen), diploma thesis, TU Dresden, 2011

[20] Wintrich A., Nicolai U., Tursky W., Reimann T.: Application Manual Power Semiconductors (german: Applikationshandbuch Leistungshalbleiter), ISBN 978-3-938843-85-7, 2015

[21] Bieler, A.; Muehlfeld, O.: Analytical Modelling of Dynamic Power Losses Inside Power Modules for 2-Level Inverters, PCIM Europe International Exhibition and Conference for Power Electronics, p. 1044-1050, Nuremberg (Germany), 2018

[22] Staudt I.: 3L NPC & TNPC Topology (AN-11001), Semikron Application Note, 2015

Autonomous Characterization of Lithium-Ion Battery Model Parameters utilizing a Mathematical Optimization Methodology

Galo D. Astudillo, Hamzeh Beiranvand, Helge Krüger, Sandra Hansen, Marco Liserre
Chair of Power Electronics, KIEL UNIVERSITY
Kaiserstraße 2, 24143 Kiel
Kiel, Germany
Phone: +49 431 880-6100
Fax: +49 431 880-6103
Email: danielastudilloheras@gmail.com,{hab,hkr,sn,ml}@tf.uni-kiel.de
URL: http://www.pe.tf.uni-kiel.de

Christian Werlig, Andreas Würsig
Fraunhoffer ISIT
Fraunhoferstraße 1
Itzehoe, Germany
Phone: +49 4821 17-0
Fax: +49 4821 17-4250
{christian.werlich,andreas.wuersig}@isit.fraunhofer.de
URL: https://www.isit.fraunhofer.de/

Acknowledgment

Funded by the European Union - European Regional Development Fund (EFRE), the German Federal Government and the State of Schleswig-Holstein (LPW-E/1.1.2/1486).

Keywords

≪State of Charge≫, ≪Optimization≫, ≪Batteries≫, ≪Electric Vehicle≫.

Abstract

Kalman filtering is commonly used for state-of-charge (SOC) estimation for lithium-ion (Li-ion) batteries owing to its simplicity, computational efficiency, and relatively precise results. However, kalman filters depend on the Li-ion battery model. Several laboratory tests such as incremental current and dynamic stress tests are required to determine battery model parameters in model-based SOC estimation. These tests such as incremental current test and dynamic stress test are time-consuming and can take multiple days. A mathematical optimization along with a battery test method, which does not need rest time for battery, are adopted to reduce the battery parameter identification time, drastically. A mathematical optimization stage is embedded prior to Kalman Filter based SOC estimation computing the battery open circuit voltage (OCV) and as well as an initial guess of the RC parameters of the battery equivalent circuit. Therefore, it reduces the required number of tests to one. Extensive numerical studies on a 2 Ah Lithium-ion cell verify the effectiveness of the proposed method by achieving a RMS error less than one percent.

I. Introduction

The battery management system (BMS) controls the charging/discharging process and guarantees the safety by permanent monitoring and parameter estimation of the electric vehicles (EVs) battery packs [1].

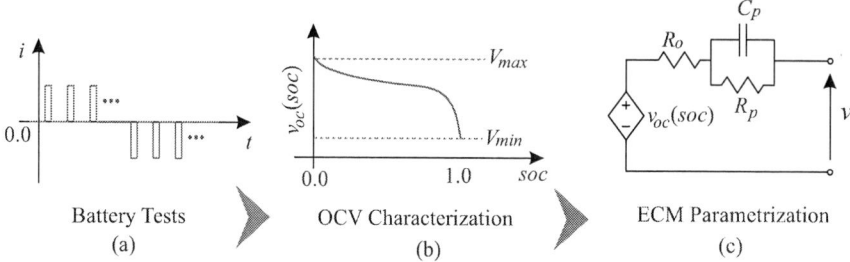

Battery Tests (a) OCV Characterization (b) ECM Parametrization (c)

Fig. 1: Conventional process for ECM parameter characterization for lithium-ion batteries v_{oc}: (a) battery test current profile, (b) OCV voltage characterization, and (c) model parameter estimation.

An accurate estimation of the battery state-of-charge (SOC) is a prerequisite to achieve the expected functions of the BMS.

Battery SOC estimation techniques have received a significant attention in both academia and industry due to its central role in the BMS. These methodologies can be classified into coulomb counting, lookup table, data-driven approaches, model-based approaches, and hybrid methods [2]. Surveys in [3, 4] show that model-based methods are the most promising solutions and are widely used in EV applications. Among the model-based approaches, equivalent circuit model (ECM) is preferred over electrochemical models (e.g. single particle model [5]) for SOC estimation owing to its computational efficiency, direct connection to coulomb counting and simple implementation of the Kalman filter family observers as the state-of-the-art solution [6]. However, efficient exploitation of the ECM model is subject to precisely estimate and calibrate the ECM parameters including open circuit voltage (OCV), RC pairs and the equivalent internal resistance.

The characterization of the variable voltage source is usually performed with an OCV test that provides pairwise values of the terminal voltage of the battery after a long relaxation time and the corresponding SOC [7]. To obtain the OCV-SOC curve, laboratory tests such as incremental current (IC) tests, low current (LC) tests at a very small C-rate (e.g. $\frac{1}{20C}$) or combined tests (CT) are required [8]. In the case of the IC tests, long relaxation times are needed to reach the near equilibrium potential. On the other hand, LC tests discharge/charge the battery at a small C-rate consuming a significant time. In addition to IC tests, a second dynamic current test is usually required to obtain the remaining parameters of the ECM. Nonetheless, these tests are time consuming and as a result autonomous characterization methods, independent from test types, are required.

The motivation behind this work is to tackle with the problem of the lithium-ion battery ECM parameterization not only for fresh cells with minimum test requirements but also to calibrate the model parameters during the operation and compensate for the SOC estimator degradation due to cell aging and sensor drifting.

A two-stage optimization process is proposed to increase the SOC estimation performance. In the fist stage, the OCV-SOC characterization is formulated as an optimization problem and solved by newton-based optimization approaches in a long time horizon. The time horizon could be tens of cycle or few months depending on the application. The second stage is to find the ECM parameters in the sample horizon to contribute to the unscented Kalman filter (UKF) based SOC estimation by providing a rel-atively precise initial guess of the ECM parameters. An advantage of using the UKF instead of the Extended Kalman Filter (EKF) is that the first one achieves at least 2nd order accuracy, while the EKF provides 1st order accuracy on the propagation of the state distribution [9]. Extensive numerical studies are carried out on lithium-ion batteries and the results are compared with the existing literature. The obtained results verify the effectiveness of the proposed mathematical optimization based approach for ECM characterization and SOC parameter estimation of the lithium-ion batteries. It is shown that ECM can be equaly parameterized by different tests such as IC test, Federal Urban Driving Schedule (FUDS), Highway Driving Schedule (US06) and Beijing Dynamic Stress Test (BJDST), and the Dynamic Stress Test (DST) utlizing the proposed autonomous characterization technique.

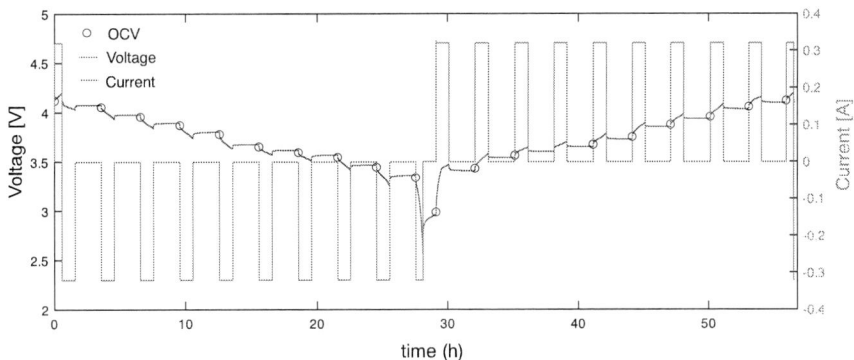

Fig. 2: A typical IC test for characterizing OCV.

II. Battery Modeling

The simplified model of the battery cell using an equivalent electric circuit is presented in Fig 1 as well as the conventional process of ECM characterization. This representation greatly reduces the complexity of the modeling of the battery by characterizing it with a RC circuit. The first order ECM is characterized with following equations:

$$\frac{d}{dt}v_c(t) = -\frac{v_c(t)}{R_p C_p} + \frac{i(t)}{C_p} \tag{1}$$

$$v(t) = i(t)R_o + v_c(t) + v_{oc}(soc(i(t))) \tag{2}$$

$$soc(i(t)) = soc_o + \frac{\eta}{C_{bat}} \int i(t)dt \tag{3}$$

where the capacitor voltage is $v_c(t)$, the terminal voltage is $v(t)$, the state of charge is *soc*. The capacity of the battery is denoted by C_{bat}. In addition, v_{oc} is the open-circuit voltage, i is the current and η is the coulomb efficiency. R_o, R_p, and C_p are the ECM parameters.

Results of [10] show that it is not necessary to perform a temperature-dependent OCV-SOC characterization if the battery capacity is "estimated dynamically and continuously". Therefore, the dependancy of OCV-SOC on temperature is neglected. Furthermore, we assume R_o, R_p and C_p to be fixed during the parametrization of the ECM at a given temperature T, which could be the nominal temperature.

III. OCV-SOC characterization

A set of experiments are carried out to show the typical IC test requirements for OCV-SOC characterization. An IC test at a current rate of C/10 with relaxation times of 2 hour is conducted on a 1865 cell with 3200 mAh capacity from LG employing a battery cycler (Biologic, VSP3e) and the results are shown in Fig. 2. Normally, OCV is defined as the equilibrium potential after setting the current to zero for a long time as can be seen from the experimental results. Theoretically, OCV can be achieved at minimum Gibbs free energy resulting from the overall electrochemical reactions in the battery cell [11]. Since the parasitic processes such as corrosion continue, OCV is interpreted as the first plateau after the relaxation of the most relevant reactions in the cell [12]. However, these kinds of tests which are associated with battery relaxation require a lot of time as demonstrated in Fig. 2. Thereby, other test methods as in [13] shall be considered in battery model calibration and characterization.

In the literature, OCV-SOC curves have been represented with look-up tables or many analytical functions (e.g. sigmoid, polynomial, exponential, chebyshev, logaritmic, etc). The OCV-SOC curve fitting procedure comparing several suitable functions can be found in [10]. In general, fitting the curve with a polynomial is one the most accurate and simplest approaches [8, 10, 14–16]. Hence, due to its simplicity and accuracy, we consider a polynomial function of the following form:

$$v_{oc}(soc) = \sum_{l=0}^{n} a_l soc^l \tag{4}$$

where n is the degree of the polynomial and a_l ($\forall l \in \{0, 1, ..., n\}$) are the coefficients. Now, it is important to notice the characteristics of $v_{oc}(soc)$ to establish a proper formulation of the fitting problem in hand. The main characteristics that will be included in the formulation are:

1. The maximum (V_{max}) and minimum (V_{min}) value of the terminal voltage at the equilibrium potential are known (i.e. when $soc = 1$ and $soc = 0$). Such parameters are usually known beforehand, and they establish the boundaries of $v_{oc}(soc)$ as $v_{oc}(0) = V_{min}$ and $v_{oc}(1) = V_{max}$.

2. The function $v_{oc}(soc)$ is monotonically increasing $\forall soc \in [0, 1]$.

3. The curvature of the $v_{oc}(soc)$ is non-positive when $soc = 0$ (e.g. $\frac{d^2}{d\,soc^2} v_{oc}(soc)|_{soc=0} \le 0$) and non-negative when $soc = 1$ (e.g. $\frac{d^2}{d\,soc^2} v_{oc}(soc)|_{soc=1} \ge 0$).

4. Due to the polarization, depending on the sign of the current $v_{oc}(soc)$ can be greater or smaller than the terminal voltage. If the current is negative (discharging) then $v_k < v_{oc}(soc_k)$ and if the current is positive (charging) $v_k > v_{oc}(soc_k)$.

The considerations presented above need to be included in the optimization problem. To do that, we first consider N samples, and separate each $k-th$ sample for $K = \{K^+ \cup K^0 \cup K^-\}$ where $K^+ = \{k \in \mathbb{Z}^+ : i_k > 0\}$, $K^o = \{k \in \mathbb{Z}^+ : i_k = 0\}$, and $K^- = \{k \in \mathbb{Z}^+ : i_k < 0\}$. The unknown variables are included in the vector $c = [a_0, a_1, \cdots, a_n]^T$. Finally, the constrained optimization problem is defined as follows:

$$\underset{c}{\text{argmin}} := f^+(c) + f^-(c) + f^o(c) \tag{5}$$

where

$$f^+(c) := \sum_{k \in K^+} \frac{1}{2} max(0, v_{oc}(soc_k) - v_k)^2 \tag{6}$$

$$f^-(c) := \sum_{k \in K^-} \frac{1}{2} max(0, v_k - v_{oc}(soc_k))^2 \tag{7}$$

$$f^o(c) := \sum_{k \in K^o} \frac{1}{2} (v_{oc}(soc_k) - v_k^*)^2 \tag{8}$$

constrained to equality and inequality constraints as:

$$\begin{aligned} h_1 &:= \sqrt{N}(v_{oc}(1) - V_{max}) = 0 \\ h_2 &:= \sqrt{N}(v_{oc}(0) - V_{min}) = 0 \end{aligned} \tag{9}$$

$$\begin{aligned} g_1 &:= -\sqrt{N}\left(\frac{d^2}{d\,soc^2} v_{oc}(soc)\Big|_{soc=1}\right) \le 0 \\ g_2 &:= \sqrt{N}\left(\frac{d^2}{d\,soc^2} v_{oc}(soc)\Big|_{soc=0}\right) \le 0 \end{aligned} \tag{10}$$

Note that the formulation of Eq. (5) is based on the penalty method. Hence, a third cost $f^0(c)$ is included in order to keep $v_{oc}(soc_k)$ in the neighborhood of v_k when $i_k = 0$. For the optimization problem, the experimental test (i.e. the current profile) is expect to have as much variation as possible. Therefore, both discharge and charge currents should be part of such test, as well as some zero currents where relaxation time is not necessarily required. An example of a profile that contains the aforesaid variations is the FUDS profile. The constraints are scaled by a factor of $\sqrt{N}$ to include a proper weight considering the number of data points available.

IV. Parameter Estimation

IV.a ECM parametrization

After that v_{oc} is determined, the second step is to obtain optimal values for $\boldsymbol{p} = [R_o, R_p, C_p]^T$. For this task, a formulation in the discrete-time domain for the future terminal voltage v_{k+1} is developed. Hence, Eq. (1) is solved for a small step $\delta = \Delta T$ and a current $i[k]$ during the k-th interval:

$$v_{ck+1} = v_{ck}e^{-\delta/C_pR_p} + i_kR_p(1 - e^{-\delta/R_pC_p}) \tag{11}$$

Next, a discrete version of Eq. (3) is obtained with the bilinear transformation:

$$soc_{k+1} = soc_k + \frac{\delta}{2C_{bat}}(i_k + i_{k-1}) \tag{12}$$

Moreover, the terminal voltage can be represented as:

$$v_k = i_kR_o + v_{ck} + v_{oc}(soc_k) \tag{13}$$

$$v_{k+1} = i_{k+1}R_o + v_{ck+1} + v_{oc}(soc_{k+1}) \tag{14}$$

Substituting Eq. (11) in Eq. (14) results in:

$$v_{k+1} = i_{k+1}R_o + v_{ck}e^{-\delta/C_pR_p} + i_kR_p(1 - e^{-\delta/R_pC_p}) + v_{oc}(soc_{k+1}) \tag{15}$$

Finally, solving for v_{ck} from Eq. (13) and replacing it in Eq. (15):

$$\begin{aligned} v_{k+1} = {} & i_{k+1}R_o + (v_k - i_kR_o - v_{oc}(soc_k))e^{-\delta/C_pR_p} + i_kR_p(1 - e^{-\delta/R_pC_p}) \\ & + v_{oc}(soc_{k+1}) \end{aligned} \tag{16}$$

Note that R_o, R_p, C_p are the only unknowns, since soc_{k+1} is obtained with Coulomb counting with Eq. (12). To find these unknown variables, the problem is formulated as the following optimization problem:

$$\underset{\boldsymbol{p}}{\text{argmin}} := \sum_{k=1}^{N} \frac{1}{2}(v_{k+1}(p) - v_{k+1}^*)^2 \tag{17}$$

where v_{k+1}^* is the future sample of the terminal voltage. It is important to mention that in this formulation there is not separating the sample points, since the N available data points are used in this task. Hence, such problem is relatively easy to solve with unconstrained optimization techniques. The proposed objective functions are solved using Sequential Least Squares Programming (SLSQP).

IV.b SOC estimation with UKF

After obtaining optimal values for the coefficients of the OCV-SOC curve and the ECM, we aim use this values in the model for the SOC estimation and terminal voltage prediction. For this task, an UKF is adopted. The UKF utilizes a discretized version of the nonlinear dynamic system of the following form:

$$\boldsymbol{x_{k+1}} = f(\boldsymbol{x_k}, i_k) + \boldsymbol{w_k}, y_k = g(\boldsymbol{x_k}) + u_k \tag{18}$$

Where x_k is the vector of states, w_k and v_k are the process and the observation noise, respectively. The function f is a mapping between the prediction of the future state $\boldsymbol{x_{k+1}}$ given the current state $\boldsymbol{x_k}$, while the function g returns the observation. The state distribution is represented by a Gaussian random variable, and it is specified by sample points that capture the statistics (e.g. mean and covariance) of the state distribution. Moreover, the propagation of this statistics through the system (i.e. $f(\boldsymbol{x_k}, i_k) + \boldsymbol{w_k}$) captures the posteriori mean and covariance with the unscented transformation (UT). Now, the space state representation of the ECM can be formulated as follows:

$$\boldsymbol{x_{k+1}} = A_k\boldsymbol{x_k} + i_k\boldsymbol{B_k} + \boldsymbol{w_k} \tag{19}$$

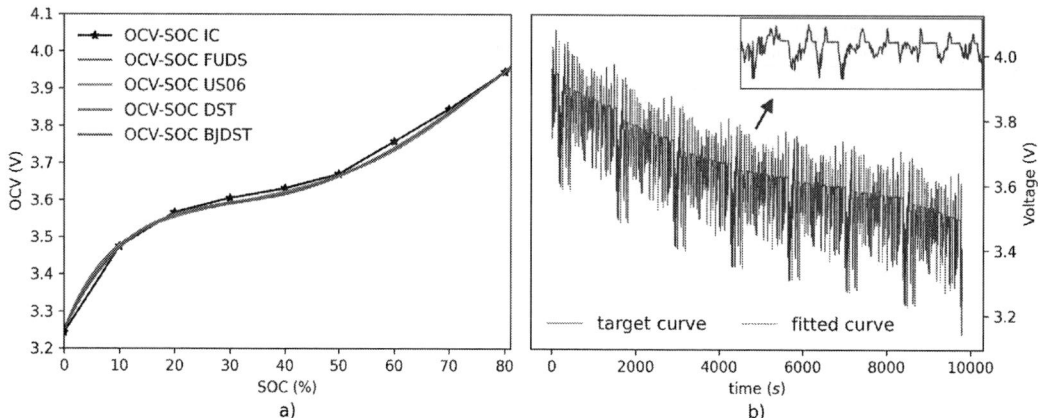

Fig. 3: a) OCV-SOC curves comparing incremental current test vs data-driven characterization at 25 °C and b) Curve-fitting ECM with data from FUDS.

with the following measurement equation:

$$y_k = i_k R_{ok} + v_{ck} + v_{oc}(soc_k) + u_k \tag{20}$$

with

$$
x_k = \begin{bmatrix} soc_k \\ v_{ck} \\ R_{ok} \end{bmatrix}, \; A_k = \begin{bmatrix} 1 & 0 & 0 \\ 0 & e^{-\delta/\tau} & 0 \\ 0 & 0 & 1 \end{bmatrix}, \; B_k = \begin{bmatrix} \delta/C_{bat} \\ R_p(1 - e^{-\delta/\tau}) \\ 0 \end{bmatrix}, \; w_k = \begin{bmatrix} w_{soc\,k} \\ w_{v_c\,k} \\ w_{R_o\,k} \end{bmatrix} \tag{21}
$$

where C_{bat} is the capacity of the battery, δ is the sampling time, and τ is the time constant ($\tau = R_p C_p$). The function v_{oc} returns the OCV of the circuit given the SOC. In addition, the vector w_k and the scalar u_k are zero-mean Gaussian stochastic processes with a known covariance Q_{w_k} and R_{u_k}, respectively.

V. Numerical Studies

To carry out a comparative study, experimental data of a LiNiMnCo-cathode Graphite-anode battery from [17] are used. This data set has been used as a benchmark in some researches [18, 19]. However, the current intention is to obtain the OCV-SOC curves at 25°C from four current stress tests, with three of them derived from driving cycles: Federal Urban Driving Schedule (FUDS), Highway Driving Schedule (US06) and Beijing Dynamic Stress Test (BJDST), and the Dynamic Stress Test (DST). The OCV-SOC characterization, ECM parametrization and UKF-based SOC estimation are explained in the followings:

OCV-SOC characterization: To evaluate the robustness, the proposed optimization method for OCV-SOC characterization is applied to all the test data (e.g. FUDS, US06, BJDST, and DST) and the results compared versus IC test as the reference value (v_{oc}^*). The obtained OCV-SOC curves are shown in Fig. 3 (a). This calibration process can be performed after few cycles autonomously without disconnecting the EV from the operation.

ECM parametrization: The second step is to utilize the obtained OCV-SOC curve for estimating temporary ECM parameters before applying UKF. Considering that FUDS could reflect the expected power variations in EV applications, it was selected to obtain the parameters of the ECM. This profile has been selected because of its complexity and it distribution of discharge/charge/zero currents: 55 %, 25 %, 20 %; respectively. The obtained parameters are Listed in Table I and compared versus the results of [19].There is a very small variation in the optimal values of the ECM parameters. The internal battery resistances R_o are very similar. Nonetheless, the parameters of the RC branch have noticeable differences. Fig. 3 (b) demonstrates the quality of the fitted terminal voltage achieved from the proposed methodology despite the considerable differences in RC paper with [19].

UKF-based SOC estimation: Once the FUDS profile has been utilized for finding the values of the

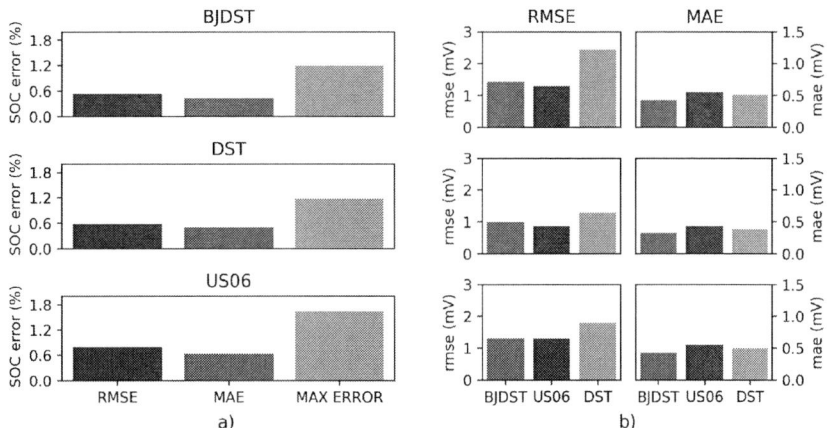

a) b)

Fig. 4: Estimation and prediction errors: a) SOC at nominal SOC_o and b) Terminal voltage: First row considers $-10\%SOC_o$ error, second row SOC_o, and third row $+10\%SOC_o$.

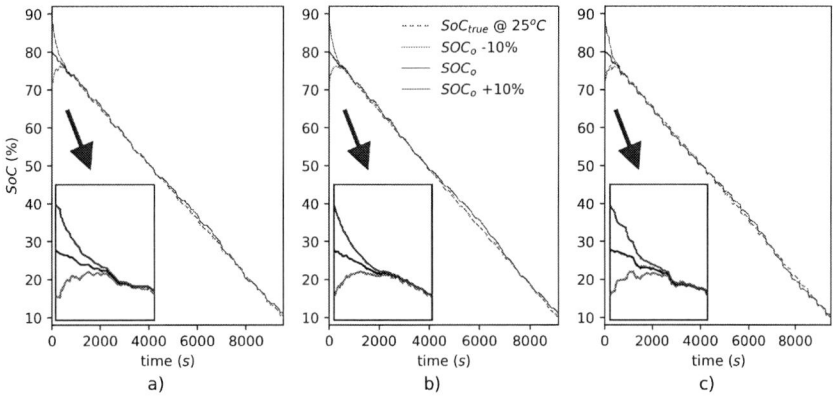

a) b) c)

Fig. 5: SOC estimation considering an initial error using the same parameters obtained from FUDS and tested on: a) BJDST b) US06 and c) DST .

OCV-SOC and the ECM, the other three tests (BJDST, US06, DST) are adopted to validate the model. This procedure is carried out with the data from each test. Moreover, an error in the initial SOC is added ($\pm 10\%$) to monitor the response of the observer (i.e. UKF). The values of the covariance for the implementation of the UKF are selected as $R_{u_k} = 1e-6$ and $Q_{w_k} = diag([1e-6, 1e-4, 1e-4])$. Fig. 4 shows a summary of the RMSE and MAE of the SOC estimation and terminal voltage prediction on each test, varying the initial SOC and considering percentage values (i.e. $0\%<SOC<100\%$). Moreover, these results are close to the results presented in [19], except for the maximum error. The maximum errors of US06 and and BJDST presented in Fig 4 a) are both $\sim 50\%$ smaller in the proposed methodology.

The SOC estimation with each stress test and the introduced $\pm 10\%$ are also shown in Fig. 5. The results show that in despite of the introduced error of $\pm 10\%$ in the initial SOC, the estimator successfully corrects the wrong initial SOC.

Table I: The optimized ECM parameters with FUDS data in comparison with IC test + DST in [19]

	Parameters	Proposed Methodology	Reference [19]
ECM Parameters	R_o (Ω)	0.07152	0.0710
	R_p (Ω)	0.01544	0.0342
	C_p (F)	881.99	1135.2
Curve-fitting Error	MAE (V)	5.424e-4	3.708e-4
	$RMSE$	8.905e-4	5.988e-4

VI. Conclusions

The conventional open circuit voltage versus state-of-charge (OCV-SOC) characterizations based on incremental current test (ICT) and dynamic stress test (DST) are time consuming and a complete process could take several weeks in laboratory environment. Moreover, the re-calibration of battery OCV-SOC after a specific number of cycles is required to compensate the cell degradation effects on the OCV-SOC curve. Therefore, autonomous and robust OCV-SOC characterization respective to the battery test method becomes an inevitable task in applications such as electric vehicles. This paper presents a methodology for autonomous equivalent circuit model (ECM) parameter characterization without the need for time-consuming ICT and DST. Characterization process is formulated as an optimization problem and integrated prior to the unscented kalman filter based SOC estimation. The methodology can be implemented in real-time in low-cost micro-controllers as the problem is solved by newton-based mathematical methods. Numerical studies are carried out to validate the effectiveness of the proposed method. In the worst case scenario the SOC estimation root mean square error (RMSE) is confined to 1% and maximum error to less than 2% when there is an error of $\pm 10\%$ in the initial SOC. Therefore, the methodology is robust and able to autonomously characterize the OCV-SOC curves and estimate the ECM parameters in real-time.

References

[1] H. Rahimi-Eichi, U. Ojha, F. Baronti, and M.-Y. Chow, "Battery Management System: An Overview of Its Application in the Smart Grid and Electric Vehicles," *IEEE Industrial Electronics Magazine*, vol. 7, no. 2, pp. 4–16, Jun. 2013.

[2] D. N. How, M. Hannan, M. H. Lipu, and P. J. Ker, "State of charge estimation for lithium-ion batteries using model-based and data-driven methods: A review," *Ieee Access*, vol. 7, pp. 136 116–136 136, 2019.

[3] Y. Zheng, M. Ouyang, X. Han, L. Lu, and J. Li, "Investigating the error sources of the online state of charge estimation methods for lithium-ion batteries in electric vehicles," *Journal of Power Sources*, vol. 377, pp. 161–188, 2018.

[4] J. Meng, M. Ricco, G. Luo, M. Swierczynski, D.-I. Stroe, A.-I. Stroe, and R. Teodorescu, "An overview and comparison of online implementable soc estimation methods for lithium-ion battery," *IEEE Transactions on Industry Applications*, vol. 54, no. 2, pp. 1583–1591, 2017.

[5] W. Li, Y. Fan, F. Ringbeck, D. Jöst, X. Han, M. Ouyang, and D. U. Sauer, "Electrochemical model-based state estimation for lithium-ion batteries with adaptive unscented kalman filter," *Journal of Power Sources*, vol. 476, p. 228534, 2020.

[6] X. Bian, Z. Wei, J. He, F. Yan, and L. Liu, "A two-step parameter optimization method for low-order model-based state-of-charge estimation," *IEEE Transactions on Transportation Electrification*, vol. 7, no. 2, pp. 399–409, 2020.

[7] I. Snihir, W. Rey, E. Verbitskiy, A. Belfadhel-Ayeb, and P. H. Notten, "Battery open-circuit voltage estimation by a method of statistical analysis," *Journal of Power Sources*, vol. 159, no. 2, pp. 1484–1487, 2006.

[8] Y. Li, H. Guo, F. Qi, Z. Guo, and M. Li, "Comparative study of the influence of open circuit voltage tests on state of charge online estimation for lithium-ion batteries," *IEEE Access*, vol. 8, pp. 17 535–17 547, 2020.

[9] E. Wan and R. Van Der Merwe, "The unscented Kalman filter for nonlinear estimation," in *Proceedings of the IEEE 2000 Adaptive Systems for Signal Processing, Communications, and Control Symposium (Cat. No.00EX373)*. Lake Louise, Alta., Canada: IEEE, 2000, pp. 153–158.

[10] B. Pattipati, B. Balasingam, G. Avvari, K. Pattipati, and Y. Bar-Shalom, "Open circuit voltage characterization of lithium-ion batteries," *Journal of Power Sources*, vol. 269, pp. 317–333, Dec. 2014.

[11] P. Vágner, R. Kodỳm, and K. Bouzek, "Thermodynamic analysis of high temperature steam and carbon dioxide systems in solid oxide cells," *Sustainable Energy & Fuels*, vol. 3, no. 8, pp. 2076–2086, 2019.

[12] D. del Olmo, M. Pavelka, and J. Kosek, "Open-circuit voltage comes from non-equilibrium thermodynamics," *Journal of Non-Equilibrium Thermodynamics*, vol. 46, no. 1, pp. 91–108, 2021.

[13] T. Mamo and F.-K. Wang, "Long short-term memory with attention mechanism for state of charge estimation of lithium-ion batteries," *IEEE Access*, vol. 8, pp. 94 140–94 151, 2020.

[14] X. Hu, S. Li, H. Peng, and F. Sun, "Robustness analysis of State-of-Charge estimation methods for two types of Li-ion batteries," *Journal of Power Sources*, vol. 217, pp. 209–219, Nov. 2012.

[15] C. Weng, J. Sun, and H. Peng, "An Open-Circuit-Voltage Model of Lithium-Ion Batteries for Effective Incremental Capacity Analysis," in *Volume 1.* Palo Alto, California, USA: American Society of Mechanical Engineers, Oct. 2013, p. V001T05A002.

[16] Q.-Q. Yu, R. Xiong, L.-Y. Wang, and C. Lin, "A Comparative Study on Open Circuit Voltage Models for Lithium-ion Batteries," *Chinese Journal of Mechanical Engineering*, vol. 31, no. 1, p. 65, Dec. 2018.

[17] "Battery Research Overview | Center for Advanced Life Cycle Engineering." [Online]. Available: https://calce.umd.edu/battery-research-overview

[18] Y. Xing, W. He, M. Pecht, and K. L. Tsui, "State of charge estimation of lithium-ion batteries using the open-circuit voltage at various ambient temperatures," *Applied Energy*, vol. 113, pp. 106–115, Jan. 2014.

[19] F. Zheng, Y. Xing, J. Jiang, B. Sun, J. Kim, and M. Pecht, "Influence of different open circuit voltage tests on state of charge online estimation for lithium-ion batteries," *Applied Energy*, vol. 183, pp. 513–525, Dec. 2016.

SOC governed algorithm for an EV Cascaded H-Bridge connected to a DC charger

Giulia Tresca[1], Andrea Formentini[2], Filippo Gemma [1], Federico Lusardi[3],
Riccardo Leuzzi[1], Pericle Zanchetta[1,4]

[1] Electrical, Computer and Biomedical Engineering Dept., University of Pavia, Pavia, Italy

[2] Telecommunications Engineering and Naval Architecture Dep., University of Genova, Genova, Italy

[3] MTA S.p.A (Power Electronics BU), Italy

[4] Electrical and Electronic Engineering Dept.,Nottingham, University of Nottingham, Nottingham, , United Kingdom

Abstract

Cascaded H- Bridge (CHB) converters have been considered valid candidates for replacing the two-level inverter in EV powertrain applications. This paper presents a State of Charge-governed algorithm for charging Li-Ion battery modules within a Cascaded H-Bridge converter for EV powertrain, connected to a DC charger. The novelty of this algorithm lies in the balanced and concurrent charge of the battery modules installed in all submodules of the three-phases, with no extra middle stage converter needed. Simulation and experimental results are shown to prove the validity of the novel architecture and the experimental setup is described.

Keywords

«**Battery modules**», «**Cascaded H- Bridge**», «**Charging algorithm**», «**Charging protocol**», «**Multilevel converter**»

I.   Introduction

Battery energy storage systems (BESS) play a strategic role for the Green Deal challenges that the world is approaching to face. One of the fields mostly affected by this challenge is the automotive one, where studies on new EV powertrain configurations are carried out.

Multilevel converters have been considered one of the main candidates for replacing and improving the classical 2-level inverter topology. The main benefits are better efficiency at partial loads, enhancement of the fault tolerant strategies and more flexibility in the design thanks to their modularity [1]-[3]. Moreover, the current trend to increase the DC link voltage for electrical powertrain requires urgently to define new solutions able to not stress the devices and improve the power quality without increasing losses: the modular structure of multilevel converters can guarantee both these requirements [4].

In [3], a Cascaded H-Bridge is compared with an IGBT and a SiC 2-level inverters, showing that the efficiency and the power density are comparable. In [5], a Modular Multilevel converter is proposed as valid topology for an electrical powertrain. In [6], the multilevel converter is mixed with a reconfigurable structure in order to optimally manage each battery cell.

Beside the motoring phase studies, a parallel research focus has been developed around the charging techniques for battery modules connected in multilevel configuration. Several research works in literature propose multilevel converters as the preferred topology to be installed within the charging stations because of their capability to face the grid high voltage thanks to their modularity [7]-[8]. A smaller number of research studies are focused on battery charging algorithms using multilevel converters within the electrical powertrain.

The authors of studies regarding charging processes for multilevel converters propose a direct connection between grid and converter [9]-[13]. In case of three-phases converters, each phase of the motor is connected to one phase of the grid and the single submodules are charged via customized modulations. In [9], the authors propose to use the power factor and the SOCs of the batteries to select the best switching state. On the other hand, [10] calculates the optimal switching states for the SOC balancing. Moreover, [11] proposes two other strategies to improve the efficiency of the charging process: the first one calculates the active power, exchanged between the grid and the converter, and uses it to predict the optimal switching state to balance the SOC of the battery modules; the second one uses the AC input current to calculate the best switching states each quarter of period. Despite the latest trend foresees to perform the charging process directly from the grid, as V2G (Vehicle tom Grid), the most common charging infrastructure is characterized by a DC charger [8], which manages the full control of the power flow to the battery pack. Therefore, it is timely to define charging strategies for three-phases multilevel converters connected to a DC source. In [12], a hybrid multilevel three-phase topology is presented and a DC source is considered to perform the charging process of the battery modules connected within the converter. However, the DC source is connected in parallel to one phase at the time; therefore the charging process is divided in three different time intervals, reducing the control complexity.

This paper proposes a State of Charge (SOC) governed algorithm for a CHB connected in parallel to a DC charger, as it is shown in Fig.1a. The real novelty of this work is the simultaneous charging process for all the battery modules installed within the three phases of the converter. The algorithm implements both the control of the charger output voltage and current and the CHB submodules in order to guarantee a balanced charging process. The paper is structured as follows. Section II gives an overview about the converter structure and charging protocol adopted. Section III explains in detail the charging algorithm implemented. In section IV and V, the simulation and experimental results are shown, respectively. Section VI concludes the paper.

II. CHB structure and charging protocol

The CHB converter considered for the charging algorithm is shown in Fig.1a. Each phase has n submodules (SM) made of one H-bridge connected to a battery module. When the CHB is in motoring phase, the switches K1, K2, K3 and K4 are turned on and K5, K6 and K7 are turned off. Vice-versa, for the charging process, K5, K6 and K7 are turned on to connect the CHB in parallel to the charger and disconnect the motor.

Charging protocol

The key points for a valid charging protocol are a good utilization of the capacity, high energy efficiency and a short charging time. In [14], different charging protocols are discussed and analysed. The reference protocol for this algorithm is the so-called Constant Current Constant Voltage (CC - CV) charging, which is divided in two main stages:

1) In the first phase, the battery cell is charged with a constant current value I_{ch}, until its voltage reaches a specified value, defined by the datasheet.
2) In the second and last phase, the cell voltage is charged with a constant voltage equal to V_{ch}. As a consequence, the charging current exponentially decreases. Usually, V_{ch} is equal to the maximum voltage of the battery cell. The CV phase stops when the charging current drops below a predefined value, defined by the datasheet, or after a pre-determined maximum charging time.

In an equivalent way, it is possible to divide the two phases of the charging process using as key parameter the State of Charge (SOC) of the battery modules. It is common practice to lead the constant current phase until the battery module reaches a SOC value equal to SOC_{th} and then continue with the Constant Voltage phase until the current value drops below a certain value. SOC_{th} is decided in order to obtain an initial charging current in the CV phase less or equal to the one employed during the CC phase. Fig.1b shows the current and voltage behaviour during the selected charging protocol.

III. Balanced Charging Algorithm

The charging algorithm proposed in this paper is the combination of a coherent utilization of the H-bridge converter configurations and a simultaneous control of current and voltage of the DC charger.

When the battery cell needs to be charged, the H-bridge is set in the active configuration (AC), shown in Fig.2a. On the other hand, if the battery cell does not need to be charged, the H-bridge converter is set to the bypass configuration (BC), shown in Fig.2b.
The voltage and current controls are carried out to fulfil the CC and CV charging phases.

Algorithm phases

The initial SOC value is calculated by measuring the initial open-circuit voltage of the batteries and inserting it within the characteristic SOC- V_{ocp} curve, provided by the battery datasheet. Once the initial SOC is known, while the evolution of the SOC of each battery module is calculated with the Coulomb counting [15]:

$$SOC(k + 1) = SOC(k) + \frac{i(k)\eta}{3600 * C_n} Ts \tag{1}$$

Where i(k) is the current flowing in the battery module at the kth instant, C_n is the capacity of the battery module, T_s is the discretized time interval between k and k+1 and η is the efficiency of the charging transformation.
For the sake of simplicity, η is considered equal to 1.

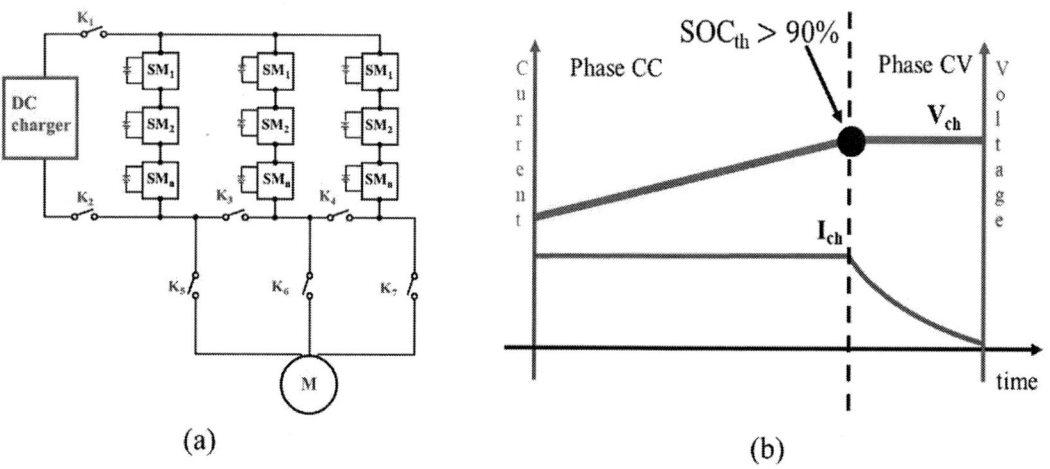

(a)

(b)

Fig. 1 – (a) CHB converter configuration for both motoring and charging phases. (b) Charging protocol CC – CV stages.

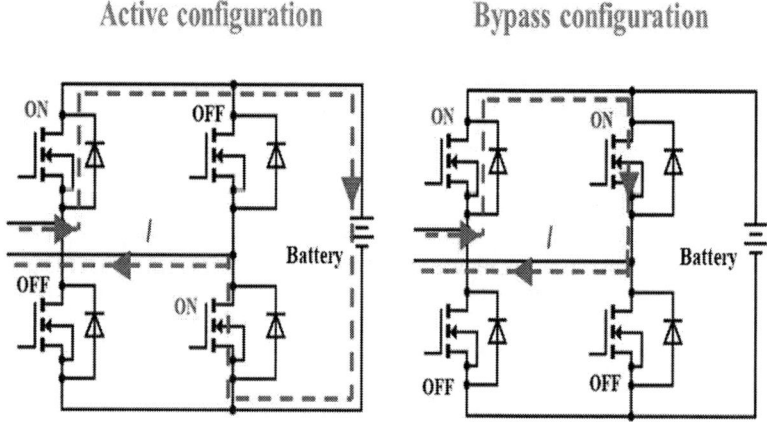

Fig. 2 – (a) shows the active configuration; (b) shows the bypass configuration. In (a), the current charges the battery module. In (b), the battery module is completely bypassed and therefore not charged.

Assuming to have n submodules connected in series per phase, the SOC value of the battery modules connected to the H-bridges determines the algorithm procedure.

1) All battery modules have SOC values below SOC_{th}

The H-bridges are set in active configuration, as shown in Fig. 3a, and all battery modules are charged. The output voltage of the charger is controlled in order to have a current flowing in the less charged phase equal to I_{ch}. Therefore, the other two phases will be charged with a smaller current – presenting a higher equivalent impedance.

2) m battery modules reach SOC_{th}

The m submodules are set in bypass configuration, as shown in Fig.3b, so that the relative battery modules are excluded from CC phase. However, an optimized current control needs an equal number of submodules per phase in active configuration. Vice versa, the current flowing in the three phases would be totally biased toward the less charged phase, provoking an unbalanced charging dynamic. Therefore, the other two phases will have a total number of submodules in active configuration equal to $n-m$. For these phases, the submodules will be set periodically in active configuration, so that the charging process can be performed in uniform way for all the battery modules.

3) All battery modules have a SOC value equal to SOC_{th}

The charger provides a fixed voltage and all submodules are set in active configuration, as shown in Fig.3c. The CV phase ends when the global current drops below a defined value.

Maximum current control

According to the phase of the charging algorithm, the implemented control is different, as it is shown in Fig.4.

When the CC phase is performed, the variable under control is the current flowing in the battery modules. However, the utilization of a unique DC voltage source makes possible the control of only one phase current. The charger voltage output needs to be optimally controlled to not create over current phenomena in any of the battery modules connected, but still allows a fast-charging process.

This result is achieved by using as control feedback variable the maximum current flowing in the three phases, obtaining two main benefits:

1) the over-current phenomena are completely avoided because the PI calculates a voltage able to guarantee the maximum current equal to the reference value. The currents flowing in the other two phases will be equal to lower values;
2) the maximum current is flowing in the less charged phase – the equivalent impedance is the lowest between the three phases. Therefore, the less charged modules will be charged faster than the other two phases, leading to a balanced charging dynamic.

The maximum current control forces only the current of the less charged phase to be constant. The other two phases perform an equivalent constant voltage charging process, with an increasing current flowing. When the CV phase is performed, the current control is replaced by the voltage control: the DC output voltage of the charger is set to the sum of the maximum voltage of the battery modules connected in one phase. Even in the case the CV would start with a minimum difference of SOC between the battery modules of the three phases, the global current would split consequently to reduce the SOCs gap between the three phases. When all phase currents drop below a certain value I_{cv} defined by the datasheet, the CV phase is terminated and the DC charger voltage is nailed to zero.

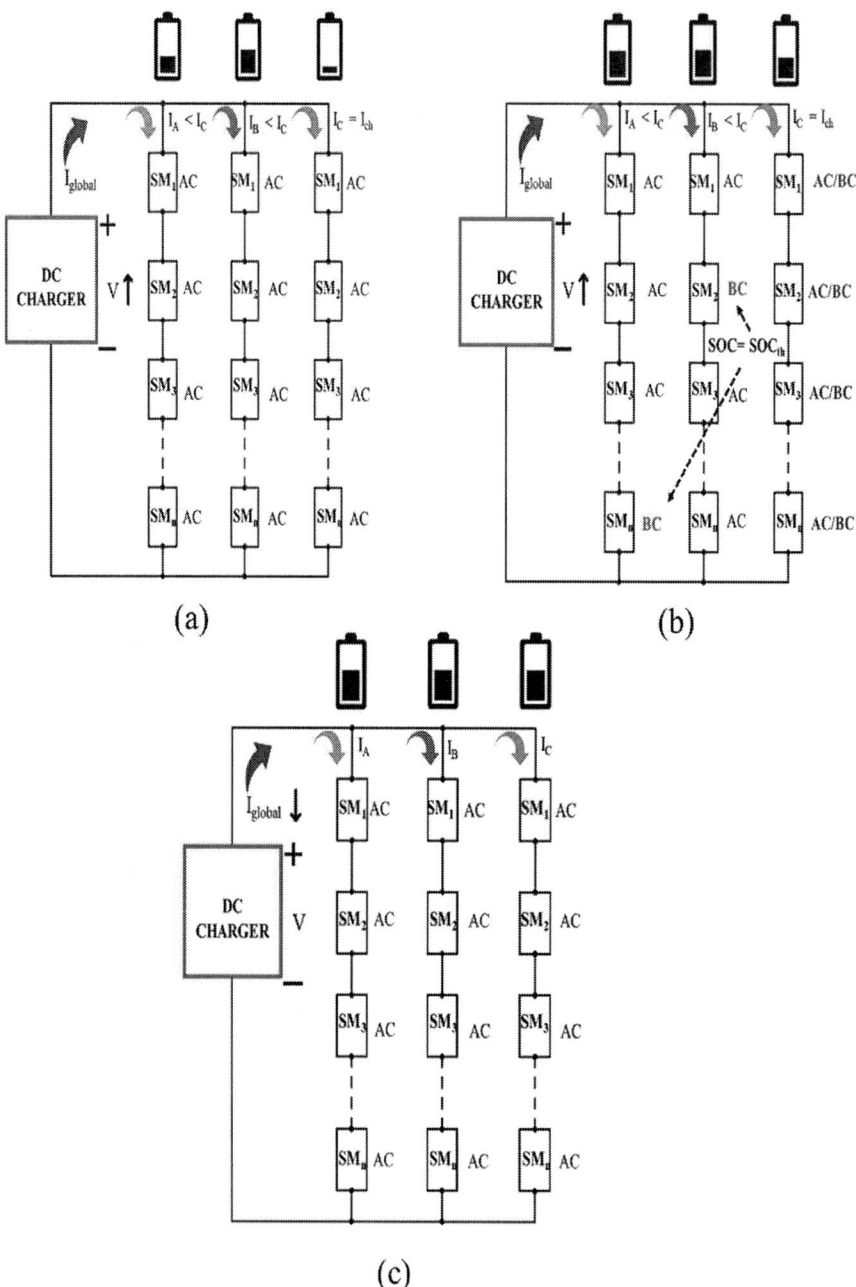

Fig. 3 – (a) All submodules are in active configuration (AC). The output voltage of the charger is controlled in order to have the nominal current flowing in the phase C, the less charged one. (b) Two submodules are set in bypass configuration (BC) to exclude their battery modules. The submodules of the phase C are periodically alternated to have always in conduction n-1 submodules. (c) - The CV phase starts: all submodules are in active configuration; the charger provides a fixed voltage.

IV. Simulations

The simulations are carried out in Simulink, MATLAB. The number of submodules per phase is chosen to be 3, as it is considered an optimized configuration for CHB in automotive application [16]. The simulated battery modules are made by serial and parallel connections of the MOLICEL battery cell. Its characteristics and the details about the battery modules simulated are shown in Table I.

The charging process of the MOLICEL has been modelled with the discharging curves given by the datasheet. The difference between the mechanical parameters of the battery cells is taken into account by varying the actual resistance and capacity with a normal distribution respect to a standard deviation.

Simulation results

Fig.5a and Fig.5b show the battery modules SOCs and the charging currents through the phases, respectively. The initial SOCs are less than SOC_{th}, therefore, the CC phase starts for all the battery modules. As soon as one battery module reaches SOC_{th}, the CC phase is eligible for only two battery modules per phase. The algorithm alternates periodically the remaining battery module per phase, so that the charging process is balanced. When the battery modules of the three phases reach SOC_{th}, the CV phase is eligible and starts for all the phases.

Despite of the significant difference between the initial SOC values – around 55%- at the end of the charging process, the maximum difference of SOC values between all the battery modules is less than 1.5%. Indeed, after the initial SOC values gaps are fulfilled, the balanced dynamic is achieved by alternating the activation of the submodules, which guarantees that the battery modules in one phase can be charged in a balanced way.

Meanwhile, the phase currents follow the maximum current control: the less charged phase is crossed by the highest charging current. When two or more submodules are alternated, the phase currents show a small ripple due to the different equivalent impedance seen by the charger, as it is shown in Figure 5b.

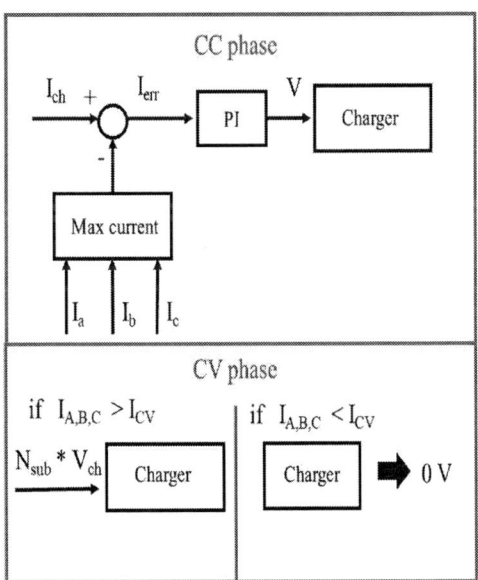

Fig. 4 – Implemented control according to the charging protocol phase.

TABLE I. SIMULATION PARAMETER

Battery cell characteristics	
Battery cell capacity	2.6 Ah
Battery cell voltage	3.6 V
Battery cell maximum voltage	4.2 V
Nominal charging current	2.6 A
Maximum internal resistance	20 mΩ
Battery module composition	
N° of battery cells in series	16
N° of battery cells in parallel	40
Capacity	104 Ah
Nominal voltage	57.6 V
Nominal charging current I_{ch}	104 A
Maximum Voltage	67.2 V
Minimum Voltage	40 V

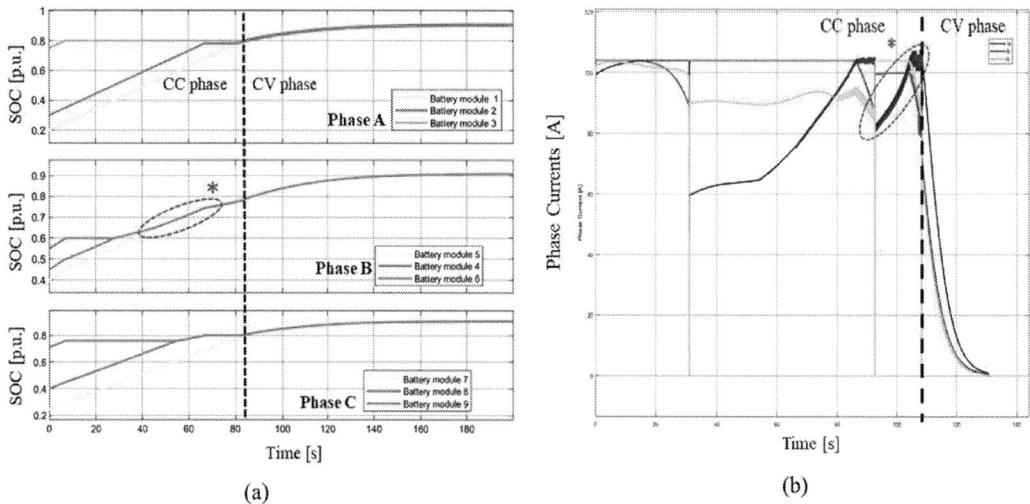

(a)   (b)

Fig. 5 – (a)The SOC trend for the battery modules is shown. *SOC balanced dynamic. (b) Charging currents through the phases A, B and C. *The phase currents show a ripple due to the alternation of the submodules.

V. Experimental results

The experimental activities are carried out with a single-phase CHB, with two submodules. Instead of battery modules, a MOLICEL INR-18650-P26A battery cell is charged with the presented protocol. Fig.6 shows the experimental setup: the DC charger is implemented with a Buck-converter connected to the two H-bridge. The algorithm is implemented through the utilization of the UCUBE platform [17]. Fig. 7a shows the estimated and measured voltages and the SOC values of both the battery cells, respectively. Fig. 7b shows the charger voltage and current, respectively.

The charging process is divided in 3 stages:

1) The initial SOC values are less than SOC_{th}: therefore, both battery cells are charged in CC phase. The charger voltage increases in order to keep a constant charging current value.
2) One battery cell reaches SOC_{th}, therefore it is bypassed. The CC phase continues for the other battery cell. As it is shown in Fig.8, the charger voltage drops to around 5 V.
3) When the second battery cell reaches SOC_{th} as well, the CV phase starts and terminates when the charging current drops below 0.1 A. The charger voltage is constant, while the current exponentially decreases.

The measurement of the battery cells has been added to demonstrate the validity of the charging algorithm. Indeed, the difference between the estimation and the calculation is completely negligible. Finally, the SOCs of the two battery cells converges at the same value, equal to 0.87%.

VI. Conclusion

This paper proposes a new perspective for EV battery charging architectures using multilevel converters. Instead of using the multilevel converter for the charging station, a CHB is employed as the power converter for the powertrain. Moreover, compared to the other research work, the three- phase CHB is connected to a unique DC charger in order to perform a simultaneous charging process for all battery modules within the converter. The novel algorithm here presented is able to control the charging process of all battery modules installed without the need of extra middle stage converters. The simulation and experimental work here presented show the feasibility and the validity of the algorithm.

Fig. 6 - Experimental setup

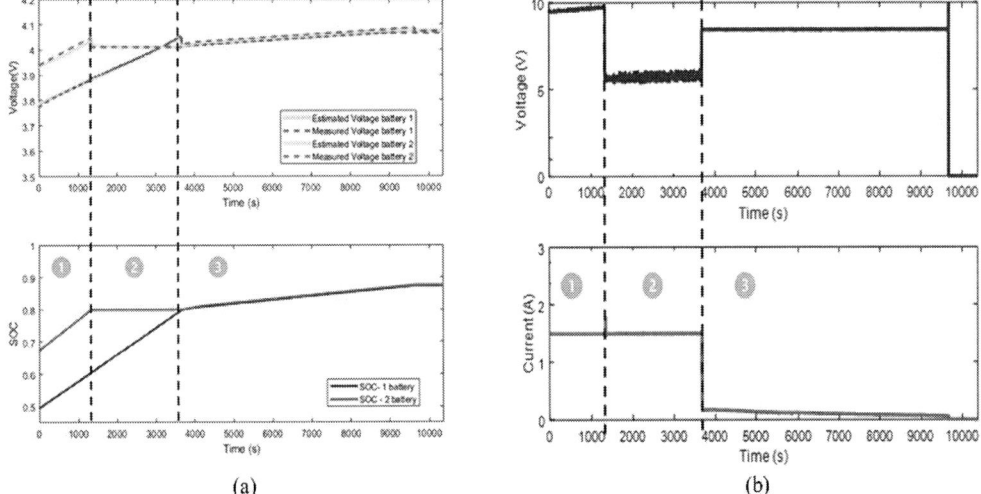

(a)   (b)

Fig. 7 – Up: the measured and estimated voltage for the two battery cells. Bottom: estimated SOC values for both battery cells. (b) Charger (buck-converter) voltage (up) and current (bottom) output .

References

[1] J. O. Estima and A. J. Marques Cardoso, "Efficiency Analysis of Drive Train Topologies Applied to Electric/Hybrid Vehicles," in IEEE Transactions on Vehicular Technology, vol. 61, no. 3, pp. 1021-1031,March 2012, doi: 10.1109/TVT.2012.2186993.

[2] E. Knischourek, K. Muehlbauer and D. Gerling, "Power losses reduction in an electric traction drive at partial load operation," 2012 IEEE International Electric Vehicle Conference, 2012, pp. 1-5, doi: 10.1109/IEVC.2012.6183196.

[3] F. Chang, O. Ilina, M. Lienkamp and L. Voss, "Improving the Overall Efficiency of Automotive Inverters Using a Multilevel Converter Composed of Low Voltage Si mosfets," in IEEE Transactions on Power Electronics, vol 34, no. 4, pp. 3586-3602, April 2019, doi: 0.1109/TPEL.2018.2854756.

[4] A. Poorfakhraei, M. Narimani and A. Emadi, "A Review of Multilevel Inverter Topologies in Electric Vehicles: Current Status and Future Trends," in IEEE Open Journal of Power Electronics, vol. 2, pp. 155-170, 2021, doi: 10.1109/OJPEL.2021.3063550.

[5]   M. Quraan, T. Yeo and P. Tricoli, "Design and Control of Modular Multilevel Converters for Battery Electric Vehicles," in IEEE Transactions on Power Electronics, vol. 31, no. 1, pp. 507-517, Jan. 2016, doi: 10.1109/TPEL.2015.2408435.

[6]   G. Tresca, R. Leuzzi, A. Formentini, L. Rovere, N. Anglani and P. Zanchetta, "Reconfigurable Cascaded Multilevel Converter: A New Topology For EV Powertrain," 2021 IEEE Energy Conversion Congress and Exposition (ECCE), 2021, pp. 1454-1460, doi: 10.1109/ECCE47101.2021.9595741.

[7]   R. Hariri, F. Sebaaly, C. Ibrahim, S. Williamson and H. Y. Kanaan, "A Survey on Charging Station Architectures for Electric Transportation," IECON 2021 – 47th Annual Conference of the IEEE Industrial Electronics Society, 2021, pp. 1-8, doi: 10.1109/IECON48115.2021.9589396.

[8]   M. Safayatullah, M. T. Elrais, S. Ghosh, R. Rezaii and I. Batarseh, "A Comprehensive Review of Power Converter Topologies and Control Methods for Electric Vehicle Fast Charging Applications," in IEEE Access, vol. 10, pp. 40753-40793, 2022, doi: 10.1109/ACCESS.2022.3166935.

[9]   A. Gholizad and M. Farsadi, "A Novel State-of-Charge Balancing Method Using Improved Staircase Modulation of Multilevel Inverters," in IEEE Transactions on Industrial Electronics, vol. 63, no. 10, pp. 6107-6114, Oct. 2016, doi: 10.1109/TIE.2016.2580518.

[10]   C. Young, N. Chu, L. Chen, Y. Hsiao and C. Li, "A Single-Phase Multilevel Inverter With Battery Balancing," in IEEE Transactions on Industrial Electronics, vol. 60, no. 5, pp. 1972-1978, May 2013, doi: 10.1109/TIE.2012.2207656.

[11]   A. Moeini and S. Wang, "The state of charge balancing techniques for electrical vehicle charging stations with cascaded H-bridge multilevel converters," 2018 IEEE Applied Power Electronics Conference and Exposition (APEC), 2018, pp. 637-644, doi: 10.1109/APEC.2018.8341079.

[12]   Z. Zheng, K. Wang, L. Xu and Y. Li, "A Hybrid Cascaded Multilevel Converter for Battery Energy Management Applied in Electric Vehicles," in IEEE Transactions on Power Electronics, vol. 29, no. 7, pp. 3537-3546, July 2014, doi: 10.1109/TPEL.2013.2279185.

[13]   S. Wang, R. Teodorescu, L. Mathe, E. Schaltz and P. Dan Burlacu, "State of Charge balancing control of a multi-functional battery energy storage system based on a 11-level cascaded multilevel PWM converter," 2015 Intl Aegean Conference on Electrical Machines & Power Electronics (ACEMP), 2015 Intl Conference on Optimization of Electrical & Electronic Equipment (OPTIM) & 2015 Intl Symposium on Advanced Electromechanical Motion Systems (ELECTROMOTION), 2015, pp. 336-342, doi: 10.1109/OPTIM.2015.7427002.

[14]   P. Keil, A. Jossen, "Charging protocols for lithium-ion batteries and their impact on cycle life—An experimental study with different 18650 high-power cells", Journal of Energy Storage, vol. 6, pp. 125-141, 2016, ISSN 2352-152X.

[15]   Wen-Yeau Chang, "The State of Charge Estimating Methods for Battery: A Review", International Scholarly Research Notices, vol. 2013, Article ID 953792, 7 pages, 2013. https://doi.org/10.1155/2013/953792

[16]   F. Roemer, M. Ahmad, F. Chang and M. Lienkamp, "Optimization of a Cascaded H-Bridge Inverter for electric vehicle applications including cost consideration", Energies, 2019, 12, 4272.

[17]   A. Galassini, G. Lo Calzo, A. Formentini, C. Gerada, P. Zanchetta and A. Costabeber, "uCube: Control platform for power electronics," 2017 IEEE Workshop on Electrical Machines Design, Control and Diagnosis (WEMDCD), 2017, pp. 216-221, doi: 10.1109/WEMDCD.2017.7947749.

Shaping the transition from Si-based power devices to SiC MOSFETs and GaN HEMTs

Dr. Gerald Deboy

Infineon Technologies Austria AG, Siemensstrasse 2, 9500 Villach / Austria

Abstract:

With an expected growth of SiC-device and module volumes to around 6 bn US$ and GaN HEMTs to 2 bn US$ in the next 5 years the transition from Si-based power devices to their corresponding Wide-bandgap technologies is now fully on its way. The advantages being created by wide bandgap devices on system level will outweigh their higher costs on device level.

These benefits are as diverse as the applications where wide bandgap power devices will be considered in the first place. SiC MOSFETs have started in the field of Photovoltaics by improving efficiency and size/weight of the PV inverter and are now penetrating the main inverter offering a range extension of up to 8%. GaN HEMTs first made an impact on chargers for mobile phones and laptops by enabling form factors up to now unachievable.

The presentation will start with an overview of key performance indicators of wide bandgap technologies in comparison to their silicon counterparts and their perspective along further generations. In a 2nd section we will discuss topologies and modulation schemes being required to reach the full system benefits of wide bandgap power devices. An outlook on future trends and applications will close the talk.

Curriculum Vitae:

Dr. Gerald Deboy received the M.S. and Ph.D. degree in physics from the Technical University Munich in 1991 and 1996 respectively. He joined Siemens Corporate Research and Development in 1992 and the Semiconductor Division of Siemens in 1995, which became Infineon Technologies later on. His research interests were focused on the development of new device concepts for power electronics, especially the revolutionary COOLMOS™ technology. From 2004 onward he was heading the Technical marketing department for power semiconductors and ICs within the Infineon Technologies Austria AG. Since 2009 he is leading a business development group specializing on new fields for power electronics. He is a Sr. member of IEEE and has served as a member of the Technical Committee for Power Devices and Integrated Circuits within the Electron Device Society. He has authored and coauthored more than 100 papers in national and international journals including contributions to three student text books. He holds currently more than 100 granted international patents and has more applications pending.

Contact Details:

Dr. Gerald Deboy
Infineon Technologies Austria AG
Siemensstrasse 2
9500 Villach / Austria
+43 51777 3541
gerald.deboy@infineon.com
https://www.infineon.com/

Reinventing Batteries Through Nanotechnology

Yi Cui
Stanford University

Abstract:

The fast growth of portable power sources for transportation and grid-scale stationary storage presents great opportunities for new battery chemistries. How to increase energy density, reduce cost, speed up charging, extend life, enhance safety and reuse/recycle are critical challenges. Here I will present how we utilize nanoscience to reinvent batteries and address many of challenges by understanding the materials and interfaces through new tools and providing new materials guiding principles. The topics to be discussed include: 1) A breakthrough tool of cryogenic electron microscopy, leading to atomic scale resolution of fragile battery materials and interfaces. 2) Materials design to enable high capacity materials: Si and Li metal anodes and S cathodes. 3) Interfacial design with polymer and inorganic coating to enhance cycling efficiency of battery electrodes. 4) New electrolyte design. 5) New battery chemistry for grid scale storage.

Curriculum Vitae:

At Stanford University, Yi Cui is the director of the Precourt Institute for Energy, co-director of the StorageX Initiative, Fortinet Founders Professor of materials science and engineering and of photon science at SLAC National Accelerator Laboratory. He has founded five companies to commercialize technologies from his lab: Amprius, 4C Air, EEnotech, EnerVenue and LifeLabs Inc. He has published more than 500 papers and is one of the world's most cited scientists (h-index 233). He is an elected fellow of the American Association for the Advancement of Science, the Materials Research Society, the Electrochemical Society, and the Royal Society of Chemistry. He is an executive editor of *Nano Letters* and co-director of the Battery 500 Consortium. His honors include Global Energy Prize (2021), Ernest Orlando Lawrence Award (2021), Materials Research Society Medal (2020) and Blavatnik National Laureate (2017). He is elected member of the US National Academy of Sciences.

Contact Details:

Yi Cui
Stanford University
476 Lomita Mall
Stanford, California 94305 USA
Phone number: +1 408-315-6762 (cell)
E-Mail address: yicui@stanford.edu
https://www.stanford.edu/

Advancing GaN Power ICs: Efficiency, Reliability & Autonomy

Dan Kinzer
NAVITAS SEMICONDUCTOR

Abstract:

Gallium Nitride (GaN) is a next-generation 'wide-bandgap' semiconductor, replacing legacy silicon chips in power electronic systems. To maximize the full potential of GaN's superior performance traits, Navitas monolithically integrates power, drive, and control to enable up to 3 times faster charging and 3 times more power in half the size and weight for mobile fast chargers, consumer electronics, solar, data centers and electric vehicles.

Integration is key with GaN power devices due to their extremely high switching speeds and sensitive gate characteristics. The next generation of GaN power ICs enable even higher efficiency, autonomy, and reliability with precision sensing of system current, voltage and temperature with real-time control and protection. Implementing integrated loss-less current sensing, external monitoring components such as large, lossy sense resistors are eliminated, reducing system power loss, complexity and system cost.

Offering GaN's superior performance and switching speed alongside the highest level of protection and sensing, GaN power ICs can be confidently used in higher power applications with stringent regulations for efficiency and reliability, such as solar inverters, motor drives, server power, EV Onboard Chargers (OBC) and DC-DC systems.

Curriculum Vitae :

For 30 years Dan has led R&D at semiconductor and power electronics companies at the VP level or higher. His experience includes developing advanced power device and IC platforms, wide bandgap GaN and SiC device design, IC and power device fabrication processes, advanced IC design, semiconductor package development and assembly processes, and design of electronic systems.

Before Co-founding Navitas, Dan served as VP R&D, VP Advanced Product Development, and Chief Technologist at International Rectifier (IR sold to Infineon for $3B), and SVP Product & Technology Development & CTO at Fairchild Semiconductor (Fairchild sold to onsemi for $2.4B).

In 2018, Dan was an inaugural inductee to the International Symposium on Power Semiconductor Devices and ICs (ISPSD) Hall of Fame.

Dan holds over 180 US patents, and a BSE degree in Engineering Physics from Princeton University.

Contact Details:

Llew Vaughan-Edmunds
Senior Director of Marketing
Navitas Semiconductor
2101 E El Segundo Blvd,
El Segundo, CA 90245, USA

Mobile: +1-310-343-8244
llew.ve@navitassemi.com
https://navitassemi.com/

Electrification Strategy of Volkswagen Group

Krick, Alexander
GROUP COMPONENTS, VOLKSWAGEN AG

Abstract:

Volkswagen AG is focusing on the world of mobility in 2030. By 2030, the global market for electric vehicles will have caught up with that of combustion engines, including in terms of sales volume.

With a view to tapping into the revenue streams offered by the new world of mobility, we are in the process of developing industry-leading platforms. The platform approach will be the key to success in the technological world of tomorrow.

These platforms form the backbone of the strategy and provide high-quality, industry leading technology at unprecedented scale and competitive cost.

The Scalable System Platform (SSP) will allow us to reduce complexity. It covers the entire product portfolio, from entry models to high-end vehicles.

The Group Components Technical Development division for E-Drive and Transmission, based at Kassel, Wolfsburg and Ingolstadt locations, is taking a leading role in the development of electric drivetrains for the SSP. One area of focus is the Group-wide responsibility to develop all future inverters. Therefore we are designing a modular system for future drivetrains on the SSP platform. Our aim is to design electric drivetrains that are the best-in-class. Inverter and software are the key components in this regard.

Curriculum Vitae:

Alexander Krick was born on May 20, 1981 in Fritzlar. After graduating from higher secondary school he studied electrical engineering at the University of Kassel, Germany and at University of Massachusetts, Dartmouth, USA. In 2009, he began his professional career as a project manager for the series development of electric drives at Volkswagen.

In 2012, Alexander Krick became the executive assistant of the plant management at Kassel site. Two years later, he was promoted to Head of the Technology Center for foundry and processing of Volkswagen Components. In 2015, Mr. Krick became in addition Head of the Steering Committee for the localization of components for New Energy Vehicles in China, where a year later he was assigned to Head of the Technical Development of Transmissions and Electric Drives at Volkswagen Automatic Transmission in Tianjin, China. From 2019 to 2020, Alexander Krick was Head of Planning and Development Steering in the Transmission and E-Drive Components business unit.

Since 2020 Alexander Krick is responsible for the Technical Development of E-Drives, Inverter and Transmissions at Volkswagen Group Components.

Contact Details:

Name:	Krick, Alexander
Company / Institution:	Group Components, Volkswagen AG
Address:	Rudolf-Leiding-Platz 1
City, Country:	Baunatal, Germany
Phone number:	+49 561 490 3049
E-Mail address:	alexander.krick@volkswagen.de

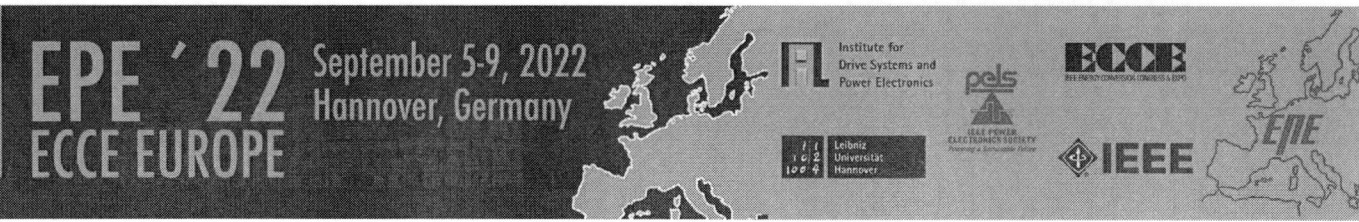

Make it Fly - The Future of Sustainable Aviation
Tanja NEULAND
AIRBUS OPERATIONS GmbH

Abstract:

Even if the impact of aviation on global warming is "only 3.5%", in absolute terms we are speaking about gigatons of CO_2, which must be massively reduced by 2050: net zero CO_2 by 2050 is the goal. To support this, Airbus wants to be a pioneer of decarbonized aviation and already in 2018, Airbus decided to take disruptive steps and continued to do so during the Corona crisis. We evaluate hydrogen powered propulsion technologies in regards to electronics & electric motors, fuel cells, liquid hydrogen storage and gas turbines. The fuel cell uses hydrogen to convert it into electrical energy. The electrical power of the fuel cell is used via power electronics to drive electric motors, which are connected to the propeller shaft via gears. Hydrogen direct combustion is the second form of drive for the propeller shaft. For this, the hydrogen (ideally in liquid form) is compressed and then sprayed into the combustion chamber. The heat generated by the ignition is used in a thermodynamic process to drive the shaft via turbine blades (similar to a classic jet engine). But on the way to this goal there are still some challenges to overcome → technically, logistically and also politically.

Curriculum Vitae:

Ms. Neuland studied aerospace engineering at the University of Applied Science in Hamburg. Since 1998 she has been working for Airbus in different positions in engineering and customer service.

Her key positions were:
- At the A380 program as Ho Attestation A380 Electric
- In the Airbus spare business as Ho Material & Logistics Engineering
- Within the Cabin Engineering as Ho Seat Architecture
- Continuing in customer service as Ho Quality SATAIR (Airbus Service Company)
- From 2020 on she is working in the hydrogen R&T area of Airbus and has overtaken her current position as hydrogen responsible (beyond ZEROe) in R&T since last October.

Contact Details:

Tanja NEULAND
Airbus operations GmbH
Hydrogen Techno IPT Leader - Propulsion of Tomorrow
+49 (0) 40 743 81017
tanja.neuland@airbus.com
https://www.airbus.com/en

The Instrumental but Extremely Challenging Role of Hydrogen Towards a Decarbonized Society

Dr. Stefan Linder
Alpiq AG, Olten, Switzerland

Abstract:

The energy transition, which in fact should be correctly named climate transition, holds unprecedented challenges that are widely underestimated. The presentation will start with a cruising altitude view that explains why all current efforts are much too slow and not orchestrated well enough to successfully meet the 2-degree target. It is shown that a successful decarbonization must be based on a few cornerstones that must be addressed swiftly, relentlessly, and in a globally coordinated manner. Hydrogen belongs to these pillars. It will be explained why hydrogen is so important, but also why there is no chance that hydrogen can develop quickly enough, unless there is both a national and a global consensus and coordinated action to overcome the barriers. The question also arises as to what role power electronics will play in the development of a hydrogen infrastructure. The presentation will show that power electronics will not be the glamorous main cast, but that it will be an indispensable and ubiquitous team member, playing its role mostly out of the limelight.

Curriculum Vitae:

Stefan Linder holds an electrical engineering and a PhD degree from the Swiss Federal Institute of Technology (ETH) Zurich. His career took him from the US Semiconductor Industry, via an 18-year employment at ABB, to Alpiq, a Swiss European electricity generation and energy trading company. At Alpiq, Stefan Linder is Head of Technology and Innovation. His main responsibilities are to conduct energy system studies and asset valuations, and to prepare strategy recommendations. Stefan Linder developed the conviction early on that hydrogen will play an instrumental role in the successful decarbonization of society. Alpiq shares this view and was among the first companies to start investing systematically in hydrogen. Alpiq is a key partner in the development of the Swiss ecosystem for the decarbonization of road freight transport with fuel cell heavy duty trucks.

Besides his role at Alpiq, Stefan Linder has engagements in several international committees in the energy and power semiconductor sectors. He is also president of the Swiss Association for Energy and Network Research.

Contact Details:

Stefan Linder
Alpiq AG
Bahnhofquai 12
Olten, Switzerland
+41 62 286 73 37
stefan.linder@alpiq.com
www.alpiq.com

Short Circuit Behavior of Dual Three-phase Permanent Magnet Synchronous Motors with Different Mutual Inductance in Electric Propulsion Application

Yinghui Yang and Georg Möhlenkamp
BRANDENBURGISCHE TECHNISCHE UNIVERSITÄT
Platz der Deutschen Einheit 1
Cottbus, Germany
Tel.: +49 (0) 355 69 5571
Fax: +49 (0) 355 69 4019
E-Mail: Yinghui.Yang@b-tu.de
URL: https://www.b-tu.de/fg-lea

Acknowledgements

The authors would like to acknowledge the helpful advice provided by Rolls-Royce Deutschland Ltd & Co KG. This research is funded by ILB - ProFIT Red Eagle under Project: P32055003

Keywords

« Permanent magnet motor », « Multiphase drive », « Mutual inductance », « Short circuit », « Fault tolerance », « Electric Propulsion »

Abstract

The dual three-phase permanent magnet synchronous motors (DTP-PMSMs) are suitable in aircraft electric propulsion unit (EPU) application because of their good reliability and fault-tolerant capability. However, this fault-tolerant capability can get weakened or even eliminated in short circuit fault due to the influence of flux coupling between the two isolated three-phase sub-systems. In this paper, a mathematical model of DTP-PMSMs with different mutual inductance between the two sub-systems is developed and analyzed. Based on the model, different short circuit protection methods are discussed and compared. Finally, a simulation of an EPU consisting of a DTP-PMSM, and a simplified constant speed propeller (CSP), is implemented to verify the short circuit behaviors.

Introduction

A typical DTP-PMSM consists of two isolated three-phase sub-systems. Compared with conventional three-phase PMSM, DTP-PMSM is more reliable due to the redundant winding configuration. For example, when there is one-point open circuit fault occurring in one sub-system, the another healthy sub-system is still able to delivery half of the nominal torque[1]. However, this benefit may not work in short circuit condition. The widely used protection method in PMSM short circuit fault is active short circuit (ASC), which creates a symmetrical short circuit in the unhealthy three-phase system with a symmetrical short circuit current flow through all three phases[2]. With two sub-systems in DTP-PMSM, there are two possible ASC options if a short circuit fault occurs internally in one sub-system. One option is only applying ASC on the unhealthy sub-system and let the healthy sub-system keep operating , which is known as active-shorted mode (ASM-ASC). Another option is applying ASC on both sub-systems, known as shorted-shorted mode (SSM-ASC)[3][4]. On the other hand, two isolated sub-systems can have flux coupling through the shared flux path, known as the mutual inductance, between the two sub-systems. With this mutual inductance, the current in one sub-system can induce EMF in another sub-system, making the two sub-systems no longer independent. Therefore, it is important to study the influence of this mutual inductance on machine operation, especially under short circuit condition, to optimize the selection of the short circuit protection method.

Machine model

In different stator winding arrangements of DTP-PMSM, the mutual inductance between two sub-systems is different. Depending on the significance of the mutual inductance, the DTP-PMSMs can be classified into two types, the one with high mutual inductance between two sub-systems (HM-DTP-PMSM) and the one with low mutual inductance between two sub-systems (LM-DTP-PMSM). Typical stator winding arrangements of both types are shown in Fig. 1 (a) and (b), where ABC represents windings of one sub-system (abc-sub-system) and XYZ represents windings of the another sub-system (xyz-sub-system). A 30-degree electrical angle shift between the two sub-systems is applied for both types[5].

Designing the stator winding arrangement is a comprehensive topic. For example, if the number of cores per slot is the same in both types of DTP-PMSM in Fig. 1, the flux linkage, and the d- and q-axis inductances will not be the same in those two types, finally resulting in different motor parameters and operation conditions, making it difficult to compare[6]. Since the influence of mutual inductance between two sub-systems is the main research object in this paper, the following assumptions are made:

1. different motors are designed in such a way that the main operation parameters are the same, as Table I shows, and only mutual inductance between two sub-systems is variable.
2. the d-axis and q-axis are identical for both sub-systems, and the inductance fluctuation is ignored with round rotor.

Table I: Parameters of DTP-PMSM

Parameter	Symbol	Value
Nominal power	$P_{(n)}$	50kW
Nominal torque	$T_{e(n)}$	205Nm
Nominal speed	$N_{m(n)}$	2320rpm
Nominal d-axis current	$i_{d(n)}$	0A
Nominal q-axis current	$i_{q(n)}$	200A
PM flux linkage	Ψ_{PM}	0.04366Wb
Phase resistance	R_s	0.01Ohm
Total d-axis inductance	L_d	300uH
Total q-axis inductance	L_q	300uH
Number of pole-pairs	P_p	8

To study the influence of mutual inductance, it is simple to use the dual dq-coordinate frame in building the mathematical model of DTP-PMSM, since the mutual inductance between the two sub-systems is a direct variable in dual dq-transformation[7]. The voltage equations are shown in (1) to (4), where number 1 and 2 represent two sub-systems, u_d and u_q are d-axis and q-axis voltages, i_d and i_q are d-axis and q-axis currents, Ψ_d and Ψ_q are d-axis and q-axis flux linkages, and ω_e is electrical rotational speed.

$$u_{d1} = R_s i_{d1} + \frac{d\Psi_{d1}}{dt} - \omega_e \Psi_{q1} \tag{1}$$

$$u_{q1} = R_s i_{q1} + \frac{d\Psi_{q1}}{dt} + \omega_e \Psi_{d1} \tag{2}$$

$$u_{d2} = R_s i_{d2} + \frac{d\Psi_{d2}}{dt} - \omega_e \Psi_{q2} \tag{3}$$

$$u_{q2} = R_s i_{q2} + \frac{d\Psi_{q2}}{dt} + \omega_e \Psi_{d2} \tag{4}$$

The flux linkage equations of the DTP-PMSM in dual dq-coordinate frame is shown in (5) to (8), where L_{dd} and L_{qq} are d-axis and q-axis self-inductances, M_{dd} and M_{qq} are the d-axis and q-axis mutual inductances.

$$\Psi_{d1} = i_{d1}L_{dd} + i_{d2}M_{dd} + \Psi_{PM} \tag{5}$$
$$\Psi_{q1} = i_{q1}L_{qq} + i_{q2}M_{qq} \tag{6}$$
$$\Psi_{d2} = i_{d2}L_{dd} + i_{d1}M_{dd} + \Psi_{PM} \tag{7}$$
$$\Psi_{q2} = i_{q2}L_{qq} + i_{q1}M_{qq} \tag{8}$$

The torque produced by the DTP-PMSM can be calculated as (9) to (11).

$$T_{e1} = 1.5 \cdot P_p\left(\Psi_{d1}i_{q1} - \Psi_{q1}i_{d1}\right) \tag{9}$$
$$T_{e2} = 1.5 \cdot P_p\left(\Psi_{d2}i_{q2} - \Psi_{q2}i_{d2}\right) \tag{10}$$
$$T_e = T_{e1} + T_{e2} \tag{11}$$

The total d- and q-axis inductances, which are assumed to be designed as constant value in different stator winding arrangements, can be calculated as (12) and (13)

$$L_d = L_{dd} + M_{dd} \tag{12}$$
$$L_q = L_{qq} + M_{qq} \tag{13}$$

A new variable k is employed in the mathematical model, to express the ratio of the mutual inductance to the self-inductance of both the d- and q-axis, as shown in (14) and (15)

$$M_{dd} = k \cdot L_{dd} \tag{14}$$
$$M_{qq} = k \cdot L_{qq} \tag{15}$$

Theoretically, the maximum value of k can reach up to 1, meaning the mutual inductance between two sub-systems is equal to the self-inductance. Therefore, the range of k is from 0 to 1. Specifically, in different stator winding arrangements, winding function (WF) approach can be used to calculate the self-inductances and the mutual inductances of all dual three-phase stator windings[8]. This inductance matrix can be transferred into dual dq-coordinate frame and then the k value can be calculated. The winding function of both DTP-PMSMs are demonstrated in Fig. 1 (c) and (d). The calculated $k_{HM-DTP-PMSM}$ is 0.86, and the calculated $k_{LM-DTP-PMSM}$ is 0.

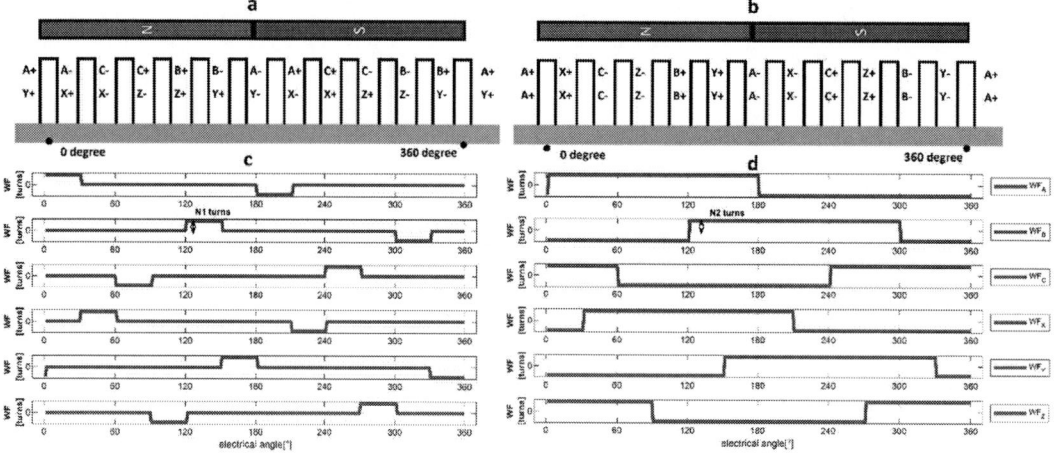

Fig.1: Two typical DTP-PMSM stator arrangements with its winding function (WF). (a) stator arrangement of LM-DTP-PMSM. (b) stator arrangement of HM-DTP-PMSM. (c) winding function of LM-DTP-PMSM. (d) winding function of HM-DTP-PMSM

Active short circuit

In SSM-ASC, the d- and q-axis voltages in both sub-systems become zero. According to machine model, the d- and q-axis currents of both shorted sub-systems can be calculated as (16) and (17).

$$i_{d_{SSM-ASC}} = -\frac{\omega_e^2 \Psi_{PM} L_q}{\omega_e^2 L_d L_q + R_s^2} \tag{16}$$

$$i_{q_{SSM-ASC}} = -\frac{\omega_e \Psi_{PM} R_s}{\omega_e^2 L_d L_q + R_s^2} \tag{17}$$

In ASM-ASC, the d- and q-axis voltages in unhealthy sub-system become zero. In the healthy sub-system, the d- and q-axis voltages are still controllable. Field oriented control (FOC) is widely used in electrical machine controls. Usually, the inner loop of the FOC is current regulator, which controls the d- and q-axis currents, respectively. In DTP-PMSM, two independent current regulators in the FOC inner loop, one for each sub-system, are used. Therefore, we can assume the d- and q-axis currents in the healthy sub-system can be still controlled as the set values. According to machine model, the d- and q-axis currents in the shorted sub-system can be calculated as (18) and (19), where i_d^* and i_q^* represent the controlled values of the d- and q-axis currents in the healthy sub-system.

$$i_{d_{ASM-ASC}} = -\frac{\omega_e^2 \Psi_{PM} L_q (k+1) - \omega_e L_q R_s i_q^* k(k+1) + \omega_e^2 L_d L_q i_d^* k}{\omega_e^2 L_d L_q + (k+1)^2 R_s^2} \tag{18}$$

$$i_{q_{ASM-ASC}} = -\frac{\omega_e \Psi_{PM} R_s (k+1)^2 + \omega_e L_d R_s i_d^* k(k+1) + \omega_e^2 L_d L_q i_q^* k}{\omega_e^2 L_d L_q + (k+1)^2 R_s^2} \tag{19}$$

According to (9) to (11) and (16) to (19), the steady state operation status of the DTP-PMSM versus motor speed after ASC can be plotted as shown in Fig. 2, where the i_d^* and i_q^* in ASM-ASC are assumed to be controlled as nominal values. It can be seen from the figure that the k value has a significant influence on motor operation if motor does ASM-ASC, while no influence if motor does SSM-ASC. When the motor speed is high, the influence of stator resistance is neglectable compared with the influence from inductance. Therefore, in a wide range of motor speed, the motor current and motor torque keep almost constant, in both ASM-ASC and SSM-ASC.

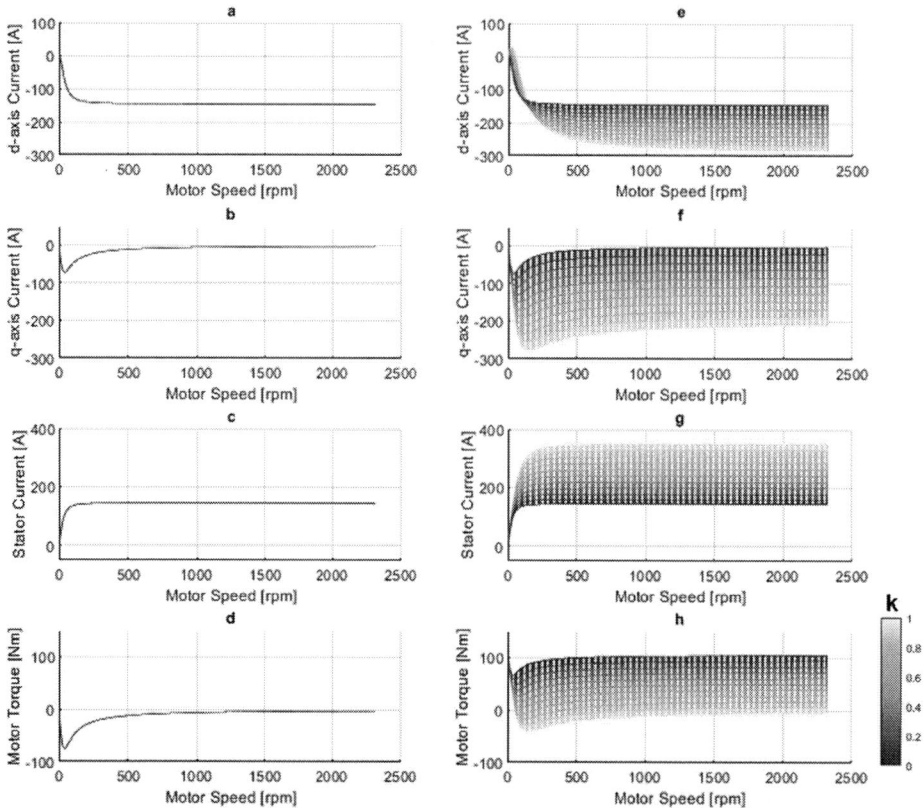

Fig. 2: Theoretical analysis of steady state operation status versus motor speed of unhealthy sub-system in DTP-PMSM during SSM-ASC (the four sub-graphs at left column) and ASM-ASC (the four sub-graphs at right column, color bar represents different k). (a) and (e) d-axis current. (b) and (f) q-axis current. (c) and (g) magnitude of stator current. (d) and (h) motor torque.

When motor speed is high, the d-axis current in the shorted sub-system in both SSM-ASC and ASM-ASC is demagnetizing current. In SSM-ASC, this value is around -145.5A for one sub-system, totally -291A for both sub-systems. In ASM-ASC, as k increases, this value increases from -145.5A to -284A. Therefore, in terms of demagnetizing effect, ASM-ASC is better for LM-DTP-PMSM.

The higher stator current usually causes higher loss in the stator windings. The symmetrical short circuit current after SSM-ASC is not necessary to be lower than motor nominal stator current, sometimes it can be higher, depending on the motor design. In the motor used in this paper , the nominal magnitude of the stator current is 200A. In SSM-ASC, the magnitude of the stator current in both sub-systems is 145.5A that is lower than the nominal value. In ASM-ASC, as k increases, the magnitude of the stator current in the shorted sub-system increases from 145.5A up to 353A. Therefore, in terms of magnitude of the state current, ASM-ASC is not a good option for HM-DTP-PMSM.

One important purpose for doing ASM-ASC is to gain some remaining torque through the healthy sub-system when short circuit fault occurs in DTP-PMSM. However, as k increases, the remaining motor torque decreases from half of the nominal torque to almost 0 Newtonmeter, even with d- and q-axis currents controlled to be as nominal values in the healthy sub-system. Due to this deceasing remaining torque, the HM-DTP-PMSM has a weaker fault tolerance capacity in terms of short circuit fault, compared to LM-DTP-PMSM.

Therefore, it can be concluded from the above discussion that ASM-ASC is more suitable for LM-DTP-PMSM, and SSM-ASC is more suitable for HM-DTP-PMSM.

Simulation

To investigate the short circuit behaviors in EPU, a simulation model is built in MATLAB/Simulink. The simulation model consists of two main parts, one DTP-PMSM connected at one end of the shaft, and one torque source with rotational speed feedbacked PI controller to simulate the simplified CSP behavior, connected at another end of the shaft. The DTP-PMSM has the same parameters as in the Table I. Two ideal switch based three-phase voltage source inverters (B6-VSIs) are used to drive the DTP-PMSM with a switching frequency 10kHz. The DTP-PMSM is controlled to operate with nominal d- and q-axis current by FOC. The PI controller in the simplified CSP is tuned to have a long reaction time that is in approximate range of 5 seconds. Shaft inertia is assumed to be 3 kilogram-meter squared. Friction is ignored. The short circuit fault occurs in abc-sub-system at 0.5 Second. Both LM-DTP-PMSM and HM-DTP-PMSM shown in Fig. 1 are simulated.

The simulation results of ASM-ASC in LM- and HM-DTP-PMSM are shown in Fig. 3. It can be observed that the DTP-PMSMs with different k values show significantly different behaviors after ASM-ASC. Since it is expected that the motor can still deliver torque in ASM-ASC, the speed controller in CSP will keep working to maintain the shaft speed to be as the nominal value. The process of ASM-ASC can be mainly divided into two parts based on shaft speed. The first part is shaft speed transient, in which the shaft speed drops and the CSP tries to recover the speed. The second part is short circuit steady state, in which the shaft speed maintains again back as nominal by CSP. In both parts of the process, the abc-sub-system in DTP-PMSM does ASC and the xyz-sub-system maintains its d- and q-axis current as nominal values by its FOC current regulator. In LM-DTP-PMSM, the healthy xyz-sub-system can still deliver approximately half of the nominal torque. The magnitude of the stator current in the shorted abc-sub-system is around 145A, and the stator current in xyz-sub-system maintains nominal value. One problem in this scenario is the d-axis current in abc-sub-system that does demagnetizing of the PM with a value of -145A. In HM-DTP-PMSM, the performance becomes even worse. The short circuit transient peak current reaches above 500A, which is more than twice higher than nominal current. With the impact of xyz-sub-system, the peak value of the steady state short circuit current in the unhealthy abc-sub-system increases to 400A. The d-axis current in the unhealthy abc-sub-system becomes -250A which can cause worse demagnetizing problem on the PM. The q-axis current of around -180A in the unhealthy abc-sub-system, together with the q-axis current in the healthy xyz-sub-system that is controlled to be as nominal value, results in a remaining torque close to 0 Newtonmeter.

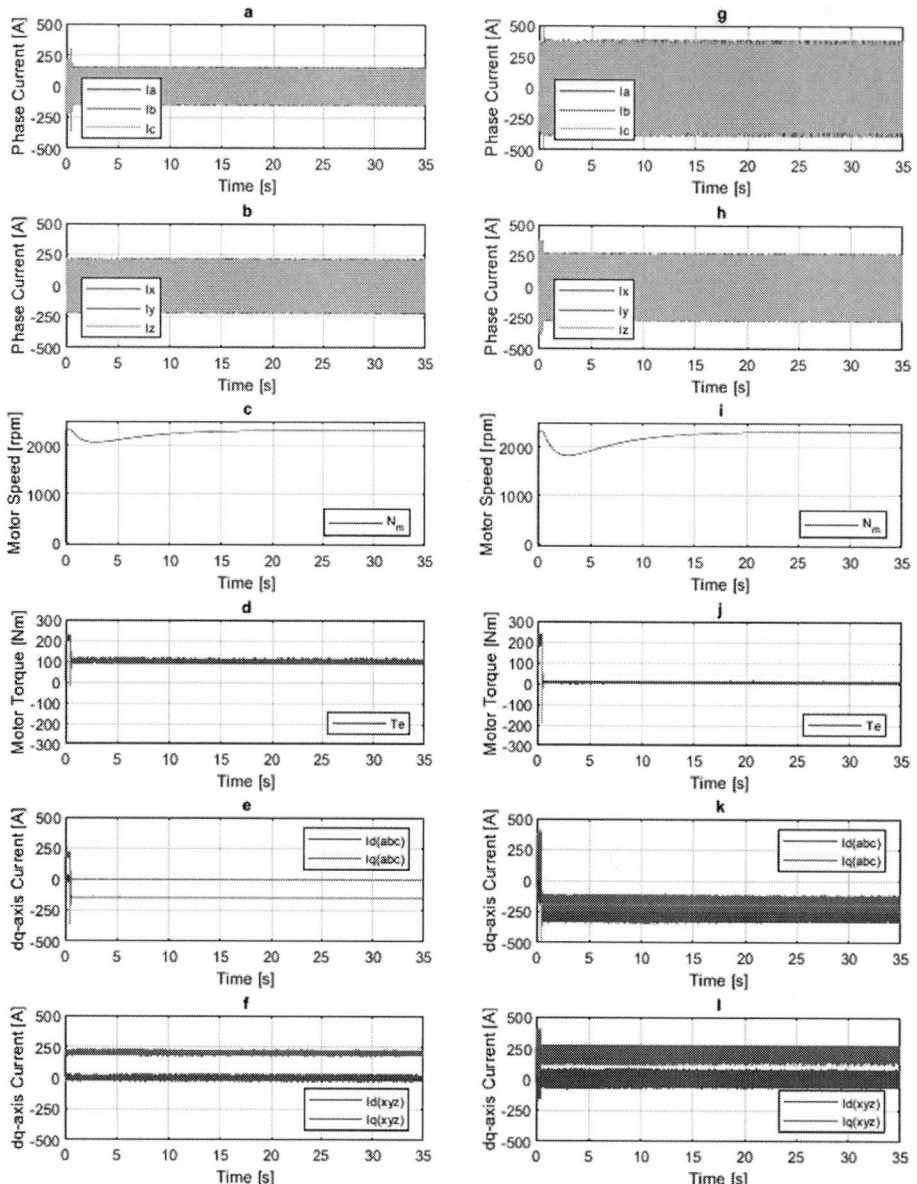

Fig. 3: Simulation results of ASM-ASC in LM-DTP-PMSM (the six sub-graphs at left column) and in HM-DTP-PMSM (the six sub-graphs at right column). (a) and (g) abc-sub-system phase current. (b) and (h) xyz-sub-system phase current. (c) and (i) motor speed. (d) and (j) motor torque. (e) and (k) dq-axis current of abc-sub-system. (f) and (l) dq-axis current of xyz-sub-system.

The simulation results of SSM-ASC in LM- and HM-DTP-PMSM are shown in Fig. 4. Since it is expected that the motor will lose all torque in SSM-ASC, simplified feathering process of the CSP is simulated after 17.5 Second. The process of SSM-ASC can be mainly divided into three parts. The first and the second parts are similar like in ASM-ASC, where the shaft speed will first drop and then get recovered. The third part is feathering, in which the shaft speed will decelerate to a very low value. It can be observed that the motors with different k values behave similarly after SSM-ASC in all three parts. Before the end of feathering, the magnitude of stator current remains around 145A, and the motor torque remains at a small negative value. When the shaft speed is low enough, the motor braking torque will increase up to -73 Newtonmeter that can even help to brake the shaft faster.

After the peak braking torque, as the shaft speed further decreases, the braking torque, as well the stator current, decreases simultaneously. In the entire process, there is no over current or over torque problem. One problem that can be observed is the demagnetizing current in d-axis, which remains around -145A in both sub-systems until the end of feathering.

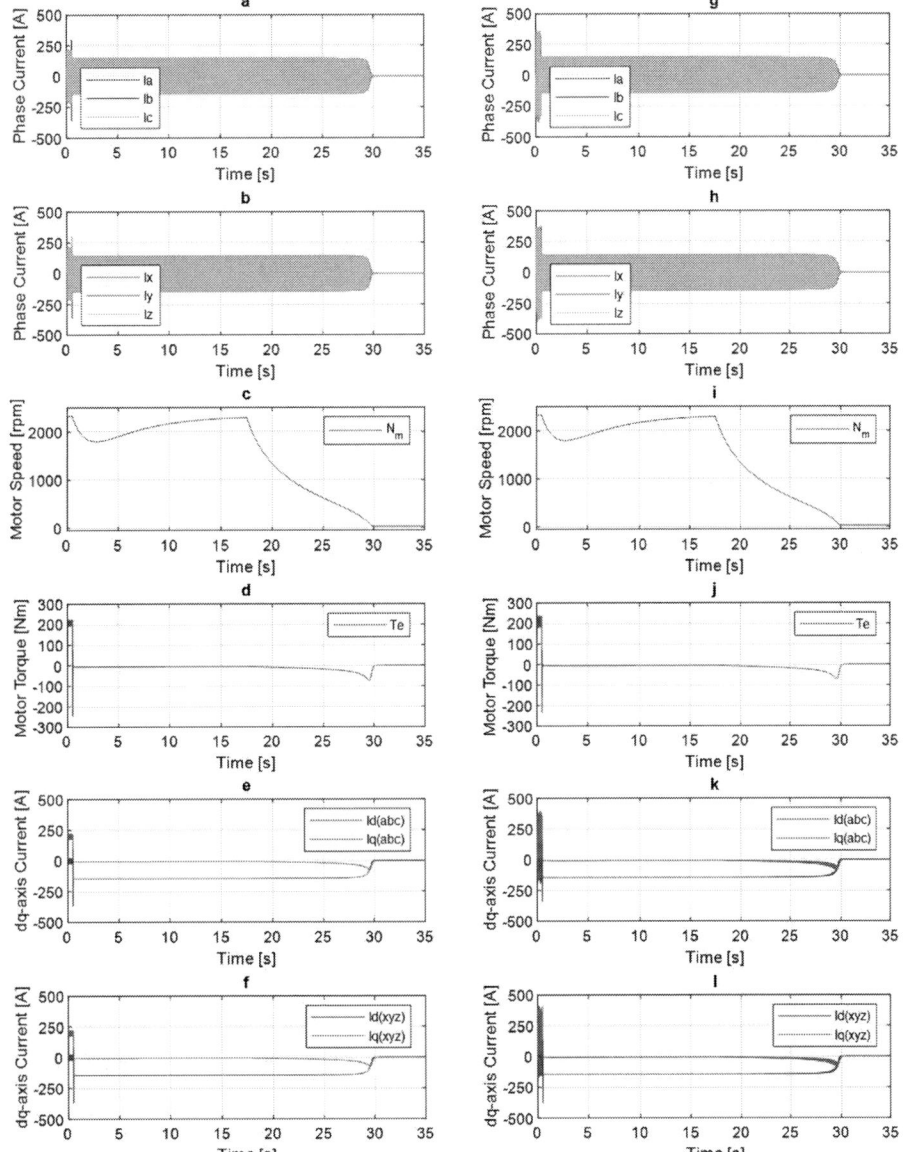

Fig. 4: Simulation results of SSM-ASC in LM-DTP-PMSM (the six sub-graphs at left column) and in HM-DTP-PMSM (the six sub-graphs at right column). (a) and (g) abc-sub-system phase current. (b) and (h) xyz-sub-system phase current. (c) and (i) motor speed. (d) and (j) motor torque. (e) and (k) dq-axis current of abc-sub-system. (f) and (l) dq-axis current of xyz-sub-system.

The simulation results of phase current and motor torque zoomed in ASC transient are shown in Fig. 5. It can be observed that both LM-DTP-PMSM and HM-DTP-PMSM can operate at similar points if the main electrical parameters are designed to be the same, even with significantly different mutual inductance between two sub-systems. With higher mutual inductance between two sub-systems, the stator current ripple becomes higher because one sub-system current can induce EMF on the another sub-system and finally creates higher ripple in the stator current. The over current during transient in HM-DTP-PMSM with ASM-ASC is the highest one in all scenarios, which is almost twice higher than others. The torque oscillation is similar in HM-DTP-PMSM with ASM-ASC ,and LM- and HM-DTP-PMSM with SSM-ASC, which is almost twice higher than in LM-DTP-PMSM with ASM-ASC. The overall transient time is around 30 milliseconds.

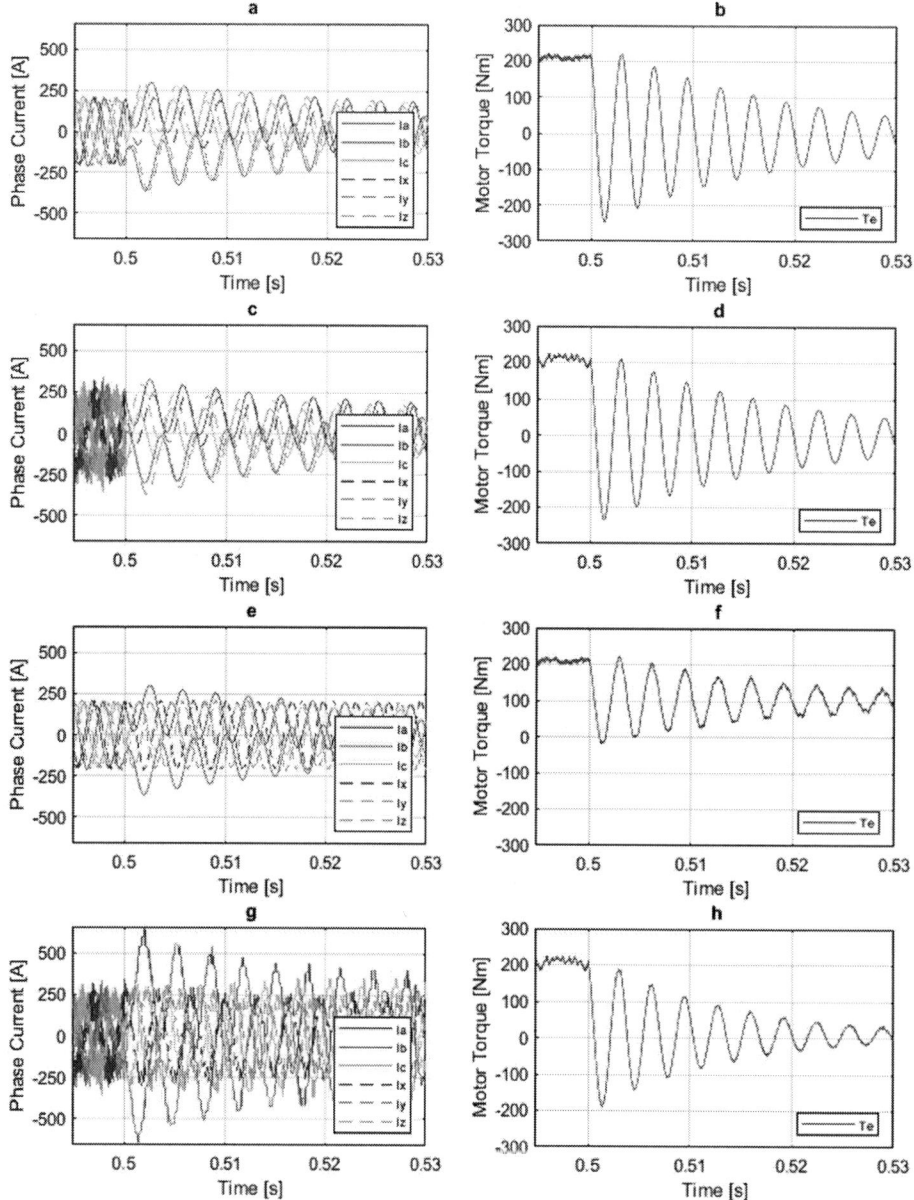

Fig. 5: Simulation results of phase current and motor torque during ASC transient. The four sub-graphs at left column are phase current. The four sub-graphs at right column are motor torque. (a) and (b) from LM-DTP-PMSM with SSM-ASC. (c) and (d) from HM-DTP-PMSM with SSM-ASC. (e) and (f) from LM-DTP-PMSM with ASM-ASC. (g) and (h) from HM-DTP-PMSM with ASM-ASC.

Conclusion

This paper focuses on the short circuit behaviors of DTP-PMSM. The influence of mutual inductance between two sub-systems is studied and analyzed under two different ASC methods, the ASM-ASC and the SSM-ASC. A mathematical model of DTP-PMSM is developed in this paper for the study. One innovative point of this mathematical model is to use a simple variable k to express the ratio of the mutual inductance to the self-inductance. The variable k represents the significance of the mutual inductance in total d- and q- axis inductance directly. Based on this mathematical model, analytical expressions of the steady state dq-axis current and the motor torque of both ASC methods are derived respectively. By changing k values, comparative analysis has been completed on demagnetizing effect, magnitude of the steady state ASC stator current, and remaining motor torque for both ASC methods. To validate the theoretical result, a simulation model of an EPU is built in MATLAB/Simulink. For each ASC method, both HM-DTP-PMSM and LM-DTP-PMSM are simulated. With dynamic DTP-PMSM and CSP models, both transient and steady state behaviors of EPU after ASC can be observed. The steady state ASC behaviors of DTP-PMSM from the simulation can verify the mathematical model. According to both theoretical analysis and the simulation result, if the mutual inductance between two sub-systems is low, the ASM-ASC performs better compared with SSM-ASC due to the sufficient remaining motor torque. Inversely, the SSM-ASC performs better if the mutual inductance between two sub-systems is high, due to less steady state stator current.

References

[1] D. Michieletto, N. Bianchi and M. Brunetti, "Dual Three-phase Motor Fault Tolerance for Modern Transport," 2021 International Aegean Conference on Electrical Machines and Power Electronics (ACEMP) & 2021 International Conference on Optimization of Electrical and Electronic Equipment (OPTIM), 2021, pp. 405-412, doi: 10.1109/OPTIM-ACEMP50812.2021.9590040.

[2] B. A. Welchko, T. M. Jahns, W. L. Soong and J. M. Nagashima, "IPM synchronous machine drive response to symmetrical and asymmetrical short circuit faults," in IEEE Transactions on Energy Conversion, vol. 18, no. 2, pp. 291-298, June 2003, doi: 10.1109/TEC.2003.811746.

[3] P. Giangrande, V. Madonna, S. Nuzzo, C. Gerada and M. Galea, "Braking Torque Compensation Strategy and Thermal Behavior of a Dual Three-Phase Winding PMSM During Short-Circuit Fault," 2019 IEEE International Electric Machines & Drives Conference (IEMDC), 2019, pp. 2245-2250, doi: 10.1109/IEMDC.2019.8785164.

[4] M. Kozovsky, P. Blaha.: Double three-phase PMSM structures for fail operational control, IFAC-PapersOnLine, Volume 52, Issue 27,2019, Pages 1-6, ISSN 2405-8963,

[5] M. Barcaro, N. Bianchi and F. Magnussen, "Faulty Operations of a PM Fractional-Slot Machine With a Dual Three-Phase Winding," in IEEE Transactions on Industrial Electronics, vol. 58, no. 9, pp. 3825-3832, Sept. 2011, doi: 10.1109/TIE.2010.2087300.

[6] A. M. EL-Refaie and T. M. Jahns, "Optimal flux weakening in surface PM machines using fractional-slot concentrated windings," in IEEE Transactions on Industry Applications, vol. 41, no. 3, pp. 790-800, May-June 2005, doi: 10.1109/TIA.2005.847312.1

[7] Y. Hu, Z. Q. Zhu and M. Odavic, "Comparison of Two-Individual Current Control and Vector Space Decomposition Control for Dual Three-Phase PMSM," in IEEE Transactions on Industry Applications, vol. 53, no. 5, pp. 4483-4492, Sept.-Oct. 2017, doi: 10.1109/TIA.2017.2703682.

[8] Z. Liang, D. Liang and S. Jia, "Inductance Calculation for the Symmetrical Non-Salient Dual Three-Phase PMSM Based on Winding Function Approach," 2018 21st International Conference on Electrical Machines and Systems (ICEMS), 2018, pp. 269-274, doi: 10.23919/ICEMS.2018.8549109.

Hybrid Silicon-SiC Inverter – Combining the Best of Both Worlds

Hans-Günter Eckel, Felix Kayser, Pham Ha Trieu To
UNIVERSITY OF ROSTOCK

Abstract:

SiC MOSFET get more and more attractive also for medium and high power inverters. There well known advantages are low switching losses and low on-state voltage at low current. But silicon IGBT still offer a better cost to chip area ratio. The combination of Si-IGBT and SiC-MOSFET will lead to a better performance than pure silicon inverters with less SiC chip area than full-SiC solutions. This might lead to a superior performance cost ratio, especially for high power inverters with large chip area. In this paper, different two level and three level hybrid topologies are investigated. The switching behavior of these Si-SiC-hybrids is experimental investigated, the maximum output power and the efficiency is compared with full Si and full SiC inverters.

Curriculum Vitae:

Hans-Günter Eckel received the Dipl-Ing. degree from the TU Braunschweig, Germany in 1989 and the Ph.D. degree from the University Erlangen-Nürnberg in 1997. He had over ten years of experience in the development of traction converters with Siemens, Nuremberg, when he joined the University of Rostock in 2008, where he is a Professor for Power Electronics and Electrical Drives.

His research activities include characterization of and gate drive for high power, high voltage semiconductors, control of multi-level inverters and the analysis of power electronic coupled loads and sources in the grid.

He has published over 200 technical papers in these areas. He is an IEEE senior member, member of the EPE ISC and the PCIM advisory board.

Contact Details:

Hans-Günter Eckel
University of Rostock
Albert-Einstein-Str. 2
18059 Rostock, Germany
+49 381 498 7110
hans-guenter.eckel@uni-rostock.de
www.iee.uni-rostock.de

Robustness of SiC Trench MOSFETs

Christian Felgemacher
ROHM Semiconductor GmbH

Abstract:

In this presentation aspects regarding the robustness of state of the art SiC Trench MOSFETs will be discussed. Results of short circuit test results on discrete devices as well as modules using multiple parallel SiC MOSFETs will be shown alongside measurements that demonstrate the timely detection and safe turn-off of both short-circuit type I and type II. As an additional aspect of device robustness results of cosmic radiation robustness tests executed on SiC MOSFETs will be presented.

Curriculum Vitae:

Christian Felgemacher received an M.Eng. (Hons.) degree in electronics and electrical engineering with management from the University of Edinburgh, Scotland, UK in 2011. From 2012 to 2017 he was a research assistant at the Centre of Competence for Distributed Electric Power Technology at the University of Kassel, Germany. During this time he worked on topics relating to reliability of power semiconductors in photovoltaic inverters and the application of modern wide-band-gap power devices to renewable energy applications. In 2018 he was awarded a doctorate in engineering (Dr.-Ing.) from the University of Kassel, Germany.

Since 2017 he works for ROHM Semiconductor GmbH where he is today leading a technical team supporting European customers in power systems related aspects on application level. He is a member of IEEE.

Contact Details:

Dr.-Ing. Christian Felgemacher
ROHM Semiconductor GmbH
Karl-Arnold-Str. 15
47877 Willich, Germany
+49 2154 921-226
christian.felgemacher@de.rohmeurope.com
www.rohm.com

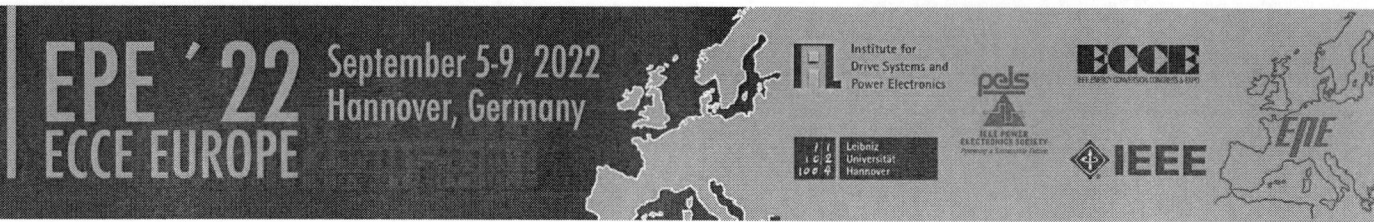

3D Predictive Fatigue Modeling of Power Modules

Ben Samples and Brandon Passmore
Wolfspeed

Abstract:

As the demand for electric vehicles continues to grow, robust fatigue modeling tools will be a key enabler in the success of power module products and technology. Predictive modeling tools can help predict performance characteristics for customers, optimize design materials/geometries/features to take advantage of the superior characteristics of SiC, estimate fatigue and lifetime, reduce time to market by minimizing testing time and resources, and reduce product cost by minimizing unnecessary over design.

The fundamental challenge of simulating SiC power module characteristics is that they multi-physics problem comprised of several different fields including electrical, mechanical, thermal, chemical, and material science as well as combinations of each. Therefore, the simulation tools must be capable of coupling multi-physics models, boundary conditions, and material properties to accurately predict power module characteristics.

In this presentation, a predictive fatigue model will be presented highlighting tradeoffs highlighting the substrate attach thermal shock life based on sweeping multiple modules parameters.

Curriculum Vitae:

Brandon Passmore completed his B.S. in Electrical Engineering from Arkansas State University in 2003, M.S. in MicroEP in 2005, and Ph.D. in MicroEP in 2008. Brandon has been with Wolfspeed since 2010 where he is a Sr. Hardware Development Engineering Manager leading the Design Engineering Team in the Power Modules BU. At Wolfspeed, his group focuses on developing new SiC power modules rated from 650 V to 1.7 kV and 50 A to 800 A and beyond. In addition, his group has been heavily involved in several projects focused on developing new packaging designs, modeling techniques, materials, and processes to fully optimize the performance of SiC power devices. Some of these technologies include electrical-thermal-mechanical multi-physics modeling, wire bondless technologies, new die attach and substrate technologies, and high heat transfer and advanced cooling technologies for the advancement of high performance SiC power modules.

Contact Details:

Ben Samples
Wolfspeed
535 W. Research Blvd.
Fayetteville, AR, United States
479 301-6017
ben.samples@wolfspeed.com

Heterogeneous Integration of Power Conversion using Power Supply on Chip and Power Supply in Package

Cian Ó Mathúna, Seamus O'Driscoll
Tyndall National Institute,

University College Cork, Ireland

Abstract:

The talk will discuss the performance opportunities and technology challenges through the emergence of heterogeneous integration of power converters (including power management ICs, power switches, magnetics and capacitors) into processor packages using PCB-embedding, 2.5D or even 3D packaging concepts to deliver significant system-level energy savings.

This talk will discuss the impact of the commercial emergence, of magnetics-on-silicon technology and associated PCB-embedded magnetics technologies which are enabling Power Supply on Chip and Power Supply in Package platforms. These technologies, along with integrated high density capacitor technologies, PMICs and power switches, are central to the ongoing development of the concept of integrated, granular power for multi-core microprocessors and other complex system-on-chip (SOC) platforms.

Curriculum Vitae:

Cian Ó Mathúna is Head of Micronamo Systems at Ireland's Tyndall National Institute. His research, over three decades, into the miniaturisation and integration of magnetics-on-silicon, has played a key role in disruptive developments in integrated power management for applications in portable electronics and high performance computing. Using semiconductor processing of thin-film magnetics, Ó Mathúna's team have made bulky power magnetic components disappear onto silicon chips. Referred to as MagIC, Tyndall's magnetics-on-silicon technology has been licensed to two of the world's leading consumer electronics companies as well as a leading semiconductor foundry.

In 2008, Cian founded the International Workshop on Power Supply on Chip (PwrSoC) which has become the highly-influential flagship workshop for the IEEE Power Electronics Society and the US-based Power Sources Manufacturers Association (PSMA). Through his leadership in PwrSoC, and his extensive collaborations with world-leading industry players in Europe, USA and Asia, Ó Mathúna has had a significant influence on the emergence of a global supply-chain that, in 2021, began delivery of high-volume production of magnetics-on-silicon for use in commercial product.

In 2013, Prof. Ó Mathúna was elevated to IEEE Fellow with the citation *"for leadership in the development of power supply using micromagnetics on silicon"*. For his impact on the industry, Ó Mathúna has, in 2021, been awarded the IEEE Power Electronics Society Technical Achievement Award for Integration and Miniaturisation of Switching Power Converters. For the development of the MagIC technology and its expected impact over the next decade, the Integrated Magnetics team has, in 2021, also received one of two 2021 EARTO (European Association of Research and Technology Organisations) Impact Innovation Awards.

Seamus O'Driscoll is currently Head of Group for Integrated Power Systems at Microelectronics Circuits Centre Ireland, Tyndall National Institute, Cork, Ireland. He is leading a number of research themes spanning ultralow-power management integrated circuits (PMIC), isolated PwrSiP converters, smart gate drivers, and point-of-load dc-dc, with all employing thin-film magnetics-on-silicon or substrate-embedded magnetic solutions. Earlier career roles have included a Power Control Silicon Systems Architect with Texas Instruments Ltd., Cork, and a Corporate Technology Staff Engineer with Artesyn Technologies Ltd., Youghal, Cork. He has released many professional power product designs to the world's leading communications and computer companies.

Contact Details:

Prof. Cian Ó Mathúna, B.E. (Elec), MEngScience, PhD, Fellow IEEE
Head of Micro/Nano Systems Centre, Tyndall National Institute,
Research Professor, School of Engineering,
University College Cork,
Lee Maltings, Dyke Parade,
Cork, Ireland.
Tel: +353 21 234 6350
www.tyndall.ie
cian.omathuna@tyndall.ie

Seamus O'Driscoll
Head of Integrated Power Systems
Microelectronics Circuits Centre Ireland,
Tyndall National Institute,
University College Cork, Ireland
www.mcci.ie
seamus.odriscoll@mcci.ie

Driving Innovations for Power Electronics with Integratable and Sustainable Magnetics

Matt Wilkowski
EnaChip

Abstract:

Magnetics are historically one of the physically largest components in power distribution networks from milliwatts to megawatts.

Consistent with the overall goal to reduce the visual presence of power converters that are necessary for continually advancing electronic devices, there is a continual push from the smallest to the largest power converters to have the magnetics blend in with other components of power delivery systems relative to their physical appearance, physical size, and cost. Power distribution systems will always require the magnetics function and accordingly the magnetics function will not disappear. This drives the term "make magnetics invisible" or more realistically become indistinguishable from other components of the overall product assembly. The reality is that magnetics become less visible by moving towards more complete integration with the semiconductor devices at the Integrated Voltage Regulator (IVR) levels or by physical size reduction at the Solid-State Transformer (SST) levels. Integration and miniaturization are the main objectives for power magnetics across all power levels

There has not been an equivalent Moore's law for magnetics to keep pace with silicon over the past fifty-plus years nor a breakthrough in magnetic materials to coincide with the more recent commercialization GaN devices. The major constituents of all magnetic materials reside in the three elements that are ferromagnetic at room temperature iron (Fe), cobalt (Co) and Nickel (Ni). Even though there has not been quantum simultaneous advancement in the key parameters of saturation flux density and power losses magnetic materials over the past decade, magnetics continues to shrink in physical size. Improved and automated manufacturing processes and innovative circuit topologies have the been key drivers to reduce the physical size of magnetics. Packaging will play a key role for closer alignment of power magnetics with improvements in semiconductor devices.

In the absence of new magnetic materials that simultaneously increase the useable flux density due to saturation constraints and reduce ac power losses, manufacturers have focused on fabrication process improvements and automation to reduce the physical size and improve the efficiency of magnetic components – effectively utilizing the available physical envelope more efficiently.

As part of the ongoing effort to either reduce the physical footprint or vertical height of magnetic components, there is the continual push to decrease volt second stress by either increasing switching frequencies at the expense of switching losses of semiconductor devices or to adopt multiple stage circuit topologies that decrease the volt second stress on magnetic components. The reduction of volt second stress leads to reducing the required value of inductance to meet peak current limitations or reduce the physical size required for a given value of inductance.

As integration of power magnetics and silicon devices continues to progress towards the application focus of this presentation Fully Integrated Voltage Regulators (FIVR) there will be a point for a cross over between off-silicon discrete magnetics that are either wire wound or assembled as PCB magnetics and magnetics on silicon. The integration of magnetics on silicon represents the ultimate integration for which magnetics are indistinguishable not only by physical visual observation but from manufacturing and reliability viewpoints as well.

Magnetics on silicon provide options to address both compatibility with silicon wafer processing and very thin profiles. Three of the technical challenges for wafer level magnetics are to provide:

1. Sufficient magnetic material cross section to achieve the values of inductance that are compatible with a switching frequency for the silicon devices to operate efficiently and with a physical footprint compatible with the physical size of the silicon devices.
2. Sufficient magnetic material cross section to support the required operating currents without saturation effects with a physical footprint compatible with silicon devices.
3. Magnetic material layer thicknesses consistent with permeability and ac power loss requirements at the specified operating frequencies

Sputtered magnetic materials applied as part of the FEOL (front end of line) process are inherently thinner per layer than electrodeposited materials but require more layers than thicker electrodeposited materials thus designs tend to have lower inductance values resulting in higher switching frequencies.

Electrodeposited materials applied as part of a BEOL (back end of line) process are inherently thicker per layer than sputtered materials and require less layers than thinner sputtered materials can have higher values of inductance and subsequently support lower switching frequency.

Manufacturing magnetics with with the same equipment and processes as the semiconductor devices they support, magnetics by default will follow the same sustainability trends for semiconductor manufacturing. As part of the development to have magnetics blend in with the silicon devices, magnetics are manufactured with the same processes and equipment as semiconductor devices. Wafer level magnetics will follow the sustainability progress of the semiconductor industry –more efficient manufacturing processes, less waste, optimal resource utilization and open innovation for end application products. Wafer level magnetics will become an integral part of the "More than Moore" silicon and packaging initiatives.

Curriculum Vitae:

Matt Wilkowski has been involved with the design and productization of power magnetic components for integration in power converters for over forty years. While at Torwico Electronics, Matt design power magnetics for aerospace and military applications. During his tenure at AT&T Bell Labs/Lucent Technologies/Tyco Electronics Power Systems from 1983 thru 2003, Matt was responsible for the technology road mapping, product design and verification, commercialization, and ongoing product support of power magnetics that that were integrated in the assembly process of power converters intended for telecommunication applications. This work entailed the development of inductor and transformer product families to support wide ranges of input and output voltages for different power ranges from 1 watt through 10KW for emerging circuit topologies to take advantage of switch frequencies from 500 kHz thru 20 MHz and accommodating physical assembly integration as well as thermal integration etc. into both ac-dc and dc-dc power converters. While part of Enpirion from 2003 through 2013, Matt became more focused on the integration of power magnetics into semiconductor device packaging leading to the commercialization of a wide product portfolio of Power System in Package (PSiP) devices. With the acquisition of Enpirion, Matt was a technology architect at Altera and transitioned to a principal engineer at Intel after the acquisition of Altera focused on the development of power magnetics for various integration levels (highly, fully, etc.,) of integrated voltage regulators (IVRs). He is currently Vice President of Technology at EnaChip and is focused on productization of wafer level magnetics to address various market applications.

Matt is a IEEE Fellow and one of the co-chairs for the PSMA Magnetics Committee and has led the organizing committee of the Power Magnetics @ High Frequency pre-APEC workshop series since the first workshop in 2016. Matt has been the past chairman of both the IEC Technical Committee 51 and the PELS Electronics Transformers Technical Committee (ETTC) and is currently leading initiatives for the development of standards for the testing and characterization of magnetic components and materials for both standards organizations.

Contact Details:

Matt Wilkowski

EnaChip
5 Crescent Avenue, Building F
Rocky Hill New Jersey USA
908-356-0056
mwilkowski@enachip.com
www.enachip.com

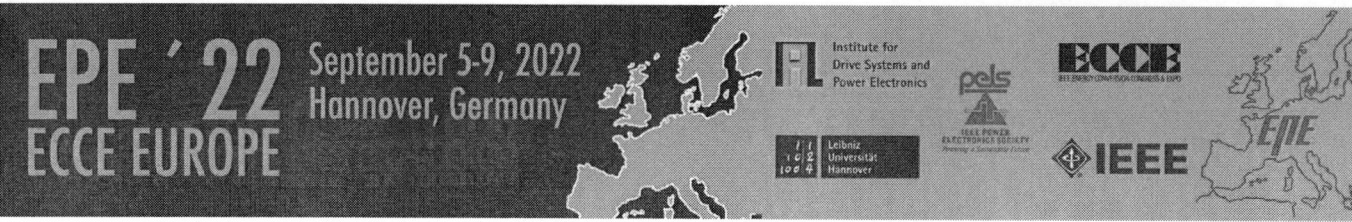

Impact of package technology on the switching behavior of high-voltage GaN FETs

Sebastian Klötzer
Nexperia

Abstract:

Package choice and assembly technique affect the performance of fast-switching high-voltage FETs.

It is well known that GaN devices in particular benefit from advanced packaging technology with reduced stray inductance, but how much difference is really to be expected under typical application conditions?

In this talk, switching waveforms for GaN FETs in a standard leaded, wire-bonded TO-type package are analyzed and compared with those for an advanced copper-clip SMD package. The GaN FETs used are nearly identical, and so the differences are solely due to packaging. Subsequently, the different fields of applications for which they are best suited are discussed.

Curriculum Vitae:

Sebastian Klötzer is a Principal Application Engineer at Nexperia. He works on application-oriented characterization of wide bandgap power semiconductors and high frequency converter design, both for industrial and automotive applications. His main interests include low-parasitic design, high-frequency magnetics design and high-performance SMD packages. He has worked as a research associate at Helmut Schmidt University Hamburg and holds a diploma in electrical engineering from the Technical University of Munich.

Contact Details:

Sebastian Klötzer
Principal Application Engineer
Nexperia Germany GmbH
Stresemannallee 101
D- 22529 Hamburg

sebastian.kloetzer@nexperia.com

Impact of power electronics on battery operation

Dirk Uwe Sauer, Prof. Dr.
Institute of Power Electronics and Electrical Drives (ISEA)
Institute for Power Generation and Storage Systems @ E.ON ERC
RWTH Aachen University

Abstract:

Lithium-ion batteries are nowadays offered in a large number of different cell designs and material combinations. This results in a wide range of performance parameters. However, in almost all cases batteries are used together with power electronic converters and a central question is the interaction between the converters and the batteries. This can for example be the influence of ripple currents or the implementation of impedance diagnostics. In the end, this is also the question of the need for smoothing elements of the power electronics on the battery side.

This presentation will give an overview of the influence of ripple currents on battery cells and their application for electrochemical impedance spectroscopy. For this purpose, the structure of the battery cell is used to show theoretically, which influences arise and results of experimental investigations are presented.

Curriculum Vitae:

Dirk Uwe Sauer if professor for Electrochemical Energy Conversion and Storage Systems at the Institute of Power Electronics and Electrical Drives (ISEA) at RWTH Aachen University since 19 years. He is a specialist for all aspects on system integration of batteries incl. testing, characterization, ageing, modeling, diagnostics, lifetime prediction and field integration into any type of mobile or stationary applications. His chair has about 75 full time employees and more than 60 students as student assistance, or in bachelor or master thesis. He is a member of the National Academy of Science and Engineering (acatech) and the Berlin-Brandenburg Academy of Sciences and Humanities (BBAW). He is also the co-founder of 4 spin-off companies.

Contact Details: Dirk Uwe Sauer, Prof. Dr.

Chair of Electrochemical Energy Conversion and Storage Systems
Institute of Power Electronics and Electrical Drives (ISEA)
Institute for Power Generation and Storage Systems @ E.ON ERC
RWTH Aachen University

Jägerstrasse 17/19, 52066 Aachen, Germany

Phone number: +49 241 80 969

E-Mail address: sr@isea.rwth-aachen.de

URL: www.isea.rwth-aachen.de

Trends in Power Electronics and Batteries

for Electrified Vehicle Infrastructure

Dr. Torsten Leifert

Volkswagen Group Charging (Elli) Hardware Platform

Abstract:

A tremendous growth in the quantity and diversity of electric vehicles is currently occurring. Different vehicle platforms have been progressively developed in recent years, and this development continues today. Among the most important questions is where and how to charge the vehicles. In Germany, it is expected that up to 15 million BEV will be on the roads by 2030.

This presentation will give an overview of the platforms, system functions and technical requirements for charging. Solutions for wall boxes and their inner circuitry will also be presented. Based on the use cases V2H and V2G, some trends in power electronics and options for storage systems will be discussed.

Curriculum Vitae:

Born in 1962 in Lower Saxony

- Diplom in electrical engineering and PhD concerning field-oriented induction machines at Leibniz University Hannover
- Lecturer at Christian Albrecht University Kiel in realtime applications for power electronics
- STIL & Linde Group: Powertrain development for forklift trucks
- Sieb & Meyer AG: Inverter for small wind turbines and fuel cells
- SMA Solar Technology AG: Development for Sunny Island inverter and in technology department
- Volkswagen / VW Group Components: Onboard charger and powertrain inverter
- VW Group Charging: Systems for bidirectional charging
- Board member of ECPE e.V.
- Member of the Wilhelm-Busch-Gesellschaft

Contact Details:

Dr. Torsten Leifert
Volkswagen Group Charging / Hardware Platform
Berliner Ring 2, D-38440 Wolfsburg
Germany
+49 01525 4954064
Torsten.leifert@volkswagen.de / Torsten.leifert@elli.eco

www.elli.eco

Impact of high frequency current pulses on battery ageing

Julia Kowal, Prof. Dr.-Ing.
Electrical Energy Storage Technology
Institute for Energy and Automation Technology
TU Berlin

Abstract:

In almost all applications, lithium-ion batteries are used in combination with power electronics. The occurring high frequency ripple currents are typically reduced by using DC link capacitors. On the other hand, it is assumed that frequencies in the kHz range do not affect ageing of batteries because it is outside the frequency range of electrochemical processes, which would mean that the capacitor could possibly also be reduced. However, still the number of investigations, especially based on automotive batteries, is quite limited.

This presentation will give an overview of published ageing tests that investigate the influence of ripple currents on battery ageing. Additionally, our own results from 18650 cells and automotive cells from the project SiCWell are presented, which show an impact of different ripple parameters on the lifetime of the cell.

Curriculum Vitae:

Julia Kowal has been professor for Electrical Energy Storage Technology at the Institute for Energy and Automation Technology at TU Berlin since 2014. Her research focus is testing, characterization, ageing, modeling, diagnostics and lifetime prediction of different battery technologies. She is managing director of the Institute for Energy and Automation Technology.

Contact Details: Julia Kowal, Prof. Dr.-Ing.

Electrical Energy Storage Technology
Institute for Energy and Automation Technology
TU Berlin

Einsteinufer 11, 10587 Berlin, Germany

Phone number: +49 30 314 25394

E-Mail address: julia.kowal@tu-berlin.de

URL: www.tu.berlin/eet

Aircraft Electrification – System-Level Potentials for Aviation Decarbonization

Kathrin Ebner, Antoine Habersetzer, Arne Seitz
BAUHAUS LUFTFAHRT e.V.

Abstract:

Aviation's ambitious climate targets have fueled the search for innovative and energy-efficient propulsion concepts. In that context, introducing electrical energy into the propulsion train opens up a vast design space. Various architectures employing battery electric and/or fuel cell electric concepts have been proposed and attract considerable attention fostered by their perspective contributions to aviation decarbonization. To understand the potential impact of these concepts, a holistic assessment regarding technological, but also system level aspects is highly relevant.

Against this backdrop, this talk will first provide an overview on aircraft electrification strategies and related technological implications, requirements and considerations. Then, their pertinence for different market segments will be examined, leading to a discussion of their possible impact on reducing aviation's overall footprint based on current flight and emission distribution. Finally, we will look at transition scenarios factoring in fleet penetration of different technological options, fuel alternatives and associated development rates. By shifting the perspective to "the bigger picture" and deconvoluting individual contributions towards the envisaged carbon-neutrality of aviation, associated potentials of the above-mentioned electrified concepts can be holistically addressed.

Curriculum Vitae:

Dr. Kathrin Ebner is leading the competence domain Energy & Fuels and acts as Strategic Research Coordinator of Hydrogen and (Hybrid-) Electric Aviation at Bauhaus Luftfahrt, Taufkirchen, Germany. Her main research focus in on electrochemical energy storage and conversion in the context of sustainable fuel production as well as propulsion applications. She holds a PhD in electrochemistry from ETH Zurich.

Contact Details:

> Kathrin Ebner
> Bauhaus Luftfahrt e.V.
> Willy-Messerschmitt-Str. 1
> 82024 Taufkirchen, Germany
> +49 89 3074-84923
> kathrin.ebner@bauhaus-luftfahrt.net
> https://www.bauhaus-luftfahrt.net

About Power Electronics Challenges in Aviation
Dr. Marco Bohllaender

Rolls-Royce Deutschland Ltd & Co KG

Abstract:

The aviation market is in change. The degree of aircraft propulsion electrification increases rapidly and requires power semiconductor and power electronic solutions to fulfil different requirements than in any other industry and application. The invited lecture will give a survey on those challenges and depict their impact on the power electronic system design in electrified aircrafts..

Curriculum Vitae:

Dr. Marco Bohllaender studied Electrical Engineering at Chemnitz University. He joined Infineon AG as development engineer and was responsible for the product development of converter power stages. In 2013 Dr. Bohllaender received his PhD in power electronics with the title "Power cycling test-based lifetime analysis methods for power semiconductors in offshore wind energy plants" at Chemnitz University. In the same year he joined Siemens AG as power electronics system architect for automotive inverters. In 2017 he took over the role of a power electronics expert at Siemens eAircraft and worked in several flight demonstrator projects. Today, Dr. Bohllaender is lead engineer in a product development project and head of the Power Electronics and Wiring team at Rolls-Royce Electrical, a business unit inside Rolls-Royce plc.

Contact Details:

Dr. Marco Bohllaender
Rolls-Royce Deutschland Ltd & Co KG
Guenther-Scharowsky-Strasse 1
91058 Erlangen
+49 152 22726816
marco.bohllaender@rolls-royce-electrical.com
www.rolls-royce.com

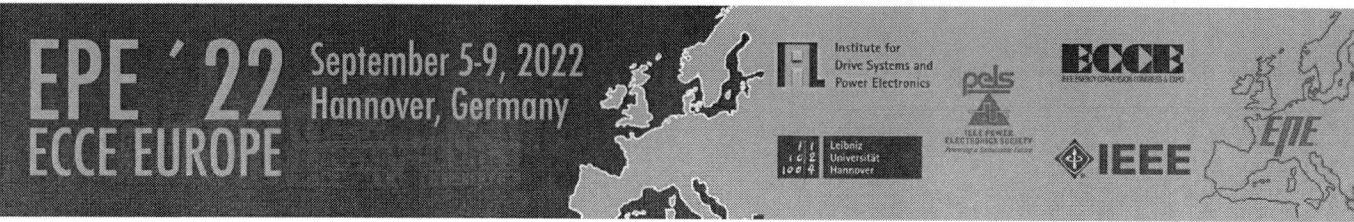

Development of electric motors for aircraft applications

Simon Wolfstädter

OSWALD ELEKTROMOTOREN GMBH

Abstract:

The trend toward the electrification of air transport is being driven by its positive contribution to a more sustainable future for aviation. Electric motors can help to increase efficiency, reduce emissions, and enable new concepts for the propulsion system of aircrafts.

The various applications in the field of aviation place different requirements on the propulsion system and the electric machine. Here, the design of the electric machines is influenced not only by the required propulsion power itself, but also by the torque and speed, as well as the overall concept of the aircraft. A general requirement for powerful, efficient, and lightweight propulsion systems places high demands on the materials in terms of electromagnetic, mechanical, and thermal behaviour. High power and torque densities of motors, even at high efficiency, lead to high power dissipation densities in the different sub-areas of the machine, which leads to special cooling challenges.

Curriculum Vitae:

2010 - 2014	Electrical engineering and information technology (B. Eng) Technical University of Applied Sciences, Aschaffenburg
2014 - 2016	Electrical engineering and information technology (M. Eng) Technical University of Applied Sciences, Aschaffenburg
Since 2017	Research and Development Engineer, Oswald Elektromotoren GmbH, Miltenberg

Contact Details:

Name	Simon Wolfstädter
Company / Institution	Oswald Elektromotoren GmbH
Address	Benzstraße 12
City, Country	63897 Miltenberg
Phone number	+49 (0) 9371 / 9719 - 0
E-Mail address	simon.wolfstaedter@oswald.de
URL (if any)	https://www.oswald.de

Powertrain trends in electric trucks

Dr. Luciana C. Afonso
Infineon Technologies AG

Abstract:

Regulations demanding reduction of CO_2 and other greenhouse gas emissions are being implemented worldwide and they are the main driver for electrification in transportation. This presentation focuses on the trends of electric powertrain for trucks, their configuration in the vehicle, their limitations and optimization potential.

Curriculum Vitae:

Dr. Luciana Caminha Afonso is system architect for the application of commercial construction and agriculture vehicles at Infineon Technologies. She has eight years of experience in the area of electric powertrain and has held positions at powertrain Tier 1 and OEMs as a development engineer, project manager, and product owner, lately with focus on trucks and buses. Dr. Afonso holds a bachelor's degree in Physics and is Doctor of science in nuclear engineering from the University of Sao Paulo, Sao Paulo, Brazil, and was a guest Ph.D. student at the Helmholtz Center Munich, Germany.

Contact Details:

Dr. Luciana Caminha Afonso
Infineon Technologies AG
Max-Planck-Strasse 5
Warstein, Germany
+49 15152542300
LucianaCaminha.Afonso@infineon.com
https://www.infineon.com/cav

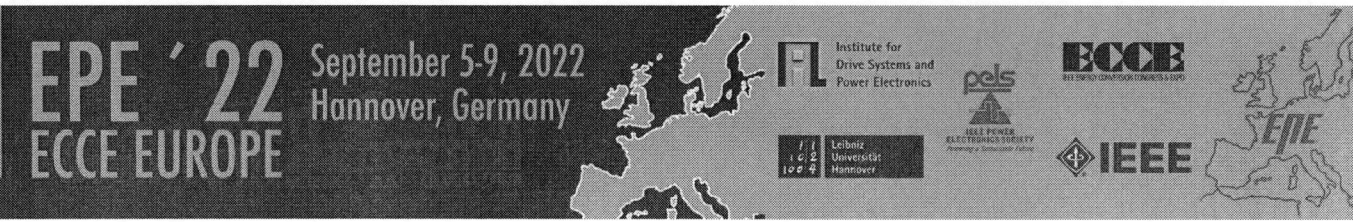

Modulation Strategy Impact of BEV Inverters on the Voltage Ripple and the High-Voltage Traction System Stability

Cornelius Rettner
Group Components, Volkswagen AG

Abstract:

Modulation strategies of traction inverters have a high impact on the efficiency and voltage utilization of the electric drivetrain. An often underestimated effect that comes along with various modulation strategies is their diverging impact on the resulting voltage ripple within the High-Voltage (HV) traction system. The voltage ripple in the first place, but also the DC-side current ripple can be considered as the stability criteria of a HV traction system, which depend upon the DC-link current and the HV system impedance.

Since the voltage ripple is the main DC-link capacitance design requirement, the DC-link current harmonics of each modulation strategy have to be analyzed in detail as well as their application range within the torque-speed diagram of an electric drive. The target is to design an ideal DC-link capacitance for each inverter within an electric platform. Therefore, we will discuss the challenge of combining the voltage ripple requirement of different modulation strategies with various HV traction systems.

Curriculum Vitae:

Cornelius Rettner is a power inverter Systems Engineer at Group Components, Volkswagen AG in Ingolstadt. He is responsible for the High-Voltage simulation and the EMI design of traction-inverters.

He was born on 30th July 1991 in Schweinfurt and received the B.Sc. and the M.Sc. degrees in electrical engineering, electronics, and information technol-ogy from the University of Erlangen-Nuremberg (FAU), Germany, in 2014 and 2016, respectively.

Cornelius Rettner started his professional career at AUDI AG, Ingolstadt as a PhD student in cooperation with the Chair of Power Elektronics (LEE), FAU Erlangen-Nuremberg in 2017. From 2020 to 2022, he worked at AUDI AG as a Research and Development Engineer in the Power Electronics Hardware Design before he joined Group Components, Volkswagen AG in 2022.

Contact Details:

Cornelius Rettner
Group Components, Volkswagen AG
Sachsstrasse 11a
Gaimersheim, Germany
+49 152 57768348
cornelius.rettner@volkswagen.de

Zero Emission Trucks & Bodies

Dipl. Ing. Martin Glaser
Daimler Truck

Abstract:

The commercial vehicle business is facing a major transformation into zero emission transportation. The presentation is about the different challenges and their possible solutions in terms of alternative powertrains for heavy duty trucks and the approach which Daimler Truck is following. Furthermore, the specifics for truck manufacturer in Power Electronics Systems will be lined out.

In some applications the body of the truck has a higher energy demand as the truck itself, e.g. construction site vehicles like concrete pumps. Some ideas about different approaches will be shared as well.

Curriculum Vitae:

After finishing his vocational education as a truck mechanics he started his studies in mechanical engineering. Mr. Martin Glaser held different positions in the entire vehicle development for Mercedes Benz Trucks and also in the powertrain development. During his career he has been working in different locations in the Daimler Truck network, besides the headquarter also in Brazil and as well in the US for some years.

After his assignment in the US he was heading a group for cost engineering in the development for truck and powertrain. In addition, from 2017 onwards, he took the technical project lead for the Actros-F. Since 2021 he is responsible for the body builder engineering for Mercedes-Benz Trucks.

Contact Details:

Martin Glaser
Daimler Truck AG
Fasanenweg 4
70771 Leinfelden-Echterdingen
+49 160 86 72 835
martin.glaser@daimlertruck.com

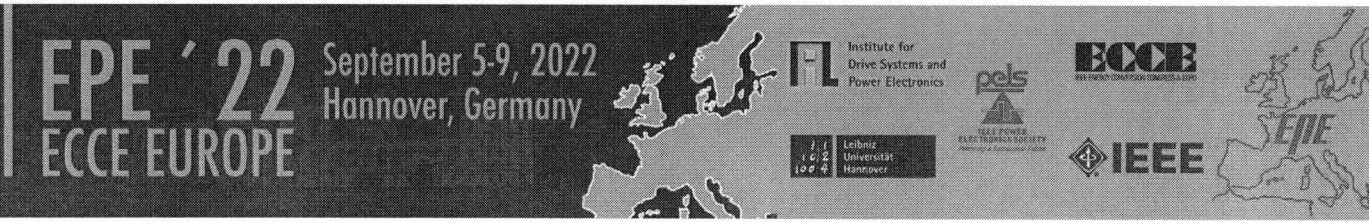

Integrating Offshore Wind & Hydrogen – An Operator's View

Dr. Florian Gremme
RWE Technology International GmbH

Abstract:

This presentation gives an overview of the ambitious goals of RWE in the future energy market. How is our core business leading the way to a green energy world? Which capacity increase in which sectors is intended to accelerate the energy transition? Besides treating this questions, the presentation shows RWE's vision and strategy how, i.e. in which project steps we want to reach commercialization of offshore hydrogen production and why do we see it as necessary to produce hydrogen offshore. Technical as well as commercial aspects will be discussed in the presentation.

Curriculum Vitae:

2007 – 2013	Studies of Mechanical Engineering with a focus on Energy and Process Technology at Ruhr University Bochum
2013 – 2019	PhD in Mechanical Engineering with focus on Energy and Process Technology, in particular Nuclear Technology as Scientific assistant at Ruhr University Bochum
2014 – 2015	Stay abroad at the Research Centre Cadarache.
2019 – today	Engineering Manager at RWE Technology International GmbH in new build projects for a CCGT, in offshore hydrogen production projects and ammonia import projects.
2020 – today	Lecturer for Project Management at Ruhr University Bochum

Contact Details:

Dr. Florian Gremme
RWE Technology International GmbH
Ernestinenstr. 60
45141 Essen
+49 152 51565870
florian.gremme@rwe.com
www.rwe.com

Status quo and future prospects of power electronic solutions for electrolysis plants

Sven Schumann

Siemens Energy

Abstract:

Power electronics plays an important role in the realization of electrolysis applications as the planned and realized plant sizes are growing rapidly. Based on the typical load profile of a PEM electrolyser, examples from realized plants are presented, considering the electrical power supply. An outlook into the future shows expected key requirements for the grid connection to increasingly challenged power networks and the connection to renewable power sources.

Curriculum Vitae:

Sven Schumann received his Master of Science degree in Electrical Power Engineering and Business Administration from the Rheinisch-Westfälische-Hochschule Aachen (RWTH Aachen) in 2012. He received his PhD degree from the same university in 2017 in the area of high voltage insulation, supervised by Prof. Schnettler from the Institute of High Voltage Technology. Since 2022 he is heading the the Primary & Secondary Electrical Engineering group at Siemens Energy, which participates in the development and erection of electrolysis plants. Before, he worked as Systems Engineer for electrolysis plants, as researcher for digitalization in manufacturing at Siemens AG, and as chief engineer at the Institute of High Voltage Technology at RWTH Aachen.

Contact Details:

Dr. Sven Schumann
Siemens Energy
Freyeslebenstr. 1
91058 Erlangen, Germany
+49 (173) 5842435
sven.schumann@siemens-energy.com

Modular power supply system for large scale water electrolyzers

Dr. Ralf Juchem, Dr.-Ing. Klaus Rigbers
SMA Solar Technology AG

Abstract:

Since the demand for green hydrogen will rapidly increase over the next decade, an enormous upscaling of water electrolysis plants in the multi 100MW range is anticipated. There are several factors challenging the development of suited power supply systems: besides the increasing and yet unknown final block sizes of the large scale electrolyzers, support of the electrical grid and restrictions on the footprint of the power supply must be considered. The presentation will describe these challenges and it will give an insight in an ongoing development project of such a large scale power supply unit using transistor technology instead of traditional thyristor technology. This approach will facilitate a much easier (turn-key) grid integration and even grid support functionality.

Curriculum Vitae Ralf Juchem:

- 1988 : Diploma in Physics of the Philipps University Marburg, Germany
- 1995 : PhD (Dr. rer. nat.) of the University of Karlsruhe (KIT), Germany
  - Focus Physical Chemistry and Electrochemistry
- 1995 – 2002 : Head of Modeling and Simulation at the Institut für Solare Energieversorgungstechnik (ISET), Kassel, Germany
- 2002 – 2005 : Head of Simplorer development, Ansoft Corp., Pittsburgh, PA, USA
- Since 2005 : Senior Expert Engineer, SMA Solar Technology AG, Niestetal, Germany

Contact Details:

Dr. Ralf Juchem
SMA Solar Technology AG
Sonnenallee 1
34266 Niestetal, Germany
+49-561-9522-4723
Ralf.Juchem@SMA.de
www.sma.de

Curriculum Vitae Klaus Rigbers

- 2002: Diploma in Electrical Engineering from Technical University Braunschweig, Germany
- 2002 – 2008: Research Assistant at Institute for Power Electronics and Electrical Drives (ISEA) at RWTH Aachen, Germany
- 2008 – 2009: Development Engineer at SMA Solar Technology AG
- 2009 – 2015: Head of Team Power Electronics in the Technology Development at SMA Solar Technology AG, Niestetal, Germany
- 2011: Dr.-Ing. in Power Electronics from RWTH-Aachen University, Germany
- Since 2015: Head of Team Inverter Technologies in the Innovation Center at SMA Solar Technology AG, Niestetal, Germany

Contact Details:

Dr.-Ing. Klaus Rigbers
SMA Solar Technology AG
Sonnenallee 1
34266 Niestetal, Germany
+49-561-9522-3238
Klaus.Rigbers@SMA.de
www.sma.de

Properties of a Lithium-Ion Battery as a Partner of Power Electronics

Alexander Blömeke[*] ⊙, Katharina Lilith Quade ⊙, Dominik Jöst ⊙,
Weihan Li ⊙, Florian Ringbeck ⊙, Dirk Uwe Sauer[*] ⊙

Chair for Electrochemical Energy Conversion and Storage Systems
Institute for Power Electronics and Electrical Drives (ISEA)
RWTH Aachen University, Jägerstraße 17-19, Aachen 52066, Germany
Jülich Aachen Research Alliance, JARA-Energy, Germany
Helmholtz Institute Münster (HI MS), IEK-12, Forschungszentrum Jülich, Germany
Phone: +49 241 80 96920
Fax: +49 241 80 92203
Email: batteries@isea.rwth-aachen.de
URL: https://www.isea.rwth-aachen.de

Acknowledgments

The authors marked with [*] contributed equally to this work.
This research received funding by:

Deutsche Forschungsgemeinschaft
(DFG, German Research Foundation)
Funding Code: GRK 1856 (mobilEM)

Federal Ministry of Education
and Research (BMBF)
Funding Code: 03X90330A (OSLiB)

Keywords

≪Battery≫, ≪Battery impedance measurement≫, ≪Filtering≫,≪Impedance analysis≫, ≪Impedance measurement≫, ≪Lithium-ion≫, ≪Lithium-ion battery≫, ≪Radio frequency (RF)≫

Abstract

Lithium-ion batteries are nowadays offered in a large number of different cell designs and material combinations. This results in a wide range of performance parameters. However, in almost all cases, batteries are used together with power electronic converters, and a central question is the interaction between the converters and the batteries. This could, for example, be the influence of ripple currents or the implementation of impedance diagnostics. But this also leads to the question of the necessity of filters of the power electronics on the battery side.

Introduction

Lithium-ion battery technology is the dominating battery technology for mobile communication devices or electric vehicles. First, this paper gives an overview of the state-of-the-art battery characteristics. Then, after the principle design of lithium-ion batteries is described, the performance of nowadays battery systems is summarised. The chapter concludes with a resume of the ageing and cell design aspects of battery applications.

Electrochemical impedance spectroscopy (EIS) is used to characterise and track the behaviour of batteries. The battery's impedance, which directly interacts with the power electronics, correlates with various internal and external conditions. Impedance data from one battery cell under different conditions is shown. Measurements at higher frequencies, up to 50 MHz, are also carried out. The impact of the temperature on the impedance is further analysed.

Lithium-ion batteries

Lithium-ion batteries are the key to far-reaching hybridisation and electrification of drives, cordless mobile devices, or stationary energy storage systems. Current forecasts predict an increase in battery cell production from a good 1000 GWh/year in 2025 to over 3000 GWh/year in 2030. By comparison, the storage capacity of all pumped storage power plants in Germany is currently around 40 GWh. While, according to the current state of knowledge, by 2030, a large number of batteries will probably be assigned to lithium-ion technology. There is now a high degree of differentiation of properties, so that a cell technology tailored to each application is available.

Principle design of lithium-ion batteries

"Lithium-ion" is a collective term for various material combinations that have in common that they contain about 3 %$_{w/w}$ lithium and that this lithium takes over the actual charge storage and charge transport between the two electrodes. Various properties can be achieved by choosing different electrode materials, cell designs and manufacturing processes. Therefore, it is generally not correct to speak of the properties of "the" lithium-ion battery. Even for a given combination of materials, there are usually products with a wide range of properties. Therefore, statements such as "material X has a longer life than material Y" are rarely correct in absolute terms. Instead, it is necessary to look very closely at the characteristics of each product in terms of capacity, charge and discharge performance, calendar and cyclic life, safety, temperature operating ranges and costs.

A distinction is made between lithium-ion batteries, in which the lithium is never metallic in regular operation, and lithium-metal batteries. The latter deposits lithium metallically on the anode when charged. This technology plays almost no role in applications yet, but will gain importance in "solid-state batteries". Here we focus on lithium-ion batteries as the technology of choice for the coming years.

Lithium-ion batteries are pure intercalation batteries (Fig. 1). This means that the lithium diffuses into or out of the existing crystal structures of the positive electrode (called the cathode) and the negative electrode (anode). While lithium is present as a positive ion when it moves in the electrolyte, it is neutralised again by the external electron flow after it has been deposited on interstitial sites in the electrodes. The cathode material used in lithium-ion batteries today is predominantly a metal oxide in which nickel, cobalt, manganese and aluminium are used in various stoichiometric ratios. The stoichiometry essentially determines all technological properties and the costs. In particular, there is an effort to reduce the cobalt content as much as possible because it is the most expensive and scarcest of the materials. For example, cathode materials with eight parts nickel and one part cobalt and manganese are increasingly being used (NMC 811). In addition, lithium iron phosphate (LFP) plays a role as a cathode material with significantly lower energy densities. However, because iron is significantly cheaper than cobalt and nickel, lower costs per kWh can be achieved. In addition, LFP is a material with different electrochemical properties that make it an inherently safer material.

The anode in commercial lithium-ion batteries consists of various modifications of graphite, whereby some products with exceptionally high energy densities replace part of the graphite with silicon. This can increase the energy density because more lithium can be incorporated per silicon mass. However, while graphite has a volume increase of about 10 % when lithium is intercalated, this is up to 300 % for silicon, which causes enormous mechanical stress and thus problems with the cycle life. Therefore, only parts of the graphite have been replaced by silicon. Another anode material is lithium titanate (LTO). Cells with this material have a significantly lower cell voltage and thus also a considerably lower energy density. At the same time, with this material, which has virtually no change in volume during cyclisation

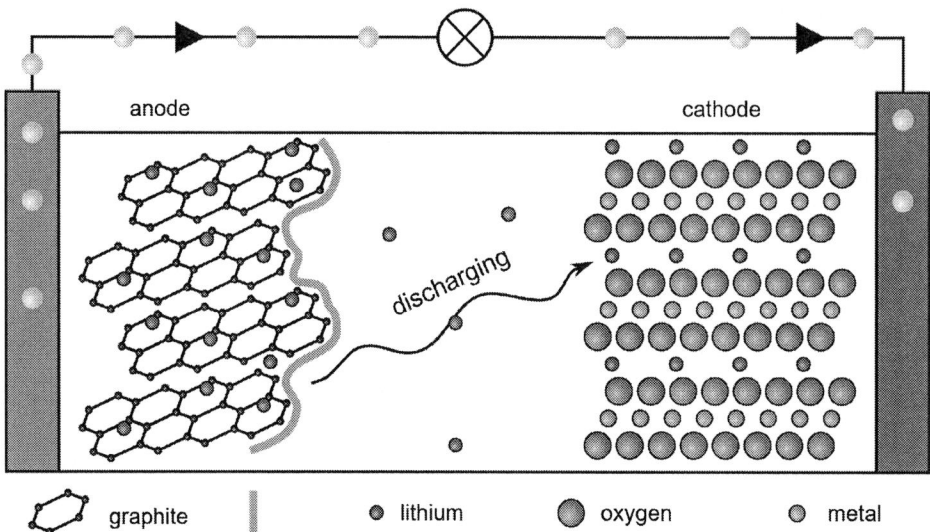

Fig. 1: Schematic of a lithium-ion battery

and top layer formation, lifetimes in the range of 100,000 cycles or very high charge rates of up to 100 C can be achieved. The C-rate expresses the current strength that can be achieved in relation to the capacity. The inverse of the numerical value indicates the duration in hours that is theoretically required for a complete charge or discharge. A typical current of 1 C thus corresponds to a duration of one hour, while 100 C corresponds to a duration of one-hundredth of an hour, i.e. 36 s.

Between the two electrodes there is a porous separator which acts as an insulator for the electronic current and reliably prevents a short circuit between the two electrodes. Today's commercial lithium-ion batteries contain an organic, water-free electrolyte ("solvent") with a conducting salt that provides sufficient lithium ions for good conductivity. The electrolyte fills not only the pore volume of the two electrodes by about 30 % so that ions can penetrate the deeper layers of the active material via the electrolyte without much resistance but also the pores of the separator and all other free spaces between the two electrolytes. If the moistening with electrolyte is incomplete due to inadequate filling or due to consumption of the solvent by ageing processes, this leads to an increase in internal resistance and a decrease in capacity. A large number of additives are usually added to the electrolyte, which has a very significant influence, for example, on the service life and safety of the battery cell. These additives' type, composition and quantity are among the best-kept secrets of battery cell manufacturers and can hardly be broken down using standard analysis methods.

Electrical performance

Today, lithium-ion batteries are offered in many modifications concerning the electrode materials, the design and the internal structure, which determines the performance. In addition, other factors such as electrolyte composition and additives play an essential role, e.g. for the maximum charging currents, low-temperature behaviour, safety or cycle and calendar life. Generalised statements about which material combinations are particularly good in terms of individual properties are often suggested in the literature and company publications. However, it can be seen in the market that there is extensive coverage of product properties for almost every material combination. Moreover, the manufacturer's specifications often only give an incomplete picture of the actual composition of the battery, as laboratory analyses show time and again. It is, therefore, worthwhile to systematically examine the products available on the market for their properties and not to automatically exclude products based on specific information from the manufacturer on the materials used.

Currently, there is an increasing differentiation of characteristics in the mobility sector. In the 2010s, vehicle manufacturers were keen to cover a wide range of different vehicle requirements with one cell to achieve an economy-of-scale effect. In the meantime, the number of units planned for the coming years is so high that optimisation can take place with regard to costs, energy density or other properties for the respective product segment. In the process, much shorter cycle lives are now being accepted. Whereas in the 2010s, traditional vehicle manufacturers set requirements for cycle life in the range of 2000 to 4000 full cycles, today, 500 to 800 full cycles are usually sufficient. On the one hand, this is a consequence of the Wöhler effect, which leads to significantly higher energy turnover and thus a higher total mileage when the battery is partially cycled. On the other hand, batteries have become much larger and, today, typically have a range of between 300 and 500 km. 800 cycles at a range of 300 km are already 240,000 kilometres (without the Wöhler effect) and sufficient for almost all passenger cars. But battery cells with a service life of 3,000 cycles and more are needed for trucks or typical stationary applications since the battery capacity is ultimately used once or even twice a day in these applications.

The volumetric energy density related to the volume is around 2 to 2.5 times higher for lithium-ion batteries than the gravimetric energy density. Nevertheless, very high power densities can be achieved, which come into the range of supercaps. The highest power densities are achieved with LTO batteries. The highest energy densities among the commercially established technologies are achieved with NMC cathodes with high nickel contents in combination with graphite anodes with an admixture of about 5 to 15 % silicon. The highest power and energy densities cannot be achieved simultaneously for any battery technology. The internal resistance must be minimal for high performance, but this can only be achieved if the ionic resistance is as small as possible. For this, the electrodes must be very thin to allow short paths for the ions. With thin electrodes, however, the ratio between the weight and volume of the active masses needed for energy storage and the passive parts of the cell such as current conductors, housing and separators is significantly lower than with thick electrodes.

For hybrid vehicles, battery cells with very high power densities are primarily used, as high power must be retrieved from small batteries. Cells with high performance can be charged and discharged very quickly. Accordingly, high numbers of cycles per day can be achieved. Therefore, high-performance cells are usually batteries with high cycle life. Even at high current rates, a high proportion of the capacity available at low currents can be used in lithium-ion batteries. This makes lithium batteries very suitable for high current loads, such as those in power tools or uninterruptible power supplies in addition to hybrid vehicles. On the other hand, high amounts of energy are needed for all-electric vehicles for corresponding ranges; therefore, cells with high energy densities are used. The performance for the drive is achieved despite the lower energy densities because the batteries are large. With 60 kW h battery capacity, 180 kW of drive power can already be served with a 3 C discharge capacity, which virtually all automotive batteries achieve.

The efficiencies of lithium-ion batteries are 90 to 95 % and are thus very high compared to all other battery technologies. This is due to the low internal resistance and the high cell voltages of 3.3 to 3.7 V for LFP and NMC types, which are thus almost twice as high as lead-acid batteries with 2.0 V and around three times as high as NiCd and NiMH batteries with 1.2 V, for example. The high voltage level thus also reduces the wiring effort at a given system voltage and positively affects the internal resistance of the battery system. In operation, the efficiency can also be lower at low temperatures, leading to accelerated heating of the battery.

Ageing

Various processes on both electrodes and the boundary layers between electrolyte and electrode materials lead to the ageing of lithium-ion batteries. However, the most important effect in all NMC and LFP variants is the formation of a boundary layer between graphite and electrolyte on the negative electrode ("solid electrolyte interphase" (SEI)). At the negative electrode potential, the electrolyte is not stable and therefore reacts spontaneously with graphite and lithium. The process is only stopped because the reaction product itself forms a separating boundary layer. This boundary layer is very dense and reduces the reaction rates considerably. This makes it comparable to verdigris on copper, for example. Copper roofs can therefore become very old. The boundary layer also allows long lifetimes in lithium-ion batteries. At the same time, however, the boundary layer must remain permeable for the lithium ions during charging and discharging. Therefore, in principle, it continues to grow depending on the voltage level, state of charge, temperature and cycle depth. Since the boundary layer contains lithium compounds that cannot be dissolved again, the growth of the boundary layer deprives the battery of free lithium necessary for charge storage. The available capacity decreases accordingly. At the same time, the internal resistance grows with increasing layer thickness. The growth of the boundary layer thus reduces the capacity and increases the internal resistance.

The performance decreases significantly towards lower temperatures. The exact temperature at which this becomes critical depends on the battery cell. Cell manufacturers can set lithium-ion batteries' "feel-good" temperature range in relatively wide ranges. Above all, it is important to prevent so-called lithium plating. This means a deposit of metallic lithium on the anode during charging. At any given temperature, there is a maximum charge current rate at which the ions have sufficient time to diffuse into the graphite structure. If it gets colder or the current is higher, a "jam" of ions forms, which are then reduced to metallic lithium on the surface. This metallic lithium then has no protective layer and is exposed to the direct reaction with the electrolyte. This produces insoluble reaction products, which on the one hand, hinder further ion transport, which leads to an increase in internal resistance, and on the other hand, active lithium is permanently bound and thus removed from charge storage. Therefore, lithium plating leads to greatly accelerated ageing. Modern charge management systems must pay particular attention to and prevent this effect, which is dependent on the respective state of charge and the ageing state. This is also a major reason why the real charging speed is often below the value possible due to the charging power of the charging station.

Cell designs

Three different cell designs are manufactured and used (Fig. 2). The electrodes and separators are wound from continuous rolls for round cells and placed in a cylindrical housing. These cells are traditionally used in consumer products and were initially used by Tesla in the 18650 size (18 mm diameter, 65 mm height) for their vehicles. Today, the 21700 format is increasingly used in vehicles, and a 46800 format has been announced for the next generation. Compared to the 18650 cell, this will have a volume around eight times larger and a correspondingly higher capacity.

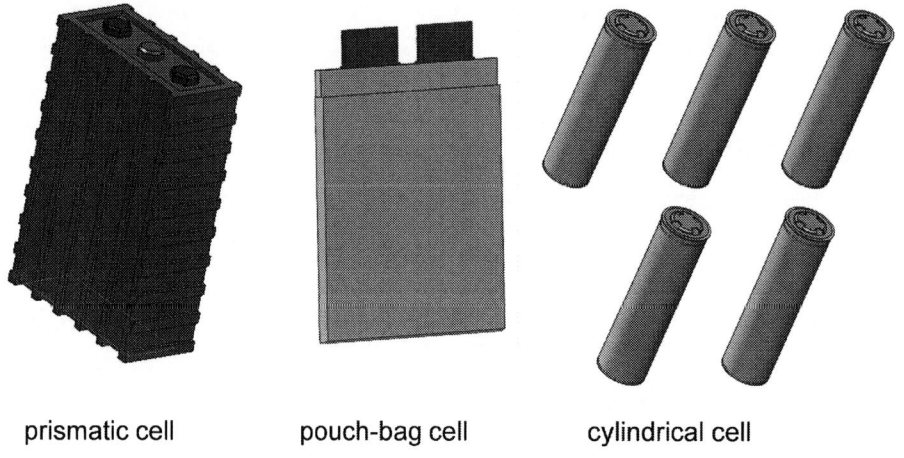

prismatic cell   pouch-bag cell   cylindrical cell

Fig. 2: Principal designs of lithium-ion batteries: prismatic, pouch-bag, and cyclindrical

The electrodes are stacked and welded in a foil in flat or pouch-bag cells. Here, very different designs are known with regard to height, width and thickness and thus also capacities from less than one to well over one hundred Ah. The contacts can be attached to one side but also to opposite sides of the cell.

In prismatic cells, cubic housings are used, which are almost always made of metal for vehicles. The electrodes are either oval-wound cell stacks or layered designs like pouch-bag cells. There is currently a trend toward layered stacks, which on the one hand, allow higher volumetric utilisation of the cell volume and, on the other hand, enable much more uniform pressure conditions and mechanical loads.

Extreme cell designs are sometimes used to house the batteries, with cells barely 10 cm high but up to a metre wide.

Due to the change in volume of the materials during charging and discharging and possible gas formation, prismatic and pouch-bag cells are usually braced in vehicles today. This makes it possible to achieve longer service lifes. However, this is not necessary for cylindrical cells, as the housing builds up this pressure due to the geometry.

The cell designs have different properties with regard to cooling in the battery pack. In principle, however, almost all electrode materials can be used in all three cell designs. So far, it is unclear whether one of the three cell designs would prevail in the automotive sector to the detriment of the others or whether one of the three designs would be eliminated in the foreseeable future.

Impedance spectroscopy

As a lithium-ion battery is an electrochemical system, electrochemical impedance spectroscopy (EIS) can be used to characterise the electrical behaviour under different circumstances. The frequency range is often between 10 mHz and 10 kHz. For slide-state batteries, this might increases up to 10 MHz or even higher, as first tests on laboratory test cells suggest [1].

Non-solid-state batteries such as lithium-ion or lead-acid batteries are used in most battery-driven applications nowadays. The higher frequencies for characterisation are not interesting, as the chemical reactions have much lower time constants. However, if the system powered by the battery applies non-constant loads at high frequencies, the battery's response is of interest. The measurements have to be done first to model the overall behaviour at high frequencies. One example of such a load might be a DC/DC converter connected to a battery. Due to improper design or filtering, the DC/DC converter might have current ripples. This noise, considered from electromagnetic compatibility (EMC) perspective, must be reduced. Here also, the frequency response of the battery is a necessary parameter.

Tested batteries and used equipment

The cylindrical 18650 battery cell Samsung 35E, with lithium nickel cobalt aluminium oxides (NCA) on the cathode and silicon-carbon (Si/C) on the anode, is further investigated in the frequency range from 10 mHz to 50 MHz, thus covering corresponding time constants of 100 s down to 20 ns. The "Vector Network Analyzer - Bode 100" by the company "OMICRON electronics GmbH" is used in the frequency range of 5 Hz to 50 MHz, and the "Digatron EIS-Meter" by the company "Digatron Power Electronics GmbH" for the frequencies from 10 mHz to 6 kHz. The tests are performed in a temperature chamber "Binder MK56" by the company "BINDER GmbH".

In sum, four different batteries of the same type are used for the measurements. Two cells, "Samsung 35E Aged - Everlast 014" and "Samsung 35E Aged - Everlast 025", are aged cells. The description of the tests and the raw data of the cyclic tests is published in [2]. The cell's nominal capacity at begin of life (BOL) is 3.35 Ah. The end of life (EOL) is often defined as a state of health (SOH) concerning the normalised capacity (SOH_C) of 80 %. For this cell, this would be 2.68 Ah. The "Samsung 35E Aged - Everlast 014" shows after 417 Equivalent Full Cycles (EFC) a state of health (SOH_C) of 91.9 % as the capacity dropped to 3.08 Ah. As also shown in Fig. 3 the capacity of the Everlast 025 cell after 899 EFCs is 2.83 Ah resulting in an SOH_C of 84.5 %.

Electrochemical impedance spectrosocpy

With the real-valued voltage and current signals, a battery's complex impedance is calculated. This data is available in the battery management system (BMS). First, the voltage (1) and the current (2) must be transformed to the frequency domain. If the measurement fulfils the linear time-invariant (LTI) criteria,

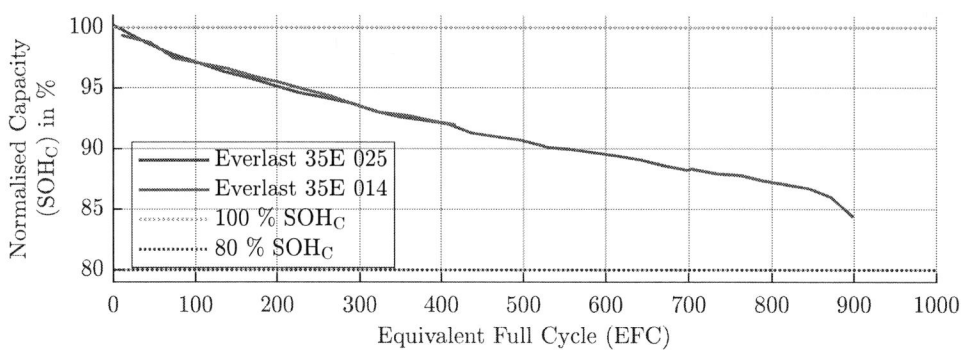

Fig. 3: Ageing of the Everlast 35E 014 and Everlast 35E 025 cell

the resulting real part of the impedance in the frequency domain possess point symmetry, and the imaginary part possesses mirror symmetry. Thus only positive frequencies are shown in the diagrams. Two representations are dominant in literature, the Bode diagram and the Nyquist diagram (3). The Bode diagrams x-axis, representing the frequency, is often scaled logarithmic, but the magnitude and the phase on the y-axis are scaled linearly. In the frequency domain of interest, often between $10\,\mathrm{mHz}$ and $10\,\mathrm{kHz}$, the behaviour of a battery is conductive. In order to have most of the data in the I Quadrant, the y-axis is flipped.

$$v(t) \circ\!\!-\!\!\bullet \underline{V}(f) \tag{1}$$

$$i(t) \circ\!\!-\!\!\bullet \underline{I}(f) \tag{2}$$

$$\frac{\underline{V}(f)}{\underline{I}(f)} = \underline{Z}(f) = \underbrace{|\underline{Z}(f)| \cdot e^{j \cdot \angle \underline{Z}(f)}}_{\text{Bode Diagram}} = \underbrace{\mathrm{Re}\{\underline{Z}(f)\} + j \cdot \mathrm{Im}\{\underline{Z}(f)\}}_{\text{Nyquist Diagram}} \tag{3}$$

In Fig. 4 and Fig. 5, impedance data of Samsung 35E cells is shown. The impedance correlates with different states. The cell's ageing here is plotted with regard to the normalised capacity (SOH$_C$), the state of charge (SOC), represented by the open-circuit voltage after one hour of rest, and the temperature affects the impedance of a battery.

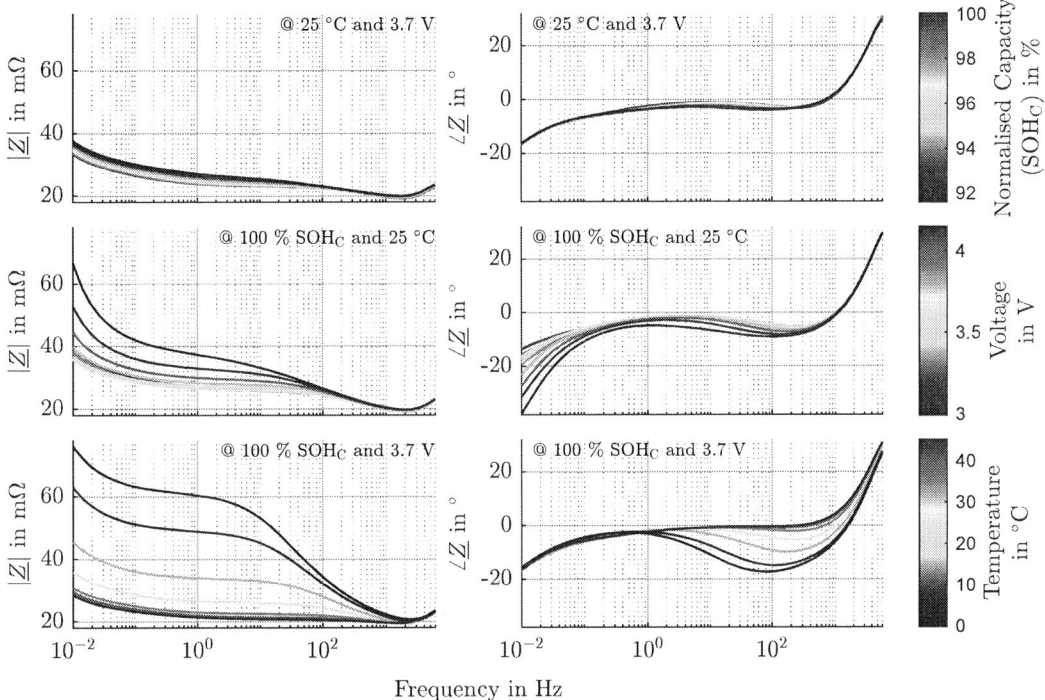

Fig. 4: EIS measurements in a Bode diagram

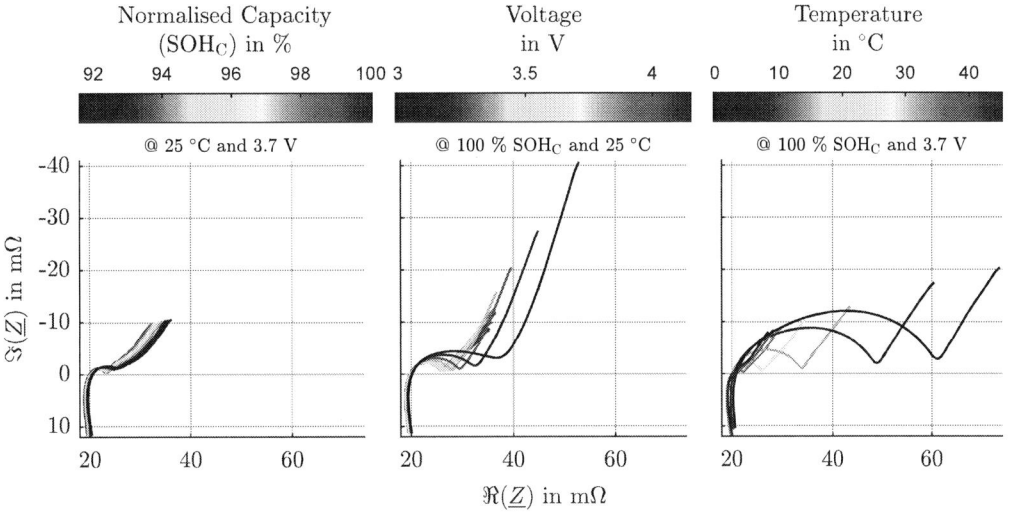

Fig. 5: EIS measurements in a Nyquist diagram

Radio frequency impedance analysis

As lately also analysed by Landinger et al. [3], the absolute value of the impedance in the radio frequencies is increasing and varies with the internal electrical connection, the state of charge and the temperature. In this work, we measure the impedance in the frequency range from 5 Hz to 50 MHz and investigate the influence of ageing, state of charge and temperature. Fig. 6 and Fig. 7 present the results. The noise in the range $|\underline{Z}(f)| < 1\,\Omega$ is relatively high, as the chosen measurement system and the connection are not designed for such low impedance. In contrast to the conventional representation in the Bode diagram for the battery sector, the y-axis is scaled logarithmic now.

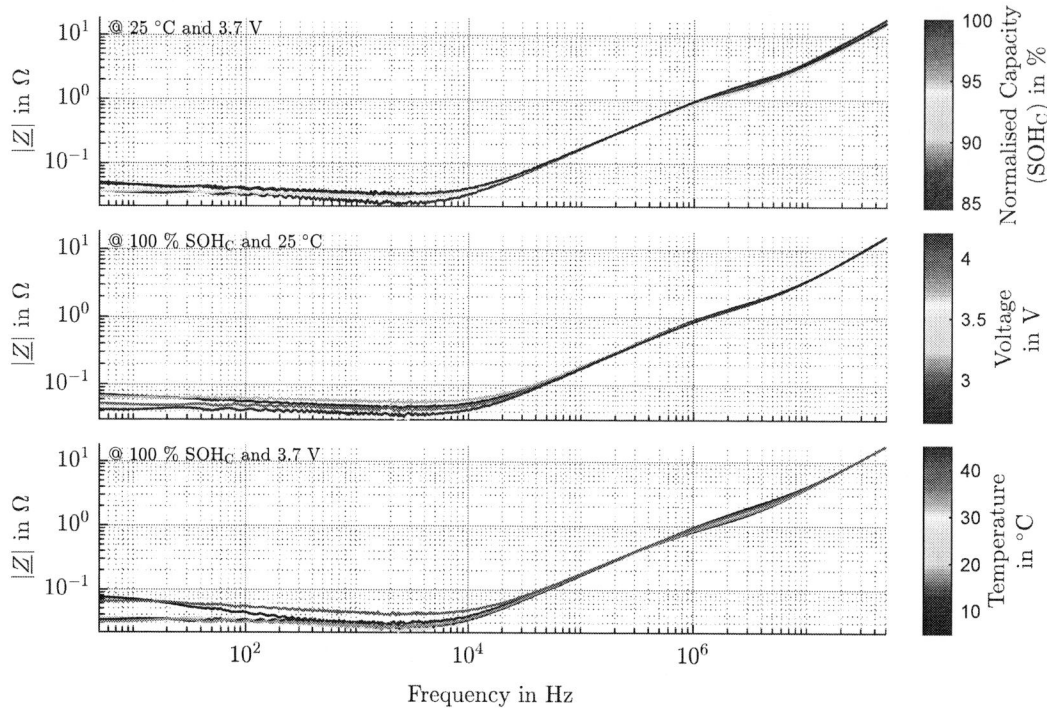

Fig. 6: Radio frequency impedance measurements in a Bode diagram

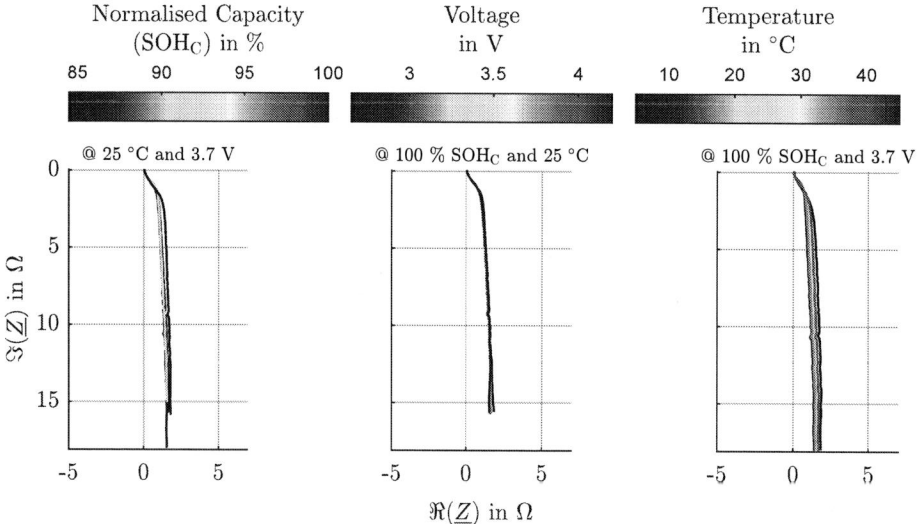

Fig. 7: Radio frequency impedance measurements in a Nyquist diagram

The measurements at different voltages and temperatures are performed with the same cell. For the measurements, at different capacities, four different cells are used. Two of them are new cells, and the other two are the aged cells from Fig. 3. For the state of charge (SOC), represented by the voltage, no significant change in the impedance curve is evident. The curves at the varying state of health (SOH$_C$) differ, but no correlation exists. This might be due to different batches of the four cells, as they were not bought together. Different characteristics of different batches of this specific cell were analysed in [4]. The temperature and the impedance show a correlation. In Fig. 8, the frequency range from 300 kHz to 13 MHz is shown with more detail. Furthermore, a 1 Ω Resistor was measured with the same setup and calibration to verify if the impact was due to the setup or the battery cell. Even though an offset at the 15 °C measurement exists, no correlation with the temperature occurs for the 1 Ω Resistor. The Arrhenius diagram endorses this. With a coefficient of determination $R^2 = 0.99862$, the suggested fit between the temperature and the absolute value of the impedance indicates a correlation.

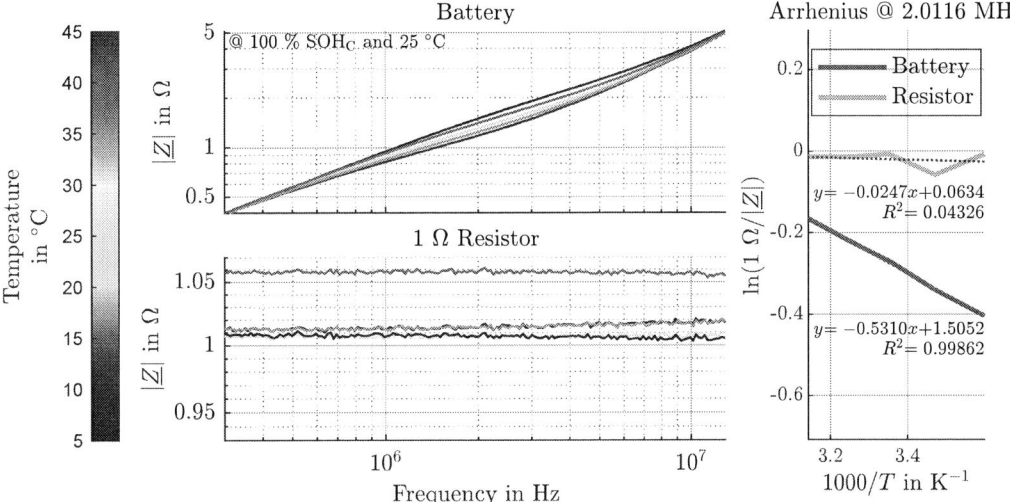

Fig. 8: Radio frequency impedance measurements at different temperatures and Arrhenius

Conclusion

As the number of battery cells on the market increases, so does the diversity. General statements about chemistries and cell formats are rarely universally applicable. The impedance of the battery characterises it, but is also an important parameter for the interconnection of the battery system and the power electronics. The absolute value of the impedance of the measured 18650 cell increases with the frequency and shows at least a correlation with the temperature. A correlation with ageing or the state of charge does not become apparent. All three conditions impact the impedance in the lower frequency range from 10 mHz to 6 kHz.

References

[1] P. Vadhva, J. Hu, M. J. Johnson, *et al.*, "Electrochemical impedance spectroscopy for all-solid-state batteries: Theory, methods and future outlook," *ChemElectroChem*, vol. 8, no. 11, pp. 1930–1947, Apr. 2021. DOI: 10.1002/celc.202100108.

[2] D. Jöst, A. Blömeke, D. U. Sauer, and F. Ringbeck, "Timeseries data of a drive cycle aging test of 28 high energy NCA/C+Si round cells of type 18650," Institut für Stromrichtertechnik und Elektrische Antriebe, RWTH-2021-02814, 2021. DOI: 10.18154/RWTH-2021-02814.

[3] T. F. Landinger, G. Schwarzberger, and A. Jossen, "High frequency impedance characteristics of cylindrical lithium-ion cells: Physical-based modeling of cell state and cell design dependencies," *Journal of Power Sources*, vol. 488, p. 229 463, 2021. DOI: 10.1016/j.jpowsour.2021.229463.

[4] M. Schindler, J. Sturm, S. Ludwig, J. Schmitt, and A. Jossen, "Evolution of initial cell-to-cell variations during a three-year production cycle," *eTransportation*, vol. 8, p. 100 102, May 1, 2021. DOI: 10.1016/j.etran.2020.100102.

AUTHOR INDEX

Abdalrahman, Adil 2241, 3282, 3757
Abdullah, Ahmed.................................. 554
Abedini, Hossein.................................. 865
Aceña, Javier Cañas............................. 484
Adabi, Jafar...................................... 2537
Addin, Ali Sharaf.............................. 1824
Afonso, Luciana C. 4018
Aganza-Torres, Alejandro................... 1328
Agarwal, Ritika................................ 3615
Agirrezabala, Eneko 3327
Aguglia, D. 1955
Ahmed, Emad M................................. 1015
Aiello, Giuseppe 2628
Aillerie, Michel................................. 315
Aizpuru, I.. 2903
Aizpuru, Iosu 325, 3327, 3574, 3750
Akuru, Udochukwu B. 2958
Al-Haddad, Kamal............................. 1025
Alaluss, Mohamed 1424
Alatise, Olayiwola 1497, 2477
Albrecht, Fabian 2726
Aldarmon, Mohamed............................ 2574
Ali, Mohammad............................ 2392, 3022
Ali, Ramy 390
Ali, Rana Asad.................................. 698
Allard, Bruno 169, 3862
Allioua, Abdelmoumin.......................... 2835
Alvarez, Asier................................... 279
Alvarez-Herault, Marie-Cecile 1147
Alves, Wendell Da Cunha...................... 1046
Alvi, Muhammad H.............................. 1692
Aly, Mokhtar.................................... 1015
Andersen, Michael A. E....................... 1561
Ando, Y. 1785
Andresen, Jan.................................... 1684
Ansari, Sajad A................................. 3440
Antonopoulos, Antonios.................... 297, 432
Anzola, J. 2903, 2967
Anzola, Jon...................................... 3574
Apostolidou, Nena.............................. 1796
Appel, Tobias.................................... 1121
Apte, Pramod.................................... 2773
Arabsalmanabadi, Bita......................... 1025
Arias, Manuel.................................... 152
Arrizabalaga, Antxon........................... 325
Arrozy, Juris.................................... 681
Arruti, Asier.............................. 3574, 3750
Artal-Sevil, J. S. 2903, 2967

Arza, Joseba..............................484, 2011
Asllani, Besar................................... 2515
Asoodar, Mohsen 2843
Atzler, Frank................................... 3391
Aunsborg, Thore Stig.......................... 825
Aviñó, Oriol.................................... 2715
Ayarzaguena, Ibán......................1765, 3336
Aztiria, Jon..................................... 325
Baars, Nico..................................... 2788
Babin, Anthony 3696
Baburske, Roman 1424
Bacha, S. 2422, 3179
Bacha, Seddik...........................3140, 3928
Bacheti, Gabriel Gaburro 421
Bachmann, Matthias........................... 3501
Badenhop, Niklas..........................145, 1939
Baek, Seung-Hyuk 2877
Bagaber, Bakr...........................3037, 3711
Baimel, D. 3254
Baimel, N. 3254
Bak, Claus Leth................................ 2504
Bakhos, Gianni 3928
Bakran, Mark-M.................805, 1036, 2744
Bakri, Reda.................................... 1046
Balachandran, Arvind 1456
Balasubramanian, Sridhar...................... 2030
Ballestín-Bernad, V...................... 2903, 2967
Banana, Shady.................................. 1064
Banavath, Satish Naik 730
Banda, Joseph............................. 187, 289
Barba, V. 1975
Barbi, Eli...................................... 3254
Barg, Sobhi 361
Barman, Subhranil............................. 2462
Barón, Kevin Muñoz 2698
Bashar, Erfan.................................. 2477
Basic, Duro.................................... 125
Basler, Michael 242
Basler, Thomas............... 1424, 1713, 1733, 3373
Bauer, Luca.................................... 971
Bauer, Pavol...........................1319, 3607, 3729
Baumann, Michael 1167
Baumann, Timm Felix 2355
Bäumler, Christian 1733
Bayer, Markus 115
Bayhan, Sertac 3518
Bayram, Islam Safak 3518
Beck, Simon...........................1434, 2038

Beckemeier, Christian..2327
Beczkowski, Szymon Michal...2661
Beineke, Stephan 3501
Beiranvand, Hamzeh............................833, 3092, 3846, 3966
Belhaouane, Mohamed Moez 582
Benchaib, Abdelkrim 3928
Bendfeld, Christian 1620
Benech, Philippe 169
Bensetti, Mohamed 3883
Bergmann, Lukas..1036
Bergveld, Henk Jan...3796
Bermejo, Jose Manuel..1765, 3336
Bernal, Carlos 3327
Bernal-Agustín, J. L..2967
Bernal-Ruiz, Carlos 3750
Bernichon, Thomas 3920
Bertilsson, Kent 361
Bertin, Matthieu 534
Beukes, Johan 3112
Beye, Mamadou Lamine..2736
Beza, Mebtu 1187
Bezerra, Vinicius Freire...2689
Bhatnagar, Pallavee 3804
Bhattacharya, Arghyadip 178
Bhoi, Sachin Kumar..3031
Biadene, Davide..865
Biela, Juergen 1402
Biela, Jürgen 651, 662, 933, 1391, 1434, 2038, 2544
Bieler, Arne 1121
Bier, Anthony 922, 2736
Billa, Laxma R...2301
Bimmel, Luc 2736
Binder, Andreas 2316
Bitsi, Konstantina 3246
Blaabjerg, Frede...............2110, 2182, 2496, 2504, 2939
Blanes, J. M. 3382, 3401
Blank, Thomas..232
Blanquez, Francisco R. 2189, 2451
Blasco-Gimenez, Ramon 2189, 2451
Blasuttigh, Nicola..3846
Blatsi, Zoe..2824, 3813
Blömeke, Alexander 4025
Böcker, Joachim 2276, 2432, 2754, 3625, 3686
Bockholt, Jan 1286
Boettcher, Norman 1128
Bohllaender, Marco 4016
Bohne, David..514
Boige, Francois...944
Boisson, Guillaume Piquet...960
Bolzoni, A..1371
Bongiorno, Massimo..1187
Bonten, Remco 634

Böorngen, Hannes 1754
Borcherding, Holger..2852
Börngen, Hannes 3362
Boroyevich, Dushan...2806
Bosch, Swen 2219
Bosga, Sjoerd G. 3246
Bouscayrol, Alain..2175
Boutleux, Emmanuel 251
Boutry, Arthur 2515
Brabetz, Ludwig..2383
Branco, Cesar Augusto Santana Castelo 2948
Braun, Gerrit 2205
Braz, Cesar 1445
Briff, Pablo 451
Bringezu, Thilo 662
Brinker, Tobias..2977
Brogioli, Doriano Constantino 833
Brommer, Volker..2726
Bronstein, S. 3254
Brooks, Michael 279
Brückner, Thomas 1824
Brulin, Pierre-Yves 3831
Brunner, Andreas 593
Brunner, Frank 3775
Brüns, Michael 474
Bruyere, Antoine 1046
Bruyere, Paul..960
Bucarey, Victor 1074
Budo, Kohei...213, 351
Bueno, Emilio José. 421
Bueno-Mariani, Guilherme 3272
Bugarski, Stevan 2334
Bünte, Andreas...380
Burgos, Rolando..1692, 2806
Burgos-Mellado, Claudio..1074, 3429
Burkart, Ralph M. 203
Burke, Richard 3696
Bushra, Rehnuma...2392, 3022
Busquets-Monge, Sergio 2715
Buticchi, Giampaolo 3014
Buttay, Cyril...2049, 2515
Byen, Byengjoo 1207
Caarls, Esin Ilhan 681
Cabrera, Michel..169
Cacciato, Mario...2628
Caillierez, Antoine 3883
Cajander, D. 1955
Cakal, Gokhan 3947
Caldognetto, Tommaso 865
Camargo, Renner Sartório 421
Camurca, Luis 3101
Can, Görkem 3092

Cano, Tania C.	335
Cao, Jingming	3215, 3225
Cao, Yongtao	2003
Cappelle, Jan	1300
Cárcamo, Alberto	1083
Carcouet, S.	843
Carpita, Mauro	1543
Carrasco, Miguel	370
Casado, P.	3382, 3401
Castellazzi, Alberto	689, 2156, 2285, 2402, 2893, 3084
Castelli-Dezza, Francesco	1476
Castro, Ignacio	335
Catalán, Pedro	2011
Catellani, Stéphane	922, 990
Ceccarelli, Lorenzo	681
Chakraborty, Sajib	2101, 3031
Chang, Che-Wei	1692
Charkaoui, Abdelmouneim	442
Chatterjee, Kishore	178, 2462
Chen, Zhe	2011
Chen, Zhu	3235
Chevalier, Florian	3582
Chida, Makoto	1580
Chinthavali, Madhu Sudhan	344
Chiumeo, Riccardo	3206
Choksi, Kushan	344
Choudhury, Soham	1966
Chub, Andrii	730
Cimetiere, Xavier	1046
Clerc, Guy	251
Clerici, Alessio	3206
Cobaleda, Diego Bernal	2581
Cogitore, Bruno	1216
Colmenero, Manuel	2189, 2451
Cosso, Simone	2919
Coumont, Martin	1966
Crovetti, Paolo	554
Cui, Yi	3986
Czerwenka, Philipp	593
Dahmen, Christopher	1824, 1855
Damian, Ioan Catalin	2266
Damm, Gilney	3590
Danielsson, Christer	2843
Dargahi, Vahid	2073
Davari, Pooya	2496
Davidson, Jonathan N.	3440
De Bernardinis, Alexandre	315
De Carne, Giovanni	3014
De Cesaris, Ivan	223
De Donato, Giulio	1569
De Doncker, Rik W.	709, 1266, 2119, 3599, 3676, 3740, 3766, 3893
De Lillo, Liliana	3450
De Matos, Jose Gomes	2948
De Oliveira, Eduardo Facanha	2441
De, Dipankar	689
Deb, Arkadeep	1497
Deblecker, Olivier	504
Deboy, Gerald	3984
Deck, Patrick	514
Deckers, Martijn	2795
Degaa, Laid	3696
Delette, Gérard	922
Deng, Kai	3235
Dennetiere, Sébastien	582
Derammelaere, Stijn	3344
Despouys, Olivier	2486
Dick, Christian P.	514
Dickmann, Stefan	758
Dieckerhoff, Sibylle	1466, 2596, 2607, 2644, 3775
Dieng, A.	2092, 2930
Dierks, Rebecca	1533
Dietrich, Tim-Hendrik	1094
Disselkamp, Simon	2912
Domae, Shinichi	3084
Domes, Daniel	2744
Domes, Konrad	1137
Dong, Chaoyu	3215, 3225
Dong, Dong	1692, 2515
Dong, Jianning	1319
Dong, Tenghui	3084
Dorner, Oscar	1177
Dos Santos, Pedro Leal	604
Dragicevic, Tomislav	2496, 2939, 3429
Drexler, Christoph	411, 1167
Driesen, J.	3655
Driesen, Johan	2795
Drimizi, Youssef	2869
Drissi, Khalil El Khamlichi	3786
Duarte, Jorge L.	681, 798
Duarte, Jorge	2788
Duchamp, Jean-Marc	169
Dujic, Drazen	2049
Dumtzlaff, Jacob	1865
Duquesne, Thierry	3582
Dürbaum, Thomas	88, 307
Duun, Sune Bro	825
Dworakowski, P.	2422
Dworakowski, Piotr	2049
Ebel, Thomas	3130
Ebner, Kathrin	4015
Eckart, Martin	3646
Eckel, Hans-Guenter	3460

Eckel, Hans-Günter...... 11, 59, 70, 980, 1294, 1703, ... 1744, 1885, 1895, 2308, 4003
Eckstein, Mattea 1277
Effenberger, Thomas 1754
Eggers, Malte 1466
Ehlich, Martin 2852
El Baghdadi, Mohamed 2101, 2293, 3031
El Sherif, Alaa 3796
El-Refaie, Ayman 719, 1692
Ellinger, Thomas 2885
Emmers, G. 3655
Emmers, Glenn 2795
Empringham, Lee 3450
Encarnação, Lucas Frizera 421
Endo, Yusuke 2285
Epping, Daniel 749
Erckrath, Tobias 1350, 1620
Eremia, Mircea 2266
Eriksson, Lars 1456
Erlbacher, Tobias 1128
Ernst, Alexander 3149, 3159
Es-Seghier, Hajar 922
Escoffier, René 990
Etoz, Burhan 1497
Faber, Samuel 307
Falchi, Daniele 2486
Faramehr, Soroush 3822
Farhangi, Shahrokh 787
Fauth, Leon 2003, 2638, 3838
Fayolle-Lecocq, Murielle 990
Fazli, Nastaran 11
Fehr, Hendrik 49, 3391
Felgemacher, Christian 442, 4004
Fernández, Arturo 152
Ferreyra, Fabio 554
Festerling, Tobias 1237
Finney, Stephen 80, 3470, 3813
Fischer, Katharina 1674, 1804
Fischer, Manuel 749
Fischer-Baeumer, Rico 1137
Fölkel, Lorandt 279
Formentini, Andrea 2919, 3975
Forouzesh, Mojtaba 1590, 1601
Forsstrom, Ville 3301
Förster, Nikolas 2432
Foster, Martin P. 3353, 3440
Foteinopoulos, Georgios 1985
Fräger, Lukas 145, 641, 1939, 2588, 2773
Frank, S. R. 3411
Franzki, Jonas 261
Frey, David 1147
Fricke, Tobias 1247

Fricke, Torben 1381
Friebe, Jens ...... 1914, 2003, 2327, 2392, 2588, 2638, ... 2655, 2689, 2773, 2977, 3022, 3059, 3545, 3838
Fritze, Eric 758
Fröhling, Sören 1674
Fuchs, Simon 1434, 2038
Fuhrmann, Jan 980
Fukunaga, Shuhei 108
Ganeshpure, Dhanashree Ashok 3729
Gao, Xiang 3014
Gaona, Daniel 2441
Garces, Santiago Ramos 3344
Garcia, Raul Murillo 2355
Garrigós, A. 3382, 3401
Gaubert, Jean-Paul 1525
Gauthier, Jean-Yves 3862
Gavelle, Mathieu 2618
Gehl, Adrian 2912
Geiss, Michael 2554
Gemma, Filippo 3975
Geng, Weiwei 3722
Geng, Xiaomeng 2596, 2644, 3775
Gennaro, Francesco 2628
Gensior, Albrecht 49, 370, 3391
Gerges, Tony 169
German, Ronan 2175
Germishuizen, J. J. 3318
Geury, Thomas 2101
Gholami, M. 3179
Gholami, Mehrdad 3140
Ghumman, Sukhjit S 2763
Gieraths, Antje 767
Gierschner, Magdalena 1294
Gierschner, Sidney 11
Gillon, Frédéric 1046
Girona-Badia, Jaume 3704
Glaser, Martin 4020
Gleissner, Michael 805, 1036
Gnärig, Lasse 370
Goetz, Stefan 1025, 1064, 1197, 3636, 3665
Gohler, Katherina 1804
Gohrmann, Kai 1137
Golev, Victor 1286
Goller, Maximilian 1733
Gomes, Lucas Vinícius De Araújo 3059
Gomes, Zariff Meira 3590
Gómez, Alexis A. 1765, 3336
Gomis-Bellmunt, Oriol 2486, 3704
Gonzalez, Jose Ortiz 1497
Gonzalez-Hernando, Fernando 3938
Gonzalez-Torres, Juan-Carlos 3928
Götz, Georg Tobias 709

Gräber, Hendrik	2977
Grabs, Volker	97
Gradinger, Thomas B.	203
Grant, Thomas	2301
Grass, Norbert	2366
Grau, Vivien	854
Gremme, Florian	4021
Griepentrog, Gerd	160, 2780, 2835
Grodnichev, Anton	624
Groke, Holger	3169
Groon, Fabian	3092
Groten, Jonas	279
Gruson, François	582
Guerrero, Bruno	944
Gui, Qiuye	49
Guillaud, Xavier	582
Günes, Ece Olcay	1361
Gupta, Kirti	2110
Gupta, Krishna Kumar	3615, 3804
Gutierrez, Alonso	2618
Haag, Felix	2726
Haake, Daniel	624
Haarer, Jörg	971, 1237, 1277
Habersetzer, Antoine	4015
Hably, A.	3179
Hably, Ahmad	3140
Hackl, Philipp	39
Haederli, Christoph	3282
Häfner, Ying-Jiang	2241, 3282, 3757
Hagedorn, Maximilian	1875
Hajar, K.	3179
Hajar, Khaled	3140
Hajian, Masood	468
Hakkila, Akseli	297
Hald, Alex	380
Hameyer, Kay	3005, 3235
Hammes, David	11
Handt, Karsten	2607
Hanf, Michael	3169
Hanisch, Lucas Vincent	261
Hanisch, Lucas	1094
Hänsel, Stefan	572
Hansen, Sandra	3966
Hanson, Alex J.	1722
Hanson, Jutta	1966
Hao, Chuantong	80, 3470
Hardan, Faysal	468
Harmand, Souad	2996
Hasan, Md. Mahamudul	3031
Hasler, J. P.	1371
Hassan, Tayssir	1466
Hatori, K.	1785

Hatori, Kenji	777
Hattori, Takato	739
Hauenschild, Philipp	1506
Haug, Martin	279, 698
Hayes, John G.	2470
Hegazy, Omar	2101, 2293, 3031
Heide, Daniel	3711
Heien, Christian	1294
Heimler, Patrick	1713
Hein, Yves	1294
Helmholdt-Zhu, Ting	97, 854
Hembel, Ahmed	3947
Henke, Markus	261, 1094, 2030
Henkenjohann, Jonas	1684
Henn, Jochen	3599
Henneberg, Dustin	2885, 3491
Herbold, Johannes	749
Hernando, Marta M.	1083, 1765, 3336
Herzog, Hans-Georg	952
Heydari, Rasool	2682
Hikihara, Takashi	108
Hiller, M.	3411
Hiller, Marc	115, 999
Hillmer, Hartmut	2383
Hilt, Oliver	2596, 2644, 3775
Himker, Niklas	1631
Himmelmann, Patrick	999
Hiraki, Eiji	2164
Hirning, David	971, 1237, 1277, 3536
Hissel, Daniel	315
Hjerrild, Jesper	2504
Hoerner, Michael	1754
Hofer, Heimo	1445
Hofer, Matthias	2251
Hoff, Bjarte	3198
Hoffmann, Klaus F.	758, 2726, 3188
Hoffmann, Madlen	3262
Hoffstadt, Thorben	1157
Hofmann, Viktor	195, 400
Hofmann, Wilfried	3957
Hofstetter, Patrick	195, 400
Hölscher, Jonas	2432
Holtje, Pauline	1665
Holzke, Wilfried	3149, 3159, 3169
Horn, Markus	2383
Hortans, Magnus	3309
Hoshi, Nobukazu	1776, 1844
Hosseinabadi, Farzad	3031
Hosseini, Elham	1025
Hou, Jingning	3722
Houwen, Simon	3344
Hridya, I	187

Hu, Anliang	651
Hu, Bin	2182
Hu, Xiaowei	3722
Huang, Jiasheng	1561
Huerta, Gabriel Ramos	1226
Huesgen, Till	2230
Huisman, Henk	634, 673, 681
Hutzler, Michael	1445
Idir, Nadir	2996, 3582, 3822
Igic, Petar	3822
Iida, Masaki	2164
Iman-Eini, Hossein	787
Imgart, Paul	1187
Incurvati, Maurizio	223, 268
Inoue, Michiko	3420
Iraola, Unai	3327
Ishihara, Mastaka	2164
Itoh, Jun-Ichi	902, 1104, 2127
Ittamveettil, Hridya	289
Izurza, Pedro	484
Jaber, Hamzeh J.	2156, 2285, 3084
Jacques, Dries	3344
Jagannath, Sriram	3362
Jahdi, Saeed	1497, 2477
Jain, Anekant	3615
Jain, Sanjay K.	3615, 3804
Jamal, Adeel	2780
Jaman, Shahid	3031
Jankovic, Marija	442
Jayathurathnage, Prasad	1947
Jena, Kasinath	3804
Jenhani, Firas	1343
Jeong, Byunghwang	1207
Jeschke, Sebina	3235
Jha, Kapil	187, 289
Jia, Hongjie	3215, 3225
Jia, Ming	1266
Joebges, Philipp	1266
Johansson, N.	1371
Johnson, C. Mark	3450
Jonsson, Tomas	1456
Jordà, Xavier	2715
Jørgensen, Asger Bjørn	825, 1641, 2661
Jöst, Dominik	4025
Jovanovic, Raka	3518
Juchem, Ralf	4023
Judge, Paul	80
Junemann, Lennart	1665
Jung, Marco	624, 1515, 1611, 1620
Junghans, Christoph	3460
Junyent-Ferre, Adria	2574
Kabbara, Wassim	3883

Kacetl, Jan	1197, 3636, 3665
Kacetl, Tomáš	1197, 3636, 3665
Kacki, Marcin	2470
Kadem, Karim	3590
Kaerst, Jens Peter	544
Kaiser, Jeremias	307
Kallfass, Ingmar	2698, 3565
Kamel, Tamer	468
Kaminski, Nando	2230, 3149, 3169
Kamm, Simon	2698
Kampen, Dennis	145, 1939, 2588
Kamper, Maarten J.	2958
Karakasli, Vefa	2835
Karamanakos, Petros	297, 1476, 1754
Karau, Fabian	3292
Karnehm, Dominic	767
Karwatzki, Dennis	195
Kasten, Henning	3501
Kayser, Felix	59, 4003
Keilmann, Robert	891
Kempchen, Malte	2912
Kemper, Philipp	749
Kennel, Ralph	1754, 2366, 3362
Kerekes, Tamas	1933
Keshavarzi, Davood	1064
Khader, Meriem	2655
Khan, Basit Ali	2537
Khan, Mohammed Ali	135
Khan, Nameer	3796
Khan, Siam Hasan	484
Khanzadeh, Babak	2344
Khenfri, Fouad	3831
Kiehnle, Philip	999
Kiffe, Axel	1157
Kikuchi, Naoto	1104
Kim, Dong-Uk	1207
Kim, Sungmin	1207, 2877
Kinzer, Dan	3987
Kirsch, Andreas	380
Kitagawa, Wataru	739
Kjærsgaard, Benjamin Futtrup	825
Klee, Matthias	1515
Klever, Severin	3676
Klötzer, Sebastian	4011
Knebusch, Benjamin	1665, 3048
Ko, Youngjong	3014
Kobayashi, Hiroyasu	1580
Kocewiak, Lukasz	2504
Koch, Jan-Niklas	2852
Koczy, Dawid	3149
Kohlhepp, Benedikt	88, 307
Kojima, Tetsuya	3740

Kondo, Keiichiro 1580
Kondratenko, Dmytro 1906
Kopp, Tobias 912
Kormska, Tomáš 1114
Körner, Patrick 2021
Korthauer, Bastian 3625
Kosesoy, Yusuf 634
Kostka, Benedikt 1649
Kostynski, Daniel 3855
Koteich, Mohamad 534
Kouro, Samir 1015
Koutroulis, Eftychios 1985
Kowal, Julia 4014
Kragl, Robert 2554
Krick, Alexander 3989
Krigar, Tim 2375
Krishnamoorthy, Harish Sarma 730
Krüger, Helge 3966
Krümpelmann, Marcel 1631
Kubulus, Pawel Piotr 2661
Kuder, Manuel 767
Kumar, Amit 451
Kumar, Kaushik Naresh 1486
Kumar, Manish 3511
Kuperman, A. 3254
Kuprat, Johannes 3067
Kuring, Carsten 2596, 2644, 3775
Kurrat, Michael 912
Kurukuru, V S Bharath 135
Kusaka, Keisuke 1104, 2127
Kusche, Stephan 3704
Kusebauch, Manuel 3491
Küster, Pierre 411
Kwak, Jaedon 2893
Kyyrä, Jorma 1947
La Mantia, Fabio 833
Labonne, A. 3179
Labonne, Antoine 3140
Labrousse, D. 843
Lacerda, Vinícius Albernaz 3704
Laclaverie, Julien 944
Laforet, David 1445
Lamar, Diego G. 335, 1083, 1765, 3336
Lange, Jarren 2276
Lange, Yannic 2644
Langfermann, Sascha 1939, 2588
Lanzarotto, D. 2564
Larrañaga, Uxue 3938
Larrazabal, Igor 1765, 3336
Larsson, Anders 1456
Lataire, Philippe 2293
Laumen, Michael 3766

Lauri, Andrea 865
Laza, Saioa Burutxaga 370
Lazkano, Markel Zubiaga 484
Le Leslé, Johan 2526
Le Métayer, Pierre 2049
Lee, Jaehong 2877
Lee, Seung-Hwan 2877
Lee, Yonghwa 2402
Lefebvre, Bruno 2515
Lefevre, Guillaume 2526
Legay, Florian 3529
Lehn, Peter W. 1995, 2084, 2145, 2763
Leifert, Torsten 4013
Lemaire-Semail, Betty 2175, 2996
Lembeye, Yves 1216
Lenz, Kevin 442
Lenzen, Patrick 2413
Leuer, Michael 3292
Leuzzi, Riccardo 3975
Lévy, PE 843
Lewicki, Arkadiusz 1906
Lexow, Daniel 1744
Li, Feifei 3235
Li, Ke 3822
Li, Marui 3215, 3225
Li, Qiang 3722
Li, Weihan 4025
Li, Xiang 2301
Li, Xupeng 3373
Li, Zheming 2744
Liang, Mincui 3786
Lichtenstein, Timo 1674
Liebfried, Oliver 2726
Liegmann, Eyke 1754, 3362
Lievre, Aurelien 2175
Lin, Siqi 1914, 2638
Lin-Shi, Xuefang 3862
Lindemann, Georg 3555
Linder, Stefan 3992
Lippold, Florian 1506
Liserre, Marco 421, 833, 3014, 3067, 3092, 3101, 3846, 3966
Liu, Chao 1561
Liu, Steven 604
Liu, Xing 1733, 3373
Liu, Yan-Fei 1590, 1601
Liu, Yining 1947
Llanos, Jacqueline 3429
Löfgren, Jonas 3920
Lombard, Philippe 169
López, Abraham 152
Lorenz, Andreas 814

Lorenz, Erwin 1167
Lorenz, Malte 1875
Lorenz, Oscar 873
Loudot, Serge 3883
Lu, Xuyang 3822
Lu, Yizhou 883
Luan, Shaokang 3309
Luckert, Franz 2706
Luecke, Stefan 3075
Luh, Matthias 232
Luo, Fang 344, 2860
Lusardi, Federico 3975
Lutsch, Michael 88
Lutz, Josef 1713
Lutzen, Hauke 2230
Ma, Wenhao 80
Maamri, Nezha 1525
Maibach, Philippe 3282
Maier, Robert W. 2744
Maitra, Abhishek 1424
Mallwitz, Regine 891, 912, 1094, 1247, 1506
Mambetow, Arthur 145
Manthey, Tobias 2655, 2689, 3059
Marca, Ygor Pereira 798
Marcaide, Inko 3920
Marcault, Emmanuel 2618
Marchesoni, Mario 2919
Margreiter, Thomas 223
Margueron, Xavier 1046
Marks, Hendrik 2030
Marquardt, Rainer 1855
Marroquí, D. 3382, 3401
Martin, Jérémy 990, 2736
Martinez, Wilmar 1914, 2197, 2581
Martinez-Garcia, Herminio 1056
Martinez-Padron, Daniel S. 1256
Martnez, Wilmar 2638
Marx, Philipp 1237, 1277, 3536
März, Martin 493, 3262
Mashaly, Aly 442
Mashayekh, Ali 767
Mathúna, Cian Ó 4006
Mattavelli, Paolo 865
Matthies, David 3159
Maussion, Pascal 2869
Maynard, X. 843
Mazuela, Mikel 325, 3327, 3574
Meddour, Aissam Riad 3696
Mehran, Kamyar 614, 3353
Mehrasa, M. 3179
Mehrasa, Majid 3140
Meier, Hans 2021

Meinert, Janus Dybdahl 825
Meissner, Michael 758, 3188
Mellor, Phil 2477
Mendoza-Araya, Patricio 1177, 1226
Meng, Qingchao 933
Menzel, Steffen 3169
Merlin, Michael M. C. 2824, 3813
Merlin, Michael 80, 3470
Mersche, Stefan 115
Mertens, Axel 641, 1350, 1533, 1631, 1649, 1665,
..1684, 1865, 1875, 2003, 2066, 2392, 2706, 3022, 3037, 3048,
3075, 3555, 3711
Miaja, Pablo F. 152
Mijatovic, Nenad 2496, 2939
Miller, T. J. E. 3318
Minami, Masataka 2285
Mir, Tabish Nazir 468
Mirza, Abdul Basit 344
Mirzadeh, Mina 1350
Mirzaeva, Galina 3903
Miskiewicz, Rafal 1486
Mistretta, C. 1975
Mita, Salvatore 2628
Mo, Wai Keung 3130
Möckel, Andreas 3391
Moench, Stefan 242
Mogorovic, Marko 203
Mohanta, MK Kharabela 689
Möhlenkamp, Georg 3993
Mohsenzade, Sadegh 614, 3353
Moldenhauer, Deniz-Heinz 2205
Mondal, Gopal 572
Mondzik, Andrzej 3804
Monmasson, Eric 1256
Mönninghoff, Sebastian 3005
Montero, E. Rodriguez 1834
Morales-Paredes, Helmo K. 1074
Morand, Julien 2526
Morel, F. 2422, 2564
Morey, Philippe 1543
Morshed, Muhammad 2301
Motte-Michellon, Denis 1216
Mouselinos, Theodoros P. 1551
Moussa, Hassan 3590
Movagharnejad, Hedieh 3048
Mu, Yunfei 3215
Müller, Jonas 2230
Müller, Tankred 474
Munk-Nielsen, Stig 825, 1641, 2661, 3309
Muñoz-Carpintero, Diego 1074, 3429
Muruaga, Endika Bilbao 3529
Musolino, Francesco 554

Mustafeez-Ul-Hassan 2860
Musumeci, S. 1975
Muyllaert, Koenraad 2383
Mysore, Madhu Lakshman 1424
Naeve, Tomasz 1445
Nagayasu, Kiwa 2164
Naghibi, Javad 614, 3353
Nahalparvari, Mehrdad 2843
Najjar, Mohammad 2682
Nakamura, Keiichi 777
Nakamura, Taketsune 3084
Nami, Ashkan 2241, 3757
Nannen, Hauke 160
Nassurdine, B. Mohamed 843
Nayak, Khirod Kumar 2241, 3757
Nayampalli, Vishwas Acharya 1703
Nazeri, Ahmad Ali 1309, 1336, 1343, 2670, 3871
Neal, Harley 2301
Nee, Hans-Peter 2843
Nehmer, Dominik 1036
Neira, Sebastian 2824, 3813
Neuland, Tanja 3991
Neumann, Christian 1895
Neumann, Ingmar 1445
Neumeister, Matthias 572
Nguyen, Allen 1722
Nguyen, Khanh-Hung 562, 1309
Nguyen, Van-Sang 922, 990
Nguyen, Xuan Viet Linh 169
Nian, Heng 2182
Niasar, Mohamad Ghaffarian 3729
Nie, Shuang 2145
Niedernostheide, Franz-J. 2744
Niedernostheide, Franz-Josef 1424
Nielebock, Sebastian 493, 2607
Niemetz, Michael 2021
Niggemann, Oliver 3545
Nikowitz, Mario 2251
Nishio, Atsushi 351
Nishitani, Yota 3420
Nishizawa, Shin-Ichi 1128
Noboru, Wakana 777
Noisette, Philippe 3910
Nooshabadi, Morteza Tadbiri 787
Nordström, Lars 883, 1006
Nymand, Morten 2682
O'Donnell, Terence 390
O'Driscoll, Seamus 4006
Obernolte, Urs 854
Odeh, Charles 1906
Okada, Ryohei 1776, 1844
Olbrich, Markus 2912

Oliveira, Hercules Araujo 2948
Orbay, Raik 3920
Orchard, Marcos 3429
Orfanoudakis, Georgios I. 1985
Örgüt, Osman 1361
Orlik, Bernd 3149, 3159, 3169
Ortega, David 1765, 3336
Ortiz-Gonzalez, Jose 2477
Orts, C. 3382, 3401
Oshnoei, Arman 2939
Ota, Ryosuke 1776, 1844
Ouyang, Ziwei 1413, 1561
Owzareck, Michael 1939, 2588
Oyarbide, Estanis 3327
Paasch, Kasper M. 3130
Pace, Loris 3582
Páez, J. D. 2422
Pagnani, Daniela 2504
Panigrahi, Bijaya Ketan 2110, 3511
Papadopoulos, Georgios 1391
Papadopoulos, Theofilos 432
Papafotiou, George 2788
Papanikolaou, Nick 1796, 2257
Papastergiou, Konstantinos 2355
Pascal, Yoann 3067
Pasquier, Christophe 3786
Passalacqua, Massimiliano 2919
Passmore, Brandon 4005
Pathmanathan, Mehanathan 1995, 2084, 2145, 2763
Patin, Nicolas 1256
Patti, Dario 2628
Patzelt, Nikolaus 1923
Paul, Arup Ratan 178
Pauls, Denis 2441
Pavone, Mario 554
Pedroso, Douglas 335
Peftitsis, Dimosthenis 1486, 2355
Pelletier, Sebastien 223
Penczek, Adam 3804
Peng, Hujun 3235
Péra, Marie-Cécile 315
Pereda, Javier 2824
Pereira, Thiago 3014, 3092, 3101, 3846
Perez, Gaëtan 960
Perez-Cebolla, Francisco Jose 3574, 3750
Peroutka, Zdenek 1114
Perpiñá, Xavier 2715
Perrin, Rémi 2526
Perrin, Remi 3272
Petritz, Andreas 279
Petzoldt, Jürgen 2885, 3491
Peyghami, Saeed 2939

Pfeiffer, Jonas 411, 1167
Pfost, Martin 2375, 2413
Phanse, Ajinkya 1722
Phulpin, Tanguy 3883
Pichon, Pierre-Yves 2526
Pickert, Phil Leon 1381
Piepenbrock, Till 2432
Pietrzak-David, Maria 2869
Pigott, John 3796
Pinheiro, José Renes 3590
Piqué, Gerard Villar 3796
Piróg, Stanislaw 3804
Placzek, Julius M. 833
Plat, Arnaud 3862
Plötz, Till-Mathis 980
Pogulaguntla, Aditya 730
Pohlmann, Sebastian 767
Polezhaev, Vladimir 2230
Ponick, Bernd 1381, 1665, 3048, 3711
Poormohammadi, Fereshteh 2795
Pöschke, Florian 3704
Pouresmaeil, Edris 2537
Pouresmaeil, Kaveh 2788
Pouresmaeil, Mobina 2537
Pramanick, Sumit 1658, 3511
Pree, Elias 1445
Prenleloup, Pierre 3529
Prieto-Araujo, Eduardo 2486, 3704
Puls, Simon 2852
Puschmann, Frank 749
Qin, Zian 3607
Quabeck, Stefan 3893
Quade, Katharina Lilith 4025
Quay, Rüdiger 242
Rabkowski, Jacek 1486, 3938
Rädel, Uwe 2885, 3491
Radha, Krishna Moorthy 344
Rafiq, Aamir 1658
Raggini, Diego 3206
Raghavendra, I Venkata 730
Rahmani, Mehdi 2496
Raison, Bertrand 1147
Rajabian, Amir Azam 614
Ramdane, Brahim 1216
Ramirez, Fernando 289
Rasekh, Navid 3120
Rasool, Haaris 2101, 2293
Raßmann, Rando 1286
Rathjen, Kai-Uwe 758
Rault, Pierre 582
Ravyts, Simon 1300
Raya, Mariana 2715

Razi, R. 3179
Razi, Reza 3140
Regnat, Guillaume 2526
Rehlaender, Philipp 2432, 2754, 3625
Reimann, René 3159
Reincke-Collon, Carsten 370, 3391
Reindl, Andrea 2021
Reiner, Richard 242
Reißenweber, Lukas 525
Reitmeier, Dominik 2211
Remón, Daniel 1083
Rettner, Cornelius 4019
Reyes-Chamorro, Lorenzo 3429
Reynaud, Jean-François 3529
Ribeiro, Luiz Antonio De Souza 2948
Richard, Lucas 1147
Rickert, Kai 115
Rigbers, Klaus 4023
Rigogiannis, Nick 2257
Ringbeck, Florian 4025
Risch, Raffael 651
Rizoug, Nassim 3696, 3831
Robinson, Jonathan 572
Rocha, Gabriel Silva 2948
Roche, Jan-Philipp 3545
Rodríguez, Alberto 335, 1083, 1765, 3336
Rodriguez, Daniel C. 3893
Rodriguez, Joan Marc 2574
Rodriguez, José 1015
Roes, Maurice G. L. 798
Roes, Maurice 2788
Roß, Tilo 3391
Rossi, Mattia 1476
Rothenburger, Max 2383
Roth-Stielow, Jörg 971, 1237, 1277, 3536
Rouphael, Rosalie 1525
Rudolph, Christian 474
Rueß, Manuel 3565
Rufer, Alfred 30
Ruppert, Lukas A. 3766
Ruthardt, Johannes 971
Rylko, Marek S. 2470
Sadarnac, Daniel 3883
Saeidi, Mahmoud 1336, 1343, 3871
Safdarzadeh, Omid 2316
Sah, Gyanendra Kumar 1885
Sahan, Benjamin 1137
Sahin, Ilker 1361
Sahoo, Subham 2110, 2182
Sahu, Malaya Kumar 2241, 3757
Sahu, Silpashree 689
Said, Nasri 2618

Saito, Wataru	1128
Sakai, J.	1785
Salehi, Navid	1056
Samples, Ben	4005
Sanchez, Juan	873
Sanchez-Ruiz, Alain	484
Santos, Francisco	3101
Sanusi, Bima Nugraha	1413
Sanz-Alcaine, José Miguel	3750
Sarlioglu, Bulent	3947
Sato, Kota	1580
Sato, Takashi	3420
Sauer, Dirk Uwe	4012, 4025
Sauerland, Henning	3159
Sawicki, Jean–paul	315
Scarcella, Giuseppe	1569
Scelba, Giacomo	1569, 2628
Schäffner, Philipp	279
Schafmeister, Frank	2432, 2754, 3625, 3686
Schanen, Jean-Luc	787
Schanen, JL	843
Schefer, Hendrik	891, 912, 1094
Schellekens, Jan	634
Schierle, Guido	3188
Schiestl, Martin	223, 268
Schillinger, Tobias	3646
Schillingmann, Henning	2030
Schlegel, Christian	1923
Schlegel, Ludwig	3957
Schmid, Markus	268
Schmidhuber, Michael	411, 1167
Schmies, Dominik	2276
Schmitz, Laurids	3599
Schnabel, Fabian	624, 1515
Scholjegerdes, Moritz	3005
Schön, André	814
Schrödl, Manfred	2251
Schueltzke, Jens	1167
Schuerhuber, Robert	39
Schuhmann, Thomas	3646
Schullerus, Gernot	593, 2334
Schulte, Horst	3704
Schulz, D.	3411
Schulze, Gerold	2383
Schulze, Hans-Joachim	1424
Schumann, Christian	2058
Schumann, Sven	4022
Schümann, Ulf	1286
Schupp, Jan	3309
Schütt, Michael	1885, 2308
Schwarz, Babette	1381
Schwendemann, R.	3411

Scohier, Martin	504
Scrimizzi, F.	1975
Sebastián, Javier	1765, 3336
Seibel, Axel	1515, 1620
Seitz, Arne	4015
Seliger, Norbert	22
Semail, Eric	2996
Sen, Paresh C.	1590, 1601
Sepehr, Amir	2537
Serdyuk, Yuriy	2344
Sergentanis, Grigorios	3450
Serra, Amiron Wolff Dos Santos	2948
Seybold, Felix	3536
Shahparasti, Mahdi	2682
Sharma, Kanuj	2698
Shawky, Ahmed	1015
Shen, Chengjun	2477
Shen, Xiaobing	1914, 2197
Shinoda, Kosei	3928
Shintani, Michihiro	3420
Shousha, Mahmoud	279, 698
Shuqin, Wang	1815
Siala, Sami	125
Siemaszko, Daniel	3910
Siemieniec, Ralf	1445
Sievers, Markus	3855
Singh, Rupam	135
Singh, Shashank Shekhawat	279
Singh, Sukhjit	2084
Skala, Aleksander	3804
Skibin, Stanislav	3301
Soeiro, Thiago Batista	1319, 3729
Solomentsev, Michael	1722
Solovyov, Vyacheslav	2860
Soltau, N.	1785
Soltau, Nils	777
Sönmez, Ertugrul	593, 2334
Soundararajan, Ajeeth Phrassanna	3729
Soupremanien, Ulrich	922
Spieler, Matthias	1692
Sprunck, Sebastian	1611
Sreekanth, T	730
Stadler, Alexander	525
Stadlober, Barbara	279
Staiger, Jochen	2219
Stala, Robert	3804
Stalleicken, Frederik	2607
Stallmann, Frederik	641
Stärz, Ronald	223, 268
Stathis, Spyridon	1402
Staubach, Christian	1137
Steckler, P. B.	2564

Stefanski, L.	3411
Steffen, Jonas	1515
Steinhart, Heinrich	2219
Štengl, Josef	1114
Stevic, Marija	2985
Stewart, Joshua	2806
Steyn, Kyle	3112
Stille, Karl Stephan	2276
Stock, Alexander	1
Stöckl, Thomas	952
Stone, David A.	3440
Strunk, Robin	1350
Stul, Koen	1300
Stutz, Christian	493
Suberski, Martin	2885, 3491
Sujeeth, Arjun	2628
Sullivan, Charles R.	2470
Svensson, Jan R.	1187
Tabrizi, Gholamreza	1611
Takamori, Taro	1128
Takayama, Hajime	108
Takeshita, Takaharu	213, 351, 739
Talla, Jakub	1114
Tang, Chengjun	2813
Tang, Zhongting	1933
Tashakor, Nima	1025, 1064, 1197, 3636, 3665
Tatakis, Emmanuel C.	1551
Tegtmeier, Bernd	1674
Teske, Peter	1466
Thiringer, Torbjörn	2344, 2813, 3920
Thoma, Jürgen	2554
Thönelt, Nick	1713
Thönnessen, André	3676
Tian, Fanghao	2581
Tillmann, Philipp	3740
Tiwari, Arvind Kumar	289
Tiwari, Arvind	187
To, Pham Ha Trieu	59, 70, 4003
Tornello, Luigi Danilo	1569
Torres, C.	3382, 3401
Torrico, Grover	361, 1815
Tournez, Florian	2175
Tran, Dai Duong	2293
Tran, Manh Tuan	2101, 2293
Tresca, Giulia	3975
Trescases, Olivier	3796
Tricoli, Pietro	468
Trochimiuk, Przemyslaw	1486
Tschepp, Andreas	279
Turrisi, Gaetano	1569
Tzanakis, Athanasios	3920
Uicich, Simon	3862
Ulbing, Alexander	3855
Ulmer, Sabrina	593, 2334
Ulrich, Burkhard	459
Umetani, Kazuhiro	2164
Unruh, Peter	1620
Unruh, Roland	3686
Urkizu, June	325
Vaccaro, Luis	2919
Vaessen, Peter	3729
Vagg, Christopher	3696
Vagnon, Eric	2515
Vahid, Sina	719
Vala, Sama Salehi	344
Valderrama, Carlos	504
Valenzuela, Rodrigo Alonso Alvarez	814
Van Cappellen, Leander	2795
Van Mierlo, Joeri	2101
Van Oosterwyck, Nick	3344
Van Tuan, Mai	351
Vandenbussche, Thomas	1300
Vanfretti, Luigi	3928
Vanwalleghem, Bart	3344
Vasiladiotis, Michail	1923
Vatamanu, Lucian	1046
Vázquez, Aitor	1083
Vázquez, Francisco	1765, 3336
Velasco-Quesada, Guillermo	1056
Velazco, Diego	251
Vellvehi, Miquel	2715
Venkataramanan, Giri	3480
Venugopal, Ravinder	2985
Verdier, Jacques	169
Vermeerch, Pierre	582
Veroni, Alessandro	3206
Vershinin, K.	2564
Viana, Caniggia	1995, 2084
Viarouge, I.	1955
Viarouge, P.	1955
Vidal-Albalate, Ricardo	2189
Videau, Nicolas	944
Videt, Arnaud	3822
Villar, Irma	3529, 3938
Vitorino, Montiê Alves	2689, 3059
Vogelsberger, M.	1834
Volzer, Benjamin	2554
Von Hoegen, Anne	3740
Wada, Keiji	1128
Wagner, Valentin	514
Wakelin, Bruce	3309
Wallart, Francois	251
Wallscheid, Oliver	2276, 2432
Waltereit, Patrick	242

Wang, Chu 3722
Wang, Jun 2136, 3120
Wang, Kangan 3014
Wang, Rui 673, 1641
Wang, Xiaoya 3722
Wang, Xin 315
Wang, Yanbo 2011
Wang, Yangang 2301
Waradzyn, Zbigniew 3804
Watanabe, Hiroki 1104
Wattenberg, Martin 873
Weicker, Martin 2316
Weires, Jonas 604
Weiser, Mathias C. J. 3565
Weiss, Xavier 1006
Wenzel, Johannes C. 2066
Werlig, Christian 3966
Weyh, Thomas 767
Wicht, Bernhard 2912
Wieczorek, Nick 3775
Wiemer, Adrian 2544
Wiesemann, Julius 1865
Wiesner, E. 1785
Wiesner, Eugen 777
Wijnands, Korneel 673, 798, 2788
Wilkowski, Matt 4008
Willer, Felix 3838
Willich, Viktor 3555
Wohlrath, Fritz 525
Wolbank, T. 1834
Wolf, Mihaela 2596, 2644, 3775
Wolfstädter, Simon 4017
Wölk, Alexander 279
Wouters, Hans 2197
Woywode, Oliver 758
Wu, Weimin 1985
Wu, Xiangqiang 1933
Wu, Yuxuan 2860
Wunsch, Bernhard 3301
Würfl, Joachim 2596, 2644, 3775
Würsig, Andreas 3966
Xia, Peizhou 3470
Xiao, Qian 3215, 3225
Xiao, Xiong 1966
Xie, Jun 2885, 3491
Xie, Lihong 2136
Xu, Huihui 709
Xu, James 3796
Xu, Qianwen 883, 1006
Xu, Wei 2136
Xu, Zhongqing 912
Xu, Zixiao 2182

Yadav, Sachin 3607
Yamaguchi, Masamichi 2127
Yamashita, Shota 213
Yamauchi, Kohei 2119
Yang, Huoming 1466
Yang, Jiajun 3014
Yang, Juefei 2477
Yang, Yinghui 3993
Yang, Yongheng 2257
Yaqoob, M. 1815
Yasuda, Takumi 902
Yeganeh, Mohammad Sadegh Orfi 2496, 2939
Yu, Guangyao 1319
Yu, Xiao 562, 1309, 2383
Yu, Xiaodan 3225
Yuan, Xibo 2136, 3120
Zacharias, Peter 411, 562, 1309, 1328, 1336, 1343,
.... 2383, 2670, 3871
Zacher, Benjamin H. 2058
Zampardi, Giorgia 833
Zanchetta, Pericle 3975
Zatocil, Heiko 160
Zdanowski, Mariusz 3938
Zhang, Bo 1733
Zhang, Shimin 709
Zhang, Yaqian 2182
Zhang, Zhe 1561
Zhang, Zhuoqi 1776
Zhang, Ziqian 39
Zhao, Hongbo 1641, 3309
Zheng, Zhixue 315
Zhetessov, Aidar 3480
Zhu, Zi-Qiang 2958
Ziani, Adel 944
Ziegler, Philipp 971, 1237, 1277, 3536
Zilic, Rufad 1336
Zocher, Markus 2366
Zolfi, Pouya 719
Zou, Zhixiang 3014
Zsurzsan, Tiberiu Gabriel 1561

IEEE
445 Hoes Lane
Piscataway, NJ 08854-4141

ISBN 978-1-6654-8700-9